W9-BQZ-529

Handbook of Flow Visualization

Handbook of Flow Visualization

Edited by

Wen-Jei Yang

The University of Michigan, Ann Arbor

HEMISPHERE PUBLISHING CORPORATION

A member of the Taylor & Francis Group

New York Washington Philadelphia London

HANDBOOK OF FLOW VISUALIZATION

1 2 3 4 5 6 7 8 9 0 HAHA 8 9 8 7 6 5 4 3 2 1 0 9

This book was set in Times Roman by Waldman Graphics, Inc. The editor was Evelyn Pettit; the designer was Bermedica Production; and the production supervisor was Bermedica Production.
Cover design by Debra Eubanks Riffe.
Arcata Graphics/Halliday was printer and binder.

Library of Congress Cataloging-in-Publication Data

Handbook of flow visualization.

Bibliography
Includes index.
1. Flow visualization. I. Yang, Wen-Jei,
TA357.H285 1989 620.1′064 88-34774
ISBN 0-89116-669-6

Contents

Contributors

Tsuyoshi Asanuma
Mechanical Engineering Department
Tokai University
Hiratsuka, Kanagawa 259-12 Japan

James F. Brymer
Cardiovascular Research
Henry Ford Hospital
2799 West Grand Boulevard
Detroit, Michigan 48202

Giovanni Maria Carlomagno
Facolta' di Ingegneria
Piazzale Tecchio, 80
80125 Naples, Italy

K.C. Cheng
Department of Mechanical Engineering
University of Alberta
Edmonton, Alberta T6G 2G8 Canada

Kunihiro Chihara
Automation Research Laboratory
Kyoto University
Uji, Kyoto 611 Japan

A. Cogotti
Industrie Pininfarina S.P.A.
Corso Stati Uniti 61
Torino, Italy 10100

J.P. Crowder
Boeing Commercial Airplane Co.
3834 East Mercer Way
Mercer Island, Washington, 98040

Luigi de Luca
Facolta' di Ingegneria
Piazzale Tecchio, 80
80125 Naples, Italy

John M. Dewey
University of Victoria
P.O. Box 1700
Victoria, British Columbia V8W 2Y2 Canada

Junta Doi
Faculty of Agriculture
University of Tokyo
Hongo, Bunkyo-ku
Tokyo 113, Japan

Shigeru Eiho
Automation Research Laboratory
Kyoto University
Uji, Kyoto 611, Japan

W. Frank
Institut für Strömungslehre und Stromungsmaschinen
Universität Karlsrughe
Kaiserstrasse 12
D-7500 Karlsruhe, West Germany

Peter Freymuth
Department of Aerospace Engineering Sciences
University of Colorado
Boulder, Colorado 80309

Ronald K. Hanson
Department of Mechanical Engineering
Stanford University
Stanford, California 94305

Lambertus Hesselink
Department of Aeronautics and Astronautics
Stanford University
Stanford, California 94305

M. Hirata
Department of Mechanical Engineering
University of Tokyo
Hongo, Bunkyo-ku
Tokyo 113, Japan

A. Jimoh
School of Chemical Engineering
Georgia Institute of Technology
Atlanta, Georgia 30332

N. Kasagi
Department of Mechanical Engineering
University of Tokyo
Hongo, Bunkyo-ku
Tokyo 113, Japan

J. Koster
Department of Aerospace Engineering Sciences
University of Colorado
Boulder, Colorado 80309

Fareed Khaja
Cardiovascular Research
Henry Ford Hospital
2799 West Grand Boulevard
Detroit, Michigan 48202

Robert A Kudlinski
NASA Langley Research Center
Hampton, Virginia 23665

Michiyoshi Kuwahara
Osaka Industrial University
Daito, Osaka 574 Japan

Thomas M. McFarland
Cardiovascular Research
Henry Ford Hospital
2799 West Grand Boulevard
Detroit, Michigan 48202

S.T. McMillan
School of Chemical Engineering
Georgia Institute of Technology
Atlanta, Georgia 30332

W. Merzkirch
Lehrstuhl für Stromungslehre
Universitat Essen
Postfach 103764
D-4300, Essen, West Germany

R.J. Moffat
Department of Mechanical Engineering
Stanford University
Stanford, California 94305

R. Monti
Universita di Napoli
80125 Naples, Italy

Thomas J. Mueller
Department of Aerospace and Mechanical Engineering
University of Notre Dame
Notre Dame, Indiana 46556

Y. Nakayama
Department of Mechanical Engineering
Tokai University
1117 Kitakaname
Hiratsuka, Kanagawa 259-12 Japan

R. Owen
Department of Aerospace Engineering Sciences
University of Colorado
Boulder, Colorado 80309

Michel Philbert
Office of National d'Etudes et de
Recherches Aerospatiales
B.P. 72
92322 Chatillon, France

F.F. Philpot
School of Chemical Engineering
Georgia Institute of Technology
Atlanta, Georgia 30332

J. R. Pincombe
School of Engineering and Applied Sciences
University of Sussex
Falmer, Brighton, Sussex BN1 9QT, United Kingdom

Armando Rapillo
Facoltá di Ingegneria
Piazzale Tecchio, 80
80125 Naples, Italy

R. Řezníček
Institute of Mechanization
Vysoke Skoly Zemedelske
16021 Praha-Suchdol, Czechoslovakia

A.J. Ridgway
School of Chemical Engineering
Georgia Institute of Technology
Atlanta, Georgia 30332

Hani Sabbah
Cardiovascular Research
Henry Ford Hospital
2799 West Grand Boulevard
Detroit, Michigan 48202

Jerry M. Seitzman
Department of Mechanical Engineering
Stanford University
Stanford, California 94305

G.S. Settles
Mechanical Engineering Department
Pennsylvania State University
University Park, Pennsylvania 16802

Paul D. Stein
Cardiovascular Research
Henry Ford Hospital
2799 West Grand Boulevard
Detroit, Michigan 48202

C.R. Smith
Department of Mechanical Engineering and Mechanics
Lehigh University
Bethlehem, Pennsylvania 18015

Robert E. Smith
NASA Langley Research Center
Hampton, Virginia 23665

H.-W. Sung
School of Chemical Engineering
Georgia Institute of Technology
Atlanta, Georgia 30332

Jean Surget
Office National d'Etudes et de Recherches Aerospatiales
B.P. 72
92322 Chatillon, France

Yoshimichi Tanida
Institute of Interdisciplinary Research
University of Tokyo
Komaba, Meguro-ku
Tokyo 153, Japan

Josef Tanny
Department of Fluid Mechanics and Heat Transfer
Tel Aviv University
69978 Tel Aviv, Israel

A.B. Tsinober
Department of Fluid Mechanics and Heat Transfer
Tel Aviv University
69978 Tel Aviv, Israel

Claude Véret
Office National d'Etudes et de Recherches Aerospatiales
B.P. 72
92322 Chatillon, France

H. Werlé
Office National d'Etudes et de Recherches Aerospatiales
B.P. 72
92322 Chatillon, France

Allen E. Winklemann
Department of Aerospace Engineering
University of Maryland
College Park, Maryland 20742

Y.-R. Woo
School of Chemical Engineering
Georgia Institute of Technology
Atlanta, Georgia 30332

Wen-Jei Yang
Department of Mechanical Engineering and Applied Mechanics
University of Michigan
Ann Arbor, Michigan 48109

A.P. Yoganathan
Cardiovascular Fluid Mechanics Laboratory
School of Chemical Engineering
Georgia Institute of Technology
Atlanta, Georgia 30332

Preface

During the era of the Renaissance, a gradual change from the purely philosophical science of the Scholastics toward the observational science of the present day finally became apparent. At this time, one individual of accomplishments in every field appeared upon the scene. Leonardo da Vinci (1452–1519) was an advocate of the experimental method who justified the necessity of observation as follows:

> I will treat of such a subject. But first of all I shall make a few experiments and then demonstrate why bodies are forced to act in this manner.

This is the method that one has to pursue in the investigation of phenomena of nature. It is true that nature begins by reasoning and ends by experience; but, nevertheless, we must take the opposite route: as I have said, we must begin with experiment and try through it to discover the reason.

"Seeing is believing" is a famous expression which is common in both the East and the West. One has a desire to see a phenomenon and understand it. Air flow is invisible, though it is evidenced by the motion of tree leaves and branches. Although water flow can be seen, its streamlines or local velocities are not visible to the naked eye. Flow visualization means to make the flow (of fluid, heat and mass) behavior that is invisible visible. After seeing the flow field, one can interpret the flow phenomena. This is the essence of Da Vinci's basic premise: "first experience and then reason."

Da Vinci has been considered by many individuals as the first to apply flow visualization. Typical phenomena that he was the first to sketch or describe are the formation of eddies at abrupt expansions and in the wakes; the profiles of free jets; the velocity distribution in a vortex; the propagation, reflection and interference of waves; and the hydraulic jump.

The first important application of the flow visualization technique was due to O. Reynolds in 1883 [1]. He investigated, in greater detail, the circumstances of the transition from laminar to turbulent flow by injecting a liquid dye into the water flowing through a long horizontal pipe. He discovered the law of similarity and defined the Reynolds number. In 1904, L. Prandtl obtained a picture of the motion of water along a thin flat plate in which the streamlines were made visible by the sprinkling of particles on the surface of the water. The traces left by the particles on the picture are proportional to the flow velocity. A very thin layer exists near the wall in which the velocity is substantially smaller than at a large distance from it. This observation led him to the concept of the boundary layer. Prandtl utilized the same method of flow visualization in several experiments on boundary layer control in 1904. Many other historically important discoveries in flow phenomena were made through the application of rather simple methods of flow visualization.

This book compiles all major techniques of flow visualization and demonstrates their applications in all fields of science and technology. It consists of four parts. Part 1 begins with a chapter introducing essence and scope of flow visualization. The reader is advised to read the chapter to become familiar with the contents of this handbook. The fundamentals of flow phenomena, e.g., the flows of fluid, heat and mass, form the core of Part 1. The understanding of basic concepts, theoretical formulation and method of solution is essential in the interpretation and judgment of results obtained by the flow visualization techniques. Part 2 presents the methods of conventional flow visualization, and the methods of computer-assisted flow visualization are the subjects of Part 3. Computer graphics are employed to display the results in colors or contours. The conventional and computer-assisted flow visualization methods can be referred to as the first- and second-generation methods, respectively, since the latter results from the evolution of the former, e.g., evolution of flow field. Part 4 presents the applications of methods of both generations to various fields in both science and technology.

This volume is the result of contributions of more than 50 specialists who should be congratulated for devoting their time and effort. Special thanks is also expressed to Hemisphere Publishing Corporation and to Alicia Hwang for her editorial assistance.

Wen-Jei Yang

Handbook of Flow Visualization

Part 1
Fundamentals

Chapter 1

Introduction

Wen-Jei Yang

Fluid flow, heat and mass transfer, and electrical flow deal with the transfer of physical properties such as momentum, energy, mass, and electricity. Called transport phenomena, they can be investigated both by experiment and by theory. Experimental flow study can be performed in the forms of flow measurement and flow visualization, while the theoretical approach includes analysis and computation as shown in Fig. 1. The task of flow visualization is to make the process of transport phenomena visible. This can be done through experiments or by simulation (theory). Flow visualization is superior to flow measurement in that the physical phenomena in the entire flow field are retrieved. However, because the information obtained is qualitative, the technique has been inherently penalized (Table 1). The advance of computing machines has led flow visualization into a new era of quantifying the flow information, and thus, the fatal shortcoming of the conventional method has been overcome.

For convenience, the methods used in flow visualization are classified as either conventional or computer assisted. The former, referred to as the first-generation method, includes all traditional techniques in use since the time of Leonardo da Vinci (1452–1519). These techniques are classified into four groups in Table 2: wall tracing, tuft, tracer, and optical methods. The applicable flow range for various methods is presented in Fig. 2. One major objective of research in flow visualization is the extension of the flow range in both directions, toward higher as well as lower velocities.

Table 1 Comparison of Flow Experiments

	Flow Measurement	Flow Visualization
Range	Local	Whole field
Information	Qualitative	Quantitative

In general, different substances are introduced into the flow field as a contrast medium or as a tracer to distinguish the object to be observed. Very often, the flow field needs to be illuminated in order to achieve the field image of high contrast and resolution. The image is then displayed in an appropriate form. This is the general practice in conventional flow visualization techniques. The image thus obtained provides only qualitative information about the flow field. In the case of computer-assisted flow visualization methods, the abovementioned process, called image for-

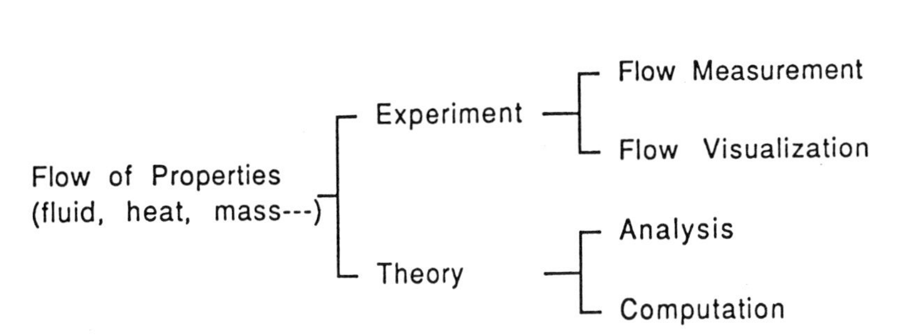

Fig. 1 Methods to study flow of properties.

Table 2 Conventional Flow Visualization Methods

Method	Type
Wall tracing	Liquid film
	Sublimation
	Thermosensible paint
	Electrolytic etching
	Soluble chemical film
Tuft	Surface tuft
	Tuft grid
	Depth tuft
Tracer	
Direct injection	Solid particles
	Liquids
	Gases
Chemical reaction	Chemical
	Electrolytic
	Photochemical
Electrical	Hydrogen bubble
	Spark
	Smoke wire
Optical	Shadowgraphy
	Schlieren
	Interferometry
	Holography
	Stream birefriengence
	Moiré
	Liquid crystal

mation, is to be followed by digital image processing. This results in better quality images being displayed. The image-acquisition system includes both the conventional flow visualization device and the imaging device, often referred to

Table 3 Flow Visualization Methods

First Generation	Second Generation
Traditional techniques	Computer-assisted techniques
Qualitative information	Quantitative information
Contrast media for visualization	Recording, display, and storage for reprocessing
Recording and display	Color as the parametric variables

as machine vision. Recent advances in computational techniques and computer color graphics have contributed to a new era in flow visualization. Results obtained by theory, measurement, and their combination can be displayed, using colors as the parametric variables. The results can also be stored in the computer for reprocessing. Figure 3 summarizes the computer-aided flow visualization techniques, while Table 3 compares the differences between the first- and second-generation methods of flow visualization.

Digital image processing and its bionics are depicted in Fig. 4. The process of identifying an object with the human eye takes the steps shown in Fig. 4a. Light rays reflected from the object pass through the crystalline lens (camera lens) to be focused on the retina (film or plate in the camera). Light rays stimulate millions of optical receptors in the retina. The stimuli are then transmitted through the optic nerves to the cerebrum, where the image of the object is identified. Figure 4b shows a schematic diagram of digital image processing. For automatic identification of an object, it is necessary to have a system similar to that of human vision. The image input is equivalent to that of the human

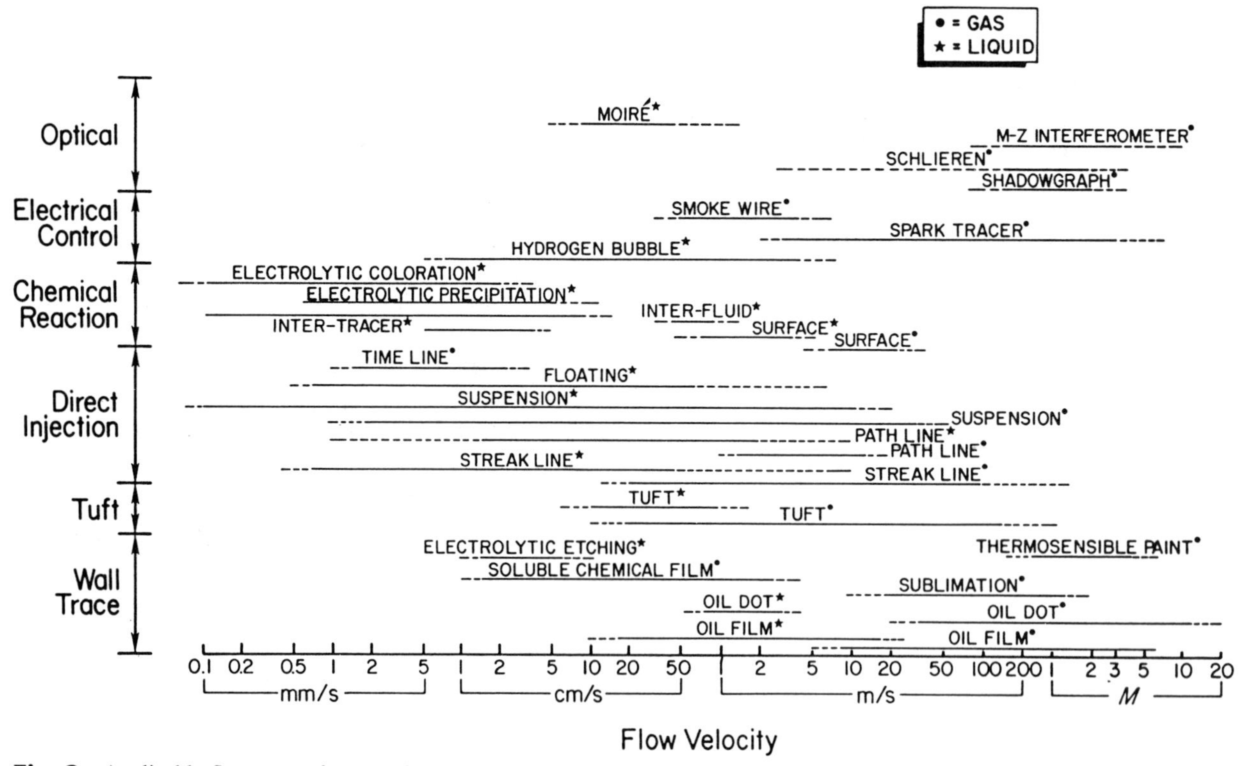

Fig. 2 Applicable flow range for each flow visualization method.

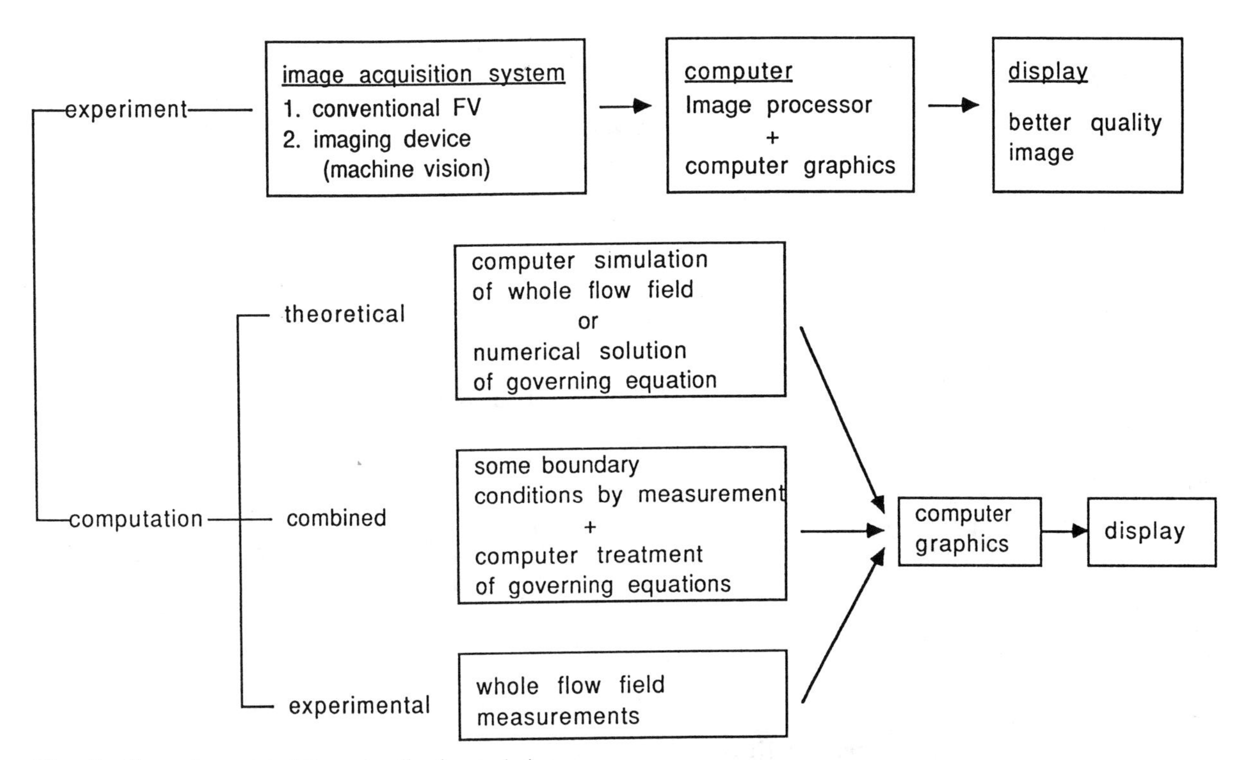

Fig. 3 Computer-assisted flow visualization techniques.

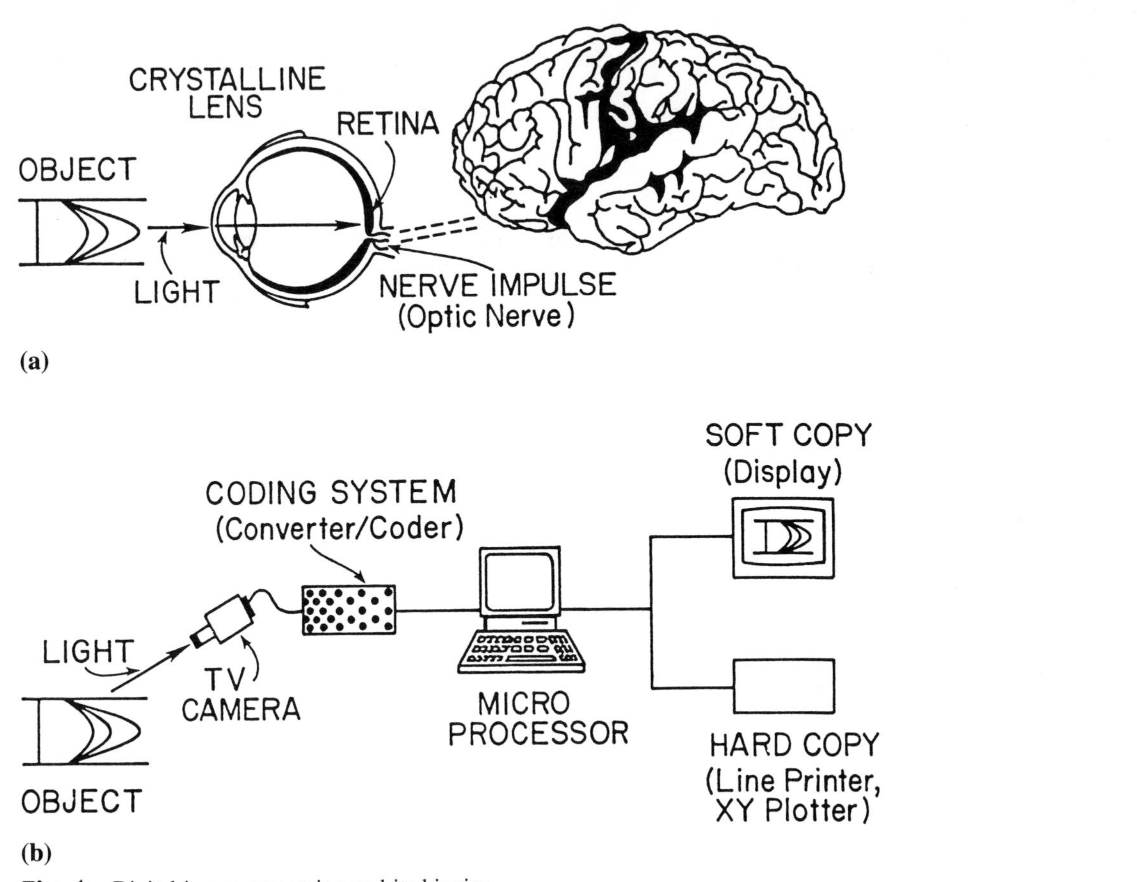

Fig. 4 Digital image processing and its bionics

IN FLOW VISUALIZATION
Three factors to be considered

Moving Particles
Position in Space
Time

P_i A Position in Space
o x
A Moving Particle Q t_i = Time Instant

1. Path of a Particule P_i
   Same particle over a time period

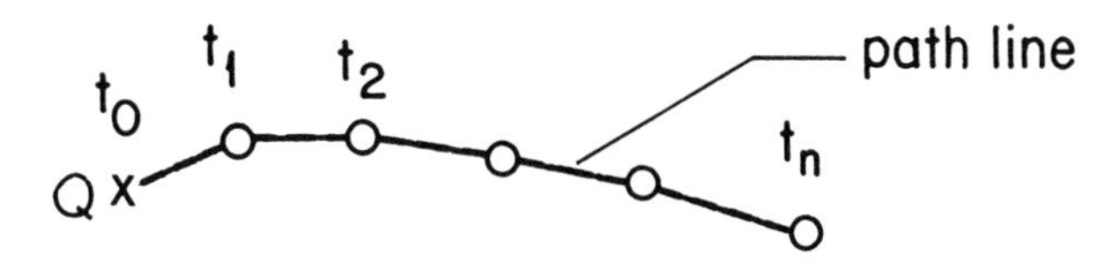

LaGrange method (Open Shutter)
Timewise Change in Particle Behavior

2. Flow Direction on Various Particles at t_i

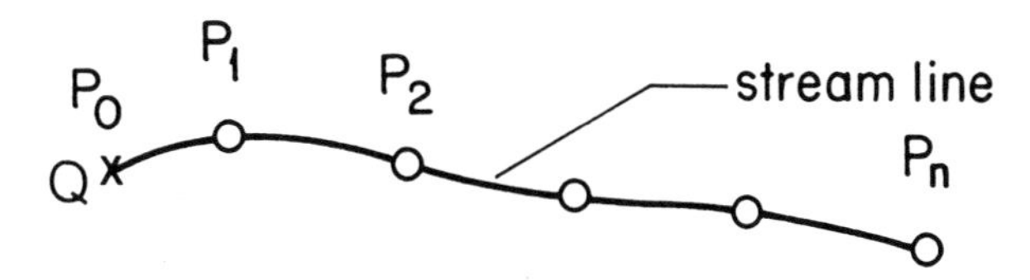

Euler Method (Fast Shutter Speed)
Time/Space Change in Flow Behavior

3. Connection of all Particles Passing Through a Point Q

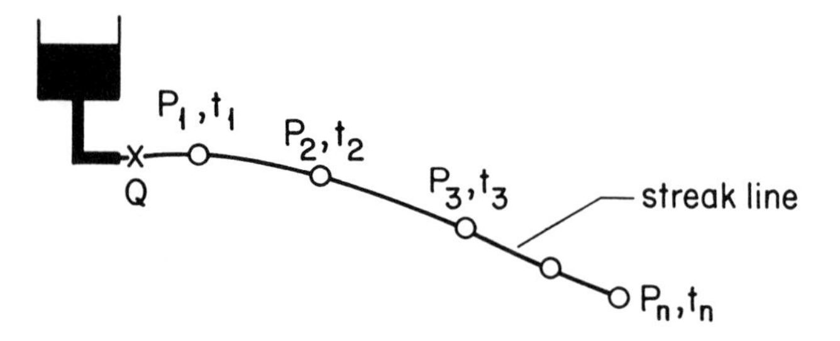

Continuous Tracer Injection at a Fixed Point Q
t_i: Time after Injection

4. Connection of Various Particles at t_i after Injection from a Fixed Line Source

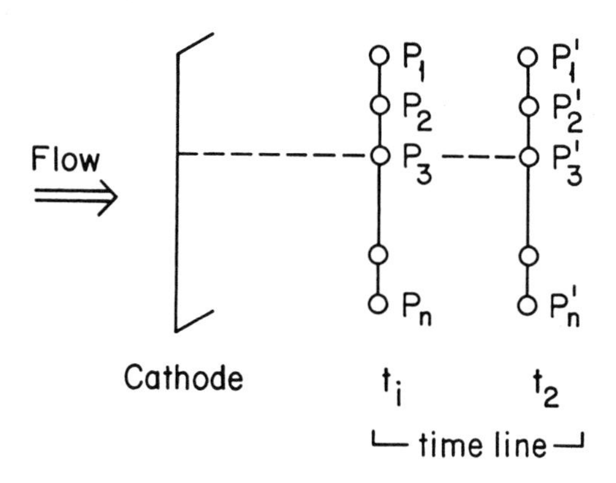

Horizontal Broken Line = Streak Line (From a Point on Cathode)

Fig. 5 Physical quantities to be observed in the behavior of tracers.

eye. The input system consists of a television camera and a coding unit (converter/coder). The device functions to convert the image of an object into a signal (equivalent to nerve impulses). An electronic computer analyzes, identifies, and judges the signal. An output device is needed for judgment of the results of the image processed by the computer. Thus, the computer and the output device together serve the functions of the cerebrum.

What do we observe in the behavior of tracers in the flow field? Three factors must be considered: moving particle (tracer) P_i, its position in space Q and time t_i. Four physical quantities of importance can be derived from the behavior of the tracer:

1 Pathline From the Lagrangian viewpoint, namely, a closed system with a fixed identifiable quantity of mass, the independent variables are the initial position with which each particle is identified and the time. Hence, the locus of the same particle over a time period from t_0 to t_n is called the pathline.

2 Streamline From the Eulerian viewpoint, namely, an open system with constant control volume, all flow properties are functions of a fixed point in space and time, if the process is transient. The flow direction of various particles at a time t_i forms a streamline.

3 Streakline Consider a continuous tracer injection at a fixed point Q in space. The connection of all particles passing through the point Q over a period of time is called the streakline. The pathline, streamline, and streakline are different in general but coincide at a steady flow.

4 Timeline When a pulse input is periodically imposed on a line of tracer source placed normal to a flow, a change in the flow profile can be observed. The tracer image is generally termed the timelines. Time lines are often generated in the flow field to aid in the understanding of flow behavior such as velocity and velocity gradient. Figure 5 summarizes the flow behavior that can be disclosed from flow visualization.

Chapter 2

Fluid Dynamics

Tsuyoshi Asanuma and Yoshimichi Tanida

I INTRODUCTION

When a body is placed in stationary fluid, it is subjected to a force normal to the element of surface (Fig. 1*a*). This normal force acting on a unit surface area is called *pressure*. If the gravity effect is neglected, the pressure is uniform everywhere. For example, a human body feels a force of 1013 mbar all over the surface, that is, the so-called atmospheric pressure.

When a body moves in stationary fluid, or when a body is placed in a flowing fluid, it feels not only the pressure of fluid but also another force that acts parallel to the surface element (Fig. 1*b*). In this case of a moving body or fluid, the pressure is not uniform in general but varies from place to place due to the effect of the displacement of fluid by the body motion. The force (tangential to the surface) is a shearing stress generated by the viscosity of fluid, which makes the fluid stick on the body surface (condition of no-slip).

Integrating these forces over the whole surface gives the lift and drag forces acting on the body. The drag is divided into profile and friction drags, which are attributed to the pressure and the shearing stress, respectively.

The concept of a perfect, i.e., inviscid and incompressible, fluid is very useful for the analysis of flow, but it leads to the erroneous result that a body moving at a constant speed in an infinite field does not experience the drag (d'Alembert's paradox).

The shearing stress may be interpreted as the transmission of the momentum of the fluid in the direction perpendicular to the body surface, which extends outside to make a velocity gradient. In many practical cases in which water or air are used as the working fluid, the influence of viscosity is often confined to a very thin layer in the immediate neighborhood of the body surface (boundary layer; see Sec. IV), and the flow outside the boundary layer can be treated

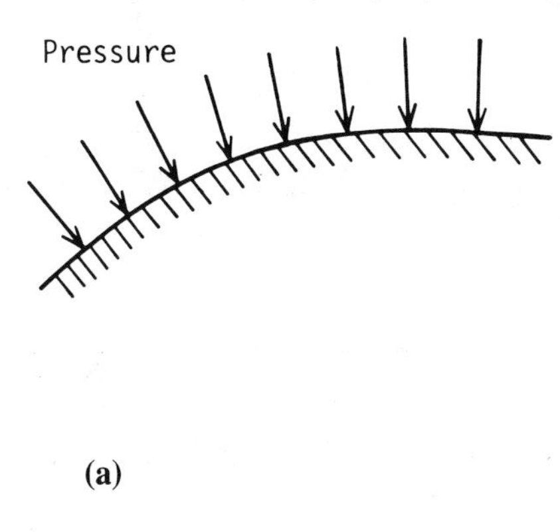

(a)

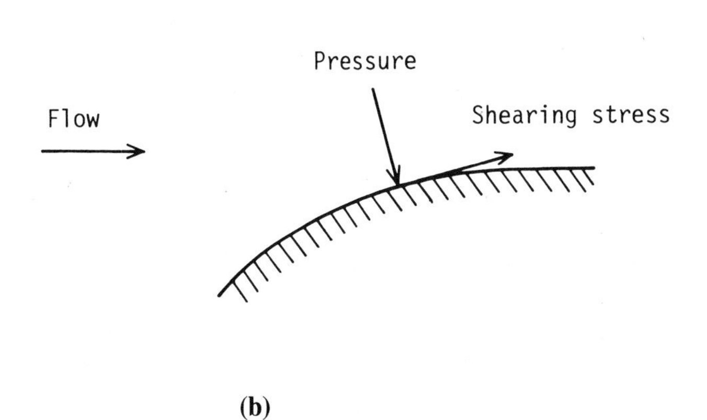

(b)

Fig. 1 Pressure and shearing stress (*a*) in stationary fluid; (*b*) in flowing fluid.

as the perfect fluid. Then, an extensive mathematical analysis should be possible.

II NATURE OF FLUID

A Viscosity

Consider a fluid contained between two large parallel plates, which are separated by a very small distance h (Fig. 2). At time $t = 0$, the upper plate is suddenly set in motion with a constant velocity V. Assuming that the pressure is constant everywhere, we can observe the velocity distribution varying with time, as shown in the figure. The fluid adheres to both plates, and as time proceeds, the region of fluid dragged by the upper plate spreads in the y direction until a final steady state of linear velocity distribution $u(y) = Vy/h$ is attained. The interpretation of this phenomenon may be that x momentum of the fluid, which is given by the motion of the upper surface, is transmitted through the fluid by the molecular transport process in the y direction.

To maintain the final state of steady motion, a tangential force must be applied to the upper plate against the frictional force in the fluid. The frictional force per unit area (shearing stress) is given by

$$\tau = \mu \frac{V}{h} = \mu \frac{du}{dy}$$

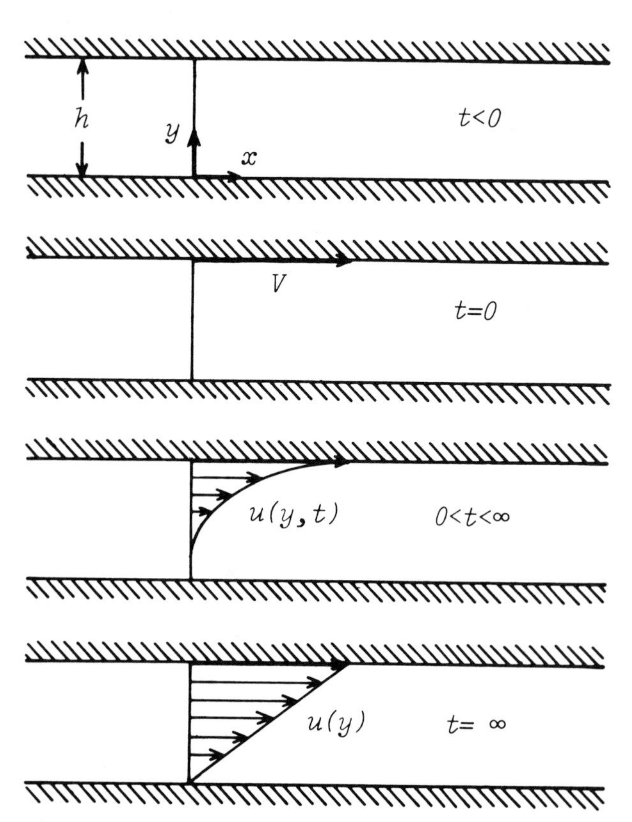

Fig. 2 Velocity distribution in viscous fluid between two parallel flat plates.

The proportionality factor μ is called the viscosity of the fluid. This relation is known as Newton's law of viscosity, and the fluid that is subject to this law is called Newtonian fluid. The viscosity μ is a property of the fluid. Generally speaking, the viscosity can be taken to be independent of pressure, but it depends on temperature. With increasing temperature, the viscosity decreases for liquids and increase for gases.

In all fluid motions, the interaction between frictional and inertial forces is important, so we define here the kinematic viscosity

$$\nu = \frac{\mu}{\rho}$$

where ρ is the density of fluid. Some kinematic viscosity data are given in Tables 1 and 2.

Recently, there have been in use many industrially important materials, which are not subject to Newton's law of viscosity, called non-Newtonian fluids. According to the way of shearing, the fluids can be classified into three categories as follows (Fig. 3):

1. Purely viscous fluid, in which velocity gradient du/dy is a function of shear stress τ only, $du/dy = \phi_1(\tau)$. Newtonian fluid is of a special case, and plastic (Bingham) fluid, pseudoplastic fluid, and dilatant fluid belong to this category.
2. Time-dependent fluid, in which du/dy is a function of τ and time t, $du/dy = \phi_2(\tau, t)$. When the fluid moves with a constant shear rate at a given temperature, the viscosity or shear stress decreses or increases as time proceeds. These are called thixotropic and rheopectic fluids.
3. Viscoelastic fluid, in which du/dy is a function of τ and internal strain s, $du/dy = \phi_3(\tau, s)$. According to the

Table 1 Kinematic Viscosity of Water and Air

	Kinematic Viscosity ν, cm²/s	
Temperature, °C	**Water**	**Air, p = 760 mm Hg**
−10	—	0.1240
0	0.01794	0.1326
10	0.01310	0.1417
20	0.01010	0.1513
30	0.00804	0.1615
40	0.00659	0.1728

Table 2 Kinematic Viscosity of Selected Liquids

Liquid	Kinematic Viscosity ν, cm²/s
Mercury	0.00114
Alcohol	0.0151
Benzol	0.0075
Glycerine	8.48

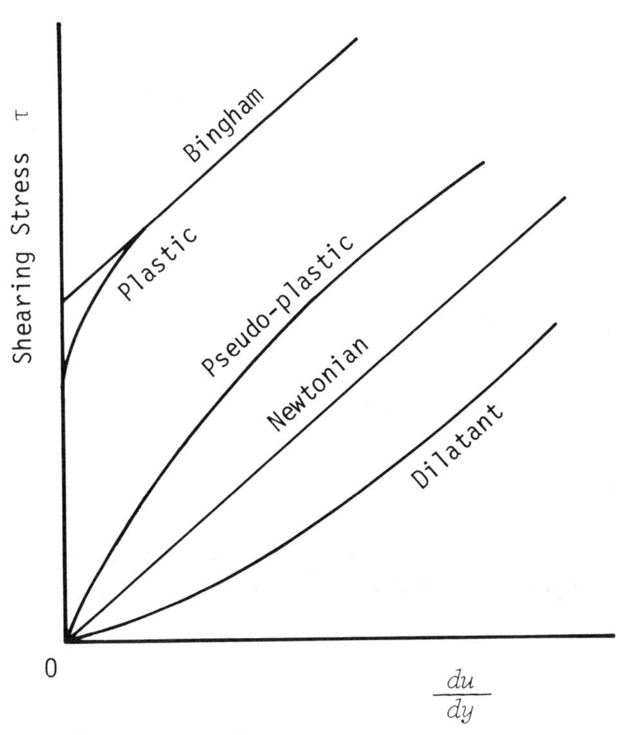

Fig. 3 Characteristics of non-Newtonian fluids.

macromolecular nature of fluid, a part of the external work done is stored as an elastic energy to give elastic recoil effect, Weissenberg effect, Borus effect, Toms effect, etc.

B Compressibility

Compressibility is a measure of the change of volume or density of a fluid under the change of pressure and temperature. The modulus of compressibility is defined as the ratio of the relative volume change of fluid to the change of pressure p.

$$\beta = \frac{1}{K} = -\frac{1}{V}\frac{dV}{dp} = \frac{1}{\rho}\frac{d\rho}{dp}$$

where K is the modulus of elasticity or bulk modulus of fluid.

The compressibility of liquids is very small, so the liquids can be treated as incompressible fluids except in a few cases such as water hammering; e.g., for water $K = 2.2 \times 10^3$ MPa. In the gases, the bulk modulus is equal to or of the same order of the pressure; e.g., for air $K = 1.4 \times 10^2$ kPa, which means that the air is compressible about 20,000 times as much as the water.

In the flowing fluid of velocity U, there may occur the pressure change Δp of the order of $\frac{1}{2}\rho U^2$, so that the relative density change is

$$\frac{d\rho}{\rho} = \frac{dp}{K} \propto \frac{\rho U^2}{K} = \frac{U^2}{a^2}$$

where a is the velocity of sound given by $a = \sqrt{K/\rho}$.

This leads to the conclusion that the compressiblity can be neglected in the flow of gases if the flow velocity is small enough as compared with the velocity of sound, or if the Mach number (see Sec. III.A) U/a is small compared with unity.

C Heat and Mass Transfer

As with the monentum transport in a viscous flow, the heat flow will occur in the direction of temperature gradient; that is, the local heat flow rate per unit area (heat flux $= q$) is proportional to the gradient of temperature T normal to this area. That is described as

$$q = -k\frac{\partial T}{\partial n}$$

where n represents the direction normal to the area considered and the proportionality factor k is the thermal conductivity of the fluid.

In three-dimensional field, the above equation can be written in a vector form as

$$\mathbf{q} = -k\,\nabla T$$

The balance of heat flux gives the local temperature change in time

$$\frac{\partial T}{\partial t} = \frac{k}{\rho C_p}\nabla^2 T$$

where the proportionality factor $\alpha = k/\rho C_p$ is called the thermal diffusivity.

Next, in a binary mixture of A and B, the diffusion of each species occurs due to its concentration gradient in the mixture. If the density of fluid is constant, Fick's law of diffusion gives for the species A

$$m_A = -D_{AB}\frac{\partial \rho_A}{\partial n}$$

where the proportionality factor D_{AB} describes the mass diffusivity of the species A against the species B. The mass diffusivity $D_{AB} = D_{BA}$ for a binary system is a function of temperature, pressure, and composition.

III FUNDAMENTALS OF FLUID MOTION

A Principle of Similarity

The performances of aircraft, ships, water turbines, and other big machines are usually predicted by laboratory experiments that use geometrically similar scaled-down models. A fundamental, important question is how we can get two dynamically similar flows about geometrically similar bodies with different fluids and different velocities.

1 Reynolds Number (Re) We now consider the important case of incompressible viscous flow, in which only frictional and inertial forces are present. The element of fluid moves under the influence of these two forces, so that the flow becomes similar when the ratio of friction and inertial forces is identical with each other.

In the steady flow, the acceleration of a fluid particle is given by $\rho v\, \partial v/\partial s$ and the friction force per unit volume by $\mu\, \partial^2 v/\partial n^2$, where $\partial v/\partial s$ and $\partial v/\partial n$ denote the changes of velocity in the direction of motion and in the direction normal to the motion, respectively. If we represent the characteristic length and velocity by L and V, then $\rho v\, \partial v/\partial s$ and $\mu\, \partial^2 v/\partial n^2$ are proportional to $\rho V^2/L$ and $\mu V/L^2$, respectively. Hence, the condition of similarity is written as

$$\frac{\text{Inertial force}}{\text{Friction force}} = \frac{\rho V^2/L}{\mu V/L^2} = \frac{VL}{\nu} = \text{const}$$

This ratio is nondimensional and known as the Reynolds number.

The pressure relative to the reference pressure and the shear stress made dimensionless by the dynamic head $\frac{1}{2}\rho V^2$ are functions of the Reynolds number only. It follows that the dimensionless lift and drag coefficients are also functions of the Reynolds number only.

2 Froude Number (Fr) A ship moves on the free water surface, generating waves that cause resistance against the ship's motion. In this case, the gravitational force is more predominant than the friction force, causing the flow to become similar when the ratio of inertial and gravitational forces is identical. Hence, the condition of similarity is given by

$$\frac{\text{Inertial force}}{\text{Gravitational force}} = \frac{\rho V^2/L}{\rho g} = \frac{V^2}{gL} = \text{const}$$

This ratio is nondimensional, and its square-root value is known as the Froude number. That is,

$$\text{Fr} = \frac{V}{\sqrt{gL}}$$

In the laboratory, it is usually impossible to make the Reynolds and Froude numbers identical simultaneously, so only the Froude number is made identical to obtain the exact wave resistance. The friction drag is estimated approximately by calculations.

3 Mach Number (M) In compressible flows, the perturbations generated by an obstacle propagates up- and downstream at a finite speed of this gas, equal to the speed of sound given by

$$a^2 = \left(\frac{\partial p}{\partial \rho}\right)_s = \kappa \frac{p}{\rho}$$

where the subscript s denotes the isentropic change of state and κ the ratio of specific heats.

Let the differences of pressure and density from the state of fluid (p_∞, ρ_∞) be p' and ρ'; then

$$a^2 = \frac{p'}{\rho'}$$

The pressure change p' in the flow is on the order of $\rho_\infty U^2$, so we can write

$$\frac{\rho'}{\rho_\infty} = \frac{p'}{\rho_\infty a^2} \sim \frac{U^2}{a^2}$$

If $\text{M} = U/a << 1$, or if the flow velocity is low enough compared with the speed of sound, the change in density of fluid can be neglected, that is, the flow is considered incompressible.

The nondimensional ratio M is known as the Mach number of flow and is an important parameter in high-speed compressible flow.

B Fundamental Flows

1 Laminar and Turbulent Flows in Pipe Osborne Reynolds carried out an experiment on flow in a circular pipe, in which he visualized the motion of fluid by injecting a dye into the stream through a thin tube (Fig. 4). At a low flow rate, the dye extended over the whole length of the pipe in a form of straight line (laminar flow). As the flow rate increases over a critical value, however, the dye begins to move irregularly and mix with the fluid. Reynolds found that the transition from laminar to irregular (turbulent) flow takes place at a definite value of the Reynolds number (critical Reynolds number). The critical Reynolds number depends on the inherent irregularity (turbulence) of the fluid entering the pipe, and its lowest value is approximately $\text{Re}_{\text{crit}} = (Ud/\nu)_{\text{crit}} = 2300$, where U is the average velocity of flow and d the inner diameter of the pipe.

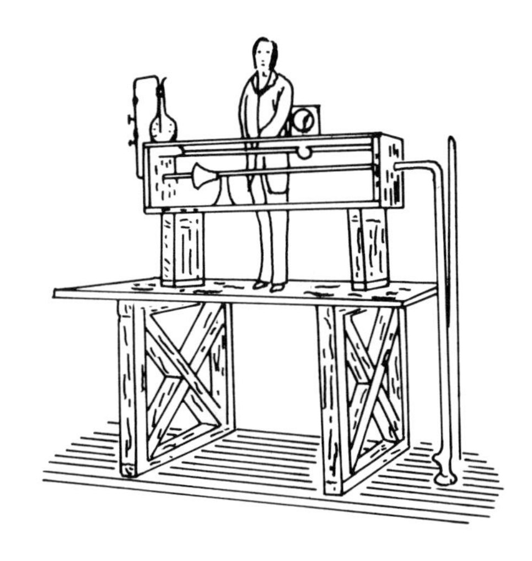

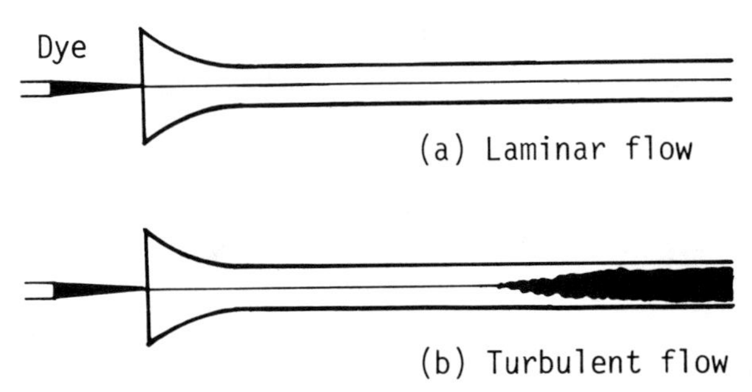

Fig. 4 Laminar and turbulent flows in pipe: (*a*) Laminar; (*b*) turbulent.

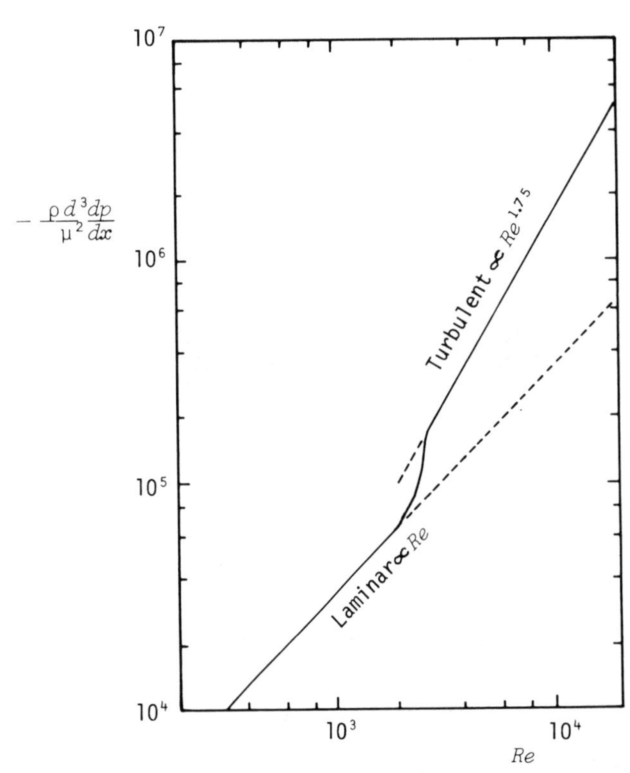

Fig. 5 Variation of pressure drop with Reynolds number for pipe flow.

Now we consider the flow of a given rate in a circular pipe, which is fully developed in the axial direction or $\partial u/\partial x = 0$. In laminar flow, known as Hagen–Poiseuille flow, the velocity profile is a paraboloid

$$u = 2U\left[1 - \left(\frac{r}{d/2}\right)^2\right]$$

and the pressure drop along the axial direction is given by

$$\frac{dp}{dx} = -\frac{\rho U^2}{2d}\frac{64}{\text{Re}}$$

In turbulent flow, the velocity profile becomes flat near the axis of the pipe, and the pressure drop is larger than in laminar flow (Fig. 5).

2 Flow Past a Circular Cylinder We will observe the flow around a circular cylinder, which looks very simple but actually holds very important physical meanings.

A cylinder of diameter d is placed in a uniform flow of velocity U, with its axis normal to the flow. Figure 6 shows how the flow pattern around the circular cylinder changes with the increase of the Reynolds number, defined as $\text{Re} = Ud/\nu$.

For the Reynolds number to be satisfactorily low, the flow pattern must be completely steady and symmetrical upstream and downstream. The viscous effect due to the presence of the cylinder spreads over long distances, where the flow velocity is appreciably different from U. As Re is increased, the viscous effect is gradually confined to a narrow region near the cylinder.

When Re exceeds about 4, the upstream–downstream symmetry of flow pattern disappears. The flow is still steady, but the fluid that comes along the cylinder surface moves away from it (separation; see Sec. IV.B) to form twin eddies attached behind the cylinder. These eddies stretch downstream more and more with Re increasing to about 40.

When Re exceeds about 40, striking changes of the flow pattern occur downstream of the cylinder. The twin eddies are no longer able to attach to the cylinder, and each of them moves off downstream alternately with a definite period. The result is that two rows of discrete vortices, known as the Karman vortex street, are produced downstream of the cylinder, as shown in Fig. 7. From the viewpoint of wake instability, von Karman [1] proved that the vortex street can exist in neutral equilibrium only when $h/a = 0.281$ (Fig. 8).

The frequency of the vortex shedding f is specified by the nondimensional parameter

$$St = \frac{fd}{U}$$

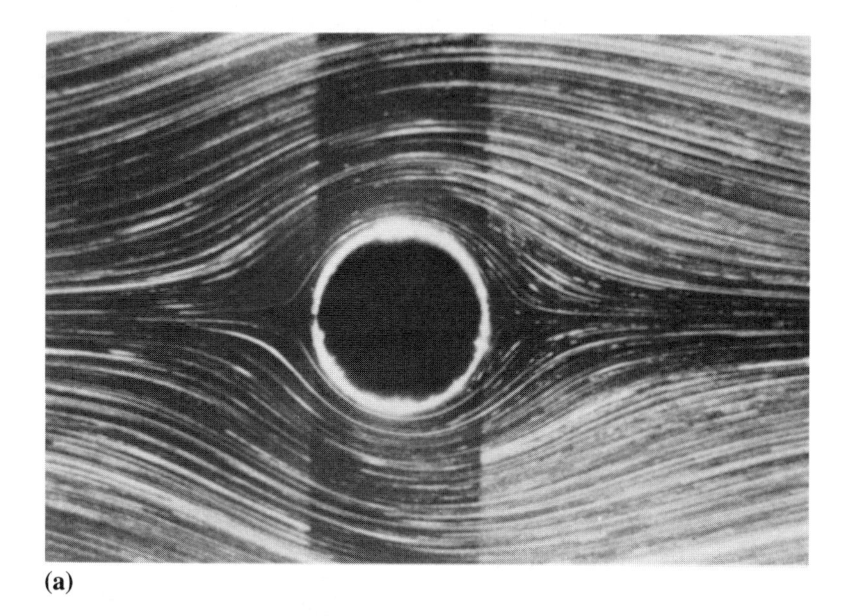

(a)

(b)

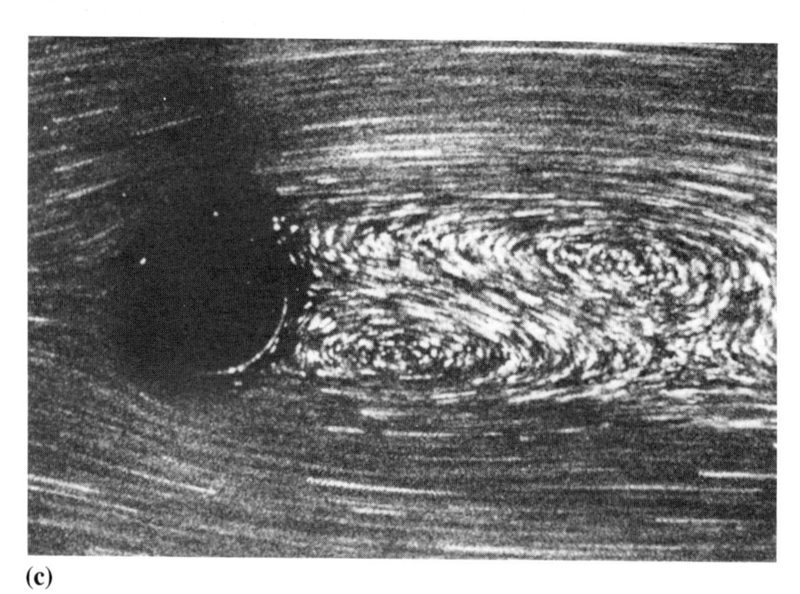

(c)

Fig. 6 Flow past a circular cylinder: (*a*) Re = 1.1; (*b*) Re = 26; (*c*) Re = 55 (courtesy of S. Taneda, Kyushu University).

Fig. 7 Karmam vortex behind a circular cylinder (Re = 140) (courtesy of S. Taneda, Kyushu University).

known as the Strouhal number. It is a function of Re, but given approximately by 0.2 in the wide range of the Reynolds number, from about 10^3 to 10^5.

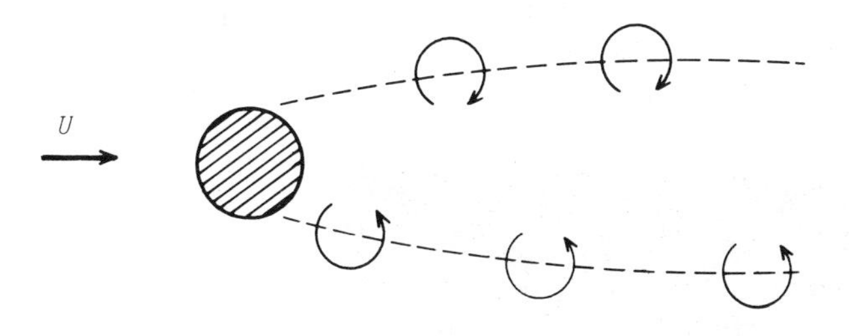

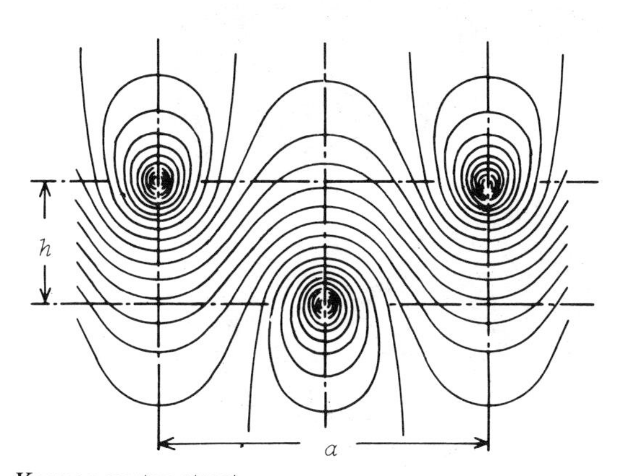

Fig. 8 Karman vortex street.

IV BOUNDARY LAYERS

A Concept

As mentioned in the case of a circular cylinder (Sec. III.B), the viscous effect of the fluid is gradually confined to a narrow region near the cylinder as Re is increased. For a sufficiently high Reynolds number, it becomes predominant only in the thin layer near the boundary, when the external flow outside it can be treated as inviscid.

This thin layer is known as the boundary layer (Prandtl, 1904), in which the velocity of fluid increases from 0 at the boundary (no-slip condition) to the value that corresponds to external frictionless flow, as shown in Fig. 9. It is useful to define the boundary layer thickness δ such that, for example,

$$u = 0.99u_0 \qquad \text{at} \qquad y = \delta$$

where u_0 is the asymptotic velocity of the external flow.

In the flow over the obstacle of characteristic length L, the inertia force is of the order $\rho U^2/L$, whereas the friction force is of the order $\mu U/\delta^2$. The equality of the order holds in the boundary layer; then

$$\frac{\delta}{L} \sim \left(\frac{UL}{\nu}\right)^{-1/2} = \mathrm{Re}^{-1/2}$$

This means that the boundary layer thickness is much smaller as compared with the scale of the obstacle for high Reynolds number.

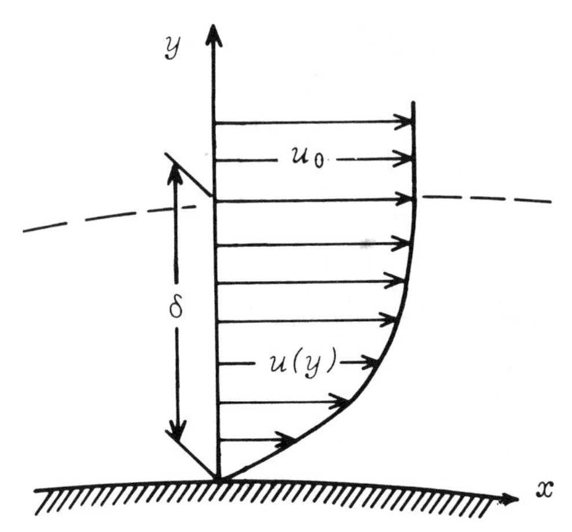

Fig. 9 Boundary layer.

The pressure in the boundary layer is practically constant in a direction normal to the boundary and given by that of the external flow.

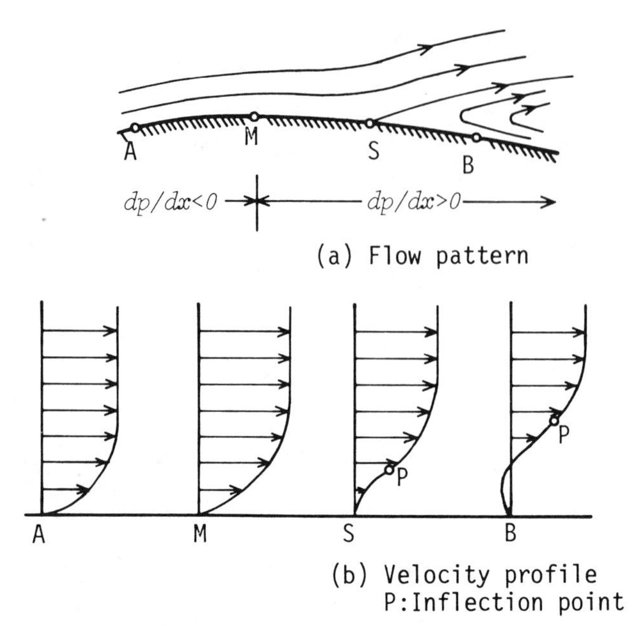

Fig. 10 Separation of decelerated flow: (*a*) Flow pattern; (*b*) velocity profile (P = inflection point).

B Separation

In the immediate vicinity of the boundary, the inertial force can be ignored as compared with the friction force, so the momentum theorem gives

$$\frac{dp}{dx} = \left.\frac{\partial \tau}{\partial y}\right|_{y=0} = \mu \left.\frac{\partial^2 u}{\partial y^2}\right|_{y=0}$$

Then, the curvature of the velocity profile there depends only on the pressure gradient, while at the edge of the boundary layer it is always negative. Hence, for a flow with decreasing pressure (accelerated flow, $\partial p/\partial x < 0$) $\partial^2 u/\partial y^2 < 0$ over the whole width of the boundary layer, whereas for a flow with increasing pressure (decelerated flow, $\partial p/\partial x > 0$) $\partial^2 u/\partial y^2$ changes its sign in the boundary layer. As a result, in a decelerated flow there may exist a point for which $\partial u/\partial y|_{y=0} = 0$, at which the fluid close to the boundary is not able to travel along the boundary and separates from it, as shown in Fig. 10. This is called separation of flow. The point of separation is defined as a point of transition from forward to reverse flow in the immediate vicinity of the boundary, or $\partial u/\partial y|_{y=0} = 0$.

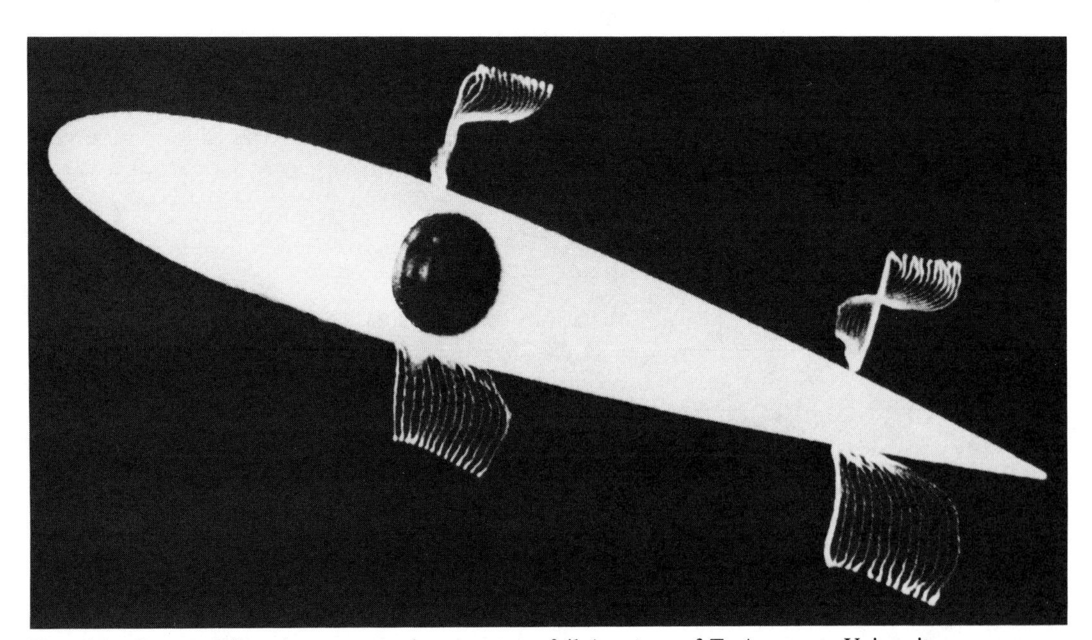

Fig. 11 Reversed flow due to separation on an aerofoil (courtesy of T. Asanuma, University of Tokyo).

The overall flow pattern depends greatly on whether the separation occurs or not. The flow upstream of the separation point is not much affected by separation, but the flow downstream of the separation point changes entirely, as seen in Fig. 6. Figure 11 gives another example, that is, the flow past an aerofoil inclined to the flow direction, showing reverse flow behind the separation point on the upper surface of the aerofoil.

C Transition of Boundary Layer

As mentioned in Sec. III.B, the flow in a circular pipe changes from the laminar to the turbulent state for a Reynolds number over 2300. This leads to an inference that the transition of the boundary layer also occurs depending on a Reynolds number $\mathrm{Re} = u_0\delta/\nu$.

The stability calculations for boundary layers show that the critical Reynolds number depends not only on the velocity profile but also on the pressure gradient. An approximate analysis gives the critical Reynolds number as a function of shape factor, as shown in Fig. 12 [2], where the Reynolds number is defined as $u_0\delta^*/\nu$, using the displacement thickness $\delta^* = \int_0^\delta (1 - u/u_0)\,dy$ and the shape factor $\Lambda = (\delta^2/\nu)(du_0/dx)$. It is noted that the critical Reynolds number increases with decreasing pressure ($\Lambda > 0$) and decreases with adverse pressure gradient ($\Lambda < 0$).

Now we consider the simplest case of the zero-pressure gradient. Suppose that a flat plate of negligible thickness is placed in a uniform flow of velocity U with no inclination. Downstream of the leading edge, the laminar boundary layer grows up along the plate surface as $\delta/x \sim (Ux/\nu)^{-1/2}$ (Sec. IV.A), and the transition to turbulent boundary layer occurs at a point x corresponding to the critical Reynolds number $\mathrm{Re}_{crit} = U\delta^*/\nu = 645$. In the developed laminar boundary layer, the velocity profile is parabolic. In the turbulent boundary layer, the momentum transfer of the fluid across it occurs to yield an additional shear force $-\rho\overline{u'v'}$ due to the turbulence (u', v'), which is known as the Reynolds stress. Then, in the turbulent boundary layer

$$\tau = \mu \frac{\partial u}{\partial y} - \rho\overline{u'v'}$$

This turbulent mixing makes the velocity profile flat such that

$$\frac{u}{u_0} = \left(\frac{y}{\delta}\right)^{1/7} \qquad \text{(1/7th Power Law)}$$

The boundary layer transition brings about a marked change in the separation as well as the pressure distribution, as seen in Fig. 13 for a circular cylinder. The boundary layer transition forces the separation point to shift downstream remarkably, causing an abrupt decrease in the drag at a definite Reynolds number (critical Reynolds number), as shown in Fig. 14.

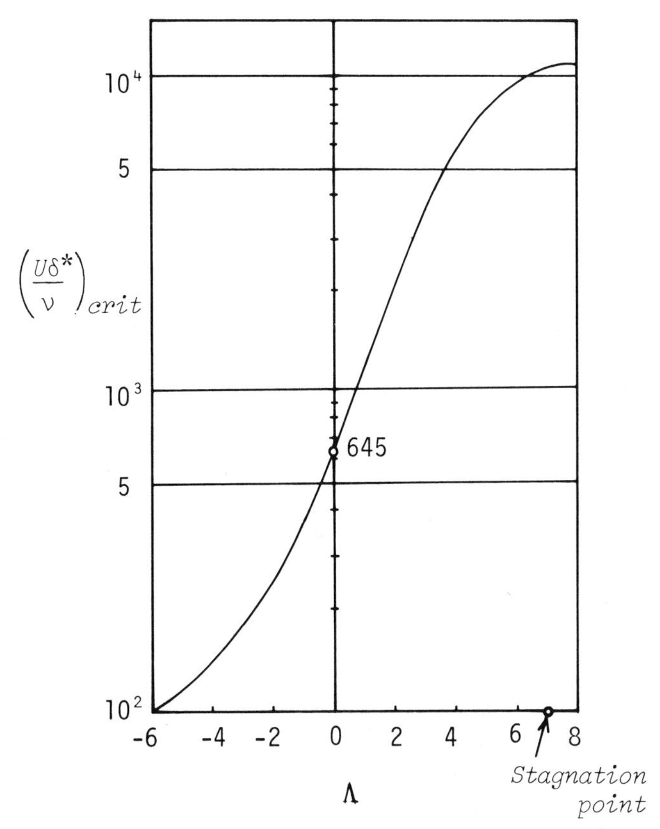

Fig. 12 Critical Reynolds number of boundary layer (after H. Schlichting [2]).

D Wakes and Jets

1 Wakes The flow past an obstacle of finite length coalesces again at its rear end and forms a region of strongly decelerated flow extending downstream (so-called wake). In the region far downstream, the velocity-defect profile of the wake becomes similar regardless of the geometry of the obstacle. When the Reynolds number is high, the velocity defect in the wake (Fig. 15) is given such that

$$\frac{u_1}{u_{1_{max}}} = f_1(\eta) \qquad \eta = \frac{y}{b}$$

$$u_{1_{max}} \propto \frac{db}{dx} \propto x^{-\alpha}$$

where b is the half width of the wake and $\alpha = 1/2$ and $2/3$ for two-dimensional and circular wakes, respectively.

This velocity profile has inflexion points, so that the wake is naturally unstable and the Karman vortex street is observed even in the wake of a flat plate.

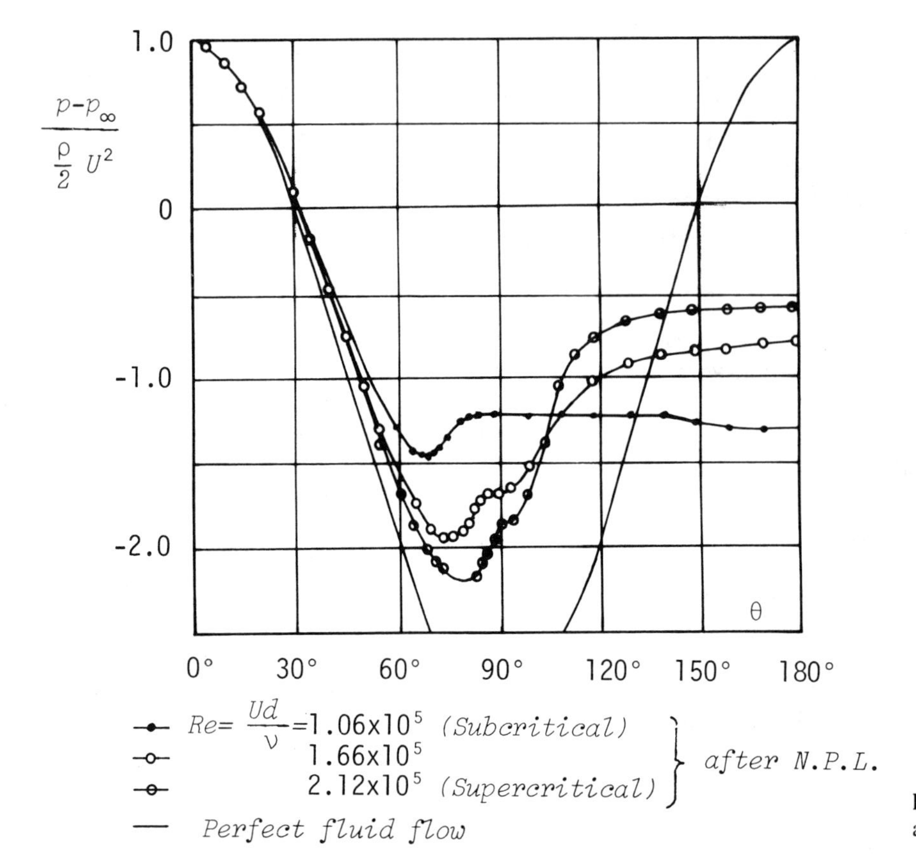

Fig. 13 Pressure distribution around a circular cylinder.

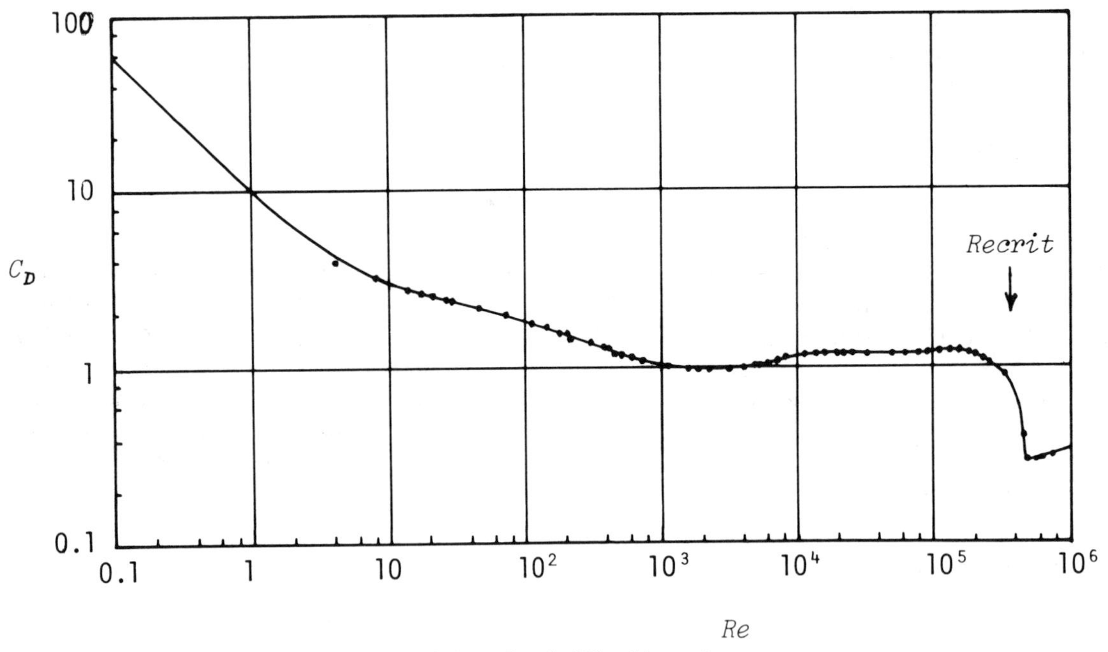

Fig. 14 Drag coefficient for a circular cylinder (after C. Wieselsberger).

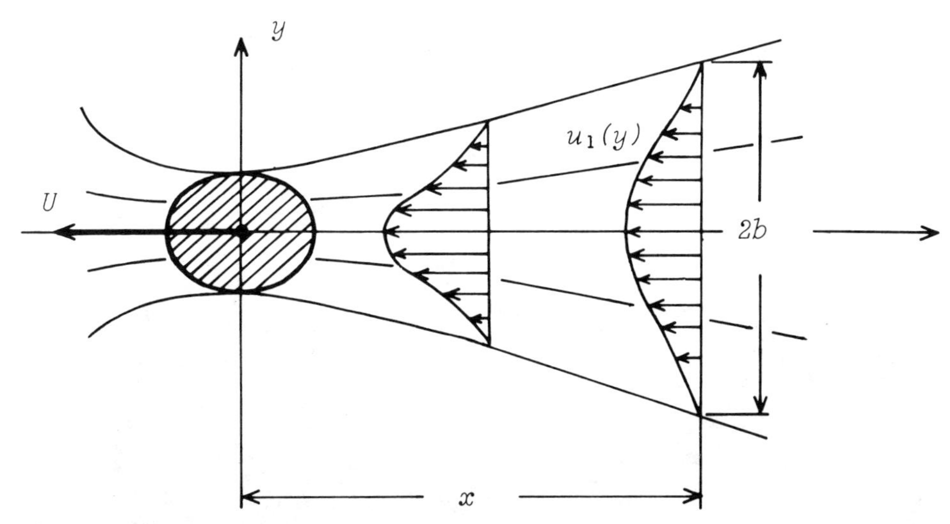

Fig. 15 Wake.

2 Jets Similarly for the case of a jet, in which fluid is ejected from an orifice into a stationary fluid of the same kind, the following relations hold.

$$\frac{u_1}{u_{1_{\max}}} = f_2(\eta)$$

$$u_{1_{\max}} \propto x^{-\beta}$$

$$b \propto x$$

where b is the half width of jet and $\beta = 1/2$ and 1 for two-dimensional and circular jets, respectively.

V COMPRESSIBLE FLOWS

A Introduction

Let us first observe a small body flying at a constant speed U in stationary air (Fig. 16). This projectile creates small disturbance continuously at each location it passes through. This disturbance propagates as a pressure pulse into all directions of the surrounding air at the local speed of sound a. When the flying speed is less than the sonic speed, the wave front does not overtake the preceding one. This is the case of subsonic flow, in which the Mach number $M = U/a$ is less than unity. Incompressible flow is a limiting case of $M = 0$, because of infinite sonic speed.

On the other hand, when the projectile flies at a speed over the sonic one, all the pressure disturbances exist only inside a confined cone and silence outside it. This cone is known as the Mach cone, and the semiangle of its apex is the Mach angle. Figure 17 shows a schlieren photograph of the Mach waves generated by a thin wing of wedge profile.

Figure 18 shows the change of the flow past a thin aerofoil, depending on the Mach number. With increasing Mach number of the free stream, a supersonic flow region appears on a part of the aerofoil surface. The Mach number at which

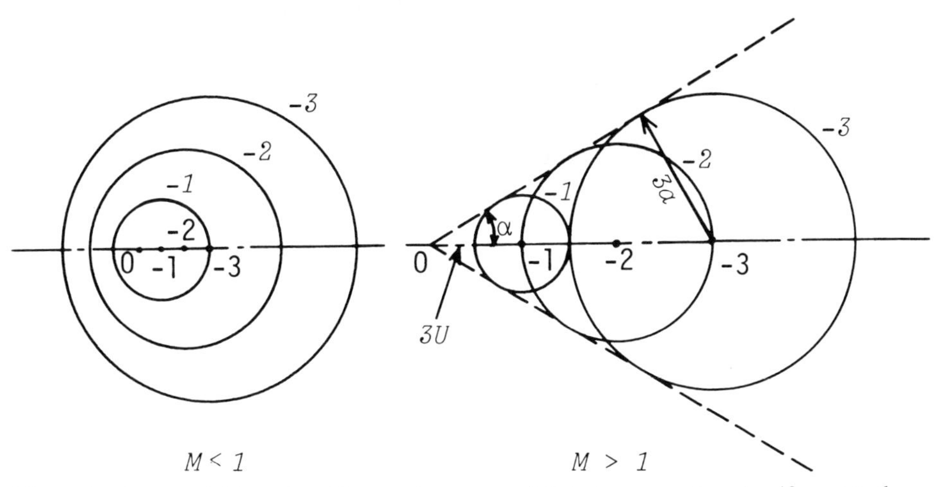

Fig. 16 Pressure field produced by a point source of disturbance moving at uniform speed.

Fig. 17 Schlieren photograph of a thin wing placed in a supersonic speed (original photo in color, courtesy of E. Ohta, Waseda University).

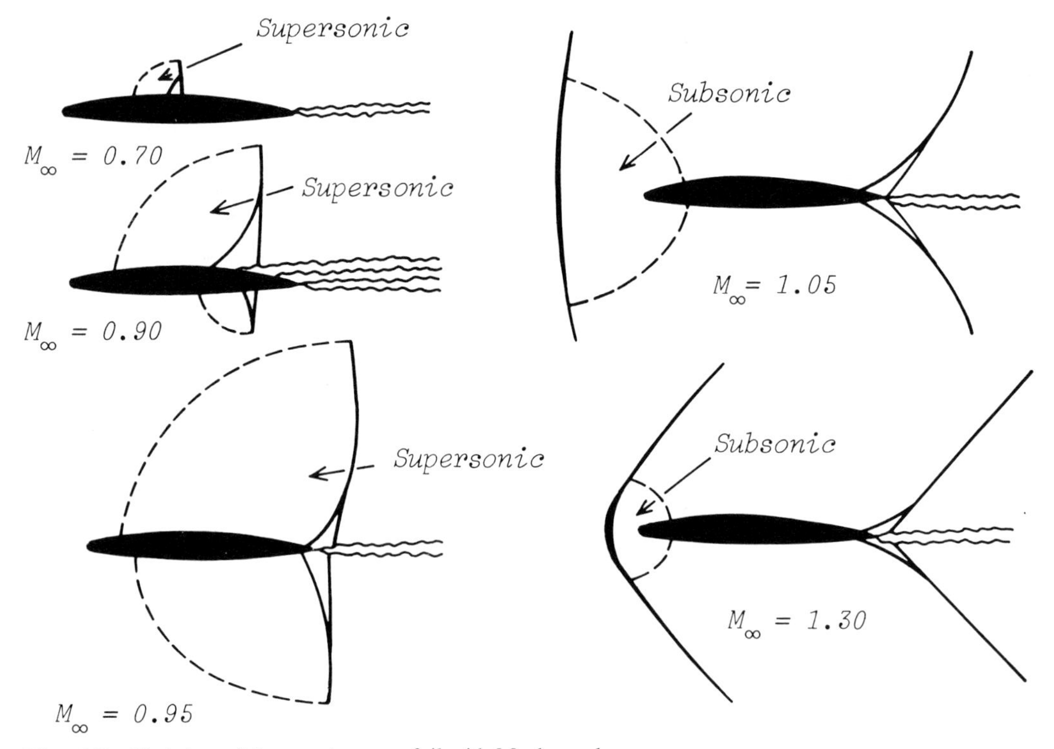

Fig. 18 Variation of flow past an aerofoil with Mach number.

the flow velocity becomes exactly sonic at a point on the aerofoil is called the critical Mach number. With further increase of the Mach number, shocks which produce discontinuous changes appear at the lee side of the supersonic region. With increasing Mach number, the shocks shift downstream, getting stronger until they reach the trailing edge of the aerofoil. Then, the flow becomes wholly supersonic. The subsonic reigon around the leading edge is caused by the thickness effect of the aerofoil, when the front shock is detached from the aerofoil (see Sec. V.C).

B Subsonic Flows

We consider a thin body, such as an aircraft wing, placed in a compressible uniform flow of velocity U. When the flow is subsonic or the Mach number is less than unity, the flow pattern is not much different from that in incompressible flow. This leads to the similarity laws which relate a compressible flow to an incompressible flow.

If an aerofoil of the same profile is placed in compressible flow at the same angle of attack, the pressure coefficient, lift coefficient, and moment coefficient are all affected by the Mach number in proportion to the factor $1/\sqrt{1 - M^2}$. For example, the pressure coefficients $C_p = (p - p_\infty)/\frac{1}{2}\rho U^2$, in which p_∞ is the pressure of the free stream, are related by

$$C_{p_M} = \frac{1}{\sqrt{1 - M^2}} C_{p_0} \qquad \text{(Prandtl–Glauert Rule)}$$

When the Mach number approaches unity, a local supersonic region appears, which is followed by the shock that brings about an unfavorable increase of drag. This effect is smaller for an aerofoil of finite span because of the three-dimensional "relief effect." On the other hand, if the aerofoil is swept back with an angle σ, the compressibility effect appears only in the direction normal to the aerofoil axis. The effective Mach number for the aerofoil is then M cos σ. These two effects result in the use of the swept-back wing having low aspect ratio for the aircraft in high subsonic flight.

C Supersonic Flows

1 Small Perturbations In two-dimensional supersonic flow, small changes of flow occur uniformly along the Mach waves, which correspond to the wave front of the Mach cone. In Fig. 19, only the momentum of the flow in the direction normal to the Mach wave is affected, so the velocity change across the Mach wave is

$$u = -v \tan \beta$$

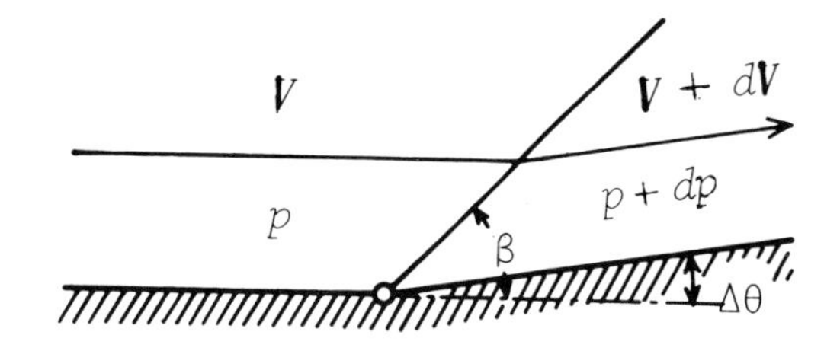

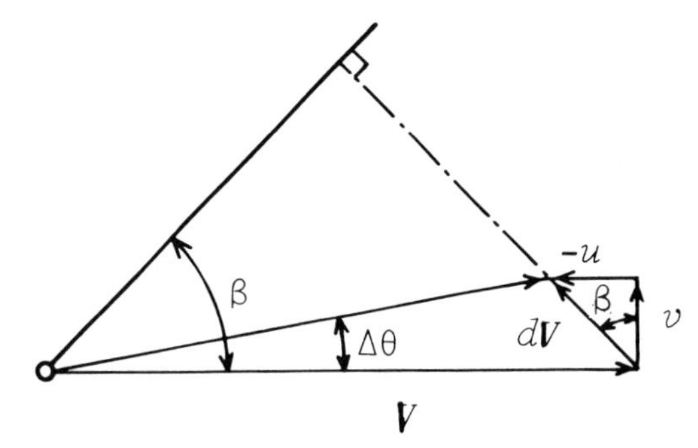

Fig. 19 Flow across Mach wave.

when the flow is deflected by an angle of

$$\Delta\theta = \frac{v}{U} \qquad |\mathbf{V}| = U$$

Because of small perturbations, it can be considered that the energy of the flow is conserved across the Mach wave, and then the Bernoulli equation gives the pressure rise as

$$p - p_\infty = -\rho U u$$

Hence, the pressure coefficient is given by

$$C_p = \frac{p - p_\infty}{\frac{1}{2}\rho U^2} = \frac{2\,\Delta\theta}{\sqrt{M^2 - 1}}$$

Example 1: Flat-plate Aerofoil When an aerofoil of flat plate is placed in uniform supersonic flow at a small angle of attack α (Fig. 20), the flow along the upper surface is deflected first downward and then upward by the angle α across the Mach waves, which emanate from the leading and trailing edges, respectively.

The surface pressure is uniform along the chord, given as

$$p - p_\infty = \mp \frac{\kappa p_\infty M^2}{\sqrt{M^2 - 1}} \alpha$$

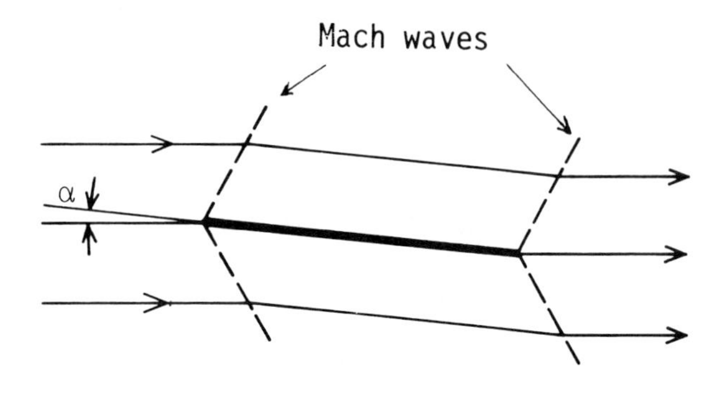

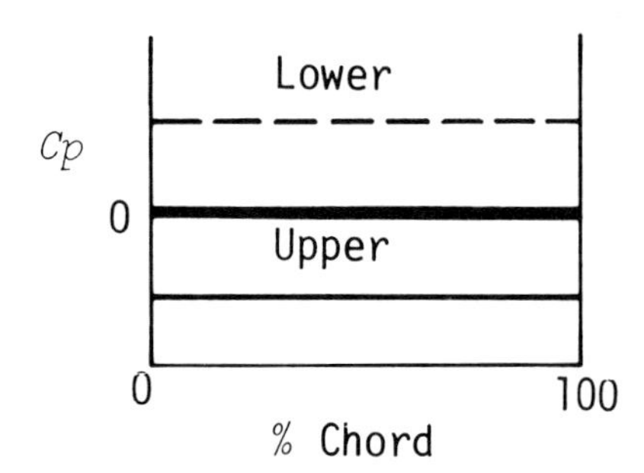

Fig. 20 Supersonic flow past a flat plate.

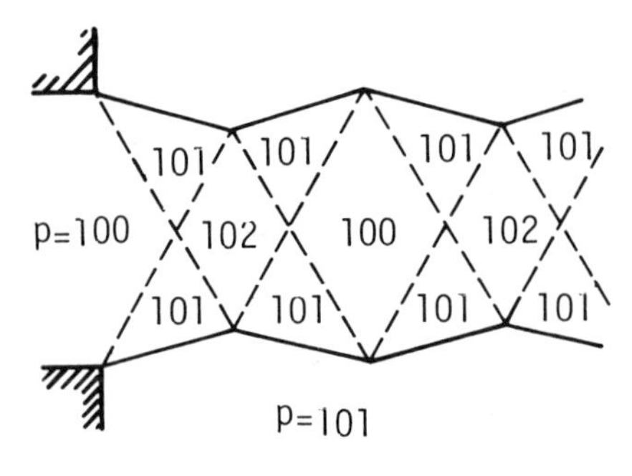

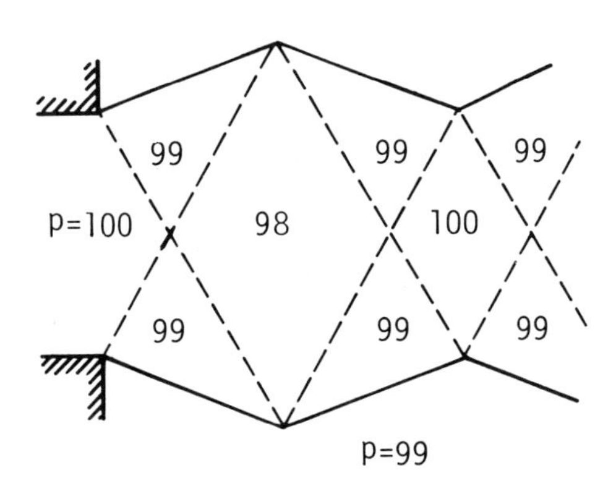

Fig. 21 Flow pattern at exit of two-dimensional nozzle: (*a*) Overexpansion; (*b*) underexpansion.

where the minus and plus signs indicate the upper and lower surfaces. The lift coefficient is then computed as

$$C_L = \frac{4\alpha}{\sqrt{M^2 - 1}}$$

Example 2: Overexpanded and Underexpanded Jets When the exit pressure of the supersonic nozzle is lower or higher than the back pressure of the exhaust chamber, it is said that the jet is overexpanded or underexpanded.

In the overexpanded jet (Fig. 21*a*), the compression waves originate at the corners of the nozzle exit to adjust the pressure difference at the jet boundaries. The compression waves are then reflected as the expansion waves at the jet boundaries, and vice versa. Hence, the wave and flow patterns undergo cyclic changes along the jet axis. Figure 21*b* shows the underexpanded jet.

2 Normal Shock When the exit pressure of the supersonic nozzle is much lower than the back pressure, the adjustments of the pressure difference (Fig. 21) cannot be made in the exhaust chamber but are done by normal shock, which stands "normal" to the oncoming supersonic flow inside the nozzle.

Across the normal shock (Fig. 22*a*) exists discontinuity in pressure, density, velocity, and entropy, whereas the stagnation temperature is conserved. The Mach number fore and aft of the normal shock are uniquely related with each other as

$$M_2^2 = \frac{M_1^2 + 2/(\kappa - 1)}{2\kappa/(\kappa - 1)M_1^2 - 1}$$

The pressure and temperature ratios are given as

$$\frac{p_2}{p_1} = \frac{2\kappa}{\kappa + 1}\left(M_1^2 - \frac{\kappa - 1}{2\kappa}\right)$$

$$\frac{T_2}{T_1} = \frac{2(\kappa - 1)}{(\kappa + 1)^2}\left(\frac{\kappa - 1}{2} + \frac{1}{M_1^2}\right)\left(\frac{2\kappa}{\kappa - 1} - \frac{1}{M_1^2}\right)$$

3 Oblique Shocks The oblique shock stands "oblique" to the oncoming supersonic flow (Fig. 22*b*). No change occurs in the tangential velocity (V_t) across it. So the above relations for the normal shock can be applied for the oblique shock by using the Mach number of the normal velocity, $M_1 \sin \sigma$, instead of the approach Mach number M_1.

For a given inlet Mach number, when the deflection angle δ of flow is moderate, either weak or strong shock may exist, depending on the back pressure, as shown in Fig. 23. When the back pressure is relatively small, the weak shock occurs with a relatively small angle, a relatively small pressure ratio, and the downstream state usually supersonic. When the back pressure is relatively large, the strong shock occurs with a relatively large angle, a relatively large pressure ratio, and the downstream state usually subsonic.

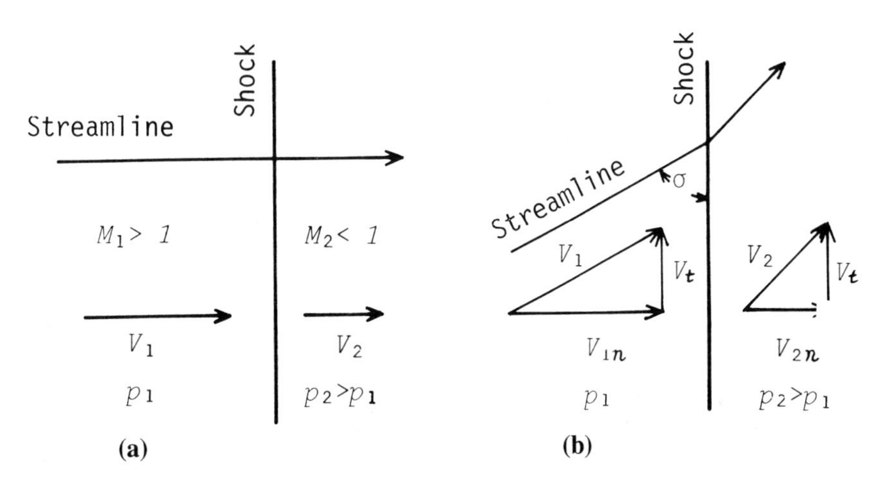

Fig. 22 Normal and oblique shocks: (*a*) Normal; (*b*) oblique.

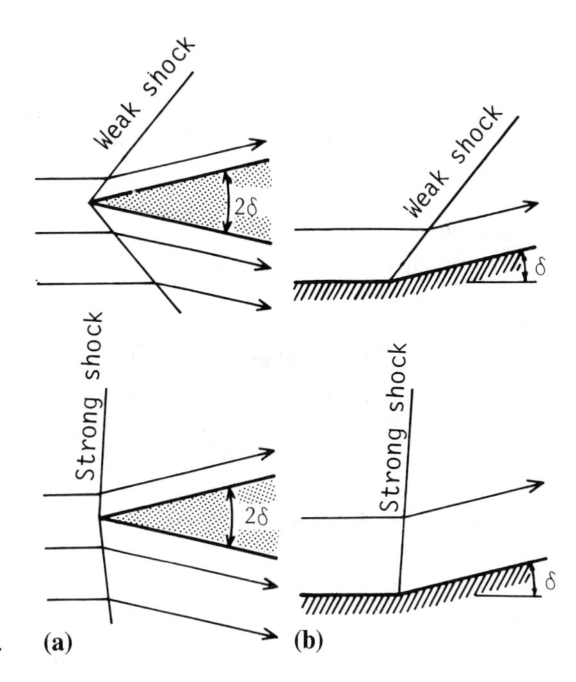

Fig. 23 Weak and strong shocks: (*a*) Weak; (*b*) strong.

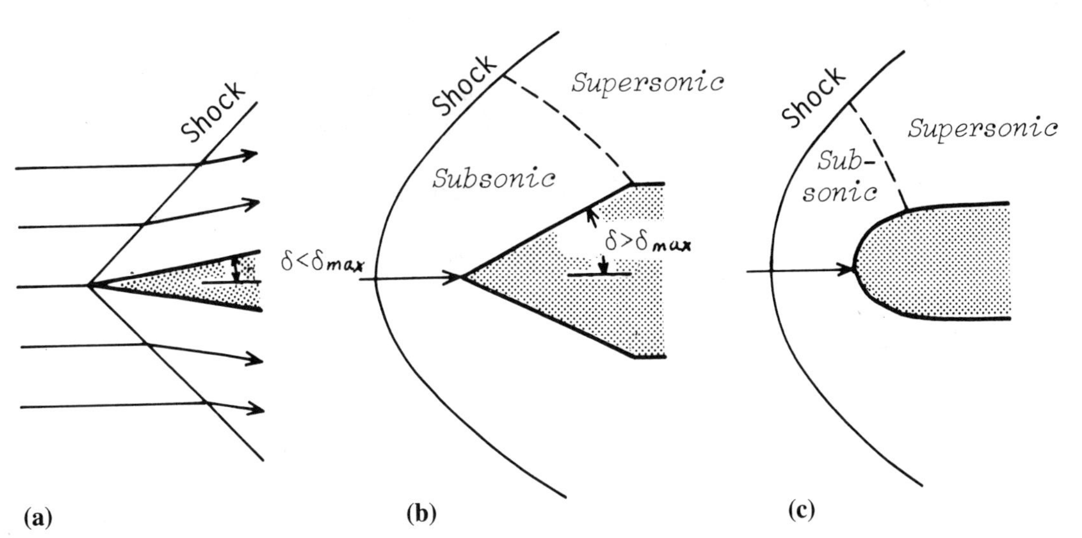

Fig. 24 Attached and detached shocks: (*a*) Attached shock on wedge; (*b*) detached shock on wedge; (*c*) detached shock on blunt body.

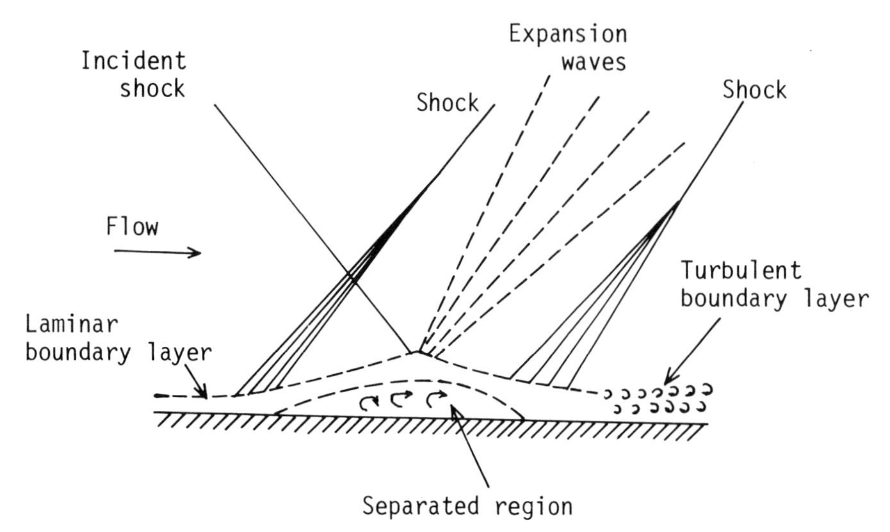

Fig. 25 Shock–boundary layer interaction.

With the deflection angle increased over a critical value, the shock cannot emanate from the corner of deflection, but it is detached upstream with the downstream state subsonic (Fig. 24).

For a vanishing deflection angle, the weak and strong shocks correspond to the Mach wave and the normal shock, respectively.

4 Shock–Boundary Layer Interaction When the shock occurs near the wall or is incident on the wall, the discontinuous pressure rise in the shock violently distorts the boundary layer flow, often to cause the transition and separation.

Figure 25 shows an example of oblique shock incident into the boundary layer on the flat plate. The incident shock is reflected as an oblique shock, accompanied by other shocks due to the thickening of the boundary layer.

D Analogy between Compressible Flow and Shallow Water Flow

An analogy exists between a two-dimensional compressible flow and a shallow water flow with free surface, as shown in Fig. 26 and Table 3. The shock wave corresponds to the hydraulic jump.

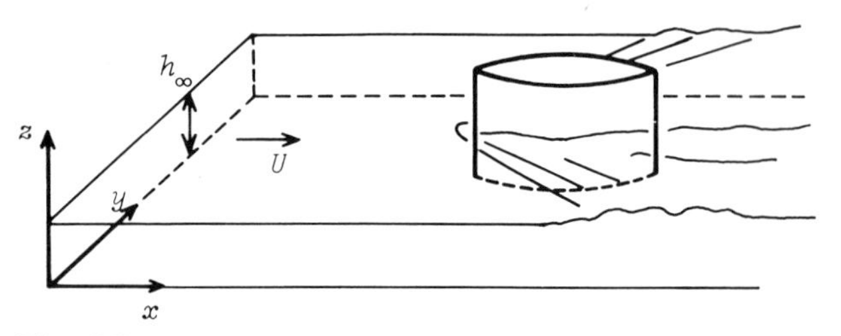

Fig. 26 Hydraulic analogy of supersonic flows.

Table 3 Hydraulic Analogy of Supersonic Flow

Shallow Water Flow		Supersonic Flow ($\kappa = 2$)	
Depth	$\dfrac{h}{h_\infty}$	Density	$\dfrac{\rho}{\rho_\infty}$
Square of depth	$\left(\dfrac{h}{h_\infty}\right)^2$	Pressure	$\dfrac{p}{p_\infty}$
Propagating velocity of long wave	$\sqrt{gh_\infty}$	Speed of sound	$\sqrt{\kappa \dfrac{p_\infty}{\rho_\infty}}$

By means of this analogy, the shallow water flow is used for the qualitative observation on a compressible flow. The depth of shallow water should be less than 6–7 mm for a good analogy [3].

Figure 27 shows an application of the hydraulic analogy for the flow through a turbine cascade, compared with a photograph taken by the schlieren method.

VI FLOWS UNDER BUOYANCY EFFECTS

A Stratified Flows

When a flow has a density distribution in a definite direction, it is called *stratified flow*. The atmosphere and the sea on the earth under the effects of gravity are typical examples.

Now, we consider a stratified flow that consists of fluids of two kinds generating the boundary between them. In such a case, where the heavier fluid lies stably beneath the lighter one, it is difficult for perturbations to occur in the vertical direction but they are liable to occur in the equidensity planes. Hence, the so-called internal wave as well as blocking effect occur at the boundary of two fluids.

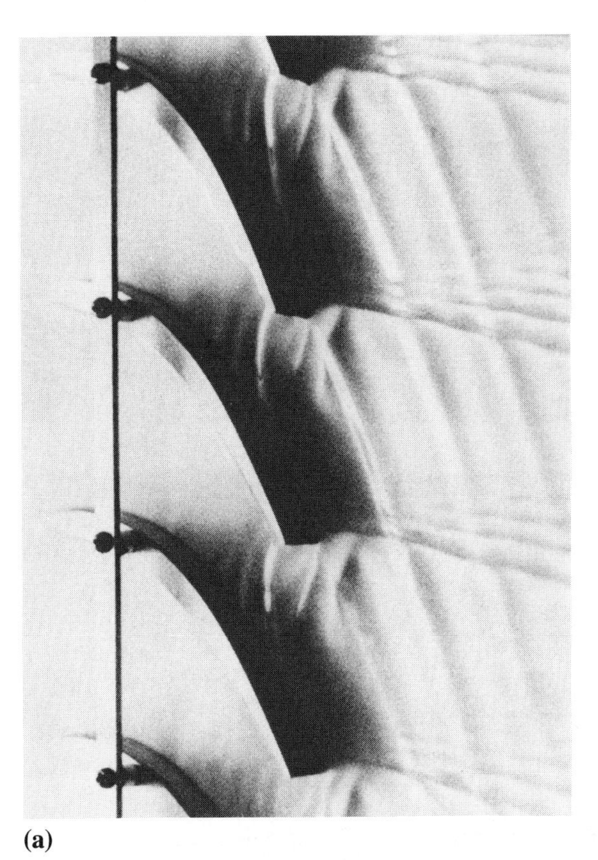

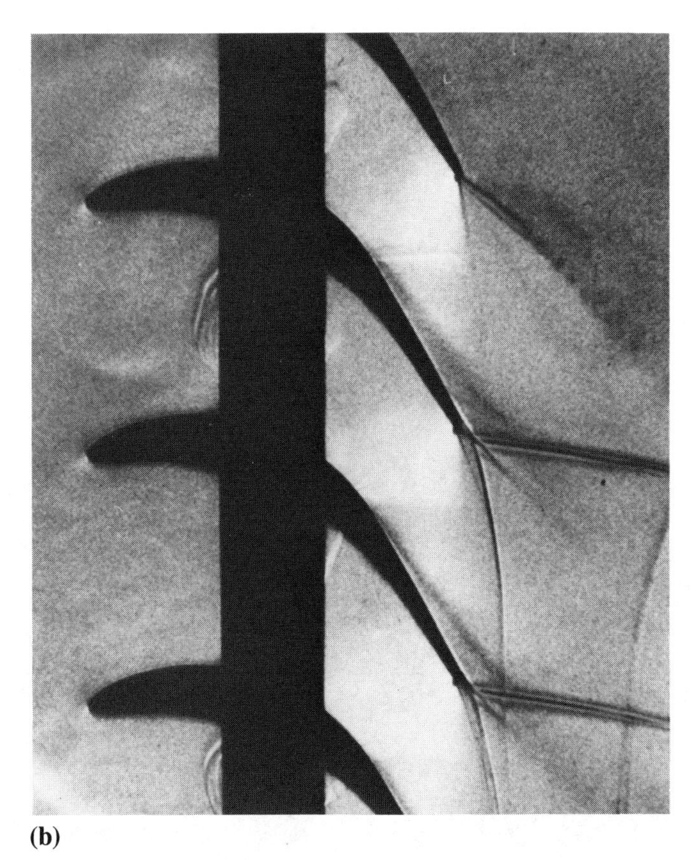

(a) (b)

Fig. 27 Application of hydraulic analogy to a flow through a turbine cascade: (*a*) Observation of flow through a turbine cascade in shallow water table (courtesy of Y. Saito and M. Honjo, Mitsubishi Heavy Industries, Ltd.); (*b*) schlieren photograph of flow through a turbine cascade (courtesy of T. Ikeda and T. Mitsuzaka, Toshiba Corporation).

The stability of the internal wave depends on the Richardson number, which is defined as the ratio of the buoyancy and inertia forces:

$$\mathrm{Ri} = -\frac{(g/\rho)(d\rho/dy)}{(dU/dy)^2}$$

For small Ri, the internal wave becomes unstable, which leads to the mixing of two fluids and the turbulent flow.

When a sphere is placed midway at the boundary, the flow pattern in the boundary plane is similiar to that of the two-dimensional circular cylinder. The blocking effect is observed typically when a cylinder is placed horizontally with the axis normal to the flow in the boundary plane. The approaching flow in the boundary plane is blocked by the cylinder in a one-dimensional manner.

B Convections

Temperature that is not uniform causes not only the flow of heat (heat transfer, see Sec. II.C) but also a flow to be generated by the buoyancy effects due to the difference of the fluid density (convection). The rate of the density change against the temperature variation is given by the coefficient of thermal expansion

$$\beta = -\frac{1}{\rho}\frac{d\rho}{dT}$$

In such a case, the motion of fluid depends on the ratio of the buoyancy and inertial forces. That is,

$$\frac{\text{Buoyancy force}}{\text{Inertial force}} = \frac{g\rho\beta\,\Delta T}{\pi U^2/L} = \left(\frac{g\beta L^3\,\Delta T}{\nu^2}\right)\left(\frac{\nu}{UL}\right)^2 = \frac{\mathrm{Gr}}{\mathrm{Re}^2}$$

where ΔT is the characteristic temperature difference. A nondimensional parameter $\mathrm{Gr} = g\beta L^3\,\Delta T/\nu^2$ is called the Grashof number. When the Grashof number is large enough for the above ratio to be of the order 1, the buoyancy effect becomes predominant in generating flows (natural convection). On the other hand, when the Reynolds number is large enough, the buoyancy effect can be neglected and the heat is conveyed by the flow itself (forced convection).

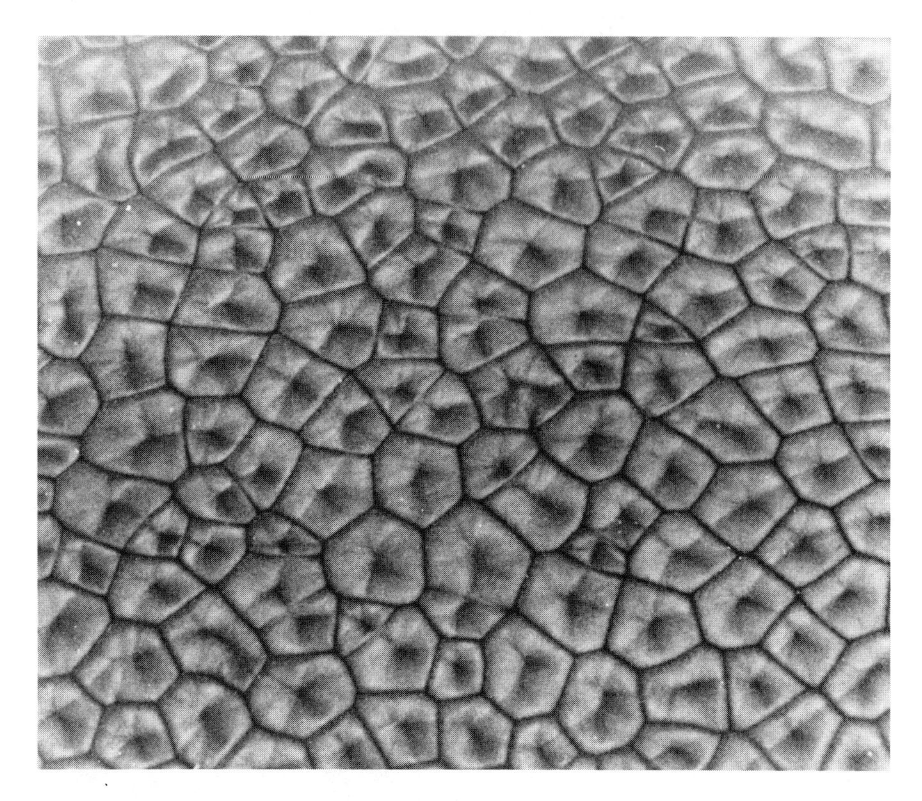

Fig. 28 Benard's cell (courtesy of R. Kimura, University of Tokyo).

Observe the heat balance in the elements of fluid, and we have

$$\frac{\text{Heat transferred}}{\text{Thermal inertia}} = \frac{k\,\Delta T/L^2}{\rho Cp\,\Delta TU/L} = \left(\frac{k}{\rho Cp}\right)\left(\frac{1}{UL}\right) = \left(\frac{\alpha}{\nu}\right)\left(\frac{\nu}{UL}\right) = \frac{1}{\text{Pr}}\frac{1}{\text{Re}}$$

where $\text{Pr} = \alpha/\nu$ is the Prandtl number, which has the magnitude of order 1 for gases. For high Re, the heat transfer is predominant only in the immediate neighborhood of the heated/cooled wall or in the thermal boundary layer. The thickness of the thermal boundary layer is given by

$$\frac{\delta_t}{L} \propto (\text{Pr} \cdot \text{Re})^{-1/2}$$

For laminar boundary layer, the velocity–boundary layer thickness δ_v is nearly proportional to $\text{Re}^{-1/2}$ (see Sec. IV.A); then

$$\text{Pr} \propto \left(\frac{\delta_v}{\delta_t}\right)^2$$

In natural convection, whether the flow is stable or not depends principally on the Rayleigh number, defined by $\text{Ra} = \text{Gr} \cdot \text{Pr}$. We now consider a fluid contained between two large parallel plates separated by a small distance. When the lower plate is heated so that $\text{Ra} > 1700$, the so-called Benard convection occurs, which generates the circulating flows in hexagonal, honeycomblike, cylindrical cells as shown in Fig. 28. The flow becomes irregular for $\text{Ra} > 4.7 \times 10^4$.

VII BASIC ITEMS RELATED TO FLOW VISUALIZATION

A Vorticity and Circulation

When the inviscid fluid that is stationary first begins to move, the flow is "irrotational" forever, and in two-dimensional flow holds

$$\zeta = \frac{\partial v}{\partial x} - \frac{\partial u}{\partial y} = 0$$

On the contrary, when ζ is not 0, the flow is "rotational." This quantity ζ is called vorticity.

Now, as shown in Fig. 29, we consider a vortex in an infinite two-dimensional field, which rotates about its axis with a velocity proportional to the radius r, or with an angular velocity ω, in the core of radius r_0, and with a velocity inversely proportional to r outside it. Then, the vorticity is finite and uniformly distributed inside the core and zero outside it. The vorticity can be observed by putting small elements in the flow, as shown in Fig. 30. The strength of this vortex is given by $\Gamma_v = 2\pi r_0^2\omega$.

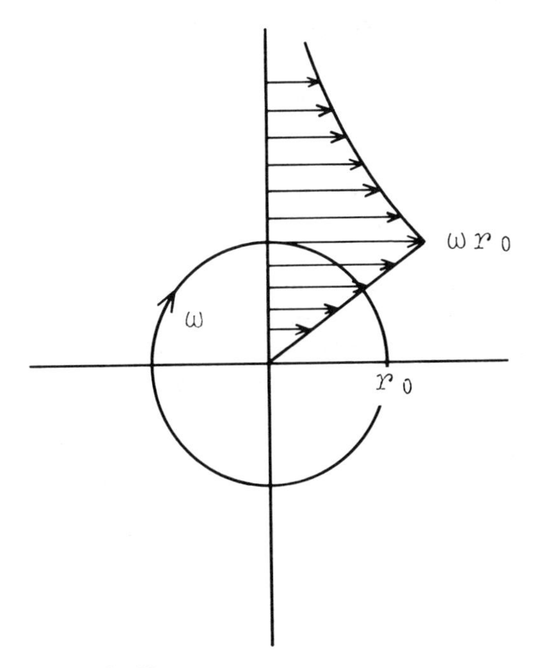

Fig. 29 Vortex.

The circulation is defined as the line integral of the velocity around any closed curve.

$$\Gamma = \oint_C \boldsymbol{V} \cdot \boldsymbol{n}\, dl = \int_S \zeta\, dS$$

The circulation around a vortex is then $\Gamma = \Gamma_v$.

In a perfect fluid flow, the vortex is always occupied by the same part of the fluid, and its strength is invariable (Helmholtz's theorem). The circulation along the closed line which moves with the fluid remains constant with respect to time (Kelvin's theorem).

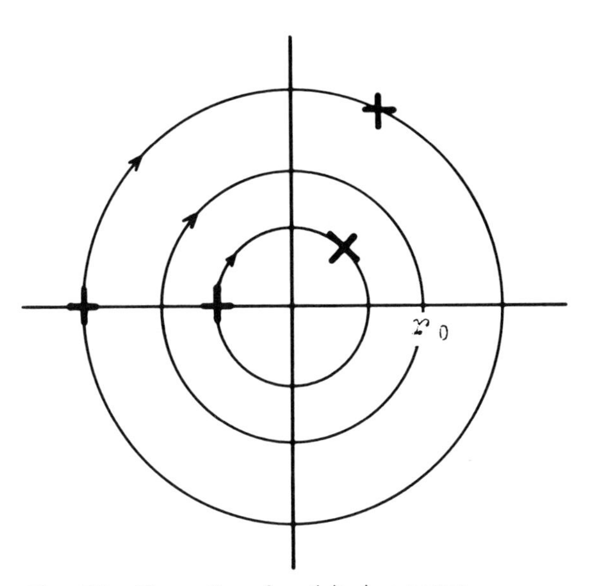

Fig. 30 Observation of vorticity in a vortex.

The viscous flow is usually rotational, and it is often investigated by using a model of inviscid rotational flow. As a typical example, a flow past a circular cylinder will be considered. As described in Sec. III.B the circular cylinder sheds the vortices of clockwise and counter-clockwise senses alternately in the range of moderately high Reynolds number. The effects of the vortex street behind the cylinder have been studied by replacing it with the rows of the inviscid discrete vortices.

B Pathlines, Streaklines, and Streamlines

Let us trace a particle of the flowing fluid, and we can draw a line in space as the locus of the particle. This trajectory is called pathline.

Next, at a given time, we look at all fluid elements that have passed through a given fixed point and draw a line connecting them. This is called streakline.

Further, at a given time, we observe the flow directions of all fluid elements in the whole flow field, put small line segments tangential to the flow directions, and then we can

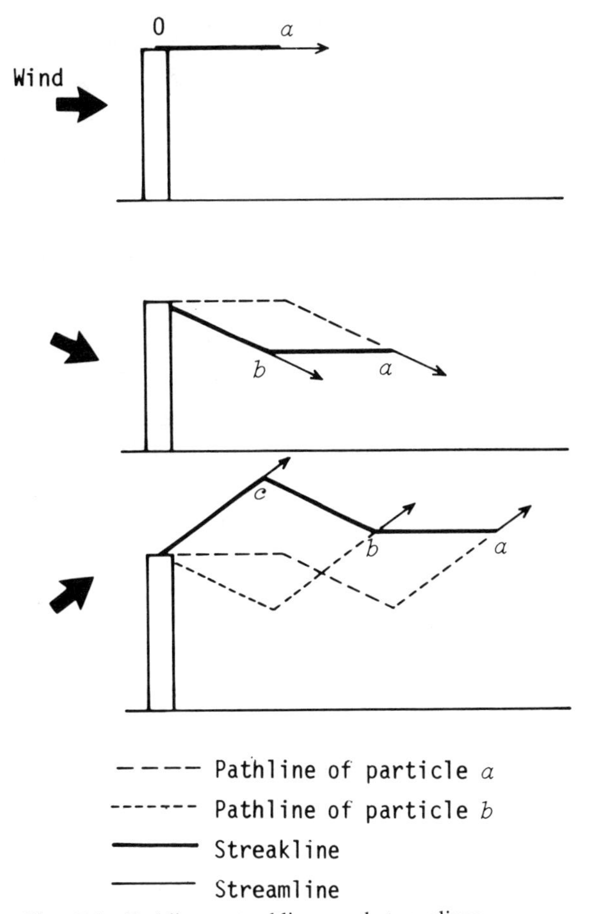

Fig. 31 Pathlines, streaklines, and streamlines.

draw lines as the envelopes for these segments, which satisfy the equation

$$\frac{dx}{u} = \frac{dy}{v} = \frac{dz}{w}$$

They are called streamlines, which can show the whole flow field at that instant.

These three are all identical in steady flow, but usually not in unsteady flow. Figure 31 illustrates the case of unsteady flow; the figure shows how the smoke emanating from a chimney is carried by the wind of changing direction with time. This suggests that caution must be taken in interpreting flow phenomena based on flow visualization techniques, as is seen in the following example [4]. We consider a case in which a parallel shear flow that corresponds to a half-jet ejected into a fluid at rest is perturbed by an unamplifying traveling sinusoidal wave. If a dye is injected near the inflection point of velocity profile ($y = 0$ in Fig. 32), the patterns of streaklines appear as if the flow develops into a row of discrete vortices (Fig. 32*b*), despite the fact that the wave itself is unamplifying and no discrete vortex exists (Fig. 32*a*). Observing the flow in the coordinate system moving with the local velocity, the per-

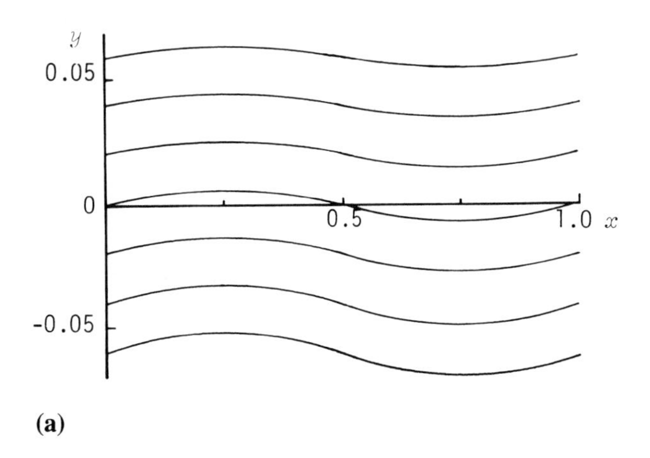

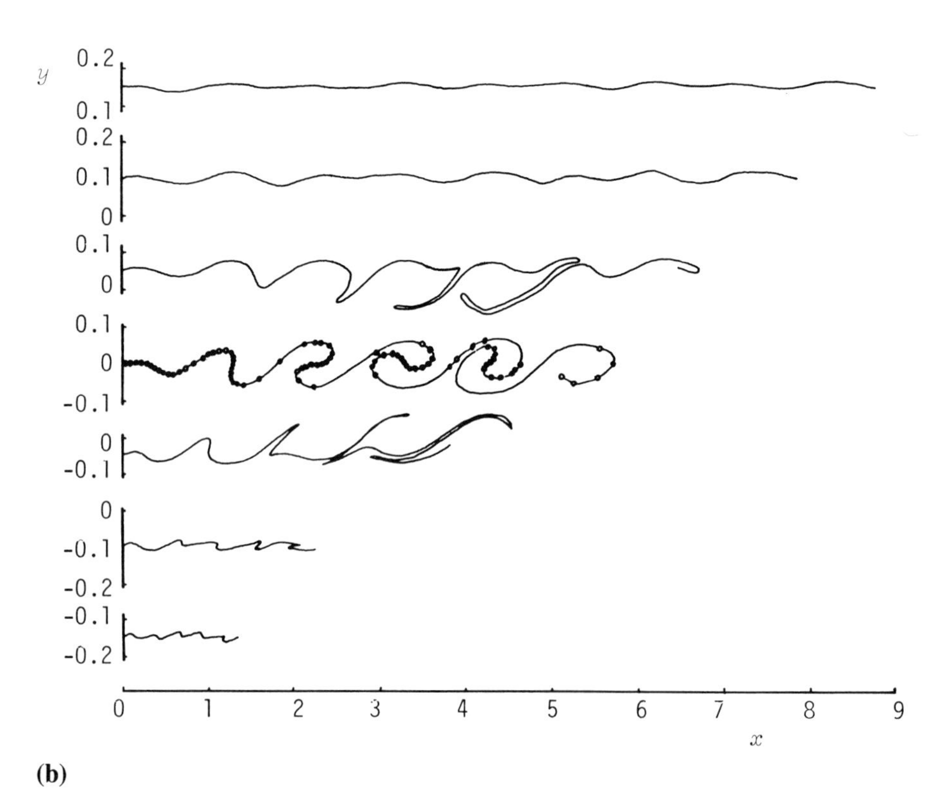

Fig. 32 Apparent vortices in a perturbed shear flow: (*a*) Streamlines; (*b*) Streaklines (after F. R. Hanna [4].

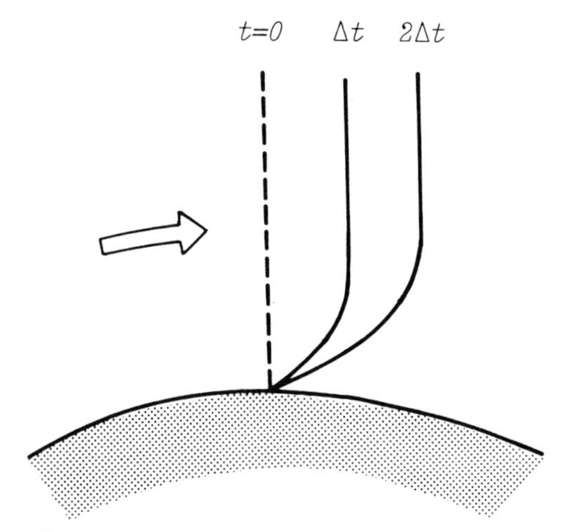

Fig. 33 Timeline.

turbation waves form the cat-eye streaklines. Hence, fluid particles trapped within these eyes keep rolling around the center of the eye so as to form the apparent discrete vortex.

There is another one called timeline, which is often used in flow visualization but is somewhat different from the above. If we draw an arbitrary line in the flow at a certain instant, it will be washed downstream with the marked fluid particles as time proceeds (Fig. 33). Observing this line with a definite time interval, we can use it as a clock marked in the fluid.

C Limiting Streamlines

The boundary layer developed on the three-dimensional body is usually skewed, that is, as shown in Fig. 34, the velocity vector changes its direction markedly in the boundary layer. The limiting streamlines are defined as the envelopes of the velocity vector in the immediate vicinity of the boundary.

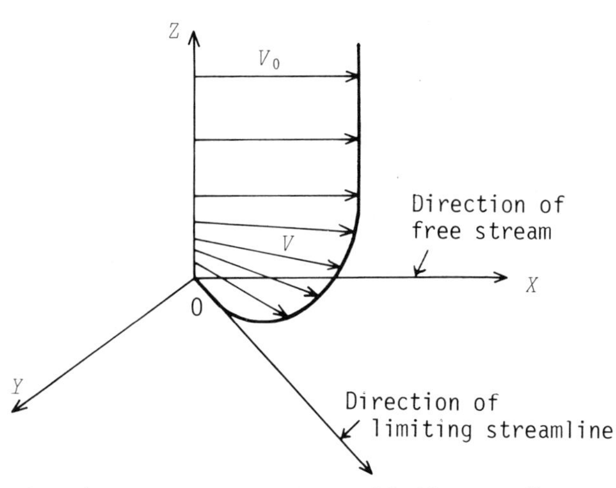

Fig. 34 Skewed boundary layer and limiting streamline.

Generally speaking, the limiting streamlines are not identical with the streamlines at the edge of the boundary layer, so that careful observation is indispensable. In the use of the oil-film method, furthermore, deviations occur between the limiting streamlines observed and those on the solid boundary, due to the flowing oil film. The flow directions are estimated as follows [5]:

At the wall surface:

$$\left(\frac{dy}{dx}\right)_{z=0} = \frac{\mu(\partial v/\partial z)_{z=0} - h(\partial p/\partial y)}{\mu(\partial u/\partial z)_{z=0} - h(\partial p/\partial x)}$$

At the oil surface:

$$\left(\frac{dy}{dx}\right)_{z=h} = \frac{2\mu(\partial v/\partial z)_{z=0} - h(\partial p/\partial y)}{2\mu(\partial u/\partial z)_{z=0} - h(\partial p/\partial x)}$$

where h is the thickness of the oil film. Figure 35 shows the deviation of the limiting streamlines along the surface of the swept-back aerofoil, either with or without oil film.

D Trajectories of Foreign Particles in Flows

Foreign particles added to the flow for flow visualization should have densities as close to that of the fluid as possible, so that their trajectories will coincide with the pathlines of the fluid particles.

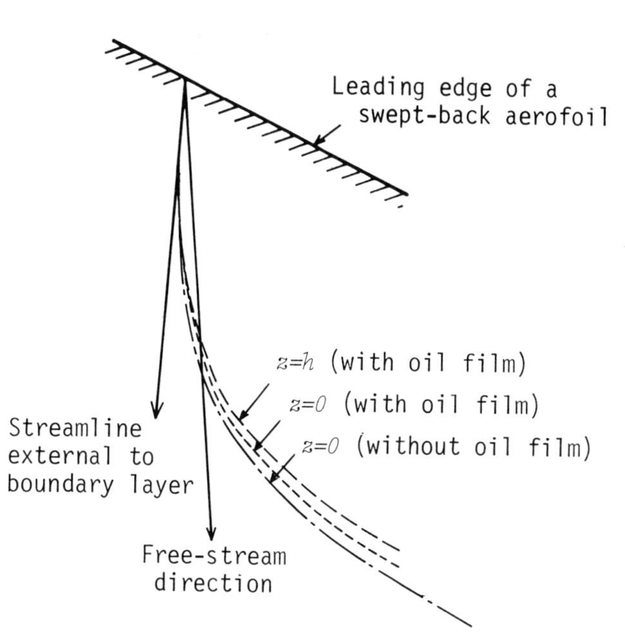

Fig. 35 Limiting streamlines observed by the oil-film method (after L. C. Squire [5]).

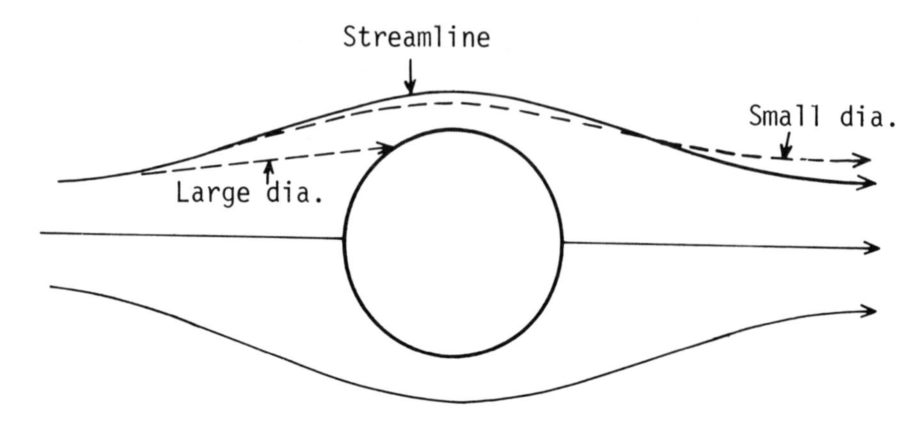

Fig. 36 Trajectories of particles of different sizes.

The motion of a particle of diameter d_p in flow can be given approximated by the following equation:

$$\frac{du_p}{dt} = a\frac{du_f}{dt} + b(u_f - u_p)$$

$$a = \frac{3}{2(\rho_p/\rho_f + 0.5)}$$

$$b = \frac{18\nu}{(\rho_p/\rho_f + 0.5)\, d_p^2}$$

where the subscripts f and p denote the quantities of fluid and particle. The first and second terms of the righthand side of the equation are attributed, respectively, to the inertial effects and the drag due to relative motion. When $\rho_p/\rho_f = 1$, then $a = 1$ and then $u_p = u_f$, that is, the motion of the particle gives the pathline of the fluid correctly.

In practice, however, the densities are not always identical with each other. Then, the size of the particle must be small enough to get a large value of the factor b, which makes $(u_f - u_p)$ satisfactorily small. Figure 36 illustrates that for a given particle density, the smaller particles give better pathlines.

The problem of varying densities must also be considered when liquid dyes are injected into the stream. In this case also, the density difference causes the deviation of the streakline observed from the real ones.

REFERENCES

1. Von Karman, Th., Über den Mechanismus des Wiederstandes, den ein bewegter Körper in einer Flüssigkeit erfahrt, *Gött. Nachricht.* vol. 12, p. 509, 1911; vol. 13, p. 547, 1912.
2. Schlichting, H., and Ulrich, A., Zur Berechnung des Umschlages Laminar-Turbulent, *Jahrb. Luftfahrtforschung*, vol. 1, p. 8, 1942.
3. Tomita, Y., A Study of High-Speed Flow by using Shallow Water Table, *Trans. JSME*, vol. 21, no. 101, p. 22, 1955.
4. Hama, F. R., Streaklines in a Perturbed Shear Flow, *Phys. Fluids*, vol. 5, no. 6, p. 644, 1962.
5. Squire, L. C., The Motion of a Thin Oil Sheet under the Steady Boundary Layer on a Body, *J. Fluid Mech.*, vol. 11, p. 161, 1962.

Chapter 3

Heat and Mass Transfer Fundamentals for Flow Visualization

Wen-Jei Yang

I INTRODUCTION

The first step in the study of transport phenomena is to define the system. There are closed and open systems. The latter is also referred to as control volume. Everything external to the system is the surroundings, and the system is separated from the surroundings by the boundary. The boundary of an open system is also called control surface. While a closed system allows only heat to cross the boundary, both heat and mass can transfer across the control surface in an open system.

Heat is transferred from one region to another by virtue of temperature difference between them. There are three distinct modes of heat transmission: conduction, convection, and radiation. Conduction is a process by which heat is transmitted by molecular diffusion, without displacement of the molecules. It takes place in both solids and fluids without internal motion. In convection, heat is transferred by the combined action of molecular diffusion, energy storage, and mixing motion. Radiation is a process of heat flow through a space or a vacuum in the form of electromagnetic waves. In most situations that occur in nature, heat flows not by one but by several of these mechanisms acting simultaneously.

II TRANSPORT PROPERTIES

The viscosity μ is the physical property that characterizes the flow resistance of simple fluids. It is defined in Newton's law of viscosity

$$\tau = -\mu \frac{\partial u}{\partial x} \qquad (1)$$

as the constant proportionality. Here, τ denotes shear stress, and $\partial u/\partial x$ is the velocity gradient. In a similar manner, the thermal conductivity k and mass diffusivity D characterize the speeds of heat and mass diffusion, respectively; k and D are the proportionality constants in Fourier's law of heat conduction and Fick's law of mass diffusion, respectively. These are discussed later in the chapter.

In addition to thermal conductivity, the thermal diffusivity a is widely used in unsteady heat transfer. It is a compound physical property defined as

$$a = \frac{k}{\rho C_p} \qquad (2)$$

Here, ρ is the density and C_p represents the specific heat under constant pressure. The specific heat is a measure of heat necessary to raise the temperature of unit mass of the substance over a small temperature range, and it depends on both the temperature and the mode of heating. The thermal diffusivity signifies the ratio of the speed of heat diffusion to the heat capacity of unit volume of the substance. Thus, it measures the change of temperature that would be produced in the unit volume by the quantity of heat that flows in unit time.

The role of mass diffusivity in unsteady mass diffusion is analogous to that of thermal diffusivity in heat diffusion.

III TRANSPORT-RATE EQUATIONS

All three modes of heat transfer are governed by the basic laws relating the rate of heat flow q (W) to the temperature difference or gradient. In conduction, the basic relation is called Fourier's law of heat conduction, which reads [1, 2]

$$q_k = -kA \frac{\partial T}{\partial x} \tag{3}$$

in the x direction, or

$$\mathbf{q}_k = kA\nabla T \tag{4}$$

in three-dimensional form. Here, the subscript k denotes conduction; A (m^2), cross-sectional area perpendicular to the direction of heat flow; and $\partial T/\partial x$ (°C/m), temperature gradient in the direction of heat flow x. Equation (1) can be expressed in a finite difference form as

$$q_k = kA \frac{\Delta T}{\Delta x} \tag{5}$$

where ΔT is the temperature difference between two locations at a distance of Δx. The definition of thermal conductivity can be derived from Eq. (5) as

$$k = \frac{q/A}{\Delta T/\Delta x} \tag{6}$$

which yields the unit W/(m · °C).

The basic equation for convection heat transfer between a surface and a fluid is called Newton's law of cooling, which states

$$q_c = h_C A(T_s - T_f) \tag{7}$$

where the subscript c represents convection; h_c (W/m² · °C), heat transfer coefficient; T_s (°C), surface temperature; and T_f (°C), fluid temperature. The rate of radiation heat transfer between two surfaces with temperatures T_1 and T_2 is derived from the Stefan–Boltzmann equation as

$$q_r = \sigma A_1 F_{12}(T_1^4 - T_2^4) \tag{8}$$

Here, the subscript r signifies radiation; σ = Stefan–Boltzmann constant equal to 5.67×10^{-8} W/(m² · K⁴); and F_{12} = shape factor accounting for the emissivities and relative geometries of the surfaces.

Both Eqs. (5) and (8) can be reduced to a simpler form like Eq. (7) by defining

$$h_k = \frac{k}{\Delta x} \tag{9}$$

and

$$h_r = \sigma F_{12}(T_1 + T_2)(T_1^2 + T_2^2) \tag{10}$$

Then, all rate equations take the form of

$$q = hA\,\Delta T = \frac{\Delta T}{1/hA} \tag{11}$$

Considering $1/hA$ as thermal resistance, Eq. (11) is analogous to Ohm's equation for current i (amp) flowing through a resistance R (ohm) with a voltage drop ΔE (V). In other words, heat flow is analogous to the flow of electricity: q and i, ΔT and ΔE, and $1/hA$ and R are analogous. The analogy between the flows of heat and electricity can be helpful in analyzing heat transfer problems. An example is the combined mechanisms of convection and radiation acting in parallel like heat transfer from the hot gas at T_g to the furnace wall at T_s. The total rate of heat transfer is

$$q = q_c + q_r = (h_c + h_r)A(T_g - T_s) \tag{12}$$

IV ENERGY BALANCE EQUATION

The equations of continuity and momentum are derived by applying the principles of conservation of mass and momentum, respectively [2]. Likewise, the law of conservation of energy is employed to obtain the equation of energy. Consider a stationary volume element through which a fluid is flowing. For the fluid contained within this volume element at any time, energy conservation reads

$$\begin{pmatrix}\text{Rate of}\\ \text{accumulation}\\ \text{of total energy}\end{pmatrix} = \begin{pmatrix}\text{rate of}\\ \text{total energy in}\\ \text{by convection}\end{pmatrix} - \begin{pmatrix}\text{rate of}\\ \text{total energy out}\\ \text{by convection}\end{pmatrix} + \begin{pmatrix}\text{net rate of}\\ \text{heat addition}\\ \text{by conduction}\end{pmatrix} + \begin{pmatrix}\text{net rate of}\\ \text{heat addition}\\ \text{by radiation}\end{pmatrix} - \begin{pmatrix}\text{net rate of}\\ \text{work done by}\\ \text{system on}\\ \text{surroundings}\end{pmatrix} + \begin{pmatrix}\text{rate of}\\ \text{internal heat}\\ \text{generation}\end{pmatrix} \tag{13}$$

This is the first law of thermodynamics written for an open system at unsteady state. The total energy E (kJ/kg) includes the internal, kinetic, and potential energy. It can be written in mathematical form as

$$\underbrace{\frac{\partial}{\partial t}(\rho E)}_{\substack{\text{Rate of gain of}\\ \text{energy per unit}\\ \text{volume}}} = \underbrace{-(\nabla \cdot \rho \mathbf{u} E)}_{\substack{\text{rate of energy}\\ \text{input per unit}\\ \text{volume by}\\ \text{convection}}} \underbrace{-(\nabla \cdot \mathbf{q}_k)}_{\substack{\text{rate of energy}\\ \text{input per unit}\\ \text{volume by}\\ \text{conduction}}} \underbrace{-(\nabla \cdot q_r)}_{\substack{\text{rate of energy}\\ \text{input per unit}\\ \text{volume by}\\ \text{radiation}}} \underbrace{-(\nabla \cdot [\boldsymbol{\tau} \cdot \mathbf{u}])}_{\substack{\text{rate of work}\\ \text{done on fluid per}\\ \text{unit volume by}\\ \text{viscous forces}}} \underbrace{-(\nabla \cdot p\mathbf{u})}_{\substack{\text{rate of work}\\ \text{done on fluid per}\\ \text{unit volume by}\\ \text{pressure forces}}} + \underbrace{q'''}_{\substack{\text{rate of internal}\\ \text{heat generation}\\ \text{per unit volume}}} \tag{14}$$

Here, t (s) denotes time; p (Pa), pressure; q''' (W/m^3), volumetric rate of internal heat generation; and $\boldsymbol{\tau}$ (Pa), shear stress tensor. The first two terms are combined to become $\rho\ DE/Dt$, in which D/Dt signifies the substantial derivative.

It is convenient to express the heat equation in terms of the fluid temperature. The transport rates $\mathbf{q}_k$ and $\boldsymbol{\tau}$ are expressed in terms of potential gradients. Then, for a Newtonian fluid conductivity, Eq. (14) becomes

$$\rho C_v \frac{DT}{Dt} = \nabla \cdot \mathbf{k}T - (\nabla \cdot \mathbf{q}_r) - T\left(\frac{\partial p}{\partial T}\right)\rho\,(\nabla \cdot \mathbf{u}) + \mu\phi + q''' \tag{15}$$

in which μ [kg/(m · s)] is the absolute viscosity and ϕ (s^{-2}) denotes the dissipation function. For incompressible fluids with constant thermal conductivity, Eq. (15) becomes

$$\underbrace{\rho C_p \left(\frac{\partial T}{\partial t}\right.}_{\text{Unsteady}} \underbrace{\left. + \mathbf{u} \cdot \nabla T\right)}_{\text{convection}} = \underbrace{k\,\nabla^2 T}_{\text{conduction}} \underbrace{- (\nabla \cdot \mathbf{q}_r)}_{\text{radiation}} + \underbrace{q'''}_{\substack{\text{internal}\\ \text{heat generation}}} \tag{16}$$

This simplified version of the energy equation is widely used. The radiation term includes the forms of energy due to electromatic and ultrasonic waves, while the internal heat generation term includes the energy from nuclear, chemical, and electrical sources.

Equation (16) is subject to the initial condition and appropriate boundary or surface conditions. The initial condition describes the state (steady temperature distribution) of the system prior to the disturbance. There are three typical boundary conditions—(1) prescribed surface temperature, prescribed heat flux across the surface; (2) no heat flux (insulation) across the surface; (3) heat transfer with the surroundings across the surface by conduction, convection, radiation or—their combinations.

V CONDUCTION

In the absence of convection and radiation, Eq. (4) for isotropic media is reduced to

$$\frac{\partial T}{\partial t} = \alpha \nabla^2 T + \frac{q'''}{\rho C_p} \tag{17}$$

This is the general expression for conduction. The simplest case is steady, one-dimensional pure conduction, for which Eq. (17) is reduced to

$$\frac{d^2T}{dx^2} = 0, \quad \frac{d}{dr}\left(r\frac{dT}{dr}\right) = 0, \quad \frac{d}{dr}\left(r^2\frac{dT}{dr}\right) = 0 \tag{18}$$

in the Cartesian, cylindrical, and spherical coordinate systems, respectively. For prescribed surface temperatures, the corresponding solution and the thermal resistance are summarized in Table 1.

The solutions for unsteady conduction problems described by Eq. (17) are often difficult to obtain. A simple approach may be used under conditions for which temperature gradients within the system are small. It is called the lumped capacitance method, which assumes a uniform temperature distribution across a system at any instant during a transient process. First, one should calculate the Biot number defined as

$$\text{Bi} = \frac{hL}{k} = \frac{L/kA}{1/hA} \qquad L = \frac{V}{A_s} \tag{19}$$

where L = characteristic length, V = volume, and A_s = heat transfer area. The Biot number signifies the ratio of the internal thermal resistance of a solid L/kA to the boundary layer thermal resistance $1/hA$. If the condition Bi $\leq$ 0.1 is satisfied, the error associated with using the lumped capacitance method is small, and the result will be valid for any system geometry.

An overall balance on the solid with a volumetric rate of internal heat generation q''' yields

$$\rho C\,V\frac{dT}{dt} = hA(T - T_\infty) + q'''V \tag{20}$$

where T_∞ denotes the new ambient temperature.

Table 1 Summary of Heat Conduction Solutions for Various Geometries with Specified Surface Temperatures[a]

Geometry	Temperature Distribution	Thermal Resistance
flat plate	$T(x) = (T_2 - T_1)\dfrac{x}{\Delta x} + T_1$	$\dfrac{\Delta x}{kA}$
hollow cylinder of length l	$T(r) = \dfrac{T_1 - T_2}{\ln (r_1/r_2)} \ln\left(\dfrac{r}{r_2}\right) + T_2$	$\dfrac{\ln (r_2/r_1)}{2\pi lk}$
hollow sphere	$T(r) = \dfrac{T_1 - T_2}{1/r_1 - 1/r_2}\left(\dfrac{1}{r} - \dfrac{1}{r_2}\right) + T_2$	$\dfrac{r_2 - r_1}{4\pi k r_1 r_2}$

[a]The mass transfer expressions may be obtained by replacing T and k by C and D, respectively.

It is subject to the initial condition that the system is at a uniform temperature T_0. The solution can be readily obtained as

$$\frac{T - T_\infty + (q'''V/hA)}{T_0 - T_\infty + (q'''V/hA)} = e^{-\text{Bi Fo}} = e^{-t/\tau} \tag{21}$$

in which

$$\text{Fo} = \text{Fourier number} = \frac{\alpha t}{L^2}$$
$$\tau = \text{thermal time constant} = \frac{\rho VC}{hA} \tag{22}$$

where Fo indicates the ratio of the heat conduction rate to the rate of thermal energy storage in a solid, and τ is the product of the resistance to convection heat transfer and the thermal capacitance of the solid.

In general, the solution of Eq. (17) without internal heat generation takes the form of

$$\frac{T - T_\infty}{T_0 - T_\infty} = f\left(\text{Bi, Fo}, \frac{x}{L}, \frac{y}{L}, \frac{z}{L}\right) \tag{23}$$

VI BOUNDARY LAYERS

The boundary layer behavior is relevant to convection transport. To introduce the concept of boundary layer, consider flow over the flat plate of Fig. 1*a*. Fluid particles in contact with the plate have zero velocity, while the fluid velocity at distance $y = \delta$ from the surface is at the freestream value u_∞. The symbol δ stands for the boundary layer thickness in which the motion of particles is retarded from u_∞ to 0 on the surface. It is formally defined as the value of y for which $u = 0.99\ u_\infty$. The boundary layer velocity profile refers to the manner in which u varies with y through the boundary layer. The surface shear stress τ_s is evaluated from the velocity gradient at the surface as

$$\tau_s = -\mu\left(\frac{\partial u}{\partial y}\right)_{y=0} \tag{24}$$

It provides the basis for determining the friction coefficient:

$$C_f = \frac{\tau_s}{\rho u_\infty^2/2} \tag{25}$$

Like the development of a velocity boundary layer, a thermal boundary layer must develop if the fluid freestream

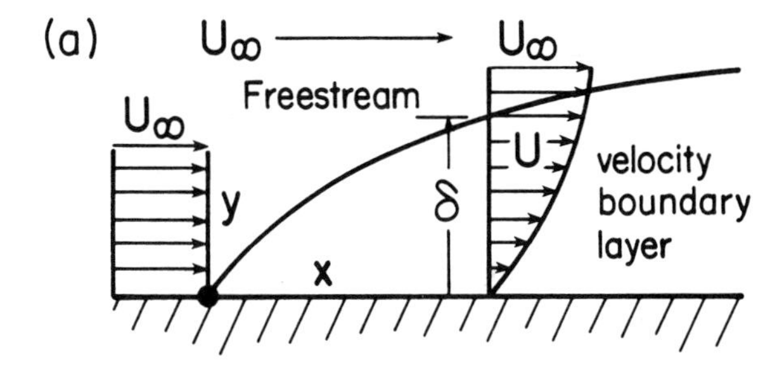

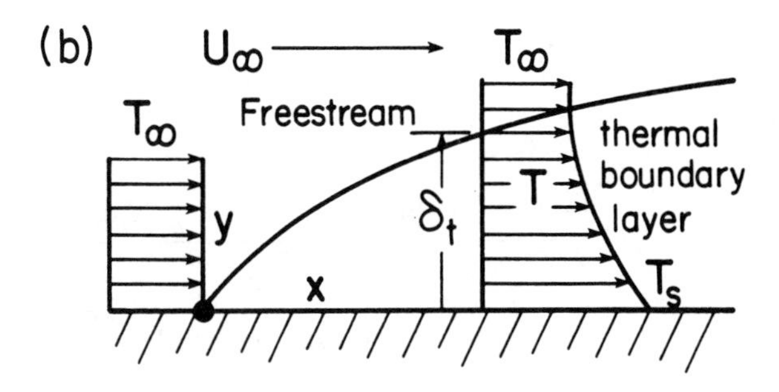

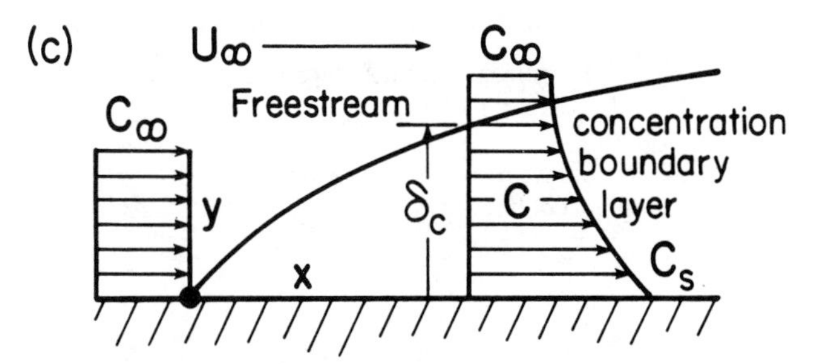

Fig. 1 Boundary layers: (*a*) velocity; (*b*) thermal; (*c*) concentration.

and surface temperatures differ (see Fig. 1*b*). The temperature profile at the leading edge is uniform, with $T = T_\infty$, while the surface is at a temperature of T_s. The fluid region in which temperature gradients exist is the thermal boundary layer. The thermal boundary layer thickness δ_t is defined as the value of y for which the ratio $(T_s - T)/(T_s - T_\infty) = 0.99$. The local heat flux in the fluid at the surface $y = 0$ may be obtained by applying Fourier's conduction law as

$$\frac{q_k}{A} = -k\left(\frac{\partial T}{\partial y}\right)_{y=0} \tag{26}$$

This amount of heat must be transferred to the freestream. By equating this expression to Newton's law of cooling, one obtains

$$h = \frac{-k(\partial T/\partial y)_{y=0}}{T_s - T_\infty} \tag{27}$$

In a similar manner, the concentration boundary layer thickness δ_c is defined as the value of y for which $(c_s - c)/(c_s - c_\infty) = 0.99$. Here, the concentration of a designated species in the fluid at the surface C_s differs from that in the freestream C_∞, as seen in Fig. 1*c*. The local mass flux at the surface is obtained by applying Fick's law as

$$N = -DA\left(\frac{\partial c}{\partial y}\right)_{y=0} \tag{28}$$

The mass transfer coefficient is found to be

$$h_D = \frac{-D(\partial c/\partial y)_{y=0}}{c_s - c_\infty} \tag{29}$$

Hence, the surface concentration (temperature) gradient will influence the mass (heat) transfer coefficient and consequently the rate of mass (heat) transfer across the boundary layer.

VII CONVECTION

In order to determine the rate of convection heat transfer using Eq. (7), one needs to know the magnitude of heat transfer coefficient h_c. One can obtain A and ΔT through measurements. It is convenient to express h_c in the dimensionless form as the Nusselt number defined as h_cL/k, where L is the characteristic length of the system and k is the thermal conductivity of the fluid. At subsonic flow with negligible aerodynamic heating, the Nusselt number is a function of the Reynolds number Re and the Prandtl number Pr:

$$\mathrm{Nu} = f(\mathrm{Re}, \mathrm{Pr}) \tag{30}$$

In natural convection, the Nusselt number depends on the Grashof number Gr and the Prandtl number as

$$\mathrm{Nu} = g(\mathrm{Gr}, \mathrm{Pr}) \tag{31}$$

Here, the dimensionless parameters are defined as

$$\mathrm{Re} = \frac{uL}{\nu} \qquad \mathrm{Pr} = \frac{\nu}{\alpha} \qquad \mathrm{Gr} = \frac{g\beta\,\Delta TL^3}{\nu^2} \tag{32}$$

where u (m/s) denotes the fluid velocity; ν (m^2/s), kinematic viscosity; α (m^2/s), thermal diffusivity; g (m/s^2), gravitational acceleration; and ΔT (°C), surface-fluid temperature difference. In general, Eqs. (30) and (31) are obtained empirically with a few exceptions that are theoretically derived. The following compound parameters often

Table 2 Summary of Convective Heat Transfer Correlations for Common Fluids[a]

(i) Forced convection, external flow (physical properties at film temperature)

Correlation	Geometry	Conditions
$Nu_x = 0.03332\ Re_x^{1/2}\ Pr^{1/3}$	Flat plate	Laminar, local
$Nu = 0.0664\ Re^{1/2}\ Pr^{1/3}$	Flat plate	Laminar, average
$Nu_x = 0.565\ Pe_x^{1/2}$	Flat plate	Laminar, local liquid metals[b]
$Nu_x = 0.0296\ Re_x^{4/5}\ Pr^{1/3}$	Flat plate	Turbulent, local, $Re_x \leq 10^8$
$Nu = (0.037\ Re^{4/5} - 871)Pr^{1/3}$	Flat plate	Mixed, average, $Re \leq 10^8$
$Nu = C Re^m\ Pr^{1/3}$	Cylinder	Average, $0.4 < Re < 4 \times 10^5$:

Re	C	m
0.4–4	0.989	0.330
4–40	0.911	0.385
$40–4 \times 10^3$	0.683	0.466
$4 \times 10^3–4 \times 10^4$	0.193	0.618
$4 \times 10^4–4 \times 10^5$	0.027	0.805

(ii) Forced convection, internal flow (physical properties at bulk temperature)

Correlation	Conditions
$Nu = 4.36$	Laminar, fully developed, constant q''_s
$Nu = 3.66$	Laminar, fully developed, constant T_s
$Nu = 1.86\ Gz^{1/3}\ (\mu/\mu_s)^{0.14}$	Laminar, constant T_s
$Nu = 3.66 + \dfrac{0.0668\ Gz}{1 + 0.04\ Gr^{2/3}}$	Laminar, constant T_s thermal entrance region
$Nu = 0.023\ Re^{4/5}\ Pr^{1/3}$	Turbulent, fully developed, $L/D_h \geq 60$, $Re \geq 10^4$
or $Nu = 0.023\ Re^{4/5}\ Pr^n$	Same, $n = 0.3$ for heating, $n = 0.4$ for cooling
or $Nu = 0.027\ Re^{4/5}\ Pr^{1/3} \left(\dfrac{\mu}{\mu_s}\right)^{0.14}$	Same
$Nu = 4.82 + 0.0185\ Pe^{0.827}$	Turbulent, fully developed liquid metals, constant q''_s
$Nu = 5.0 + 0.025\ Pe^{0.8}$	Turbulent, fully developed, liquid metals, constant T_s

(iii) Free convection (physical properties at film temperature)

Correlation	Geometry	Conditions
$Nu = 0.59\ Ra^{1/4}$	Vertical plates and cylinders	Laminar, $Ra \leq 10^9$
$Nu = 0.10\ Ra^{1/3}$	Same	Turbulent, $Ra \geq 10^9$
$Nu = 0.54\ Ra^{1/4}$	Horizontal plates, heating surface facing upward, or cooling surface facing downward	Laminar, $10^5 \leq Ra \leq 10^7$
$Nu = 0.15\ Ra^{1/3}$	Same	Turbulent, $10^7 \leq Ra \leq 10^{10}$
$Nu = 0.27\ Ra^{1/4}$	Horizontal plates, heating surface facing downward, or cooling surface facing upward	Laminar, $10^5 \leq Ra \leq 10^{10}$
$Nu = 0.59\ Ra^{1/4} \cos\theta$	Inclined plates and cylinders, θ = angle of inclination	Laminar, $Ra \leq 10^9$
$Nu = 0.10\ Ra^{1/3} \cos\theta$	Same	Turbulent, $Ra \geq 10^9$
$Nu = C\ Ra^n$	Horizontal cylinders	(see below)
$Nu = 2 + 0.43\ Ra^{1/4}$	Spheres	$1 \leq Ra \leq 10^5$, $Pr \simeq 1$

Horizontal cylinders:

Ra	C	n
$10^{-10}–10^{-2}$	0.675	0.058
$10^{-2}–10^2$	1.02	0.148
$10^2–10^4$	0.850	0.188
$10^4–10^7$	0.480	0.250
$10^7–10^{12}$	0.125	0.333

[a]For forced convection, the mass transfer correlations may be obtained by replacing Nu, Gr and Pr by Sh, Gr_D, and Sc, respectively in cases (i), (ii), and (iii).

[b]The mass transfer analogy is not applicable.

appear in the convective equations (30) and (31):

$$\begin{aligned}
\text{Stanton number (St)} &= \frac{\text{Nu}}{\text{Re} \cdot \text{Pr}} \\
\text{Peclet number (Pe)} &= \text{Re} \cdot \text{Pr} \\
\text{Rayleigh number (Ra)} &= \text{Gr} \cdot \text{Pr} \\
\text{Graetz number (Gz)} &= \frac{\text{Re} \cdot \text{Pr}\, D_h}{L}
\end{aligned} \tag{33}$$

where D_h is the hydraulic diameter. In addition to Re and Pr, heat transfer in high-speed flow involves other dimensionless parameters such as

$$\begin{aligned}
\text{Mach number (M)} &= \frac{u}{u_s} \\
\text{Knudsen number (Kn)} &= \frac{l_m}{L} \\
\text{Specific heat ratio} &= \frac{C_p}{C_v} \\
\text{Recovery factor} &= \frac{T_{as} - T_f}{u^2/2C_p}
\end{aligned} \tag{34}$$

Hence, u_s (m/s) represents the sonic velocity; l_m (m) = molecular mean free path; T_{as} (°C) = adiabatic surface temperature; and C_p and C_v [kJ/(kg · K)] = specific heat at constant pressure and constant volume, respectively.

Table 2 presents a summary of convective heat transfer correlations for common fluids. Those correlation equations have been in common use. Once the heat transfer coefficient h_c is evaluated using Eq. (30) or (31), the rate of convective heat transfer can be readily determined by Eq. (7).

VIII RADIATION

Infrared (IR) vision, namely, infrared thermography, is a process used to produce the thermal image of a body surface from which a heat flow situation on the surface can be visualized. Radiation possesses dual characteristics of both waves and particles photons.

A wave has its frequency ν and wavelength λ, which are related by

$$\lambda\nu = c \tag{35}$$

where c (m/s) is the speed of propagation that is the speed of light (3×10^9 m/s in vacuum) for electromagnetic (EM) waves and the sonic velocity in case of acoustic waves. Figure 2 depicts the spectrum of IR and EM waves. The IR wave is one part of the EM waves. Its wavelength varies between 0.7 and 1000 μm.

Sometimes, it is useful to think of EM radiation from a corpuscular point of view. That is, an EM wave of frequency ν is associated with a photon, which is a particle with no mass and no charge but with an energy e given by

$$e = h\nu \tag{36}$$

Here, h is Planck's constant, which is equal to 6.624×10^{-27} erg·s. From Eqs. (35) and (36), together with the spectrum of EM radiation (Fig. 2), one can see that decreasing λ implies increasing e. This is observed in the infrared radiation, through which relatively small energies are released as a result of a molecule decreasing its speed of rotation. On the other hand, γ- or X-ray radiation is associated with a release of relatively large energies re-

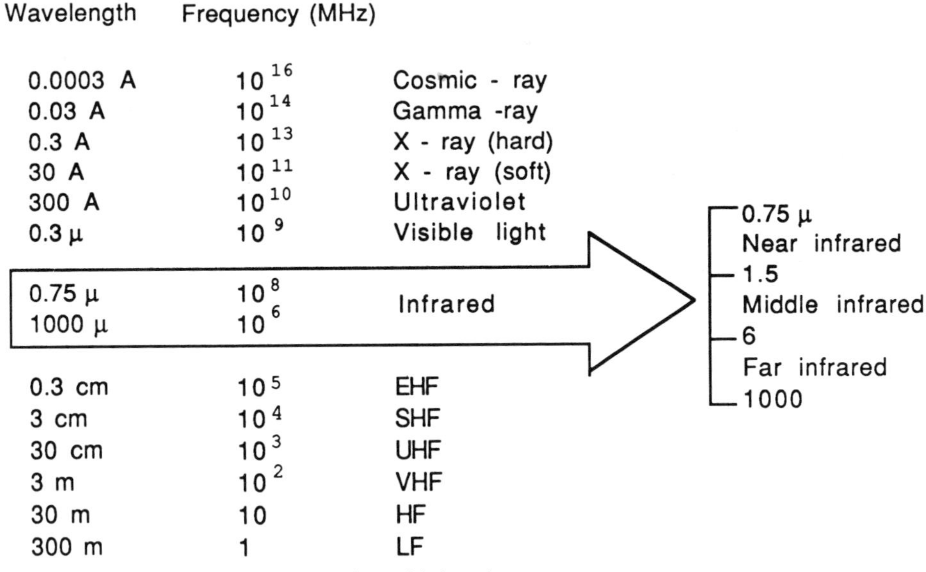

Fig. 2 Spectrum of electromagnetic and infrared waves.

sulting from an atomic nucleus going from a high energy state to a lower one. Therefore, as the surface temperature is raised, the emitted radiant energy tends toward smaller λ or larger e. When a molecule or atom undergoes a change in state from a high to a low level, the emission of radiant energy or photons occurs. Conversely, absorption occurs when radiant energy is added to a molecule or atom, resulting in a change from a low to a high energy state.

All surfaces at a temperature higher than absolute zero radiate heat, called emission. When a radiation beam strikes a surface, reflection, absorption, and transmission occur. A surface is called blackbody, or black surface, if it absorbs all radiation irradiated on it (called a perfect absorber). A black surface is also a perfect radiator, which emits the maximum possible radiation at a given temperature. Figure 3 shows the spectral distribution (E_{b_λ} versus λ) of a black surface at a certain temperature T_s. The term E_{b_λ} is the rate of emissive heat transfer per unit surface area for a wave with wavelength λ and is called monochromatic emissive power. The curves, called spectroradiometric curves, are characterized by three laws: Planck's law for the shape, Wien's displacement law for the peak and the Stefan–Boltzmann law for the area between the curve and the abscissa. Equations for these laws follow:

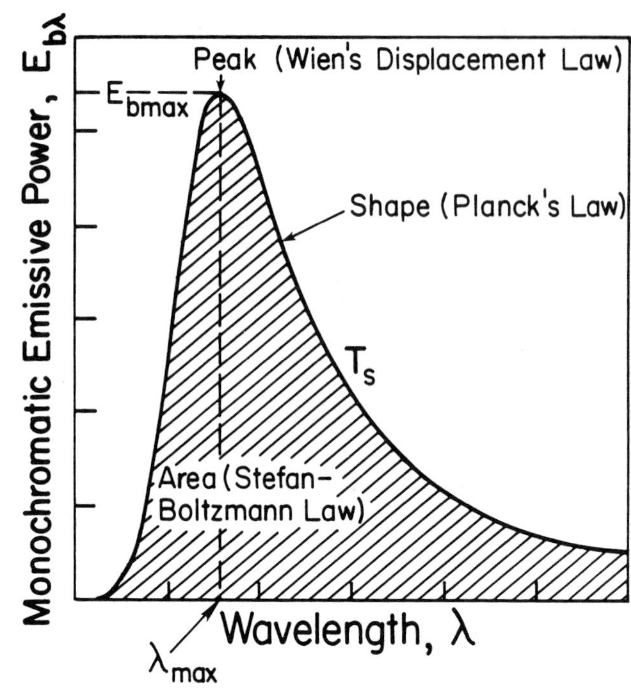

Fig. 3 Spectral distribution of a blackbody at temperature T_s.

Planck's law:

$$E_{b_\lambda} = C_1\lambda^{-5}\left[\exp\left(\frac{C_2}{\lambda T}\right) - 1\right]^{-1} \quad (37)$$

Wien's displacement law:

$$\lambda_{\max}T = 2896\ (\mu\text{K}) \quad (38)$$

Stefan–Boltzmann law:

$$E_b = \int_0^\infty E_{b_\lambda}\, d\lambda = \sigma T^4 \quad (39)$$

Here, E_b is the total emissive power of a black surface that is equal to the area in Fig. 3; $C_1 = 3.7415 \times 10^{-4}$ μW^2; $C_2 = 1.4387 \times 10^4$ μK; $\sigma = 5.6724 \times 10^{-8}$ W/(m^2 · K^4); σ represents the Stefan–Boltzmann constant.

A real surface has the emissive power E, which is lower than that of a blackbody E_b at the same temperature. Their ratio ε is called emissivity or emissive coefficient:

$$\varepsilon = \frac{E}{E_b} \quad (40)$$

When a surface intercepts an irradiation with heat flux G, heat balance requires

$$G = R + A + T$$

where R, A, and T denote the heat fluxes of reflection, absorption, and transmission, respectively. Dividing both sides of the equation yields

$$1 = \rho + \alpha + \tau \quad (41)$$

$$\rho = \frac{R}{G} \qquad \alpha = \frac{A}{G} \qquad \tau = \frac{T}{G} \quad (42)$$

where ρ, α, and τ represent reflectivity, absorptivity, and transmissivity, respectively. Usually τ is negligible.

Kirchhoff's law states that at thermal equilibrium, the ratio of the emissive heat flux E of a surface to its absorptivity is the same for all bodies in the same enclosure:

$$\frac{E_1}{\alpha_1} = \frac{E_2}{\alpha_2} = \cdots = \frac{E_n}{\alpha_n} \quad (43)$$

From Kirchhoff's law, the ratio of E/E_b is equal to the absorptivity:

$$\varepsilon = \alpha \quad (44)$$

Lambert's cosine law leads to

$$E_b = \pi I_b \qquad E_{b_\lambda} = \pi I_{b_\lambda} \quad (45)$$

Thus, the total emissive heat flux of a diffuse surface is related to its intensity I_b by a factor of π. Similarly, the monochromatic emissive heat flux E_{b_λ} is π times its intensity I_{b_λ}.

When two surfaces at different temperatures T_1 and T_2 ($T_1 > T_2$) are situated in a space devoid of any matter, radiant heat exchange takes place, resulting in a net loss of radiant heat from the surface at T_1 to the surface at T_2. The simplest case is where both surfaces are black, infinite in dimension, and parallel. Since radiant energy from one surface is completely absorbed by the other, the net heat exchange can be obtained by applying the Stefan–Boltzmann equation as

$$q_r = (E_{b_1} - E_{b_2})A_1 \tag{46}$$

Here, the observer is stationed on surface 1 with area A_1. If the surfaces are not both black, finite in dimension, and at any orientation, then a correction factor F_{1-2} is needed in Eq. (46), to read

$$q_r = F_{1-2}A_1(E_{b_1} - E_{b_2}) \tag{47}$$

which is Eq. (8). The factor F_{1-2} is zero when both surfaces do not see each other. When it takes the maximum value of unity, Eq. (47) is reduced to Eq. (46). It is a modulus to account for the emissivities and relative geometries of the actual surfaces.

IX MASS TRANSFER

Heat transfer will occur whenever there exists a temperature difference in a system. Similarly, whenever there exists a difference in the concentration or density of some species in a mixture, mass transfer must occur. Hence, just as a temperature gradient constitutes the driving potential for heat transfer, a concentration gradient for some species in a mixture acts as the driving potential for transport of the species. There are two modes of mass transfer, mass diffusion and convective mass transfer, which are analogous to the conduction and convection modes of heat transfer. However, there is no mass transfer mechanism which is similar to radiative heat transfer.

Fick's law of diffusion describes the rate of mass diffusion as

$$N = DA \frac{dC}{dx} \tag{48}$$

in the x direction, or

$$\mathbf{N} = -DA\, \nabla C \tag{49}$$

in three-dimensional form. Here, C denotes the concentration of a designated species. It can be written in finite difference form as

$$N = DA \frac{\Delta C}{\Delta x} \tag{50}$$

in which D (m^2/s) is the proportionality factor between the mass flux N/A and the concentration gradient $\Delta C/\Delta x$ or dC/dx. Table 1 summarizes the concentration distribution and the diffusion resistance.

In mass transfer by convection, fluid motion combines with diffusion to promote the transfer of a species for which a concentration gradient exists. The mass transfer rate of a designated species may be expressed as

$$N = h_D A(C_s - C_f) \tag{51}$$

where h_D (m/s) is the convection mass transfer coefficient. The dimensionless form of h_D is called the Sherwood number, defined as

$$\text{Sh} = \frac{h_D L}{D} \tag{52}$$

The Sherwood number is to the concentration boundary layer what the Nusselt number is to the thermal boundary layer. The functional dependence of Sh is

$$\text{Sh} = f(\text{Re}, \text{Sc}) \tag{53}$$

for forced convection and

$$\text{Sh} = g(\text{Gr}_D, \text{Sc}) \tag{54}$$

for natural convection. Here,

$$\begin{aligned} \text{Sc} &= \text{Schmidt number} = \frac{\nu}{D} \\ \text{Gr}_D &= \text{mass transfer Grashof number} \\ &= \frac{g\beta_D\, \Delta X L^3}{\nu^2} \end{aligned} \tag{55}$$

where β_D is the concentration coefficient of volumetric expansion, defined as $-(\partial\rho/\partial X)/\rho$, and X is the mass function of the species in migration. Other dimensionless parameters pertinent to convection mass transfer include

$$\begin{aligned} \text{Le} &= \text{Lewis number} = \frac{\alpha}{D} \\ \text{Pe}_D &= \text{mass transfer Peclet number} = \text{Re Sc} \\ \text{St}_D &= \text{mass transfer Stanton number} = \frac{\text{Sh}}{\text{Re Sc}} \end{aligned} \tag{56}$$

The mass transfer correlations may be obtained from Table 2 for the heat transfer correlations by replacing Nu and Pr by Sh and Sc, respectively. Once the value of Sh is calculated, the mass transfer coefficient can be determined by Eq. (52), followed by evaluating the mass flux using Eq. (51).

The mass balance equation can be readily derived by applying the heat and mass transfer analogy, Eq. (16) without the radiation term, as

$$\underbrace{\frac{\partial C}{\partial t}}_{\text{Unsteady}} + \underbrace{\mathbf{u} \cdot \nabla C}_{\text{convection}} = \underbrace{D\, \nabla^2 C}_{\text{diffusion}} + \underbrace{R'''}_{\text{internal mass generation}} \tag{57}$$

The initial and boundary conditions are also analogous to those in heat transfer problems.

REFERENCES

1. Kreith, F., and Bohn, M. S., *Principles of Heat Transfer*, 4th ed., Harper, New York, 1986.
2. Bird, R. B., Stewart, W. E., and Lightfoot, E. N. *Transport Phenomena*, Wiley, New York, 1960.

Part 2
Methods

Chapter 4

Liquids

H. Werlé

The methods using tracers for visualization in water or in other liquids are manifold and of different natures. The interested reader may consult the various review papers [1–9] that describe the state of the art. Only the most commonly used processes will be reviewed here.

It should be noted that these methods, based on the use of elements in suspension within the fluid, are in some cases limited by the discrepancies between the trajectories of the particles and of the fluid elements they represent. This mainly concerns particles not having the same density and viscosity as the fluid. As they have different diameters, they are subjected to averaging effects. They are also affected by centrifugal or centripetal forces and gravity effects.

Let us recall that if the trajectories and emission lines of correct tracers coincide with the streamlines of a steady flow, at least in the laminar regime, this is no longer true for an unsteady flow or in the turbulent regime. However, as will be seen later, these tracers may, provided that some precautions are taken, give a correct picture of the instantaneous or averaged flow with the separation structures.

The main liquid tracers used in water are as follows:

- Diluted milk (mixture of milk, alcohol, and dye whose density and viscosity are the same as those of water), used in the ONERA tunnels up to 1979 [10] (Fig. 1*a*, *f*, *g*).
- Diluted rhodorsil (stable white mixture replacing milk), used since 1979 (Fig. 1*b–e*)
- Fluorescent dyes [11]
- Such substances as ink, commercial dye, solution of carbon tetrachloride and benzine, and potassium permanganate [4]

The dye streaks are emitted from upstream ramps (Fig. 1*a*) or from the models themselves at isolated and suitably chosen points on the surface (Fig. 1*b*, *d*, *g*), with a wall flow passage being obtainable by a progressive reduction in the flow of the dyes (Fig. 1*f*).

In a steady flow, the streaks reveal the streamlines (Fig. 1*a*), the vortex axis and structure (Fig. 1*g*), and the wall flow with all the singularities (e.g., separation line, partition point, and focus) characterizing the separation.

In turbulent flow, this process preserves a certain effectiveness despite the rapid diffusion of the dye (Fig. 1*c*). When the flow becomes unsteady, the dyes then form only emission lines, which nonetheless reveal the instantaneous pattern of the flow, in particular of the wake (Fig. 1*b*), jet (Fig. 1*c*), and vortex (Fig. 1*d*). To illustrate the more random phenomena such as the formation of turbulence spots under transient conditions, a variation of the process consists of emitting through a slot a continuous colored wall layer covering all or part of the model (Fig. 1*e*).

The other main tracers used in water are the following:

- The solid tracers, which are usually introduced by a probe upstream of the model, include glass balls, perspex powder, and aluminum particles, for example [4], and spherical polystyrene beads of a density close to that of water and textile tufts fixed on the model surface (Fig. 2*a*, *c*) [12].
- The most widely used gas tracers are the hydrogen bubbles obtained by electrolysis, a method created by Geller [13], which is employed extensively [14, 15]. These tracers are emitted by a cathode either fixed upstream (Fig. 2*d*), revealing the flow shape with the vortex sheddings, also in unsteady regime [15], or placed on the model, for instance, on the wing leading edge (Fig. 2*e*), bringing to light the scroll vortex sheet and only this sheet. This can be seen in Fig. 2*e* as a

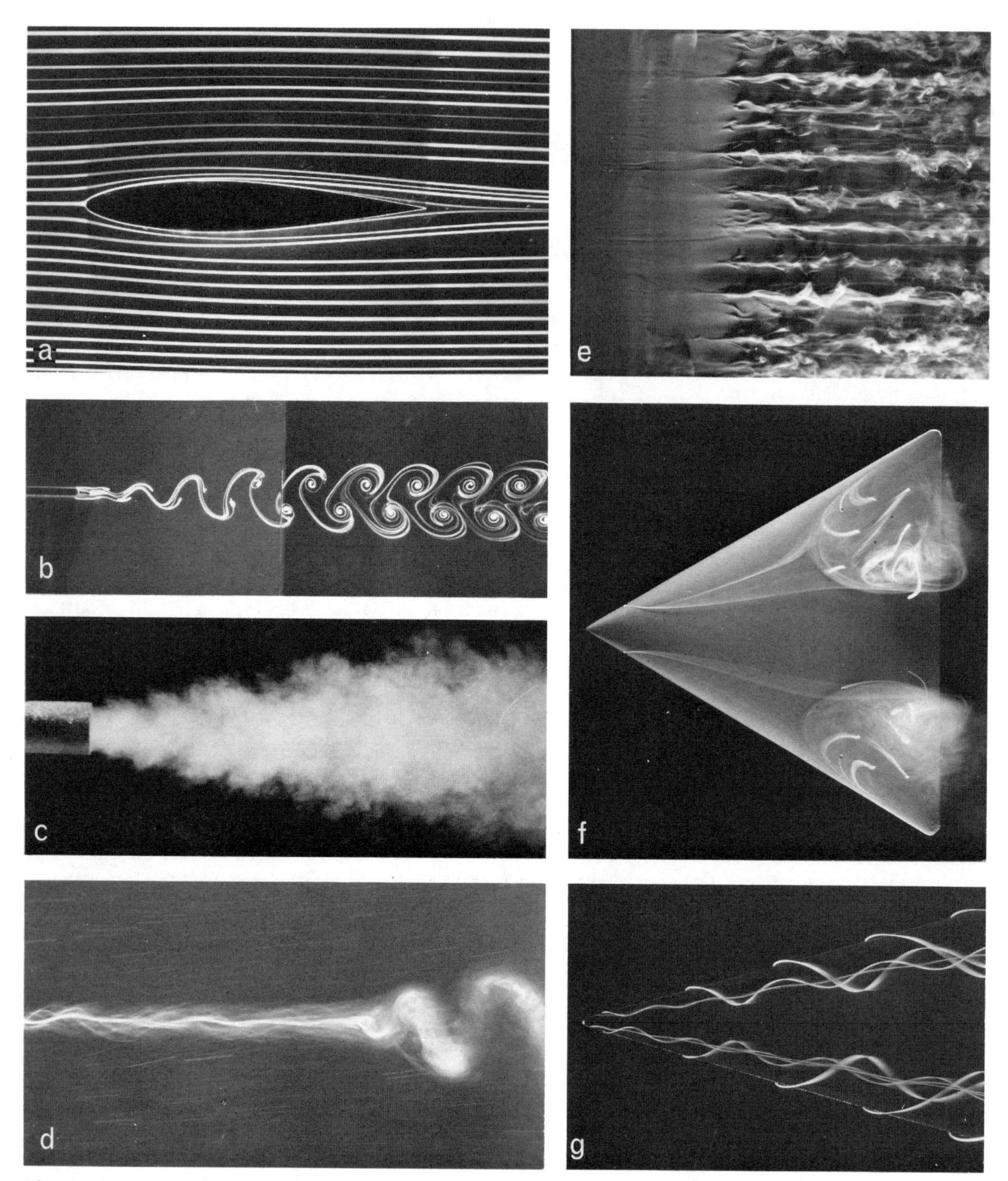

Fig. 1 Flow visualizations by dye emissions (examples of fundamental studies); (*a*) Streamlines around a profile without incidence; (*b*) von Karman vortices behind a flat plate without incidence; (*c*) discontinuous structure of an isolated colored jet; (*d*) spiral breakdown of an isolated vortex; (*e*) boundary layer transition on a long flat plate without longitudinal pressure gradient; (*f*) three-dimensional free rear separation over a thick delta wing with incidence; (*g*) vortex structure of the fixed separation over a thin delta wing with incidence.

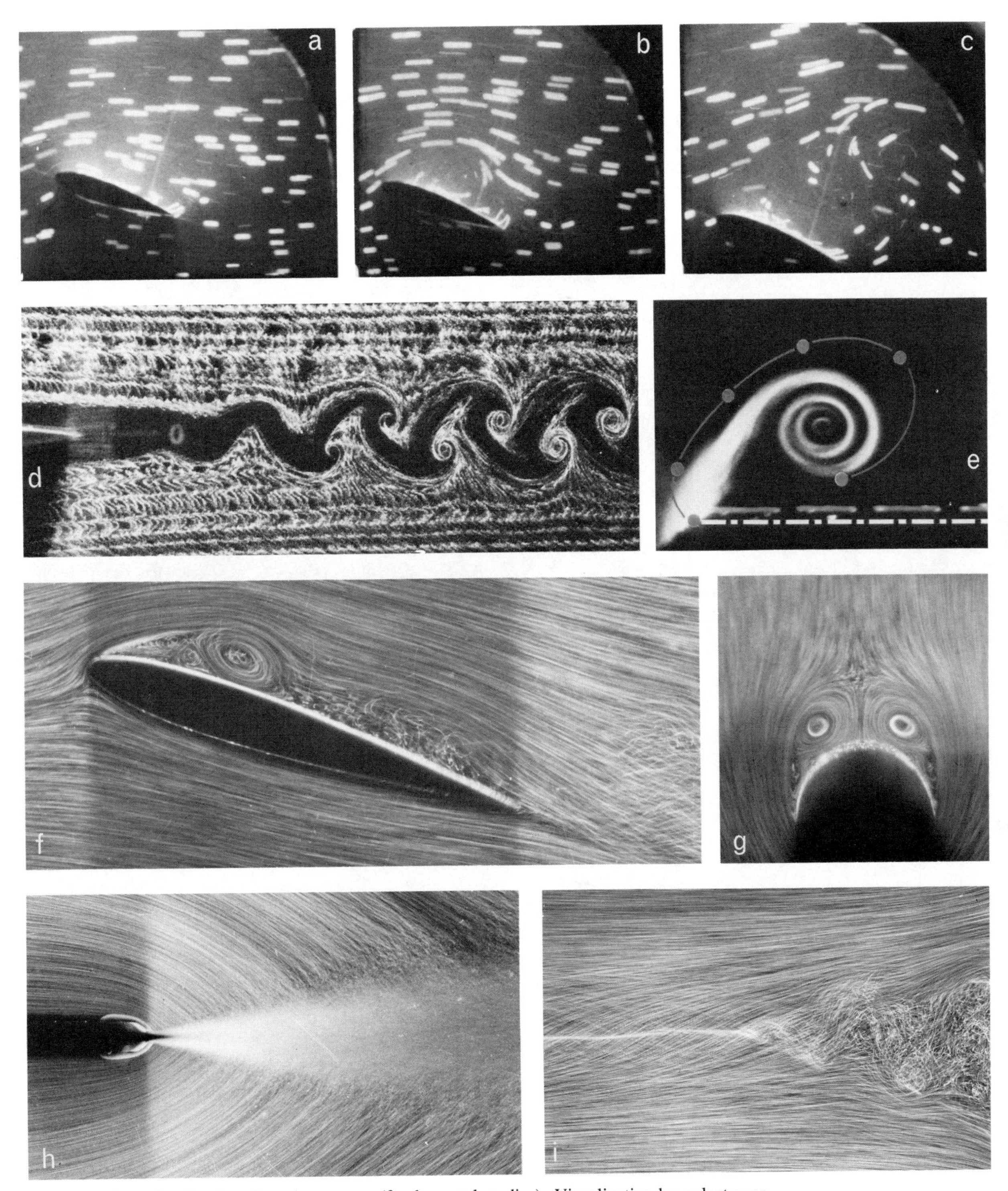

Fig. 2 Flow visualizations by other tracers (fundamental studies). Visualization by polystyrene beads and tufts: (*a*, *b*, *c*) flow around a profile with flapping motion. Visualization by hydrogen bubbles: (*d*) Flow and wake behind a profile without incidence (McAlister and Carr [15]); (*e*) scroll vortex sheet issued from the sharp leading edge of a thin delta wing (cross section at mid-chord with theoretical points). Visualization by air bubbles: (*f*) Instantaneous flow around an oscillating profile (medium section); (*g*) symmetric vortices over a cylindrical fuselage with incidence (cross section at half length); (*h*) isolated jet in two-dimensional flow (mean flow); (*i*) spiral breakdown of an isolated vortex (section along the vortex axis).

cross-section at mid-chord, and it is compared with results of a numerical calculation [16].

The other commonly used gas tracers are the air bubbles (ONERA process) obtained by means of an emulsifying agent introduced during the filling of the tank. These small tracers suspended in the water bring out the shape of the flow in longitudinal sections (Fig. 2*f*, *h*, *i*) and cross-sections (Fig. 2*g*), illuminated by a sheet of light. In particular, the air bubbles facilitate visualization of flow with jet (Fig. 2*f*) or vortices in steady regime (Fig. 2*g*) and in unsteady regime (Fig. 2*f*), and finally the axis and the breakdown (Fig. 2*i*) of a vortex with sufficient intensity, as the air bubbles accumulate along the said axis under the effect of centripetal forces.

REFERENCES

1. Balint, E., Techniques of Flow Visualization, *Aircr. Eng.*, vol. 25, pp. 161–167, 1953.
2. Werlé, H. Méthodes d'Étude, par Analogie Hydraulique, des Écoulements Sub-, Super- et Hypersoniques, AGARD Rep. 399, 1960; The Hydraulic Analogy Laboratory ONERA, ONERA Publ. 103, 1961.
3. Merzkirch, W., Making Flows Visible, *Int. Sci. Technol.*, vol. 58, pp. 46–56, 1966.
4. Macagno, E., Flow Visualization in Liquids, Univ. of Iowa, Des Moines, IIHR Rep. 114, 1969.
5. Werlé, H., Hydrodynamic Flow Visualization, *Ann. Rev. Fluid Mech.*, vol. 5, pp. 361–382, 1973.
6. Asanuma, T., ed., *Flow Visualization*, Hemisphere, Washington, D.C., 1979.
7. Merzkirch, W., ed., *Flow Visualization II: Proceedings of the Second International Symposium*, Hemisphere, Washington, D.C., 1982.
8. Yang, W. J., ed., *Flow Visualization III: Proceedings of the Third International Symposium on Flow Visualization*, Hemisphere, Washington, D.C., 1985.
9. Véret, C., ed., *Flow Visualization IV: Proceedings of the Fourth International Symposium*, Hemisphere, Washington, D.C., 1987.
10. Werlé, H., Le Tunnel Hydrodynamique au Service de la Recherche Aérospatiale, ONERA Publ. 156, 1974.
11. Campbell, D. R., ARL Rep. 73-005, 1973.
12. Werlé, H., Les Mouvements Plans Non Stationnaires à Circulation Constante. Comparaison entre Résultats Théoriques et Expérimentaux, *Rech. Aeronaut.*, vol. 26, pp. 13–18, 1952.
13. Geller, E., An Electrochemical Method of Visualizing the Boundary Layer, *Aeronaut. Sci.*, 22, pp. 869–870, 1955.
14. Schraub, F., Kline, S., Henry, J., Runstadler, P., and Littel, A., Use of Hydrogen Bubbles for Quantitative Determination of Time Dependent Velocity Fields in Low Speed Water Flows, Stanford Univ. Rep. MD-12, 1964.
15. McAlister, K. W., and Carr, L. W., Water Tunnel Visualizations of Dynamic Stall, *J. Fluids Eng.*, vol. 10, no. 3, 1979.
16. Rehbach, C., Étude Numérique de Nappes Tourbillonnaires Issues d'une Ligne de Décollement Près du Bord d'Attaque, *Rech. Aerosp.*, vol. 6, pp. 325–330, 1973.

Chapter 5

Gases: Smokes

Thomas J. Mueller

I INTRODUCTION

Throughout the history of aerodynamics there has been a great deal of interest in development of techniques that can be used to help visualize a given flow phenomenon. This visualization of flow patterns has played a singularly important role in the advancement of our physical understanding of fluid mechanics. Flow visualization has led to the discovery of flow phenomena and has helped in the development of mathematical models for complex flow problems. It is also useful in the verification of existing theories, and has been an important tool in the development of complicated engineering systems. Most visualization methods allow for detailed study of a problem without the introduction of probes that can influence the character of the flow.

Flow visualization in wind tunnels closely followed the development of such tunnels, and can be traced to Dr. Ludwig Mach of Vienna in 1893 [1]. Mach's in-draft tunnel had a cross section of 180 × 250 mm, and was driven by a centrifugal fan that could produce a speed of 10 m/s. A piece of wire mesh over the inlet was used to straighten the flow. One side of the test section was clear glass and the others were painted black. The flow was observed and photographed by using silk threads, cigarette smoke, and glowing iron particles. The only smoke-flow photographs presented in Ref. [1] are for the flow past a plate perpendicular to the flow. The smoke is faint and difficult to make out; nevertheless, it was a beginning.

In France, about 1899, E. J. Marey, who was famous for his photographic studies of animal locomotion, turned his attention to photographing air in motion [2]. Marey, cognizant of Mach's work, used a vertical wind tunnel with a 200 × 300 mm cross section. The front and sides of the test section were glass, and the back was covered with black velvet. Air was drawn into the tunnel by a small suction fan after passing through fine silk gauze at the inlet. Smoke obtained by burning wood shavings entered the wind tunnel upstream of the gauze straighteners through a row of fine tubes. Excellent smoke-flow photographs were obtained using a magnesium flash.

Although the interest in smoke visualization continued during the first thirty years of the twentieth century, significant improvements in the technique and results were not achieved until the 1930s [3, 4]. The best of the early two-dimensional smoke tunnels was developed by A. M. Lippisch in Germany in the mid-1930s [5]. He obtained a large number of good smoke photographs for the flow around plates, cylinders, and airfoils, including the Lippisch rotor wing. Many of these photographs indicted that the turbulence level was higher than desirable. Lippisch also began the development of an intermittent smoke-delivery system.

In this same time period, F. N. M. Brown at the University of Notre Dame began his research in flow visualization. His first smoke tunnel, developed primarily for classroom demonstrations, became operational in 1937 [3, 4]. An in-draft tunnel with a large inlet-contraction section, it produced speeds up to 3 m/s. The test section was 305 mm wide by 1219 mm long by 1.58 mm deep. Photographs were obtained at about 3 m/s using titanium tetrachloride for smoke. Brown also developed a three-dimensional smoke tunnel that was operational in 1940. This in-draft tunnel had a single 6.30 mesh/cm screen followed by

The author would like to thank Drs. S. M. Batill, R. C. Nelson, and P. F. Dunn for their comments and suggestions during the preparation of this manuscript.

a 9 : 1 contraction in area. The test section, which had a 610-mm-square cross section, was about 914 mm in length, and speeds up to 12.2 m/s could be attained using a 1-hp DC motor. Smoke, produced by coking wheat straw, was introduced upstream of the antiturbulence screen through a row of tubes, i.e., a rake. A new three-dimensional smoke tunnel was designed in the early 1940s, but little progress was made during World War II.

Work began in 1947 on a research three-dimensional smoke tunnel, which was operational at the University of Notre Dame in 1948 [3, 6]. As a result of the lessons learned earlier, this smoke tunnel had five 5.51 × 7.09 mesh/cm bronze screens at the inlet of the 12 : 1 contraction section. The test section was maintained at 610 × 610 mm in cross section and 914 mm long. Useful speeds of 10.7 m/s could be attained using a centrifugal fan driven by a 5-hp motor. Coked wheat straw was used with a total of 12,000 W of steady lighting to obtain photographs of the flow. This three-dimensional smoke tunnel evolved slowly into the ones used today. In conjunction with his smoke tunnel development, Brown developed a movable smoke rake and the first easy-to-use kerosene smoke generator that could produce large quantities of smoke. He was also the first to take three-dimensional and stereo photographs of smoke flows. In fact, most of the progress and refinement of smoke-visualization techniques are credited to Brown [in 7].

Expanding the techniques of Brown, V. P. Goddard (also at the University of Notre Dame) was able to produce the world's first smoke-visualization supersonic tunnel in 1959 [8, 9]. An in-draft supersonic wind tunnel with seven screens and a large inlet contraction to the nozzle throat was designed and built. A modified Schlieren system permitted the simultaneous photographing of both smoke and shock-wave patterns. Using the same smoke-generation and -injection techniques as in the subsonic tunnels, smoke photographs were taken at speeds up to 404 m/s (i.e., a Mach number of 1.38). Most researchers in this field still do not believe this is possible until they actually see it. Batill et al. [10–12] have developed quantitative procedures for the design of three-dimensional wind tunnel inlets. The inlet and other flow-conditioning devices (e.g., screening and honeycombs) are of primary importance in achieving the uniform low-turbulence flow in the test section of a subsonic or supersonic smoke-visualization wind tunnel.

II EQUIPMENT AND TECHNIQUES

It was clear by 1959 [13, 14] that the smoke tunnel was an important research tool. From publications since 1959, it is evident that the state of the art established by Brown, Goddard, and Lippisch has not changed to the present day. It is also evident from the work of Brown and Goddard that the equipment necessary to develop smoke visualization must be considered and designed as a complete system. The system consists of a low-turbulence wind tunnel, smoke-generation and injection apparatuses, and lighting and photographic equipment. The arrangement and use of these system components must be coordinated to obtain good results.

A Low-Turbulence Wind Tunnels

The smoke wind tunnels developed by Brown and his colleagues represent a compromise. Because all smoke is toxic to some degree, an in-draft tunnel that exhausts outside the laboratory is highly desirable. Furthermore, because the best position for the smoke tubes is upstream of the inlet, the distance the smoke travels from injector to test section must be as small as possible to avoid excessive diffusion. This requires that the inlet (i.e., contraction section and screens) be as short as possible while producing a turbulence intensity less than 0.1% in the test section. The latest version of Brown's subsonic smoke tunnel, shown in Fig. 1, uses an inlet with 12 screens followed by 24 : 1 contraction in area to the test section. The shape of the contraction was obtained from the two-dimensional studies of Smith and Wang [15]. These tunnels have operated successfully for more than 30 years.

Choosing the shape of the contraction section has been simplified by the development of design criteria for three-dimensional (square or rectangular cross-section) inlets. A rational design procedure for high-contraction-ratio subsonic wind tunnel inlets has been developed by Batill and his colleagues [10–12] as a result of extensive experimental and numerical research. A set of design charts was developed for a family of wall contours defined by matched cubic polynomials. These charts can be used for the aerodynamic design of high-contraction-ratio inlets in the range of 10 : 1 to 40 : 1. A subsonic inlet with a contraction ratio was designed using these charts and evaluated in conjunction with various screen configurations [10]. The results of these experiments will be discussed later.

B Smoke-Generation and -Injection Techniques

The word *smoke* is used in a very broad sense, and includes a variety of smokelike materials such as vapors, fumes, and mists. The smoke must be generated in a safe manner and must possess the necessary light-scattering qualities so that it can be readily photographed. It is also important that the smoke not adversely affect either the wind tunnel or the model being studied. Another desirable, but not necessary, qualification is that the smoke be nontoxic, in the unlikely

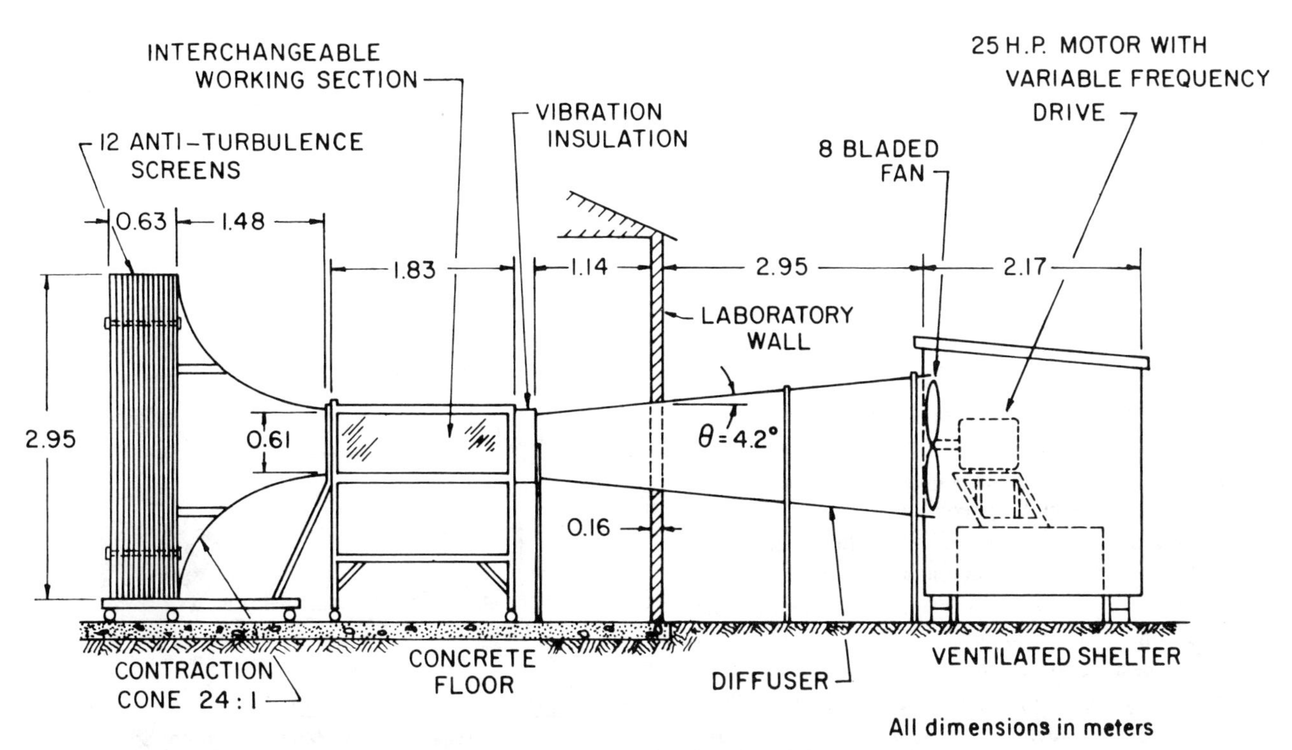

Fig. 1 Low-turbulence subsonic smoke wind tunnel (courtesy of University of Notre Dame).

event that the experimenters are exposed to it. Finding a smokelike substance that meets all these criteria is not easy.

Many methods have been used to generate smoke. Examples are the combustion of tobacco [1], rotten wood [2], wheat straw [3]; the products of reactions of various chemical substances such as titanium tetrachloride and water vapor; and the vaporization of hydrocarbon oils, to name a few [16]. The smoke so generated is referred to as an aerosol, which is defined as either a fine liquid or a fine solid suspension in gas. A great deal of interest has been focused on the properties and generation of aerosols because of their close relationship to meteorology, air pollution, cloud chambers, smokes, combustion of fuels, colloid chemistry, etc. [16].

Two very practical items must be carefully examined before the smoke-generating material is chosen for flow visualization. The smoke or aerosol particles must be small enough to follow closely the flow pattern being studied [17], but large enough to scatter enough light so that photographs of the smoke pattern can be obtained. These typically are particles having diameters on the order of the wavelength of the incident light, from approximately 0.1 to 1.0 μm. Although many materials and substances have particle sizes below 1 μm, the most practical ones for flow visualization are tobacco smoke (0.01–1.0 μm), rosin smoke (0.01–1.0 μm), carbon black (0.01–0.03 μm), and oil smoke (0.03–1.0 μm). Particles of this size usually have particle Stokes numbers much less than unity, implying that they follow the flow. Note that water-vapor droplets (fog) are generally much larger than 1 μm, in the range of 2 to 70 μm. If one considers the light-scattering ability of the particles used, another constraint becomes apparent. According to the laboratory methods of measuring particle size, particles should be larger than about 0.10 μm to scatter enough light to be readily seen [17]. This light-scattering criterion indicates that tobacco smoke and carbon black particles are mostly lower than 0.1 μm and are therefore more difficult to photograph.

Resin is a semisolid organic substance exuded from various plants and trees or prepared synthetically, whereas rosin is the hard, brittle residue remaining after oil of turpentine has been distilled. Maltby and Keating [18] describe an electrically fused, pyrotechnic resin smoke generator that was developed for use in wind tunnels with speeds up to 16 m/s. This smoke generator was manufactured by Brocks Fireworks Company, Ltd., in England and was based on the vaporization of resin. Maltby and Keating [18] also mention that some ammonium chlorate is present in the resin canisters, which must be stored away from heat since this substance is unstable. The device is simply a smoke bomb adapted for wind tunnel use. This appears to be the only mention of a resin or rosin smoke generator used for wind tunnel flow visualization.

There are, of course, a large number of possible hydrocarbon mixtures, i.e., oils, which undoubtedly could be used to produce smoke by combustion or vaporization. From the point of view of laboratory safety, it would be desirable to use vaporization rather than the combustion technique. Table 1 shows properties of interest for three common hydrocarbon mixtures [19]. It would also be safer

Table 1 Physical Properties of Hydrocarbons Used for Smoke Generation (adapted from Ref. [16])

Hydrocarbon	Boiling Point, °F (°C)	Flash Point, °F (°C)	Auto Ignition, °F (°C)
Mineral oil	600 (315.5)	275–500 (135–260)	500–700 (260–371)
Kerosene	350–550 (176.6–287.7)	110–130 (43.3–54.4)	440–560 (226.6–293.3)
Coal tar oil	96 (35.5)	60–77 (15.5–25)	

to use an oil that would vaporize at the lowest possible temperature and be the least flammable. Figure 2 presents data for the temperature versus percentage distilled (vaporized) for six oils. Mineral oil requires the highest temperature vaporization, while charcoal lighter fluid requires the lowest. The second lowest temperature for vaporization is for kerosene. Since kerosene is less flammable than charcoal lighter fluid, it is the obvious choice. Kerosene seems to offer the best compromise when particle size, light-scattering ability, low vaporization temperature, and low flammability are considered. It should also be mentioned that particles smaller than 1 μm in diameter can be readily inhaled into the lungs and should therefore be used with care.

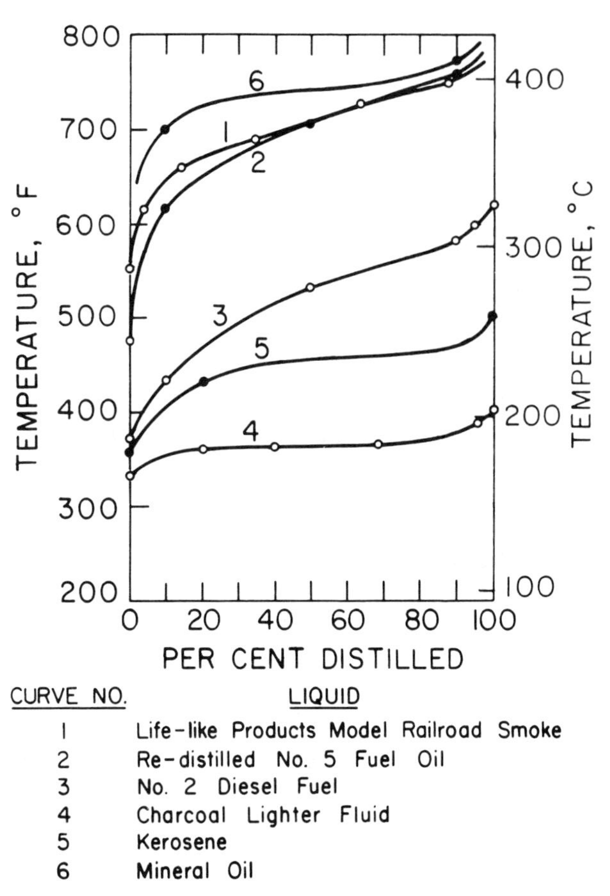

Fig. 2 Distillation curves for hydrocarbon mixture [16].

C Smoke-Tube Method

Although the first oil smoke generator was developed by Preston and Sweeting [20], one of the most successful oil smoke generators was designed by F. N. M. Brown in 1961. Large quantities of dense kerosene smoke were produced quickly and safely with this generator. This four-tube kerosene generator is shown schematically in Fig. 3.

A flat-electric-strip heater is shown inside a square thin-wall tube. The entire unit is set at a convenient angle (about 60°) and a sight-feed oiler is mounted on the unit at the upper end of each tube so that oil drips on the upper end of the strip heater. It has been determined that a drip rate of approximately two drops per second is enough to produce an adequate amount of smoke for the applications shown later. Faster rates result in inefficient, wasteful operation. Furthermore, an extremely fast drip rate can result in backfiring of the unit. A centrifugal blower mounted at the low end of the unit is used to force the smoke through the system. The centrifugal blower is more or less mandatory in the event of backfiring; the sudden increase in pressure is easily transmitted through the rotor.

Before entering the smoke rake, the smoke passes through a heat exchanger made of 42-mm-dia. pipe, as shown in Fig. 4. The primary purpose of this heat exchanger is to cool the smoke to room temperature. The entire system has drain cocks conveniently located; one is at the bottom of each tube of the generator to remove excess oil not converted to smoke, and others are at the bottom of the heat exchanger to remove any condensed oil. After passing through the heat exchanger condenser system, the smoke flows into a 97-mm-dia. manifold, and is passed through an absorbent cloth filter. This filter serves a dual purpose: it removes most of the remaining lighter tars, and it helps distribute the smoke uniformly into the evenly spaced tubes that extend from the manifold. These tubes determine the initial smokeline spacing. Such an array of tubes with the manifold has a rakelike appearance and is therefore referred to as a smoke rake.

Appropriate measures should be taken to guard against leaks in the system, and to provide an outside exhaust from the tunnel, for the inhalation of oil smoke is a health hazard. Various types of oil have been tried. It has been found that kerosene produces the best quality of smoke. Oil smoke generators of this type have been constructed in single, double, and quadruple units. This grouping of units is basically a method of increasing the volumetric output of smoke. This type of smoke generator has been modified by the Lockheed–Georgia Company to include flame detectors and a nitrogen purge system to satisfy industrial safety requirements [14].

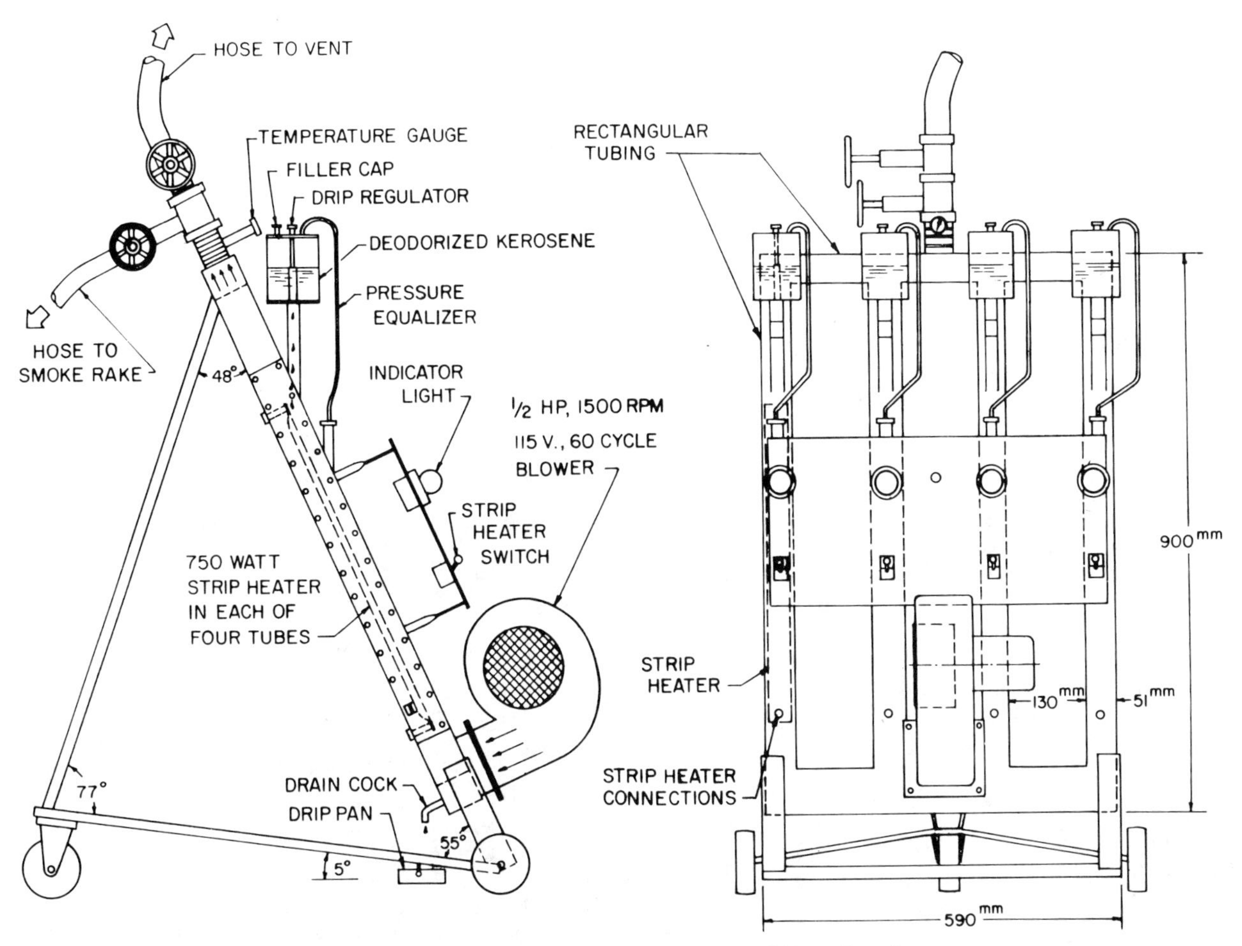

Fig. 3 F. N. M. Brown kerosene smoke generator (courtesy of University of Notre Dame).

D Alternative Methods

The only reason for trying to replace the kerosene or other oil smoke-generation techniques would be to produce a less toxic and nonflammable smokelike substance. All products of combustion, reactions of chemical substances, oil and paraffin, vapors, and aerosols are toxic to some degree [21–23]. Many of these substances are flammable, and some are corrosive or chemically active. The only technique available that does not have one or more of these undesirable properties appears to be the steam–liquid nitrogen method [24, 25]. However, generating large quantities of steam in some type of boiler presents a different safety hazard. After generation, the steam is mixed with liquid nitrogen and introduced into the wind tunnel. Although the apparatus used to generate the steam is somewhat different in Refs. [24] and [25], the end product is the same. The MIT, Iowa State, and Notre Dame University laboratories have used pure steam with reasonable success. The steam or steam–liquid nitrogen mixtures have been used in both in-draft and closed-circuit subsonic wind tunnels.

Although the use of steam for flow visualization has the advantage of being clean and nontoxic, it does have disadvantages. For example, the system temperature must be controlled carefully if a neutrally buoyant fog is to be obtained. As pointed out earlier, water vapor, on the average, is composed of much larger particles than is oil smoke. Another disadvantage is that steam condenses to water on cold model surfaces, walls, and other protrusions in the test section. The steam does photograph reasonably well, and it can produce usable visualization records.

Griffin and Votaw [21] developed an aerosol generation system for low-speed wind tunnel flow visualization. This system used a polydisperse liquid aerosol of di(2-ethylhexyl)-phthalate, referred to as DOP. This flow visualization method worked well at the relatively low speed used (maximum tunnel speed was 8 m/s). One of the reasons this aerosol system was developed was to avoid the toxic nature of oil smokes. Although DOP was originally thought to be nontoxic, later studies found it to be toxic, and this limits its use.

Although many substances have been used to produce smoke for flow visualization in wind tunnels, kerosene smoke appears to have a significant advantage over all other substances studied. However, it must be used with great care because of its toxic nature.

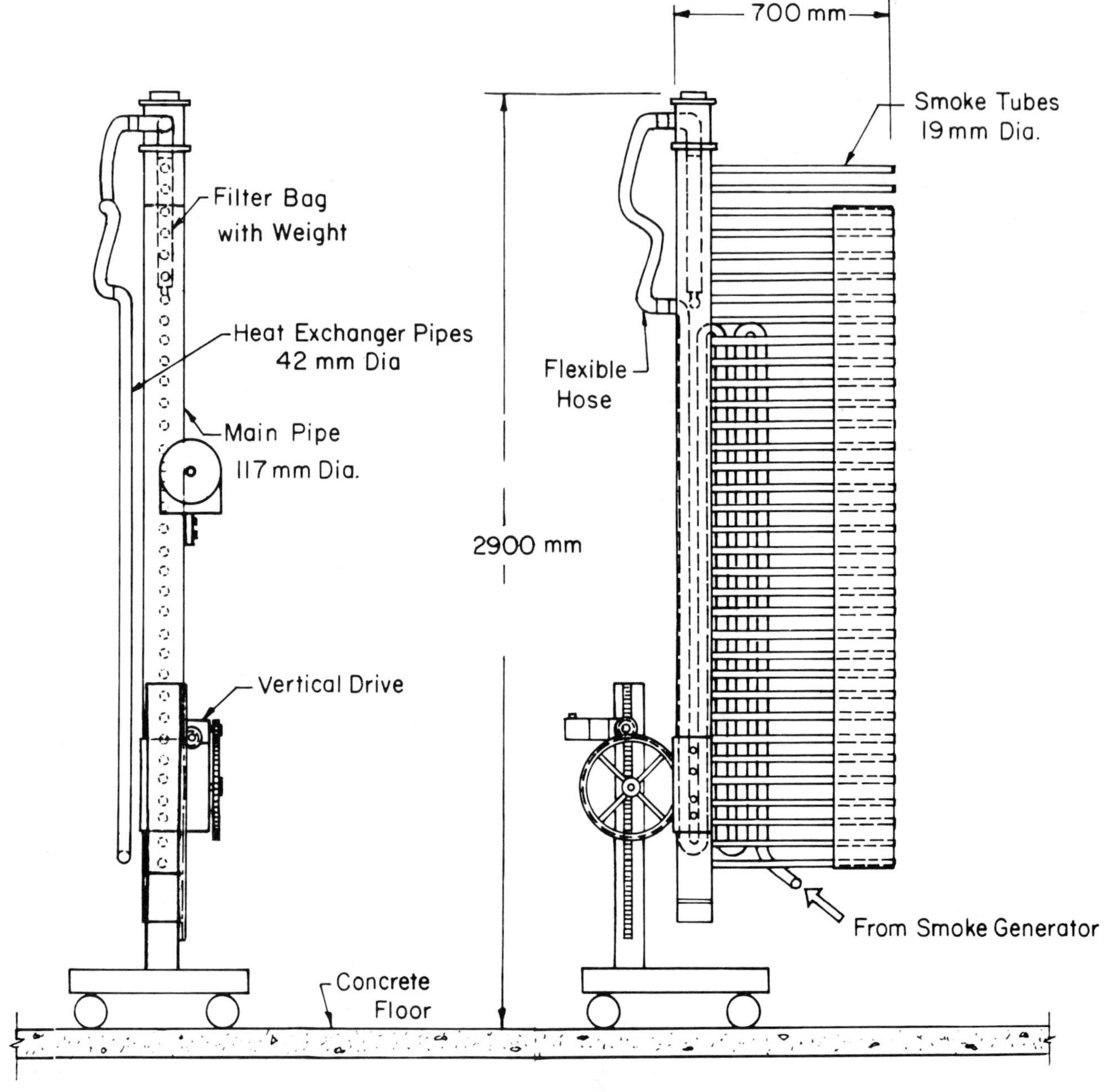

Fig. 4 F. N. M. Brown smoke rake (courtesy of University of Notre Dame).

E Smoke-Wire Method

For very-low-speed experiments (i.e., test section velocities less than about 4 m/s), the smoke wire may be used where the smoke is generated and introduced directly in the test section [13]. This technique does not require a low-turbulence environment in the test section as does the smoke-tube method.

The smoke-wire technique, developed by Raspet and Moore at Mississippi State University in the early 1950s and subsequently improved and extended [26–30], is capable of producing very fine smoke filaments, and it can be used to study the detailed structure of complex flow phenomena. The "smoke"[1] is produced by vaporizing oil from a fine (≈0.1 mm) wire by the use of resistive heating. As the wire is heated, it expands and sags, which is not desirable; The wire should therefore be prestressed. The technique was initially applied to the measurement of velocity profiles in a boundary layer. Applications have included the investigation of the large eddy structure in tur-

[1]Although the product is not smoke in a strict definition of the term, it will be referred to as such.

bulent shear flows [30–32], as well as flow over airfoils [33], wings [34], and circular cylinders [35].

The smoke-wire technique is limited but fortunately is ideally suited to applications where the Reynolds number based on wire diameter is low (<40). Practical applications require wind tunnel speeds on the order of 4–6 m/s. The method consists of a fine wire positioned in the flow field, coated with oil, and heated by passing an electrical current through the wire [13]. As the wire is coated with the oil, small beads of the oil form on the wire, and at each of these beads the smoke filaments originate when the wire is heated. Cornell [16] conducted an extremely thorough study of materials used in smoke generation for flow visualization; he indicated that particles formed in this manner should be classified as a vapor-condensation aerosol, so they are actually small liquid particles ($\simeq$1 μm dia.), not solid particles or products of a combustion process.

A number of different liquids have been used to coat the smoke wire to produce the smoke filaments. These include several types of lubricating and mineral oils and a commercially available product, Life-Like Model Train Smoke, produced by Life-Like Products, Baltimore, Maryland. The results were similar to those achieved in Ref. [40], which indicated that the model train smoke produced the best smoke filaments. This product is composed of a commercial-grade mineral oil to which a small amount of oil of anise and blue dye have been added. The oil is very easy to work with, and such small quantities are used that the commercial product is ideal and inexpensive.

There would be obvious safety problems associated with large quantities of the smoke, but since small amounts are produced, safety problems are minimized. Reference [29] indicates that since the amount of smoke is so small, the method is quite suitable for limited use in closed-circuit tunnels.

As the coated wire is heated, fine smoke streaklines are formed at each droplet on the wire ($\simeq$8 line/cm for the 0.076-mm wire). Depending upon the current through the wire, the beads can be vaporized very rapidly or, if a lower current is used, continuous filaments of adequate density will emanate from the wire for as long as 2 s. With lower current, the streaklines become fainter and cannot be photographed. For example, a 0.076-mm-dia. 302 stainless wire of 0.040-m length was heated using a power supply setting of approximately 0.7 amp. Both AC and DC power supplies can be used, but a DC power supply is usually used because it is easier to control and because steady-state current results in smoother and better defined smokelines. Because of the relatively short duration of the smoke generation for a single wire coating, it is important that the event being photographed, the lighting, the camera, and the smoke be properly controlled and synchronized. To accomplish this, a timing circuit should be designed and used. Examples of such timing circuits are given in Refs. [13, 29].

F Smoke-Flow Photography

One of the most critical aspects of flow visualization photography is getting enough light on the subject of interest. In wind tunnel applications this problem is often the most difficult to overcome because of space constraints. The main problem is the large quantity of light required for proper illumination of the smoke streaklines while maintaining high contrast between the smoke and the background. As most wind tunnels have very limited viewing areas, one is generally forced to use less light than desirable to minimize unwanted reflections or adverse heating problems. Space constraints often restrict the placement of lights and cameras.

Figure 5 illustrates the three most commonly used lighting arrangements. When the lights are positioned on the opposite side of the tunnel and out of the direct field of view of the camera, the arrangement is called *back lighting*. Here, the forward-scattered light from the particles is photographed. Glare from the model can be minimized using the back-lighting arrangement. If the lights are positioned on the same side of the tunnel as the camera, the arrange-

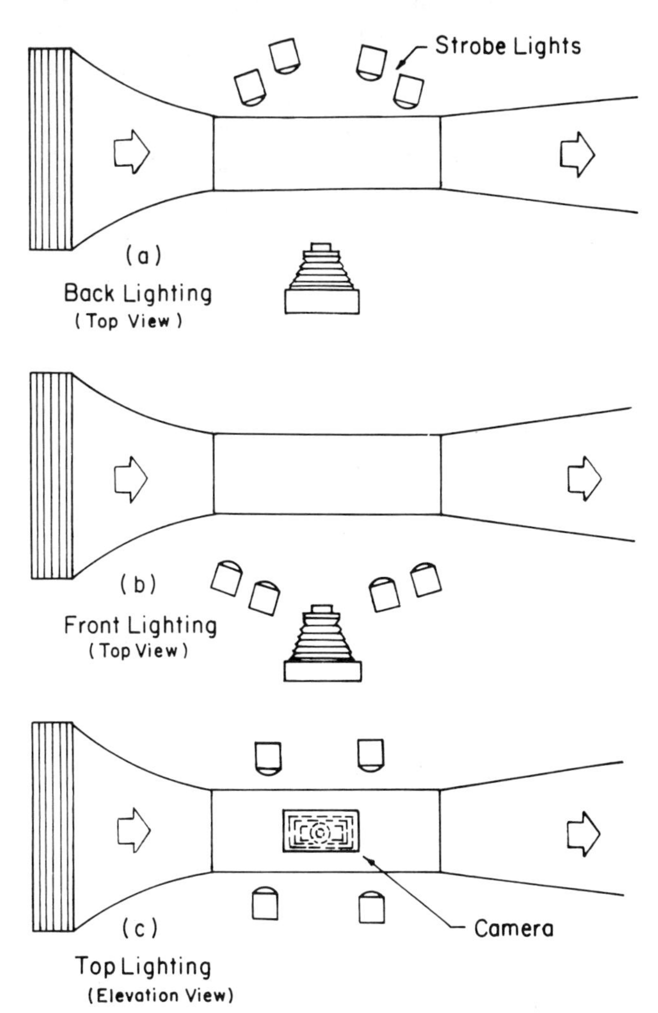

Fig. 5 Lighting and camera arrangements.

ment is called *front lighting*. For this arrangement, the back-scattered light from the particles is photographed. Major difficulties with front lighting are the reflections and glare from the model and tunnel windows. These problems can be minimized by proper positioning of the lights and camera. When there is a limited viewing area and space is restricted, however, it may not be possible to position the lights properly. The final lighting arrangement is called *top* or *bottom lighting*. This arrangement can be used in any test section where the flow can be illuminated in a plane perpendicular to the normal viewing direction.

The type of light source required for smoke-visualization tests depends upon the particular test being conducted. The two most widely used light sources are stroboscopic and high-intensity continuous lamps. Short-duration flash lamps such as the GenRad type 1540 strobolumes or their equivalent have been found to produce satisfactory results. These lamps operate from an electrical source of 120 V and 60 cycles. The power consumed varies up to 500 W, depending upon the flash rate. The rated duration of the flash is approximately 30 μs. For still photographs, the strobolumes are synchronized with the camera shutter, and they operate similarly to any camera using an electronic flash. With movies, the flash rate is adjusted to the framing speed. The number of lamps required depends on the nature of the test being conducted. For continuous lighting, high-intensity spot lamps are required. One of the most significant developments in the field of smoke visualization in the past 10 years has been the introduction of laser light to illuminate the smoke. Although other light sources have been used for many years to produce narrow sheets of light, the minimal spreading of laser sheets together with the variety of power levels available have had a dramatic effect on smoke visualization. The sheet of laser light is usually produced by passing the laser beam through a cylindrical lens or a glass rod, as illustrated in Fig. 6.

G Photographic Procedures

Over the years, many universities and both government and private research laboratories have attempted to obtain flow visualization data at subsonic speeds. Most of these efforts have ended in failure for a variety of reasons. Even in tunnels where quality smoke is available, only poor smoke-flow photographs were obtained. One can only speculate on the reasons for such failure. One difficulty is probably the ability of the experimenters to see the flow pattern with their eyes, while being unable to obtain good photographic records. This is usually because, in such installations, the photographic work is left to a photographic technician. To achieve quality smoke photographs, the experimental aerodynamicist should also be a photographer. An experimental aerodynamicist is familiar with the phenomena and their influences on the smoke pattern in the flow field. This is not to say that a photographic technician cannot be used; however, the aerodynamicist must be willing to spend enough time with the photographer for the technician to know what is required [19].

Processing the photographic data presents another challenge. Due to the difficulties in lighting, the photographic plates are usually underexposed and the exposure is gen-

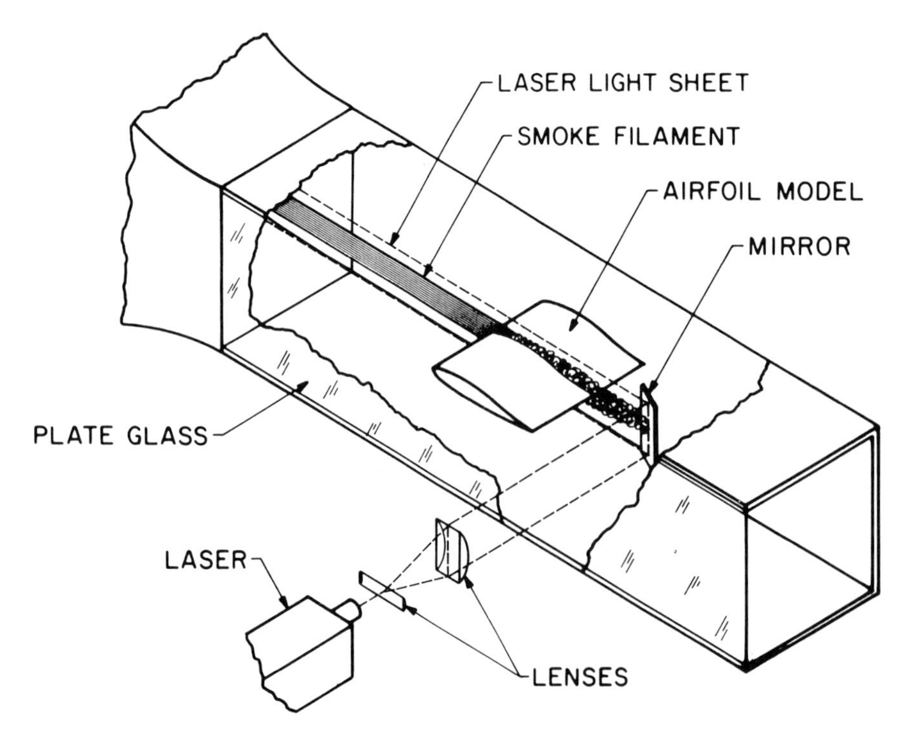

Fig. 6 Low-speed wind tunnel test section showing a laser-light arrangement for smoke visualization.

erally uneven. To produce a quality print, an uneven exposure of the negative is therefore required. This procedure, referred to as *photographic dodging*, enables one to enhance the photographic image, i.e., to capture all the data on the print. Again, if the person processing the film does not know what to look for in the print, much, if not all, useful data will be lost.

It should be remembered that the use of flow visualization techniques in the study of complicated flow patterns is an much as art as it is a science. Obtaining good flow visualization data requires a great deal of patience and time; each new application will present the researcher with new photographic difficulties. However, with patience and some effort, meaningful flow visualization data can be acquired. The interest in obtaining quantitative information from smoke photographs has led to use of digital computers to analyze the global and local flow patterns [35]. The use of digital image processing to analyze flow visualization images from photographs or video frames is increasing very rapidly [36–41]. Examples of the results obtained from a study of the interaction of turbulent scales in boundary layer flows are described by Corke in Refs. [39, 42].

III EXAMPLES OF SMOKE VISUALIZATION

As indicated earlier, the inlet design along with the associated turbulence-management devices such as screens play a critical role in development of a smoke-visualization wind tunnel. During an experimental evaluation of a 30 : 1 contraction ratio inlet, designed matched cubic polynomial wall geometry, the influence of the number of screens was also studied by Batill et al. [10]. The inlet was fabricated with clear Lexan side walls for flow visualization studies. Five screens were made of multistrand nylon mesh with 0.051-mm-dia. strand in a 10.24/cm mesh. The 6 screens placed upstream of the nylon screens were made of aluminum with a 0.228-mm-dia. wire in 7.08/cm mesh. The inlet entrance was 1193.8 mm square, the exit was 217.9 mm square, and the distance between the entrance and exit was 1437.6 mm. The flow uniformity, angularity, and turbulence intensity were measured for cases with 0–11 screens. The effect of the number of screens on the coherence of the smokelines, or smoke tubes, injected at the beginning of the inlet or at the first screen was also studied. Figure 7 shows the smoke tubes in the inlet for 2, 4, and 11 screens, using the back-lighting method. For 2 screens the smoke tube began to break up within the first 20% of the length of the inlet, and was almost completely diffused before reaching the test section. Increasing the number of screens to 4 improved the integrity of the smoke tubes, but there were still signs of flow unsteadiness. The use of 11 screens resulted in well-defined, coherent smoke tubes throughout the length of the inlet. Antiturbulence devices must be thought of as a single system and therefore designed together.

The use of the smoke-tube method for an airfoil using top lighting (i.e., the arrangement shown in Fig. 5*c*) is shown in Fig. 8. If the smoke tube that impinges on the nose of the model is too high (i.e., Fig. 8*a*), a large amount of smoke goes into, and adjacent to, the boundary layer. In this situation, the details of free-shear layer transition after separation on the upper surface are masked by the excessive amount of smoke. However, if the impinging smoke tube is lowered so that a very small amount of smoke goes into the boundary layer (i.e., Fig. 8*b*), more detail is obtained of the free-shear layer undergoing transition. The unsteady character of the airfoil wake is evident if one compares Figs. 8*a* and 8*b*, which, although taken for the same airfoil and wind tunnel settings, were taken at different times in the unsteady-flow cycle.

The smoke photograph shown in Fig. 9 was taken during a study of boundary-layer characteristics of the Lissaman 7769 airfoil at low Reynolds numbers. The boundary layer on the upper surface of this airfoil separates at about 35% chord, and some indication of the transition from laminar to turbulent flow after separation is observed. On the lower surface, a leading-edge separation bubble is present, which acts as a boundary-layer trip. Thus, the boundary layer on the lower surface is turbulent from about the 15% chord location to the trailing edge. These observations obtained from smoke photographs are very useful in explaining and interpreting data obtained from force, pressure, and hot-wire anemometer measurements. When the free-stream turbulence level of the basic wind tunnel (i.e., <0.1%) is raised to 0.3% by placing a screen at the beginning of the test section, the smoke tubes appear as shown in Fig. 10. A comparison of Figs. 9 and 10 clearly indicates the importance of a low turbulence level in smoke-visualization wind tunnels.

The increased interest in improving the aerodynamic characteristics of land vehicles offers another opportunity to use smoke visualization to study flow patterns [43, 44]. Typical results from three-dimensional studies of the flow around a car body are shown in Fig. 11 [44]. This study is of a 1/5-scale model of a hardtop car, using a hydrocarbon vapor. Figure 11*a* shows the side view with 0°-angle yaw. For this case, the flow appears to be attached until it reaches the rear end of the roof. What appears to be a thick smokeline flowing from slightly above the bumper along the side of the model is actually a group of parallel smokelines in a horizontal plane. Two views of the hardtop model at a yaw angle of 20° are shown in Figs. 11*b* and 11*c*. These photographs show vortices produced by the car geometry at this condition. These visual results correlate with the aerodynamic characteristics of the car model.

The smoke-tube method combined with a laser-light sheet have provided quantitative data concerning the vortex trajectories produced by bodies at high angle of attack [45,

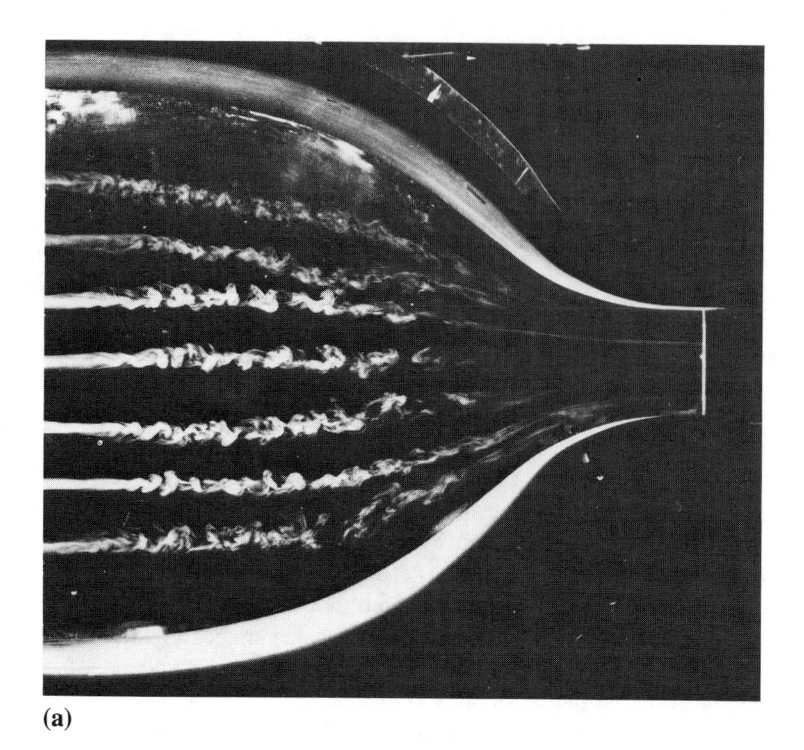

(a)

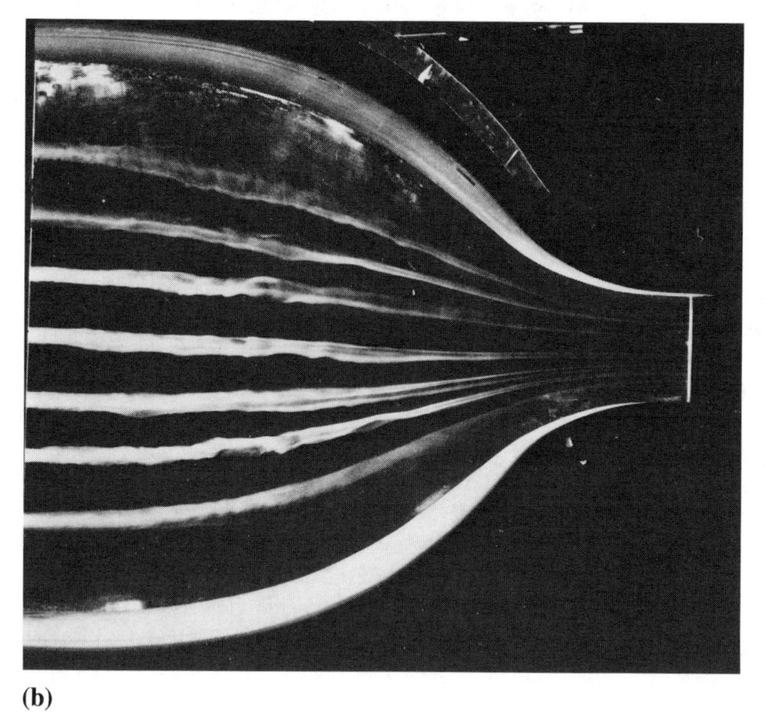

(b)

Fig. 7 Flow visualization of screen influence: (*a*) 2; (*b*) 4; (*c*) 11 screens (courtesy of S. M. Batill, University of Notre Dame).

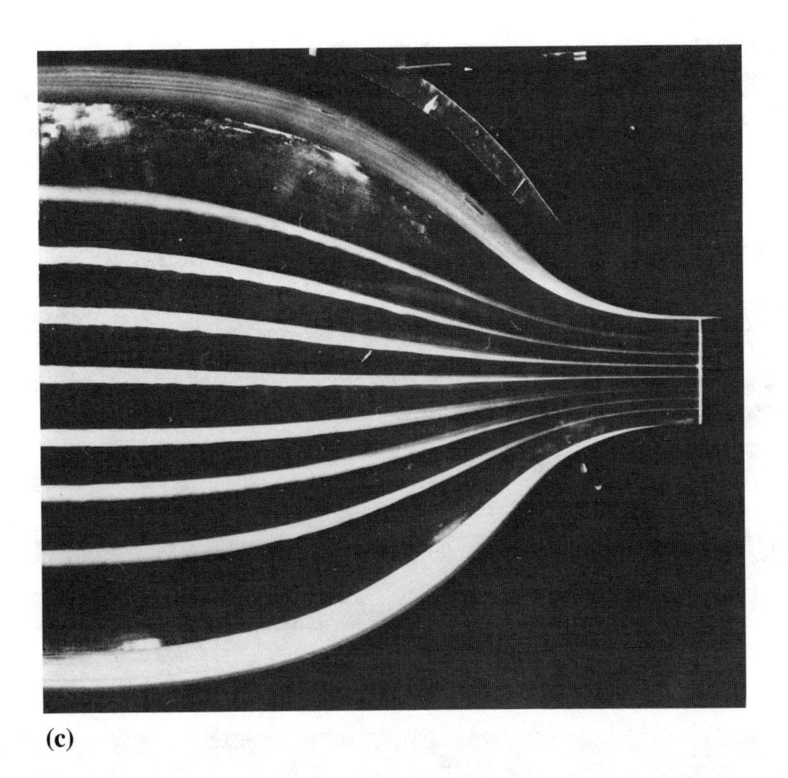

Fig. 7 (continued) **(c)**

46]. The experimental arrangement in a low-speed smoke tunnel is shown in Fig. 12. The laser-light sheet illuminates a cross section of the flow field. The intense light sheet can be titlted to any angle and moved along the longitudinal axis of the model. The vortex pattern from a finned-cone model at 25% angle of attack using an Argon–Ion laser with an output of about 3 W is shown in Fig. 13. The smoke photograph was obtained by using multiple exposures with the laser-light sheet at different longitudinal stations along the model. For this condition, the vortex pattern is symmetrical, and the trajectories can be measured directly from the photograph [46].

The difficulties associated with transonic flow experiments are well known. The opportunity to visualize transonic flows with smoke provides an additional tool that complements the well-established optical techniques. It appears that the only transonic smoke tunnel is the one developed about 1980 at the University of Notre Dame in conjunction with the U.S. Air Force Wright Aeronautical Laboratories [19]. This facility is of the in-draft continuous-flow type. The inlet has 11 screens followed by a contraction ratio of 150 to the test section, which is 101.6 mm square. The upper and lower walls of the test section have slots feeding into suction plenums. The two side walls were designed to include glass for schlieren and/or smoke photography. Figure 14 shows an example of the photographic results obtained in a study of the wake behind a circular cylinder, using the back-lighting method. The quality of the visualization data is good since the change in wake structure between model numbers of 0.75 and 1.08 is clearly seen.

In a study of the three-dimensional flow over finite wings at low Reynolds numbers, Bastedo and Mueller [34] used the smoke-wire technique and top lighting. Figure 15 shows the smoke-wire visualization of the wing-tip vortex at a chord Reynolds number of 80,000. The smoke wire was located at a position 87% of the distance from wing root to wing tip. The wing is at an angle of attack of 15°. Visualization results of this type were helpful in interpreting static pressure measurements at various spanwise stations.

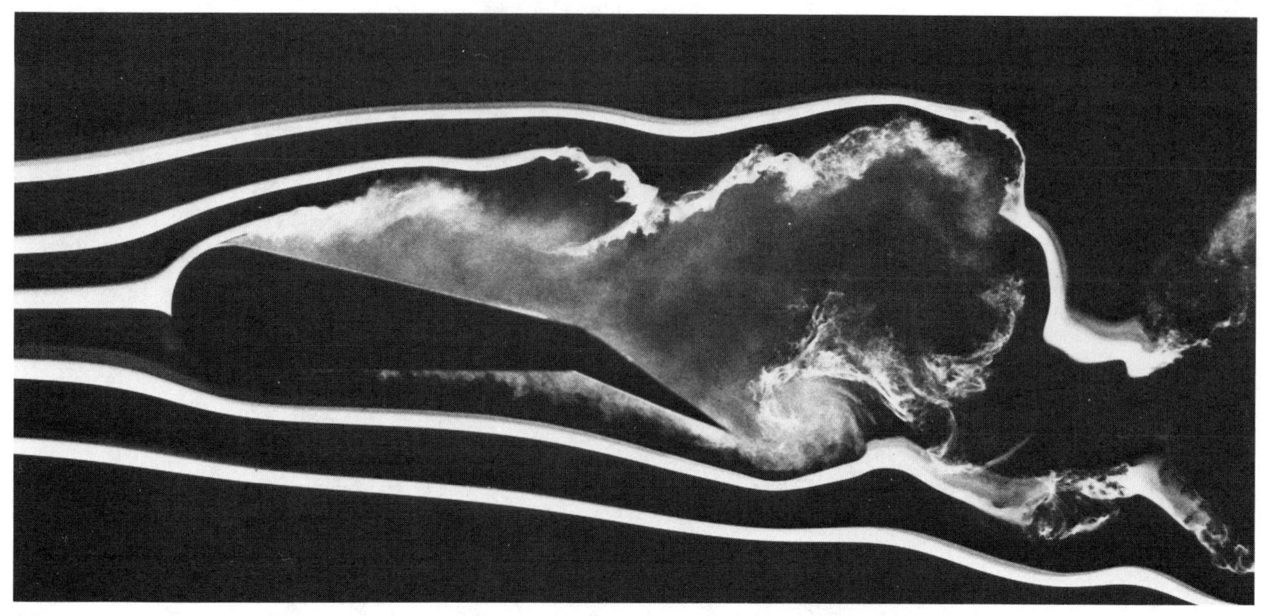

(a)

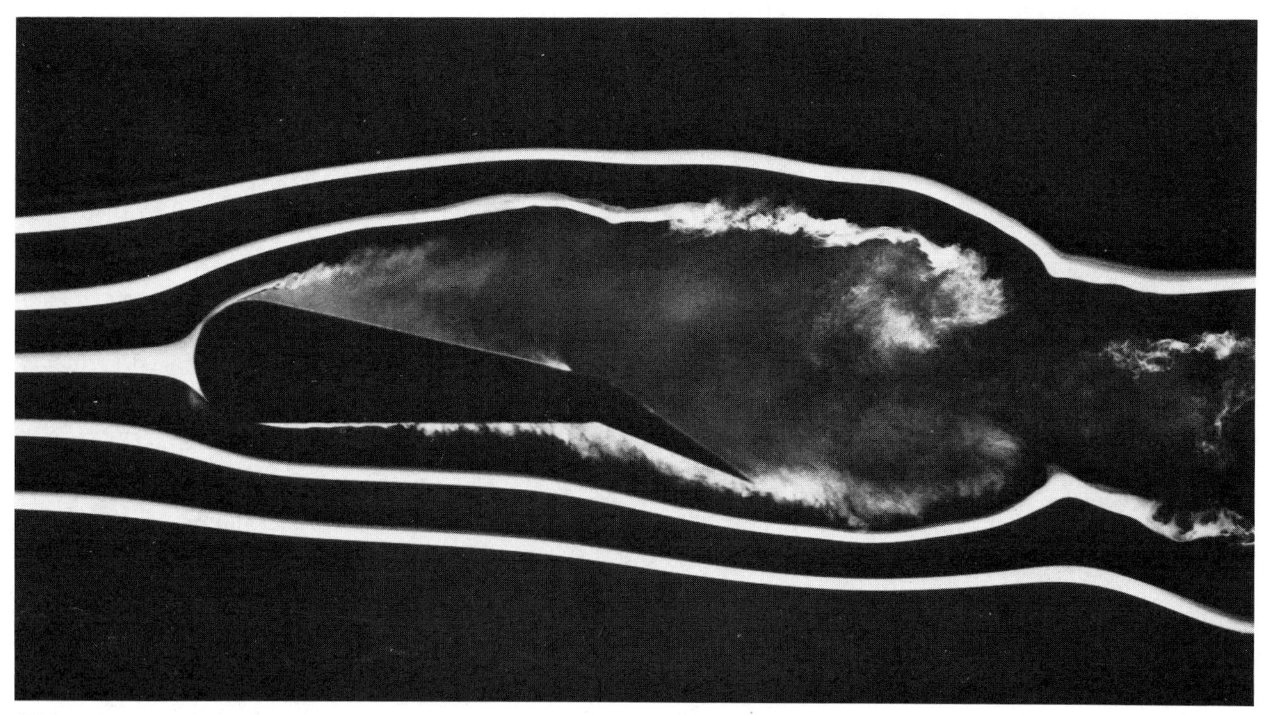

(b)

Fig. 8 Cylindrical leading-edge model at an angle of attack of 5°, a flap angle of 20°, and a chord Reynolds number of 235,000: (*a*) Large amount of smoke masks details of leading-edge separation and subsequent free-shear layer transition; (*b*) small amount of smoke allows the separation and the early stages of free-shear layer transition to be seen (courtesy of T. J. Mueller, University of Notre Dame).

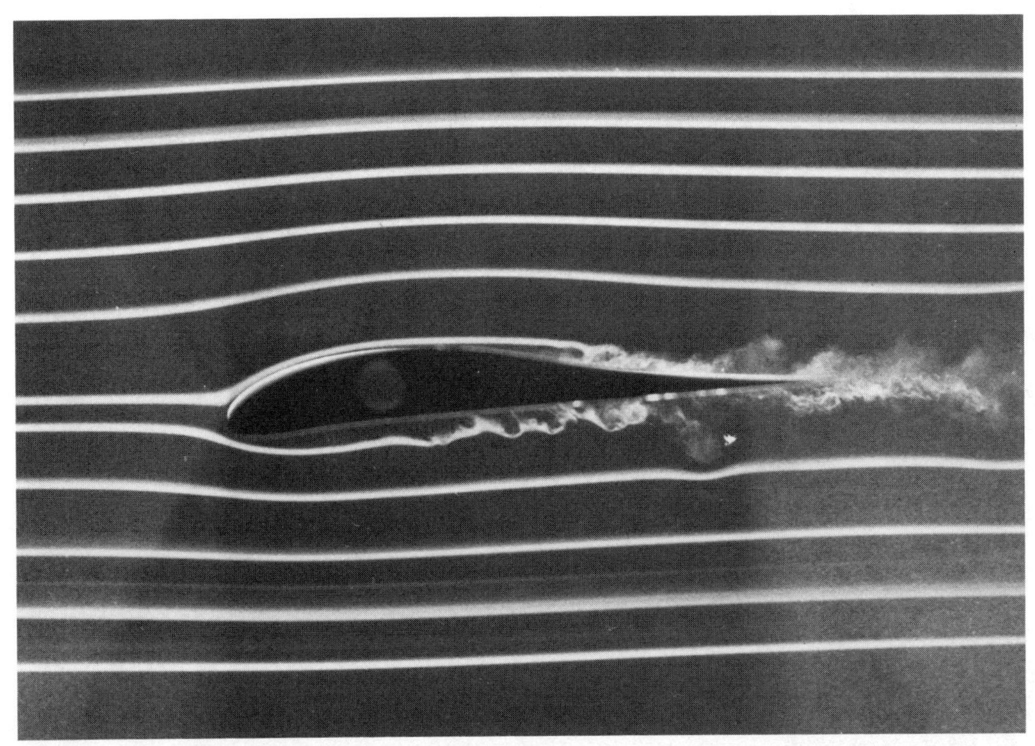

Fig. 9 Smoke photograph of the Lissaman 7769 Airfoil at a $-2.5°$ angle of attack and a chord Reynolds number of 150,000. (courtesy of T. J. Mueller, University of Notre Dame).

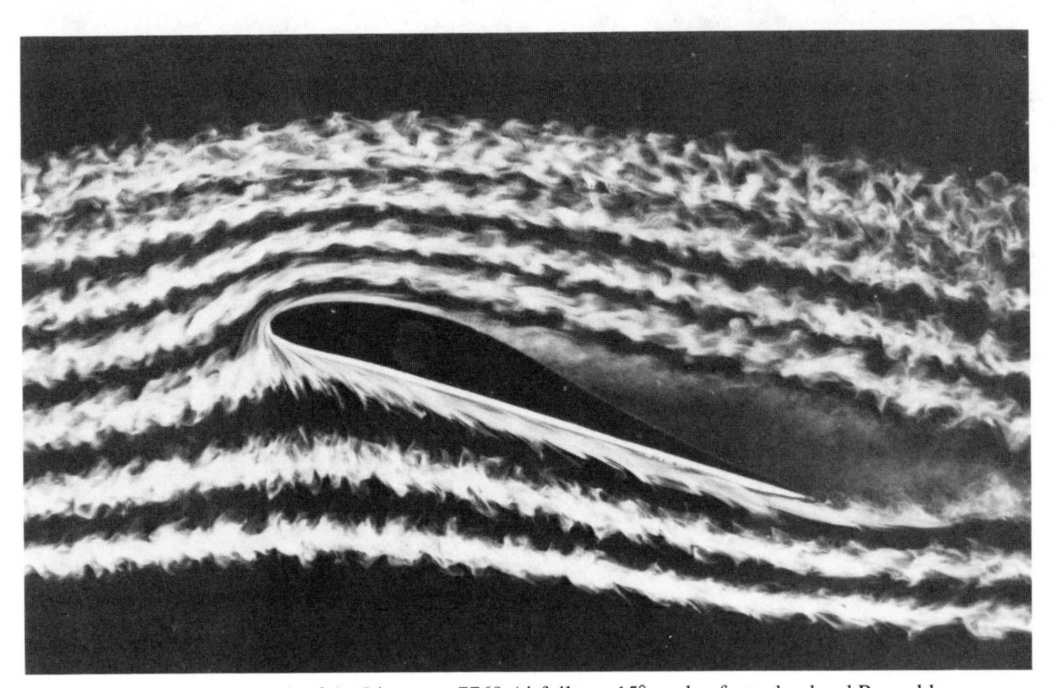

Fig. 10 Smoke photograph of the Lissaman 7769 Airfoil at a 15° angle of attack, chord Reynolds number of 150,000, and a free-stream turbulence level of 0.3%, produced by a 7.09-mesh/cm screen at the upstream end of the test section (courtesy of T. J. Mueller, University of Notre Dame).

(a)

(b)

Fig. 11 Three-dimensional smoke tunnel with 1/5-scale hardtop model, 5 m/s, liquid paraffin, mist: (*a*) Side view, $\beta = 0°$; (*b*) plan view, $\beta = 20°$; (*c*) rear view, $\beta = 20°$ (courtesy of M. Takagi, Nissan Motor Co., Ltd.).

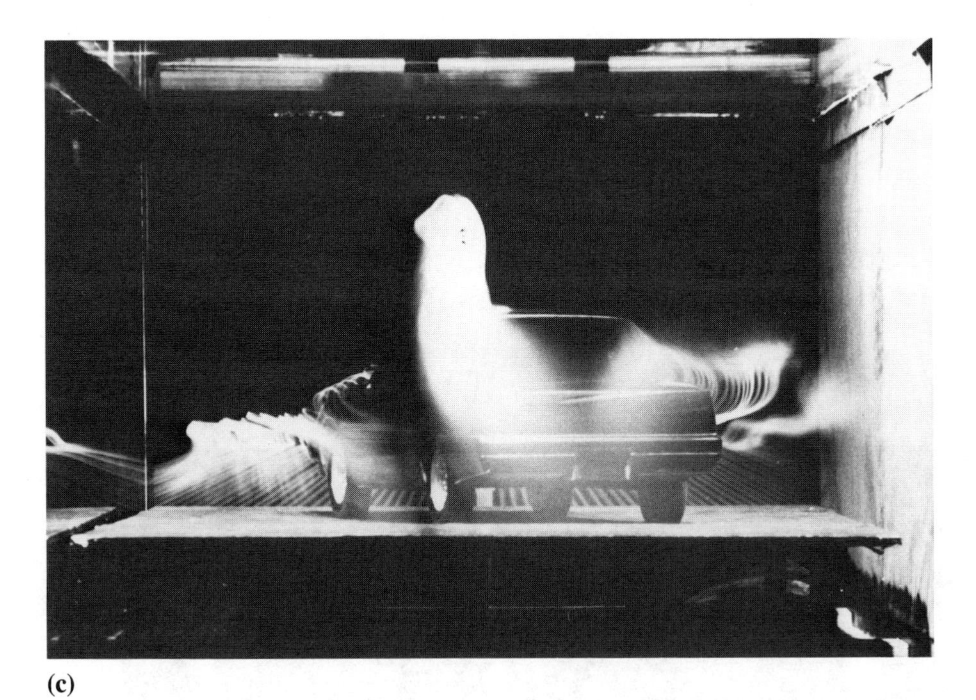

(c)

Fig. 11 (continued)

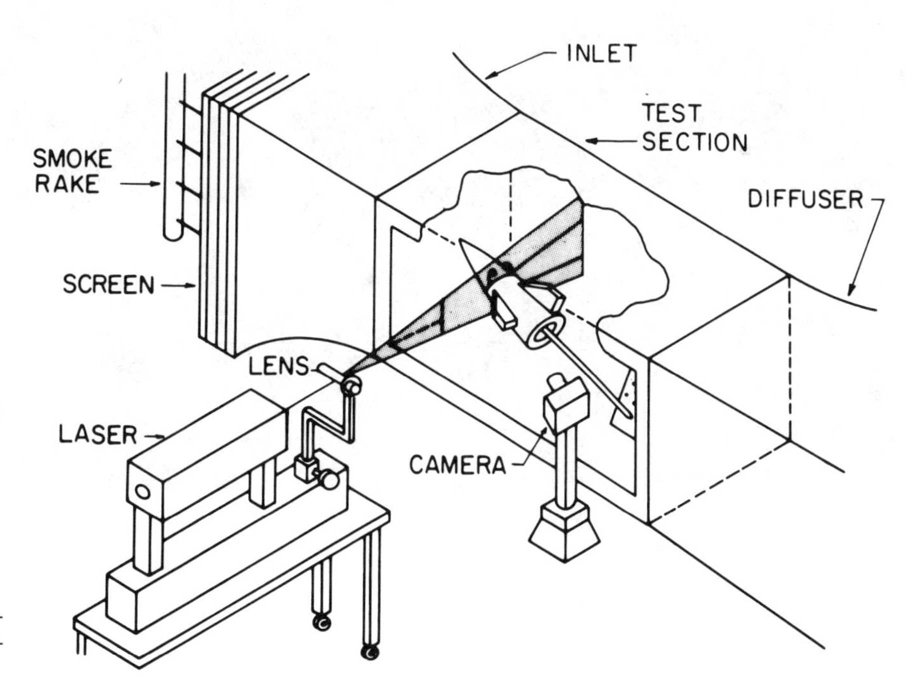

Fig. 12 Laser-light sheet arrangement (courtesy of R. C. Nelson, University of Notre Dame).

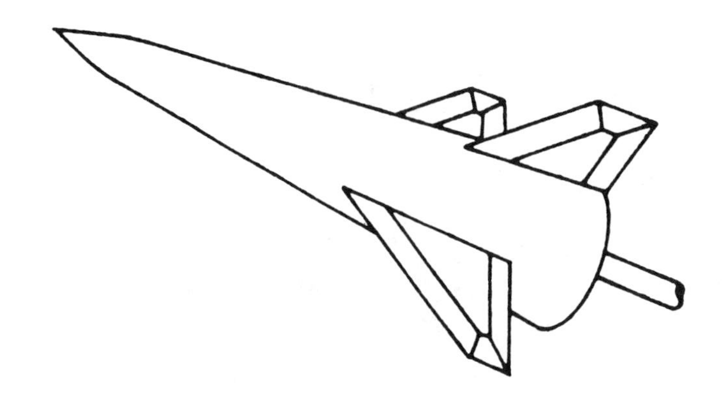

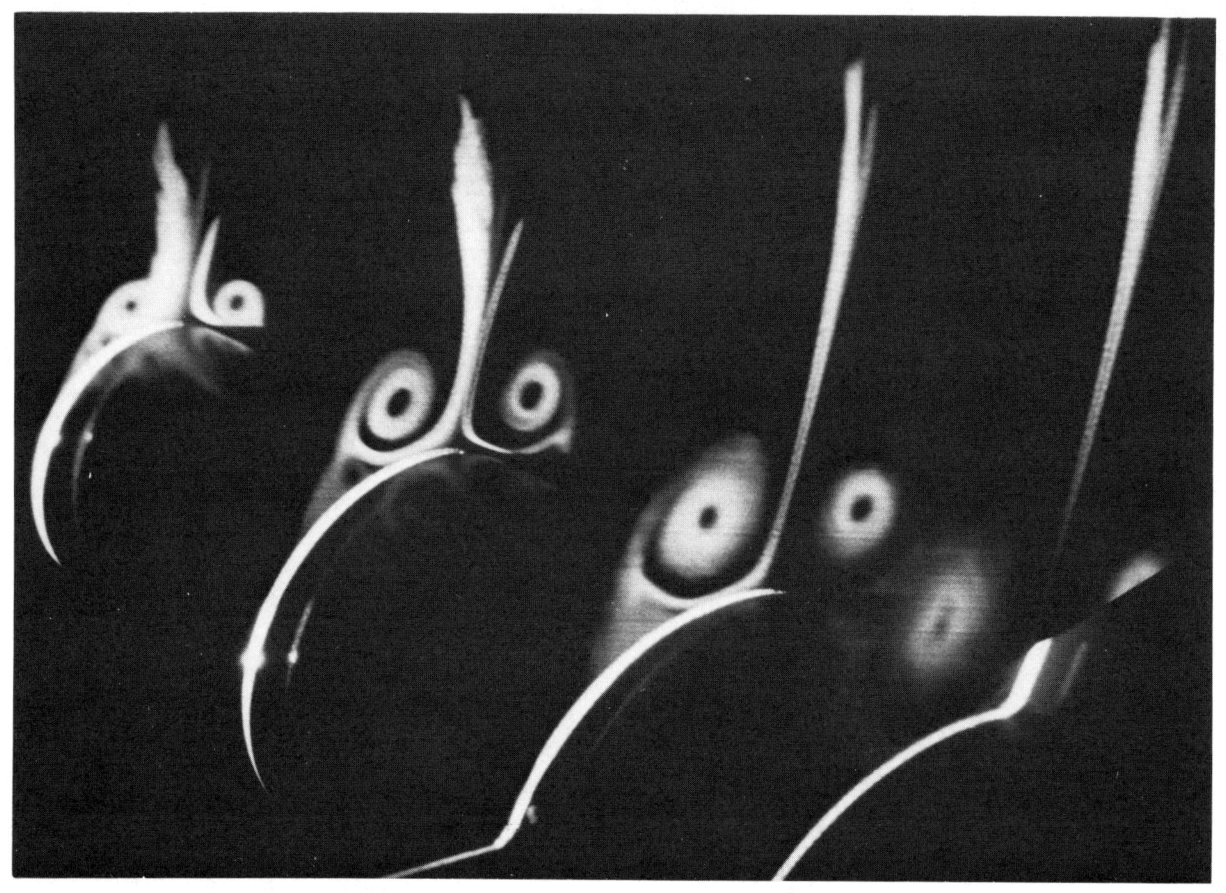

Fig. 13 Symmetric-body vortex pattern on a finned-cone research model (courtesy of R. C. Nelson, University of Notre Dame).

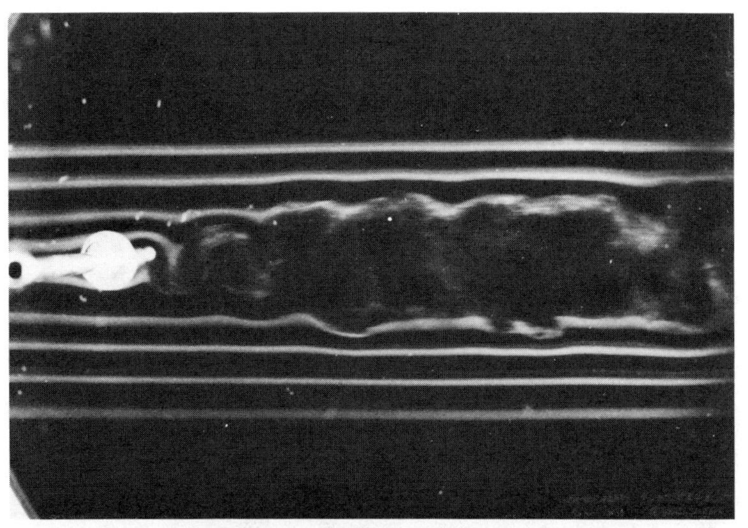

$M_\infty = 0.5$, $R_D = 32{,}000$

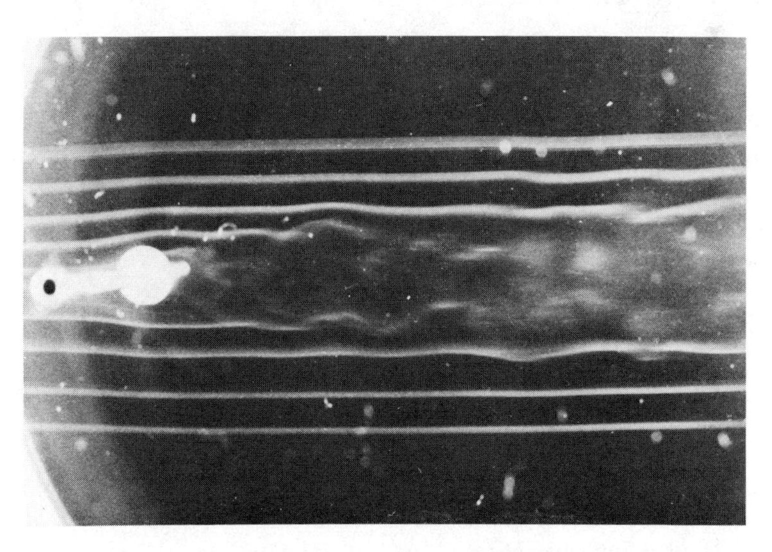

$M_\infty = 0.75$, $R_D = 42{,}000$

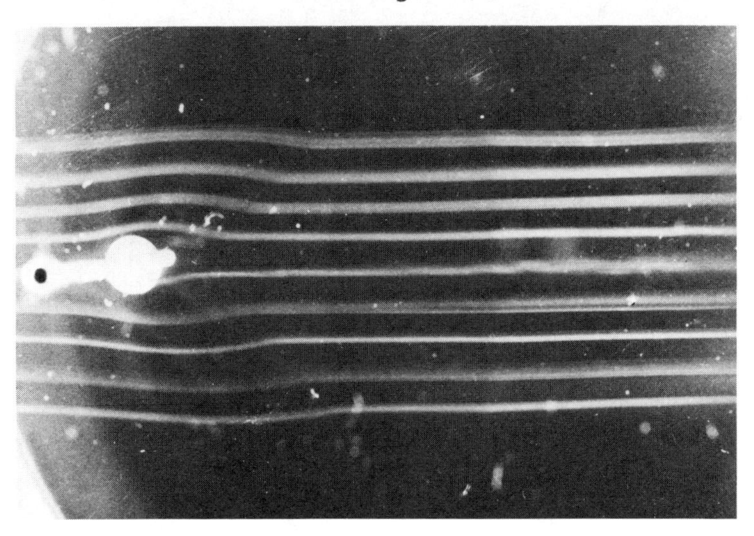

$M_\infty = 1.08$, $R_D = 49{,}000$

Fig. 14 Transonic smoke flows visualization of a circular cylinder, using a contraction ratio = 150 inlet and 11 screens (courtesy of S. M. Batill, University of Notre Dame).

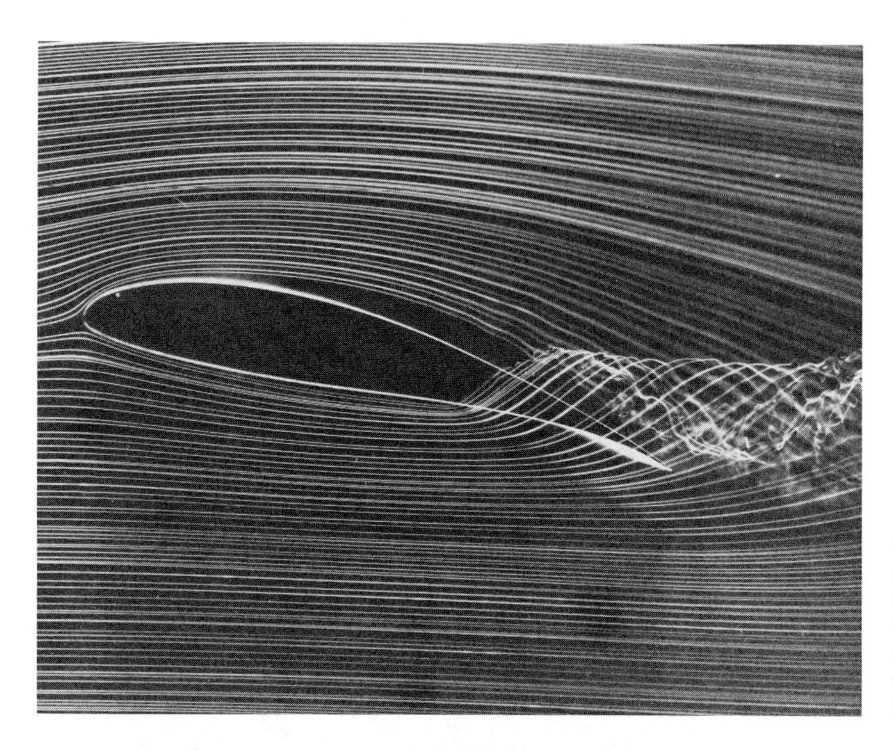

Fig. 15 Smoke-wire photograph of finite wing at an angle of attack of 14°, a chord Reynolds number of 80,000, with the smoke introduced at a location 87% of the distance from wing root to tip (courtesy of T. J. Mueller, University of Notre Dame).

IV CONCLUDING REMARKS

Smoke visualization has developed to the point where it can produce a global view of very complex flow fields. This global picture is useful in understanding complex flow phenomena as well as indicating specific locations where quantitative measurements should be taken, and in evaluating possible probe interference effects. If the flow of interest is steady, the smoke streaklines are identical to streamlines and capable of being compared with analytical or numerical results for the same flow. Even when the flow is unsteady, the streakline pattern obtained is helpful in understanding complicated flow phenomena.

REFERENCES

1. Mach, L., Über die Sichtbarmachung von Luftstromlinien, *Z. Luftschif. Phys. Atmosph.*, vol. 15, no. 6, pp. 129–139, 1986.
2. Marey, E. J., Des mouvements de l'air lorsqu'il recontre des surfaces de differentes formes, *C. R. Séances Acad. Sci.*, vol. 131, pp. 160–163, July 16, 1900.
3. Mueller, T. J., On the Historical Development of Apparatus and Techniques for Smoke Visualization of Subsonic and Supersonic Flows, AIAA 11th Aerodyn. Test. Conf., Colorado Springs, AIAA Paper 80-0420, March 18–20, 1980.
4. Mueller, T. J., Smoke Visualization in Wind Tunnels (An Historic Sketch), Astronaut. Aeronaut., vol. 21, pp. 50–54 and p. 62, 1983.
5. Lippisch, A. M., Results from the Deutsche Forschungsanstalt für Segel-flug Smoke Tunnel, J. R. Aeronaut. Soc., vol. 43, pp. 653–672, September 1939.
6. Brown, F. N. M., *See the Wind Blow*, F. N. M. Brown, Notre Dame, 1971.
7. Merzkirch, W., *Flow Visualization*, Academic Press, New York, 1974.
8. Goddard, V. O., McLaughlin, J. A., and Brown, F. N. M., Visual Supersonic Flow Patterns by Means by Smoke Lines, *J. Aerosp. Sci.*, vol. 26, no. 11, pp. 761–762, 1959.
9. Goodard, V. P., Development of Supersonic Streamline Visualization, Rep. to NSF on Grant 12488, March 1962.
10. Batill, S. M., Caylor, M. J., and Hoffman, J. J., An Experimental and Analytic Study of the Flow in Subsonic Wind Tunnel Inlets, AFWAL-TR-83-3109, 1983.
11. Batill, S. M., Hoffman, J. J., Aerodynamic Design of High Contraction Ratio Subsonic Wind Tunnel Inlets, AIAA 22d Aerosp. Sci. Mtng., Reno, Nev., AIAA Paper 84-0416, January 9–12, 1984.
12. Batill, S. M., Hoffman, J. J., Aerodynamic Design of Three-Dimensional Subsonic Wind Tunnel Inlets, *AIAA J.*, vol 24, no. 2, pp. 268–269, 1986.
13. Mueller, T. J., Flow Visualization by Direct Injection, in *Fluid Mechanics Measurements*, ed. R. J. Goldstein, pp. 307–375, Hemisphere, Washington, D.C., 1983.
14. Mueller, T. J., Recent Developments in Smoke Visualization, in *Flow Visualization III: Proceedings of the Third International Symposium on Flow Visualization*, ed. W. J. Yang, pp. 30–40, Hemisphere, Washington, D.C., 1985.
15. Smith, R. J., and Wang, C., Contracting Cones Giving Uniform Throat Speeds,'' *J. Aeronaut. Sci.*, vol. 1, pp. 356–360, October 1944.
16. Cornell, D., Smoke Generating for Flow Visualization, Mississippi State Univ., Aerophysics Res. Rep. 54, November 1964.
17. Stanford Research Institute, Characteristics of Particles and Particle Dispersoids, copies available from SRI Publications Dept., Menlo Park, Calif. 94025.

18. Maltby, R. L., and Keating, R. F. A., Smoke Techniques for Use in Low Speed Wind Tunnels, AGARDograph 70, pp. 87–109, 1962.
19. Batill, S. M., Nelson, R. C., and Mueller, T. J., High Speed Smoke Visualization, AFWAL-TR-81-3002, March 1981.
20. Preston, J. H., and Sweeting, N. E., An Improved Smoke Generator for Use in the Visualization of Airflow, Particularly Boundary Layer Flow at High Reynolds Numbers, ARC R&M 2023, ARC 711, October 1943.
21. Griffin, O. M., and Votaw, C. W., The Use of Aerosols for the Visualization of Flow Phenomena, *Int. J. Heat Mass Transfer*, vol. 16, pp. 217–219, 1973.
22. Griffin, O. M., and Ramberg, S. E., Wind Tunnel Flow Visualization with Liquid Particles Aerosols, in *Flow Visualization*, ed. T. Asanuma, pp. 65–73, Hemisphere, Washington, D.C., 1979.
23. Sellburg, L. D., Smoke Generator Type Lorinder for Flow Visualization in Low Speed Wind Tunnels, Dept. of Aeronautical Engineering, Royal Institute of Technology, Stockholm, Sweden, KTH AEOA TN 55, 1966.
24. Bisplinghoff, R. L., Coffin, J. B., and Haldeman, C.W., Water Fog Generation System for Subsonic Flow Visualization, *AIAA J*, vol. 14, no. 8, pp. 1133–1135, 1976.
25. Brandt, S. A., and Iversen, J. D., Merging of Aircraft Trailing Vortices, *AIAA J. Aircr.*, vol. 14, no. 12, pp. 1212–1220, 1977.
26. Cornish, J. J., "Device for the Direct Measurements of Unsteady Air Flows and Some Characteristics of Boundary Layer Transition, Mississippi State Univ., Aerophysics Res. Note 24, p. 1, 1964.
27. Sanders, C. J., and Thompson, J. F., An Evaluation of the Smoke-Wire Technique of Measuring Velocities in Air, Mississippi State Univ., Aerophysics Res. Rep. 70, October 1966.
28. Yamada, H., Instantaneous Measurements of Air Flows by Smoke-Wire Technique, *Trans. JSME*, vol. 39, p. 726, 1973.
29. Corke, T., Koga, D., Drubka, R., and Nagib, H., A New Technique for Introducing Controlled Sheets of Smoke Steaklines in Wind Tunnels, *Proc Int. Cong. Instrum. Aerosp. Simul. Facil.*, IEEE Publ. 77 CH 1251-8 AES, p. 74, 1974.
30. Nagib, H. M., Visualization of Turbulent and Complex Flows using Controlled Sheets of Smoke Streaklines, in *Flow Visualization*, ed. T. Asanuma, pp. 257–263, Hemisphere, Washington, D.C., 1979.
31. Kasagi, N., Hirata, M., and Yokobori, S., Visual Studies of Large Eddy Structures in Turbulent Shear Flows by Means of Smoke-Wire Method, in *Flow Visualization*, ed. T. Asanuma, pp. 245–250, Hemisphere, Washington, D.C., 1979.
32. Torii, T., Flow Visualization by Smoke-Wire Technique, in *Flow Visualization*, ed. T. Asanuma, pp. 251–256, Hemisphere, Washington, D.C., 1979.
33. Batill, S. M., and Mueller, T. J., Visualization of Transition in the Flow over an Airfoil using the Smoke-Wire Technique, *AIAA J.*, vol. 19, no. 3, pp. 340–345, 1981.
34. Bastedo, Jr., W. G., and Mueller, J. T., Spanwise Variation of Laminar Separation Bubbles on Wings at Low Reynolds Numbers, *AIAA J. Aircr.*, vol. 23, no. 9, pp. 687–694, 1986.
35. Nagib, H. M., Corke, T. C., and Way, J. L., Computer Analysis of Flow Visualization Records Obtained by the Smoke-Wire Technique, Dynamic Flow Conf., Baltimore, September 1978.
36. Hernan, M. A., and Jimenez, J., The Use of Digital Image Analysis in Optical Flow Measurements, *Proc. 2d Symp. Turb. Shear Flows, London*, Paper 702, pp. 7.7–7.13, July 1979.
37. Giamati, C. C., Application of Image Processing Techniques to Fluid Flow Data Analysis, NASA Tech. Memo 82760, February 1981.
38. Borleteau, J. P., Concentration Measurement with Digital Image Processing, IEEE, ICIASF 83 Record, pp. 37–42, 1983
39. Corke, T. C., Two-dimensional Match Filtering as a Means of Image Enhancement in Visualized Turbulent Boundary Layers, AIAA Paper 83-379, January 1983.
40. Corke, T. C., and Guezennec, Y., Discrimination of Coherent Features in Turbulent Boundary Layers by the Entropy Method, AIAA Paper 84–534, January 1984.
41. Utami, T., and Ueno, T., Visualization and Picture Processing of Turbulent Flow, *Exp. Fluids*, vol. 2, pp. 25–32, 1984.
42. Corke, T. C., A New View on Origin Role and Manipulation of Large Scales in Turbulent Boundary Layers. Ph.D., dissertation, Illinois Institute of Technology, Chicago, 1981.
43. SAE-RVAC, Aerodynamic Flow Visualization Techniques and procedures, SAE Inform. Rep. HS J1566, January 1986.
44. Takagi, M., Hayashi, K., Shimpo, Y., Uemura, S., Flow Visualization Techniques in Automotive Engineering, *JSAE Rev.*, pp. 77–84, March 1982.
45. Nelson, R. C., Flow Visualization of High Angle of Attack Vortex Wake Structures, AIAA 23d Aerospace Sci. Mtng., Reno, Nev., AIAA-85-0102, January 14–17, 1985.
46. Nelson, R. C., The Role of Flow Visualization in the Study of High-Angle-of-Attack Aerodynamics, in *Tactical Missile Aerodynamics*, ed. M. J. Hemsch and J. M. Nielsen, pp. 43–88, AIAA, New York, 1986.

Chapter 6

Electric Sparks and Electric Discharge

Y. Nakayama

I SPARK-TRACING METHOD

The flow visualization method using electric sparks is called the spark-tracing method [1]. This method can obtain the velocity distribution of gas flow by putting electrodes of suitable shape into the gas, supplying high-voltage pulses, and creating the spark lines that move downstream with the gas flow. The generation of sparks is controlled electrically. This method can accurately measure both steady and unsteady flow. It is therefore an excellent method of measuring quantitatively a flow field, together with the hydrogen-bubble method.

This spark method was the first with which Townsend observed the flow of air heated by generating a single electric discharge perpendicular to the wall and applying the schlieren method [2], which gave the velocity distribution in the boundary layer on a flat plate, using multiple sparks [3]. Saheki tried to light a spark path by generating discharges repeatedly and by controlling spark intervals [4].

Later, the method of using a high-voltage, high-frequency generator was developed, and many kinds of flow were visualized [5]. As pulse generators were improved, the field utilizing this method was extended [6, 7]. The effects of measuring conditions, accuracy, and acceleration were also examined [8–13].

Next, a spark-line-straightening method [14], making the first spark line on a fine wire, and a method of simultaneously visualizing streamlines [14, 15], by putting fine particles into gas, were developed. The spark-tracing method thus became easy to apply, thanks to the efforts of many researchers.

When an electric spark travels in air, it makes an instantaneous path of ionized air. This path has a very low resistance for a very short period (0.1–1 ms), during which detectable ionization exists. When a pair of electrodes are placed in the air flow to be measured and high-voltage pulses are applied to them, the first electric spark connects these electrodes through the shortest distance, making an ionized path. This ionized path moves together with the air flow, and the second electric spark travels along this moving path with very low electrical resistance. Subsequent electric sparks travel along the moving ionized path one after another, tracing the timelines of the air flow as shown in Fig. 1. By taking a photograph of these sparks, the timelines of the air flow can be obtained.

The relation between flow velocity u and pulse frequency f is given by the following formula:

$$u = \frac{\Delta b}{c} f \tag{1}$$

where Δb is the distance between timelines on a film or a printed picture, and c is its magnification factor. Since the life of an ion generated by a discharge is less than 1 ms, the repetition frequency must be more than 1 kHz.

On the other hand, when flow velocity is calculated from the measurement of spark distance, it is necessary to make the spark distance on a film more than 1 mm to assure the accuracy of measurement within 1%, since the resolution of film particles is 0.01 mm.

From this, we can see that the lower limit of the flow velocity to be measured is about 1 m/s. It is thus possible

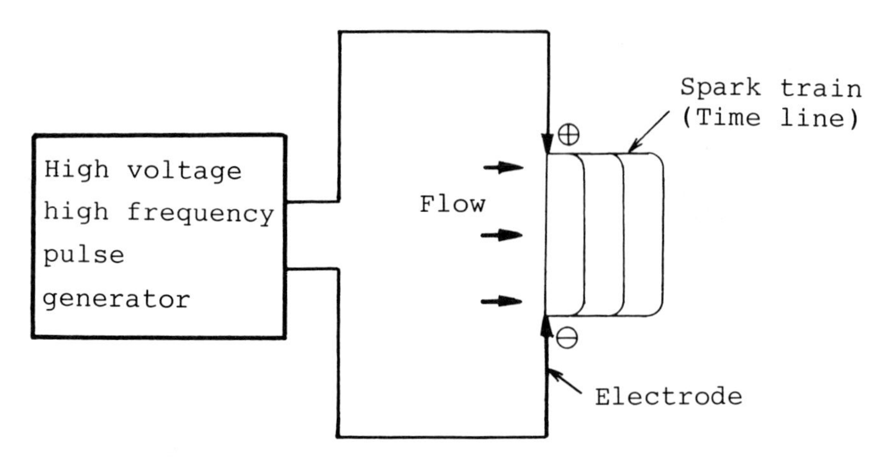

Fig. 1 Principle of spark-tracing method.

to measure up to supersonic flow, but the upper limit is determined by the characteristics of a pulse generator.

To make clear timelines by discharge in the air flow, a high-voltage, high-frequency, short-width pulse generator must be used. The high-speed flashing device (Strobokin) [13] developed as a stroboscopic device and other devices manufactured for laboratories were used as power supply sources.

A device manufactured for this exclusive use is now available, making it easy to apply this method. Figure 2*a* shows the whole view, and Fig. 2*b* shows the block diagram of this device [16, 17]. In the pulse-generating circuit, arbitrary frequency pulses varying from 1 μs to 1 s are generated through the frequency division circuit, using 1-MHz pulses from a crystal oscillator. With a presetting counter, a number of pulses varying from 1 to 200 can be set; the pulses up to the preset number are sent to the trigger pulse circuit through the gate circuit when a start signal is given to the presetting counter. These trigger pulses are amplified, and trigger the high-voltage switching circuit. In this circuit, electric charge and discharge are synchronized with the trigger pulses, and the output is sent to a pulse transformer. The boosted output from this transformer is applied to the electrodes; the pulse width is 0.5 μs.

The form of electrodes determines discharge condition, straightness of the first spark line, and thickness of a spark line. The form thus affects the accuracy of measurement when the velocity of gas flow is measured quantitatively. A sewing needle, a bead needle, a tungsten wire whose end was ground to a needle shape, and an electrode made of a fine metallic wire were inserted into a thin glass tube. Each was fixed with acryl resin, after making the ends of the wire flush, and the tube was used as a pointed electrode. Figure 3 shows the difference in spark discharge caused by various forms of electrodes; electrodes were made by forming the ends of sewing needles (0.85 mm dia.) into various shapes and inserting a fine tungsten wire (20 μm dia.) into a glass tube (No. 5) [18, 19]. Pointed electrodes (No. 3) and tungsten electrodes (No. 5) worked best because of the stability of a spark line and the straightness of the first spark.

It is necessary to arrange the electrodes according to the type of flow, as shown in Fig. 4. Pointed electrodes are suitable for measuring the velocity distribution of uniform flow, wake flow, pipe flow, and so on (Fig. 4*a*). But in order to see the whole view of inside flow in a long channel (for example, the inside flow of a centrifugal blower impeller), one must stretch line electrodes along the walls of the channel (Fig. 4*b*1).

In the case of outside flow, such as the flow around a model made of metal, the line electrodes are set so as to be enlarging toward downstream, as shown in Fig. 4*b*2. To visualize a boundary layer, the model is made of an insulator; L-shaped and pointed electrodes are set at the measuring points facing the flow direction. When sparks are generated, the velocity distribution in the boundary layer on the model can be obtained accurately. When reverse flow exists in the vicinity of the tail edge of a wing model, it is necessary to use a T-shaped electrode (Fig. 4*c*). To generate the first spark from a fixed point, it is better to put a projection of 0.5–1 mm at that fixed point [20, 21]. If a plate is used as the opposite electrode, as shown in Fig. 4*d*, it is possible to create a three-dimensional flow pattern.

II SPARK-LINE-STRAIGHTENING METHOD

It is difficult to make the first spark line straight. If the point of the electrode is sharpened, the spark line is straight to some extent, but good results are not always obtained. Straight spark lines are necessary to get accurate velocity distributions of each part. It was found that the first spark line could be made straight by stretching a fine platinum wire between electrodes [14]. Figure 5*a* shows a straight spark line generated on a 10-μm platinum wire stretched between electrodes, and Fig. 5*b* is a photograph of uniform velocity flow visualized by the same method [14].

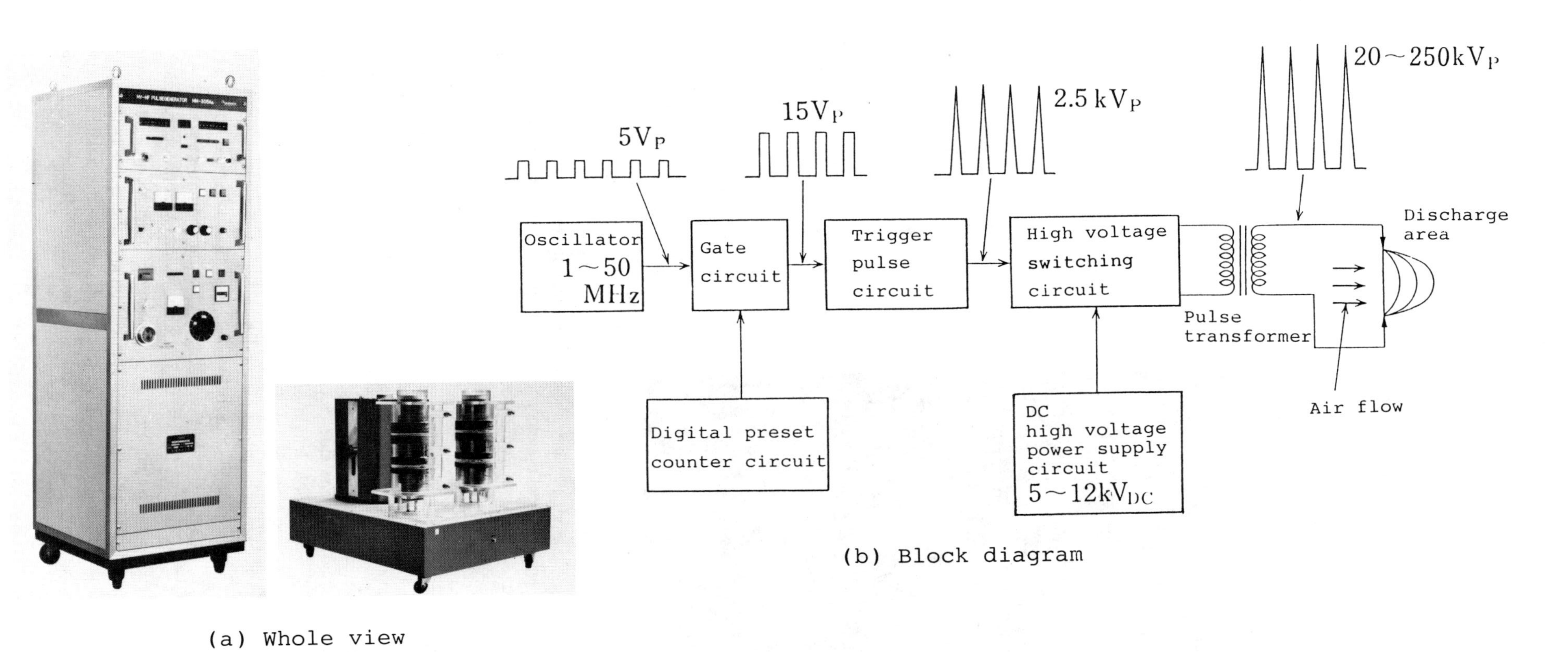

(a) Whole view

Fig. 2 High-voltage, high-frequency pulse generator having a maximum supply voltage of 250 kV, a pulse width of 0.5–1 μs, and a maximum frequency of 100 kHz [16].

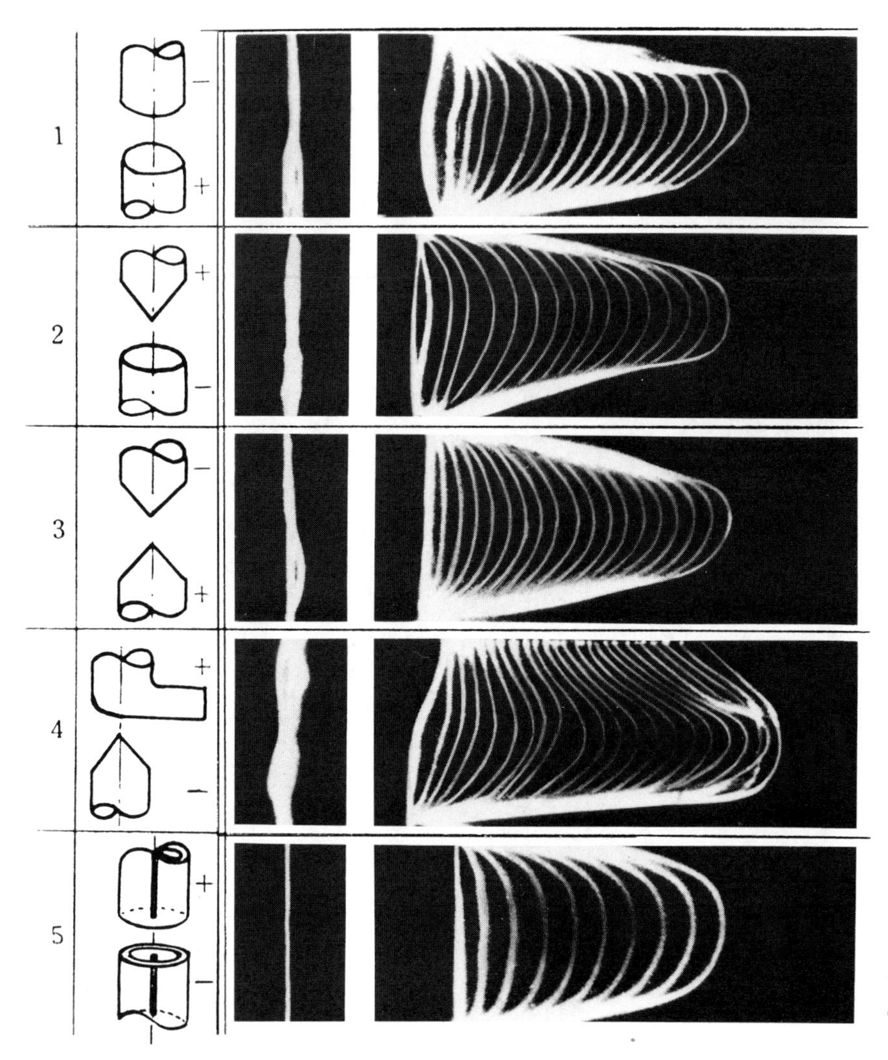

Fig. 3 Difference in spark discharge caused by various forms of electrodes [18, 19].

III GENERATION OF SPARKS

To measure flow velocity by the spark method, electric discharge must be made through the ionized path moving together with the flow, and the timeline must be shown. The minimum required voltage was measured by changing the shape of the electrodes, the spacing or interval of the electrodes, and the flow velocity [22]. The results are shown in Fig. 6. As the interval between the electrodes becomes larger, the required discharge voltage becomes greater. The effects of electrode shape and flow velocity could not be recognized. The relation between electrode interval H and the required discharge voltage is shown by Eq. (2):

$$E = 3.45H^{0.77} \tag{2}$$

The lifetime of an ion is 0.1–1.3 ms. It becomes longer as flow velocity becomes lower and frequency becomes higher, but it becomes independent of frequency to some extent when the frequency becomes high [23].

As an ionized path moves downstream with air flow, it gradually becomes longer, and its electric resistance increases; the electric spark then flies along a line that is the shortest distance between two electrodes, instead of along the ionized path.

Figure 7 shows the relation between frequency f and the ratio of the maximum attainable spark distance Lmax to the gap of electrodes H [18, 19]. This distance changes according to the gap of electrodes, applied voltage, and frequency.

If one side electrode is of the line type, the maximum attainable length of sparks increases. If electrodes on both sides are of the line type, the spark train can cover the whole length of electrodes.

The distance between two successive spark lines, Δb, cannot be extended beyond the maximum spark interval shown in Fig. 8. Figure 8 shows the relation of the maximum value Δb to flow velocity U observed in a wind tunnel. It changes with applied voltage, distance between electrodes H, and flow velocity U, and is also related to the diffusion time of ions and the length of a discharge path [18, 19].

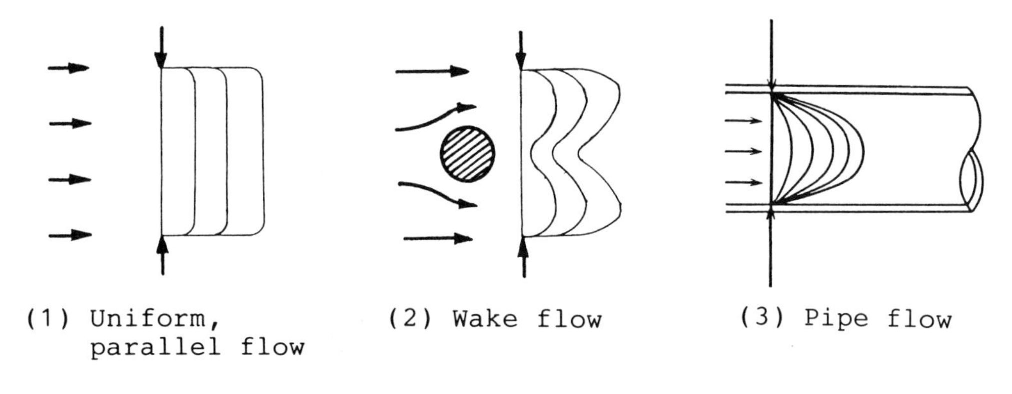

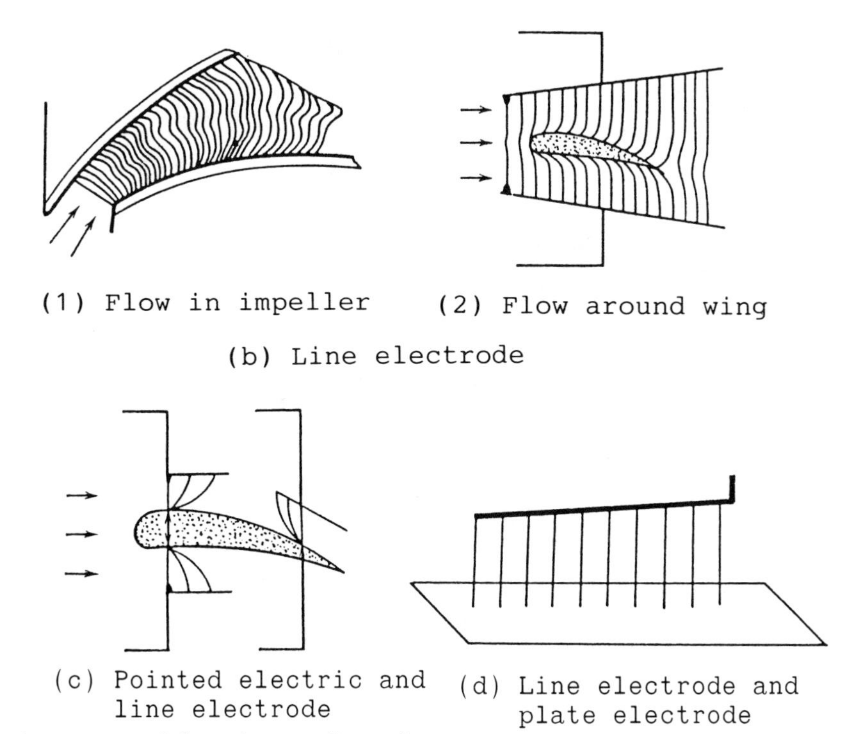

Fig. 4 Arrangement of electrodes according to flow type.

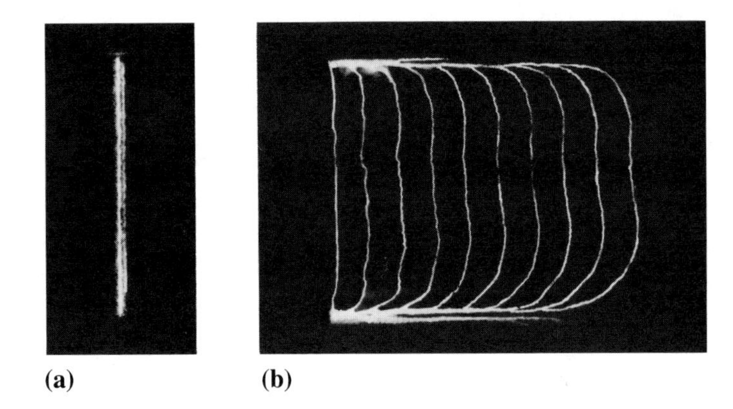

Fig. 5 Spark-line-straightening method [14]. (*a*) Straight spark line; (*b*) uniform velocity flow.

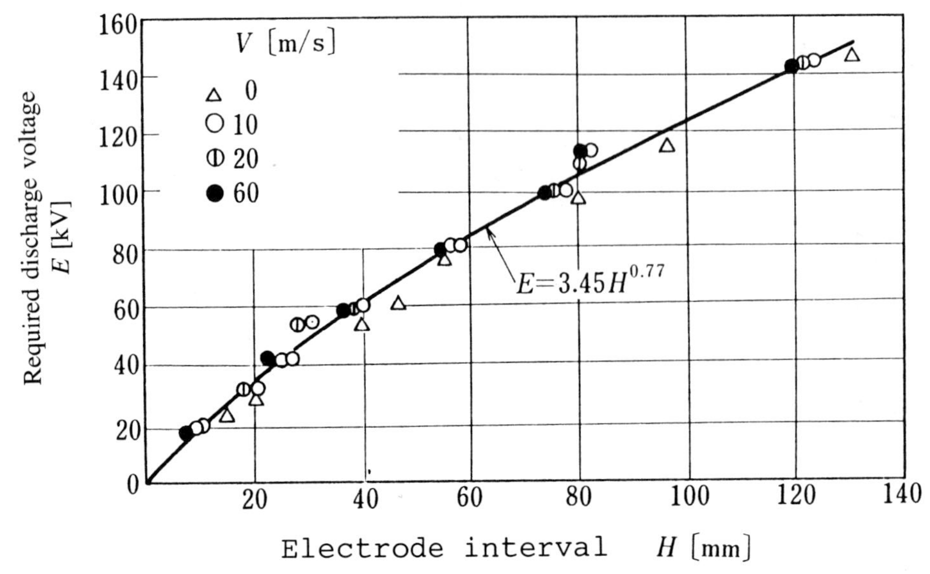

Fig. 6 Required discharge voltage and electrode interval [22].

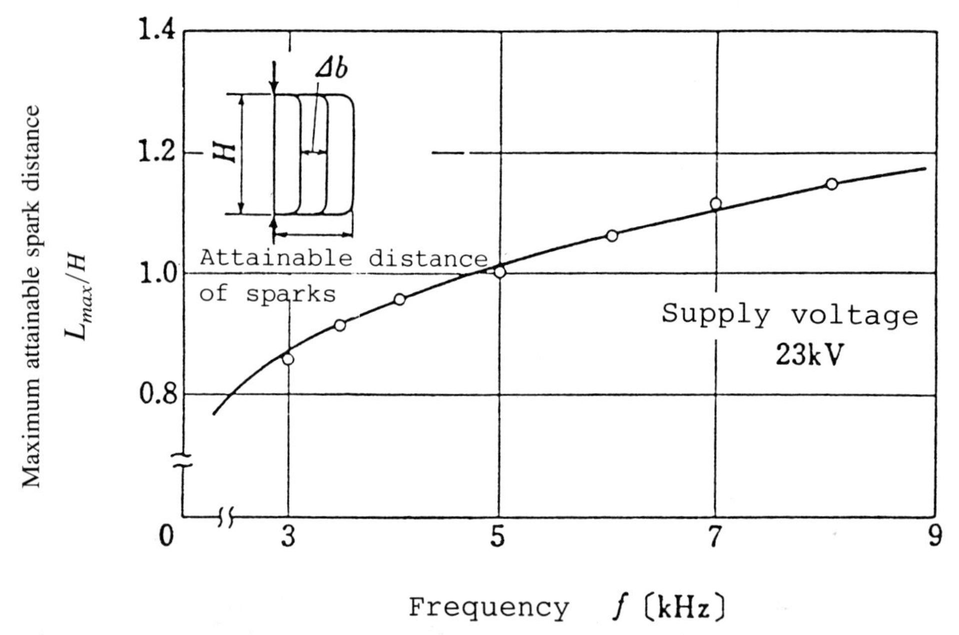

Fig. 7 Maximum attainable spark distance (E = supply voltage) [18, 19].

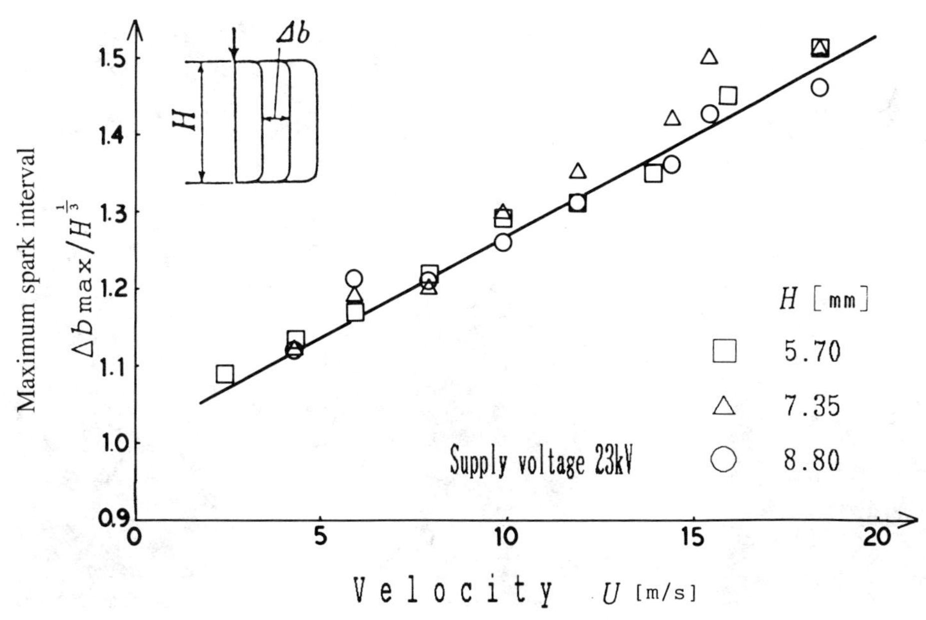

Fig. 8 Maximum interval between two spark lines (H = distance between electrodes) [18, 19].

IV METHOD OF SIMULTANEOUSLY VISUALIZING STREAMLINES

By the spark-tracing method, only timelines are obtained, and flow direction cannot be visualized. It is necessary to apply the tuft and tracer methods together to determine flow direction. It is, however, difficult to apply these methods simultaneously and to accurately measure flow velocity, especially in cases where the flow direction is unclear, e.g., divergent flow and unsteady flow. It is much more difficult to get the flow velocity by the spark-tracing method for divergent flow with rotation (e.g., the flow in an impeller), since the effect of centrifugal force increases. To surmount this problem with the spark-tracing method, a method of deciding flow direction was developed, in which fine particles are put into flow and a light-emitting tail is formed by spark lines [14, 15]. When aluminum nitride (AlN) particles (about 1 μm) are put into the air and sparks are generated, the particles on a spark line burn instantaneously, and the flow direction is indicated by the light-emitting tail of the particles. With this method, flow direction becomes clear, and a highly accurate velocity distribution can be obtained with photographs. Figure 9 shows photographs of a divergent channel visualized using this method [15].

V RELIABILITY

In order to obtain the stable spark line of the first discharge with a pointed electrode, it is necessary to sharpen the end of the electrode. When a pointed and a line electrode are used, it is best to use the pointed electrode as a cathode [4]. When trying to obtain a straight spark positively, it is better to use the spark-line-straightening method as described in Sec. II.

When a spark is generated, the gas expands because heat is generated, and the flow is disturbed. This disturbance can be minimized by reducing the discharge energy to the minimum level necessary for photographing, increasing spark-discharge frequency, and shortening pulse width.

When measuring gas flow of higher than atmospheric pressure, higher discharge voltage proportional to absolute pressure is required; accordingly discharge current becomes large, and more disturbance occurs. It is therefore better to use this method with gas flows less than 3 times atmospheric pressure [5].

Spark lines generated successively cause distortion of their path due to deflection of the initial spark line or to ions generated by spark discharge. Thus, a spark train deviates from a current timeline. These errors are divided into random and systematic ones [5, 11, 13].

Figure 10 shows random error related to the spark-tracing method when the flow velocity at the center of a free jet was measured. The ordinate is the ratio of flow velocity u_n^* found from nth and $(n + 1)$th spark lines to flow velocity u measured with a Pitot tube. The index n on the abscissa is the number of reference spark lines. This figure shows that errors can be reduced considerably if the measurement is carried out with reference to n = 3–5 spark lines.

Systematic error is defined as that originating from short-circuit spark lines through an actual flow with velocity distribution having curvature. Figure 11 shows the relation between the number of reference spark lines and the ratio

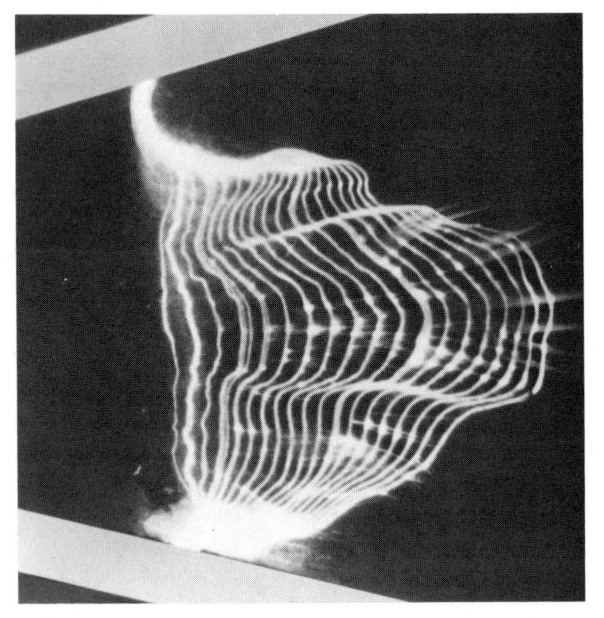

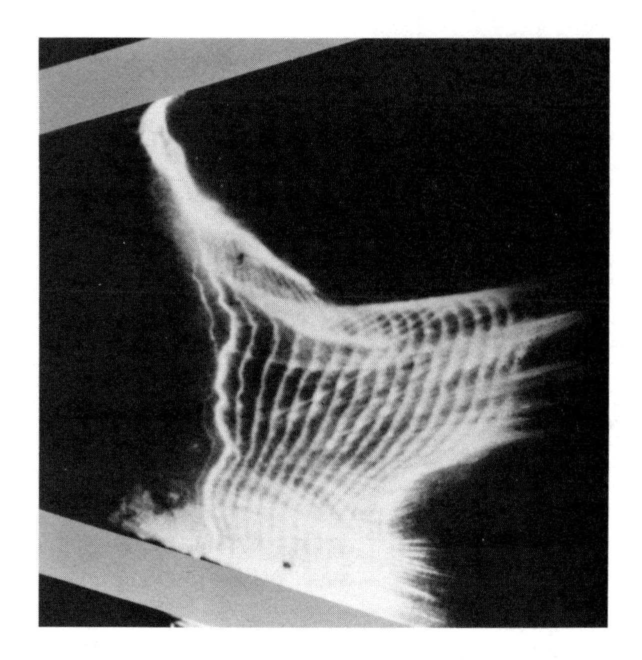

Fig. 9 Method of simultaneously visualizing streamlines [15].

of flow velocity u_n or u_n^* to real velocity, assuming straight fly of the initial spark, and parabolic velocity distribution. In the figure, u_n is the velocity obtained from the first to the nth spark lines and u_n^* is the velocity obtained from the nth and $(n + 1)$th spark lines. It is apparent that errors increase with an increase of n and a decrease of frequency.

The temperature of the air heated by discharge rises at about 0.136 m/s, and remains almost constant for 10 ms, even after a spark has disappeared [5]. The value depends on discharge energy and spark frequency.

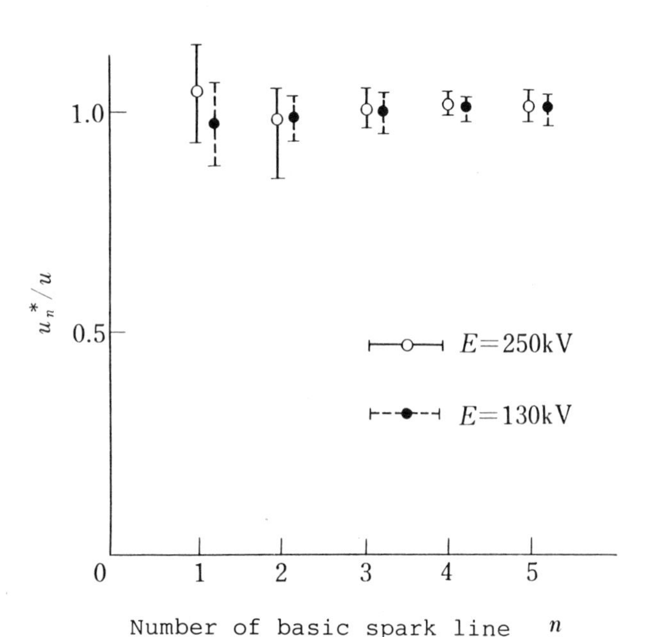

Fig. 10 Random error related to the spark-tracing method (experiment) [11].

For fundamental flow, Fig. 12 shows the velocity distributions in the inlet and the fully developed regions of laminar and turbulent flows in a rectangular channel [18]. It is clear from these photographs that the measured results coincide with the theoretical values; the error is 5% except near the wall. Figure 13 shows one more example of discharge successfully achieved with an electrode interval of 150 mm in the flow in a duct; the velocity distribution was

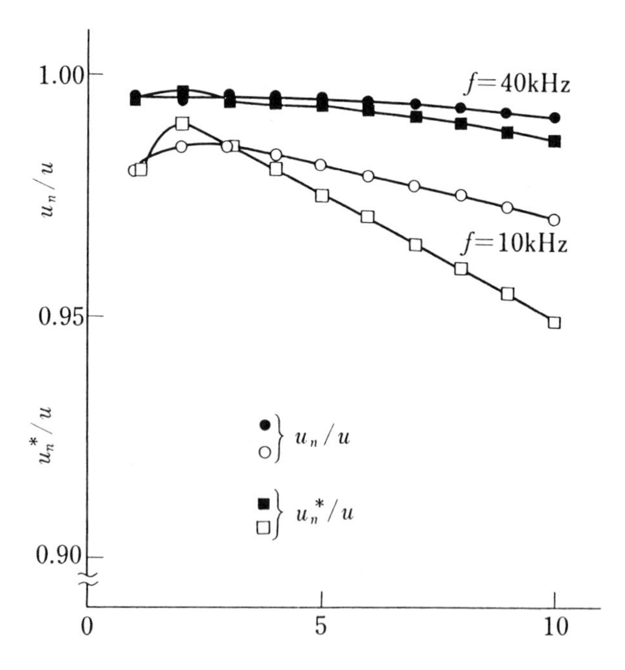

Fig. 11 Systematic error related to the spark-tracing method (calculation) [11].

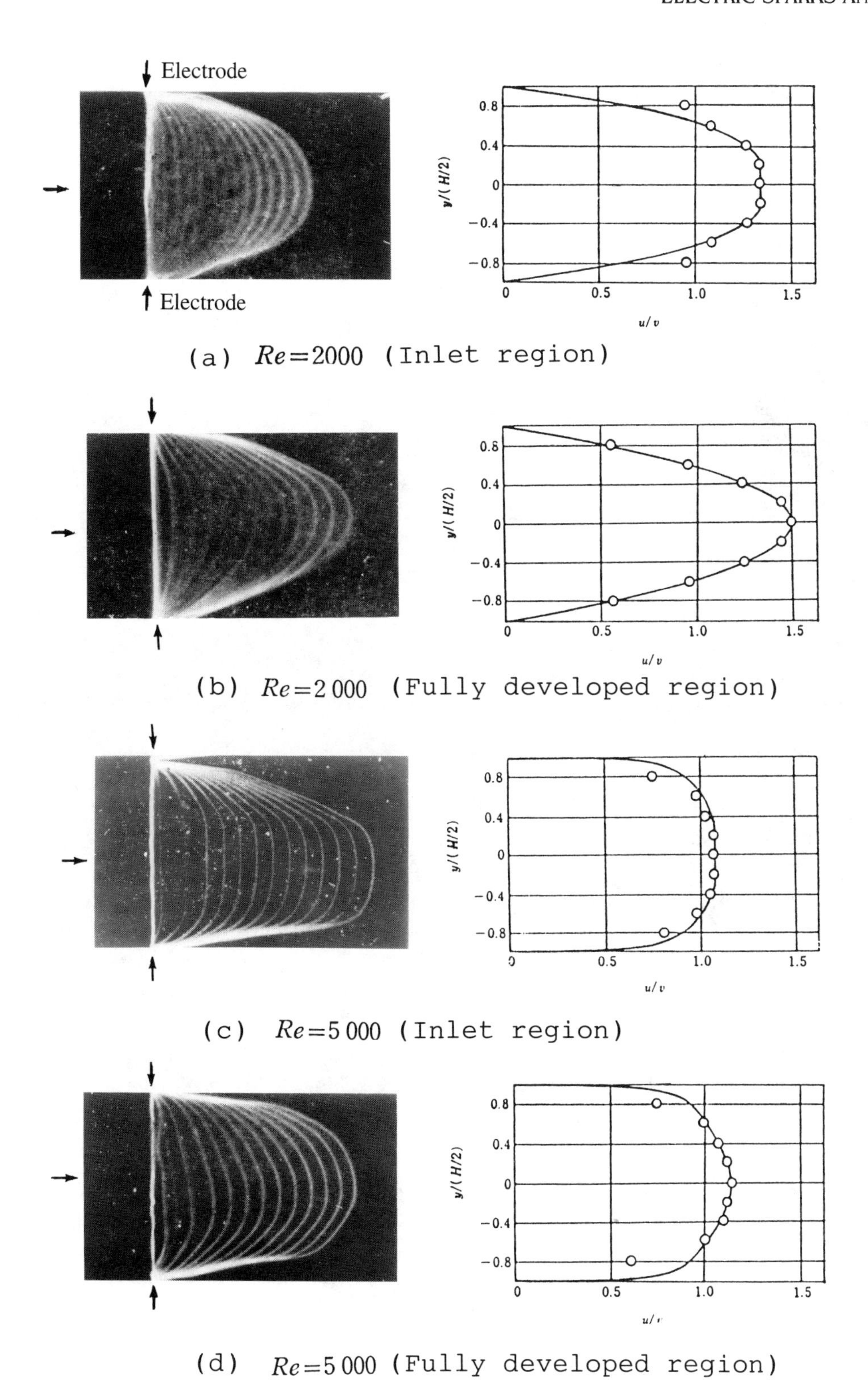

Fig. 12 Velocity distribution in rectangular channel [19].

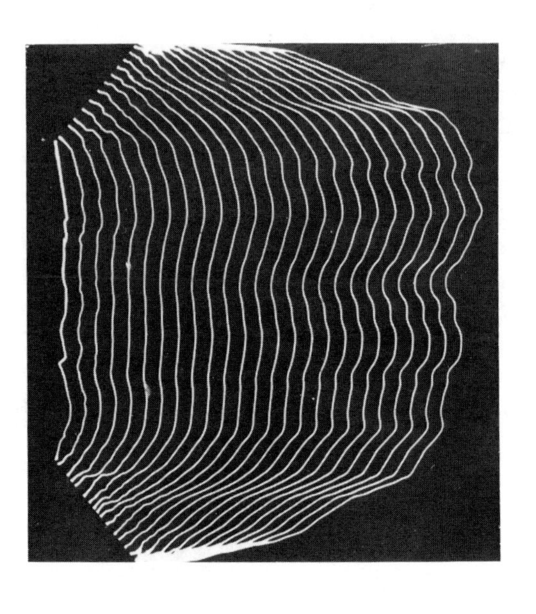

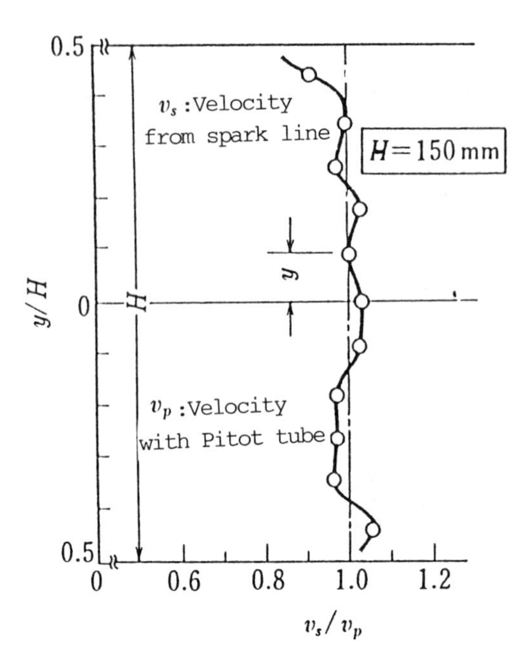

Fig. 13 Flow in rectangular duct (150 mm × 75 mm) [9].

visualized [24]. The figure on the right shows a comparison with values measured with a Pitot tube. These values agree within ±3% except in the neighborhood of the wall.

Figure 14 shows the relation between the average acceleration $\bar{a}$ and the error of flow velocity Δu obtained by spark lines, varying the acceleration in a pipe of 20° convergent angle. In this figure, the error increases in proportion to the increase of acceleration. This relation can be expressed by the equation for spark frequency f = 20–30 kHz [13]:

$$\Delta u = 0.016 \left(\frac{\bar{a}}{g}\right)^{0.94} \tag{3}$$

Cal. $(d_t = 0.8\text{mm})$ $\rho_t/\rho_f = 1/2$ × , $= 1/5$ +, $= 1/10$ *

d_t = Diameter of tracer

ρ_t : Density of tracer

ρ_f : Density of air

Exp. ● f=10kHz, ○ f=20kHz, ⊕ f=33kHz

Velocity error Δu [m/s]

Average acceleration $\bar{a}/g$

Fig. 14 Relation of velocity error to average acceleration [13].

From Fig. 14 it appears that the energy given to sparks in unit time is small in the case of low frequency (10 kHz), so the density of ionized air becomes large, and error becomes quite small.

According to other research [12], $\Delta u_{\max}$ can be expressed by the next equation.

$$\Delta u_{\max} = 4.24 \times 10^{-3} \left(\frac{a}{g} - 3000\right)^{0.733} \tag{4}$$

VI APPLICATION EXAMPLES

This method is favorable for high rather than low velocity; it is possible to measure from 1 m/s to supersonic velocity. Only because of the development of excellent high-frequency, high-voltage pulse generators has the visualization of high-velocity flow become possible; the utilization of these generators is very promising.

A Inside Flow

In order to confirm the effect of a new rectifier developed to shorten the length of an upstream straight pipe for a flowmeter, a pipe having no rectifier was compared with

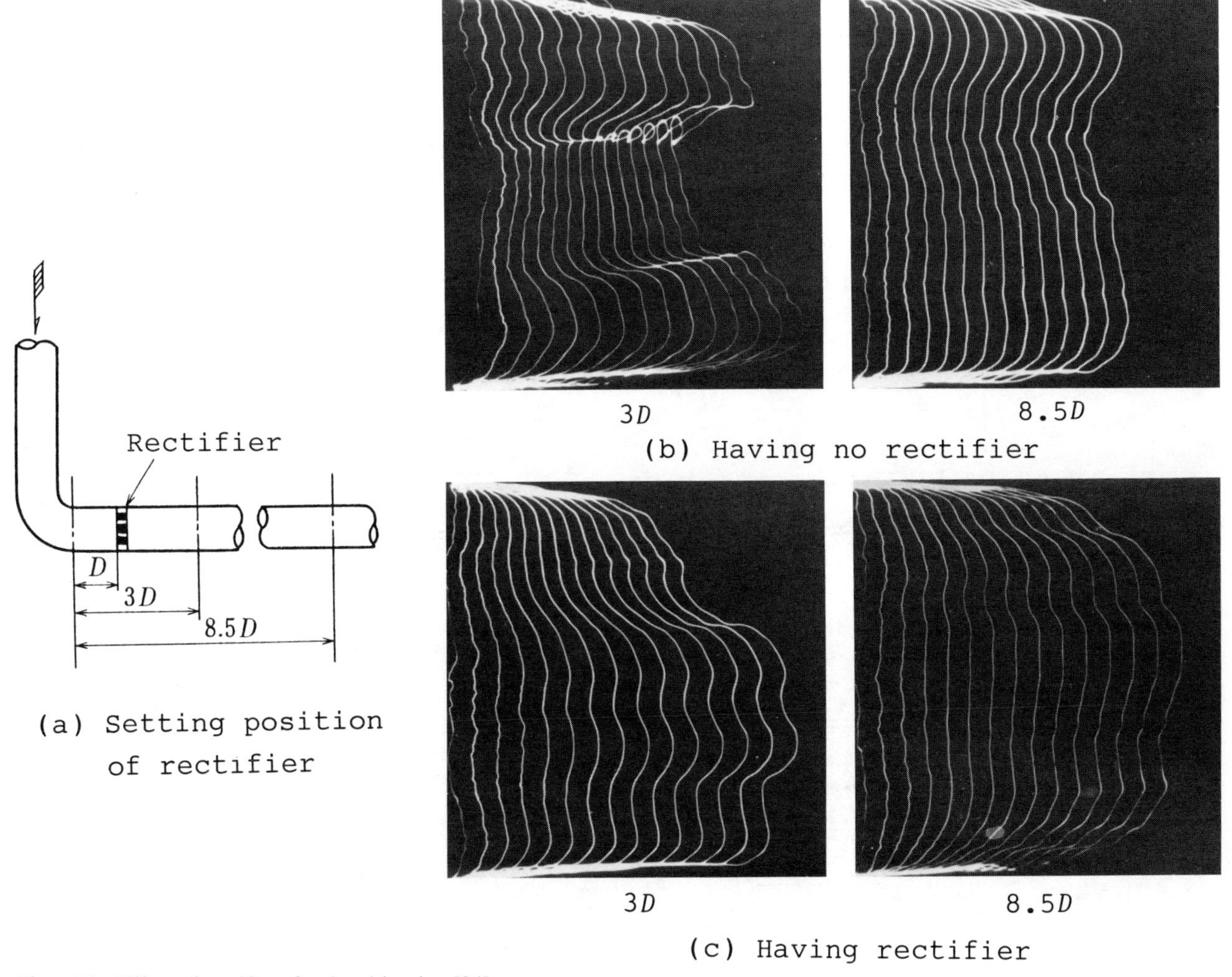

Fig. 15 Effect of rectifier after bend in pipe [24].

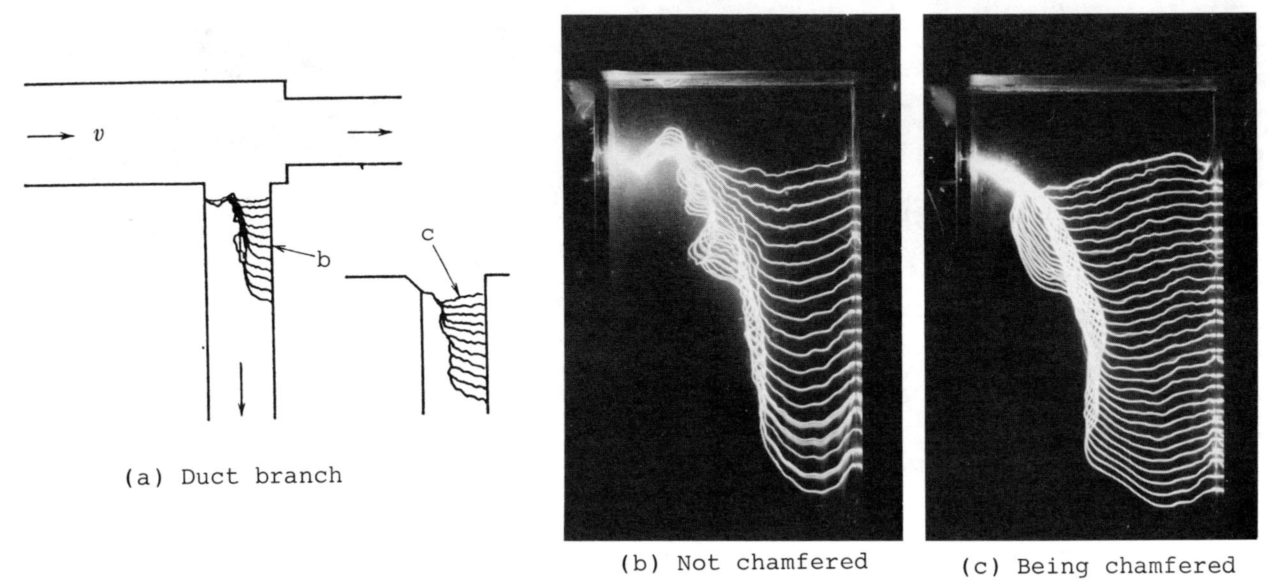

Fig. 16 Flow in duct branch [25, 26].

one having a rectifier after the bend [24]. Some results of this experiment are shown in Fig. 15. Velocity distribution is much different from turbulent distribution at the position of $8.5D$ in the no-rectifier case, but it shows fairly close velocity distribution at $8.5D$ in the rectifier case. The visualization result of a duct branch is also reported [25, 26]. Some examples of this experiment are shown in Fig. 16. In Fig. 16*b* the upstream side of the branch duct was not chamfered; in Fig. 16*c* the duct was chamfered at the same position. It is obvious that chamfering reduces the separation zone.

This spark-tracing method was applied to research on swirling flow in the continuous combustor of a gas turbine [27–29]. Figure 17*a* shows the model of a cylindrical combustor; Figs. 17*b*, *c* are photographs of the swirling flow patterns of this combustor model, taken simultaneously from the front and side, using two cameras 60 mm downstream from the inlet point. Swirling phenomena and the reverse flow occurring in the middle part are well observed. Tests were conducted on an internal combustion engine with the cylinder and part of the cylinder head made of transparent acryl resin; the flow pattern of the air sucked into the cylinder and the flow direction were made visible by light-emitting AlN find particles [30]. Figure 18 shows the flow pattern in the cylinder in two photographs taken simultaneously from the top and the side, using two cam-

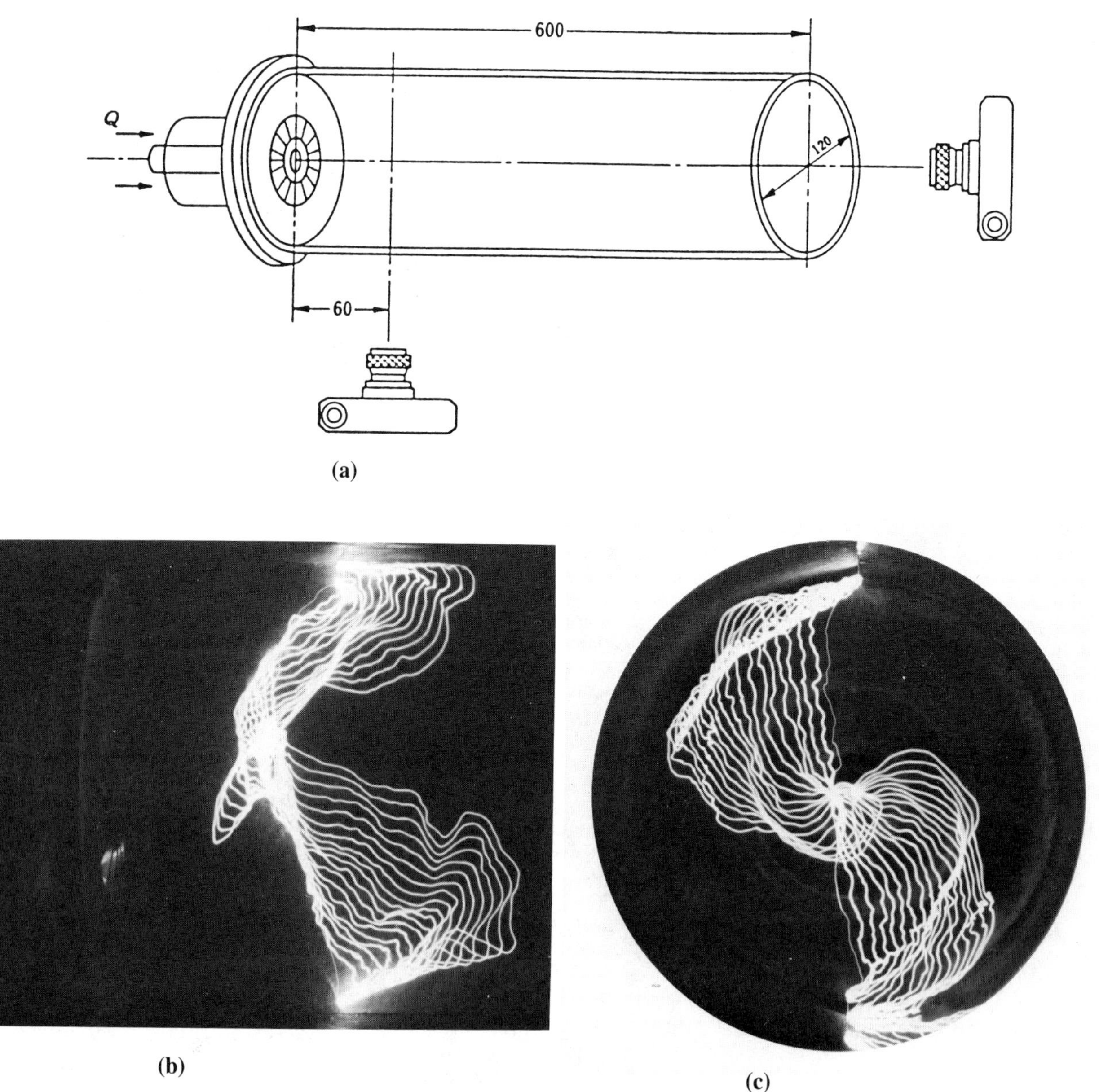

Fig. 17 Flow in cylindrical combustor [27–29]. (*a*) Cylindrical combustor model; (*b*) side view; (*c*) front view.

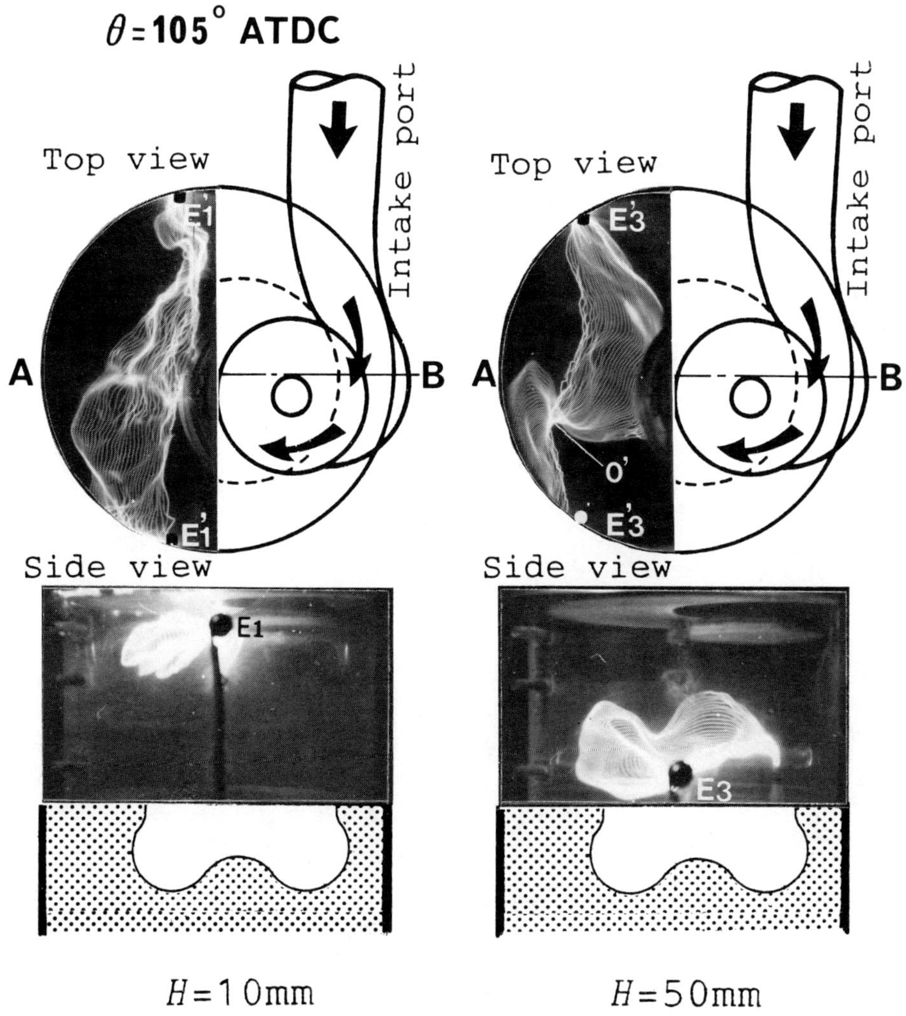

Fig. 18 Behavior of swirling flow induced into cylinder of motored model engine with high swirl port (N = 600 rpm) [31].

eras. The engine was driven at 600 rpm, and the photographs of the flow pattern in the cylinder were taken at crank angle 105° after the upper dead point on the suction stroke was reached. It is clear that considerable swirling flow occurs near the piston head. The steady flow near the air suction port in a diesel engine cylinder was also visualized [31].

Visualization of internal flow within turbomachines has been tried by various methods [32]. However, it is not easy to observe continuously the relative flow within centrifugal impellers driven at high speeds. The method employed by Fister to stop and grasp the visualized images of the spark-tracing method by using a rotating prism is considered to be the most effective [33]. A rotating image-stopping processor (Stroborotator), developed later, is superior in dynamic balancing because it uses a triangular prism [14, 34]. The principle and setting are shown in Fig. 19. The inner flow in turbomachines and reverse flow at a low flow rate were observed using this method [14, 33–35]. One example of this experiment is shown in Fig. 20 [34]. This method was also applied to the inner flow in a centrifugal pump impeller [35, 36]. The visualized results of the main flow are shown in Fig. 21*b*, and the results of local flow near the *Z* point (Fig. 21*a*) are shown in Fig. 21*c*. This rotating image-stopping processor made the quantitative measurement of inner flow in impellers more accurate when coupled with the application of the method of simultaneously visualizing streamlines using the light-emitting tails of mixed fine particles. Figure 22 shows a drawing of an apparatus used to visualize flow in a certrifugal blower impeller. The line electrodes installed inside the impeller are as shown in Fig. 23*b*; these electrodes are fitted on the hub side, intermediate part, and shroud side of the blade surface. Figure 23*c*1–3 shows the visualized results at design flow-rate point; a quasi–three-dimensional velocity distribution was formed, as shown in Fig. 23*c*4 [14]. The acceleration flow by sweepback shockwave was visualized using this method [37, 38]

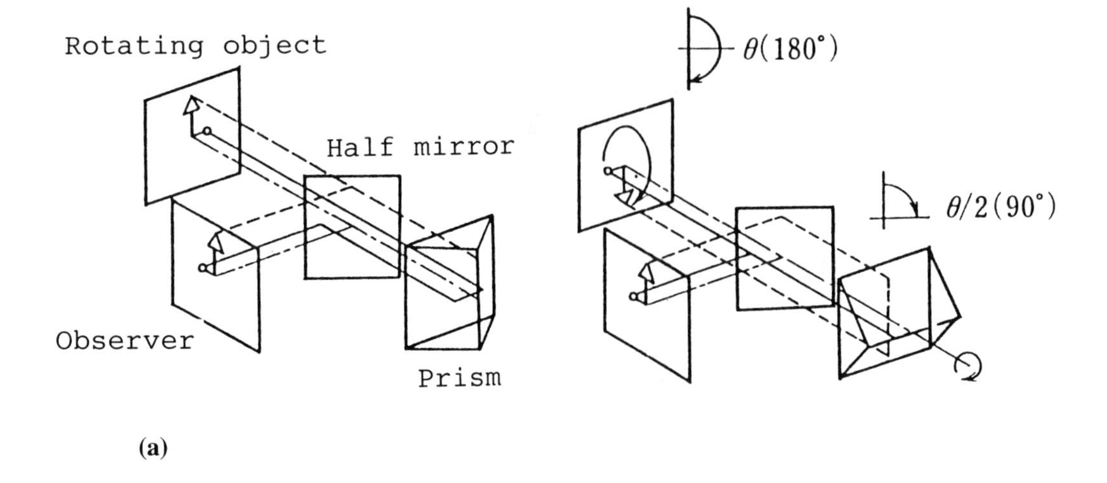

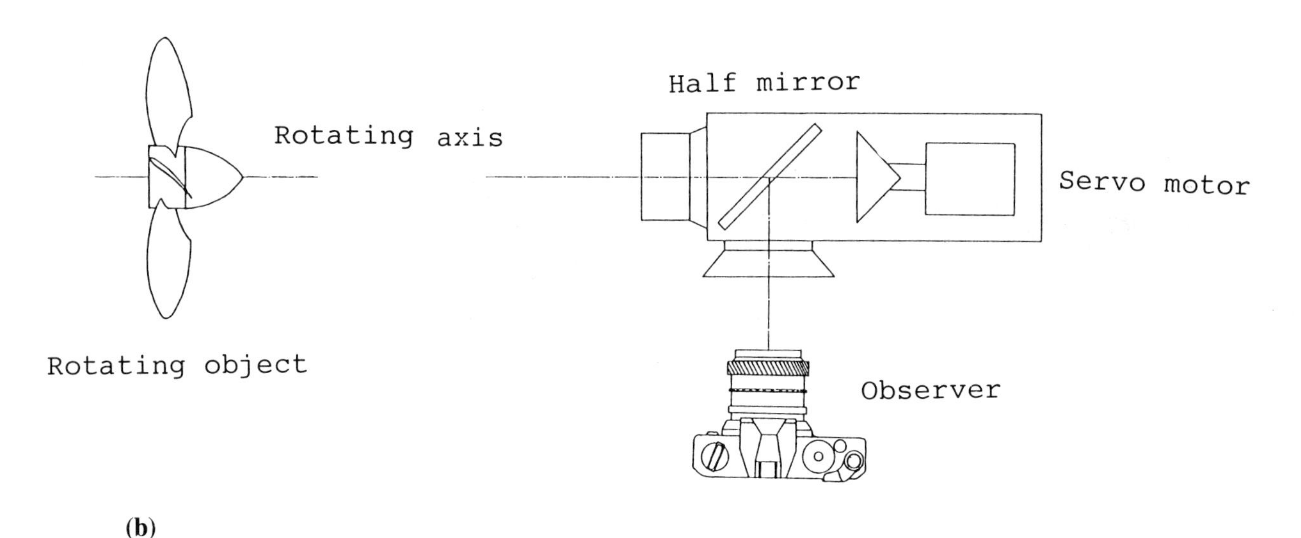

Fig. 19 Rotating image-stopping processor [14, 34]. (*a*) Principle of triangular rotating prism; (*b*) setting of rotating image stopping processor.

B Outside Flow

To observe the flow around a cylinder, a tungsten wire electrode is buried in an acryl resin cylinder in the direction of diameter, and each end of this wire is opposed to a tungsten wire electrode on the outside. High voltage is supplied to both electrodes. The flow patterns of the spark trains are shown in Fig. 24 [19]. When the rotating angle of this cylinder is changed, measurement can be done and the separation point found. Figure 25 shows the visualized result of the flow pattern when a cylinder of 20 mm dia. is put in a parallel flow [39]. It is clear from this figure that flow velocity decreases near the front stagnation point and increases over the upper and lower surfaces of the cylinder. Thereafter, separation occurs, and complex vortices arise in the wake.

The spark-tracing method is used for flow visualization around flying golf balls [40, 41]. The starting velocity of a golf ball (about 40–60 m/s) is too high for the smoke method (as smoke diffuses) and too low for the optical method (density change is small). The spark-tracing method is, therefore, one of the most suitable methods. Figure 26 shows the flow visualization results around dimpleless and dimpled golf balls, in cases of rotation and no rotation.

Figure 27 shows the visualization result of the peripheral part of the jet. It is clear that rolling up occurs on the

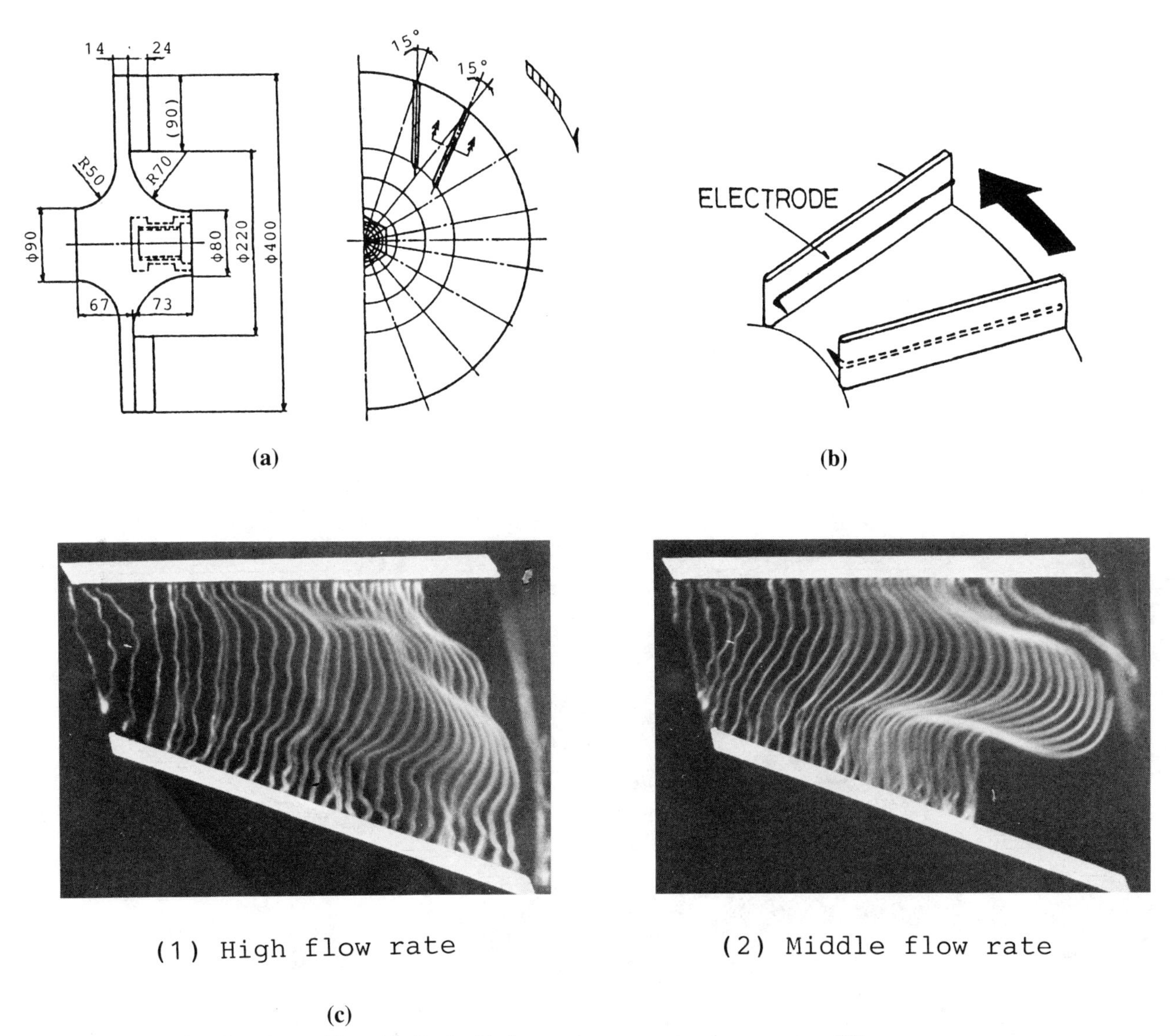

Fig. 20 Inner flow in turbomachine visualized with the rotating image-stopping processor [34]. (*a*) Impeller; (*b*) electrodes; (*c*) results of flow visualization.

peripheral part of the jet with flow, unevenness arises in the circumferential direction, and the jet flows while expanding [42].

Figure 28 shows the flow pattern around a brass wing (NACA-0020), using line electrodes [39]. By changing the attack angle continuously, the separation phenomena on the wing can be observed continuously. Figure 29 is one example of flow visualized around a wing (NACA-0015) using multielectrodes. Pointed electrodes are arranged on a wing made of an insulator such as acryl, and L-shaped electrodes are set outside the wing. The boundary layer at the point aimed at on the wing is visualized. Flow separation reaches the neighborhood of the leading edge at an attack angle $\alpha = 20°$, and it can be observed that the flow rolls up from the front to the back of the wing [20]. Figure 30 is an example observing flow around an oscillating wing; the frequency is 5 Hz, average attack angle, $\alpha = 15°$, and angular amplitude $\Delta\alpha = \pm 1.75°$. The flow around this wing was visualized by the multielectrode spark-tracing method, using pointed electrodes plugged into the wing, and L-shaped electrodes outside the wing; the photographs were taken with a high-speed camera [11].

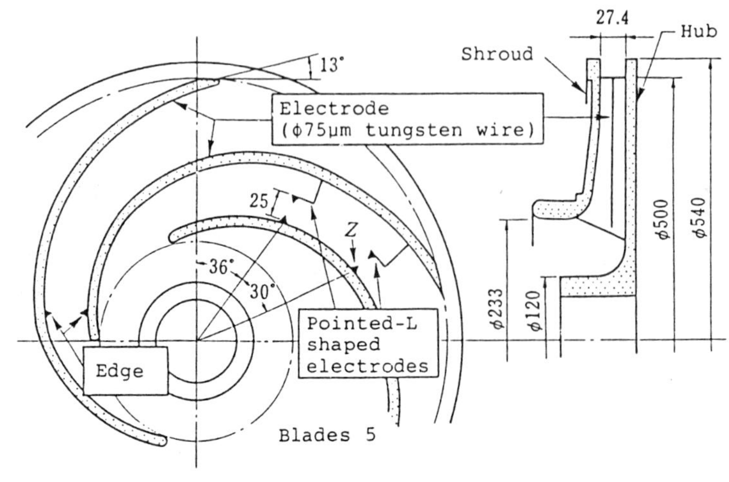

(a) Impeller

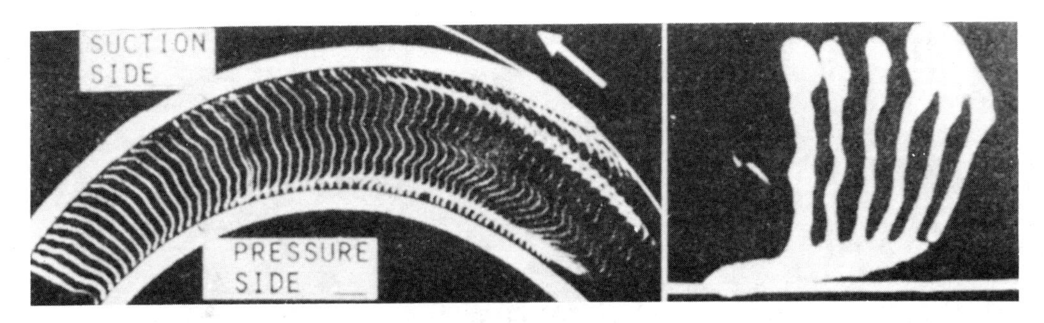

(1) Large flow rate

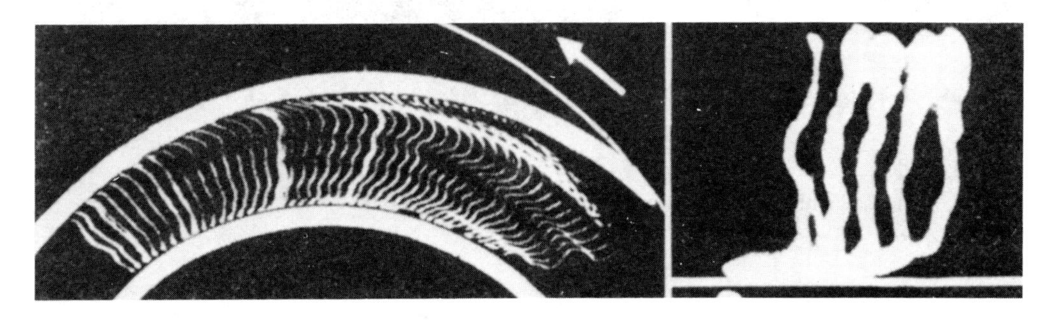

(2) Design flow rate

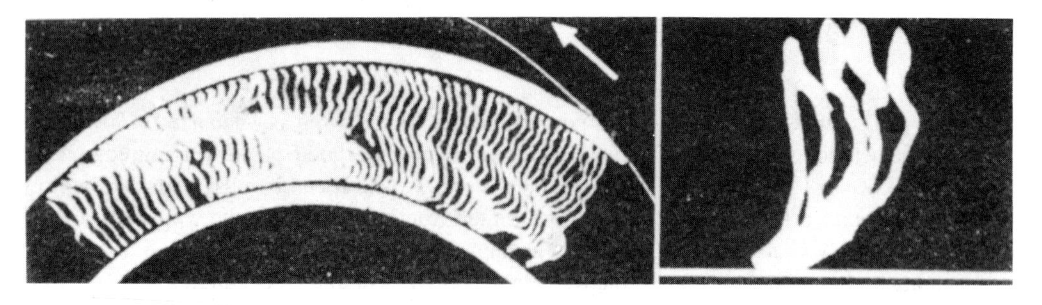

(3) Small flow rate

(b) Main flow (c) Local flow near Z point

Fig. 21 Inner flow in centrifugal pump impeller visualized with the rotating image-stopping processor [35, 36].

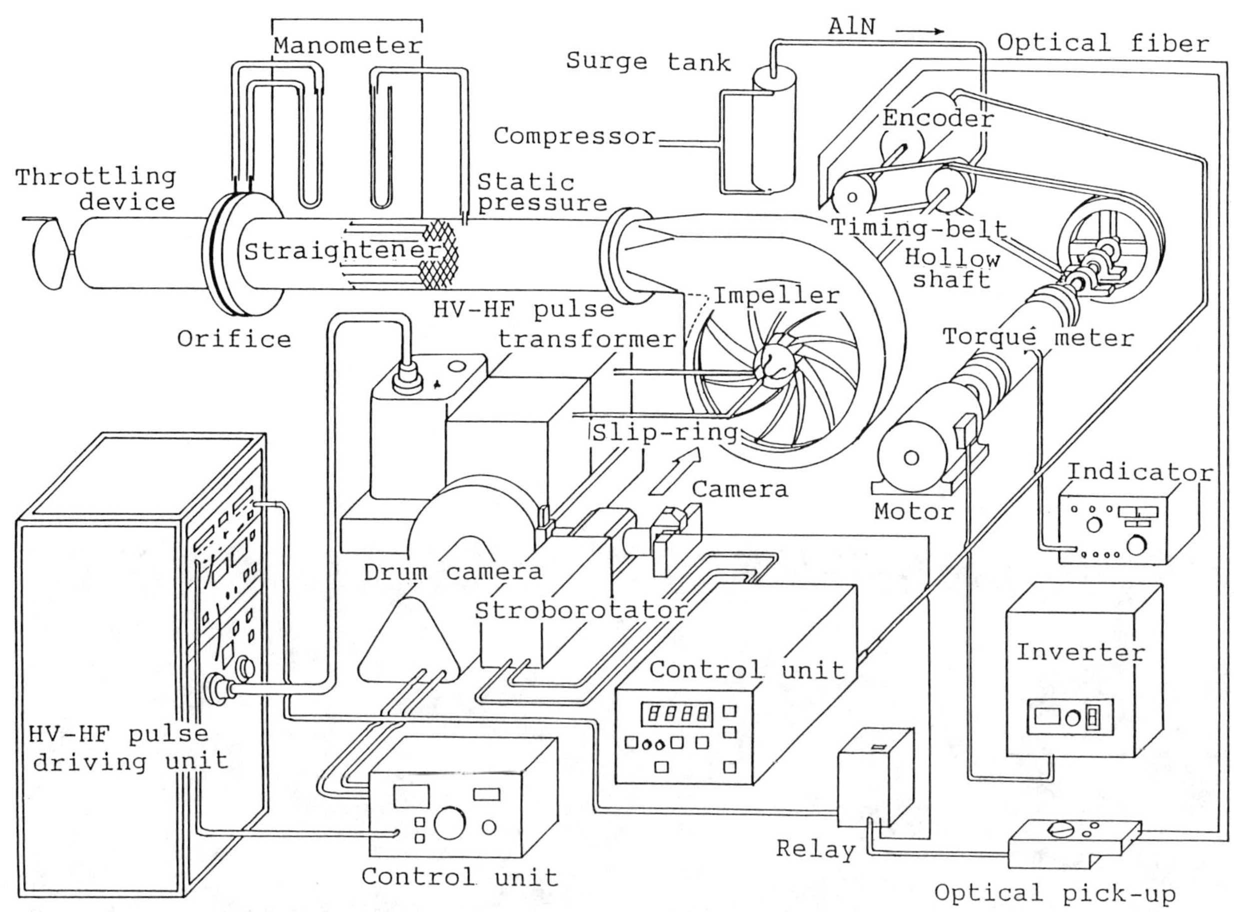

Fig. 22 Experimental apparatus used to visualize flow in a centrifugal blower impeller.

VII ELECTRIC-DISCHARGE METHOD

When discharge occurs in a gas, the intensity of light emission varies according to the density of the gas. The electric-discharge method utilizes this principle [1]. Although visualization of the cross-sectional shape of shock wave using optical systems, such as the schlieren method, has been considered difficult, the cross-sectional shape of the shock wave can be successfully observed by this method. The discharge form of electricity changes in various ways, depending on such variables as gas pressure and the shape of the electrodes. The electric-discharge methods are thus classified according to these forms as the spark-discharge method, the glow-discharge method, the electron-beam method, etc.

The spark-discharge method is generally used for the gas flow from atmospheric pressure to a pressure of 10^{-2} mmHg. There are examples visualizing the shock wave on a wedge model in a gun tunnel by the spark-discharge method [43, 44]. Figure 31 shows the electric circuit. The anode is buried in a Bakelite model surface, and the cathode is placed in free stream. When a thyratron is operated by a trigger signal, the electric circuit is closed, a high voltage is applied to the electrodes, and an electric discharge occurs.

If a pointed electrode is installed in a model surface, the discharge shape is linear, and one point of a crossed shock-wave plane visualized [43]. If a line electrode is installed in a model surface, the shape of the electric discharge becomes planar [44]. If a line electrode is placed on a model in the direction of flow, as shown in Fig. 32*a*, the side view of an oblique shock wave is visualized, as shown in Fig. 32*b*. If a line electrode is placed on a model perpendicular to the flow, as shown in Fig. 33*a*, the sectional shape of the shock wave is visualized, as shown in Fig. 33*b*. The characteristics of a hypersonic gun tunnel were Mach number, 10; static pressure, 67 Pa; running time, 10 ms; Reynolds number, 2×10^5; interval of electrodes, 6 cm; and field intensity, 150 V/cm.

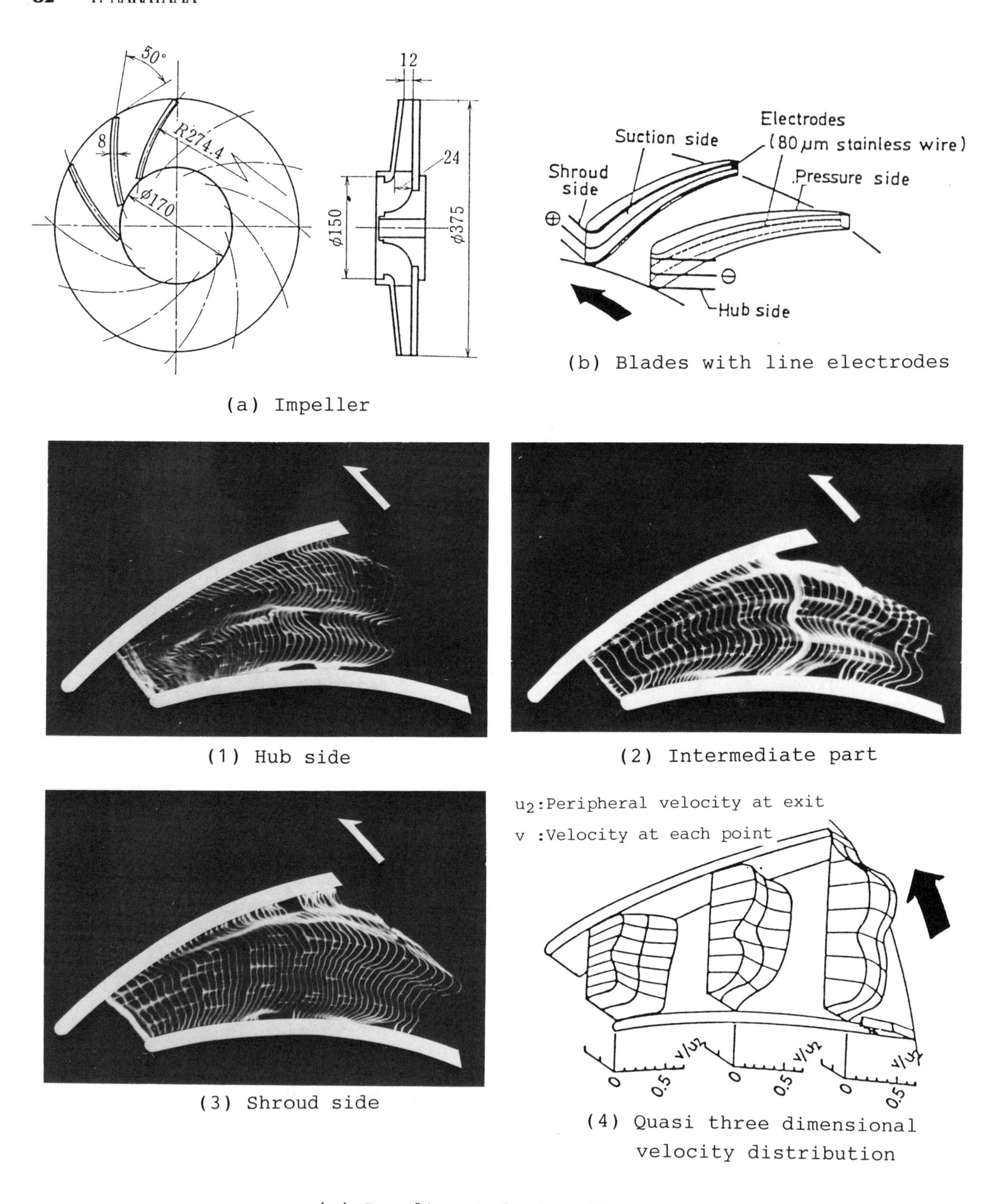

Fig. 23 Relative velocity distribution in centrifugal blower impeller [14].

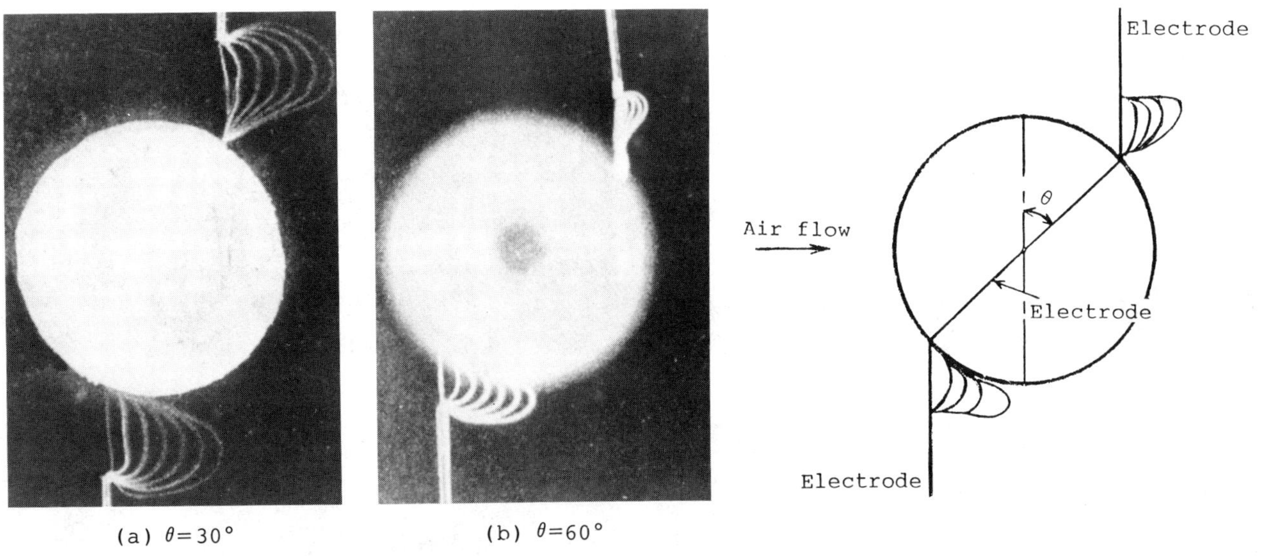

Fig. 24 Flow around cylinder (pointed electrodes) [19].

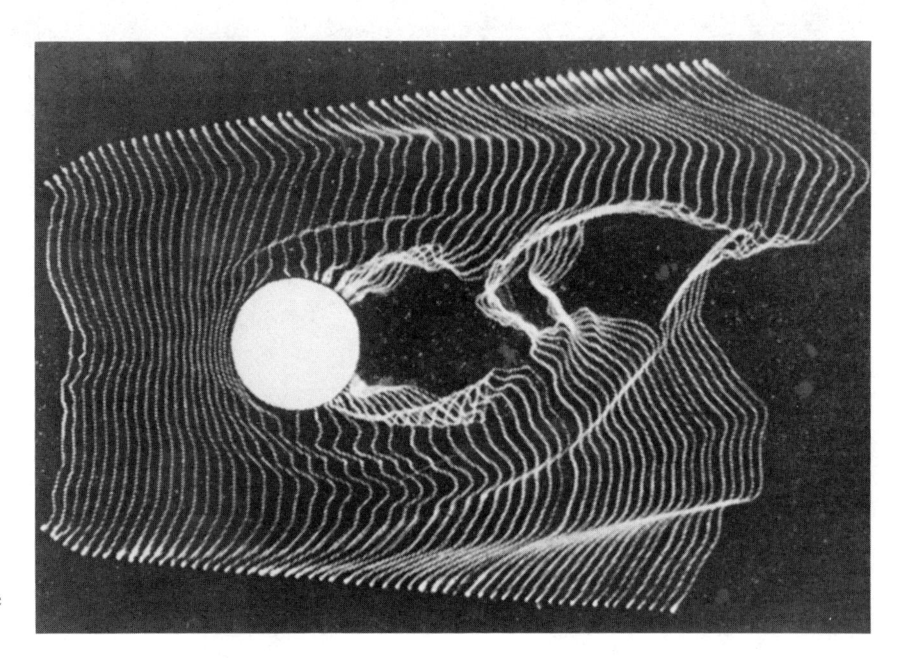

Fig. 25 Flow around cylinder (line electrodes) [39].

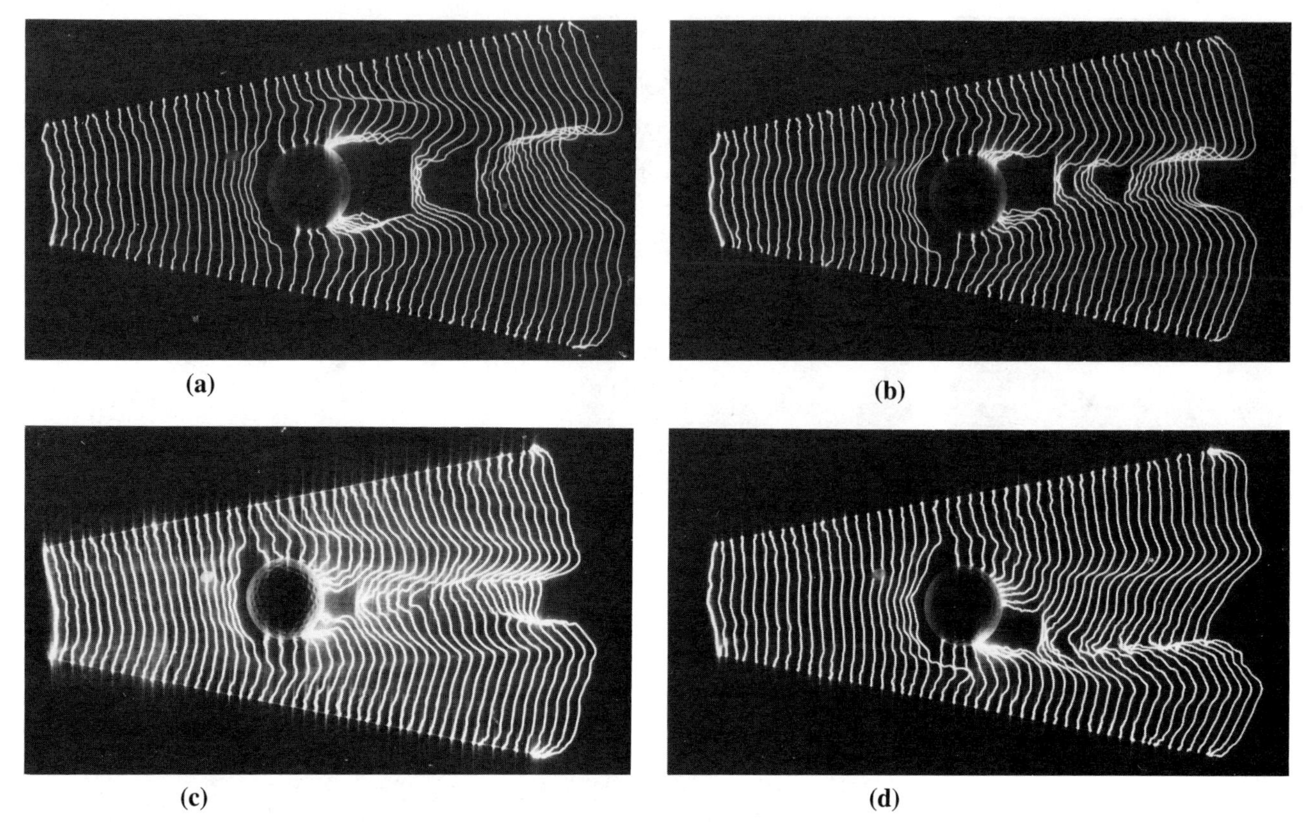

Fig. 26 Flow visualization around dimpleless and dimpled golf balls [40, 41]. (*a*) Dimpleless on rotating; (*b*) dimpleless 2500 rpm; (*d*) dimpled no rotation; (*d*) dimpled 2500 rpm.

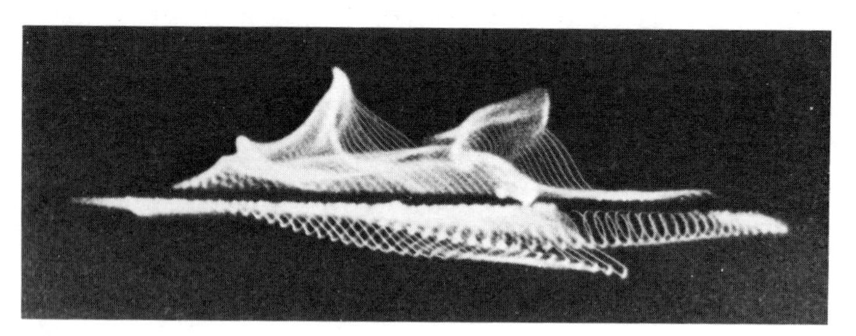

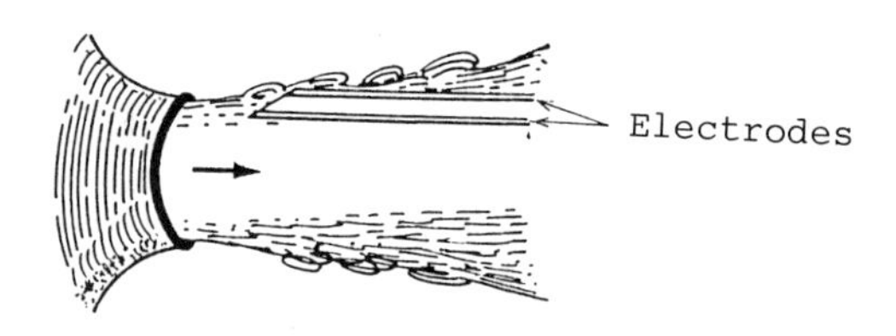

Fig. 27 Rolling up in peripheral part of jet [42].

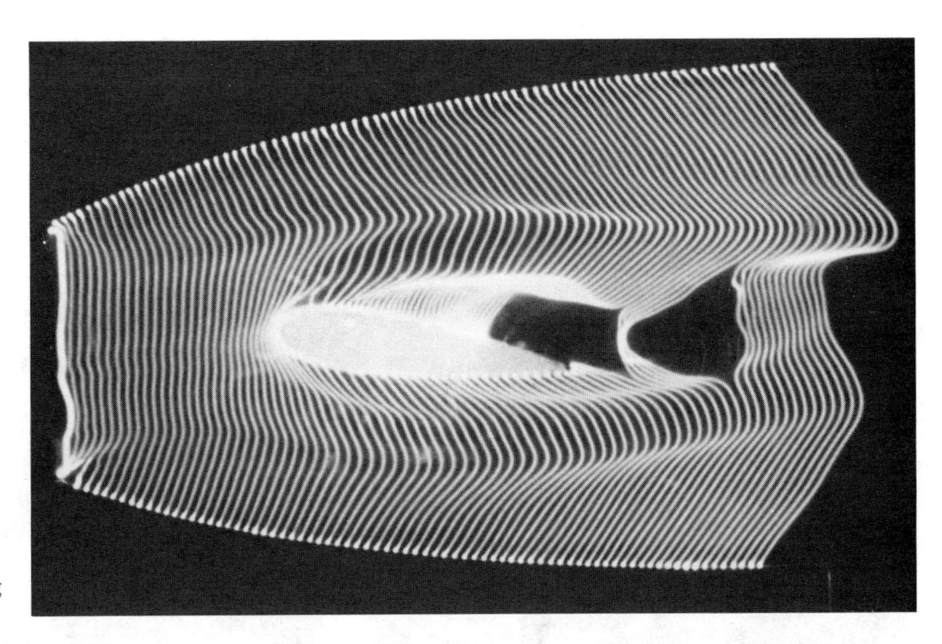

Fig. 28 Flow pattern around a wing (NACA-0020) [39].

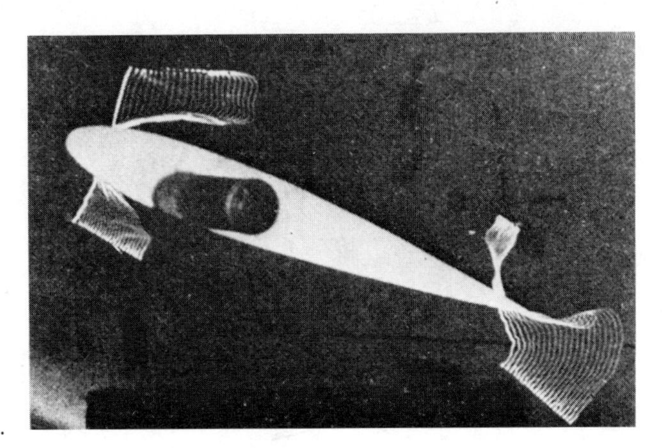

Fig. 29 Flow pattern around a wing (NACA-0015) [20].

The glow-discharge method is used for visualization of a very low-pressure field (10^{-1}–10^{-4} mm Hg) [45]. As shown in Fig. 34, by letting the glow discharge cover the whole flow field and observing the change of light-emission intensity, the whole field is visualized.

A flow visualization method for even lower pressure fields (under 10^{-4} mm Hg) is the electron-beam method [46]. As shown in Fig. 35, when the electron beam passes through a flow field, the reduced degree of density appears on a fluorescent screen as an image. From this image, the flow pattern can be obtained [47].

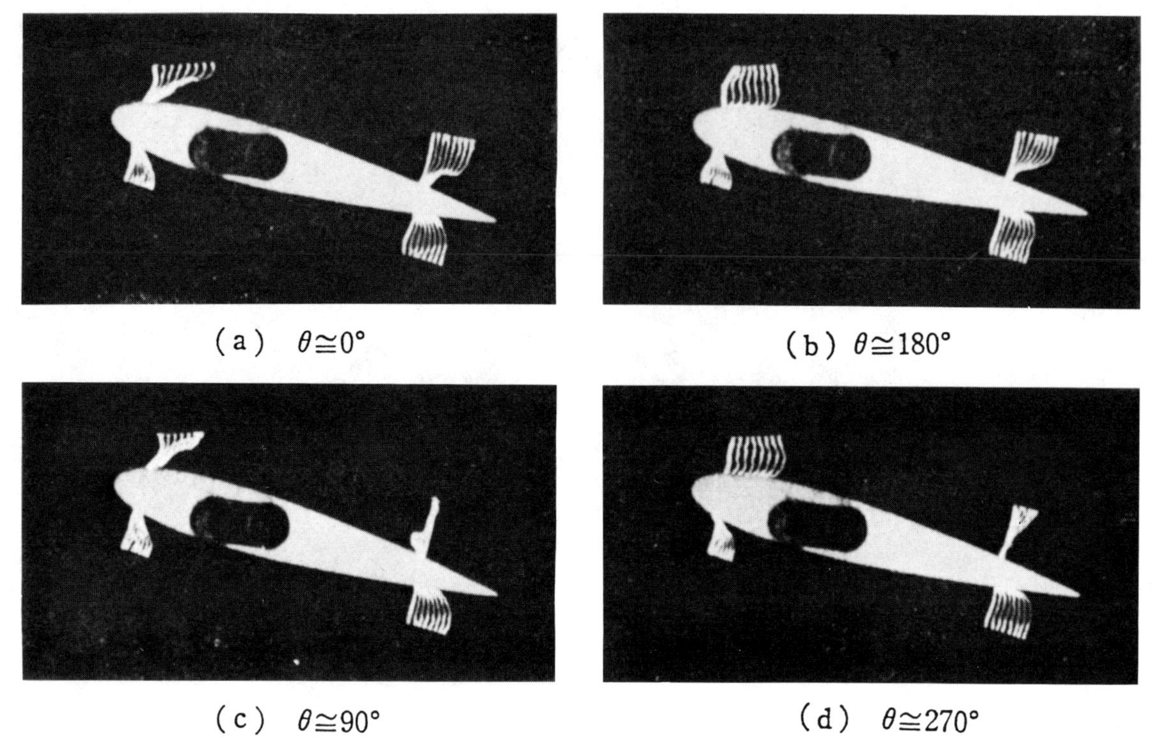

Fig. 30 Flow around oscillating wing [11].

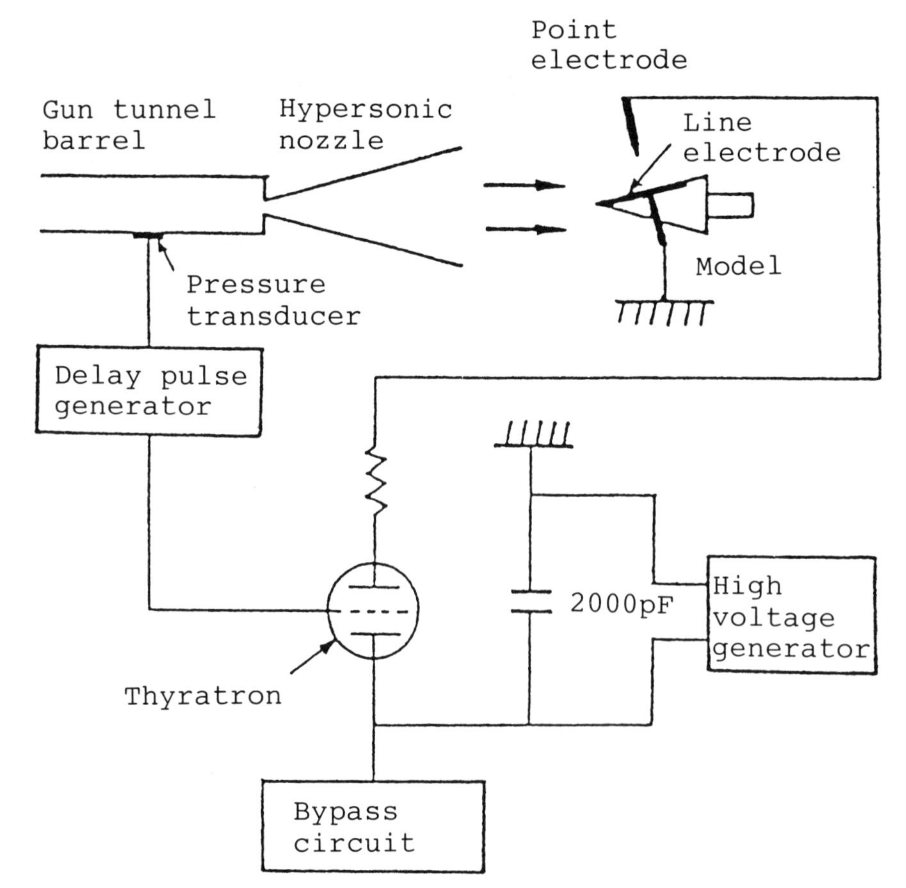

Fig. 31 Electric circuit used for visualizing the shock wave on a wedge model in a gun tunnel [41].

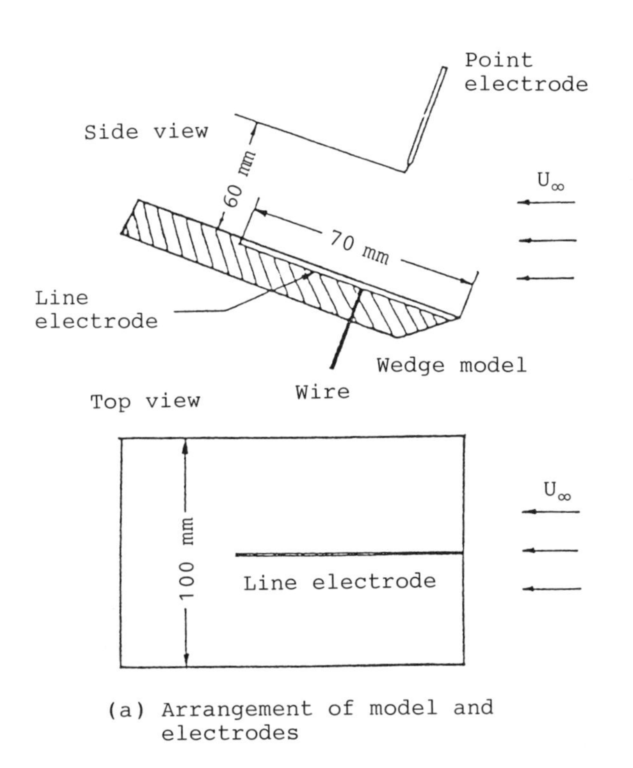

(a) Arrangement of model and electrodes

(b) Side view

Fig. 32 Visualization of lateral shock shape over wedge model [44].

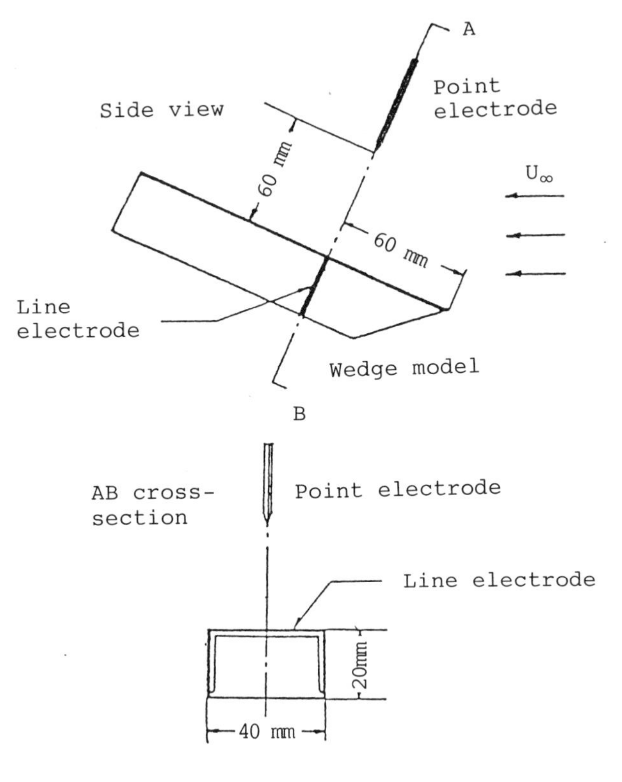

(a) Arrangement of model and electrodes

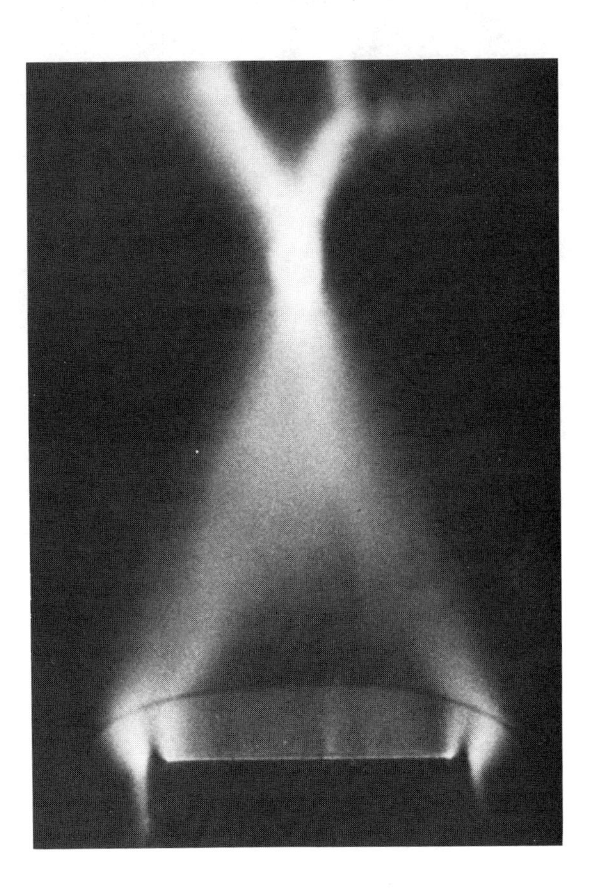

(b) Cross section view

Fig. 33 Visualization of cross-sectional shock shape over wedge model with finite width [44].

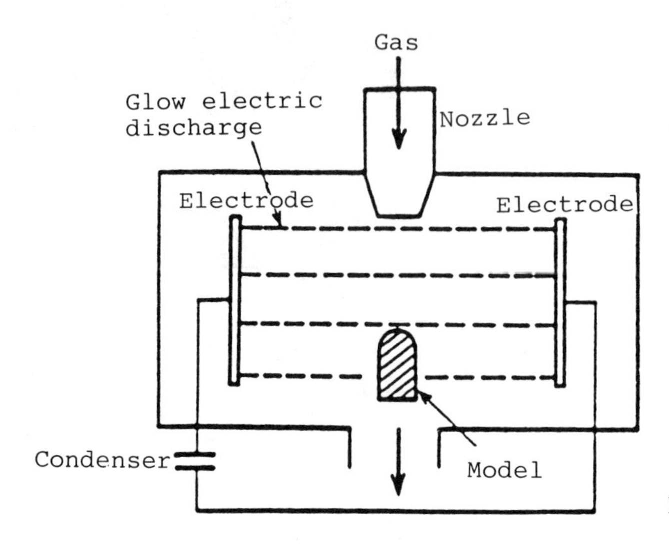

Fig. 34 Principle of glow-discharge method.

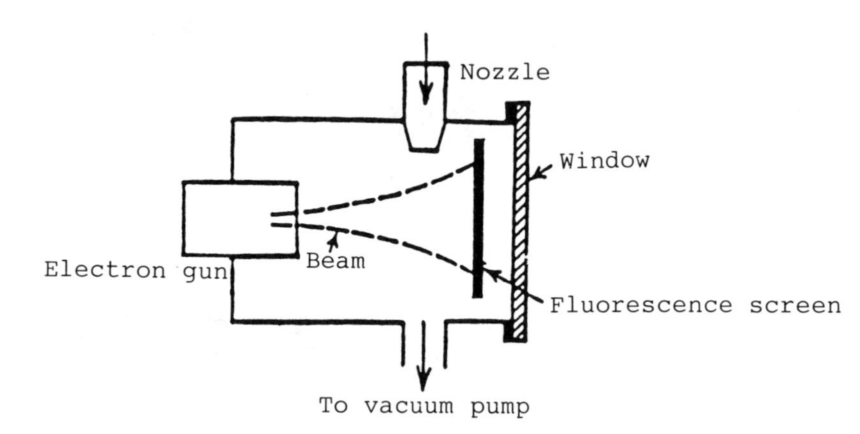

Fig. 35 Principle of electron-beam method.

REFERENCES

1. Flow Visualization Society of Japan, *Flow Visualization Handbook,* pp. 255–272 and 356–357, Asakura, Tokyo, 1986.
2. Townsend, H. C. H., Hot-Wire and Spark Shadowgraphs of the Air Flow through an Air-Screw, *Philos. Mag.*, vol. 7, no. 14, p. 700, 1932.
3. Townsend, H. C. H., Visual and Photographic Methods of Studying Boundary Layer Flow, *Rep. & Mem. Aeronaut. Res. Counc.*, no. 1803 p. 39, 1973.
4. Saheki, Y., About Measuring the Flow Velocity using Electric Sparks, *Bull. Fac. Eng. Univ. Hokkaido*, no. 8, pp. 185–190, 1947.
5. Bomelburg, H. J., Herzog, J., and Weske, J. R., The Electric Spark Method for Quantitative Measurements in Flowing Gases, *Z. Flugwiss.*, vol. 7, no. 11, pp. 322–329, 1959.
6. Früngel, F., Bewegungsaufnahmen rasher Luftströmungen und Strosswellen durch Hochfrequente Hochspannungsfunken, *Jahrb. Wiss. Ges. Luft-Raunfahrt*, p. 175, 1960.
7. Früngel, F., and Thorwat, W., Spark-Tracing Method Progress in the Analysis of Gaseous Flows, *Proc. 9th Int. Cong. High-Speed Photogr.* Denver, pp. 166–170, 1970.
8. Rudinger, G., and Somers, L. M., Behavior of Small Regions of Different Gases Carried in Accelerated Gas Flow, *J. Fluid Mech.*, vol. 7, no. 7, pp. 161–176, 1960.
9. Takahashi, N., Kato, T., Flow Visualization by Electric-Spark Method, *Oyo Buturi*, vol. 38, no. 7, pp. 709–713, 1969.
10. Nagao, F., Ikegami, J., Miwa, M., Okazaki, T., Sugimoto, N., and Nanba, T., Measurement of Air Flow using Repeated Discharge, *Trans. JSME* vol. 37, no. 296, pp. 797–806, 1971.
11. Asanuma, T., Tanida, Y., and Kurihara, T., On the Measurement of Flow Velocity by Means of Spark Tracing Method, *3d Symp. Flow Visual., ISAS Univ. of Tokyo,* pp. 33–38, 1975.
12. Matsuo, K., Ikui, T., Setoguchi, T., and Yamamoto, Y., The Error in Measuring an Accelerated Flow Velocity by a Spark Tracer Method, *6th Symp. Flow Visual., ISAS Univ. of Tokyo,* pp. 59–62, 1978.
13. Asanuma, T., Tanida, Y., and Kurihara, K., On the Measurement of Flow Velocity by Means of Spark Tracing Method, in *Flow Visualization*, ed. T. Asanuma, pp. 151–156, Hemisphere, Washington, D.C., 1979.
14. Nakayama, Y., Yamamoto, T., Aoki, K., and Ohta, H., Measurement of Relative Velocity Distribution in Centrifugal Blower Impeller, *Bull. JSME*, vol. 28, no. 243, pp. 1978–1985, 1985.

15. Nakayama, Y., Okitsu, S., Aoki, K., and Ohta, H., Flow Direction Detectable Spark Method, in *Flow Visualization*, ed. T. Asanuma, pp. 163–168, Hemisphere, Washington, D.C., 1979.
16. Nakayama, Y., Flow Visualization using Hydrogen Bubble and Spark Tracing Method, *J. Turbomach.*, vol. 3, no. 3, pp. 41–51, 1975.
17. Sugawara, S., The Development of High Voltage High Frequency Pulse Generator for Flow Visualization, *J. Flow Visual. Soc. Jpn.*, vol. 7, no. 26, pp. 315–316, 1987.
18. Nakayama, Y., and Hayashi, M., Flow Visualization using Spark or Hydrogen Bubble Method, *Proc. 2d Int. JSME Symp. Fluid Mach. Fluidics, Tokyo*, vol. 3, pp. 131–140, 1972.
19. Nakayama, Y., Hayashi, M., and Endo, H., Quantitative Measurement of Air Flow using Spark Method, *1st Symp. Flow Visual., ISAS Univ. of Tokyo*, pp. 57–60, 1973.
20. Asanuma, T., Tanida, Y., Kurihara, T., and Tanikatsu, T., Measurement of Flow Velocity around an Airfoil by Spark Tracing Method, *1st Symp. Flow Visual., ISAS Univ. of Tokyo*, pp. 61–64, 1973.
21. Asanuma, T., Tanida, Y., Kurihara, T., and Tanikatsu, T., Visualizing Studies on Nonsteady Flow around an Aerofoil, *Bull. Inst. Space Aeronaut. Sci. Univ. Tokyo*, vol. 7, no. 2, pp. 491–551, 1971.
22. Nakayama, Y., Aoki, K., Ohta, H., and Satoh, T., Basic Characteristics of the Spark Tracing Method, JSME Paper 800-14, pp. 175–177, 1980.
23. Ikui, T., Matsuo, K., and Yamamoto, Y., A Basic Study of Spark Tracing Method, *4th Symp. Flow Visual., ISAS Univ. of Tokyo*, pp. 101–104, 1976.
24. Akashi, K., Tanaka, T., and Koga, K., Effect of Nonuniform Flow and Rectifier after Bend, *5th Symp. Flow Visual., ISAS Univ. of Tokyo*, pp. 51–54, 1977.
25. Ohta, H., Nakayama, Y., Hagiwara, A., Aoki, K., Satoh, K., Satoh, T., Kohzu, J., Kohno, M., and Shinohara, H., Flow Visualization of Branch Region of Duct Line, *J. Flow Visual. Soc. Jpn.*, vol. 1, no. 2, pp. 149–152, 1981.
26. Nakayama, Y., Hagiwara, A., and Ohta, H., The Relationship between Shape and Energy Loss of Rectangular Duct Branches Suitable for Production Rationalization, *Bull. JSME*, vol. 29, no. 252, pp. 1752–1758, 1986.
27. Nakayama, Y., Aoki, K., Ohta, H., and Wakatsuki, M., Spark Method Visualization of the Swirling Flow Mixing Process, in *Flow Visualization III: Proceedings of the Third International Symposium on Flow Visualization*, ed. W.-J. Yang, pp. 707–711, Hemisphere, Washington, D.C., 1985.
28. Ferrel, G. B., Aoki, K., and Lilley, D. G., Flow Visualization of Lateral Jet Injection into Swirling Crossflow, AIAA 23d *Aerosp. Sci. Mtng. Reno, Nev.*, pp. 1–13, 1985.
29. Aoki, K., Nakayama, Y., and Wakatsuki, M., Study on the Cylindrical Combustor Flow with Swirling Flow, *Trans. JSME*, vol. 51, no. 468, pp. 2759–2766, 1985.
30. Ryu, H., Yumino, T., Asanuma, T., Kajiyama, K., and Moro, N., Visualizing Study on Swirling Flow Induced into Cylinder of Motored Model Engine, *J. Flow Visual. Soc. Jpn.*, vol. 5, no. 18, pp. 255–260, 1985.
31. Kajiyama, K., Moro, N., and Kano, S., Flow Visualization in Diesel Engine Cylinder by Spark Tracing Method, in *Flow Visualization II: Proceedings of the Second International Symposium*, ed. W. Merzkirch, p. 153, Hemisphere, Washington, D.C., 1982.
32. Nakayama, Y., Visualization of Flow in a Rotating Impeller, *J. Turbomach.*, vol. 11, no. 3, pp. 173–184, 1983.
33. Fister, W., Eikelman, J., and Witzel, U., Expanded Application Programs of the Spark Tracer Method with Regard to Centrifugal Compressor Impellers, in *Flow Visualization II: Proceedings of the Second International Symposium*, ed. W. Merzkirch, p. 107, Hemisphere, Washington, D.C., 1982.
34. Yamamoto, T., Aoki, K., Ohta, H., and Nakayama, Y., Visualization of Internal Flow in Turbomachine, *J. Flow Visual. Soc. Jpn.*, vol. 2, suppl., pp. 19–22, 1982.
35. Ohki, H., Yoshinaga, Y., and Tsutsumi, Y., Visualization of Flow Patterns in Rotating Centrifugal Impellers, *J. Flow Visual. Soc. Jpn.*, vol. 3, no. 11, pp. 330–337, 1983.
36. Ohki, H., Yoshinaga, Y., and Tsutsumi, Y., Visualization of Relative Flow Patterns in Centrifugal Impellers, in *Flow Visualization III: Proceedings of the Third International Symposium on Flow Visualization*, ed. W.-J. Yang, pp. 723–727, Hemisphere, Washington, D.C., 1985.
37. Matsuo, K., Ikui, T., Yamamoto, Y., and Setoguchi, T., Measurement of the Unsteady Flow using a Spark Tracing Method, 5th *Symp Flow Visual., ISAS Univ. of Tokyo*, pp. 43–46, 1977.
38. Matsuo, K., Ikui, T., Yamamoto, Y., and Setoguchi, T., Measurements of Shock Tube Flows using a Spark Tracer Method, in *Flow Visualization*, ed. T. Asanuma, pp. 157–162, Hemisphere, Washington, D.C., 1979.
39. Nakayama, Y., Aoki, K., and Ohta, H., Flow Visualization by Spark Method (in the Case of Line Type Electrodes), *4th Symp. Flow Visual., ISAS Univ. of Tokyo*, pp. 105–108, 1976.
40. Nakayama, Y., and Wada, K., Visualizing Study on the Flow around Golf Ball using Spark Tracing Method, *J. Flow Visual Soc. Jpn.*, vol. 5, suppl., pp. 99–104, 1986.
41. Nakayama, Y., Aoki, K., Kato, M., and Okumoto, T., Flow Visualization around Gold Balls, in *Flow Visualization IV: Proceedings of the Fourth International Symposium*, ed. C. Veret, pp. 191–198, Hemisphere, Washington, D. C., 1987.
42. Nakayama, Y., Davies, P. O. A. L., McConachie, P. J., and Baxter, D. R. J., Analysis of Shear Layer Structure using Spark Method, ISVR Univ. of Southampton, ISVR Memo 608, pp. 1–11, 1981.
43. Kimura, T., Nishio, M., Fujita, T., and Maeno, R., Visualization of Shock Wave by Electric Discharge, *AIAA J.* vol. 15, no. 5, pp. 611–612, 1977.
44. Kimura, T., and Nishio, M., A Method of Shock Visualization using Electric Discharges, *J. Flow Visual. Soc. Jpn.*, vol. 2, no. 7, pp. 661–666, 1982.
45. Kimura, T., and Nishio, M., Visualizing Method of Shock Wave by Electric Discharge Using Point-Line Electrodes, in *Flow Visualization III: Proceedings of the Third International Symposium on Flow Visualization*, ed. W.-J. Yang, pp. 473–477, Hemisphere, Washington, D.C., 1985.
46. Gowen, F. E., and Hopkins, V. D., A Wind Tunnel using Arc-Heated Air for Mach Numbers from 10 to 20, *Proc. 2d Symp. Hypervel. Techn., Univ. of Denver*, p. 27, 1962.
47. Venable, D., and Kaplan, N. Electron Beam Method of Determining Density Profiles across Shock Waves in Gases at Low Densities, *J. Appl. Phys.*, vol. 26, no. 5, pp. 639–640, 1955.

Chapter 7

Surface Tracing Methods

R. Řezníček

I INTRODUCTION

This chapter reviews methods where visualization is provided by means of a suitable coating of the surface of bodies in the flow; the fluid flowing around the body changes the coating, allowing determination of certain flow characteristics. These changes are either observed by the naked eye or recorded by photographic, cinematographic, or videogenic recording techniques. These methods, reviewed in [1–4], may be divided into three groups:

1. **Chemical methods**, where a chemical reaction occurs between the flowing fluid and a suitable substance applied in a thin layer on the surface of the body in flow. Sites on the surface of the body where the reaction rate or intensity is higher change tint or color, compared to the rest of the surface. The greatest reaction occurs in turbulent flow, so that the turbulent flow and transition boundaries are visualized.
2. **Physical methods**, based on sublimation, evaporation, or dissolution of the surface coating on the body in the flow. Due to a more intense mixing process in turbulent flow, the physical processes occur more rapidly, and laminar and turbulent flow may be differentiated from each other on the basis of different optical characteristics of the surface (different index of refraction, different color tint).
3. **Mechanical methods**, which are based on mechanical processes such as change of position or shape of small particles on the surface of the body in the flow, change of thickness of suitable liquids on the surface, or deposition of particles in the flow on the coated surface of the body.

The materials for coatings used in these methods must not be toxic or harmful to the human organism; they must not cause corrosion or other damage to the model surface coating or to other experimental devices.

In general, these methods visualize characteristics of the flow on the body surface in models as well as in actual objects. It is then possible to determine the local direction of the flow on the body surface, laminar and turbulent flow regions, and transient regions, eddies, and shock waves.

It should be emphasized that the pattern obtained by these methods is determined by the flow inside a boundary layer. The pattern indicates, e.g., flow direction in the boundary layer. Flow direction beyond the boundary layer may be essentially different. Cases of secondary flow in the boundary layer are well known. In a region of flow separation, flow in the boundary layer may be in a direction opposite to that beyond this layer. A photo of the visualization layer on the body is itself frequently insufficient to yield detailed data about flow direction. It is necessary either to follow visually the time development of the layer conversion or to take several photos in the course of the layer conversion. For this reason it is advantageous to use cinematographic or videogenic recording techniques. These procedures are used more frequently where the patterns on the surface are produced by free flow around a stationary body.

In certain cases, it is necessary to take into account the effects of gravitation and possibly inertial forces (e.g., during rotation) on the visualization layer. This effect may sometimes be very important. The simultaneous action of gravity and shear stress on the visualization layer may also be used to determine changes of shear stress on the surface of nonhorizontal walls of a body in the flow, e.g., for a more precise determination of the position of the boundary layer transition. The substance used to produce the visualization layer should be chosen for each case on the basis of preliminary experiments. Cases will be presented in the following discussion, and relevant references are given. It is impossible to recommend a universal mixture. When choosing the mixture, it is necessary to take into account the flow rate, the duration of flow, the effect of gravity and inertial forces, as well as the toxicity of chemical substances used, their corrosive action, and availability. These methods represent a simple, cheap approach to obtaining important data about flow around the surface of bodies.

In the case of liquids, these methods may be used for flow velocities of 1 m/s and above. The electrolytic method is also suitable for lower velocities, 0.01–0.1 m/s. For gas flow, velocities of 10 m/s and above, up to supersonic velocities, should be considered.

II THE USE OF SURFACE-TRACING METHODS FOR VISUALIZATION OF LIQUID FLOW

A Chemical Methods

Methods based on chemical reactions are used commonly for visualization of liquid flow. The liquid used is usually water, and it is hard to find a substance that would react suitably with water. This problem is solved by adding to the water a chemical substance that reacts with a reagent in the coating of the body in flow [5, 6]. Depending on the choice of chemical substances, it is possible to achieve changes in color of the visualization layer on the surface in flow. These changes occur quickly or more intensely at sites of greater contact of the two chemical substances, due either to higher concentration or to a more intense mixing process. The case of a homogeneous spread of chemical reagent in flowing water indicates a region of turbulent flow around the body. The pattern obtained on the surface of the object in flow is usually not permanent. Once the coating has been used, it normally cannot be reused. To repeat the experiment, it is necessary to give the surface a new coating. The chemical reagent may also be supplied to the water flow by tubes with openings situated at desirable sites close to the surface in flow [7]. The chemical reagent can also be applied through openings in the surface. It is thus possible to examine the direction of flow on the surface of bodies in flow. This method was used successfully for visualization of flow on the surface of models of ship hulls [8]. The color streaks obtained are sharp and long in the region of laminar flow. During turbulent flow, the color of the streaks is paler, the streaks are shorter, with a pearlike shape.

The chemical reagent used for visualization on the surface of a body in flow is usually added, finely spread, in a suitable binder in a ratio of about 1 : 1 by mass; the mixture is applied to the surface. Common varnishes, paints, glues, and gels may be used as binders [4]. After drying, solidification, and grinding (e.g., emery grinding) of the coating, the surface is ready for the experiment. It is possible to use the following combination of reagents in the surface coating and reagents in flowing water [3, 8]:

1. Lead white (basic lead carbonate)–ammonium sulfate (saturated solution); brown color
2. Calomel (mercurous chloride)–ammonia liquor; black color
3. Ferric chloride–pyrogallol (inorganic gallic acid); black color
4. Potassium iodide–sodium thiosulfate; blueish violet color

Most of these substances are toxic, and relevant safety rules should be adhered to when handling them.

B Physical Methods

Physical methods are used more frequently than chemical methods during the flow of water around a body. The following substances were tested as a base of the coating: acetoacetanilide, exalgine, acetanilide, phenacetine, hydroquinone diacetate, and benzoine [9]. The surface is coated dry with these substances, using a spray gun; the solvent has evaporated to a smaller or larger extent before striking the body surface. The layer applied is as thick as 5–12 mm, and it must be smooth. The model surface should be dark colored; the coating applied is usually pale. In turbulent flow around the body, the layer is dissolved more rapidly in comparison to laminar flow; the laminar region of flow is shown by a pale layer on the surface. By contrast, in the region of turbulent flow, a dark color of the model surface is prevalent, since there is a more rapid dissolution of the layer by action of a more intense mixing.

Acetone or light petroleum fractions are most suitable as solvents for the surface coating with 4–5% concentrations of the tracer. Benzoic acid is easy to use. [8, 10]; available in the form of white crystalline powders, it dissolves well in acetone. The suitable concentration for application with a spray gun is 150 g of benzoic acid in one liter of acetone. The solubility of the layer in water at 24°C is 3.4 g/one liter of water. The necessary time of water flow is several tens of seconds. After the object is removed from the water, it should be dried, using a suitable texture and a hot-air stream.

The surface coating of the body in flow should be perfect. It is necessary to use cement, primer, and fine emery grinding to achieve this. These layers, particularly dull-black paint, have to be sufficiently resistant to acetone, the solvent of the visualization coating. Two-component paints and cements based on epoxide are suitable. For finding the transient region, a 5% solution of hydroquinone diacetate in acetone or a 4–5% solution of acetanilide in a mixture of the same volumes of acetone and petroleum are suitable. Local inequalities of the surface of the body in flow induce local disturbances of the flow wakes, creating dark zones in the region of laminar flow which may serve for determining the direction of flow on the body surface [9, 11, 12]. These inequalities may be produced artificially. For example, it is possible to fix into the surface stainless steel wires that protrude about 1 mm above the surface.

C Mechanical Methods

When using mechanical methods of visualizing liquid flow on the surface of bodies, preparation of the surface is very simple. The body may, e.g., be painted with a strong layer of oil paint and then, without allowing the paint to dry, put into the liquid flow. Fine striae are produced in the paint, determining the direction of local flow on the surface. It is also possible to use water-soluble glue-containing wall paints. This method makes it possible to establish regions of separation, since the striae disappear in this region. In a similar approach, called the oil-dot method, the body is covered before the experiment with drops of oil paint, varnish, oil, or glue, and it is put wet into the flow, which spreads the drops in the direction of the stream. A suitable consistency of drops must be used, since they should spread only after a certain time and not immediately. The lines obtained in this way on the body surface yield good information about the flow on the body surface (Figs. 1 and 2). This method was also elaborated on for obtaining quantitative results [13, 14]. The length of the track obtained by spreading of the drop is directly dependent on the surface shear stress caused by the flowing liquid. The drop size is of no essential importance in this case.

An interesting new concept [13] is the use of ferromagnetic wafers instead of drops. These wafers are, for example, circular, 5 mm dia., 0.2–0.5 mm thick, and cut from a sheet. After being magnetized, they are applied onto the model surface, which is also made of ferromagnetic material. The initial location of the mark is recorded (e.g., with the help of a photo or by drawing on the surface), and the track of the wafer is measured after a certain period of flow. In this case, the magnitude of the shear stress depends on the length of the wafer shift for a given time.

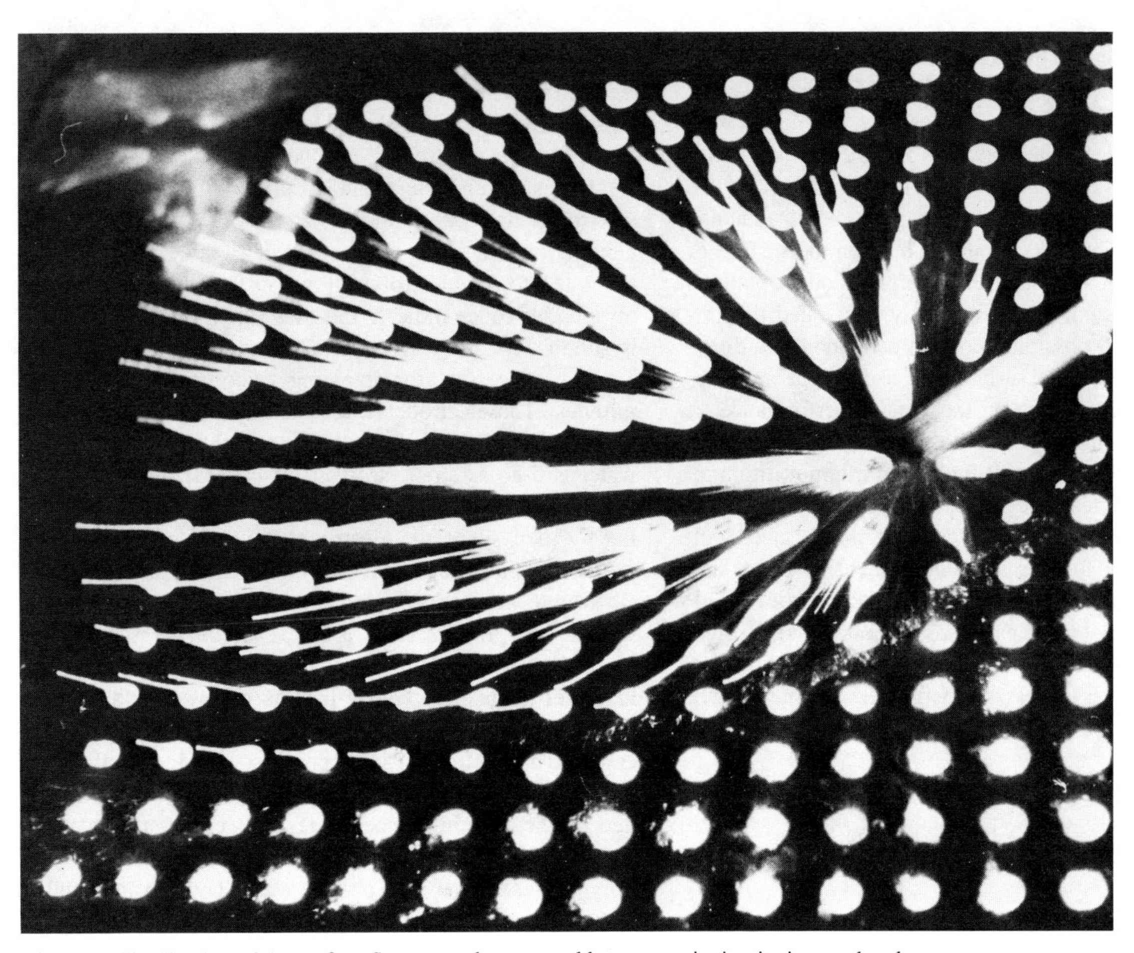

Fig. 1 Visualization of the surface flow on a plane caused by a water jet impinging on the plane. The oil-dot method was used, with varnish forming the dots [58].

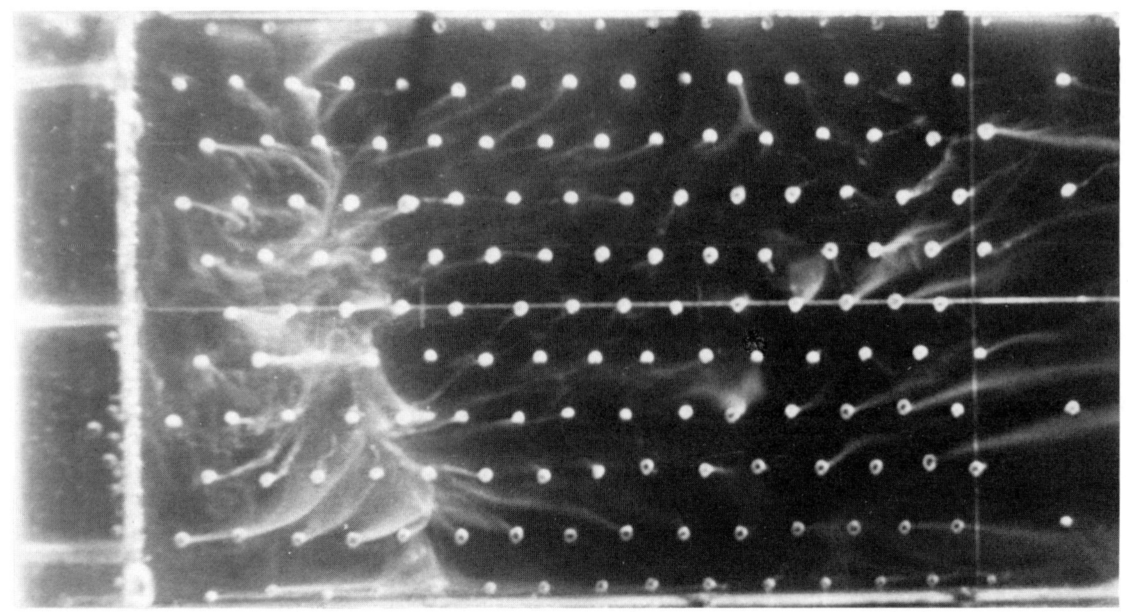

Fig. 2 Surface flow of water behind a backward-facing step. The oil-dot method was used, with water-soluble adhesive forming the dots [57].

A method using oil film containing powdered pigment is also used, most frequently in the flow of gases; however, it can also offer good results in water flow. Oil mixed with pigment is spread on the body by brush or by spraying. When water flows around the body, the appearance of the oil film is altered; the motion of the particles and the change of their location yield data about the flow on the surface of the body [15, 16]. Although the interpretation of patterns obtained still does not have a perfectly elaborated theoretical basis, it is still possible to obtain data about the direction of local flow on the surface and in the separation regions [17, 18]. Depending on the type of performance, it is possible to use motor as well as transmission oils, paraffin, silicone oils, and diesel fuel, with minium, titanium dioxide, soots, or lead chromate as pigments. The choice of the pigment depends on the color of the surface; with white pigment the surface is coated with black paint, and with dark pigment white paint is used. Addition of a small amount (about 1% of the paint suspension mass) of oleic acid reduces the tendency of particles to coagulate. For each particular case, it is necessary to find the optimum layer thickness.

D Electrochemical Methods

Electrochemical methods based on the use of electrolysis should also be mentioned [19–22]. In these methods the surface tested is metallic and serves as one electrode. The medium flowing is the electrolyte in which a second electrode is situated. When an electric voltage source is connected to the electrodes, changes on the surfaces of the electrodes occur. The metallic surface of the body in the electrolyte flow either releases atoms or deposits them, depending on whether it is an anode or cathode. This process is related to the electrolyte flow, so that tracks remaining on the electrodes indicate the flow direction. This method is suitable for low velocities of about 10^{-2} to 10^{-1} m/s. The time necessary for obtaining a good pattern is several minutes. The electrode represented by the surface investigated may be, e.g., of tin, and it serves as an anode. Tap water is used as an electrolyte; the cathode material is unimportant. Striae are formed on the anode surface, indicating by their shape the electrolyte flow. Using the surface tested as a cathode, metal will be deposited on it from the electrolyte in streaks, corresponding in shape to the electrolyte flow. In this case, a copper electrode is used, the electrolyte being a solution of 50 g H_2SO_4 and 150 g $CuSO_4$ in 1 liter of distilled water. It is necessary to add into the solution an organic compound to prevent the formation of crystalline structure in the copper deposited, thus facilitating the involvement of the shape of the metal layer deposited by the streaming electrolyte. This compound may be, e.g., 100 mg polyvinylpyrolidone in 1 liter of electrolyte. Polypropylethylene may also be used. The arrangement of electrodes should provide a uniform distribution of electric current density on the electrode of about 10 to 20 mA/cm.

III THE USE OF SURFACE-TRACING METHODS FOR VISUALIZATION OF GAS FLOW

A Chemical Methods

The first chemical method we will consider is based on providing the body in flow with a coating containing lead compounds and, through an opening in the surface, applying a hydrogen sulfide stream close to the surface. Hydrogen sulfide induces blackening during contact with the coating; at the sites of turbulent flow, hydrogen sulfide diffuse very rapidly, so blackening does not occur. The blackened coating cannot be reversed to the original white color. To repeat the experiment, it is necessary to remove the blackened coating and to recoat the surface. As this process is rather tedious, it is sometimes easier to soak the chemical agent into a filtration paper or suitable texture (e.g., mousseline) and then fix this paper or texture onto the surface to be investigated. The paper or texture used may then be easily replaced. However, if the body has a complicated shape, it is impossible to cover the surface properly in this way.

The coating necessary to coat such a shape is prepared by mixing 10 g of lead carbonate, 5 ml of water, and 0.5 ml of glycerol. The paste thus obtained is applied to the surface with a brush. By adding acetone (~10 ml), we obtain a solution that may be sprayed onto the surface [24]. Hydrogen sulfide may be prepared in a Kipp device by a reaction of hydrochloric acid with iron sulfide. For flows at a velocity of 25 m/s, it is satisfactory to use a hydrogen sulfide flow rate of 15 ml/min through an inlet opening or tube [6]. These chemicals are of course harmful, and it is better to use mercuric chloride instead of lead carbonate and to bring a thin stream of ammonia and air near to the surface. This method can be used for studying the boundary layer [23, 24] at even higher velocities than with a method based on smoke filaments. As a matter of fact, at higher velocities a spread of the filament occurs, making observation impossible in the case of smoke filaments. At higher velocities this spread is also manifested in the case of the ammonia filament; however, in this case it does not disturb the observation too much, since the reaction is induced even by an ammonia filament of a rather low concentration. The mixture of air and ammonia may be obtained, e.g., by bubbling air through ammonia liquor.

Another chemical method begins by covering the body to be tested with ozalide paper (paper for collotype) and, with the help of thin nozzles, inserting filaments of a mixture of ammonia and air in front of the body in flow, which cause dark streaks to appear on the surface of the ozalide paper in the direction of the local flow [25].

A further chemical method is based on using a coating of potassium iodide + starch. A small amount of gaseous chlorine is added to the flowing air. The chlorine reacts with iodide in the surface coating, mainly in the region of turbulent flow, since turbulent flow, as compared to laminar flow, creates more intense mixing and more frequent contact of the chlorine with the surface coating. One consequence of this reaction is a change of the coating color to purple. Chlorine releases iodine from potassium iodide, and free iodine yields a color reaction with starch. This method is not suitable for work in a tunnel because chlorine is harmful; it has, however, been used during an aircraft flight. In this case the aircraft flies through a cloud of air enriched in chlorine that has been released from other aircraft, from factory chimneys, or from nozzles on the aircraft investigated [1, 3]. The coating is prepared by mixing thoroughly 10 g of titanium white, 0.5 g of potassium iodide, 1 ml of 1% aqueous starch solution, 0.5 ml of glycerol, and 1 ml of normal sodium thiosulfate. The mixture is applied with a brush onto the surface to be investigated.

Indication agents can also be used [3, 32]. For the coating, it is possible to use

1. **chromocresolpurple:** To a solution of chromocresolpurple, prepared by dissolving 0.1 g of chromocresolpurple in 10 ml of alcohol, are added 10 g of titanium white, 4 ml of water, 1 ml of sulfuric acid, and 0.5 ml of glycerol, mixed to obtain a paste, which is spread with a brush onto the surface. Another possibility lies in spraying a mixture of 0.5 g of chromocresolpurple, 50 ml of alcohol, 50 ml of water, 3 ml of glycerol, 0.5 ml of phosphoric acid (specific mass of 1.75), and 15 g of titanium white or alumina onto the test surface. The coating is yellow, but its contact with gaseous ammonia changes it to violet. The mixture of air and ammonia may be diluted. After completing the experiment, the flow of ammonia is stopped, and the object is exposed to flowing air causing the violet spots to be reconverted to yellow. Diethylamine may also be used instead of ammonia.
2. **Congo red:** To a solution prepared by dissolving 0.1 g of Congo red in 10 ml of alcohol are added 5 ml of water, 1 ml of sulfuric acid, 5 ml of glycerol, and 10 g of titanium white; a paste is prepared, which is then applied with a brush onto the test surface. The coating is blue, but its contact with gaseous ammonia causes it to change to red with a very good contrast. If the object is exposed to the effect of a gaseous inorganic acid, the red color returns to blue. Carbon dioxide contained in the air does not yield this effect.
3. **Bromothymol blue (BTB):** This dye is frequently used as an indicator because it is easy to observe and record the course of the color change. From 5 g of BTB and 500 g of ethyl alcohol, 700 ml of yellow mixed solution

are prepared. To provide a suitable viscosity for producing a proper coating on the object surface, a solution of 300 g of wheat flour is prepared in 500 ml of water; after dilution with 8 liters of hot water, it is thoroughly stirred. To this mixture, 700 ml of mixed BTB solution are added; after thorough mixing, the paste obtained is spread onto the test surface, preserving its chemically neutral character. The object is situated into a flow field of air, and a gaseous mixture of ammonia and air is applied. The yellow coating changes to blue upon contact with gaseous ammonia. After the evaluation, the coating is washed off with water.

Fig. 3 Visualization of the air flow on the surface of a wing model. A method of naphthalene coating sublimation was used [1].

B Physical Methods

Of the physical methods, we will first consider the sublimation method, which is based on the fact that a coating of a suitable material on the surface of the body in flow is sublimated more strongly in the region of the turbulent boundary layer than in the region of the laminar boundary layer. A basic requirement is that the color of the coating be different from that of the body in the flow. It is then possible to establish the transition boundary between the surface free of the coating, i.e., that portion of the surface where the coating substance is passed into the flowing medium by action of the sublimation, and the surface where the coating remained unaffected. In certain cases the difference between these two portions of the surface is insufficiently distinct, and is properly visible only after intense oblique illumination [26]. The following substances are most frequently used for coatings: hexachloroethane, naphthalene, diphenyl, acenaphthene, hydroquinone, diethylether, fluorene, camphor, borneol [9]. The test surface is coated by dry sparying, using a solvent, e.g., of acetone, light petroleum fractions, benzene, xylene, whenever necessary. The solvent chosen must not be damaging to the body surface being tested.

In practice, for a rapid laboratory investigation of transient regions and wakes, hexachloroethane is the most suitable coating. It may be prepared for spraying in the form of a 10% solution. The coating obtained is dull white, so the surface of the test body should be black. In this case, at velocities of 50–130 km/h, a distinct boundary is obtained in 30–120 s, and it remains visible for about 2–3 min after the model is removed from the tunnel. Naphthalene, diphenyl, and acenaphthene are good indicators at velocities of 150–700 km/h (Fig. 3). Acenaphthene may also be used to indicate the transition at supersonic velocities; e.g., at 1900 km/h a distinct boundary was produced in 10 min. At velocities of 2000 km/h, a satisfactory pattern was obtained in 9 min with a thick azobenzene coating. For experiments with aircraft in flight, diethylether was used for low velocities, i.e., 200 km/h. The pattern was developed after 10–45 min, depending on the atmospheric temperature and the altitude. Acenaphthene and fluorene were used for velocities of 400–800 km/h, and the time of the pattern development was 8–30 min [9, 24, 27–30].

Other physical methods similar in the way of observation and possibilities of marking the transition boundaries are the so-called evaporation methods. Good results are obtained particularly with a simple method using kaolin coatings [26, 31]. In this method, the surface of the test body is painted black and then covered by a layer of kaolin, e.g., by spraying with a kaolin suspension. A suitable suspension is obtained from 100 ml of colorless frigilene, 100 ml of butyl acetate, 100 ml of butyl alcohol, 50 ml of xylene, and 100 g of kaolin. When the layer has dried, it should be finely ground with emery to remove inequalities in the kaolin layer. Then the layer is sprayed with a volatile liquid; for low velocities, about 100 m/s, it is sprayed with nitrobenzene, for high subsonic velocities with ethylbenzoate, methylsalicylate, and isosaphrol. In this way the original kaolin layer, which was dull white due to diffused light reflection, becomes bright; i.e., the light reflection on the layer is now regular, and the layer seems to be dark when illuminated in a suitable direction. Undesirable reflected light may be eliminated by using a polarization filter on the camera. When a model prepared in this way is placed in flow, the volatile liquid is evaporated more intensely in the turbulent flow, as compared to the laminar flow, region as a result of the intense mixing process. Thus, the part of the surface exposed to turbulent flow becomes dull due to the diffused reflection of light, whereas the laminar region is still bright and thus the surface is dark in suitable illumination [23, 26]. The advantage of this approach is that the basic kaolin layer may be used several times. However, this layer may sometimes affect (reduce) inequalities of the surface.

Another evaporation method—the method of liquid layers—is similar and as a matter of fact simpler in comparison to the method of kaolin coatings; however, its results offer less contrast [33]. In this method, the test surface is given a thin layer of volatile liquid by spraying or painting. Some of the volatile liquids used were already mentioned in the description of the preceding method. The volatile liquid is evaporated during flow around the body in the turbulent region of the boundary layer much more rapidly than in the laminar layer, so that behind the transition boundary the liquid layer disappears; it is evaporated completely. To increase the contrast, it is desirable to paint the body surface with a dull black coating. The advantage of this method is that the liquid layer, being very thin, does not affect inequalities of the surface. The method was also used for supersonic velocities of the flow [3, 33–37].

C Mechanical Methods

In yet another, mechanical method for the visualization of flow, the oil-film method, the body surface is given a coating film of a mixture of soots and petroleum. In the flow, petroleum evaporates and flows in the stream direction [38, 39]. By the petroleum flow the soots are arranged on the surface in the flow direction. This simple method works very well for determining transition regions as well as separation regions. Certain oils besides petroleum are also suitable [3]. It is advantageous to give the surface a pale, mostly white coating. Traces of soot on the body surface may be recorded using photographic processes. They can also be printed on white paper pressed to the body surface. This is easily done when most of the liquid phase of the coating has evaporated. By painting the body surface with a black coating, it is possible to work in a manner analogous to the use of a pale powdered pigment and black paper. For powdered pigment it is possible to use magnesia alba (basic magnesium carbonate), titanium dioxide, lead chromate, minium, zinc oxide, kaolin. To prevent coagulation of the pigment particles, it is recommended that 1% of oleic acid mass be added to the oil suspension. The viscosity, density, and surface tension of the oil, in addition to its gravity and inertia [45], affect the pattern induced by the flowing fluid. On the basis of experience, it should be mentioned that for each case of flow, it is necessary to find the most suitable type of oil and pigment and their mutual ratio, which is usually 1 : 1 to 1 : 4 by volume. Pigments of various colors may be used advantageously. Depending on the type of flow, suitable oils may be transmission and motor oils, paraffin, silicone, and linseed oil, as well as diesel fuel and kerosene (fuel for jet aircraft). The oil suspension may be applied onto the surface with a brush, by spraying, or by a texture soaked with the suspensions [3, 40]. In applying the suspension, thickness of the layer and method of its equalization should also be tested, since characteristics of oils and pigments as well as conditions of production of the pattern in the layer are variable [41]. Sometimes, better results may be achieved by preliminary formation of a very thin layer of light oil before applying the visualization oil layer, e.g., spreading it with a texture soaked with light oil. An oil film on the test surface may be used for low as well as supersonic velocities [42–45, 49, 59].

Figure 4 shows a visualization obtained by this method on the surface behind a backward-facing step. This case of flow is shown as an example of using methods of surface treatment not only for interesting characteristics from the standpoint of complicated flow, but also for observation simultaneously of several characteristic phenomena, such as the region of flow separation and the region of boundary layer reattachment. It is possible to follow the formation of eddies, the effect of side walls, the direction of the flow, and other phenomena.

Figures 5 and 6 show two other examples of the use of this method. As in the case of visualizing liquid flow, drops placed on the test surface of a body may also be used to study flow of gases; this is known as the oil-dot method. The liquid used may be, as in the oil-film method, a mixture of oil and pigment. The drops are applied with a brush or tube onto the test surface. It is also possible to construct a device to apply a large number of drops simultaneously. The distribution of drops must be such that observation of the tracks of oil drops yields proper evaluation of the character of the flow on the surface. The droplets on the test surface are affected by the shear stress of the flowing fluid and by gravity and inertial forces. The height of drops should not exceed the boundary layer. This method produces a pattern similar to that obtained by using fiber indicators (tufts). The direction of the oil streaks flowing from the drops shows the direction of flow at these sites, and their length provides information about the value of the shear stress on the test surface. As a matter of fact, the length of the streaks is not essentially dependent on the diameter of the drops on the surface [13, 14, 46–48]. A procedure was also tested whereby the visualization liquid is driven to the surface of the body through small openings [14].

In another method, the body surface is coated with oil and sprayed rather strongly with fine powder, e.g., powdered sulfur or the above mentioned pigments. Then, using very moist air as the flow medium, condensed water is deposited on the body; the flow forms streaks on the surface in the direction of the flow [1].

The body can also be coated with oil, and finely ground chalk or other powdered pigment may be added into the flowing gas. The chalk is deposited on those sites on the surface where intense eddies are present. In this way the region of turbulent boundary layer and the region of turbulent separation are determined; in the latter case the chalk is introduced into a wake and comes to the body surface by back flow [49–51]. It is also possible to use methods based on chemical reactions of substances in the surface layer and in the air [4].

Fine powder, e.g., lycopodia, which is a pale yellow powder with a linear grain size of about 0.04 mm, may be used for visualization as follows: lycopodium is dried and

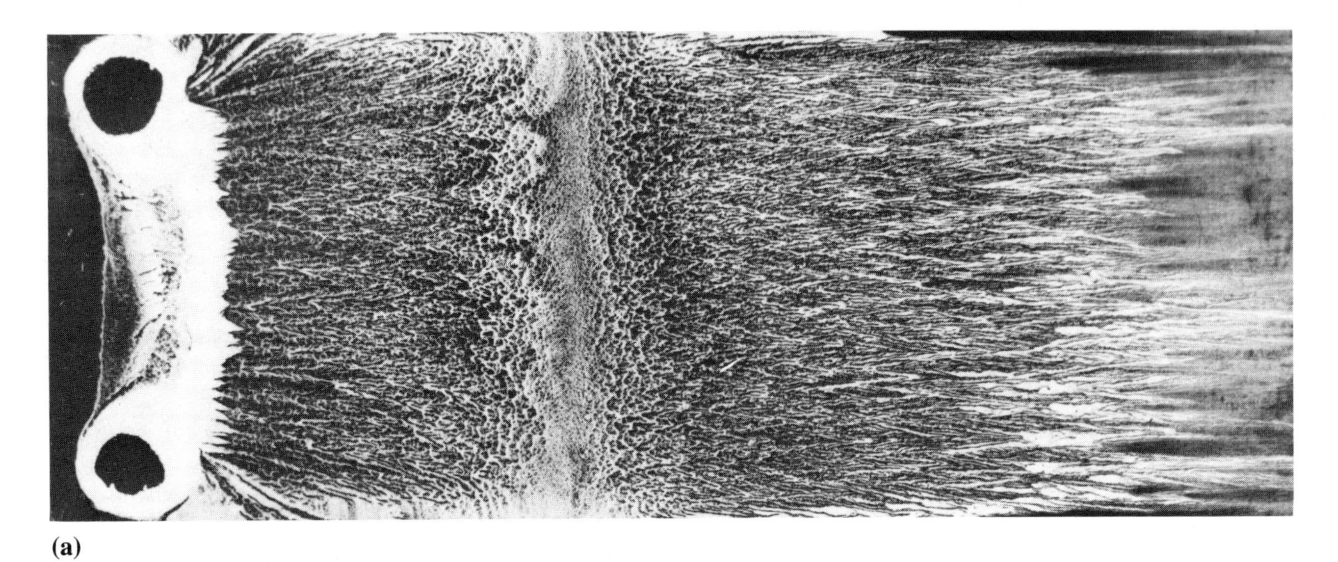

(a)

(b)

Fig. 4 Visualization of air flow on the surface behind a backward-facing step. The method used was spreading a thin layer of the visualization oil layer on the surface in the flow. Visualization liquids used: (*a*) Kerosene + magnesia alba 3 : 1 by volume; (*b*) kerosene + lampblack (soots) 3 : 1 by volume [57, 58].

applied to the model surface, which has first been degreased, polished, and given a dark coating. The ventilator of the tunnel is then put into operation. The powder remains on the surface of the model at those sites where the velocity of the air in the layer adjacent to the model, of a thickness corresponding to the grain size, is not sufficiently high to remove the powder. The powder thus adheres in the thicker laminar boundary layer. However, it is removed from the entering edge, where the laminar boundary layer is thin, and also from sites where the boundary layer is turbulent, which also means a high-velocity gradient at the surface. The method is satisfactory for velocities to about 30 m/s; for higher velocities the laminar boundry layer is too thin [23].

In a method that met with success, the test surface was coated with an oil that exhibited fluorescence in ultraviolet rays. When flow occurred around the body, the thickness of the oil film was affected by tangential stress, induced on the body surface by the flow of air. The oil on the one hand is shifted along the surface, and on the other hand is evaporated, so that the shape of the oil layer is related to the magnitude and direction of the tangential stress on the surface. The difference in the thickness of the oil layer may be easily recognized during its exposure to ultraviolet rays, inducing fluorescence in the oil, since sites where the oil layer is very thick emit more intense light. Since the thickness of the oil film affected by the flow persists for a certain time, dependent on the viscosity of the oil, after interrupting

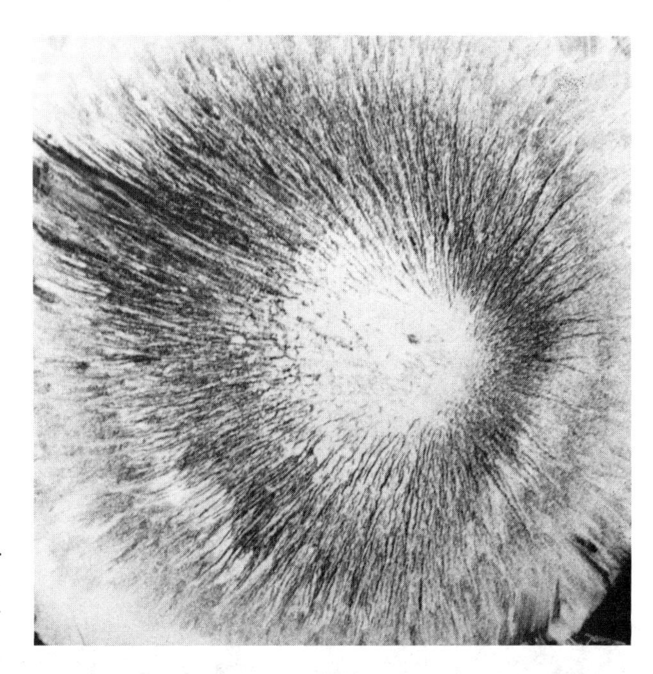

Fig. 5 Visualization of the flow on the surface of a planar horizontal plate on which an air jet impinges at an angle of 70°. The thin oil-film method was used, where the oil containing kerosene + magnesia alba 3 : 1 by volume [58].

Fig. 6 Oil-film method used to visualize the surface flow near the tip of a wing [58].

the flow the model may be removed from the tunnel and a photo made during exposure to ultraviolet rays.

For successful use of this method, the model surface must have a nonfluorescent white color and should not absorb the oil used. A white or bright surface is necessary to reflect the incident ultraviolet radiation as well as the induced fluorescent radiation that amplifies the effect observed.

The method is especially suitable for visualization of the region of laminar and turbulent flow, back flow, and flow separation. It is possible to locate pressure maxima and shock waves. This method can also be used at high velocities; Fig. 7 shows visualization of an eddy series in a boundary layer in supersonic flow around a wing ($M = 3$).

To examine transverse flow in a boundary layer, it is more suitable to distribute oil on the surface of the body in small, separate droplets or in longitudinal streaks. On the basis of changes of shape of the droplets or streaks, it is possible to ascertain the direction of the tangential stress on the surface. In selecting the oil we consider its viscosity, which affects the ability of the oil layer to adjust to the

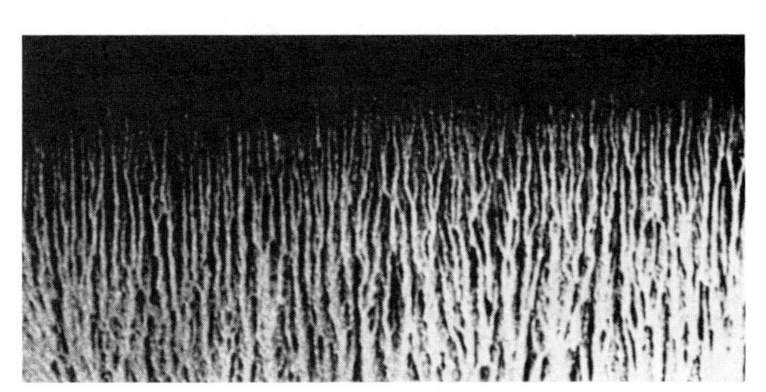

Fig. 7 Visualization of an eddy series on the wing surface with the use of an oil film and photography in ultraviolet rays [1].

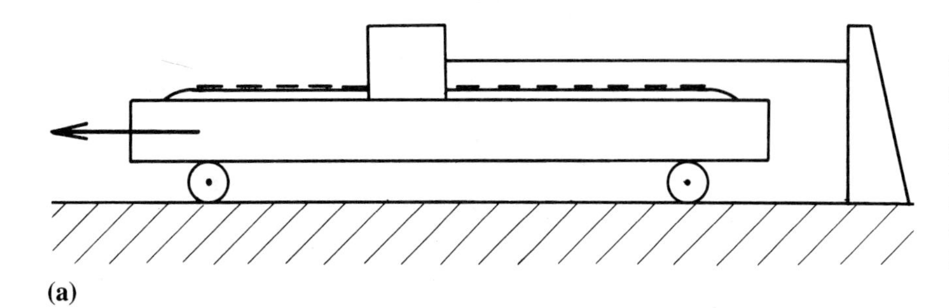

(a)

Fig. 8 Visualization of the flow in a thin liquid layer, using a water layer about 1 mm thick and visualization particles of lycopodium on the surface of the water layer; (*a*) Schematic diagram of the equipment; (*b*) flow around a cylinder; (*c*) flow around a wavy surface [1].

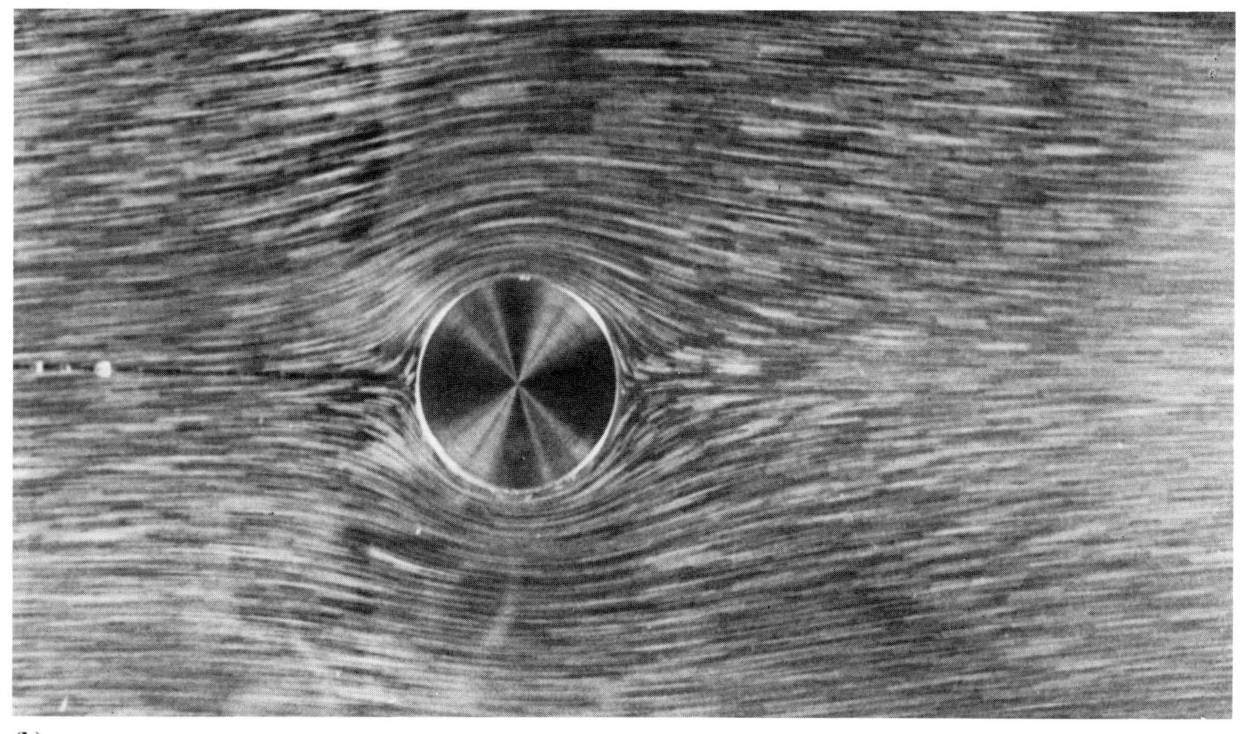

(b)

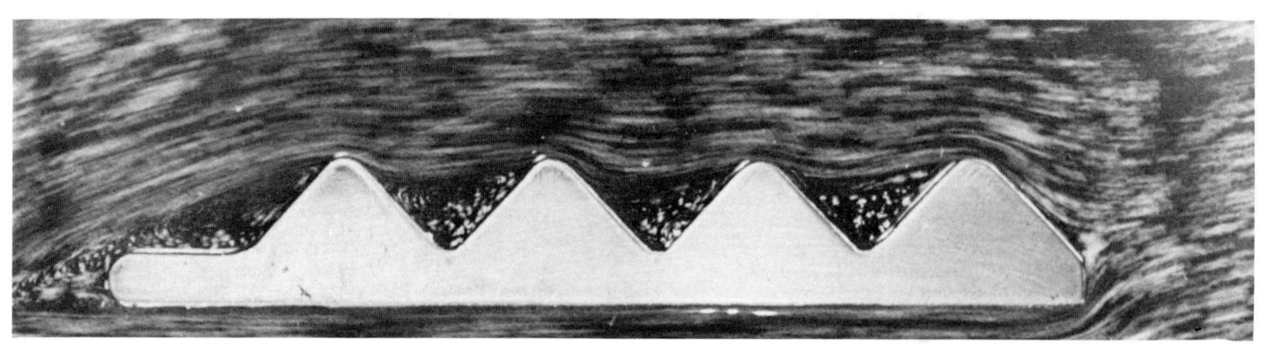

(c)

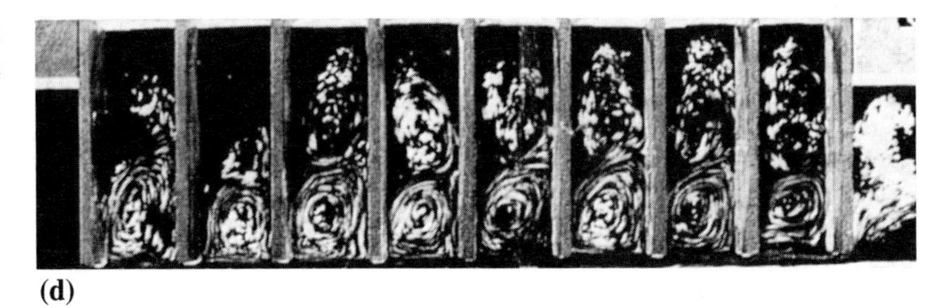
(d)

Fig. 8 (continued) (*d*) Flow in a labyrinth box.

flow, the time the oil adheres in a sufficiently thick layer on the model, and the time the oil layer maintains its altered shape after flow interruption. Since these effects of the oil's viscosity are crucial from the standpoint of the method considered, it is necessary in each particular case to select oil of a suitable viscosity on the basis of a preliminary test. During this test it is necessary to consider that the oil chosen has to easily wet the model surface. At low flow velocities, oils with a lower viscosity are satisfactory; at higher velocities oils with higher viscosities should be used. Common motor oils may be used [52–54], and fluorescein may be added to the oil. For the photographic process it is necessary to select a suitable oil, source of ultraviolet radiation, and photographic film with sufficient sensitivity to the fluorescence radiation obtained.

The effect of the air flow on the thickness of a liquid film on the body surface may be measured by light interference in this thin layer [55, 56]. During the reflection of light by the thin layer, interference occurs between the rays reflected by the upper surface of the layer and those reflected by the bottom surface. At sites on the layer where its thickness conforms to the condition for the disappearance of the light by the interference, dark streaks are formed. Those are sites where one ray is delayed by passage through the layer in such a way that it is out of phase with other rays reflected from the upper surface of the layer.

When the layer thickness varies with time, the interference streaks move; the number of interference streaks passing a site on the surface of the body in a unit of time is proportional to the time change of the layer thickness.

The methods mentioned may also be used for quantitative determination of the surface shear stress. For this purpose it is assumed that the time change (reduction) of the layer thickness is proportional to the local value of the surface shear stress [55, 56].

Thin layers of liquid adhering to the surface of a horizontal plate may also be used for the visualization of flow around models of bodies in other ways. For example, on a thin liquid film (e.g., 0.5–2 mm of water) visualization particles are applied (lycopodium, aluminum dust). For visualization observation, the liquid layer may be stationary while the model is moved on a plate with the help of a string. In making photos, it is possible for the plate to be situated on a carriage that moves; the liquid layer also moves and flows around the model, which is fixed by the string to the stationary base of the carriage. The camera is situated on a stationary base above the model (see Fig. 8*a*). With a liquid layer that is sufficiently viscous and thin, this method makes it possible to obtain a model of potential flow (see Fig. 8*b*, showing the flow of water around a cylinder), the thickness of the water layer is about 1 mm, visualized with lycopodium. Other results obtained in this way are shown in Fig. 8*c*, visualized flow around a wavy surface and flow in a labyrinth box Fig. 8*d* [1].

Other important methods of visualization on the surface of bodies in flow are fiber indicators (tufts), heat-sensitive coatings, liquid crystals, and thermovision. They are discussed elsewhere.

REFERENCES

1. Řezníček, R., *Visualisace Proudění*, Academia, Praha, 1972.
2. Merzkirch, W., *Flow Visualization*, Academic, New York, 1974.
3. Asanuma, T., *Handbook of Flow Visualization*, Asakura Shoten, Japan, 1977. In Japanese.
4. Kottke, V., Verfahren zum Sichtbarmachen von Gasströmungen, *Chem. Ing. Tech.*, vol. 52, no. 9, pp. 687–695, 1980.
5. Clayton, B. R., and Massey, B. S., Flow Visualization in Water, *J. Sci. Instrum.*, vol. 44, no. 1, pp. 2–11, 1967.
6. Street, P. J., and Twamley, C. S., A Technique for Flow Visualization using Chemical Indicator Solutions, *J. Sci. Instrum.*, vol. 44, pp. 558–571, 1967.
7. Tagori, T., Masunaga, K., Okamoto, H., and Suzuki, M., Visualization of Longitudinal Vortex near the Wall by Various Methods, in *Flow Visualization*, ed. T. Asanuma, pp. 201–206, Hemisphere, Washington, D.C., 1979.
8. Takosato, T., and Ouchi, H., Streamlines near the Bow of Large Tankers, Symp. Res. Rep. Eng. Coll., Ibaraki Univ. 12, 1965. In Japanese.
9. Main-Smith, J. D., Chemical Solids as Diffusible Coating Films for Visual Indication of Boundary–Layer Transition in Air and Water, Rep. Mem. Aeronaut. Res. Counc. 2755, 1950.
10. Takosato, T., The Effect of Turbulent Flow Stimulation, *Symp. Ship Build. Assoc.*, pp. 109–115, 1961. In Japanese.
11. Walker, W. P., Detection of Laminar Flow on Ship Models, *Trans. Inst. Nav. Arch.*, vol. 91, no. 4, pp. 220–228, 1949.
12. Itsui, T., Experiments in the Water Tank using Special Coatings, *Symp. Ship Build. Assoc.*, no. 92, 1957. In Japanese.

13. Atraghji, E., More Than Meets the Eye: The Oil Dot Technique, in *Flow Visualization II: Proceedings of the Second International Symposium*, ed. W. Merzkirch, pp. 619–628, Hemisphere, Washington, D.C., 1982.
14. Bisgood, P. L., The Application of a Surface Flow-Visualization Technique in Flight, Rep. Mem. Aeronaut. Res. Counc. 3769, 1974.
15. Ishihara, T., Kobayashi, T., and Iwanaga, M., Visualization of Laminar Separation by Oil Film Method, in *Flow Visualization II: Proceedings of the Second International Symposium*, ed. W. Merzkirch, pp. 283–287, Hemisphere, Washington, D.C., 1982.
16. Ishihara, T., Koya, S., and Mori, W., The Flow inside the Hydraulic Clutch, *Symp. Jpn. Eng. Soc.*, pp 39–45, 1968. In Japanese.
17. Murai, H., Ihara, A., and Narasaka, T., Visual Investigation of Formation Process of Oil-Flow Pattern, in *Flow Visualization II: Proceedings of the Second International Symposium*, ed. W. Merzkirch, pp. 629–633, Hemisphere, Washington, D.C., 1982.
18. Arakawa, C., and Tagori, T., Fundamental Experiments of Oil Film on a Rotating Disc, in *Flow Visualization II: Proceedings of the Second International Symposium*, ed. W. Merzkirch, pp. 727–737, Hemisphere, Washington, D.C., 1982.
19. Cognet, G., Mallet, J., and Wolff, M., Wall Streamline Visualization by Electrochemical Method, in *Flow Visualization II: Proceedings of the Second International Symposium*, ed. W. Merzkirch, pp. 227–237, Hemisphere, Washington, D.C., 1982.
20. Jaksic, M. M., and Tobias, C. W., Hydrodynamic Flow Visualization by an Electrochemical Method, in *Flow Visualization II: Proceedings of the Second International Symposium*, ed. W. Merzkirch, pp. 647–651, Hemisphere, Washington, D.C., 1982.
21. Bousgarbies, J. L., and Renaud, A., Study of the Wall Streamlines inside an Annular Cavity with Two Rotating Walls, in *Flow Visualization II: Proceedings of the Second International Symposium*, ed. W. Merzkirch, pp. 641–645, Hemisphere, Washington, D.C., 1982.
22. Taneda, S., Ishii, K., Takada, S., and Izumi, K., New Flow Visualization Method using Electrolysis, Rep. Appl. Mech. Lab., Kyushu Univ. 42, 1975. In Japanese.
23. Jaňour, Z., Experimental Examination of the Boundary Layer Development on the Airfoil, Rep. Aircr. Res. Inst., Prague 6, 1949. In Czech.
24. Preston, J. H., and Sweeting, N. E., Experiments on the Measurement of Transition Position by Chemical Methods, Rep. Mem. Aeronaut. Res. Counc. 2014, 1945.
25. Johnston, J. P. A Wall-Trace Flow Visualization Technique for Rotating Surface in Air, *Trans. ASME, Ser. D,* vol. 86, no. 4, 1964.
26. Richards, E. J., and Burstall, F. H., The China-Clay Method of Indication Transition, Rep. Mem. Aeronaut. Res. Counc. 2126, 1945.
27. Owen, P. R., and Ormerod, A. O., Evaporation from the Surface of a Body in an Airstream, Rep. Mem. Aeronaut. Res. Counc. 2875, 1951.
28. McCormack, P. D., Welker, H., and Kellcher, M., Taylor–Goertler Vortices and Their Effect on Heat Transfer, *J. Heat Transfer*, vol. 92, no. 1, 1970.
29. Diep, G. B., Le Fur, B., and Brun, E. A., Heat and Mass Transfer on Sweptback Circular Cylinder in Supersonic Flow, *Proc. 3d Int. Heat Transf. Conf., Chicago*, vol. 2, 1966.
30. Ginoux, J. J., Streamwise Vortices in Laminar Flow, Proc. Recent Developments in Boundary Layer Research, Part I, AGARDograph 97, 1966.
31. Holder, D. W., Transition Indication in the National Physical Laboratory 20 in. × 8 in. High Speed Tunnel, Rep. Mem. Aeronaut. Res. Counc. 2079, 1945.
32. Sadeh, W. Z., Brauer, H. J., and Durgin, J. R., A Dry-Surface Coating Method for Visualization of Separation, in *Flow Visualization II: Proceedings of the Second International Symposium*, ed. W. Merzkirch, pp. 635–639, Hemisphere, Washington, D.C., 1982.
33. Gray, W. E., A Simple Visual Method of Recording Boundary-Layer [Liquid Film], RAE Tech. Note Aero. 1816, 1946.
34. Persoz, B., and Germier, G., Enouits de Visualisation a Base de Corps Gras, *Rech. Aeronaut.*, no. 60, 1957.
35. Garner, H. C., and Bryer, D. W., Experimental Study of Surface Flow and Part-Span Vortex Layer on a Cropped Arrowhead Wing, Rep. Mem. Aeronaut. Res. Counc. 3107, 1957.
36. Tigue, J. G., Two Techniques for Detecting Boundary-Layer Transition in Flight at Supersonic Speeds and at Altitudes above 20,000 Feet, NASA Tech. Note D-18, 1959.
37. Haines, A. B., *Some Notes on the Flow Patterns Observed over Various Sweptback Wings at Low Mach Numbers*, Her Majesty's Stationery Office, London, 1960.
38. Abbot, I. H., and Sherman, A., Flow Observation with Tufts and Lampblack of the Stalling of Four Typical Airfoil Sections in the NACA Variable Density Tunnel, NACA Tech. Note 672, 1938.
39. Settles, G. S., Flow Visualization Techniques for Practical Aerodynamic Testing, in *Flow Visualization III: Proceedings of the Third International Symposium on Flow Visualization*, ed. W.-J. Yang, pp. 306–315, Hemisphere, Washington, D.C., 1985.
40. Murai, H., Flow Analysis by the Oil Film Method, *J. Jpn. Eng. Soc.*, vol. 74, p. 634, 1971.
41. Haines, A. B., and Rhodes, C. W., Tests in the Royal Aircraft Establishment 10 ft × 7 ft High Speed Tunnel on Three Wings with 50-deg Sweepback and 7.5 percent Thick Section, Rep. Mem. Aeronaut. Res. Counc. 3043, 1954.
42. Hall, I. M., and Rogers, E. W. E., The Flow Pattern on a Tapered Sweeptback Wing at Mach Numbers between 0.6 and 1.6, Rep. Mem. Aeronaut. Res. Counc. 3271, 1960.
43. Maxworthy, T., Experiments on the Flows around a Sphere at High Reynolds Numbers, *J. Appl. Mech.*, vol. 36, no. 3, pp. 598–611, 1969.
44. Raithby, G. D., and Eckert, E. R. G., The Effect of Support Position and Turbulence Intensity on the Flow near the Surface of a Sphere, *Wärme Stoffübertrag.*, vol. 1, pp. 87–94, 1968.
45. Taneda, S., and Tomonari, Y., Accelerating Flow Boundary Layer Transition, Rep. Appl. Mech. Lab., Kyushu Univ. 41, 1974.
46. Meyer, R. F., A Note on a Technique of Surface Flow Visualization, Nat. Res. Counc. Can. Aero-Rep., LR-457, 1966.

47. Lambourne, N. C., Pusey, P. S., Some Visual Observations of the Effects of Sweep on the Low-Speed Flow over a Sharp-Edged Plate at Incidence, *Rep. Mem. Aeronaut. Res. Counc.*, no. 3106, 1958.
48. Peake, D. J., The Flows about Upswept Rear Fuselages of Typical Cargo Aircraft, Nat. Res. Counc. Can. Rep. DME/NAE, 3, 1968.
49. Maltby, R. L., and Keating, R. F. A., The Surface Oil Flow Technique for Use in Low Speed Wind Tunnels, AGARDograph 70, 1962.
50. Hignett, E. T., *The Use of Dust Deposition as a Means of Flow Visualization*, Her Majesty's Stationery Office, London, 1963.
51. Hignett, E. T., Surface Flow Pattern as Visualized by Dust Deposits on the Blades of a Fan, *J. R. Aeronaut. Soc.*, vol. 67, no. 633, pp. 127–127, 1963.
52. Hopkins, E. J., Photographic Evidence of Streamwise Arrays of Vortices in Boundary-Layer Flow, NASA, Tech. Note D-328, 1960.
53. Loving, D. L., and Katzoff, S., The Fluorescent-Oil Film Method and Other Techniques for Boundary-Layer Flow Visualization, NASA Memo 3-17-59L, 1959.
54. Stalder, J. R., The Use of a Luminescent Lacquer for the Visual Indication of Boundary-Layer Transition, NACA Tech Note 2263, 1951.
55. Murphy, J. S., Measurement of Wall Shearing Stress in the Boundary-Layer by Means of an Evaporating Liquid Film, *J. Appl. Phys.*, vol. 27, no. 9, pp. 232–239, 1956.
56. Tanner, L. H., Surface Flow Visualization and Measurement by Oil Film Interferometry, in *Flow Visualization II: Proceedings of the Second International Symposium*, ed. W. Merzkirch, pp. 613–617, Hemisphere, Washington, D.C., 1982.
57. Jezek, J., and Reznicek, R., Visualization of a Backward-Facing Step Flow, in Flow Visualization IV: Proceedings of the *4th Int. Symp. on Flow* Visualization, ed. C. Véret, pp. 365–370, Hemisphere, Washington, D.C., 1987.
58. Reznicek, R., Hevrova, L., and Hevr, M., Surface Tracing Flow Visualization Methods, Res. Proj., Physics Dept., VSZ, 165 21 Praha, 1988.
59. Stanbrook, A., The Surface Oil Flow Technique for Use in High-Speed Wind Tunnels, AGARDograph 70, 1962.

Chapter 8

Liquid Crystals

N. Kasagi, R. J. Moffat, and M. Hirata

I INTRODUCTION

Visualization of the thermal field in a fluid may be as important as visualization of the velocity field for those working in mechanical, civil, aerospace, chemical, and many other engineering and scientific fields related to fluid mechanics and heat transfer.

Knowledge of the temperature distribution in a flow field is very useful for understanding not only the convective heat transfer mechanism but also the structure of the flow field itself. In forced convection, when the temperature difference is small enough for temperature to be regarded as a passive scalar, the flow field is unaffected by the thermal field; temperature can therefore act as a good tracer of the fluid motion. It is also important to obtain information about the temperature distribution in natural convection, where the temperature difference creates a density gradient in the fluid and, through interaction with the gravitational field, drives the flow. If information about the local wall temperature distribution is available, along with the local wall heat flux, the local heat transfer coefficient can be calculated, and the local friction factor can be estimated based upon the analogy between momentum and heat transfer.

Various temperature-measuring devices, such as mercury thermometers, thermocouples, resistance thermometers, and infrared thermometers, have already been developed and applied in a variety of situations. Problems often arise, however, when one must know the surface or spatial temperature distribution at a given moment. Among the devices just mentioned, thermocouples are used most often, and offer the added advantage of being inexpensive. In the case of surface temperature measurement, however, thermocouples are not particularly advantageous, because a large number of these point sensors are needed for good coverage of an area. In addition, the raw data must be reduced and plotted before one can visualize the physical situation. The infrared thermometer, on the other hand, does not require direct contact with the surface, and isothermal contours can be obtained in a few seconds. However, the surface emissivity may not be well known, and these thermometers are usually expensive.

Recently, a number of researchers in the heat transfer and fluid mechanics community, using temperature-sensitive liquid crystals as a temperature-measuring tool, have reported promising results in both surface and spatial temperature measurements. Liquid crystals have been well known since their existence was first observed in 1888 by the Austrian botanist Friedrich Reinitzer [1], but their industrial application and commercial availability were established only two decades ago with the advent of electronic display devices. In the last decade, with progress in liquid crystal manufacturing as well as application techniques, the literature on liquid crystal applications in fluid mechanics and heat transfer has increased rapidly, researchers appear to have accepted the liquid crystal technique as a legitimate temperature-measuring technique. Liquid crystals have the following advantages:

1. They are easily handled and inexpensive.
2. They have a small and predictable effect on the flow and thermal field when proper care is taken.
3. Measurement is nearly instantaneous.
4. They offer satisfactory accuracy and resolution.

Liquid crystals are particularly useful for the preliminary study of an unknown thermal field before precise local temperature measurements are undertaken, and also as a one-

Table 1 Classification of Liquid Crystal Applications [2, 3]

Item		Classification	Description
A	(1)	Three-dimensional (spatial) visualizations	Liquid crystal microcapsules suspended in a liquid with lighting in an arbitrary plane
	(2)	Two-dimensional (surface) visualizations	Liquid crystal paint coated on or liquid crystal sheet bonded to a wall surface
B	(1)	Display of time-dependent temperature distribution	Time scale of phenomenon much larger than that of the liquid crystal
	(2)	Display of time-averaged temperature distribution	Time scale of phenomenon much smaller than that of the liquid crystal
C	(1)	Qualitative observation	Visual interpretation or documentation by color still photo, color motion picture, or video system
	(2)	Quantitative measurement	Measurement by monochromatic illumination, narrow band-pass filtering, or digital color-image processing of still photograph, motion picture, or video data

or two-dimensionality check of a test flow field, which is otherwise simply assumed in most experimental investigations.

Applications of liquid crystals to fluid flow and heat transfer visualizations are classified by several different criteria in Table 1 [2, 3]. The first criterion is the number of dimensions in the measurement. In the case of three-dimensional measurements in water, the specific density of the liquid crystal capsules is usually 1.02–1.03, and only a small amount of the capsules, e.g., 0.1 wt. %, is required; hence, the properties of the test liquid should not be affected. In the case of two-dimensional measurements, the additional heat conduction resistance of the liquid crystal layer, which may be as thin as 100 μm, must be taken into account. The second criterion is the time scale of the phenomenon under study; the measurement is either time dependent or time averaged. Typically, time scales in liquids tend to be larger than in liquid crystals; thus, such applications can be classified as time dependent. To the best of the authors' knowledge, the frequency response of the liquid crystal sheet in water is flat up to between 1 and 10 Hz, depending on the structure. Time scales in gases tend to be smaller than in liquid crystals; thus, measurements in air are generally time averaged. The third criterion divides the applications into qualitative and quantitative observations. In the case of quantitative observations, spectral information on the liquid crystals' selective scattering must be used to obtain information on the temperature distributions.

II PHYSICAL AND CHEMICAL PROPERTIES

The following is a general description of liquid crystals, with an emphasis on cholesteric liquid crystals; details of their physical and chemical properties are given elsewhere, e.g., Refs. [4–6].

An organic compound is usually optically nonisotropic in its crystallized, solid phase but optically isotropic in the liquid phase at temperatures above its melting point. A particular group of organic compounds, however, exhibit behavior midway between that of an isotropic liquid and a nonisotropic crystalline solid; these compounds are generally termed *liquid crystals* or *mesophases*. The special term *thermotropic liquid crystal* denotes a compound that displays a nonisotropic liquid character at a temperature between its pure crystalline and isotropic liquid states. On the other hand, a *lyotropic liquid crystal* is a compound that shows optical anisotropy when dissolved in a particular solvent, e.g., soapy water. A liquid crystal phase can also be created by mixing several compounds [7], such as different pure liquid crystal compounds, liquid crystal and nonliquid crystal compounds, or even different nonliquid crystal compounds.

Liquid crystals are generally classified in three groups, based upon molecular structure and optical characteristics as first proposed by Friedel [8]:

1. Smectic
2. Nematic
3. Cholesteric

Brief descriptions and schematics of the molecular arrangements of these liquid crystals are summarized in Table 2 [3, 9] and Fig. 1 [10, 11], respectively. Each liquid crystal group has its own molecular anisotropy. In particular, the cholesteric liquid crystal group has a peculiar helical molecular orientation, in which the pitch of the helix depends on certain physical and chemical conditions.

The term *cholesteric* liquid crystals applies when the molecular structure is characteristic of a large number of compounds that contain cholesterol. (Cholesterol by itself does not have a liquid crystal phase.) As shown in Fig. 1, cholesteric liquid crystals consist of very thin, molecular microlayers (approximately 3 Å in thickness); the long axis of the molecules is parallel to the plane of the layers, and

Table 2 Classification of Liquid Crystals [3, 9]

Type	Origin of word (meaning)	Characteristics
Smectic	Smectos (purifying detergent)	Viscous and muddy Soaplike Birefringent Optically positive Nonrotary polarized
Nematic	Nematos (threadlike)	Muddy Birefringent Optically positive Nonrotary polarized
Cholesteric	Cholesterol	Muddy fluidity Birefringent Optically negative Rotary polarized Circular dichroism

the direction of the molecular axis in adjacent layers appears to rotate as one moves in a direction normal to the layers. This angular displacement, which averages about 15 min of arc per layer, is present in each successive layer so that the overall displacement traces out a helical path.

These layers have peculiar optical characteristics. If linearly polarized light is transmitted perpendicularly through the molecular layers, the direction of the light's electric vector will be rotated progressively to the left along a helical path. Thus, the plane of polarization, which is determined by the electric vector and the direction of propagation, will also be rotated to the left, through an angle proportional to the thickness of the transmitting material. Crystalline substances that rotate the plane of polarization of light in this manner are called *rotary polarized* or *optically active*. While certain types of highly active quartz

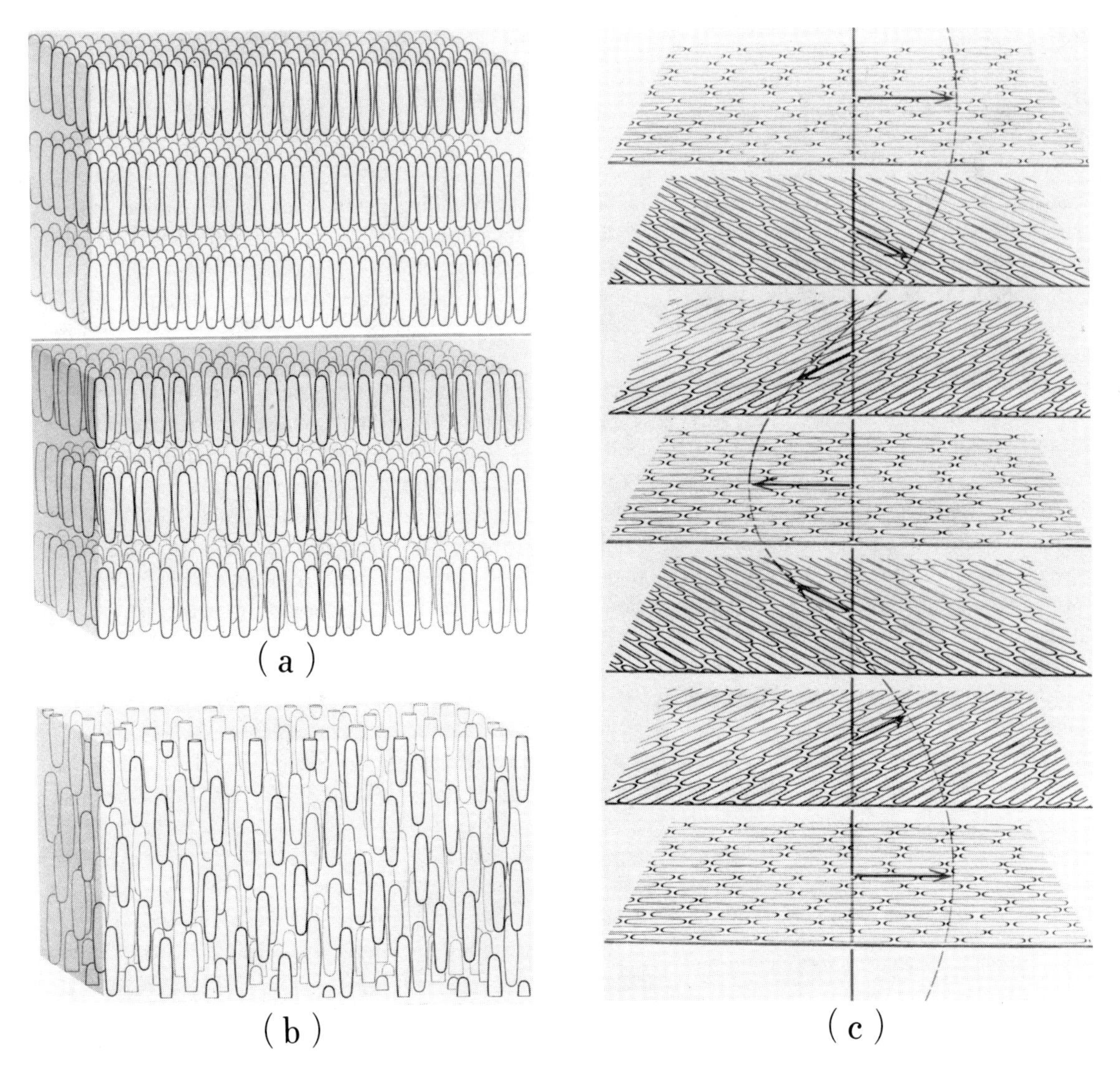

Fig. 1 Molecular arrangements of liquid crystals: (*a*) Smectic; (*b*) nematic; (*c*) cholesteric [10, 11].

rotate the plane of polarization about 20°/mm, an optically active cholesteric substance rotates the polarization plane through an angle of as much as 18,000°, or 50 rotation/mm [10]. Such liquid crystals are clearly the most optically active substances known.

Another peculiar characteristic exhibited by cholesteric liquid crystals is circular dichroism. When ordinary white light is directed at these substances, the light's electric vector is separated into clockwise and counterclockwise rotating components. Depending on the material, one component is transmitted and the other is reflected. This property gives the cholesteric phase its characteristic color changes when illuminated by white light. The characteristic wavelengths a given cholesteric liquid crystal scatters generally depend on the compound ingredients in the liquid crystal, temperature, imposed electric and magnetic fields, pressure, shear stress, chemical vapors in the environment, light angles of incidence and reflection, and other factors that can change the pitch of the molecular microlayers. Among these influential factors, liquid crystal temperature-measurement techniques utilize the temperature dependency. Cholesteric liquid crystals usually reflect long wavelengths (red) at lower temperatures and short wavelengths (blue) at higher temperatures.

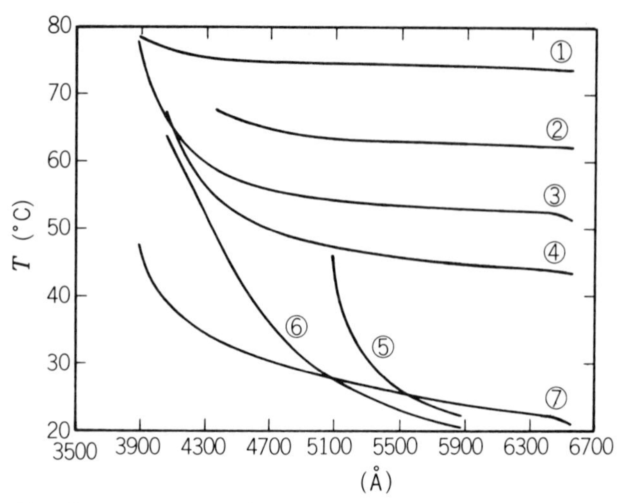

Fig. 2 Wavelength of maximum scattering as a function of temperature for mixtures of cholesteric esters and cholesteric nonanoate: (1) Pure cholesteric nonanoate; cholesteric nonanoate (2) with 20% cholesteryl hydrocinnamate; (3) with 20% cholesteryl butyrate; (4) with 20% cholesteryl propionate; (5) with 20% cholesteryl chloride, (6) with 20% cholesteryl acetate; (7) with 20% cholesteryl methyl carbonate [7].

Thermotropic liquid crystals can be separated into two categories: the enantiotropic mesophase and the monotropic mesophase [6]. The former, which is used in temperature measurements, changes phase reversibly in both heating and cooling processes, while the latter enters the liquid crystal phase only in heating or cooling processes. It has been reported that the color change of a certain kind of enantiotropic mesophases becomes irreversible at temperatures above 100°C [12], but this effect can be avoided by choosing the proper liquid crystal compounds. A pure liquid crystal has a fixed temperature-sensitive range, which is called the *event temperature range*. One can, however, create a mixture that has a desired event temperature by mixing appropriate liquid crystals. Figure 2 shows the temperature dependency of the wavelength of maximum scattering for cholesteryl nonanoate and cholesteryl ester mixtures [7]. The temperature level and the temperature range of color change depend on the mixture. These factors therefore result in a variety of sensitivities of liquid crystals. It is said one can obtain liquid crystals that have event temperature ranges roughly between 1 and 100°C and that tolerate temperatures from -20 to $+350$°C [13]. However, the encapsulated liquid crystals described later do not well tolerate temperatures above 100°C.

III FUNDAMENTALS OF LIQUID CRYSTAL APPLICATIONS

Since cholesteric liquid crystals are chemically unstable, their temperature characteristics may change through reaction with surrounding gases, such as oxygen and carbon monoxide. In addition, liquid crystal mixtures tend to crystallize under certain environmental conditions and thus lose their ability to change color. Moreover, the liquid crystal itself is not easy to handle because of its fluidity. In order to avoid chemical contamination and physical damage in practical use, liquid crystals are usually enclosed in microcapsules, about 5–30 μm dia., which are made of a dense colloid containing gelatine and Arabian gum. Micrographs of these liquid crystal capsules are shown in Fig. 3 [2, 3]. These microcapsules can be suspended in a liquid for temperature-distribution measurement, as described later. Encapsulated liquid crystals are said to have a spatial resolution of about 0.02 mm in temperature measurements [13].

A Applying Liquid Crystals

Since a cholesteric liquid crystal phase reflects a certain component of circularly polarized light and allows the other components to pass through it, the background beneath the liquid crystals must be completely black; otherwise, the color change will not be visible. Any black paint can be used, as long as the paint and its solvent do not damage liquid crystal capsules chemically. Water-based paints are less apt to interfere with liquid crystals than are hydrocarbon-based paints.

Liquid crystal paint is commercially available as a slurry of liquid crystal microcapsules in a water-soluble binder that is transparent when dry. A thin layer of this liquid

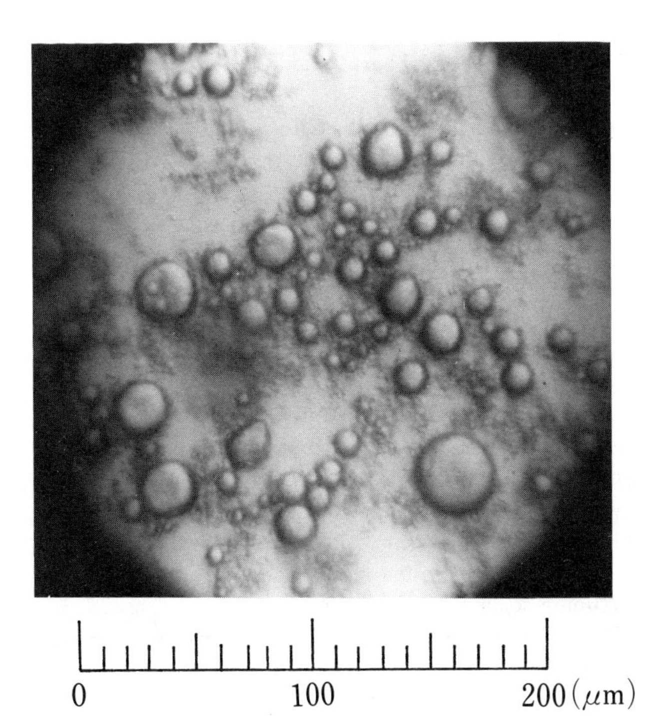

Fig. 3 Microcapsules of cholesteric liquid crystals [2, 3].

crystal slurry can be applied to the background surface with a brush, a paint sprayer, an artist's air brush, by silk screening, or by blading. Dry layers about 30–50 μm thick generally produce good colors.

The surface finish of the dried paint can affect the intensity of the image. Spray-coated layers sometimes have rough surfaces which scatter white light from the asperities on the surface and reduce the saturation of the colors from the image.

Thicker layers are recommended when two or more paints are mixed to form a multiple-event coating. As a general rule, the final thickness should be 30–50 μm for each component.

For surface temperature measurements in water, a liquid crystal sheet is preferable to a paint. A typical assembly, about 100 μm thick, is a thin sheet of acetate or polyester film on which cholesteric microcapsules and black backing have been uniformly coated. The sheet must be in good thermal contact with the surface, to minimize the thermal disturbance, and sealed, to prevent water from contacting the liquid crystal.

B Calibration and Stability

In-place evidence of calibration can be obtained by installing fine-wire thermocouples on the test structure and applying the liquid crystal paint directly over the thermocouples.

Samples of paints can be calibrated using a strip of material heated at one end and cooled at the other, properly designed to produce a linear temperature gradient from end to end. Alternatively, a block of material can be ''swept'' slowly through the event temperature range in a transient mode, with color and temperature observed as function of time. Constant-temperature water baths have also been used.

The perceived color at a given temperature is affected by both the viewing angle and the illuminating angle. Both the lighting and the observation angles should remain fixed throughout the experiment. When color photographs are used, a color calibration strip and the test image should be recorded on the same roll of film in order to account for differences in processing procedures. When liquid crystals are calibrated in this fashion, their accuracy and reproducibility can be maintained to within ±0.1°C.

Under good storage conditions, liquid crystals will remain accurate for a year or more, though the brilliance of the colors may fade with time. This applies even though the paints have been chemically stabilized by encapsulation. One must avoid exposing liquid crystals to ultraviolet light or fluorescent light or to high temperatures, since all of these hasten the loss of color.

C Interpreting the Image

Liquid crystal images can be interpreted directly (by the experimenter watching the experimental event) or indirectly (using conventional photography or video recording), or using digital image processing (from either a video image or a photograph). In many cases, the visual interpretation is all that is required, and no calibration is necessary. Where precise quantitative measurements of temperature are required, image-processing techniques (discussed in a later section) have significant advantages.

Quantitative interpretation is complicated by the fact that liquid crystals do not change color abruptly at a single precise temperature, but sweep through the spectral range from red to blue over a range of temperatures called the *ambiguity band*. (For spectral distributions of liquid crystals' light scattering, see [14].) Commercially encapsulated paints are sold by two criteria: the event temperature, and the ambiguity-bandwidth. Ambiguity-band widths range from 1 to 15°C, and can be specified at the time of purchase.

In some situations, a human observer can understand the pattern displayed by a liquid crystal surface without assigning specific temperatures to the perceived images. If quantitative temperature information is needed, the simplest approach is monochromatic visual interpretation. In this mode of interpretation, one particular color (usually the yellow or green) is identified as the ''event color,'' and the ambiguity band is interpreted as an uncertainty in temperature. For such applications, narrow ambiguity bands are preferred. Again, for best accuracy, the lighting and viewing angles must be kept constant during the calibration and the test.

Monochromatic lighting has been used in some experiments to reduce the ambiguity band. If a narrow-band monochromatic light is used [15, 16], the isothermal contours are sharply defined. A 5890-Å sodium lamp is recommended, because in that region the temperature-versus-wavelength curve of liquid crystals has a nearly horizontal slope, as shown in Fig. 2, and good temperature resolution is obtained.

If white light is used in conjunction with a number of narrow band-pass optical filters, photographs of multiple isotherms can be made and a contour diagram assembled by superimposing the band-pass photographs [14, 17]. In [14], 18 band-pass filters were used to obtain the calibration curve relating the filter wavelengths to the peak-brightness temperatures. The measurement uncertainty was estimated to be about ± 0.1°C in the range of 29.5–31.4°C.

Images that span a wide range of temperatures pose a special problem. Narrow-band paints can be mixed to produce a multiple-event paint, which will produce several isotherms in the same image [18, 19]. As long as the ambiguity bands of the paints do not overlap, the calibrations of the paints are not affected. Each isotherm is an independent event, from a different component of the paint. Up to 10 different slurries can be mixed to form a single, multiple-event slurry. As long as the event temperatures of the components are separated by an interval wider than their ambiguity bands (i.e., as long as the component color bands do not overlap), each component acts independently. With multiple-event liquid crystals, one can map the distribution of temperature on a surface without having to adjust conditions between tests.

Images from multiple-event paints can be interpreted monochromatically, either directly or by black-and-white or color photography. The monochromatic interpretation is unambiguous when the specimen has a monotonic distribution of temperature, of known slope, and the identity of one of the isotherms can be determined. When the temperature distribution has an interior maximum or minimum, the sequence of colors around each event color must be observed. For most cholesteric paints, the sequence red–yellow–blue represents increasing temperature. By noting the sequence on each event isochrome, the slope of the temperature field can be identified in the vicinity of each isotherm.

Wide-band paints offer another approach to the wide-range situation. If the experiment can be scaled to operate within the ambiguity band of an existing paint, then the entire temperature range within the image can be mapped with one paint. Wide-band paints require chromatic rather than monochromatic interpretation, however, and this can only be done with precision using digital image processing, discussed in a later section.

D Time Response

The intrinsic time response of liquid crystal material has been reported in only a few works [3, 20–23]. As temperature changes, the helical pitch of the molecular layers of cholesteric liquid crystals changes with a finite time constant, which depends mainly upon its constituents. Fergason [20] has proposed the following formula for the transient wavelength of scattering:

$$\lambda = \lambda_0 - |\lambda_1 - \lambda_0| \exp(-\beta t) \tag{1}$$

where λ, λ_0, and λ_1 are the transient, steady-state, and initial wavelengths of maximum scattering, and β is the inverse of the time constant. In addition, he has determined the values of the time constant for cholesteryl nonanoate as 0.1 s, and for cholesteryl oleyl carbonate as 0.2 s, at a maximum scattering of 5640 Å.

The time constant of a liquid crystal measuring system depends on more than just the intrinsic time response of the liquid crystal material. The system response depends also on the thickness of the liquid crystal layer, its thermal properties, the heat transfer coefficient, and other thermophysical properties. The time response should be checked for each experimental condition [3, 23].

One method for evaluating the time response of a liquid crystal measurement [3] is shown schematically in Fig. 4. This is a typical composite-layer arrangement used for the

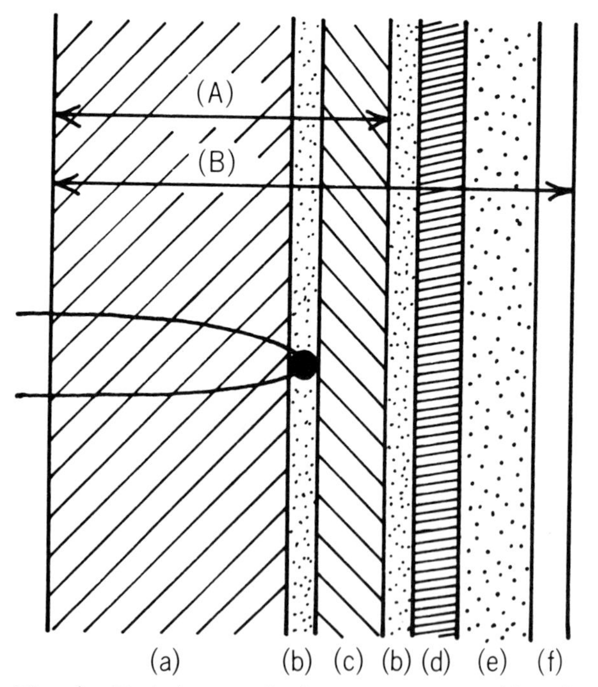

Fig. 4 Typical composite layer arrangement with a thermocouple for visualization of wall temperature distribution: (*a*) Adiabatic substrate; (*b*) glue; (*c*) stainless steel foil: (*d*) black paint: (*e*) liquid crystals; (*f*) acetate or polyester [3].

qualitative and quantitative observation of a surface temperature distribution. An adiabatic substrate, a stainless steel foil heater, and a liquid crystal sheet are glued together to form the test surface. The liquid crystal sheet is composed of a layer of black paint, a layer of cholesteric liquid crystals, and a thin transparent film. This film, made of acetate or polyester, protects the liquid crystals from water erosion or dust particles in air. The transient temperature (i.e., color) change in the liquid crystal layer is caused by a change in the surface heat transfer coefficient. Since the total thickness of these layers above the adiabatic substrate is very thin, on the order of 0.1 mm, this experimental situation can be reduced to the one-dimensional physical model shown in Fig. 5. The model consists of three layers: a stainless steel layer of uniform volumetric heat generation with one adiabatic surface; a composite layer of glue, black paint, and liquid crystals; and a surface film.

This one-dimensional unsteady heat-conduction problem has been analyzed using a finite difference scheme. The thermal properties of the middle layer (2) in Fig. 5, however, were not available due to the difficulty in measuring the individual thicknesses of the glue, black paint, and liquid crystals, and to the lack of basic data on the thermal properties of liquid crystals. Hence, values of the overall thermal conductivity and diffusivity of this composite layer were measured experimentally for the actual assembly.

To measure the thermal conductivity, the test plate was immersed in a uniform flow, and the laminar boundary layer heat transfer measured with and without the liquid crystal layer [segments (A) and (B) in Fig. 4] with the same wall heat flux. The heat flux of the system in Fig. 5 is given as

$$\frac{\dot{Q}}{A} = K(T_0 - T_\infty) = \frac{T_0 - T_\infty}{1/h + \delta_1/2\lambda_1 + \delta_2/\lambda_2 + \delta_3/\lambda_3} \tag{2}$$

where $\dot{Q}$ = volumetric heat generation
A = surface area
K = overall thermal conductance
T_0 = temperature of the adiabatic surface
T_∞ = fluid temperature
h = heat transfer coefficient
δ_i = thickness of the ith layer
λ_i = thermal conductivity of the ith layer

If the thermal conductances with and without the liquid crystal layer are denoted by K_b and K_a, respectively, then the thermal conductivity of the layer (2) in Fig. 5 is obtained by the following formula:

$$\frac{1}{\lambda_2} = \frac{1}{\delta_2}\left(\frac{1}{K_b} - \frac{1}{K_a} - \frac{\delta_3}{\lambda_3}\right) \tag{3}$$

Using the values of the known thermal properties in Table 3 and the measured thickness of the middle layer (2), δ_2 = 68 μm, the following typical value of the thermal conductivity was obtained [3]:

$$\lambda_2 = 0.223 \pm 0.032 \quad (20:1) \quad (\text{W/mK}) \tag{4}$$

For the measurement of thermal diffusivity, the electric power input was varied sinusoidally at different frequencies in the same flow field as above. The time-dependent color change was recorded along with the phase signal of the power input by a video system, and the phase lag of the color change was measured for each frequency. Figure 6 shows the results of such a measurement [3]. The experimental data were plotted and compared with the analytical curves for various thermal diffusivities, obtained by the finite-difference calculation for the system shown in Fig. 5. This comparison gave an approximate value of the liquid crystal layer's thermal diffusivity:

$$\kappa_2 = 2.5 \times 10^{-8} \pm 0.5 \times 10^{-8} \quad (20:1) \quad (\text{m}^2/\text{s}) \tag{5}$$

The frequency response of the color change could then be calculated with the values given in Eqs. (4) and (5) and in Table 3, when the test plate in Fig. 4 is used for observation of the wall temperature fluctuation in a turbulent boundary layer [26]. In this case, it was assumed that the average heat transfer coefficient was 1100 (W/m^{2}K) with a fluctuating component of ±20% of the mean.

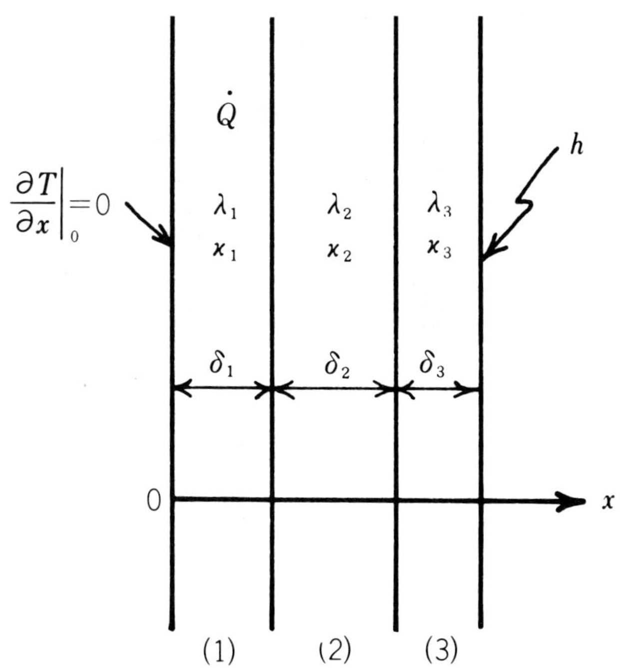

Fig. 5 One-dimensional heat-conduction model of the composite layer system with time-dependent surface heat transfer coefficient $h = \bar{h}(1 + \alpha \sin \omega t)$: (1) Stainless steel foil: (2) glue, black paint, and liquid crystals; (3) acetate or polyester film [3].

Table 3 Physical Properties of the Composite Layer Used in the Unsteady Heat-Conduction Calculation [3].

	Thermal Conductivity λ (W/m^{2}K)	Thermal diffusivity κ (m^2/s)	Thickness δ (m)
Acetate (cellulose-diacetate)	0.201[a]	0.73–1.25 $\times$ 10^{-7}[a]	3 $\times$ 10^{-6}, 50 $\times$ 10^{-6}
Stainless steel foil	14.7[b]	3.87 $\times$ 10^{-6}[b]	30 $\times$ 10^{-6}

[a]Data from [24].
[b]Data from [25].

The results for two types of surface films are represented in Fig. 7; note that the two layers differ in thickness. Curve (A) is the ratio of the amplitude of the wall temperature fluctuation to the global temperature difference between the wall and the fluid. Curves (B) and (C) are the amplitude ratios and the phase differences of the temperature fluctuations, which occur at the center of the liquid crystal layer and on the wall surface. Thus, Curve (A) illustrates the surface temperature fluctuation on the composite wall shown in Fig. 5 under the fluctuating heat transfer coefficient, while Curves (B) and (C) illustrate the traceability of the liquid crystal. The results in Fig. 7 indicate that the traceability is better with a thinner surface film, as expected. As an example, consider the case where the film thickness is δ_3 = 3 μm and the global temperature difference is 10°C. According to Fig. 7, at 1 Hz, the amplitude of wall temperature fluctuation is about 1.2°C, while the temperature fluctuation in the liquid crystals reaches 90% (1.1°C) of the surface fluctuation with a phase lag of 15°. If the temperature resolution of the observing system is higher than the temperature fluctuation in the liquid crystal (here, 1.1°C), the wall temperature fluctuation can be observed through the color change. Since the time response is dependent upon the flow field and the experimental conditions, the correct physical interpretation of the liquid crystal's transient behavior can be attained only through knowledge of the time response behavior. Thus, time-dependent calibration measurements, as in the example above, are vital if quantitative measurements are to be made under unsteady conditions.

Another assembly that has been used in water studies [23] consists of a transparent electrical heating film (10–20 Å of vapor-deposited gold on 0.1-mm-thick polycarbonate), painted with liquid crystal and sealed against a black insulating layer. The assembly was made with the polycarbonate protecting both the heating film and the liquid crystal from contact with water. The time constant of this system was calculated using finite-difference analysis and confirmed by cyclic variation of the applied power, with constant flow conditions on the water side. The results showed that changes in color could be observed at frequencies up to 2 Hz when the variations in power were ±15% of the mean value needed to bring on the color change event. No attempt was made to correct the time-varying indicated tem-

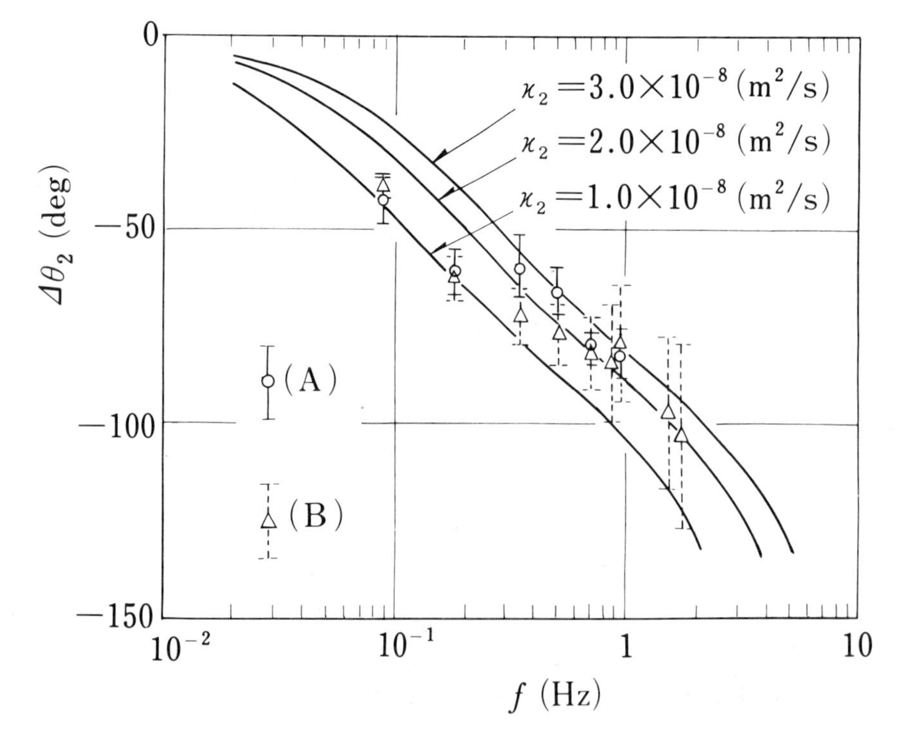

Fig. 6 Comparison of the measured and numerical phase lags in color response at various frequencies. (A) and (B) indicate repeated measurements [3].

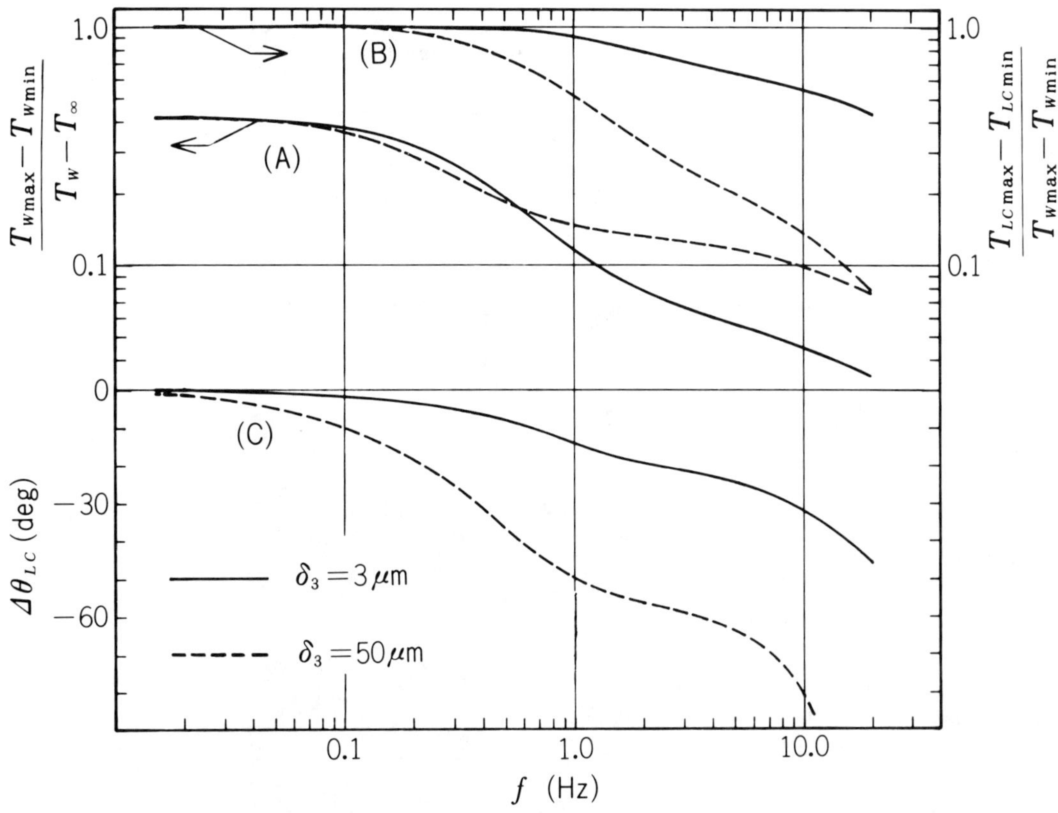

Fig. 7 Frequency response of the color change of liquid crystals in the composite layers with two kinds of surface films: (A) Ratio of the wall temperature amplitude to the mean wall–fluid temperature difference; (B) ratio between the temperature amplitudes at the liquid crystal layer and at the wall surface; (C) phase lag in color change of liquid crystals relative to the wall surface temperature fluctuation [3].

perature for the response lag of the sensing system; the objective was simply to determine whether or not the fluctuations were discernible.

IV APPLICATIONS

The optical characteristics of the cholesteric phase, and their variation with the various physical factors previously noted have led to many possible applications, such as display devices [10] and thermal non-destructive testing [20, 27]. The following are typical examples of the liquid crystal applications to flow, temperature, and heat transfer measurements [2, 3, 18, 28].

Steady wall temperature distributions on surfaces have been visualized and used for qualitative and quantitative investigations of fluid flow and heat transfer in both air and water flows. With aerodynamic heating, the locations of laminar-to-turbulent transition and of weak shock waves [29] and the heat transfer distribution on the surface of aircraft models [30] have been visualized in wind tunnels. From visualization of the wall temperature distribution around a heated circular cylinder in a cross flow [31], the local Nusselt numbers have been obtained, along with qualitative information on the effects of the flow separation, the separation bubble region, the turbulent boundary layer, and the turbulent wake on the surface heat transfer. Liquid crystals have also been used to visualize the secondary flow patterns on the walls of curved ducts [32, 33].

The first use of liquid crystals to measure convective heat transfer appears to have been by Venneman and Butesfisch [34], who used a transient technique, observing the time to change following a stepwise onset of flow. This technique has since been used by others [35], and two other techniques have emerged: steady-state mapping of surface temperature on a wall whose back face was maintained at constant temperature [36] and mapping the surface temperature of an electrically heated film of uniform heat release [23, 37].

The heat transfer coefficient distribution has been measured on the surface opposite a single round, impinging jet [36] as well as with multiple impinging jets [38, 39], and qualitatively observed on a surface equipped with heat transfer augmentation devices opposite a round impinging jet [40]. In addition, film-cooling heat transfer on gas turbine blades has been studied with liquid crystals [15, 16,

41]. A specially prepared composite-surface element, which utilizes a vapor-deposited gold layer on a polyester film, has also been used for the measurement of local convective heat transfer coefficients in various flows [23, 42, 43]. The gold layer is so thin that light can pass through the layer, which permits observation of color change on the back side of a test plate.

As an example of steady-state measurement with a multiple-event paint, Fig. 8 shows an image from a 6-event liquid crystal paint used to map the temperature distribution on an otherwise adiabatic wall around and above a heated cube in mixed convection [18]. The yellow isochrome is most visible and is often used as the event marker. The six visible isotherms in this photograph correspond to 30, 35, 40, 45, 50, and 55°C. The air temperature was 25°C; the outer isotherm is identified as 30°C. The temperature distribution is monotonic.

Figure 9 is an example of quantitative temperature measurement illuminated by a 5980-Å sodium lamp [15]; these measurements are part of a basic study of full-coverage film cooling (FCFC) for future high-temperature gas turbines. A flat-plate model with 30°-slant injection holes arrayed in a staggered manner was fixed at the bottom of a wind tunnel test section. A liquid crystal paint 100 μm thick was applied uniformly to the FCFC plate. The injected air was heated to a prescribed temperature, while the main stream was maintained at room temperature. The bright line in the photograph corresponds to a certain absolute temperature; this line therefore represents an isotherm. Changing the temperature of the heated secondary air shifted the position of the isotherm. Hence, by superimposing successive photographs, an isothermal contour diagram could be created.

Transient temperature fluctuations on the walls have also been studied in the turbulent boundary layers on a flat wall [26] as well as on a concave wall [44]; geometries vary from a natural convection boundary layer along a vertical

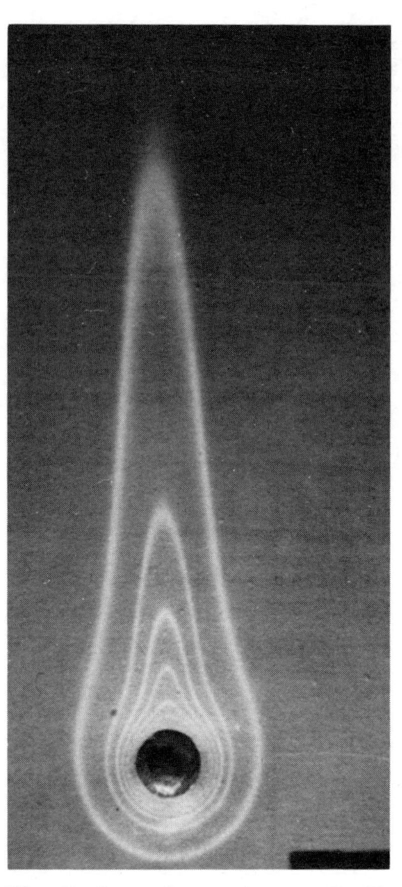

Fig. 8 Image from a six-event liquid crystal paint on the wall behind a heated cube: 30, 35, 40, 45, 50, and 55°C isotherms [18].

flat plate [45] to a two-dimensional impinging jet [46]. All of these tests have been devoted to understanding the characteristic turbulence structures, particularly large-scale structures, and associated heat transfer mechanisms.

Figure 10 shows the combined hydrogen bubble–liquid crystal technique in a fully developed turbulent boundary

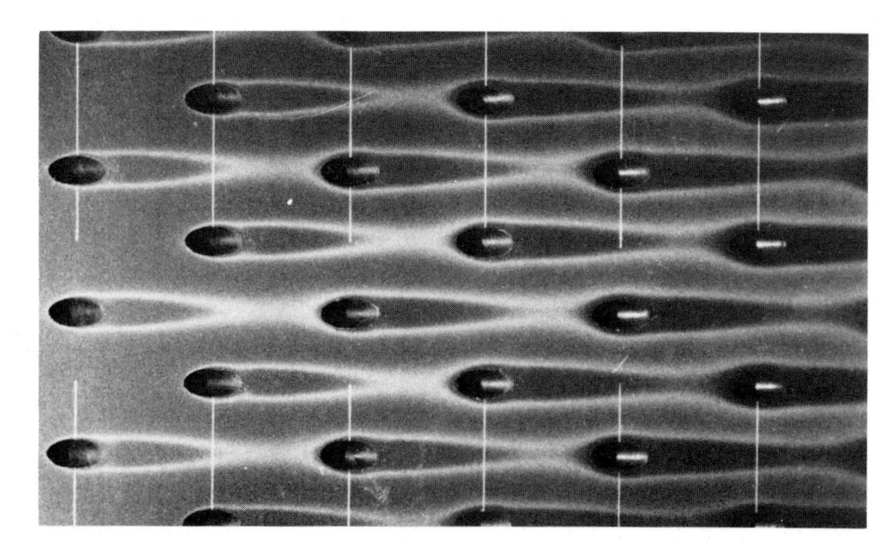

Fig. 9 Surface temperature measurement in a full-coverge film-cooled flat wall with a monochromatic light [15].

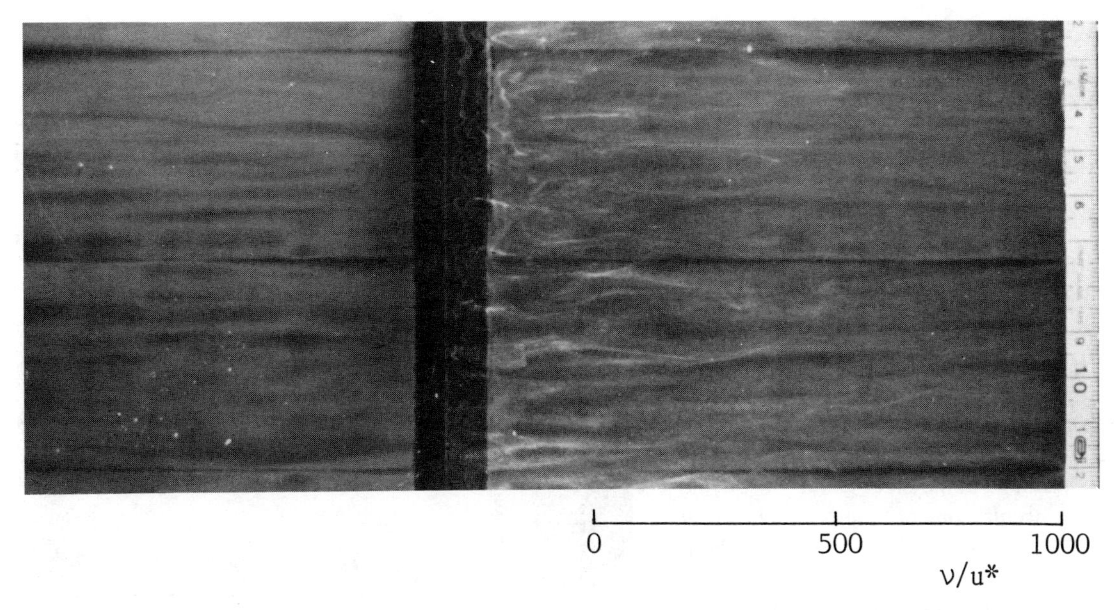

Fig. 10 Simultaneous visualization of wall temperature fluctuation and near-wall flow field in a turbulent boundary layer along a flat plate: free stream velocity, 0.20 m/s; momentum thickness Reynolds number, 990; hydrogen bubble wire at $yu^*/\nu = 6$; y, distance from the wall; u^*, friction velocity; ν, kinematic viscosity [26].

layer along a flat plate [26]. The aim of this study was to investigate the three-dimensional, time-dependent temperature field associated with organized fluid motions near the wall, for better understanding of turbulent convective heat transfer. The boundary layer was developed along a flat plate in a 2000-mm-long water channel with a cross section of 300 × 300 mm. In this instantaneous photograph, H_2 bubbles show the low-speed streaky structures near the wall, they lift up and eventually break up after a certain streamwise distance. In the liquid crystal mapping, the temperature field also appears to have a streaky structure, with a streamwise length scale much longer than its spanwise length scale.

A similar investigation has been carried out on natural convection; the result is shown in Fig. 11 [45]. This photograph was taken in turbulent natural convection along a vertical plate, which was heated with a constant heat flux in water. The streamwise position was 2.1 m from the leading edge, and the modified Rayleigh number reached an order of 10^{14}. The instantaneous wall temperature pattern again suggests that the large-scale turbulence structures characteristic of this type of flow field play an important role in turbulent transport, although they are somewhat different from those found in forced convection (see Fig. 10).

Qualitative measurements of three-dimensional temperature fields in fluid flows have been carried out in several experimental works; these measurements include temperature distributions in experimental simulations of a heated island [47], mixing in a chemical reactor tank [48], flow pathlines and temperature distributions in a cavity flow [49], and natural convection inside a cylindrical vessel [50].

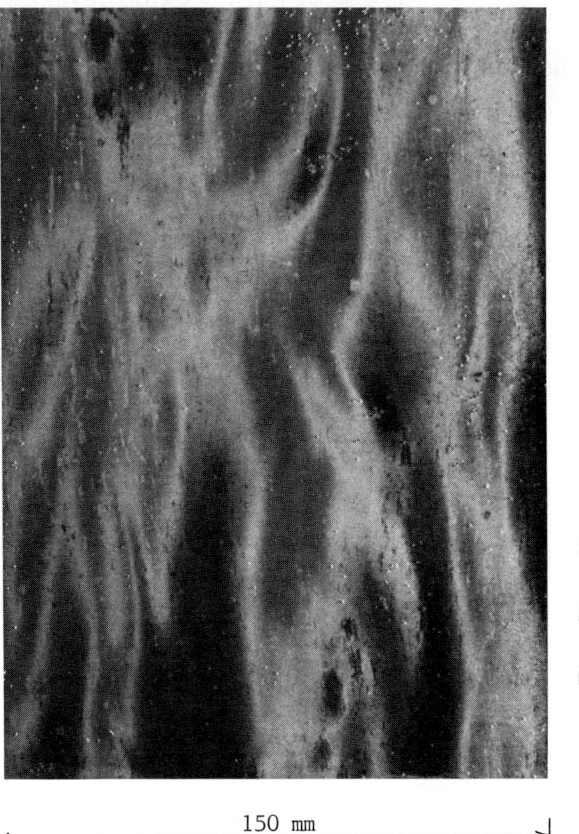

Fig. 11 Visualization of wall temperature fluctuation in a turbulent natural convection along a vertical plate heated with a constant heat flux in water: modified Rayleigh number, 2.7×10^{14}; distance from the leading edge, 2.1 m [45].

In these experiments, the test fluid was either water or silicone oil with 0.1 wt. % of liquid crystal microcapsules; the orientation of observation is shifted 45° from the incidence of a slit light sheet. Measurement in an arbitrary cross section of the flow field is made possible by traversing the slit light sheet.

Figure 12, from [49], shows the temperature and flow pattern in a shear-driven cavity flow. The top surface was maintained 4°C hotter than the bottom until a stably stratified temperature field was established. The top surface was then put into uniform motion. The shear-driven flow carried warm fluid down the wall, before buoyancy began to show its effect. The photograph was made with 0.01% liquid crystal slurry in water, illuminated by a sheet of white light. The photograph was taken from the side, about 45° off the lighting axis. Higher concentrations of liquid crystal caused diffuse scattering. Neutrally buoyant liquid crystal material was selected by flotation separation.

Figure 13 shows the visualization of the natural convection in a cylindrical container [50]. The bottom surface was maintained at a higher temperature than the top surface, while the cylindrical side wall was kept nearly adiabatic. The test fluid was silicone oil, with a Prandtl number of 250 and specific density of 1 at 25°C. The Rayleigh number based on the container height was 1.2×10^7. The instantaneous pathlines and temperature distributions visualized by the liquid crystal microcapsules show clearly the large-scale recirculating flow patterns and associated heat transport.

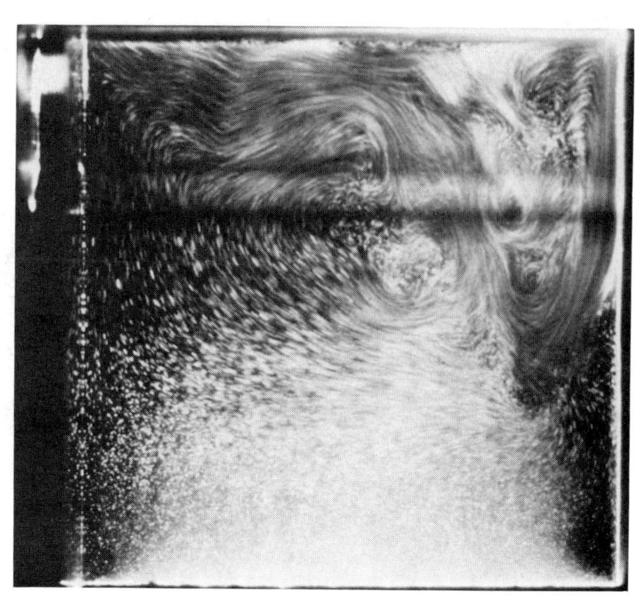

Fig. 12 Temperature distributions and flow patterns observed in a shear-driven flow in a rectangular cavity, shortly after the impulsive start of top surface motion; the cavity was stably stratified, 4°C hotter at the top, before the motion started [49].

Viewing angle is particularly important when using suspended liquid crystals. Although colors can be observed over a range of angles, the best color is obtained with a viewing angle 30 to 45° off the lighting axis, in partial back scatter.

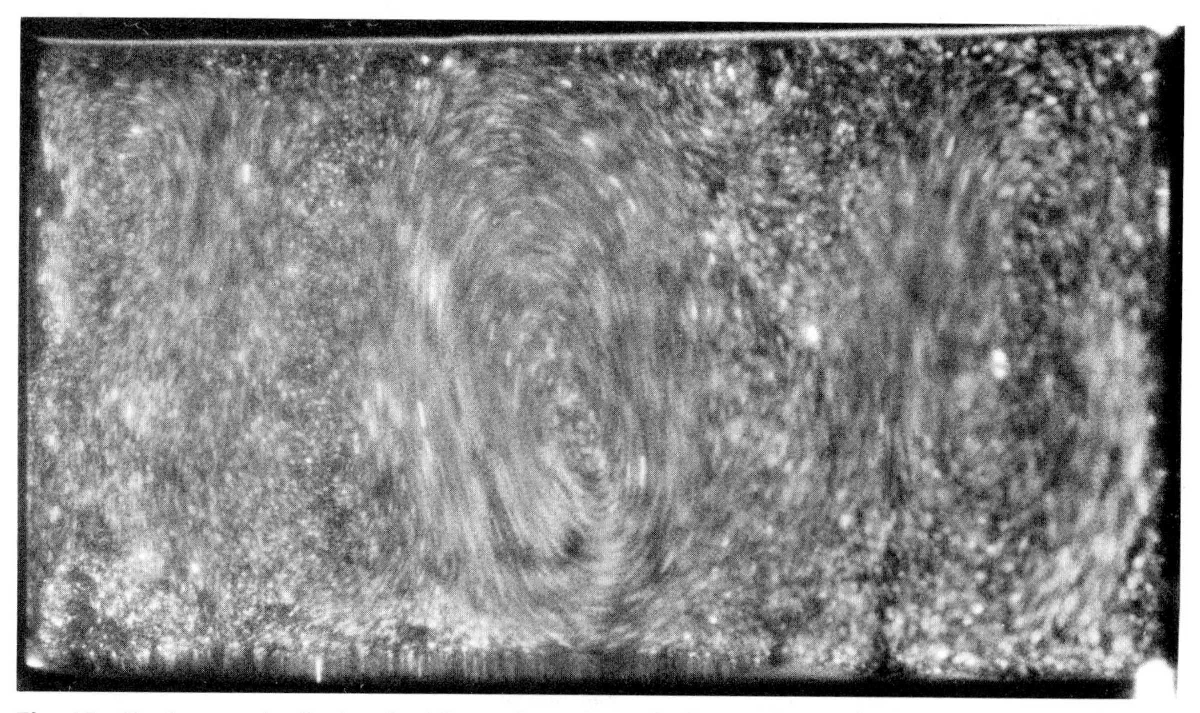

Fig. 13 Simultaneous visualization of pathlines and temperatures in the natural convection in a cylindrical container heated from below: $H/D = 0.5$ where H = height and D = 93 mm dia.: Rayleigh number, 1.2×10^7; exposure time, 0.5 s [50].

V DIGITAL IMAGE PROCESSING

Liquid crystal images can be interpreted by either monochromatic intensity (at one [16] or several [14, 17] wavelengths) or by chromatic interpretation [51–53]—directly associating the color to temperature, over the full range.

Figure 14 [51] shows the results of successive digital processing of the image from Fig. 8 based on monochromatic intensity (i.e., brightness of the yellow line). In Fig. 14*a* the digital image is shown as first derived, while in Fig. 14*b* the centroids of the areas have been identified and lines have been drawn through the centroids to reveal "single-line" representations of the isotherms. Digital processing, which includes image enhancement, thresholding, aberration correction, and coordinate transformation, has also been applied to monochromatic images similar to that in Fig. 9, superimposing multiple images into a single isothermal diagram [16].

In Fig. 14*c* the image is further processed and shown in expanded (R, θ) coordinates. The angle θ is measured from the center of the generating body, with (R, θ) in the downward direction with respect to gravity. The radius R is the distance from the center of the image to the isotherm in question. The temperature field can be reconstructed from Fig. 14*b* using an orthogonal ray-tracing technique, and interpolating. Figure 14*c* is a useful form for evaluating moments as indicators of shape and size. The system used in preparing these figures is also usable in chromatic mode, and is discussed in the following material.

Chromatic interpretation has some advantages in terms of freedom from the effects of nonuniform lighting and reduced ambiguity-band width, as well as offering significantly wider useful range from a single paint.

Chromatic processing systems can be assembled from readily available components, and are compatible with moderate sized (i.e., desktop or personal) computers. A typical image-acquisition and -processing system contains the following elements:

- Event to be imaged
- Lighting system
- Camera and lens system
- Analog storage processor (e.g., film or videotape)
- Analog-to-digital converter
- Computer and color monitor

The following discussion is generally descriptive of any true-color digitizing system. A schematic is shown in Fig. 15. Light from the event is collected by a color video camera, which generates a signal containing both intensity and color information. For a time-varying phenomenon, a sequence of video frames (individual images) can be recorded on video tape for later analysis. For a steady-state phenomenon, the signal is passed directly to a color decoder. This instrument decodes each incoming video frame into three frames containing the intensities of the red, green, and blue light that, when combined, produced the original color frame. This decoding, called RGB decomposition, is central to the concept of color-image processing and reproduction. These three signals are digitized by three analog–digital convertors, pixel by pixel.

The intensity of each pixel is assigned an integer between 0 and 255 (for an 8-bit digitizer) so that each decoded

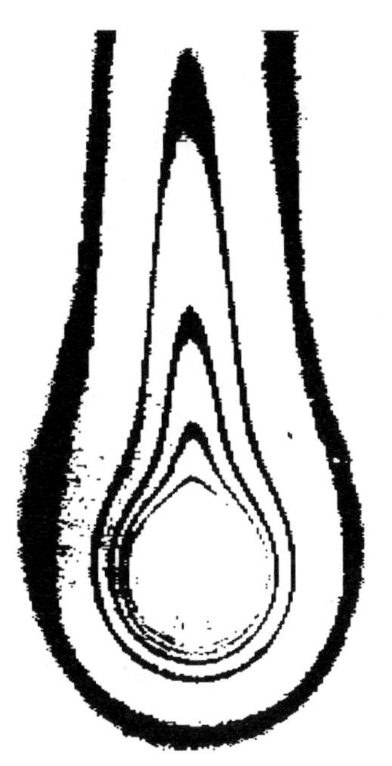

(a)

Fig. 14 Successive digital processing of the monochromatic image in Fig. 8: (*a*) Digital reproduction as generated. (Figure continues.)

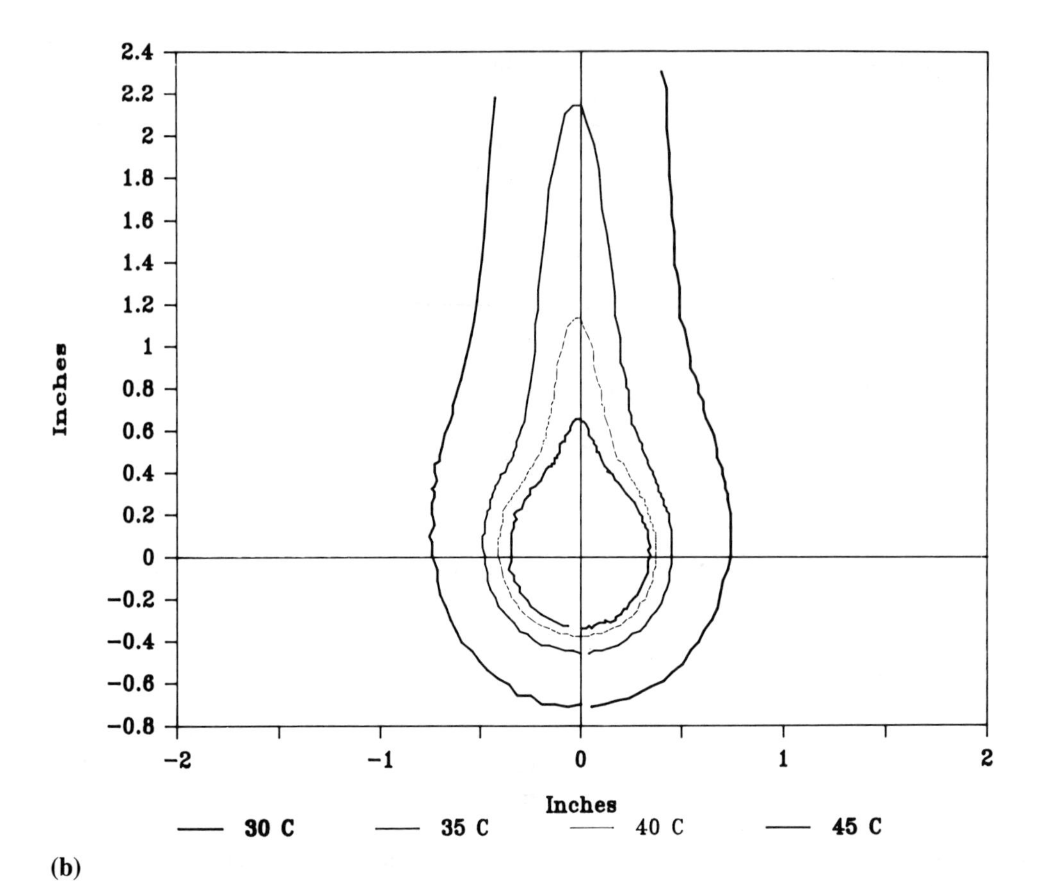

(b)

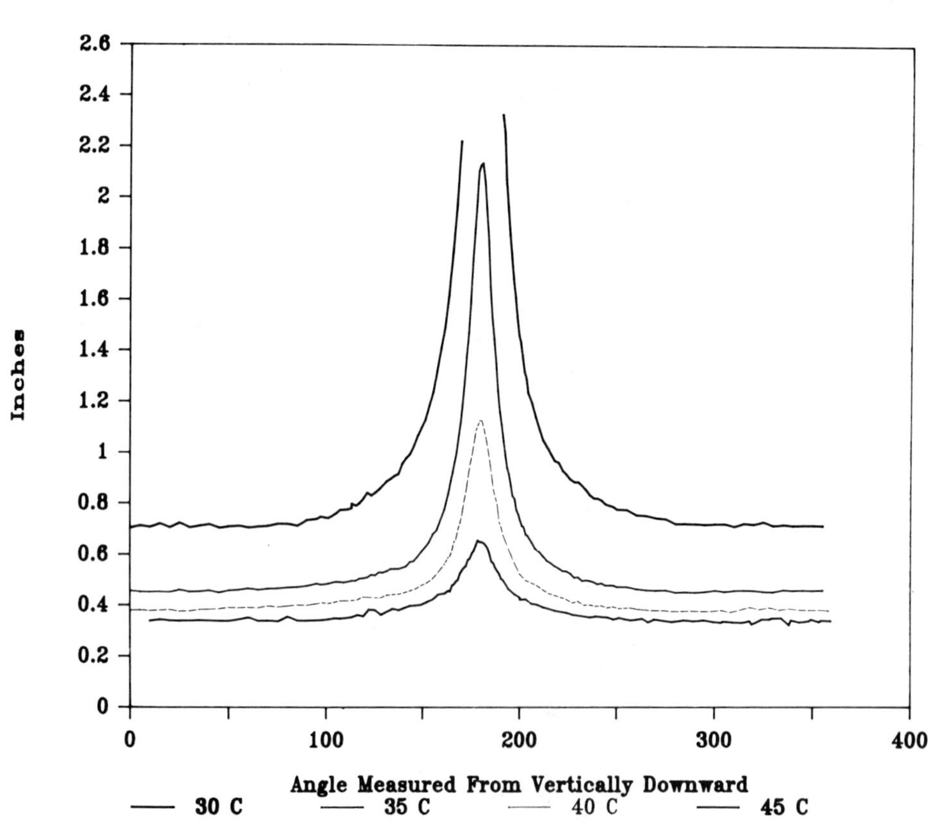

(c)

Fig. 14 (continued) (*b*) single-line representation of the image, using the centroidal points of each band to define the line; (*c*) expanded R–θ representation of the image, ready for moment analysis [51].

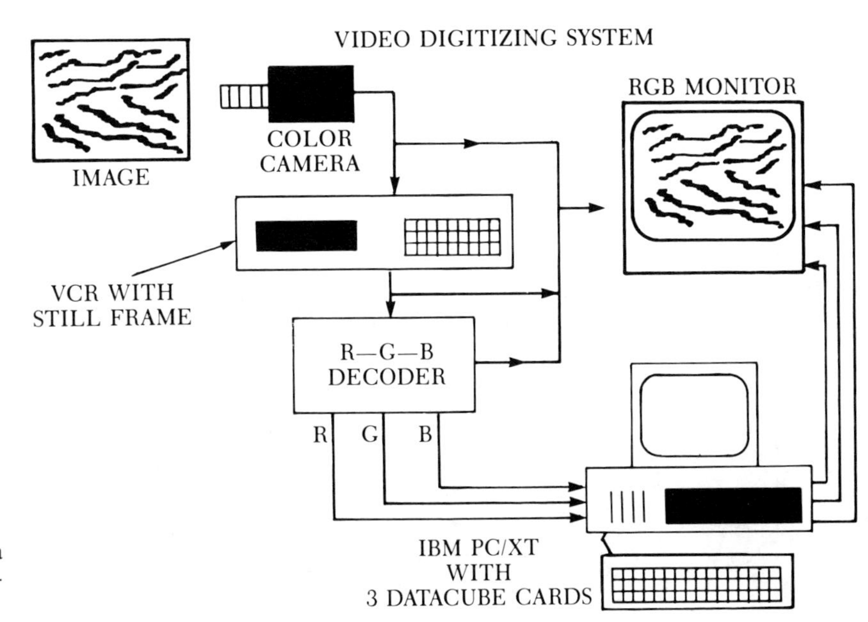

Fig. 15 Schematic diagram of a video system for chromatic interpretation of liquid crystal images [51].

frame is represented by a matrix of integers. The original image, the sum of the three decoded frames, is thus represented by three matrices (red, green, and blue). At the standard video rate of 30 frames/s, even a small system (384 × 485) acquires more than 16.7 million bytes per second of event time.

The three RGB matrices for one image are stored in computer memory (or on disk) until the image is processed. Processing consists of calculating the value of each scalar of interest at every pixel. These calculations require a calibration curve relating the three RGB intensities to the value of the measured quantity. The processed result is stored as a fourth matrix, the measured quantity. This final matrix is also recognizable as an image and can be displayed and further processed into pseudocolor images, if desired.

In the present instance, true-color interpretation can be obtained, from which surface temperature is to be obtained, pixel by pixel, using a chromatic calibration of the liquid crystal material. This requires describing each perceived color by a set of three scalars, its RGB components.

A standard mathematical representation of color is used by the television and color-printing industries [54]. Experiments on the response of the human eye to colored light have established that any given color can be matched by some combination of intensities from three reference light sources. The procedure is simply to adjust the three intensities until a perceived match is found. The three source lights are called "primaries"; while there are many established sets of primaries, the colors comprising most sets are generally recognized as red, green, and blue of standard wavelengths. The two most important sets of primaries are the NTSC RGB system used in color video transmission and the XYZ tristimulus values used as the historical basis for color calculation. Color video systems use the RGB decomposition, but color-matching experiments have established a linear transformation between the RGB system and the XYZ system.

$$\begin{bmatrix} a_{11} & a_{12} & a_{13} \\ a_{21} & a_{22} & a_{23} \\ a_{31} & a_{32} & a_{33} \end{bmatrix} \begin{bmatrix} R \\ G \\ B \end{bmatrix} = \begin{bmatrix} X \\ Y \\ Z \end{bmatrix} \tag{6}$$

The **RGB** and **XYZ** vectors contain the intensities of the light from the three primaries. The **RGB** vector is obtained for each pixel from the digitally stored RGB matrices. The fraction of the total amount of light attributable to each of the tristimulus primaries is given as

$$x = \frac{X}{X + Y + Z} \tag{7a}$$

$$y = \frac{Y}{X + Y + Z} \tag{7b}$$

$$z = \frac{Z}{X + Y + Z} \tag{7c}$$

The values x, y, and z are *chromaticity coordinates,* and can be used to produce a graph of all possible colors called a *chromaticity diagram*. The diagram is a two-dimensional surface displayed as y versus x, shown in Fig. 16. This is a sufficient description; since the three components must sum to 1.0, z can be determined if x and y are known. The perimeter of the surface, represents the pure (single-wavelength) colors of the spectrum; these are the colors of the rainbow. The wavelengths of the spectral colors vary from about 400 to 700 nm; wavelength is an unambiguous descriptor of the position of the color on the diagram. White light, a uniform mixture of all wavelengths, is located at

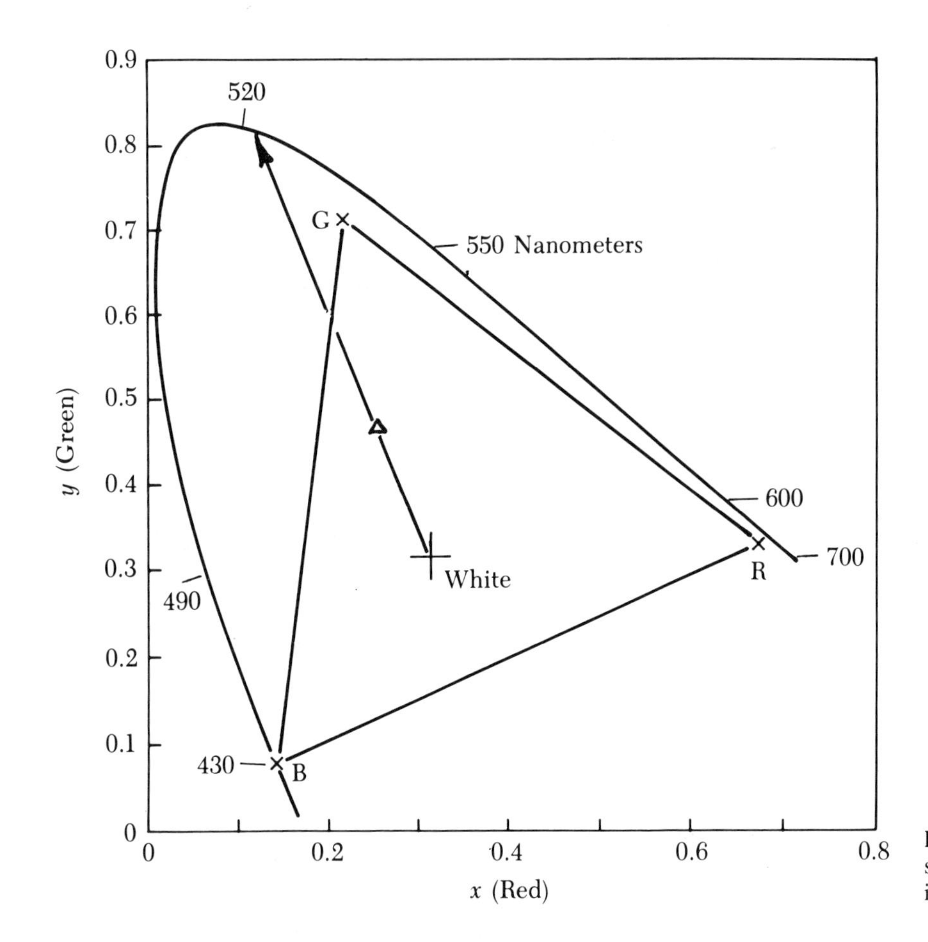

Fig. 16 A chromaticity diagram showing the identification of the dominant wavelength of a data point [51].

the point $x = 0.33$, $y = 0.33$, near the center of the diagram. The colors that are reproducible with the RGB system are contained within the triangle formed by the RGB coordinates and are a subset of all possible colors.

Any point on this surface can be described with the chromaticity coordinates (x, y) or with polar coordinates using a radius and an angle. For the present analysis, the angle is defined in terms of the "dominant wavelength" of the point in question. The dominant wavelength of a particular data point is found by extending to the edge of the diagram a line connecting the white point with the data point. The dominant wavelength is the wavelength of the spectral color that, when mixed with an appropriate intensity of white light, would produce a match to the perceived color rep-

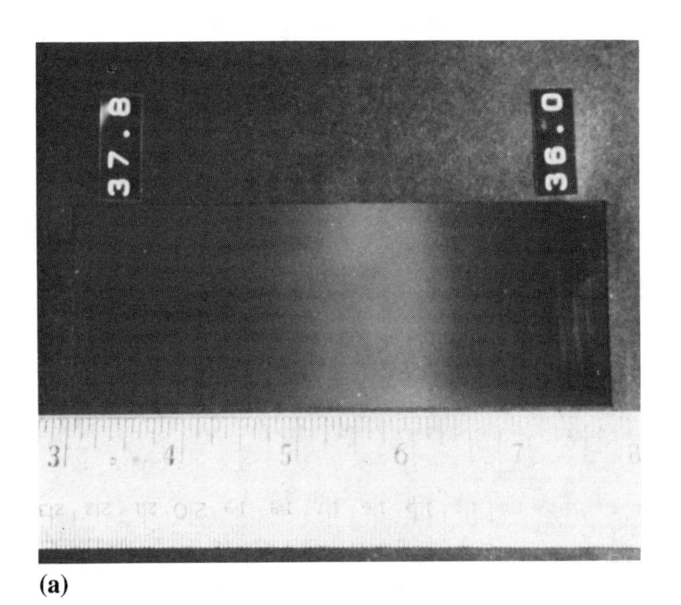

(a)

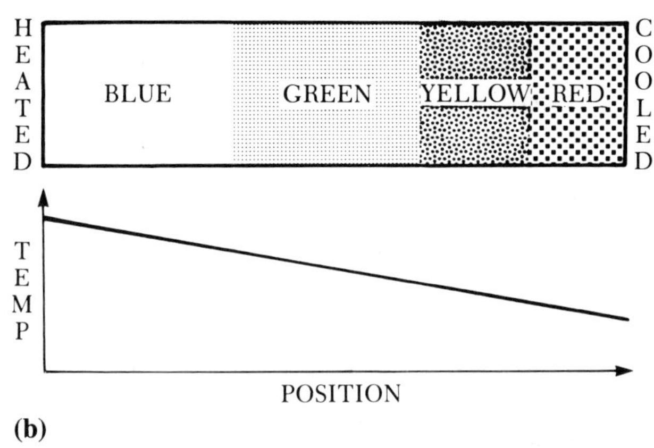

(b)

Fig. 17 Linear gradient calibration of liquid crystal sheet: (*a*) photograph of a liquid crystal calibration using a linear temperature gradient test device [39]; (*b*) schematic of a linear gradient calibration result, in perceived color versus temperature coordinates [51].

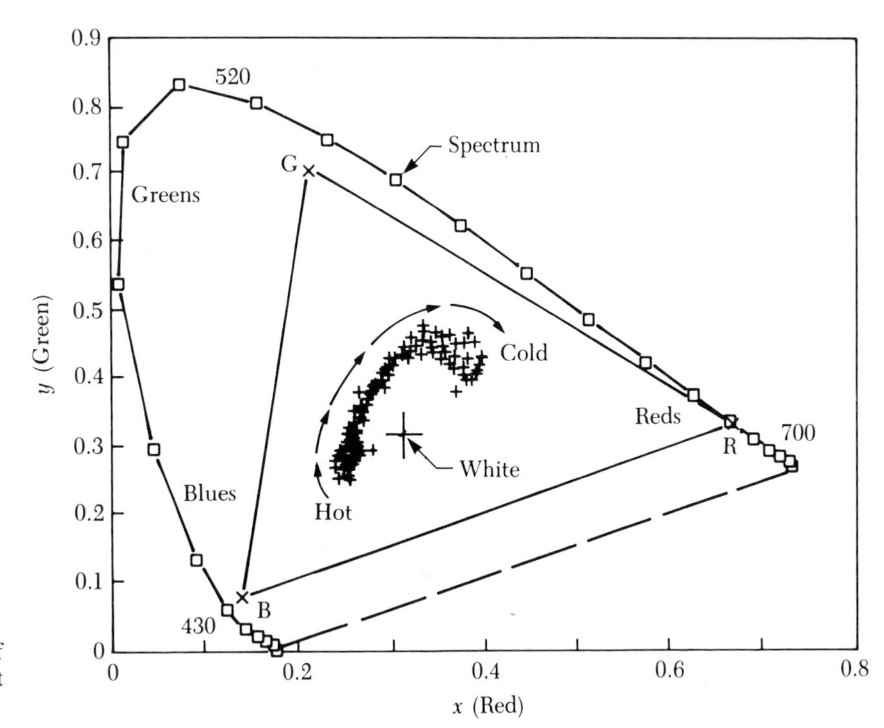

Fig. 18 Chromatic interpretation of an image similar to that in Fig. 17, but at a lower temperature [51].

resented by the data point. The "hue" of a color is related to the dominant wavelength; its saturation is related to its distance from the perimeter.

Experiments have shown that a calibration relationship can be produced between the temperature of a liquid crystal surface and the color descriptions in the chromaticity diagram. A Mylar sheet coated with liquid crystals was placed on an aluminum bar. The crystal had a qualitative operating range of about 5°C. A linear temperature gradient was produced in the bar so that the temperature was linearly related to position along the liquid crystal sheet. A color distribution was formed in the crystal and allowed to reach a steady state. The temperature was monitored using several thermocouples embedded in the aluminum bar.

Figure 17*a* [39] shows a photographic image of such a calibration; a schematic of the calibration is shown in Fig. 17*b* [51]. An RGB color image was acquired with the imaging system, and chromaticity coordinates were calculated along lines parallel to the edge of the strip. The chromaticity coordinates deduced from pixels along these lines are shown in Fig. 18. These data all lie within the RGB triangle since they were obtained through a video system. They show a monotonic increase in wavelength as the temperature drops. A calibration curve was produced relating temperature to dominant wavelength, as shown in Fig. 19. The calibration has a useful range of about 2.5°C and can be implemented by a computer look-up table or curve fit.

To test the measurement system, a different temperature distribution was produced in the calibrator and another image acquired. The calibration curve from the first image was used to calculate the temperature along the bar; the results are compared with thermocouple measurements in Fig. 20. The solid line in the figure represents approximately 140 measurement points (pixels) along the bar. The agreement is very good except at the low-temperature end of the bar. Near the extremes of the useful range the color intensity was not high enough for accurate determination of the dominant wavelength. Subsequent experiments have shown the uncertainty in the temperature measurement to be about 5% of the useful (i.e., quasi-linear) calibration range, in this case 5% of 2.5°C, or 0.125°C.

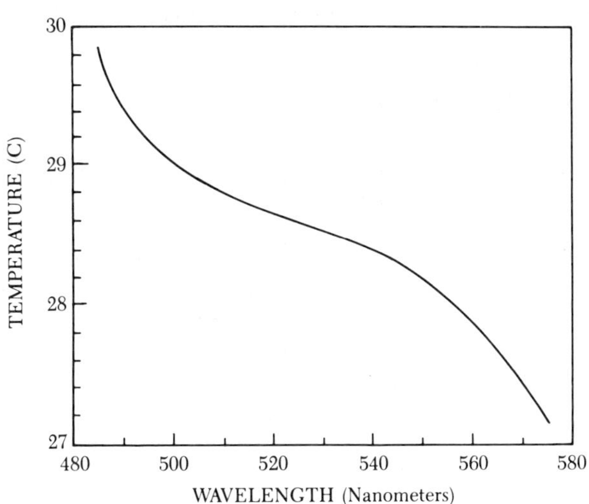

Fig. 19 Chromatic calibration of a liquid crystal showing dominant wavelength versus temperature [51].

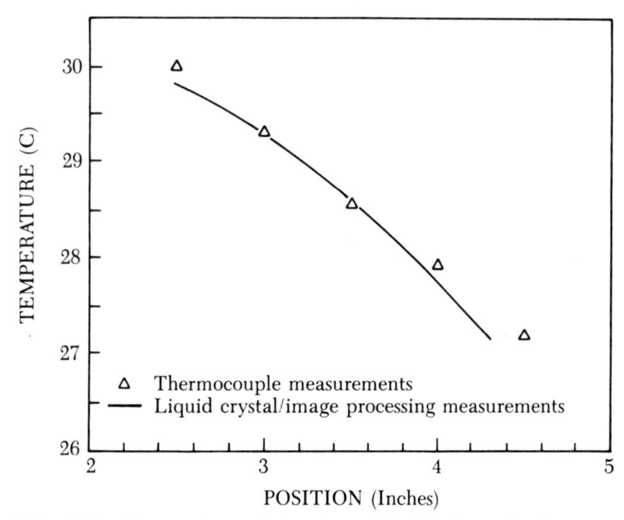

Fig. 20 Comparison of the chromatically determined temperature with thermocouple measurements on the specimen [51].

VI CONCLUSION

A brief description has been given of liquid crystal properties, with an emphasis on the cholesteric liquid crystals, and on the fundamentals of practical applications of thermal visualizations. In addition, liquid crystal measurements have been classified in several categories, and examples of these measurements have been given. Liquid crystals offer significant advantages over conventional thermometers, particularly with respect to spatial and surface temperature measurements in both steady- and unsteady-state conditions. When used properly, liquid crystals perform very well with respect to temperature resolution, accuracy, reproducibility, and time response. Thus, there is a promising future for the development of liquid crystal measurement techniques. Image-processing technology continues to advance, liquid crystal techniques will certainly contribute to advances in heat transfer and fluid mechanics research.

REFERENCES

1. Reinitzer, F., Beiträge zur Kenntniss des Cholesterins, *Monatsh. Chem. Wien*, vol. 9, pp. 421–441, 1888.
2. Kasagi, N., Liquid Crystal Applications in Heat Transfer Experiments, Thermosci. Div., Mech. Eng. Dept., Stanford Univ., Stanford, Calif., Rep. IL-27, 1980.
3. Kasagi, N., Temperature Measurements with Liquid Crystals, in *Temperature Measurements in Heat Transfer Research*, ed. I. Tanasawa, pp. 102–118, Yoken-do, Tokyo, 1985. In Japanese.
4. Gray, G. W., *Molecular Structure and the Properties of Liquid Crystals,* Academic, New York, 1969.
5. Brown, G. H., *Advances in Liquid Crystals*, vols. 1–3, Academic, New York, 1975–1978.
6. Kobayashi, S., ed., *Liquid Crystal: Its Properties and Application*, Engineer's Practical Library, vol. 29, Nikkan-Kogyo Shinbunsha, Tokyo, 1970. In Japanese.
7. Fergason, J. L., Goldberg, N. N., and Nadalin, R. J., Cholesteric Structure: II. Chemical Significance, *Mol. Cryst.*, vol. 1, pp. 309–323, 1966.
8. Friedel, G., Les États Mésomorphes de la Matiére, *Ann. Phys. Paris*, vol. 19, pp. 273–474, 1922.
9. Furuhata, Y., Toriyama, K., and Nomura, S., Liquid Crystal and Its Application (I), *Solid Phys.*, vol. 4, no. 5, pp. 242–251, 1969. In Japanese.
10. Fergason, J. L., Liquid Crystals, *Sci. Am.*, vol. 211, no. 2, pp. 76–85, 1964.
11. Fergason, J. L., Cholesteric Structure: I. Optical Properties, *Mol. Cryst.,* vol. 1, pp. 293–307, 1966.
12. Dixon, G. D., and Scala, L. C., Thermal Hysteresis in Cholesteric Color Responses, *Mol. Cryst. Liq. Cryst.*, vol. 10, pp. 317–325, 1970.
13. Lauriente, M., and Fergason, J. L., Liquid Crystals Plot the Hot Spots, *Electron. Des.*, vol. 10, pp. 71–79, 1967.
14. Akino, N., Kunugi, T., Ichimiya, K., Mitsushiro, K., and Ueda, M., Improved Liquid-Crystal Thermometry excluding Human Color Sensation: Part 1, Concept and Calibration, in *Pressure and Temperature Measurements*, ed. J. H. Kim, and R. J. Moffat, pp. 57–62, ASME Fluids Eng. Div., vol. 44, and Heat Transfer Div. vol. 58, 1986.
15. Kasagi, N., Hirata, M., and Kumada, M., Studies of Full-Coverage Film Cooling: Part 1. Cooling Effectiveness of Thermally Conductive Wall, ASME Paper 81-GT-37, 1981.
16. Kasagi, N., Hosoya, K., Hirata, M., and Suzuki, Y., The Effects of Free Stream Turbulence on Full-Coverage Film Cooling, *Proceedings of the 1987 Tokyo International Gas Turbine Congress, Tokyo, October 1987*, vol. 3, pp. 217–222, 1987.
17. Akino, N., Kunugi, T., Ichimiya, K., Mitsushiro, K., and Ueda, M., Improved Liquid-Crystal Thermometry Excluding Human Color Sensation: Part 2. Application to the Determination of Wall Temperature Distributions, in *Pressure and Temperature Measurements*, ed., J. H. Kim, and R. J. Moffat, pp. 63–68, ASME Fluids Eng. Div., vol. 44, and Heat Transfer Div., vol. 58, 1986.
18. Moffat, R. J., Experimental Methods for Air Cooling of Electronic Components, Keynote Paper, *Int. Symp. Cooling Technol. Electron. Equip.*, Honolulu, 1987.
19. Abe, Y, Sano, T., Tamaki, M., and Iritani, Y., Visualization of a Thermal Stratification in a Two-Dimensional Cavity by using Multiple Liquid Crystals, Photo *J. Flow Visual.*, no. 3, pp. 2–7, 1986. In Japanese.
20. Fergason, J. L., Liquid Crystals in Nondestructive Testing, *Appl. Opt.*, vol. 7, no. 9, pp. 1692–1737, 1968.
21. Parker, R., Transient Surface Temperature Response of Liquid Crystal Films, *Mol. Cryst. Liq. Cryst.*, vol. 20, no. 1, pp. 99–106, 1973.
22. Grodzka, P. G., and Facemire, B. R., Tracking Transient Temperatures with Liquid Crystals, *Lett. Heat Mass Transfer*, vol. 2, pp. 169–178, 1975.
23. Simonich, J. C., and Moffat, R. J., New Technique for Mapping Heat-Transfer Coefficient Contours, *Rev. Sci. Instrum.*, vol. 53, no. 5, pp. 678–683, 1982.

24. Murahashi, S., Oda, R., and Imoto, M., eds., *Plastic Handbook,* Asakura, Tokyo, 1972. In Japanese.
25. Touloukian, Y. S., ed., *Thermophysical Properties of Selected Materials*, vol. 3, pp. 596–597, Purdue Univ., Lafayette, Ind., 1976.
26. Iritani, Y., Kasagi, N., and Hirata, M., Heat Transfer Mechanism and Associated Turbulence Structure in a Near-Wall Region of a Turbulent Boundary Layer, in *Turbulent Shear Flows 4*, ed. L. J. S. Bradbury, F. Durst, B. E. Launder, F. W. Schmidt, and J. H. Whitelaw, pp. 223–234, Springer-Verlag, Berlin, 1984.
27. Woodmansee, W. E., Cholesteric Liquid Crystals and Their Application to Thermal Non-Destructive Testing, *Mater. Eval.*, vol. 24, pp. 564–572, 1966.
28. Hirata, M., and Kasagi, N., Studies of Large-Eddy Structures in Turbulent Shear Flows with the Aid of Flow-Visualization Techniques, in *Studies in Heat Transfer: A Festschrift for E. R. G. Eckert*, ed. J. P. Hartnett, T. F. Irvine, Jr., E. Pfender, and E. M. Sparow, pp. 145–164, Hemisphere, Washington, D.C., 1979.
29. Klein, E. J., Liquid Crystals in Aerodynamic Testing, *Astronaut Aeronaut.*, vol. 6, pp. 70–73, 1968.
30. Ardasheva, H. M., and Ryzhkova, M. V., The Use of Liquid Crystals in an Aerodynamic Heating Test, *Fluid Mech. Sov. Res.*, vol. 6, pp. 128–136, 1977.
31. Cooper, T. E., Field, R. J., and Meyer, J. F., Liquid Crystal Thermography and Its Application to the Study of Convective Heat Transfer, *J. Heat Transfer*, vol. 97, pp. 442–450, 1975.
32. Durao, M. D., Investigation of Heat Transfer in Straight and Curved Rectangular Ducts using Liquid Crystal Thermography, M.S. thesis, Naval Postgraduate School, Monterey, Calif., 1977.
33. Kelleher, M. D., Flentie, D. L., and McKee, R. J., An Experimental Study of the Secondary Flow in a Curved Rectangular Channel, ASME Paper 79-FE-6, 1979.
34. Venneman, D., and Butesfisch, K. A., The Application of Temperature Sensitive Crystals to Aerodynamic Investigations, European Space Agency, ESRO-RR-77, translated from DLR-FB-73-121, 1973.
35. Ireland, P. T., and Jones, T. V., Detailed Measurements of Heat Transfer on and around a Pedestal in Fully Developed Passage Flow, in *Heat Transfer 1986: Proceedings of the Eighth International Heat Transfer Conference, San Francisco, August 1986*, ed. C. L. Tien, V. P. Carey, and J. K. Ferrell, vol. 3, pp. 975–980, Hemisphere, Washington, D.C., 1986.
36. den Ouden, C., and Hoogendoorn, C. J., Local Convective-Heat-Transfer Coefficients For Jets Impinging on a Plate; Experiments using a Liquid Crystal Technique, in *Heat Transfer 1974: Proceedings of the Fifth International Heat Transfer Conference, Tokyo, September 1974*, vol. 5, pp. 293–297, Hemisphere, Washington, D.C., 1974.
37. McComas, J. P., Experimental Investigation of Ground Effects on a Heated Cylinder in Crossflow, M. S. thesis, Naval Postgraduate School, Monterey, Calif., 1974.
38. Goldstein, R. J., and Timmers, J. F., Visualization of Heat Transfer from Arrays of Impinging Jets, *Int. J. Heat Mass Transfer*, vol. 25, pp. 1857–1868, 1982.
39. Abuaf, N., Urbaetis, S. P., and Palmer, O. F., Convection Thermography, General Electric Research Labs Rep. 85CRD168, September 1985. Public distribution allowed.
40. Ali Khan, M. M., Kasagi, N., Hirata, M. and Nishiwaki, N., Heat Transfer Augmentation in an Axisymmetric Impinging Jet, in *Heat Transfer 1982: Proceedings of the Seventh International Heat Transfer Conference, München, Federal Republic of Germany, September 1982*, ed. U. Grigull, E. Hahne, K. Stephan, and J. Straub, vol. 3, pp. 363–368, Hemisphere, Washington, D.C., 1982.
41. Brown, A., and Saluja, C. L., The Use of Cholesteric Liquid Crystals for Surface-Temperature Visualization of Film-cooling Processes, *J. Phys. E: Sci. Instrum.*, vol. 11, pp. 1068–1072, 1978.
42. Hippensteele, S. A., Russell, L. M., and Stepka, F. S., Evaluation of a Method for Heat Transfer Measurements and Thermal Visualization using a Composite of a Heater Element and Liquid Crystals, *J. Heat Transfer*, vol. 105, pp. 184–189, 1983.
43. Baugh, J. W., Hoffman, M. A., and Makel, D. B., Improvements in a New Technique for Measuring and Mapping Heat Transfer Coefficients, *Rev. Sci. Instrum.*, vol. 57, no. 4, pp. 650–654, 1986.
44. Simonich, J. C., and Moffat, R. J., Local Measurements of Turbulent Boundary Layer Heat Transfer on a Concave Surface using Liquid Crystals, Thermosci. Div., Mech. Eng. Dept., Stanford Univ., Stanford, Calif., Rep. HMT-35, 1982.
45. Kitamura, K., Koike, M., Fukuoka, I., and Saito, T., Large Eddy Structure and Heat Transfer of Turbulent Natural Convection along a Vertical Flat Plate, *Int. J. Heat Mass Transfer*, vol. 28, no. 4, pp. 837–850, 1985.
46. Yokobori, S., Kasagi, N., Hirata, M., and Nishiwaki, N., Role of Large-Eddy Structure on Enhancement of Heat Transfer in Stagnation Region of Two-dimensional Submerged Impinging Jet, in *Heat Transfer 1978: Proceedings of the Sixth International Heat Transfer Conference, Toronto, August 1978*, vol. 5, pp. 305–310, Hemisphere, Washington, D.C., 1978.
47. Kimura, R., Visualization of Temperature Fields of Water by Liquid Crystals, *2d Symp. Flow Visual., ISAS Univ. of Tokyo*, pp. 99–102, 1974. In Japanese.
48. Kuriyama, M., Ohta, M., Yanagawa, K., Arai, K., and Saito, S., Heat Transfer and Temperature Distributions in an Agitated Tank Equipped with Helical Ribbon Impeller, *J. Chem. Eng. Jpn.*, vol. 14, pp. 323–330, 1981.
49. Rhee, H. S., Koseff, J. R., and Street, R. L., Flow Visualization of a Recirculating Flow by Rheoscopic Liquid and Liquid Crystal Techniques, *Exp. Fluids*, vol. 2, pp. 57–64, 1984.
50. Akino, N., Kunugi, T., Seki, M., Shiina, Y., and Okamoto, Y., Natural Convection in a Horizontal Silicone Oil Layer in a Circular Cylinder Heated from Below, *23d Nat. Heat Transfer Symp. Japan,* Sapporo, pp. 382–384, 1986. In Japanese.
51. Hollingworth, K., and Moffat, R. J., Digital Image Processing Applied to Liquid Crystal Images, Doctoral Qualifying Seminar, October 1985, Also Thermosci. Div., Mech. Eng. Dept., Stanford Univ., Stanford, Calif., Rep. IL-85, 1986.
52. Akino, N., Kunugi, T., Ueda, M., and Kurosawa, A., Im-

provement in Automatic and Quantitative Liquid-Crystal Thermometry; Temperature Evaluation using Regression Equation of RGB Tristimulus Values, *Fall Mtng. Atomic Energy Soc. Japan*, Fukuoka, p. 117, 1986. In Japanese.
53. Akino, N., Kunugi, T., Ueda, M., and Kurosawa, A., Liquid-Crystal Thermometry and Visualization of Heat Transfer; I. Quantitative Evaluation of Liquid-Crystal Color by Hue, *Ann. Mtng. Atomic Energy Soc. Japan*, Nagoya, p. 35, 1987. In Japanese.
54. Sproson, W. N., *Color Science in Television and Display Systems*, Adam Hilger, Ltd., Bristol, England, 1983.

Chapter 9

Tufts

J. P. Crowder

I INTRODUCTION

The tuft method is one of the earliest techniques of surface flow visualization. In its simplest form, the method can be implemented with almost no special effort, using readily available materials and equipment. Tufts are point indicators of the local flow direction. They are most often used to distinguish between attached or separated flow. Each tuft can depict the flow conditions at only one location. Visualizing the flow over a surface requires that an array of tufts be applied. The spatial resolution of the visualization experiment is determined by the number density of the tuft array.

In this discussion, tuft techniques are described in the somewhat arbitrary terms of *conventional tufts* and two recent adaptations of the tuft technique known as *fluorescent minitufts* and *flow cones*. These three categories form the bulk of tuft applications for surface flow visualization experiments. A wider variety of other applications extending the method to special cases, including flow field applications, are also described in somewhat less detail.

A Conventional Tufts

Conventional tufts are such simple and obvious devices that it may seem that the technique needs no particular description. We shall see, however, that this apparent simplicity is somewhat deceiving.

Tufts are short pieces of flexible string or yarn attached to an aerodynamic surface in such a way that they can move freely under the influence of the flow. They can range in size from hairlike fibers to thick ropelike yarn. They are equally effective in gas or liquid flows.

The effectiveness of tufts in indicating the flow properties depends on the distinctive behavior of the tuft material under the various conditions. In steady attached flow, tufts remain relatively stationary and become aligned with the flow direction. When they are installed in a dense array, their indicated direction is very well correlated with neighboring tufts. In separated flow, tufts generally develop distinctive unsteady motions and show large deviations in the indicated flow direction from the free-stream direction and from neighboring tufts.

Figure 1 shows a typical tuft image on a wind tunnel model. The model is of a jet transport airplane with the landing flaps extended, tested in a low-speed wind tunnel at Mach 0.2. The regions of separated flow are distinctively marked, compared to the attached flow, by large deviations in the apparent tuft direction, and by blurred tuft images indicating rapid motion.

The details of these effects depend strongly on a wide range of factors including the flow properties, tuft materials, installation procedures, and data-recording methods.

One of the fundamental considerations in conventional tuft techniques is the requirement that their behavior be visually observed. Every tuft experiment has a basic dimensional scale that is best expressed by the distance from which the tufts must be observed. The greater the observation distance the larger the tuft material must be. This factor influences all the other factors important in selecting the size relationship of the tufts to the aerodynamic surface, i.e., tuft length, diameter, spacing, and the resulting intrusiveness of the experiment.

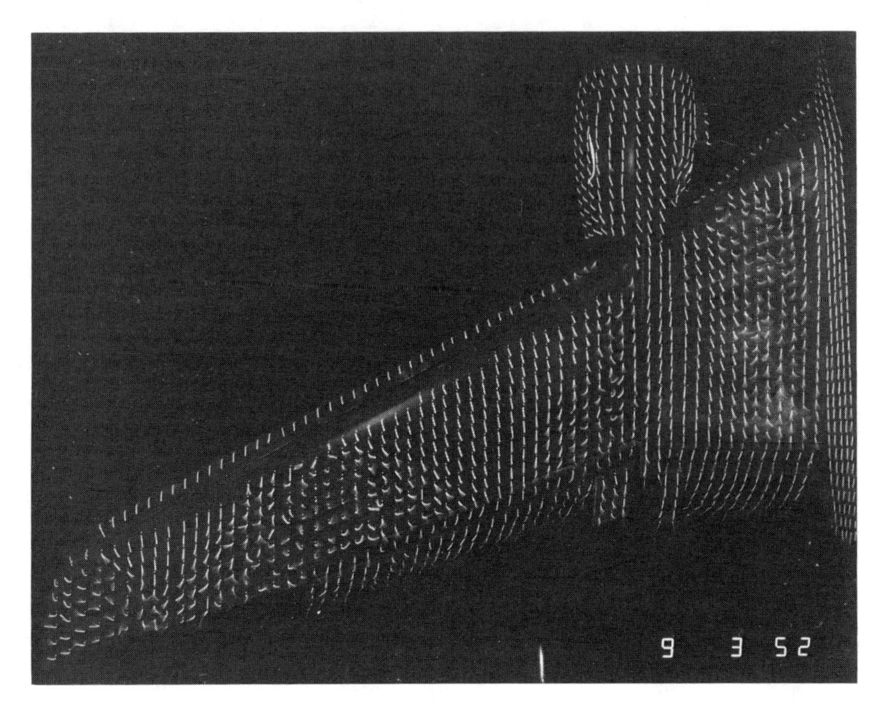

Fig. 1 Typical tuft indications on wind tunnel model.

These considerations are described in some detail in the following sections.

B Fluorescent Minitufts

Fluorescent minitufts are an adaptation of conventional tufts that resulted from trying to reduce the tuft size as much as possible in order to minimize the intrusiveness of tufts on small wind tunnel models. The key element in the fluorescent minituft technique is the elimination of the size of the tuft as a factor in making the tuft visible. This is accomplished by making the tuft optically bright. In photographing any small object, if its brightness can be increased significantly more than the surrounding background, its photographic image can be effectively magnified compared to its geometrically scaled image size.

In the fluorescent minituft technique, the tuft material is treated with a fluorescent dye and the image is recorded with fluorescence photography. With this process, the scene is illuminated with light at a wavelength that is absorbed by the fluorescent dye. It then radiates light at a longer wavelength. A barrier filter is placed over the camera lens to block the reflected light at the original excitation wavelength while allowing the light at the fluorescence wavelength to pass. The result of this process is that the illumination energy can be increased almost without limit, while still maintaining an effectively dark background. In this way, practically any size of tuft material can be made easily visible just by providing sufficient illumination energy.

Besides the fluorescent dye and the fluorescence photography, the other fundamental consideration is a practical method of attaching large numbers of minitufts to small model surfaces in a way that avoids disturbing the flow. All these considerations are described in detail in the following sections.

C Flow Cones

Flow cones are a further recent adaptation of the tuft technique aimed specifically at large-scale, high–Reynolds number flow visualization applications usually associated with flight testing of airplanes. Conventional tufts of a size required for visibility under these conditions frequently develop an unsteady whipping motion that is apparently not a consequence of the flow behavior, but more likely due to the flexible nature of the tuft material. This motion is dynamically similar to the flag-waving motion of flexible sheets.

Flow cones are tuftlike indicators that were developed specifically to avoid the unstable motion of conventional tufts. They are small, rigid, conically shaped elements attached to the aerodynamic surface with a string flexure or short chain. They remain perfectly stationary in steady, attached flow, even when nearby conventional tufts are observed to be undergoing large flagging motion. As a result, flow cones provide much greater fidelity of indication, especially for separation onset indications.

A secondary benefit of flow cones is their relatively large and smooth surface area, which makes it possible to employ special reflective tape as a means of making them extremely optically bright under certain lighting conditions. By this means, very small flow cones can be photographed at large distances in bright daylight, as is usually required for air-

Fig. 2 Reflective flow cones on AT-6 airplane at stall.

to-air photography, or with a very wide-angle view to permit imaging of a large area from an on-board camera.

Figure 2 shows a typical reflective flow cone result. In this example, flow cones on the wing upper surface of an AT-6 airplane at the stalling point are photographed with a very wide-angle lens to cover the entire wing area from root to tip.

D Other Adaptations

Many other adaptations of the tuft method, developed and reported over the years, fall outside the categories just described. These techniques are briefly described here, but are not discussed in detail later.

1 Tuft Grid The tuft grid method [1] is a technique for indicating the flow direction in the flow field, off the model surface. It is normally used in a cross plane, perpendicular to the flow direction. A typical example is shown in Fig. 3. The tufts are attached to wires stretched between the edges of a frame surrounding the flow field. A camera positioned downstream of the tuft grid provides a view that graphically indicates large deviations in the lateral velocity components as well as regions of increased turbulence.

Other orientations of tuft grids are possible, such as in a streamwise plane [2]. In that case, however, there is a greater likelihood that the wakes of the tufts and their support members will disturb the flow significantly.

2 Streamers Streamers are generally much longer tufts that are employed singly or in arrays, such as in Fig. 4, to indicate flow direction and quality in the flow field, off the surface. This example shows the flow at the exit of a 6-in. dia. jet. The exit Mach number was 0.6. The streamers were used to depict the swirl angle of the flow, although the swirl vanes were not installed for the condition shown.

The behavior of streamers is often dominated by a tendency for dynamic instabilities to develop in them. These are caused by traveling waves that propagate along the length of the streamer. This usually shows up as a rapid whipping motion at the free end of the streamer.

The greatest limitation to the use of streamers in arrays, as in Fig. 4, is the problem of tangling. If the flow is unsteady or separated and the streamers are closer together than their length, they can easily become tangled, thereby destroying their effectiveness. For essentially steady flows, as in this example, that is seldom a problem; however, care must be taken to keep the streamers combed and untangled during the installation and while the flow is being started.

Streamers can also be used on an aerodynamic surface, as shown in Fig. 5, but are generally less satisfactory than arrays of short tufts. Probably the only advantage to streamers, as in this example, is ease of installation. One long streamer could possibly indicate the same thing as a streamwise array of many short tufts. In this example it was surprising, upon comparison of the lengthwise distribution of surface flow angle with the adjacent individual tufts, to see the generally good agreement. The tendency for tangling is

Fig. 3 Tuft grid in wind tunnel (photograph courtesy of NASA).

Fig. 4 Flow field streamers in exit of 6-in. jet.

readily apparent in Fig. 5. This illustrates what amounts to a general principle in tuft applications: To avoid tangling, the tufts must be no longer than the spacing between adjacent members of the array.

3 Tuft Wand An individual streamer is often attached to the end of a slender wand for hand-held surveys of the flow field. Around leading-edge regions of a model, such a device can graphically depict stagnation zones, as indicated by very high spatial gradients in the flow direction. Around aft regions of a model, separations can be readily marked, usually by the highly unsteady motion of a streamer and by large deviations in the flow direction. When the flow is separated, its direction is frequently completely reversed in the wake and a streamer will blow forward.

With streamers made of material having a very small diameter (e.g., 0.002 in.) and relatively long lengths, it is surprising how well the streamer apparently follows a highly curved streamline. Of course, the streamer shape must reflect the balance of forces over its entire length. However, it seems as if the aerodynamic cross-flow forces, which are responsible for producing the streamer curvature, predominate over the axial forces, which tend to straighten the streamer out.

Since a tuft wand is most often used as a hand-held device, it is necessarily limited to rather low speeds. Because of the probably large wake from the wand, it must

(a)

(b)

Fig. 5 Surface streamers on wind tunnel model: (*a*) Angle of attack = 14°; (*b*) angle of attack = 16°.

be used as a qualitative device with a high probability of significant disturbances to the flow field.

4 Pin Tufts There is widespread recognition that the flow direction on the surface of a model within the boundary layer can be substantially different from that just outside the boundary layer. Tufts mounted at the tip of short pins are occasionally used to examine this effect [3]. As in the case of streamwise oriented tuft grids, caution must be exercised to ensure that the wake of the support pin does not adversely affect the behavior of the tuft. Ideally the pin diameter should be at a subcritical Reynolds number (Re < 100), but in practice the choice of mounting pin is dictated by handling loads, manufacturing procedures, and most often by whatever material is readily at hand.

5 Hinged Tufts Because of concern about the stiffness of tuft materials possibly interfering with their indicating behavior, some workers have developed configurations utilizing a mechanical hinge or swivel joint [4]. The relative merit of such a device must be weighed against the possible difficulty of manufacturing and installing large numbers of them and the likelihood that such a device would present a relatively large disturbance to the flow.

6 Mass Balanced Hinged Tufts In considering the use of tufts on rotating surfaces such as propellers and wind turbines, it is generally recognized that conventional tufts may be adversely affected by the centrifugal acceleration field, and give a biased indication of flow direction. One attempt to overcome this factor involved the use of a rigid tuftlike indicator made from cylindrical plastic soda straws supported on a single-axis hinge with a mass balance. These devices were employed on a large-diameter wind turbine [5] and were reported to be very effective in indicating separation. The relative complexity of such a device would tend to limit the number of them that could be practically manufactured and installed, however.

II HISTORY

A Conventional Tufts

Tufts are seemingly ubiquitous and no doubt have been used for flow studies long before the modern era of aeronautics. However, the earliest reference found in a cursory literature search was the 1928 British R & M by Haslam [6]. Judging from the several contemporary references [7, 8] and the relatively advanced methodology in use at that time, the technique was well known and regarded as a standard tool.

B Fluorescent Minitufts

Fluorescent minitufts were developed by the author [9–11] as a deliberate attempt to provide for concurrent flow visualization testing on small high-speed wind tunnel models. The first applications of the method were undertaken during the DC-10 wing development program at Douglas Aircraft Company starting in 1969.

The technique, as eventually developed and reported entailed three aspects: (1) preparation of the tuft material; (2) an application method practical enough for routine use in industrial wind tunnel programs; and (3) fluorescence photography for recording the data.

The answer to the original question of "How small can a tuft be and still be visible?" was quickly found to be "Smaller than anyone would care to use."

Eventually, it was recognized that another major benefit of very small tufts, besides their nonintrusiveness, was their relative insensitivity to centrifugal acceleration on rotating propellers and fans. The first demonstrations of this principle were conducted in 1982 as a development program of the X-Aero Company. The first applications of fluorescent minitufts on a propeller were conducted in flight during the development of the Machen Superstar airplane [12, 13]. Applications of fluorescent minitufts to a rotating propeller in water is reported in Ref. [14].

C Flow Cones

Flow cones [15] were developed as an outgrowth of a brief participation in flight test flow visualization studies on the DC-10 airplane at Douglas Aircraft Company in 1974. At that time the unstable behavior of large tufts typical of flight testing on large airplanes was observed firsthand. The flow cone shape evolved in fairly short order and proved very effective in initial demonstrations. However the development of a practical device required investment in mass production tooling and was not accomplished until 1983. Flow cones are now marketed by the X-Aero Company [16] and are protected by a U.S. patent [17].

III FUNDAMENTAL CONSIDERATIONS

In this section the fundamental considerations of tuft methodology are discussed in some detail. Many of these considerations apply to any of the several adaptations covered, whether conventional tufts, fluorescent minitufts, or, in some aspects, flow cones.

A Spatial Resolution

An important factor to consider in tuft flow visualization methodology is that each tuft is an individual indicator of flow conditions at one point (as explained above, some streamer applications can overcome this limitation). It is usually desirable to depict the flow over a surface in two dimensions, and tufts are therefore employed in arrays. The spacing between adjacent tufts can be no closer than the tuft length to avoid tangling. This arrangement defines the characteristic spatial resolution of the technique. This factor is probably the most serious limitation of tuft techniques. Small-scale features of the surface flow can be resolved only down to the tuft-spacing dimension.

This principle is illustrated in Fig. 6. For comparison, Fig. 7 shows the same flow visualized using the fluorescent oil technique, permitting substantially greater spatial resolution of small-scale features.

The model is a large bump on the wall of a 14- by 18-in. low-speed wind tunnel. It is arranged to produce a massive separation on the afterbody, in order to study the effects of vortex generators in suppressing the separation. The tufts in this example are fluorescent minitufts with a diameter of 0.0019 in. Data recording is by fluorescence

(a)

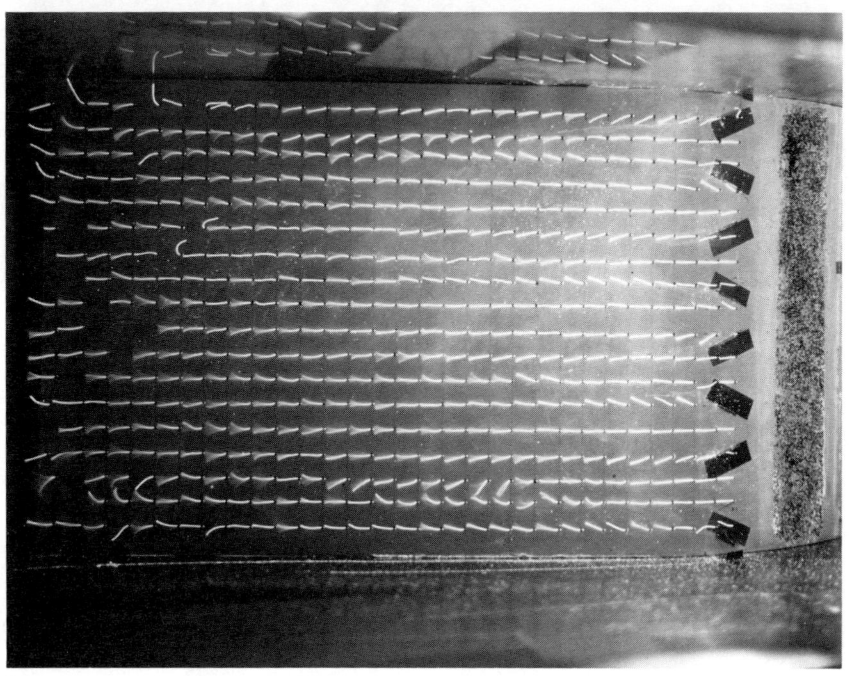

(b)

Fig. 6 Tufts on wind tunnel wall bump: (*a*) Basic configuration; (*b*) vortex generators installed.

photography using an ultraviolet flash lamp having a duration of 4 ms.

The image of Fig. 6*a* shows the massive separation on the basic configuration. The distinctive cues for separated flow of blurred tufts and large deviations in the indicated flow direction are well illustrated. The tufts clearly show the highly unsteady nature of the separated flow. Figure 7*a* shows the same condition visualized by fluorescent oil. Much the same extent to the separated region is evident, although the unsteadiness of the flow is not so well marked. In this case, because the model surface is vertical, the oil has developed a marked assymetry, which is probably due to the effects of gravity on the oil indicator material. Overall, the tufts give a much better indication of the nature of the flow.

In Figs. 6*b* and 7*b*, for the configuration with vortex

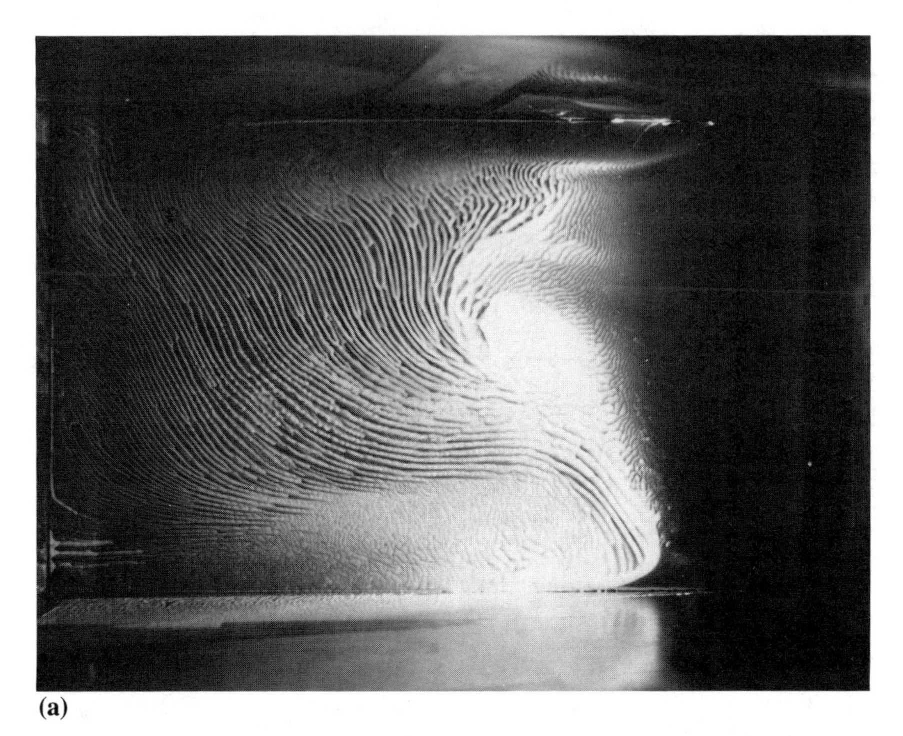

(a)

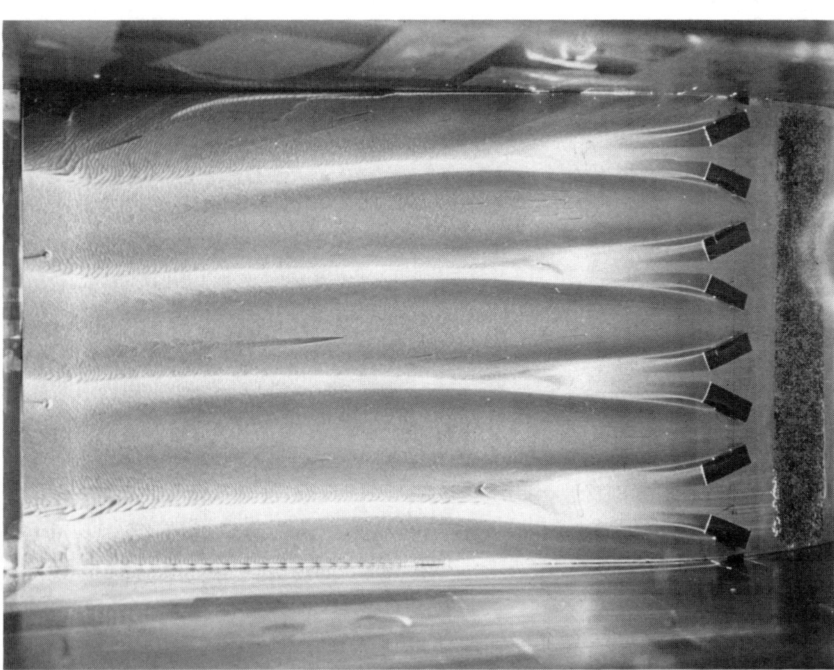

(b)

Fig. 7 Fluorescent oil on wind tunnel wall bump: (*a*) Basic configuration; (*b*) vortex generators installed.

generators installed, the afterbody separation is largely suppressed. However, the vortex generator wakes are small-scale features that are not very well depicted by the tufts. It is clear that the massive separation is not present, but it is nearly impossible to resolve the structure of the vortex generator wakes. The oil flow, on the other hand, shows many more details of the wake structure and some indications of local shear-stress variations from the variations in indicator thickness. Gravitational effects on the oil indicator material are still evident.

B Tuft Materials

Tufts can be made—and no doubt have been—of any material that meets the basic requirements, i.e., flexibility and relative durability. Beyond that, the requirements tend to be dictated by the details of the experiment. In many cases the choice of tuft material is determined by whatever the experimenter can locate within arm's reach.

1 Conventional Tufts Conventional tufts, i.e., those viewed and recorded in white light, must be large enough to be visible from whatever viewing distance is dictated by the facility. For a typical low-speed wind tunnel situation with a viewing distance of 4 to 6 ft, materials such as knitting yarn or crochet yarn are frequently prescribed. These tufts might be made in lengths of 1 in. or so to provide enough flexibility for them to respond to the flow.

In larger scale installations, such as on wing surfaces in flight, where the tufts are viewed from on board the aircraft at distances of 10 to 30 ft, twine or heavy yarn might be employed. For even larger distances, a light rope or heavy twine might be used. One popular material suggested for jet-transport-wing flow visualization applications is known as parachute riser cord. This is a woven nylon rope composed of a tubular sheath having several separate strings running through the interior. The internal strings are often removed to make the tubular sheath flattened and more flexible.

In the larger scale installations, where tuft whipping occurs, the tips of the tufts suffer considerable shredding damage. Often the tufts will be dipped in resin or fused by melting in an attempt to prolong their life.

One unique material reportedly [18] used for tufts in medium-scale flight-testing applications is a flat ribbon made from $\frac{1}{4}$-in.-wide magnetic recording tape, as found in tape cassettes. In using this material, the tufts must be installed with an orientation close to the flow direction. Flat ribbon material has considerable lateral stiffness that resists bending until the loads become large enough to buckle the ribbon. It is reported that tufts made of this material tend to remain more stable prior to separation, but snap out of position when the flow separates. Thus they give an exaggerated indication of flow separation. In addition, such tape tufts have a much larger surface area, which makes them relatively easier to observe.

2 Fluorescent Minitufts The fluorescent minituft technique employs the optical brightness of the tuft fluorescence to provide visibility, thereby permitting any size of tuft material, provided enough illumination energy can be supplied.

The material used for fluorescent minitufts is usually nylon monofilament. This material is available in several sizes, but the smallest monofilament that is readily available seems to be what is known as 15 denier. (The denier is a unit of measure employed in the textile industry. It corresponds to the weight in grams of a 9000-m length of the material.) This material has a diameter of 0.0019 in. A smaller nylon filament is available in a 30-denier, 10-strand yarn. The yarn is untwisted so that it is relatively easy, if not tedious, to unravel individual strands of up to 10 ft in length. A single strand of this material has a diameter of 0.0007 in.

Use of these extremely thin materials may seem like overkill in striving for small tufts, but some applications definitely require this smallest size. Fortunately, this smallest size is seldom required, as it is difficult to work with. The 15-denier monofilament is the size most often utilized for wind tunnel model applications. At first, even this size might seem too small to be practical for an industrial wind tunnel testing environment, but with practice one can adapt to it without much trouble.

As the size of the aerodynamic surface increases, the size of the minitufts can be increased correspondingly, while still maintaining their nonintrusive nature. For large wind tunnel models or medium-sized airplanes, fluorescent minitufts can be made from sewing thread or 2-lb test nylon fishing line with a diameter of 0.006 in. This larger material is much easier to handle and very much brighter, making it much easier to photograph.

3 Fluorescent Minituft Preparation In its basic form, the nylon filament used for the smaller sizes of fluorescent minitufts is not fluorescent. It must be dyed, much like any other textile, and fortunately, very effective fluorescent dyes are available. Known as optical brighteners, these dyes are used in most white textiles. The fluorescence, which is weakly stimulated by sunlight, makes the fabric seem "whiter than white."

For minitufts made from larger material, starting with 0.006-in-dia. sewing thread, ordinary white textile material is already brightly fluorescent when purchased from any fabric store.

Fluorescent minituft material in the two smaller sizes, 15 denier, and 3 denier, 0.0019- and 0.0007-in.-dia. respectively, is available from the X-Aero Company with the fluorescent dye already incorporated [19]. For those who prefer to have their own supply, the preparation process is described below.

The first step in making fluorescent minituft material is to acquire the bare nylon stock. It is available as a 15-denier monofilament and a 30-denier, 10-strand yarn. One of the 10 strands must be unraveled prior to use. This material is not commonly available at the retail level. This writer was able to acquire sample quantities of the material from the manufacturer* just by explaining the application and mentioning Douglas Aircraft Company or Boeing Airplane Company. Judging from this experience, it seems that the smallest available unit of such material is a 2-lb. bobbin. Two pounds of 15-denier nylon monofilament has a length

*E. I. DuPont de Nemours and Company, Textile Fibers Department, Wilmington, Delaware, phone (302) 999-4363.

of 544 km. Needless to say, this is essentially an infinite quantity of fluorescent minituft material.

There are apparently many types of optical brightener fluorescent dyes suitable for this process. The material used in this instance is available from Sandoz, Inc., and is known as Leucophor EFR Liquid. It is used a concentration of 1% in water with 2% acetic acid added. The nylon material should be first wound onto an open wire reel. The reel is immersed in the solution for 15 min. at a temperature of 180°F with frequent stirring.

After the dye bath, the reel of material should be rinsed in fresh water, dried for several hours, and then wound onto small spools. During this step, the freshly dyed minituft material should be wiped with tissue pads to remove the loose fluorescent powder that adheres to the filament. Otherwise, this powder may transfer to the model surface in irregular patterns and obscure the flow visualization data.

It is possible to have the nylon dyed by a commercial dye company, but these establishments are accustomed to 1000-lb. quantities. It is difficult to find a company willing to dye material in 2-lb. quantities.

4 Flow Cones Flow cones are a product of the X-Aero Company and are available in several sizes, materials, and finishes [16]. The most common type is made from injection-molded plastic with a short piece of string extending out of the apex for attachment to the surface. Figure 8 shows several different styles of flow cones. The one at the top of the figure is made from thin steel sheet with a short chain attached to a small steel plate. This style is for use in high-temperature jet exhaust flow. The plate is bonded to the airplane skin with high-temperature adhesive.

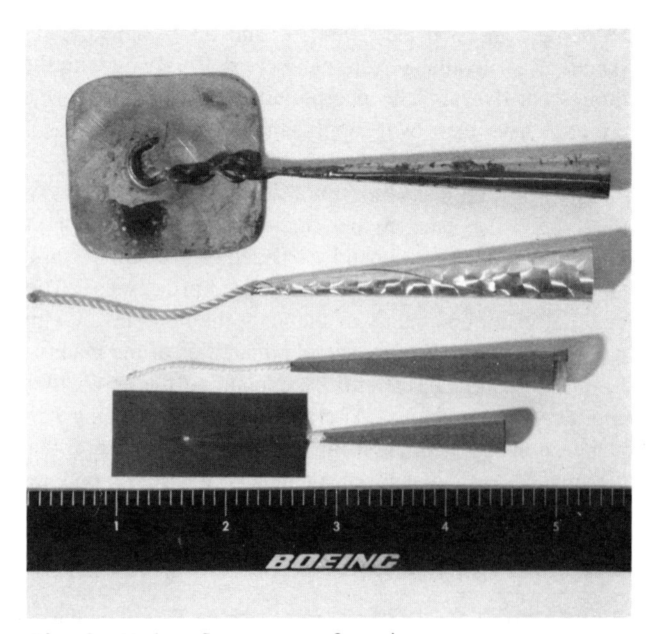

Fig. 8 Various flow cone configurations.

C Tuft Behavior

Tufts indicate certain things about the flow as a result of forces that develop on the tufts and the response of the tufts to those forces. Some of the more important aspects of tuft behavior are discussed below.

1 Stiffness The stiffness of the tuft material can have a significant influence on the design of an experiment. This is a potential factor in any tuft experiment, but is particularly important when the tufts are made of monofilament, as in the fluorescent minituft method. As the tufts are made smaller in diameter, the airloads responsible for bending them in the desired directions become correspondingly smaller.

In general, there is a minimum tuft length for a given tuft material and test condition, based on the fidelity with which the tufts will indicate the flow. Too many factors are at work to permit specifying this minimum length for all cases, but the conditions for which this is most often a controlling factor are found in a typical low-speed atmospheric wind tunnel flow with a Mach number of 0.1. For such a flow, a rough guide would be as shown in Table 1.

Other factors that affect the minimum length include the flow dynamic pressure and the orientation of the tufts at their attachment point on the model surface. Although it is desirable to attach them to the model surface oriented in the direction of the flow, other considerations, such as the attachment procedures and surface geometry, may dictate an orientation that requires the tuft to bend through a large angle in order to achieve alignment with the flow.

2 Motion Rate Tufts indicate the flow condition by developing a characteristic rapid motion involving rotation about the attachment point. This is the most obvious cue to indicate the unsteadiness associated with separated flow. Tufts in separated flow usually have such rapid motion that they appear, both to the eye and in photographs, as a blur. It is important to have a feeling for the speed of this tuft motion in order to select exposure times of the photographic data-recording system.

It has sometimes been stated that an exposure time of $\frac{1}{25}$ or longer is required to register this tuft motion. That would be difficult to accomplish with electronic flash lamps, which are required for the intense ultraviolet illumination used in the fluorescent minituft method. However,

Table 1 Nominal Minimum Tuft Length for Low-Speed Wind Tunnel Applications

Tuft Diameter (in.)	Minimum Length (in.)
0.0007	0.4
0.0019	0.6
0.006	0.8

recent experience has indicated that the typical tuft motion in a low-speed wind tunnel is still usually sufficiently fast to provide blurred images with exposure times as short as 1 ms.

On the other hand, other experience has amply demonstrated that exposure times of 10 μs will almost always freeze the tuft motion. In that case there can be some loss of information in the data record. However, it is not too serious since, even if the tuft motion is not depicted by image blur, the separation can still be detected by the random orientation of the tufts. This aspect is especially prominent when tufts are installed in a dense array, so the neighboring tufts enhance the random nature of their orientation. Also, several photographs at the same condition will have differing random orientations of the same tufts at different times.

3 Stability Under certain conditions tufts have a tendency to develop a self-excited instability that produces a whipping motion at the free end. This condition does not reflect anything about the flow, but is a consequence of the flexible nature of the tuft material. This is similar to the whipping behavior of flags or towlines. This behavior has been well described in the literature [20–22].

In tuft installations on small-scale wind tunnel models, tuft whipping does not occur very often. Most often it is restricted to a few individual tufts that perhaps have some particular orientation that result in a peculiar loading condition. More frequently it can be seen at the free ends of long streamers.

However, for large tufts typical of installations on large-scale airplanes at high Reynolds numbers, tufts frequently develop a continuous whipping motion that is perceived by the eye as a continuous blur. The main body of the tuft usually shows an excursion of about 20° while the last 10–20% of the tuft length has, superimposed on it, an angular excursion of nearly 60°. This motion is sometimes seen to commence and die out randomly, and adjacent tufts may or may not be moving at any given time.

Figure 9 shows the typical appearance of conventional tufts and flow cones on the wing of a large jet transport airplane. The photograph was taken with a short exposure so the tufts are frozen in various orientations; several of them are lifted off the surface. By contrast, all the flow cones visible in the picture are oriented in the same direction and show no evidence of motion.

The unstable tuft motion is observed on aerodynamic surfaces that experience would suggest are entirely free of separation, such as a wing upper surface near the leading edge at a high-speed cruise condition, an engine fan cowl at cruise power, or a fuselage center section. Experienced flight test personnel seem to recognize this tuft whipping behavior as something commonplace and of no particular concern. If it is sometimes commented on, it might be explained as a response to the boundary layer turbulence.

Fig. 9 Behavior of conventional tufts on jet transport wing.

This is very unlikely since, although the boundary layer is no doubt turbulent for these flight conditions, there is no obvious difference in the tuft motion between locations having widely different Reynolds numbers; an example of such a location is between the wing leading edge and the fuselage center section.

A more direct argument against the idea that this tuft whipping motion is an indication of flow properties can be made by comparing the response of flow cones in the same flow condition. When flow cones are installed on a wing where whipping tufts are located, the flow cones remain perfectly still while the surrounding tufts continue their unstable motion.

This tuft whipping motion can usually be distinguished from the typical tuft response to large-scale separation. In that case the tufts usually cease the rapid whipping motion and undergo a slower, large-scale flopping motion, oftentimes blowing forward against the freestream flow direction. However the self-excited tuft motion greatly obscures indications of separation onset. This is especially true with high-speed separation, which is usually associated with smaller eddy sizes.

The other difficulty caused by tuft whipping, besides the poor fidelity of flow indication, is the very high reaction loads produced on the tufts and the attachments to the aircraft surface. These large, unsteady reaction loads lead to disintegration of the tufts and require extra effort to be attached to the airplane surface.

4 Response to Turbulence It has sometimes been claimed [3, 23] that behavior of tufts on small-scale wind tunnel models is different in a turbulent boundary layer than in a laminar boundary layer. In other words, they can act as boundary layer transition indicators. In the experience of this author, this does not happen with any reliability. It is true that there are sometimes observed differences in the steadiness between individual tufts when the flow is nominally nonseparated, but these slight differences are probably due to minor environmental factors.

The fact that significant differences due to boundary layer state have not been observed does not rule out the possibility that they can, under some conditions, actually occur. Moreover, the detection of boundary layer transition is an extremely important experimental task for current wind tunnel testing of transonic transport aircraft. Therefore the slightest possibility that tufts could serve that function deserves careful examination.

Figure 10 shows the result of a wind tunnel test designed to look for such effects. The model was a two-dimensional wing panel with an NACA 0018 section shape. The wind tunnel test section had dimensions of 14 $\times$ 18 in. The model had a chord length of 6 in. and was installed spanning the 14-in. dimension. The angle of attack was close to 0, but was adjusted to produce a slightly proverse pressure gradient to promote laminar flow to 80% chord. The test was conducted at Mach 0.25, providing a dynamic pressure of 93 lb/ft^2 and a unit Reynolds number of 1.7 million per foot.

Fluorescent minitufts with a diameter of 0.0007 in. were installed over the right half-span, while minitufts with a diameter of 0.0019 in. were used on the left half-span. A trip strip was applied to the model surface at 20% chord running from 25 to 75% of the model span. Thus, both minituft installations were exposed to regions of laminar and turbulent boundary layer flow. The trip strip consists of an array of circular disks with a height of 0.012 in., a diameter of 0.05 in., and a spacing of 0.1 in.

The model was spray painted with a thin layer of naphthalene, and the boundary layer transition location was observed by the sublimation technique [24]. With this process the naphthalene indicator material evaporates at a greater rate in regions of turbulent boundary layer than in the laminar boundary layer. At the start of the run the model is coated with a uniform layer of the indicator. After a few minutes of exposure to the flow, naphthalene in the turbulent region has completely evaporated, leaving a detectable thickness of the material in the laminar region. The transition edges are clearly marked with typical turbulent wedges emanating from each of the tufts in the front row.

In order to emphasize any tuft motion that may have been present, but frozen by the ultraviolet flash exposure duration of 4 ms, the three separate flash exposures were superimposed on the same photograph.

The figure clearly shows no distinguishable tuft behavior differences between the laminar and turbulent boundary layer regions. Furthermore, the presence of the tufts in the laminar boundary eventually causes transition to a turbulent boundary layer. Thus, even if tufts did show some subtle effect of turbulence, they would not be useful as transition indicators, as their presence would alter the flow phenomenon being studied.

5 Effects of Centrifugal Acceleration Rotating surfaces, such as those found on propellers and turbines, are particularly difficult subjects for flow visualization experiments because of the effects of centrifugal acceleration on the indicator materials. The fluorescent minituft technique has been successfully applied to propellers and turbines because of the unique properties of very small tufts in their response to the centrifugal acceleration.

Tufts indicate certain features of the flow in response to the forces acting on them. On nonrotating models, the forces are chiefly aerodynamic. On rotating models, the centrifugal acceleration imposes substantial inertial loads, in addition to the aerodynamic loads. Simple dimensional analysis suggests that the aerodynamic forces on a tuft depend on the tuft diameter, while the inertial forces depend on the cross-section area, or the diameter squared. Therefore, as the tuft diameter is reduced, the inertial force de-

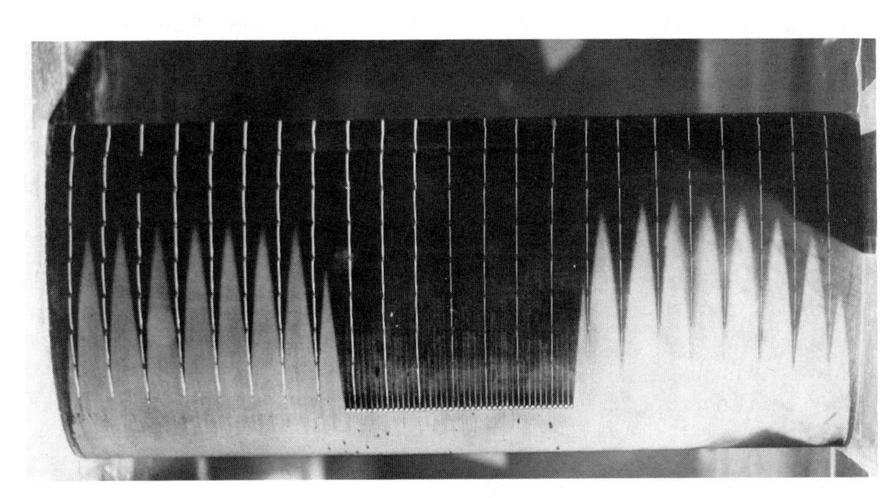

Fig. 10 Minituft behavior in laminar and turbulent boundary layers.

creases at a greater rate than the aerodynamic force. Consequently, at some particular tuft size, the centrifugal loading will become insignificant.

Furthermore, the tuft airloads depend on the dynamic pressure, the rotational component of which is proportional to the rotational rate squared times the radius squared Ω^2R^2. The inertial loads due to rotation depend on Ω^2R. Thus, for a given rotation rate, the ratio of aerodynamic to inertial loads should increase as the radius is increased.

From this reasoning, it seemed probable that for some combination of tuft diameter, rotational rate, radius, and Reynolds number, very small tufts such as those used in the fluorescent minituft method would indicate the surface flow properties without excessive biasing due to centrifugal acceleration. These effects are difficult to determine analytically, so actual experimental trials were required to demonstrate the hypothesis.

Confirmation of the expected minituft response on small-scale propellers was demonstrated in simple experiments by comparing the behavior of tufts of two different sizes on the same-scale propeller. If both tuft sizes indicated the same response, and the reasoning that the smaller tuft would have a more accurate response were correct, then the larger of the two tufts must be following the flow correctly as well.

An exhaustive calibration of size versus radius effects has not yet been established, but sufficient experience is in hand to permit some tentative recommendations. For a propeller diameter of 10 in., tuft material having a diameter of 0.0007 in. seems to be required. Tuft material having a diameter of 0.0019 in. appears to be an accurate indictor for a propeller diameter of 30 in. or more. Propellers larger than 60 in. can employ tuft material in the 0.006-in.-dia. size range.

Results of an attempt to calibrate these effects for larger tuft sizes are presented in Fig. 11. A range of tuft sizes were installed on a 13-ft-dia. CV-440 airplane propeller and examined during a ground runup. The six different tuft materials and sizes are noted on the figure alongside each of the three propeller blades. All the tuft materials, except for the $\frac{1}{2}$-in. tape, appear to show similar behavior at the propeller tip. At the one-half radius station, however, only the smaller of the two materials, the sewing thread and knitting yarn, are indicating the same, and presumably correct, flow direction.

The results of these trials led to the conclusion that for diameters greater than 13 ft, any of the evaluated tuft materials would be relatively unaffected by the effects of rotation. However the $\frac{1}{4}$-in.- and $\frac{1}{2}$-in.-wide tape materials exhibited excessive self-excited whipping behavior.

6 Static Electric Charges A peculiar problem with the smaller fluorescent minitufts is caused by the presence of static-electric charges that interfere with the free response of the minitufts to the flow. These charges apparently develop as a consequence of handling the model and minituft material. The air loads on the minitufts are not enough to overcome the attractive forces associated with these

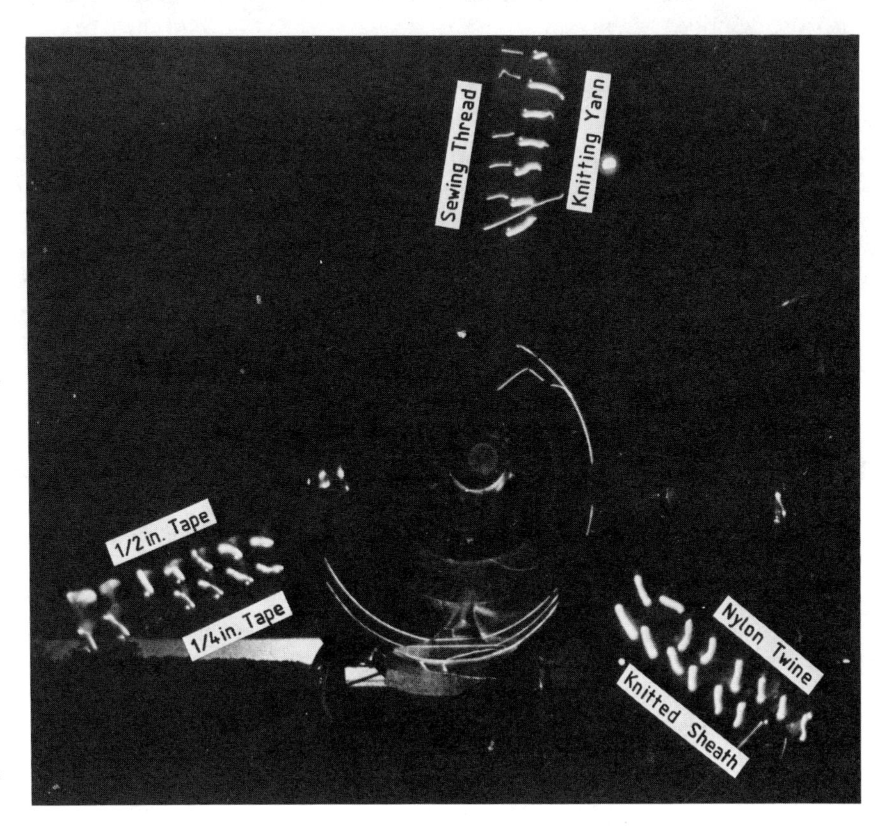

Fig. 11 Behavior of various-size tuft materials on rotating propeller.

charges. They are not always present, but can be easily detected by blowing on the tufts by mouth. If static charges are present, the minitufts will remain stationary on the surface. The attractive forces associated with these charges are large enough to resist minituft airloads at dynamic pressures as large as 100 lb/ft^2.

The easiest way to eliminate the static-electric charges is to treat the minitufts and the model surface with any of several home laundry antistatic products. These are paper sheets impregnated with special chemicals, and they are ordinarily used in clothes dryers. For minituft applications, the antistatic sheets are rubbed lightly over the minitufts and the model surface after the tufts have been installed and cut free. The response of the minitufts to light blowing by mouth is observed as the surface is lightly rubbed with the antistatic sheets. At first the minitufts will stay stuck to the surface, but gradually they will become more mobile. The treatment should be continued until the minitufts are seen to be standing up off the surface and responding to light puffs of air.

D Attachment Procedures

Attachment of tufts to a model surface is that aspect of the method which requires the greatest effort. It is also the most tedious. The sad fact remains, that if high spatial resolution is important, then literally thousands of individual tufts must somehow get attached to the test surface. There are, however, some procedures that can help ease the effort, and they are described here.

1 Individual Flow Cones and Tufts The most obvious attachment method, which could be called the "brute-force" method, is to attach each individual tuft with an individual piece of adhesive tape. This is the procedure usually used with flow cones, since these devices are, by nature, individual elements. Tufts, on the other hand, usually are formed, in one way or another, from a continuous length of material. Whatever the method, it is perhaps difficult to appreciate the significance of this consideration until one has personally attached several thousand tufts or flow cones to an airplane surface, or until one has impatiently watched a crew of bored mechanics doing the task.

The required strength of the flow cone attachments depends on the test conditions. For general aviation airplanes and slower vehicles such as cars and trucks, it is usually sufficient to attach each flow cone to the surface with a small piece of vinyl adhesive tape. The only important factor is that the tape should leave only a minimal distance between the apex of the cone and the edge of the tape. If too long a free length of string is exposed, the flow cone might develop an unstable oscillatory motion.

For high-speed airplanes, especially where high-speed separation is present, as on a maneuvering fighter, a much more substantial attachment is called for. This might require that a double layer of tape be used with the attachment string folded over the first tape before the second tape layer is placed. Also a strong tape, such as 0.008-in.-thick aluminum tape, might be required to resist tearing at the trailing edge of the tape attachment when the flow cones are pulled forward in regions of strongly separated reversed flow.

Large conventional tufts, as used on full-size airplanes, are usually attached in the same manner. This procedure can require a substantial investment of man-hours for a complete installation.

2 Tuft Board As the tuft materials and test surfaces have become smaller, several stategies have been developed to speed up tuft installations. One frequently used method is to make up lengths of adhesive tape with the tufts attached to them, using a fixture known as a tuft board. The tape can then be applied to the model surface much more efficiently.

A tuft board is basically a large, flat surface, such as a sheet of plywood, with a row of pins or small nails spaced along two opposite edges. The tuft material is stretched back and forth between these pins as on a loom. The tape either is placed, adhesive side down over the tuft material, or is fastened, adhesive side up to the tuft board, prior to stringing the tuft material. After the tuft material is fastened to the tape adhesive surface, the tufts are cut with a sharp razor blade.

With tuft tape, the tufts might stick out along only one edge of the tape, in which case the tape would be attached to the model surface in a wing-spanwise direction. An alternate arrangement is for the tufts to stick out on both sides of the tape and have the tape installed in a streamwise direction. The tufts would then have to bend approximately 90° to become aligned with the flow direction.

3 Adhesive Tape Attachment Another installation technique is to temporarily lay continuous lengths of the tuft material over the test surface at regular intervals with the desired tuft spacing, and then place the adhesive tape over the tuft material, sticking it to the surface. With wing surfaces, it is sometimes convenient to wrap the wing with a continuous length of tuft material in a helical fashion. After the adhesive tape is applied, the tufts must be cut free. Since the surface might be too delicate to permit cutting with a razor, this step might require use of small scissors, causing the work to be somewhat tedious.

One problem with attaching tufts using continuous lengths of tape, especially if the tape is oriented perpendicular to the flow, is that if the tape is not well adhered to the test surface, it can develop local regions where the flow peels it away from the surface in progressively larger amounts. In any case, use of tape for attaching tufts to a model surface entails a disturbance due to the thickness of the tape. For large-scale installations, as on a full-size airplane, this is usually insignificant, unless the tape were to peel off. On small wind tunnel models, the thickness of the tape, which is seldom less than 0.004 in., can be a substantial disturbance.

4 Fluorescent Minitufts For the fluorescent minituft method, a special attachment procedure was developed using liquid adhesive. This process permits attachment of minitufts with a very small disturbance but, in some cases, requires a very careful preparation of the model surface to achieve proper adhesive bonding. The adhesive material used is a nitrocellulose lacquer with a relatively large volatile solvent fraction. It can be applied over the minituft material in a conveniently large drop that will, upon drying, shrink to a thin patch, much thinner than the minituft diameter itself.

The bonding chemistry is basically the same as that in the adhesion of paint. If a surface can be painted, then minitufts can be applied. If the surface has already been painted with good adhesion, the minituft glue can be applied over the paint with almost no preparation beyond a simple solvent wipe. However, bare metal surfaces can sometimes require a much more rigorous cleaning procedure than required for painting. This is probably because of the very small surface bonding area of each minituft attachment drop. Some stainless steel alloys, in particular, have proven to be extremely resistant to minituft glue adhesion.

Because of possible problems in this process, a detailed procedure for the most difficult case is presented below in a step-by-step fashion. Not all these steps are usually required, but that should be determined for a particular case before deciding to bypass the procedure.

1. Wipe the model surface with solvent to remove grease and oil. Use clean paper wipes or tissues, rather than shop rags. Preferred solvents are the chlorinated hydrocarbons such as trichloroethylene or trichloroethane. Methyl ethyl ketone, acetone, or Freon are less suitable because of their rapid evaporation rate; evaporation makes it more likely that the contaminates will be redeposited on the surface. Do not use alcohol.
2. If possible, lightly abrade the surface with Scotch-Brite abrasive pad or aluminum oxide paper. Do not use silicon carbide paper on aluminum. Care should be taken to not compromise the aerodynamic smoothness requirements of the model.
3. After abrading the surface, thoroughly rinse with solvent and clean tissues. Wipe the area one time only with a clean section of the tissue. Wiping more than once with the same tissue redeposits contaminants. Continue wiping until the tissues show no stain. Avoid letting the solvent evaporate completely before it can be wiped.
4. Apply Alodine solution* to the surface with a swab or brush. Allow it to remain wet for 3 min. Any areas that resist wetting should be scrubbed with an abrasive pad wet with the Alodine solution.
5. Rinse the Alodine from the surface with clean water. Dry the surface with clean tissues. Be careful to limit skin contact with the Alodine and to wash thoroughly after use.
6. Cover the cleaned area with paper to avoid contamination until after the minituft adhesive is applied.

After the surface preparation is complete, the minituft material can be placed across the surface with a regular spacing corresponding to the minituft length. This can be as separate lengths of the material or one continuous length wrapped around the surface, as shown in Fig. 12. Care should be taken to avoid stretching the monofilament material. Illumination of the work area with ultraviolet light is helpful in working with the small-diameter material.

Simple fixtures can be very helpful in speeding up the

*The Alodine material mentioned in step 4 is a chromate conversion solution that promotes adhesion of paint to the surface. It is available from aircraft paint suppliers. In this application a small quantity of wetting agent such as Kodak Photo-Flo should be added to make approximately a 1% solution concentration.

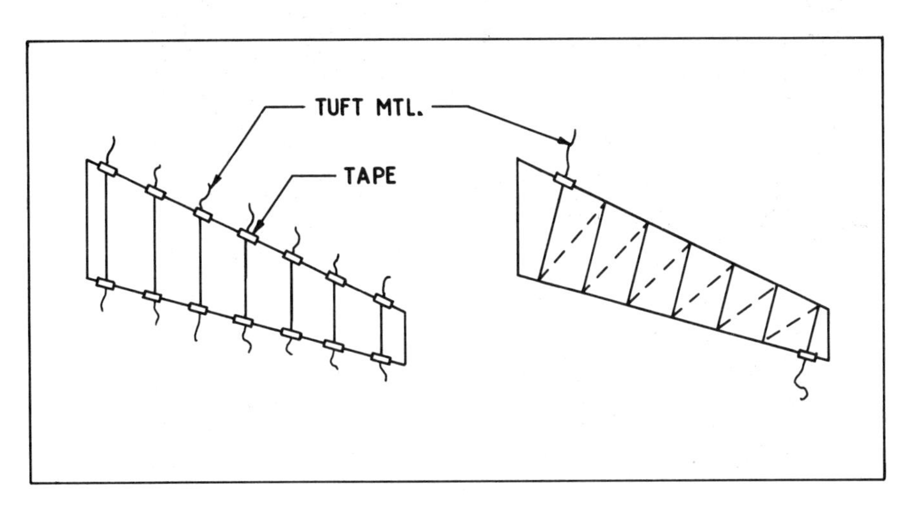

Fig. 12 Sketch of minituft installation procedures.

layout process. For example, Fig. 13 shows a tail panel with notched plates temporarily attached to the leading and trailing edges with double-back adhesive tape. The notches are cut in the plates with an interval equal to the minituft spacing divided by the tangent of the sweep angles.

Figure 14 shows another fixture for layout of minitufts on a propeller blade surface. In this case, the minituft material is strung between the pins on the two posts before the model surface is placed in the fixture.

The minituft adhesive is prepared as a solution of nitrocellulose cement, solvent, and a small quantity of paint pigment. The cement can be any commonly available model airplane construction cement or DUCO Household Cement. It should be mixed in proportions of 1 : 1 with solvent such as acetone or methyl ethyl ketone.

A small quantity of paint pigment should be added to the solution to make it more easily seen during the installation process and to obscure the minituft material where it is attached to the model surface. Without the pigment, the fluorescent minitufts show through the adhesive drop when photographed, usually with an orientation different from the minituft direction. This tends to add a confusing element to the data record.

The adhesive should be applied over the minituft material using a hypodermic syringe. The syringe should be half filled with the adhesive solution. Then a small bubble of air should be pulled into the syringe. By compressing the air bubble slightly with the syringe plunger, the flow of the adhesive can be set to a convenient rate to permit rapid drop placement over the minitufts.

A 22-gauge needle is recommended for the glue-dispensing syringe. The sharpened tip must be broken off with pliers for safety as well as to avoid cutting the minituft material. If a needle having a small enough diameter is not available, it will be difficult to control the adhesive flow rate. In that case, the needle can be closed off by bending it through two sharp 120° bends to form a zigzag. This collapses the needle about the right amount.

After the adhesive has dried, the minitufts should be cut free with a sharp razor or melted apart with a hot soldering iron. The adhesive ordinarily will dry to the point where it can be cut in about 30 min.

On some occasions a peculiar difficulty has been experienced with the adhesive solution described above. If the solution is used more than several days after it is mixed, a subtle chemical change seems to occur. The material looks and smells the same, but it dries very much slower. Sometimes more than 24 h have been required for the adhesive to become hard enough for cutting the minitufts. On this account, it is good practice to discard the mixed adhesive solution at the end of an application period.

E Viewing

The mechanism by which the tuft response is interpreted is visual. Above all else, the tufts must be visible. The limitations imposed by this requirement pertain to all the types of tufts discussed here and all the recording schemes em-

Fig. 13 Minituft installation fixture on wind tunnel model tail panel.

Fig. 14 Minituft installation fixture for wind tunnel model propeller blade.

ployed. Three basic categories of viewing systems are discussed in the following sections.

1 White Light Imaging This is the viewing system most familiar to us. It entails illuminating the scene with light at a visible wavelength and observing the patterns of reflected light. Those objects that appear dark are the ones that absorb more light than their surroundings. Conversely, bright-appearing objects are more effective in reflecting light than surrounding objects. The illumination energy must be in the visible wavelengths, and it can be natural sunlight or artificial light.

The property that makes the tufts visible is contrast. The tufts must appear to have a brightness different from the surrounding parts of the scene. Therefore the first requirement of visibility is that the tufts have a color different than the surface to which they are attached.

There is a logical question about whether it is better to have dark tufts on a light surface or vice versa. For wind tunnel models, the background color can usually be chosen for whatever arrangement is decided as optimum. For large vehicles, such as airplanes, it is much more awkward to have to paint the surface for a special purpose. Therefore the choice of tuft color is usually dictated by the existing test surface color.

Tufts are usually observed from such a long distance that they are perceived as small objects. Each tuft occupies a very small fraction of the total field of view. In this situation, it is usually advantageous for the small object to be brighter than the background. The observer then perceives light coming from the tuft and compares it to a relative absence of light coming from the surrounding area.

In a photograph, the original image is formed as a negative where the bright parts of the image act to darken the negative material. As the exposure time, or brightness, is increased, the dark regions on a negative tend to grow larger as the light falling on it fogs the adjacent negative material. In this way, bright object images can be made larger than their geometrically scaled size on the negative.

For a white light imaging situation, there is a definite optimum level of illumination. Too much or too little light degrades the visibility. Many other factors can interfere with the image quality. Chief among them is glare. Wind tunnel models are frequently made of aluminum or steel and left unpainted. Glare can be a severe problem with such material as a background surface, especially since the model has a large range of surface slopes. This makes it likely that at some condition a specular image of the light source will appear on the surface.

All these considerations are responsible for making this aspect of flow visualization potentially troublesome.

2 Fluorescence Imaging The easiest way to make a small object visible is to make it bright. The best example of this principle is the experience of observing stars. These objects are truly small in that their size, expressed as a fraction of the field of view, is below the limit of resolution for any sensor. Yet stars are extremely visible because their brightness effectively magnifies the image size well beyond their geometrically scaled size.

Fluorescence imaging employs this principle by using ultraviolet illumination to increase the brightness of the objects that are fluorescent, compared to the surrounding area that is not fluorescent. The scene is illuminated at an ultraviolet wavelength that is absorbed by the fluorescent material and then re-emitted at a longer wavelength in the visible portion of the spectrum. Human vision is not very sensitive to the usual ultraviolet illumination wavelength, but black-and-white photographic film is very sensitive to it. A filter must be employed on the camera to block the illumination energy that is reflected from the scene and to pass light at the fluorescence wavelength.

This selective filtration process permits the illumination energy to be increased, almost indefinitely, without increasing the background brightness. Thus, fluorescent minituft material of any size can be made visible, with high contrast against a nonfluorescent background surface, merely by increasing the illumination energy.

This is in distinct contrast to white light imaging, where increasing the illumination energy increases the brightness of all parts of the scene, although not necessarily at the same rate. Fortunately, the required level of ultraviolet illumination energy is conveniently available from any number of lighting systems, as discussed in Sec. III.G.

The absolute brightness of fluorescent minitufts tends to be less important than their relative brightness, compared to the surrounding nonfluorescent background. The process is easiest if the ambient light level can be made completely dark. Inside the laboratory, as in a medium-sized wind tunnel, it is quite easy to merely turn out the lights. In large wind tunnels that is not always easy to do. In flight, of course, the ambient light level is controlled by the time of day as well as the extent of cloud cover.

In most wind tunnels, even with all the lights turned off, there is often some residual lighting, such as from building skylights, that cannot be completely extinguished. It is seldom necessary to provide complete darkness, however. In fact, some low level of background illumination is often very desirable to help provide a dim image of the model surface.

In flight, it is usually impractical to provide sufficient ultraviolet illumination to make fluorescent minitufts visible against a surface exposed to strong sunlight. However, on shaded portions of the aircraft and at short distances, it may be possible. Usually fluorescent minituft imaging on any vehicle outside the laboratory entails nighttime operations.

The background color of the surface on which fluorescent tufts are to be installed plays an important role that may seem surprising at first glance. The best background color for fluorescent tufts is white. Bare metal is almost as good in terms of fluorescent brightness, but has the disadvantage of producing specular glare highlights. There is a tendency to assume a black background is required because that is the appearance of the data images. With fluorescence photography, however, a black background signifies only a nonfluorescent background and a low level of ambient light.

The effect of a black background on a fluorescent tuft image is to reduce the brightness of the tufts by a factor of 2, or one f stop. As the tufts absorb ultraviolet radiation and convert it to fluorescent radiation, they emit light in all directions. The fluorescence that is directed toward the surface is either absorbed or reflected, depending on the color of the surface.

3 Reflective Surface Imaging Flow cones offer another way of providing enough object brightness to make them visible against a relatively bright background. The smooth and rigid surface of the flow cone can be coated with a special reflective material, an adhesive film manufactured by the 3M Company, similar to the material employed in highway signs. This film has the optical property that any light falling on it is reflected back in the direction of the source within a small angle range.

The reflective coating of flow cones makes it possible to make them bright enough to be strongly visible in bright daylight, even at distances of hundreds of feet or with very wide-angle optics. This process requires only that an illumination source with enough intensity be located immediately adjacent to the camera. This allows some of the light returned to the source to spill over into the camera lens.

This arrangement of the light source presents a problem when viewing the scene through a window. It is difficult to avoid glare from the light source reflecting into the camera lens. If the scene must be viewed through a window, such as from on board a pressurized-cabin airplane, a special window with two separate transparent panes situated close together should be installed. Another approach would be to place the light source outside the window so that only the camera looks through the window.

Sunlight can be used as an illumination source for air-to-air viewing of reflective flow cones. The chase airplane is positioned in line between the sun and the target airplane. Reflective flow cones are extremely visible with the eye, as well as with film or video cameras, under these conditions.

F Recording

The results of flow visualization experiments are data, just as much as balance readings or pressure measurements. As such, they should be recorded with the same standards of accuracy and reportability. Because the data are in a graphic form, more thought must be given to this aspect of the experimental process.

1 Memory The human eye–brain combination is an extremely powerful interpretive visual system. It can discern subtle details that the most sensitive photographic imaging is hard pressed to record. Unfortunately, it is not so well suited to recording such information. As a result, the data tend to be rather ephemeral, and they can also be subject to prejudical interpretation. One tends to see what one is expecting to see.

2 Sketches Sketches often become the long-term storage medium for data originally recorded by human memory. Needless to say, there are many opportunities for data to be lost or distorted in this process. It also depends on a certain amount of artistic skill, something many of us are lacking.

Sketches are also used to interpret other visual data records that are of too poor a quality to be presented by themselves. In other words, when the photographs did not come out, but the results were obvious anyway, a simple sketch should be able to get the point across.

However it is rationalized, resorting to this medium is most often an admission of failure in some aspect of the experiment.

3 Photography Photographic recording of flow visualization data must be considered the baseline recording technique. It is accurate, sensitive enough for most applications, and reasonably convenient.

The factors that must be considered in setting up a photographic data-recording system include the choice of camera size, film type, logging the test conditions, presenting the results, and archiving the data. Selection of all these parameters is beyond the scope of this discussion, but this writer has developed some opinions that might serve as a useful starting point.

For conventional tufts or nonreflective flow cones, the photographic record must provide the maximum possible resolution because the method tends to suffer from an inherent lack of contrast. Therefore, the choice of camera becomes quite important in terms of the film size. It is probably true to say that the best resolution is achieved with the largest film size, but anything larger than a 4- by 5-inch camera is very impractical. However, because of operational convenience, a medium-sized roll-film camera of 70-mm-square or 60- by 90-mm format is probably the largest size convenient for routine use.

Fluorescent minitufts and reflective flow cones produce images of much greater contrast than conventional tufts. As a consequence, 35-mm cameras offer more than sufficient image quality with the additional advantage of a vast selection of available equipment.

One particular camera parameter that sometimes is very important is the maximum flash-synchronization shutter speed. Most 35-mm cameras use a focal plane shutter with an operation time of 1/60 s, or 17 ms. That is the fastest shutter speed for which the shutter curtain is ever fully open. Therefore, if a flash illumination system is used, even though the flash duration may be 1 ms or less, the shutter must remain open for the full 17-ms period. If the photograph is being taken at a high ambient light level, such as outdoors in the daytime, it may be difficult to control the background exposure.

Black-and-white film has until recently been the obvious choice for most applications. It provides the highest sensitivity and the most convenient processing. However, modern color films are nearly as fast, and the recent phenomenon of 1-h film-processing has made it possible to greatly shorten the turn-around time for photographic data acquisition.

4 Fluorescence Photography Fluorescence photography [25] is a distinct process absolutely vital to the fluorescent minituft technique, so it is appropriate to describe it in substantial detail. Only through this process can the very small minitufts be made visible.

The process uses selective light filtration. The scene is illuminated with light at ultraviolet wavelength. This is usually centered on a wavelength of 360 nm. Most often the light source is a xenon flash lamp with a wide-spectrum output. The lamp output is passed through a narrow bandpass filter known as the exciter filter. It has the appearance of a dense black glass, and it filters out most of the lamp output in the visible part of the spectrum.

The ultraviolet illumination excites the fluorescent dye in the minituft material to radiate in the visible part of the spectrum between 400 and 500 nm. The light reaching the camera is composed of both the visible fluorescence radiation and the reflected ultraviolet radiation. Panchromatic black-and-white film is very sensitive to ultraviolet light so the reflected ultraviolet must be blocked at the camera lens with a low-pass filter known as the barrier filter. This has a pale amber color.

Figure 15 summarizes the optical properties of some of the available exciter and barrier filter materials. The exciter filters shown were produced by Corning Glass Works [26] as molded blanks or polished sheets in various sizes. For illuminating purposes, it is not usually necessary to go to the additional expense of a polished filter.

In the final stages of preparing this chapter, it was discovered that Corning Glass Works no longer manufactures this material. Similar exciter filter material is available from Kopp Glass, Inc. [27] and Hoya Corporation [28]. The recommended filter material, Corning 9863, appears to be most similar to Hoya U-330.

The barrier filters are from the Kodak Wratten series [29] and are available from commercial photographic supply dealers as 3-inch-square unmounted gelatin sheets.

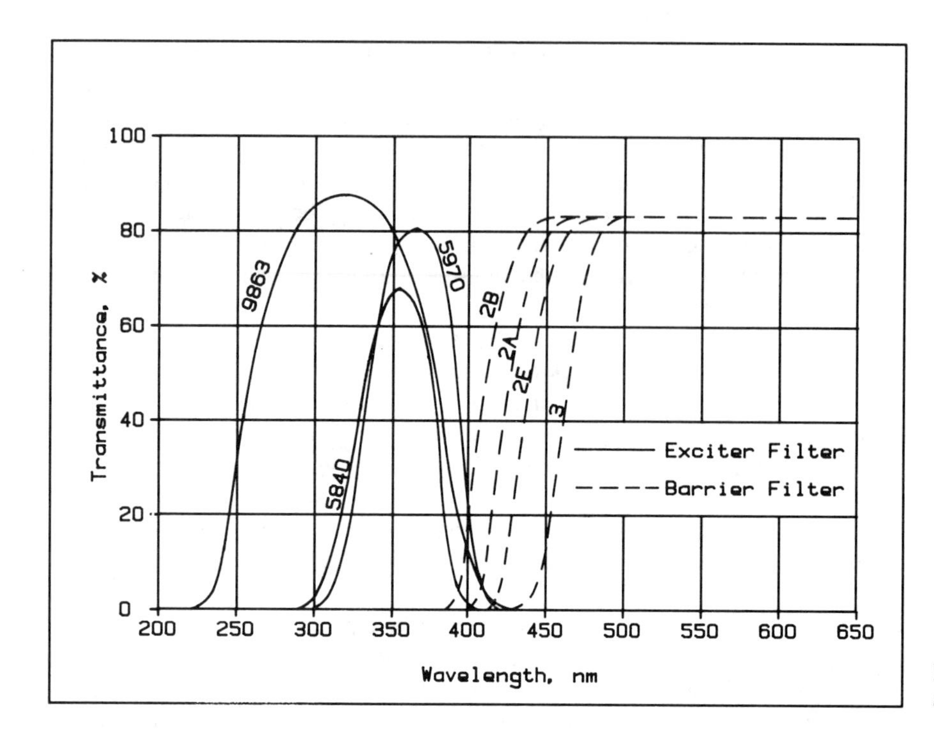

Fig. 15 Optical properties of exciter and barrier filters.

The choice of filter combinations is difficult to specify beforehand since it depends on many factors specific to a particular application. The choice of exciter filter depends on the illumination intensity available, while the barrier filter should be selected to attenuate the minituft material fluorescence spectrum as little as possible. The extent of the overlap between the two filters controls the amount of visible glare in the photographs.

The fluorescence spectrum of the minituft material is shown in Fig. 16. When excited with ultraviolet radiation centered on a wavelength of 360 nm, it displays a peak fluorescence radiation at 435 nm. This is close enough to the exciter filter band-pass wavelength to require care in selecting a barrier filter without excessive attenuation of the fluorescence.

For example, an exciter filter of Corning 5970 glass has the highest transmission at 360 nm, but the transmission curve extends relatively far into the visible spectrum. This

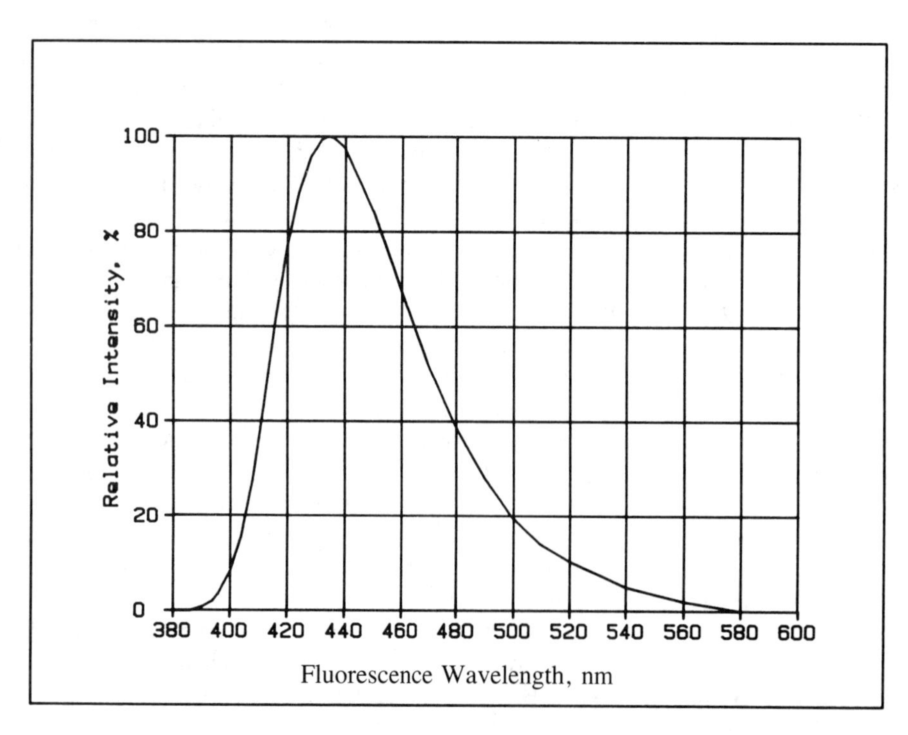

Fig. 16 Minituft dye fluorescence spectrum.

can result in substantial reflected light at the fluorescence wavelength. The properties of this exciter filter make it more difficult to separate the excitation wavelength from the fluorescence, even though this exciter filter can produce the greatest tuft brightness. For tufts applied to a bare metal surface, there would likely be substantial glare that would interfere with the tufts images. This exciter might be best for illuminating tufts against a black background.

By contrast, an exciter filter made of Corning 5840 has a greatly reduced peak transmittance at 360 nm, but a much narrower width to the transmittance curve so that it would produce very little interference with tuft fluorescence. This material might be best for illuminating tufts on a shiny metal surface.

Figure 17 shows visual results of several filter combinations. The best choice tends to be a tradeoff between excitation brightness and glare control.

Selection of an optimum combination is best done with trial photographs for a particular set of circumstances. The combination found best for most applications utilizes an exciter filter of Corning 9863 glass and a Wratten 2E barrier filter.

5 Still versus Cine Photography Sometimes motion pictures are considered for recording tuft experiments, based on the notion that the main indication of separation is the tuft motion. A cine record is nothing more than a time series of still photographs, however. Unless the frame rate is fast enough, moving tufts appear blurred in the motion picture frames, just as they do in still photographs.

Another type of tuft-imaging consideration might justify a cine record, however. That is to depict the unsteady onset of separation or the intermittent nature of the flow. However, that kind of behavior tends to be random. Cine recording would show the correct time history of the separation process, but those details might not be significant. Still images could just as well depict such behavior by recording a number of images at random times.

The greatest difficulty in making cine records is that the lighting must be continuous. This is not particularly difficult for imaging conventional tufts or flow cones, but becomes very awkward for the intense ultraviolet illumination required for fluorescent minitufts. The other disadvantages of a motion picture record include limited resolution due to the small image size, inconvenient film processing, the need for projection viewing, and the difficulty of making hardcopy images for report documentation.

6 Video Enough recent progress in electronic imaging has been made to anticipate that video imaging will continue to make irresistible progress toward displacing photographic recording. That is not to say that it will ever completely replace photographic recording. More likely, one application after another will succumb as the properties of video are progressively improved. Industrial imaging, such as that involved in flow visualization is a natural application for video because of the tremendous potential advantage of the instant turnaround for the data records.

The principal disadvantage of video imaging, compared to photographic film is, of course, the reduced spatial resolution of the medium. The types of images produced by fluorescent minitufts and reflective flow cones, however, are very tolerant of the reduced image resolution experienced with the video format. These images tend to be very high in contrast and to require only enough spatial resolution to resolve the orientation of the indicators. There is no great loss of visual information if the minitufts appear as fuzzy blobs, as long as the blobs can be discerned to be pointing in some direction.

In applying the principles of fluorescence imaging to video recording, the unique properties of video photosensors need to be understood. First, most illumination sources are very rich in infrared radiation and most exciter filter materials are relatively transparent to those wavelengths. Second, most video photosensors are very sensitive to infrared radiation. Therefore the appropriate barrier filter must include an infrared blocking filter. Such material is not readily available from the traditional photographic equipment sources, but can be produced by modern interference-filter manufacturing techniques, although at considerably greater expense. Modern industrial video cameras are frequently supplied with such a filter, usually called an infrared cut filter.

There may be a tendency to think of video imaging only in terms of moving picture, or cine, imaging. That certainly is the more prominent method available with the medium. However, the criticisms of cine imaging offered in Sec. III.F.5 still pertain when a video rather than a film cine camera is utilized. It still requires a continuous light source, and the data must be viewed as a real-time history presentation. Unless the actual temporal dimension is of significance, this medium may be much less convenient than still imaging.

Video images can, however, be recorded in a still format. In addition, the still images can be acquired with a short-duration flash illumination system. This process requires a still video recorder. Two, basically different, still video-recording systems are available, analog and digital. Analog video still recorders utilize a 12-in.-dia. rotating magnetic disk medium and are able typically to record 100 video frames per side. The images can be accessed instantly in any order. Video imaging in this manner is sometimes called video snapshot recording.

Experience with fluorescent minituft recording using a video disk recorder and flash illumination system has shown how the system can be very convenient. For example, a typical wind tunnel run might consist of taking data at a series of angles of attack from 0 lift to beyond the maximum lift angle. This might entail 20 or 30 test points and require several minutes of real-time data taking. With a video snap-

(a)

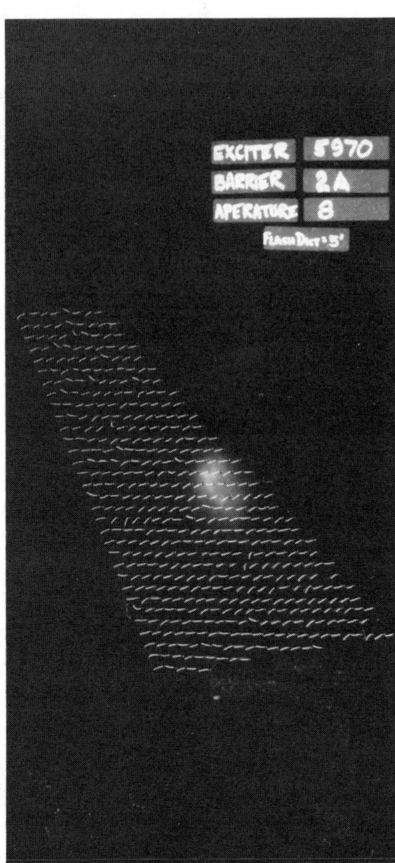

(b)

Fig. 17 Effects of exciter and barrier filter combinations on minituft visibility: (*a*) 5970 Exciter, 2B barrier filters; (*b*) 5970 exciter, 2A barrier filters; (*c*) 5970 exciter, 2E barrier filters; (*d*) 9863 exciter, 2B barrier filters; (*e*) 5840 exciter, 2B barrier filters.

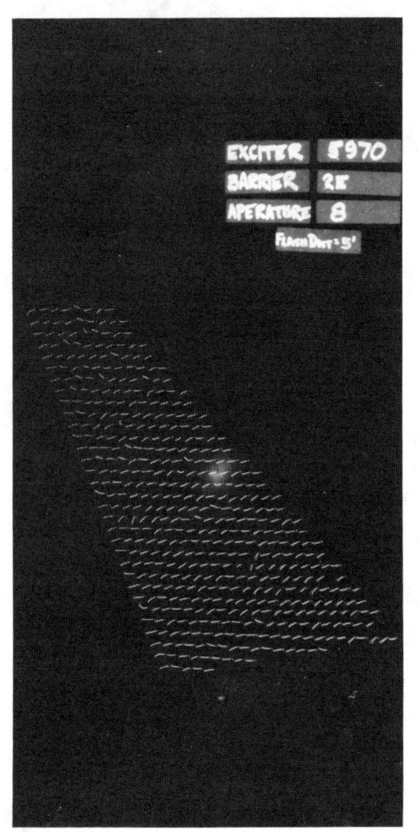

(c)

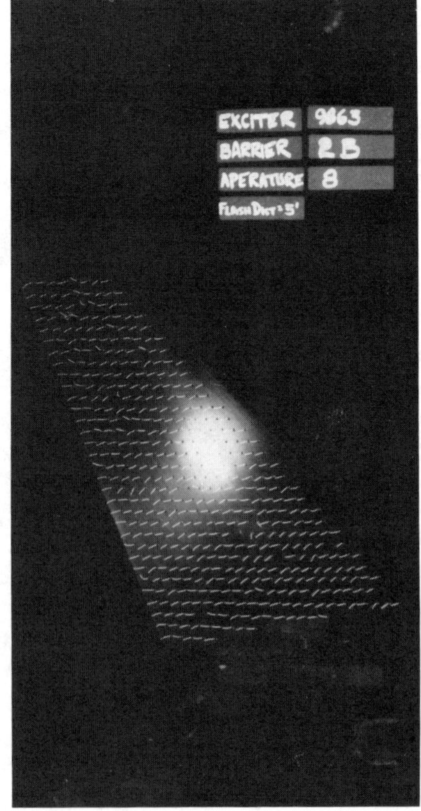

(d)

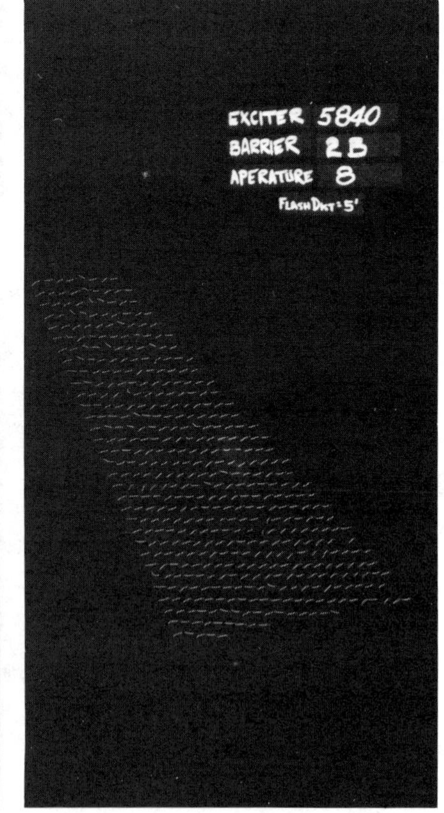

(e)

shot recorder, one image at each angle of attack can be taken and stored on the disk recorder. After the entire sequence is recorded, the series can be played back in a very rapid sequence. In this way, the series can be replayed in a much shorter time than if a real-time video tape had to be played. In addition, the few frames that are the most interesting can be displayed for indefinite periods while the remaining images can be rapidly bypassed.

7 Computer Graphics Systems The second video still-recording system, just now becoming available, is based on a digital frame grabber. This type of system digitizes the video image and stores it in a solid-state memory that is interfaced to a computer. In particular, such systems are now available that are based on microcomputers such as the IBM PC and other compatibles. The cost of digital video frame-storage systems are substantially less than the equivalent analog recorders, but tend to be much slower, in terms of time required to access a stored image.

The really exciting potential for such digital video frame-storage schemes is realized when image processing is incorporated into the system. This holds the prospect of significantly improving the quality of video images by software procedures for contrast enhancement and low-light-level image accumulation. Simple procedures can be used for taking the difference between two images to detect, for example, only those features that show motion or are stationary. As development of microcomputer and digital video camera hardware continues to progress at an accelerating pace, these approaches to flow visualization imaging are certain to become more common, if not predominant.

G Ultraviolet Illumination Systems

Ultraviolet illumination is the key to fluorescence imaging of minitufts. Until recently, there has been a relative lack of industrial applications for ultraviolet illumination systems. Certainly, experimental aerodynamic applications have not been sufficiently important to support a market in this area. Recently, however, the use of ultraviolet illumination for the rapid curing of adhesives, resins, and inks has been expanded. These applications are more widespread throughout the general industry, and they offer the potential for a favorable fallout in the aerodynamic testing industry.

It should be noted that the ultraviolet region of the spectrum is generally considered to extend from a wavelength of less than 200 nm to about 380 nm. The visible blue spectrum starts at about 400 nm. The ultraviolet portion of the spectrum is further subdivided into the short-wave region, extending up to a wavelength of 300 nm, and the long-wave, or black light, region. Fluorescent minituft dyes absorb radiation in the blacklight region, centered on 360 nm, as do most other common fluorescent dyes. This is fortunate, since shorter wavelength ultraviolet radiation is rather hazardous. It causes skin burns, promotes skin cancer, causes painful retinal damage, and generates ozone. Black light, by comparison, is rather innocuous. Nevertheless, any unusual radiation should be avoided as much as possible, especially when it is of high intensity, as is often the case for flow visualization imaging applications.

Figure 18 shows several different illuminating systems for fluorescent minituft applications. Sitting on the desk top are, respectively from left to right, a high-power xenon flash lamp with an electrical pulse energy of 2000-W-s and a duration of 4 ms, a 400-W continuous high-pressure metal-halide arc lamp, and a 400-W continuous mercury arc lamp with standard screw base and black glass envelope. Standing against the front of the desk is a fixture with four 40-W BLB-40 fluorescent tubes. The two systems on the floor in front of the desk are, respectively left to right, a xenon flash lamp with 500-W-s pulse energy and 10 μs duration, and a continuous medium pressure mercury arc lamp with 5000-W power. These lamps are covered in detail in the following sections.

1 Continuous Ultraviolet Illumination Continuous sources of ultraviolet radiation are almost all based on some cycle of mercury vapor. The most common source uses a low-pressure mercury arc similar to standard fluorescent 40-W tubes. The gas-discharge plasma inside the tube generates considerable short-wave ultraviolet radiation, which is converted to the final output wavelength by a phosphorescent coating on the walls of the tubes. Standard white light fluorescent tubes use a phosphor that generates light in the visible part of the spectrum, while the black light tubes use a different phosphor that produces ultraviolet radiation in the 340- to 380-nm range.

These lights, designated BLB-40, are very convenient to purchase and use as they can be installed in standard fluorescent tube fixtures and are readily available in most lighting supply stores. Since this cycle generates black light by converting a shorter wavelength radiation with a chemical phosphor, it is relatively efficient. However, the intensity is rather low, being spread out over a large surface area. It is somewhat difficult to concentrate enough light to make the fluorescent minitufts visible.

Four BLB-40 bulbs in a reflector fixture will produce, at a distance of 4 ft, a radiation intensity at 360 nm of about 100 $\mu W/cm^2$. This is about the minimum intensity needed to produce a usable image of 0.0019-in. dia. minitufts using a sensitive video camera. It is about $\frac{1}{10}$ the intensity required to produce a comfortable image for direct eye viewing. A photograph would require an exposure of several seconds' duration.

Various medium-pressure mercury arc lamps are produced commercially in sizes up to about 5000 W. These lamps are gas-discharge arcs with a broad-band spectrum extending from the ultraviolet through the visible and into

Fig. 18 Various ultraviolet illumination systems.

the infrared region. Flow visualization imaging requires that the visible radiation be largely blocked with an exciter filter, as discussed in Sec. III.E.2.

A 5000-W medium-pressure mercury arc lamp like that shown in the bottom front of Fig. 18 will produce an intensity of about 2000 $\mu W/cm^2$ at a distance of 4 ft. This is a very useful intensity for video imaging of the 0.0019-in.-dia. minitufts, and it permits very comfortable direct eye viewing. Photographic exposure times are still over 1 s.

A newly available illuminating system is shown on top of the desk in the center of Fig. 18. It uses a high-pressure metal-halide lamp cycle. The model shown uses 400 W of electrical power and, with the separate black glass exciter filter shown, produces an illumination intensity of 1500 $\mu W/cm^2$ at a distance of 4 ft. This compares favorably with the much larger and more expensive 5000-W medium-pressure mercury arc lamp also shown in Fig. 18.

2 High-Power Flash The most useful light source for fluorescent minituft imaging is a high-power photographic flash lamp of the type developed for commercial studio applications. These systems generally consist of a power supply the size of a small suitcase and one or more flash tubes in a lamp fixture having a reflector. The light they produce covers the entire visible spectrum and extends relatively far into the ultraviolet and infrared regions. The radiation is produced by an electric discharge through an arc tube filled with xenon gas. An exciter filter is required to absorb the visible radiation produced by this type of lamp.

Many different flash lamp systems are commercially available with electrical energy levels in the range of 1000 to 5000 W-s. Exposure times for these lamps usually lie in the range of 1 to 5 ms, and they often allow for fast recycle times as short as once per second. The lamp system shown on top of the desk on the left side of Fig. 18 is a typical example.

The photographic sensitivity when using a flash lamp is usually expressed as a flash guide number. The guide number, divided by the distance from the lamp to the subject, gives the camera aperture f stop. The guide number is specific to a constant set of parameters, including film speed, developing process, exciter filter, barrier filter, and minituft size. Table 2 summarizes some typical flash guide numbers for a range of different-sized minituft materials.

3 Short-Duration Flash The use of fluorescent minitufts on rapidly moving objects, such as rotating propellers, requires a flash lamp system with much shorter exposure times than the high-power flash lamps just described in Sec. III.G.3. Based on the typical motion rates and the amount of blur that could be tolerated without loosing too much

Table 2 Typical Flash Guide Numbers for Two Flash System Types[a]

	Guide Number	
Tuft Material	**High Power (5000 W-s)**	**Short Duration (360 W-s)**
Minituft, dia. = 0.0007 in.	125	45
Minituft, dia. = 0.0019 in.	250	90
Sewing thread, dia. = 0.006 in.	500	180

[a]Film speed, ASA 400; developer, Diafine; exciter filter, Corning 9863; barrier filter, Wratten 2E.

definition of minituft orientation, a short-duration flash lamp system should have exposure times of from 1 to 10 μs.

It is much more difficult to provide such short flash durations at high energy levels. The highest flash energy that is readily available is about 500 W-s at 10 μs. Because of these factors, the size, weight, and cost of such systems are several times larger then those of typical high-power flash lamp systems. The unit shown at the left side on the floor in Fig. 18 is a typical short-duration flash lamp system.

4 Windows It may be surprising to realize that the windows through which the illumination radiation must pass are an important component of the overall illumination system. Because ultraviolet radiation is invisible to human vision, our senses tend to fail us. Many of the common transparent materials found in wind tunnel and airplane windows are partially or completely opaque to ultraviolet radiation. Just because we can see through them does not mean that the excitation illumination radiation will make it through.

Most glass is almost completely transparent to the ultraviolet radiation of interest to the fluorescent minituft method. This includes quartz, borosilicate, i.e., Pyrex, and the soda-lime glass used in ordinary building windows. However, most so-called safety glass, which incorporates a laminated film of plastic material, is nearly opaque to these wavelengths. Also most transparent acrylic and polycarbonate plastics, such as Plexiglas and Lexan, are partially opaque.

The nominal optical properties of Plexiglas brand transparent plastics are well documented in Ref. [30]. This product bulletin lists several grades of available material having varying ultraviolet transmittance. The most commonly available material is "G" grade, which has a relatively sharp cutoff in transmittance at 360 nm. At that wavelength the nominal transmittance is about 80% for each thickness of 0.25 in. The "UVT" grade of Plexiglas has a transmittance of about 89% at the same wavelength and thickness, but it is much less commonly available. Another grade, "UF-4," is almost completely opaque to wavelengths below 400 nm. All these materials look the same to the unaided eye.

In addition, as described in Ref. [31], the ultraviolet transmittance of this material is strongly affected by exposure to sunlight. An outdoor exposure of 2 years is sufficient to render G grade Plexiglas completely opaque at a wavelength of 360 nm.

The safest recommendation for designing an ultraviolet illumination-system installation is to use glass windows. If plastic window material must be used, try to locate material with enhanced ultraviolet transmittance. Direct measurement of the window properties should be done because of the unstable nature of the material.

H Durability

The effort required to install tufts can sometimes represent a serious shortcoming of the method. If only one test point is under study, a liquid indicator might be much faster to use, as well as providing better spatial resolution. Most test programs, however, encompass a wide range of conditions that require study at many test points. The installation effort is easily justified in most cases. However, for the tuft-installation effort to be a practical investment, the tufts must be durable enough to survive as long as the test program requires. The three categories of tuft applications tend to have different durability problems, as discussed in the following sections.

1 Conventional Tuft Durability Conventional tufts, as commonly used in wind tunnel model-scale applications, are generally made from a relatively large and light-weight yarn, such as knitting yarn. The greatest durability problem with this material occurs in heavily separated flow, where the tufts tend to gradually shred and shorten. Since these tufts are often applied just for a short test series, because of their intrusiveness, they are usually durable enough to accomplish the test goals.

In large-scale applications, such as on flight test airplanes where self-excited whipping behavior is prevalent, tuft durability is a much more serious problem. The flagging motion of the tufts acts to cause shredding of the free end and also produces very large reaction loads on the attachment. These large reaction loads dictate that the method of attaching the tufts to the aircraft surface be especially durable. Usually the tufts are secured to the airplane surface with two layers of tape, with the tuft folded back on itself. Heavy aluminum tape, which can resist tearing, is often used for this.

The tuft whipping motion is a consequence of disturbance waves that travel along the length of the tuft. These waves present segments of the tufts at a large local angle to the flow such that the cross-flow forces feed energy into the tuft. As the waves reach the ends of the tufts, considerable stored energy must be released, resulting in strong local stresses in the tuft material. At high dynamic pressures the tufts can be very quickly destroyed. This behavior is difficult to predict, sometimes seeming to be an almost random occurrence that is not necessarily triggered by flow separation. In fact, local separations tend to ease the tuft stresses by reducing the local dynamic pressure.

2 Fluorescent Minituft Durability Fluorescent minitufts are very durable and are able to survive long periods of testing with negligible degradation in most applications. However, in a significantly large number of cases they develop a severely destructive type of motion in which they survive no more than a few seconds of test time. This is usually not a random event, but depends on encountering a certain type of flow condition. It seems that certain types of separation with high local loadings produce a strong and rapid rotational motion in minitufts. This is distinct from the typical trailing-edge wing separation and often is associated with high transonic Mach number shock wave–induced separations or interactions with strong vortex flows. When this happens, the minitufts are observed to rotate in a full circle in a plane parallel to the model surface at a very high rate of speed. After some large number of cycles, they simply break off at the point where they enter the glue drop.

It has been difficult to predict those situations that induce this premature failure in minitufts, but when such a flow condition has been identified, the minituft failure is always very repeatable.

In an attempt to study the problem, a test fixture has been devised that seems to simulate the failure mode very accurately. Figure 19*a* shows a photograph of this test fixture, which consists of a rapidly rotating shaft with its end set flush with a flat surface. A compressed air jet impinges on the end of the shaft, and a minituft attached to the end of the shaft is blown by the jet in one direction. As the shaft rotates, the minituft experiences the same type of bending as a naturally rotating minituft. Minituft failure with this fixture is very repeatable.

Figures 19*b* and 19*c* show close-up details of the minituft specimen. The appearance in Fig. 19*b* is typical of minitufts before they have been stressed. The minituft filament takes on a relatively large radius of curvature as it bends into the flow direction. Figure 19*c* shows the appearance of the filament after exposure to the test flow, just before the failure point.

It was hoped that this device would help to identify a material more durable than the nylon presently used, but so far this has not been accomplished. The only progress toward curing this problem has been to establish that making the minituft very short, about $\frac{1}{4}$ in., is helpful. This seems to reduce the loads on the minituft to the point where it folds back on itself with a relatively large radius of curvature. Premature failure seems to depend on the minituft's folding back on itself with a sharp bend. The fracture apparently is caused by fatigue in the minituft after it has been bent a large number of times.

3 Flow Cone Durability Compared to conventional tufts, flow cones are much more durable for typical flight test applications. Because they do not develop the rapid motion in attached flow that the flexible tufts experience, the reaction loads remain relatively small in those conditions. However, in strongly separated flows, especially at transonic flight conditions with high local dynamic pressures, destructive loads on flow cones can be developed in two different modes.

Flow cones most often fail by gradually shredding the trailing edge under those test conditions where they develop motion that causes them to repeatedly strike the aircraft surface. Flight testing at high altitudes subjects the flow cones to very low temperatures. The material from which they are fabricated obviously has a large influence on their durability. Most common plastic materials become very brittle at such low temperatures, which aggravate their tendency to break up under these conditions. Special plastic formulations have been relatively effective in producing low-temperature ductility to resist this failure mode.

The other failure mode for flow cones involves the attachment of the flexure string to the aircraft surface in conditions of strong separation with reverse flow. This results in the string's tending to tear the edge of the tape as the flow cones are pulled forward. Resistance to this failure mode requires a more durable tape attachment, and is relatively easy to accomplish with double layers of a thicker tape.

One style of flow cone developed for very high dynamic pressure and temperature, as experienced in jet exhaust flow, is made from thin steel sheet material and attached to the aircraft with a short chain brazed to a steel plate. The plate is bonded to the aircraft surface with high-temperature adhesive. This material fabrication and attachment method is very durable, but much more expensive because of the greater effort of manufacture.

I Intrusiveness

It has generally been recognized that conventional tufts large enough to be visible on small wind tunnel models tend to be relatively intrusive. This intrusiveness affects the data most directly in terms of an increase in drag and a reduction in the maximum lift coefficient. Such tufts are usually employed in dedicated flow visualization experiments and are removed during most of the data runs. The results must be used with caution because of these considerations.

(a)

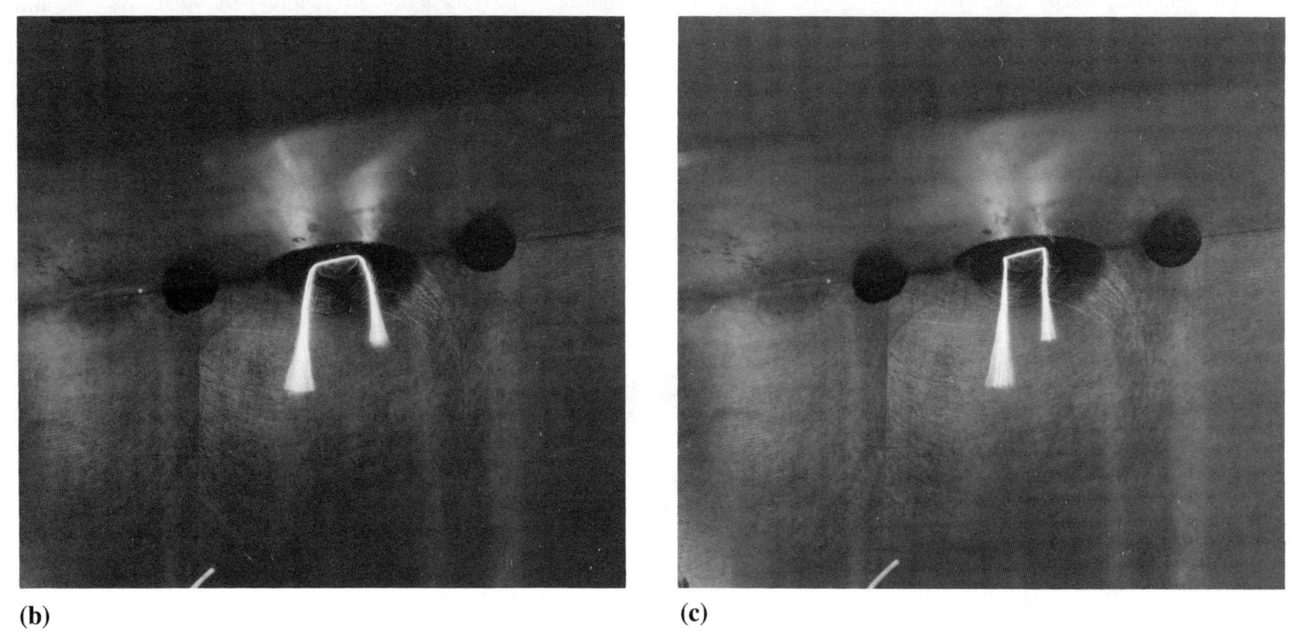

(b) (c)

Fig. 19 (*a*) Minituft durability test fixture; (*b*) minituft durability specimen at start of test; (*c*) minituft durability specimen just before failure.

The fluorescent minituft technique was specifically developed to permit large numbers of tufts to be employed for flow visualization experiments without the adverse effect of conventional tufts. The simple reasoning was that if the tufts were small enough, their aerodynamic effects would become insignificant. It is unlikely that even fluorescent minitufts will have no effect on the flow, but it is only necessary that the intrusiveness be insignificant compared to the data resolution sought in the baseline experiment.

This simple assertion of insignificant intrusiveness has been relatively difficult to demonstrate. The first reaction of observers to the brightly visible appearance of fluorescent minitufts is to assume that their physical size corresponds to their appearance. A more reasoned assessment of minituft effects still must recognize that the minitufts are a much larger disturbance than would ordinarily be tolerated on a wind tunnel model surface.

There is no convenient way to assess potential minituft interference effects except by direct experimental trials. As the interference effects that do exist get smaller, and therefore further submerged in the data noise, the difficulty in resolving the effect becomes greater. It is a rather difficult experimental problem to resolve vanishingly small data increments from large, and sometimes noisy, signals.

This problem calls for a carefully controlled test program, with minituft on-and-off configurations tested back to back, with as many repeat runs as feasible in order to establish the data resolution.

1 A High Lift Wind Tunnel Model The highest quality set of minituft intrusiveness data yet obtained were acquired as part of a cooperative test program between the Boeing Aerodynamics Laboratory and the University of Washington Aerodynamics Laboratory (UWAL). This entailed the use of a typical jet transport model with landing flaps extended. The model was a derivative of a Boeing 707 configuration, maintained by UWAL in a permanently rigged configuration for use as a calibration model in the 8- by 12-ft low-speed UWAL wind tunnel. This wind tunnel operates at a static pressure of 1 atm. The data were taken at a dynamic pressure of 36 lb/ft^2. This test condition corresponds to a unit Reynolds number of 1.04 million per foot.

The model had a span of 88 in., an aspect ratio of 7.5, and a wing-sweep angle at the one-quarter chord line of 35°. The landing flaps were a double-slotted type with a total deflection angle of 40°. A slotted leading-edge slat was fitted to the wing.

The test consisted of measuring the lift and drag properties over a range of angle of attack from the angle for zero lift to beyond the maximum lift condition with closely spaced points at approximate intervals of one degree.

The configurations tested included the baseline without tufts, and installations of minitufts made from 0.0019-in.-dia. monofilament attached with glue drops, 0.006-in.-dia sewing thread tufts attached with glue drops, the same sewing thread tufts attached with tape, and 0.015-in.-dia. tufts made from crochet yarn, as in a typical conventional tuft installation. For the configurations with the glue attachment, approximately 900 tufts were installed on the port wing upper surface from the leading edge to the trailing edge. For the cases with tape attachment, only about 400 tufts were installed on the same wing panel.

Figure 20 shows the typical appearance of the fluorescent minitufts on the model at an angle of attack of 26°, the angle of attack for maximum lift.

Each configuration was tested with 5 repeat runs. The baseline clean configuration was tested with 5 repeat runs at the beginning of the test series and another 5 runs at the end of the test series. A least-squares curve-fitting procedure was employed to derive specific parameters that allowed many data points over a relatively wide range of angles of attack to contribute to the particular parameter.

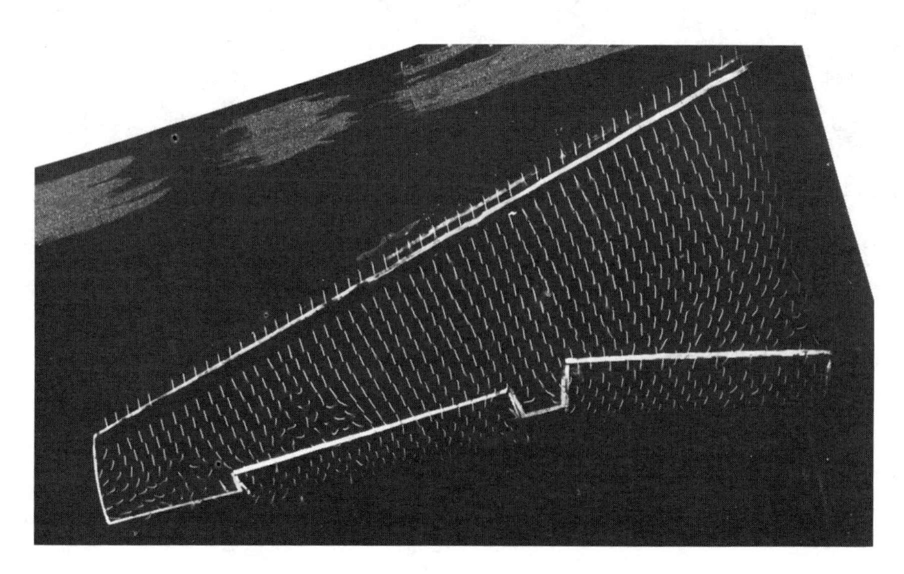

Fig. 20 Fluorescent minitufts on high Lift Model in UWAL 8- by 12-ft low-speed wind tunnel (photo courtesy of the University of Washington).

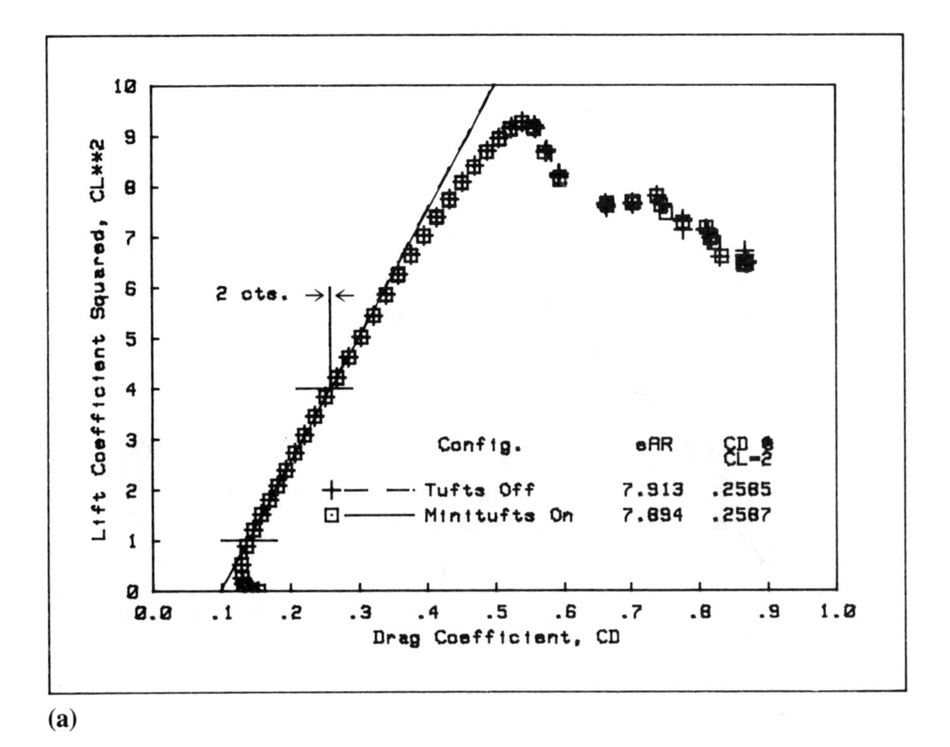

(a)

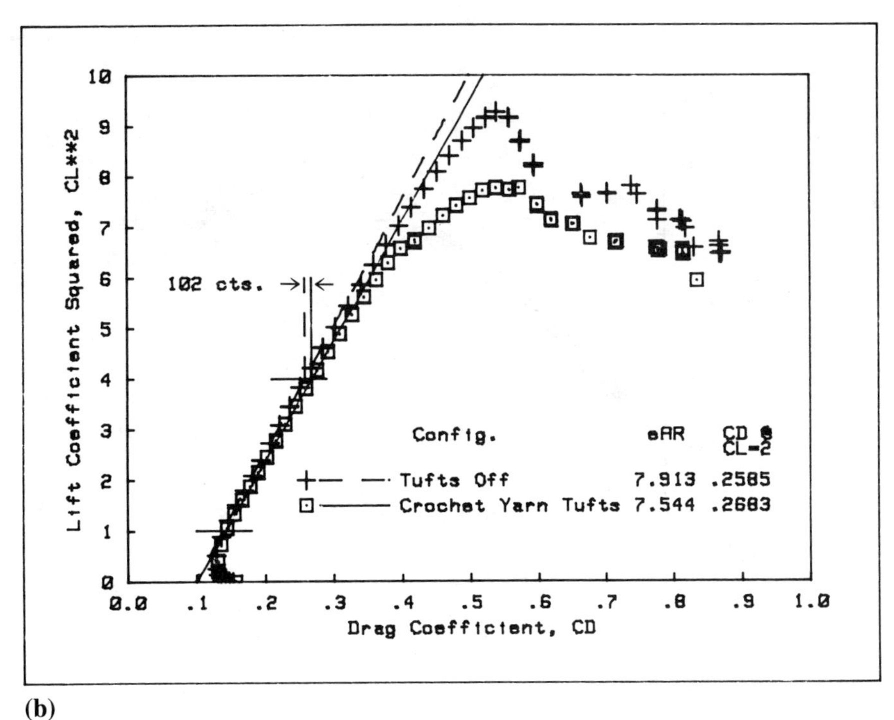

(b)

Fig. 21 Effect of tufts on drag curves for high lift wind tunnel model: (*a*) Minitufts installed; (*b*) crochet yarn tufts installed.

The drag data were analyzed by plotting the drag coefficient versus lift coefficient squared. In this form the curve is linear over the range of lift coefficients between 1.0 and 2.0. A linear least-squares curve fit to the data in this range was performed. The specific parameters derived from these data were the drag curve slope and the drag coefficient at a lift coefficient of 2.0. The drag curve slope is expressed as the effective aspect ratio,

$$\text{eAR} = \frac{1}{\pi}\frac{d(C_D)}{d(C_L^2)}$$

Over this angle of attack range, about 45 data points contributed to the curve fit for each configuration. Figure 21 presents typical drag data, comparing the baseline configuration with minitufts in Fig. 21*a* and with the crochet yarn tufts in Fig. 21*b*. The more interesting comparison is

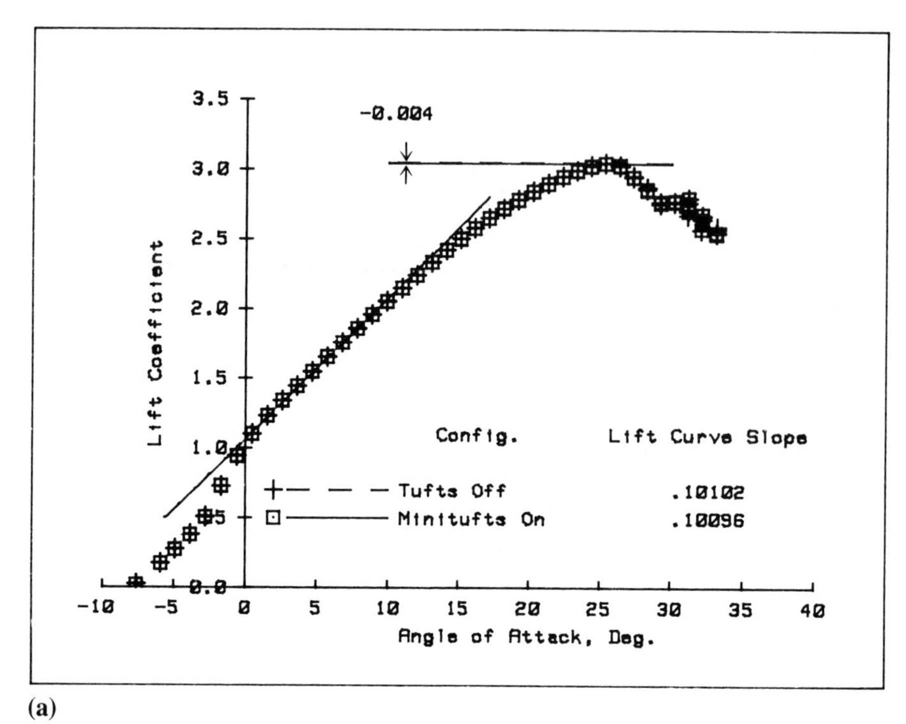

(a)

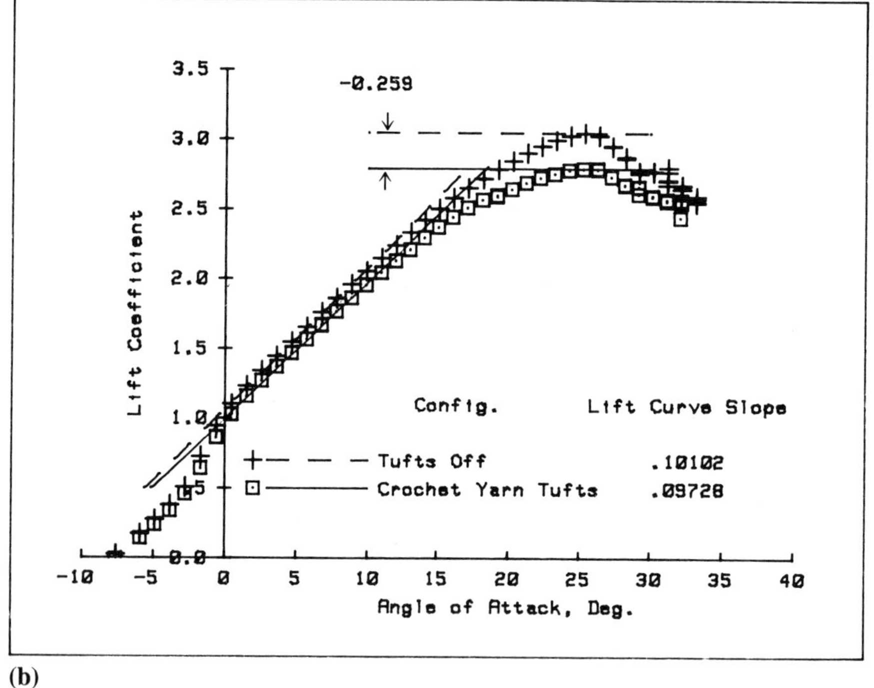

(b)

Fig. 22 Effect of tufts on lift curves for high lift wind tunnel model: (*a*) Minitufts installed; (*b*) crochet yarn tufts installed.

that of Fig. 21*a*, but the effect is too small to be depicted in this figure. All the data points fall on top of each other. Figure 21*b* illustrates more graphically the interference effects of large tufts typical of conventional wind tunnel applications. Each symbol plotted on the graph actually represents 5 data points, but they all have less variability than can be shown here.

Typical lift curve data are shown plotted in Fig. 22*a* and 22*b* for the same configurations. A linear least-squares curve fit was performed to the data points over the same interval of lift coefficient from 1.0 to 2.0 to derive the lift curve slope. The other parameter derived from these data was the maximum lift coefficient, taken as the highest single point in the data array.

Table 3 Tuft Effects Summary Parameters

Configuration	eAR	%	*Cd* at $C_L = 2$	ΔC_D, counts	$\frac{dC_L}{d\alpha}$	%	$C_{L_{max}}$	$\Delta C_{L_{max}}$
Baseline	7.913	100.0	0.2585	0	0.1010	100.0	3.047	0
Baseline, repeat	7.917	100.1	0.2585	0	0.1012	100.2	3.055	+0.008
Minitufts	7.894	99.8	0.2587	2	0.1010	100.0	3.043	−0.004
Sewing thread + glue	7.858	99.3	0.2599	14	0.1007	99.7	3.016	−0.031
Sewing thread, repeat	7.835	99.0	0.2606	21	0.1003	99.3	2.985	−0.062
Sewing thread + tape	7.839	99.1	0.2603	18	0.1003	99.3	2.992	−0.055
Crochet yarn + tape	7.544	95.3	0.2683	98	0.0973	96.3	2.788	−0.259

Table 3 summarizes the specific parameters derived from these wind tunnel data for all the configurations.

As can be seen from these summary parameters, the minitufts show a very small interference effect while the larger tufts show progressively greater interference effects. Whether the minituft effects are truly insignificant for this case would depend on the test objectives and the quality of the data obtained throughout the test program. It is worth pointing out that the minitufts installed on the slat and leading-edge region of the wing probably have the greatest interference effect. Therefore it is to be expected that limiting the minituft installation to the aft half-chord of the wing, for example, would result in a substantially reduced interference effect.

2 A Transonic Wind Tunnel Model Intrusiveness data on a transonic wind tunnel model were acquired in the 8- by 12-ft Boeing Transonic Wind Tunnel (BTWT). This wind tunnel runs at 1 atm total pressure and a temperature of about 100°F. At Mach 0.85 the corresponding unit Reynolds number is 4.05 million per foot.

The model used for this study was the wind tunnel calibration model with a span of 5.6 ft, a sweep angle at the one-quarter chord line of 35°, and an aspect ratio of 8.0. The wing had an uncambered section and was mounted on the center plane of the axisymmetric body. This model was carefully maintained in good condition and tested at regular intervals to document the wind tunnel behavior.

A total of 2800 minitufts were installed on the starboard wing on both the upper and lower surface from just behind the boundary layer trip at 10% chord to the trailing edge. The minitufts were made from 0.0019-in. dia monofilament with a length of 0.5 in. This installation was somewhat more extensive than would ordinarily be used in order to exaggerate any interference effects. A photograph of the minitufts installed on the wing is shown in Fig. 23.

Comparison runs were conducted for both the minitufts off baseline and the minitufts installed condition. The data are presented in Fig. 24 as plots of drag coefficient versus lift coefficient squared for 3 Mach numbers. At each Mach number 3 repeat runs were made. A total of about 45 data points are plotted within the lift coefficient range of −0.3 to 0.3. The drag data plotted in this fashion form a linear curve over this lift coefficient range, thereby permitting a linear least-squares curve fit to be established. In this way all the data in this range can contribute to define the drag coefficient at 0 lift and the drag curve slope, expressed as the effective aspect ratio.

The results show a decrease in the drag level of about 2 drag counts due to the minitufts. It was not possible to conduct the test with the minitufts on and off as a contiguous test series. There were several other runs between the two series. Also, it was not possible to repeat the minitufts-off data to establish the long-term repeatability of the data. Therefore, it remains somewhat uncertain whether the minitufts actually reduce the drag or whether that effect is just an instrumentation shift.

The main point to be made from these data is that for a minituft installation of much greater density than would ordinarily be used, and on both upper and lower surfaces of the wing, the drag increment is essentially within the data scatter. Presumably a minituft installation with fewer

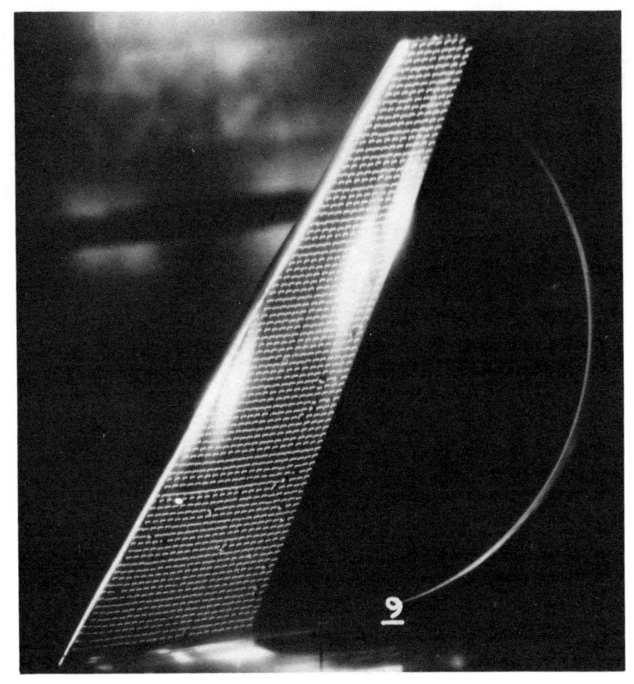

Fig. 23 Fluorescent minitufts on transonic wind tunnel model.

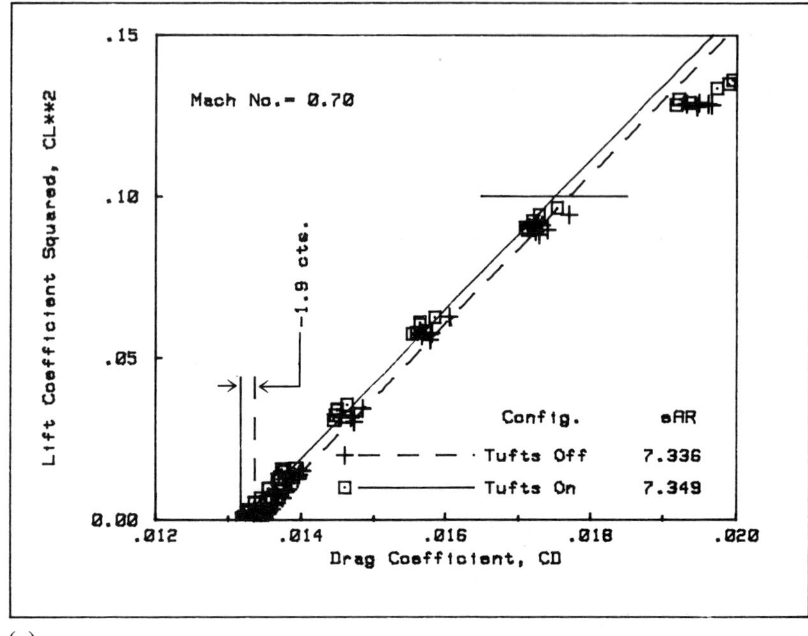

(a)

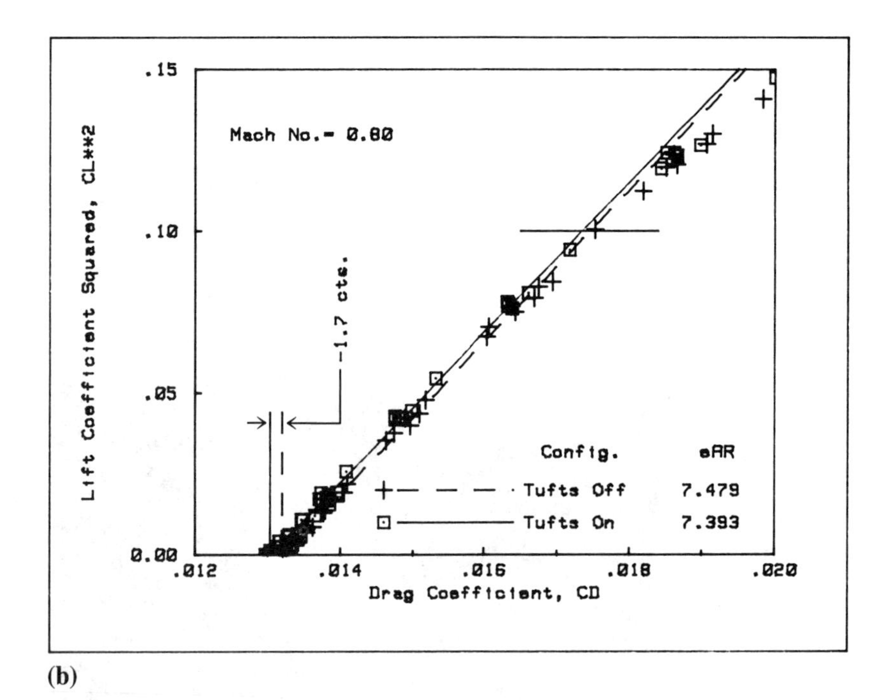

(b)

Fig. 24 Effect of minitufts on drag curves for transonic wind tunnel model: (*a*) M = 0.70; (*b*) M = 0.80. (Figure continues.)

elements and only on the upper surface would have had less than one half the observed interference.

3 Supersonic Transport Wind Tunnel Model Data from a minituft installation on a supersonic transport model tested at Mach 1.6 to 2.4 are presented here. These data were taken in the 9- by 7-ft leg of the NASA Ames Unitary wind tunnel as part of a cooperative Douglas Aircraft/NASA supersonic transport development program. They are not quite as extensive as the examples just discussed since no repeat runs were made and the number of data points is much smaller. Figure 25 shows the typical appearance of the minitufts at an angle of attack somewhat higher than the cruise condition.

The drag data are plotted in a fashion similar to that just given, with one exception. Because of the wing camber, the linearity of the drag curve was enhanced by plotting the square of lift coefficient minus the minimum drag lift coef-

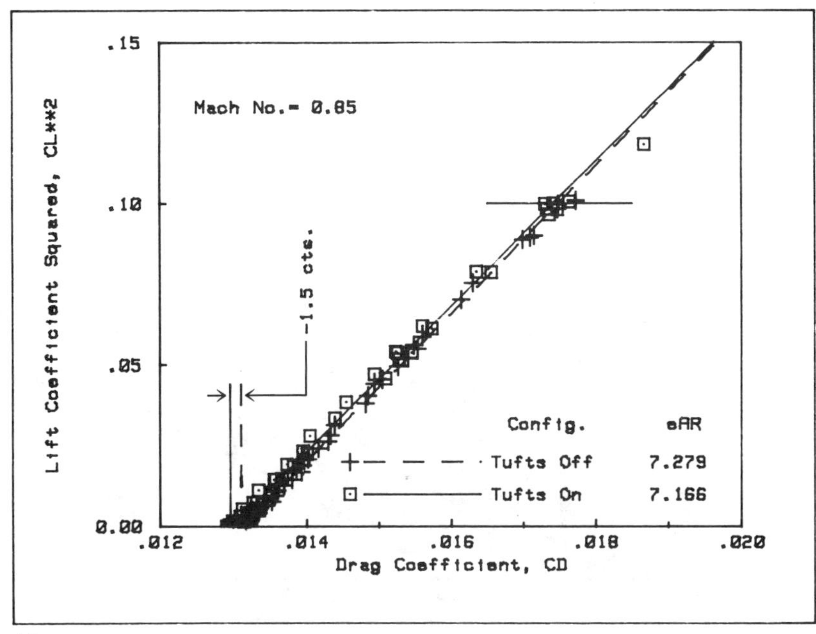

Fig. 24 (continued) (*c*) M = 0.85. **(c)**

ficient. This is a constant value of $Clm = 0.15$. These data are presented in Fig. 26 for five Mach numbers ranging from 1.6 to 2.4. It is more difficult to make a definitive judgment on the minituft interference effects from these data because of the sparse data points and lack of repeat runs. Nevertheless, the general impression is that the minituft interference effects lie within the data scatter of the test, about ± 2 drag counts.

4 Two-Dimensional Transonic Airfoil Model Minituft interference effects were studied in detail with a small two-dimensional wing in the Boeing Model Transonic Wind Tunnel (BMTWT). This is a continuous running facility with a test section of 4.6- by 7.2-in. cross section. It runs at a total pressure of 1 atm and a temperature of about 85°F, so that at Mach 0.8 the unit Reynolds number is 4.08 million per foot. The model has a chord length of 6 in. and

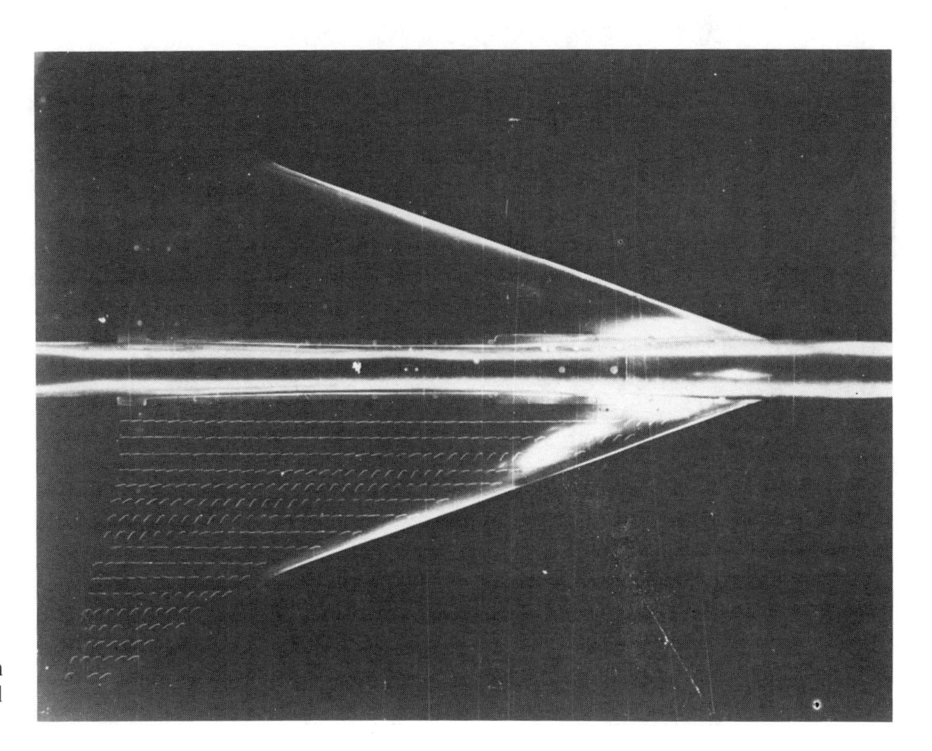

Fig. 25 Fluorescent minitufts on supersonic transport wind tunnel model.

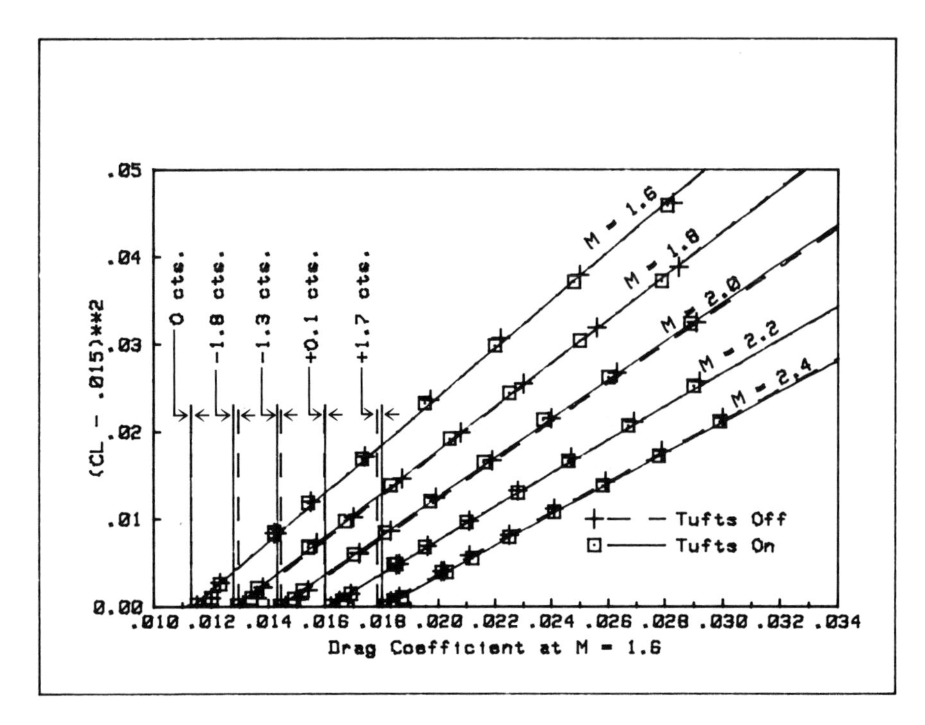

Fig. 26 Effect of minitufts on drag curves for supersonic transport wind tunnel model.

spans the 4.2-in. dimension across the test section. The airfoil section shape was specially designed so that in the presence of the confined test section, the chordwise pressure distribution was similar to a realistic wing in flight.

The test was run at Mach 0.8 and at one angle of attack. The profile drag was measured by traversing a total pressure probe through the wake and integrating the sectional drag coefficient. A boundary layer trip was installed at a chordwise station of 10% to establish a turbulent boundary layer.

Minitufts made from 0.0019-in.-dia monofilament with a length of 0.38 in. were installed from a chordwise station of 18% to the trailing edge in 13 rows. Each row had 12 minitufts spread along the span of the model. The wake profile drag was measured as each row of tufts was removed, starting with the most forward row.

Figure 27 presents the results of the test in terms of the drag coefficient versus the number of minituft rows on the model. Five repeat runs were made for the initial condition

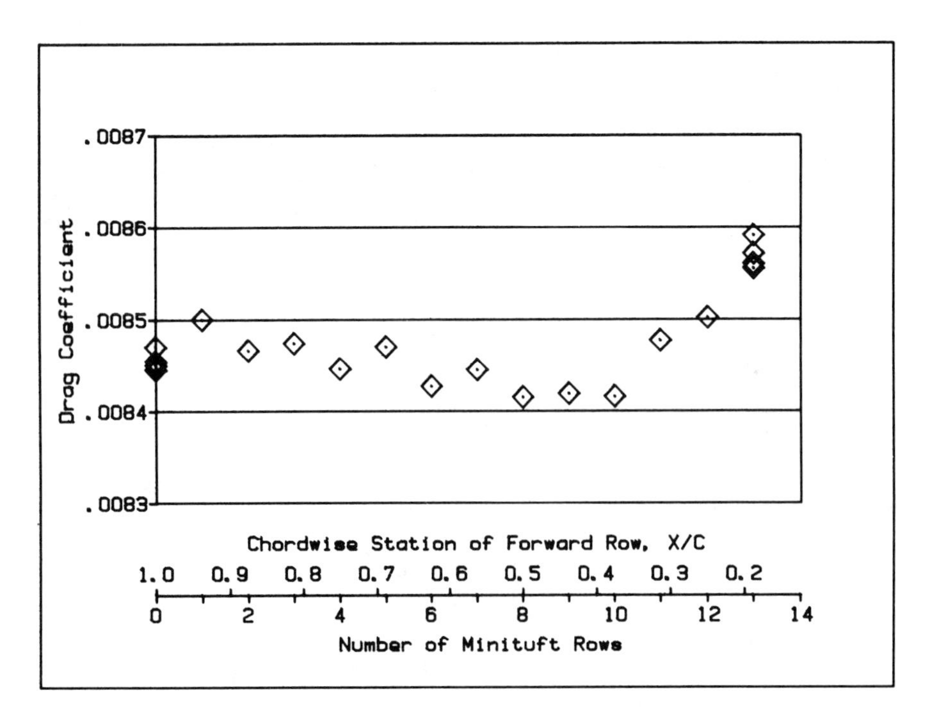

Fig. 27 Effect of minitufts on drag of transonic two-dimensional airfoil wind tunnel model.

with all 13 rows in place, and 5 repeat runs were made when all the minitufts had been removed.

The results show a data-scatter level of ±0.2 drag counts. As the first 5 rows of minitufts were removed, back to a chordwise station of 50%, the sectional drag coefficient dropped by about 1.5 drag counts, out of a level of 85 counts. As the remaining minituft rows were removed, the drag remained essentially constant, or possibly rose slightly. This confirms the hypothesis that the farthest forward minitufts produce the greatest interference and that behind a certain chordwise location they have negligible effect.

5 Other Minituft Intrusiveness Data Two other sources of minituft intrusiveness data have been published in Ref. [32,33]. The first of these presents the results of an extensive evaluation of the method in the RAE, Farnborough, 5-m pressurized low-speed wind tunnel. In that report it was concluded that the minitufts were sufficiently intrusive to make it inadvisable to conduct force testing with the minitufts in place. A drag increment due to minitufts of 20 drag counts was cited, and an increment in maximum lift coefficient of 0.2 out of a baseline level of 1.75 was shown.

Without knowing more details of the test, it is difficult to critique these reported results. The drag increment could well be an artifact of a very sparse data set with no repeat runs, but the minituft effect on the maximum lift coefficient is unlikely to be so easily explained. The reference does not specify the unit Reynolds number of the test, but because the wind tunnel can presumably operate at 3 atm, the test unit Reynolds number may have been sufficiently greater than that in the UWAL wind tunnel to account for the increased interference.

The reference notes that the hypothetical action of minitufts in increasing the drag might be either acting to cause premature transition, if they are installed in regions of normally laminar boundary layer, or acting as a local roughness to increase the drag in an already turbulent boundary layer.

It has always been recommended practice to avoid installing minitufts in regions of laminar boundary layer. Indeed, there seldom is a need for the visualization of separation in these regions. On occasion, however, it has been possible to compare the tripping effectiveness of individual minitufts with typical boundary layer trips. Minitufts usually cause transition, but not very effectively.

It should be noted that typical boundary trip heights in the transonic BTWT are in the range of 0.004 to 0.006 in. Trip heights for the 1-atm low-speed UWAL wind tunnel range from 0.01 to 0.02 in. The actual height of minitufts made from 0.0019-in.-dia monofilament is between 0.002 and 0.003 in., depending on the thickness of the attachment glue drop.

In Ref. [33] an evaluation of fluorescent minitufts was conducted on a fighter airplane wind tunnel model at transonic conditions. It was concluded that "tufts caused significant changes in the force data." In particular, a drag increase of 30 counts was reported for one configuration at a lift coefficient of about 0.8 and a drag level of 1000 counts. Another configuration indicated a "significant drag reduction."

Again, it is difficult to critique such results based on the few details provided in the reference, but the question of minituft nonintrusiveness is central to the utility of the technique and deserves careful consideration. It would be unfortunate for the method to be limited by a perception of greater interference than actually exists. Three points deserve to be made with respect to these results.

First, 30 counts of drag increase might be very significant at conditions near the cruising point on an airplane wing, where the drag level may be in range of 200 to 400 drag counts. However such an increment is much less significant at conditions of high lift where the drag level and the rate of change of drag is becoming very large as is the case with these results. Such conditions usually include a considerable extent of local separation and associated flow unsteadiness, making it very difficult to maintain data quality.

Second, the reported data from which these conclusions are drawn are extremely sparse. No indications of data repeatability are provided. Under these conditions, such conclusions are difficult to sustain.

Finally, the report mentions that the adhesive used for attaching the minitufts to the model was a water-soluble glue. Implementation of the fluorescent minituft technique required the development of a practical method of installing tufts. As originally reported, this called for the use of an adhesive with a large fraction of volatile solvent. This is a very important requirement in order to permit placement over the tuft material of a reasonably large drop of adhesive that will shrink to the size of a thin patch over the minituft attachment point. Microscopic examination of a nominal minituft attachment shows the adhesive thickness to be much less than the diameter of the minituft material. If this practice is not followed, the resulting size of the attachment glue drop can be quite large and therefore intrusive.

This discussion is not intended to uncritically defend the technique. It is clear that there will always be some model locations or test condition where minitufts are liable to be significantly intrusive. On the other hand, the other data presented here clearly show that there are very many conditions where the presence of a large number of minitufts have an insignificant effect.

The important point is to recognize the value of nonintrusive, concurrent flow visualization data acquisition and to develop an understanding of those conditions that permit it. For critical situations, direct determination of minituft interference will remain the only certain manner of ascertaining those effects. This must, however, entail carefully designed experimental trials with plentiful data and many repeat runs.

IV APPLICATION EXAMPLES

The best way to assess the overall performance of any flow visualization technique is to examine the visual data records. Several recent examples of tuft flow visualization experiments are presented in this section. These examples are categorized in terms of conventional tufts, fluorescent minitufts, and flow cones. They encompass results from low-speed and high-speed wind tunnels and full-scale vehicles in flight or on the road. With the exception of the examples of conventional tufts, these results are from the author's recent experience.

A Conventional Tufts

1 AA-1X Light Airplane Figure 28 shows a view of conventional tufts on the starboard upper wing panel of an AA-1X airplane at the stall-onset point during an attempted spin entry. This wing is specially modified with a leading-edge droop on the outboard part of the span. This photograph was provided by the Flight Applications Branch of NASA Langley Research Center.

2 T-38 Jet Trainer Conventional tufts on the wing of a T-38 jet trainer are shown in Fig. 29. The flight condition is unspecified except that the purpose of the flight was to investigate wing rock. The view is from the rear cockpit. This is a good example of self-excited tuft whipping motion. Several of the tufts in the first and second row show the typical appearance of unsteady waves propagating along the length of the tufts. This example, and the following one, was provided by NASA/Ames Dryden Flight Research Facility.

3 AD-1 Research Airplane An air-to-air photograph of conventional tufts on the wing of the AD-1 skewed wing research airplane is shown in Fig. 30. The outboard portion of the port wing is indicating a leading-edge separation, which is probably associated with a rolled-up leading-edge vortex because of the high wing sweep angle.

B Fluorescent Minitufts

In addition to the examples presented below, several other examples of fluorescent minitufts have been presented elsewhere in this discussion to illustrate specific points. The photograph shown in Fig. 1 is perhaps the most representative example of a wind tunnel model fluorescent minituft flow visualization experiment. It was taken in the 5- by 8-ft Boeing Research Wind Tunnel (BRWT). The model was photographed and illuminated from a ceiling window about 48 in. above the model. The illumination source was a 2000 W-s flash lamp having a flash duration of 4 ms. The minitufts are made from 0.002-in.-dia. monofilament with lengths of about 0.63 in.

The model is a half-span model of a twin engine jet transport mounted on the right-hand wall turntable. The half-span was 50 in. Leading-edge slats and trailing flaps are deployed in a landing configuration. The wind tunnel Mach number is 0.2.

Distinctive features in the minituft patterns include local regions of separation behind the outboard end of the lead-

Fig. 28 Conventional tufts on AA-1X wing upper surface during spin (photograph courtesy of NASA).

Fig. 29 Conventional tufts on T-38 wing (photograph courtesy of NASA).

ing-edge slat and forward of the outboard end of the trailing-edge flaps. The disturbed tufts on the wing upper surface behind the engine nacelle show regions of interaction with rolled-up vortices shed from strakes on the engine nacelle. Streamers are installed on the tips of these strakes.

1 Two-Dimensional Wing in Blow-Down Wind Tunnel A particular advantage of tufts for surface flow visualization is the very fast response time. This is especially useful in blow-down wind tunnels, where the run time is measured in seconds. In the example shown in Fig. 31, three data images are presented from a series of six angle-of-attack conditions that were photographed in a run time of 10 s.

The two-dimensional wing model had a span and chord length of 12 in. The camera and ultraviolet flash lamp were installed in the ceiling, about 18 in. above the model. Fluorescent minitufts having a diameter of 0.002 in. and a

Fig. 30 Conventional tufts showing leading-edge separation on NASA AD-1 (photograph courtesy of NASA).

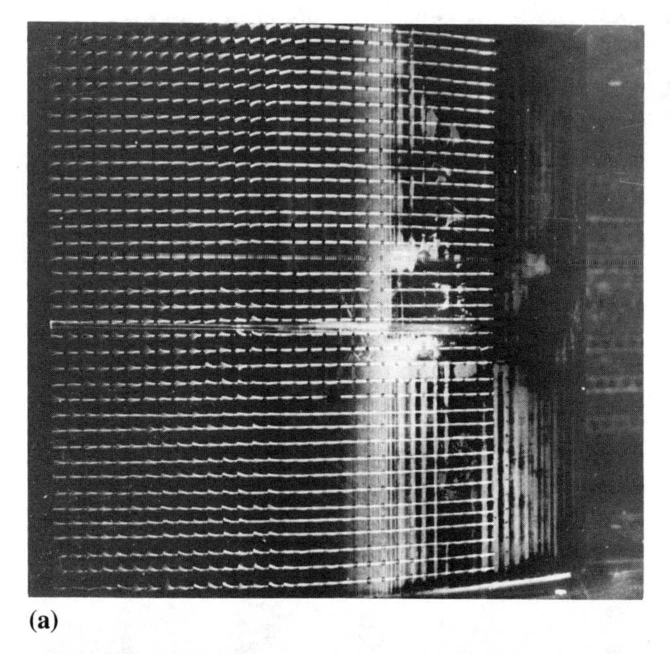

(a)

(b)

(c)

Fig. 31 Fluorescent minitufts on two-dimensional model in blowdown transonic wind tunnel: (*a*) Angle of attack = 1°; (*b*) angle of attack = 3°; (*c*) angle of attack = 6°.

length of about 0.63 in. were installed from a chordwise station of 20% to the trailing edge. The test Mach number was 0.74.

At the first angle of attack of 1°, shown in Fig. 31*a*, the wing is unseparated, except for small regions along the walls near the trailing edge. As the angle of attack is increased to 3°, shown in Fig. 31*b*, a shock wave–induced separation starts in the center of the span. This separation grows laterally with increasing angle of attack. At the 6° angle of attack, Fig. 31*c*, the separation still has not completely extended to the walls. Even though the model is two-dimensional, the flow does not seem to have gotten the message.

2 A Generic Fighter Wind Tunnel Model with Simultaneous Smoke Figure 32 shows a generic fighter wind tunnel in the 5- by 8-ft BRWT. The model span is 20 in. and is imaged from a distance of 48 in. The minitufts are made from 0.002-in.-dia. monofilament. The illumination system is a 2000-W-s flash lamp with 4-ms duration.

The rolled-up vortices emanating from the strake leading edge are visualized with two different procedures. The starboard vortex has been seeded with condensed glycol vapor smoke dispensed from a traversing wand positioned upstream of the model. The port vortex is marked by natural water vapor condensation. The test condition shown in the

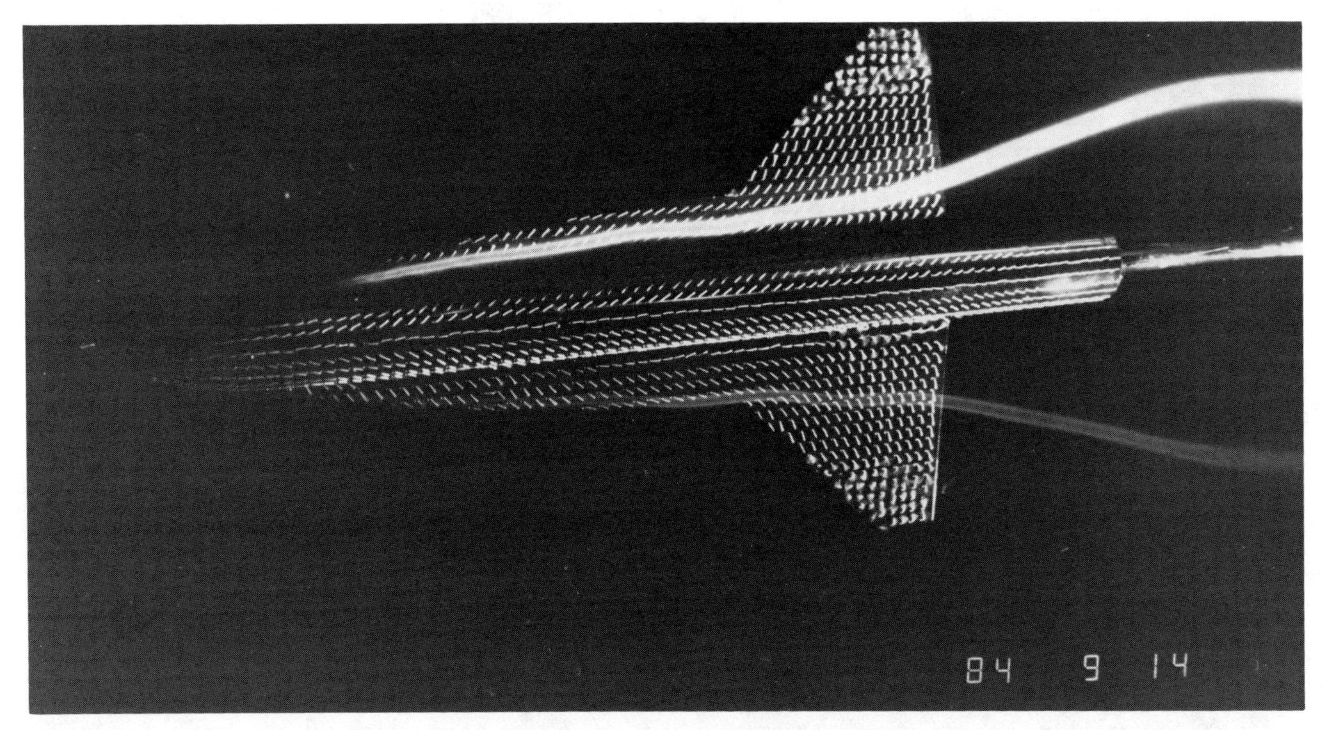

Fig. 32 Fluorescent minitufts of generic fighter model in low-speed wind tunnel.

figure is a Mach number of 0.2 and an angle of attack of 30°. This is just before the angle for spontaneous vortex-core bursting.

The very high outflow angle of the flow on the wing upper surface is evident. Local regions of vigorous separation near the wing tip are well marked by the blurred tufts even though the exposure time of the flash lamp was 4 ms.

3 Cessna 206 Airplane Figure 33 shows an application of fluorescent tufts to the situation of an airplane in flight. The photograph was taken at night. The tufts are made from cotton-covered polyester sewing thread with a diameter of 0.006 in. No additional fluorescent dye treatment beyond that employed by the thread manufacturer for white textile material was used. The tufts on the flap surface were 1 in. long and held in place with drops of lacquer adhesive. On the main wing surface, the tufts are 3 in. long and attached with individual pieces of vinyl tape. The illumination system consisted of a 2000-W-s flash lamp power supply driving a single flash lamp. The lamp was located on the tip of the vertical tail, along with the camera.

In Fig. 33 the airplane was in a steady glide. The flaps deflected to the 40° landing configuration experience extensive separation. Each of the separation cells appears to originate at a flap bracket or actuator push rod cutout. The main wing is unseparated.

In Fig. 33 the airplane has been pulled up into a full stall. The photograph was taken just at the point where the stall break occurs. The main wing is completely separated while on the flap surface the flow has almost completely reattached. This illustrates the fast response time of tufts. The flash lamp power supply has a cycle time of 1 s, thereby permitting a fast time series of data images.

4 Piper Seneca Light Twin-Engine Airplane Another example of in-flight fluorescent tufts is presented in Fig. 34. The photograph shows a view of the wing trailing edge, nacelle afterbody, and inboard wing of a Piper Seneca light twin-engine airplane. The camera and illumination system were installed in the aft cabin, and the flight condition in this case was at cruise.

A region of separation in the center of the nacelle afterbody extends back to the wing trailing edge. A more extensive separation area originates at the inboard side of the nacelle and extends aft and inboard out of the view of the camera.

5 Machen Superstar Propeller The first application of fluorescent minitufts to a rotating propeller was accomplished during the development of the Machen Superstar, a re-engined version of the Piper Aerostar airplane. This airplane was experiencing high-speed performance problems attributable to propeller performance. This flow visualization application was instrumental in developing a revised propeller configuration resulting in a significant increase in high-speed performance.

Figure 35*a* presents the flow visualization result for the baseline configuration at an economy cruise condition,

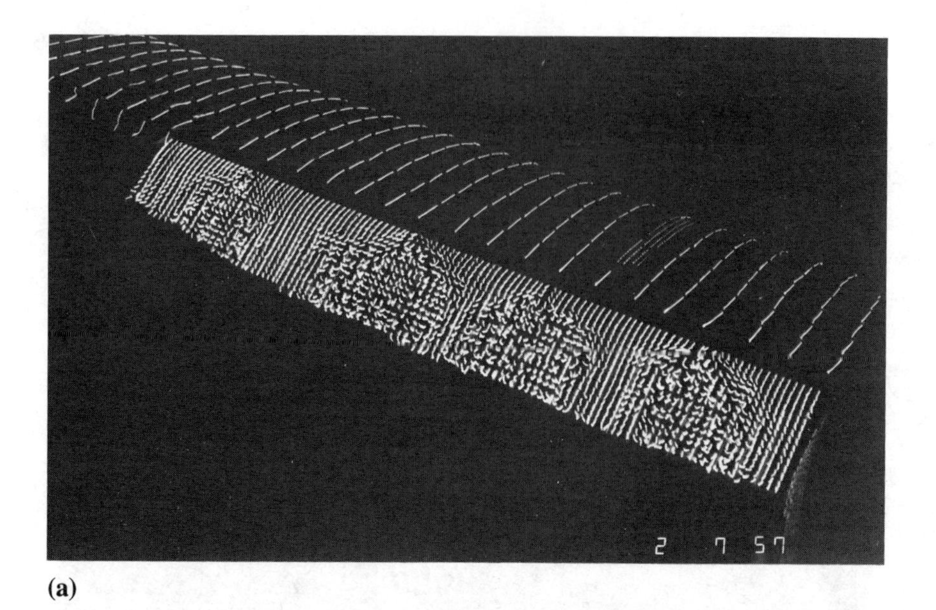

(a)

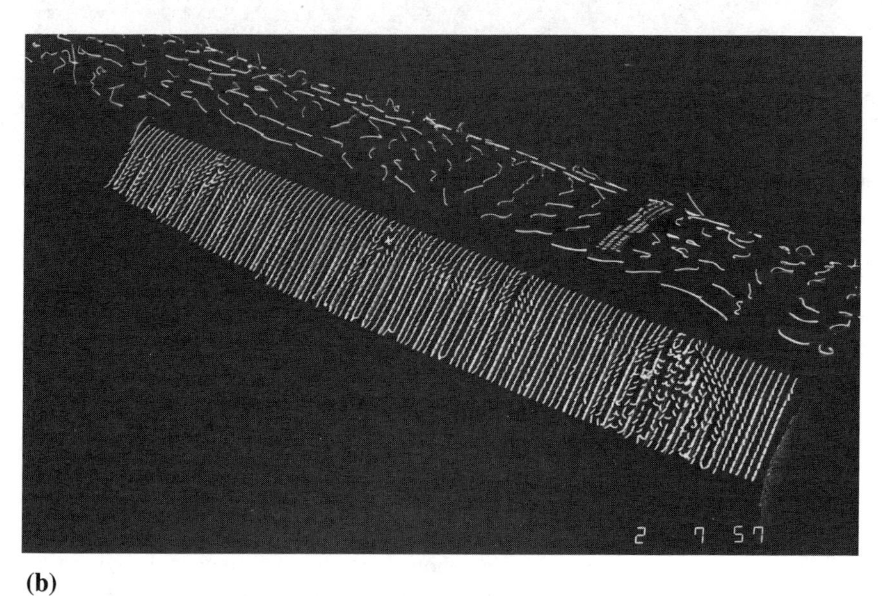

(b)

Fig. 33 Fluorescent tufts on Cessna 206 airplane in flight: (*a*) During steady glide, 70 knots indicated air speed; (*b*) at stall break point.

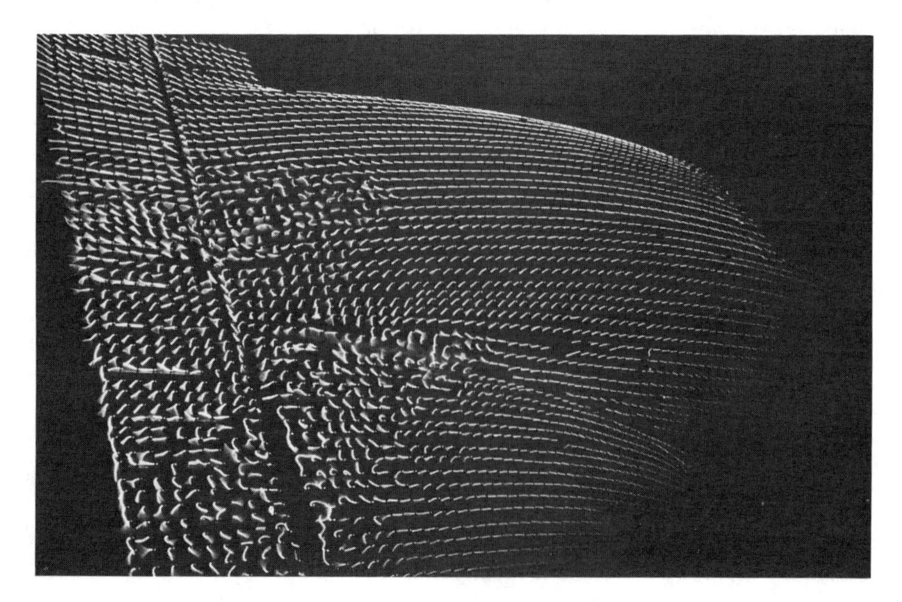

Fig. 34 Fluorescent tufts on Piper Seneca in flight.

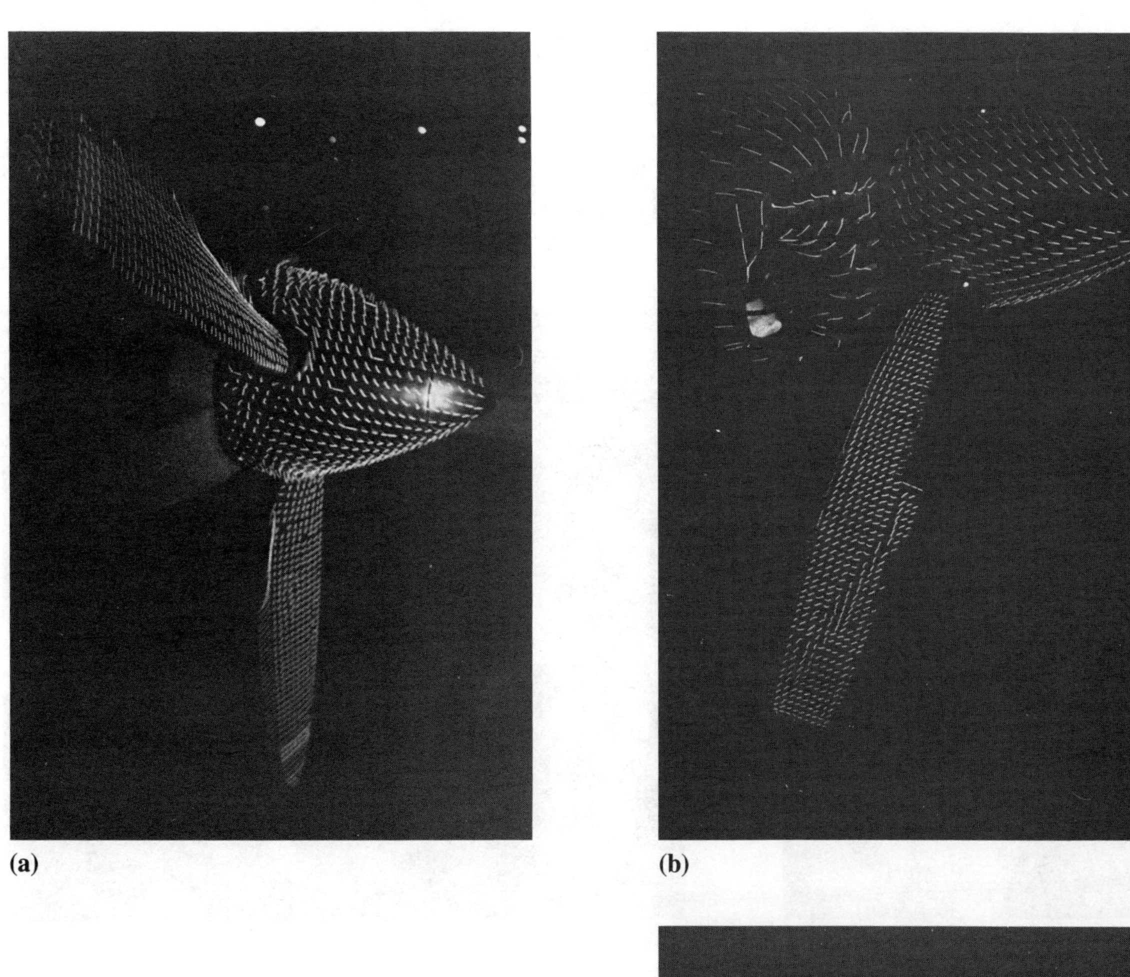

(a) (b)

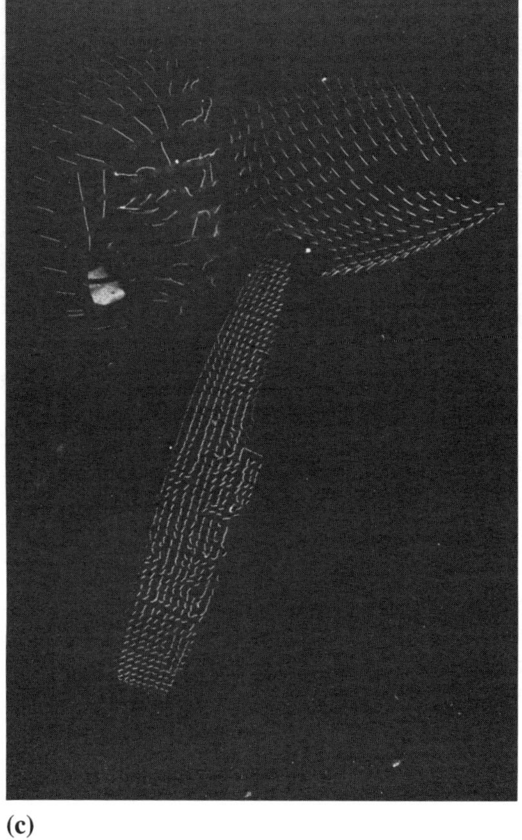

(c)

Fig. 35 Fluorescent minitufts on rotating propeller, Machen Superstar airplane in flight: (*a*) Economy cruise; (*b*) high-speed cruise; (*c*) climbing near ceiling.

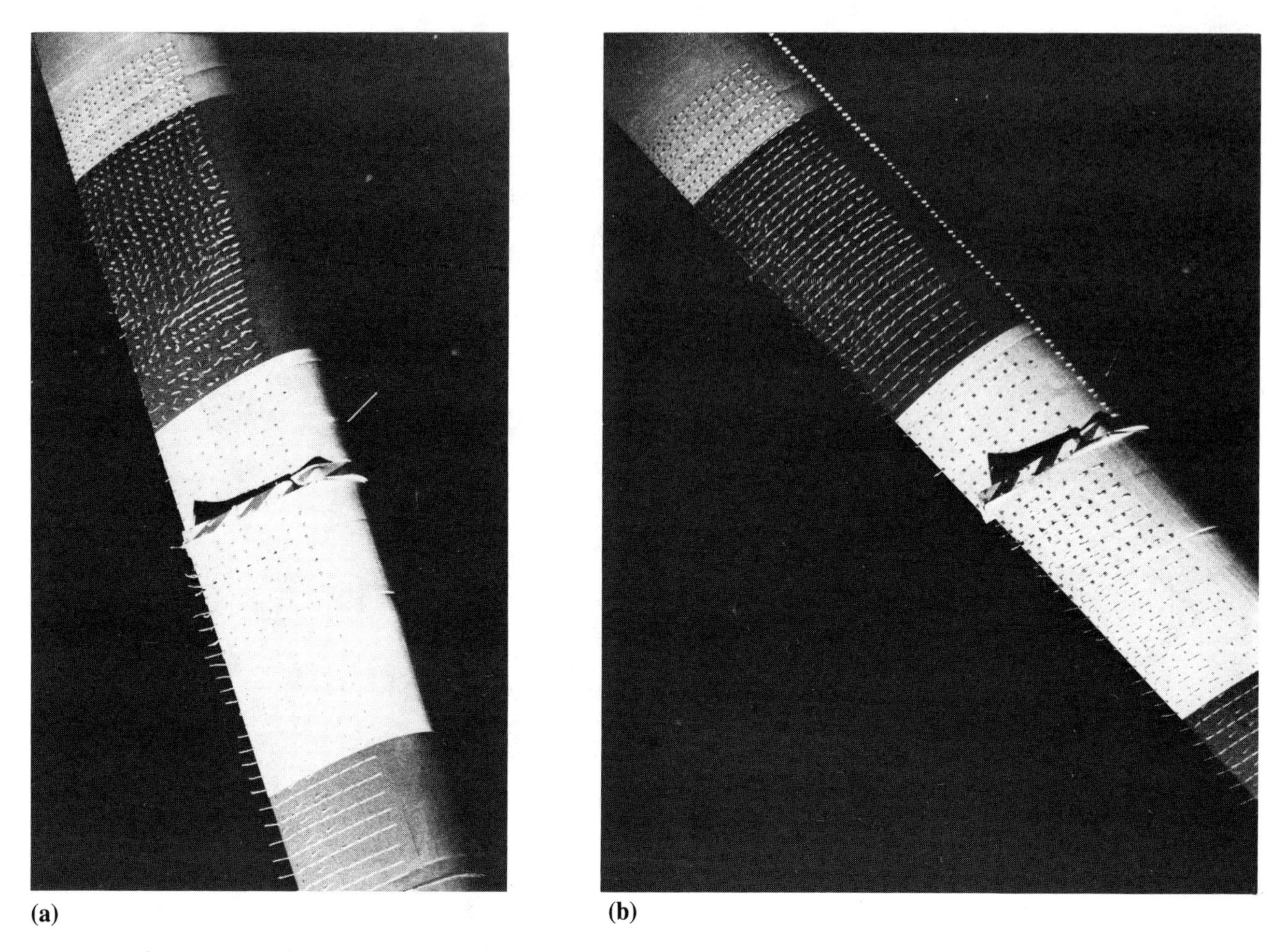

Fig. 36 Tufts on NASA/Boeing MOD-II 300-ft-dia. wind turbine: (*a*) Baseline configuration; (*b*) with leading edge vortex generators.

where the propeller performance was good. The minitufts show essentially no separation. In Fig. 35*b* the airplane was at a high-speed cruise condition at an altitude of 25,000 ft. The propeller-tip Mach number was 0.93, and the airplane was suffering a sudden loss of performance. The minituft image shows a distinctive separation extending diagonally across the outer half-radius that is believed to be caused by a shock wave.

The final example from this test program shows, in Fig. 35*c*, the separation pattern on the propeller surface while climbing at a low air speed near the airplane ceiling. As would be expected under these conditions, almost the entire propeller surface is separated except for a small region of attached flow near the tip.

Further details of this application are presented in Refs. [12,13].

6 NASA/Boeing MOD-II Wind Turbine Figure 36 presents two photographs of tufts on the NASA/Boeing MOD-II wind turbine. This machine has a rotor diameter of 300 ft. The view is from the ground close to the rotor disk, looking up along the blade, which is greatly foreshortened. Figure 36*a* shows the baseline configuration at a wind speed of 25–30 mph, while Fig. 36*b* shows the rotor with vortex generators installed near the leading edge at a slightly higher wind speed of 28–33 mph.

The design of this wind turbine employs a fixed-pitch teetering two-blade rotor out to 70% of the blade radius. That point is marked in the photograph by the striped panel near the center of the picture. From the 70% radius station and beyond, the rotor blade has controllable pitch. By design, this type of rotor blade experiences increasing separation at wind speeds greater than the design speed on the fixed-pitch portion of the rotor. The outer panel beyond the 70% station remains unseparated by controlling the blade-pitch angle. Further details of this wind turbine are provided in Ref. [34].

The tufts are made from light nylon twine having a diameter of 0.1 in. and length of about 6 in. They are attached to the blade surface with small pieces of aluminum tape. The photographs were taken at night with white light illumination provided by two 500-W-s 10-μs duration flash lamp systems. The flash lamps were located on the ground and situated at a shallow grazing angle to the blade surface.

As shown in Fig. 36*a*, extensive separation was present on the fixed-pitch portion of the rotor at wind speeds somewhat below the design speed. Even though this wind turbine had been in service for several years, this separation had not been previously detected. Indirect evidence from control system discrepencies led to the flow visualization study that documented this behavior.

The result in Fig. 36*b* shows the effect of an array of leading-edge vortex generators in suppressing the separation at greater wind speeds. This simple configuration change resulted in alleviating the control system problems and increasing the annual energy production by about 15%. This beneficial effect could not be easily documented except by the flow visualization experiment. Because of the highly intermittent nature of local wind behavior, determination of wind turbine performance requires an elaborate statistical analysis of energy production history over several months of operation. The tuft flow visualization technique was the only practical means of confirming the vortex generator effects.

7 Porsche 944 Automobile Fluorescent minituft images on an automobile tested on the road are shown in Fig. 37. This experiment is potentially useful for two particular applications.

The first pertains to validation of wind tunnel results. Most automobile wind tunnels do not employ ground-plane boundary layer control or moving wheels. There is therefore a reasonable likelihood of modeling discrepencies for that part of the flow near the ground or around the wheels. This on-the-road experiment offers a convenient and economical means of acquiring data that can be correlated with similar wind tunnel experiments.

The other application of this technique is for development programs of automobile aerodynamics when a wind tunnel is not available. Many race cars utilize extensive aerodynamic effects on small wing panels used for generating down force to help in increasing tire traction. Probably most of these configurations are developed from on-the-road testing using indirect measures of aerodynamic performance. This flow visualization technique allows the separation-onset conditions to be directly determined.

Fluorescent minitufts on an automobile can be imaged with either one of two different techniques. The one that requires the least investment in equipment involves photographing the test car from a chase car traveling alongside at the same speed. This permits the use of a high-power, long-duration ultraviolet flash. However, it also requires a long section of highway where the test speeds can be reached and presents a potentially significant safety problem.

The second method, the one presented here, requires only a short section of roadway. The results shown in Fig. 37 were acquired on an airport taxiway. The car was driven past a stationary camera and photographed with a short-duration ultraviolet flash lamp system.

Figure 37*a* shows the front quarter view of a Porsche 944 running at a speed of 100 mph past a stationary camera. The ultraviolet illumination system provided a flash energy of 500 W-s with a duration of 10 μs. Figure 37*b* shows a rear-quarter view of the same setup. The tufts on the body were made from white sewing thread with a length of about 1 in. They were attached to the car body with long strips of vinyl tape made up on a tuft board as described in Sec. III.D.2. The tufts on the rotating wheel were made from 0.002-in.-dia. monofilament, attached with liquid adhesive.

When these tufts are photographed with such a short-duration flash lamp, all the tuft motion is frozen, thereby eliminating one of the cues for separated flow. The rapid, random tuft motion associated with separated flow is still graphically evident in the image, however. The tuft orientations are randomly distributed. It is evident that the entire side of the car body is essentially separated downstream of the front wheel well.

8 Wind Tunnel Stilling Chamber and Nozzle Contraction One of the more unusual applications of flow visualization, presented in Fig. 38, is the stilling chamber and test section nozzle contraction in the 5- by 8-ft Boeing Research Wind Tunnel (BRWT). The view, taken from the upper corner with a wide-angle lens, shows fluorescent tufts on the opposite lower wind tunnel floor and wall. The dimensions of the wind tunnel at this point are 15 ft wide and 19 ft high. This wind tunnel runs at a test section static pressure of 1 atm and a maximum test section Mach number of 0.2. The nozzle contraction ratio is 7.2 so the flow in

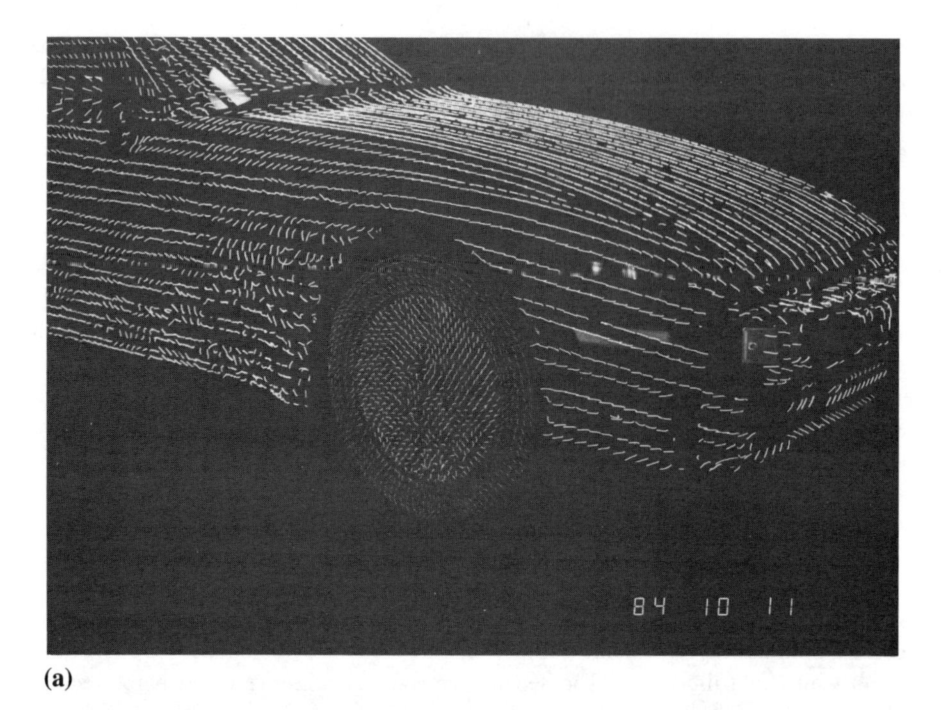

(a)

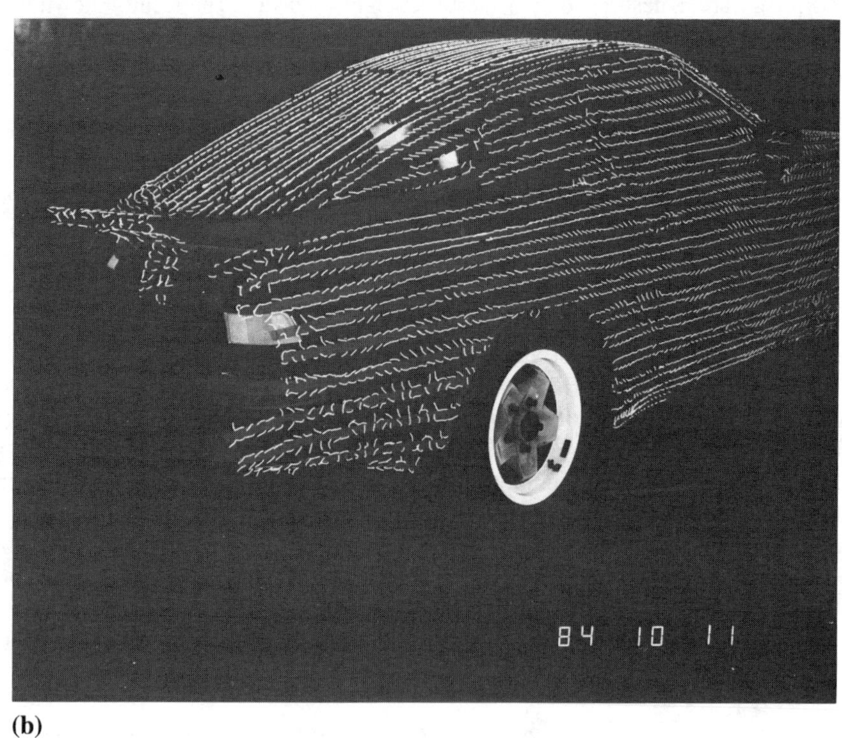

(b)

Fig. 37 Fluorescent tufts on Porsche 944 automobile running past stationary camera at 100 mph: (*a*) Front quarter view; (*b*) rear quarter view.

the stilling chamber has a Mach number of 0.028. The tufts in this example are made from sewing thread and are attached to the wind tunnel walls with vinyl tape.

The flow visualization results show a large separation on the floor of the stilling chamber at the entrance to the contraction nozzle.

C Flow Cones

Recent experience with flow cones in flight test programs has demonstrated their superior performance in terms of fidelity of indication and greatly enhanced stability, compared to conventional tufts. Most of the applications, how-

Fig. 38 Fluorescent tufts showing separation in Bellmouth of 5- by 8-ft low-speed wind tunnel.

ever, have utilized nonreflective flow cones with contrasting colors. These applications still suffer imaging problems due to limited camera resolution of very small objects at large distances. The full potential of the reflective flow cones has not been well utilized, mainly because of the slight additional effort required to implement the reflective lighting system.

Results from these nonreflective flow cone applications are not represented here, mainly because of difficulties in providing enough graphical quality to resolve these small objects. The following flow cone examples are chosen to illustrate the enhanced contrast and visibility that can be achieved with the reflective flow cones.

1 Air-to-Air Imaging: Thrush Agricultural Airplane Figure 39 shows a good example of the high visibility available with reflective flow cones. This photograph was taken at a distance of 500 ft in bright daylight. The picture has been deliberately printed dark to enhance the contrast of the bright flow cones with the dark background.

The camera had an image size of 60 × 90 mm and used a moderately wide-angle lens. The shutter speed was

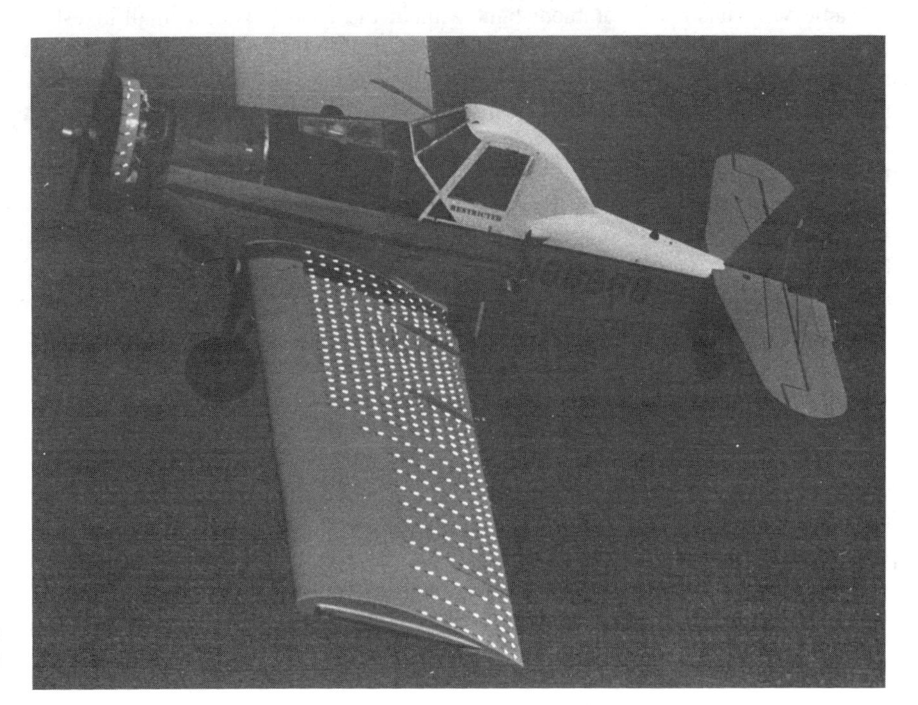

Fig. 39 Reflective flow cones on agricultural airplane photographed at distance of 500 ft in bright daylight.

1/500 s. The film type was Kodak Panatomic-X, a fine-grain film with very low light sensitivity. Such a film was needed to maintain a dark background in the high ambient light environment of flight testing in the daytime. The illumination system employed a 2000-W-s flash lamp with no filters.

Because the camera field of view had a moderately wide angle, the image of the airplane occupied only a small portion of the entire negative, about 10%. This feature permitted the scene to be photographed without careful aiming of the camera. It was just pointed in the general direction of the target airplane. This can be a very important consideration for air-to-air photography with a single-seat chase airplane.

2 On-board Imaging: North American AT-6 Airplane The photograph presented earlier, in Fig. 2, shows a good example of the potential of the reflective flow cones for producing high-contrast images with a very wide-angle view. In this case the root section of the wing where it intersects the fuselage is just out of view behind the cockpit sill at the bottom of the picture. The photograph was taken with a 35-mm camera using a lens having a focal length of 17 mm. Placed alongside the camera was a 2000-W-s flash lamp. In this photograph the airplane is just at the force break point in a stall maneuver, and the flow cones over a major portion of the wing are indicating complete separation characterized by random and widely scattered orientations of the flow cones.

In Fig. 40 the same airplane is shown at a condition prior to the stall, where the flow is not separated. The flow cones are all oriented in the same direction with no evidence of any unsteadiness. By way of comparison, Fig. 40*b* shows the same view taken with ambient light without the flash lamp so that the reflective feature of the flow cones is not utilized. This picture illustrates how rapidly the visibility of the flow cone images decreases as the distance from the camera increases when using such a wide-angle lens. In this picture the cockpit canopy is closed, thereby obstructing more of the inboard part of the wing panel.

For the images in Fig. 2 and 40*a* the canopy was open to avoid problems of glare from the window surface when a flash lamp is placed close to the camera lens. This aspect of the lighting setup is one of the difficulties that must be overcome when implementing the reflective flow cone procedure.

3 STOL Transport Wing Mockup Figure 41*a* shows a mockup of a reflective flow cone installation on the wing of a large STOL jet transport airplane. The purpose was to demonstrate how, in a single wide-angle photograph, the flow conditions over an entire wing panel from root to tip could be documented. The outline of the wing planform was laid out on an airport taxiway, and reflective flow cones were distributed over the entire surface. The camera with a wide-angle lens was located on top of a tall ladder corresponding to the location of the tip of the vertical tail on the actual airplane.

In this picture, the photographic print was exposed lighter than an optimum print to make it easier to see the test setup. Figure 41*b* shows a magnified segment of the original picture, printed darker for maximum contrast of the flow cones. This illustrates the resolution with which the reflective flow cones can be imaged, even at the leading edge of the wing planform.

4 NASA/Boeing MOD-II Wind Turbine The flow visualization study conducted on the NASA/Boeing MOD-II 300-ft-dia. wind turbine, already discussed in Sec. IV.B.6, had been planned with the intention of using reflective flow cones. This approach was selected because of the convenience afforded by the possibility of conducting the experiment in daytime. It was recognized from the outset that the relatively massive flow cones would experience bias in their indicated direction due to the centrifugal acceleration of the rotor blade. For the region of interest on the rotor, the acceleration was about 10 g.

A quick test in the wind tunnel with the flow cones ballasted to 10 times their ordinary weight showed that the deviation of the flow cone from the flow direction would be no more than about 20° due to the centrifugal effects. This was initially considered to be a tolerable deviation, and the flow cones could still indicate a distinctive difference between attached and separated flow.

Accordingly, the wind turbine blade was prepared with several thousand reflective flow cones distributed from a chordwise station of 25% to the trailing edge. This required not a small investment in effort, considering the large area involved and the difficulty of access to the rotor blade surface several hundred feet above the ground.

Regrettably, it was not until all the flow cones had been installed that the aerodynamic effects of an array of flow cones oriented at 20° to the local flow direction were tested in the wind tunnel on a model of the rotor blade. This configuration, it turns out, makes a very effective vortex generator system, substantially altering the flow behavior under study. The flow cones had to be removed to the 70% chord station, where their effects were shown to be negligible.

Figure 42 shows the results of the abbreviated reflective flow cone flow visualization experiment conducted on the wind turbine. Figure 42*a* shows the overall view of the rotor blade, and Fig. 42*b* shows a magnified portion of the same photograph.

The photograph was taken with an image size of 60- by 90-mm, using fine-grain film. The reflective flow cones were illuminated with a white light flash lamp having an electrical energy of 400 W-s and a duration of 10 μs. The

(a)

(b)

Fig. 40 Flow cones on AT-6 airplane: (*a*) Reflective lighting; (*b*) ambient lighting.

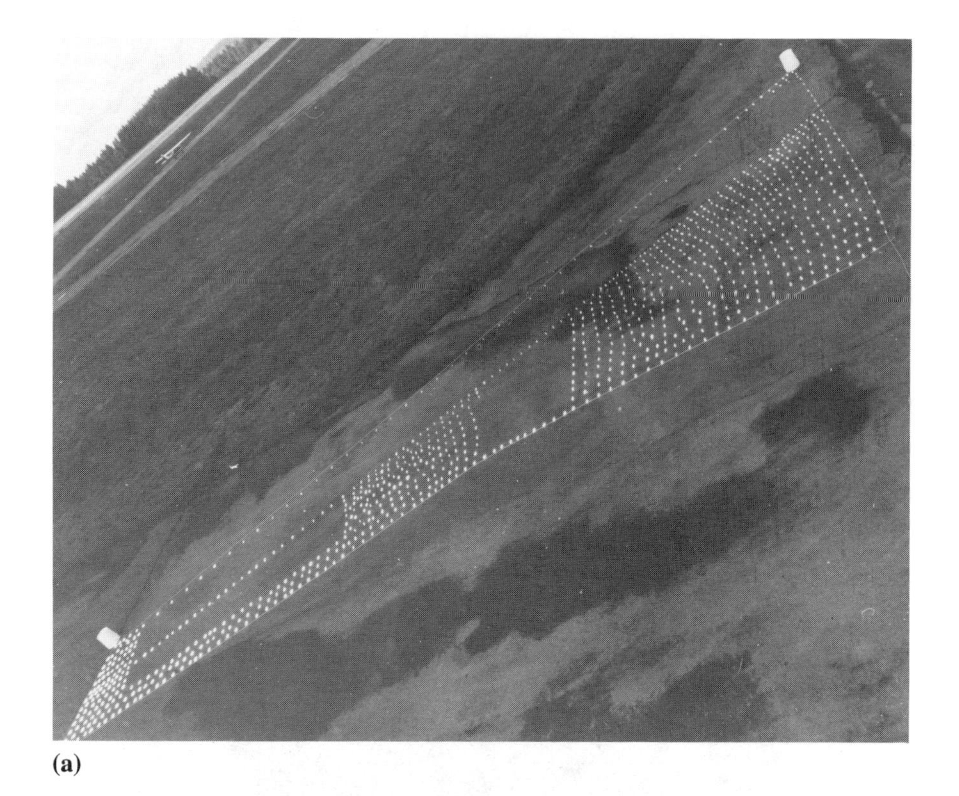

(a)

(b)

Fig. 41 Reflective flow cone demonstration on STOL transport mockup: (*a*) Overall view; (*b*) close-up view.

camera shutter duration was 2 ms, and the camera and lamp were located on the ground about 400 ft downwind of the wind turbine.

The wind turbine rotor tip speed is about 240 ft/s. In the 2-ms shutter duration, the rotor tip moves almost 6 in. This results in a large blur in the image of the rotor blade. However, the reflective flow cones are made visible by the 10-μs-duration flash lamp which freezes their motion with negligible image blur.

The flow cones on the controllable-pitch outer rotor blade panel, beyond the striped panel visible in Fig. 42*b*, apparently indicate attached flow except along the trailing edge. They appear to be deflected outward about 20° due to the centrifugal acceleration. The flow cones on the fixed-pitch inboard rotor blade, however, indicate a mostly radial orientation, consistent with separated flow.

The utility of flow cones for this application is considerably limited by their relatively high mass. The reflective

(a)

(b)

Fig. 42 Reflective flow cones on NASA/Boeing MOD-II wind turbine: (*a*) Overall view; (*b*) close-up view.

imaging procedure, however, provides a convenient means of recording data in the daytime. Perhaps the mass balanced indicator described in Sec. I.D.6 could be combined with this process to yield a more practical system.

This example does serve to illustrate another application, which should be considerably more practical. Reflective flow cones could be attached to an airplane surface imaged from a stationary camera and flash lamp on the ground. Flow visualization on the lower side of an airplane could be easily visualized by flying the airplane over a stationary camera at an altitude of several hundred feet. With some additional care, upper-facing surfaces on an airplane could possibly be similarly imaged from a tower.

5 Toyota Automobile Figure 43 presents a final reflective flow cone example to illustrate the image quality available with a small flash illumination system at a distance of several tens of feet. The automobile was photographed from another car driving alongside at close to the same speed. The camera was operated with a shutter duration of

Fig. 43 Reflective flow cones on automobile.

2 ms. The flash lamp was an Olympus T32, a small battery-powered unit that ordinarily slips on top of a 35-mm camera.

V CONCLUSIONS

The principle purpose of tuft surface flow visualization is to detect separation. Efficient, high speed vehicles all have one thing in common: a requirement to avoid separation. For airplanes, any approach to the operating limits, the so-called "flight envelope," is always accompanied by a dramatic increase in the extent of flow separation. Therefore methods of detecting separation assume extra importance in such instances.

There are many ways to detect separation, but few so easy and cost effective as tufts. Tuft methods are unique in offering such a wide range of applicability and sophistication. The tuft is perhaps the most intuitively obvious flow visualization indicator, but developing a detailed understanding of tuft response can entail considerable effort. The simplest tuft experiments can be implemented with very little effort, but in many other situations a rather high level of preparation is required for best results.

Tufts can be equally effective at speeds of only inches per second, or at hypersonic Mach numbers. They can be utilized in air, water, or any other transparent fluid. Tuft experiments can be conducted on tiny wind tunnel models, as well as on the largest airplane wings.

Tufts may have been one of the earliest flow visualization techniques ever employed, but they are very likely to be one of the latest, as well.

REFERENCES

1. Bird, J. D., Visualization of Flow Fields by Use of a Tuft Grid Technique, *J. Aeronaut. Sci.*, vol. 19, no. 7, pp 481–485, 1952.
2. Petrov, A. V., Certain Types of Separated Flow over Slotted Wings, *Fluid Mech. Sov. Res.*, vol. 7, no. 5, 1978.
3. Pankhurst, R. C., and Holder, D. W., *Wind Tunnel Technique,* Academic Press, Sir Isaac Pitman and Sons, Ltd., London, 1965, pp. 156–157.
4. Shapiro, A. H., Design of Tufts for Flow Visualization. *AIAA J.*, vol. 1, no. 1, pp. 213–214, 1963.
5. Nyland, T. W., and Savino, J. M., Wind Turbine Flow Visualization Studies, *Proc. Wind Power '85,* SERI/CP-217-2902, pp. 559–564, 1985.
6. Haslam, Flt. Lt., Wool Tufts: A Direct Method of Visually Discriminating between Steady and Turbulent Flow over the Wing Surfaces of Aircraft in Flight, Rep. Mem. Aeronaut. Res. Counc. 1209, 1928.
7. Gough, M. N., and Johnson, E., Methods of Visually Determining the Air Flow around Airplanes, NACA Tech. Note 425, July 1932.
8. Clark, K. W., Methods of Visualizing Air Flow, Rep. Mem. Aeronaut. Res. Counc. 1552, May 1932.
9. Crowder, J. P., Fluorescent Minitufts for Non-Intrusive Flow Visualization, Douglas Aircraft Co. Rep. MDC-J7374, 1976.
10. Crowder, J. P., Hill, E. G., and Pond, C. R., Selected Wind Tunnel Testing Techniques at the Boeing Aerodynamics Laboratory, AIAA 11th Aerodyn. Test. Conf., Colorado Springs, AIAA Paper 80-0458, March 18–20, 1980.
11. Crowder, J. P., Add Fluorescent Minitufts to the Aerodynamicist's Bag of Tricks, *Astronaut. Aeronaut.*, Vol. 18, no. 11, pp. 54–56, 1980.
12. Crowder, J. P., In-Flight Propeller Flow Visualization using

Fluorescent Minitufts, NASA Conf. Publ. 2243, pp. 91–95, March 1982.
13. Crowder, J. P., Fluorescent Minitufts for Flow Visualization on Rotating Surfaces, in *Flow Visualization III: Proceedings of the Third International Symposium on Flow Visualization,* ed. W.-J. Yang, pp. 55–59, Hemisphere, Washington, D.C., 1985.
14. Stinebring, D. R., and Treaster, A. L., Water Tunnel Flow Visualization by the Use of Fluorescent Minitufts, in *Flow Visualization III: Proceedings of the Third International Symposium on Flow Visualization,* ed. W.-J. Yang, pp. 65–70, Hemisphere, Washington, D.C., 1985.
15. Crowder, J. P., and P. R. Robertson, Flow Cones for Airplane Flow Visualization, in *Flow Visualization III: Proceedings of the Third International Symposium on Flow Visualization,* ed. W.-J. Yang, pp. 60–64, Hemisphere, Washington, D.C., 1985.
16. Flow Cone Product Bull., X-Aero Co., Mercer Island, Wash.
17. Crowder, J. P., Flow Direction and State Indicator, U.S. Patent no. 4,567,760, February 4, 1986.
18. Lissaman, P. B. S., Aerovironment, Inc., Monrovia, Calif., private conversation.
19. Fluorescent Minitufts Product Bull., X-Aero Co., Mercer Island, Wash.
20. Phillips, W. H., Theoretical Analysis of Oscillations of a Towed Cable, NACA Tech. Note 1796, 1949.
21. Paidousis, M. P., Dynamics of Flexible Slender Cylinders in Axial Flow: Part 10 Theory; Part 2. Experiments, *J. Fluid Mech.,* vol. 26, no. 4, pp. 717–751, 1966.
22. Datta, S. K., and Gottenberg, W. G., Instability of an Elastic Strip Hanging in an Airstream, *J. Appl. Mech.,* vol. 45, no. 1, pp. 195–198, 1975.
23. Merzskirch, W., *Flow Visualization,* pp. 18–19, Academic, New York, 1974.
24. Main-Smith, J. D., Chemical Solids as Diffusible Coating Films for Visual Indications of Boundary Layer Transition in Air and Water, RAE R&M 2755, February 1950.
25. Ultraviolet and Fluorescence Photography, Kodak Publ. M-27, Eastman Kodak Co., Rochester, N.Y., rev. February 1972.
26. Color Glass Filters, Corning Laboratory Products Catalog CF-3, Corning Glass Works, Corning, N.Y., 1965.
27. Color Filter Glasses, Kopp Glass, Inc., Pittsburgh, Pa.
28. Hoya Color Filter Glass, Hoya Optics, Inc., Fremont, Calif.
29. Kodak Wratten Filters for Scientific and Technical Use, Kodak Publ. B-3, Eastman Kodak Co., Rochester, N.Y., rev. March 1968.
30. Ultraviolet Filtering and Transmitting Formulations of Plexiglas Acrylic Plastic, PL-612d, Rohm and Haas Co., Philadelphia, 1979.
31. Transmittance and Exposure Stability of Colorless Plexiglas Cast Sheet, PL-53i, Rohm and Haas Co., Philadelphia, 1979.
32. Dobney, D. G., Hanson, P., and Fiddes, S. P., The "Minituft" Surface Flow Visualization Method. Experience of Use in the RAE 5 m Pressurised Low-Speed Wind Tunnel, *Aeronaut. J.,* vol. 90, number 891, pp. 10–17, 1986.
33. Mann, M. J., Huffman, J. K., Fox, C. H. Jr., and Campbell, R. L., Experimental Study of Wing Leading Edge Devices for Improved Maneuver Performance of a Supercritical Maneuvering Fighter Configuration, NASA Tech. Paper 2125, March 1983.
34. McMasters, J. H., Crowder, J. P., and Robertson, P. R., Recent Applications of Vortex Generators to Wind Turbine Airfoils, *3d Appl. Aerodyn. Conf.,* Colorado Springs, October 1985.

Chapter 10

Streaming Birefringence

W. Merzkirch

A number of liquids or liquid solutions exhibit the optical effect of streaming birefringence, i.e., these liquids become birefringent under the action of shear forces in a flow. Light propagation in such a medium is directionally dependent. An incident light wave separates in the birefringent liquid into two linearly polarized components whose planes of polarization are perpendicular to one another, and which propagate at different phase velocities. Different values of the refractive index are therefore assigned to the two components. They are out of phase when leaving the birefringent liquid, and this difference in optical phase can be visualized and measured by interferometric means. Many attempts have been made to use the effect of streaming birefringence, known for more than one century [1, 2], for determining shear rates or deformation velocities, or just for the visualization of the flow of such liquids.

Birefringence can be generated in fluids consisting of elongated and deformable molecules (polymers) or having elongated, solid, crystal-like particles in solution (coloidals). If the fluid is at rest, these particles or molecules are randomly distributed, and the fluid is optically isotropic and thus not birefringent. The shear forces in a flow cause the particles or long molecules to align in a preferential direction; the fluid is then anisotropic or birefringent. Theoretically, any fluid consisting of nonspherical particles or molecules should show this effect. Boyer et al. [3] verified experimentally that the effect, though weak, even exists in air flows with strong shear. For flow visualization or measurement, however, high optical sensitivity is necessary, which restricts the discussion to fluids whose particles (or molecules) are much larger than the molecules of normal gases or liquids.

Flow visualization by streaming birefringence depends on the behavior of the fluid's refractive index; optical interference is the principal technique for visualization. For quantitative evaluation there must be a relationship between the observable refractive index field and the state of the flow, a so-called flow–optic relation. The relevant theory is not totally developed, and determining the flow quantities often generates difficult numerical problems. The theory was not developed primarily for performing flow measurements; instead, the interest was to determine, from a simple flow pattern, characteristic molecular constants or physicochemical properties of the fluid. The double refraction in polymers has been analyzed by Philippoff [4], while Wayland [5] investigated the effect of streaming birefringence in colloidal solutions.

An aqueous solution of Milling Yellow dye (chemical nomenclature: H5G) has high optical sensitivity and has been used therefore in many visualization experiments. This commercial dye consists of the crystals of the pure dye and certain additives, e.g., Na_2SO_4 and NaCl. Dissolving the powdered dye in water requires the addition of energy, in the form of heat or ultrasound waves. The dye crystals, about 1–2 μm long, are birefringent. When the crystals suspended in the solution align under the action of flow shear, the gross effect is that the whole solution becomes birefringent to a certain degree. The properties of a Milling Yellow solution have been described by Swanson and Green [6] and by Pindera and Krishnamurthy [7]. This colloidal liquid exhibits strong non-Newtonian behavior; only for very low deformation velocities can the liquid be considered to behave like a Newtonian fluid. This viscous behavior is sensitive to temperature and dye concentration. Schmitz and Merzkirch [8] have shown that the flow curve of a particular concentration of Milling Yellow is, for a wide range of deformation velocities, exactly proportional to the flow curve of human blood.

The oldest and most widely used technique for visualizing the flow of a birefringent liquid is an apparatus called a polariscope, which is also used for photoelastic experiments (Fig. 1). A linearly polarized beam of light is directed through the flow; after traversing a second polarizer (the "analyzer"), the two beams resulting from light separation due to double refraction interfere with one another. In most cases it can be assumed that the optical axis of the birefringent fluid is perpendicular to the direction of light propagation (the z direction). The two beams are not separated in space; they coincide but propagate at different speeds in the fluid, and are out of phase when they leave the flow field. The optical phase difference of two interfering light rays in a two-dimensional test field of constant thickness b (in the z direction) is

$$\frac{\Delta\psi}{2\pi} = \Delta n \frac{b}{\lambda}$$

where λ is the light wave length, and $\Delta n(x,y)$ is the difference of the refractive indices assigned to the two separated rays. Fringes ("isochromates") appear in the field of view at positions where the quantity $\Delta\psi/2\pi$ is equal to an integer. The isochromatic fringe pattern (Fig. 2) yields the distribution of the experimental data $\Delta n(x,y)$.

Wayland [5] has derived a flow–optic relation that expresses Δn as a power series of the quantity $\dot{\varepsilon}_{max}/D$, where $\dot{\varepsilon}_{max}$ is the maximum deformation velocity of a plane (two-dimensional) flow and D is the coefficient of rotary diffusion counteracting the aligning mechanism of the shear forces. Attempts at quantitative evaluation have been made by accounting only for the linear term of the power series:

$$\Delta n(x,y) \frac{b}{\lambda} = c \frac{\dot{\varepsilon}_{max}}{D} \tag{1}$$

where c is a constant that has to be determined by calibration. This linear approximation restricts the applicability to small values of $\dot{\varepsilon}_{max}$, or to very low flow velocities or low Reynolds numbers (creeping flow). In this range, the flow curve of the Milling Yellow solution is almost linear; it follows that Eq. (1) applies to flow conditions under which this liquid behaves like a Newtonian fluid. The quantity $\dot{\varepsilon}_{max}$ in Eq. (1) is related to the velocity components u,v of the plane flow by

$$\dot{\varepsilon}^2_{max} = 4\left(\frac{\partial u}{\partial x}\right)^2 + \left(\frac{\partial u}{\partial y} + \frac{\partial v}{\partial x}\right)^2 \tag{2}$$

Peebles and Liu [9] introduce a stream function for determining the velocity field with the aid of Eq. (2). Solutions for u and v in the flow through channels of varying cross sections, including stepwise expansions, have been found with satisfactory accuracy by Prados and Peebles [10], Peebles and Liu [9], and Horsmann et al. [11], while Durelli and Norgard [12] combined this method with the hydrogen bubble technique.

The pattern visualized with the polariscope may include a second, different fringe system ("isoclines"), which carries additional information. The data density of the isoclinic fringes, however, is much lower than that of the isochromates, and it has not been used for flow-measuring purposes.

A technique that enables one to use streaming birefringence without being restricted by the amount of the flow velocity or by the viscous behavior of the fluid has been described by Schmitz and Merzkirch [13]. The two refractive indices, n_1 and n_2, are measured separately by means of a Mach–Zehnder interferometer (Fig. 3). Use is made of the facts that this interferometer has two exits, and that the two waves separated due to birefringence have different linear polarization. Each exit is equipped with a polarizer

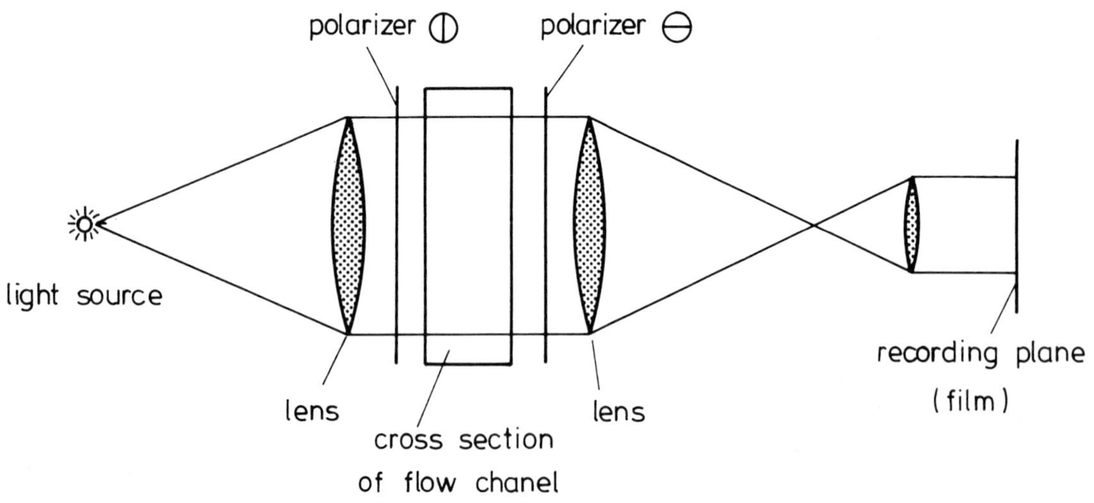

Fig. 1 Polariscope with crossed polarizers.

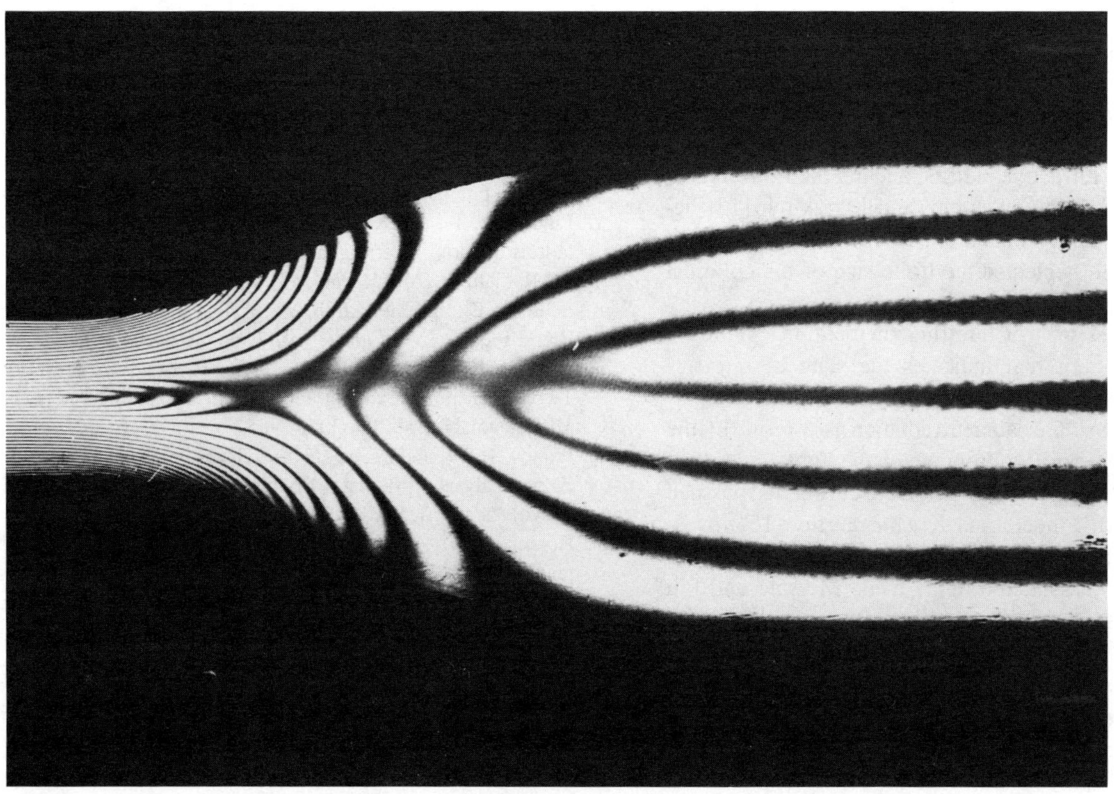

Fig. 2 System of isochromates in the flow through a divergent channel. Fluid: Milling Yellow.

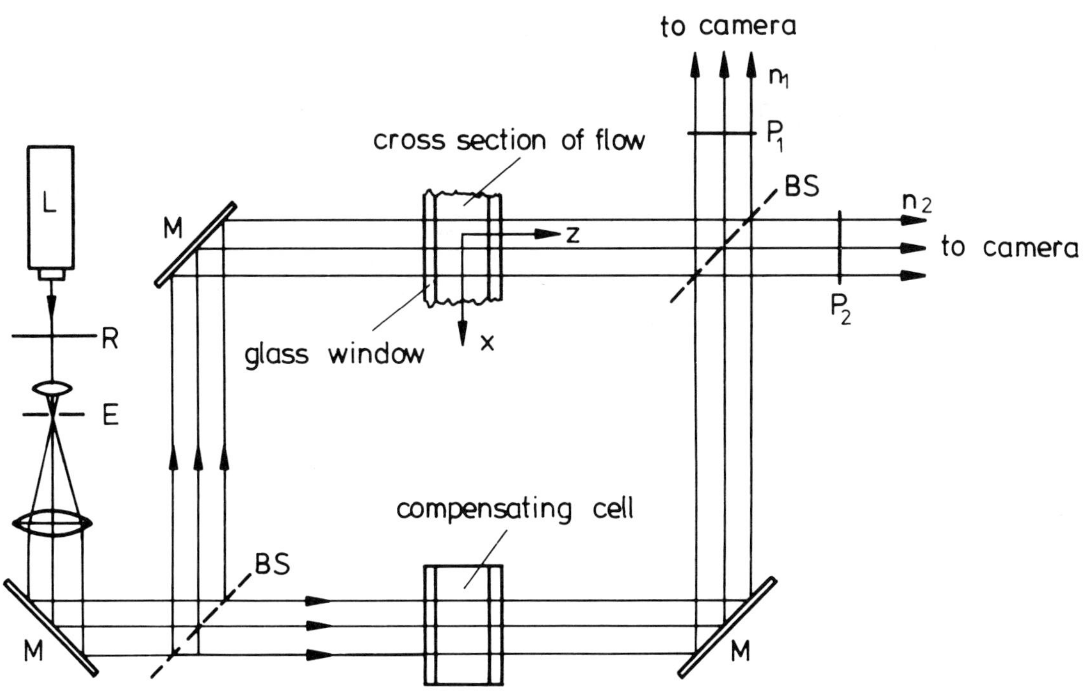

Fig. 3 Mach–Zehnder interferometer applied to streaming birefringence: L, laser; R, phase plate providing circular polarization; E, beam expander and spatial filter, M, mirror; BS, beam splitter; P, polarizer; *n*, refractive index (after [13]).

with a direction of polarization for passing only one of the two waves. This separate measurement of n_1 and n_2 delivers more information than the classical polariscope with which one determines only the difference $\Delta n = n_1 - n_2$. The quantitative evaluation of the measured data requires equations relating the refractive indices, and not their difference, to the state of the flow. Schmitz and Merzkirch [13] developed such a theory on the basis of the early work of Boeder [14], who neglected the finite size of the colloidal particles causing the birefringence.

Light scattered from a birefringent medium exhibits a characteristic pattern that indicates the state of birefringence. As in other applications, it is assumed that the pattern observable in the scattered radiation is a result of the state of flow at the position from which the light is scattered and that this radiation is not affected during its passage through the fluid. Pindera and Krishnamurthy [15] investigated some fundamental optical parameters of a Milling Yellow solution in scattered light, while McAfee and Pih [16, 17] demonstrated the possibility of determining quantities of the flow through simple geometries from the observation of scattered light. A theoretically derived flow–optic relation necessary for quantitative evaluation is not available for streaming birefringent light scattering. McAfee and Pih [17] therefore proposed an empirical relationship that was tested for relatively simple flow cases. Horsmann and Merzkirch [18] extended these investigations and developed, on an empirical basis, a flow-optic relationship that applies to the general three-dimensional case.

In experiments that have been reported using one of the three optical techniques (polariscope, Mach–Zehnder interferometer, scattered-light technique), an aqueous solution of Milling Yellow dye was used. Other birefringent fluids for visualization experiments have been described (e.g., polyvinyl alcohol [19], and vanadium pentoxide [20]), but none of them is as sensitive as Milling Yellow.

REFERENCES

1. Mach, E., *Optisch-akustische Versuche, Calve,* Prague, 1873.
2. Maxwell, J. C., Double Refraction of Viscous Fluids in Motion, *Proc. R. Soc. London, Ser. A,* vol. 22, pp. 46–47, 1873.
3. Boyer, G. R., Lamouroux, B., and Prade, B. S., Atmospheric Birefringence under Wind Speed Gradient Shear, *J. Opt. Soc. Am.,* vol. 68, pp. 471–474, 1978.
4. Philippoff, W., Streaming Birefringence of Polymer Solutions, *J. Polym. Sci., Part C,* vol. 5, pp. 1–9, 1964.
5. Wayland, H., Streaming Birefringence as a Rheological Research Tool, *J. Polym. Sci., Part C,* vol. 5, pp. 11–36, 1964.
6. Swanson, M. M., and Green, R. L., Colloidal Suspension Properties of Milling Yellow Dye, *J. Colloid Interface Sci.,* vol. 29, pp. 161–163, 1969.
7. Pindera, J. T., and Krishnamurthy, A. R., Characteristic Relations of Flow Birefringence: Part 1. Relations in Transmitted Radiation, *Exp. Mech.,* vol. 18, pp. 1–10, 1978.
8. Schmitz, E., and Merzkirch, W., A Test Fluid for Simulating Blood Flows, *Exp. Fluids,* vol. 2, pp. 103–104, 1984.
9. Peebles, F. N., and Liu, K. C., Photoviscous Analysis of Two-dimensional Laminar Flow in an Expanding Jet, *Exp. Mech.,* vol. 5, pp. 299–304, 1965.
10. Prados, J. W., and Peebles, F. N., Two-dimensional Laminar Flow Analysis, Utilizing a Double Refractive Liquid, *AIChE J.,* vol. 5, pp. 225–234, 1959.
11. Horsmann, M., Schmitz, E., and Merzkirch, W., The Application of Streaming Birefringence to the Quantitative Study of Low–Reynolds Number Pipe Flow, in *Flow Visualization,* ed. T. Asanuma, pp. 369–375, Hemisphere, Washington, D.C., 1979.
12. Durelli, A. J., and Norgard, J. S., Experimental Analysis of Slow Viscous Flow using Photoviscosity and Bubbles, *Exp. Mech.,* vol. 12, pp. 169–177, 1972.
13. Schmitz, E., and Merzkirch, W., Direct Interferometric Measurement of Streaming Birefringence, *Rheol. Acta,* vol. 22, pp. 75–80, 1983.
14. Boeder, P., Über Strömungsdoppelbrechung, *Z. Phys.,* vol. 75, pp. 258–278, 1932.
15. Pindera, J. T., and Krishnamurthy, A. R., Characteristic Relations of Flow Birefringence: Part 2. Relations in Scattered Radiation, *Exp. Mech.,* vol. 18, pp. 41–48, 1978.
16. McAfee, W. J., and Pih, H., A Scattered Light Polariscope for Three-dimensional Birefringent Flow Studies, *Rev. Sci. Instrum.,* vol. 42, pp. 221–223, 1971.
17. McAfee, W. J., and Pih, H., Scattered Light Flow-Optic Relations Adaptable to Three-dimensional Flow Birefringence, *Exp. Mech.,* vol. 14, pp. 385–391, 1974.
18. Horsmann, M., and Merzkirch, W., Scattered Light Streaming Birefringence in Colloidal Solutions, *Rheol. Acta,* vol. 20, pp. 501–510, 1981.
19. Nakatani, N., Yamada, T., and Soezima, Y., Development of a New Aqueous Solution Highly Sensitive to Flow Birefringence, *Jpn. J. Appl. Phys.,* vol. 10, pp. 1034–1039, 1971.
20. Liepsch, D., Moravec, S., and Zimmer, R., Visualization of stationary and pulsating flow in artery models, in *Flow Visualization II: Proceedings of the Second International Symposium,* ed. W. Merzkirch, pp. 587–591, Hemisphere, Washington, D.C., 1982.

Chapter 11

Speckle Photography

W. Merzkirch

Speckle photography is an experimental technique for measuring in-plane deformations of rough, solid surfaces. The application of speckle photography to the measurement and visualization of fluid flows is twofold: first, to measure the velocity of tracer particles in a plane of a flow field (''speckle velocimetry''); second, with a different optical setup, to measure the refractive index gradient in flows with varying density or concentration. In both cases, the information is available in a whole field of view at a given instant of time.

I SPECKLE VELOCIMETRY

This method was developed independently in a number of laboratories; interestingly, the first results were all reported in the same year [1–4]. A review of the technique has been given by Landreth et al. [5]. The method is explained here by considering an object point imaged on a recording plane (Fig. 1). The object point is displaced a small distance ds; assuming an imaging ratio of 1 : 1, the image point is displaced in the same direction and by the same amount ds. With photographic double exposure, one exposure taken before and a second after the displacement, the image point appears twice, separated by ds (Fig. 1). In fluid flow, the object point represents a single tracer particle moving along the flow with a velocity w. The displacement of the particle between the two photographic exposures is $ds = w \cdot dt$, where dt is the time interval between exposures. Knowing dt, determination of velocity w requires measurement of ds.

After developing the double exposure, the two image points appear on the plate as two bright spots on a dark background. These two points are illuminated from behind with a thin laser beam (Fig. 2). For the propagation of light in the space to the right of the plate, the two illuminated spots can be considered as two point sources; they emit light into two cones (''halo''). In the overlapping regime of the two cones the light from the two sources can interfere. Parallel and equally spaced interference fringes (''Young's fringes'') can be observed in a plane normal to the optical axis and at a distance l from the two sources. The fringes are normal to the direction of ds and separated by the spacing Δ:

$$\Delta = \frac{l \cdot \lambda}{ds}$$

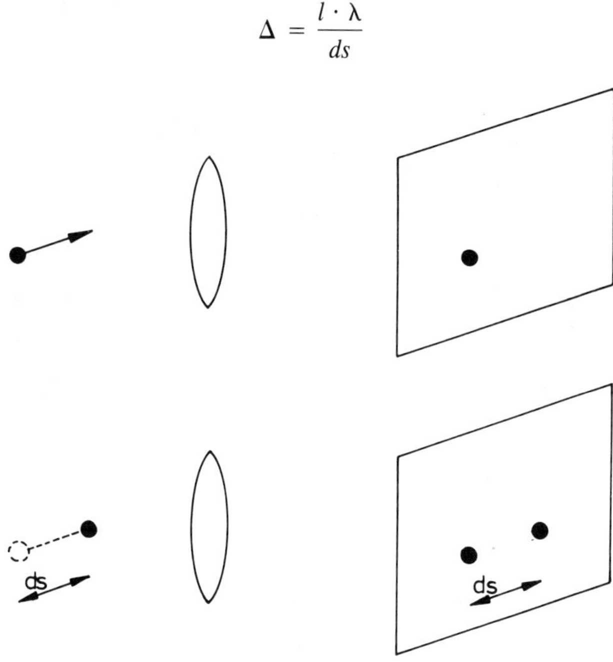

Fig. 1 Displacement ds of an object point and its image in the recording plane.

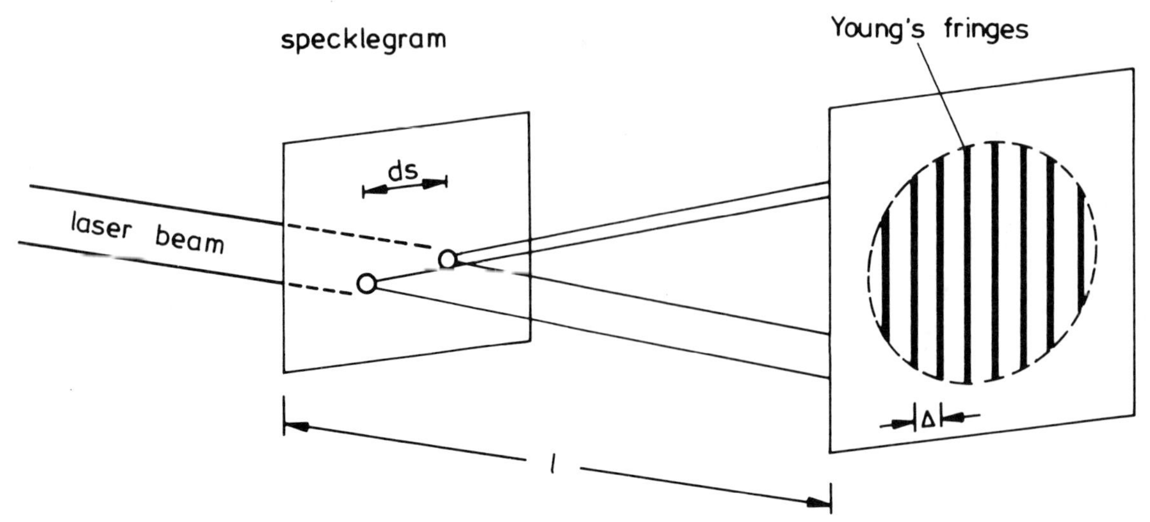

Fig. 2 Illumination of the specklegram by a laser beam, and the generation of Young's interference fringes in a plane at distance l from the specklegram.

where λ is the laser wave length. Measurement of the vector ds is thereby reduced to measurement of the fringe direction and fringe spacing in a system of Young's interference fringes (Fig. 3).

The laser beam in the setup shown in Fig. 2 appears on the developed plate as a small circular area of diameter $\leqq 1$ mm. If only one pair of particle images is present in the circular area, the flow obviously is seeded at a low seeding rate, so it is possible to identify or to image individual tracer particles. This is not a common situation. At higher seeding rates the circular area may include a number of pairs of particle images. The recording on the double exposure might not even be an imaging process, but a process of multiple interference caused by scattering of light from the tracer particles. Then the result of the recording is a real speckle pattern, a granular structure consisting of a distribution of "speckles." The differences between these two modes of speckle velocimetry have been explained by Adrian [6]. In the following discussion the term *speckle* is understood as a particle image when the seeding rate was low.

Fig. 3 System of Young's interference fringes.

The small circular area illuminated by the laser beam includes a number of pairs of speckles distributed randomly in that area. If all these speckles are displaced by the same amount and in the same direction, the resulting system of Young's fringes will be more or less the same as if only one pair had been illuminated. But if the illuminated speckles have not all experienced the same displacement, different systems of Young's fringes will be generated, which overlap each other. The pattern is to a certain degree decorrelated; the fringe system is of reduced contrast or visibility. If it is still possible to identify a dominating frequency in the fringe pattern, i.e., to measure a mean value of the fringe spacing Δ, then one attributes a mean value of the displacement ds to all speckles appearing in the circular area. Reduction in size of this area, e.g., by focusing the laser beam with a lens, is desirable for improving the local resolution. Single speckles illuminated near the edge of the circular area contribute to the noise level of the signal, i.e., they reduce the fringe visibility.

Discussion of a single pair of speckles in the field can serve to explain some limitations of the method. Each speckle has a certain diameter d_p whose value can be derived from diffraction theory; d_p is equivalent to the particle image diameter when the method is explained in terms of an imaging process. The lower limit for the displacement ds that can be measured in this way is given by $ds \geqq d_p$. There exists also an upper limit when the fringe spacing Δ becomes too small, i.e., when the fringe system becomes too narrow. A rule of thumb is that this upper limit is reached for $ds \approx 20d_p$. These two values give the range of tracer velocities that can be measured in an experiment with a given time interval of the double exposure.

Many of the experiments using speckle velocimetry are performed with a light sheet generated by expanding a laser beam in one plane. Use of a double-pulsed ruby laser as the light source is very suitable for taking the double exposure. The width of the single pulses, as well as the time interval between pulses, can be made short enough to allow for a tolerable value of the displacement to be measured.

Particle concentration and particle size affect the amount of light available for taking the double exposure; this problem is probably the most severe restriction to the application of speckle velocimetry. The light intensity distribution is quite different from laser Doppler velocimetry, where all the light is concentrated in the focal point, i.e., the measuring volume. When the laser light is expanded in a sheet, however, the incident intensity scattered from an individual particle is by orders of magnitude lower than in laser Doppler systems. The low sensitivity of photographic material to radiation in the red part of the spectrum (particularly the ruby-laser wavelength) contributes to this intensity problem. Consequently, reported applications of speckle velocimetry are either cases with relatively large tracer particles or flows of low velocity, so that the duration of each light pulse is relatively long.

The intensity problem can be improved by taking multiple exposures instead of a double exposure [7]. This also improves the clarity of the Young's fringes, but the method is then restricted to flow fields with smooth and moderate velocity changes. Another possibility is to use the method in forward scattering [8], but this is possible only with plane flow fields. Illumination for the recording can even be provided with a white source [9, 10]. Since such a source excludes the possibility of interference, the recording is then strictly a particle-imaging process.

In analyzing the developed double exposure (''specklegram'') by the method described in Fig. 2, the specklegram is scanned point by point with the laser. For each illuminated point one obtains a pattern of Young's fringes. From the measured fringe direction one determines the two velocity components of the respective point on the illuminated plane. The information stored in the specklegram is a very dense distribution of data on displacement vectors or velocity. Rational analysis of such an extended data field requires an automated processing system. Meynart [11] describes an interactive system requiring an operator. A fully automated system for point-by-point scanning has been developed by Erbeck [12]: Each system of Young's fringes is digitized; fringe spacing and direction are determined by a computer, which also controls scanning by the laser beam. In this system, optical elements are incorporated that compensate for the radial decrease of mean light intensity in the fringe system (halo) and for formation of secondary speckles. Erbeck and Keller [13] describe an improved form of the system.

From the fringe pattern it is not possible to determine the particle direction (positive or negative). This difficulty is overcome by an additional displacement of the recording plate between exposures; see, e.g., [14]. This technique of image shifting is equivalent to the frequency shift in laser Doppler anemometry.

As with any light-sheet method, the velocity component normal to the illuminated plane is not recovered. If a particle moving in this third direction is in the plane of the sheet at the instant of one of the two exposures, it will appear as a single image (or speckle) in the double exposure. The light scattered from such particles contributes to the ground noise and reduces the visibility of the fringes. Signal noise or decorrelation is also generated by turbulent fluctuations in the velocity vector, as well as by a nonuniform distribution in particle size and concentration. This noise does carry a certain information, and attempts have been made to decode this information to obtain, in addition to the measurement of mean velocity, data on turbulence characteristics [15] or particle size distribution [16]. A technique for compressing the two-dimensional Young's fringes to a one-dimensional pattern has been described by Yao and Adrian [17].

In addition to measurement of the velocity vector, the speckle technique allows direct visualization of the velocity field. This analysis is performed by spatial filtering of the developed specklegram. A schematic representation of this reconstruction process is shown in Fig. 4. The specklegram is illuminated with an expanded laser beam, and it is imaged by means of a lens in a plane (''Fourier plane'') where, in the aforementioned point-by-point analysis, the pattern of Young's fringes had been observed. An opaque screen placed in this plane has a transparent hole at an off-axis location. A small amount of light can pass through the hole, and a second lens forms an image of the specklegram in the final image plane. Since every point of the specklegram is illuminated by the expanded laser light, the Fourier plane is covered by an infinite number of overlapping fringe systems. Light can pass through the hole if the hole, in the Fourier plane, is located on a bright interference fringe. This requirement is met by fringe systems that are normal to the radius combining the hole and the axis, and for which the ratio of the distance from the hole to the optical axis and the fringe spacing is an integer. As a consequence, only such points appear illuminated in the (final) image plane for which the flow velocity in the respective object points of the flow plane is of a certain value and direction. The points illuminated in the image plane lie on fringes or contours of equal velocity component. The velocity difference from fringe to fringe (not to be confused with the Young's fringes) is constant. The amount and the direction of the velocity component, visualized by means of these

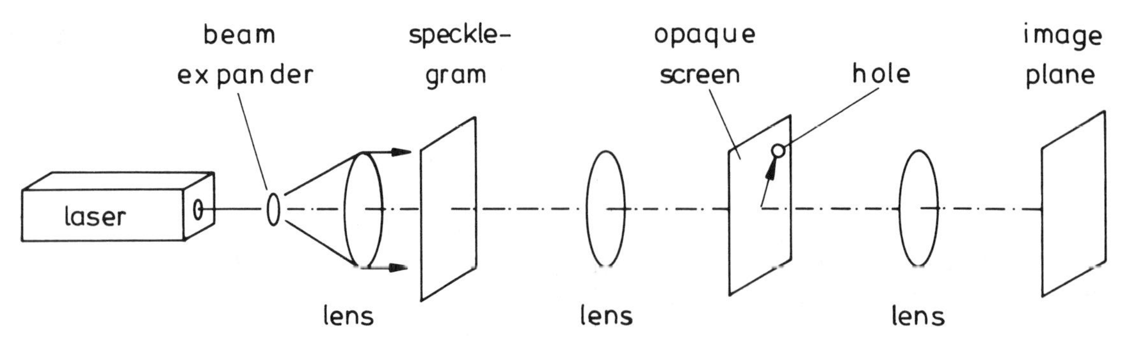

Fig. 4 Optical setup for reconstruction of a specklegram by the method of spatial filtering.

"equi-velocity fringes," depends on location of the hole in the Fourier plane; see, e.g., [8, 10, 18].

Figure 5 shows a system of equi-velocity contours in the field of a plane internal gravity wave, which has been produced by a cylinder moving upward through density-stratified salt water [8]. The observable velocity contours are at the same time curves of equal phase in the wave system. Locating the hole at a different position in the Fourier plane results in visualizing a different velocity component. This reconstruction by spatial filtering suffers from the low light intensities that have to be managed.

Application of speckle photography to velocity measurements is not yet widespread. An apparent limitation is the availability of an ultrashort and intensive light source for generating the double exposure. This problem is reduced when the flow velocity to be measured is low; this explains successful application of the method to Bénard convection [18, 19], the low-speed flow in a towing tank [20, 21] or a pipe [22]. However, the general suitability of speckle velocimetry for gas flows with higher velocities has also been verified [23–25].

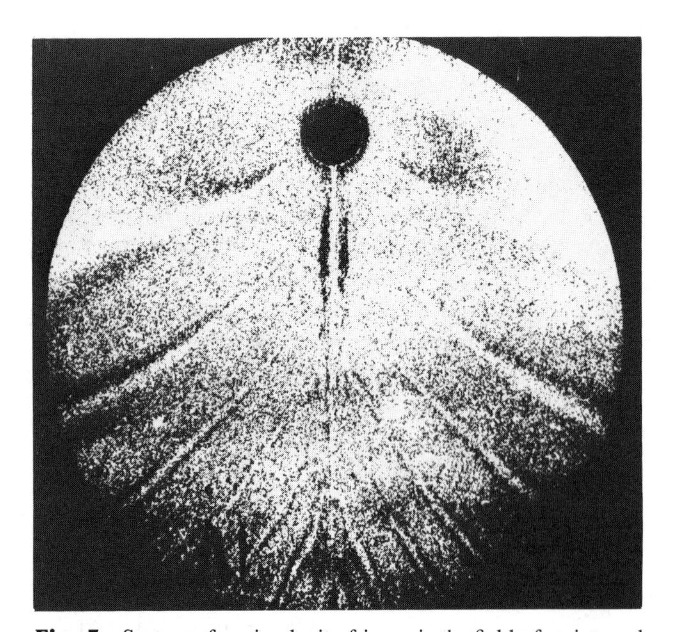

Fig. 5 System of equi-velocity fringes in the field of an internal gravity wave behind a circular cylinder moving upward in density-stratified salt water; see also [8].

II SPECKLE PHOTOGRAPHY OF FLOWS WITH VARYING FLUID DENSITY

This application of speckle photography is analogous to the schlieren and other "classical" methods, in which a light beam is transmitted through a flow with varying fluid density or concentration. However, the signal obtained from a schlieren record is not very suitable for quantitative evaluation, because it is necessary to measure the shades of gray of the schlieren photograph, and this cannot be done with high precision even using image-processing methods. Many attempts have been made to modify these systems so that the signal form is more appropriate for quantification; e.g., a form is desired that is equivalent to the fringes known from an interferogram. The results of these attempts are the moiré methods. Here a more or less regular reference pattern is produced in the undisturbed field of view, while the pattern is distorted in the presence of a flow field with refractive index changes. A comparison of the disturbed pattern with the undisturbed reference pattern permits measuring and mapping of the deflection of light in the field of view. While these moiré patterns are large scale (of a dimension known from interference methods), the method discussed in this section relies on an arbitrarily oriented micro-sized pattern, produced by speckle photography. The first descriptions of this method have been given by Köpf [26] and Debrus et al. [27]. A micro-sized reference pattern is generated by a plate of ground glass in the light path (Fig. 6). The recorded speckle pattern is determined by the scattering characteristics of the ground glass. After an exposure of this reference pattern has been taken (in the absence of the flow), the refractively disturbing (compressible) flow is turned on, and a second exposure is taken on the same photographic plate. Due to the deflection of the light in the flow, this second pattern is distorted with respect to the reference pattern. The displacement of indi-

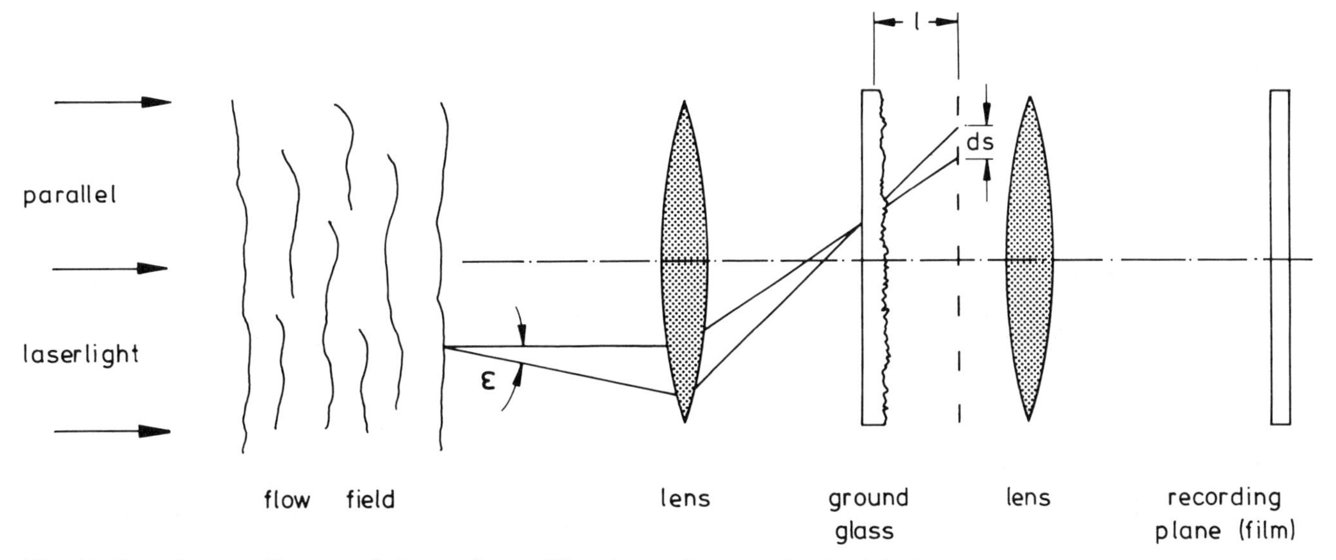

Fig. 6 Optical setup with a ground glass as the speckling element for measuring the deflection of light in a compressible flow.

vidual speckles is a measure of the local deflection angle $\varepsilon(x,y)$; (x,y designate the coordinates in the field of view or the recording plane, while z is the direction of the optical axis of the optical setup).

The system shown in Fig. 6 (see [28]) is an improvement of the setup originally used by Köpf [26]. Due to imaging of the test field on the ground glass, a light ray, deflected by an angle ε in the test flow, arrives at the ground glass at the same point as the corresponding undisturbed ray of the reference exposure. The two rays form an angle ε if a 1 : 1 imaging ratio is chosen. In the plane at distance l from the ground glass, the two rays are separated by $ds = \varepsilon \cdot l$. At the same time, ds is the displacement of the recorded speckles in the double exposure, and the distribution $ds(x,y)$ can be determined by the point-by-point reconstruction technique, via an evaluation of the respective systems of Young's fringes (Fig. 2). The distance l must be such that the recorded speckle patterns are always correlated; i.e., l must be decreased if the deflection angle ε increases. It is thus possible to investigate extended flow fields producing large values of ε, which was not possible with the original arrangements used by Köpf [26] and Debrus et al. [27].

In contrast to the schlieren method, this speckle photographic system permits quantitative measurement of the distribution of the deflection angle $\varepsilon(x,y)$. In one experiment, i.e., in one double exposure, the displacement ds is measured as a vector; the two components of the deflection angle, ε_x and ε_y, are therefore measured simultaneously. They can be converted into the respective derivatives of the refractive index or fluid density. The quantitative results are comparable to those obtained from a shearing or schlieren interferogram. However, the number of data points obtainable per square unit of the speckle record is much higher than with an interferogram. In the latter case, the derivation of data is restricted to the existence of an interference fringe, whereas the micro-sized speckle pattern allows quantitative evaluation in practically any point of the field of view. The usefulness of the speckle method has been proven by applications to flames [29, 30], convective heat, and mass transfer [31, 32].

While the latter references are applications to laminar flow problems, it has been demonstrated by Wernekinck et al. [33] that the method can even be used for resolving the light deflection in turbulent flow with fluctuating density.

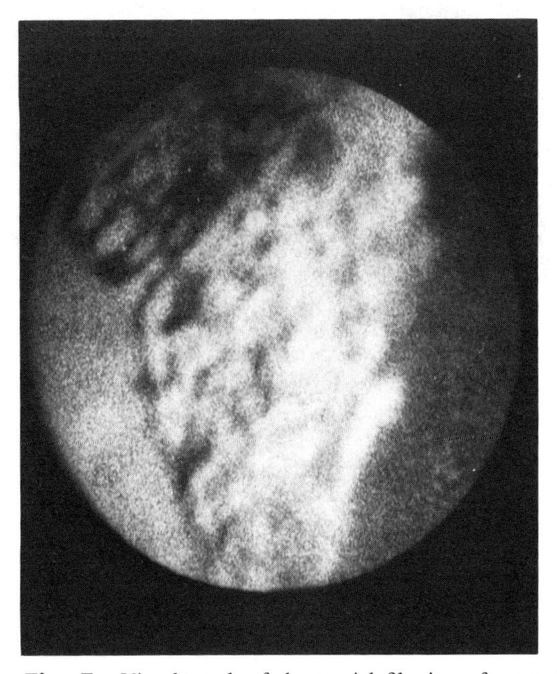

Fig. 7 Visual result of the spatial filtering of a specklegram taken of a turbulent jet of helium exhausting into still air.

The system then must be operated with a pulsed light source, e.g., a ruby laser. It is thus possible to map the randomly distributed deflection angles and to provide quantitative information from a turbulent flow field; in most cases, this is not possible with an interferometer due to the usual blurring of the interference fringes. In addition to the availability of quantitative data, a very dense distribution of information, i.e., values of the deflection angle $\varepsilon(x,y)$, can be obtained from the micro-sized speckle pattern. An automated evaluation system, mentioned in the previous section, is of great advantage. It is possible to analyze the plane distribution $\varepsilon(x,y)$ by means of appropriate statistics. Erbeck [34] (see also [35]) has developed an algorithm that permits deriving statistical properties of a flow with homogeneous turbulence by statistical analysis of the plane distribution of $\varepsilon(x,y)$.

The specklegram taken from a refractive index field can also be analyzed by the spatial filtering technique (Fig. 4). The result compares to a schlieren pattern (Fig. 7). Improvement of the quality of such a direct visualization should be the object of further research.

REFERENCES

1. Barker, D. B., and Fourney, M. E., Measuring Fluid Velocity with Speckle Patterns, *Opt. Lett.*, vol. 1, pp. 135–137, 1977.
2. Dudderar, T. D., and Simpkins, P. G., Laser Speckle Photography in a Fluid Medium, *Nature*, vol. 270, pp. 45–47, 1977.
3. Grousson, R., and Mallick, S., Study of the Distribution of Velocities in a Fluid by Speckle Photography, *SPIE J.*, vol. 136, pp. 266–269, 1977.
4. Lallement, J. P., Desailly, R., and Froehly, C., Mésure de Vitesse dans un Liquide par Diffusion Cohérente, *Acta Astronaut.*, vol. 4, pp. 343–356, 1977.
5. Landreth, C. C., Adrian, R. J., and Yao, C. S., Double Pulsed Particle Image Velocimeter with Directional Resolution for Complex Flows, *Exp. Fluids*, vol. 6, pp. 119–128, 1988.
6. Adrian, R. J., Scattering Particle Characteristics and Their Effect on Pulsed Laser Measurements of Fluid Flow: Speckle Velocimetry vs. Particle Image Velocimetry, *Appl. Opt.*, vol. 23, pp. 1690–1691, 1984.
7. Iwata, K., Hakoshima, T., and Nagata, R., Measurement of Flow Velocity Distribution by Multiple-Exposure Speckle Photography, *Opt. Commun.*, vol. 25, pp. 311–314, 1978.
8. Gärtner, U., Wernekinck, U., and Merzkirch, W., Velocity Measurements in the Field of an Internal Gravity Wave by Means of Speckle Photography, *Exp. Fluids*, vol. 4, pp. 283–287, 1986.
9. Bernabeu, E., Amare, J. C., and Arrojo, M. P., White-Light Speckle Method of Measurement of Flow Velocity Distribution, *Appl. Opt.*, vol. 21, pp. 2583–2586, 1982.
10. Suzuki, M., Hosoi, K., Toyooka, S., and Kawahashi, M., White-Light Speckle Method for Obtaining an Equi-velocity Map of a Whole Flow Field, *Exp. Fluids*, vol. 1, pp. 79–81, 1983.
11. Meynart, R., Digital Image Processing for Speckle Flow Velocimetry, *Rev. Sci. Instrum.*, vol. 53, pp. 110–111, 1980.
12. Erbeck, R., Fast Image Processing with a Micro-computer Applied to Speckle Photography, *Appl. Opt.*, vol. 24, pp. 3838–3841, 1985.
13. Erbeck, R., and Keller, J., Image Processing of Young's Fringes in Laser Speckle Velocimetry, in *The Use of Computers in Laser Velocimetry*, eds. H. J. Pfeifer and B. Jaeggy, pp. 26/1–26/5, ISL, St. Louis, France, 1987.
14. Adrian, R. J., Image Shifting Technique to Resolve Directional Ambiguity in Double-pulsed Velocimetry, *Appl. Opt.*, vol. 25, pp. 3855–3858, 1986.
15. Hinsch, K., Schipper, W., and Mach, D., Fringe Visibility in Speckle Velocimetry and the Analysis of Random Flow Components, *Appl. Opt.*, vol. 23, pp. 4460–4462, 1984.
16. Genceli, O. F., Schemm, J. B., and Vest, C. M., Measurement of Size and Concentration of Scattering Particles by Speckle Photography, *J. Opt. Soc. Am.*, vol. 70, pp. 1212–1218, 1980.
17. Yao, C. S., and Adrian, R. J., Orthogonal Compression and 1-D Analysis Techniques for Measurements of Particle Displacements in Pulsed Laser Velocimetry, *Appl. Opt.*, vol. 24, pp. 44–52, 1984.
18. Meynart, R., Equal Velocity Fringes in a Rayleigh-Bénard Flow by a Speckle Method, *Appl. Opt.*, vol. 19, pp. 1385–1386, 1980.
19. Simpkins, P. G., and Dudderar, T., Laser Speckle Measurements of Transient Bénard Convection, *J. Fluid Mech.*, vol. 89, pp. 665–671, 1978.
20. Lourenco, L., Krothapalli, A., Buchlin, J. M., and Riethmuller, M. L., Noninvasive Experimental Technique for the Measurement of Unsteady Velocity Fields, *AIAA J.*, vol. 24, pp. 1715–1717, 1986.
21. Lourenco, L., and Krothapalli, A., The Role of Photographic Parameters in Laser Speckle or Particle Image Displacement Velocimetry, *Exp. Fluids*, vol. 5, pp. 29–32, 1987.
22. Barakat, N., El-Ghandoor, H., Merzkirch, W., and Wernekinck, U., Speckle Velocimetry Applied to Pipe Flow, *Exp. Fluids*, vol. 6, pp 71–72, 1988.
23. Meynart, R., Instantaneous Velocity Field Measurements in Unsteady Gas Flow by Speckle Velocimetry, *Appl. Opt.*, vol. 22, pp. 535–540, 1983.
24. Meynart, R., Speckle Velocimetry Study of Vortex Pairing in a Low-Re Unexcited Jet, *Phys. Fluids*, vol. 26, pp. 2074–2079, 1983.
25. Hinsch, K., Mach, D., Schipper, W., Air Flow Analysis by Double-Exposure Speckle Photography, in *Flow Visualization III: Proceedings of the Third International Symposium on Flow Visualization*, ed. W.-J. Yang, pp. 576–580, Hemisphere, Washington, D.C., 1985.
26. Köpf, U., Application of Speckling for Measuring the Deflection of Laser Light by Phase Objects, *Opt. Commun.*, vol. 5, pp. 347–350, 1972.
27. Debrus, S., Françon, M., Grover, C. P., May, M., and Roblin, M. L., Groundglass Differential Interferometer, *Appl. Opt.*, vol. 11, pp. 853–857, 1972.

28. Wernekinck, U., and Merzkirch, W., Speckle Photography of Spatially Extended Refractive Index Fields, *Appl. Opt.*, vol. 26, pp. 31–32, 1987.
29. Farrell, P. V., and Hofeldt, D. L., Temperature Measurement in Cases using Speckle Photography, *Appl. Opt.*, vol. 23, pp. 1055–1059, 1984.
30. Shu, J. Z., Li, J. Y., Speckle Photography Applied to the Density Field of a Flame, *Exp. Fluids*, vol. 5, pp. 422–423, 1987.
31. Sivasubramanian, M. S., Cole, R., and Sukenek, P. C., Optical Temperature Gradient Measurements using Speckle Photography, *Int. J. Heat Mass Transfer*, vol. 27, pp. 773–780, 1984.
32. Wernekinck, U., and Merzkirch, W., Measurement of Natural Convection by Speckle Photography, in *Heat Transfer 1986: Proceedings of the Eighth International Heat Transfer Conference, San Francisco, August 1986*, ed. C. L. Tien, V. P. Carey, and J. K. Ferrell, vol. 2, pp. 531–535, Hemisphere, Washington, D.C., 1986.
33. Wernekinck, U., Merzkirch, W., and Fomin, N. A., Measurement of Light Deflection in a Turbulent Density Field, *Exp. Fluids*, vol. 3, pp. 206–208, 1985.
34. Erbeck, R., Die Anwendung der Speckle-Photographie zur Statistischen Analyse Turbulenter Dichtefelder. *VDI-Fortschritt Ber.*, ser. 8, no. 112, 1986.
35. Erbeck, R., and Merzkirch, W., Speckle Photographic Measurement of Turbulence in an Air Stream with Fluctuating Temperature, *Exp. Fluids*, vol. 6, pp. 89–93, 1988.

Chapter 12

Shadowgraph and Schlieren

Michel Philbert, Jean Surget, and Claude Véret

I LIGHT PROPAGATION IN A HETEROGENEOUS MEDIUM

When light propagates in a vacuum or in a transparent homogeneous medium, it moves along straight lines defined as light rays. Due to the electromagnetic nature of the radiation, we can also define hypothetical surfaces, called wavefronts, which are the spatial loci where the electromagnetic field vibrates in phase. Light rays of a beam are everywhere normal to the wavefronts.

If we consider a point source (i.e., light source about the size of the oscillation wavelength), light is emitted radially along straight rays (Fig. 1) and the wavefronts are spheres centered on the light source S. When the light source is very far from a wavefront (considered to be at infinity), the light rays are parallel lines and the wavefront is a plane perpendicular to the rays (Fig. 2).

A medium is optically heterogeneous when the refractive index is not the same everywhere. This optical heterogeneity has either a chemical origin (mixture of different materials), a physical origin (distribution of pressure or temperature within the same material), or both together. Through such a medium, the light rays are no longer straight lines, and the wavefronts have surface shapes other than spherical or planar.

Let us consider, for example, a parallel light beam falling on a heterogeneous medium located between two parallel planes P_1 and P_2 (Fig. 3). The medium outside these planes is homogeneous. An incident light ray propagates along a sinuous path and emerges in a direction different from that of incidence. The deviation angle α is due to the heterogeneities met during propagation. However, if we consider a plane wavefront Σ corresponding to propagation in a homogeneous medium between P_1 and P_2, and the distorted wavefront Σ' due to the heterogeneities, the change between the points can be characterized by the distance Δ at a given point M.

As the light rays are perpendicular to the wavefronts, the deviation angle α is related to the change of wavefront shape Δ by

$$\alpha = \frac{d\Delta}{dM} \tag{1}$$

where dM is the infinitesimal change of position of point M on the wavefront and $d\Delta$ is the corresponding infinitesimal change of distance Δ.

The deviation angle is then the derivative of distance Δ with regard to the coordinates of point M.

II OPTICAL VISUALIZATION METHODS

The aim of optical visualization methods is to make visible the light ray deviations (or wavefront deformations) due to the refractive index heterogeneities in a transparent medium. To become visible, these deviations or wavefront deformations must involve changes in illumination or color on an observation screen. Such transformations may be obtained with the three main optical techniques: shadowgraphy, schlieren, and interferometry.

These techniques are not new. Around 1860, L. Foucault [1] in France and A. Toepler [2] in Germany published papers concerning the use of schlieren technique, one concerning control of the optical quality of mirror surfaces,

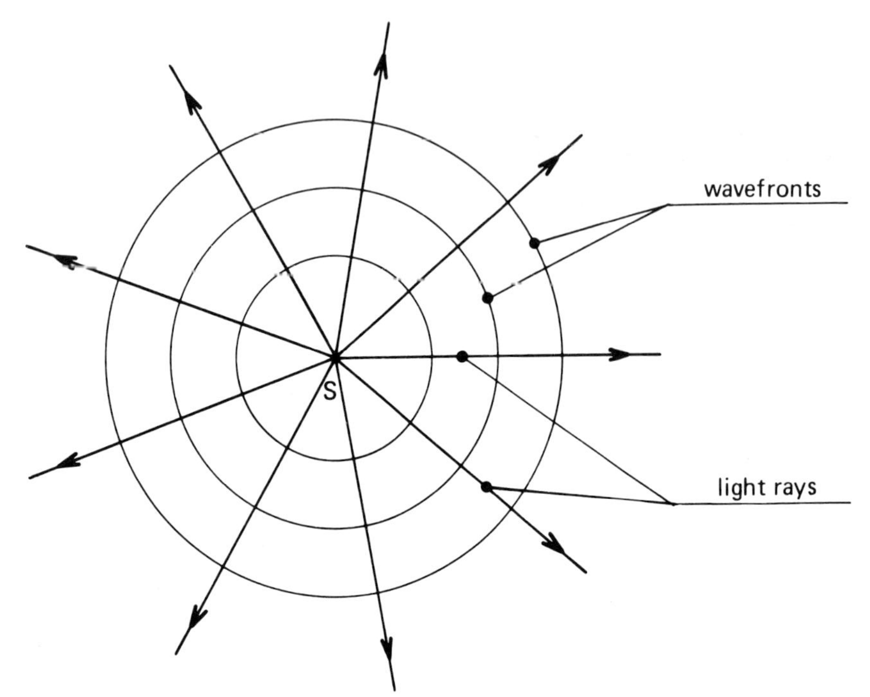

Fig. 1 Rays and wavefronts from a light source at a finite distance.

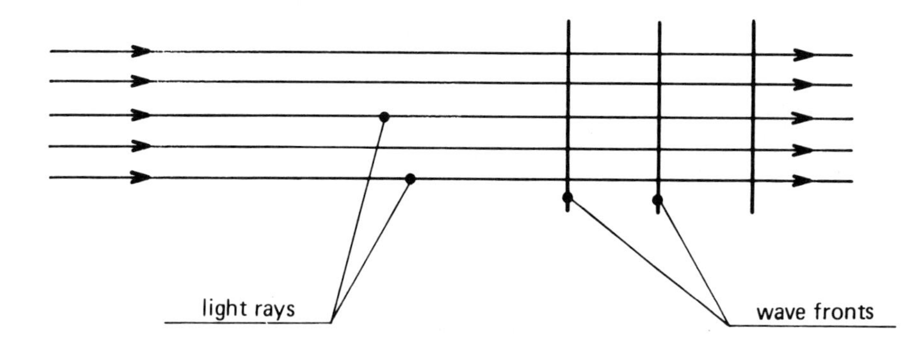

Fig. 2 Rays and wavefronts from a light source at infinity.

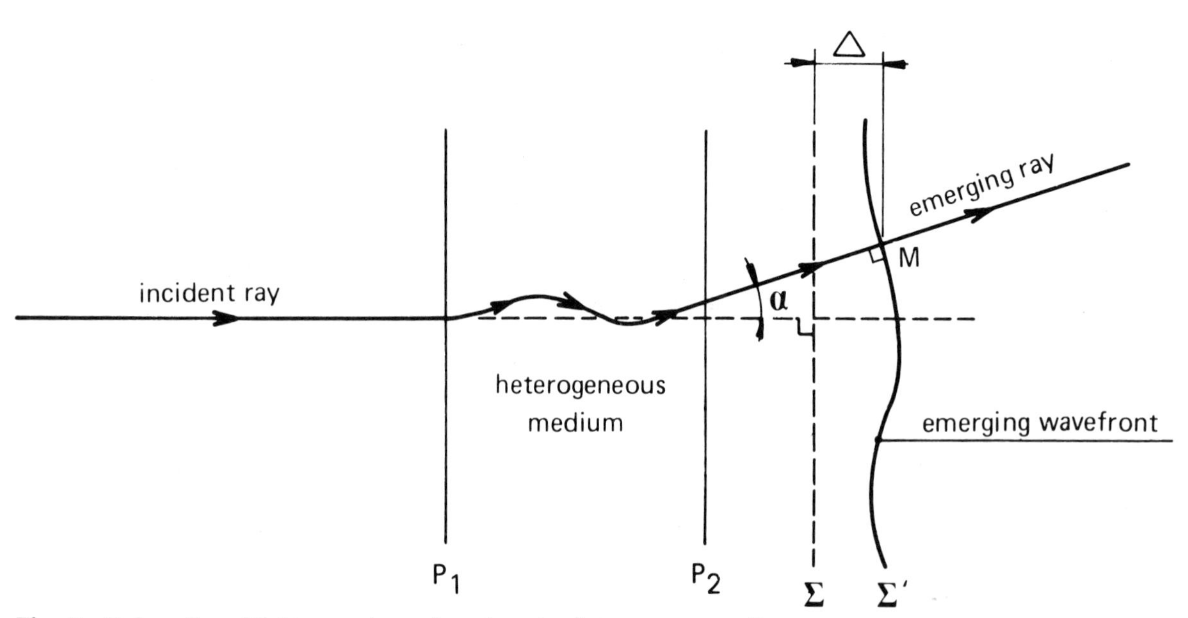

Fig. 3 Deformation of light ray and wavefront through a heterogeneous medium.

and the other concerning glass homogeneity control. Toepler was the first scientist to develop the technique for observation of liquid or gaseous flow.

During the developmental period of aerodynamics and hydrodynamics optical techniques were used and improved with the experimentation in wind tunnels and water tanks, taking advantage of other technical developments such as more powerful and convenient light sources and better recording materials. A new step has now been taken using data-acquisition and -processing systems, which allow comparisons between experimental and calculated results presented in the shape of synthetic images.

Visualization methods are often used only qualitatively. Even so, thanks to the large number of qualitative continuous data, an image can contribute to the understanding of complex physical phenomena in a flow field and clarify the data acquired discontinuously by other metrological means.

In certain particular cases, however, it is also possible to get quantitative data of the density distribution in a given field, as we shall see.

A Shadowgraphy

Shadowgraphy is the simplest technique for implementing flow visualization. A common observation illustrates the principle of the shadowgraphic technique. On a sunny day, observe on a wall or floor the projected shadow of a window. Dark lines and bands appear, even if the window is clean. Open the window, and all bands disappear; the illumination becomes uniform.

Observe also the projected shadow of a heat source, a lighted lamp or heater. Over the shadow, moving dark lines are visible, as though the hot object were smoking.

Such phenomena arise from heterogeneities within the window glass, in the first case, and from the air refractive index distribution due to the heat, in the second case. The traces observed on the screen (wall or floor) constitute a shadowgram.

To reproduce this experiment in a laboratory, we use a small-diameter light source S (e.g., concentrated filament incandescence lamp, high-pressure mercury or xenon lamp, or electroluminescent diode) and a screen E (Fig. 4). Introducing a burner near the plane P, midway between source and screen, we see moving shadows on the screen, as temperature-gradient fluctuations in the heated air over the burner generate density and refractive index heterogeneities. Without the burner, the screen E would have been uniformly illuminated, as all light rays between the source S and the screen E are straight lines. A light ray SM, which would have reached the screen at a point A along the path MA, is deflected by a density gradient at point M to path MA', the angle between directions MA and MA' being equal to α. As the screen is less illuminated at A and more illuminated at A' than without a density gradient at M, there are illumination changes on the screen related to density gradients in the space between the source and the screen.

When the density gradient is the same in all points of the plane P, all light rays are deflected by the same angle, and the screen remains uniformly illuminated. Illumination changes appear only when there are density-gradient changes (producing deflection changes) near a point M. The

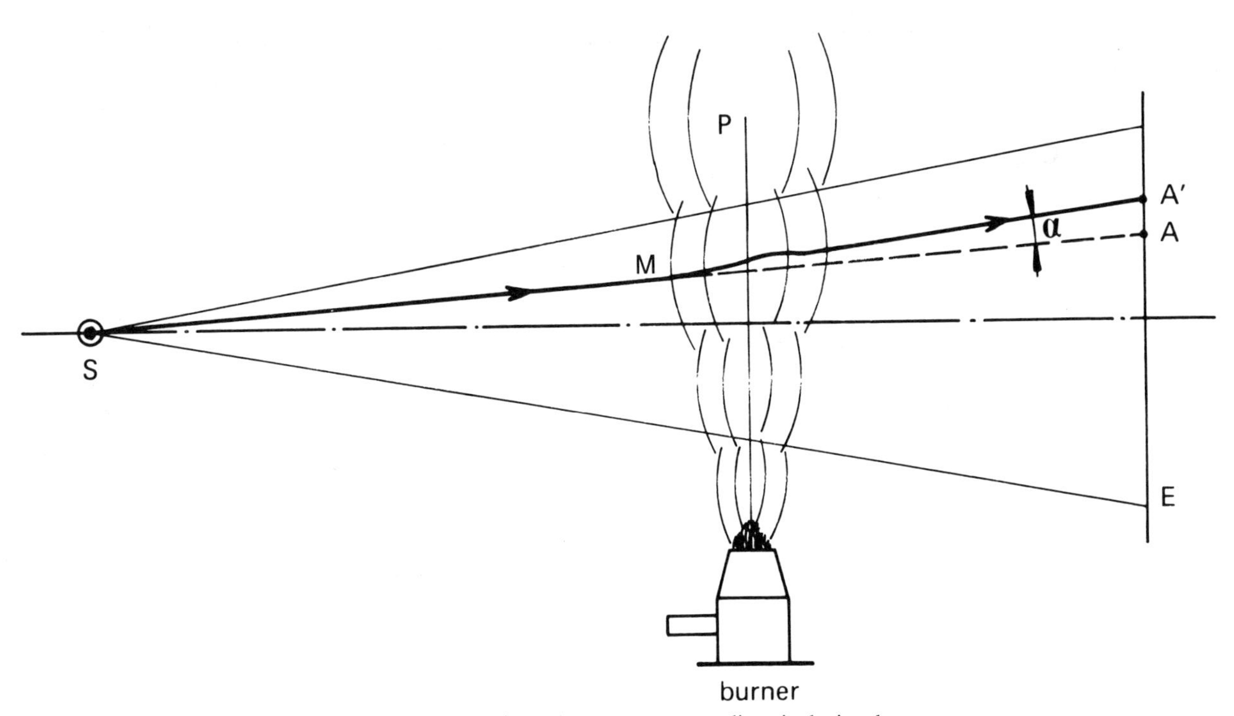

Fig. 4 Shadowgraph principle: A light ray crossing a heterogeneous medium is deviated.

shadowgraph method sensitivity is then proportional to $d\alpha/dM$, the derivative of deflection α with regard to the coordinates of point M.

For recording purposes or remote observation, any type of photographic, cinematic, or videogenic camera can be focused on the screen. A white screen is, however, not very efficient regarding the luminosity because a large part of the diffusely reflected light is lost in directions other than that of the camera lens.

A more efficient screen utilizes a retroreflective material like Scotchlite, in which small glass balls are dispersed on the surface of an adhesive sheet. For every incidence angle of illumination on such a screen, the reflected light is concentrated, in a direction opposite to that of the illumination, within a cone whose angle depends on the material gradation. When the eye or the camera lens is very close to the light source, the screen appears very bright (Fig. 5*a*). To collect the reflected light, one can use a beam splitter *B*

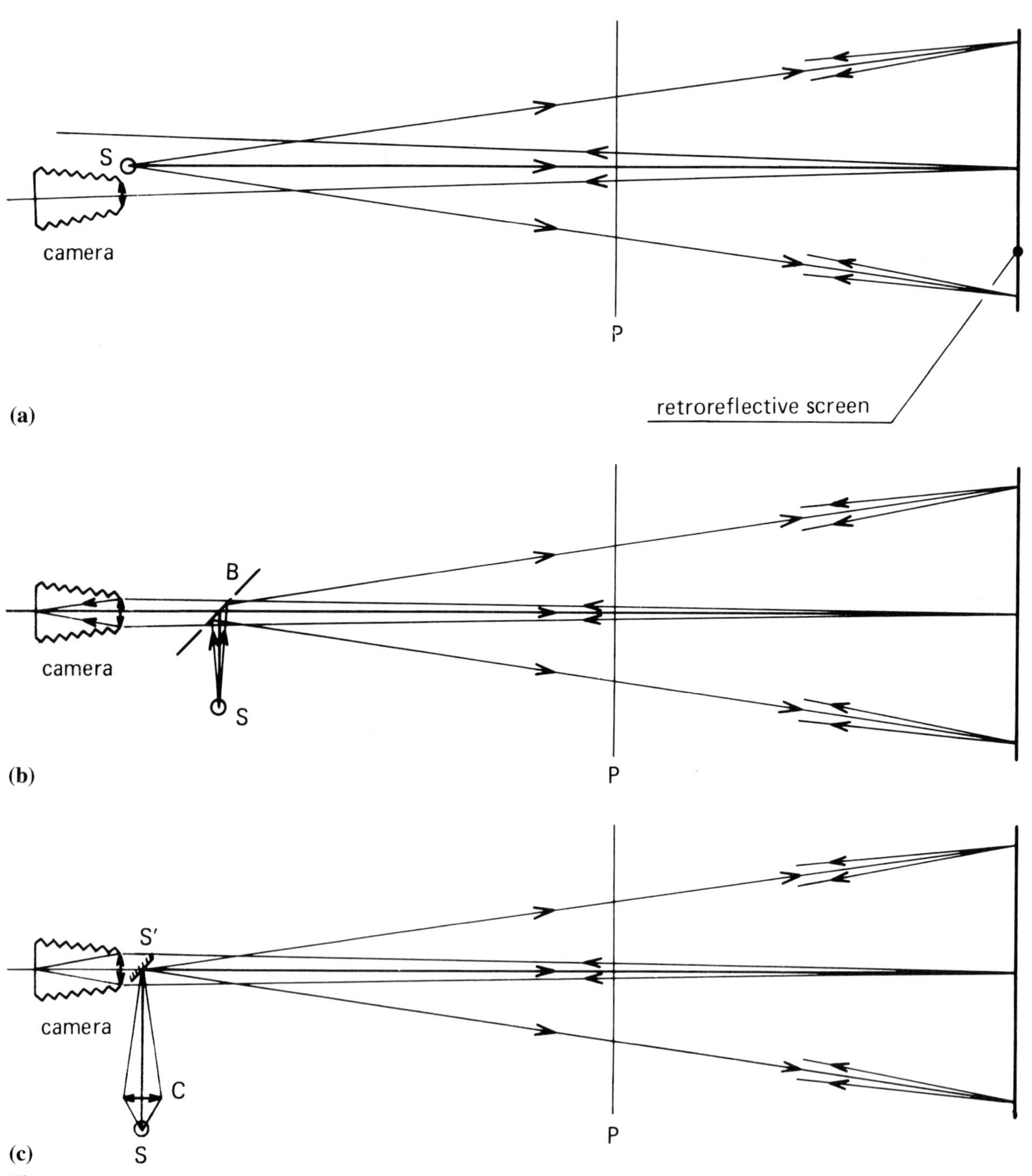

Fig. 5 Shadowgraph with a retroreflective screen.

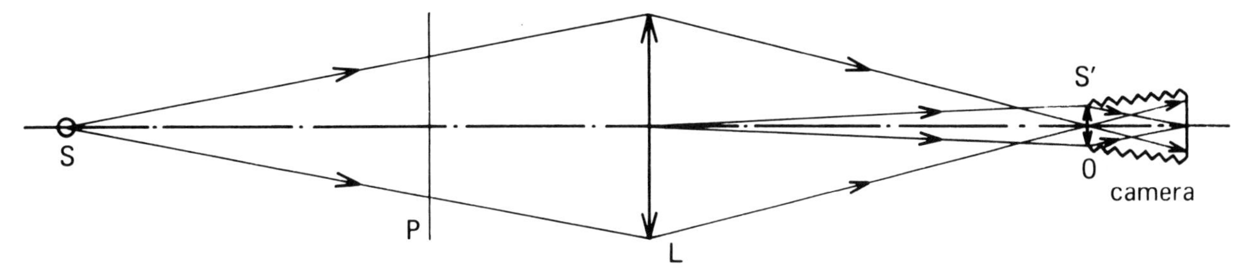

Fig. 6 Shadowgraph arrangement by transmission.

(Fig. 5*b*) or a small plane mirror placed at the center of the camera lens, on which is formed, by the condensing lens *C*, an image *S′* of the light source *S* (Fig. 5*c*). Arrangements 5*b* and 5*c* have this advantage: when there is an object in the field, like a wind tunnel model, the perspective from the camera lens is the same for this object and its shadow on the screen. The image observed with the camera is thus not disturbed by the object itself.

Shadow images may also be acquired in direct observation with maximum light efficiency. As shown in Fig. 6, a lens *L* is substituted for the screen. This lens forms an image of the light source *S* in *S′*, where the recording camera lens *O* is placed. The diameter of lens *O* has to be large enough to admit all rays deviated by the heterogeneities in the field. The recording plane has to be conjugated, not with the object plane *P*, but with all other planes, e.g., the middle plane of lens *L*. In fact, in conjugating the object and the recording plane, the shadowgraphic sensitivity is cancelled: a change in ray direction from an object point *A* has no effect on the position of image point *A′*; the ray reaching *A′* is likewise deviated, but the position of *A′* is unchanged (Fig. 7).

There is a variety of arrangements with lenses and/or mirrors for direct observation shadowgraphy. As they are basically the same for schlieren arrangements, they will be reviewed later.

B Schlieren Principle

Consider the optical arrangement in Fig. 8. A condensing lens *C* images a light source *S* at the object focus of a first field lens L_1. A second field lens L_2, placed downstream from the object field *P*, concentrates the light in its image focus, which is conjugated with the light source (*O* is a camera lens and *E* the observation screen).

This arrangement is like a shadowgraph: when the screen is conjugated with the object field, there is no sensitivity with regard to the index heterogeneities in the object field *P*. To achieve sensitivity, we introduce in the focal planes of the field lenses so-called schlieren diaphragms; the first D_1 is the primary, or entrance, diaphragm, and the second D_2 is the secondary, or exit, diaphragm.

There is a large variety of possible diaphragm couples. Some will be described later. For now, consider a square hole as entrance diaphragm D_1 and a knife edge as exit diaphragm D_2 (Fig. 9). The light source must be large enough to fill the square as uniformly as possible.

Consider the hatched ray pencil through point *A* of the object field *P* (Fig. 8). Generated by source *S*, it crosses square D_1, passes through the image of this square on the knife edge D_2, and converges on the screen in *A′*, conjugate of *A*. As shown in Fig. 10*a*, the position of the knife edge is adjusted so that the rays through half of the square image

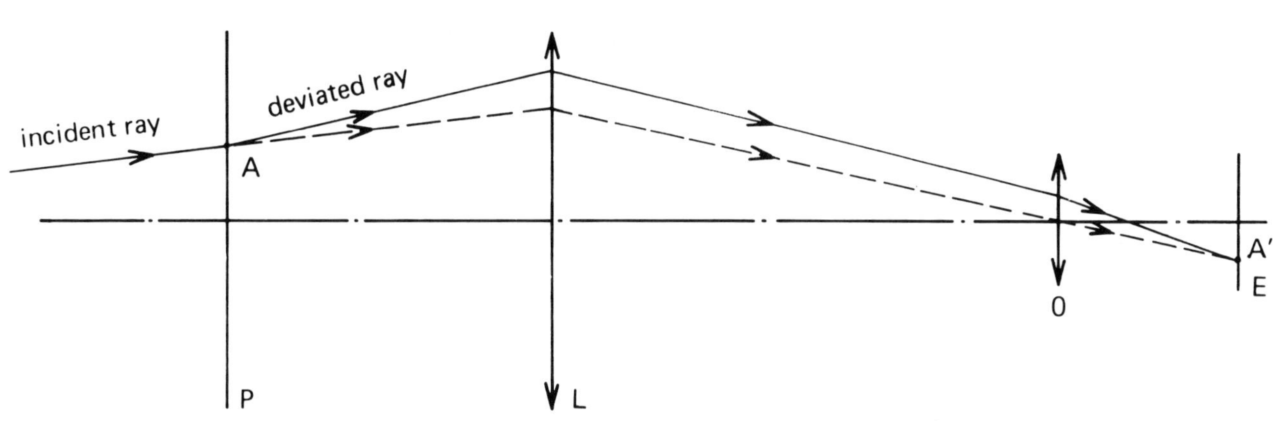

Fig. 7 The shadowgraphic sensitivity is null for a couple of conjugated points such as *A* and *A′*.

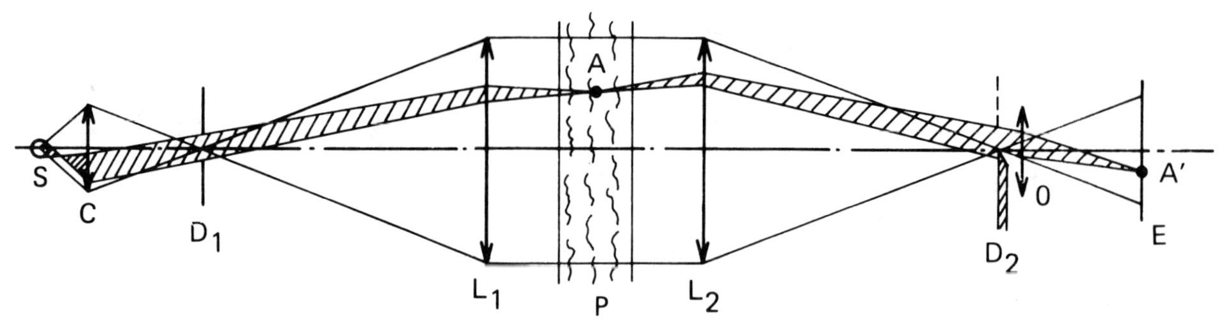

Fig. 8 Optical diagram of a schlieren system.

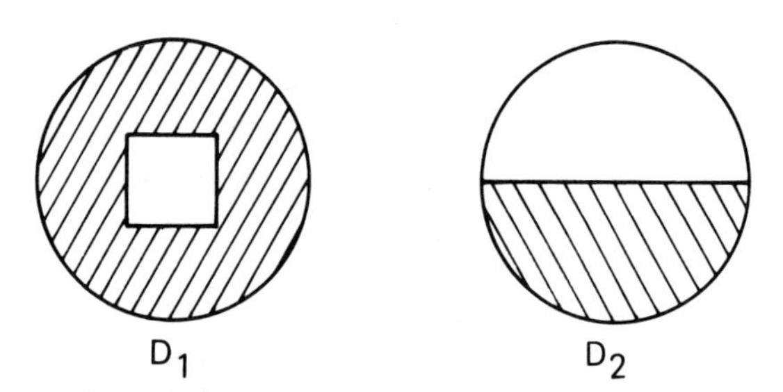

Fig. 9 Couple of schlieren diaphragms: Entrance one D_1 (square hole) and exit one D_2 (knife edge).

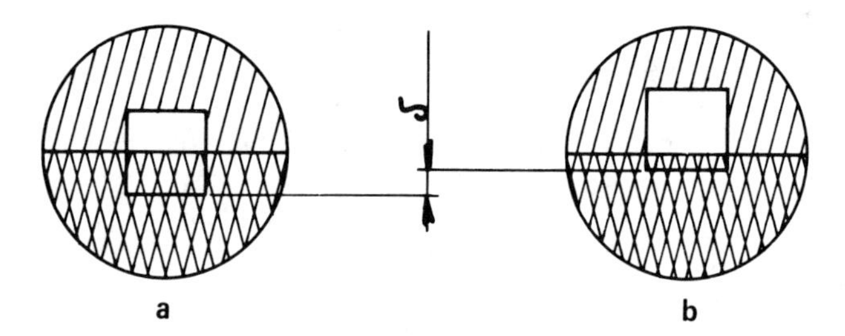

Fig. 10 Image of the entrance diaphragm D_1 in the plane of the exit one D_2: (*a*) Adjustment before experiment; (*b*) shift δ of the square image due to a deviation ε_x of the ray pencil.

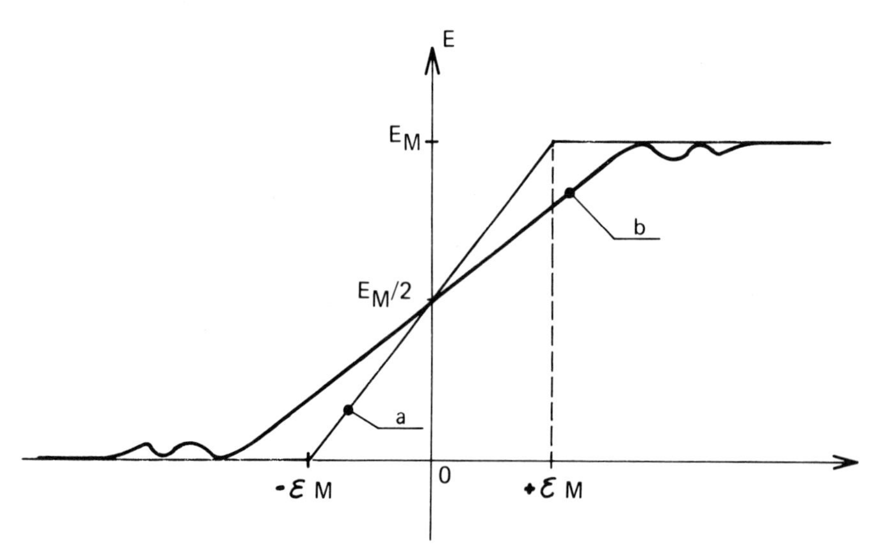

Fig. 11 Responsitivity curve $E(\varepsilon_x)$ of schlieren for a diaphragm couple square hole–knife edge: (*a*) Negligible diffraction; (*b*) when diffraction effect becomes more important.

are stopped; consequently, the illumination in A' is reduced by one half. An index heterogeneity in A deviates the pencil by a small angle ε and shifts the square on knife edge D_2 by a quantity δ:

$$\delta = f \cdot \varepsilon \tag{2}$$

where f is the focal length of lens L_2.

The change in illumination ΔE in A' is proportional to the free surface change on D_2 (Fig. 10*b*):

$$\Delta E = k \cdot a \cdot \delta = k \cdot a \cdot f \cdot \varepsilon \tag{3}$$

where a is square side length, k is proportionality coefficient, which depends on the source brightness.

The illumination changes in A' arise only when the deviation ε has a component ε_x in a direction x perpendicular to the edge, so that the schlieren sensitivity for this diaphragm couple (square and knife edge) is directional. A line of equal illumination in the image corresponds to the locus of equal deviation ε_x at the output of the perturbed medium in the object field. Unlike shadowgraphy, when there is a constant refractive index gradient in the field, all rays are equally deviated and the illumination in the image on the screen is uniformly changed.

Figure 11 shows the responsivity curve $E(\varepsilon_x)$ for this diaphragms couple (square and knife edge), with a primary adjustment so that the background illumination on the screen is $E_m/2$ (half of the maximum illumination E_m). The responsivity is linear between two limit values of ε_x, $-\varepsilon_M$ and $+\varepsilon_M$. The sensitivity s, defined as $\Delta E/\varepsilon_x$, is a constant in the interval between the limit values, and becomes null outside this interval. From relation (3), we can write

$$\begin{aligned} s &= k \cdot a \cdot f && \text{for} \quad -\varepsilon_M < \varepsilon_x < \varepsilon_M \\ s &= 0 && \text{for} \quad |\varepsilon_x| > \varepsilon_M \end{aligned} \tag{4}$$

with

$$\varepsilon_M = \frac{a}{2f} \tag{5}$$

The sensitivity, corresponding to the slope of the responsivity curve, is proportional to side a of the square and to focal lens f of the collimator for a constant value of k, i.e., for a constant source brightness. Relation (4) is, however, only a first-order approximation for relatively small values of a and f. Actually, increasing the square side at constant f requires an increase in the size of the light source image without changing brightness; there is a practical limit to this increase.

Further, when increasing f, at constant a, the diffraction effects become more and more important in the image of the square. The diffraction pattern enlarges the image with regard to the geometrical image so that the responsivity curve is changed (Fig. 11*b*). The slope of this curve, i.e., the sensitivity, is reduced.

1 Resolving Power in a Schlieren Device The resolving power of the image field on the screen depends on the aperture of camera lens O, placed near the exit diaphragm, which is used to conjugate the object field and the screen. Due to diffraction effects, reducing this aperture, the sharpness at the edge of the images is reduced, and fringes appear along the edges.

With regard to schlieren sensitivity, to exchange the entrance and the exit diaphragms would have no effect, but using a square hole as the exit diaphragm limits the aperture. On the other hand, using the knife edge increases the lens aperture: only half of this lens aperture is optically used, but it can be larger than the square hole and the resolving power is better.

2 Variety of Schlieren Diaphragms A large variety of schlieren diaphragms may be used. Some of them provide black-and-white images, as with the square–knife edge couple; with other couples, including colored patterns, the images observed on the screen are colored. The lines of equal color in these images correspond to the loci of equal deviation at the exit of the object field.

Another diversity relates to directionality of the observed deviations. As noted above, the square–knife edge couple is directional; the illumination variations are proportional to the component of the deviation perpendicular to the knife edge. However, with a circular hole as entrance diaphragm and a circular mask as exit diaphragm, the deviation is undirectional; the sensitivity is the same for every deviation direction.

Figure 12 shows several kinds of commonly used schlieren diaphragms. With a directional entrance, such as a hole, a slit, or a knife edge, may be associated for black-and-white observation, either a knife edge, a band, or a grid; for colored observation, a set of colored bands, a rainbow, or a bicolor diaphragm may be used. For undirectional sensitivity, an entrance hole may be used with bicolor or multicolor rings or a circular rainbow.

Many other devices can be imagined, each giving a sensitivity curve different from Fig. 11.

3 Interferential Schlieren Colored observations are also obtained using differential interferometers in place of the entrance and exit diaphragms. A device often used as interferometer is a polarization biprism made of two quartz prisms of the same angle glued one over the other (Fig. 13). One of the prisms has the crystal axis parallel and the other perpendicular to the edge. Between crossed polarizer and analyzer, we observe a fringe pattern parallel to the prism edges, the fringe spacing of which is inversely proportional to the prism angle.

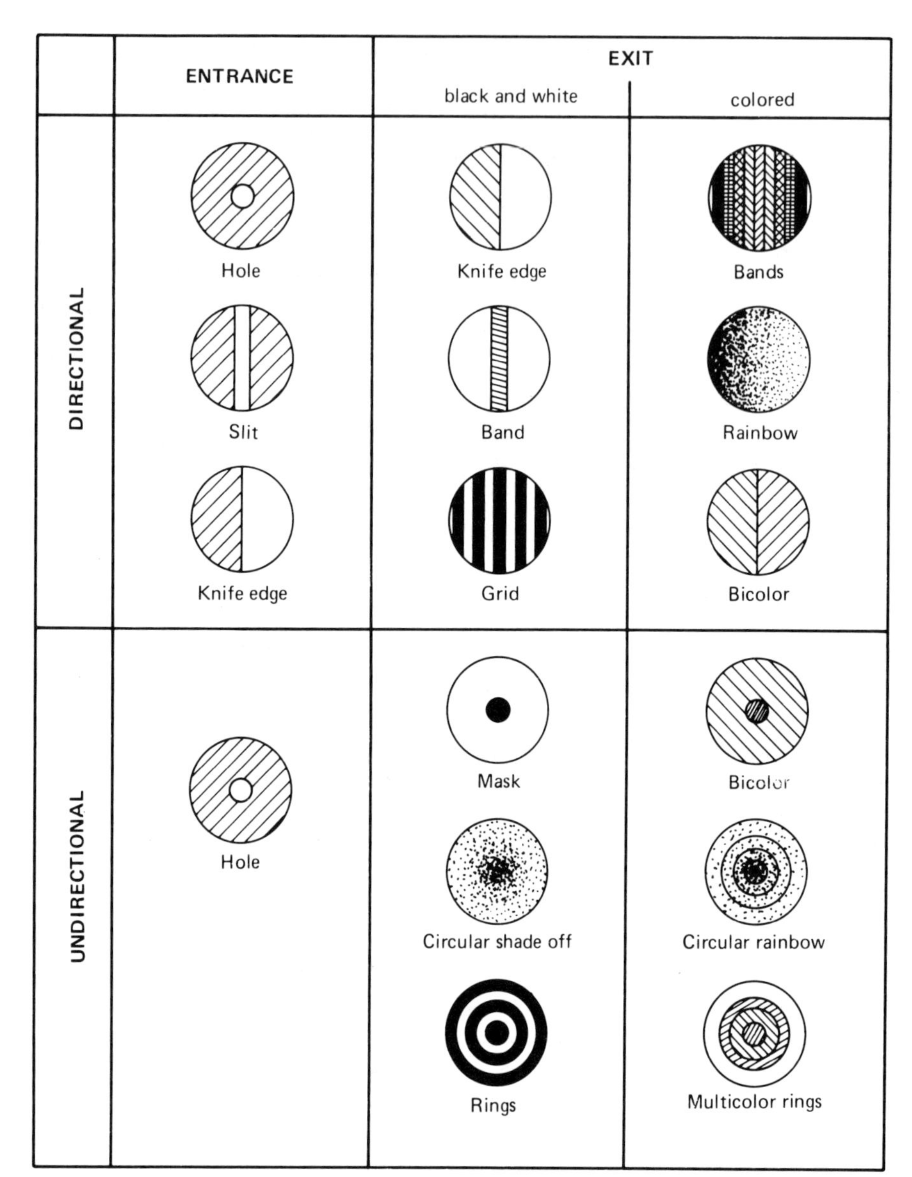

Fig. 12 Couples of schlieren diaphragms.

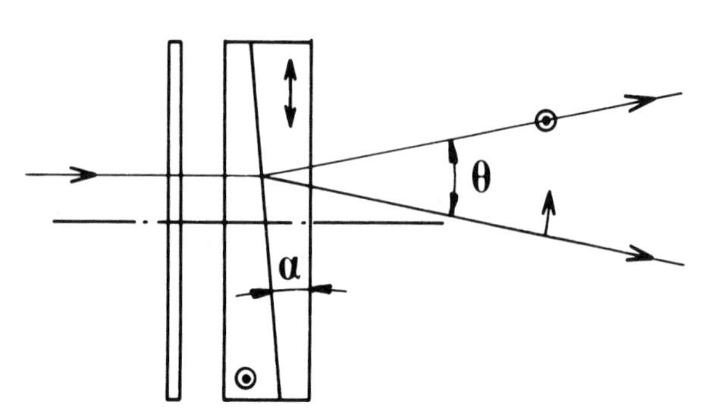

Fig. 13 Birefringent biprism: Two crystal prisms of the same angle cut with their crystal axis perpendicular to each other. Angular separation between the two emerging rays: $\theta = 2\alpha\,\Delta n$, with Δn crystal birefringence.

One biprism, preceded by a polarizer oriented at 45° of the edge direction, is placed as entrance diaphragm D_1; an identical biprism, followed by an analyzer crossed with the polarizer, is exit diaphragm D_2, the prism edges of both biprisms being parallel to one another.

Translating the prism D_2 on the image of the prism D_1 in a direction perpendicular to the prism edges and the localized fringe pattern changes the background color on the screen E when a white light source is used. With heterogeneities in the object field, a multicolored image is observed on the screen, for which the equicolor lines correspond to the loci of equal ray deviations at the exit of the field. The sensitivity is directional, as when we used multicolor bands as exit diaphragms.

When there is an object such as a model within the field, the edge of the image on the screen is doubled. The doubling distance between images is proportional to the prism angles of the biprisms. For quantitative evaluation, we can consider that the observed fringe pattern in the image corresponds to the interference between a light wave, distorted in crossing the object field, and the same distorted wave translated by a quantity equal to the doubling distance.

C Phase Contrast

Phase contrast observation is also possible using specific diaphragm couples. With a hole or a slit as entrance diaphragm and a band or an edge phase plate as exit diaphragm, the sensitivity is directional. With a hole entrance diaphragm and a phase circular mask exit diaphragm, the sensitivity is undirectional.

The phase plate or mask is a thin coating on a transparent plate. This coating is both absorbing and dephasing. The absorption corresponds to an optical density of 2 or 3, the dephasing is a quarter wave.

For small wave distortion through the object field, the contrast in the image on the screen may be some 4 to 10 times higher than with schlieren. However, the phase contrast has to be reserved for small-wave distortion visualization because, with the same light source, the image is darker and perturbations due to diffraction at the object edges are more important.

III MAIN OPTICAL ARRANGEMENTS USED FOR FLOW VISUALIZATION IN WIND TUNNELS

The test section of a wind tunnel is closed by two plano-parallel transparent windows made of good optical-quality glass. Two main categories of optical arrangements are used: reflection and transmission. Three kinds of arrangements by reflection are shown in Fig. 14.

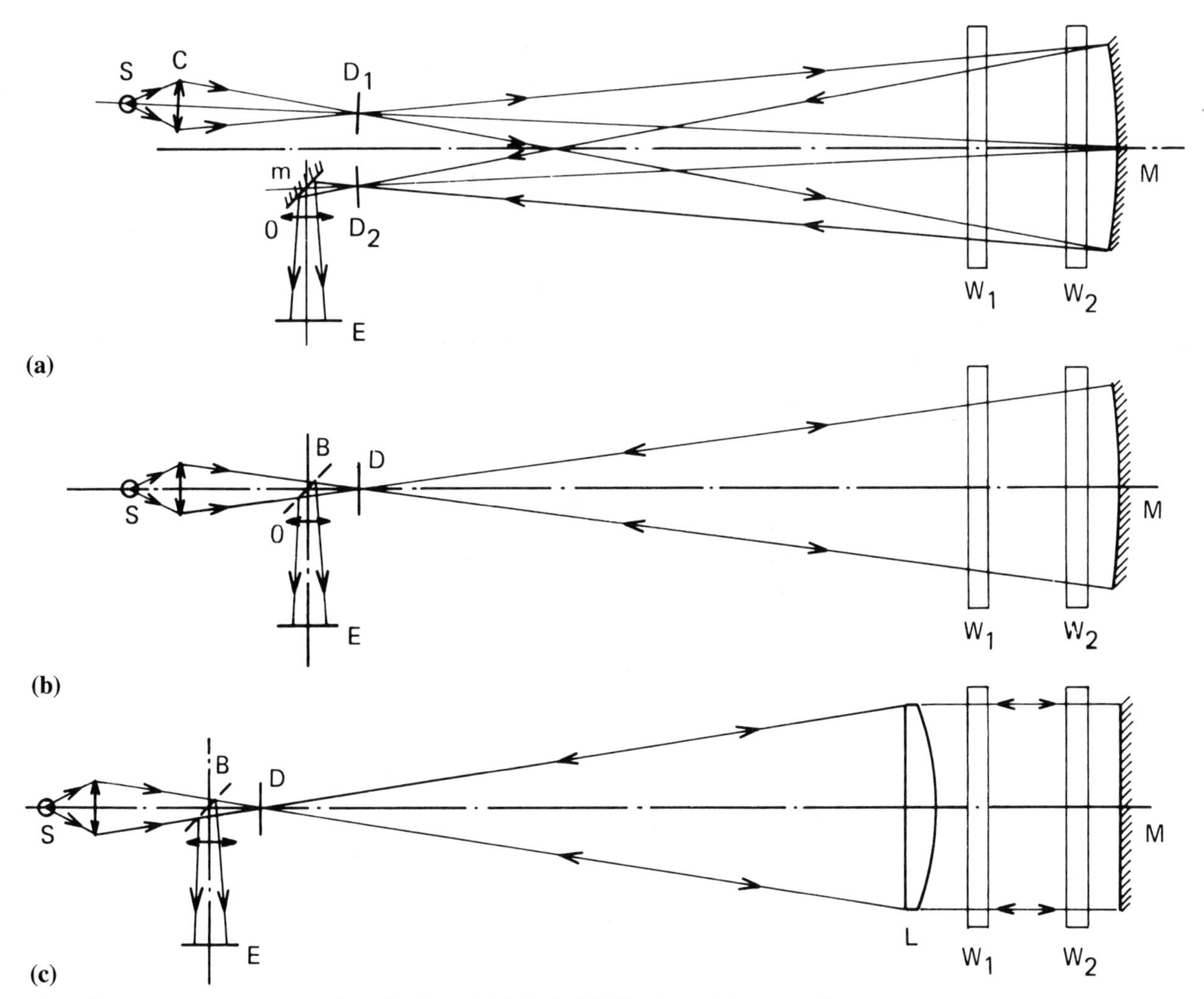

Fig. 14 Schlieren arrangements by reflections: (*a*) Spherical field mirror with separated entrance and exit diaphragms, (*b*) spherical field mirror with the same diaphragm for entrance and exit placed at the curvature center; (*c*) field optics: lens and plane mirror.

In arrangement *a*, a spherical mirror *M* is placed near the test section closed by the windows W_1 and W_2. Entrance and exit diaphragms D_1 and D_2 are placed symmetrically with regard to the curvature center of mirror *M*. Diaphragm D_1 is illuminated by the image of the light source *S*, formed by the condensing lens *C*. Behind diaphragm D_2, camera lens *O* forms an image of the test section on screen *E*, after reflection on plane mirror *m*.

This arrangement has a small inconvenience in that a ray does not cross the field at the same point before and after reflection by mirror *M*, so that the image on the screen is slightly doubled.

To avoid this doubling, arrangement *b* (Fig. 14) may be used. In this arrangement a single diaphragm *D* is used for entrance and exit, placed at the center of curvature of mirror *M*. In this case, a diaphragm such as a knife edge, a grid, or even a biprism must be used. Illumination and record are separated by means of a beam splitter *B*, with the loss of half the light from the source.

In arrangements *a* and *b*, the field is crossed in diverging light so that we get a perspective deformation on the screen. Arrangement *c*, using a field lens *L* and a plane mirror *M*, allows the crossing in parallel light. Because of the chromatic aberration of the lens, a quasimonochromatic light source has to be used in this case.

Some examples of arrangements by transmission are shown in Fig. 15, with collimating mirrors, and Fig. 16, with collimating lenses.

In Fig. 15*a* the collimating mirrors M_1 and M_2 are spherical. They have the same focal length and the same incidence angle on both sides of the observation axis, in order to cancel the coma aberration in the image of entrance diaphragm D_1 on exit diaphragm D_2. The spherical aberration effect in this image is negligible if the aperture number (focal length/mirror diameter) is not less than 10.

The entrance diaphragm is illuminated by means of light source *S*, condensing lens *C*, and plane mirror m_1. Parallel light goes through the observation field; the image is formed

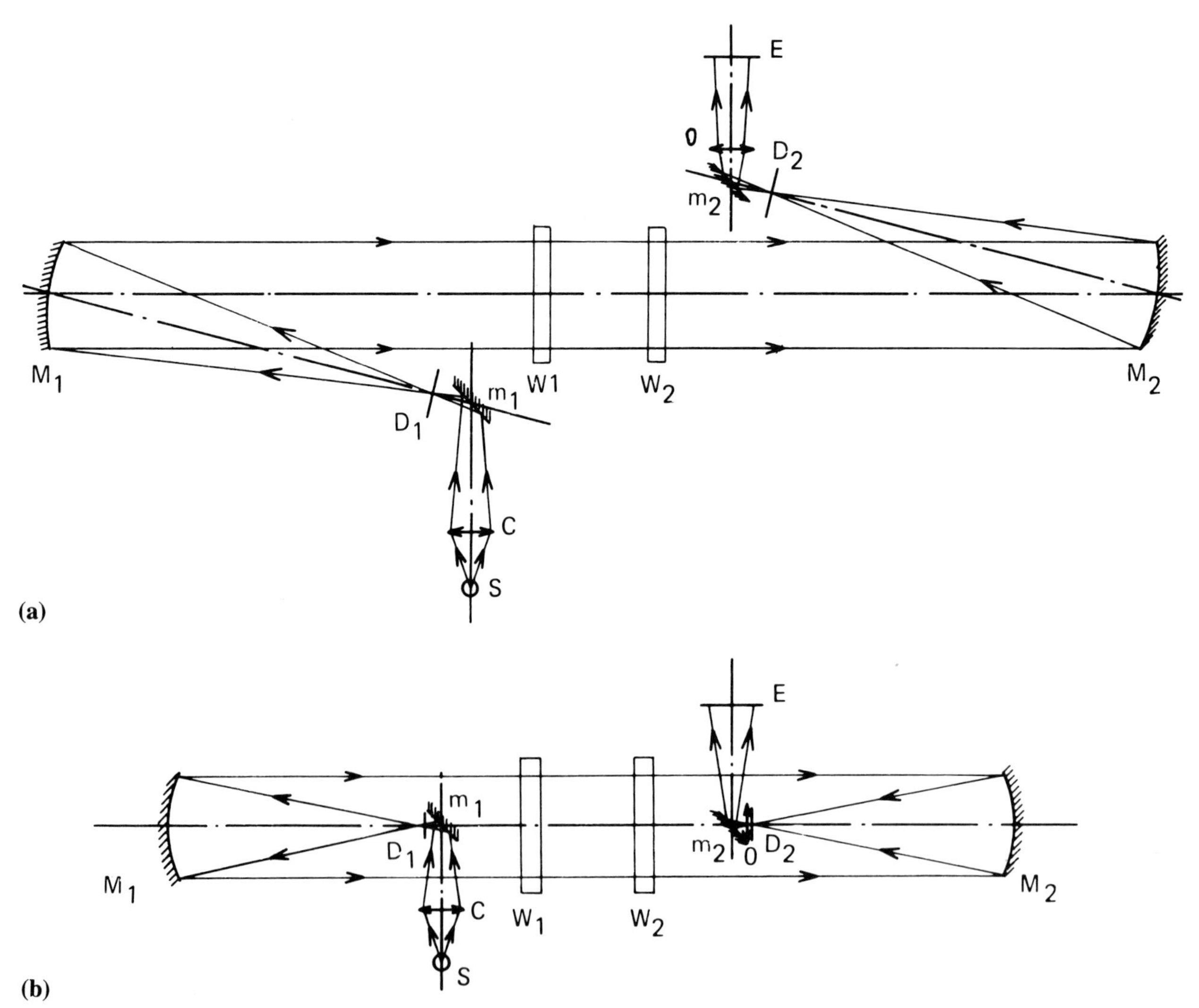

Fig. 15 Schlieren arrangements by transmission with mirror collimators: (*a*) *Z* Type; (*b*) coaxial type.

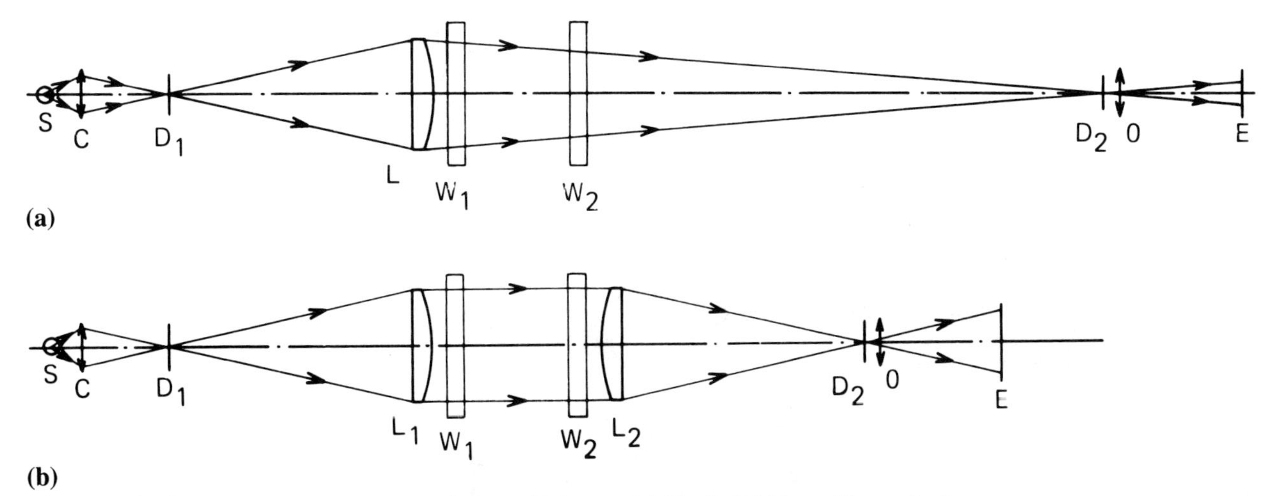

Fig. 16 Schlieren arrangements by transmission with lenses: (*a*) Single field lens; (*b*) two field lenses for observation in parallel light.

on screen *E* by camera lens *O*, after reflection on plane mirror m_2.

For very large observation fields, the aperture number conditions (≧10) imply very long focal length and the equipment would be cumbersome. A more convenient arrangement is the coaxial one shown in Fig. 15*b*. Spherical aberration of the collimating mirrors is suppressed by using parabolic surfaces and placing the entrance and exit diaphragms D_1 and D_2 on the common observation axis. Thus, the aperture number may be reduced to 5, and the focal length reduced by half. The diaphragms D_1 and D_2, the mirrors m_1 and m_2, and their supports introduce occultations within the field, which however have no consequence when they are masked by the model placed in the test section.

Two other arrangements, using lenses as field optics, are shown in Fig. 16. The first (Fig. 16*a*) includes a single lens *L* conjugating the entrance diaphragm D_1 on the exit one D_2. The second (Fig. 16*b*) includes two similar lenses L_1 and L_2, each on one side of the test section so that the field is crossed by parallel light.

To correct spherical and chromatic aberrations, every lens can be a doublet, including two lenses of different glass types (e.g., crown and flint). As for the windows, their optical qualities have to be very good. Such a solution is possible, but, for diameters less than 200 or 300 mm; it is expensive for larger diameters.

Singlet lenses with spherical surfaces may be used when the aperture number of the incident and emerging beams is more than 10. For smaller aperture numbers, one face of the singlet lens must be aspherical to reduce the spherical aberration. With singlet lenses, chromatic aberration is not corrected, so quasimonochromatic or filtered light sources have to be used.

IV USE OF OPTICAL METHODS IN FLUID MECHANICS

The optical techniques provide a general view of a wide field in the shape of *images*. In some cases, such as two-dimensional or axisymmetric flows, the results may be used quantitatively to restore the density distribution within the flow. However, they are most often used only qualitatively, thanks to their ability to contribute very efficiently to the understanding of the complex physical phenomenon. Viewing an image gives synthetic knowledge that allows judicious choice of a limited number of sites for performing precise measurements: pressure, temperature, velocity, and so forth.

A less subjective way of using images is possible with the development of fast data-acquisition and -processing hardware and software. With such equipment, geometrical data included in the image, such as location and shape of shock waves, boundary layers, wakes, and vortices, may be quantified and systematically processed.

Modern large computers and sophisticated software give aerodynamicists especially apt tools for flow numerical simulation in a large field around an in-flight structure. At the computer output, results are often presented in the shape of colored or monochromatic images. A comparison between this computer simulation and lab experiments is necessary to validate the computational codes. This comparison may be efficiently performed among other means, between computational and flow visualization images.

The pictures in Figs. 17–19 illustrate the contribution of the flow visualization techniques. Figure 17 is a shadowgram taken with the shadowgraph equipment of the Modane ONERA Center S2MA wind tunnel. The flow is

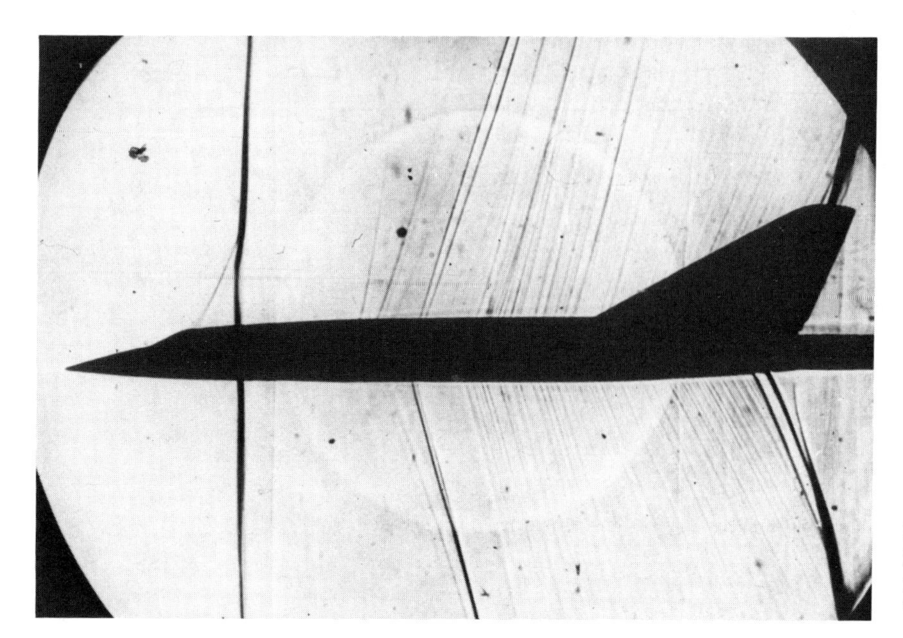

Fig. 17 Shadowgram of a transonic flow around an airplane model taken in the Modane ONERA Center S2MA Wind Tunnel.

Fig. 18 Schlieren picture of a hypersonic flow around a model of rocket taken in the Chalais–Meudon ONERA Center R2 wind tunnel.

transonic, and we see the shock waves originating from the fuselage, the wings, and the drift, as well as the trace of the boundary layer along the fuselage. Figure 18 shows a monochrome schlieren picture taken in the Chalais–Meudon ONERA Center R2 wind tunnel. The flow around a rocket model is hypersonic (about Mach 5). Figure 19 is an interferential schlieren picture, reproduced in monochrome from a colored one, taken in the Chalais–Meudon ONERA Center S8 wind tunnel. The flow around the leading edge of a two-dimensional model is supersonic (about Mach 2). In this two-dimensional case, the iso-illumination curves on the image correspond to the iso-density curves in the flow. Such an image can be processed in order to restore the density distribution in the flow.

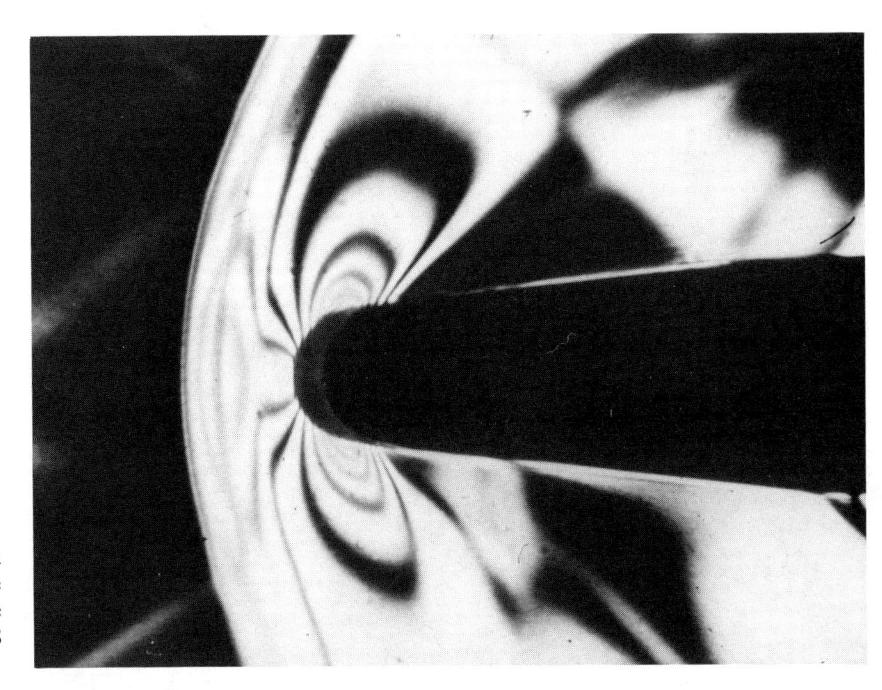

Fig. 19 Interferential schlieren of a supersonic flow near the leading edge of a two-dimensional model taken in the Chalais–Meudon ONERA Center S8 wind tunnel.

REFERENCES

1. Foucault, L., *Mémoire sur la Construction des Télescopes en Verre Argenté. Ann. Observ. Imp. Paris,* vol. 5, pp 197–237, 1859.
2. Toepler, A., *Beobachtung nach Einer Neuen Optischen Methode,* M. Cohen and Sohn, Bonn, 1864.
3. Schardin, H., Die Schlierenverfahren und Ihre Anwendungen, *Ergeb. Exakten Naturwiss.*, vol. 20, pp. 303–439, 1942.
4. Trolinger, J. D., Laser Instrumentation for Flow Field Diagnostics, AGARDograph 186.
5. Asanuma, T., ed., *Flow Visualization,* Hemisphere, Washington, D.C., 1979.
6. Merzkirch, W., ed., *Flow Visualization II: Proceedings of the Second International Symposium,* Hemisphere, Washington, D.C., 1982.
7. Yang, W.-J., ed., *Flow Visualization III: Proceedings of the Third International Symposium of Flow Visualization,* Hemisphere, Washington, D.C., 1985.
8. Véret, C., ed., *Flow Visualization IV: Proceedings of the Fourth International Symposium,* Hemisphere, Washington, D.C., 1987.

Chapter 13

Interferometry

Michel Philbert, Jean Surget, and Claude Véret

I INTRODUCTION

Propagating through a heterogeneous transparent medium, a light wavefront suffers deformations (see Chap. 12, Sec. I).

Interferometry is a technique for comparing the shape of a wavefront to that of a reference wavefront and showing the difference in wavefront shapes by means of dark and bright lines: the interference fringes. A fringe (black or white) appears every time the optical path difference between the distorted and reference wavefronts is an integer of the light wavelength.

The optical path along a ray between two wavefronts Σ and Σ' (Fig. 1) is given by

$$L = \int_{A}^{A'} n(X,Y,Z)\, dl \tag{1}$$

where $n(X,Y,Z)$ = refractive index at a point $M(X,Y,Z)$
dl = elementary geometrical path along a ray

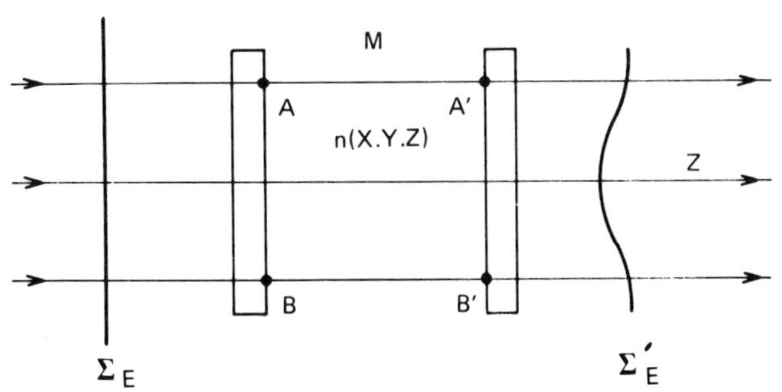

Fig. 1 The plane light wave Σ_E suffers deformations passing through the heterogeneous transparent medium *M*.

When the medium is homogeneous, n is a constant independent of the point M location, so that the optical path between A and A' is

$$L(A \rightarrow A') = ne \tag{2}$$

where e = the distance between wavefronts Σ and Σ'.

In a heterogeneous medium, n is a function of the location of point M, and the light ray is no longer a straight line. The optical path has to be evaluated with the integral equation (1) all along the ray.

In some practical cases, the changes in refractive index within the medium are small enough to permit neglecting changes of the ray shape. This is most often the case in aerodynamic flows. Thus, when a two-dimensional model in a wind tunnel test section is placed between two parallel plane windows, the refractive index may be considered constant along a light ray crossing the flow perpendicular to the windows. The refractive index gradient is perpendicular to the light rays. Through the test section, these rays are curved slightly, but this change in relation to a straight line is negligible, and the optical path can be considered to be the product of the local refractive index and the distance between the two windows.

II DESCRIPTION OF INTERFEROMETERS

A Conventional Interferometers

Conventional interferometers are those that provide interferograms, using an incoherent light source. Fringes arise from the optical path difference between two wavefronts originating simultaneously from the same light source.

We shall consider two types of interferometers primarily used in fluid dynamics: the Michelson and the Mach–Zehnder.

These devices both include two light beams propagating along separate paths: a test beam crossing the test section C, and a reference beam outside the test section.

At the instrument output, the interferogram due to the superimposition of beams E and R (Fig. 2) reveals the difference between the wavefront Σ_E' and Σ_R. This difference is equal to the difference between Σ_E' and Σ_E due to the aerodynamic phenomenon studied on condition that the profiles of Σ_E and Σ_R are identical.

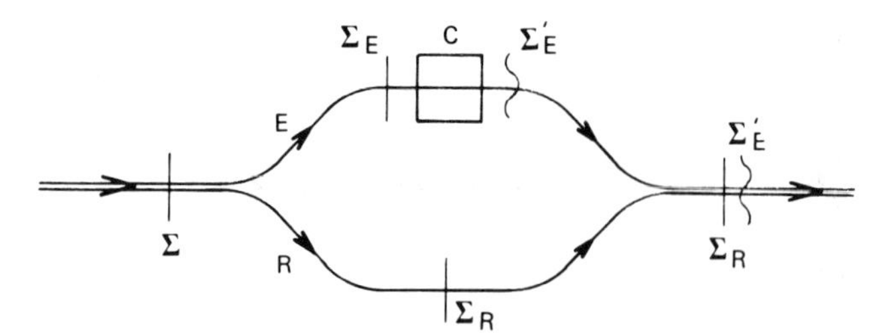

Fig. 2 The interferogram reveals the difference in optical path distribution after light travels through two different paths *E* and *R*, one of them being heterogeneous within a cell C.

1 Michelson Interferometer The beam of parallel rays (Fig. 3) supplied by lens O_1 from light source S is split by beam splitter G_1 inclined at 45° into a test beam and a reference beam. The incident beam and the split beams correspond to wavefronts Σ, Σ_E, and Σ_R. The plane mirrors M_1 and M_2, placed symmetrically with regard to the semireflective face S_1 of beam splitter G_1, reflect the test and reference beams. The test beam transmitted by G_1 and the reference beam reflected by G_1 are mixed and able to interfere. The corresponding wavefronts will have kept the profiles Σ_E and Σ_R (Fig. 4) if all the optical surfaces are perfectly plane and the transparent media (glass and air) are homogeneous. The superposed beams meet at lens O_2, which forms, on observation screen E, an image of test section C close to mirror M_1 and windows H_1 and H_2.

For the reference beam, H_3, H_4, and G_2 are compensating plates, identical to H_1, H_2, and G_1. They are used to equalize the optical paths of both interferometer arms, taking into account especially the fact that beam splitter G_1 is crossed two times more by Σ_E than by Σ_R.

When image M_2' of mirror M_2 in plate G_1 is parallel to M_1, Σ_E and Σ_R are also parallel, entering lens O_2. The fringe pattern is identical with that of an air plate of thickness e corresponding to the distance between M_2' and M_1. Annular fringes appear in focal plane S' of lens O_2, using a wide source S and uniform illumination on screen E, using a pinhole source S (infinite fringe pattern).

When M_2' and M_1 form angle α, the angle between Σ_E and Σ_R is 2α, and the fringe pattern on screen E is in the

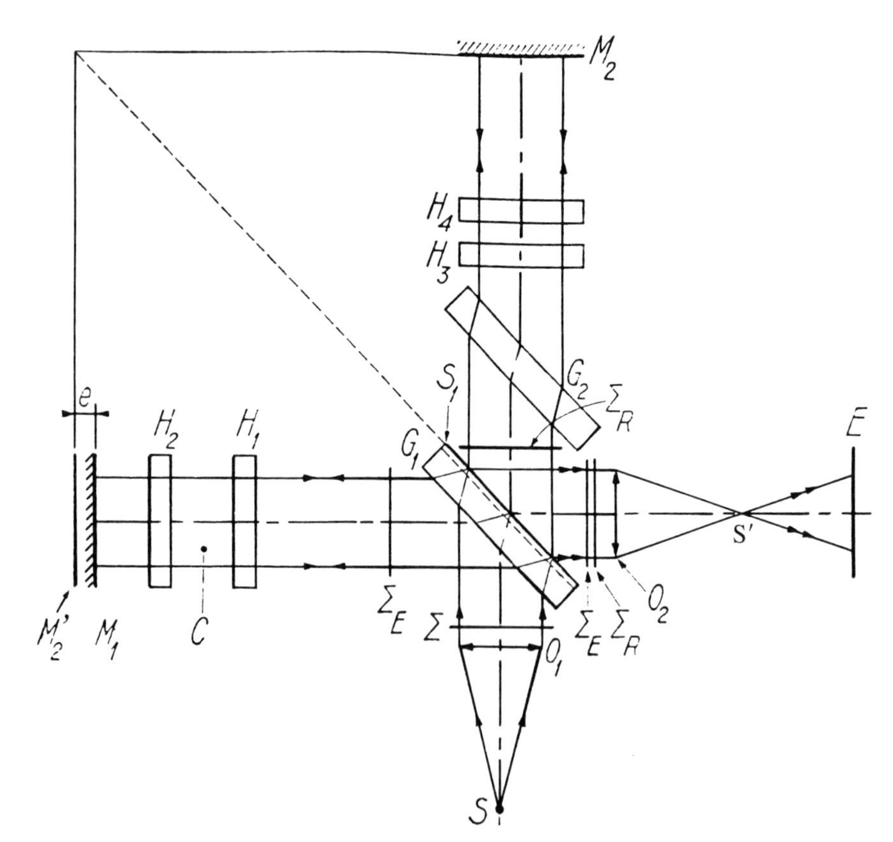

Fig. 3 The Michelson interferometer, optical components: S, light source; O_1, O_2, collimating lenses; G_1, beam splitter; G_2, compensating plate; M_1, M_2, plane mirrors; H_1, H_2, windows closing test section C; H_3, H_4, compensating plates; E, observation screen; Σ, Σ_E, Σ_R, wavefronts.

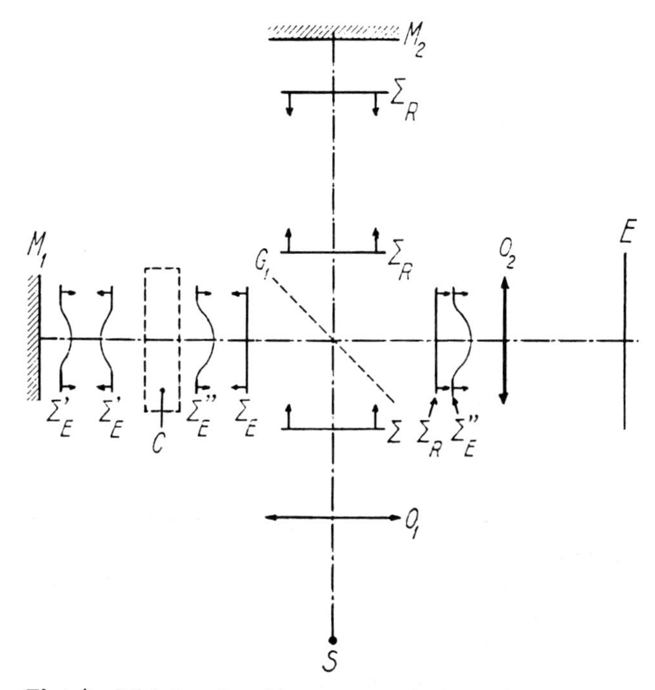

Fig. 4 Michelson interferometer, wave propagation.

shape of equidistant straight parallel fringes (finite fringe pattern).

The interferometer, initially adjusted for infinite fringe, works as follows: when the wind tunnel is running, test wavefront Σ_E is distorted into profile Σ'_E after the first crossing of the test section, and into profile Σ''_E after the second crossing. The reference wavefront does not cross the flow and keeps the same profile Σ_R all along its path; the interferogram therefore gives difference Δ between the wavefronts as

$$\Delta = \Sigma''_E - \Sigma_R \tag{3}$$

2 Mach–Zehnder Interferometer As in the Michelson interferometer, the illumination beam, collimated by lens O_1 with the corresponding incident wavefront Σ, is split by beam splitter G_1 inclined at 45° into a test beam and a reference beam (Fig. 5). The test beam reflected by coated face S_1 of G_1 is reflected by mirror M_1 in order to cross the test section closed by windows H_1 and H_2. The beam is then transmitted through the second beam splitter G_2 before reaching lens O_2, forming a test section image on observation screen E.

At the same time, the reference beam, transmitted through G_1, crosses compensating plates H_3 and H_4, identical to the test section windows, and is successively reflected by plane mirror M_2 and coated face S_2 of G_2 before being superimposed on the test beam.

Beam splitters G_1 and G_2 and mirrors M_1 and M_2 stand at the vertices of a rectangle, their faces parallel each other. Beam splitter G_2 imposes the same optical path delay on the test beam that G_1 does on the reference one. Consequently, when the 14 faces of the mirrors and plates are perfectly plane, and the 6 crossed glass materials are homogeneous, the test and reference wavefronts keep plane profiles Σ_E and Σ_R and are mixed beyond G_2. Their interference then provides, on observation screen E, an infinite fringe pattern with uniform illumination. When one of the elements, such as G_2, is tilted slightly, Σ_E and Σ_R are no longer parallel, and a straight parallel equidistant fringe pattern appears, localized near G_2. Tilting a second element, such as M_1, changes the location of the fringe pattern localization plane, placed at the intersection of the beams propagating beyond G_2, making it possible to adjust the position of this localization plane within the test section. This adjustability is an appreciable advantage of this type of interferometer, as is the fact that the test section is crossed only once. Double crossing, as in the case of the Michelson interferometer, introduces bias as soon as the flow refractive index gradients are large enough in order that the refracted light rays do not cross the flow for the second passage at the same location.

When the wind tunnel is running, the test wavefront Σ_E (Fig. 6) takes on a distorted profile like Σ'_E after crossing

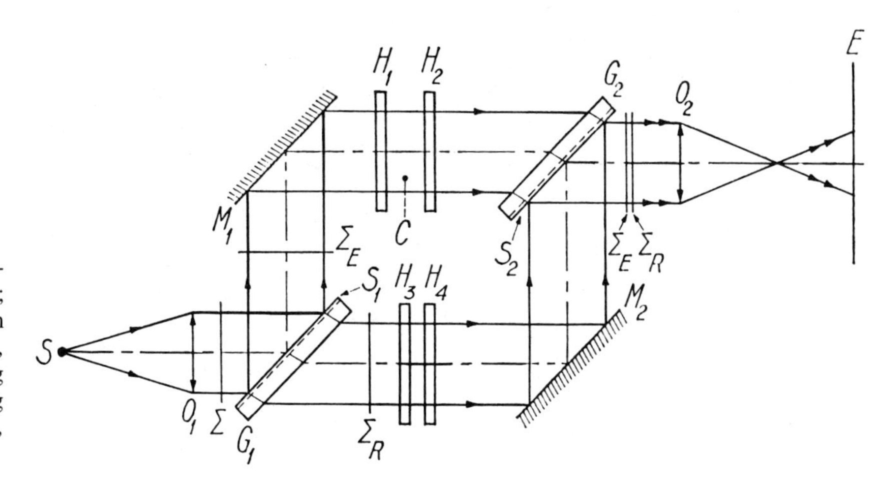

Fig. 5 Mach–Zehnder interferometer, optical components: S, light source; O_1, O_2, collimating lenses; G_1, beam splitter; G_2, beam splitter; M_1, M_2, plane mirrors; H_1, H_2, windows closing test section C; H_3, H_4, compensating plates; E, observation screen; Σ, Σ_E, Σ_R, wavefronts.

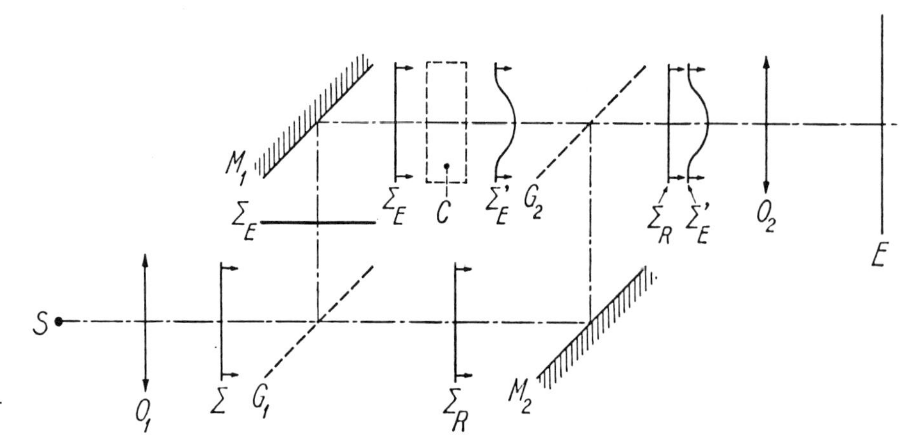

Fig. 6 Mach–Zehnder interferometer, wave propagation.

the test chamber, and the observed fringe pattern corresponds to the optical path difference Δ between wavefronts Σ_E' and Σ_R

$$\Delta = \Sigma_E' - \Sigma_R \tag{4}$$

Figures 7 and 8 show two examples of interferograms taken with a Mach–Zehnder interferometer.

B Holographic Interferometers

Since the advent of lasers, thanks to the highly coherent light they emit, a new interferometer type has arisen; the holographic interferometer.

Holography has the remarkable property of allowing a light wavefront recording in view of a subsequent restitution. Thus, in contrast with classical interferometry, in which both interfering wavefronts have to travel simultaneously along separate paths before being mixed, it has become possible to make interfere wavefronts, traveling the same path at different times. The use, with the holographic method, of a single path has the advantage of self-compensating for the optical components' eventual defects and misadjustments, as they act likewise on both interfering wavefronts. Thus, the holographic interferometers are much easier to use and cheaper to build than the classical ones. Because of these advantages, the holographic interferometer is now used almost exclusively in fluid dynamic research.

The optical arrangement of such an interferometer implemented for an ONERA wind tunnel is shown in Figs. 9 and 10. Two examples of interferograms acquired with this device are shown in Figs. 11 and 12.

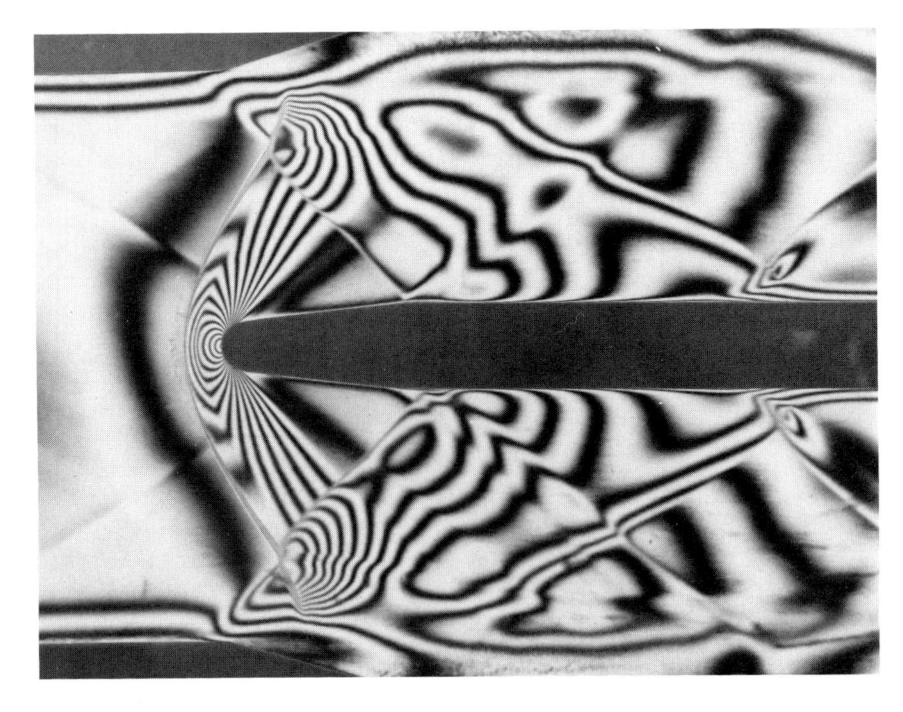

Fig. 7 Mach–Zehnder interferogram showing two-dimensional supersonic flow with an infinite fringe adjustment.

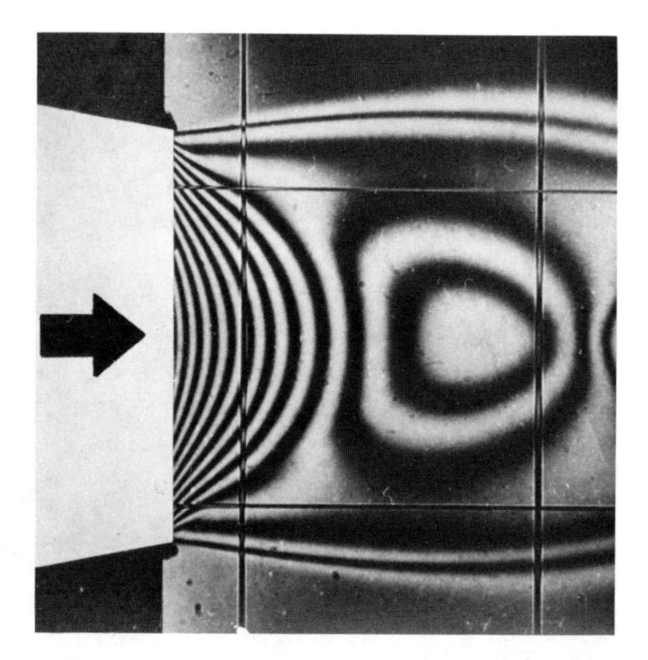

Fig. 8 Mach–Zehnder interferogram showing axisymmetric flow at the exit of a blast pipe with an infinite fringe adjustment.

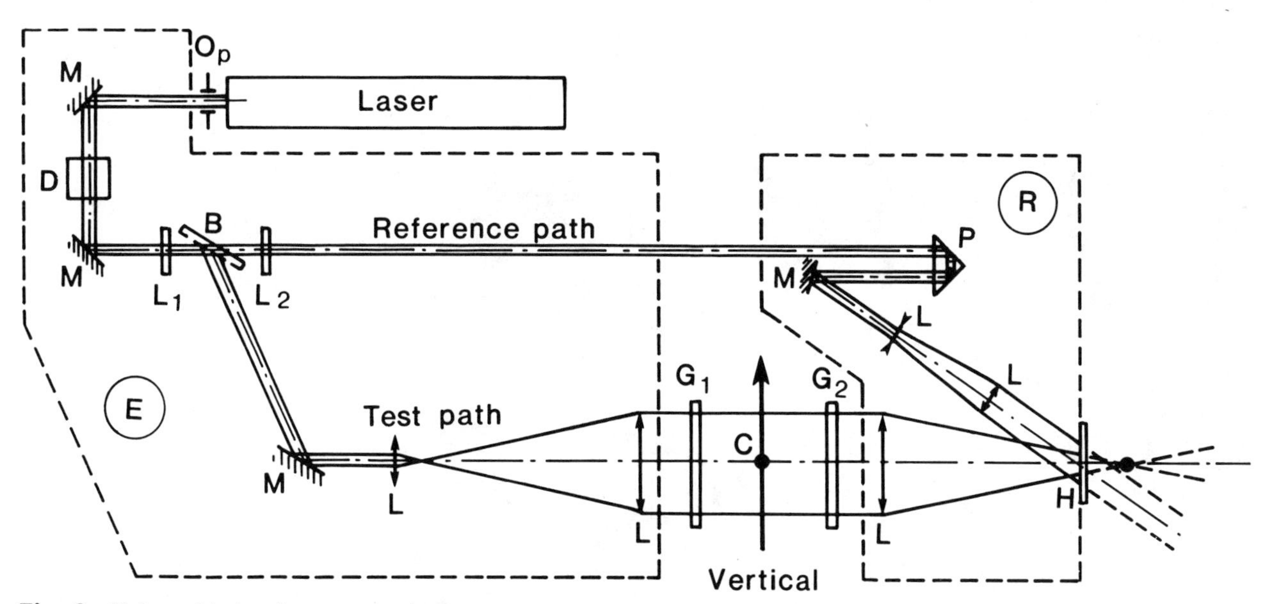

Fig. 9 Holographic interferometer, optical components: *E*, emitting block; *R*, receiving block; O_p, shutter; *M*, plane mirrors; *D*, Bragg cell; L_1, L_2, half wave plates used to adjust the intensity ratio between the transmitted and the reflected beams by the beam splitter *B; L*, lenses; *P*, movable prism to adjust equality of reference and test paths; G_1, G_2, windows closing the test section *C; H*, holographic plate.

Fig. 10 Holographic interferometer in place on the S8 wind tunnel of the ONERA Chalais–Meudon Center.

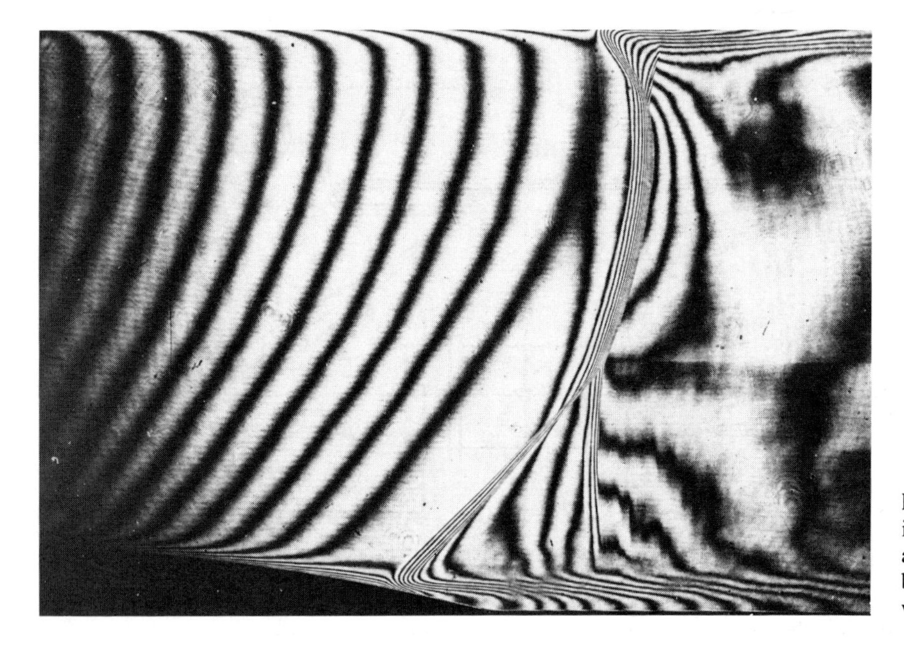

Fig. 11 Infinite fringe holographic interferogram of a transonic aerodynamic flow near the leading edge of a bump in the wall in the S8 ONERA wind tunnel (exposure time 52 μs).

III IMPLEMENTATION OF INTERFEROMETRY IN WIND TUNNELS

The implementation of interferometry in wind tunnels requires great care in proportion to the high sensitivity of the process not only for the studied phenomenon, but also to control ambient conditions such as air convection flows through the light beams as well as sonic waves and ground vibrations during the run.

Besides their high cost, the classical interferometers' main disadvantage lies in their bulk, which makes them particularly sensitive to vibrations. This bulk is due to the requirement that the reference beam, which goes around the test section, have the same diameter as the test beam.

By contrast, in a holographic interferometry set up, the reference beam diameter, reduced to some millimeters, makes an easier way to the outer test chamber, or even within it, in a small region of uniform flow. Thus, the size of the equipment is considerably reduced so that it is less sensitive to ambient perturbations. Further, the simplicity

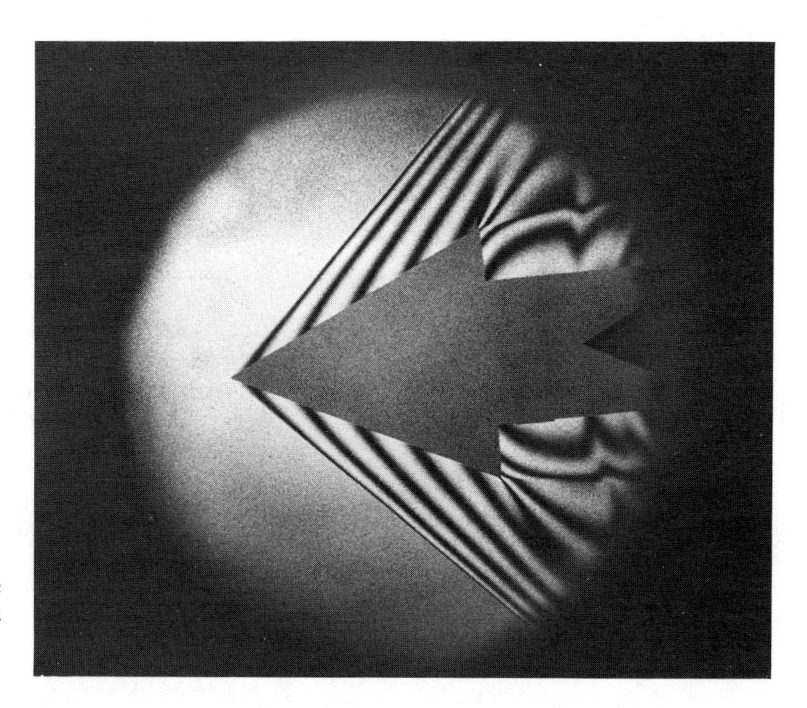

Fig. 12 Infinite fringe holographic interferogram of a cone model in a supersonic aerodynamic flow (M = 1.45) in the S8 ONERA wind tunnel.

of the optical arrangement, allied with the large tolerance acceptable to the optical component quality, leads to a cost lower than that of a classical interferometer, without reduction in performance.

For conventional interferometry and holographic interferometry applying, refer to [1–5] bibliographical references.

IV INTERFEROGRAM DATA PROCESSING

The aim of the data processing is the determination of the gas density ρ distribution in a whole field. The operating mode is the same for interferograms obtained by classical or holographic interferometers. It includes the photometric analysis of the photographic recording plate, then calculations on the data acquired.

The recording-plate analysis consists of acquisition of the fringe middle locations and model outlines in a coordinate system connected to the field. The fringe numbering may be arbitrary with respect to the zero-order fringe, which is not necessarily numbered. In fact, the only thing of consequence is the difference between the interference states at the various considered locations with regard to the interference state at a reference location R where the density ρ_R is provided by pressure and temperature measurements. The line-by-line scanning of the recording plate is performed with an optical microdensitometer and may be automated.

For a given point A in the field, the calculation, which may likewise be automated, consists first of determining the difference Δn_A between the gas refractive indices at two points A and R.

In the case of a two-dimensional flow, the optical path difference Δ_A for the rays passing respectively through A and R is determined, making use of the radiation wavelength λ, from interference states N_A and N_R at points A and R provided by the recording-plate analysis

$$\Delta_A = (N_A - N_R)\lambda = dL_A + \delta \tag{5}$$

$$dL_A = L_A - L_R = e\,\Delta n_A \tag{6}$$

$$\delta = (N'_A - N_R)\lambda \tag{7}$$

where e = distance between the window faces within the test section (two-dimensional flow); dL_A = optical path difference for the rays passing through the flow at points A and R during a run of the wind tunnel; δ = optical path difference for the same rays passing through A and R, just before a run; and N'_A = interference state at A before a run, measured from a reference recording plate taken just before the run. It may also be extrapolated from a uniform flow region of the interferogram. Difference δ is nul for an infinite fringe adjustment for, then, N'_A and N_R are equal.

Taking into account the preceding relations, the refractive index difference at the considered field point, before and during a run is

$$\Delta n_A = \frac{(N_A - N'_A)\lambda}{e} \tag{8}$$

The second part of the calculation introduces the Gladstone–Dale relation connecting refractive index n to density

ρ, through molecular refractivity k

$$n - 1 = k\rho \tag{9}$$

Then, for point A, writing down $\Delta\rho_A$ the difference of densities ρ_A and ρ_R, respectively at points A and R is

$$\Delta n_A = k\,\Delta\rho_A = k(\rho_A - \rho_R) \tag{10}$$

The density at point A, related to density ρ_{io} in the part of the flow where the generating conditions are defined, is given, from (8) and (10), by

$$\frac{\rho_A}{\rho_{io}} = \frac{\lambda}{ke\rho_{io}}(N_A - N'_A) = \frac{\rho_R}{\rho_{io}} \tag{11}$$

The ratio ρ_R/ρ_{io} may be calculated from the generating pressure P_{io}, the pressure P_R at reference point R and the ratio γ of the specific heats at constant pressure and volume:

$$\frac{\rho_R}{\rho_{io}} = \left(\frac{P_R}{P_{io}}\right)^{1/\gamma} \tag{12}$$

REFERENCES

1. Trolinger, J. D., Laser Instrumentation for Flow Field Diagnostics, AGARDograph 186.
2. Asanuma, T., ed., *Flow Visualization,* Hemisphere, Washington, D.C., 1979.
3. Merzkirch, W., ed., *Flow Visualization II: Proceedings of the Second International Symposium,* Hemisphere, Washington, D.C., 1982.
4. Yang, W.-J., ed., *Flow Visualization III: Proceedings of the Third International Symposium on Flow Visualization,* Hemisphere, Washington, D.C., 1985.
5. Véret, C., ed., *Flow Visualization IV: Proceedings of the Fourth International Symposium,* Hemisphere, Washington, D.C., 1987.
6. Colloque National de Visualisation et de Traitement d'Images, Institut National Polytechnique de Lorraine, Nancy, France, 1985.

Chapter 14

Light Sheet Technique

Michel Philbert, Jean Surget, and Claude Véret

I INTRODUCTION

The light sheet technique is a process complementary to the previously described optical method for a flow diagnostic. Actually, as it requires the flow seeding, this technique is easier to implement in low-velocity incompressible flows than in high-velocity flows. Also, the use of a thin light slice allows in-depth localization of the details to be visualized and consequently restoration of phenomena in their three dimensions with an appropriate exploration of the volume to be analyzed.

The light sheet technique is commonly used in water tanks in conjunction with air-bubble seeding. Thanks to the development of lasers, new applications of the technique have been extended to aerodynamic flow visualization. In fact, the high luminous power and weak beam divergence of the laser light source make it possible to obtain much thinner and more intense light sheets than with the previously used systems that relied on slit and incandescent light sources. The use of pulse lasers with very short duration and high energy appears particularly well adapted to the detailed visualization of the flows' unsteady features (turbulent boundary layer separation, wakes, and vortex breakdown).

II DESCRIPTION

Application of the light sheet technique to flow visualization requires simultaneous implementation of the following items: realization of the light sheet, flow seeding, contingent auxiliary illumination, and choice and location of the camera.

A Realization of the Light Sheet

Implementation of the light sheet technique in a wind tunnel test section is shown in Fig. 1. An argon ionized laser, having a power of some watts, is placed over the test section with the light beam parallel to the flow direction. This beam is reflected downward by a 90° prism to meet a glass rod that acts as a highly powerful lens in a direction perpendicular to the cylinder axis so that the light is spread in a plane, within a sector. The sector angle depends on the rod and incident beam diameters. To ensure a light sheet as thin as possible (~1 mm) within the observation field (Fig. 2), a low-power convergent lens is sometimes placed downstream of the rod for the beam's natural divergence compensation.

The whole system, including prism, rod, and lens, is placed on a cart movable along a longitudinal axis in order to allow successive illumination of different flow slices. The rod is also mobile in rotation around the incident beam axis, making it possible to adjust the light sheet orientation with regard to the flow direction (from perpendicular to parallel to the flow). The rod technique is often used because it is cheap, easy to implement, and not cumbersome. Its main advantage is provision of simultaneous illumination within the whole visualization field. However, it sometimes does not allow perfect uniformity within the field without loss of light.

Another way to generate the light sheet is fast scanning of the laser beam with the use of a system that has a rotating or oscillating mirror. In this case, an image is obtained using time integration during one or several beam scannings of the observed phenomenon. Thus, the illumination of all

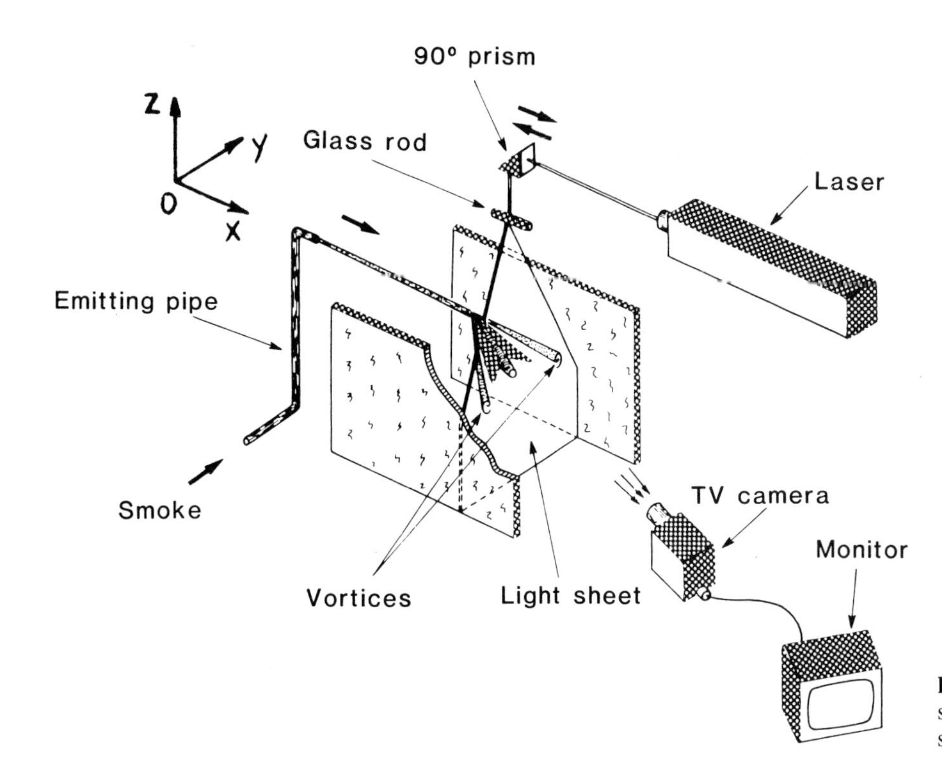

Fig. 1 Implementation of the laser sheet technique in a wind tunnel test section.

the points is not simultaneous, but with linear scanning, uniformity of illumination is obtained. This device is not recommended for observation of fast evolution phenomena, and it is not usable with pulse lasers.

Shapes other than a plane light slice, such as conical surfaces, may also be realized using a double mirror scanning system [1]. They are sometimes used with a stroboscopic illumination technique to study particular flow types.

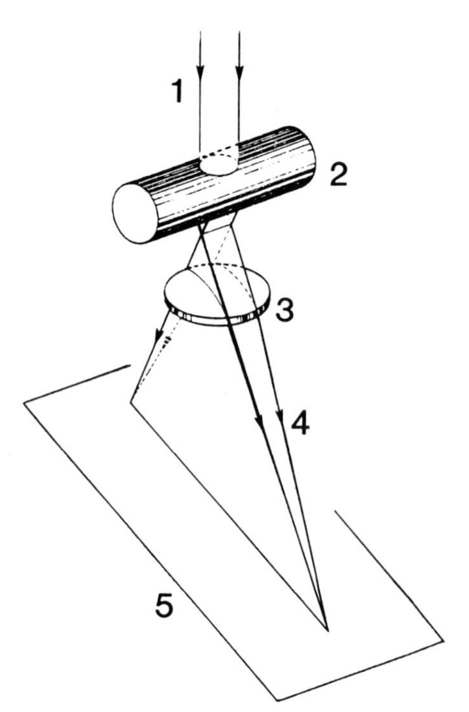

Fig. 2 Spreading of a laser beam by means of a cylindrical glass rod: 1, Laser beam; 2, glass rod; 3, converging lens; 4, light sheet; 5, screen.

B Flow Seeding

The light sheet visualization technique requires convenient flow seeding. This seeding may be done either by tracer injection from the model wall or by injection within the flow by means of a seeding pipe. In both cases, the emission point has to be precisely localized in order for the tracer to effectively seed the whole part of the flow that is of interest. This is shown in Fig. 1 concerning the wing vortices on a plane with incidence; the end of the smoke emission pipe has to be localized in y and z for the tracer stream to reach the model apex zone, where both vortices originate. The adjustment is sometimes uneasy enough, and the emitted plume cross section has to be adapted in a dimension in relation to that of the vortex. Indeed, a plume too thin seeds only a part of the phenomenon to be observed and gives only a partial view. A plume too wide seeds zones external to the vortex, possibly masking its structure. In proportion to increase in the flow velocity, aerodynamic perturbations arise in the wake of the smoke emission pipe. These perturbations may modify the flow around the model and generally generate the tracer stream breakdown so that the smoke is spread. These perturbations may be prevented either by placing a profiled box around the pipe or by placing this pipe upstream in the wind tunnel, where the flow velocity is reduced. Another very efficient solution is injecting the tracer through the model wall.

The tracer species used most frequently in large wind tunnels are oil or kerosene smokes. Also used is steam generated from condensation of humid air meeting cold nitrogen in a liquid or gaseous phase. This last tracer has an advantage as compared to oil smoke in being nonpolluting, but its persistence in a flow is very dependent on humidity in the air; its use in dry air is very problematic.

C Auxiliary Illumination

It is often useful to illuminate the model with an auxiliary light source in order to be able to locate the light sheet with regard to the model. This auxiliary illumination may be obtained by means of several adjustable projectors placed around the test section. If the auxiliary light is a color contrasting well with that of the light sheet, observation of the test field will be improved.

D Recording

Photographic, cinematic, and/or video cameras are placed either outside the test section or, sometimes, within the flow itself. In these cases, the cameras are placed inside protective boxes with windows, their shape being profiled to limit the flow perturbation. The direction of observation is generally oriented perpendicular to the light sheet in order to prevent distortion of perspective in the recorded images. Unfortunately, it is in this direction that intensity of the light scattered by the tracer particles is generally the lowest.

III APPLICATIONS TO FLOW VISUALIZATION

Some examples of applications of the light sheet technique to flow visualization are shown in Figs. 3–7.

In Fig. 3 we see the swirling flow on a delta plane model with incidence 12.5°, in a low-velocity flow (20 m/s). The laser cross section perpendicular to the flow is located downstream of the model. The flow was seeded with steam.

Figure 4 shows the longitudinal evolution of the apex vortices over a delta wing with incidence 20° and flow velocity 20 m/s. Several successive cross sections are recorded on the same picture in order to follow development of the vortex from the apex of the model.

Figure 5 shows the same vortex, in the same aerodynamic conditions, but with a light sheet oriented along the

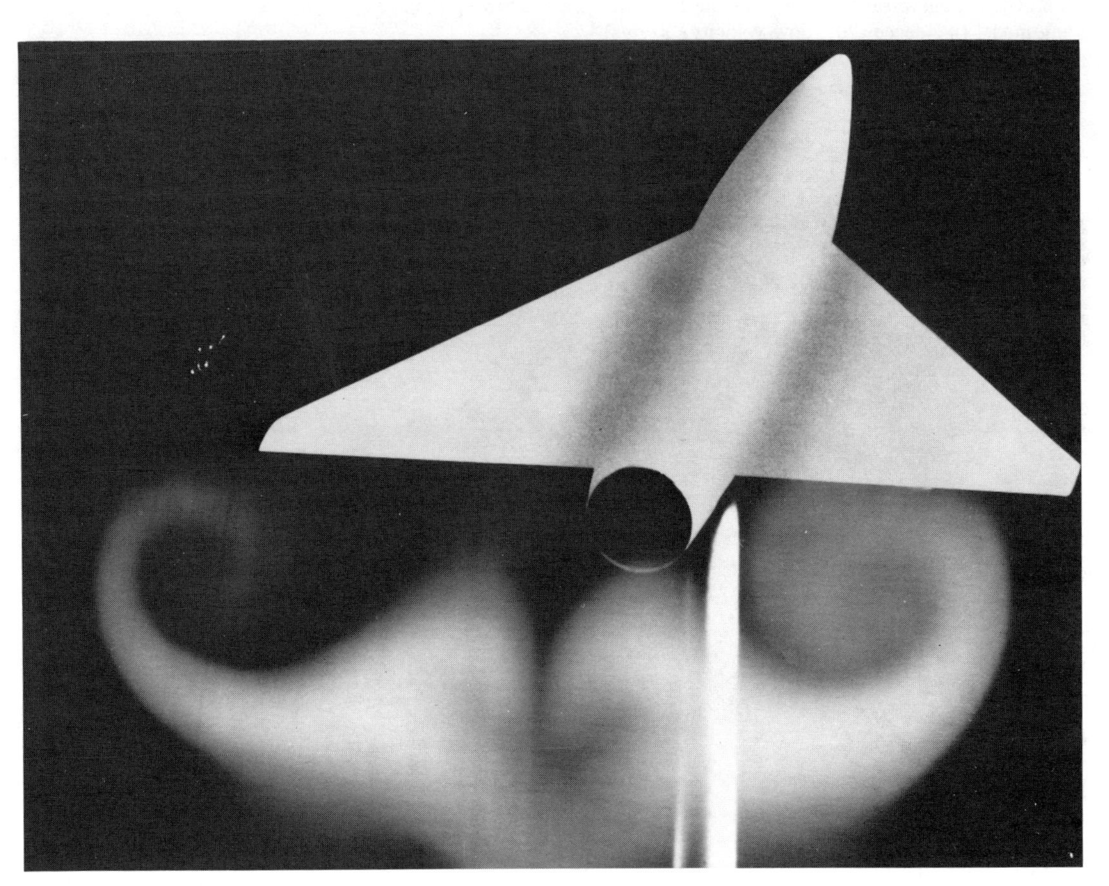

Fig. 3 Light sheet cross section of the seeded flow in the wake of a delta plane model (incidence, 12.5°; flow velocity, 20 m/s).

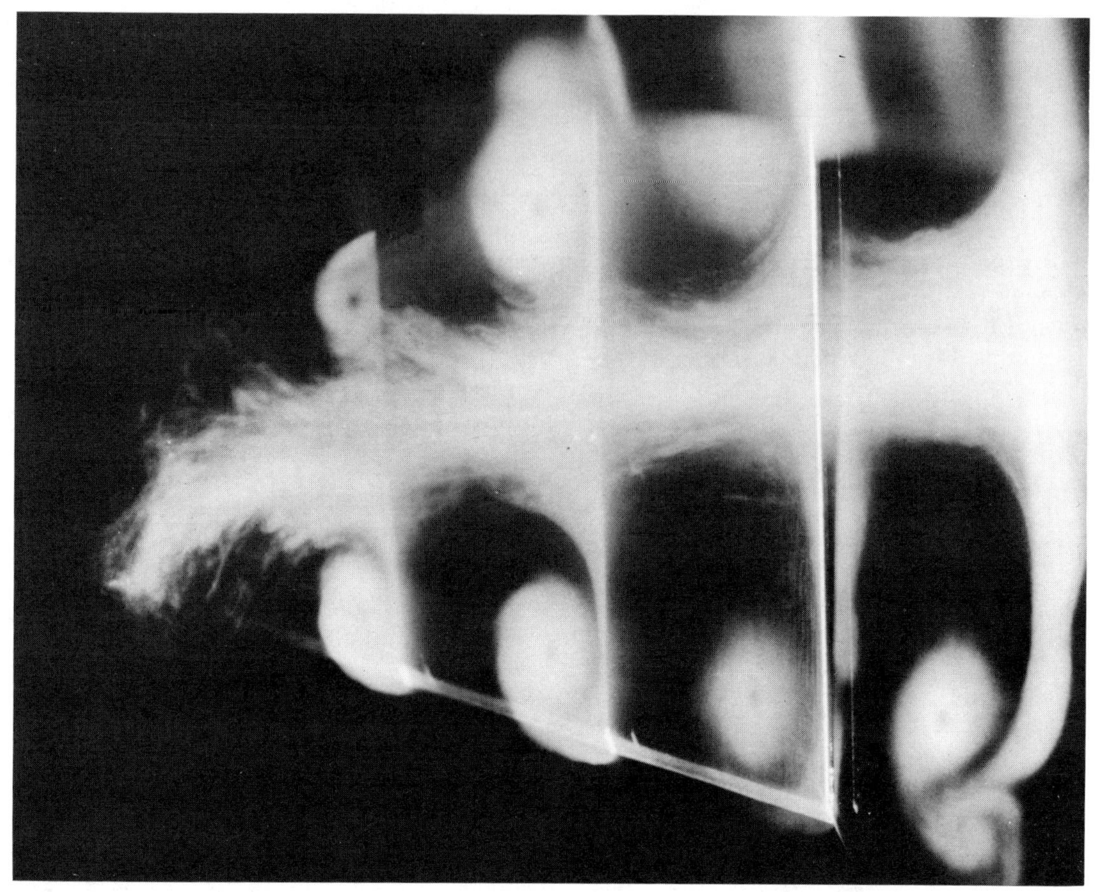

Fig. 4 Superimposition of four transversal cross sections on the same picture showing the apex vortices over a delta wing model (incidence, 20°, flow velocity, 20 m/s).

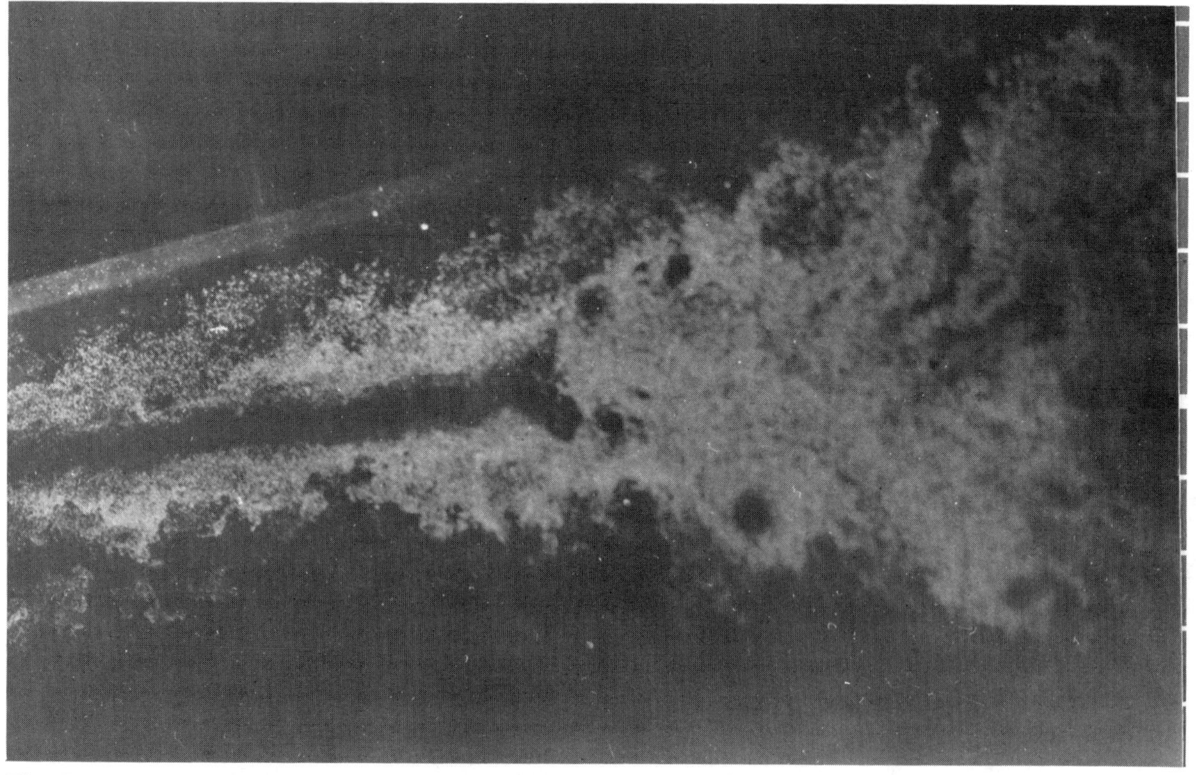

Fig. 5 Longitudinal cross section of the apex vortex over a delta wing model showing the vortex core and the breakdown point downstream of the trailing edge. The picture was taken with a Yag pulse laser (pulse length, 15 ns).

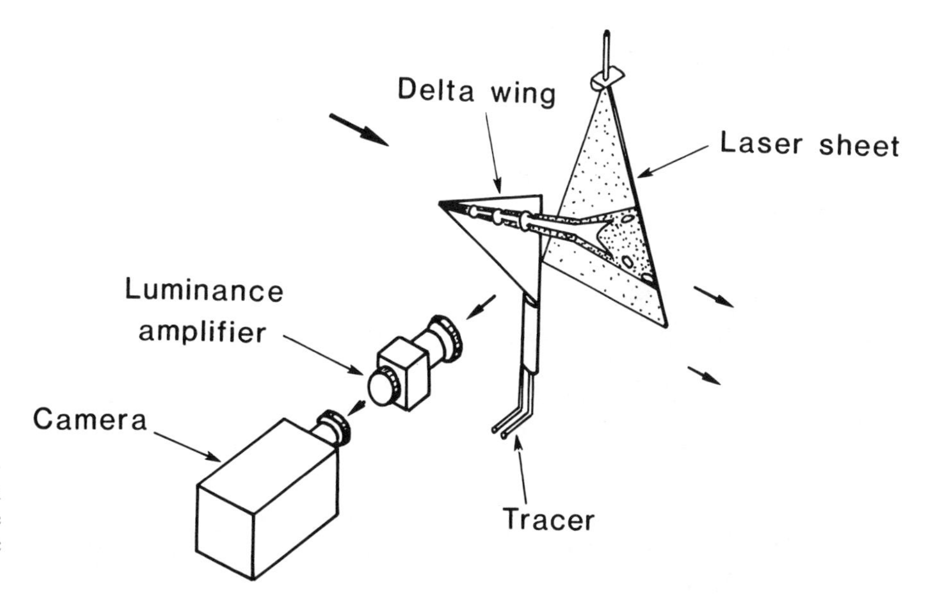

Fig. 6 Arrangement for the high-speed cinematic recording of the vortex breakdown by means of a luminance amplifier associated with a cinematic camera.

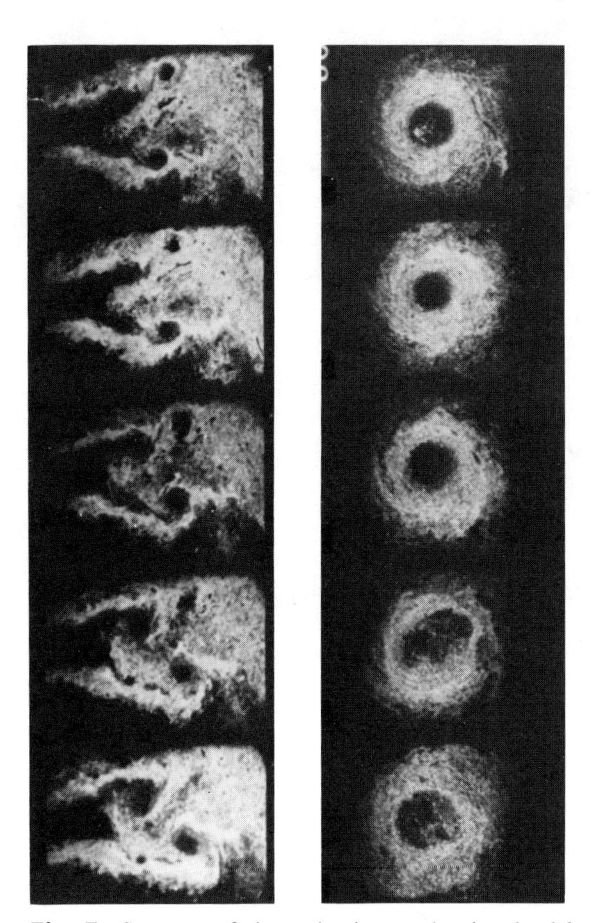

Fig. 7 Sequence of cinematic pictures showing the delta wing vortex evolution: (*Left*) Longitudinal cross section, 130 × 90 mm; (right) transversal cross section, 90 × 70 mm (exposure time, 50 μs; camera framing rate, 1000 frame/s; film type, 4X Negative 7224).

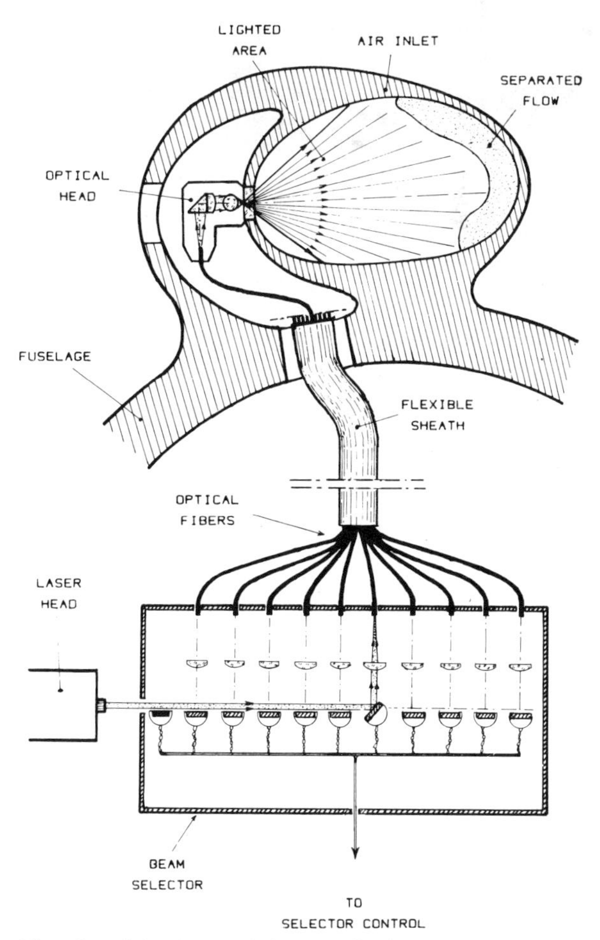

Fig. 8 Light sheet technique application to air intake flow visualization.

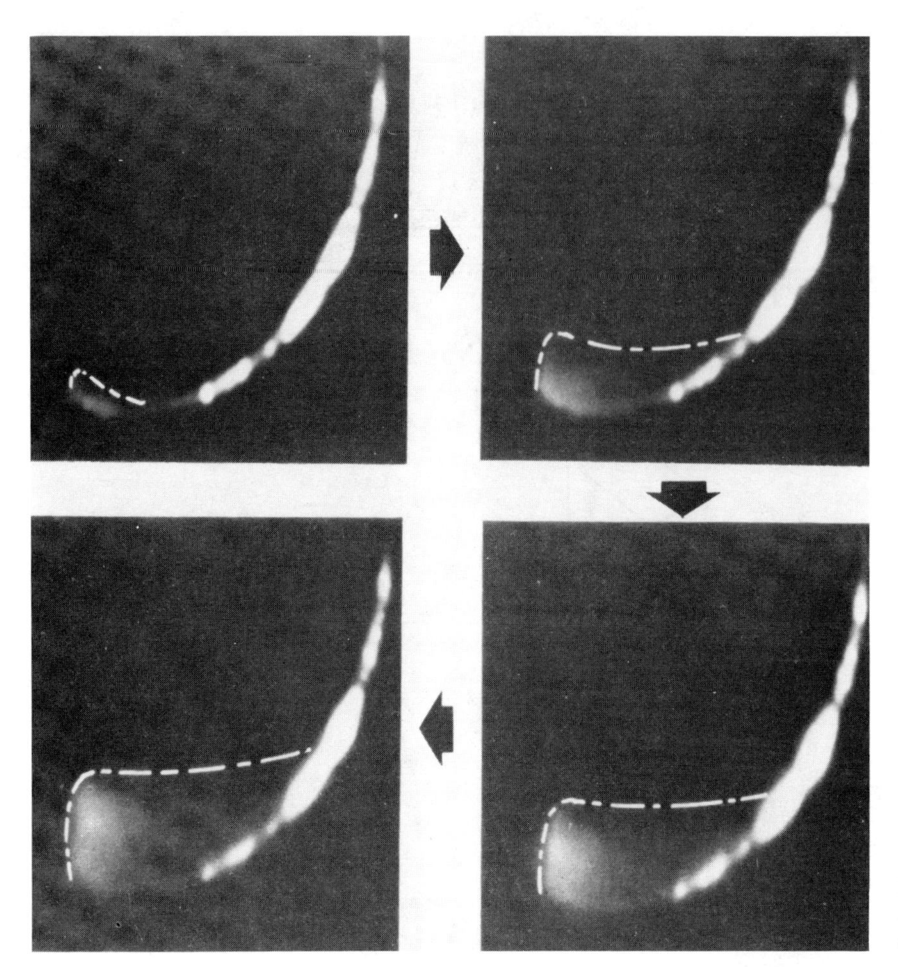

Fig. 9 Coming out of the airstream separation inside the air intake, angle of attack increasing.

vortex axis in order to see a longitudinal cross section. For this experiment, the cw laser was replaced by a Yag pulse laser having a pulse length of 15 ns. The vortex core containing no seeding particles appears as a black channel. The vortex breakdown is seen from the central channel dislocation in alternate dark nuts.

For cinematic recording with a cw laser, the camera is associated with a luminance amplifier tube [2] as shown in Fig. 6. With this device, Fig. 7 shows a sequence of pictures from a 16-mm film showing evolution of the vortex breakdown. The framing rate is 1000 frame/ps, and the exposure time for every image is 50 μs. A 15-W ionized argon laser was used, and the tracer was ammonium sulfite injected through holes in the model wall.

The light sheet technique was also used to study the internal flow in plane engine air intake [3]. Figure 8 shows an experimental arrangement [4], making use of optical fibers and a rotating mirror selector to introduce the laser beam sequentially. Figure 9 presents the evolution of the separated flow contour when the angle of attack increases.

The light sheet technique is still used for the implementation of other flow diagnostic techniques such as the velocity field measurements by either double laser pulses [5, 6], Doppler effect [7], or induced fluorescence [8, 9].

REFERENCES

1. Porcar, R., Prenel, J. P., Diemunsch, G., and Hamelin, P., Visualizations by Means of Coherent Light Sheets: Applications to Various Flows, in *Flow Visualization III: Proceedings of the Third International Symposium on Flow Visualization,* ed. W.-J. Yang, pp. 123–127, Hemisphere, Washington, D.C., 1985.
2. Philbert, M., and Faléni, J. P., Applications de l'Amplification de Luminance à la Visualisation par Tomoscopie Laser, Colloque National de Visualisation et de Traitement d'Images, Institut National Polytechnique de Lorraine, pp. 49–57, Nancy, France, 1985.
3. Stanislas, M., Ultra High Speed Smoke Visualization of Unsteady Flows Using a Pulsed Ruby Laser, in *Flow Visualization III: Proceedings of the Third International Symposium on Flow Visualization,* ed. W.-J. Yang, pp. 41–44, Hemisphere, Washington, D.C., 1985.
4. Philbert, M., Faléni, J. P., and Deron, R., Air Intake Flow Visualization, in *Flow Visualization IV: Proceedings of the Fourth International Symposium,* ed. C. Véret, pp. 753–758, Hemisphere, Washington, D.C., 1987.
5. Haertig, J., and Smigielski, P., Visualisation d'Ecoulements par Tranche Laser et Mesure de Vitesses Instantanées, Colloque National de Visualisation et de Traitement d'Images,

Institut National Polytechnique de Lorraine, pp. 59–67, Nancy, France, 1985.

6. Meynart, R., Instantaneous Velocity Field Measurements in Unsteady Gas Flow by Speckle Velocimetry, Appl. Op., vol. 22, pp. 535–540, 1983.
7. Seiler, F., and Oertel, H., Visualization of Velocity Fields with Doppler Pictures, in *Flow Visualization III: Proceedings of the Third International Symposium on Flow Visualization,* ed. W.-J. Yang, pp. 454–459, Hemisphere, Washington, D.C., 1985.
8. McDaniel, J. C., and Hanson, R. K., Quantitative Planar Visualization in Gaseous Flow Fields using Laser Induced Fluorescence, in *Flow Visualization III: Proceedings of the Third International Symposium on Flow Visualization,* ed. W.-J. Yang, pp. 113–117, Hemisphere, Washington, D.C., 1985.
9. Rhee, S. R., Koseff, J. R., and Street, R. L., Simultaneous Flow and Temperature Field Visualization in a Mixed Convection Flow, in *Flow Visualization IV: Proceedings of the Fourth International Symposium,* ed. C. Véret, pp. 659–664, Hemisphere, Washington, D.C., 1987.

Chapter 15

Planar Fluorescence Imaging in Gases

Ronald K. Hanson and Jerry M. Seitzman

I INTRODUCTION

Although flow visualization is regarded as a well-developed field, with a long history of important contributions to fluid mechanics and combustion, remarkable advances in measurement capability have occurred in the past few years. Particularly noteworthy has been the development of spectroscopy-based techniques, which combine species specificity with the capability for simultaneous, spatially resolved measurements at a large number of points within a plane, in essence providing two-dimensional (2-D) images of flow-field properties. These new techniques generally utilize sheet illumination from a tunable, pulsed laser source with right-angle detection of scattered light using a solid-state array detector. The detector may be coupled directly to a laboratory computer for fast storage, processing and display of planar image data. Techniques based on laser-induced fluorescence as well as Raman, Rayleigh, and Mie scattering have been reported, yielding measurements of species concentration or mole fraction, temperature, density, velocity, and even pressure. Temporal resolution is generally controlled by the laser source, with 5 to 20-ns pulse lengths being typical. Spatial resolution is set by the array detector, with 100 × 100 arrays (10,000 pixels) now typical, though arrays with several hundred thousand and even 1 million pixels are seeing initial use. The extension of two-dimensional imaging to three dimensions using a rapidly scanned illumination plane has recently been reported.

The history of this field is quite brief, with nearly all of the progress having been made since Eckbreth's 1981 review [1] of spatially resolved, single-point laser diagnostics. Several research laboratories have made important contributions, but the pioneering work on the use of planar array detection was done at Yale University with initial emphasis on Mie [2] and Rayleigh [3] scattering. The important extension to imaging based on laser-induced fluorescence (LIF) was made nearly simultaneously by groups at Lund Institute of Technology [4], SRI International [5], and Stanford University [6]. More recently, imaging based on Raman scattering has been demonstrated at Sandia Laboratories [7], and LIF imaging from evaporating spray droplets has been reported [8] by collaborators from United Technologies Research Center and the University of Texas at Dallas. Finally, the extension of LIF imaging to sense velocity has been reported by researchers at Princeton University [9] and Stanford University [10]. The continued high level of activity at these and other laboratories promises to yield rapid progress in this new field, although significant research challenges remain.

The potential of imaging diagnostics in studies of nonreacting and reacting flows is readily apparent. The techniques should find extensive use for qualitative purposes, for example, in surveying complex flows and in guiding development of realistic flow models, as well as for quantitative flow-field analysis. Detailed study of large- and small-scale features of flow fields will be enabled, for both

---

The authors are pleased to acknowledge the important contributions made by several colleagues, particularly G. Kychakoff, R. Howe, P. Paul, M. Allen, B. Hiller, M. Lee, and J. McDaniel. Our research has been sponsored by the U.S. Air Force Office of Scientific Research.

reacting and nonreacting cases, as well as fundamental studies aimed at developing improved reacting-flow turbulence models. Although optical access may be a problem in some cases, imaging diagnostics may be expected to reveal previously unobserved phenomena in the complex flow fields of practical devices. Reexamination of flow fields previously studied with traditional line-of-sight visualization methods and single-point diagnostics should also lead to observations of previously masked flow features and, ultimately, to improved understanding of fluid mechanics.

Progress in the development of flow-field imaging is tightly coupled to advances in laser sources and, particularly, in array detector technology. In the case of lasers, the need is for higher repetition rates and broader wavelength tunability, for example, providing convenient access to ultraviolet wavelengths needed to excite chemically stable compounds. As regards array detectors, the need is for improved noise performance, broader wavelength coverage (again to facilitate ultraviolet operation), and of course faster and larger arrays (more pixels). The anticipated availability of large arrays with capability for rapid, on-chip transfer of image data will also be important in establishing high-repetition-rate imaging. Since image intensification is commonly employed, there is also a requirement for improved microchannel plate (MCP) intensifiers. Finally, reductions in the cost of imaging equipment are critical if these techniques are to be employed in more laboratories.

The aim of this chapter is to provide an overview of LIF-based approaches to planar imaging diagnostics, which currently offer higher sensitivity for species measurements and greater versatility for measuring multiple flow-field parameters than other light-scattering techniques. Example results are presented with the objective of illustrating both current capability and the future potential of these methods.

II LASER-INDUCED FLUORESCENCE

A Basic Principles

Laser-induced fluorescence is a well-established, sensitive technique for detecting population densities of atoms and molecules in specific quantum states. Although LIF was used initially in studies of spectroscopy and chemical analysis, it is now also recognized as a powerful fluid mechanics diagnostic with the potential for monitoring flow-field parameters such as mixture mole fractions, density, temperature, velocity, and pressure. The development of LIF methods has, until recently, been driven by single-point measurements, but the underlying concepts carry over directly to multiple-point planar imaging.

The principles of LIF are well known (see [11–13] and articles cited therein for an up-to-date review) and need not be repeated in detail here. In brief, a laser source is tuned to excite (usually from the ground electronic state) a specific electronic absorption transition in the species of interest. Following the absorption process, collisional redistribution in the electronically excited state may occur prior to either collisional quenching or radiative de-excitation (fluorescence) of the molecule back to a lower electronic state. The fluorescence, which occurs over a range of wavelengths, is usually collected at right angles and filtered spectrally at the photodetector. For a given species, the variables in the LIF process are the transition pumped, the detection spectral band pass, the spectral intensity of the laser, and, in the case of a narrow-band-width laser, the location of excitation within the absorption line profile. Laser-induced fluorescence may be thought of as spatially resolved absorption, with the intersection of the illumination and collection beam paths controlling the spatial resolution of the measurement. In the case of sheet beam illumination and detection with an array detector, the sheet thickness and the area of a pixel image (set by the size of the detector pixels and the magnification of the collection optics) control the size of the measurement volume at each image "point" [14].

The governing equation for the LIF signal S, based on a simple two-level model with weak (unsaturated) excitation, for a single detector pixel, is [11–16]

$$S = C\,E\,V\,N_s\,B\,F_{vj}(T)\,\frac{A}{A+Q} \tag{1}$$

Where C is a group of constants specific to the experimental setup, E is the laser energy per pulse per unit area per unit frequency, V is the measurement volume for the detector element, N_s is the number density of the absorbing species, F_{vj} is the population fraction for the pumped state, B is the Einstein coefficient for absorption, and A is the appropriate (for the transitions monitored) Einstein coefficient for spontaneous emission. The parameter Q represents the sum rate of all other transfer processes which eliminate emission into the detection band width. In the most common case, with broad-band collection and no chemical or predissociative removal from the upper electronic state, $Q(\chi_i, P, T)$ is simply the electronic quench rate, which in general depends on temperature, composition through the mole fractions χ_i, and pressure. Procedures for dealing with system calibration and variations in the quench rate are discussed in [14–16].

B Equipment Requirements

A schematic diagram of a typical experimental setup for planar LIF (PLIF) imaging is shown in Fig. 1. Though the exact setup may vary, all PLIF measurement systems consist of two basic subsystems: (1) a laser illumination source

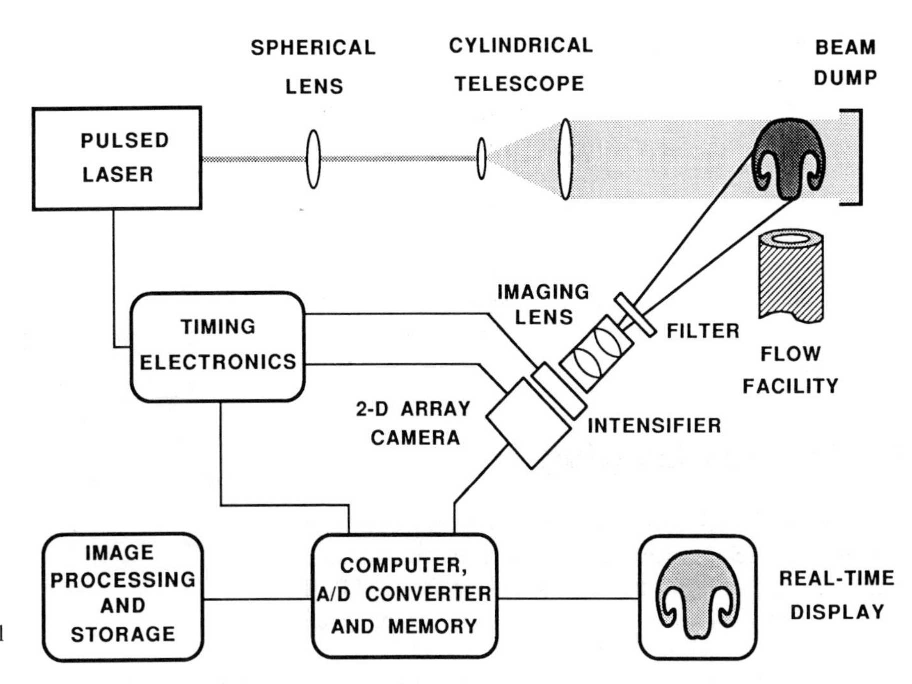

Fig. 1 Schematic diagram of typical experimental setup for PLIF imaging.

and associated sheet forming optics; (2) an imaging system consisting of collection optics, a 2-D photodetector (often image intensified), and control electronics.

In most PLIF applications, pulsed lasers are the preferred illumination source, as their short pulse lengths (typically 5–20 ns) provide the ability to freeze motion in the flow field and their high spectral irradiance increases the signal-to-noise ratio in flow fields with significant background luminosity. Commercially available Nd:YAG- and excimer-pumped dye laser systems are typically used as they provide high energy (10–100 mJ/pulse) output and, in conjunction with frequency-conversion techniques such as frequency mixing in birefringent crystals and Raman shifting, provide continuous tuning throughout the range (200–1000 nm) where many species of interest have absorption bands. For discrete excitation wavelengths in the ultraviolet region (150–400 nm), the direct or Raman-shifted output of tunable, narrowed (1 cm^{-1} band width) excimer lasers may be suitable. Whereas Nd:YAG systems are presently limited to repetition rate of a few tens of Hertz and the excimer systems can be operated at rates up to 200–500 Hz, visible-wavelength, metal vapor lasers provide higher repetition rates (5–15 kHz) suitable, in some cases, for acquiring 3-D data sets and for measurements in turbulent gaseous flows. In particular, metal vapor lasers may be useful for imaging species with broad-band visible absorption transitions. For production of the laser sheet, three simple lenses are typically required. The laser beam is loosely focused with a long-focal-length spherical lens, producing a sheet of slowly changing thickness (typically 200–500 μm at the focus) across the imaged region. The sheet itself is formed and collimated by a cylindrical telescope, consisting of two cylindrical lenses.

Fluorescent light from a segment of the illuminated plane is collected at right angles, passed through a spectral filter, and imaged either directly onto an array detector or onto the front face of an image intensifier [14]. Simple lenses can be used to image the flow, though in visible wavelength and low f-number applications, aberration-corrected commercial lenses provide the best image quality. Image intensification is needed when the fluorescence intensity is low relative to the detection limit of the array detector, with proximity-focused MCP intensifiers providing the best choice for imaging applications, as they exhibit less geometrical distortion and nonuniform response than field-focused devices [14]. Additionally, MCPs are more easily time gated (reducing the noise associated with background luminosity), with minimum gate widths on the order of 1 ns obtainable. The intensifiers are normally coupled to the detectors by fiber optic bundles (as opposed to lenses), which provide good transmission efficiency and minimize the weight and size of the imaging system. The types of detectors currently employed are photodiode (PD) arrays [14–15], low-light-level silicon-intensified-target (SIT) Vidicon cameras [16], and charge-coupled device (CCD) arrays [17; see also 29]. While SIT devices have been successfully used in PLIF experiments, they are generally inferior to the solid-state PD and CCD arrays in terms of blooming, framing rates, weight, size, and durability, with CCD arrays offering superior performance in terms of noise, linear dynamic range, and spatial resolution. A significant potential advantage of CCD systems is the increasing availability of low-cost CCD cameras employing standard video formats, which can be interfaced to inexpensive microcomputers with off-the-shelf hardware (e.g., frame grabbers) and software. In some cases though, the low cost may be outweighed by the disadvantages associated with the fixed framing rates (60 Hz, 50 Hz, and their subharmonics) of many of these standard CCD systems. For high-speed imaging, improvements in the recording rate of solid-state cameras are required. One possibility is to employ on-chip storage of multiple images on large detector arrays. If intensification is employed, special short-lived phosphors will be required on the output face of the intensifier. Finally, the array is read, and the data are digitized, displayed, and stored, if desired, for subsequent image processing.

C Species Imaging

The bulk of fluorescence imaging activity has been concerned with species concentration or mole fraction measurements. In reacting flows, the species-specific nature of PLIF provides direct information on the possible coupling between chemistry and fluid motion. Examples of PLIF species imaging to probe the burning and burned gases in a small-scale spray flame [18, 19] are provided in Figs. 2 and 3. These single-shot images of OH and CH were obtained with an intensified, 100 × 100 photodiode array camera (Reticon MC521). A Nd:YAG-pumped dye laser was used in both cases, with energies per pulse in the range 10–15 mJ. In order to discriminate against elastic scattering from the droplets present, OH was pumped at the $Q_1(6)$ transition of the (1,0) band, at 283 nm, and the fluorescence from the (1,1) band was collected. The diagonal nature of CH transitions required excitation (426 nm) and detection (431.5 nm) within the same (0,0) vibrational band using a custom three-cavity filter with 1 nm half-width.

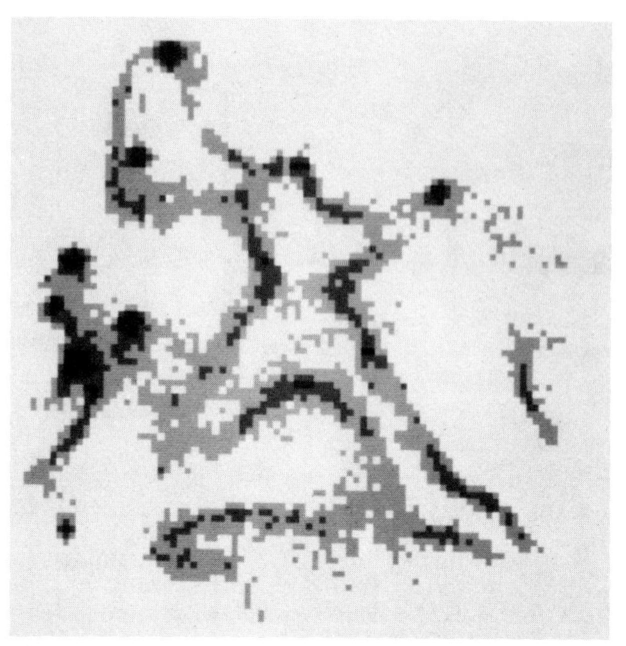

Fig. 3 PLIF image of CH and droplet Mie scattering in a spray flame. The 1.2 × 1.2 cm image is shown with four gray scales, with black denoting the highest signal [19].

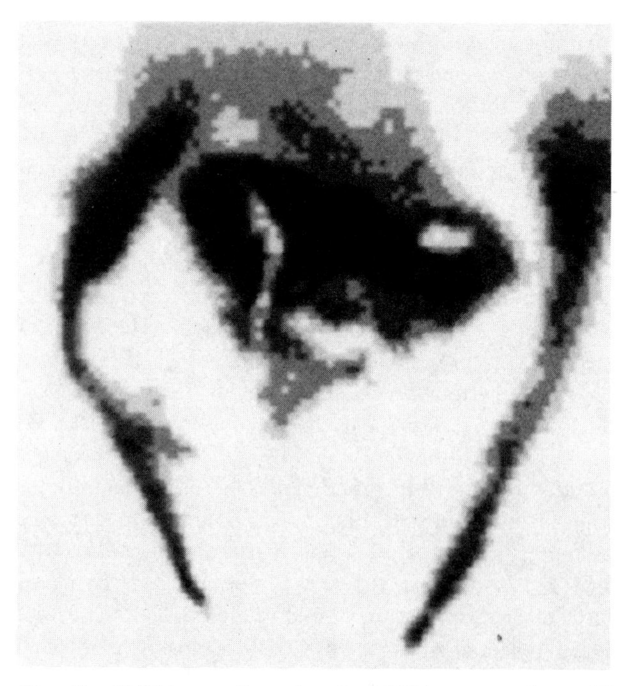

Fig. 2 PLIF image (8 × 8 cm) of OH in a spray flame. The image, digitized to 256 levels, is displayed with a five-level gray scale, with white representing the lowest signal and black the highest signal [19].

The results shown in Fig. 2 provide information on the instantaneous distribution of OH in the central plane of the turbulent heptane–air spray flame. The air-atomizing siphon nozzle is just below the 8 by 8-cm field of view. Peak concentrations of OH (about 1000 ppm) are found in two distinct zones of combustion: a thin sheath flame at the periphery of the spray, and an interior, distributed reaction zone located above the primary fuel vaporization zone. The results for CH shown in Fig. 3 are for a smaller sized zone, 1.2 × 1.2 cm, located at the tip of the primary reaction zone. A portion of the elastically scattered light has been transmitted to enable the simultaneous display of droplet positions within the burning region marked by CH. The instantaneous flame zone (displayed as gray) is seen to have a thin filamentary structure that encompasses most of the local droplets (shown in black); the maximum CH level is estimated at 100 ppm. Similar results have also been reported in gaseous flames using C_2 and CH imaging of instantaneous flame zones [20]. An important point to be made about these measurements is that the OH and CH distributions provide different and complementary information about the flame flow field. The OH is an indicator of where already-burned gases are located, i.e., where combustion has occurred, while CH marks the location of the thin, instantaneous reaction zone. These observations are also evident in a set of composite results of CH and OH in a small glass-blowing torch [21]. Since only a single laser source was available in both of these studies, the imaging of CH and OH (or C_2) was not performed simultaneously, but two-laser, dual-species (OH and C_2) imaging has been reported using 1-D [22] and 2-D [23] diode arrays.

The advantages of PLIF imaging over more traditional techniques have been clearly demonstrated in other studies, e.g., [24, 25], in which laser schlieren visualization was contrasted with PLIF imaging of NO_2 and OH in a combustion-torch flow field. In this experiment, a lean methane–air mixture, near the flammability limit, was first ignited in a small cylindrical prechamber. The high-temperature combustion products exited from the prechamber through a circular orifice as a jet, leading to ignition and flame propagation in the main chamber. While the line-of-sight schlieren technique is useful, in this symmetrical flow, for following the high-temperature interface between unburned reactants and the burned combustion products in the torch, it provides little direct information on the combustion chemistry. Further, the spatial resolution of schlieren visualization is limited in highly turbulent flows. The observation of OH by PLIF, on the other hand, serves as a clear indicator of the chemical reaction front of the flame and the subsequent effect of fluid motion on the state of the burned gases, and this method retains its spatial resolution in turbulent flow fields. The objective of the NO_2 imaging was to track the nonreacted gases, and this was accomplished by seeding the air with a low, chemically nonperturbing level (700 ppm) of NO_2, which decomposes rapidly (primarily by H-atom attack) in the flame zone. Thus the NO_2 imaging provides a complement to OH imaging, useful up to about 700 K, where dissociation begins to occur [25].

The use of NO_2 as a flow tracer is quite attractive because of the range of possible visible excitation wavelengths (400–650 nm), and hence the relative ease of laser selection. For example, it should be possible to use a high-repetition-rate (5–15 kHz) metal vapor laser together with PLIF imaging for monitoring the real-time evolution of flow phenomena. By comparison, imaging of OH requires a near-UV laser, typically at 285–310 nm, and has thus far been limited to a 10–30Hz repetition rate for a Nd:YAG/pump/laser and 250 Hz with an excimer laser (either for direct excitation at 308 nm or for dye laser pumping).

In the results cited thus far, and in nearly all past work to develop LIF, the species have been chemical intermediates or otherwise reactive and hence not suited for many fluid mechanics imaging applications. Exceptions that have seen some use are NO [26, 27] and biacetyl [28, 29], and, quite recently, O_2 [30]. The extension of LIF imaging to O_2 is of course quite important, for both combustion and fluid mechanics applications, and has recently been demonstrated using both 2-D [31] and 1-D [32] array detectors. The laser excitation of O_2 is carried out in the UV spectrum (Schumann–Runge band system, $B \leftarrow X$) and is highly temperature sensitive [33] as a result of the much larger Franck–Condon factors associated with absorption transitions from excited vibrational levels of ground state O_2. The fact that the B state of O_2 is predissociated leads to a reduction in the fluorescence yield [30, 33], but, more importantly, it also removes the usual dependence of the fluorescence signal on the mixture- and temperature-dependent quenching rate. An argon fluoride (ArF) laser at 193 nm seems to be a particularly attractive excitation source [31], owing to the large energy per pulse (hundreds of millijoules) and high repetition rates available (500 Hz), and to system simplicity, though fairly complex spectroscopic calculations are needed to predict the fluorescence when a standard broad-band laser is used [33]. Raman shifting of a line-narrowed ArF or KrF (248 nm) laser would allow optimum excitation over a range of O_2 temperatures.

An example result of O_2 imaging in a combustion flow using single-pulse broadband ArF laser excitation is shown in Fig. 4 [31]. The 5 by 5-cm region imaged was located on the central vertical plane of a fuel-rich (CH_4–air) Meker-burner flame. The laser beam is incident from left to right, with sufficient attenuation to give an apparent asymmetry to the flow field. As expected, the image shows a high level of O_2 upstream of the inner conical flame, since the premixed fuel and air have not yet burned, and a second zone of high signal at the outer diffusion flame, where air contacts the rich products of the inner flame. It is important to realize that the signal level is a function of both O_2 mole fraction and temperature and that a separate measurement would be needed to deconvolute these two effects. The preferred approach would appear to be to employ either a tunable, narrow-line-width ArF laser or a two-laser/two-detection-channel arrangement. The latter should enable simultaneous determination of O_2 concentration and temperature.

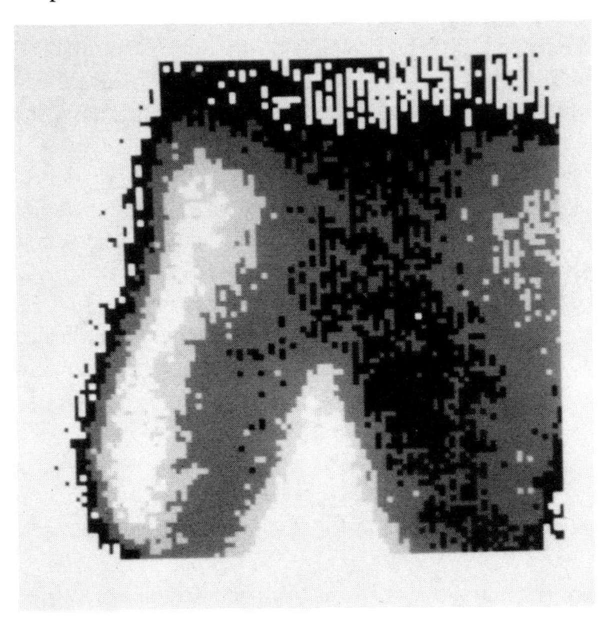

Fig. 4 PLIF image (5 × 5 cm) of O_2 in a fuel-rich Meker burner flame. This is a four-level gray scale display, with white representing the largest signal and black the smallest [31].

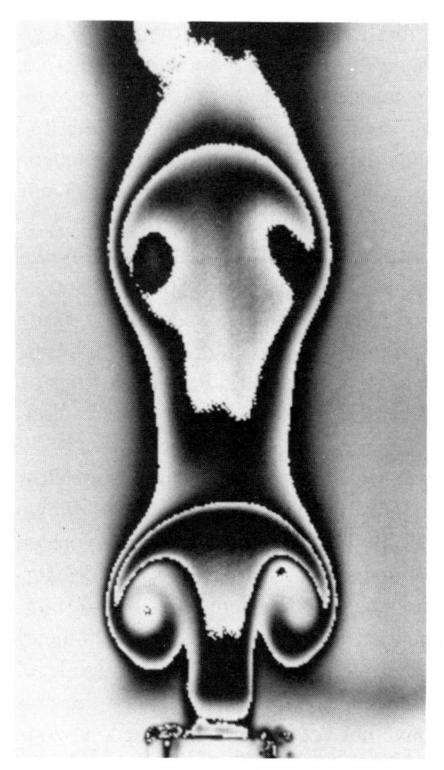

Fig. 5 PLIF image of axially forced nitrogen jet seeded with biacetyl. Here the biacetyl emission, digitized to 4096 levels, has been divided into three sublevels, each displayed with a continuous gray scale. Black denotes the highest signal in each sublevel and white the lowest [29].

An example of PLIF applied to a nonreacting fluid mechanics experiment is illustrated in Fig. 5, which is a PLIF image of an axially forced, room-temperature nitrogen jet issuing from a 6-mm-dia. nozzle at a mean exit flow rate of 85 cm/s [29]. The flow has been seeded with biacetyl (also known as diacetyl or 2-3-butanedione) to enable fluorescence imaging at visible wavelengths. In this case, the light source was a pulsed xenon fluoride (XeF) laser, at 351 nm, and the broad-band emission, which appears green to the eye, was detected with a low-noise, high-resolution (Thomson-CSF 576 × 384) CCD array camera. The vortex structure is clearly visible, and the interface between the mixed and unmixed fluid is well defined. Results such as these illustrate the potential of PLIF imaging in studies of fluid mechanical phenomena such as mixing, entrainment rates, and coherent structures.

D Species Imaging (Multiphoton Excitation)

Recently multiphoton LIF has been adopted as a means of exciting species not accessible with single-photon excitation, particularly those that have their resonance transitions in the vacuum UV region. Although the absorption process is weaker than with allowed single-photon processes, satisfactory LIF signals can be generated, in some cases, using intense laser sources together with multipass optical arrangements. There are some disadvantages to multiphoton excitation, associated primarily with the nonlinear dependence of the signal on illumination intensity and some uncertainties in the critical processes that must be included in a proper LIF model, but these are outweighed by the importance of gaining access to critical atomic and molecular species. Thus far, this approach has been demonstrated in combustion flows for atomic O [34, 35] and H [35], using intensified 1-D (linear) array detection, and for CO, using both 1-D [36] and 2-D [37, 38] arrays for imaging.

Sample results [37] of single-shot 2-D imaging in the central vertical plane of a CO–air diffusion flame are shown in Fig. 6. The extension of multiphoton LIF imaging of CO is important owing to its significance as a combustion intermediate or product species, as well as its use as a simple fuel in basic experiments. In the imaging experiments reported thus far, a common pumping scheme was employed that involved two-photon excitation (at 230.1 nm) from the $X^1\Sigma^+$ to the $B^1\Sigma^+$ state. Subsequent emission to various vibrational levels of the $A^1\Pi$ state, at wavelengths from 451 to 725 nm, was monitored through an appropriate filter on a 2-D [37, 38] intensified diode array. The 4 by 4-cm region image shown in Fig. 6 is located just above the exit of the 6-mm-dia. fuel tube. The laser energy was 2 mJ/pulse at the entrance to the multipass cell, which served both to

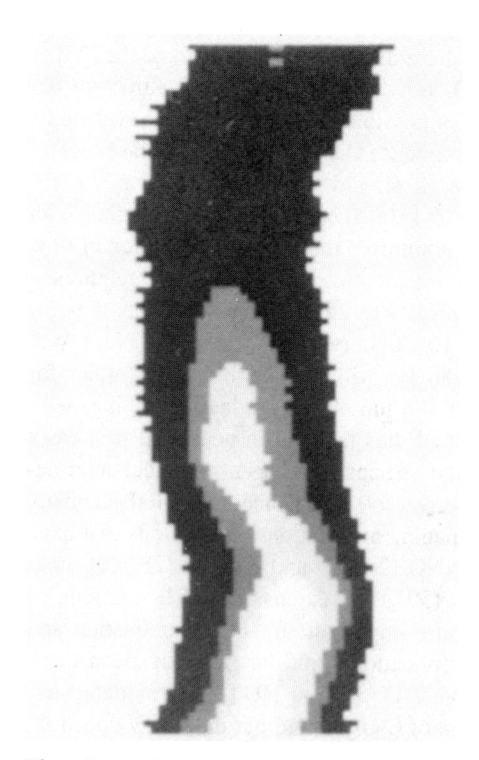

Fig. 6 PLIF image of CO in a CO–air diffusion flame using two-photon excitation. The largest CO concentration is indicated by white in this 4- by 4-cm example [37].

maintain the beam intensity over a large region and to provide a more uniform intensity distribution than obtained in a single-pass configuration. The low-speed flame was unstable, and the resulting low-frequency oscillation in flame position is apparent in this figure. Modeling of the two-photon LIF process, needed to render the imaging quantitative, is still under development [38], but the sensitivity demonstrated in this and other hydrocarbon–air flames suggests that CO imaging will become a useful tool in studies of combustion aerodynamics.

Another variation of multiphoton excitation that is emerging for imaging applications is planar multiphoton dissociation, in which multiple photons are used to photolyze an initial species, leaving one of the reaction products in an excited state that subsequently emits fluorescent light. In the one current example of this method [19], an ArF laser (193 nm) is used to fragment C_2H_2, via a two-photon process, leading to strong CH ($A \rightarrow X$) emission at 431 nm. Imaging this emission serves to mark the location of C_2H_2, and hence the hot unburned fuel regions, in combustion flow fields.

E Liquid/Vapor Imaging

In an important extension of gas-phase imaging, it has been shown [8] that organic exciplexes (**exci**ted-state com**plex**) can be used to enable simultaneous imaging of vapor and liquid distributions present in evaporating sprays. These exciplexes are formed through a reversible reaction involving the parent fuel and an additive organic compound with the effect of shifting the liquid-phase fluorescent emissions by up to 150 nm from the peak of the gas-phase emission [39]. Thus the fluorescence from both phases can be visualized simultaneously using a single-wavelength laser source. Work is in progress to incorporate a 2-D diode array camera for recording these images [40]. Although there are difficulties in quantifying this technique and in extending it to combusting flows [41], the method is likely to be of considerable use in studies of evaporating sprays relevant to combustion. Finally, it should be noted that the developers of this exciplex approach hope to extend the method to yield droplet temperature [42].

A related fluorescence-based imaging diagnostic for fuel spray characterization has also been reported [43]. In this technique, the sprayed liquid is doped with a fluorescent dye and illuminated with a sheet of cw He–Cd laser light (10 mW at 421.6 nm). The resulting fluorescence is imaged at right angles onto an intensified television camera, and the images are subsequently processed digitally. When the absorption is low, the fluorescent emission is directly proportional to the mass of fuel instantaneously present in the imaged volume. This new diagnostic thus gives direct information on the spray nozzle distribution pattern needed in developing improved spray nozzles and models of fundamental spray processes.

F Temperature Imaging

Temperature is a critical parameter in many flow fields and so the development of temperature imaging is particularly important. Successful 2-D imaging results have already been reported based on both PLIF [27, 44] and planar Rayleigh scattering [45], though the latter is more correctly viewed as a density measurement. There are two current strategies for LIF-based temperature imaging: two-laser (i.e., two-line) excitation schemes in which the ratio of two fluorescence signals is used to infer the relative population in two absorbing levels, and hence the temperature through the Boltzmann relation; and one-laser excitation schemes using either a spectrally narrow or broad laser source and either broad-band collection or ratios of narrow-band signals.

The two-line scheme has been evaluated [44] in the post-flame region of a premixed flat-flame burner using the $Q_1(5)$ lines in the (1,0) and (1,1) vibrational bands of OH ($A \leftarrow X$ system). The broad-band fluorescence was detected with an intensified Vidicon camera, and the ratio of the LIF signals for the two excitation pulses, corrected for laser pulse energy, was formed at each pixel in the 100×100 array. Since the transitions pumped have a common upper state, the fluorescence yield is the same and the LIF signal ratio depends directly on the relative populations in the vibrational levels $v'' = 1$ and 0, and hence the temperature. Although successful results were obtained, with a stated precision of 10%, the method is limited to regions with a sufficiently high temperature to produce significant populations of OH. Thus an alternative compound such as NO may be preferred. The requirement for two lasers is a serious disadvantage in some cases.

Two variations of the single-laser approach have been demonstrated [27, 46], both involving use of a stable tracer molecule, which can be seeded into the flow at a constant mole fraction χ_s. In this case the LIF signal [see Eq. (1)] is proportional to a simple function of temperature [27]:

$$S(T) \propto \chi_s F_{vj}(T) \frac{1}{T^{1/2}} \tag{2}$$

where the specific temperature dependence can be dominated by the Boltzmann fraction in the absorbing state (or states) F_{vj} and only weakly affected by the assumed temperature dependence of the electronic quench rate.

An example of single-shot results obtained in a rod-stabilized CH_4-air flame, with NO as the tracer, and using an intensified 100×100 photodiode array camera and exciting the $Q_1(22)$ line at 225.6 nm, is shown in Fig. 7 [27]. The temperature contours are spaced by 200K, and the

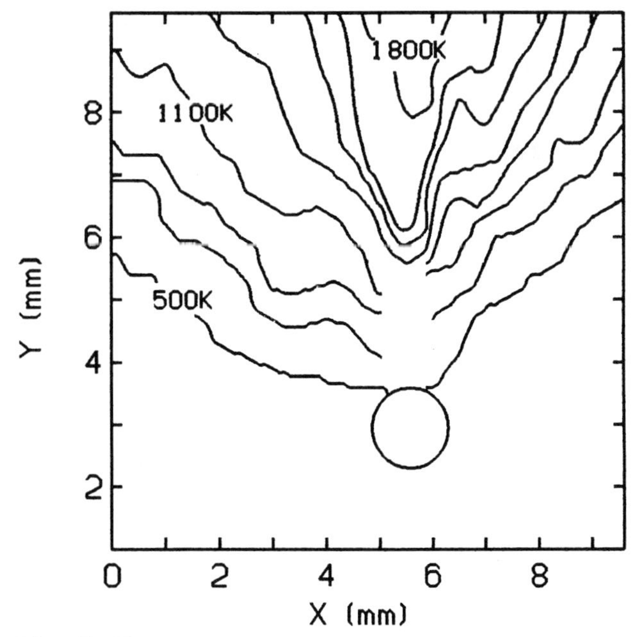

Fig. 7 Temperature contours in a rod-stabilized methane–air flame using PLIF of NO. The lowest seven contours represent 200K increments; the last increment is 100K (1700K to 1800K) [27].

asymmetric pattern results from the unstable nature of the flame. The optimum choice of the transition pumped is a function of the desired sensitivity and temperature range of interest. This particular line enabled a self-calibrating procedure. The level of NO seeding was about 2000 ppm, and there was no evidence of significant NO reaction occurring in the fuel-lean flows investigated. The estimated accuracy of these initial measurements was 100–200K, or about 5–10%, but improvements in the experimental system, including the laser energy, should enable a significant reduction in measurement error. In addition to its primary advantages of requiring only a single laser source and producing an "instantaneous" temperature map, this method is applicable to nonreacting flows.

An important second example of a single-laser approach was the demonstration of temperature imaging using a broad-band argon fluoride laser (at 193 nm) to excite O_2 in a heated air stream [46]. The broad spectral width (0.5 nm) of the laser results in excitation of a large number of O_2 transitions in several vibrational bands, with a resulting strong dependence on temperature. Although limited in its present form to gas mixtures with a constant mole fraction of O_2, the method is ideally suited for numerous practical applications involving flows of nonreacting air. For low-temperature flows, Raman shifting of the excimer laser to shorter wavelengths can be employed to increase sensitivity.

A third single-laser technique has recently been suggested [47] that combines the advantages of single-laser excitation with the two-line excitation scheme just described. In this case, the temperature measurement is based on the overlap of two absorption transitions within the line width of a single narrowed excimer laser (a KrF laser at 248 nm in [47]). The UV laser promotes the molecules to two separated upper states, both with predissociation rates that dominate the collisional transfer rates. The ratio of the spectrally separated fluorescence signals from the two upper states can be interpreted to yield the rotational Boltzmann temperature as discussed above. This approach may be suited to several gases, including OH, O_2, and H_2O (a two-photon transition for the latter).

G Velocity and Pressure Imaging

Variations of PLIF also have been developed for imaging 2-D fields of velocity in subsonic [48, 49] and supersonic [9, 10, 50] flows and extended to include pressure in one case [50], with the applications thus far having been limited to nonreacting flows. The velocity measurements are based on Doppler-shifted absorption of a narrow-line-width laser source, with the broad-band collection of the subsequent fluorescence serving to indicate the relative absorption. This approach is attractive in that it is nonintrusive and avoids the use of particles, though it requires the presence of a suitable tracer species. Work thus far has been carried out in nitrogen flows seeded with trace levels of iodine (I_2) or sodium vapor, which absorb at readily accessible cw laser wavelengths. The method is quite general, however, and work is in progress to enable measurements with more convenient species such as O_2 and NO [51].

The current strategy for imaging velocity makes use of two laser frequencies, fixed at separate locations within an absorption line of the tracer species, and requires four successive PLIF images to infer two velocity components in the plane of illumination. Sketches of the experimental arrangement and the relationship of the laser frequencies (and directions) to the absorption line profile are shown in Figs. 8 and 9; details are available in [50]. The excitation was carried out using a tunable, single-mode argon ion laser operating at 514.5 nm, which is coincident with the overlapping $P(13)/R(15)$ lines of the (43,0) band of the $B \leftarrow X$ system of I_2. Four overlapping sheets of light were generated and selected sequentially using a chopper. A piezo-driven intracavity etalon was employed to rapidly change the laser frequency of the fourth beam by about 750 MHz (nine axial mode spacings). The four successive fluorescence images were recorded on separate frames of an intensified 100 × 100 diode array camera, requiring a total measurement time of about 250 ms for the 200-mW power level available in each laser sheet. The size of the region imaged was 16 × 16 mm, yielding individual measurement volumes 160 × 160 × 80 μm. The measured signals were corrected for beam intensity variations and then input to

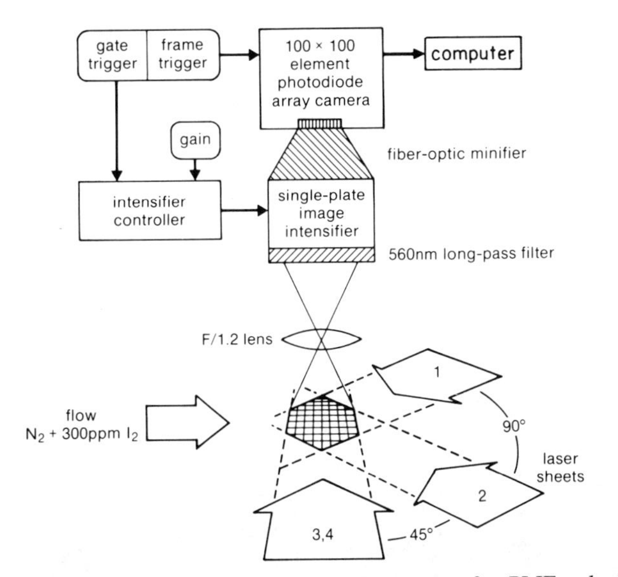

Fig. 8 Illumination scheme and apparatus for PLIF velocity imaging scheme [50].

simple relations [50] to infer the velocity components. An important advantage of the two-frequency scheme is that it is self-calibrating and requires no prior knowledge of the line-shape function. Further, the measurements also yield the slope of the line that may be used to infer pressure when pressure broadening is sufficiently large.

Example images obtained in an underexpanded Mach 1.5 supersonic jet, with a background pressure of 125 Torr, are shown in Figs. 10 and 11 for velocity (two components) and pressure respectively [50]. The results successfully replicate the major features expected within the shock cells imaged. The appearance of the images is compromised by

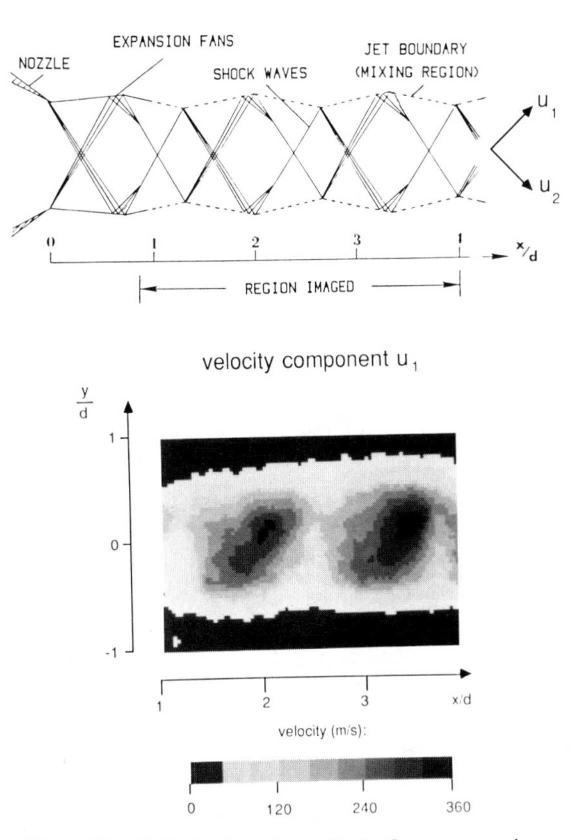

Fig. 10 Velocity imaging with I_2 fluorescence in an underexpanded supersonic jet. This is a gray scale display of one of the measured velocity components, with the gray-scale-to-velocity conversion shown below the figure [50].

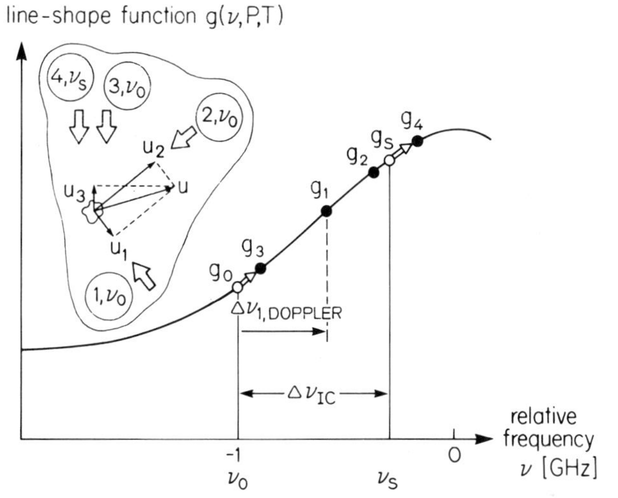

Fig. 9 Relative frequencies with 4-beam velocity/pressure imaging scheme; υ_0 and υ_s are the incident laser frequencies with values of the absorption line-shape function g_0 and g_s. The line-shape values g_1 through g_4 correspond to the Doppler-shifted frequencies of laser sheets 1 through 4 [50].

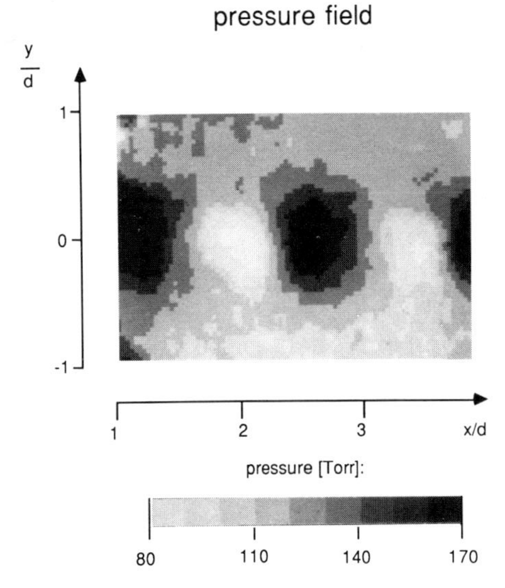

Fig. 11 Pressure imaging in an underexpanded supersonic jet. The gray-level-to-pressure mapping is indicated in the horizontal scale [50].

the need to display only a modest number of gray scales for clarity, though the data are actually recorded to 8-bit resolution. The estimated error in the velocity measurement is 8% (at the maximum velocities) and is about 15% for pressure, but improved accuracy as well as temporal resolution may be expected in future work. Additionally, it may be possible to incorporate simultaneous determinations of temperature and density.

Although PLIF imaging of velocity is still in an early stage of development, with significant challenges yet to be resolved, the potential of the method is clear. If sufficient improvements can be made, then the velocity data can be used in novel ways, for example, to calculate "images" of various velocity correlations or the vorticity distribution and their evolution. When considering fluorescence-based approaches, however, one should be aware of the progress being made with competitive imaging techniques, such as those based on particle-tracking or laser-marking concepts. For example, stereoscopic visualization using scattered light from tracer particles in liquid flows has been coupled with advanced digital image processing to yield 3-D velocity information at a much improved rate [52]. Visualization based on laser-induced phosphorescence [53], Mie scattering from laser-induced aerosols (laser "snow") [54], and even Raman excitation with subsequent LIF tracking [55] are other possibilities for extracting velocity field information, though these methods currently appear to have less potential than LIF for high-speed and low-density flows.

H Three-dimensional Imaging

A recent extension of planar imaging is the observation and measurement of 3-D structures in gases, achieved by scanning the recording plane employed in 2-D imaging. The laser sheet can be rapidly scanned across the flow field by electro-mechanical, electro-optic, or acousto-optic devices. Initial results based on this approach are now available for both Rayleigh [56] and fluorescence [57] imaging. As an example of 3-D data acquisition and display, Fig. 12 shows three slices or cuts through the flow field of an axially forced jet mixing with a surrounding slow-flowing stream [57]. The nitrogen jet was seeded with biacetyl to enable fluorescence imaging, and the 3-D data set was compiled by sequentially recording 24 individual vertically oriented planes of data spaced across the width of the jet. Although each plane of data (100 × 100 pixels) was recorded with a single laser pulse, thus freezing the flow, the 24 planes were recorded in separate cycles of the highly reproducible, cyclic flow. The locations of the three planes shown in this single-frame display are set by movable cursors (see the lines visible on the sides of the display "box"). The display program [58] utilized allows wide flexibility both in the orientation of the planes displayed and in the rate of change of the display, so that the planes may be scanned with time to provide a "3-D movie."

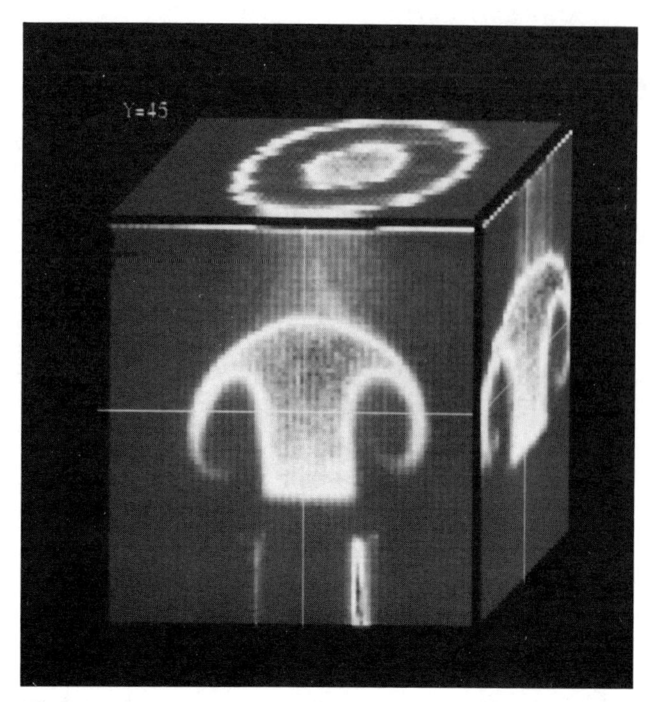

Fig. 12 Display of 3-D image data for a forced jet, seeded with biacetyl, mixing with a coflowing stream of N_2. Each face of the cube display depicts an orthogonal slice through the 3-D data set, as indicated by the white lines. In this display, white represents the highest biacetyl concentration [57].

III TRENDS IN PLANAR IMAGING OF GASES

Planar imaging techniques have advanced rapidly, to a stage at which selected methods are in use as tools in application-oriented studies. Further developments in imaging diagnostics can be anticipated, however, based on current research, though the field is new enough to leave room for creative surprises.

Imaging diagnostics developments are driven by measurement needs, which are relatively fixed, and at the same time constrained by the fast-evolving capabilities of laser sources and solid-state cameras. This situation enables prediction of some trends with high confidence, for example, the recording of planar images with increased repetition rates and spatial resolution (more pixels per camera frame). These improvements will enable detailed study of small- and large-scale flow-field features and their temporal evolution. Steady progress can also be anticipated in extending the list of species accessible by PLIF imaging, not only by single-photon excitation but also through use of multiphoton and photofragmentation schemes. In this connection, current work on fluorescence imaging of O_2 is of special importance owing both to the significance of this species and to its potential for imaging temperature [46] and velocity [51]. With expected simplifications in imaging apparatus, experiments involving simultaneous imaging of multiple quantities, for example, imaging multiple species or temperature together with species, will become more common. With improved availability of tunable, narrow-line-width laser sources, both cw and pulsed, variations in PLIF imaging based on absorption line-shape concepts may be expected. In contrast, use of spectrally broad sources offers prospects for instantaneous particle sizing by quantitative planar imaging of Mie-scattered light [59].

On a more speculative level, if velocity imaging schemes can be improved sufficiently, it should be possible to convert such images to maps of vorticity [50] (and the evolution of vorticity), or to correlate velocity/vorticity with other imaged properties. Similarly, refinements in species and temperature data will allow image processing on a higher level, enabling determination of scalar gradients and various spatial and temporal correlations. And, in time, these parameters will be measured with full 3-D data sets.

The impact of planar imaging on combustion and fluid mechanics modeling is not yet clear, though the future of imaging diagnostics and advanced modeling are surely intertwined; it is likely that the changed nature of data available through imaging will lead to changes in modeling perspectives and methods.

The area of camera technology is central to imaging, and it is encouraging that impressive advances are being made in the performance and convenience of solid-state detector array systems. The trend toward high-resolution, low-noise CCD arrays is particularly notable, with large, 1000×1000 pixel arrays now available commercially. These arrays have architecture that will enable very flexible use of pixels. For example, it is possible to use a large array as a piece of "electronic film," that is, to record several separate images on a reduced area of the array and to transfer the image data electronically on the chip for short-term storage. In this way, the array can be used as a high-speed digital camera with frame-acquisition rates of several kilohertz. Alternatively, multiple images may be recorded at separate locations on the array, so that the array functions as an "electronic plate," much like a rotating-prism camera. With electrostatic deflection in an image-converter tube for example, images could be sorted and acquired at tens of megahertz.

Research with high-resolution image sensors is already in progress in the flow-imaging community [29]. An early example of such work is illustrated in Fig. 13 which is a PLIF image of an unsteady, room-temperature nitrogen jet issuing from a 5-mm-dia. nozzle at an initial velocity of about 10 cm/s [29]. The solid-state camera was a commercial unit with a Thomson-CSF 576×384 CCD array. This sensor is operated at reduced temperature ($-125°C$) to reduce dark current, and the circuitry is specially de-

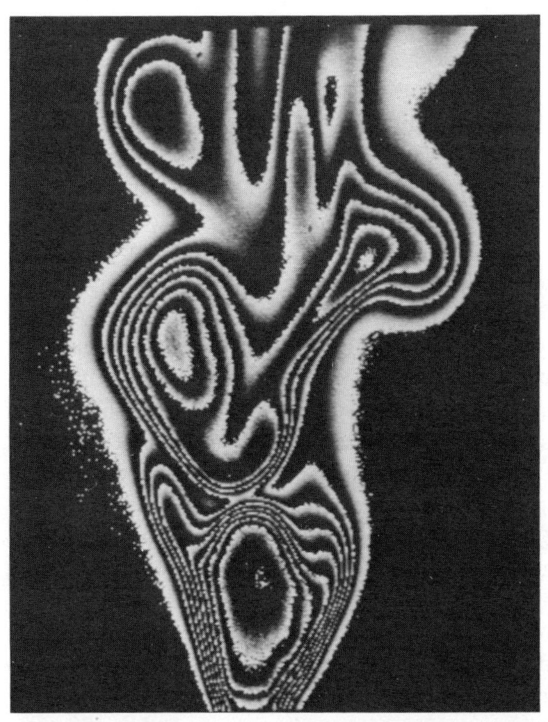

Fig. 13 PLIF image of unsteady jet obtained with a high-resolution (576×384 pixel), high-dynamic-range CCD camera [29].

signed for low-noise performance; the resulting measured background noise level is about 10 photoelectrons per pixel. The key performance advantages of this array, relative to currently common 100×100 and 128×128 photodiode arrays, are (1) a higher fill factor for the active portion of the imaging sensor (93%), which leads to more efficient use of light; (2) a high quantum well depth, about 5×10^5 photoelectrons, which together with the low noise level provides for a linear dynamic range (defined as the ratio of the largest detectable signal to the signal level where the signal-to-noise ratio is 1 : 1) of more than 10^4; (3) a much larger number of pixels, which will allow observation of a greater range of scale sizes in a single image and also will enable development of electronic-film schemes for recording multiple images between readout cycles; and finally, (4) the use of a cooled CCD lowers the detection limit and eliminates the need for a complex image intensifier, when the scattered light is in the sensitive region of the array. The superior dynamic range and spatial resolution are already apparent [17, 29]. For example, in Fig. 13 the intensity has been divided into nine major ranges (corresponding to the dark contours in the figure), and each range has been further subdivided into 6 gray levels. Thus more than 50 distinct levels are apparent when the data are displayed on a high-resolution monitor. Unfortunately that level of resolution is difficult to convey on a small black-and-white print such as Fig. 13. Controlled tests of the system have verified that 12-bit resolution of the signal is possible, in contrast with the 6- to 7-bit capability of the array cameras in current use.

Another important trend in flow-imaging research is to establish the capability for 3-D imaging. The effort in this area of imaging research is presently directed toward two main issues: (1) the need to record multiple planes of image data in a short time, so that the data approach the instantaneous 3-D distribution; (2) questions of how to display and to use 3-D data. Presently, recording rates of up to about 500 Hz have been employed, as both laser sources (e.g., excimer-pumped lasers) and solid-state cameras (e.g., photodiode array systems) are available with such capability. The long-term strategy may involve, on the laser side, use of metal vapor laser systems capable of operation up to about 20 kHz, systems based on amplification of a high-repetition-rate train of short pulses from a mode-locked laser source, or application of a long-pulse flash-lamp-pumped dye laser, with the 3-D set acquired as frames taken by short exposures within the 1- to 5-μs pulse. On the recording side, the most promising schemes seem to be those indicated above, based on electronic-film and -plate concepts. The latter presents some challenges but will enable imaging into the MHz regime.

As to the question of display of 3-D image data, several possibilities are suggested by the state of the art in fields where image data are already in use. There are, however, important differences between imaging solid objects with distinct boundaries and imaging flow properties, which have more diffuse distributions, and these differences complicate predictions of the future [57, 60, 61]. It may be that the value of 3-D data is less likely to accrue, in the long term, from the visual use of 3-D data displays than from the need for 3-D data to infer quantities, such as property gradients and spatial correlations, that have an inherent 3-D rather than a 2-D character. Nonetheless, there is a need for 3-D displays of flow-field properties for at least qualitative use in studying flow fields, and the future will surely lead to exciting developments in this area of research. The quantity of information stored in a single 3-D data set (2.4×10^5 pixels with the 24-plane-deep 100×100 pixel images used for Fig. 12; and 22 megapixels for 100 planes from a 576×384 pixel camera) serves as an impressive indicator of the potential of planar and 3-D imaging.

REFERENCES

1. Eckbreth, A. C., Recent Advances in Laser Diagnostics for Temperature and Species Concentration in Combustion, *18th Int. Symp. Combust.*, pp. 1471–1488, Combustion Institute, Pittsburgh, Pennsylvania, 1981.
2. Long, M. B., Webber, B. F., and Chang, R. K., Instantaneous Two-dimensional Concentration Measurements in a Jet by Mie Scattering, *Appl. Phys. Lett.*, vol. 34, no. 1, pp. 22–24, 1979.
3. Escoda, M. C., and Long, M. B., Rayleigh Scattering Measurements of the Gas Concentration Field in Turbulent Jets, *AIAA J.*, vol. 21, no. 1, pp. 81–84, 1983.
4. Alden, M., Edner, H., Holmsted, G., Svanberg, S., and Hogberg, T., Single-Pulse Laser-induced OH Fluorescence in an Atmospheric Flame, *Appl. Opt.*, vol. 21, no. 7, pp. 1236–1240, 1982.
5. Dyer, M. J., and Crosley, D. R., Two-dimensional Imaging of OH Laser-Induced Fluorescence in a Flame, *Opt. Lett.*, vol. 7, no. 8, pp. 382–384, 1982.
6. Kychakoff, G., Howe, R. D., Hanson, R. K., and McDaniel, J. C., Quantitative Visualization of Combustion Species in a Plane, *Appl. Opt.*, vol. 21, no. 18, pp. 3225–3227, 1982.
7. Long, M. B., Fourguette, D. C., Escoda, M. C., and Layne, C. B., Instantaneous Ramanography of a Turbulent Diffusion Flame, *Opt. Lett.*, vol. 8, no. 5, pp. 244–246, 1983.
8. Melton, L. A., and Verdieck, J. F., Vapor/Liquid Visualization in Fuel Sprays, *Combust. Sci. Technol.*, vol. 42, pp. 217–222, 1985.
9. Cheng, S., Zimmermann, M., and Miles, R. B., Supersonic-Nitrogen Flow-Field Measurements with the Resonant Doppler Velocimeter, *Appl. Phys. Lett.*, vol. 43, no. 2, pp. 143–145, 1983.
10. McDaniel, J. C., Hiller, B., and Hanson, R. K., Simultaneous Multiple-Point Velocity Measurements using Laser-induced Iodine Fluorescence, *Opt. Lett.* vol. 8, no. 1, pp. 51–53, 1983.
11. Lucht, R. P., Applications of Laser-induced Fluorescence Spectroscopy for Combustion and Plasma Diagnostics, in *Laser Spectroscopy and Its Applications,* ed. L. J. Radziemski, R. W. Solarz, and J. A. Paisner, pp. 623–676, Marcel Dekker, New York, 1986.
12. Bechtel, J. H., Dasch, C. J., and Teets, R. E., Combustion Research with Lasers, in *Laser Applications,* ed. J. F. Ready, and R. K. Erf, vol. 5, pp. 129–212, Academic, New York, 1984.
13. Crosley, D. R., Laser-induced Fluorescence Measurements of Combustion Intermediates, *High Temp. Mater. Process.*, vol. 7, pp. 41–54, 1986.
14. Kychakoff, G., Howe, R. D., and Hanson, R. K., Quantitative Flow Visualization Technique for Measurements in Combustion Gases, *Appl. Opt.*, vol. 23, no. 5, pp. 704–712, 1984.
15. Kychakoff, G., Hanson, R. K., and Howe, R. D., Simultaneous Multiple Measurements of OH in Combustion Gases using Planar Laser-induced Fluorescence, *20th Int. Symp. Combust.*, pp. 1265–1272, Combustion Institute, Pittsburgh, Pennsylvania, 1985.
16. Cattolica, R. J., and Vosen, S. R., Two-dimensional Measurements of the [OH] in a Constant Volume Combustion Chamber, *20th Int. Symp. Combust.*, pp. 1273–1282, Combustion Institute, Pittsburgh, Pennsylvania, 1985.
17. Kychakoff, G., and Hanson, R. K., Digital Flowfield Imaging, *Int. Wkshp. Phys. Eng. Computerized Multidimens. Imag. Process.*, ed. O. Nalcioglu, Z. H. Chu, and T. F. Budinger, Proc. SPIE, vol. 671, pp. 72–80, 1986.
18. Allen, M. G., and Hanson, R. K., Planar Laser-induced Fluorescence Monitoring of OH in a Spray Flame, *Opt. Eng.*, vol. 25, no. 12, pp. 1309–1311, 1986.
19. Allen, M. G., and Hanson, R. K., Digital Imaging of Species

Concentration Fields in Spray Flames, *21st Int. Symp. Combust.*, Combustion Institute, Pittsburgh, Pennsylvania, 1987.

20. Allen, M. G., Howe, R. D., and Hanson, R. K., Digital Imaging of Reaction Zones in Hydrocarbon-Air Flames using Planar Laser-induced Fluorescence of CH and C_2, *Opt. Lett.*, vol. 11, no. 3, pp. 126–128, 1986.
21. Dyer, M. J., and Crosley, D. R., Fluorescence Imaging for Flame Chemistry, *Proc. Int. Conf. Lasers '84*, pp. 211–218, San Francisco, California, 1985.
22. Alden, M., Edner, H., and Svanberg, S., Simultaneous Spatially Resolved Monitoring of C_2 and OH in a C_2H_2/O_2 Flame using a Diode Array Detector, *Appl. Phys. B*, vol. 29, no. 2, pp. 93–97, 1982.
23. Allen, M. G., and Hanson, R. K., Simultaneous Imaging of Species Distributions in Two-Phase Reacting Flowfields, Proc. Int. Laser Sci. Conf., Seattle, 1986.
24. Cattolica, R. J., and Vosen, S. R., Two-dimensional Fluorescence Imaging of a Flame-Vortex Interaction, *Combust. Flame*, vol. 48, no. 1, pp. 77–87, 1986.
25. Cattolica, R. J., Combustion-Torch Ignition: Fluorescence Imaging of NO_2, *21st Int. Symp. Combust.*, Combustion Institute, Pittsburgh, Pennsylvania, 1987.
26. Kychakoff, G., Knapp, K., Howe, R. D., and Hanson, R. K., Flow Visualization in Combustion Gases using Nitric Oxide Fluorescence, *AIAA J.*, vol. 22, no. 1, pp. 153–154, 1984.
27. Seitzman, J. M., Kychakoff, G., and Hanson, R. K., Instantaneous Temperature Field Measurements using Planar Laser-induced Fluorescence, *Opt. Lett.*, vol. 10, no. 9, pp. 439–441, 1985.
28. Itoh, F., Kychakoff, G., and Hanson, R. K., Flow Visualization in Low-Pressure Chambers using Laser-induced Biacetyl Phosphorescence, *J. Vac. Sci. Technol. B*, vol. 3, no. 6, pp. 1600–1603, 1985.
29. Paul, P. H., van Cruyningen, I., Hanson, R. K., and Kychakoff, G., High Resolution Digital Flowfield Imaging of Jets, *J. Exper. Fluids*, 1988, in press.
30. Massey, G. A., and Lemon, C. J., Feasibility of Measuring Temperature and Density Fluctuations in Air using Laser-induced O_2 Fluorescence, *IEEE J. Quantum Electron.*, vol. QE-20, no. 5, pp. 454–457, 1984.
31. Lee, M. P., Paul, P. H., and Hanson, R. K., Laser-Fluorescence Imaging of O_2 in Combustion Flows using an ArF Laser, *Opt. Lett.*, vol. 11, no. 1, pp. 7–9, 1986.
32. Goldsmith, J. E. M., and Anderson, R. J. M., Laser-induced Fluorescence Spectroscopy and Imaging of Molecular Oxygen in Flames, *Opt. Lett.*, vol. 11, no. 2, pp. 67–69, 1986.
33. Lee, M. P., and Hanson, R. K., Calculations of O_2 Absorption and Fluorescence at Elevated Temperatures for a Broadband Argon-Fluoride Laser Source at 193 nm, *J. Quant. Spectrosc. Radiat. Transfer.*, vol. 36, no. 5, pp. 425–440, 1986.
34. Alden, M., Hertz, H. M., Svanberg, S., and Wallin, S., Imaging Laser-induced Fluorescence of Oxygen Atoms in a Flame, *Appl. Opt.*, vol. 23, no. 19, pp. 3255–3257, 1984.
35. Goldsmith, J. E. M., and Anderson, R. J. M., Imaging of Atomic Hydrogen in Flames with Two-Step Saturated Fluorescence, *Appl. Opt.*, vol. 24, no. 5, pp. 607–609, 1985.
36. Alden, M., Wallin, S., and Wendt, W., Applications of Two-Photon Absorption for Detection of CO in Combustion Gases, *Appl. Phys. B*, vol. 33, no. 4, pp. 205–212, 1984.
37. Haumann, J., Seitzman, J. M., and Hanson, R. K., Two-Photon Digital Imaging of CO in Combustion Flows using Planar Laser-induced Fluorescence, *Opt. Lett.*, vol. 11, no. 12, pp. 776–778, 1986.
38. Seitzman, J. M., Haumann, J., and Hanson, R. K., Quantitative Two-Photon LIF Imaging of Carbon Monoxide in Combustion Gases, *Appl. Opt.*, vol. 26, no. 14, pp. 2892–2899, 1987.
39. Melton, L. A., Spectrally Separated Fluorescence Emissions for Diesel Fuel Droplets and Vapor, *Appl. Opt.*, vol. 22, no. 14, pp. 2224–2226, 1983.
40. Dobbs, G., United Technologies Research Center, unpublished.
41. Melton, L. A., and Verdieck, J. F., Vapor/Liquid Visualization in Fuel Sprays, *20th Int. Symp. Combust.*, pp. 1283–1290, Combustion Institute, Pittsburgh, Pennsylvania, 1985.
42. Murray, A. M., and Melton, L. A., Fluorescence Methods for Determination of Temperature in Fuel Sprays, *Appl. Opt.*, vol. 24, no. 17, pp. 2783–2787, 1985.
43. Brown, G. M., and Kent, J. C., Fluorescent Light Section Technique for Fuel Spray Characterization, in *Flow Visualization III: Proceedings of the Third International Symposium on Flow Visualization*, ed. W.-J. Yang, pp. 118–122, Hemisphere, Washington, D.C., 1985.
44. Cattolica, R. J., and Stephenson, D. A., Dynamics of Flames and Reactive Systems, *Prog. Astronaut. Aeronaut.*, vol. 95, pp. 714–721, 1985.
45. Long, M. B., Levin, P. S., and Fourguette, D. C., Simultaneous Two-dimensional Mapping of Species Concentration and Temperature in Turbulent Flames, *Opt. Lett.*, vol. 10, no. 6, pp. 267–269, 1985.
46. Lee, M. P., Paul, P. H., and Hanson, R. K., Quantitative Imaging of Temperature Fields in Air using Planar Laser-induced Fluorescence of O_2, *Opt. Lett.*, vol. 12, no. 2, pp. 75–77, 1987.
47. Andresen, P., Bath, A., Groger, W., and Lulf, H. W., Laser-induced Fluorescence with Tunable Excimer Lasers as a Possible Method for Instantaneous Temperature Field Measurements at High Pressures, Appl. Opt., vol. 27, no. 2, pp. 365–378, 1988.
48. Hiller, B., McDaniel, J. C., Rea, E. C., Jr., and Hanson, R. K., Laser-induced Fluorescence Technique for Velocity Field Measurements in Subsonic Gas Flows, *Opt. Lett.*, vol. 8, no. 9, pp. 474–476, 1983.
49. Hiller, B., and Hanson, R. K., Two-Frequency Laser-induced Fluorescence Technique for Rapid Velocity-Field Measurements in Gas Flows, *Opt. Lett.*, vol. 10, no. 5, pp. 206–208, 1985.
50. Hiller, B., and Hanson, R. K., Simultaneous Planar Measurements of Velocity and Pressure Fields in Gas Flows using Laser-induced Fluorescence, *Appl. Opt.*, vol. 27, no. 1, pp. 33–48, 1988.
51. Paul, P. H., Lee, M. P., and Hanson, R. K., Molecular Velocity Imaging of Supersonic Flows Using Pulsed Planar Laser-Induced Fluorescence of No, *Opt. Lett.*, 1988, in press.
52. Chang, T. P., Wilcox, N. A., and Tatterson, G. B., Application of Image Processing to the Analysis of Three-dimen-

sional Flow Fields, *Opt. Eng.*, vol. 23, no. 3, pp. 283–287, 1984.

53. Hiller, B., Booman, R. A., Hassa, C., and Hanson, R. K., Velocity Visualization in Gas Flows using Laser-induced Phosphorescence of Biacetyl, *Rev. Sci. Instrum.*, vol. 55, no. 12, pp. 1964–1967, 1984.
54. Hassa, C., and Hanson, R. K., Fast Laser-induced Aerosol Formation for Visualization of Gas Flows, *Rev. Sci. Instrum.*, vol. 56, no. 4, pp. 557–559, 1985.
55. Miles, R., Cohen, C., Connors, J., Howard, P., Huang, S., Markovitz, E., and Russell, G., Velocity Measurements by Vibrational Tagging and Fluorescent Probing of Oxygen, Appl. Opt., vol. 12, no. 11, pp. 861–863, 1987.
56. Yip, B., and Long, M. B., Instantaneous Planar Measurement of the Complete Three-dimensional Scalar Gradient in a Turbulent Jet, *Opt. Lett.*, vol. 11, no. 2, pp. 64–66, 1986.
57. Kychakoff, G., Paul, P. H., van Cruyningen, I., and Hanson, R. K., Movies and Three-dimensional Images of Flowfields using Planar Laser-induced Fluorescence, *Appl. Opt.*, vol. 26, no. 13, pp. 2498–2500, 1987.
58. Ottolini, R., Sword, C., and Claerbout, J. F., On-line Movies of Reflection of Seismic Data with Description of a Movie Machine, *Geophysica.*, vol. 49, no. 2, pp. 195–200, 1984.
59. Allen, M. G., and Hanson, R. K., Digital Imaging in Spray Flames, Univ. of California, Davis, Western States Section Paper 85-13, 1985.
60. Russell, G., and Miles, R. B., Display and Perception of 3-D Space-filling Data, *Appl. Opt.*, vol. 26, no. 6, pp. 973–982, 1987.
61. Stickland, R. N., and Sweeney, D. W., Digital Imaging Sequence Analysis for Computing Optical Flow in Flame Propagation Visualization, SPIE Tech. Symp. S.E. Optics, Electro-Optics, and Sensors, Orlando, Fla., 1987.

Part 3

Image Processing and Computer-Assisted Methods

Chapter 16

Digital Processing of Interferograms

Giovanni Maria Carlomagno and Armando Rapillo

I INTRODUCTION

Interferometric techniques are widely used for quantitative measurements of flow fields and/or properties of transparent fluids. In order to make such an evaluation, however, an accurate analysis of interferograms has to be done.

The simplest case, in which an interferogram gives straightforward, explicit information, is the plane two-dimensional laminar flow field visualized by means of a reference beam interferometer with infinite fringe width alignment. In this case each fringe represents the locus of constant refractive index of the fluid (which, depending on the specific problem, may be related to such variables as density, temperature, or mole fraction), and thus only the knowledge of fringe orders and fringe positions is required to describe the flow field quantitatively. In all other cases the reconstruction of a flow field from interferograms requires the use of a more or less complex evaluation procedure to decode the integrated information they contain [1].

The starting points in interferogram analysis are the accurate location of fringe positions or shifts within the image and the identification of fringe orders. The information available in an interferogram can be expressed, according to Merzkirch [2], by the data function $D(x,y)$, where x and y are the coordinates in the recording plane; z is the coordinate in the direction of the undisturbed light beam. The data function may be related to the refractive index distribution of the fluid in the test section (phase object) by

$$D(x,y) = \int_{z_1}^{z_2} R(x,y,z)\, dz \qquad (1)$$

where $R(x,y,z)$ is a function of the refractive index of the fluid and the type of interferometer, and $z_2 - z_1$ is the length of the test section. The data function $D(x,y)$ may be interpreted as phase distribution of the incoming light on the interferogram plane (fringe order).

In particular, for a homogeneous gas, one has

$$R(x,y,z) = K\,\frac{\rho(x,y,z) - \rho_\infty}{\lambda} \qquad (2)$$

$$R(x,y,z) = Kh\,\frac{\delta\rho/\delta y}{\lambda} \qquad (3)$$

respectively for a reference beam and a shearing interferometer. In (2) and (3), ρ and K are respectively the mass density and the Gladstone–Dale constant of the gas, λ is the wavelength of the light, h is the distance between the two interfering light beams in the test section, the y direction is assumed to be perpendicular to the undisturbed fringes and the subscript ∞ refers to undisturbed conditions.

For the case of the diffusion of an isothermal binary gas mixture [3] one has respectively for the two types of interferometers

$$R(x,y,z) = (n_1 - n_2)\,\frac{X_1(x,y,z) - X_{1\infty}}{\lambda} \qquad (4)$$

$$R(x,y,z) = (n_1 - n_2)\,\frac{h\,\delta X_1/\delta y}{\lambda} \qquad (5)$$

where n_1 and n_2 are the refractive indexes of the two components of the mixture, both evaluated at mixture temperature and pressure, and X_1 is the mole fraction of one component of the mixture.

In a plane two-dimensional flow only one interferogram is needed. Integration of (1) is easily performed, and the fringe shifts represent either density (or mole fraction) values or values of the gradient component in the y direction of the same quantities, respectively, for a reference beam and a shearing interferometer. Therefore, for the latter one, an integration of the fringe shift data is necessary in order to obtain density (or mole fraction) profiles.

For an axisymmetric two-dimensional flow [2] only one interferogram is needed to compute the field of refractive index. In this case, if the radial coordinate r lies in the plane y–z, the coordinate x being the axis of symmetry, for a cross section x = const, Eq. (1) can be rewritten as

$$D(y) = \int_{r=y}^{r} R(r)(r^2 - y^2)^{-1/2}\, d(r^2) \tag{6}$$

The refractive index function for a homogeneous gas is

$$R(r) = K\,\frac{\rho(r) - \rho_\infty}{\lambda} \tag{7}$$

for the reference beam interferometer and

$$R(r) = Kh\,\frac{\delta\rho/\delta(r^2)}{\lambda} \tag{8}$$

for the shearing one.

For an isothermal binary mixture one has

$$R(r) = (n_1 - n_2)\,\frac{X_1(r) - X_{1\infty}}{\lambda} \tag{9}$$

for the reference beam interferometer and

$$R(r) = (n_1 - n_2)h\,\frac{\delta X_1/\delta(r^2)}{\lambda} \tag{10}$$

for the shearing one. It has to be recalled [2] that, in the finite fringe width alignment, for a reference beam interferometer $D(x,y) = \Delta s/s$, where $\Delta s/s$, is the relative fringe shift, whereas for a shearing one $D(x,y) = (\Delta s/s)/y$. In the infinite fringe width alignment, for a reference beam interferometer $D(x,y) = \Delta 1/\lambda$, where $\Delta 1$ is the optical path length difference between interfering rays, whereas for a shearing one $D(x,y) = (\Delta 1/\lambda)/y$.

By applying the Abel inversion to the integral appearing in (6), the following relationship is obtained:

$$R(r) = -\left(\frac{1}{\pi}\right)\left[\frac{d}{d(r^2)}\right]\int_{r}^{r\infty} D(y)(y^2 - r^2)^{-1/2} d\,(y^2) \tag{11}$$

Equation (11) shows that, in applying the Abel inversion to an interferogram obtained with a reference beam interferometer, the derivative of fringe shift with respect to the lateral distance must be used. Therefore, considerable errors may arise if the fringe shift measurement is not accurate enough. With a shearing interferogram a further integration must be performed in order to compute density profiles.

For the case of three-dimensional flows [2], the evaluation procedure to obtain the flow field requires a full tomographic reconstruction technique from several interferograms, each taken from a different viewing direction. The phase object under investigation is subdivided into N finite elements, and the refractive index function $R(x,y,z)$ of Eq. (1) is assumed to be constant within each element. Then, for each light ray, the integral in (1) is approximated by a summation. By taking a number of interferograms from different viewing angles, a set of N (or more, if generalized inversion methods are used) equations must be obtained in order to solve the R_i unknowns ($i = 1, \ldots, N$).

As pointed out in Ref. [2], the precision of the method increases with the number of elements N, and the major problems in performing the evaluation procedure are finding an appropriate subdivision of the object field, replacing the integral in Eq. (1) by a summation or a series of suitable analytic functions, and solving the respective set of equations. It is also true that a fast, accurate method of acquiring the various data functions given by the interferometric images (which constitute the known terms in the system of equations) is a very important step in performing the three-dimensional evaluation procedure.

Thus, the procedure of evaluating interferograms ranges from integration of density gradients as visualized by the shearing interferometer, to Abel inversion for axisymmetric flow fields, to tomographic reconstruction for three-dimensional flow fields.

The classical processing required to obtain the data function $D(x,y)$ from an interferogram generally consists of a long, careful procedure that includes acquisition of the image in form of a photographic transparency (positive or negative), measurement of fringe center positions by means of an optical comparator, and numerical computation of the fringe shift distribution curve. This kind of operation is generally a very time consuming process that partially depends on the sensitivity and ability of the operator. In order to improve accuracy, eliminate human judgment, and, in particular, handle large amounts of data, an automated fringe-reading and evaluation procedure constitutes a very powerful tool for the interferometric techniques.

When there are few fringes on the interferogram, it may be necessary, in order to increase the amount of measurements, to interpolate among fringe position data by picking up additional points at arbitrary positions within the fringe pattern. In this case, automatic evaluation of the fringe order is badly needed since it would be very inaccurate to perform such an interpolation manually.

Also the evaluation of such techniques as speckle interferograms [4, 5], electronic speckle pattern interferometry (ESPI) [6, 7], and holographic interferometry [8, 9] can be successfully automated.

The increasing availability of low-cost electro-optical sensing devices, microprocessors, and frame memories has made the automated evaluation of interferograms by digital analysis very appealing.

The scope of this chapter is to determine some of the design criteria for the different parts of a system that can perform digital processing of interferograms (DPI) and to give a brief review both of the options which have been available for each of them and of the usual methods for interferogram evaluation. A number of applications will also be reviewed in the last section.

In the field of thermal fluid dynamics the design of a DPI system requires, besides experience in flow visualization, the knowledge of some basic concepts of electronics and computers. In the following the most important of them will be evidenced from a user point of view, leaving the deeper questions of specific subjects to specialized handbooks.

Schematically a system for DPI may be divided in the following, logical or physical, parts: (1) image acquisition and digitization; (2) interferogram evaluation including (a) preprocessing and/or image enhancement, (b) fringe segmentation, and (c) fringe order recognition; (3) oriented processing. The chapter is developed according to this scheme.

II IMAGE ACQUISITION

When a DPI system is used in an experiment, it should be regarded as a measuring instrument and designed accordingly. The configuration of the system is strictly related to the application to which it is oriented. In fact, costs may easily rise by a factor of ten or a hundred with increasing image resolution, but not gaining for that increase much measurement accuracy while generally increasing acquisition and computing times.

Image-acquisition systems can be broadly classified as *real time* or *static,* depending on their working speed. Real-time systems should be intended to sense, digitize, and transmit an image in a time that is a fraction of the characteristic experimental time scale. In practice, systems that make use of a video camera, which has at least a standard video refresh rate (25 or 30 frame/s), are commonly considered as real-time systems. Higher refresh rates are generally difficult to achieve. When the video rate is insufficient, the alternative is, first, to acquire images by means of photographic media, then to convert them into digital data with a slower (static) system such as a microdensitometer [10] (accurate but very slow and expensive), a graphic tablet (cheap but slower than a microdensitometer and very sensitive to human judgment), or else the video camera itself.

It has to be emphasized that photography has been up to now the best means of acquiring and storing images and that it is still unmatched by electronic devices. No electronic digital system can yet acquire data with such density of information and speed as, for example, a standard 16-mm movie camera, let alone any of the available high-speed photographic systems. A choice in terms of cost, image-storage capability, and image-acquisition speed is always in favor of photographic media, while for a quantitative analysis the need for digital processing is even more necessary. In the following, only real-time systems are discussed.

A schematic layout of the main components of a standard DPI system is shown in Fig. 1. The image given by an interferometer is sensed by a black-and-white video camera connected both to a video processor, mainly providing image acquisition as well as other optional functions, and to a monitor. The monitor can visualize either the video signal directly from the camera itself for monitoring purposes or the signal from the video processor reconverted by means of a digital-to-analog converter. The video processor is interfaced with the host computer to transmit image data and receive instructions. It may have, besides the fundamental blocks (e.g., synchronisms and timings, sample and hold, analog-to-digital converter, and frame memory), look-up tables and/or an arithmetic logic unit in order to carry out in real time a preprocessing of the incoming image (thresholding, filtering, and addition or subtraction of images).

A Sensor Requirements

The digitization of an image starts by transducing point by point the light intensity into an electrical signal by means of an electro-optical sensing device. A typical response of such a sensor to continuous illumination is shown in Fig. 2, where I is the intensity of the light impinging on the sensor and V is the corresponding output signal which is generally a voltage or an electrical current intensity. At low light levels the signal V is due mainly to the sensor noise, called *dark current,* while above a certain light level the sensor signal is clipped, as the sensor is in *saturation.* Between these two limits (dark current and saturation) the relationship between the logarithms of input and output is approximately linear, and the slope of the line is generally referred to as Γ. Sensors with Γ greater than unity show a nonlinear response favouring high light intensity, while Γ less than unity is desired at low light levels. A sensor having a Γ equal to 1 exhibits a linear response to light-intensity variations and is normally preferred in quantitative work.

The output analog signal V is generally degraded with respect to light intensity I in terms of spatial resolution, noise, dynamic range, geometric distortion, lag, and spectral response.

The electro-optical sensing devices can be divided into two groups: tubes (Vidicon, Plumbicon, Tivicon, Chalnicon, etc.) and solid-state imagers (charge coupled devices

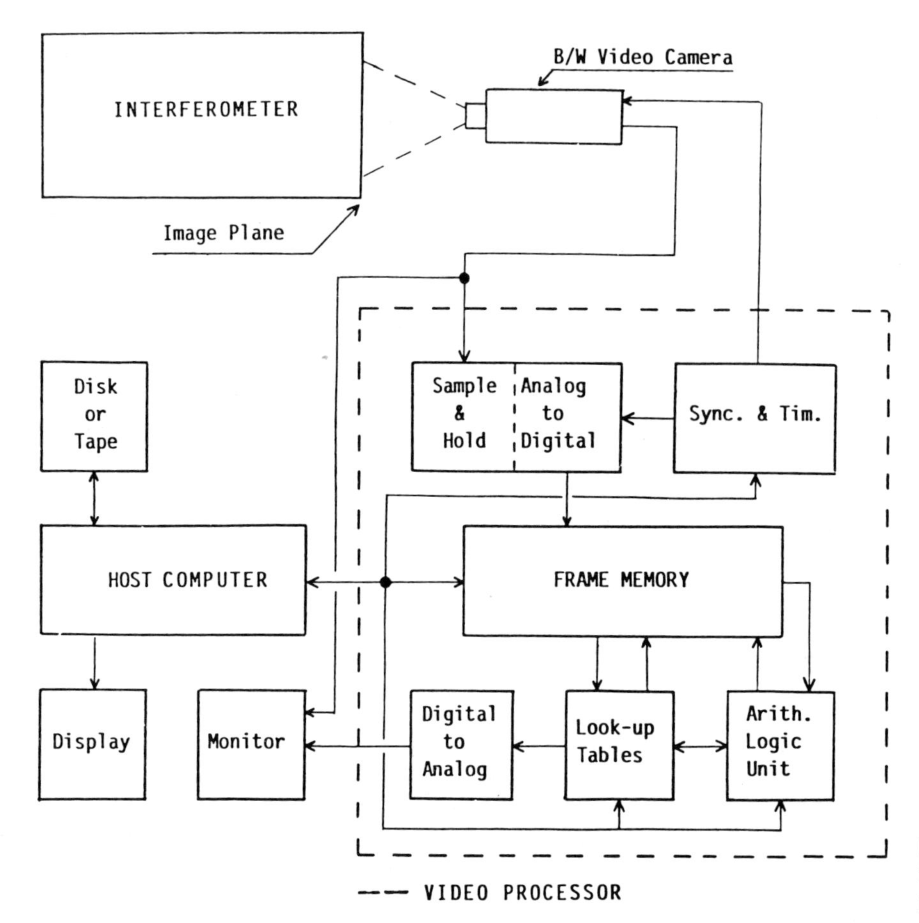

Fig. 1 Schematic representation of an image-processing system for digital processing of interferograms (DPI).

CCD, charge injection devices CID, metal oxide semiconductors MOS, etc.) Table 1 compares a tube and an SSI sensor. The following discussion concentrates on solid-state imagers (SSIs) because their costs are currently comparable to tubes, and they are best suited for use in digital image processing. Furthermore, they are relatively resistant to burn in, which may occur when the sensor saturates.

The sensing area of an SSI consists of a matrix (linear or bidimensional) of discrete photodiodes that sense the light irradiance in fixed points (pixels) in the image plane. Thus, these intensity data are naturally compatible with the digital computer-processing techniques that work on a fixed grid of data values. The practical absence of geometric distortion, peculiar to their solid-state construction, is a

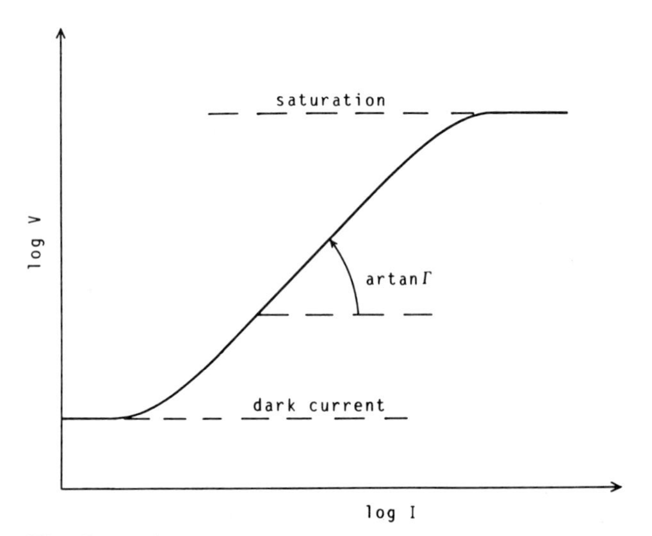

Fig. 2 Typical response curve of an electro-optical sensor.

Table 1 Comparison of Two Good and Typical Sensors

Characteristic	Tube	Solid-state Imager (SSI)
Spatial resolution (pixel/side)	900	1024
Geometric distortion (%)	1	0
Gamma	0.8	1
Lag (%)	15	0
Signal-to-noise ratio (dB)	40	48
Relative sensitivity	0.5	1
Minimum operational life (hr)	3000	40,000
Burn in	Yes	No
Magnetic field influence	Yes	No
Relative power consumption	High	Low
Relative camera dimensions	Large	Small
Voltage requirements (V)	500	15

significant advantage for image recognition and for accurate measurement on interferograms. It is not necessary to perform any spatial calibration within the field of view, and the evaluation of other optical characteristics, e.g., image magnification, becomes much easier.

The most important characteristic in interferogram acquisition is the linear spatial resolution of the sensor d, defined as number of pixels per unit length. This is limited both by technology and by the amount of information to be stored, transferred,and computed by machines.

Comparing photographic media and solid-state sensors, the linear resolution of a high-quality film is typically 500 mm^{-1} as compared with 100 mm^{-1} for a very good commercially available sensor. In terms of area resolution (number of pixels per unit area), the ratio between films and SSIs is therefore about 25. Actually, comparing standard films (250 mm^{-1} linear resolution) to commercial sensors (70 mm^{-1}), said ratio can be fixed to about one half. It should be further evidenced that large format (60 $\times$ 60 mm) is standard for films, while it is generally difficult to find sensors with sensing area larger than about 100 mm^2.

On the other hand, photographic films show a nonlinear response to light intensity, practically having $\Gamma \approx 0.7$, while SSIs have a unity Γ.

In Fig. 3 the number of 8-bit bytes necessary to store a single frame is reported as a function of both the needed spatial resolution, expressed as number of pixels per side of a square sensor (*image format*), and the needed light-intensity resolution, expressed as number of bits of the converted light signal (*image depth*) [11]. The figure also shows the main areas of application of digital image processing including interferometry. The isolines of constant total storage (broken lines) indicate the largest amount of storage that can be treated directly in the central memory, e.g., a processor that has 20-bit addressing can manage up to 1 Mbyte in the fast mode. The use of virtual memory-management systems or mass-storage units requires more complex software and/or longer computing times.

While determining the needed spatial resolution of the SSI to be used in a specific DPI application, two aspects must be considered. The first is the accuracy required to locate objects and fringes in the test section. This is related directly to the number of pixels N_p along the measuring direction and to the image magnification M. The linear spatial resolution in the test section s for a width of observed area W_o is given by

$$s = \frac{N_p}{W_o} = d \cdot M$$

being the linear spatial resolution on the image plane d fixed by the pixel center-to-center spacing of the sensor. By increasing magnification of the object, the spatial resolution in the test section can be increased. A greater spatial resolution, however, implies that smaller areas of the image must be digitized at one time, and, in order to patch up the entire field of view, a pixel-matching program has to be introduced into the processing procedure.

The other limiting factor for the spatial resolution of sensor d is the fringe density on interferograms. Expressing this parameter in terms of spatial frequency f_o as number of fringe pairs (light and dark) per unit length on the image plane, the theoretically detectable upper limit frequency is given by the Nyquist sampling theorem:

$$f_{\mathrm{Nyq}} = \frac{1}{2d}$$

The effective response of a sensor [12] to a pattern of black-and-white strips changes when the spatial frequency

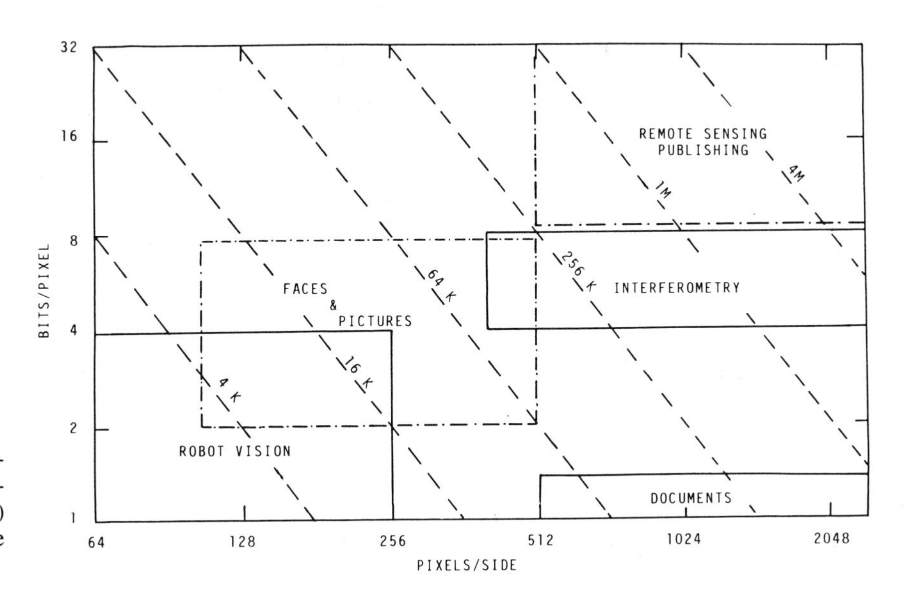

Fig. 3 Memory requirements (number of 8-bit bytes) as a function of sensor spatial resolution (image format) and A/D conversion resolution (image depth) [11].

of the strips on the image is varied; this characteristic is called modulation transfer function (MTF). The MTF of a sensor should ideally be defined only by the size and shape of photosites. In practice, this ideal MTF is degraded by charge transfer inefficiency and lateral photocharge diffusion, which cause in the sensor an image smearing in the direction of charge transfer. Lateral photocharge diffusion occurs when photogenerated charge diffuses sideways into adjacent photosites before being collected. This is a function of the depth of penetration of the photons into the substrate, which is, in turn, wavelength dependent. For silicon devices, the near infrared photons penetrate more deeply, producing electron-hole pairs deeper in the material. These can diffuse sideways before being collected at a potential well near the surface. For general visible imaging applications an infrared-absorbing filter is often used to improve MTF (Fig. 4*a*,) but in interferometry, where only a specific wavelength is used, a filter is not needed. The degradation of the MTF with the wavelength of the impinging light (Fig. 4*b*) must be taken into account.

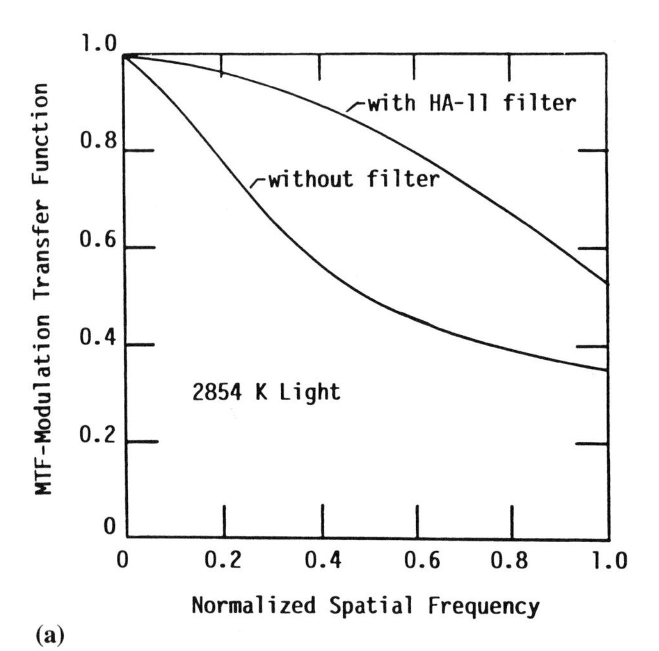

(a)

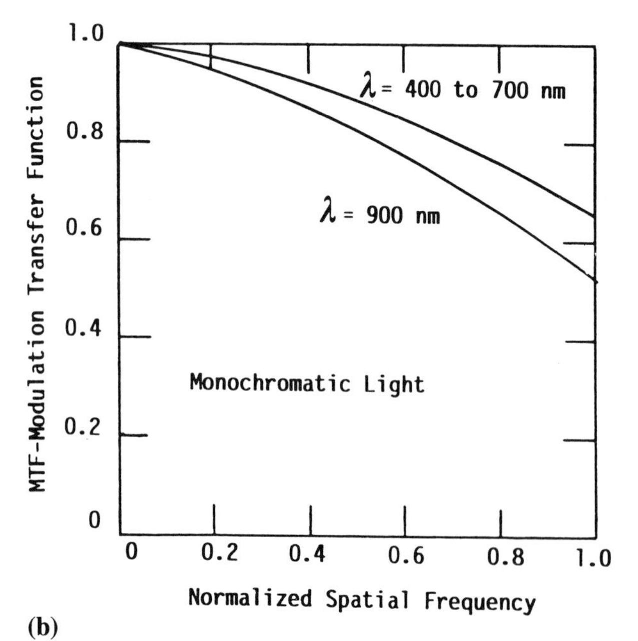

(b)

Fig. 4 Modulation transfer function versus normalized spatial frequency (f_o/f_{Nyq}) of a linear array of 2592 photodiodes (TEXAS Instruments TC106-1). (*a*) White light; (*b*) monochromatic light.

For an area array it is important to consider the linear-spatial-resolution MTF and Nyquist frequency limitations separately in the horizontal and vertical directions, since the pixel size and separation are very often different in the two dimensions.

There are many applications [14, 15] where a high linear spatial resolution is needed in only one direction, e.g., when density profiles are measured in few sections of the interferogram. In these cases a very simple solution is to use a linear array of photodiodes mounted on a translation stage driven by a stepping motor. Then it is possible to have a very high spatial resolution (up to 4096 pixel/line) without treating the large amount of data (16 Mbyte) of an equivalent square area sensor.

This solution, however, while considerably reducing costs inherent to resolution, is limited by the slower speed required for scanning the entire image (as it is done mechanically), and it is therefore not generally useful for study of transient phenomena. On the other hand, the elapsed time between the acquisition of two lines could be used to process the previous line signal.

Another characteristic that has to be considered when selecting a sensor is the signal-to-noise ratio, or the dynamic range of the sensed light. The dynamic range of a sensor is usually given as the ratio between maximum peak signal and rms noise; a typical value for SSIs at room temperature is about 500 : 1.

The noise in an image sensor is generally expressed as the current generated by the sensor when no light impinges on it, i.e., the dark current. There are three major components to the dark output signal:

1. Dark leakage current, which is directly proportional to the integration time (elapsed time between two frames) and shows a very strong function with temperature, approximately doubling every 7°C. Cooling of the sensor may raise the dynamic range up to 10,000 : 1.
2. Fixed pattern noise, which is caused by the incomplete cancellation of the clock switching transients of the pixel scanning circuit, and is capacitatively coupled into the video line. It can be removed from the signal by means of a differential readout with an obscured dummy array.
3. Random pixel noise, due to the nonrepetitive fluctuations superimposed on the dark level; it cannot be removed by signal processing except by filtering.

Since SSIs operate in the charge storage mode, they respond to the total light energy that is absorbed during exposure. The maximum output signal (white) is reached at saturation. The single cell of an array generally consists of a photodiode with an associated storage capacitor. During this integration (exposure) time, the charge stored on each cell is gradually removed by the photocurrent, that is directly proportional to diode sensitivity and to the light irradiance (radiant flux per unit area). For a given sensor, the charge output of each diode (below saturation), is proportional to the product of the radiant flux impinging on the sensor itself by the integration time. Thus, there is an obvious tradeoff between scanning speed and the light irradiance required to saturate the sensor.

Longer integration times are limited by dark current. Shorter ones, on the contrary, are limited by sensor sensitivity, by the band width of the sensor electronic-scanning circuit, or by the rate of data output. For low-light-level applications it is often desirable to provide cooling to reduce the dark current. Because SSIs typically dissipate on the order of 100 mW and their shape is similar to that of integrated circuits, their cooling (Peltier, liquid nitrogen) is much easier than in tubes.

An important characteristic of solid-state devices that has to be emphasized is the absence of lag (usually defined as the percentage of signal remaining after the third frame scanning). Since the photodiodes are reset after every frame scanning, the lag is practically null, i.e., the sensor has no memory of the previous image. This characteristic of SSIs is very useful when imaging transient phenomena; by comparison, the lag of tubes can reach values as high as 30%.

The main operational problem of SSIs is the lack of uniform sensitivity among the photodiodes. These disuniformities can reach values of 15% in very-high-pixel-density sensors. Fortunately, the larger disuniformities are in the peripheral area, and selected sensors with smaller disuniformities are available at higher costs. However, since systems must generally be calibrated because of nonuniform illuminated backgrounds, the differences in sensitivity among photodiodes can be automatically compensated for.

The early 1980s saw a veritable explosion in development of solid-state imaging devices. The basic types of such imagers now available fall into four categories: charge coupled devices (CCD), charge injection devices (CID), charge prime devices (CPD), and metal oxide semiconductors (MOS) [16].

Particular mention must be made of the CID technology (General Electric); these sensors have an architecture amenable to random accessing small image areas and/or selected lines within the overall array [17]. When applicable, this capability can be very useful for limiting the image acquisition only to areas of interest, thus saving computer memory and/or increasing the frame rate.

In addition to individual SSIs, complete video cameras and imaging systems are currently available [18]. In many cases the complete camera may be a good choice for a DPI system, saving a certain amount of circuit design work. However, these cameras are often optimized to produce TV-compatible video signals (interlaced, smoothed, etc.), which is not necessarily the ideal choice for a DPI application.

B Analog to Digital Conversion

The image, acquired by a sensor, has to be digitized before it is processed. Digitization of video signals is achieved by means of special hardware whose core is the analog–digital converter (ADC). The fundamental characteristics of an ADC that have to be considered are the resolution of conversion (number of bits) and the speed at which the video signal must be sampled.

The resolution of an ADC represents the number of intensity quantization levels, in other words, the number of gray levels ranging from dark to white, detected in an image. The resolution of conversion is expressed by the number of bits b of the ADC *(image depth),* with the relationship between the number of detectable gray levels N_g and b being $N_g = 2^b$; for example, an 8-bit ADC gives a resolution of 256 gray levels.

The number of reliable bits that can be extracted from a digitization system is a function of the signal-to-noise ratio (SNR) of the sensor, with the digitization itself being an extra source of noise. If the number of levels N_g of the digitization system is chosen so that the quantization noise is equivalent to the noise introduced by the sensor, it can be shown [11] that the number of bits b of the ADC should be given by

$$b = 0.17(\mathrm{SNR}) - 1.8$$

where SNR is measured in dB.

It should be noted that the optical noise (such as speckle) is often larger than that due to the sensor, and thus, a lower SNR has to be taken into account. Furthermore, a high image depth is not needed in most interferometric analysis, because only minima and maxima of the signal (fringe centers) must be detected. This means that in designing a DPI system, one can often accept an ADC resolution lower than that of the sensor. In some cases, however, such as direct phase-measurement techniques [12, 19], the accuracy in measuring light irradiance is directly related to the measurement of phase.

A smaller number of bits for an ADC, besides needing less memory capacity (see Fig. 3), means lower cost and higher speed. This is because, when flash converters are used as ADCs, the complexity of such integrated circuits doubles every bit more.

The minimum speed at which the video signal has to be converted is practically given by the ratio between the total number of pixels of the sensor and the integration time or, more simply, by the clock frequency. For a frame rate of $\frac{1}{30}$ s and an array of 512 × 512 pixels the conversion time should not be greater than about 100 ns.

In carrying out the A/D conversion, the video signal has to be sampled and held. The purpose of a sample-and-hold S/H circuit is to sample the pixel video output and "freeze" it during the A/D conversion period, until the next video output is available for sampling. The electronic circuit, which performs this function, is often already included in the sensor chip.

Most SSIs provide some pixels covered with an opaque mask, referred to as *black reference pixels*. The video output voltage from these pixels is always the zero level, representing black. If the video voltage excursions were referred to the black reference, it would be necessary to sample this black reference and offset the rest of the video signal with respect to this voltage. This technique, which can be carried out either manually or automatically is called *DC restoration* [12].

Once the data are obtained by the ADC, they must be stored in the computer memory. Three techniques for performing computer data input are (1) programmed input/output (I/O), (2) interrupt-driven I/O, and (3) direct memory access (DMA). The first two techniques consist of the computer's reading an input port either on a programmed command or on an interrupt request by an external device. Both require several software steps for each read-data item, and thus they need either a slow data rate or a very fast computer. Generally, the incoming data from a video sensor is too fast for either technique to be used, and DMA is required. The DMA technique allows the input device to take control of the computer memory and to store the data directly without any action by the computer [12].

In many cases it is convenient to use oriented video processors instead of a general-purpose ADC to interface the computer to the video camera. They offer, besides the S/H and ADC circuits, large memory buffers to store one or more images as well as an oriented processor that can carry out several preprocessing works (look-up tables, addition or subtraction of two images, etc.) in shorter times. Furthermore, they also generally have a digital–analog converter that allows, with only one frame delay, the visualization on a monitor of the digitized preprocessed image.

C Computer Requirements

Computer requirements are strictly related to the size of the image. The image size includes its format (number of pixels), its depth (bits needed to code each pixel), and the number of operations needed for the desired process. Format and depth define the amount of storage; format and number of operations per pixel define the total number of operations needed in the processing and, thus, the speed of the processor required for a fixed computing time.

The choice of the image format is the limiting factor when designing a DPI system. If a square format is assumed, the requirements grow with the square of the number of pixels per side; the maximum capacity of the central memory of the computer is easily reached with a small number of images. Also the transferring time to peripherals begins to be sensible if several million bits of data have to be transferred.

Whenever possible, reduction of resolution in one of the two scanning directions can sensibly reduce the computer requirements without affecting measurement accuracy. Accuracy in locating fringes is related to the pixel spacing and number, but description of the flow field generally depends on fringe density. E.g., if the average fringe spacing on the interferogram is 0.5 mm, and if no discontinuities such as shock waves are present, then a uniform description of the flow field can be obtained with a $\frac{1}{2}$-mm grid. Thus, it is possible to take on the sensor a line every $\frac{1}{2}$ mm. If the sensor side is 10 mm long, only 20 sections on the interferogram are needed to uniformly describe the flow field.

It must be emphasized that a DPI system should have great flexibility of data acquisition in order to handle different applications well by optimizing accuracy, speed, computing time and memory.

As already mentioned, a simple DPI system can be realized by means of a linear array that scans the image. This solution is not expensive and gives the maximum available spatial resolution (up to 4096 pixels) and line-scan speed (up to 128 MHz) but is relatively slow in image scanning.

Classical DPI systems are realized by area arrays, and currently the largest area array available (Tektronix) has 2048 × 2048 pixels. If one skips a fixed number of horizontal lines in the A/D conversion step to reduce the amount of stored data, the frame rate is not increased because all the horizontal lines of a frame will always be electronically scanned.

A sensor that matches the advantages of the first and second system is the CID area array. In fact, this array has the possibility of skipping electronically the scan of a predetermined number of horizontal lines, so as to achieve a proportional increase of the frame rate.

Figure 5 [11] can give a first estimation of computing time requirements, i.e., computer speed needed. For different processing activities, there are reported the isolines of the total number of operations that have to be performed as a function of the image format and of the number of operations per pixel. The vertical right-hand axis is marked with a few selected algorithms. The number of operations given here for these routines represents the inner loop of a careful assembler implementation; complete programs would probably have higher operation counts, especially if they are written using a high-level language. In general, with increasing capability of hardware, the upper limits of the different processing activities tend to increase.

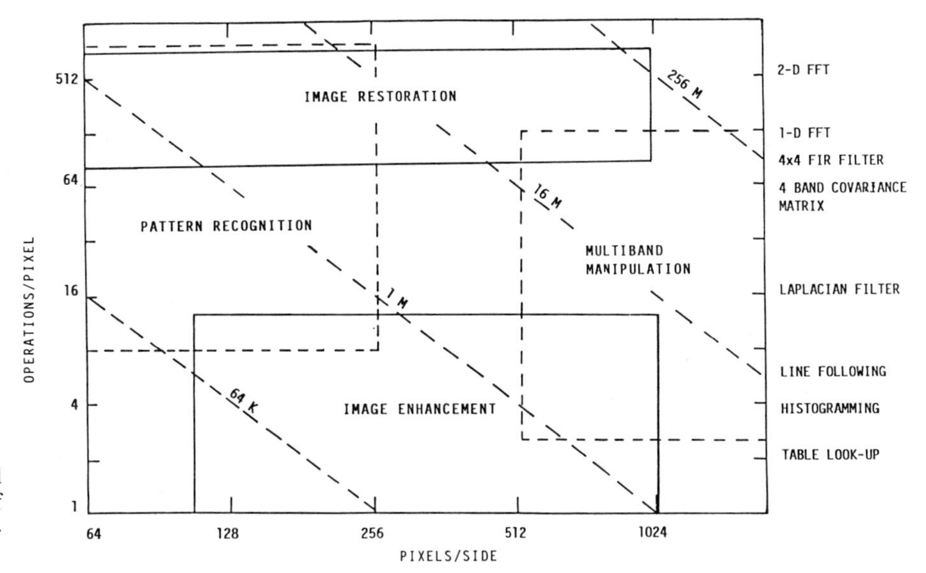

Fig. 5 Computing requirements (total number of operations) as a function of spatial resolution and number of operations per pixel [11].

III INTERFEROGRAM EVALUATION

In order to quantitatively evaluate flow-field patterns, loci of equal phase (equal optical path difference between two interfering rays) have to be recognized on the interferograms.

The intensity $I(x,y)$ on the interferogram plane may be written as

$$I(x,y) = I_o(x,y) + I_c(x,y) \cos [I(x,y)] + I_n(x,y) \quad (12)$$

where $I(x,y)$, which is related to the data function introduced in Sec. I, and represents the phase distribution of the wavefront to be determined; $I_o(x,y)$ is the background illumination which gives the variations due to uneven light intensity in the beam and reflections; $I_c(x,y)$ is the fringe amplitude or contrast as light-intensity variation from dark to light due to interference (this term may also vary over the field of view for the same reasons mentioned above); $I_n(x,y)$ is a light-intensity variation, unrelated to the quantity being measured, which is referred to as optical noise.

The light-intensity distribution given by (12), after transduction by the sensor and digitization by the ADC, is further degraded because of nonlinearities due to the sensor and converter. This is due mainly to dark current, electronic noise, Γ value, modulation transfer function, and saturation. When considering a digitized signal that can be modeled by relation (12), all these nonlinearities may be considered as additional contributions to the optical noise.

The fringe maxima and/or minima, or the fringe sides, are easily detectable lines of equal phase. Thus, in evaluating an interferogram, each fringe has to be recognized and those pixels that represent the constant-phase line have to be picked up. In this way the interferogram can be binarized, set bits representing loci of equal phase.

The next step in interferogram evaluation is to assign the fringe order to each fringe, so as to have the total phase shift and thus the optical phase difference between two interfering rays.

At this point the reduction of the image in engineering data can be carried out by using the relationship between the optical path difference being measured and the physical property (density, mole fraction, etc.) to be determined.

While developing this procedure automatically, several problems that may occur more or less in all the applications must be overcome. The nature of the interferogram and its quality will tailor the right procedure.

A Filtering

Several sources of noise contribute to degradation of the signal, and they may be divided essentially in two classes, electronic and optical noises.

The electronic noise is due to the sensing device and to the electronic circuits; in order to reduce this noise much care must be paid to the circuit design and, if needed, to cooling the sensor.

The optical noise, in most cases, is greater than the electronic one, so greater effort is needed to minimize it. The optical noise is due to speckles of the laser light and to impurities in the optics or along the light path from the laser to the image plane. Much of the laser speckle can be removed by a procedure of spatial filtering [20]. Spatial filtering depends on the fact that scattered light propagates in directions different from the direct laser light and, hence, arrives at the lens focal plane at points different from that

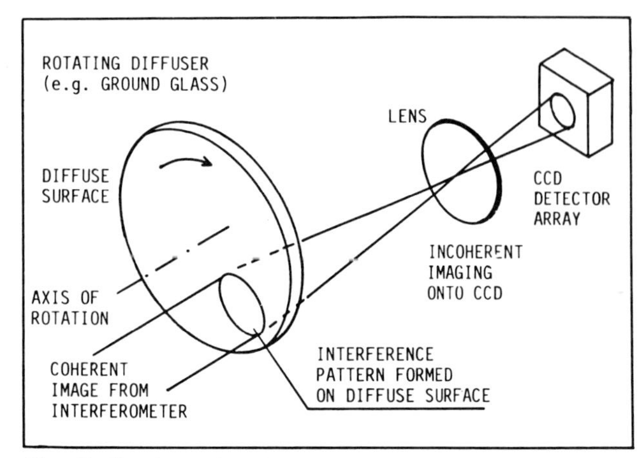

Fig. 6 Incoherent imaging system to remove laser speckle [12].

at which the direct beam is focused. By centering a small aperture (pinhole) around the focal spot of the direct beam, it is possible to block the scattered light while allowing the direct beam to pass unscathed. The result is a light with a very smooth irradiance distribution.

Another technique reported by Prettyjohns [12] to reduce the speckle noise is to form the image onto a rotating diffuser and then to incoherently image it onto the sensor (Fig. 6). Rotating the diffuser causes a time-varying random speckle pattern that averages out during the integration period of the sensor.

In most cases it is convenient to smooth the light irradiance data by means of digital filtering. This can be done in a simple way by averaging small groups of data, taken along a line or within a small area [21–24] or, with a more extensive processing work, by attenuating high frequencies in the Fourier transform image [25, 26].

Low-pass filters, based on data averaging, are currently used to smooth a noisy image. Essentially they consist of replacing the center pixel value of a small segment or area with a weighted average of the pixel values within it. Some typical convolution masks to assign different weights to the neighborhood pixels for a 3- by 3-pixel area are shown in Fig. 7.

$$H = 1/9 \begin{vmatrix} 1 & 1 & 1 \\ 1 & 1 & 1 \\ 1 & 1 & 1 \end{vmatrix} \qquad H = 1/8 \begin{vmatrix} 0 & 1 & 0 \\ 1 & 4 & 1 \\ 0 & 1 & 0 \end{vmatrix}$$

$$H = 1/10 \begin{vmatrix} 1 & 1 & 1 \\ 1 & 2 & 1 \\ 1 & 1 & 1 \end{vmatrix} \qquad H = 1/16 \begin{vmatrix} 1 & 2 & 1 \\ 2 & 4 & 2 \\ 1 & 2 & 1 \end{vmatrix}$$

Fig. 7 Typical convolution masks for digital filtering of noise images.

A very powerful low-pass filter is the median filter, which consists of the center pixel value in a moving window of odd pixels being replaced with the middle value (the median) of the series of such sorted values; this causes pulses having widths less than half the window size to be suppressed. The advantage of this filter, aside from the absence of arithmetical operations (being the median operator essentially a sort algorithm), is to preserve the edges.

If the image is stationary (e.g., steady flows or digitizing transparencies), there is also the possibility of averaging several images [11, 27] so that the amplitude of averaged noise decreases proportionally to the square root of the number of images. This improvement can be fairly expensive; e.g., reducing the noise by an order of magnitude implies averaging 100 consecutive images. The amount of core storage needed to accumulate the intermediate results is larger than that required for the final image. If the image is converted with an 8-bit/pixel resolution, the addition of 100 images requires at least a 15-bit/pixel memory. Also note that a real-time system that has to be averaged 100 times is not real time anymore.

There are cases in which the video signal is noisy and the number of fringes on the interferogram is relatively small. Besides filtering the signal, in order to increase the number of measurements on an interferogram, it is also necessary to find the phase distribution encoded in the fringes at arbitrary positions within the fringe pattern. In these cases a piecewise approximation [28] of sinusoidal function or polynomials of the video line signal can be used for precise phase measurements even in the presence of speckle noise and few fringes in the cross section. This method generally requires considerable computing time, and its convergence and stability depend on a priori estimates of fringe positions.

A simple algorithm to filter the signal as well as to accurately estimate the period and the phase of truncated sinusoidal fringe patterns in the presence of noise is described by Snyder [29]. The essence of its method is first to smooth and reduce the data by processing with a simple adaptive digital filter that locates the symmetry points of the fringe pattern (i.e., the extrema of the underlying sinusoid). A straight line is then fit to the set of symmetry point positions by the methods of the least square. The period (or frequency) and the phase of the fringe pattern are simply related to the slope and intercept of the straight line fit to the reduced data.

A simple technique [14] to reduce the effect of the noise consists of working with a high-intensity light so as to saturate the sensor in the white fringes. In this way the fringe contrast amplitude is much greater than the dynamic range of the sensor, and thus the noise amplitude becomes negligible with respect to the fringe contrast amplitude. A limitation of this method consists of the possibility of picking up only light intensity minima; in fact, information is lost on both the maxima and the fringe sides, even if digital filtering is applied on the signal.

B Fringe Recognition

The key step in DPI is to recognize fringes by picking up pixels that have the same position on a fringe (minima, maxima, sides, etc.), say, fringe segmentation. The algorithms to be implemented for this aim must be chosen depending on kind and complexity of interferograms to be evaluated.

For example, the video signal along a scanning line ξ of the interferogram of Fig. 8*a* is plotted in Fig. 8*b* [14]. The effects of the terms I_o and I_c of Eq. (12), as variation of fringe contrast, and of I_n, as noise superimposed on the true signal, can be seen. Notice that most of the white pixels work in saturation.

The first effect, when it is due to uneven illumination, can be corrected by means of a reference image, e.g., an image acquired without fringe patterns or shifts. In this way, besides a background correction, the differences in sensitivity among pixels are also corrected. If, however, the fringe contrast variation is due to high fringe density of the interferogram, i.e., related to the MTF of the sensor, no correction can be made unless MTF and approximate fringe density distribution are known a priori.

The most important fringe-segmentation methods are analyzed in the following discussion.

1 Fixed Threshold This is the simplest way to carry out fringe segmentation. It requires an image with low noise and constant fringe contrast. If a threshold, i.e., the mean gray value, is fixed, the pixels that have a light intensity greater than the threshold are set as white pixels (level 1); the others are considered as black pixels (level 0). In this way the original image is transformed into a binary image. It is then easy to find lines of constant phase as edges between dark and light regions (fringe sides) or to find fringe minima or maxima by means of skeletonization techniques [30]. Whenever possible (even illuminated and noise-free images), it may be more convenient to perform thresholding before digitizing the image either by an analog device or, as reported by Chen [31], by means of an opticRAM directly in the computer memory. In this way the ADC can be eliminated because the video signal is already binarized. Furthermore, by using the opticRAM the data are promptly available in the computer memory. Since the threshold is fixed, the method fails when the fringe contrast is considerably variable along the scan line (e.g., the sides of the plot of Fig. 8*b*). As will be seen later, use of an adaptive threshold can bypass this limitation with a little more computing effort.

2 Fringe Extrema This method consists of searching the extrema (minima and/or maxima) of the light irradiance signal along a scan line of the interferogram [5, 9, 14, 32]. The noise and the variable amplitude of fringe signal make necessary more tests during the extrema research. The basic procedure consists of setting a threshold and, when the signal crosses this value, storing the minimum or maximum that has been previously found. Then searching for the next extremum is continued until the threshold is again crossed. The signal modulations due to noise, if in the vicinity of the threshold value, may cause a false extremum detection. For this reason another test (noise-rejection threshold) has to be done in order to make sure that the difference between the extremum and the threshold is greater than a fixed noise amplitude σ (Fig. 9). The image is then binarized by setting the positions corresponding to pixel extrema at a high level.

In the presence of a strong dark current and/or signal saturation, more than one pixel can exist as the fringe extremum (e.g., the center of the plot of Fig. 8*b*). In this case the first and the last pixel of the extremum interval have to be stored, and the central position between them can be taken as the position of the extremum along the fringe.

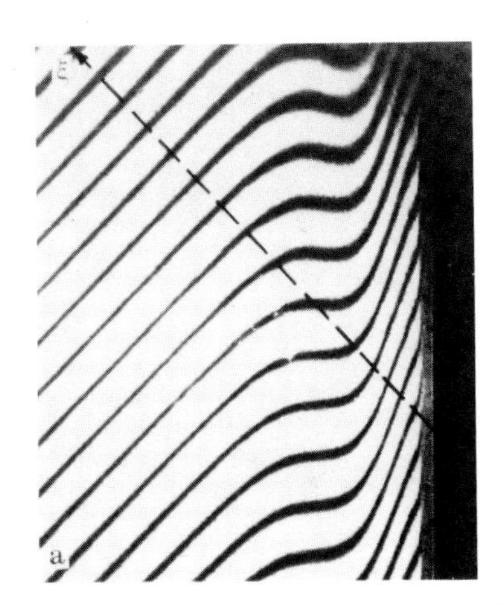

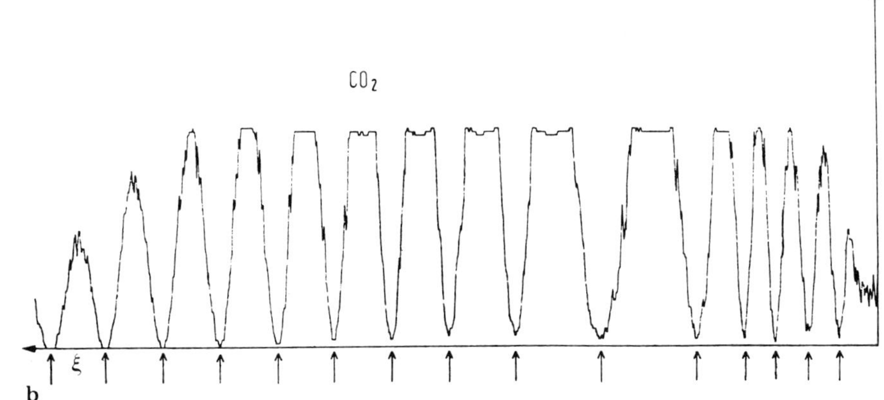

Fig. 8 Convective wall plume induced by a linear mass source: (*a*) Finite-fringe Wollaston interferogram near the wall; (*b*) plot of the line video signal along the broken line of (*a*) [13].

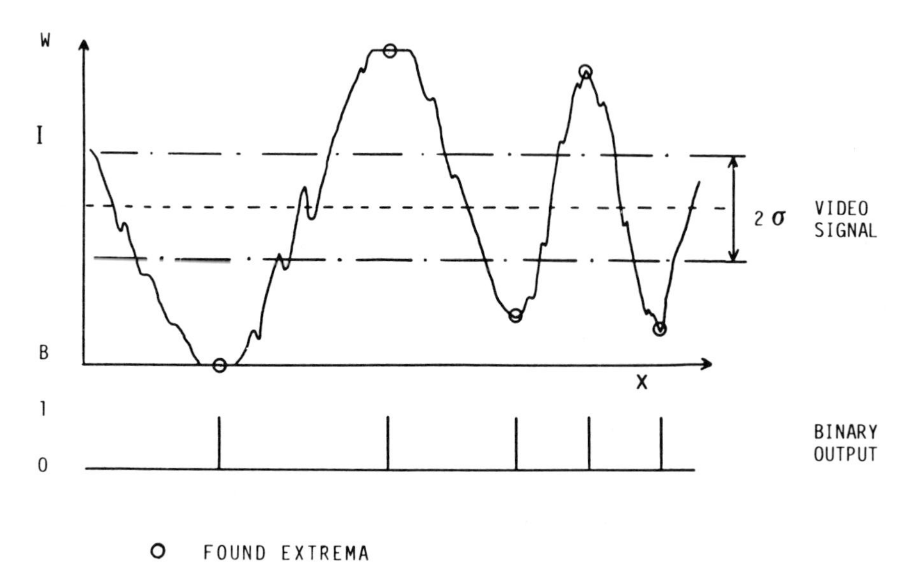

Fig. 9 Fringe segmentation. Extrema detector.

3 Adaptive Thresholding Most of these methods were developed initially for document scanning and optical character recognition. Since the nature of the signal coming from these images (steplike) is different from the interferometric one (which is approximately sinusoidal), they may sometimes fail at fringe segmentation. Three major algorithms must be mentioned: peak and valley, transition-refreshed peak and valley, and robust two-dimensional adaptive thresholder.

The peak and valley algorithm [33] stores the whitest (peak) *W* and the blackest (valley) *B* values into two buffers during a time period of one scan line. At the beginning of each scan line, the first pixel value is stored in both buffers. These two buffers are then updated during the line scan when a higher or lower pixel value, respectively, occurs. The threshold used throughout the line scan is determined by averaging *W* and *B*. For the next line, the buffers are reset and the procedure is repeated.

The transition-refreshed peak and valley detector [34] works in a similar manner with the main difference being that the black buffer is refreshed every white-to-black transition and the white buffer is refreshed every black-to-white transition. The transitions are detected by introducing two parameters α and β respectively for black-to-white and white-to-black transitions (Fig. 10). At the beginning of each line, *B* and *W* are set at predefined values; the thresh-

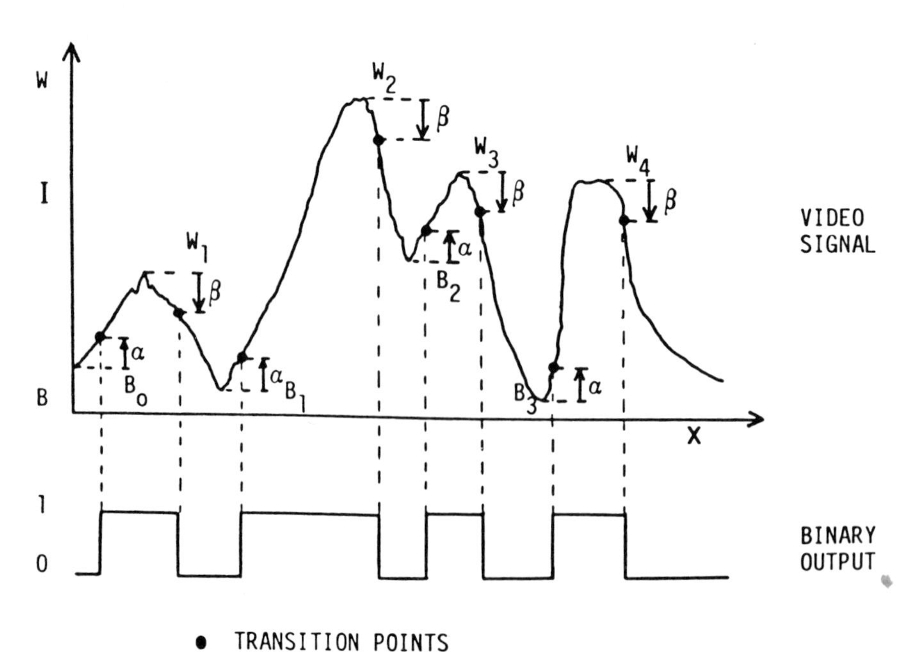

Fig. 10 Fringe segmentation. Transition-refreshed peak and valley detector [33].

old to detect transition is calculated with respect to each extremum (peak or valley) of the line signal by adding α to each valley B and by subtracting β from each peak W.

The robust two-dimensional adaptive thresholder [33] is substantially a transition-refreshed peak and valley detector in two dimensions. It therefore has the advantage of a lower anisotropy, which is typical of one-dimensional algorithms. The method consists of a two-dimensional adaptive threshold-setting algorithm with an updating memory procedure. It can dynamically update the nominal black-and-white reference levels to provide more adaptive capability. Local neighboring information is gathered by including the adjacent pixels along the vertical and horizontal directions. The flowchart relative to this algorithm is fully described in [33].

Since all adaptive thresholding methods transform the original image into a binary image, as does the fixed-threshold method, the iso-phase lines can be obtained in the way already described.

4 Floating Threshold Introduced by Becker et al. [35, 36], this is a two-pass hysteresis detection scheme for one-dimensional segmentation. In the first pass the scan line is searched for extrema, by using a noise-rejection thresholding. In the second pass, between two adjacent extrema, a threshold is calculated as the average of such extrema and the side of corresponding fringe is recognized when the signal crosses this floating threshold (Fig. 11).

This method works well even on fringe fields having abrupt changes in image brightness, fringe density, and contrast. The advantage of the floating threshold method over the fringe extrema method is that the fringe sides are not affected as much by noise as are the fringe extrema, due to the maximal slope of the signal there. As compared with the adaptive methods, the floating threshold method computes the adapted threshold by averaging the actual extrema of the local signal, for which a double scan of the line data is needed. By contrast, the adaptive methods compute the threshold by using the previously found extrema so that it is necessary to scan the line data only one time.

5 Fourier Transform Introduced by Takeda et al. [37] and extended by Macy [38] and Bone et al. [39] to two-dimensional schemes, this method determines the phase distribution on the interferogram without searching for fringe positions and orders. The problem with this approach is the amount of operations to be done on the entire image.

This method consists of superimposing on the interferogram phase distribution an additional phase that is a linear function of the coordinates, e.g., by tilting one of the mirrors of the interferometer (finite fringe width alignment). This technique separates, in the transform plane, the positive and negative frequency components of the sinusoidal intensity from each other and from some spurious intensity contributions. The underlying phase can be retrieved by filtering the transformed image so as to retain only the positive frequency components.

It should be noted that Fourier transform methods are not well suited to fringe fields that have fringe frequencies varying much across the scan line.

In most of the methods described, the algorithms for fringe recognition on the interferogram are based on analysis of individual one-dimensional cross sections of the fringe pattern. This procedure works well if the cross section is almost perpendicular to the fringe direction, but it

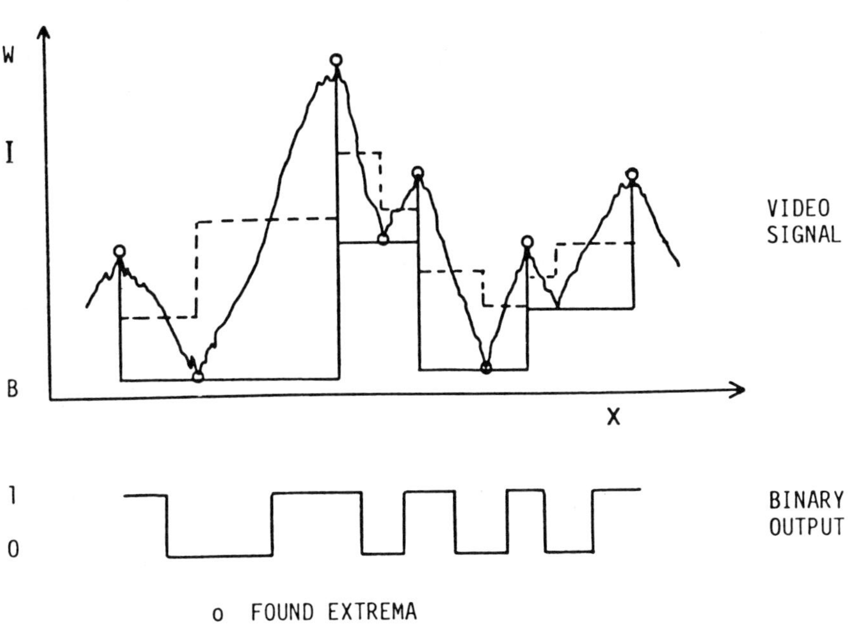

Fig. 11 Fringe segmentation. Floating threshold detector [35].

may fail if the fringes have significant components parallel to the scan line. In many applications, for example, the fringes may be rings and the scanning of the image in only one direction may cause a deficiency of fringe points and poor accuracy wherever the scan line is tangent to the rings. In these cases, a double scan of the image, e.g., horizontally and vertically [8], may overcome this inconvenience.

Another technique, developed by Robinson [5], consists of processing the image by random accessing the frame memory. It means that an algorithm, not restricted to a line and/or column analysis, has to be developed to suit a specific problem (parallel fringes, concentric rings, etc.). For example, speckle interferograms can be usefully processed. A speckle interferogram shows a pattern of parallel and equally spaced fringes whose direction and spacing must be determined. The procedure proposed by Robinson, shown in Fig. 12, can be written schematically as follows:

1. Search for the most illuminated pixel near the pattern center, referred to as image center.
2. Compute the sums of the intensity values along a series of small radial vectors passing through the center. The vector angle is varied in small increments over a range of 180°, and the vector giving the largest summed value is taken as the nearest to the fringe direction.
3. Repeat this procedure using a smaller range of angular scan, a smaller angular increment between the radial vectors, and a longer vector. This step is repeated using an iterative approach until the best fitting vector is found within the limit of 1°.
4. Average the data along a direction parallel to the fringe direction. Thus, the two-dimensional pattern is reduced to a line pattern whose wavelength gives the fringe spacing.

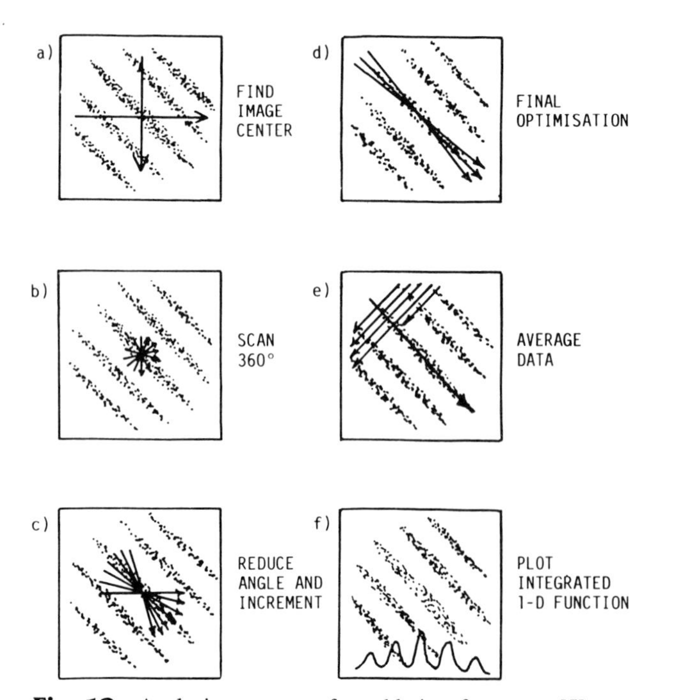

Fig. 12 Analysis sequence of speckle interferograms [5].

C Fringe Order Determination

After the fringe recognition procedure, a binary image whose set bits represent selected points of the fringe field is in the frame memory. The first step in determining the fringe order is to join all those pixels belonging to the same fringe. This can be achieved by implementing a sequential tracking procedure as reported by Rosenfeld and Kak [21], Seguchi et al. [40], and Funnell [32]. A sequential tracking procedure generally consists of testing the neighborhoods of the pixel on a curve that is being tracked and then picking the best candidate for the next point. If the curve being tracked can branch, it may be necessary to pick more than one next point; in that case, all but one of the chosen points are stored for later investigation, and the tracking proceeds with the one remaining point as next point. The tracking procedure for each fringe must terminate either on the boundaries or on a pixel of the same curve (closed fringe); if it does not, an error (disconnection) has occurred.

An interesting scheme for the numbering of erroneously extracted line fields, proposed by Becker et al. [36], is described in some detail. The idea of the scheme is based on the fact that in general there are only a few locations where the fringe lines are falsely extracted, while the line field in most parts of the interferogram is represented correctly. The interferogram is divided into rectangular segments by a grid with suitably sized meshes (Fig. 13). Each mesh is then numbered individually, starting at the mesh with the maximum number of line segments and with a minimum number of inversion points of fringe counting (e.g., a mesh containing parallel lines). The algorithm tries to continue the numbering into the adjacent meshes, using the already-known numbers of the common grid line. The mesh with the best conditions is always chosen as the next to be numbered. If the line numbering of the actual mesh is inconsistent with the numbering of the adjacent meshes,

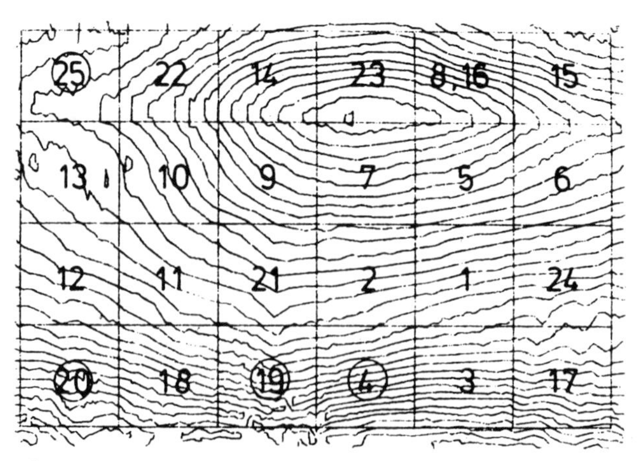

Fig. 13 Correction scheme for fringe numbering [36].

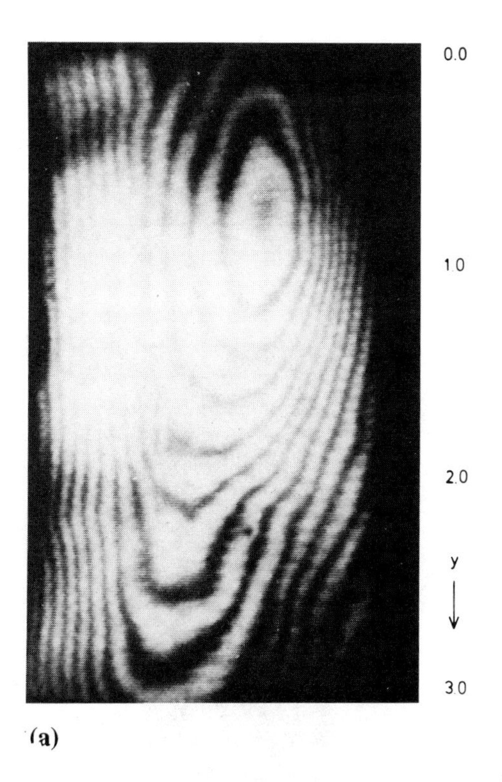

(a)

it is put back and handled later. In this way the meshes containing disconnections are processed after all other meshes have been numbered. The numbering scheme now tries to number the line segments inside the erroneous meshes, using the numbers already known in the surrounding meshes. If a line must have different numbers, it is divided into segments. The cutting position may be determined at the minimum angle between the polygon segments. In areas where the fringe number function is well behaved, the fringe disconnections may be completely corrected with this approach. In areas with discontinuities, however, the line cutting and connection problem may be ambiguous. In Fig. 13, for example, a line field with disconnections is shown. The order in which the meshes are processed by the numbering scheme is shown by the numbers inside the meshes. Encircled numbers mark the meshes that contain lines erroneously connected, as localized by the algorithm.

Once the lines are recognized, a "best fitting" algorithm could be applied to each of them in order to smooth the interferogram [36, 41]. This is very useful since it reduces the amount of information to be stored; only the coefficients of the polynomial are needed to plot fringe contours, and it allows interpolation of the discrete data.

For example, the polygonal line map of the holographic interferogram of Fig. 14*a* is shown in Fig. 14*b*, while a contour line map of the polynomial surface derived from the numbered fringe polygons is seen in Fig. 14*c*.

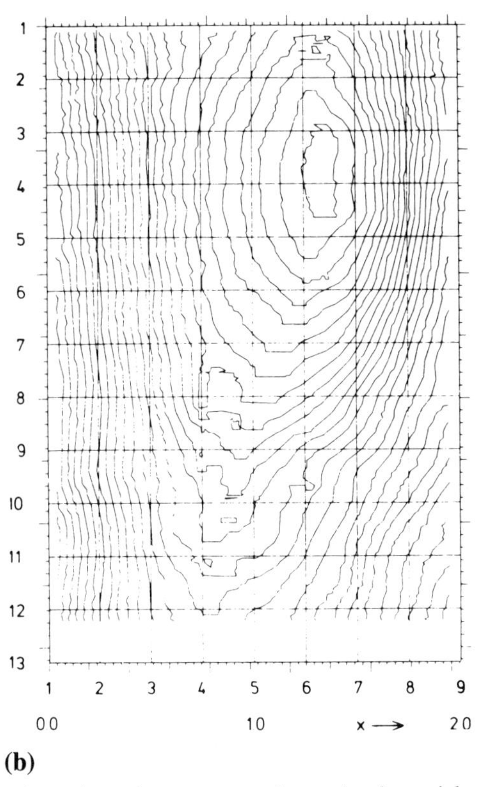

(b)

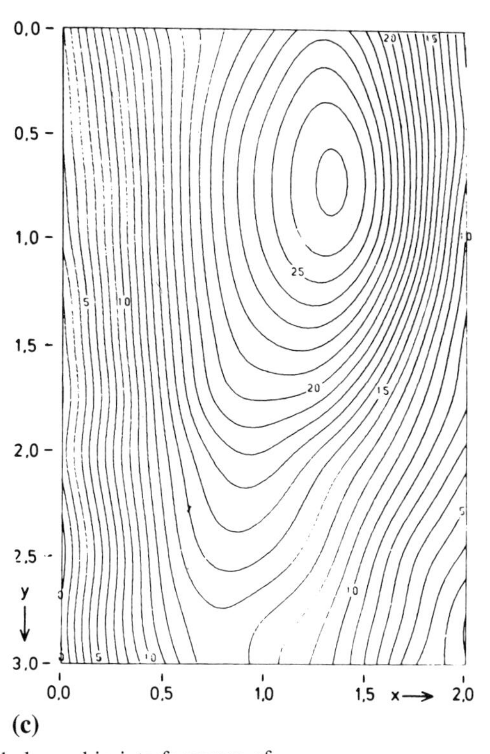

(c)

Fig. 14 Fringe segmentation and polynomial approximation of a holographic interferogram of a car tire: (*a*) Holographic interferogram; (*b*) polygonal representation of the fringe field of (*a*); (*c*) contour line plot of the surface computed from (*b*) by polynomial least-square approximation [36].

The last step in interferogram evaluation, and one of the most critical, is to assign the fringe order to each fringe. This procedure cannot be generalized because the fringe order distribution is strictly related to the phase distribution, which depends both on the particular experiment under investigation and on the interferometric technique being used. If the trend of phase field (monotonous, symmetric, etc.) is known a priori, it can be easy to find a procedure that automatically gives the fringe orders.

According to Becker et al. [35], it is possible to establish some criteria that are always valid in fringe order recognition: (1) the difference in fringe order of two adjacent fringes can be only ± 1 or 0; (2) fringes of unequal order cannot touch or intersect each other; (3) a fringe can never end except at the boundary of the field of view unless the fringe is closed; (4) the sum of the fringe order differences derived along a closed line through the interferogram always has to be 0; (5) if unsteady processes are studied by taking a series of interferograms and the interframe time is chosen properly, the difference between the fringe number function at time t and $t + dt$ is less than 1 in most sections of the interferogram.

Since several tests must generally be performed to make the right decision in assigning fringe order, it is preferable to implement in the computer program an interactive procedure [32]. Once the fringe lines have been correctly recognized, it is sufficient to code only one point on each fringe so as to give the order to the whole fringe.

Finally, when the fringe order is assigned to each fringe, the quantitative evaluation of the physical properties visualized on the interferogram can be carried out by means of the relationship between this property and the phase shift of the light (see Sec. I).

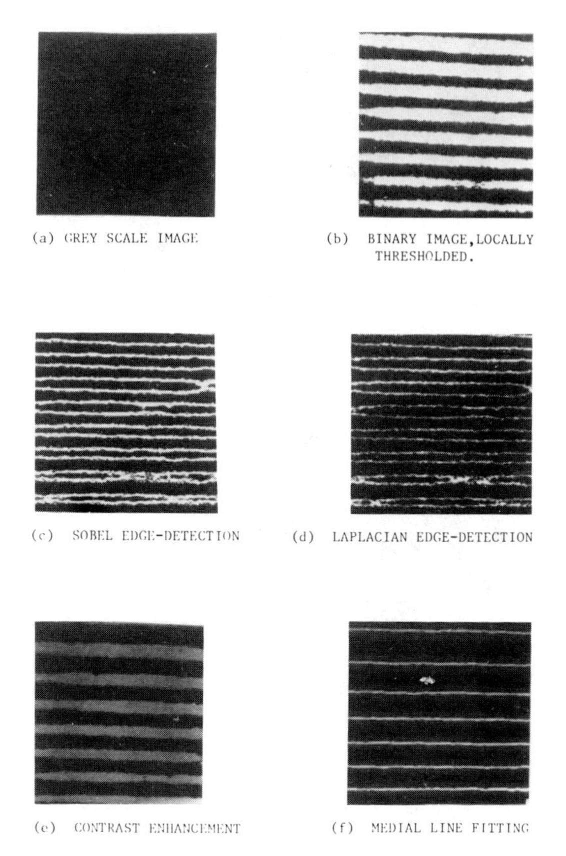

Fig. 15 Examples of interferogram processing: (*a*) Gray-scale image; (*b*) binary image, locally thresholded; (*c*) Sobel edge detection; (*d*) Laplacian edge detection; (*e*) contrast enhancement; (*f*) medial line fitting [47].

IV APPLICATIONS

While studying an automated fringe-reading procedure, one can pick up many ideas by looking for similar systems that have been developed in other interferometric visualizations, such as surface reconstruction [42] or stress [9, 19, 43], displacement [44], and vibration [45] analysis. In these cases interferograms are generally obtained by means of holographic or moiré techniques. A few studies utilizing DPI systems have been carried out for quantitative flow-field visualization even though some of them only give evidence to the capability of such systems without extensive applications.

An application to the four-dimensional nature of turbulence (three coordinates and time dependency) was carried out by Lockett and Collins [46] by means of holographic interferometry. The automatic evaluation of the corresponding interferograms was started by Hunter and Collins [47]. They acquired holographic interferograms by means of either photographic media (high-speed film, motor-driven SRL camera) or video camera. Interferographs were digitized by using a linear CCD camera suspended above a motor-driven movable bed. Figure 15*a–f* shows examples of interferogram processing, as adaptive (local) thresholding, high-pass filtering (edge detection), contrast enhancement, and fringe center (medial line) fitting [47].

A simple application of a DPI system to an unsteady heat transfer problem, in which the position of a single fringe is measured as a function of time, is presented by Carlomagno et al. [48]. They studied the delay time before the onset of natural convection in a fluid near a suddenly heated horizontal wire (line heat source). A new type of reference beam interferometer [49] was used, and a linear CCD array of 1024 photodiodes sensed the light. The interferometer was aligned in the infinite fringe mode so that the fringes on interferograms represent isotherms of the temperature field around the heated wire. For the phenomenon being studied, the fringes appear initially as concentric circles (Fig. 16*a*) indicating that heat transfer from wire to medium is by conduction only. After a period of time, the isotherms become asymmetrical with respect to the horizontal plane (Fig. 16*b*) as those above the wire are being convected upward while those below the wire remain nearly stationary. Finally, a more pronounced natural convection

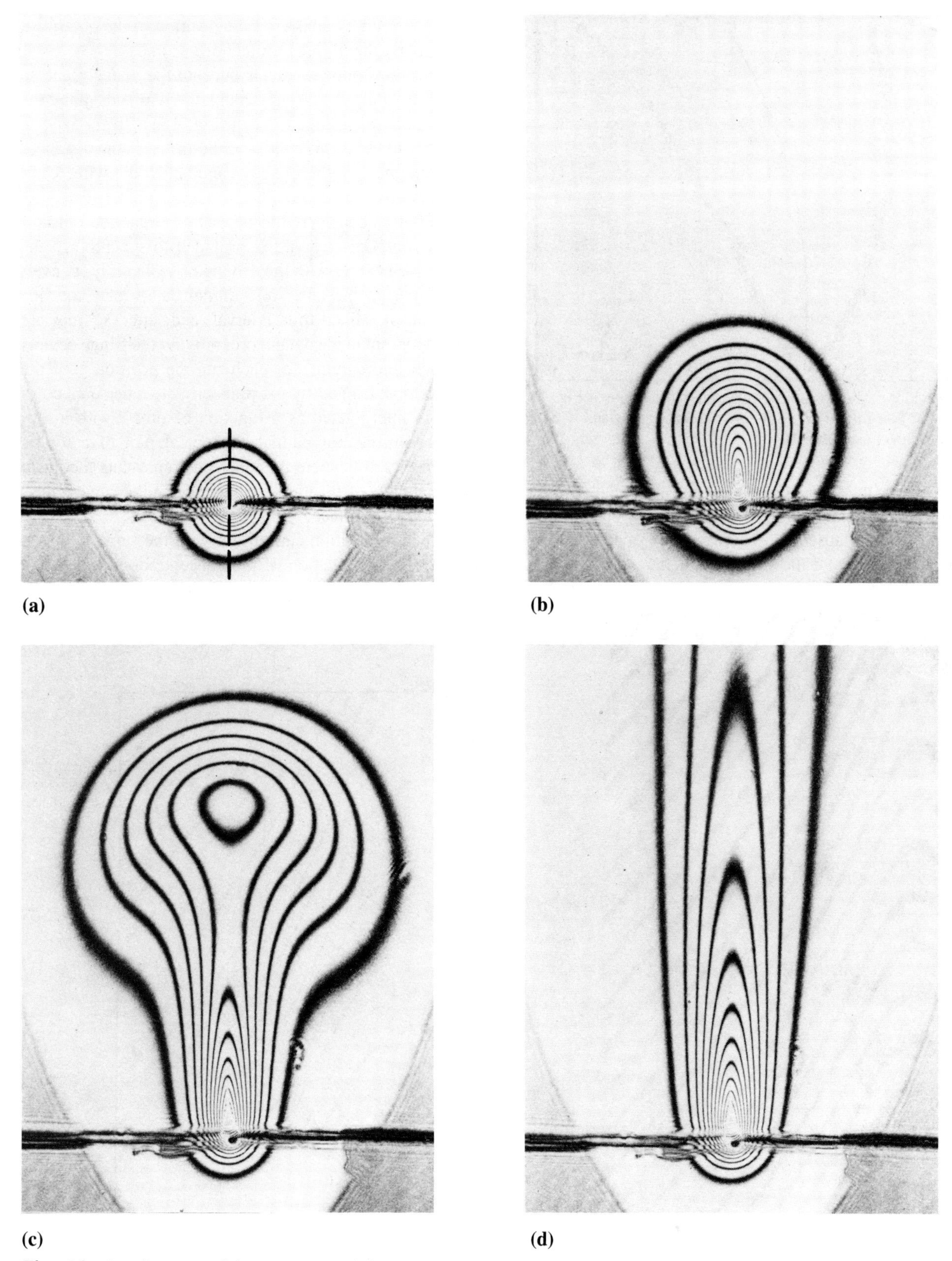

Fig. 16 Interferogram of the temperature field around a suddenly heated fine wire in distilled water: (*a*) $t = 4$ s; (*b*) $= 15$ s; (*c*) $t = 29$ s; (*d*) steady-state regime.

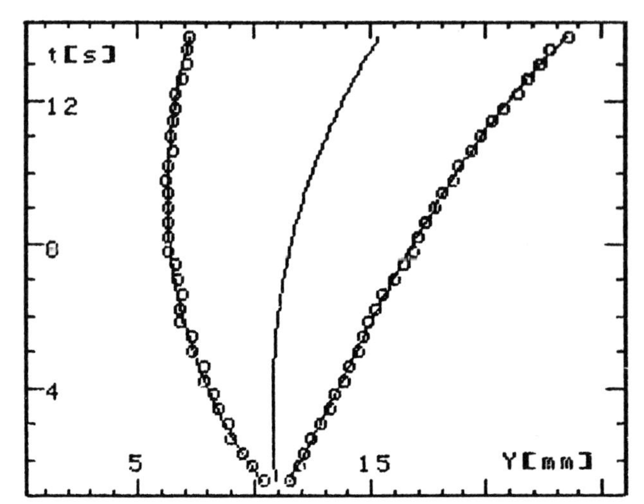

Fig. 17 Plot of the positions of the outermost fringe of Fig. 16 as given by a sequence of 32 interferograms. The elapsed time between each acquisition is 0.4 s.

takes place, with a heated "cap" (Fig. 16*c*) that breaks away from the wire and rises to form the steady-state plume (Fig. 16*d*). The elapsed time between the application of current to the wire and the starting of appreciable convection is measured by analyzing the position of the outermost fringe around the wire as a function of time.

The array of photodiodes is put into the image plane so as to sense the light passing through the wire axis along the vertical direction (broken line in Fig. 16*a*). In this way it is possible to measure, as a function of time, the distances of the outermost isotherm above and below the wire. Until these distances are equal, the heat transfer between wire and medium is considered to occur by conduction only.

A microprocessor board is provided to synchronize the start of wire heating with the driving of CCD array, convert the signal from the latter into digital data, delay the line acquisition of a fixed time interval, and, after 32 lines are acquired, compute the data in order to find the fringe centers from each line datum. For each run the position Y of the two outermost fringe centers, transferred to a personal computer, are interpolated as a function of time t with a 4th-order best-fitting polynomial. A typical plot of a test is shown in Fig. 17, where the inner line represents the mean value of the two Y positions, which does not change unless convection takes place.

Another application carried out with the same DPI system [1, 13] is evaluation of the concentration field of the

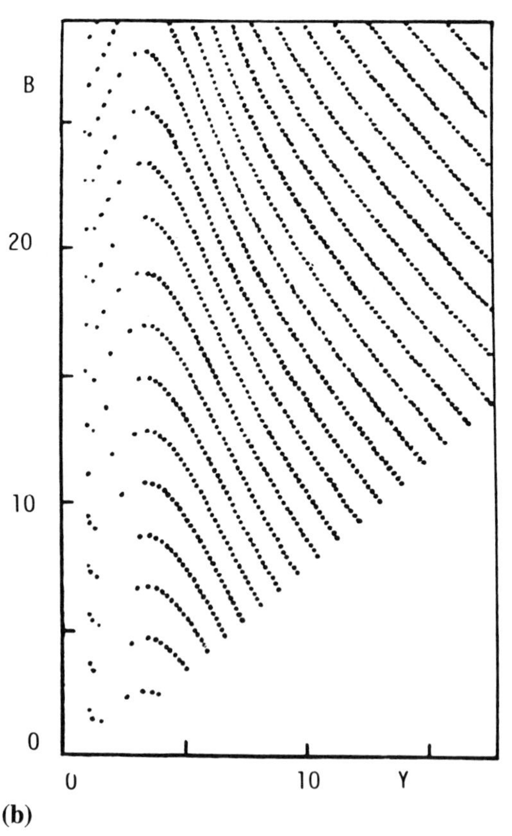

(a) (b)

Fig. 18 Convective wall plume induced by a linear mass source: (*a*) Finite fringe Wollaston interferogram near the wall; (*b*) segmentation of the interferogram of (*a*).

steady, buoyancy-driven laminar-flow about an impermeable, vertical flat plate induced by a linear mass source (convective wall plume).

The experimental model is fed with a helium or carbon dioxide stream that mixes with ambient air. A Wollaston prism interferometer is used to visualize the flow. A digitization of the interferogram of Fig. 18*a* relative to a CO_2 wall plume is represented in Fig. 18*b*. Flow is from top to bottom, and the plate is located at the left side of the interferogram. In this case the array, oriented normal to the undisturbed fringes, is moved vertically by means of a translational stage, driven by a stepping motor. Notice the loss of data points when the fringes are almost parallel to the array.

Eight plots of the array signal taken at different distances *B* from the injecting nozzle are shown in Fig. 19 [1]. In this case the plate is located at the right side of each diagram, where the signal clearly shows the double image of the wall, typical of a Wollaston prism interferometer. Some of the diodes, which are in the most illuminated zones, work in saturation, so that the output signal variations, from light to dark, are much higher than those induced by noise. Therefore, the dark fringe center recognition which is performed by the fringe extrema method becomes much easier. The diode output signal is unfiltered, and the vertical lines represent the computed positions of the fringe centers. Far away from the plate, i.e., and in the left part of diagrams, the fringes are not displaced and the distances among them are constant, since no diffusion takes place there. Large displacements of the fringes are instead evident near the wall, where mole fraction gradients are present.

The procedure for evaluating mole fraction distribution is as follows: (1) fringe displacements are measured along the direction that is normal to the undisturbed fringes (i.e., array direction); (2) these data are computed in order to have mole fraction gradients; (3) starting from a point at which concentration is known (e.g., outside the boundary layer), an integration is carried out so as to obtain absolute mole fraction values of CO_2; see Eq. (5). These three steps are repeated for each scanned line, and data are interpolated in order to plot lines of constant mole fraction, as shown in Fig. 20 [1]. For the present case, recognition of the fringe order is rather easy since mole fraction increases monotonically from zero to the maximum value reached at the plate wall.

An axisymmetric flow field was studied by means of

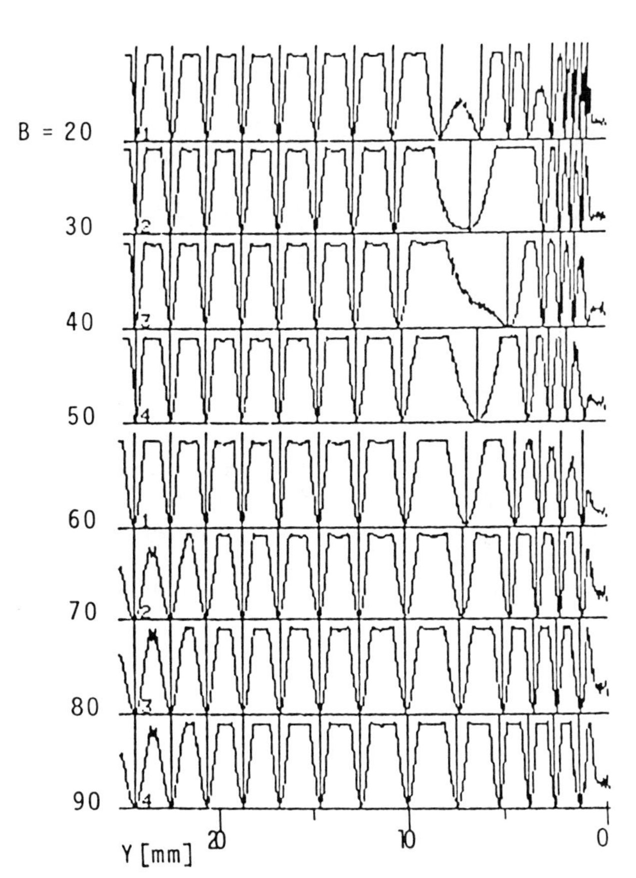

Fig. 19 Convective wall plume. Eight plots of line video signal from a linear array of 1024 photodiodes at various distances *B* from the injecting nozzle [1].

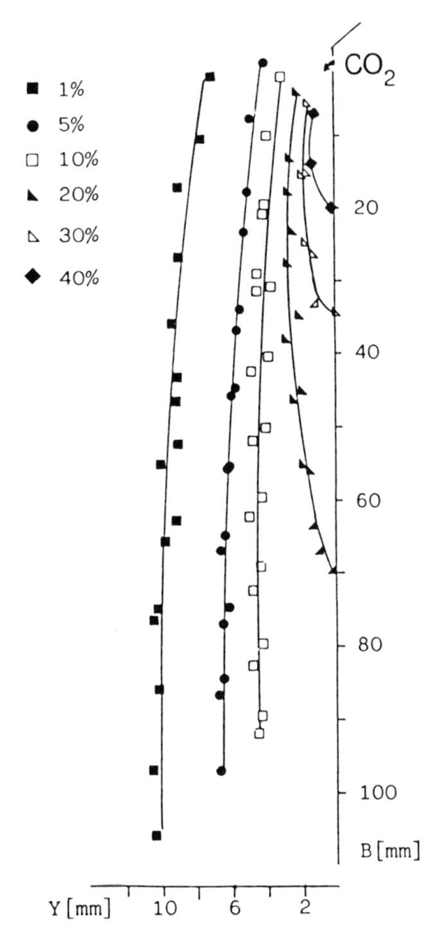

Fig. 20 Convective wall plume. Reconstructed mole fraction distribution from the interferogram relative to the analysis of Fig. 19 [1].

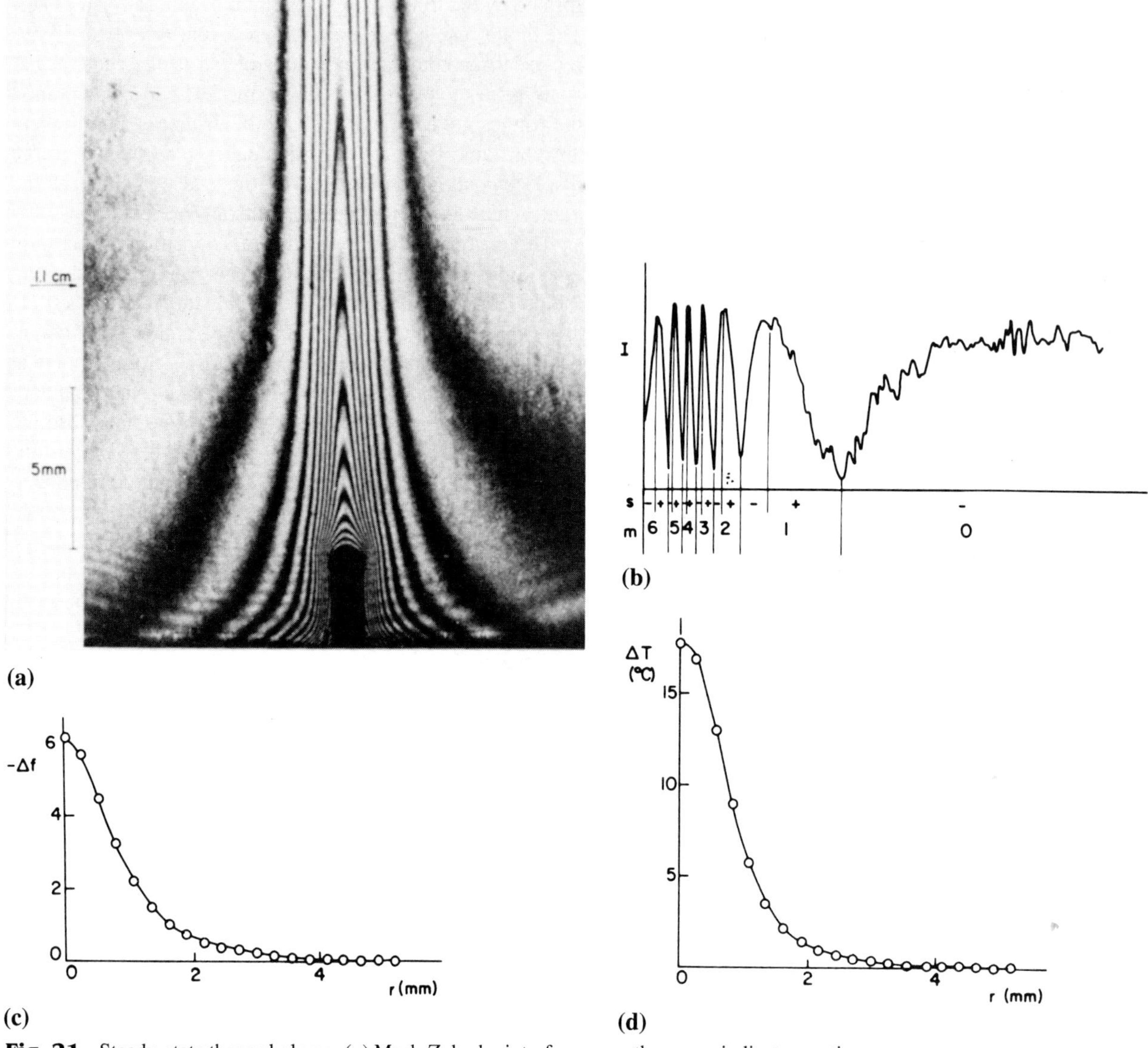

Fig. 21 Steady-state thermal plume: (*a*) Mach-Zehnder interferogram; the arrow indicates section at a height of 1.14 cm from the base chosen for analysis; (*b*) right side of microdensitometer plot with fringe extrema numbering; (*c*) fringe shift distribution; (*d*) temperature distribution as a function of radius [10].

DPI, first, by Boxman and Shlien [10], who used a microdensitometer and later, by Chay and Shlien [15], who used a solid-state linear array. They measured the scalar field (temperature or species concentration) of the plume induced in a liquid by a point source in both the starting and the steady regimes, and they took interferograms by means of a Mach–Zehnder interferometer. The evaluation procedure consisted of searching for the fringe extrema, evaluating the fringe orders, and determining the axisymmetric temperature field from the fringe shift field.

Identification of the fringe orders is straightforward in the steady-state case, where for physical reasons it is known that the temperature decreases monotonically with increasing radius (Fig. 21). The situation is more complicated in the case of the starting plume because of the presence of three "eyes," one on the axis and the other two symmetrically disposed with respect to the axis (Fig. 22). The latter two represent local extrema in the temperature field. In both figures three graphs are plotted next to the interferogram [10]. The first (Fig. 21*b* and Fig. 22*b*) is the output signal of the microdensitometer relative to a horizontal section of the respective interferogram; the second (Fig. 21*c* and Fig. 22*c*) gives the fringe shift distribution evaluated from the previous graph; and the third (Fig. 21*d* and Fig. 22*d*) is the temperature distribution as a function of radius, which is derived by the fringe shift data by means of the Abel inversion (see Sec. I). The authors indicate that the Abel inversion tends to exaggerate inaccuracies, since the fringe shift data are entered in Eq. (11) in the form of a derivative, and dictate, therefore, the absolute need of precise evaluation of data.

Speckle photography is a relatively new technique in

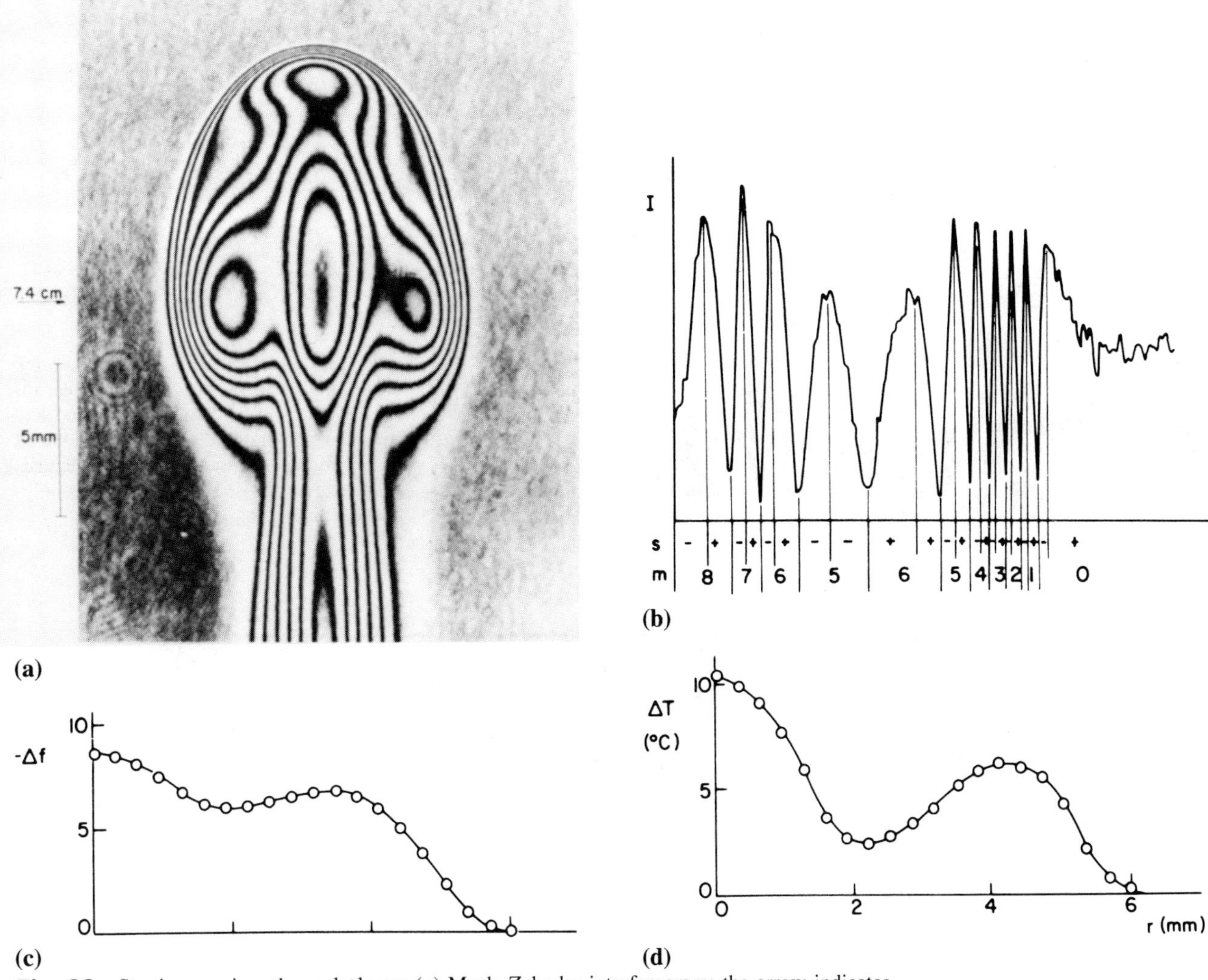

Fig. 22 Starting cap in a thermal plume: (*a*) Mach–Zehnder interferogram; the arrow indicates section at a height of 7.4 cm from the base chosen for analysis; (*b*) right side of microdensitometer plot with fringe extrema numbering; (*c*) fringe shift distribution; (*d*) temperature distribution as a function of radius [10].

experimental fluid dynamics [50–52], while several studies have been performed on deformation and stress analysis [5–7, 25, 43, 44]. The data are produced in the form of equally spaced interference fringes (similar to Young's fringes). For data reduction it is necessary to determine direction and spacing of the fringes (see Sec. III.B). Since for a single speckle photograph one must evaluate as many fringe patterns as measurement points needed, the use of a DPI system is highly recommended.

A digital processing system of speckle interferograms, applied to the study of laminar and turbulent jets, was proposed by Erbeck [4]. The system consists basically of a microcomputer interfaced to a video camera and to an *x–y* positioning stage that allows speckle scanning. Much attention is paid to preparing for acquisition of the image. A rotating ground glass is used as a screen in order to suppress additional speckle, and a special compensation plate that has a decreasing transmission toward its center is used to improve the typical uneven illumination of speckle interferograms. In this way it is possible to threshold electronically the image that is memorized with only 1-bit resolution (black and white). Speckle analysis is carried out by means of a normalized correlation function of the light intensity, which is calculated along several lines in Young's pattern. As it is the correlation function of statistical character, it does not depend on the absolute coordinates of the intensity distribution but on the separation of two points of interest. The wavelength of the correlation function is a measure of the fringe spacing in the respective direction. The fringe pattern is scanned in four directions (horizontal, vertical, and both diagonals), each along two lines. The direction nearly parallel to the fringes is then disregarded. The procedure selects the two best correlation functions to determine the spacing and direction of the fringes. The advan-

(a)

Fig. 23 Interaction of a Karman vortex street and an airfoil: (*a*) Mach–Zehnder interferogram of two vortices (initial and secondary) passing the airfoil; (*b*) evaluated interferogram of (*a*); automatic fringe reading provides the bounds of fringes as polygons. (Figure continues.)

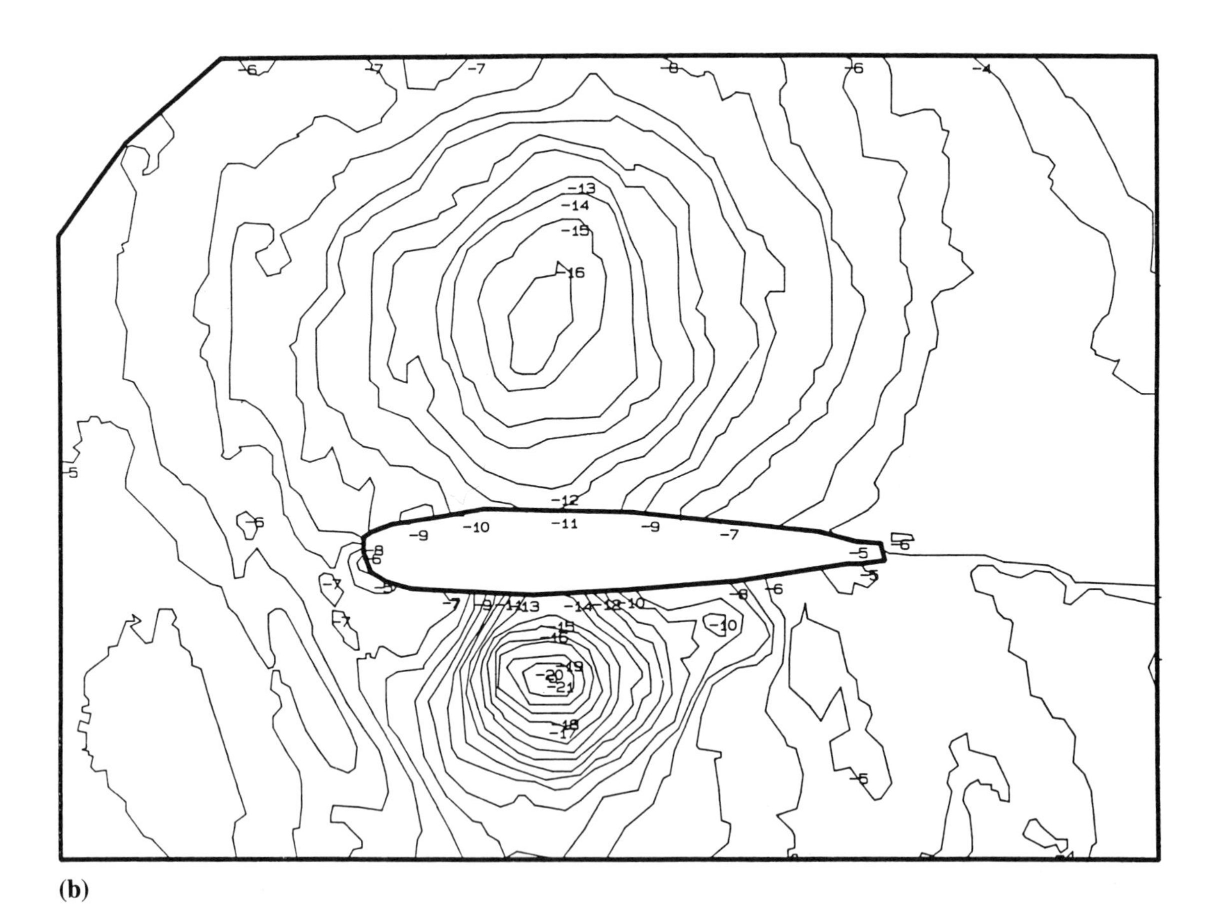

(b)

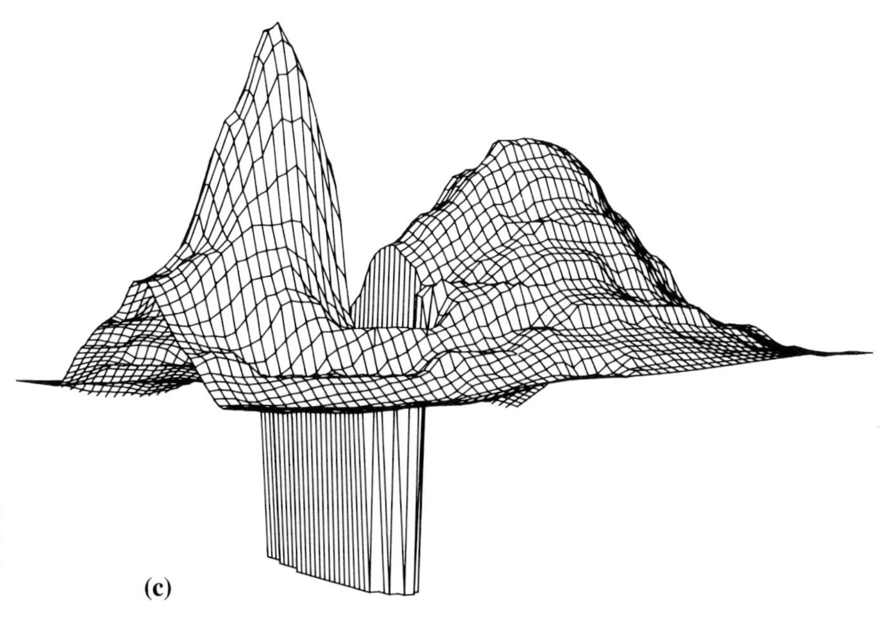

Fig. 23 (continued) (*c*) density plot in a view from upstream on the stagnation point of the airfoil; the surface of the airfoil is marked with straight lines [53].

tage of this method is that it treats a binary image so that multiplications while computing the correlation function are reduced to logical operators, thereby consistently increasing computing speed.

Becker, Meier, and co-workers [8, 35, 36, 53–55] described a system for digital interferogram evaluation applied to several aerodynamic studies. Their DPI system is basically composed of (1) a video camera with a spatial resolution of 512 × 512 pixels converted with 8-bit intensity resolution and (2) an image processor that provides signal digitization and storage and includes an arithmetic logic unit. This unit allows real-time addition, subtraction, or comparison of one or more image planes.

Preprocessing work consists of (1) averaging of multiple frames of the same picture; (2) spatial smoothing by averaging each pixel with the eight neighboring ones (3) removal of the uneven illumination by averaging the signal in a grid of 32 × 32 (or 64 × 64) pixels; (4) construction of a smooth surface by bilinear interpolation between the gray values in every four adjacent mesh points, which is then subtracted from the original interferogram.

The fringe segmentation is carried out by means of the floating threshold method (see Sec. III.B), implemented both horizontally and vertically. The fringe coordinates are determined at the transition from white to black (and vice versa), giving the middle of the fringe sides as the equal phase lines. A sequential tracking procedure is used to represent the fringe field as a polygonal data structure. To prevent fringe polygons from becoming connected at the boundary of the field of view or to prevent background objects, such as airfoils, from being interpreted as fringes, the boundary coordinates are used to generate a mask that specifies the input area for the coordinate extraction.

Hereafter, two examples of application of the system to the interaction of a Karman vortex street and to an airfoil, and the measurement of the three-dimensional transonic flow field around a helicopter blade are reported.

The first application was carried out in a transonic wind tunnel in which a NACA 0012 profile was placed in the wake of a cylinder in order to study the vortex–profile interaction [54]. The pressure field was evaluated by means of transducers and digitally processed by fast digital-signal-processing hardware to get spectra and correlations. The density field was visualized by means of a Mach–Zehnder interferometer and recorded by a high-speed movie camera. The automatic interferogram analysis was used to evaluate densities, pressures, and vortex traces from interferograms.

A typical sequence of the automatic interferogram analysis is shown in Figs. 23*a–c,* [53]. The interferogram of Fig. 23*a* shows two vortices passing over the airfoil. The interferogram is subsequently digitized and processed so as to have the polygonal representation of the fringe field given by Fig. 23*b,* where fringe numbers are also indicated. A perspective view (from upstream of the stagnation point) of the density field around the airfoil is shown in Fig. 23*c,* where the airfoil surface is marked with straight lines.

The second investigation carried out with basically the same DPI system concerns evaluation of the transonic three-dimensional flow field around a rotating helicopter blade [8, 55, 56]. A double-exposure holographic technique was used to visualize the flow field, and a filtered convolution back-projection method, proposed by Shepp [57], was implemented for the three-dimensional tomographic reconstruction.

Figure 24*a* shows a reconstruction of the flow field around a hovering 1/7 geometrically scaled model UH-1H helicopter rotor blade tip, at a tip Mach number of 0.9, in a plane 8% chord above the blade. The interferograms were taken every 2° for the angular intervals from 8 to 40° and from 140 to 186°. The resolution in the plane was

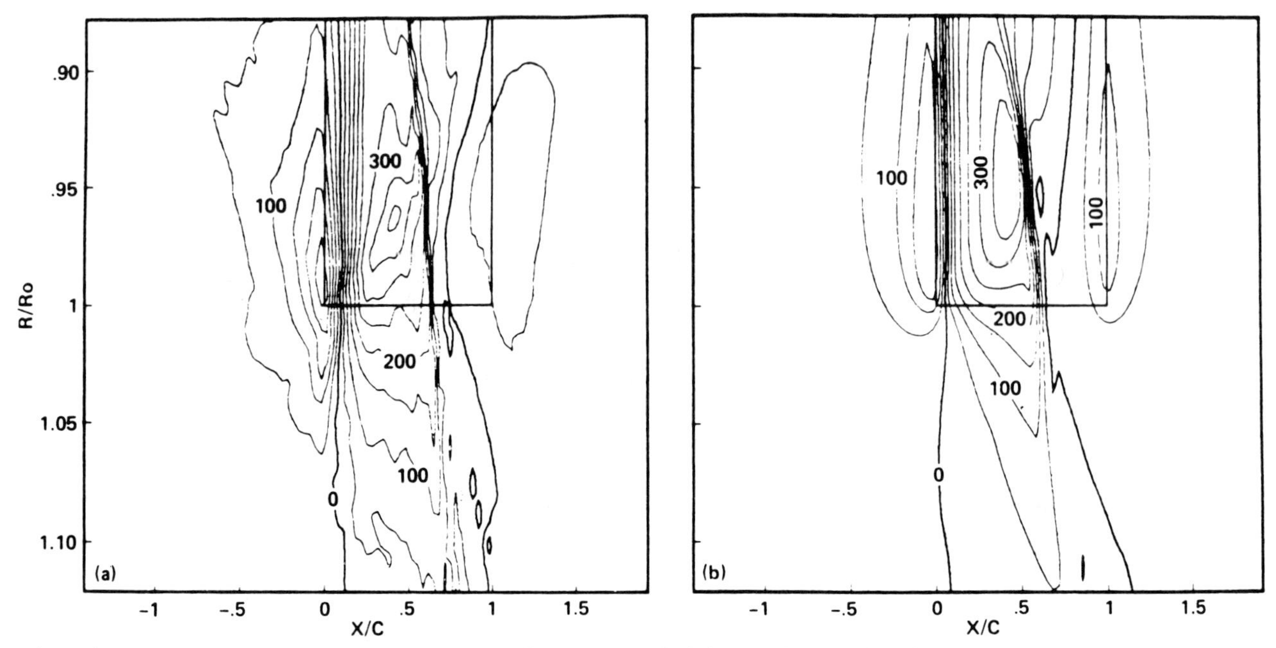

Fig. 24 Perturbation velocity contours around a helicopter rotor blade in a horizontal plane 8% chord above the blade: (*a*) Reconstruction by means of computer-assisted tomography; (*b*) numerical solution obtained by small perturbation method [56].

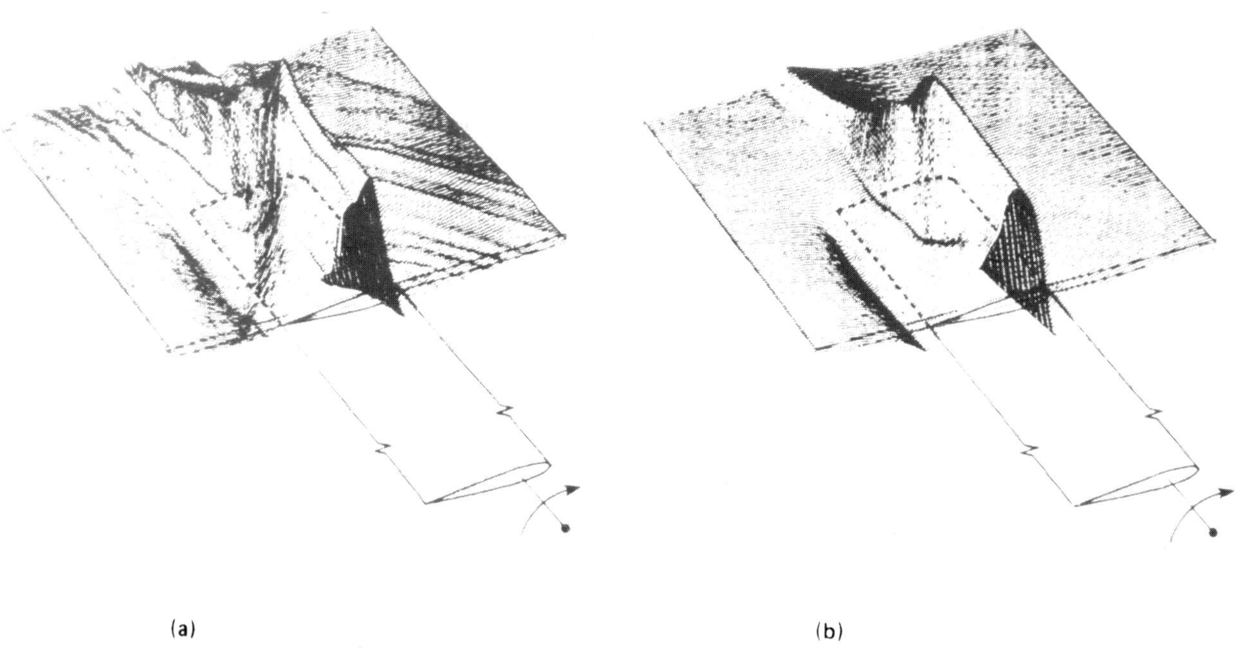

Fig. 25 Perturbation velocity contours: (*a*) Perspective view of Fig. 24*a;* (*b*) perspective view of Fig. 24*b* [56].

101 × 101 points, and 141 rays were used in the projection data at each view. In Fig. 24*b* the numerical small perturbation potential solution of the same flow field is reported. The general flow features appear very similar: a low-velocity region was present near the leading edge, which was followed by a high-velocity region over the blade surface containing a shock at approximately 60% of the chord near the blade tip.

Perspective views of the velocity contours of Figs. 24*a* and 24*b* are reported, respectively, in Figs. 25*a* and 25*b* [56]. Again the general flow shapes appear very similar. The major differences between the two results (i.e., the minor ridges in the tomographically reconstructed flow) is attributed by the authors to reconstruction artifacts caused by noncontinuous data (interferograms recorded in 2° azimuthal increments) or to noise (erroneous fringes) in the interferogram data caused by optical-component motion.

REFERENCES

1. Carlomagno, G. M., and Rapillo, A., Interferogram Analysis by Means of a Photodiode Array, in *Optics in Engineering Measurements,* ed. W. F. Fagan, *Proc. SPIE,* vol. 599, pp. 160–165, 1985.
2. Merzkirch, W., Density Sensitive Flow Visualization, in *Methods of Experimental Physics,* vol. 18A: *Fluid Dynamics,* ed. R. J. Emrich, pp. 392–395, Academic, New York, 1981.
3. Carlomagno, G. M., Schlieren Interferometry in the Mass Diffusion of a Two-dimensional Jet, *Exp. Fluids,* vol. 3, pp. 137–141, 1985.
4. Erbeck, R., Fast Image Processing with a Microcomputer Applied to Speckle Photography, *Appl. Opt.,* vol. 24, no. 22, pp. 3838–3841, 1985.
5. Robinson, D. W., Automatic Fringe Analysis with a Computer Image-processing System, *Appl. Opt.,* vol. 22, no. 14, pp. 2169–2176, 1983.
6. Nakadate, S., Yatagai, T., and Saito, H., Computer-aided Speckle Pattern Interferometry, *Appl. Opt.,* vol. 22, no. 2, pp. 237–243, 1983.
7. Oreb, B. F., Sharon, B., and Hriharan, P., Electronic Speckle Pattern Interferometry with a Microcomputer, *Appl. Opt.,* vol. 23, no. 22, pp. 3940–3941, 1984.
8. Becker, F., and Yu, Y. H., Digital Fringe Reduction Techniques Applied to the Measurement of Three-dimensional Transonic Flow Fields, *Opt. Eng.,* vol. 24, no. 3, pp. 429–434, 1985.
9. Tichenor, D. A., and Madsen, V. P., Computer Analysis of Holographic Interferograms for Nondestructive Testing, *Opt. Eng.,* vol. 18, no. 5, pp. 469–472, 1979.
10. Boxman, R. L., and Shlien, D. J., Interferometric Measurement Technique for the Temperature Field of Axisymmetric Buoyant Phenomena, *Appl. Opt.,* vol. 17, no. 17, pp. 2788–2793, 1978.
11. Jimenez, J., Planning an Experiment for Digital Image Processing, *Digital Image Processing in Fluid Dynamics,* Proc. Lect. Ser., von Karman Institute for Fluid Dynamics, Rhode-Saint-Genése, Belgium, March 1984.
12. Prettyjohns, K. N., Charge-coupled Device Image Acquisition for Digital Phase Measurement Interferometry, *Opt. Eng.,* vol. 23, no. 4, pp. 371–378, 1984.
13. Carlomagno, G. M., and Lombardi, P., Convective Wall Plume Induced by a Linear Mass Source, *Heat Technol.,* vol. 4, no. 2, pp. 80–94, 1986.
14. Carlomagno, G. M., and Rapillo, A., A Wollaston Prism Interferometer Implemented with a Digitizer, *Exp. Fluids,* vol. 4, pp. 332–336, 1986.
15. Chay, A., and Shlien, D. J., Scalar Field Measurements of a Laminar Starting Plume Cap using Digital Processing of Interferograms, *Phys. Fluids,* vol. 29, no. 8, pp. 2358–2366, 1986.
16. Beynon, J. D. E., and Lamb, D. R., *Charge Coupled Devices and Their Applications,* McGraw-Hill, Maidenhead, England, 1980.
17. Carbone, J., A Camera for All Systems, *Robot. World,* June 1986.
18. *Laser Focus/Electro-Optics Buyer's Guide,* Pennwell Publishing Co., Tulsa, Okla., 1987.
19. Sciammarella, C. A., and Ahmadshahi, M., An Optoelectronic System for Fringe-Pattern Analysis, *Exp. Technol.,* vol. 10, no. 7, pp. 8–13, 1986.
20. Trolinger, J. D., Laser Instrumentation for Flow Field Diagnostic, AGARD-AG-186, 1974.
21. Rosenfeld, A., and Kak, A. C., *Digital Picture Processing,* Academic, New York, 1976.
22. Uang, T. S., Image Enhancement: A review, *Opto-electronics,* vol. 1, pp. 49–59, 1969.
23. Martelli, A., and Montanari, U., Optimal Smoothing in Picture Processing: An Application to Fingerprints, *Proc. IFIP Cong.,* 71, Booklet TA-2, 86–90, 1972.
24. Pratt, W. K., *Digital Image Processing,* Wiley, New York, 1978.
25. Linfoot, E. H., *Fourier Methods in Optical Image Evaluation,* Focal Press, New York, 1964.
26. Chambless, D. A., and Broadway, J. A., Digital Filtering of Speckle Photography Data, *Exp. Mech.,* vol. 19, pp. 286–289, 1979.
27. Kohler, R., and Howell, H., Photographic Image Enhancement by Superimposition of Multiple Images, *Photogr. Sci. Eng.,* vol. 7, pp. 241–245, 1963.
28. Schemm, J. B., and Vest, C. M., Fringe Pattern Recognition and Interpolation using Nonlinear Regression Analysis, *Appl. Opt.,* vol. 22, no. 18, pp. 2850–2853, 1983.
29. Snyder, J. J., Algorithm for Fast Digital Analysis of Interference Fringes, *Appl. Opt.,* vol. 19, no. 8, pp. 1223–1225, 1980.
30. Hilditch, C. J., Linear Skeleton in Square Cupboards, in *Machine Intelligence,* ed. B. Meltzer and D. Michie, vol. 6, pp. 403–420, Univ. Press, Edinburgh, 1969.
31. Chen, T. H., and Christiansen, W. H., Semiautomated Method for Determination of Density Profiles from Mach Zehnder Interferograms, *Rev. Sci. Instrum.,* vol. 56, no. 8, pp. 1619–1622, 1985.
32. Funnell, W. R. J., Image Processing Applied to the Interactive Analysis of Interferometric Fringes, *Appl. Opt.,* vol. 20, no. 18, pp. 3245–3250, 1981.
33. Hsing, T. R., Techniques of Adaptive Threshold Setting for Document Scanning Applications, *Opt. Eng.,* vol. 23, no. 3, pp. 288–293, 1984.
34. Takahashi, S., A Video Signal Processing System, Japanese Patent no. 11625, 1979.
35. Becker, F., Meier, G. E. A., and Wegner, H., Automatic Evaluation of Interferograms, in *Applications of Digital Image Processing IV,* ed. G. Tescher, *Proc. SPIE,* vol. 359, pp. 386–393, 1982.
36. Becker, F., Zur automatischen Auswertung von Interferogrammen, *Mitt. Max-Planck-Inst. Strömungsforsch.,* Göttingen, Nr.74, pp. 1–106, 1982.

37. Takeda, M., Ina, H., and Kobayashi, S., Fourier Transform Method of Fringe-Pattern Analysis for Computer-based Topography and Interferometry, *J. Opt. Soc. Am.*, vol. 72, pp. 156–162, 1981.
38. Macy, W. W., Jr., Two-dimensional Fringe-Pattern Analysis, *Appl. Opt.*, vol. 22, no. 23, pp. 3898–3901, 1983.
39. Bone, J. D., Bachor, H. A., and Sandeman, J., Fringe-Pattern Analysis using a 2-D Fourier Transform, *Appl. Opt.*, vol. 25, no. 10, pp. 1653–1660, 1986.
40. Seguchi, Y., Tomita, Y., and Watanabe, M., Computer-aided Fringe-Pattern Analyzer: A Case of Photoelastic Fringe, *Exp. Mech.*, vol. 19, pp. 362–370, 1979.
41. McLain, D. H., Drawing Contours from Arbitrary Data Points, *Comput. J.*, vol. 17, no. 4, pp. 318–324, 1974.
42. Cline, H. E., Holik, A. S., and Lorensen, W., Computer-aided Surface Reconstruction of Interference Contours, *Appl. Opt.*, vol. 21, no. 24, pp. 4481–4488, 1982.
43. Muller, R. K., and Saackel L. R., Complete Automatic Analysis of Photoelastic Fringes, *Exp. Mech.*, vol. 19, pp. 245–251, 1979.
44. Nakadate, S., Yatagai, T., and Saito, H., Digital Speckle-Pattern Shearing Interferometry, *Appl. Opt.*, vol. 19, no. 24, pp. 4241–4246, 1980.
45. Nakadate, S., Yatagai, T., and Saito H., Electronic Speckle Pattern Interferometry using Digital Image Processing Techniques, *Appl. Opt.*, vol. 19, no. 11, pp. 1979–1883, 1980.
46. Lockett, J. F., and Collins, M. W., Problems in using Holographic Interferometry to Resolve the Four-dimensional Character of Turbulence: Part I. Theory and Experiment, *Int. J. Opt. Sens.*, vol. 1, no. 3, pp. 211–224, 1986.
47. Hunter, J., and Collins, M. W., Problems in using Holographic Interferometry to Resolve the Four-dimensional Character of Turbulence: Part II. Image and Data Processing, *Int. J. Opt. Sens.*, vol. 1, no. 3, pp. 225–234, 1986.
48. Carlomagno, G. M., de Luca, L., and Rapillo, A., A New Type of Interferometer for Gas Dynamics Studies, *Proc. 12th Int. Cong. Instrum. Aerosp. Simul. Facil., IEEE Aerosp. Electron. Syst. Soc., Williamsburg,* 1987.
49. Carlomagno, G. M., A Wollaston Prism Interferometer Used as a Reference Beam Interferometer, in *Flow Visualization IV: Proceedings of the Fourth International Symposium,* ed. C. Véret, pp. 105–110, Hemisphere, Washington, D.C., 1987.
50. Wernekinck, U., Merzkirch, W., and Fomin, N., Measurement of Light Deflection in a Turbulent Density Field, *Exp. Fluids,* vol. 3, pp. 206–208, 1985.
51. Kopf, U., Application of Speckling for Measuring the Deflection of Laser Light by Phase Objects, *Opt. Commun.*, vol. 5, pp. 347–350, 1972.
52. Meynart, R., Instantaneous Velocity Field Measurements in Unsteady Gas Flow by Speckle Velocimetry, *Appl. Opt.*, vol. 22, no. 4, pp. 535–540, 1983.
53. Bartels, H. H., Bretthauer, B., and Meier, G. E. A., personal communication, 1987.
54. Meier, G. E. A., Timm, R., and Becker F., Transonic Noise Generation by Duct and Profile Flow, Max-Planck-Inst. Strömungsforsch., Final Tech. Rep., 1–100, 1984.
55. Becker, F., and Yu, Y. H., Application of Digital Interferogram Evaluation Techniques, AIAA 23rd Aerosp. Sci. Mtg., Nev., AIAA Paper 85-0037, Reno, 1985.
56. Kittleson, J. K., and Yu, Y. H., Reconstruction of a Three-dimensional Transonic Rotor Flow Field from Holographic Interferogram Data, Reno, AIAA 23rd Aerosp. Sci. Mtg., AIAA Paper 85-0370, Nev., 1985.
57. Shepp, L. A., and Logan, B. F., The Fourier Reconstruction of a Head Section, *IEEE Trans. Nucl. Sci.*, vol. NS-21, pp. 21–43, 1974.

Chapter 17

Optical Image Processing

Lambertus Hesselink

I INTRODUCTION

Optical image processing refers to manipulation of images by optical means. Just as time-dependent electrical signals can be decomposed into Fourier frequency components, images can be decomposed into spatial frequency (or Fourier) components, for instance, by using an optical processor. An optical processor in its simplest form consists of an information-carrying (laser) beam that is incident on a positive lens. Such a system produces in the back focal plane of the lens a field distribution of the light that is equal to the spatial Fourier transform decomposition of the object in the front focal plane. The back focal plane is therefore often designated as the Fourier transform plane of the lens. Placing a recording medium such as a detector or film at this location allows the intensity of the light to be measured, representing the spectrum of spatial frequencies.

In analogy to processing of electrical signals, the spatial frequency components of an image may be altered by filtering techniques. For example, low-pass filtering is achieved by placing an opaque disk with a centered hole in the back focal plane of the lens, and high-pass filtering involves positioning an opaque disk on axis at that location. Band-pass filtering is achieved by using transparent rings. The phase of Fourier spectral components may also be modified using, for instance, half or quarter wave plates. Furthermore, several lenses may be placed in series to carry out sequential transforming operations.

The attractiveness of optical image processing stems from the speed with which the signals are processed. Fourier transforms are computed in the time it takes for the light to propagate from the object to the back focal plane of the lens. Massive parallel processing occurs, because all spatial frequency components are computed simultaneously. Additionally, by proper design of the processor, spatial derivatives of the image may be obtained with good accuracy. For example, schlieren systems measure the gradient of the index of refraction distribution and shadowgraph systems record the Laplacian averaged along the path length. In essence, we may think of an optical processor as an analog computer. Computations are carried out very fast, but accuracy is rather limited, often less than 256 levels, because scattered light and noise tend to contaminate the images. Thus, in terms of speed, optical processors are far superior to digital computers. In terms of accuracy, however, digital processors have the edge. Nevertheless, in many applications, it is desirable to perform hybrid operations. For example, optical processors may be used to determine spatial derivatives of the index-of-refraction field, and these images may then be further processed digitally. Of course, we could also directly record the phase of the optical wave interferometrically, followed by digital differentiation to produce the equivalent of a schlieren image in a 2-D flow. In this case, however, accuracy tends to be less than when using optical processors, because digital differentiation of noisy data can produce substantial errors.

In this Chapter we discuss the basic principles underlying optical processors. The framework for the discussion is linear systems theory, often referred to as Fourier optics in this context. Finally, flow applications of optical processing are discussed.

II CONCEPTS OF MODERN OPTICS

A Introduction

The theory of geometrical optics provides, in many cases, adequate information regarding the overall performance of an optical system. Often it is not necessary to analyze the configuration in any more detail, and the sophistication that is introduced by diffraction theory does not justify the extra effort. In some problems, however, such as holography, the propagation of Gaussian beams, and pattern-recognition schemes (using, for instance, van der Lugt filters), the finesse of diffraction theory is absolutely necessary. Furthermore it adds another dimension to our point of view of optical systems. In fact, this new look, called Fourier optics, provides insight into optical systems that is not only very useful, but very hard to obtain any other way. The use of Fourier transform techniques in the analysis of optical systems was largely stimulated by the French scientist P.M. Duffieux in the late 1930s. Today, Fourier optics is a powerful tool in the analysis of systems involving coherent as well as incoherent sources.

Diffraction theory deals with solutions to the wave equation and appropriate boundary conditions. The exact solution is often difficult to calculate, and two useful and realistic approximations are commonly made. These are called the Fresnel, or near field, and the Fraunhofer, or far field, approximation. The diffraction field is determined as a superposition of elementary waves which originate from the object under consideration, in accordance with Huygen's principle. In the Fraunhofer region the relationship between object field and diffracted field simplifies to such an extent that the two fields are related by a Fourier transformation.

The theory is linear, and a fruitful analogy between electrical and optical systems can be made. For electrical systems the output is computed using linear systems theory in terms of a convolution between the impulse response of the system and the input signal. In analogy, for optical systems we wish to calculate the field in the output plane in terms of the field at an input plane, and we need to know the impulse response (or transfer function) that relates the two fields (or Fourier transforms of the fields).

The transfer function is related to the impulse response of the system (for optical systems, a spatial impulse) by a Fourier transform operation, and arbitrary signals are synthesized from elementary impulses. The analogy between optical and electrical systems is formalized by replacing time in electrical systems with a length scale in optics problems. This analogy is similar to the correspondence between geometrical optics and classical mechanics (the Hamiltonian analogy).

Referring to Fig. 1, we now address the following problem: Given the field distribution in plane I, can we calculate the field distribution in plane II, assuming an energy flow that is chosen to be directed along the positive z axis.

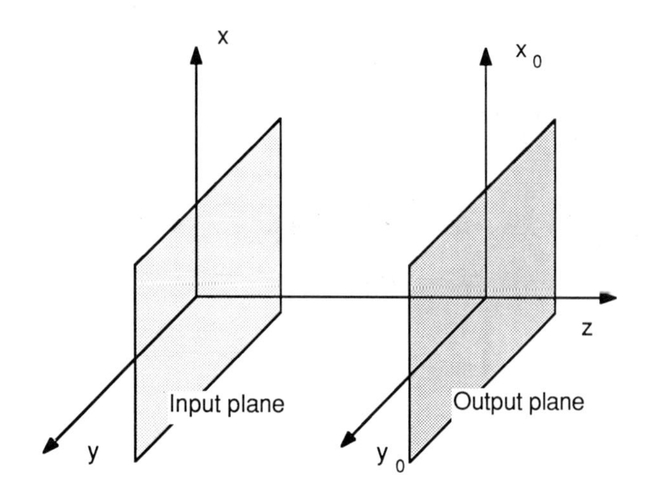

Fig. 1 Statement of the problem.

First, the solution to the wave equation is calculated for the appropriate boundary conditions. Then two approximations are introduced, the Fresnel and Fraunhofer approximations, and examples of diffraction problems are treated. Subsequently, the properties of lenses and optical processors are examined using Fourier optics theory.

B Wave Equation and Boundary Conditions

The Maxwell equations and appropriate boundary conditions form the starting point of the discussion. In all of the following analyses on diffraction theory it is assumed that the electromagnetic waves depend harmonically on time, unless otherwise stated. Mathematically this can be expressed as follows:

$$\phi = \phi(x, y, z)e^{-i\omega t} \tag{1}$$

where ϕ denotes a scalar component of the vector field.

The Maxwell equations (formulated in mks units) for free space propagation may be reduced to:

$$\nabla^2\mathbf{E} = \mu\varepsilon \frac{\partial^2\mathbf{E}}{\partial t^2}$$
$$\nabla^2\mathbf{H} = \mu\varepsilon \frac{\partial^2\mathbf{H}}{\partial t^2} \tag{2}$$

where μ and ε are the permeability and dielectric constants, respectively, and E_i and H_i represent the electric and magnetic field components. Now let $\mathbf{E} = E_1\mathbf{i} + E_2\mathbf{j} + E_3\mathbf{k}$ and represent the scalar components $E_{1,2,3}$ by $\phi_{1,2,3}$. Substitution of $\mathbf{E}$ into (1) and (2) then gives the result

$$(\nabla^2 + k^2)\phi = 0 \tag{3}$$

where the wave number $k = \omega\sqrt{\mu\varepsilon}$. This equation, called the Helmholtz equation, is obeyed by the scalar components of the electromagnetic field. Thus, the amplitude of any monochromatic optical wave that propagates through an isotropic, nonconducting medium satisfies this equation. Unique solutions are given in terms of the boundary conditions.

C Definition of the Problem and a Heuristic Approach to the Solution

Suppose that boundary conditions are prescribed on surface S. Without loss of generality this surface is taken to be plane, as shown in Fig. 2. The central problem we want to solve then can be stated as follows:

> For a given set of boundary conditions on S can we calculate the field at some other surface S_0?

Points on surface S are indicated by coordinates x, y, and $z = 0$, and points on surface S_0 by x_0, y_0, z_0. Point P_1 on S is connected to point P_0 on S_0 by vector $\mathbf{R}$. On S the field is represented by $\phi(x, y, 0)$.

A simple heuristic argument is given here to determine the form of the solution to this boundary value problem. According to Huygen's principle, the wave front at S_0 can be determined from the wave front at S in terms of a superposition of elementary secondary wavelets. Suppose that a wavelet emitted from point P_1 on S is mathematically represented by

$$K(x_0, y_0, z_0; x, y, 0) \tag{4}$$

This function, often referred to as a kernel, for free space propagation (eq. 4) has the form of a spherical wave:

$$K(x_0, y_0, z_0; x, y, 0) \sim \frac{e^{ikR}}{R} \tag{5}$$

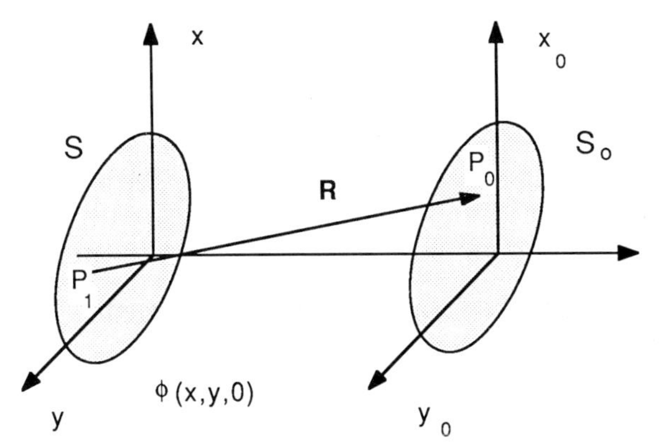

Fig. 2 Geometry for solving optical propagation problems.

and the time dependence is again suppressed here. Because each point on S will radiate with some characteristic amplitude given by $\phi(x, y, 0)$, we have to multiply the kernel by this prescribed field distribution to determine the diffracted field. Therefore, the field at P_0, which according to Huygens' principle is a superposition of all elementary waves, can be expressed mathematically as follows:

$$\phi(x_0, y_0, z_0) = \text{const} \int_S K(x_0, y_0, z_0; x, y, 0)\phi(x, y, 0)\, dS \tag{6}$$

The objective now is to determine the detailed representation of the kernel, which we expect to have the form of Eq. (5), except perhaps for additional geometric factors involved. In linear systems theory, K is referred to as the impulse response of the system. Furthermore, since Eq. (6) is a superposition integral, the linear system is completely determined by its response to unit impulses, i.e., the response of the system to several stimuli acting simultaneously is identically equal to the sum of the responses that each of the component stimuli would produce individually.

The solution to the Helmholtz equation under appropriate boundary conditions has been derived in Goodman [5], and will not be repeated here. Instead, we will use as a starting point for our discussion this solution, referred to as the Rayleigh–Sommerfeld integral:

$$\phi(x_0, y_0, z) = -\frac{1}{2\pi} \iint_{z=0} \phi(x, y, 0) \cos(\mathbf{n}, \mathbf{r}_{01}) \left(ik - \frac{1}{r_{01}}\right) \frac{\exp(ikr_{01})}{r_{01}}\, dx\, dy \tag{7}$$

where the vectors $\mathbf{n}$ and $\mathbf{r}_{01}$ are defined in Fig. 3 and $z = z_0$ for simplicity.

D Simplifying Assumptions for Solving the Rayleigh–Sommerfeld Integral

Referring to Eq. (7) and the optical configuration shown in Fig. 3, the boundary conditions are specified on S. The point $P_0(x_0, y_0, z)$ is located on $S_0(z = z_0 = \text{const})$ and connected to the origin by vector $\mathbf{R}_0$, and $P_1(x, y, 0)$ resides on S. The vector linking P_0 and P_1 is $\mathbf{r}_{01}$ (from P_0 to P_1).

In order to reduce the complexity of Eq. (7), we make the following simplifying assumptions:

1.

$$|ik| >> \frac{1}{r_{01}} \quad \text{or} \quad r_{01} >> \frac{\lambda}{2\pi} \tag{8}$$

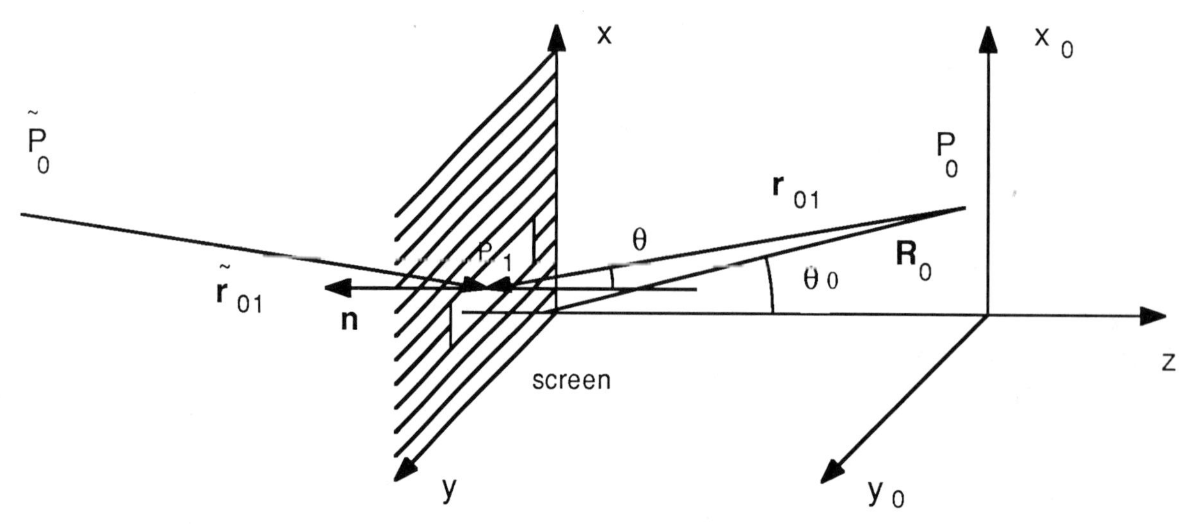

Fig. 3 Geometry for solving the Rayleigh–Sommerfeld diffraction integral.

In practical cases this condition is almost always satisfied and for $\lambda = 5000$ Å, r_{01} has to be much larger than 1 μm.

2. For large $z >> 1$, $\mathbf{r}_{01} \approx \mathbf{r}_0$ and

$$\cos(\mathbf{n}, \mathbf{r}_{01}) \approx \cos(\mathbf{n}, \mathbf{R}_0) \approx \cos\theta_0 \approx \cos\theta \approx \frac{z}{r_{01}} \tag{9}$$

which is called the *inclination factor*.

3. Replace r_{01} by z in the denominator, but *not* in the exponent, because a small error in r_{01} will cause a very large phase change as a result of the large value of k.
4. Expand r_{01} in the exponent in terms of x, x_0, y, y_0, and z.

$$\begin{aligned} r_{01} &= \sqrt{z^2 + (x_0 - x)^2 + (y_0 - y)^2} \\ &= z\sqrt{1 + \left(\frac{x_0 - x}{z}\right)^2 + \left(\frac{y_0 - y}{z}\right)^2} \\ &\approx z\left(1 + \frac{1}{2}\left(\frac{x_0 - x}{z}\right)^2 + \frac{1}{2}\left(\frac{y_0 - y}{z}\right)^2 \right. \\ &\quad \left. + 0\left\{\left[\frac{(x_0 - x)^2}{z^z} + \frac{(y_0 - y)^2}{z^z}\right]^2\right\}\right) \end{aligned} \tag{10}$$

and

$$\frac{x_0 - x}{z}, \frac{y_0 - y}{z} << 1$$

Neglect the higher order terms.

The implications of these assumptions are manifested in a simpler form of integral (7)

$$\phi(x_0, y_0, z) = \frac{e^{ikz}}{i\lambda z} \cos\theta_0 \iint_{z=0} \phi(x, y, 0) \exp\left\{\frac{ik}{2z}[(x_0 - x)^2 + (y_0 - y)^2]\right\} dx\, dy \tag{11}$$

This is called the Fresnel diffraction formula. The assumption to neglect the higher order terms in (10) amounts to requiring that

$$[(x_0 - x)^2 + (y_0 - y)^2]^2_{max} \frac{kz}{8z^4} << 1 \text{ radian}$$

in the exponent of Eq. (11), or

$$z^3 >> \frac{k}{8}[(x_0 - x)^2 + (y_0 - y)^2]^2_{max} \tag{12}$$

This condition is referred to as the Fresnel approximation.

To give an example of the limit imposed on z, let $\lambda = 0.5 \times 10^{-3}$ mm and

$$[(x_0 - x)^2 + (y_0 - y)^2]^2_{max} = 1 \text{ mm}^4$$

Then $z >> 2.5$ mm. Furthermore, in almost all practical cases when (12) is satisfied, the inclination factor $\cos\theta_0 \approx 1$, further simplifying Eq. (11).

E Interpretation of the Fresnel Diffraction Formula

It is interesting to compare Eq. (11) with (6), which was derived using a heuristic argument. More precisely, the comparison shows that

$$K(x_0, y_0, z; x, y, 0) = \frac{e^{ikz}}{z} \exp\left\{\frac{ik}{2z}[(x_0 - x)^2 + (y_0 - y)^2]\right\} \quad (13)$$

and the constant $= \cos\theta_0/i\lambda$. But the right-hand side is a quadratic approximation for a spherical wave, under the assumption that (12) is satisfied. Thus, Huygen's principle indeed provides the physical interpretation for (11), namely, that the field at $\phi(x_0, y_0, z)$ can be thought of as a superposition of elementary waves given by (13), and where the unit amplitude of each source is multiplied by the prescribed value of the field amplitude at the boundary.

Yet another interpretation of (11) can be given in terms of linear systems analysis. Equation (11) has the form of a convolution integral. The Fresnel diffraction is then interpreted to be the convolution of $\phi(x, y, 0)$ and $h(x, y, z)$

$$h(x, y, z) = \exp\left[\frac{ik}{2z}(x^2 + y^2)\right] \quad (14)$$

where $h(x, y, z)$ is called the impulse response of the system. In other symbols

$$\phi(x_0, y_0, z) = \phi(x, y, 0) ** h(x, y, z) \quad (15)$$

where the ** denotes a convolution. It should be remembered that this relationship (15) holds true only for space-invariant systems, that is, systems for which

$$h(x_0, y_0, z; x, y) = h(x - x_0, y - y_0, z) \quad (16)$$

Now if we take the Fourier transform of both sides of (15) we get

$$F[\phi(x_0, y_0, z)] = F[\phi(x, y, 0)]F[h(x, y, z)] \quad (17)$$

The function $F[h(x, y, z)]$ is called the transfer function:

$$F[h(x, y, z)] = \iint_{-\infty}^{\infty} h(x, y, z) \exp[-i2\pi(f_x x + f_y y)]\, dx\, dy \quad (18)$$

where f_x and f_y denote spatial frequency variables. In the next section we will derive an expression for them.

F Another Form of the Fresnel Integral

Alternatively, the quadratic form in the Fresnel integral (11) may be expanded to give the following result:

$$\phi(x_0, y_0, z) = \frac{\cos\theta_0}{i\lambda z}\exp(ikz) \exp\left[i\frac{k}{2z}(x_0^2 + y_0^2)\right] \iint_{-\infty}^{\infty} \phi(x, y, 0)\exp\left[\frac{ik}{2z}(x^2 + y^2)\right] \exp\left[-i\frac{2\pi}{\lambda z}(x_0 x + y_0 y)\right] dx\, dy \quad (19)$$

This integral represents a Fourier transform with frequencies

$$f_x = \frac{x_0}{\lambda z} \qquad f_y = \frac{y_0}{\lambda z} \quad (20)$$

of the function

$$\phi(x, y, 0)\exp\left[\frac{ik}{2z}(x^2 + y^2)\right] \quad (21)$$

The quadratic term in (21) denotes a phase factor across the input aperture. If this term is unity across that aperture, then the observed field distribution can be found directly from a Fourier transform operation of the aperture field distribution itself. In order for the approximation to be valid, it is necessary that

$$z >> \frac{1}{2}k(x^2 + y^2)_{\max} \quad (22)$$

This is called the *Fraunhofer*, or far-field, approximation. Notice that indeed condition (22) is more stringent than (12).

The far field can then be calculated from

$$\phi(x_0, y_0, z) = \frac{\cos\theta_0}{i\lambda z}\exp(ikz) \exp\left[\frac{ik}{2z}(x_0^2 + y_0^2)\right] \iint_{-\infty}^{\infty} \phi(x, y, 0) \exp\left[-i\frac{2\pi}{\lambda z}(x_0 x + y_0 y)\right] dx\, dy \quad (23)$$

or, simpler yet,

$$\phi(x_0, y_0, z) = \frac{\cos\theta_0}{i\lambda z}\exp(ikz) \exp\left[\frac{ik}{2z}(x_0^2 + y_0^2)\right] F[\phi(x, y, 0)] \quad (24)$$

and with spatial frequencies

$$f_x = \frac{x_0}{\lambda z} \qquad f_y = \frac{y_0}{\lambda z}$$

To give an example of the severity of condition (22) let

$$(x^2 + y^2)_{max} = 1 \text{ mm}^2$$
$$\lambda = 0.5 \times 10^{-3} \text{ mm}$$

then

$$z >> 6.3 \text{ m}$$

for the Fraunhofer approximation to be valid. Far-field diffraction patterns, however, may be observed at distances much closer than those implied by the Fraunhofer condition provided that

1. The aperture is illuminated by a spherical wave converging toward the observer or measurement location.
2. A converging lens is properly positioned between the observer and the aperture.

This notion will be elaborated on when considering the Fourier transforming properties of a lens. Notice here that the Fraunhofer diffraction formula is not space invariant anymore and that no transfer function exists for Fraunhofer diffraction. This is caused by the simplifying assumptions we have made. But of course the Fresnel approximation is still valid in the far field as well; it is just that Fresnel integrals are much more difficult to compute than Fourier integrals for practical problems. Thus, the far field can still be computed by a linear superposition approach, but the concept of transfer function is no longer applicable.

G Diffraction Problems in Optics

A commonly used configuration in optical diffraction problems is shown in Fig. 4. An object, for instance, a transparency, is located in the x–y plane and is illuminated with a monochromatic plane wave of unit amplitude and normal incidence. The field inside the aperture is then prescribed, and we wish to determine the diffraction pattern, either in the near or in the far field. In accordance with Kirchhoff's ideas, at the plane $z = 0$ the following boundary conditions are applied:

1. Edge effects are neglected at the plane of the aperture. Across Σ, the field inside the aperture $\phi(x, y, 0)$ has exactly the same value as it would have in the absence of the aperture.
2. In the geometric shadow of the aperture, the field is identically 0 at the plane $z = 0$.

These conditions agree well with experiment as long as the dimensions of the aperture are large compared with the wavelength λ. To compute the diffracted field the Fraunhofer or Fresnel formula are applied to each component of the electromagnetic field.

The effect of the object on the incoming field is generally expressed in terms of a transmission function $T(x, y)$. This function is defined as follows. If an object with a transmission function $T(x, y)$ is illuminated by a polarized (for instance in the x direction) plane wave of normal incidence and unit amplitude, then the field $\phi(x, y, 0)$ is given by

$$\phi(x, y, 0) = T(x, y)E_{x,\text{incoming}}(x, y, 0) = T(x, y) \qquad (25)$$

just downstream of the object.

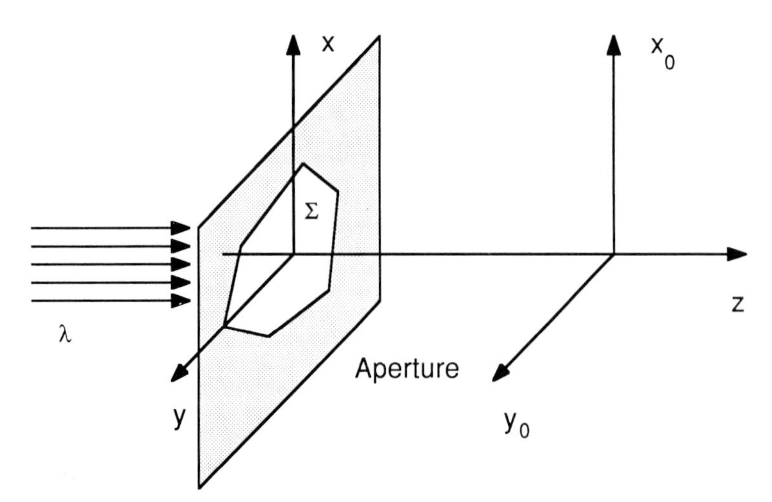

Fig. 4 Optical configuration for obtaining diffraction pattern.

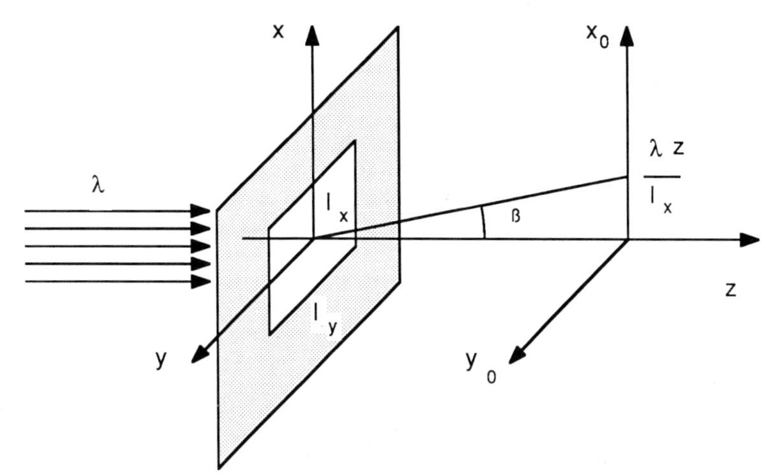

Fig. 5 Diffraction by a rectangular aperture.

In case the aperture is illuminated by a wave of amplitude $A(x, y)$, which may be complex, then the resulting boundary condition becomes

$$\phi(x, y, 0) = A(x, y)T(x, y) \tag{26}$$

Since we can detect intensities only directly, we often wish to compute the intensity by multiplying the field by its complex conjugate.

$$I(x_0, y_0, z) = \phi(x_0, y_0, z)\phi^*(x_0, y_0, z) \tag{27}$$

where the * denotes complex conjugation and $\phi(x_0, y_0, z)$ denotes the field in the output plane.

III EXAMPLES OF FRAUNHOFER DIFFRACTION

To illustrate applications of the Fraunhofer and Fresnel diffraction formulae, we will closely examine some specific diffraction configurations. The mathematical calculations furnish us with an analytic description of the diffraction patterns, but it will be shown in the following sections that very useful information can be obtained from simple heuristic arguments. This kind of reasoning is attractive in those instances when we are not interested in all the details, but only want to devise a quick scheme to obtain simple answers. The following examples are designed to develop and expand one's intuition by interpreting the results of the formal calculations using physical notions.

A Rectangular Aperture

Suppose that a rectangular aperture of dimensions ℓ_x and ℓ_y is illuminated by a normally incident, monochromatic, plane wave, with unit amplitude as shown in Fig. 5.

The transmission function is given by

$$T(x, y) = \text{rect}\left(\frac{x}{\ell_x}\right)\text{rect}\left(\frac{y}{\ell_y}\right)$$

where rect $(x/\ell_x) = 1$ for $|x| \leq \ell_x/2$ and 0 otherwise, and

$$\phi(x, y, 0) = T(x, y) \tag{28}$$

Now substitute (28) into the Fraunhofer diffraction formula (23) and calculate the intensity pattern according to (27). The result of this computation is

$$I = \phi\phi^* = \frac{\ell_x^2\ell_y^2}{\lambda^2 z^2}\,\text{sinc}^2\left(\frac{\ell_x x_0}{\lambda z}\right)\text{sinc}^2\left(\frac{\ell_y y_0}{\lambda z}\right) \tag{29}$$

The details can be found in Goodman [5], and the pattern is displayed in Fig. 6.

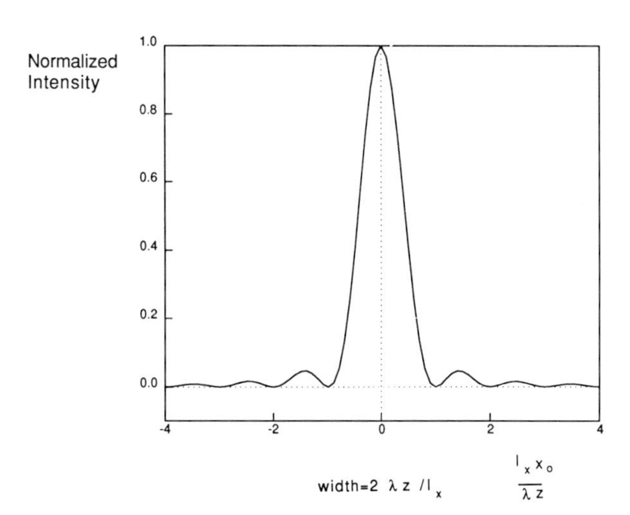

Fig. 6 Diffraction pattern from a rectangular aperture.

The result can again be interpreted in terms of a superposition of elementary waves according to Huygens' principle. At any point in the far field, elementary waves coming from the object are coherently superimposed to form an interference pattern. The peaks correspond to constructive interference and the nodes to destructive interference. The distance from the origin to the first zero in the diffraction pattern is given by $\frac{\lambda}{\ell_x}$. The quantity λ/ℓ_x corresponds to diffraction angle β as shown in Fig. 5. The angle β represents the direction to the first zero, measured from the z-axis and by multiplying α by z we can calculate the distance from the origin to the first dark fringe. Therefore, if we ignore the side lobes, we find that the far-field diffraction pattern has a width equal to

$$w = 2\,\frac{\text{wavelength} \times \text{propagation distance}}{\text{aperture size}} \tag{30}$$

This result holds in general for diffraction from objects characterized by a length scale L, ($L \sim \ell_x$ or ℓ_y) such as a flow eddy of size L. The exact value of the constant may vary from case to case, but it always attains a value of order one.

B Fourier Transform Operations Performed by a Lens

Referring to Fig. 7, a monochromatic normally incident plane wave of unit amplitude impinges upon a converging lens with focal length f. Suppose an aperture of size L limits the spatial extent of the beam a distance f in front of the lens. In principle we could have chosen the object distance to assume any value, but the distance f simplifies the mathematics without obscuring any of the principal physics.

To compute the far-field, or Fraunhofer, pattern we use a cascaded approach. Since we do not know a priori that the Fraunhofer condition is satisfied, we use the Fresnel propagation formula. Subsequently, we show that in the back focal plane of the lens, the Fraunhofer, or far-field, diffraction pattern is found. First we propagate the field from plane I to plane II. At plane I the field amplitude, say the scalar electric field (for the moment we do not worry about the coupling between electric and magnetic fields), has unit amplitude

$$\begin{aligned} E = 1 \quad &\text{for} \quad x \leq \frac{L}{2} \quad \text{or} \quad T = 1 \\ &\phantom{\text{for}} \quad y \leq \frac{L}{2} \\ E = 0 \quad &\text{for} \quad x > \frac{L}{2} \quad \text{or} \quad T = 0 \\ &\phantom{\text{for}} \quad y > \frac{L}{2} \end{aligned} \tag{31}$$

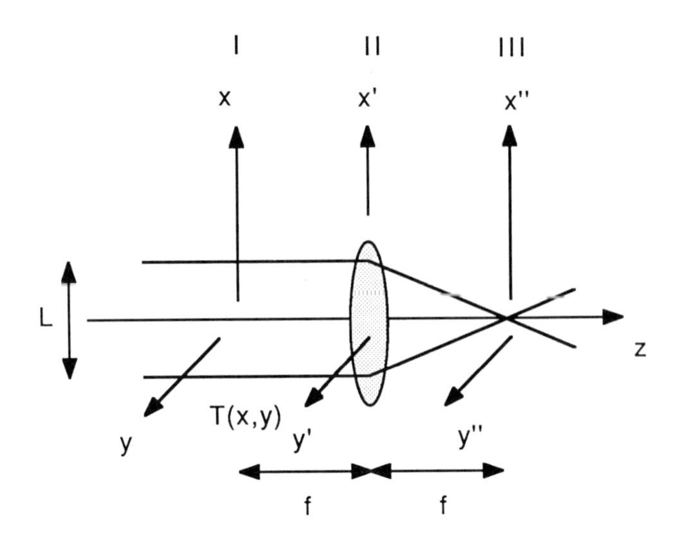

Fig. 7 Optical Fourier transforming geometry for a lens.

We describe the aperture by $T(x, y)$, referred to as the amplitude transmittance function.

Using Eq. (19) (the Fresnel approximation in the near field), the field distribution is calculated just in front (upstream) of the lens:

$$E_u(x', y') = \frac{\exp(ikf)}{i\lambda f} \exp\left[i\frac{k}{2f}(x'^2 + y'^2)\right] \iint_{-\infty}^{+\infty} T(x, y) \exp\left[i\frac{k}{2f}(x^2 + y^2)\right] \exp\left[-i\frac{2\pi}{\lambda f}(x'x + y'y)\right] dx\, dy \tag{32}$$

The propagation through the lens is described by a transmission function $T(x', y')$. Assume that the lens is infinitesimally thin and therefore the field distribution just behind the lens (downstream) can be calculated by multiplying $T(x, y)$ under the integral by $T(x', y')$ to give

$$E_d(x', y') = \frac{\exp(ikf)}{i\lambda f} \exp\left[i\frac{k}{2f}(x'^2 + y'^2)\right] \iint_{-\infty}^{+\infty} T(x, y)T(x', y') \exp\left[i\frac{k}{2f}(x^2 + y^2)\right] \exp\left[-i\frac{2\pi}{\lambda f}(x'x + y'y)\right] dx\, dy \tag{33}$$

Finally the propagation of the field from the lens to the back focal plane is calculated according to

$$E(x'', y'') = \frac{\exp(ikf)}{i\lambda f} \exp\left[i\frac{k}{2f}(x''^2 + y''^2)\right]$$

$$\iint E_d(x', y') \exp\left[i \frac{k}{2f}(x'^2 + y'^2)\right] \exp\left[-i \frac{2\pi}{\lambda f}(x'x'' + y'y'')\right] dx'\, dy' \quad (34)$$

The amplitude transmission function for the lens can be derived from geometrical consideration [5] and is given by:

$$T(x', y') = P(x', y') \exp\left[-i \frac{k}{2f}(x'^2 + y'^2)\right] \quad (35)$$

where $P(x', y')$ denotes the pupil function of the lens. In physical terms, this means that a positive lens converts a plane wave into a converging spherical wave, as is expected. Assume now that the lens diameter is large compared with L, and therefore $P(x', y') = 1$ in (35). Upon substitution of (35) into (34) the quadratic phase factor under the integral disappears and

$$E(x'', y'') = \frac{\exp(ikf)}{i\lambda f} \exp\left[i \frac{k}{2f}(x''^2 + y''^2)\right] F[E_d(x', y')]\Big|_{\left(\frac{x''}{\lambda f}, \frac{y''}{\lambda f}\right)} \quad (36)$$

where $F[E_d(x', y')]$ denotes the Fourier transform of $E_d(x', y')$ evaluated at frequencies $x''/\lambda f$, $y''/\lambda f$.

Equation (33) is of the convolution type, as indicated in (15). Upon taking the Fourier transform of both sides of Eq. (33) evaluated at frequencies $x''/\lambda f$, $y''/\lambda f$ we get

$$F[E_d(x', y')] = e^{ikf} F[T(x, y)] \exp\left[-i\pi\lambda f\,(f_x^2 + f_y^2)\right] \quad (37)$$

It is now evident that upon substitution of (37) into (36) the quadratic phase factor cancels:

$$E(x'', y'') = F[T(x, y)] = \frac{\exp(2ikf)}{i\lambda f} \iint_{-\infty}^{+\infty} T(x, y) \exp\left[-i \frac{2\pi}{\lambda f}(x''x + y''y)\right] dx\, dy \quad (38)$$

The phase factor in front of the integral represents the overall phase retardation that any spectral component of the signal undergoes as it propagates between the two planes that are separated by a distance $2f$. This factor is constant and independent of x'' and y''. Usually one is interested in the intensity or squared modulus of the amplitude, and this phase factor disappears.

Equation (38) shows the important result that for an object placed in the front focal plane of the lens the amplitude distribution at the back focal plane is given by its Fourier transform. The transform must be evaluated at frequencies $f_x = x''/\lambda f$, $f_y = y''/\lambda f$ to provide the proper scaling in the observation plane. In case the object is not located a distance f in front of the lens, we still obtain a Fourier transform of the field in the back focal plane of the lens, but multiplied by a spherical wave phase factor.

IV MODERN INTERPRETATION OF CLASSICAL INSTRUMENTS AND OPTICAL PROCESSORS

A Optical Processor

The generic optical processor configuration shown in Fig. 8 forms the basis for understanding modern optical processors such as the schlieren system. The system consists of two convex lenses that image a point light source from the focal plane of the first lens onto the back focal plane of the

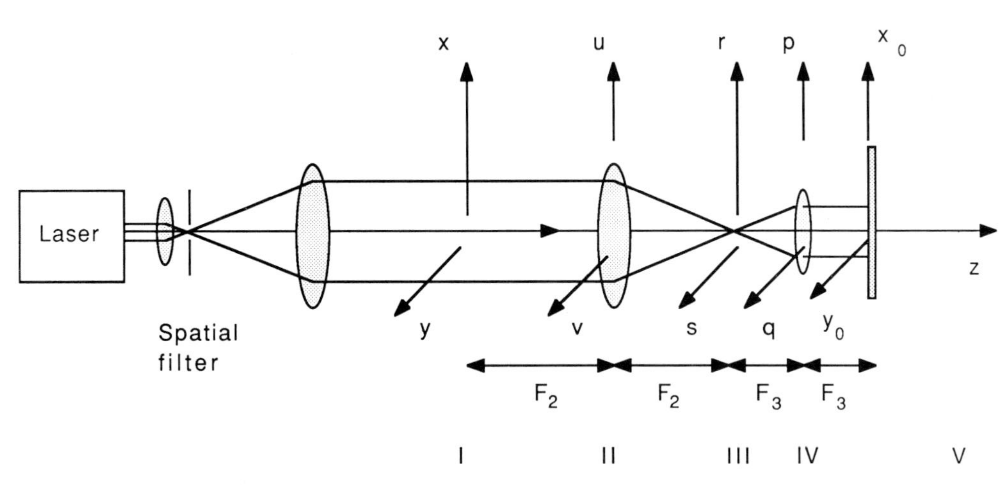

Fig. 8 Generic optical processor.

second one, referred to as the optical Fourier transform (OFT) plane. A third lens images the object onto film. For the purpose of describing the system, without sacrificing the essential physics in the problem, the discussion here is restricted to monochromatic sources. In principle, the results can be extended to polychromatic radiation by superimposing the various temporal frequency components, but the mathematics describing this configuration becomes considerably more involved.

Consider the geometry of Fig. 8 with a rectangular coordinate system (x, y), (u, v), (r, s), and (p, q) attached to the planes I, II, III, and IV, respectively. The object, located a distance F_2 in front of the lens, is illuminated by a normally incident plane wave of unit amplitude. The transmittance of the object is denoted by $T_I(x, y)$, where the harmonic phase factor $e^{-i\omega t}$ is omitted. In plane III, the electric field amplitude is computed using diffraction theory, Eq. (33), and is given by

$$E_{III}(r, s) = \frac{e^{2ikF_2}}{i\lambda F_2} \iint_{-\infty}^{+\infty} T_I(x, y)\, P\,(x + r, y + s) \exp\left[-i\frac{k}{F_2}(xr + ys)\right] dx\, dy \tag{39}$$

where P is the pupil function for lens 2, i.e.,

$$P\left(|x| > \frac{D_2}{2}, |y| > \frac{D_2}{2}\right) = 0$$

and

$$P\left(|x| \le \frac{D_2}{2}, |y| \le \frac{D_2}{2}\right) = 1$$

or for this square lens

$$P(x, y) = \text{rect}\left(\frac{x}{D_2}\right) \text{rect}\left(\frac{y}{D_2}\right)$$

Applying this same formula again to propagate the field distribution from plane III to plane V, $E_V(x_0, y_0)$ is found to be

$$E_V(x_0, y_0) = \frac{e^{2ikF_3}}{i\lambda F_3} \iint_{-\infty}^{+\infty} E_{III}(r, s)\, P\,(r + x_0, s + y_0) \exp\left[-i\frac{k}{F_3}(rx_0 + sy_0)\right] dr\, ds \tag{40}$$

where P is again the pupil function for lens 3. In this configuration the imaging condition is satisfied.

Combining Eqs. (39) and (40), $E_V(x_0, y_0)$ can be expressed in terms of $T_I(x, y)$ and geometrical parameters. Thus, we may compute the schlieren image of a flow by first determining the field $T_I(x, y)$ in terms of the overall phase and amplitude of the optical wave after it has propagated through the flow. Then the resulting field in plane III is computed using Eq. (39), subsequently filtered, and propagated to plane V using Eq. (40) with an appropriate modification to account for spatial filtering. This calculation is carried out in the following section.

B Schlieren System

The schlieren system is identical to that of the optical processor described in Sec. IV.A, except that the frequency content of the signal in the OFT plane is altered. This is accomplished, for instance, by inserting a knife blade into the focal spot or by blocking the low frequencies with a dot. A knife blade with the cutting edge oriented parallel to the s coordinate and passing through the origin changes the limits of integration in (40) as follows:

$$E_V(x_0, y_0) = \frac{e^{2ikF_3}}{i\lambda F_3} \int_0^{+\infty} \int_{-\infty}^{+\infty} E_{III}(r, s)\, P\,(r + x_0, s + y_0) \exp\left[-i\frac{k}{F_3}(rx_0 + sy_0)\right] dr\, ds \tag{41}$$

A dot with radius R centered around the origin modifies Eq. (40) to

$$E_V(x_0, y_0) = \frac{e^{2ikF_3}}{i\lambda F_3} \int_R^{+\infty} \int_{\sqrt{R^2 - s^2}}^{+\infty} E_{III}(r, s)\, P\,(r + x_0, s + y_0) \exp\left[-i\frac{k}{F_3}(rx_0 + sy_0)\right] dr\, ds \tag{42}$$

where $r^2 + s^2 \ge R^2$. Equations (39) and (41) are not altered as long as it is kept in mind that (39) is valid an infinitesimal distance upstream of the stops. Just behind these filters, Eq. (39) is applicable only for either $r > 0$, in case of the knife blade, or $r^2 + s^2 > R^2$, for the dot with radius R. The schlieren image can now be computed by substituting the appropriate expression for the field $E_{III}(r, s)$ and carrying out the integration. For the case of homogeneous and isotropic turbulence, Thompson [12] found a solution to this problem.

As a simple example consider an object given by

$$E_I(x, y) = \exp\,[i\Phi(x, y)] \tag{43}$$

where $\Phi(x, y)$ denotes the phase of the optical wave and $\Phi << 1$. In this example, the flow produces only a phase variation of the incident plane wave and no amplitude change. For a system with unity magnification the image intensity may then be approximated by

$$I_V(x_0, y_0) \approx \frac{1}{4}\left[1 - \frac{2}{\pi}\int_{-\infty}^{\infty} \frac{\Phi(\chi, y_0)}{x_0 - \chi}\, d\chi\right] \qquad (44)$$

Similarly, a shadowgraph system can be analyzed by computing the Fresnel diffraction pattern of the flow. As a general approach, we assume that the flow introduces a phase variation in the optical wave equal to the optical pathlength of the waves divided by the wavelength of light, i.e.,

$$\Phi(x, y) = \frac{2\pi}{\lambda}\int_0^D n(x, y, z)\, dl$$

where l denotes the pathlength through the medium of thickness D. The medium thus acts as a phase grating, and the diffraction pattern is computed by substituting $\Phi(x, y)$ into the Fresnel diffraction formula to obtain an explicit expression for (44).

C Optical Correlator for Turbulent Flow Measurements

Shadowgraph and schlieren pictures can also be used to obtain space-correlation functions, as first described by Uberoi and Kovasznay [13]. The method consists of making two identical transparencies of the original negative. The two copies are then placed back to face in an optical processor between the first two lenses. A photodetector is substituted for the camera and is located in the OFT plane, as shown in Fig. 8. The two transparencies are matched, and the transmitted intensity is measured as a function of the lateral displacement of one with respect to the other. Using the assumption of homogeneous and isotropic turbulence, the correlation function of the three-dimensional density field may be recovered.

The analysis for correlation functions obtained from schlieren pictures is given by Thompson [12].

Uberoi and Kovasznay [13] first proposed the technique for obtaining three-dimensional space-correlation functions of a density field using two-dimensional space-correlation functions acquired from shadowgraph pictures. The method principally relies on the assumption of a statistically homogeneous and isotropic flow. The correlation function measured in the optical correlator as described above is in general a peaked curve. The peak in the curve corresponds to perfect alignment of the two transparencies and is denoted by 1, the normalized correlation coefficient. The position of the origin, which signifies zero correlation, cannot be directly measured. It can be shown, however, that the first and third moments of the correlation function have to be zero for isotropic turbulence, when the light path through the medium is long compared with the integral scale

$$\int_0^{+\infty} \beta(\xi)\xi\, d\xi = \int_0^{+\infty} \beta(\xi)\xi^3\, d\xi = 0 \qquad (45)$$

where $\beta(\xi)$ is the measured picture correlation function. The correlation function for the flow can then be computed according to

$$R(r) = \frac{1}{8\pi D}\left(\frac{\rho_0}{kd}\right)\int_r^{+\infty} \beta(y)\left[y\,\frac{2r^2 + y^2}{r}\cos^{-1}\left|\frac{r}{y}\right| - 3y(y^2 - r^2)^{1/2}\right] dy \qquad (46)$$

where ρ_0 signifies the mean flow density, D denotes the thickness of the slab of turbulence, k is a constant (of order 1) associated with the photographic process, d is the distance between the density field and the photographic plate, and r is the displacement over which the correlation function is computed. In many cases though, the sole purpose of the correlation function is the determination of scales of turbulence (Hesselink [7]). For that purpose Eq. (46) does not have to be computed. The integral scale L_ρ, and the microscale, λ_ρ, can be expressed immediately in terms of $\beta(r)$ using Eq. (46):

$$L_\rho = \int_0^{+\infty} \frac{R(r)\, dr}{R(0)} = -\frac{\pi/8 \int_0^{+\infty} \beta(\xi)\xi^3 \log \xi\, d\xi}{\int_0^{+\infty} \beta(\xi)\xi^2\, d\xi} \qquad (47)$$

$$\lambda_\rho^2 = -2R(0)R''(0) = -\frac{6\int_0^{+\infty} \beta(\xi)\xi^2\, d\xi}{\int_0^{+\infty} \beta(\xi)\, d\xi} \qquad (48)$$

From (47) it can be seen that L_ρ depends principally on the behavior of $\beta(\xi)$ for large values of ξ; however, the correlation function is the least accurate there, mainly because the shadowgraph technique deemphasizes the low-wave-number contribution that determines L_ρ. Therefore, it is to be expected that using this method, the microscale is more accurately determined than the integral scale. A major difficulty arises with this approach when the density gradients in the flow cause rays to cross. In that case, the procedure outlined here is no longer valid, as discussed by Hesselink and White [8].

D Calculation of a Length Scale for a Density Field Using Optical Fourier Transform Photography

A rather simple method for measuring the integral scale of a turbulent medium with a known form for the correlation function involves photographing the spectrum of the light in the back focal plane of the lens.

The equations, presented in Sec. II to describe the properties of an optical processor, are based on diffraction theory. Using Eq. (19) as a starting point, the effective radius of the focal spot in a system with a slab of turbulence present between the two lenses may be computed. This calculation is lengthy. In many instances the wavelength of light is small compared with the smallest scale in the turbulence, $\lambda << \lambda_\rho$, and also $(\lambda D)^{1/2} << \lambda_\rho$, where D is the thickness of the slab. In that case geometrical optics may be used to obtain approximate results. The calculation then becomes relatively simple and will be presented here. Consider the optical arrangement of Fig. 7 where the aperture is replaced by a slab of turbulence of thickness D. The scattering medium is characterized by a fluctuating index of refraction $n = 1 + \mu$, where $\mu << 1$, and a space-correlation function N. The mean angle of deviation of a ray from its direction is denoted by $\bar{\theta}$.

$$\bar{\theta} = 4K_1 D \tag{49}$$

where K_1 is a constant (Chernov [2]). In the case of isotropic scattering, K_1 takes on the value

$$K_1 = -\frac{1}{2}\overline{\mu^2}\int_0^{+\infty}\frac{1}{2}\frac{\partial}{\partial r}\left(r^2\frac{\partial N}{\partial r}\right)dr \tag{50}$$

The mean radius of the focal spot in the presence of turbulence R_{turb} can now be computed simply using

$$\bar{\delta} = \bar{\theta}\cdot\bar{F} \tag{51}$$

Thus by measuring $\bar{\delta} = R_{\text{turb}} - R_{\text{noturb}}$, the effect of the scattering medium on the impulse response of the optical system can be ascertained. The length scale L_ρ, for instance, may be computed by assuming a correlation function $N = A\exp(-r^2/L_\rho)$ and using Eqs. (49), (50), and (51).

V PRINCIPLES OF SPECKLE VELOCIMETRY

Referring to Fig. 9, a particle-laden flow is illuminated with a sheet of laser light. Scattered radiation out of the illumination plane is recorded by a camera, which images the laser sheet onto a suitable photo-sensitive medium. In practice, high-resolution film is most widely used, but recently Collicott and Hesselink [4] have demonstrated good recording with photorefractives, allowing real-time velocimetry. Solid-state recording devices may also be used (at least in principle) but the high spatial resolution and bandwidth required for speckle recording currently have limited practical use in flow applications. Electronic speckle recording, however, has been used successfully in solid mechanics applications that require less spatial resolution. Moreover, recent advances in solid-state cameras hold promise for applications in laser speckle velocimetry too.

A speckle pattern results when the scattering particles are much smaller than the resolution cell of the imaging system and there are many particles in each cell. According to Huygens' principle, all scatterers within that small region

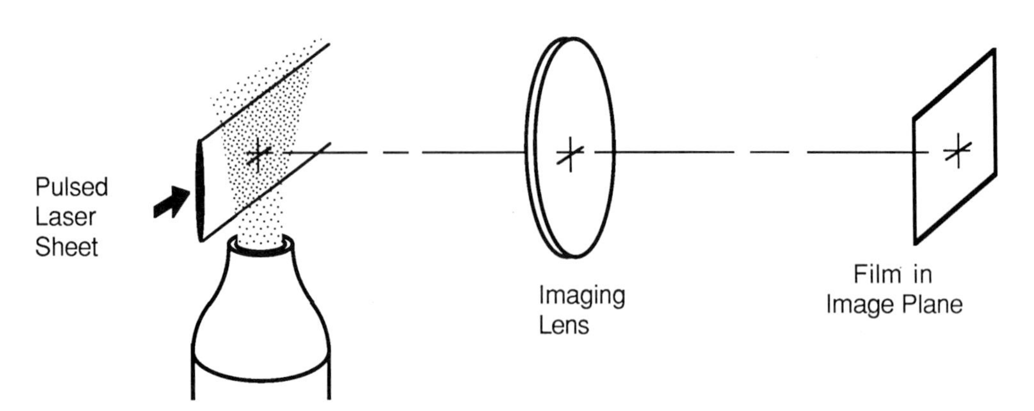

Fig. 9 Optical configuration for laser speckle velocimetry.

emit secondary spherical wavelets, which are coherently superimposed on the film at the image point. The superposition of all these waves gives rise to a net amplitude and phase of the optical wave. The value of the phase and amplitude depends on the laser wavelength and the geometry of the imaging system, and varies randomly from point to point in the image, referred to as a speckle pattern. Mathematically, we may describe this imaging system as follows:

Let $f(x, y)$ denote the field distribution of the scattered light from the particle-laden flow at the location of the illuminating plane. This field distribution is then imaged onto a recording medium at two instances in time. The recorded patterns are correlated and translated according to the local fluid motion. Assume that the first speckle pattern is denoted by $s_1(\mathbf{r})$ and the second speckle pattern by $s_2(\mathbf{r})$. The position in the image plane is $\mathbf{r}$, and $\Delta\mathbf{r}$ denotes the shift in position caused by the velocity vector $\mathbf{v}$ in the object space that acts on the particles during a time Δt between exposures. The imaging magnification is M, and

$$\Delta\mathbf{r} = \mathbf{v}\,\Delta t M \tag{52}$$

In general $\mathbf{r}$ and $\Delta\mathbf{r}$ are functions of position and time. We assume, however, that the exposure time is short enough so that the fluid motion is frozen during exposure. Assume now that $\Delta\mathbf{r} = \delta x\mathbf{l} + \delta y\mathbf{m}$, where $\mathbf{l}$ and $\mathbf{m}$ are unit vectors in the x and y direction, respectively. Then upon recording and development of the film we have a transmission function $T(x, y)$ for the double-exposure speckle pattern, which is represented by

$$T(x, y) = A(x, y) + B(x, y)s_1(\mathbf{r}) + C(x, y)s_2(\mathbf{r} - \Delta\mathbf{r}) \tag{53}$$

where the constants A, B, and C depend on the exposure characteristics.

The local velocity vectors may be obtained from optical analysis of these speckle patterns, using three distinct approaches.

A Optical Processor for Measuring 2-D Velocity Vectors

In Fig. 10 the two superimposed speckle patterns recorded on the same medium (usually film) are illuminated (after development) with a small pencil of light from a probe laser. The transmitted probe light is collected in the back focal plane of a positive lens where the Fourier transform of the double-exposed pattern is recorded. We assume that the intensity of the probe laser is uniformly equal to 1 and its beam cross section is small enough so that the two patterns are highly correlated, but displaced by the fluid motion. The field distribution in the back focal plane is then equal to the Fourier transform of the two patterns s_1 and s_2. Since we can record only intensities we have to multiply this expression by its complex conjugate. The result is a fringe pattern superimposed on a noisy background.

$$I(x', y') = A(x', y') + B(x', y')\cos^2\left(\frac{\pi}{\lambda f}M(\delta x x' + \delta y y')\right) \tag{54}$$

The bias term $A(x', y')$ is referred to as the pedestal and $B(x', y')$ denotes the diffraction halo. The cosinusoidal modulation gives rise to fringes superimposed on a noisy diffraction halo. The fringes carry the velocity information, because the spacing is inversely proportional to the local velocity magnitude and the orientation is perpendicular to the local velocity vector at the point of interrogation by the laser beam

$$|\mathbf{r}| = \frac{\lambda f}{\Lambda M\,\Delta t} \tag{55}$$

where Λ denotes the fringe spacing. By scanning the probe beam in a raster fashion, the complete velocity field may be measured. The fringe pattern for each probe beam location is usually detected by a CCD camera coupled to a computer, and the pattern may be analyzed using a variety

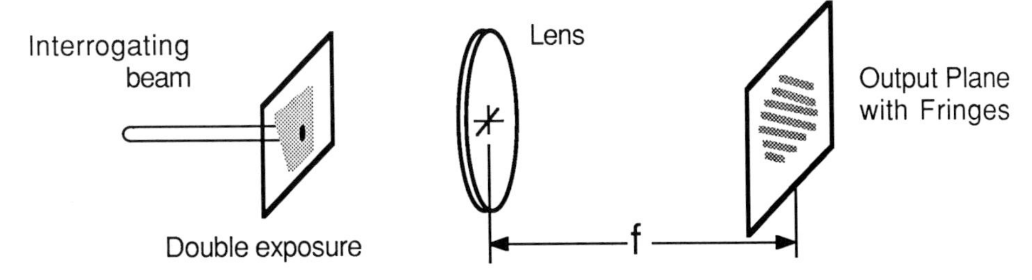

Fig. 10 Optical processor for measuring local velocity vector.

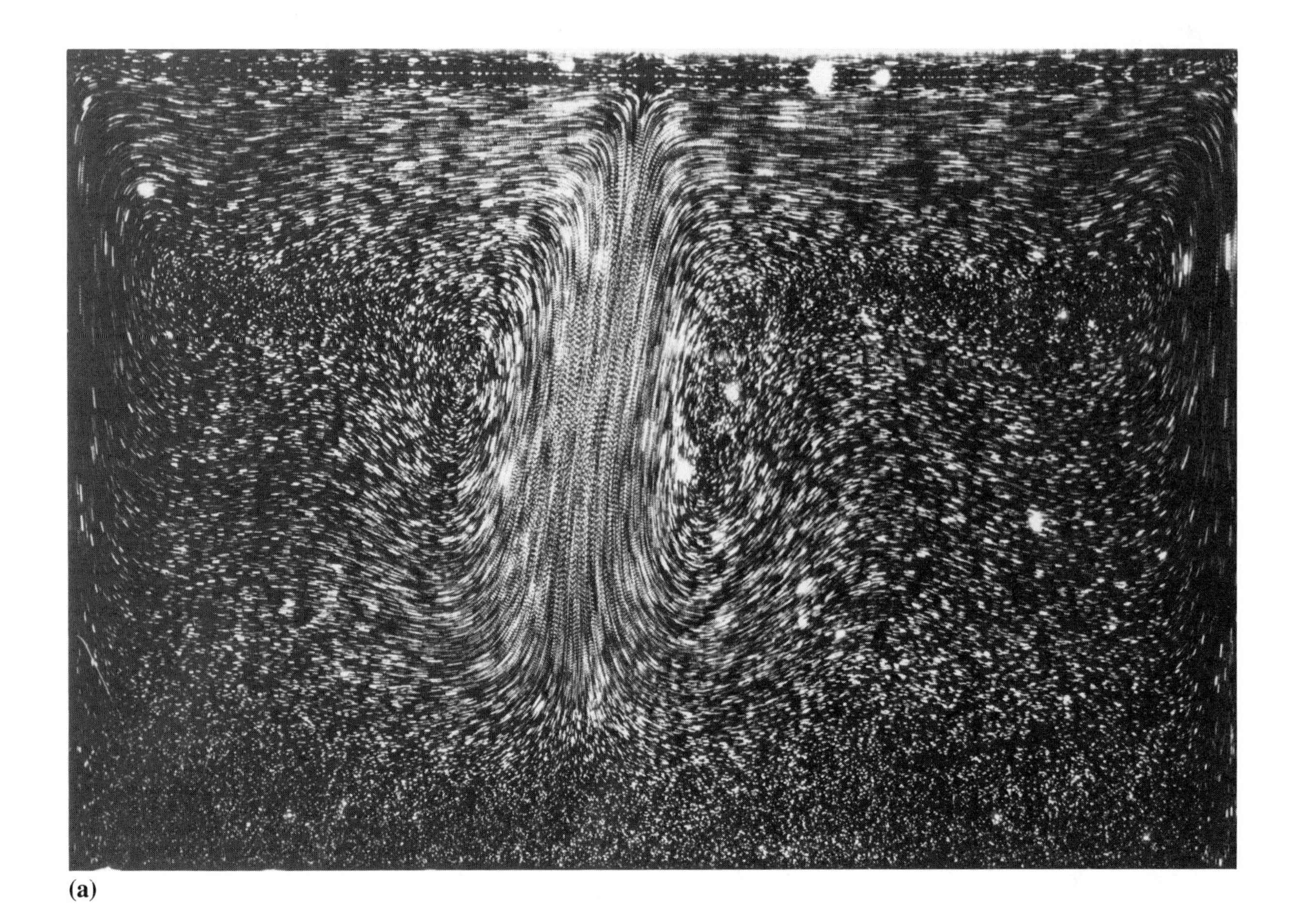

(a)

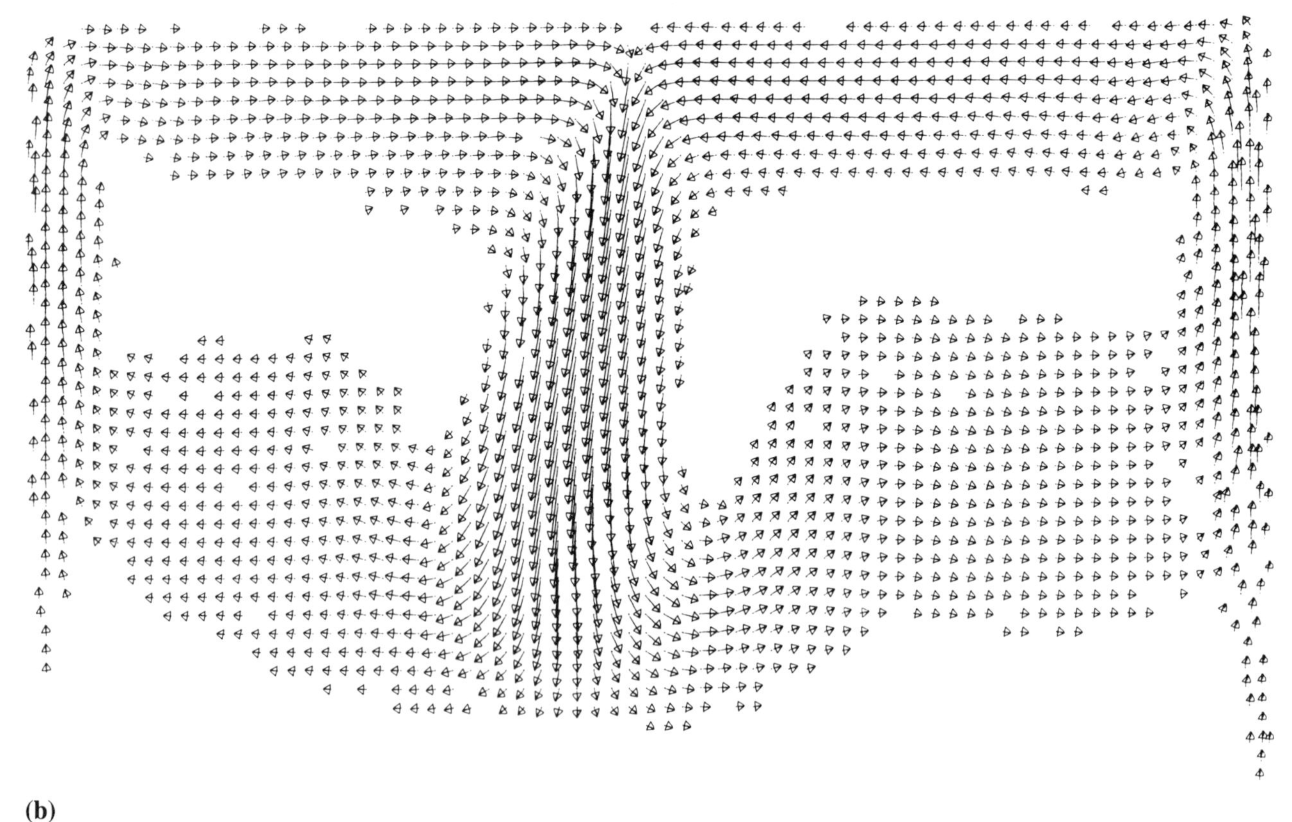

(b)

Fig. 11 Example of a velocity field measured with laser speckle velocimetry using the optical processor shown in Fig. 10 (*b*) the corresponding velocity-vector field (courtesy of Meynart [9], with permission from Cambridge University Press).

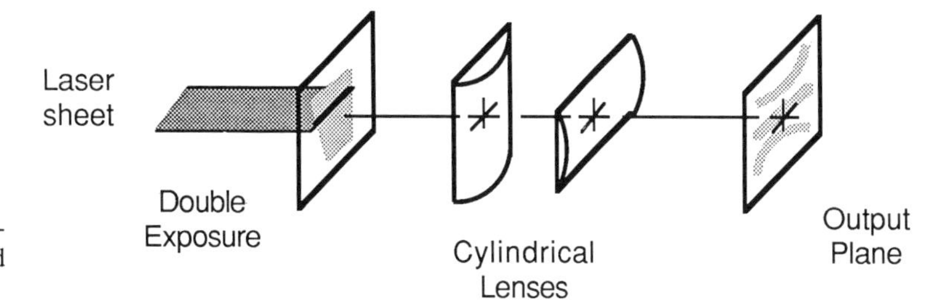

Fig. 12 Anamorphic optical processor for obtaining the velocity profile from a double-exposed speckle pattern.

of fringe readout techniques (Hesselink [6]). This approach has been successfully applied to Rayleigh–Benard convection (Meynart et al. [9]) and vortical flow fields (Smith et al. [11]). An example of the quality of results obtainable with this technique is shown in Fig. 11.

B Anamorphic Optical Processor for Measuring 1-D Velocity Profiles

A one-dimensional velocity profile may be measured from a double-exposed speckle pattern by combining a Fourier transforming and imaging operation in the optical processor. Referring to Fig. 12, the double-exposed speckle pattern is illuminated with a light sheet, and a combination of cylindrical lenses oriented horizontally and vertically simultaneously perform a 1-D imaging and 1-D Fourier transforming operation.

The expression for the intensity pattern in the output plane $I_3(u, y_3)$ is now

$$I_3(u, y_3) = A\,\delta(u, y_3) + B\,|\tilde{S}(u, y_0) + \tilde{S}[u, y_0 + \delta_y(y_0)]|^2 \cos^2[\pi\,\delta_x(y_0)u] \tag{56}$$

where $y_0 = -y_3/M_y$, A and B are real constants, and $\tilde{S}(u, y_0) = F_x[S(x_0, y_0)]$ and $\tilde{S}[u, y_0 + \delta_y(y_0)]$ are essentially speckle patterns [3]; F_x denotes the one-dimensional Fourier transform. The first term on the right-hand side of (56) is simply a zero-order line, and the second term contains the x-velocity data in the argument of the squared cosine. Fringes result from the squared cosine multiplying $|\tilde{S}|^2$, and fringe spacing in the x_3 direction varies as a function of y_3 because δ_x depends on y_3. Note that δ_x and not δ_y determines the fringe spacing and that δ_y merely shifts one of the background patterns.

After measuring the period of the fringes, $\Lambda(y_3)$, the x-velocity component as a function of y_0 can be computed as:

$$V_x(y_0) = \frac{\lambda f_x}{\Lambda(y_3)\,\Delta t\,M_{\text{im}}} \qquad y_3 = -y_0 M_y \tag{57}$$

where Δt is the time between exposures, M_{im} is the imaging magnification of the speckle recording optics, and λ is the wavelength of the laser used in the optical processor.

This technique has been applied to measurement of the velocity profile in a particle-laden jet. The fringe pattern no longer consists of straight lines, but is generally curved, as shown in Fig. 13. From this pattern the velocity profile

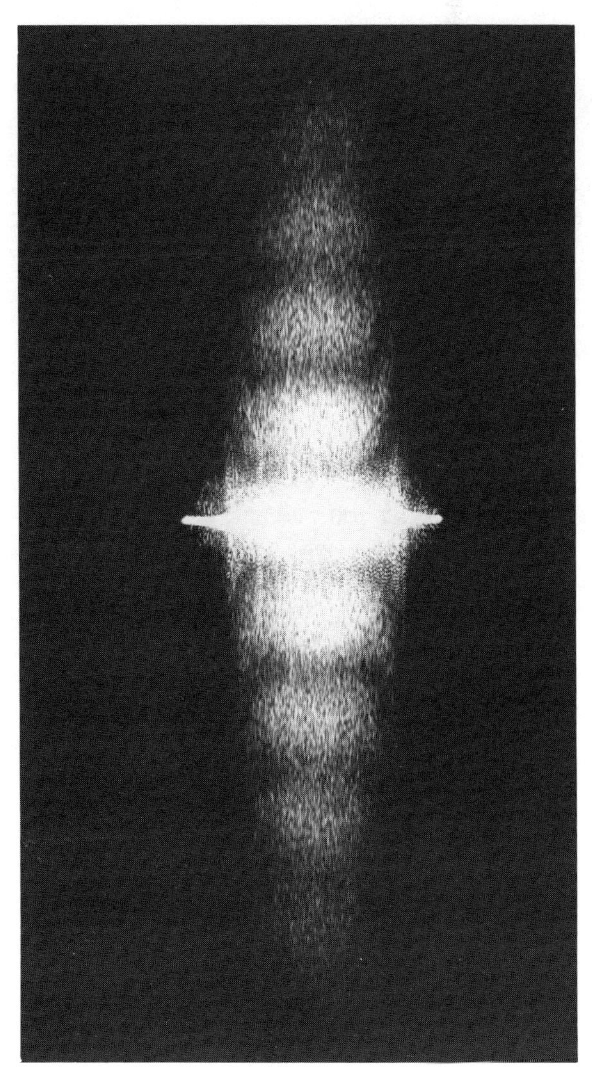

Fig. 13 Example of a fringe pattern obtained with the anamorphic optical processor shown in Fig. 12.

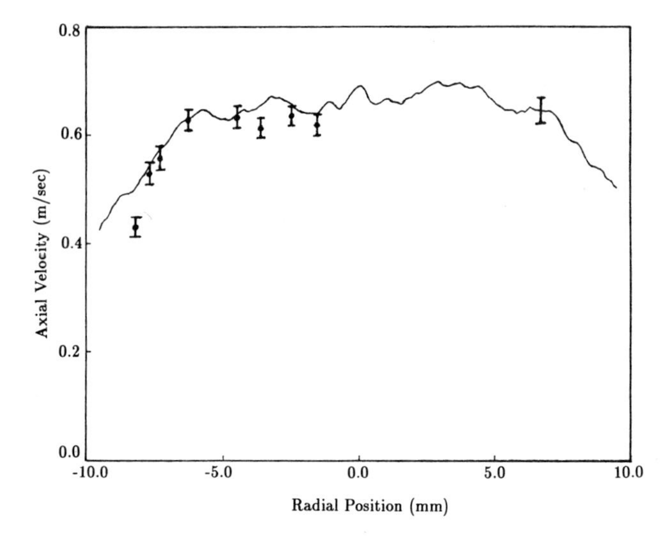

Fig. 14 Velocity profile of a gaseous jet obtained with the anamorphic optical processor shown in Fig. 12 using the fringe pattern shown in Fig. 13.

is computed using Eq. (57), and the results are shown in Fig. 14. The complete velocity profile for all y_0 stations may now be obtained by a one-dimensional scan of the laser sheet in the y_0 direction.

C Fourier Filtering Processor for Measurement of Isovelocity Contours

Isovelocity contours may be measured by Fourier filtering of multiple-exposed speckle patterns. Though, in principle, double-exposed speckle patterns could be analyzed by spatial filtering, the signal-to-noise ratio is usually too small to provide meaningful results. By superimposing several images, the fringe width is reduced and the S/N ratio is enhanced.

Referring to Fig. 15, and using a cylindrical coordinate system ρ, θ, we find that the output image in plane II is the convolution of the object field distribution and the impulse response of the system, or

$$\begin{aligned} g(r_i) &= \iint T(\mathbf{r}_0) \cdot P_{\text{filter}}(\mathbf{r}_i + \mathbf{r}_0)\, d\mathbf{r}_0 \\ &= \iint P_{\text{filter}}(\mathbf{r}_0) \cdot T(\mathbf{r}_0 - \mathbf{r}_i)\, d\mathbf{r}_0 \end{aligned} \tag{58}$$

where $\mathbf{r}_0$ and $\mathbf{r}_i$ denote the radial coordinates in the object and image plane, respectively, $P_{\text{filter}}(\mathbf{r}_0)$ is the impulse response of the filter, $T(\mathbf{r}_0)$ denotes the transmission function of the speckle pattern, and $g(\mathbf{r}_i)$ represents the image field distribution.

If we consider now the object to be a superposition of N speckle patterns, then

$$T(\mathbf{r}_0) = \sum_{p=0}^{N-1} s(\mathbf{r}_0 - p\,\delta\mathbf{r}) \tag{59}$$

where $\delta\mathbf{r}$ denotes the displacement between exposures. For a circular aperture of diameter d, with its center a distance $\mathbf{D}$ from the optical axis, the impulse response is

$$D_{\text{filter}}(\mathbf{r}_0) = \exp\left(-2\pi i \frac{\mathbf{r}_0 \cdot \mathbf{D}}{\lambda f}\right) \frac{J_1(\pi r_0\, d/\lambda f)}{\pi r_0\, d/\lambda f} \tag{60}$$

and J_1 denotes the Bessel function of the first kind.

Thus we find for $g(r_i)$, upon substitution of (59) and (60) into (58), assuming that the total displacement $N\,\delta\mathbf{r}$ is smaller than the impulse response of the system $\lambda f/d$,

$$\begin{aligned} g(r_i) = \sum_{p=0}^{N-1} \exp\left[-2\pi i p \frac{D \cdot \delta\mathbf{r}}{\lambda f}\right] \iint s(\mathbf{r}_0 - \mathbf{r}_i) \\ \exp\left[-i2\pi \frac{\mathbf{r}_0 \cdot \mathbf{D}}{\lambda f}\right] \frac{J_1(\pi r_0\, d/\lambda f)}{\pi r_0\, d/\lambda f}\, d\mathbf{r}_0 \end{aligned} \tag{61}$$

This expression denotes the product of the image of the object multiplied by the fringe pattern of the first term. The fringe pattern attains a maximum whenever

$$\frac{\delta\mathbf{r} \cdot \mathbf{D}}{\lambda f} = q \tag{62}$$

where $q = 1, 2, 3, \ldots$.

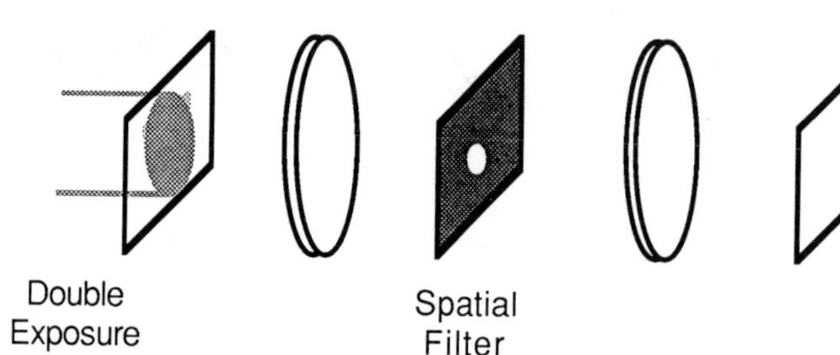

Fig. 15 Optical processor using spatial filtering to obtain isovelocity contours.

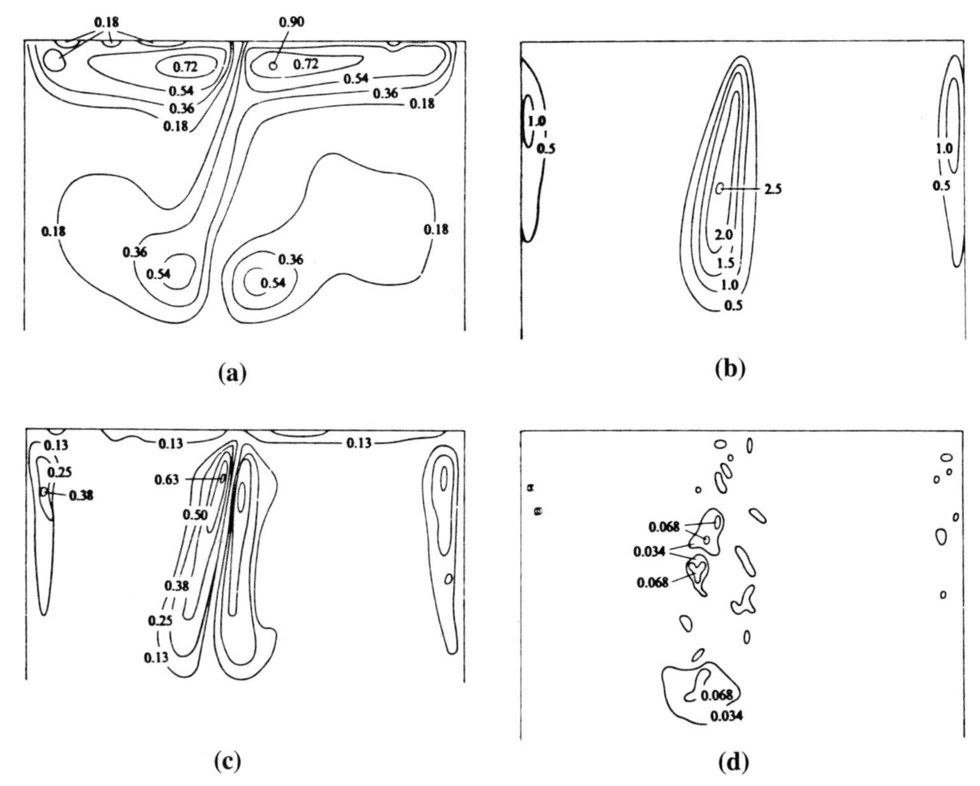

Fig. 16 Isovelocity contours in a Rayleigh–Bénard cell obtained with the optical processor of Figure 15 (*a*) Horizontal-velocity-component distribution; $u_{max} \approx 0.9$ mm s^{-1}. (*b*) Vertical-velocity-component distribution; $v_{max} \approx 2.5$ mm s^{-1}. (*c*) Vorticity distribution; $\zeta_{max} \approx 0.63$ s^{-1}. (*d*) Estimates for the gradient of the out-of-plane component. (courtesy of Meynart [9] with permission from Cambridge University Press).

Thus to apply this technique to a fluid flow application, the impulse response of the filter must be made larger than the total displacement of the particles during the superposition of N speckle patterns. By making the impulse response wider (or the aperture smaller), the fringes also become less well defined and tend to smear. This is the reason why it is difficult to obtain good fringes in a region of high velocity, which implies large particle displacements. There is thus a tradeoff between the number of speckle patterns that are superimposed (we like to have as many superpositions as possible to increase fringe contrast) and the smallest diameter we can make the filter, which tends to smear the fringes. Typical values for N are on the order of 6 [9].

Excellent results have been obtained by Meynart [9, 10] in convective flow systems. An example of results obtainable with this technique is shown in Fig. 16.

We conclude now by noting that the three methods described here for readout of superimposed speckle patterns are complementary. The two-dimensional scanning method is the most elaborate, but it provides the full two-dimensional velocity field. The anamorphic processor provides the one-dimensional velocity profile without the need for scanning, and allows the full field to be analyzed by one-dimensional scanning of the laser sheet. The Fourier filtering technique requires that several speckle patterns be superimposed, but the isovelocity contours are obtained without any scanning of laser beams. The directionality of the velocity measurements, however, has been lost.

REFERENCES

1. Baker, P. L., Ninio, F., Troup, G. J., and Bryant, R., Speckle Pattern Photographic Measurements of Protoplasmic Streaming Velocities, *Atti Fond. Giorgio Ronchi,* vol. 36, 1981.
2. Chernov, L. A., *Wave Propagation in a Random Medium.* McGraw-Hill, New York, 1960.
3. Collicott, S., Hesselink, L., Anamorphic Optical Processing of Multiple-Exposure Speckle Photographs, *Opt. Lett.,* vol. 11, pp. 410–412, 1986.
4. Collicott, S., and Hesselink, L., Real Time Speckle Velocimetry with Recording in Photorefractive Crystals, Conference AIAA-81-1376, June 6–10, Honolulu, Hawaii, 1987.
5. Goodman, J. W., *Introduction to Fourier Optics,* McGraw-Hill, New York, 1968.
6. Hesselink, L., *Digital Image Processing in Flow Visualization,* Annual Reviews, Palo Alto, Calif., 1988.

7. Hesselink, L., Propagation of Weak Shocks through a Random Medium, *J. Fluid Mech.*, vol. 196, pp. 513–553, 1988.
8. Hesselink, L., and White, B. S., Digital Image Processing of Flow Visualization Photographs, *Appl. Opt.*, vol. 22, no. 10, pp. 1454–1461, 1983.
9. Meynart, R., Simpkins, P. G., and Dudderar, T. D., Speckle Measurements of Convection in a Liquid Cooled from Above. *J. Fluid Mech.*, vol. 182, pp. 235–254, 1987.
10. Meynart, R., Instantaneous Velocity Field Measurements in Unsteady Gas Flow by Speckle Velocimetry, *Appl. Opt.*, vol. 22, no. 4, pp. 535–540, 1983.
11. Smith, C. A., Lourenco, L. M. M., and Krothapalli, A., The Development of Laser Speckle Velocimetry for the Measurement of Vortical Flow Fields, *AIAA 14th Aerodyn. Test. Conf.*, West Palm Beach, Fla., 1986.
12. Thompson, L. L., and L. S. Taylor, Analysis of Turbulence by Schlieren Photography, *AIAA*, vol. 7, no. 10, pp. 2030–2031, 1969.
13. Uberoi, M. S., and Kovasznay, L. S. G., Analysis of Turbulent Density Fluctuations by the Shadow Method, *J. Appl. Phys.*, vol. 26, pp. 19–24, 1955.

Chapter 18

Ultrasonic Image Processing

Michiyoshi Kuwahara, Shigeru Eiho, and Kunihiro Chihara

I INTRODUCTION

Ultrasound cardiography (UCG), or echocardiography, has been widely used as a noninvasive, safe, and reliable diagnostic tool in clinical medicine. A part of the ultrasound (over 2 MHz) pulses produced by a piezoelectric transducer is reflected at several interfaces within the heart and returned to the transducer, where the reflected echoes are transformed into electrical impulses.

An M-mode (time–motion display) echocardiogram presents the time course of the changing position of the echo due to the motion of tissues [1–4]. By changing the direction of ultrasound beam, we can get a 2-D cross-sectional shape of organs. The sector-scan method is used for the heart. By using digital image-processing methods for these echo data, we can evaluate various kinds of cardiac functions quantitatively [5–13].

Blood-flow information is also obtainable by analyzing Doppler signal in the frequency domain. With the rapid growth of VLSI (very large scale integrated circuits), the use of real-time flow-mapping devices to observe an intracardiac blood-flow velocity profile on an echocardiogram is spreading over clinical fields [14–20].

II AUTOMATIC ANALYSIS OF M-MODE ECHOCARDIOGRAMS

By clarifying the internal structure of the heart through UCG signals of an M-mode display, we can obtain various numerical values relating to the left ventricle, e.g., the length of the short axis (diameter), the thickness of the posterior wall, the velocity of circumferential fiber shortening, and the left ventricular volume [2]. Figure 1 shows an example of an original digitized M-mode echocardiogram. In this figure, one can see the right ventricle, the interventricular septum, the left ventricle, the endocardium and the epicardium of the posterior wall, and the lung tissue. The curve on the right side of this figure shows the A-mode (amplitude) display of the echo signal on the last

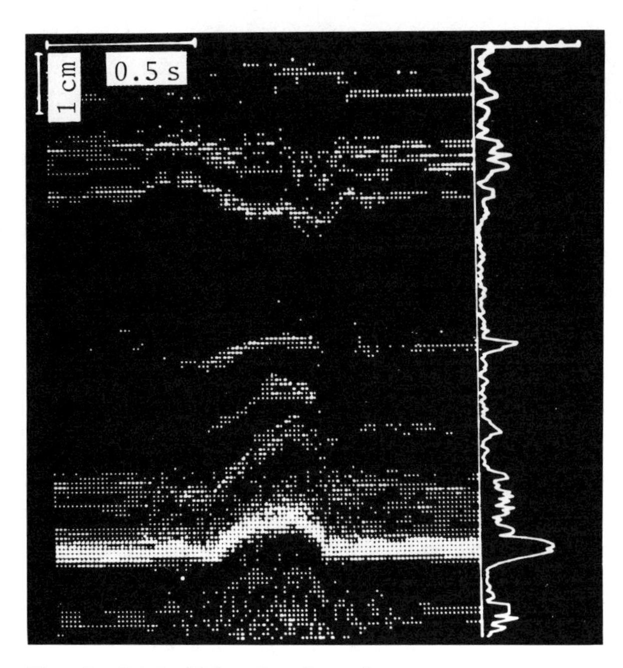

Fig. 1 Original M-mode echocardiogram.

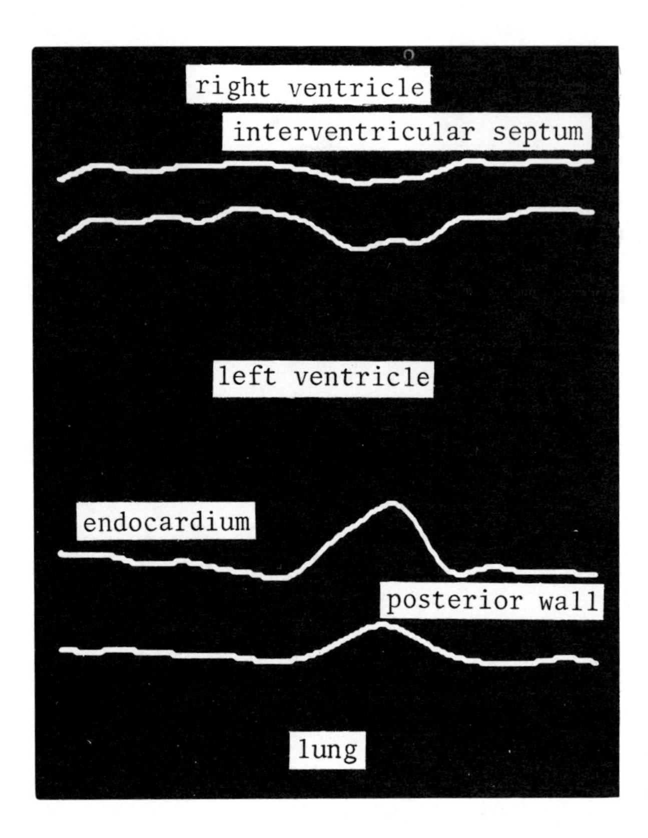

Fig. 2 Detected cardiac structures on the M-mode echocardiogram.

line of the M-mode echocardiogram. The M-mode display gives the depth of each tissue from the chest wall as a function of elapsed time.

The epicardium of the posterior wall of the left ventricle usually gives the highest value among echo signals from all tissue boundaries, and it is detectable automatically. The other edges are detected by reference to this epicardial edge. Figure 2 shows these edges detected: The upper two lines show the edges of the interventricular septum, and the lower two show the movement of endocardium and epicardium of the posterior wall. Figure 3 shows the detected edges superimposed on the original M-mode echocardiogram. From these detected lines we can obtain various information on cardiac functions as shown in Fig. 4: (a) the diameter of the left ventricle (DIM) and the posterior wall thickness (PW), (b) the volume change of the left ventricle, and (c) the velocity of the circumferential fiber shortening.

III AUTOMATIC ANALYSIS OF 2-D ECHOCARDIOGRAMS

Two-dimensional echocardiograms, transferred directly from ultrasound diagnostic equipment or through a videotape recorder to a minicomputer system, are processed digitally, and certain cardiac functions can be obtained from them.

It is most important to detect left ventricular boundaries in a cardiac cycle. Many algorithms to detect the left ventricular boundary on 2-D echocardiograms have been designed, but it is too difficult to detect the left ventricular boundary with clinically applicable quality by fully automated methods. Operator's guide is needed for stable detection of the left ventricle such as fixing some points of

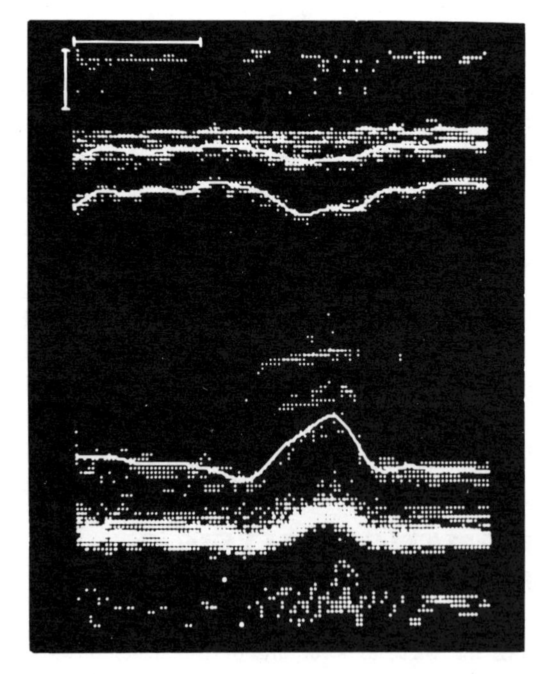

Fig. 3 Detected cardiac structures superimposed on the original M-mode echocardiogram.

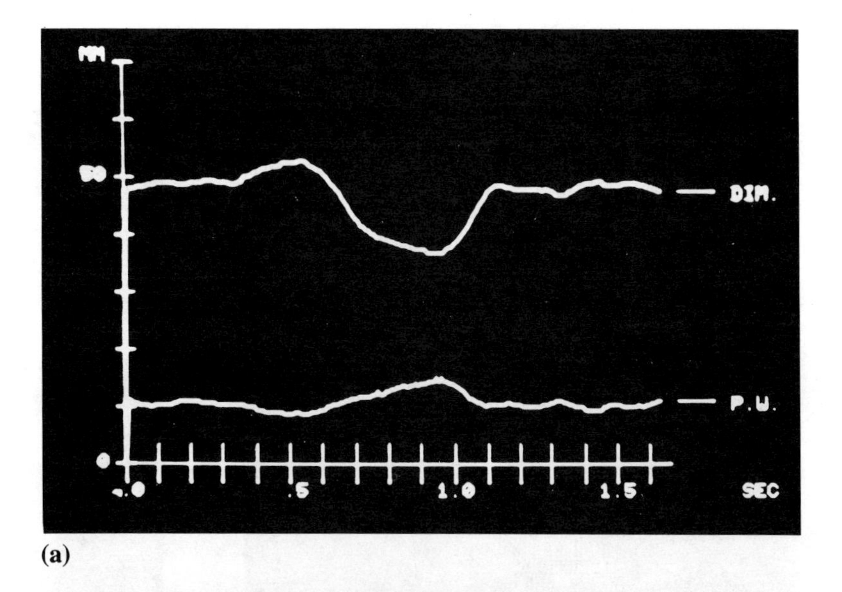

(a)

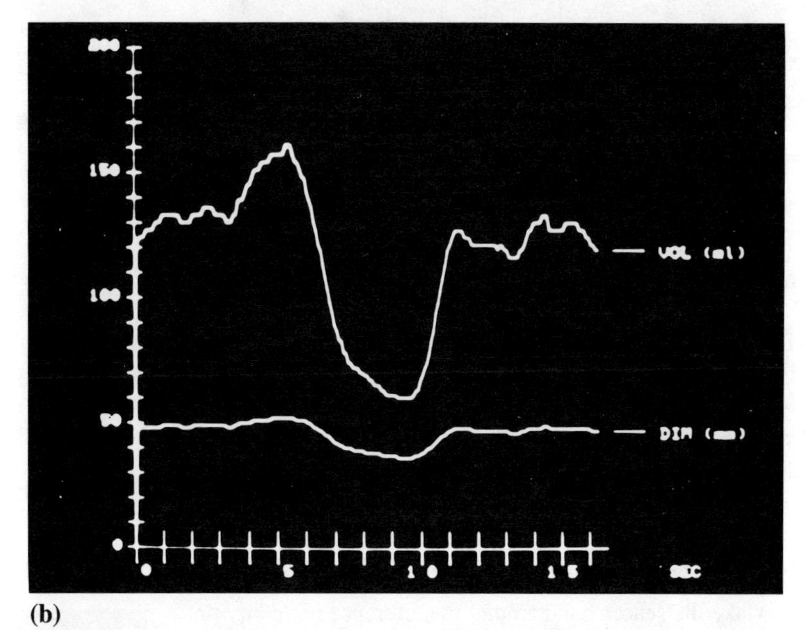

(b)

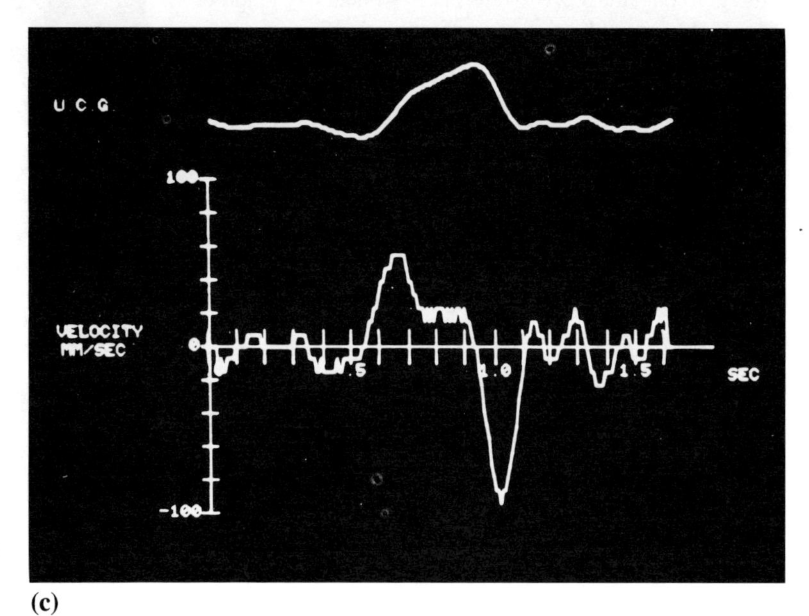

(c)

Fig. 4 Display of some cardiac functions: (*a*) Diameter of the left ventricle; (*b*) volume; (*c*) velocity of the circumferential fiber shortening.

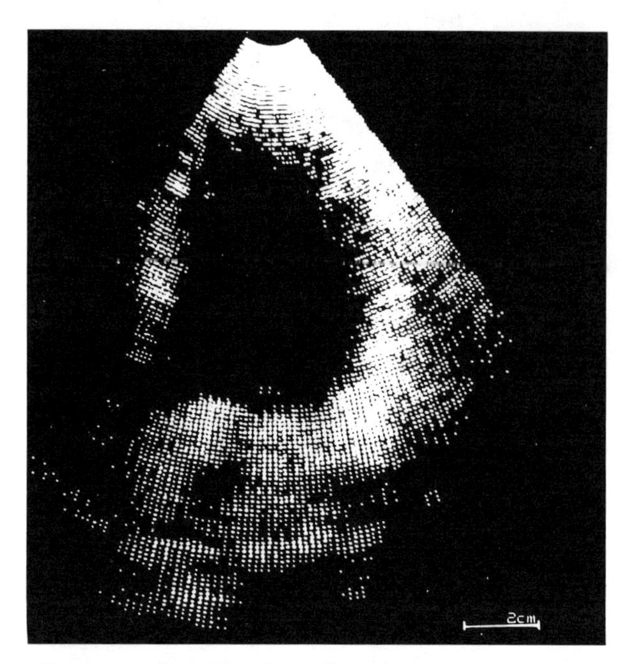

Fig. 5 Original 2-D echocardiogram.

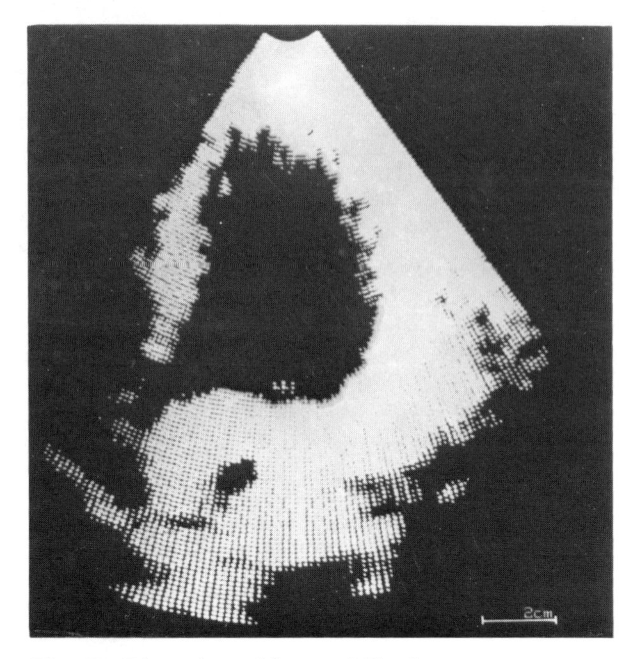

Fig. 7 Edge-enhanced image of Fig. 5.

the left ventricular boundary on the image [10, 11]. Figure 5 shows a digitized echocardiogram at the end diastolic (ED) phase. The operator gives several boundary points on this image by using a pointing device such as a cursor, a lightpen, or a joystick. We can enhance the edge by expanding the dynamic range of gray level near the boundary, shown as Fig. 6. Figure 5 is changed to Fig. 7 by this gray-level transformation, and Fig. 8 is the gradient image of Fig. 5. The left ventricular boundaries are automatically traced frame by frame as detailed in the following steps:

1. Make the search band around the reference boundary on the gradient image. The reference boundary is made of points indicated by the cursor for the first image and is made of the boundary curve just obtained on the previous frame.
2. Trace the points that have the local maximum value in the band. Points for the start and the goal and each direction for tracing are calculated from the reference boundary.
3. Smooth the boundary thus obtained and fix some points for cardiac structure, such as the apex and the aortic valve.

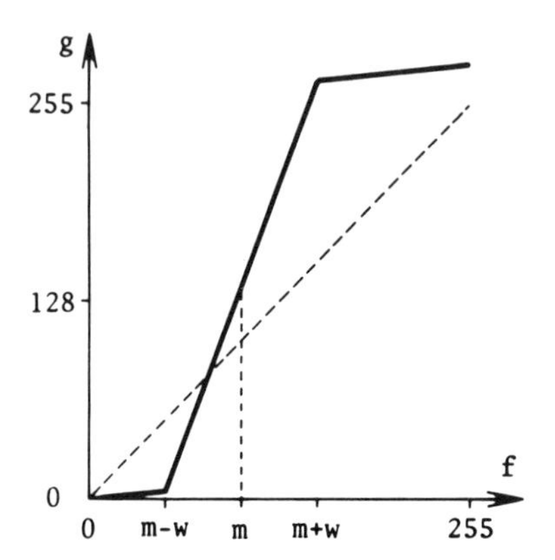

Fig. 6 Gray-level expansion near the boundary.

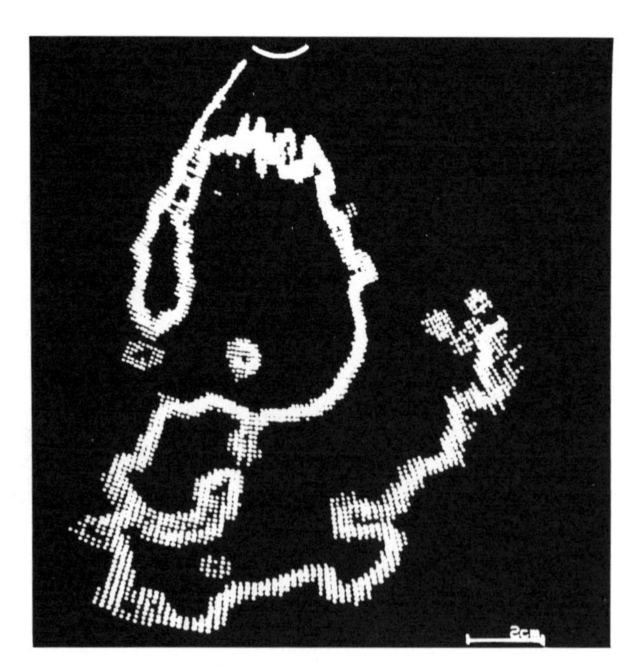

Fig. 8 Gradient image of Fig. 7.

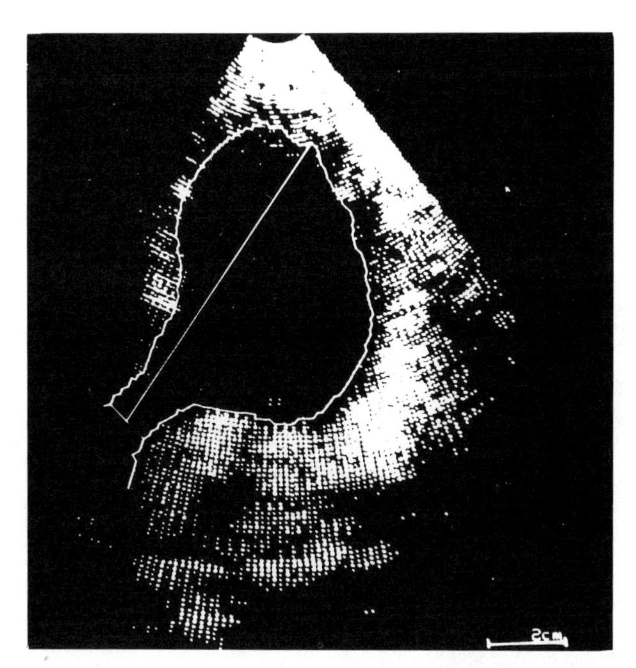

Fig. 9 Detected left ventricular boundary superimposed on the original 2-D echocardiogram.

Fig. 10 Detected boundaries in a cardiac cycle.

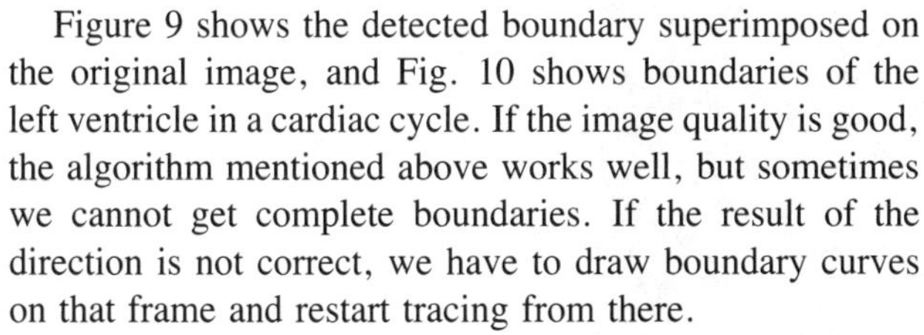

Figure 9 shows the detected boundary superimposed on the original image, and Fig. 10 shows boundaries of the left ventricle in a cardiac cycle. If the image quality is good, the algorithm mentioned above works well, but sometimes we cannot get complete boundaries. If the result of the direction is not correct, we have to draw boundary curves on that frame and restart tracing from there.

To reduce the possibility of erroneous boundary detection, we use our knowledge about the left ventricular shape and motion: The boundary curves can be assumed to be closed and to have a periodical property such as the cardiac cycle [12]. Thus a modified algorithm is shown as follows:

1. Pick up about 10 points of the left ventricular boundary on a digitized image at the first ED phase. These points are connected by spline curves, and the center of gravity of the closed curve thus obtained is fixed. The closed curve is used as the principal reference boundary in the following boundary detection.
2. Every echocardiogram in the sector format of the following frames is radially resampled at 64 points on each line of 64 directions from the center of gravity of the reference boundary.
3. The boundary point on each radial line is automatically obtained, mainly by a thresholding method for endocardial boundary and a maximum-value detecting method for the epicardium. The range for each search is restricted within 5 points both sides of the left ventricular boundaries on the previous frame.
4. The boundary points are smoothed both spatially and temporally by using present, previous, and principal reference boundary points.
5. Repeat step 3 and step 4 for frames within two thirds of the cardiac phase.
6. Repeat step 5 from the next ED frame to the frame at one third of the cardiac cycle backward.
7. Average two boundaries in the middle one third of the cardiac phase.

Figure 11 shows a digitized echocardiogram of the apical long axis view, with the detected left ventricular boundary superimposed on it.

We can obtain various kinds of information related to the cardiac function from the detected left ventricular boundaries [11–13]. Figure 12 shows volume change of the left ventricular inner cavity calculated by the area length method over 64 frames with an ECG signal. Figure 13 is an example of regional wall motion of the endocardium where the change of the length from the center of gravity of the endocardial boundary at the ED phase to each point on the wall is shown. A stroke excursion from ED to end systolic (ES) corrected by the length at ED is expressed as a percentage shortening, as shown in Fig. 14, which gives us useful information for evaluating the regional contractility of the left ventricle [11].

Figure 15 shows an example of the left ventricular boundaries (endo- and epicardium) on a short axis view. Figures 16 to 19 show the volume change calculated by Gibson's method [21], wall motion of the endocardium, change of wall thickness, and percentage shortening, respectively, obtained from the left ventricular boundaries on the short axis view. In Fig. 19 the percentage values are displayed on the line segments giving stroke excursion from ED to ES by using pseudo colors.

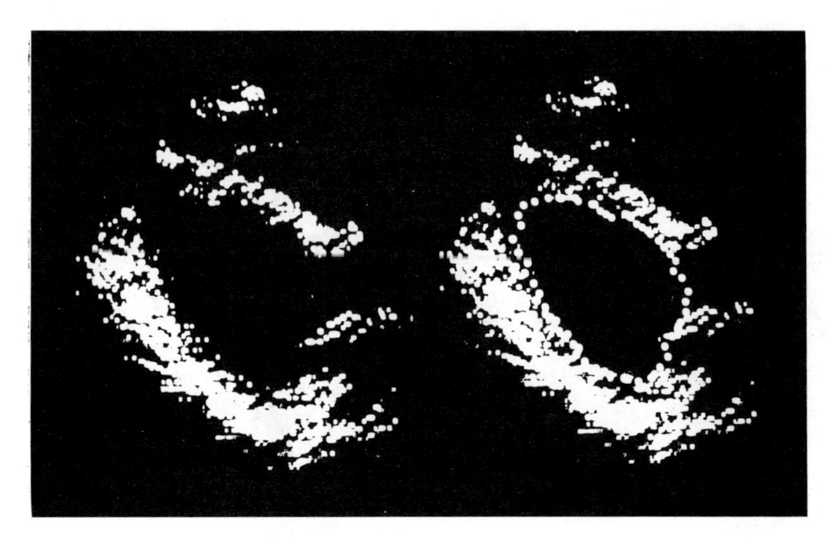

Fig. 11 Digitized echocardiogram and detected edge.

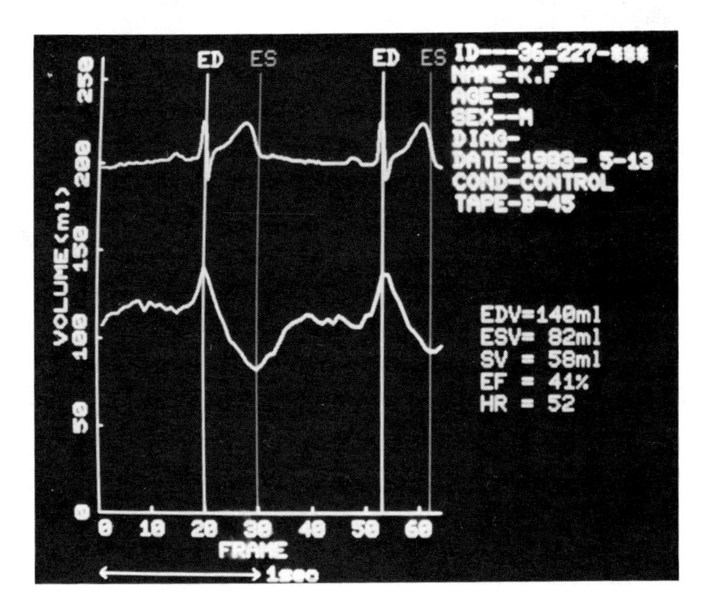

Fig. 12 Volume change with ECG.

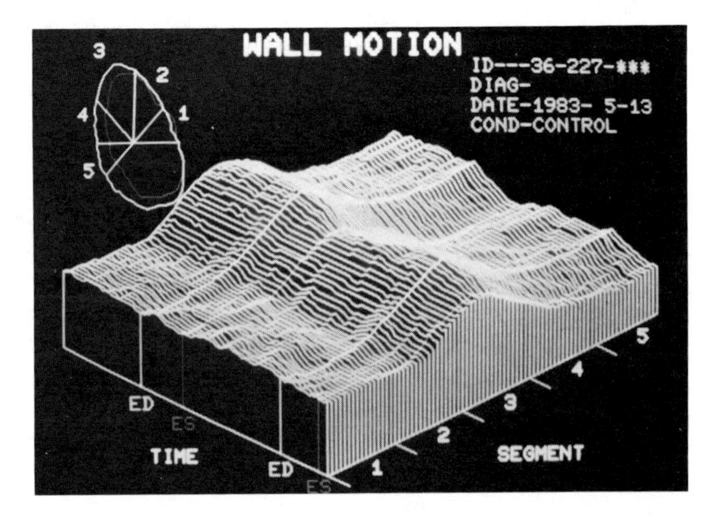

Fig. 13 Wall motion.

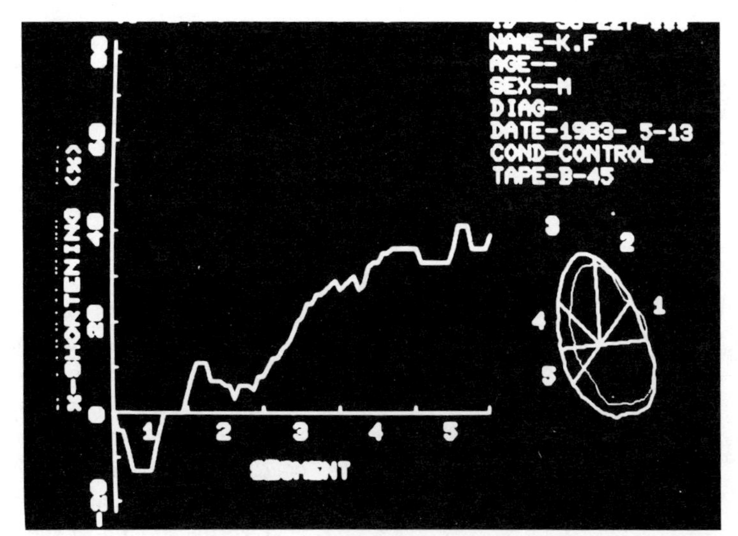

Fig. 14 Precentage shortening.

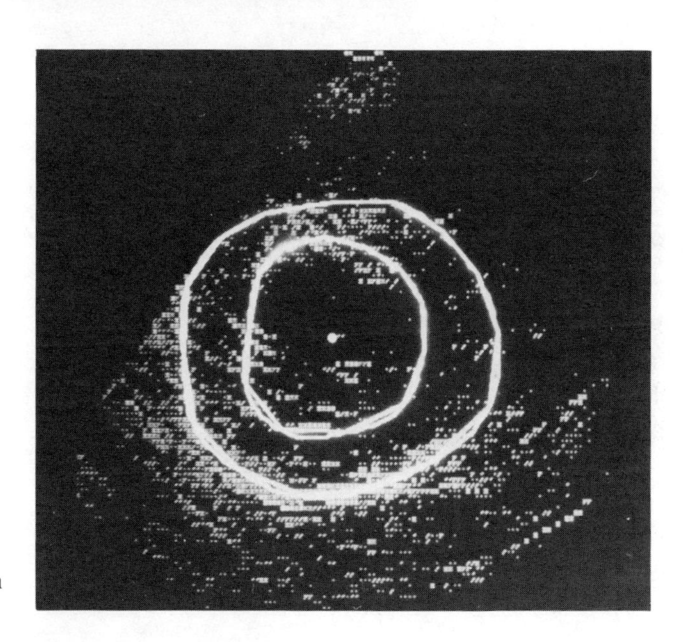

Fig. 15 Left ventricular boundaries (endo- and epicardium) on a short axis view.

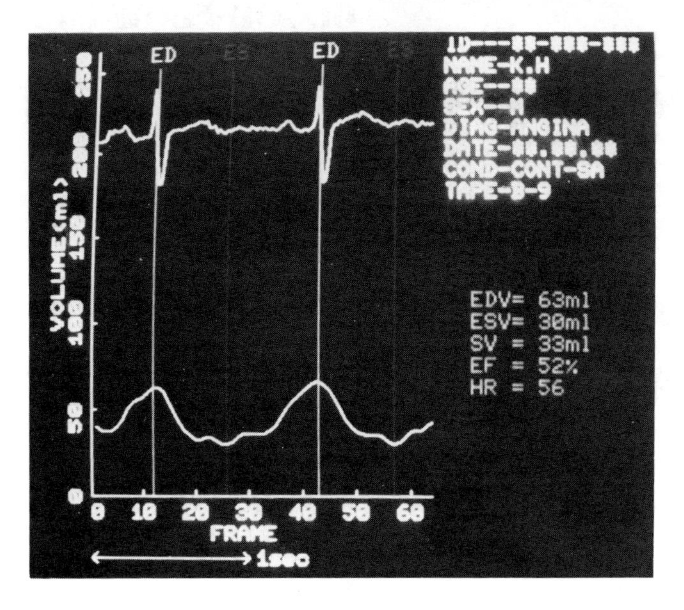

Fig. 16 Volume change.

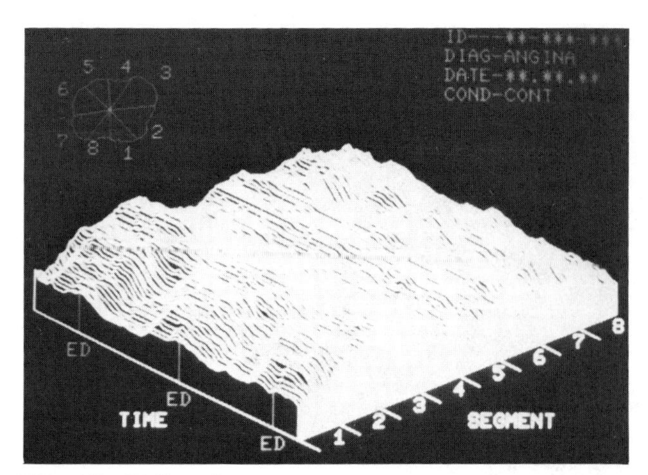

Fig. 17 Wall motion.

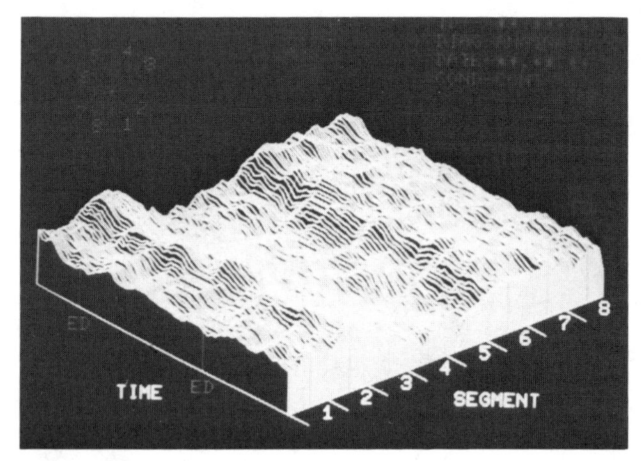

Fig. 18 Wall thickness.

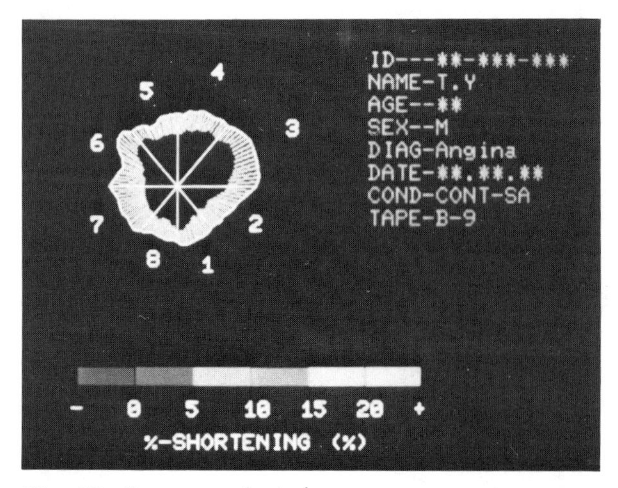

Fig. 19 Percentage shortening.

IV A MICROCOMPUTERIZED IMAGE PROCESSOR FOR 2-D ECHOCARDIOGRAPHY

Most of the systems in these applications of image processing of echocardiograms are based on minicomputer systems, and data processing is usually done off line. However, it is desirable to make tests on line and in real time for more convenient clinical examinations.

Figure 20 shows a microcomputerized image-processing system for sector-scan echocardiograms [12, 13]. For on-line image processing and analysis of the motion of the heart in a short time, we need real-time digitization of echocardiograms and a sufficient amount of storage of digitized data. Images over at least one cardiac cycle are needed to analyze the motion of the heart. The system digitizes 64 consecutive frames of sector-scan echocardiograms, which correspond to data in two or three cardiac cycles. Echo data on each frame are sampled to 128 × 128 digitized data, each 8-bit resolution, i.e., the whole size of the data is 1-Mbyte. Two 1-Mbyte IC memories (video memories) are used for on-line data storage and display of various kinds of data processed. Those video memories are accessible from the CPU and VIOC (video I/O controller) through the bus selector, as shown in Fig. 21.

The 16-bit microprocessor MC68000, with 768K-byte memories, I/O interface, and character and graphic display memories, controls the whole system and executes processing. The system works under the operating system CP/M 68K.

Using this system, we can get several kinds of cardiac information, as shown in Figs. 12–19, within a few minutes.

V AN INTRACARDIAC BLOOD-FLOW IMAGE

This section covers a flow-mapping technique and results of flow mapping in the left ventricle.

A Flow-Mapping Technique

In ultrasonic Doppler method the blood-flow velocity is measured as the Doppler frequency f_d:

$$f_d = 2\left(\frac{f_c v}{C}\right)\cos\theta$$

where f_c is a carrier frequency, v is blood-flow velocity, C is a sound velocity, and θ is a Doppler angle between the beam axis and the blood-flow direction. If it is possible to separate the Doppler signal back scattered from different

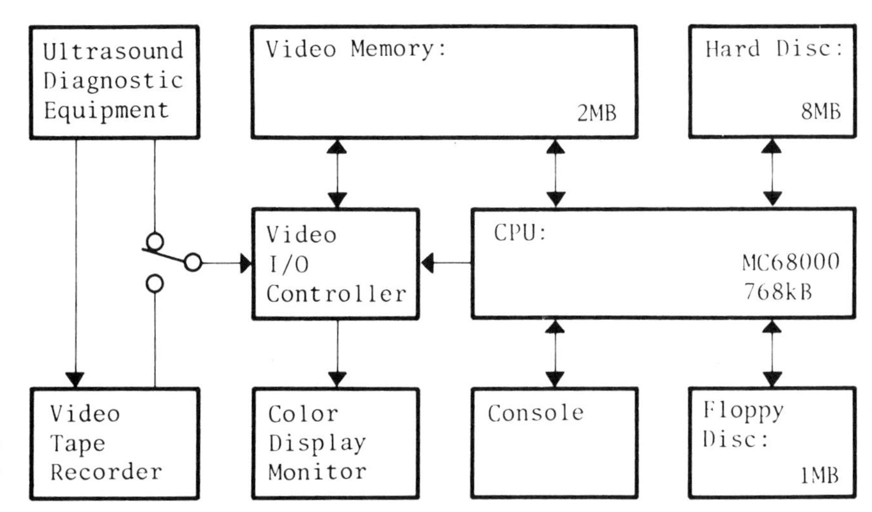

Fig. 20 Microcomputerized image-processing system for sector-scan echocardiograms.

ranges, then the blood-flow information can be arranged and displayed on the echocardiogram. In short, the basic technique of flow mapping is a multichannel ranging technique of the echo signal and an imaging technique of the blood-flow information.

In the multichannel ranging technique, a pulsed beam of pulse width T_w is transmitted at a pulse repetition period T_p, and the echo signal $x(t)$ from the blood vessel is sampled after the time $\tau_i = 2L_i/C$, where L_i is the distance between the probe and the center of the measured range $R_i = L_i - \lambda/4, L_i + \lambda/4$. As $x(nT_p + \tau_i)$ contains the Doppler signal only from the range R_i, we can easily acquire the Doppler signal along the measured beam axis by sampling the echo signal at I times during T_p.

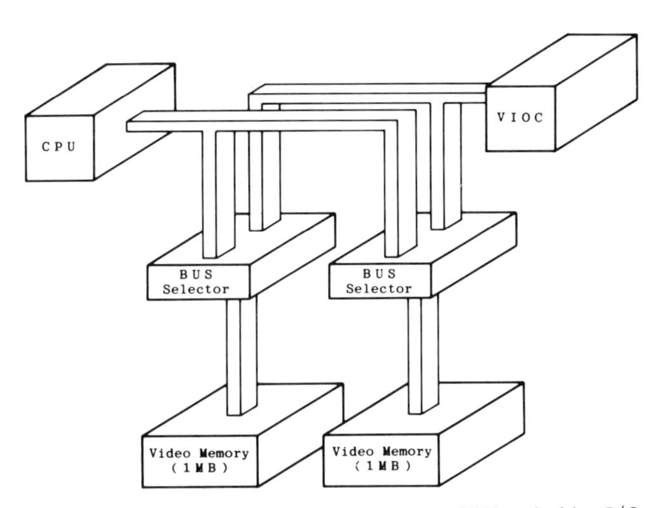

Fig. 21 Data flow from video memory to CPU and video I/O controller through bus selector.

In the imaging technique, it is very important to estimate the appropriate parameter reflecting the blood-flow state. Because many cells are distributed over the range of velocities and space in actual measurements, the Doppler frequency contained in echo signal $x(\tau_i)$, $x(T_p + \tau_i)$, $x(2T_p + \tau_i)$, . . ., where i is the measured channel number, is not a simple value. Therefore, we use the mean Doppler frequency calculated by either the auto correlation (AC) method or the fast Fourier transform (FFT) method. In FFT method, the mean Doppler frequency f_d and the variance S are calculated from the power spectrum as follows:

$$f_d = \frac{\Sigma[f_k P_t(f_d) - f_k P_a(f_k)]}{\Sigma[P_t(f_k) + P_a(f_k)]}$$

$$S^2 = \frac{\Sigma(f_k - f_d)^2[P_t(f_k) + P_a(f_k)]}{\Sigma[P_t(f_k) + P_a(f_k)]}$$

where $P_t(f)$ is the power spectrum of velocity component toward the probe and $P_a(f_k)$ is one away from the probe. Thus, the toward flow is $f_d > 0$ and the away flow is $f_d < 0$.

Generally, as the length N of the time series needed in the auto correlation (AC) method is shorter than the one in the FFT method, the AC method can select a fast beam-scanning rate. On the other hand, the spread of the Doppler power spectrum reflects the fact that the blood-flow state is not uniform in time and space; for example, there is a flow velocity distribution and an accelerated motion of red cells. The FFT method is therefore superior to the AC method for observation of the flow state in detail. The AC method is usually applied in a real-time processing device, and the FFT method is applied to observe the disturbed flow.

Now, with $T_p = 2L_{\max}/C$, where $L_{\max}$ is the maximum detectable range, according to Shannon's sampling theorem

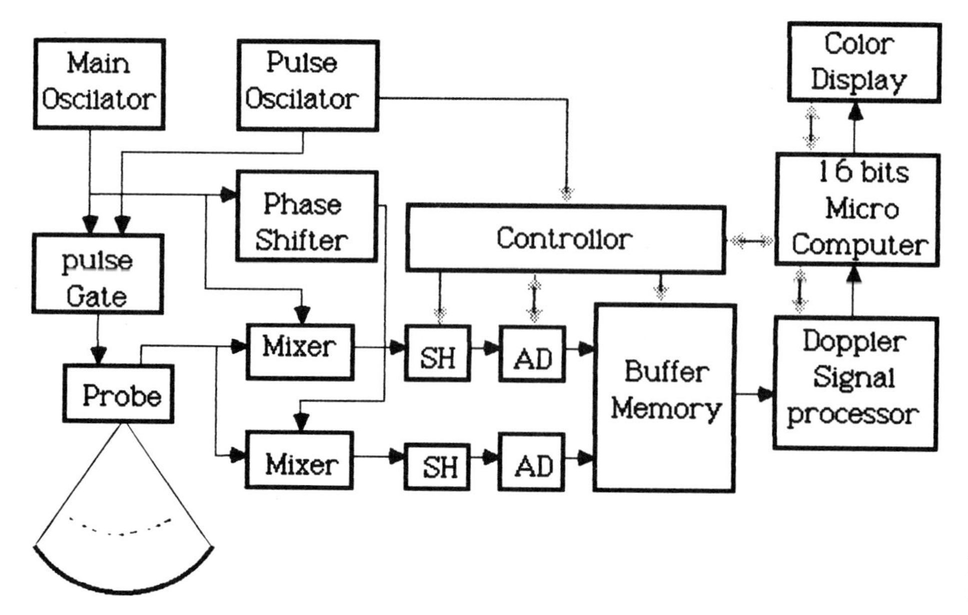

Fig. 22 Block diagram of a flow-mapping system.

the maximum detectable velocity $v_{\max}$ is restricted to

$$v_{\max} = \frac{C}{2f_c 2T_p \cos\theta} = \frac{C^2}{8f_c L_{\max} \cos\theta}$$

In the case of a rapid flow measurement, e.g., a shunt flow, a valve regurgitation flow, or a valve stenosis flow, a frequency higher than $1/2T_p$ contained within the echo signal is shifted to a lower frequency band 0, $1/2T_p$, and an aliasing frequency is displayed as the incorrect velocity information.

Figure 22 shows a block diagram of our flow-mapping system, which consisted of the following units:

- An ultrasonic Doppler unit (f_c = 2.3 MHz, T_p = 200 μs, T_w = 0.89 μs) with a probe of an electric sector scanning type and a quadrature phase-detection circuit

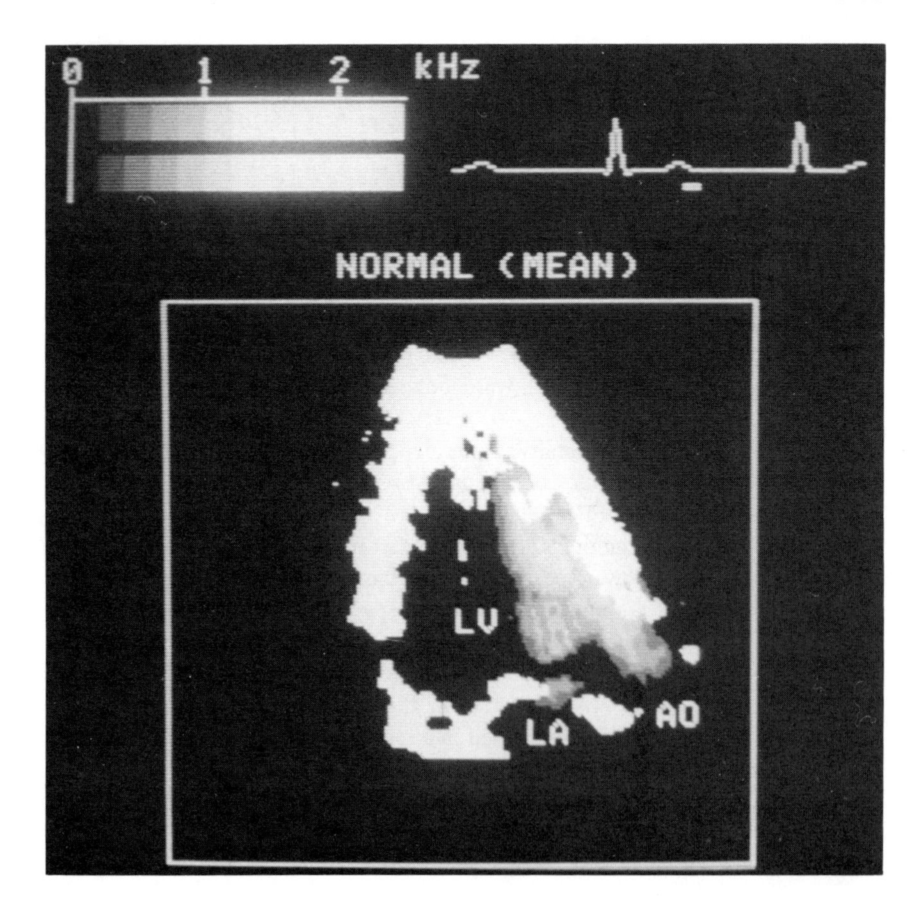

Fig. 23 Blood-flow velocity pattern within the left ventricle of normal subject at systole.

- A multichannel data-acquisition unit with two ultra-high-speed sample holders (sampling pulse width: 50 ns), two high-speed 8-bit analog–digital convertors (conversion time: 800 ns) and a 1.2-MB buffer memory.
- An image-processing unit with a 16-bit microcomputer that has two 1.2-MB floppy disks and a prototype Doppler signal processor
- A color display with 640 × 400 pixels and 16 levels/pixel

B Some Results of Clinical Application

As bones and lungs are impervious to ultrasonic beams, the probe-setting point is restricted within the range from the second intercostal to the fifth one on the left chest surface. Further, because the Doppler frequency is proportional to the cosine component of the velocity, θ should be small. Therefore, in the blood-flow measurement within a left ventricle, the ultrasonic beam is normally transmitted from the apex.

Figure 23 shows the flow mapping of f_d measured at the systole in a normal subject. Generally, this image is suited for observation of a flow profile in a left ventricle. Because each measured beam direction has a different Doppler angle, we need careful observation to estimate a real flow direction. But, as the beam direction approaches that of the ejection flow, the ejection flow is seen to slowly accelerate from the apex toward the aorta.

Figure 24 shows the flow mapping of the variance S measured at the diastole in an aortic valve regurgitation subject. This is suited for the observation of a disturbed flow. In this case, because the aortic valve does not close completely at diastole, the reversed flow from the aorta disturbs the inflow from the left atrium to the left ventricle. Thus the image is not displayed in the case of normal subjects, because the Doppler spectrum is a narrow band.

Finally, a new flow-mapping image is shown in Fig. 25. This is the two-dimensional flow-velocity vector estimated by the flow-continuous equation on the assumption that flow exists within the measured plane only [20]. The conventional flow-mapping device images only the beam component of the flow velocity, while this new technique has the ability to display the amplitude and direction of the intracardiac blood-flow velocity.

As we have shown in this chapter, we can easily observe the blood-flow state by the real-time Doppler flow-mapping device, in which the directional blood-flow velocity image is displayed within the echocardiogram by the red and/or blue color series. It is not too much to say that a state-of-the-art ultrasonic Doppler device is indispensable for clin-

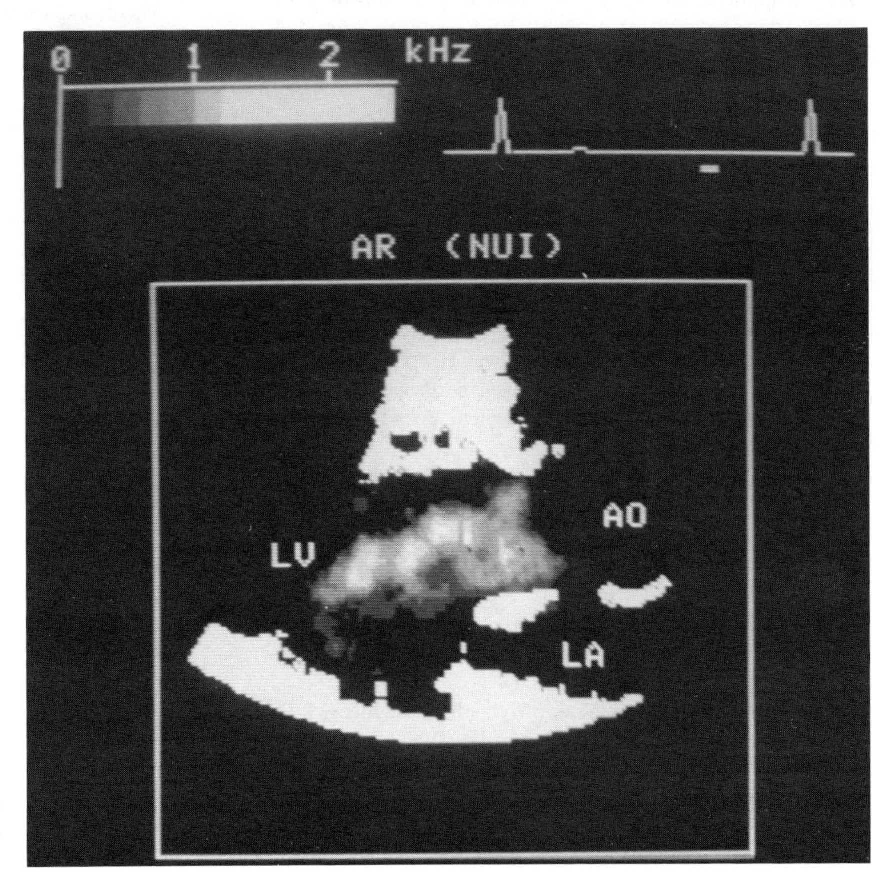

Fig. 24 Blood-flow nonuniformity pattern within the left ventricle of aortic regurgitation subject at diastole.

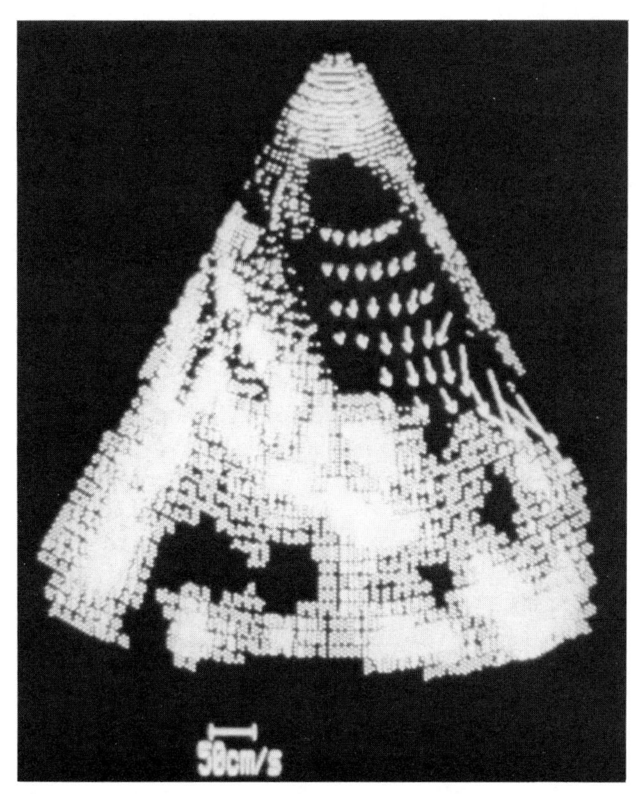

Fig. 25 Two-dimensional blood-flow velocity pattern within the left ventricle of a normal subject at systole.

ical diagnosis of an intracardiac diseased case. Since it is very difficult to understand what the displayed information means, however, careful insight and adequate knowledge of flow dynamics are essential to successful clinical application.

REFERENCES

1. Hirsh, M., Sanders, W. J., Popp, R. L., and Harrison, D. C., Computer Processing of Ultrasonic Data from the Cardiovascular System, *Comput. Biomed. Res.*, vol. 6, no. 4, pp. 336–346, 1973.
2. Kuwahara, M., Eiho, S., Kitagawa, H., Minato, K., and Miki, N., in *Real-Time Medical Image Processing*, ed. M. Onoe, K. Preston, Jr., and A. Rosenfeld, pp. 41–52, Plenum, New York, 1980.
3. Hostetler, M. S., Roemer, L. E., Malindzak, G. S., Caufield, E. J., and Petrovick, M. L., A Microprocessor-controlled Echocardiographic Tracking System, *IEEE Trans. Biomed. Eng.*, vol. BME-27, no. 5, pp. 249–254, 1980.
4. Inari, T., Usami, T., Jitsumori, A., Ohno, E., Eiho, S., and Kuwahara, M., Microcomputerized M-Mode Echocardiography with VTR Recording, *Comput. Cardiol. 1981, Florence*, pp. 557–560, 1982.
5. Robison, E. P., Pryor, T. A., Wellard, S. J., Jones, D. S., and Ridges, J. D., Recognition of Left Ventricular Borders using Two-dimensional Echocardiographic Images, *Comput. Biomed. Res.*, vol. 9, no. 3, pp. 247–261, 1976.
6. Joskowicz, G., Klicpera, M., Pachinger, O., Probst, P., Mayr, H., and Kaindl, F., Computer Supported Measurements of 2-D Echocardiographic Images, *Comput. Cardiol. 1981, Florence*, pp. 13–17, 1982.
7. Jenkins, J. M., Qian, G., Basozzi, M., Delp, E. J., and Buda, A. J., Computer Processing of Echocardiographic Images for Automated Edge Detection of Left Ventricular Boundaries, *Comput. Cardiol. 1981, Florence*, pp. 391–394, 1982.
8. Delp, E. J., Buda, A. J., Swastek, M. R., Smith, D. N., Jenkins, J. M., Meyer, C. R., and Pitt, B., The Analysis of Two-dimensional Echocardiograms using a Time-varying Image Approach, *Comput. Cardiol. 1982, Seattle*, pp. 391–394, 1983.
9. Garcia, E., Ezekiel, A., Levy, R., Zwehl, W., Ong, K., Corday, E., Areeda, J., Meerbaum, S., and Corday, S., Automated Computer Enhancement and Analysis of Left Ventricular Two-dimensional Echocardiograms, *Comput. Cardiol. 1982, Seattle*, pp. 399–402, 1983.
10. Kuwahara, M., Eiho, S., Kitagawa, H., and Ishimi, K., Automatic Analysis of Two-dimensional Echocardiograms, in *MEDINFO 80*, ed. D. A. B. Lindberg and S. Kaihara, pp. 210–213, North-Holland, Amsterdam, 1980.
11. Eiho, S., Kuwahara, M., Asada, N., Sasayama, S., Takahashi, M., and Kawai, C., Reconstruction of 3-D Images of Pulsating Left Ventricle from Two-dimensional Sector Scan Echocardiograms of Apical Long Axis View, *Comput. Cardiol. 1981, Florence*, pp. 19–24, 1982.
12. Kuwahara, M., Eiho, S., Kitagawa, H., Asada, N., Ohtsuyama, K., and Ogawa, H., A Microcomputerized On-line Image Processing System for Sector Scan Echocardiography, *Comput. Cardiol. 1983, Aachen*, pp. 105–108, 1984.
13. Eiho, S., Kuwahara, M., Asada, N., and Yamashita, K., A Microcomputerized Image Processor for 2-D Echocardiography and 3-D Reconstruction of Left Ventricle, *Comput. Cardiol. 1985, Linköping*, pp. 269–272, 1986.
14. Baker, D. W., Pulsed Ultrasonic Doppler Blood Flow Sensing, *IEEE Trans. Sonics Ultrason.*, vol. SU-7, no. 3, pp. 170–185, 1970.
15. McLeod, F. D., Multichannel Pulsed Doppler, in Cardiovascular Applications of Ultrasound, ed. R. S. Reneman, pp. 85–107, North Holland, Amsterdam, 1974.
16. Brandestini, M., Topoflow: A Digital Full Range Doppler Velocity Meter, *IEEE Trans. Sonics Ultrason.*, vol. SU-25, no. 5, pp. 287–293, 1978.
17. Chihara, K., Inokuchi, S., Hirayama, M., Sakurai, Y., Matsuo, H., Asao, M., Mishima, M., Tanouchi, J., Inoue, M., and Abe, H., Development of Computer-based Ultrasonic Multichannel Pulsed Doppler Flowmeter, *Proc. 12th Int. Conf. Med. Biol. Eng., Israel*, pp. 28/1–2, 1979.
18. Angelsen, B. A., Instantaneous Frequency, Mean Frequency, and Variance of Mean Frequency Estimation for Ultrasonic Blood Velocity Doppler Signals, *IEEE Trans. Biomed. Eng.*, vol. BME-28, no. 11, pp. 733–741, 1981.
19. Kasai, C., Namekawa, K., Koyano, A., and Omoto, R., Real-Time Two-dimensional Blood Flow Imaging using an Autocorrelation Technique, *IEEE Trans. Sonics Ultrason.*, vol. SU-32, no. 3, pp. 458–463, 1985.
20. Chihara, K., and Shirae, K., Two-dimensional Bloodflow Velocity Vector Imaging System, *Proc. 8th Ann. Conf. IEEE Eng. Med. Biol. Soc., Fort Worth*, pp. 105–108, 1986.
21. Gibson, D. G., Estimation of Left Ventricular Size by Echocardiography, *Bri. Heart J.*, vol. 35, pp. 128–134, 1973.

Chapter 19

Cardiac Image Processing

Shigeru Eiho and Michiyoshi Kuwahara

I INTRODUCTION

Medical imaging and image-processing techniques have recently been of great interest, especially since the development of X-ray computed tomography (CT) machines. Images obtained from these machines have triggered research to produce new imaging machines and to process medical images by using digital computers. Medical image processing is expected to be one of the most useful applications of image processing. Various kinds of images are used by doctors in clinical medicine to diagnose diseases, according to their experience and knowledge. However, few methods can produce from the medical images usable quantitative data for clinical diagnosis.

In the first half of this chapter, we focus on left ventricular image processing from X-rays as an example of two-dimensional image processing. (Image processing of ultrasound echocardiograms was discussed in Chap. 18.)

In the second half of this chapter, three-dimensional reconstruction methods using X-ray cineangiocardiograms, echocardiograms, and MRI-CT are discussed.

II IMAGE PROCESSING OF CINEANGIOCARDIOGRAMS

X-ray cineangiocardiography has been a very helpful technique in examining cardiac functions. A pulsed X-ray exposure starts a little before the injection of contrast material through a catheter inserted into the left ventricle or the coronary artery through the aorta; a silhouette of the organ enhanced by the contrast material is recorded on a cine film having a film rate of 30–150 frames per second. Quantitative information, such as volume change and wall motions of the left ventricle in a cardiac cycle, can be obtained by detecting the left ventricular boundary on each frame of the cineangiocardiogram. This information can also be used to diagnose cardiac diseases.

Manual techniques, such as tracing the left ventricular boundary and planimetry, have often been used by cardiologists to measure volume change of the left ventricle. However, manual methods are very troublesome and time consuming, especially when a large number of images are to be processed.

Automatic or semiautomatic analysis of cineangiocardiograms using a digital computer system has been performed. Heintzen organized the international workshop on ''Roentgen-, Cine- and Videodensitometry'' in 1969; in the proceedings, published in 1971 [1], many contributors discussed computer application for blood flow and heart volume determination. Chow and Kaneko proposed a dynamic threshold method to detect boundaries in cineangiocardiograms [2]. Slager, Reiber, and their co-workers [3, 4] developed a hard-wired system designed for operator-interactive automated outlining of left ventricular contrast angiograms. There were also many contributions on left ventricular image processing at the international conferences on ''Computers in Cardiology'' chaired by Cox and Hugenholtz since 1974 [5] and the international symposium on ''Ventricular Wall Motion'' organized by Sigwart and Heintzen in 1983 [6].

Eiho, Kuwahara, are their co-workers [7–12] have been reporting the methods of left ventricular boundary detection

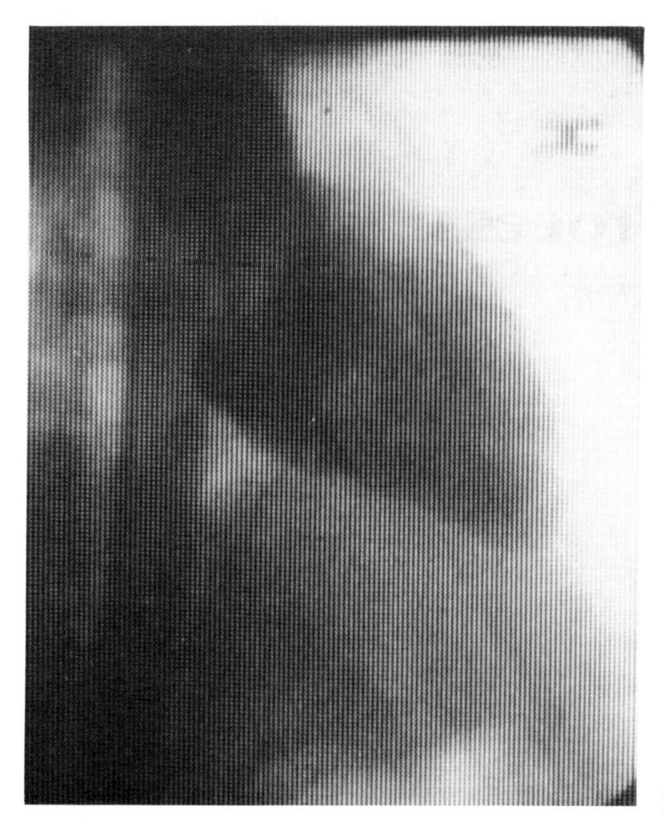

Fig. 1 Left ventricular image of cineangiogram.

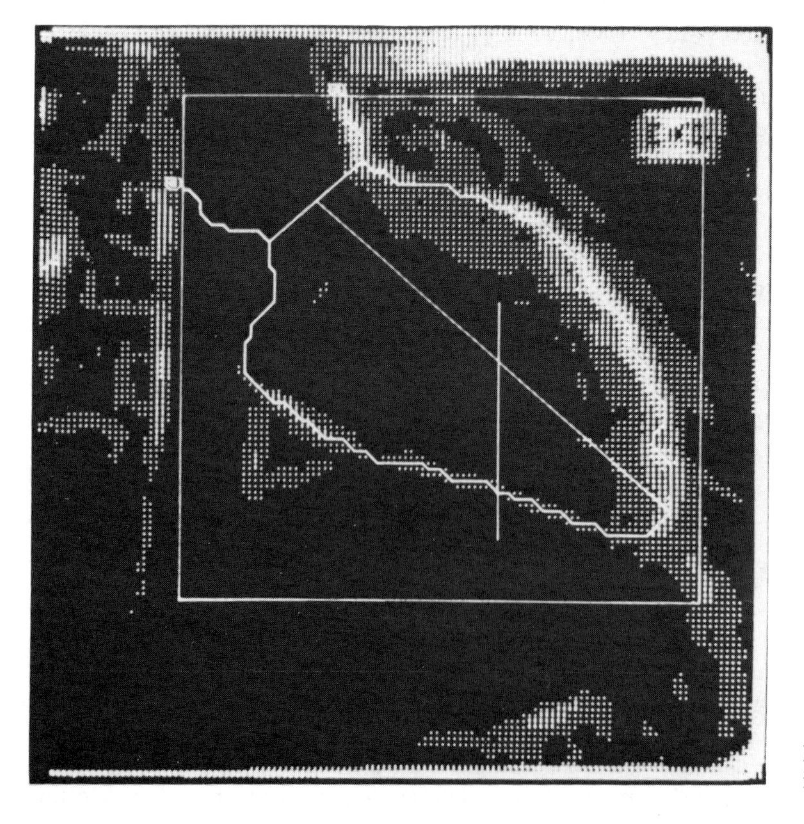

Fig. 2 Detected left ventricular boundary superimposed on gradient image.

from cineangiocardiograms, 3-D reconstruction of left ventricle, and displays of cardiac functions. Their fundamental steps for detecting the 2-D left ventricular boundary are based on automated heuristic tracing of the boundary, using the spatial gradient image of gray levels of the cineangiocardiograms.

Figure 1 shows an example of digitized left ventricular image, and Fig. 2 shows the example of detected boundary superimposed on the gradient image. Figure 3 shows 100 consecutive boundary curves obtained from cineangiocardiograms at 60 frames per second; the dotted line superimposed on each boundary curve is the boundary at the end of a diastolic phase. These consecutive boundary curves of the left ventricle can yield various pieces of quantitative information related to the cardiac function.

An example of left ventricular volume calculated by the area-length method is shown in Fig. 4. Cross marks in this figure give the volumes obtained from boundaries manually traced by a cardiologist. This volume curve is, as a whole, very closely positioned among volume values obtained by the manual method.

The wall motion of the left ventricle can be measured as the change of length from the center of gravity of end diastolic boundary to each boundary curve on every radial

Fig. 3 Detected 100 consecutive boundaries of the left ventricle.

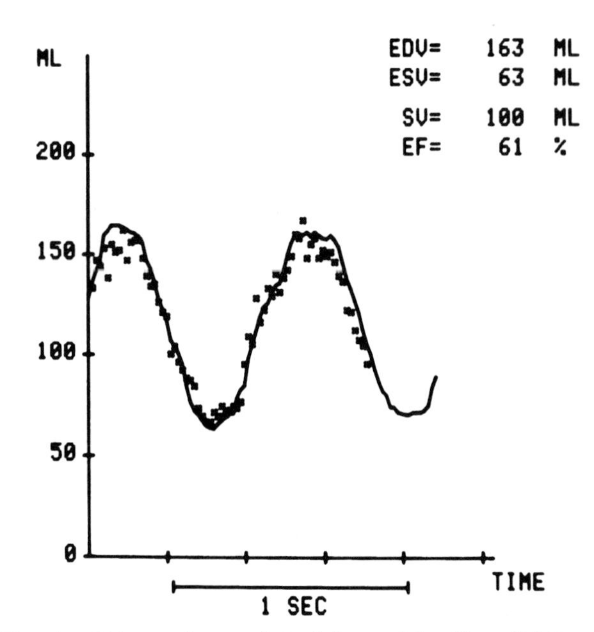

Fig. 4 Volume change of the left ventricle. The solid line is the volume curve obtained from the boundaries detected by the computer, and the cross marks are volumes obtained from the curves drawn by a cardiologist. EDV, End diastolic volume; ESV, end systolic volume; SV, stroke volume; EF, ejection fraction.

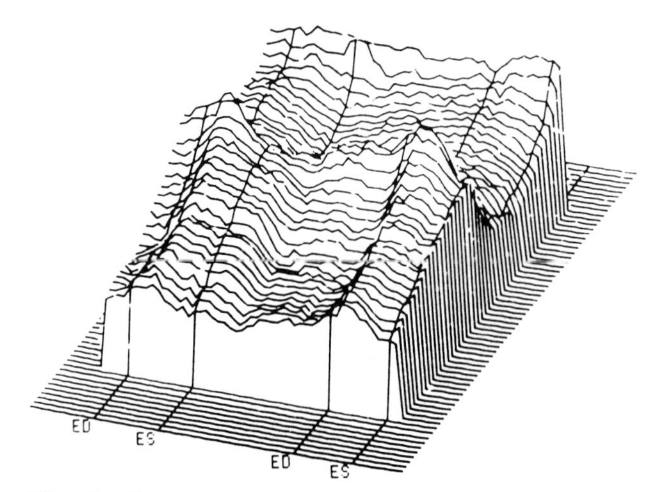

Fig. 6 Three-dimensional display of the wall motion.

direction. Figure 5 gives the motions of three arbitrarily chosen points on the left ventricular wall. Three curves on the left-hand side show the left ventricular wall segments in rest condition (control); those on the right-hand side show the movements after administering nitroglycerine (NTG) to extend capillaries. Figure 6 shows the wall motion at every point around the left ventricle, from the anterior wall to the inferior wall through the apex. By choosing two boundary curves at the end diastole (ED) and end systole (ES), one can observe contractile wall movements. Figure 7 indicates the wall movements of the left ventricle by the lengths of line segments; the left side shows rest condition (control), and the right side shows movement after an anginal attack induced by rapid pacing of the right atrium. Figure 8 gives the percentage shortening in the same patient, calculated by dividing the length of each line segment by the length from the center of gravity of the left ventricle to the point on the wall at the ED phase.

By using a high-fidelity micromanometer-tipped catheter, we can measure the ventricular pressure simultaneously during angiography. The pressure–volume loop provides a convenient framework for understanding global contractile functions of the myocardium [13].

The length of each radial grid plotted against time (Fig. 6) can be handled in the same way as an overall

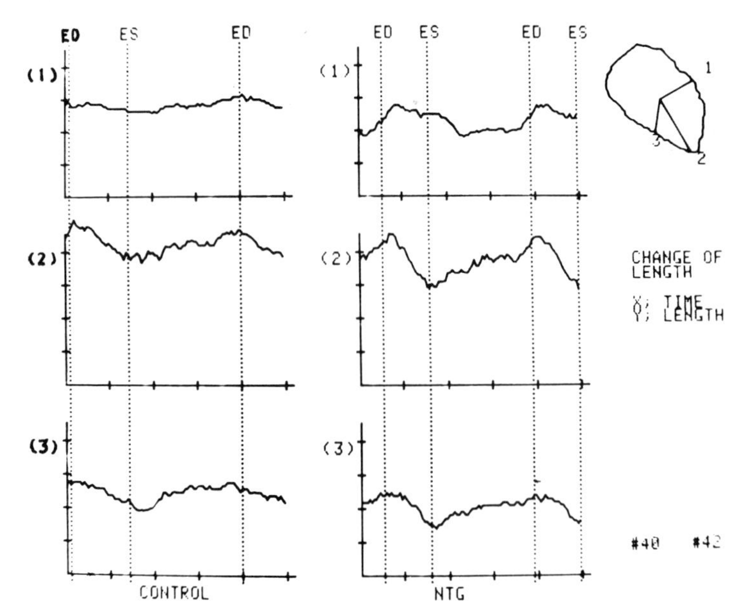

Fig. 5 Motions of three points on the left ventricular wall.

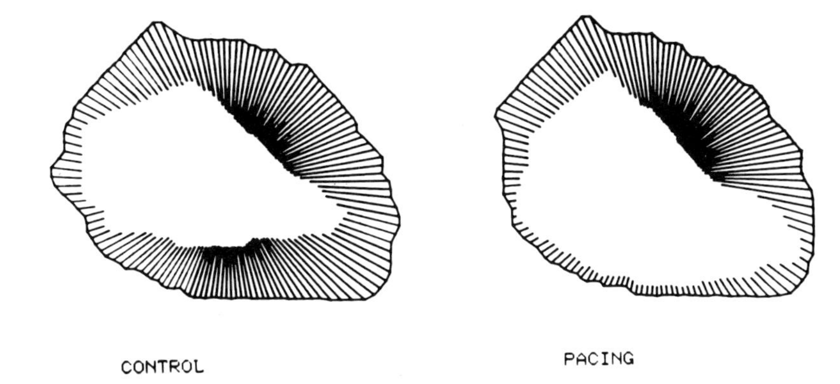

Fig. 7 Wall movements between ED and ES phases.

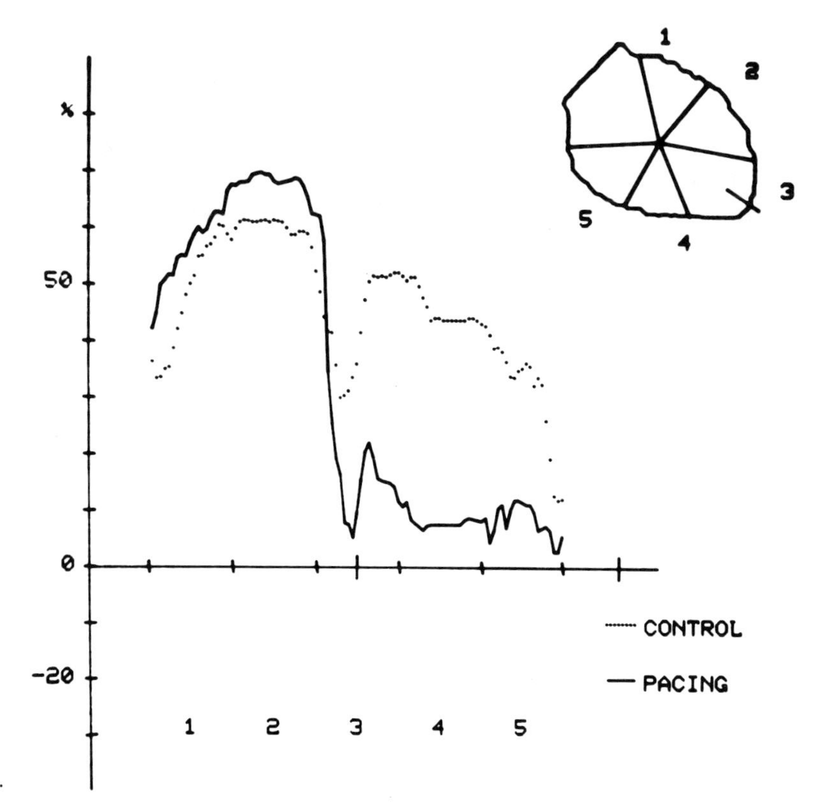

Fig. 8 Percentage shortening.

volume–time curve. The pressure–length loops are obtainable [11], as shown in Fig. 9.

Figure 9*a* shows data obtained in the rest state. The apical segment loops in the central ischemic area are deformed considerably. The anterior wall is perfused by well-developed collateral circulation from the right coronary artery; active segment shortening is maintained, but coronary reserve is critically limited. Loops in the anterior wall exhibit an inclination toward the right, suggesting delayed relaxation due to potential ischemia during the stressed state. The loops in the normally perfused inferior wall show an inclination toward the left, with lengthening of the myocardium during isovolumic relaxation. There is also an ischemic area at the inferior wall between the infarcted and the normal areas, but the loops are not a typical configuration, presumably due to a tethering effect.

Figure 9*b* shows data obtained during an anginal attack induced by rapid cardiac pacing. The deformation of the pressure–length loop is now expanded over the entire anterior wall. The loops in the inferior wall exhibit a more marked inclination toward the left, associated with augmentation of systolic excursion.

Figure 9*c* shows the result of sublingual administration of 0.3 mg nitroglycerin after an adequate time for recovery from the effect of repeated angiography. All the loops obtained after nitroglycerin injection are rectangular in configuration and have a counter-clockwise rotation.

III THREE-DIMENSIONAL RECONSTRUCTION OF THE LEFT VENTRICLE

To measure the cardiac function of the heart, we need to observe its motion, especially that of the left ventricle. We can observe some performance of the left ventricle by using 2-D images such as X-ray cineangiocardiograms or echo-

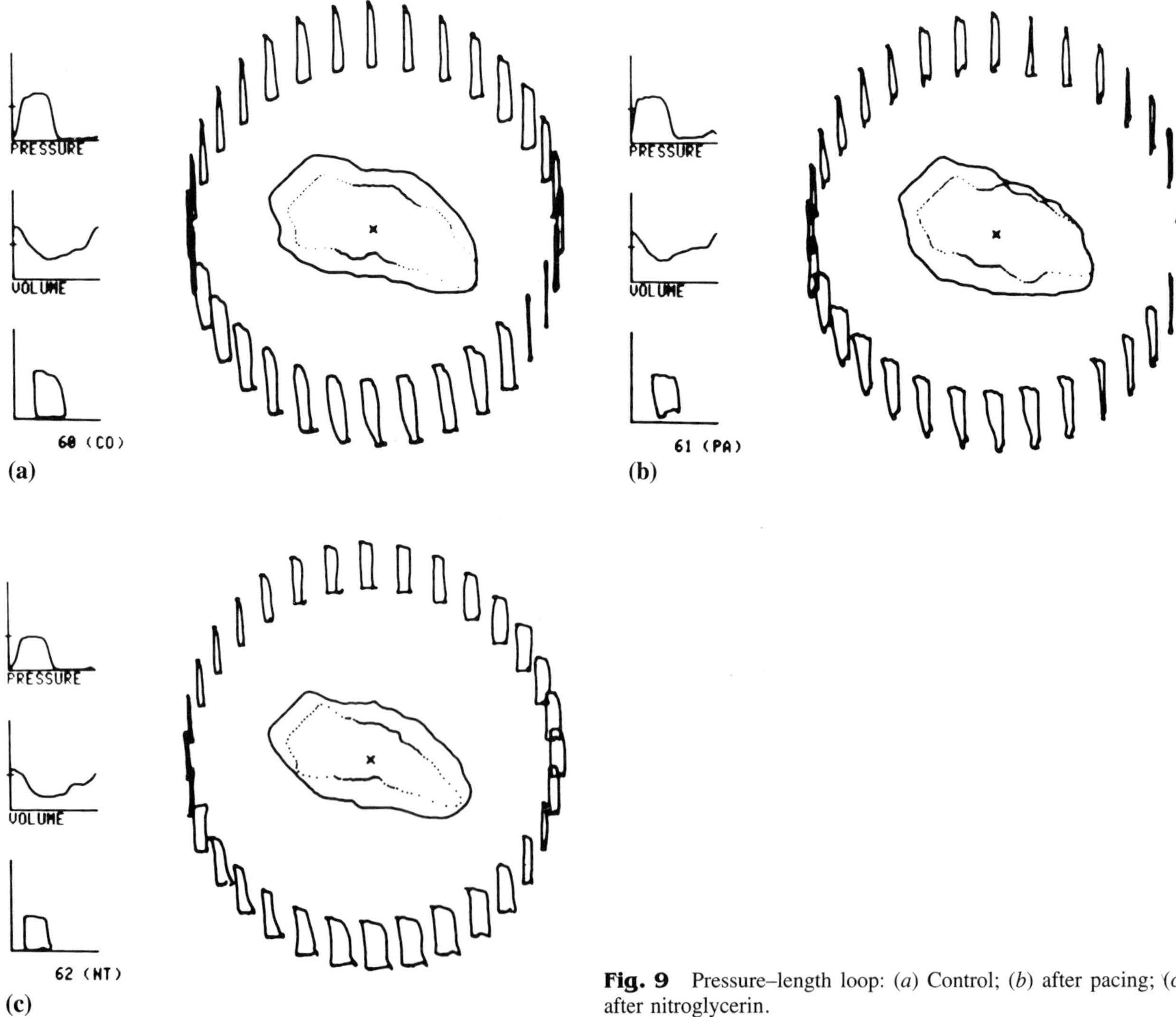

Fig. 9 Pressure–length loop: (*a*) Control; (*b*) after pacing; (*c*) after nitroglycerin.

cardiograms. But it is impossible to learn the regional function of the wall motion on the whole area of the left ventricle from these images. If 3-D shapes of the left ventricle are observable, then the left ventricular wall motion will be obtained more easily and more precisely.

Recently, various kinds of cross-sectional images were obtained by using CT imaging techniques. But conventional CT methods have some disadvantage for use with a quick-acting organ such as the heart. The dynamic spatial reconstruction (DSR) method has been developed for reconstruction 3-D heart images [14, 15], but DSR requires a special expensive system of imaging.

By using several cross-sectional shapes with the ECG signal or by using the ECG-gated method, we can obtain several cross-sectional shapes or silhouettes of the left ventricle; we can then reconstruct 3-D shapes of the left ventricle from these cross-sectional images.

In this chapter we discuss several methods of 3-D reconstruction of the left ventricle, from X-ray cineangiocardiograms, ultrasound echocardiograms, and magnetic resonance images (MRI).

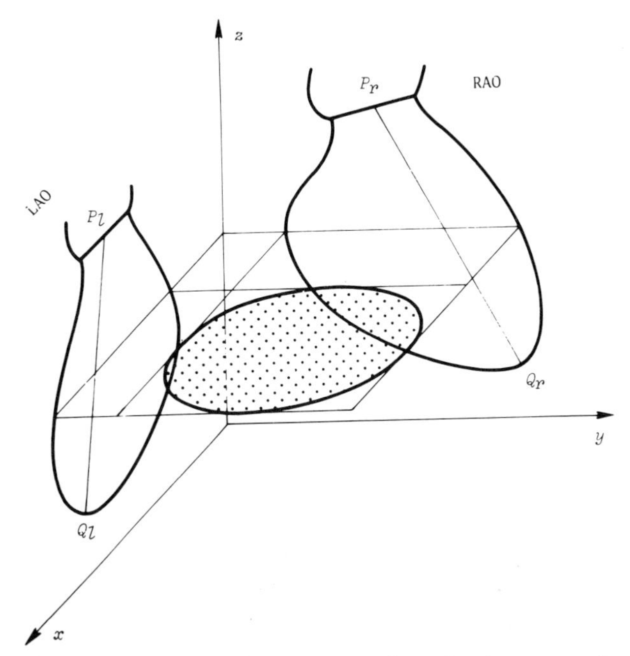

Fig. 10 Schematic diagram of three-dimensional reconstruction of left ventricle from biplane cineangiocardiograms.

A 3-D Left Ventricular Image from Biplane Cineangiocardiograms

Some conventional methods have been developed to reconstruct 3-D images of the heart using biplane cineangiocardiograms [16, 17]. One conventional way is to detect left ventricular boundaries of two orthogonal silhouettes and to reconstruct 3-D images from these boundaries.

Figure 10 shows a schematic diagram of the reconstruction method, using two boundary curves on biplane cineangiocardiograms at RAO 30° and LAO 60°. When one slices the left ventricle at a plane horizontal to the body axis, one obtains two line segments cut by the two boundary curves on the RAO and LAO planes. The cross section of the left ventricular cavity on this horizontal plane is assumed to be an ellipse, which is inscribed in the rectangle made by the two line segments of the RAO and LAO projections. The left ventricle is reconstructed by stacking these ellipses from the bottom (near the apex) to the top (the aortic valve).

After the corrections of scale and position on the RAO and LAO planes, the shape of the left ventricle is reconstructed from the ellipses on the horizontal planes. Sixteen horizontal planes are used for reconstruction; each ellipse is sampled at 32 points with an equal angle from the center of gravity of each ellipse. Thus, the 3-D shape of the left ventricle is reconstructed by $16 \times 32 = 512$ points.

Projecting 512 points on the plane orthogonal to a line of vision and removing invisible points yields a wire-framed 3-D left ventricular image. Figure 11*a* shows a 3-D shape of a wire-framed left ventricle from near RAO 30°. The left-hand side of the figure is at the ED phase, the right-hand side at the ES phase. By lightening rectangles on the wire-framed 3-D left ventricle, one can obtain a shaded 3-D left ventricle, as shown in Fig. 11*b*. Figure 12 shows 3-D left ventricular shapes in a cardiac cycle.

B Reconstruction of Three-dimensional Shape of the Left Ventricle from Cineangiocardiograms with a Rotating Arm [18]

As detailed in the previous section, we can obtain silhouettes of the left ventricle at any angle of view with a conventional cineangiographic device, though the angle is usually fixed at, e.g., RAO 30° or LAO 60°.

The more angles of silhouettes used to reconstruct the left ventricle, the more accurate images we can obtain. By continuously rotating the X-ray source and the detector of the monoplane cineangiographic device from 0 to 180° for several seconds, we can obtain silhouettes of the left ventricle at various angles during several cardiac cycles (usually a one-dose injection).

The boundaries of the left ventricle are traced automatically or semiautomatically by a computer, and the cardiac phase of each silhouette of the left ventricle is estimated by referring the ECG signal marked on the cine-pulse chart.

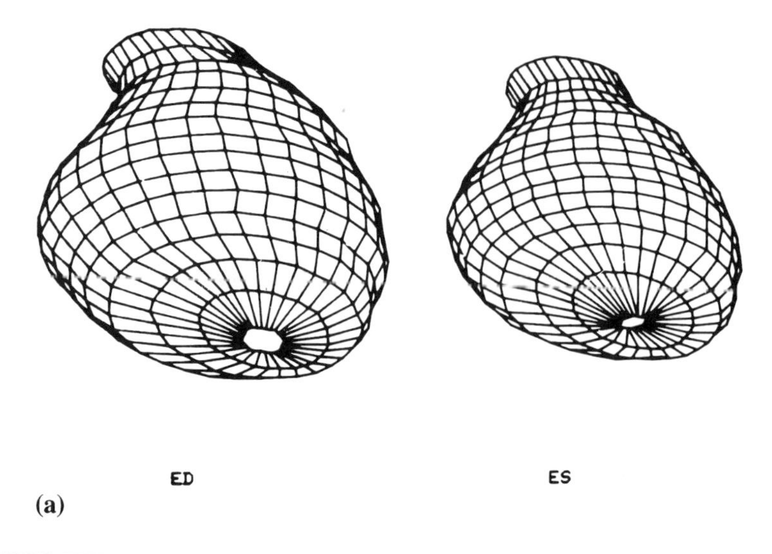

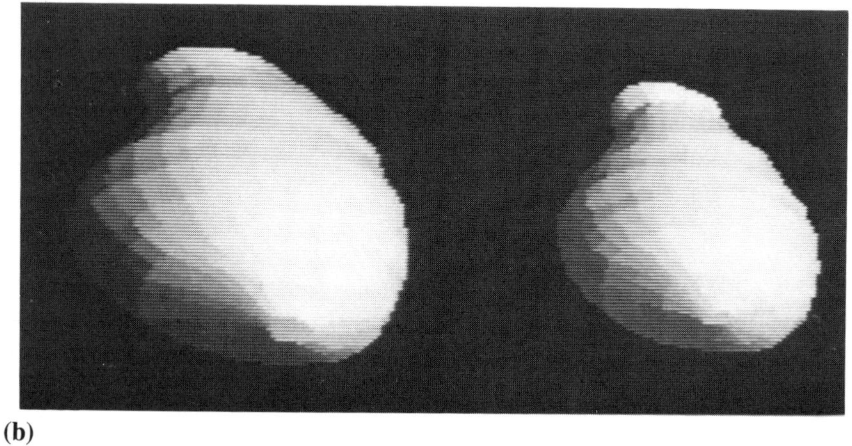

Fig. 11 Three-dimensional left ventricle reconstructed from biplane cineangiocardiogram: (*a*) Wire-framed images; (*b*) shaded images.

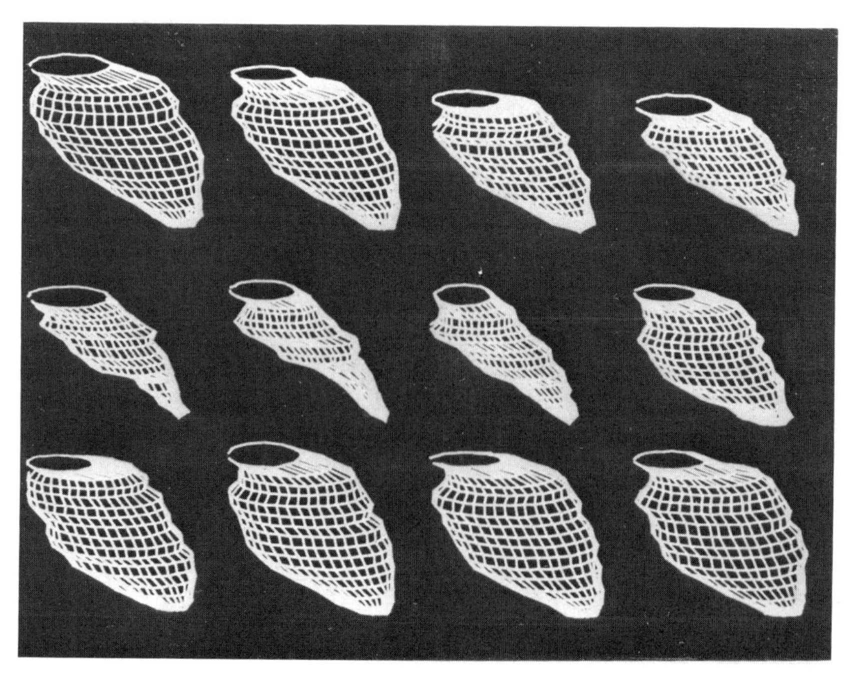

Fig. 12 Three-dimensional left ventricle shapes in a cardiac cycle.

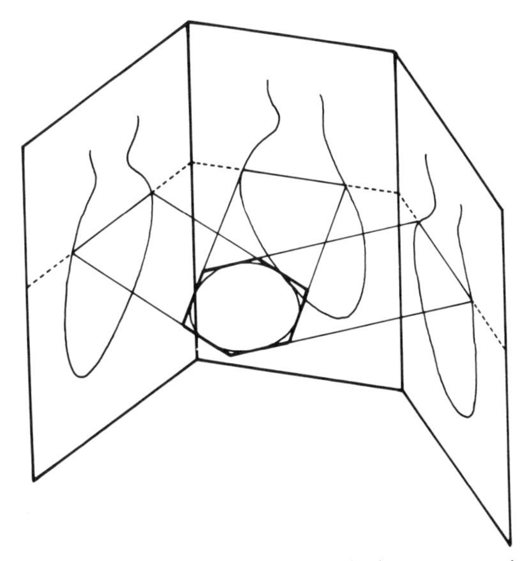

Fig. 13 Schematic diagram to obtain a cross-sectional shape of left ventricle on a horizontal plane.

Figure 13 shows a schematic diagram of the reconstruction of a cross-sectional shape on a (horizontal) plane perpendicular to the axis of rotation. Three boundary curves shown in this figure correspond to the 3 different angles of projection at the same cardiac phase. By assuming that the X-ray radiation is done in parallel beam, one can obtain the left ventricular shape on the horizontal plane by the following steps:

1. Assume an inscribed point on each side of the polygon (i.e., at the center of each side).
2. Draw a smooth curve connecting these points. The spline curve of the third order is used.

Table 1 Calculated Volume of Each Reconstructed Model[a]

Number of Projections	Angle of Arm, °	Volume, ml
2	30, 120	33.4
3	30, 90, 120	35.4
4	30, 60, 90, 120	36.1

[a]True volume is 37 ml.

The 3-D shape is reconstructed as a set of these curves. We usually use 16 planes for cross-sectional images and sample the curve on each plane at 32 points, thus building a 3-D shape of 512 points.

Figure 14 shows a model of the left ventricle, one of its X-ray images, and a reconstructed image from three silhouettes. Figure 15 shows the reconstructed 3-D shapes of the model. In these two figures, (*a*) is reconstructed from two silhouettes (30 and 120°), (*b*) from three silhouettes (30, 90, and 120°), and (*c*) from four silhouettes (30, 60, 90, and 120°). Table 1 shows volume of the reconstructed images calculated from this equation:

$$V = d \sum_{i=1}^{16} S_i$$

where S_i is an area of each cross-sectional shape and d is the distance between two consecutive horizontal planes. The true volume of the model was 37 ml.

Cineangiocardiograms of a patient who had tetralogy of Fallot were taken by rotating the arm of the X-ray device

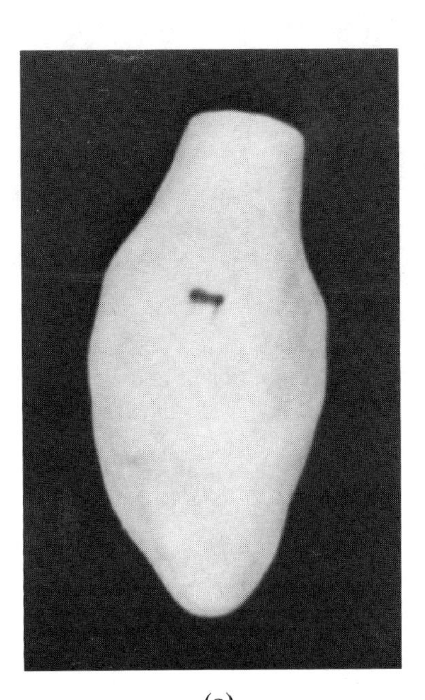

(a)

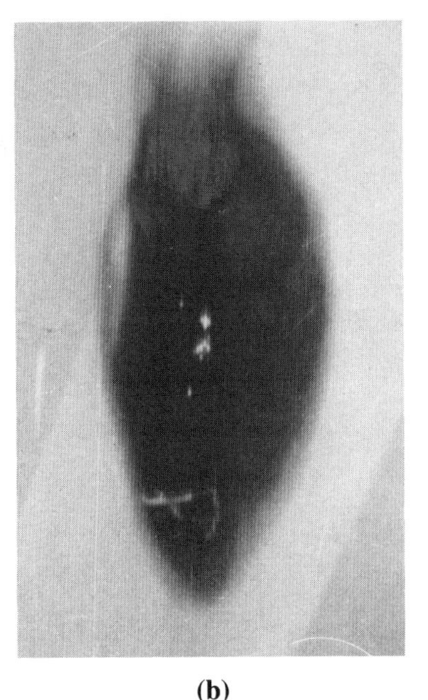

(b)

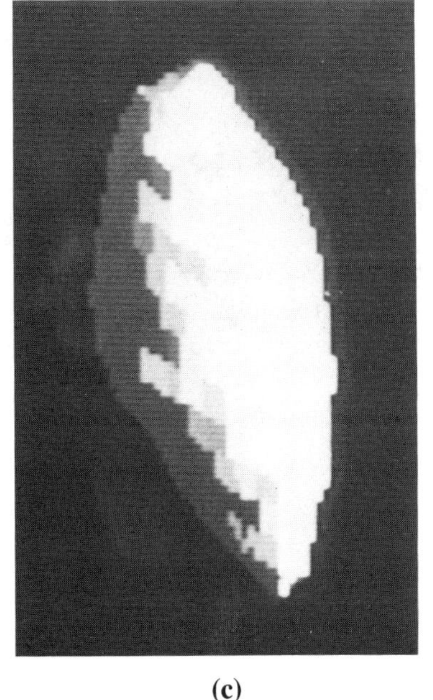

(c)

Fig. 14 Left ventricle: (*a*) Model; (*b*) X-ray image; and (*c*) reconstructed image.

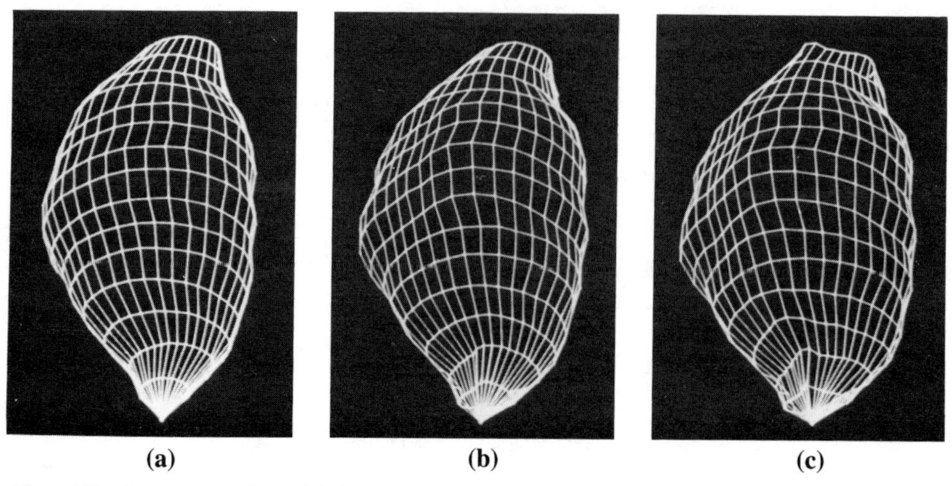

Fig. 15 Reconstructed model from (*a*) two projections, (*b*) three projections, and (*c*) four projections.

for about 4.5 s (film rate = 60 frames per second). More than 270 images for about seven cardiac cycles from angles 0 to 180° were obtained. Figure 16 shows part of the detected boundaries for every five cineangiocardiograms. Figure 17 shows shapes of the left ventricle in a cardiac cycle.

This method is used to reconstruct 3-D shapes of the left ventricle with a conventional monoplane cineangiocardiographic device by appropriate modifications.

There are, of course, many problems to be solved for clinical use, e.g., the angle of view of each frame should be recorded. The method of estimating the cross-sectional shape inscribed in the polygon should be modified; a more reasonable method uses the shapes of the boundary at successive phases and/or uses the projected intensity of dose [19].

C Reconstruction from Echocardiograms

One can observe various cross-sectional images of the left ventricle by echocardiography, and 3-D shapes of the left ventricle can be reconstructed by using a series of 2-D echocardiograms. We can use two methods for this reconstruction: One is to obtain the short axis images by changing the position of the ultrasound probe on the chest wall; the other is to take long axis images at several angles by rotating the probe whose tip is fixed at the same direction and position on the chest wall. We have been using the latter method because 3-D reconstruction may easily be obtained without measuring the position of the probe precisely [20]. Two-dimensional echocardiograms on several (usually six) cross-sectional planes are taken by rotating the ultrasound probe at the same position on a chest wall, as shown in Fig. 18. A 3-D left ventricular shape is reconstructed by the following steps:

1. Choose the left ventricular boundary curves at the same cardiac phase.
2. Rearrange them at their angles.
3. Cut these curves by 16 planes perpendicular to the axis of rotation of the probe.
4. Connect the 12 cross points on the plane with a spline curve and sample the curve at 32 points.
5. Store these 16 × 32 = 512 points as the 3-D shape of the left ventricle.

Figure 19 shows the six boundary curves and the spline curve with the 32 sample points mentioned above. Figure 20 gives an example of a wire-framed 3-D shape of the left

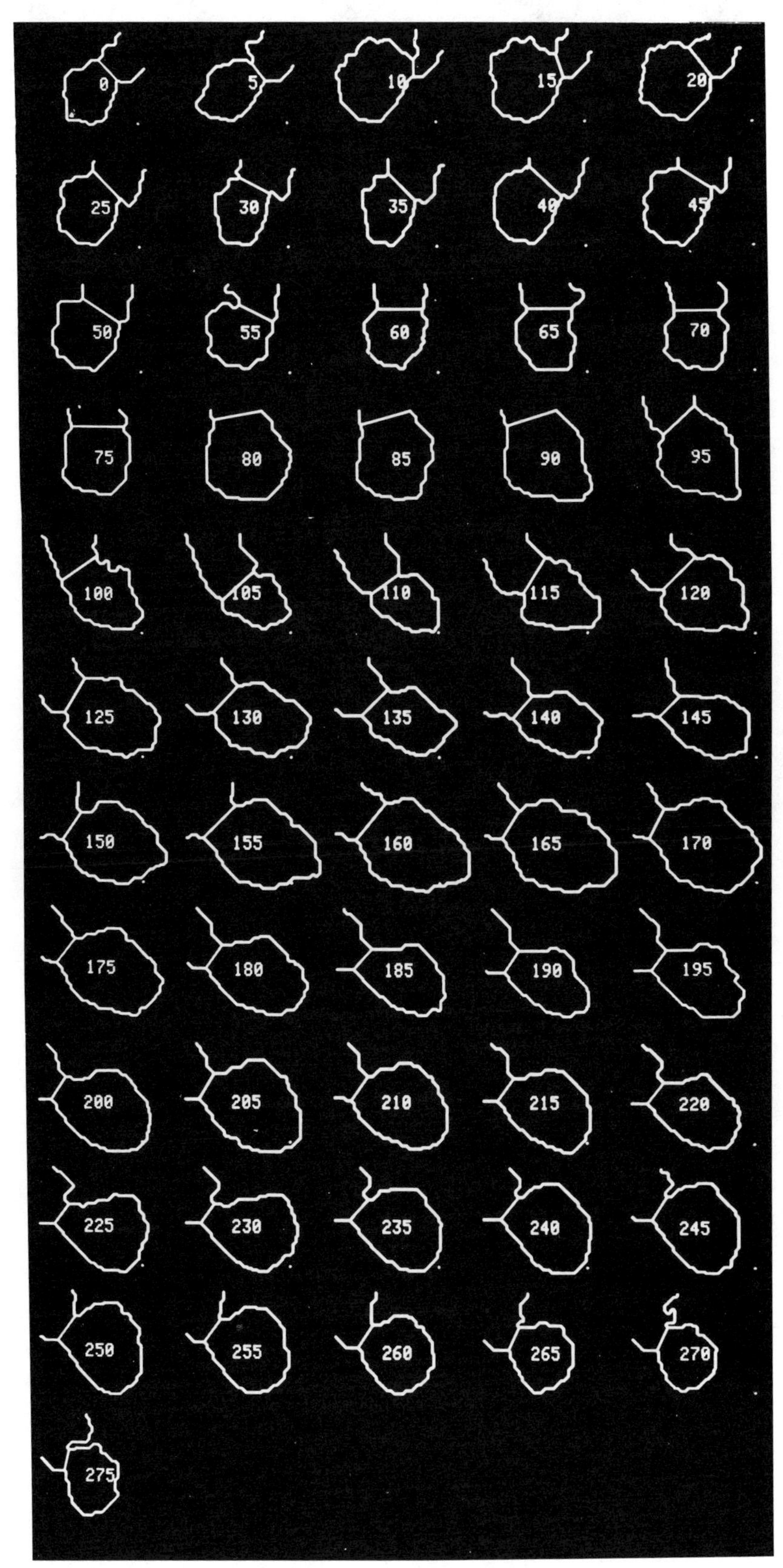

Fig. 16 Part of the boundaries detected by a computer.

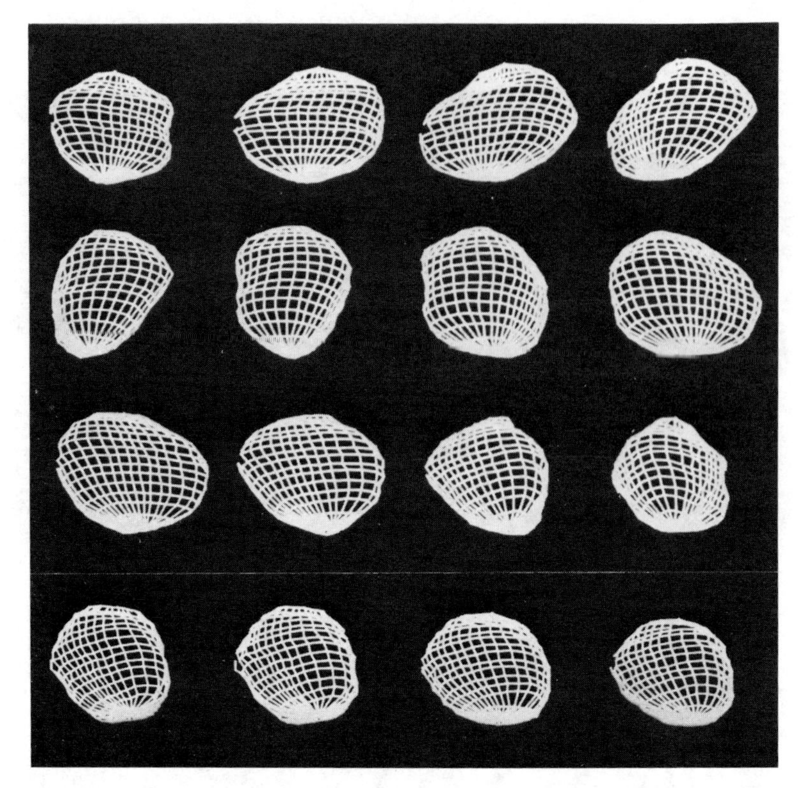

Fig. 17 Reconstructed left ventricle in a cardiac cycle.

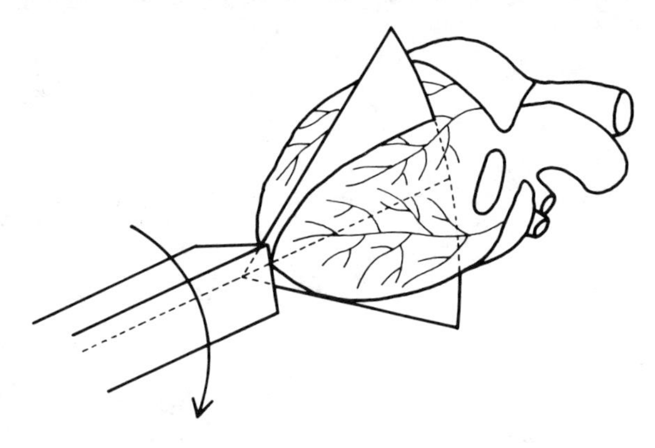

Fig. 18 Three-dimensional reconstruction from several long axis views of two-dimensional echocardiograms.

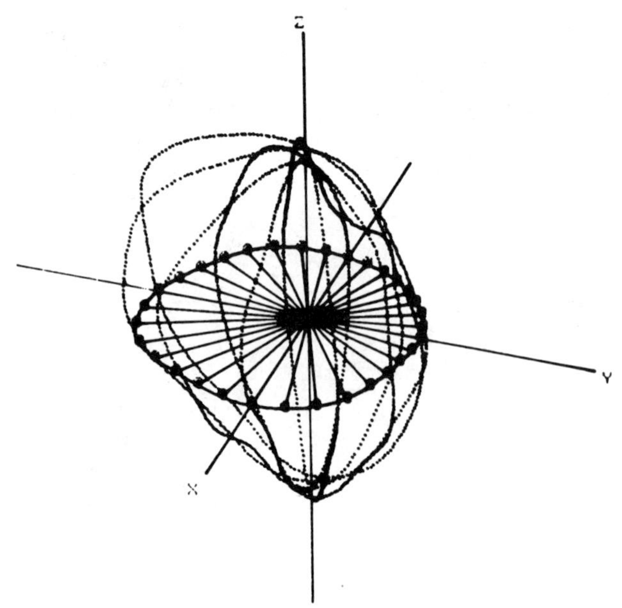

Fig. 19 Six boundary curves and cross section with 32 sample points on a spline curve.

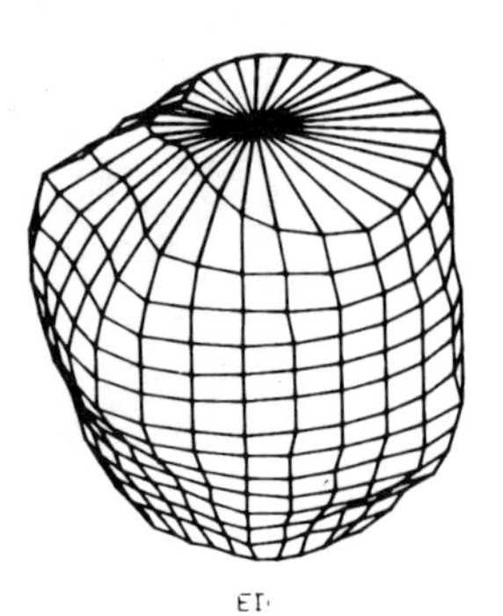

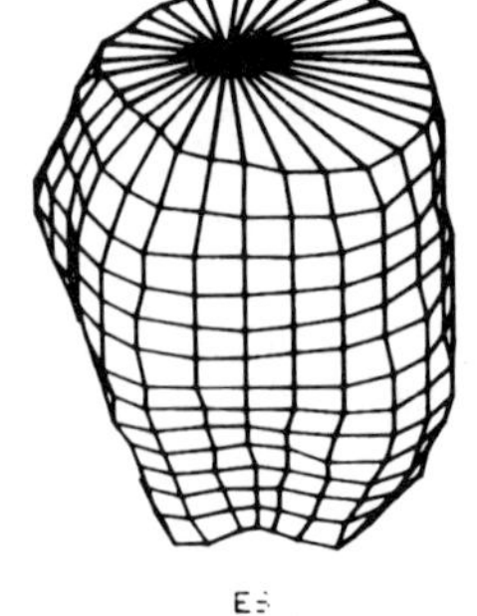

Fig. 20 Wire-framed three-dimensional shapes of left ventricle at ED and ES phases.

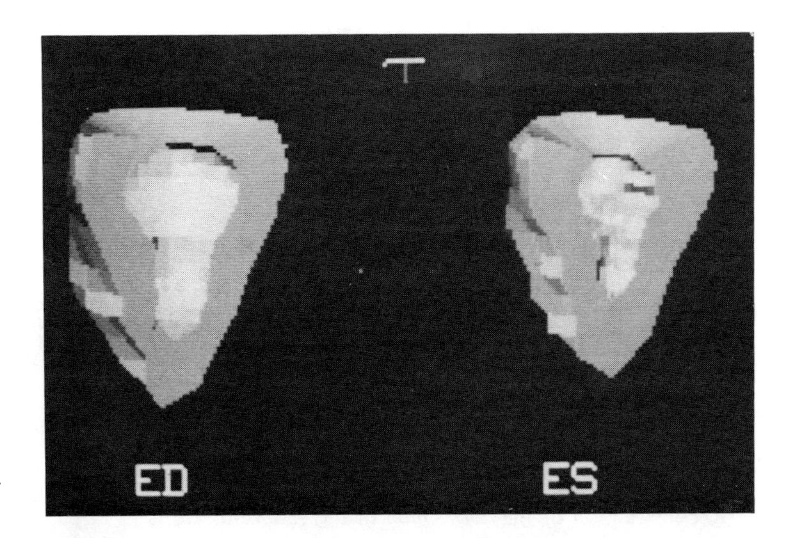

Fig. 21 Longitudinal cross sections of the myocardium at ED and ES.

ventricle. The left-hand side of this figure is at the ED phase, and the right-hand side is at the ES phase.

By using two sets of 3-D boundary data of endocardium and epicardium, we can observe cross sections of the left ventricular myocardium. Figure 21 shows longitudinal cross sections of the myocardium at ED and ES.

D Reconstruction from Magnetic Resonance Images (MRI) [21]

The magnetic resonance imaging method gives us several cross-sectional images similar to X-ray CT. Moreover, MRI is a noninvasive method, and we can set a cross-sectional plane more freely than with an X-ray CT. The quality of the images has been improved in recent years. Several sets of images in a cardiac cycle are taken by the multiphase MRI method gated by ECG.

Figure 22 shows an example of six cross-sectional images of MRI at ED phase: two transverse (z plane), two coronal (y plane), and two sagittal (x plane) images. We can get these sets of images at four cardiac phases by 6-time scanning with a 0.2 Tesler registive magnet MRI. The lines on each image show the planes of the other two cross-sectional directions, i.e., the lines on sagittal images show two coronal and two transversal planes. The left ventricular boundaries (inner and outer) on each image are traced manually by using a track ball. A boundary curve traced on a plane intersects at two points another plane perpendicular to this plane. Thus a boundary traced on a plane intersects with four cross sections at eight points. By displaying these points as bright dots on the images on four cross sections, we can easily check the correctness of the traced boundary. These dots are useful also for drawing the boundary on a rather unclear image. By using these six inner and outer boundaries, the 3-D shape of the left ventricle is reconstructed in 3-D space with 32 × 32 × 32 elements, where each element is called a ''voxel'' (volume element).

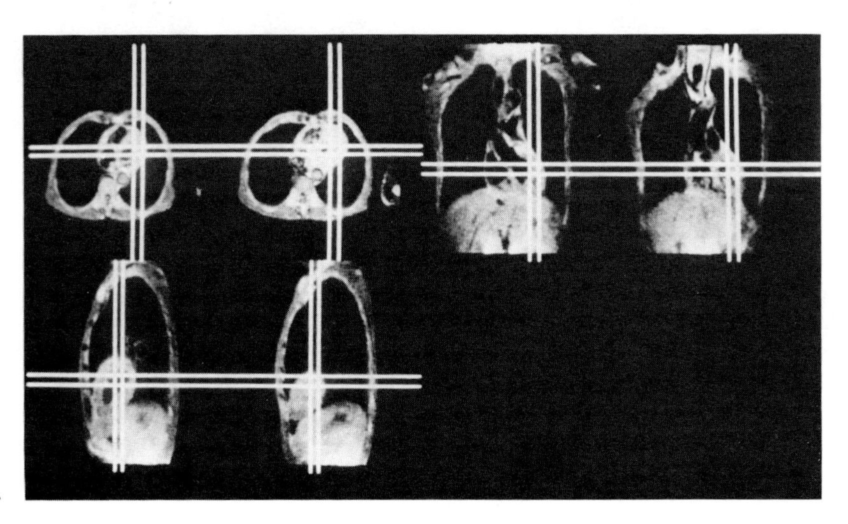

Fig. 22 Six cross-sectional images of MRI.

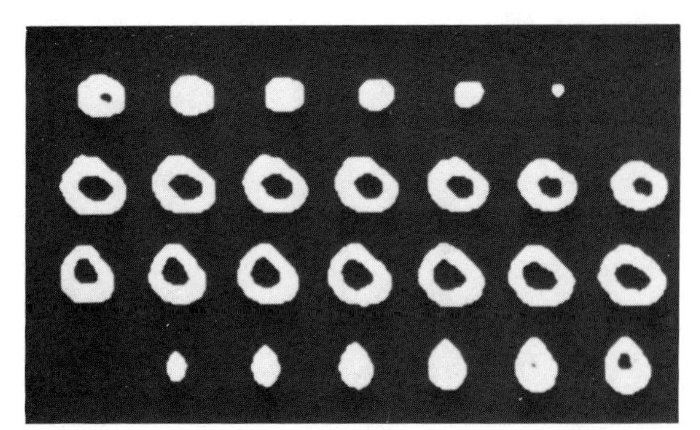

Fig. 23 Three-dimensional voxel data of myocardium on 32 z planes.

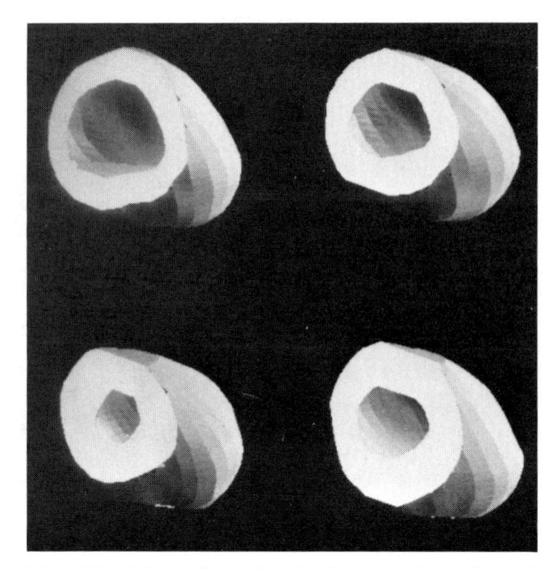

Fig. 25 Three-dimensional left ventricle cut by a plane at four cardiac phases.

Reconstruction of the 3-D shape of the left ventricle is done by choosing voxels which correspond to the inner side of the left ventricle. The steps for 3-D voxel reconstruction are as follows:

1. Draw boundary curves in the voxel space. If we cut the voxel space in the z direction, we get 32 planes and almost every plane has eight boundary points.
2. Connect these points with spline curves and digitize these curves at each voxel element.
3. Fill the inner part of the boundary. Thus we get several cross-sectional shapes of the left ventricle on every z planes.
4. Execute the same procedure to x and y planes.
5. Smooth these x, y, and z plane 3-D shapes in their 3-D space (size for smoothing is $3 \times 3 \times 3$), sum up in one 3-D voxel space, and cut by a threshold value. The result is a 3-D voxel shape of the left ventricle.

Figure 23 shows an example of the voxel data of myocardium on every z plane thus obtained. The voxel data on each z plane are changed to a smooth curve and sampled at 32 points.

Various kinds of 3-D shapes are obtainable from these 3-D data. Figure 24 is a myocardium cut by a horizontal plane with inner and outer parts of the left ventricle where the wire framed epicardium is superimposed.

Figure 25 shows 3-D left ventricles cut by a plane at four cardiac phases, and Fig. 26 shows the left ventricle and the aorta. Figure 27 shows 3-D left ventricle, left atrium, and aorta superimposed on the original MRI.

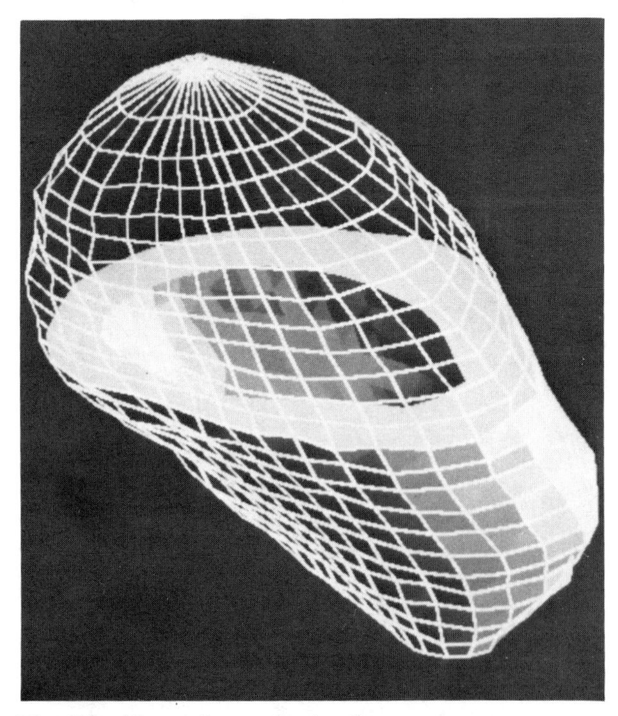

Fig. 24 Myocardium and epicardium.

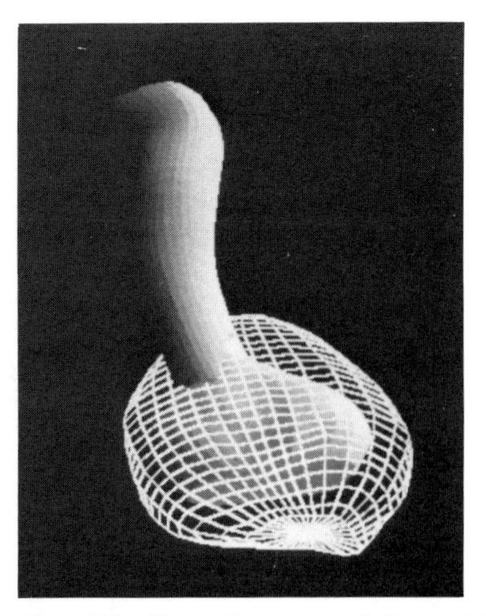

Fig. 26 Three-dimensional left ventricle and aorta.

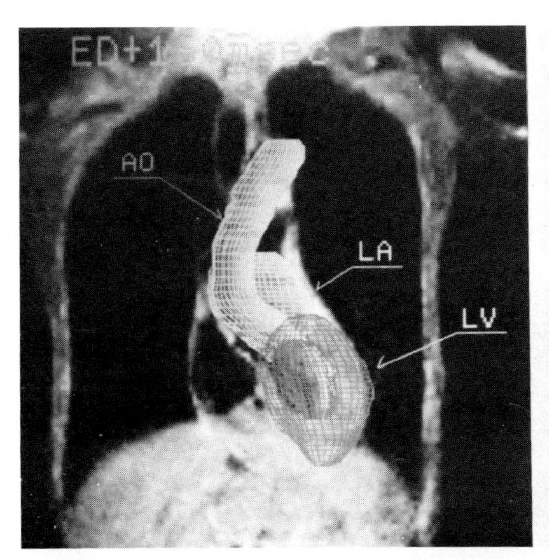

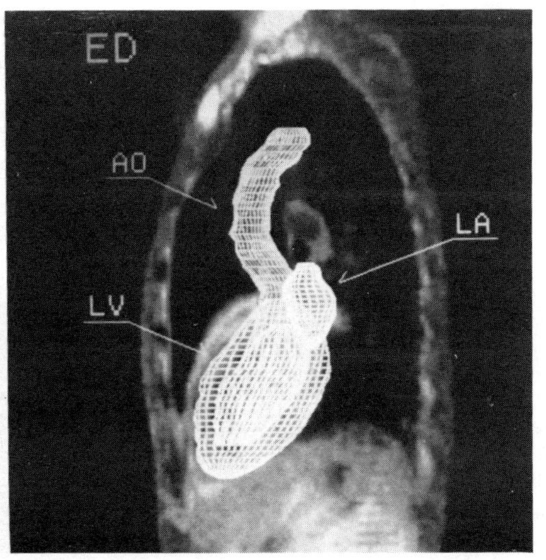

Fig. 27 Three-dimensional left ventricle, left atrium, and aorta superimposed on original MRI: (*a*) Coronal view; (*b*) sagittal view.

E Display of Functional Images

We can observe the motion of a 3-D left ventricle in the cardiac cycle by displaying 3-D sequential shapes in a cardiac cycle repeatedly, and we can check the abnormality of the motion of the regional wall from such a motion display. We can also obtain various kinds of cardiac functions from these 3-D data, which can be shown as 3-D functional images.

By comparing two 3-D images at the ED and ES phases, one can calculate a percentage shortening for the 3-D left ventricle [17]. The distances from the center of gravity of the ED image to two points on the 3-D left ventricular wall at the ED and ES phases are measured along the same direction, and named l_d and l_s. Three-dimensional percentage shortening can be calculated by the following equation:

$$\% \text{ Shortening} = \frac{l_d - l_s}{l_d} \times 100\ (\%)$$

One can display these values by pseudocolor on the silhouette of the 3-D image. The lower left image of Fig. 28 (see colorplate between pages 388 and 389) shows an example of pseudocolor display of the percentage shortening. The upper two images are 3-D left ventricle at ED and ES phases; the lower right image is the ED shape with the boundary shape of the silhouette of the ES shape superimposed. Color codes are divided every 5%, and red and black show negative percentage values, which correspond to parts for dyskinesis during a systolic period.

The values of wall thickness are also obtainable for 3-D inner and outer left ventricular shapes, as shown in Fig. 21.

Figure 29 (see colorplate between pages 388 and 389) shows an example of functional images obtained from echocardiograms. The upper two images of Fig. 29 show wall thickness at ED *(left)* and ES *(right)*. The values of wall thickness are put on the 3-D endocardium, where each 2-mm increase in the wall thickness is coded by a different color starting from an offset thickness of 8 mm. The lower left side of Fig. 29 shows 3-D percentage wall thickening, and the right side shows percentage shortening.

Color-mapping display of the regional wall motion in contraction and relaxation states, using the radial wall displacements between cardiac phases on adjacent cine frames, visualizes propagation of regional wall motions during a cardiac cycle, as shown in Fig. 30 (see colorplate between pages 388 and 389). In the figure the length of each radial line is normalized by the maximum and minimum lengths [22].

REFERENCES

1. Heintzen, P. H., ed., Roentgen-, Cine- and Videodensitometry, Georg Thieme Verlag, Stuttgart, 1971.
2. Chow, C. K., and Kaneko, T., Automatic Boundary Detection of the Left Ventricle from Cineangiograms, *Comput. Biomed. Res.*, vol. 5, no. 4, pp. 388–410, 1972.
3. Slager, C. J., Reiber, J. H. C., Schuurbiers, J. C. H., and Meester, G. T., Contouromat—A Hard-wired Left Ventricular Angio Processing System: I. Design and Application, *Comput. Biomed. Res.*, vol. 11, no. 5, pp. 491–502, 1978.
4. Reiber, J. H. C., Slager, C. J., Schuurbiers, J. C. H., and Meester, G. T., Contouromat—A Hard-wired Left Ventricular Angio Processing System: II. Performance Evaluation, *Comput. Biomed. Res.*, vol. 11, no. 5, pp. 503–523, 1978.

5. IEEE Computer Society, *Proc. Comput. Cardiol. 1974–1987.*
6. Sigwart, U., and Heintzen, P. H., eds., *Ventricular Wall Motion,* Georg Thieme Verlag, Stuttgart, 1984.
7. Eiho, S., Kuwahara, M., Fujita, M., Sasayama, S., and Kawai, C., Automatic Processing of Cineangiocardiographic Images of Left Ventricle, *Proc. 4th Int. J. Conf. Patt. Recog., Kyoto,* pp. 740–742, 1978.
8. Eiho, S., Kuwahara, M., Fujita, M., Sasayama, S., and Kawai, C., Boundary Detection of Left Ventricle from Cineangiocardiograms and Analysis of Regional Left Ventricular Wall Motion, *Proc. 6th Conf. Comput. Applic. Radiol. Comput./Aid. Anal. Radiolog. Images,* Newport Beach, pp. 221–227, 1979.
9. Fujita, M., Sasayama, S., Kawai, C., Eiho, S., and Kuwahara, M., Automatic Processing of Cineventriculograms for Analysis of Regional Myocardial Function, *Circulation,* vol. 63, no. 5, pp. 1065–1074, 1981.
10. Sasayama, S., Nonogi, H., Kawai, C., Fujita, M., Eiho, S., and Kuwahara, M., Automated Methods for Left Ventricular Volume Measurement by Cineventriculography with Minimal Doses of Contrast Medium, *Am. J. Cardiol.,* vol. 48, pp. 746–753, 1981.
11. Sasayama, S., Nonogi, H., Fujita, M., Sakurai, T., Wakabayashi, A., Kawai, C., Eiho, S., and Kuwahara, M., Analysis of Asynchronous Wall Motion by Regional Pressure-Length Loops in Patients with Coronary Artery Disease, *J. Am. Coll. Cardiol.,* vol. 4, no. 2, pp. 259–267, 1984.
12. Sasayama, S., Fujita, M., Nonogi, H., Kawai, C., Eiho, S., and Kuwahara, M., in *Ventricular Wall Motion,* ed. U. Sigwart and P. H. Heintzen, pp. 62–73, Georg Thieme Verlag., Stuttgart, 1984.
13. McLaurin, L. P., Grossman, W., Stefadouros, M. A., Rolett, E. L., and Young, D. T., A New Technique for the Study of Ventricular Pressure–Volume Relations in Man, *Circulation,* vol. 48, pp. 56–64, 1973.
14. Gilbert, B. K., Robb, R. A., and Krueger, L. M., in *Real-Time Medical Image Processing,* ed. M. Onoe, K. Preston, Jr., and A. Rosenfeld, pp. 23–40, Plenum, New York, 1980.
15. Heffernan, P. B., and Robb, R. A., Towards Improved Reconstruction of the Heart from Small Numbers of Projections, *Comput. Cardiol. 1982, Seattle,* pp. 149–152, 1983.
16. Onnasch, D. G. W., and Heintzen, P. H., A New Approach for the Reconstruction of the Right or Left Ventricular Form from Biplane Angiocardiographic Recordings, *Comput. Cardiol. 1976, St. Louis,* pp. 67–73, 1977.
17. Eiho, S., Yamada, S., and Kuwahara, M., 3-Dimensional Display of Left Ventricle by Biplane X-Ray Angiocardiograms and Assessment of Regional Myocardial Function, in *MEDINFO 80,* ed. D. A. B. Lindberg and S. Kaihara, pp. 1093–1097, North-Holland, Amsterdam, 1980.
18. Eiho, S., Kuwahara, M., Shimura, K., Wada, M., Ohta, M., and Kozuka, T., Reconstruction of the Left Ventricle from X-Ray Angiocardiograms with a Rotating Arm, *Comput. Cardiol. 1983, Aachen,* pp. 63–67, 1984.
19. Lindenau, J., Onnasch, D. G. W., Bursch, J. H., and Heintzen, P. H., Spatial Reconstruction of the Opacified Myocardium from a Small Number of Protections, *Comput. Cardiol. 1985, Linköping,* pp. 351–354, 1986.
20. Eiho, S., Kuwahara, M., Asada, N., Sasayama, S., Takahashi, M., and Kawai, C., Reconstruction of 3-D Images of Pulsating Left Ventricle from Two-dimensional Sector Scan Echocardiograms of Apical Long Axis View, *Comput. Cardiol. 1981, Florence,* pp. 19–24, 1982.
21. Eiho, S., Kuwahara, M., Fujita, Y., Matsuda, F., Sakurai, T., and Kawai, C., 3D Reconstruction of the Left Ventricle from Magnetic Resonance Images, *Comput. Cardiol. 1986, Boston,* pp. 51–56, 1987.
22. Asada, N., Tanaka, M., Hasegawa, Y., Yamashita, K., Ishii, Y., Eiho, S., and Kuwahara, M., 2-D and 3-D Analysis of Regional Wall Motion of Left Ventricle, *Comput. Cardiol. 1986, Boston,* pp. 567–570, 1987.

Chapter 20

Optical Tomography

Lambertus Hesselink

I INTRODUCTION

A Background

Classical flow visualization techniques provide information about physical observables averaged along the pathlength of the probe radiation. Space-resolved measurements cannot be obtained from a single picture, except for special cases such as axial symmetric or two-dimensional flows. Additionally, for homogeneous and isotropic turbulent flows, the spectrum of the fluctuations can be computed from a single shadowgraph as discussed by Hesselink and White [21] or schlieren photograph [45]. In general, interior flow information cannot be obtained from a single line-of-sight measurement, and we have to resort to other techniques such as laser scattering of particle-laden flows employing sheet illumination.

Tomography extends the domain of flow visualization to new applications because it provides a framework for obtaining quantitative flow measurements from a set of line-of-sight images. From classical techniques such as shadowgraphy, interferometry, and absorption, three-dimensional distributions of physical observables can be reconstructed, provided a number of images of the flow are taken from different directions. These images are commonly referred to as projections, representing line integrals of index of refraction or absorption along the pathlength (or spatial derivatives thereof). By image reconstruction the dimensionality of the imagery is increased by 1; a two-dimensional cross section is reconstructed from a set of line projections and a volume from a set of planar images.

Radon showed in 1917 [34] that from an infinite series of noiseless projections the object can be reconstructed unambiguously. In practice, of course, one can never afford such luxury, and it is easy to show that perfect image reconstruction is no longer possible. Approximate solutions, however, may be obtained from a finite set of projections.

The reconstructions are approximate in the sense that errors are introduced in the image when only a limited number of projections are available. The errors are commonly referred to as artifacts and are often streaky patterns superimposed on the image. The specifics of these artifacts depend on the object, the reconstruction technique used, the quality of the projections, and the distribution of angles over which the projections are measured. As a general observation, artifacts are introduced because there are an insufficient number of projections available to fully describe fine detail such as rapid spatial variation of the index of refraction of the object. Examples of such features include shocks, sharp density gradients, and vortices. Additional artifacts may be introduced when the projections are not uniformly spaced about the object, for instance, because projections are missing. It is then necessary to interpolate between projections, or to use iterative reconstruction schemes to obtain good results.

At present several reconstruction schemes are available[1] that can be classified into direct and indirect methods. Di-

It is a pleasure for me to acknowledge Ray Snyder's excellent contributions to optical tomography and his significant influence on our research efforts in this area. This work was supported by the U.S. Department of Energy under contract DE-FG03-85ER13312.

[1]Lawrence Berkeley Laboratory provides algorithms at a nominal cost [23].

rect methods reconstruct the image from their projections by using each projection only once. Examples in this category include convolution back projection, Fourier transform, simple and filtered back projection, as well as polynomial techniques. Iterative approaches pass through the data set several times and try to find the best reconstruction under a set of constraints. Examples of such techniques include algebraic reconstruction, simple back projection, and the maximum-entropy method.

These algorithms rely on the assumption that rays propagate along straight lines in the medium. This assumption is usually good for X-ray tomography because the wavelength of the probe radiation is shorter than any physical dimension of interest, and consequently diffraction effects can be neglected to a good approximation. The wavelength of optical waves, on the other hand, is long enough so that rays may not be straight for flows with strong density or temperature gradients. In that case, reconstructions can be severely contaminated by artifacts. These artifacts can be reduced either by using an iterative approach that corrects the projections to account for ray bending or by using post-image processing. The problems associated with ray bending are particularly severe for acoustic waves, because their wavelength is several orders of magnitude longer than the wavelength of optical waves. For this application, Devaney [11] has developed the method of back propagation.

Radio astronomers were the first to apply tomography to image microwave emission from the sun [6]. Subsequently, Hendee [17], Barrett and Swindell [2], and Robb [36] used tomography for radiologic imaging.

Applications of tomography in other areas include electron micrography [22], geology [12], oceanography [32], and industrial imaging (special issue on AO 24, 1985). In addition to radiology, tomography has been successfully applied to industrial and scientific applications. Maldanado and Olsen [29, 33] reconstructed a plasma flow from emitted intensities in an asymmetric argon arc using a series expansion, and Rowley [38] and Wolf [50] used Fourier transforms to reconstruct transparent phase objects from interferometry data. Wolf's method solves an inverse problem using the Born approximation. Shtein [39] and Vest [47] used holographic interferometry for reconstruction of flows, and Sweeney and Vest [44] determined the temperture field above a heated plate submerged in water. Dandliker and Weiss [10] made early contributions to high-resolution inversion techniques, and Iwata and Nagata [24] discussed noniterative approaches. Other fluid mechanical applications include the work by Alwang et al. [1] on slot flames.

Fundamental noise aspects of fan beam tomography have been discussed by Bennett et al. [3–5] in relation to concentration measurements in iodine-seeded flows. Hertz [18] measured temperature in a flame using interferometry and an asymmetric electric field distribution using the Kerr effect [19] in dielectric liquids.

Other applications include aerosol measurements by Williams [49], interferometric density in a Rayleigh–Benard convection cell by Kirchartz et al. [25], and interferometric measurements in a boundary layer by Cha and Vest [9]. Infrared emission from a flame was used by Uchiyama et al. [46] to measure temperature.

Tomographic measurements discussed so far have been applied to time-averaged or stationary flow fields. Time resolution is limited by data-acquisition schemes requiring mostly serial acquisition of projections. These approaches are not suitable for turbulent flow studies. For these applications Snyder and Hesselink [42] have developed an optical data-acquisition scheme for making time-resolved tomographic measurements in turbulent flows using 36 viewing angles obtained in less than 300 μs, using a 1-W argon laser.

In this chapter, the principles of image reconstruction from projections are discussed; the approach is inspired by that of Macovski. We assume that rays propagate along straight lines, but the modifications to account for the effects of ray bending are discussed as well. Examples are given of model studies that investigate the effects of the number of projections used, the noise in the data, and the angular spacing of the projections on the quality of reconstruction.

Following the discussion on reconstruction algorithms various implementations are presented. Since image reconstruction can be achieved from a variety of line-averaged data sets, examples of a few techniques are considered, including holographic interferometry, absorption, and ray-deflection tomography.

B Principles of the Method

Referring to the recording geometry of Fig. 1, the distribution of the physical property of interest is described by $g(x, y)$, which for the moment we assume to be two-dimensional; extension to three dimensions is achieved by stacking planes.

The result of measurements are estimates of path integrals along lines of known location, referred to as projections. The line along which $g(x, y)$ is measured is denoted by u, which is expressed in terms of R, the distance from the origin and the angle θ. In terms of R and θ we compute the line integral along u as follows:

$$\begin{aligned} p_\theta(R) &= p(R, \theta) \\ &= \int_{-\infty}^{\infty} g(R\cos\theta - u\sin\theta, R\sin\theta + u\cos\theta)\, du \\ &= \int_{-\infty}^{\infty} g(x, y)\,\delta(x\cos\theta + y\sin\theta - R)\, dx\, dy \end{aligned} \tag{1}$$

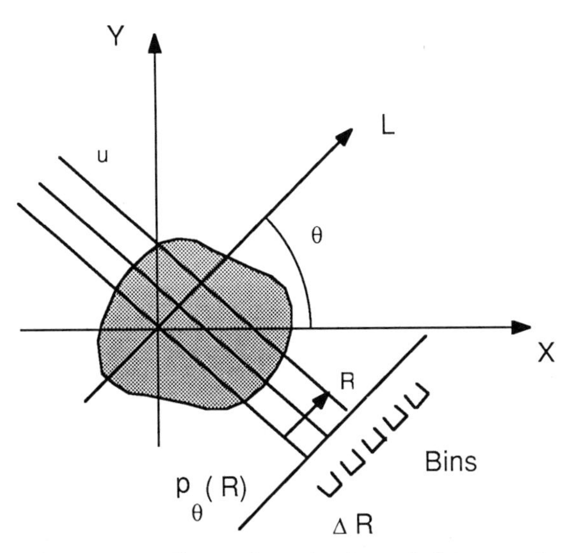

Fig. 1 Recording configuration for optical tomography.

The aim of tomography is to recover $g(x, y)$ from projection data $p_\theta(R)$.

The simplest case of reconstruction of axial symmetric objects was solved by Abel in 1826. Radon, in 1917, provided an analytical result in the general case and showed that the object can be recovered from the projections, provided that an infinite number of projections are known with arbitrary accuracy. In practice, of course, this is never possible, and approximate reconstruction methods are desired.

The physical parameters represented by $g(x, y)$ may be density, absorption coefficient, or temperature of fluids or solids. For example, in medicine we may be interested in reconstructing interior details of the brain, or in nondestructive testing applications we may wish to find cracks at the interior of a solid boule. In fluid mechanics the density or tempeature distribution inside a combusting gaseous jet may be desired.

If case measurements of the absorption coefficients are made, the procedure has to be modified slightly by taking the logarithm of the measured data prior to reconstruction.

$$I = I_0\, e^{-\int \mu(z)\, dz}$$

and

$$\mathrm{Ln}\left(\frac{I_0}{I}\right) = \int \mu(z)\, dz$$

To carry out the reconstruction several techniques have been developed, including the algebraic reconstruction technique (ART), back projection, filtered back projection, convolution back projection, iterative approaches, series expansion methods, and maximum entropy technique. The underlying principle of these approaches is to find a distribution of the measured physical observable consistent with all projections. In general, we are dealing with a restoration problem in which we wish to reconstruct an image of $N \times N$ pixels from p projections of n elements each.

II ALGEBRAIC RECONSTRUCTION TECHNIQUE

The ART technique is a simple iterative approach by which elements g_{ij} of the image matrix shown in Fig. 2 are computed from the projection data p_k, where k denotes the measured projection. The bin width in each projection is ΔR, and the values g_{ij} are assigned at points in the center of each pixel. The separation between points on the rectangular grid is denoted by Δx and Δy.

The procedure starts by setting all values of $g_{ij} = 0$ and using the known projection data p_k. The iteration values for g_{ij} are computed by determining the sum off all pixel values locted along the path of the projection (the points located inside the strip denoted by P_k shown in Fig. 2). This sum is subtracted from the measured projection value, and the difference is evenly distributed among these pixels. The procedure is repeated until the image elements g_{ij} are consistent with the measured projection data. That is to say, the sum of all g_{ij} values in a particular direction should be equal to the corresponding measured sum p_k.

Mathematically, the algorithm is defined as

$$g_{ij}^{q+1} = g_{ij}^{q} + \frac{p_k - \sum_{i=1}^{N} g_{ij}^{q}}{N} \tag{2}$$

where q denotes the iteration and the pixels g_{ij} are those that correspond to the projection P_k (the points inside the tilted rectangle).

As a simple example consider the following matrix and projections taken along the horizontal, vertical, and diagonal directions.

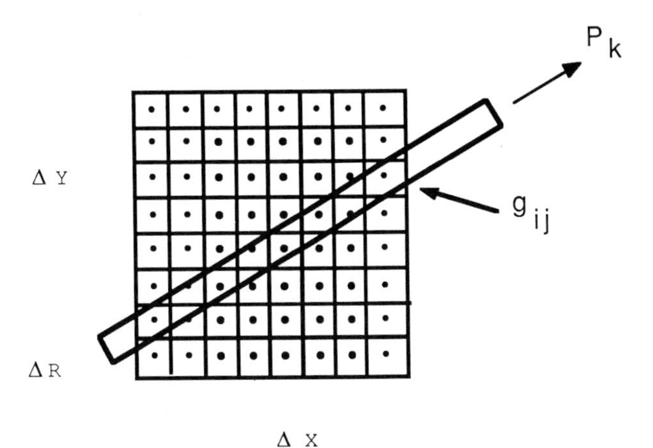

Fig. 2 Principle of the algebraic reconstruction technique (art).

$$\begin{bmatrix} 12 & 5 & 16 & 9 \\ \cdots & g_{11} = 3 & g_{21} = 7 & 10 \\ \cdots & g_{12} = 2 & g_{22} = 9 & 11 \end{bmatrix}$$

The process starts with setting all $g_{ij} = 0$ and computing $g_{11}^1 = g_{21}^1 = 2.5$ and $g_{12}^1 = g_{22}^1 = 8$ using the vertical projection values 5 and 16, respectively. These values are then substituted into the matrix, and the calculation is repeated using the horizontal projections 10 and 11. In the third iteration the two diagonal projections 12 and 9 are used to compute the final result. For larger sized images many iterations may be needed.

A variation on this scheme includes multiplicative ART

$$g_{ij}^{q+1} = \frac{p_k}{\sum_{i=1}^{N} g_{ij}^q} g_{ij}^q$$

For high-resolution images many iterations may be required. Noise in projections tends to reduce convergence and can make the reconstruction problem ill posed. Consequently, many reconstructions may be consistent with the data. The question is which one to use, because it can be shown in general that the number of feasible images is very large. Another approach to selection of the best image is the maximum entropy method.

The maximum entropy method selects that image of all possible images that has the greatest entropy

$$S = -\sum P_i \log \frac{P_i}{m_i}$$

where P_i denotes the proportion of intensity originating at pixel i, and m_i represents the initial estimate for that pixel [15]. This iterative procedure selects the image that is the least dependent on unmeasured parameters in the projections. To improve results, a priori information about the object should always be incorporated. For details of the method and numerical implementations, the reader is referred to Mohammad-Djafari and Demoment [31].

To avoid extensive computing, direct methods such as the Fourier transform approach are preferred.

III FOURIER TRANSFORM APPROACH

The starting point is Eq. (1), which can be rewritten

$$\begin{aligned} p_\theta(R) &= \iint g(x, y)\, \delta(x \cos\theta + y \sin\theta - R)\, dx\, dy \\ &= \int_0^{2\pi} \int_0^{\infty} g(r, \phi)\, \delta[r \cos(\theta - \phi) - R] r\, dr\, d\phi \end{aligned} \tag{3}$$

where $p_\theta(R)$ denotes the projection information in the θ direction. In this calculation, integration is carried out along $x \cos\theta + y \sin\theta = R$ using polar coordinates r, ϕ, and $r \cos(\theta - \phi) = R$. The delta function notation is convenient here, because $\delta[r \cos(\theta - \phi) - R]$ determines the line along which integration takes place. The expression of (3) is directly related to the Fourier transform in polar coordinates.

Consider $g(x, y)$ and let $G(f_x, f_y) = FT[g(x, y)]$.

$$G(f_x, f_y) = \iint g(x, y) e^{-i2\pi[f_x x + f_y y]}\, dx\, dy$$

or in polar coordinates

$$G(f_x, f_y) = G(\rho, \beta) \qquad f_x = \rho \cos\beta \qquad f_y = \rho \sin\beta$$

$$\begin{aligned} G(\rho, \beta) &= \iint g(x, y) e^{-i2\pi\rho(x \cos\beta + y \sin\beta)}\, dx\, dy \\ &= \iiint g(x, y)\, \delta(x \cos\beta \\ &\qquad + y \sin\beta - R) e^{-i2\pi\rho R}\, dx\, dy\, dR \end{aligned} \tag{4}$$

Comparing (3) and (4) we get

$$G(\rho, \beta) = \int p_\beta(R) e^{-i2\pi\rho R}\, dR \tag{5}$$

Thus the Fourier transform of a projection at angle β forms a line in Fourier space at exactly the same angle $\beta = \theta$, as shown in Fig. 3.

From this analysis we can now deduce the reconstruction procedure. The projections measured in object space are one-for-one Fourier transformed to provide corresponding lines in Fourier space. Once this space is filled, a 2-D Fourier inverse transformation reveals the reconstructed image.

$$\begin{aligned} g(x, y) &= \iint G(f_x, f_y) e^{i2\pi(f_x x + f_y y)}\, df_x\, df_y \\ &= \int_0^{2\pi} d\theta \int_0^{\infty} G(\rho, \theta) e^{i2\pi\rho(x \cos\theta + y \sin\theta)} \rho\, d\rho \end{aligned} \tag{6}$$

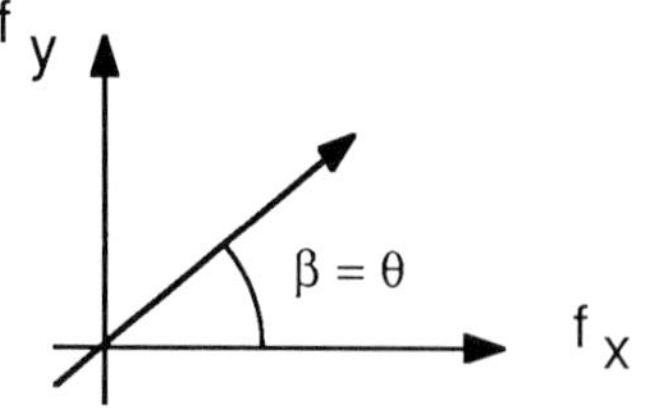

Fig. 3 Relationship between projection and Fourier space.

The problem with the Fourier approach is the need for several coordinate transformations that make computations costly. Image quality, however, is often quite good.

Alternatively, simpler approximate approaches provide faster reconstruction algorithms. A straightforward and fast technique is convolution back projection.

IV BACK PROJECTION

The idea behind back projection is to assign the measured value of the projection to each element on the line along which the integration is carried out. In other words, since the distribution of the physical variable along the line is unknown, we assume, as a first guess, that the projection value is equally distributed, similar to the starting point for ART. By summing all contributions from the projections at a given point in the image, we can find the reconstruction.

As an example consider an impulse located at the origin as a representation of a single point in a more complicated image. By following the back projection procedure we determine the impulse response. This impulse response—the reconstructed image point—is always wider than the input spot. Once the impulse response is known, the image is enhanced by filtering techniques. Mathematically, we may express this idea as follows:

Consider a δ function at the origin (a bright spot in a dark background). The corresponding projections are then also delta functions. The back projection technique proceeds by assigning the measured value along the entire averaging line. We know the point is located somewhere along the line but we do not know precisely where it is:

$$b_\theta(x, y) = \int_{-\infty}^{\infty} p_\theta(R)\, \delta(x \cos\theta + y \cos\theta - R)\, dR$$

where $b_\theta(x, y)$ denotes the back-projected density due to projection $p_\theta(R)$ at angle θ. Now, add up the contributions at all angles to get

$$\begin{aligned} g_b(x, y) &= \int_0^{\pi} b_\theta(x, y)\, d\theta \\ &= \int_0^{\pi} d\theta \int_{-\infty}^{\infty} p_\theta(R)\, \delta(x \cos\theta + y \sin\theta - R)\, dR \end{aligned} \quad (7)$$

where $g_b(x, y)$ represents the reconstructed point at the origin. By analyzing (7), it can be shown (see, for instance, Macovski [28]) that the response function is a $1/r$ blur, implying that

$$g_b(x, y) = g(x, y) ** \frac{1}{r} \quad (8)$$

where the ** denotes convolution. Thus high spatial frequency components in the image are smeared out and the image is fuzzy.

The correction constitutes a well-known approach in signal processing, namely, inverse filtering. We take the Fourier transform of $g_b(x, y)$, multiply it by ρ, and then take the inverse transform to obtain an improved image.

$$G_b(\rho, \theta) = \frac{G(\rho, \theta)}{\rho}$$
$$G(\rho, \theta) = \rho G_b(\rho, \theta)$$

The problem with this approach is that two FFT operations are required.

V FILTERED BACK PROJECTION

In an improved approach we like to retain the attractive features of back projection and filtering, but we wish to carry out only 1-D Fourier transform operations.

To illustrate the procedure we start with (7)

$$g_b(x, y) = \int_0^{\pi} d\theta \int_{-\infty}^{\infty} p_\theta(R)\, \delta(x \cos\theta + y \sin\theta - R)\, dR \quad (9)$$

and, using (5), we get

$$\begin{aligned} g_b(x, y) &= \int_0^{\pi} d\theta \int_{-\infty}^{\infty} F_1^{-1} G(\rho, \theta)\, \delta(x \cos\theta + y \sin\theta - R)\, dR \\ &= \int_0^{\pi} d\theta \int_{-\infty}^{\infty} \left[\int_{-\infty}^{\infty} G(\rho, \theta) e^{i2\pi\rho R}\, d\rho\right] \delta(x \cos\theta + y \sin\theta - R)\, dR \\ &= \int_0^{\pi} d\theta \int_{-\infty}^{\infty} G(\rho, \theta) e^{i2\pi\rho(x \cos\theta + y \sin\theta)}\, d\rho \end{aligned}$$

or

$$g_b(x, y) = \int_0^{\pi} d\theta \int_{-\infty}^{\infty} \frac{F_1[p_\theta(R)]}{|\rho|} e^{i2\pi\rho(x \cos\theta + y \sin\theta)} |\rho|\, d\rho \quad (10)$$

which produces blurred images.

We can see this by comparing (10) with the expression for the 2-D Fourier transform technique, which provides good results:

$$\begin{aligned} g(x, y) &= \int_0^{2\pi} d\theta \int_0^{\infty} G(\rho, \theta) e^{i2\pi\rho(x \cos\theta + y \sin\theta)} \rho\, d\rho \\ &= \int_0^{\pi} d\theta \int_{-\infty}^{\infty} G(\rho, \theta) e^{i2\pi\rho(x \cos\theta + y \sin\theta)} |\rho|\, d\rho \end{aligned} \quad (11)$$

Thus to get good results with the back projection method we have to multiply the integrand of (10) by $|\rho|$. Equation (10) now becomes

$$g(x, y) = \int_0^{\pi} d\theta \int_{-\infty}^{\infty} \frac{F_1[p_\theta(R)]|\rho|}{|\rho|} e^{i2\pi\rho(x \cos \theta + y \sin \theta)}|\rho| \, d\rho \tag{12}$$

We may interpret this result as follows. The 1-D Fourier transform of the back projection is multiplied by the frequency variable $|\rho|$ to eliminate $1/|\rho|$ blurring. The procedure now consists of taking the 1-D transform of each projection, filtering them by $|\rho|$, and 1-D inverse transforming the result.

$$g(x, y) = \int_0^{\pi} d\theta \int_{-\infty}^{\infty} \{F_1[P_\theta(R)]|\rho| e^{i2\pi\rho R}\} \delta(x \cos \theta + y \sin \theta - R) \, dR$$

and

$$g(x, y) = \int_0^{\pi} d\theta \int_{-\infty}^{\infty} F_1^{-1}\{F_1[p_\theta(R)]|\rho|\} \delta(x \cos \theta + y \sin \theta - R) \, dR \tag{13}$$

Alternatively, we do not have to use Fourier transforms, but may perform convolution operations because

$$F_1^{-1}\{F_1[p_\theta(R)]|\rho|\} = p_\theta(R) * F_1^{-1}[|\rho|] \tag{14}$$

Now the projections are convolved with the inverse transform of $|\rho|$ to obtain the reconstructed image. This process is computationally very attractive and the most widely used reconstruction algorithm.

The problem with this approach is that $F_1^{-1}|\rho|$ does not exist, and we need to use approximations because practical systems are always band limited. One such choice is ρ multiplied by a limiting rectangle function [16], as shown and defined in Fig. 4.

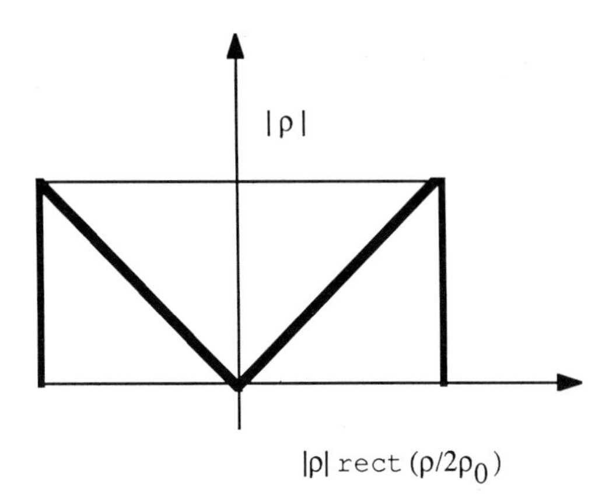

Fig. 4 Rectangular filter for convolution back projection.

$$|\rho| \text{ rect } \frac{\rho}{2\rho_0} = \text{rect } \frac{\rho}{2\rho_0} - \Lambda\left(\frac{\rho}{\rho_0}\right)$$

and

$$F_1^{-1}\left[|\rho| \text{ rect } \frac{\rho}{2\rho_0}\right] = \rho_0(2 \text{ sinc } 2\rho_0 R - \text{sinc}^2 \rho_0 R) \tag{15}$$

The convolution back projection method is now interpreted as follows. Take $p_\theta(R)$, which is band limited, convolve it with (15) and use back projection. For this purpose several filters have been designed [23, 28], and the choice of a particular filter is usually determined by trial and error.

VI IMPLEMENTATION

A Modeling Studies

Modeling of the tomographic process is useful for support of experiments. In practice, projection data are always contaminated with noise, and it is not feasible to measure an infinite number of projections. Since the quality of the reconstructed image depends on both the projection data and the reconstruction algorithm, model studies can provide answers about image quality as a function of the number of projections, their angular position about the object, noise effects, and ray bending.

To carry out modeling studies, a priori information about the object is required. This knowledge may be available from numerical simulations of the flow field or other diagnostics. To simulate the actual experiment, projections are computed at various angles through the flow by carrying out line integration of the index of refraction or the absorption coefficient along each ray direction. From these projections the image is reconstructed using any one of the available reconstruction codes. The effects of noise on the recording and data acquisition may be simulated by adding noise to the projections. The effect of ray bending on the reconstruction may be studied using an iterative approach.

First, the image is reconstructed, assuming that the rays are straight. Once the flow is known, rays are propagated through the image using exact ray-tracing formulas allowing ray paths to bend. Subsequently, a comparison is made between the projections thus computed and the original projections used for the reconstruction. Inconsistencies between the two sets of projection data are attributed to ray bending. To account for these effects, the old projections are modified by shifting projection values laterally; the amount of displacement is determined by the difference between the position of a straight ray and the actual ray location in the projection. These modified projections are then used for a second iteration to reconstruct the object. Several iterations may be needed before significantly improved results are obtained. Unfortunately, in general one cannot prove that the iterative procedure converges to the best solution. The procedure is usually halted when the difference in the mean squared sense between two images reconstructed in consecutive iterations is below a threshold value.

Examples of modeling studies have been reported by Stuck [43], Byer and Shepp [7], Snyder and Hesselink [40], Vest [48], and Modarress et al. [30]. Stuck as well as Byer and Shepp studied remote sensing of air pollution using absorption of probe radiation for the projection data. No ray-tracing calculations were carried out in these studies.

As an example we will discuss the work of Snyder and Hesselink [41], who studied tomographic reconstruction of the flow over a revolving helicopter rotor blade using holographic interferometry for measurement of the projections and convolution back projection as the reconstruction technique. Similar studies were subsequently carried out by Modarress et al. [30] using the algebraic reconstruction technique.

The flow around a revolving helicopter rotor blade is transonic and three-dimensional, including shocks producing strong acoustic radiation. Due to the rotating geometry it is difficult to make conventional probe measurements, and remote optical techniques such as holographic interferometry are preferred. Since the flow about the rotor in blade-fixed coordinates is stationary for hovering flight conditions, tomography is particularly suitable for making space-resolved density measurements.

In a coordinate system affixed to the rotating blade, a coherent beam is passed through the region of interest and pivoted about an axis in this region. Interferograms are recorded for each orientation, typically at a few degrees' spacing and over a 180° arc. In practice, the optical apparatus is not rotated, but translated laterally in repeated experiments to record projections at varying angular positions of the blade.

Referring to Fig. 5 for a single angle θ the region near the tip of the blade is probed with a collimated laser beam. The exiting beam with optical path information

$$P_\theta(B) = \int_s n \, ds \tag{16}$$

(where s is the pathlength along a ray, n the index of refraction, and B denotes the bin number associated with the ray) is recorded interferometrically, and the resulting phase distribution is referred to as the projection. Two-dimensional projections are recorded for a range of θ. To reconstruct a volume of data a plane of information is required for each projection, but a cross section may be reconstructed from single projection lines (all in the cross-sectional plane). The projection data thus acquired can be reconstructed into an index-of-refraction map by use of any of the discussed algorithms. Due to practical limitations all projections may not be available. As an example, the

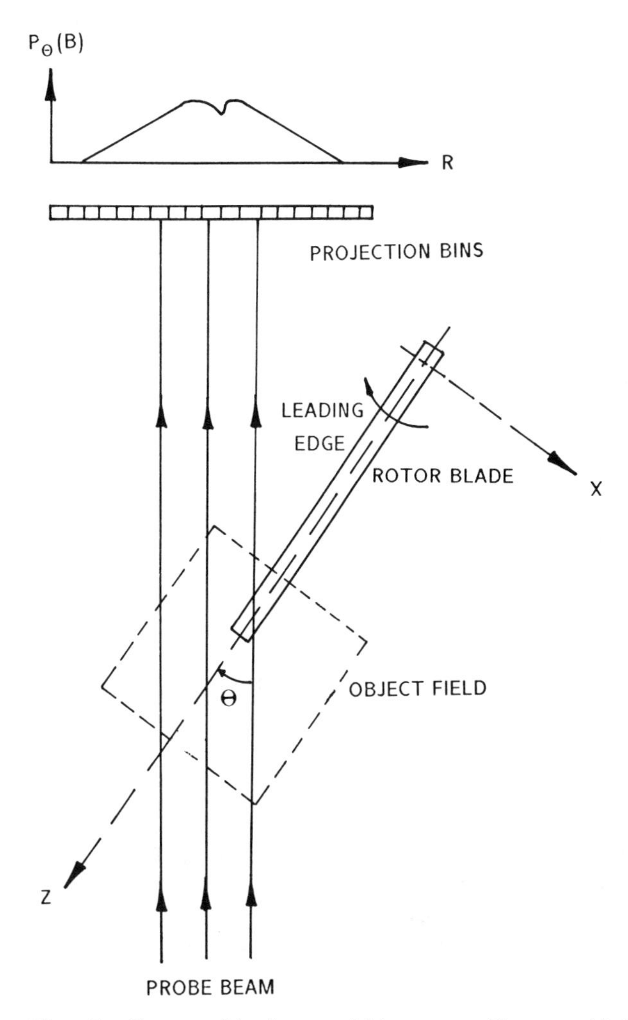

Fig. 5 Tomographic data-acquisition setup. The rotor blade rotates through the probe beam as shown. The region of the flow field that is imaged and reconstructed is outlined by the dashed line. The pathlength information in the probe beam is recorded in the projection bins away from the object.

AVRADCOM facility at NASA Ames allows projection data to be collected over only a limited range of angles, and the probe beam is smaller than the extent of the disturbed flow about the tip. Since the tomographic procedure reconstructs only that region in space where the probe beams overlap, perturbations in the flow outside this region are considered noise and reduce image quality. In the Ames facility the recording plane is located approximately 9 m from the rotor blade so that small variations in the ray direction give rise to substantial ray displacements at the measuring location.

In the modeling study the amount of data necessary for accurate reconstruction is determined using only limited views, and the effect of limited probe beam size on image quality is considered. The effect of distorted data due to beam wander is also investigated, and an iterative approach for improvement of image quality is shown to produce

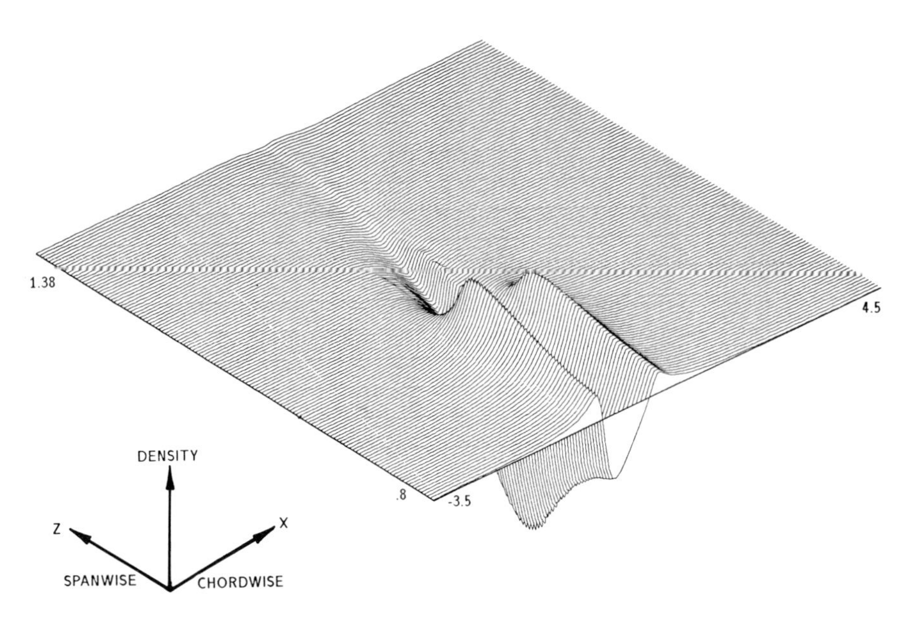

Fig. 6 Test-bed density field. Three-dimensional isometric showing the density distribution about the tip of the rotor blade in a single plane above the blade. The ridges in the plot correspond roughly to the leading and trailing edges of the blade, respectively.

higher quality results. The chief findings of the study are briefly discussed.

The test-bed density field used for this study is shown in Fig. 6. These data were kindly provided by Caradonna [8], who computed this flow making assumptions that needed experimental verification. Inspection of the data shows that density gradients are mainly directed perpendicular to the blade direction. Computations of the projections confirmed that projections perpendicular to the blade show very little structure (they are constant), but those in only a limited angular range about the rotor blade are significantly affected by the flow, as shown in Figs. 7 and 8. This leads to the conclusion that the flow can be reconstructed using only a limited number of projections centered about the rotor blade and estimating the remaining ones from simple geometry.

Reconstructed images using either all or 40 angles are shown in Figs. 9 and 10, respectively. The effect of using only a limited number of projections is manifested as the streaky artifacts extending from the shock. Results shown in Fig. 11 indicate that the position of the shock is always accurately reconstructed, but the magnitude may not be, depending on the number of iterations used. From this calculation one concludes that accurate reconstructions are possible with only 30 evenly spaced projections at angles of 2° intervals.

The effect of limited probe-beam size and variations in index of refraction along the beam path outside the region of interest is to reduce image quality; the nature of the artifacts depends on the specific geometry employed [41].

Artifacts caused by ray bending are the most influenced by views parallel to the rotor blade, and rays propagating perpendicular to this direction are essentially straight. Even along the blade direction, ray bending is small, but the effects are magnified by the long pathlength between the flow and the recording medium. In essence we assume that the rays in this direction are pivoted about a point near the tip by an amount equal to that of the angular ray deviation integrated along the pathlength through the region of interest. Displacement is very small at that location, but is mag-

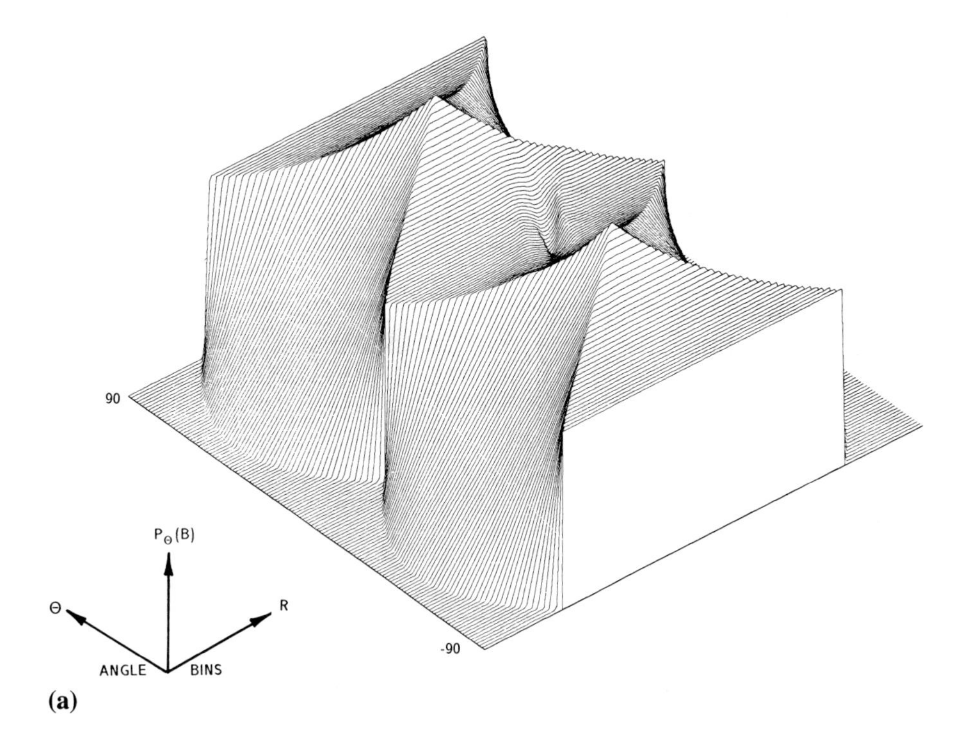

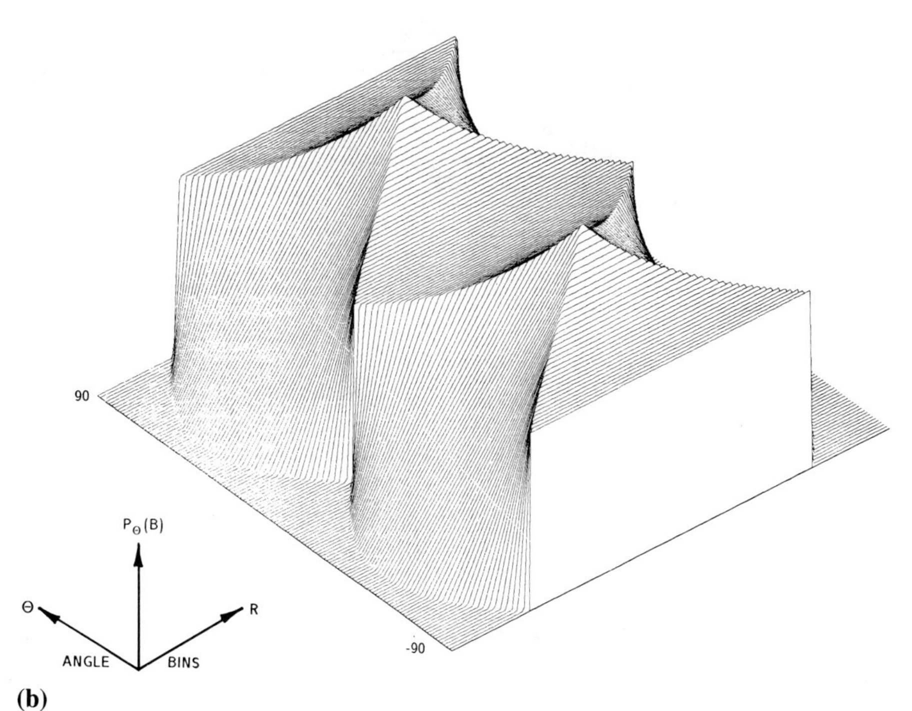

Fig. 7 Unbiased projections: (*a*) Projections for all angles through the test-bed density field; (*b*) projections through a uniform density field. Note the rectangle–trapezoid–triangle features. The dimple in the center of (*a*) contains most of the information needed to reconstruct the test bed.

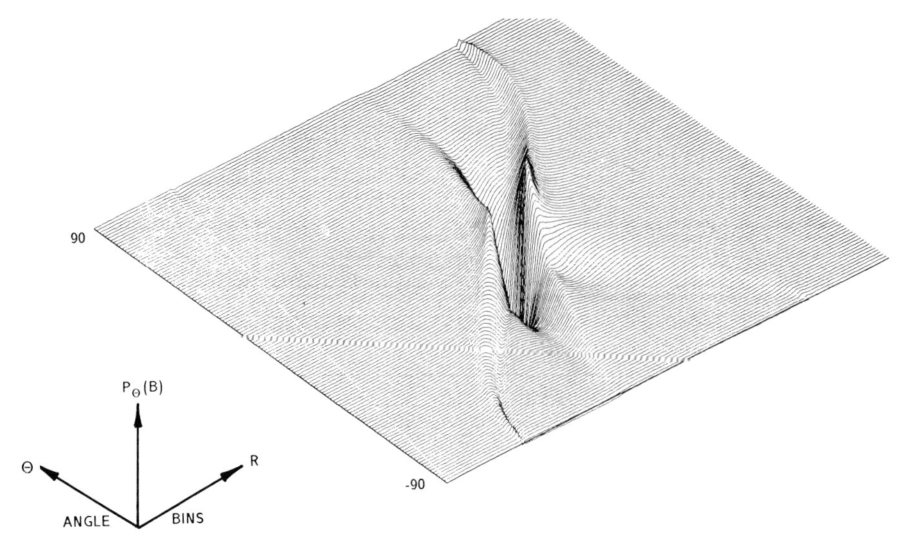

Fig. 8 Projections through test bed. Projections through density field with bias removed.

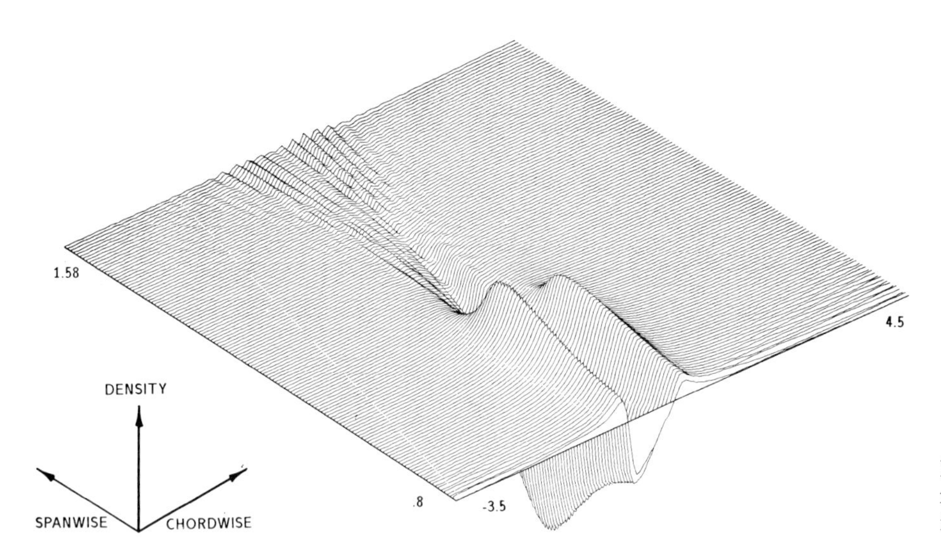

Fig. 9 All-angle reconstruction. Direct reconstruction of projections shown in Fig. 7.

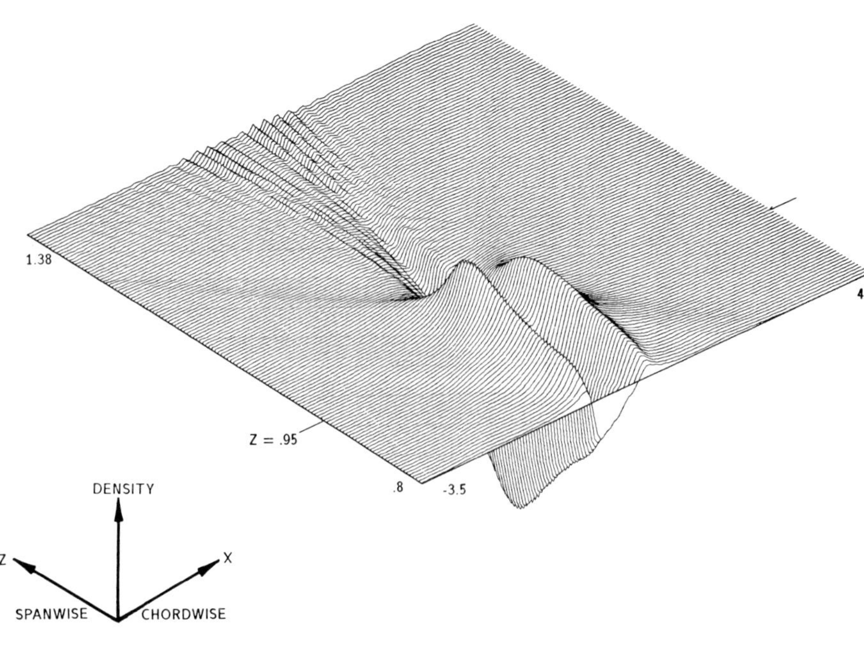

Fig. 10 40-Angle reconstruction. Reconstruction of projections in Fig. 7 with data for first and last 25 angles (first and last 50° of view) set to 0. Line marked 0.95 shows section that is displayed in cross-section graphs.

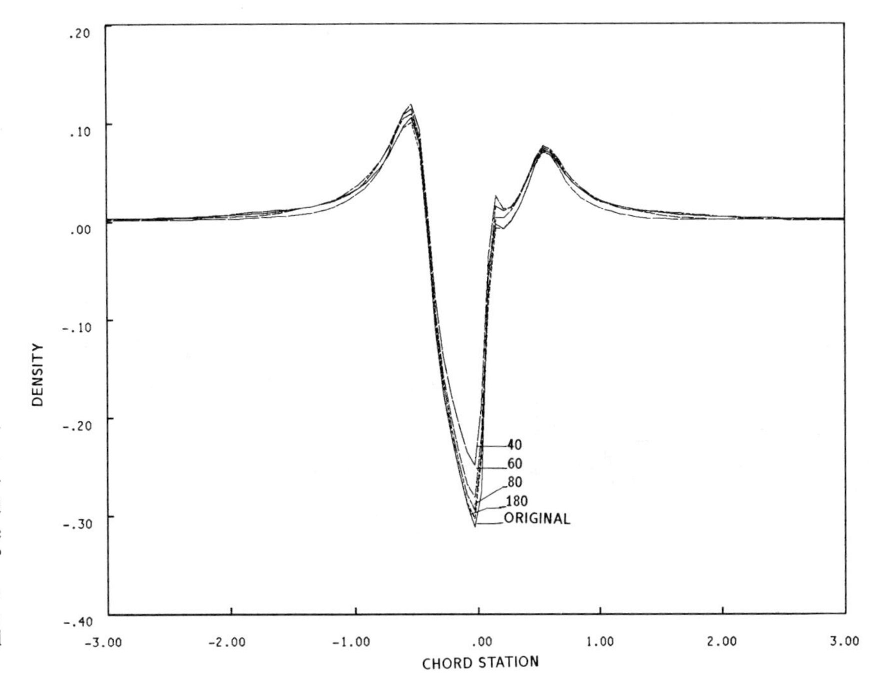

Fig. 11 Cross sections of reconstructions for varying number of views. 95%-Span cross section of reconstructed density field. Line marked *original* is a cross section through the test bed. Other lines correspond to 180° included view (90 angles as in Fig. 9), 80° included view (40 angles as in Fig. 10), 60° included view (30 angles), and 40° included view (20 angles).

nified by free space propagation outside the shock as shown in Fig. 12. When the ray is deflected over a distance larger than one bin width, substantial artifacts may arise when using the convolution back projection for reconstruction. The effects of ray bending can be reduced by an iterative procedure, which is as follows:

1. Discrete rays, spaced initially one bin width apart, are traced through the test bed and 120 chordlengths beyond, through free space, to the recording plane.
2. The sequence of ray positions at the recording plane (B_i, $i = 1, 2, \ldots, 200$) is scanned. Ray crossings are detected when $B_i < B_{i-1}$, in which case the ray with the largest phase is discarded. (We never had a case where more than two rays crossed.)
3. The ray phases P_i are interpolated linearly onto an evenly spaced array. This array constitutes the simulated projection data $P_\theta(B)$.
4. These projections are smoothed and then reconstructed to produce the 0*th* iteration.

$$P_\theta^s(B) = H ** P_\theta(B)$$
$$D_0(x, y) = \mathcal{R}\{P_0^s(B)\}$$

where ** denotes convolution, $\mathcal{R}\{\ \}$ is the reconstruction operation, and $H = [0.25 \quad 0.5 \quad 0.25]$.

5. The 0*th* iteration is smoothed

$$D_0^s = H ** [H ** D_0(x, y)]$$

where

$$H = \begin{bmatrix} 0 & 0.2 & 0 \\ 0.2 & 0.2 & 0.2 \\ 0 & 0.2 & 0 \end{bmatrix}$$

and rays are traced through eliminating crossing rays as in procedure 2 above. Straight projections are also made through the smoothed reconstruction. Thus for each bent ray there is a corresponding straight ray; a correction function can be built up consisting of a spatial displacement defined as

$$\Delta = i \times (\text{bin width}) - B_i$$

and a phase correction defined as

$$\delta = \int_{\substack{\text{straight}\\ \text{path}}} n\,ds - \int_{\substack{\text{bent}\\ \text{path}}} n\,ds$$

at each point that maps the bent projection into a straight projection.

$$\hat{P}_\theta(B_i + \Delta_i) = P_\theta(B_i) + \delta_i$$

6. The correction function is applied to the simulated projection data $P_\theta(B)$ from procedure 3 above. This field is then interpolated linearly onto an evenly spaced array.

(a)

(b)

Fig. 12 Ray traces through test bed, for 90° projection for clarity, only every fourth ray is shown: (*a*) Traces with refractive effects magnified by 50; (*b*) actual traces.

7. The corrected projections are smoothed and then reconstructed as in procedure 4 above to produce the 1*st* iteration.

$$D_1(x,\ y) = \mathcal{R}\{\hat{P}^s_\theta(B)\}$$

This process is repeated; in each case the correction determined from the ray traces through the latest reconstruction is applied to the originally determined projections to determine a new $\hat{P}_\theta(B)$.

In this study three iterations are performed with limited projection data (40 angles). Cross sections through 95% span for the zero*th* (bent) and third (corrected) iterations are compared with a reconstruction from 40 straight projections in Fig. 13. Note that the positions of the qualitative

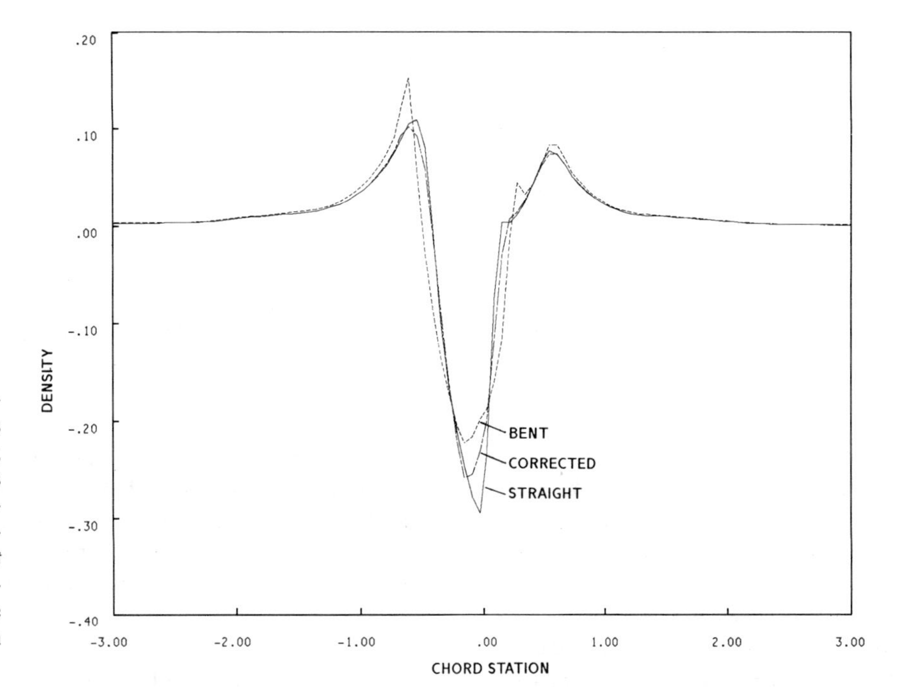

Fig. 13 Cross sections of reconstructions with corrections. 95% Span cross section of reconstructed density field; all reconstructions are from 40 angles only. Solid line marked *straight* corresponds to projections with bending suppressed; dashed line marked *bent* corresponds to uncorrected bent projections (0*th* iteration); and dashed line marked *corrected* is section through third iteration.

features are wholly restored and that the quantitative features are significantly improved. The errors remaining are mostly caused by the loss of information in ray crossings or caustics (double-valued points were discarded).

In the general case of strongly refracting objects Cha and Vest [9] have shown that imaging greatly reduces the effect of ray bending. In that case a lens images the central region of the flow of interest onto the film or hologram plane. For imaging systems Fermat's principle applies, which states that the optical pathlengths for all rays emanating from the same object point and converging to a common image point should be equal. Using this configuration, the potentially serious lateral shifts of the rays are greatly reduced. In this respect imaging methods provide similar results as the back propagation techniques discussed by Devaney [11].

Modarress et al. [30] subsequently also carried out modeling studies for the flow over a revolving helicopter rotor blade using the same test-bed data from Caradonna. They too found that accurate reconstructions could be obtained with limited projections, but using a modified iterative ART technique. This technique provides images with accuracies similar to or slightly better than those provided using the convolution back projection technique studied by Snyder and Hesselink [41], but at the expense of considerable increase in processing time.

Using the iterative ART approach, Modarress et al. [30] and Kittleson and Yu [26] produced the first experimental reconstruction of the flow about a rotating helicopter rotor blade under hovering flight conditions.

B Applications

Projections can be measured in a variety of ways. Holographic interferometry provides information about the index of refraction along the pathlength, absorption measurements naturally about the absorption coefficient, ray deflection methods about gradients in the index of refraction, and emission techniques about emitted radiation. In this section four representative techniques are discussed, including an absorption technique applied to a combusting flow, holographic interferometry in a wind tunnel experiment, deflection tomography in a flame, and a fast optical technique for making time-resolved measurements in a mixing flow using holographic interferometry.

1 Concentration and Temperature Measurements in Reacting flows Temperature and density in a reacting flow can be measured by detecting the intensity of a transmitted probe beam and comparing it with the intensity of a reference beam. Using spectroscopic techniques, the amount of absorption can be related to physical flow parameters using the Voigt function [36].

Ray and Semerjian [35] studied a flat premixed flame of methane seeded with sodium atoms, and they measured absorption by tuning to a sodium D line. In this study, a numerical simulation was carried out to analyze tomographic parameters such as the number of bins required in each projection and the effect of noisy projections on image quality. The object is axially symmetric, allowing the field to be reconstructed from a single projection. Noise consid-

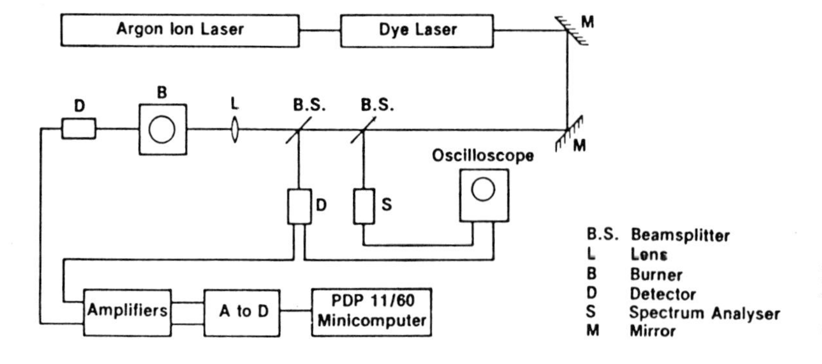

Fig. 14 Experimental configuration for single beam concentration or temperature measurements (by Ray and Semerjian [35], with permission).

erations, however, require that additional angular views are obtained in order to improve the quality of the reconstruction. They found that in their facility, reconstructions with an uncertainty of 3% of the true value could be obtained using a single projection.

The apparatus of this experiment is shown in Fig. 14. An argon-pumped dye laser beam is divided into two branches, one of which is incident on a flat flame burner and the other of which is used as a reference. The intensities of the transmitted and reference beams are detected, and the values are processed in a computer to obtain absorption coefficients. The flame is translated through the flow to make measurements at several locations in the projection, because the spatial extent of the probe beam is less than the flow cross section. Alternatively, a sheet of light allows simultaneous recording of a single projection, and this approach yields results that are quite acceptable when using filtering for noise reduction. In a related study, Emmerman et al. [13] also measured axial symmetric steady flames and found good agreement between computer simulations and experimental results.

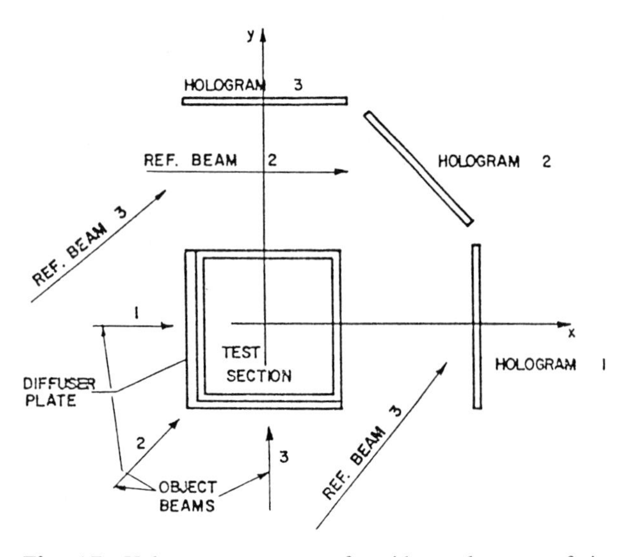

Fig. 15 Hologram arrangement for wide angular range of views (by Kosakoski and Collins et al. [27], with permission).

2 Holographic Interferometry of Three-dimensional Flows Kosakoski and Collins [27] have studied applications of holographic interferometry to several flow fields ranging from an axisymmetric but tilted round jet with respect to the pivot axis of the tomographic data acquisition system to supersonic flow over a lifting body at an angle of attack. To achieve full field coverage over an arc of 180°, three holograms are recorded simultaneously, each covering an angular range of bout 8° as shown in Fig. 15. The

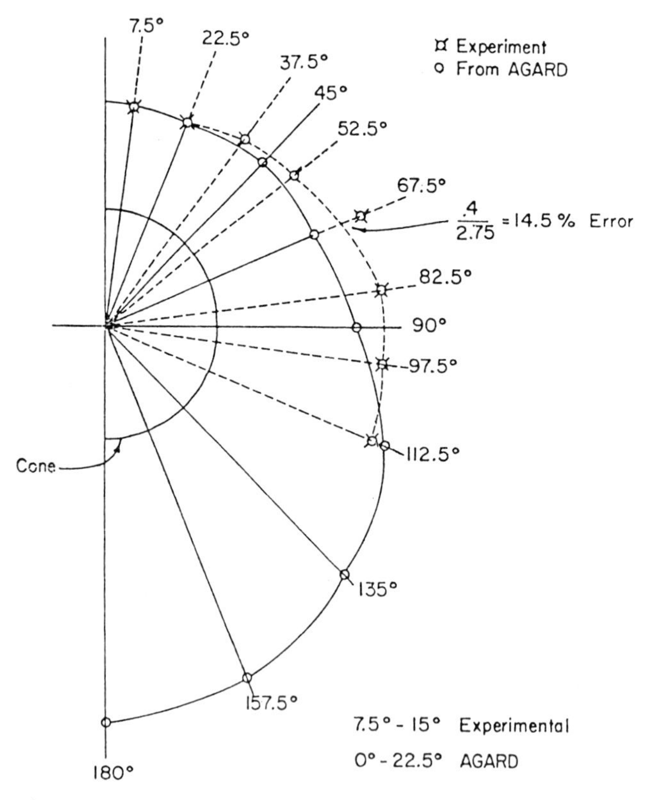

Fig. 16 Experimental results obtained with optical arrangement of Fig. 15. Comparison between the position of the shock wave on a conical body of revolution obtained experimentally and from AGARD 167 for M = 2.84; the flow is made asymmetric by tilting the cone at an angle of attack (by Kosakoski and Collins [27], with permission).

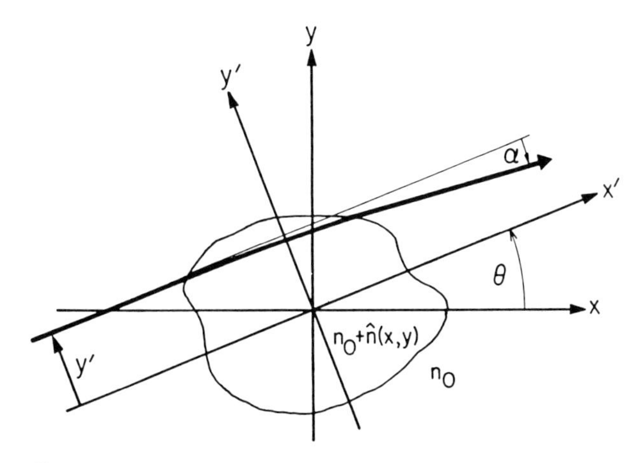

Fig. 17 Geometry for beam-deflection tomographic reconstructions (by Faris and Byer [14], with permission).

ground-glass plate allows reconstruction of the hologram from several vantage points. In this case, scattering properties of the ground-glass plate dictated that several sets of experiments had to be repeated to cover the complete angular range. Since the flow is steady, projections are acquired sequentially instead of instantaneously. With this technique, reasonable agreement between theory and experiment was achieved [27], as shown in Fig. 16. An interesting innovation in this work was the use of a transparent model to allow full coverage of the flow under investigation.

3 Deflection Tomography Faris and Byer [14] described an approach using beam-deflection tomography and an extension of convolution back projection for reconstruction of flows from such projections. The geometry for this approach is depicted in Fig. 17. Light incident on the flow is deflected from a straight path, and the angular variation along the pathlength is measured for each projection. From these projection data the index of refraction field can be reconstructed

$$n(x, y) = n_0 \sum_{p=1}^{P} \frac{\pi}{P-1} \sum_{q=1}^{Q} \alpha\left(n\,\Delta y', \frac{p\pi}{P-1}\right) \kappa(-x \sin\theta + y\cos\theta - p\,\Delta y')\,\Delta y' \tag{17}$$

using P projections of Q points each. Here α denotes the beam-deflection angle and n_0 represents the unperturbed index of refraction of the fluid. In this equation the expression for κ is

$$\kappa(p\,\Delta y') = 0 \qquad for \qquad p = even \tag{18}$$

or

$$\kappa(p\,\Delta y') = \frac{1}{\pi^2 p\,\Delta y'} \qquad for \qquad p = odd \tag{19}$$

and $\Delta y'$ denotes the step size in the y' direction perpendicular to the beam.

With this technique and the optical configuration of Fig. 18, experimental investigations of a supersonic jet, a flame, and a subsonic diffusion jet were made. In these experiments the flow was assumed to be stationary and was rotated during data acquisition. Typical results are displayed in Fig. 19.

4 High-speed, High-resolution Optical Tomography Until recently, the optical architectures used for data acquisition of the projections have not been capable of supplying time-resolved simultaneous projections from a large number of directions. Such data are required in order to apply tomography to turbulent flows. In that case a large number of projections are needed to achieve good spatial resolution, and all projections must be acquired in such a short time that the flow is essentially frozen.

Snyder and Hesselink [41] have reported a novel data-acquisition system capable of making such measurements. A schematic of the setup revealing one path through the system is shown in Fig. 20. A laser beam is divided into two waves. One is used as a diverging spherical reference

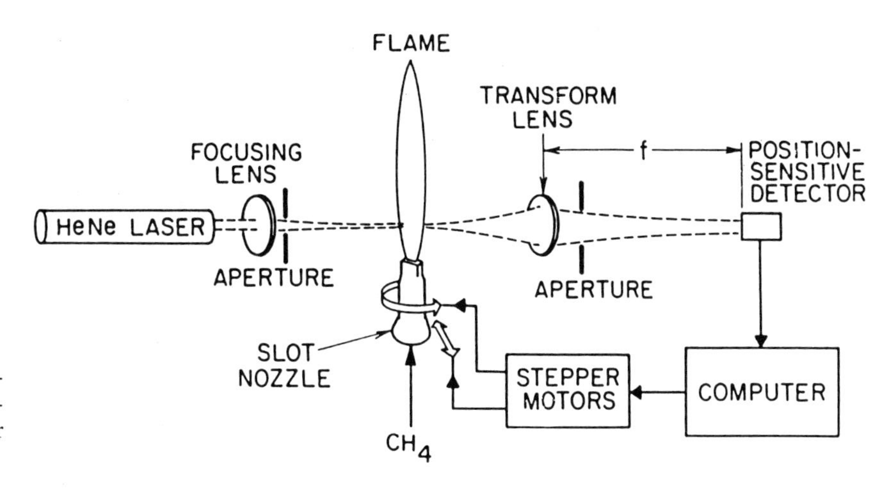

Fig. 18 Diagram of experimental apparatus for beam-deflection optical tomography of a flame (by Faris and Byer [14], with permission).

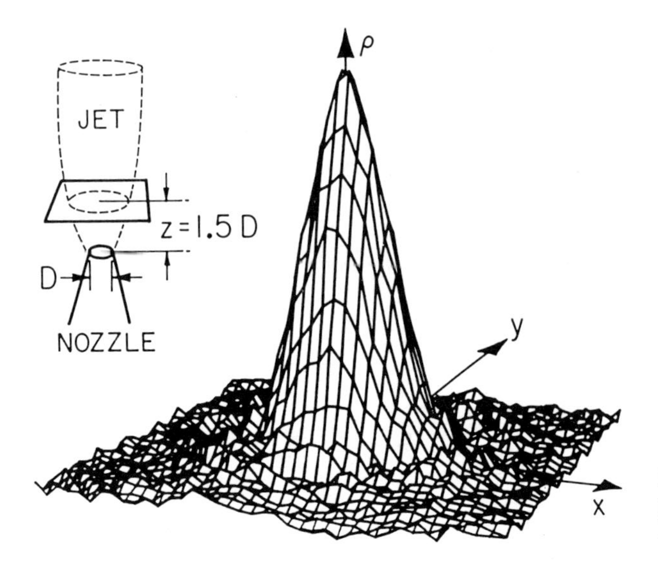

Fig. 19 Reconstruction of chlorine gas density in a horizontal plane at 1.5 nozzle diameters from the nozzle tip (by Faris and Byer [14], with permission).

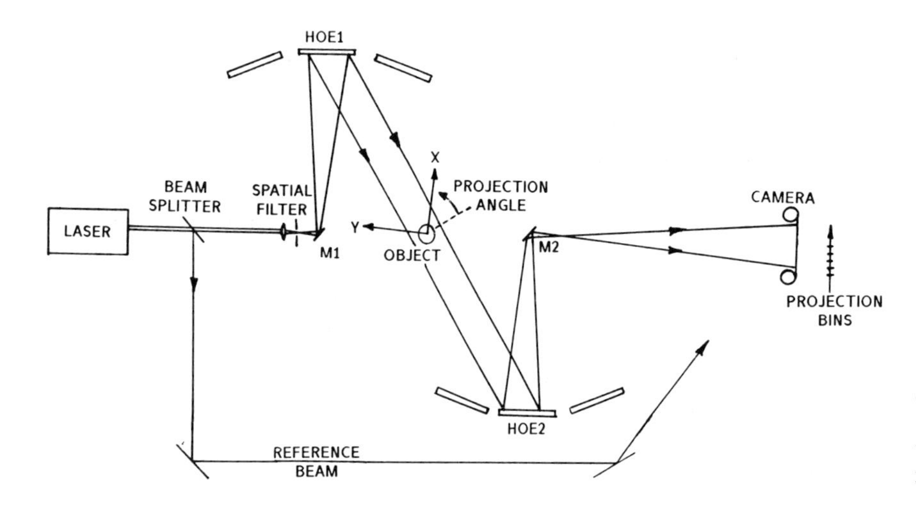

Fig. 20 Schematic of the experimental configuration. For the tests described, only the beam path shown is employed. The coordinates shown are fixed to the rod, and the rod is rotated to obtain projections at angles measured relative to the beam as shown.

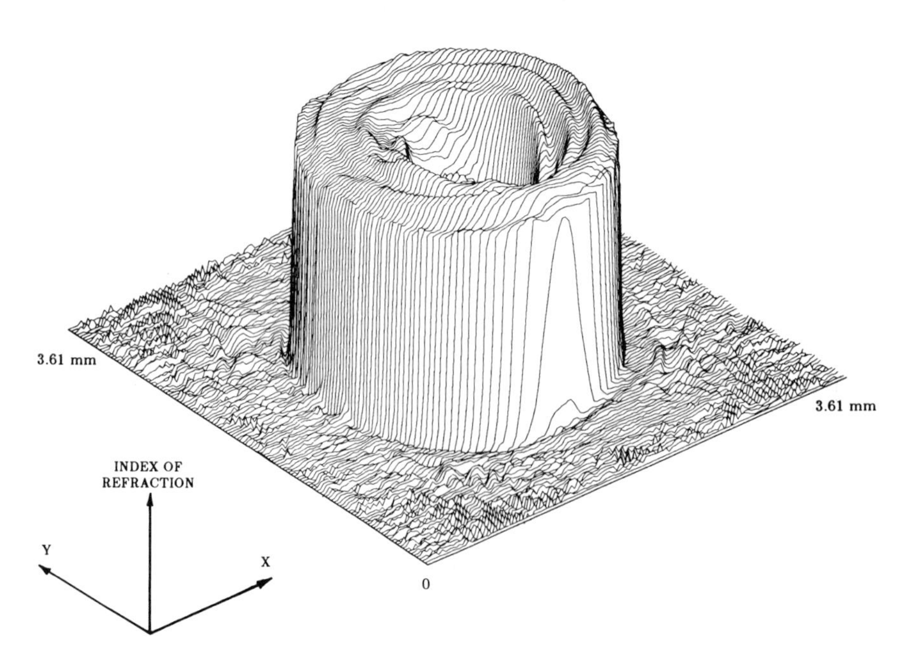

Fig. 21 Reconstruction of the index of refraction in the rod.

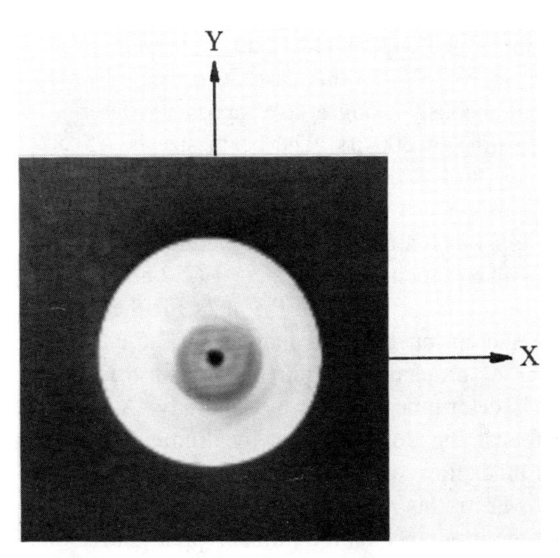

Fig. 22 Gray scale image representation of the reconstruction of the index of refraction in the rod. Lighter values correspond to a higher index of refraction. Note the asymmetry and the hole in the center.

beam whereas the other beam is spatial filtered and subsequently collimated by a holographic optical element (HOE). The first, HOE1, performs beam shaping as well as beam steering in the direction of the desired projection. The second, HOE2, is used to refocus the beam and provide an image of the object on the recording film. Two holograms are recorded, one with no flow present and the other with flow. In this way, using standard readout techniques, a double-exposure interferogram is made. Information about the optical phase of the wave is extracted by using a Fourier transform fringe-interpretation scheme as discussed by Hesselink [20]. The measured projection data are then used to reconstruct the image.

As an example the index of refraction distribution in a

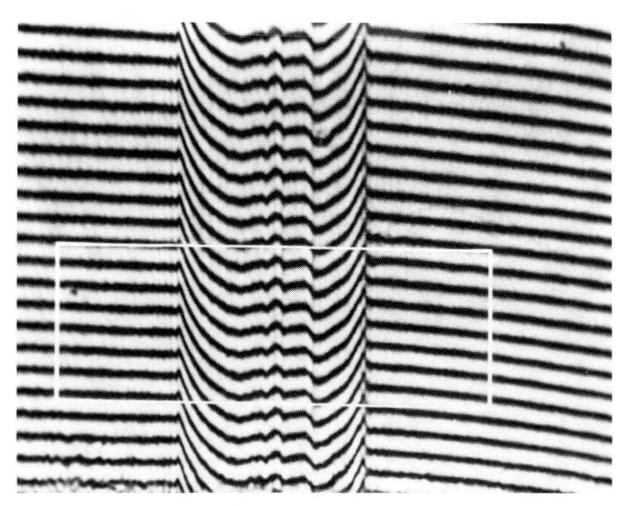

Fig. 24 Interferogram representing one projection about the rod.

glass rod is reconstructed from projections, and the result is shown in Figs. 21, 22, and 23. The reconstruction method is the convolution back projection technique, and for this particular application only five projections were used. The interferograms are analyzed with a Fourier transform technique, since the fringe pattern is essentially one-dimensional, as shown in Fig. 24. It is interesting to note here that this fringe pattern is obtained by using two holographic optical elements and no conventional imaging optics. The resolution achieved with this system is approximately 20 line pairs/mm, and the quality of reconstruction is good as shown by comparison with an actual cross section of the rod (Fig. 25).

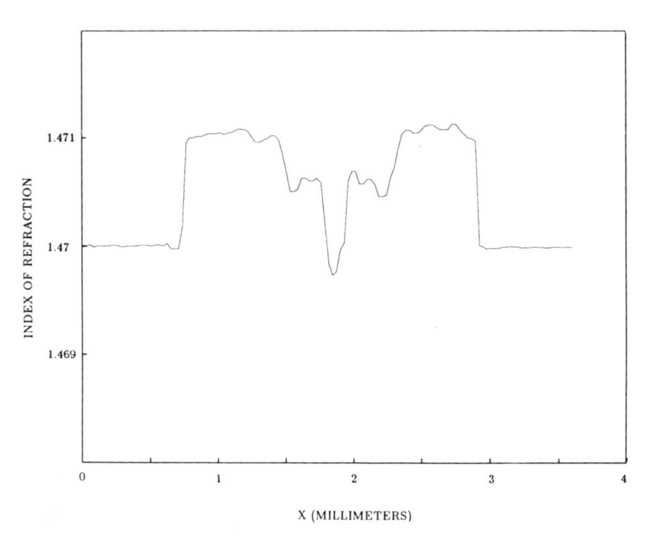

Fig. 23 Cross section of the index profile through center of the reconstruction. Radial structure is shown as well as hole in the center of the rod.

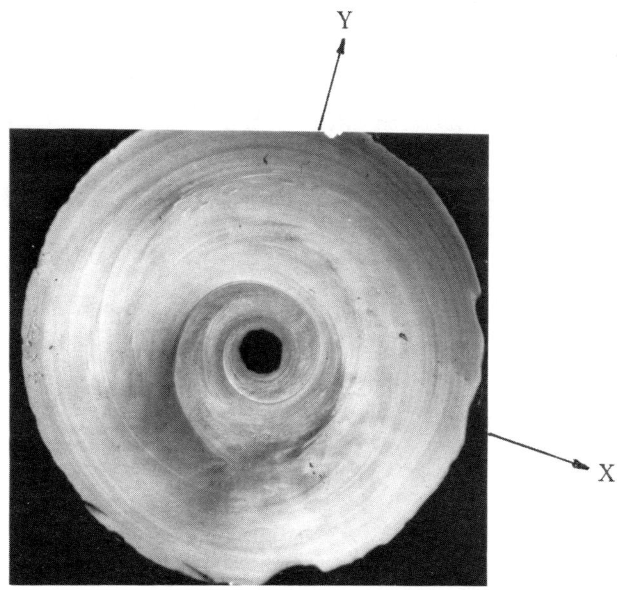

Fig. 25 Transmittance photograph of a 6-mm-long slice of the rod, obtained after the tomography experiment was completed. Note the correspondence with the results of the reconstruction in Fig. 22. The edge of the rod was chipped as it was being cut.

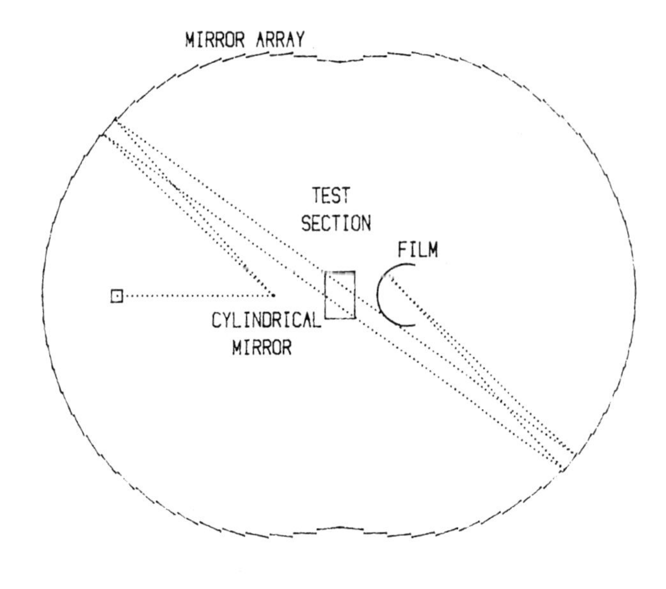

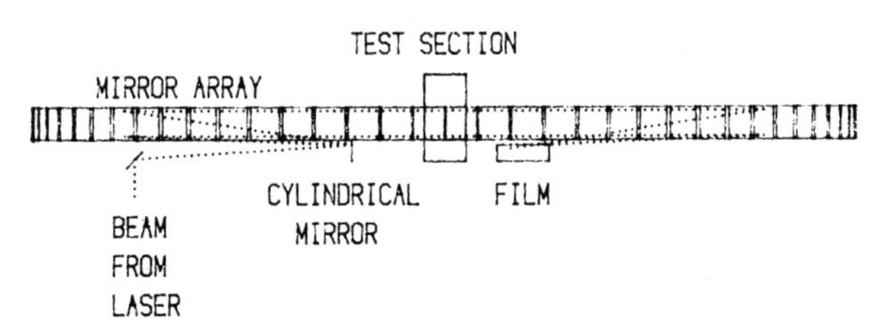

Fig. 26 Optical apparatus for measuring 36 holographic interferogram projections simultaneously.

A more complicated apparatus based on the same principle is shown in Fig. 26. In this case 36 projections are measured simultaneously using a 1-W argon ion laser as the source in a time of 300 μs. The beam size is 7.5 × 7.5 cm, allowing full three-dimensional reconstruction of the flow at an instant in time short enough to freeze fluid motion. This apparatus is used to image a coflowing jet of helium in air with a velocity ratio of 2.5 : 1 (2.2 m/s average velocity for the inner flow and 0.9 m/s for the outer flow velocity) and an inner nozzle diameter of 2.5 cm.

The number of projections required to reconstruct the coflowing jet is determined from a model study. A cross-sectional image of the flow obtained by illuminating the particle-laden inner flow with a sheet of laser light is used as the test bed. From this image, projections are computed assuming that the irradiance on the film is representative of density variations in the flow. Noise is added to the projections to simulate experimental conditions. Results of this simulation are shown in Fig. 27. The original test-bed image (obtained by seeding the inner flow with particles and photographing a cross section using laser sheet illumination) appears on the right, and along each row the noise level is kept constant, but the number of angles used for the reconstruction is varied from 20 to 30 and 40, respectively, going from left to right. From row to row the noise level is increased from a level of 1% in the top row to 5% in the middle and bottom rows, but in the last row median filtering has removed isolated noise spikes. From this study it was determined that 36 projections should provide sufficient detail to obtain images with sufficient spatial and temporal resolution to study turbulent or transition flow regimes.

An example of an interferogram obtained with this apparatus is shown in Fig. 28 and a processed version in Fig. 29. In the latter image extraneous noise has been removed

Fig. 27 Model study of a coflowing jet, using 20, 30, and 40 projections and the convolution back projection technique. Noise is added to the projections to simulate experimental conditions; 1% noise is added to the projections corresponding to the top row reconstructions, 5% to the middle row and last row. The results of the bottom row are median filtered to remove isolated noise spikes.

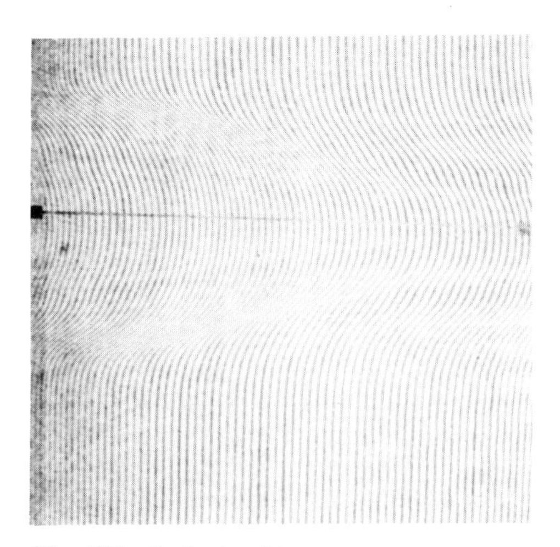

Fig. 28 Projection fringe pattern.

Fig. 29 Fourier transform of fringe pattern.

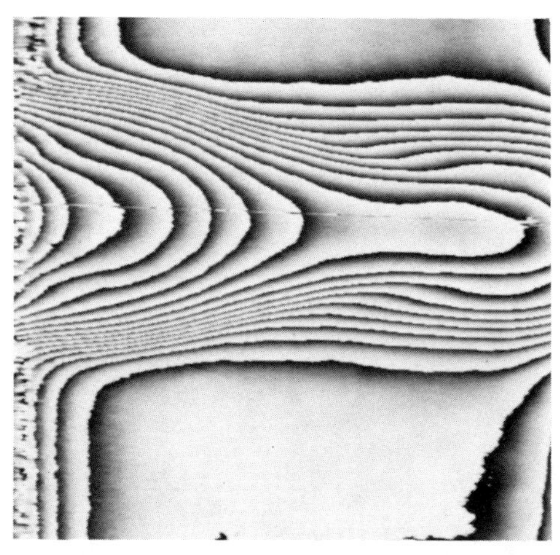

Fig. 30 Zero fringe pattern computed from pattern of Fig. 28.

by digital image processing. The interferogram used here has a bias fringe pattern due to a beam offset introduced between exposures. From this interferogram the zero fringe pattern as shown in Fig. 30 is computed using Fourier techniques. The Fourier transform of the pattern is centered around the spatial frequency corresponding to the background bias fringe pattern. By filtering, the bias is removed and upon inverse Fourier transforming a zero-fringe pattern results. Using the convolution back projection, a cross section of the flow can be reconstructed using a single line in the projections as shown in Fig. 31. The central peak in the graph corresponds to helium in the core. A density profile is denoted in Fig. 32 and a contour plot in Fig. 33. The error in the quantitative density values is less than 3%. At 15 cm farther downstream of the nozzle, in the transition region of the flow, the cross sections exhibit more structure as can be seen in Fig. 34. Using the complete set of 2-D projections, a full reconstruction of the 3-D flow is obtained shown in Fig. 35 [41]; these results represent the first snapshot tomographic reconstructions of an unsteady and complex flow structure.

VII EVALUATION OF OPTICAL TOMOGRAPHY IN FLOW APPLICATIONS

The quality of tomographic reconstructions depends largely on the reconstruction algorithm, the number of views used, the noise level and resolution of the projections, and flow topology. Geneally speaking, image quality improves with increasing number and quality of projections. In particular, for reconstruction of flows with fine detail and sharp discontinuities a large number of projections may be needed. For example, sharp discontinuities such as shocks and thin interfaces between two different fluids tend to cause streaky patterns in the reconstructed flow. These artifacts can be estimated by carrying out a modeling study in which the quality of the reconstruction is studied as a function of the number of views and resolution as well as noise level of each projection. The artifacts caused by the reconstruction can be further reduced by postreconstruction image processing such as median filtering and image enhancement often without significant loss in detail or fidelity.

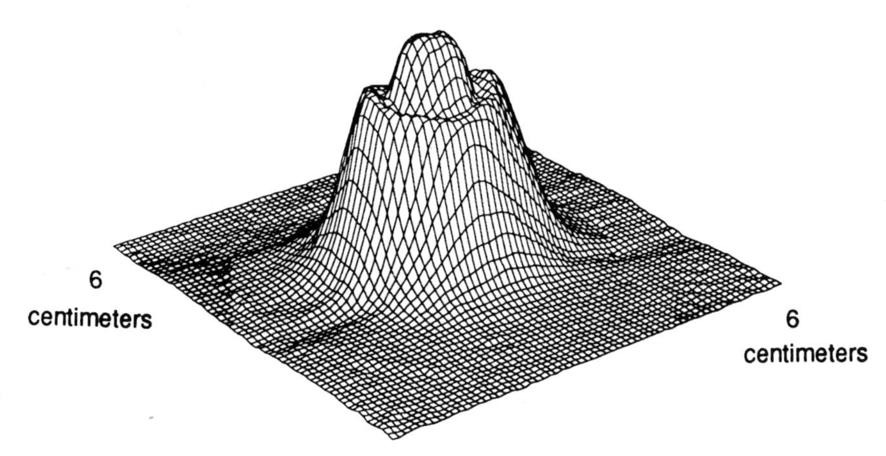

Fig. 31 Reconstructed cross section of a coflowing jet. Reconstruction error is less than 3%.

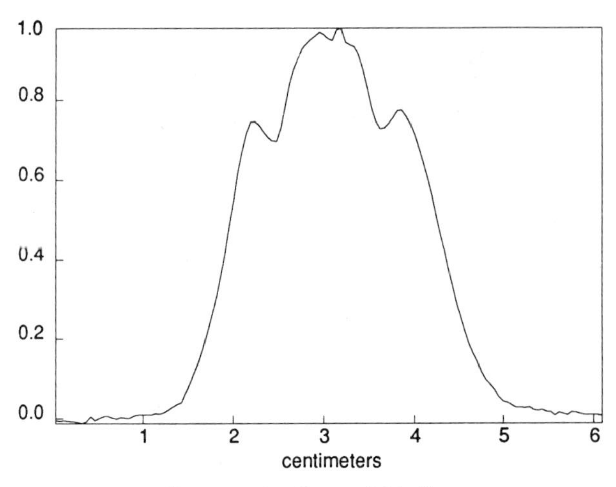

Fig. 32 Density profile of helium concentration in coflowing jet. Note that pure helium is found at the center of the jet; the plot is not normalized.

Fig. 34 Cross section of the transition region in a coflowing jet. The intensity values represent helium concentration; black denotes air and white represents pure helium.

A peculiarity of currently developed reconstruction schemes is that all projections must lie in a plane, usually along an arc of at least 180°. This need for optical access implies that walls containing the flow must be optically transparent and should not distort the probe beam. Thus optical windows should be made thin to reduce ray bending. Because of this requirement, measurements of flows in round ducts are difficult to achieve unless optical elements are introduced into the probe beam to correct for ray refraction caused by the curved walls. It is, however, not necessary to probe the flow with collimated light, and fan beam tomography can provide excellent results as well. The tradeoff is usually one of determining the optimum geometry of the optical apparatus under a set of space constraints.

In most flow applications, holographic interferometry has been used as the probing technique for measuring temperature, density, and concentration. Although interferometry is a well-developed diagnostic tool and is widely used for qualitative flow visualization, data reduction is often a complicated and time-consuming task, in particular, for turbulent flows with considerable structure. Other approaches,

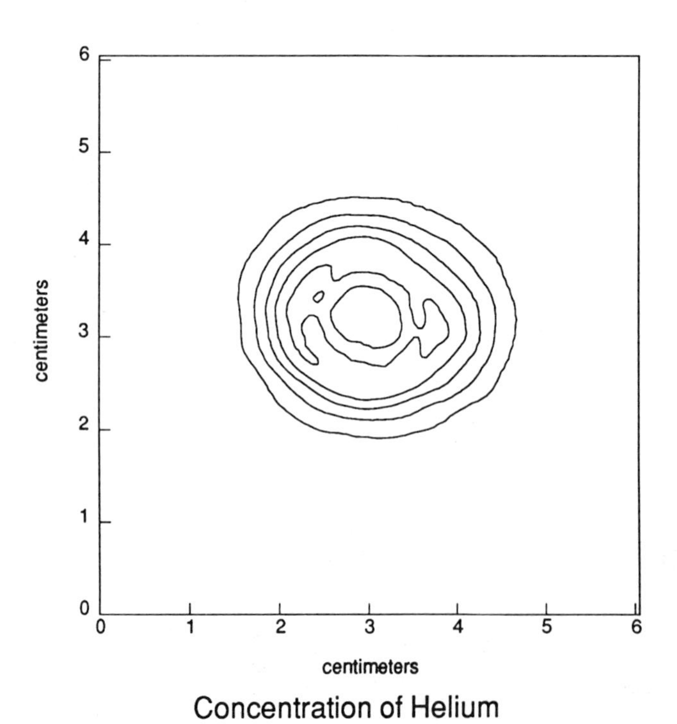

Fig. 33 Contour plot of helium concentration in a coflowing jet.

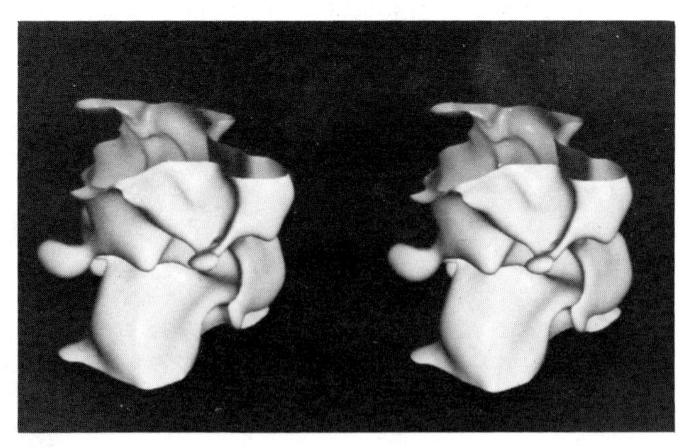

Fig. 35 Stereo pair of the 3-D reconstruction of an isodensity contour surface in the transition region of a coflowing jet. The inner fluid is a mixture of helium and argon and the outer flow is air. The flow direction is from bottom to top. Approximately half the core fluid is retained inside the volume bounded by the surface.

such as absorption and deflection tomography, should therefore be considered viable alternatives, because they require much simpler data-reduction schemes. For example, absorption tomography is very well suited to combustion research because, by tuning a laser to a particular absorption line of an intermediate species or reactant, species-specific measurements of concentration or temperature can be made, and data reduction is not nearly as complicated and time consuming as for holographic interferometry. Emission tomography is also an attractive method for combustion research because no external probe beam is required to obtain the projections. The flow can be reconstructed from the projections in a variety of ways. The most widely used reconstruction algorithm is convolution back projection. It is a fast and easy-to-implement approach that, in this author's opinion, should be tried first. In those applications where the projections are not uniformly spaced about the object, or projections are missing, convolution back projection can still provide good results, provided the missing projections are generated, for example, by interpolation of available data. Alternatively, ART or the maximum-entropy method tend to produce good results for these reconstructions, but at the expense of considerably more computation time. We have always achieved comparable results with direct and iterative approaches when using postreconstruction image-enhancement techniques such as median filtering to remove isolated noise spikes.

In addition to complications arising from incomplete data, difficulties arise when the rays are bent. All reconstruction algorithms considered here are based on the assumption that the rays are straight. In combusting and two-phase flows, or in flows with large density gradients, this may not be true as a result of refraction. In that case, the reconstruction algorithms need to be modified. As an example of an approach in which the effects of ray bending are reduced, the image is first reconstructed under the assumption that the rays are straight. Then rays are traced through the medium and the projections are computed, including the effect of ray bending. The measured projections are subsequently compared with the computed ones, and are updated to account for the differences. The updated projections are then used to compute a new reconstruction. This process is repeated and stopped when the differences between two sequential reconstructions is below a threshold value. Using this approach, much improved results have been obtained for reconstruction of the flow around a revolving helicopter rotor blade. In that application, rays nearly parallel to the shock are bent rather strongly, because the gradients in the density (and therefore index of refraction) are perpendicular to the ray paths. Unfortunately, in general one cannot be assured of convergence of this approach, and each application needs to be investigated to determine the effects of ray bending. Alternatively, the effects of ray bending can be reduced substantially by using a lens that images the object onto the measurement plane for each projection. This approach should always be tried first, if an imaging configuration can be implemented. As an example of such an approach, for interferometry the object is imaged onto the recording plane, where it is interfered with a reference wave to obtain the fringe pattern.

Difficulties also arise when the probe beam does not cover the full flow field cross section. In that case severe artifacts may contaminate the reconstruction. In addition, evaluation of interferograms becomes very difficult because reference information about density or temperature in the unperturbed region is missing. Since tomography reconstructs the full flow field, a smaller interior region cannot be reconstructed by probing only that section of the flow. However, from a coarse representation of the complete flow a smaller subsection can be reconstructed with higher resolution by a technique referred to as *onion peeling*.

Finally, it is interesting to note here that tomography provides a means of measuring cross sections or three-dimensional representations of a flow by making in-plane observations. In that respect tomography differs from scattering techniques using laser sheet illumination, in which the recorded image is obtained by collecting scattered radiation outside the plane of illumination. The tomographic approach is thus attractive for measurement of hot or environmentally adverse flow fields, such as exhausts of jet engines.

REFERENCES

1. Alwang, W. L., Cavanaugh, L., Burr, R., and Hauer, A. Some Applications of Holographic Interferometry to Analysis of the Vibrational Response of Turbine Engine Components. Electro-Opt. Syst. Des. Conf., New York, Tech. Rep. A70-33134, 1970.
2. Barrett, H., and Swindell, W., *Radiologic Imaging*, Academic, New York, 1981.
3. Bennett, K., and Byer, R. L., Optical Tomography: Experimental Verification of Noise Theory, *Opt. Let.*, Vol., 9, 270–272, 1984.
4. Bennett, K. E., Faris, G. W., and Byer, R. L., Experimental Optical Fan Beam Tomography, *Appl. Opt.*, vol. 23, pp. 2678–2685, 1984.
5. Bennett, K. E., and Byer, R. L. Fan-Beam-Tomography Noise Theory, *J. Opt. Soc. Am. A*, vol. 3A, no. 5, pp. 624–633, 1986.
6. Bracewell, R. N., Strip Integration in Radio Astronomy, *Aust. J. Physi.*, vol. 9, pp. 198–217, 1956.
7. Byer, R. L., and Shepp, L. A., Two-dimensional Remote Air Pollution Monitoring by Tomography, *Opt. Lett.*, vol. 4, pp. 75–77, 1979.
8. Caradonna, F. X., Transonic Flow on a Helicopter Rotor, Ph.D. dissertation, Stanford Univ., Stanford, Calif., 1978.
9. Cha, S., and Vest, C. M., Tomographic Reconstruction of Strongly Refracting Fields and Its Application to Interfero-

metric Measurement of Boundary Layers, *Appl. Opt.*, vol. 20, no. 16, pp. 2787–2794, 1981.
10. Dandliker, R., and Weiss, K., Reconstruction of the Three-dimensional Refractive Index from Scattered Waves, *Opt. Commun.*, vol. 1, pp. 323–328, 1970.
11. Devaney, A. J., A Computer Simulation Study of Diffraction Tomography, *IEEE Trans. Biomed. Eng.*, vol. BME-30, no. 7, pp. 377–386, 1983.
12. Dines, K. A., and Lytle R. J., Computerized Geophysical Tomography, *Proc. IEEE*, vol. 6, no. 7, pp. 1065–1073, 1979.
13. Emmerman, P. J., Goulard, R., Santoro, R. J., and Semerjian, H. G., Multiangular Absorption Diagnostics of a Turbulent Argon–Methane Jet, *J. Energy*, vol. 4, pp. 70–77, 1980.
14. Faris, G. W., and Byer, R. L., Quantitative Optical Tomography Imaging of a Supersonic Jet, *Opt. Lett.*, vol. 11, pp. 413–415, 1986.
15. Frieden, B. R., and Zoltani, C. K., Maximum Bounded Entropy: Application to Tomography Reconstruction, *Appl. Opt.*, vol. 24, no. 23, pp. 3993–3999, 1985.
16. Goodman, J. W., *Statistical Optics*, Wiley-Interscience, New York, 1985.
17. Hendee, W. R., *The Physical Principles of Computed Tomography,* Little, Brown, Boston, 1983.
18. Hertz, H. M., Experimental Determination of 2-D Flame Temperature Fields by Interferometric Tomography, *Opt. Commun.*, vol. 54, pp. 131–136, 1985.
19. Hertz, H. M., Kerr Effect Tomography for Nonintrusive Spatially Resolved Measurements of Asymmetric Electric Field Distributions, *Appl. Opt.*, vol. 25, pp. 914–921, 1986.
20. Hesselink, L., Digital Image Processing in Flow Visualization, *Ann. Rev. Fluid Mech.*, vol. 20, pp. 421–485, 1988.
21. Hesselink, L., and White, B. S., Digital Image Processing of Flow Visualization Photographs, *Appl. Opt.*, vol. 22, no. 10, pp. 1454–1461, 1983.
22. Hoppe, W., and Hegerl, R., Computer Processing of Electron Microscope Images, *Top. Curr. Phys.*, vol. 13, pp. 127–185, 1980.
23. Huesman, R. H., Gullberg, G. T., Greenberg, W. L., and Budinger, T. F., Donner Algorithms for Reconstruction Tomography, Lawrence Berkeley Laboratory, Tech. Rep. Pub. 214, 1977.
24. Iwata, K., and Nagata, R., Calculation of Three-dimensional Refractive-Index Distribution from Interferograms, *J. Opt. Soc. Am.*, vol. 60, pp. 133–135, 1970.
25. Kirchartz, K. R., Muller, U., Oertel, H., Jr., and Zierep, J., Axisymmetric and Nonaxisymmetric Convection in a Cylindrical Container, *Acta Mech.*, vol. 40, pp. 181–194, 1981.
26. Kittleson, J. K., and Yu, Y., Transonic Rotor Flow Measurement Technique using Holographic Interferometry, *J. Am. Helicopt. Soc.*, vol. 30, no. 4, pp. 3–10, 1985.
27. Kosakoski, R. A., and Collins, D. J., Application of Holographic Interferometry to Density Field Determination in Transonic Corner Flow, *AIAA J.*, vol. 12, no. 6, pp. 767–770, 1974.
28. Macovski, A., *Medical Imaging Systems,* Prentice-Hall, Englewood Cliffs, N.J., 1983.
29. Maldonado, C. D., and Olsen, H. N., New Method for Obtaining Emission for Emitted Spectral Intensities: Part II. Asymmetrical Sources, *J. Opt. Soc. Am.*, vol. 56, pp. 1305–1313, 1966.
30. Modarress, D., Tan, H., and Trollinger, J. D., Tomographic Reconstruction of Flow over Air Foils, Paper AIAA-85-0479, 1985.
31. Mohammad-Djafari, A., and Demoment, G., Maximum Entropy Fourier Synthesis with Application to Diffraction Tomography, *Appl. Opt.*, vol. 26, no. 9, pp. 1745–1754, 1987.
32. Munk, W., and Wunsch, C. Ocean Acoustic Tomography: A Scheme for Large Scale Monitoring, *Deep-Sea Res. Pt. A. Oceanograph. Papers,* vol. 26A, p. 123, 1979.
33. Olsen, H. N., Maldonado, C. D., and Duckworth, G. D., A Numerical Method for Obtaining Internal Emission Coefficients from Externally Measured Spectral Intensities of Asymmetrical Plasma, *J. Quant. Spectrosc. Radiat. Transfer,* vol. 8, pp. 1419–1430, 1968.
34. Radon, J., Uber die bestimmung von funktionen durch ihre integralwerte langs gewisser mannigfaltigkeiten. *Berichte über die Verhandlunger Sachsische Akademie der Wissenschagren (Leipzich) Math-Phys KL*, vol. 62, pp. 262–277, 1917.
35. Ray, S. R., and Semerjian, H. G., Laser Tomography for Simultaneous Concentration and Temperature Measurement in Reacting Flows, *Prog. Astronaut. Aeronaut.* vol. 92, pp. 300–324, 1983.
36. Regnier, P. R., and Taran, J. P., On the Possibility of Measuring Gas Concentration by Stimulated Anti-Stokes Scattering, *Appl. Phys. Lett.*, vol. 23, pp. 240–242, 1973.
37. Robb, R. A., *Three-dimensional Biomedical Imaging*, CRC Press, Boca Raton, Fla., 1985.
38. Rowley, P. D., Quantitative Interpretation of Three-Dimensional Weakly Refractive Phase Objects Using Holographic Interferometry, *J. Opt. Soc. Am.*, vol. 59, no. 11, pp. 1496–1498, 1969.
39. Shtein, I. N., The Use of Radon Transform in Holographic Interferometry, *Radio Eng. Electron. Phys.* (USSR), vol. 17, no. 11, pp. 2436–2437, 1972.
40. Snyder, R. and Hesselink, L., High Speed Tomography for Flow Visualization, *Appl. Opt.*, vol. 24, pp. 4046–4051, 1985.
41. Snyder, R., and Hesselink, L., Optical Tomography for Flow Visualization of the Density Field around a Revolving Helicopter Rotor Blade, *Appl. Opt.*, vol. 23, pp. 3650–3656, 1984.
42. Snyder, R., and Hesselink, L., Measurement of Mixing Fluid Flows with Optical Tomography, *Opt. Lett.*, vol. 13, no., pp 87–89, 1988.
43. Stuck, B. W., A New Proposal for Estimating the Spatial Concentration of Certain Types of Air Pollutants, *J. Opt. Soc. Am.*, vol. 64, pp. 668–678, 1977.
44. Sweeney, D. W., and Vest, C. M., Reconstruction of Three-dimensional Refractive Index Fields from Multidirectional Interferometric Data, *Appl. Opt.*, vol. 12, pp. 2649–2664, 1973.
45. Thompson, L. L., and Taylor, L. S. Analysis of Turbulence by Schlieren Photography, *AIAA J.*, vol. 7, no. 10, pp. 2030–2031, 1969.
46. Uchiyama, H., Nakajima, M., and Yuta, S., Measurement of Flame Temperature Distribuiton by IR Emission Computed Tomography, *Appl. Opt.*, vol. 24, pp. 4111–4116, 1985.

47. Vest, C. M., Formation of Images from Projections: Radon and Abel Transforms, *J. Opt. Soc. Am.,* vol. 64, pp. 1215–1218, 1974.
48. Vest, C. M., Tomography for Properties of Materials That Bend Rays: A Tutorial, *Appl. Opt.,* vol. 24, pp. 4089–4094, 1985.
49. Williams, I., A Measurment Procedure for Acquisition of Spatial Inhomogeneous Aerosol Concentrations, in *Aerosols in Science, Medicine, and Technology:* Schmallenberg, Germany F.R.C., 1980.
50. Wolf, E., Three-dimensional Structure Determination of Semi-transparent Objects from Holographic Data, *Opt. Comm.,* vol. 1, pp. 153–156, 1969.

Chapter 21

Thermography

R. Monti

I INTRODUCTION

A thermograph (TG) is an instrument able to take almost instantaneous infrared pictures of IR-opaque solid or liquid surfaces. When the emissivity of the solid or liquid surface is known, the surface temperature distributions of the observed object, projected onto a plane orthogonal to the TG "optical" axis, can be evaluated.

In its most common arrangement (single detector) the thermograph can be described as a fast-scanning pyrometer that observes finite-dimension surfaces and reconstructs the thermal image on a monitor. An image is then formed similar to that of a conventional TV set in which the brightness (or the gray levels) of each picture element (pixel) is related to the energy, radiated by each object point within the instrument wavelength range, collected by the lenses of the TG and focused on the TG sensor (typically cooled at rather low temperatures).

Apart from the different applications, in general one can say that a TG system can be operated in two different modes, direct or reflection, depending on whether it collects mainly the thermal power radiated by the observed target surface or if it collects mainly radiation emitted by other sources reflected on the target surface.

The information that can be extracted from the TG signals may need either single thermal pictures or a number of pictures taken at different instants of time; in general, quantitative measurements (e.g., surface temperatures) are obtained by means of appropriate peripherals (e.g., A/D converters, computers).

The following capabilities have recently been demonstrated with the help of appropriate software:

1. Measurement of space and time distributions of surface temperature
2. Measurement of time and space derivatives of surface temperatures
3. Evaluation of the symmetry conditions of temperature fields (or of heat transfer processes)
4. Identification of contours and/or shapes of interfaces exhibiting radiation contrast over the background
5. Assessment of steady-state conditions for temperature and for heat transfer processes
6. Measurement of the heat transfer coefficients over models in wind tunnels
7. Visualization of flow separation regions over airfoils
8. Evaluation of temperature gradients along an axis orthogonal to a model surface
9. Evaluation of thermophysical properties of the test object during heating or cooling

The advantages of a TG system, either in its own right or in comparison with other measuring devices, accrue from its typical features:

Noninvasive measurements, a feature that is essential when the specimen is highly reactive, at very high temperatures, or only "optically" accessible (e.g., remote sensing)

Continuous time- and spacewise surface temperature monitoring

What was said above indicates that thermographs have a number of capabilities that include flow visualization: extensive use of image-processing techniques, to extract the appropriate information, is made in most recent systems.

Computer-assisted methods are examined in detail in this chapter, with reference to the digital modeling that allows TG data elaboration, needed in specific applications. A number of different types of TG equipment are commercially available, differing according to

- Type of IR sensor
- Number of sensors
- IR wavelength range
- Sensor cooling system
- Signal presentation
- Peripherals for the data elaboration

These different characteristics are discussed in the following sections.

II BASIC PRINCIPLES OF NONINVASIVE SURFACE TEMPERATURE MEASUREMENT SYSTEMS

The problem of noninvasive surface temperature measurement can be assessed in the following way.

A blackbody emits radiation according to Planck's law:

$$J(T, \lambda) = \frac{C_1}{\lambda^5} \frac{1}{\exp(C_2/\lambda T) - 1}$$
$$C_i = 3.68 \times 10^{-16} \text{ (Wm}^2\text{)}$$
$$C_2 = 1.438 \times 10^{-2} \text{ (mk)} \tag{1}$$

where $J(T, \lambda)$ is the total (in the half space) power emitted by the surface per unit area and per unit wavelength λ.

Integrating over the entire spectrum one obtains

$$\int_0^\infty J(T, \lambda)\, d\lambda = \sigma T^4 \tag{2}$$

being the Stefan–Boltzman constant, $\sigma = 5.67 \times 10^{-2}$ [W/(m^{2}K^4)].

Wien's law relates the wavelength λ_{max} at which the maximum J_{max} of the emitted radiation power occurs and the emitting blackbody temperature T:

$$\lambda_{max} T = 2880 \text{ } (\mu\text{K}) \tag{3}$$

The normalized power $J^+ = J/J_{max}$ can be plotted versus the normalized wavelength $\lambda^+ = \lambda/\lambda_{max}$ and is represented by a unique curve independent of T (Fig. 1).

In general, a nonblackbody emits at each wavelength a fraction of the radiation emitted by blackbody. The surface emissivity ε depends on the surface type (and conditions), on the surface temperature, on the emission wavelength and on the two angles that identify the viewing direction with respect to the surface. For isotropic surface:

$$\varepsilon = \varepsilon(\text{surface}, \ T, \ \lambda, \ \beta)$$

where β is the angle between the normal and the TG axis.

An infrared detector is basically a converter that absorbs IR energy and converts it into an electrical signal (usually a voltage or a current). The two main types are thermal and photon detectors. Thermal detectors are absorbing receivers, like the thermocouple, the bolometer, and the pyroelectric (capacitor) detector.

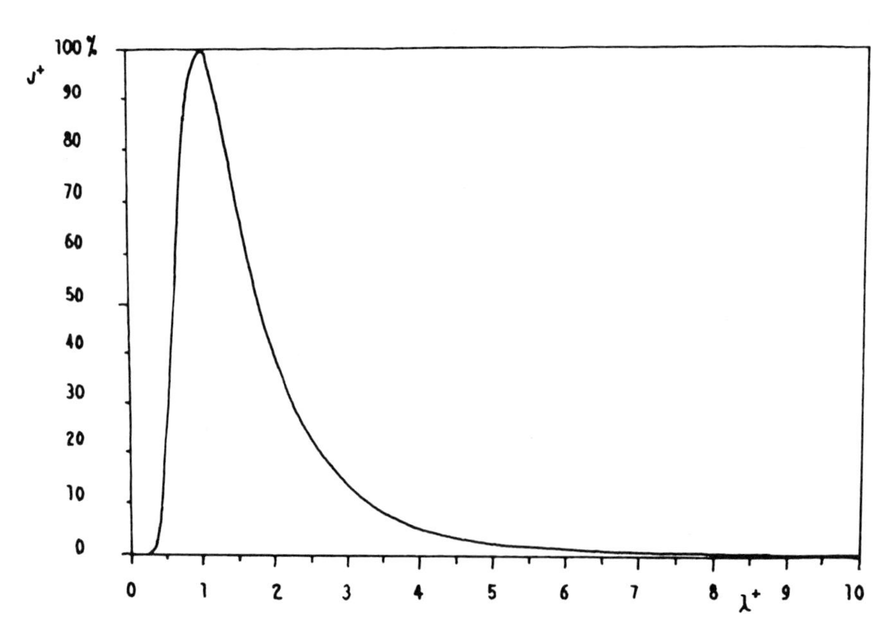

Fig. 1 Normalized spectral distribution of blackbody radiation as a function of normalized wavelength.

The most important thermal detector today is the thermistor bolometer, which utilizes the change in resistance of a semiconductor film when it is heated by radiation. Typical of thermal detectors is their "flat" spectral response. If they have been properly blackened, their output signal remains practically constant over a very wide range of wavelengths. Most thermal detectors exhibit a comparatively slow response time, due to the heating processes involved and to the finite heat capacity of the sensor. The pyroelectric thermal detector has instead a relatively fast response, owing to its use of the ferroelectric effect of certain crystals.

There is a variety of different photon (or quantum) detectors. They show distinctly different spectral responses (characterized by a sharp cutoff in the long-wavelength range). All photon detectors are made of semiconductor materials, in which the release or transfer of charge carriers (e.g., electrons) is directly associated with photon absorption. Since the energy of the photon is inversely proportional to the wavelength associated with it, the photoelectric activity disappears at wavelengths longer than the "cutoff" wavelength, i.e., when the energy of the photons is insufficient to set electrons free. To be effective, the impinging photons must exceed the so-called forbidden energy gap in the semiconductor material that strongly depends on the detector temperature (this is the reason why the detectors must be cooled at low temperatures).

Two principal types of photon detector are of particular interest today: the photoconductive and the photovoltaic detectors [1].

In a photoconductive detector the gap energy is determined by the nature of the material itself, and the effect of photon absorption is to "free" the electrons and thereby increase the detector's conductivity.

In a photovoltaic detector the gap energy is also determined by the material, but the radiation-generated charge carriers are swept away by the electric field in a *p–n* junction, thereby directly producing a voltage rather than a change in conductivity.

High-speed scanning requires a detector with a very short response time. The advantages of photon detectors are that they are more sensitive and have a much shorter response time than thermal detectors. The drawbacks are that they have a limited spectral response and require cooling for optimum sensitivity. Generally they are cooled to the temperature of liquid nitrogen, 77 K; some of them must even be cooled to the temperature of liquid helium, 4.2 K.

Depending on what sensor is employed, the thermographic signal is then proportional either to the power emitted by the observed surface (thermal detectors) or to the number of "active" photons emitted by the surface (photo detectors) in a defined spectral band.

In the first case each surface point of the emitting body (liquid or solid) will generate a signal P proportional to the total radiative power arriving at the detector of the instrument:

$$P(T) = \int_{\lambda_c}^{\lambda_c+\Delta\lambda} J(T, \lambda)\, \varepsilon(T, \lambda, \beta)\, R(\lambda)\, d\lambda \qquad (4)$$

where $R(\lambda)$ is the relative response of the entire instrument. Often the emissivity ε is assumed to be independent of T and of λ (gray-body assumption) [2].

In the applications discussed in this chapter, which deal with a limited range of surface temperatures and of wavelengths, this assumption appears reasonable and Eq. (4) reads

$$P(T) = \varepsilon(\beta) \int_{\lambda_c}^{\lambda_c+\Delta\lambda} J(T, \lambda)\, R(\lambda)\, d\lambda = \varepsilon(\beta)\, P_B(T) \qquad (5)$$

where $P\beta(T)$ is the detector signal that would be obtained when looking at a surface point of a blackbody at the same surface temperature. Knowledge of and inversion of the function $P(T)$, (by means of the calibration curve of the instrument), eventually yields the value of the surface point temperature T.

Independent of the instrument used, the surface temperature can be evaluated only if ε is known at the instrument wavelength range λ; this information can be obtained by laboratory calibration (for instance, looking at the surface and at a blackbody surface both at the same temperature).

In the second case (IR photo detectors), with the usual thermographic equipment the signal generated by the system is proportional to the number of photons emitted by the surface and collected by the sensor. If $Q = hc/\lambda$ is the energy of a photon (inversely proportional to the wavelength), the number of emitted photons (for a blackbody) is

$$N_{\lambda B} = \frac{J(T, \lambda)}{Q} = \frac{2\pi c}{\lambda^4} \left[\exp\left(\frac{hc}{\lambda T}\right) - 1\right]^{-1} \qquad (6)$$

When integrating over the entire spectrum

$$N_B = \int_0^{\infty} N_{\lambda B}\, d\lambda = \frac{0.37}{K} \sigma T^3 \text{ [photons/(cm}^2\text{/s)]} \qquad (7)$$

This is equivalent to the Planck's law expressed in terms of the number of photons emitted by the surface (instead of the radiated power). The "active" photons collected by the sensor are, however, those confined within the characteristic instrument wavelength and capable of being computed by an integral similar to that of Eq. (5).

III TYPES OF THERMOGRAPHS

The most usual TG systems commercially available today are of two basic types:

Scanning mirror infrared camera (mechanical scanning with a single detector or with a detectors array)

Vidicon-type infrared camera (electronic scanning with pyroelectric detectors)

Other systems (less common) use matrices of detectors that will mainly reduce the problems associated with the fast mechanical scanning of the first type of system. Advantages and disadvantages of these systems are illustrated in the following sections.

A Photon Detector Thermographs

Existing equipment (Hughes, AGA, Philips, Barnes, UTI) has been evolving and has yielded a large body of experience toward integrated systems with a number of accessories and peripherals. These systems have many similar characteristics but differ in temperature range, sensitivity, frame frequency, and available optics. The main drawback of some of these instruments is the liquid nitrogen–cooling system (contained in a Dewar-type tank).

Other instruments, used in field applications, utilize bottles of compressed gas. Systems have been developed that use thermoelectrically cooled sensors that avoid the liquid nitrogen (or gas bottle) refill before the measurements and are able to be used on a continuous basis whenever a power supply is available.

B Pyroelectric Detector Thermocameras

These devices are designed to eliminate two of the major drawbacks of the photon detector TG systems mentioned in the previous paragraph, namely, the cryogenic cooling of the detector and the mechanical scanning by means of rotating and/or oscillating mirrors. Typically these objectives are achieved, together with a full compatibility with TV systems, at the expense of (1) thermal image quality, (2) dynamic temperature range, and (3) accuracy of measurements.

The basic difficulty with these devices is that the detecting target is sensitive to a temperature–time variation (pyroelectric effect) so that

1. "Still" objects at a constant temperature cannot be detected; this calls for the use of a chopper, which introduces mechanically moving parts.
2. The quality of the thermal image, which depends strongly on the properties of the detecting target, is rather poor.

Pyroelectric TG systems have shown a number of difficulties precluding their immediate use in quantitative application:

1. They have no easy selection of temperature-range sensivity; this could be done only by changing the lenses or by changing the lens aperture and the radiation input on the target.
2. There is no possible positioning of the temperature window (or slot) for a given sensitivity.
3. The accuracy of temperature measurement is limited (the temperature reading is somewhat dependent on the object size).
4. The contour temperature measurement is poor. Contrast on the contour is lacking, due to thermal diffusion within the detecting target (thermal diffusion correction is at the moment necessary for a correct thermal image representation).

IV MODERN SYSTEMS AND TRENDS

Thermographic instruments were originally intended to yield qualitative surface temperature information by displaying "thermal" images on monitors. Subsequently the need for more accurate and complex diagnostic protocols and for refined thermal analysis led to digitization and computerization of those systems. At that point the displayed image quality is not of great interest for the TG users.

Once an A/D conversion of the thermographic signal is accomplished, then a number of operations can be performed at the software level to ameliorate some of the instrument characteristics (signal : noise ratio, linearity, accuracy) before extracting the significant parameters (e.g., temperature, temperature gradient, temperature time derivative, heat fluxes).

Development of software to eliminate some of the hardware deficiencies and to extract the relevant parameters is essential for implementation of the TG systems as quantitative measurement devices. Hardware and software are now being developed with the aim of arriving at a systems approach in which the thermograph is one of the subsystems of a chain of equipment that will be able to output elaborated data of immediate use for specific problems.

A Sensor Hardware

The field of TG hardware is expanding very rapidly; a number of devices under development will eventually improve the characteristics of the existing devices. The main trends are to

1. Avoid the cryogenic liquid cooling of the sensors (for scanning devices), and thus improve the TG portability
2. Avoid mechanically moving parts, both for reasons of reliability and of instrument life (at the moment all the existing TG systems include some moving parts)

3. Make the systems compatible with TV components (monitors, recorders, broadcasting system)
4. Extend the dynamic temperature range and sensitivity to cover low-, medium-, and high-temperature applications
5. Improve the signal : noise ratio

For instance, if a linear array of detectors is available, scanning in only one direction (by means of an oscillatory mirror) is needed. A matrix of detecting elements could eliminate any mechanical scanning process, simplify the system, and attain an excellent signal : noise ratio; these systems are being designed for special applications (e.g., military) but pose some problems (due to the different behavior of the sensors) when quantitative measurements are to be performed.

B Systems Configuration

Modern TG systems have been realized by a number of electronic modifications of the basic instrument to allow digitization, acquisition, and recording of the signals, and interfacing of these systems with dedicated computers for the elaboration of the parameters of interest.

One of the first systems, the one implemented at the University of Naples [3], uses the Thermograph AGA 680. The signal from the detector is digitized by an analog–digital convertor (Digimen Avioradio) connected through a multiplexer to the sensor element. The A/D conversion time is of the order of few microseconds; the digital signal (8 bits) is stored in a buffer RAM holding 64- by 128-matrix pixels. Each time an image is digitized, it is displayed on a color TV monitor. The RAM is subsequently read by a microcomputer (Apple II), transferred into the computer dynamic memory, and then recorded on a floppy disk; alternatively the images can be stored on a digital magnetic tape. The rate of acquisition can be chosen at will (for continuous storage on a magnetic tape an image every 2 s is required). When two subsequent images are to be acquired the time interval between them can be chosen by the operator.

In order to improve the signal : noise ratio, a feature exists that averages the signals over a number of subsequent images (averages can be performed over 2, 4, or 8 images); this can be performed whenever the evolution of the thermal process is sufficiently slow (say of the order of 10 s) compared with the acquisition time. The schematic of the hardware of the thermographic system is shown in Fig. 2.

Other recent commercial systems achieve similar features by different hardware arrangements. Fig. 3 illustrates the functional description of a modern commercial system (Probeye 3000 by Hughes) that utilizes an array of 10 IR sensors (indium antimonide), a single rotating mirror wheel, and a gas-bottle cooling device. A picture of this apparatus is shown in Fig. 4 and is described in detail in Ref. [4]. The main features of this apparatus are shown in Fig. 5; the presence of an A/D convertor makes it possible to use a microcomputer for image elaboration and for quantitative measurements, similar to the ones discussed below.

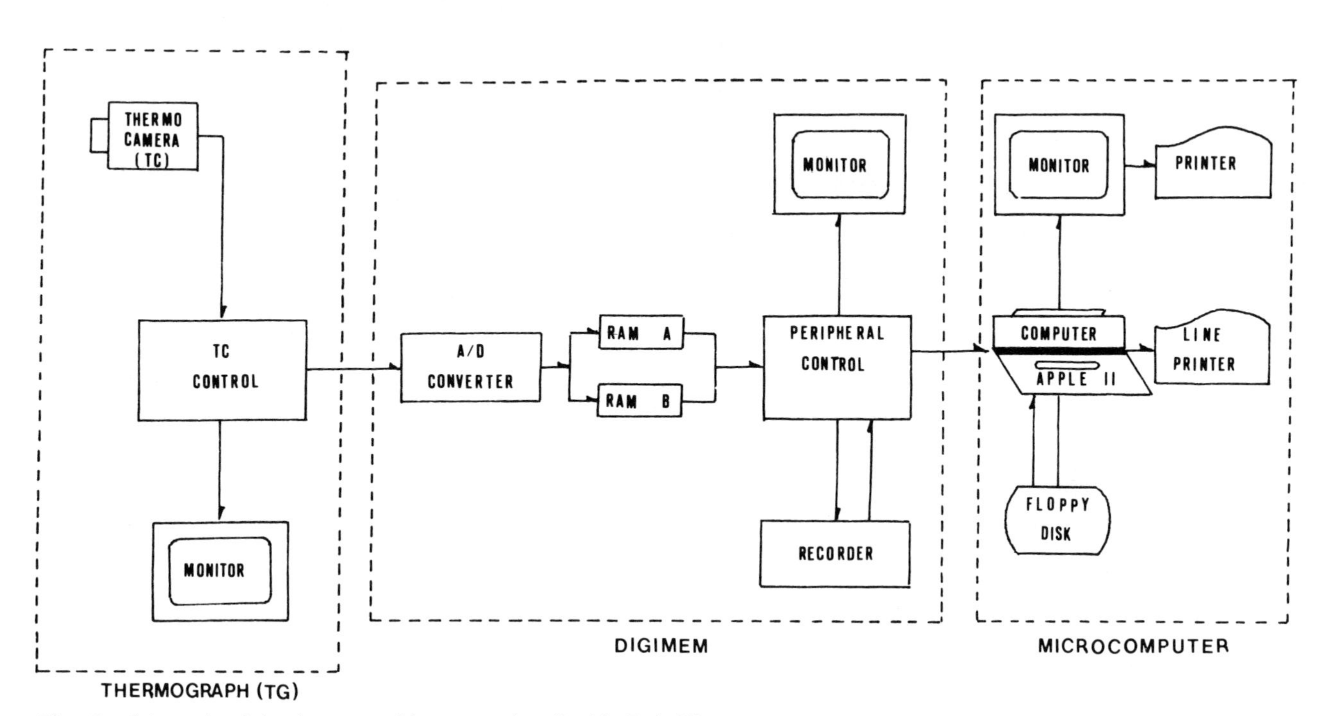

Fig. 2 Schematic of the thermographic system described in Ref. [3].

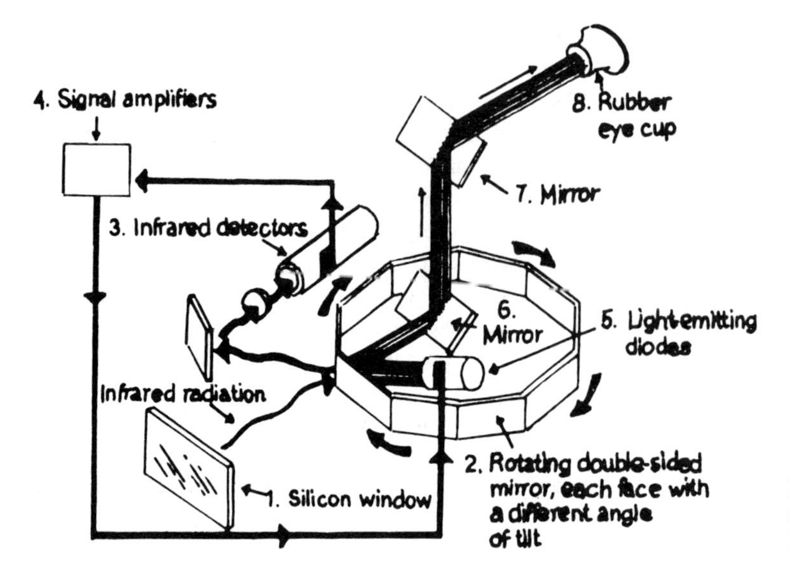

Fig. 3 Functional description of the IR detectors and optics of the TG system shown in Fig. 4 (courtesy of Hughes Aircraft Co.).

C Software

The success of thermographic techniques is mainly due to the possibility of digitization and subsequent elaboration of the thermographic images, and to the development of the appropriate thermal modelings. The most relevant results are being obtained by the implementation of

1. Systems software, for appropriate data acquisition
2. Applications software, oriented to the management and the data elaboration for specific problems
3. Software for presentation of the results on appropriate peripherals

In the process of extracting quantitative information out of digitized images it is appropriate to recall the differences that exist between a TG system and a conventional TV system:

1. The number of pixels that form a TV video image is typically larger than that provided by TG systems (say 625×625, or more, against 256×256).
2. The information content of a TG pixel is much higher than that of a TV pixel (say, 10–12 bits against 4–6 bits).
3. The acquisition frequency of TV images is typically higher than in the TG system (say, 25 images per second

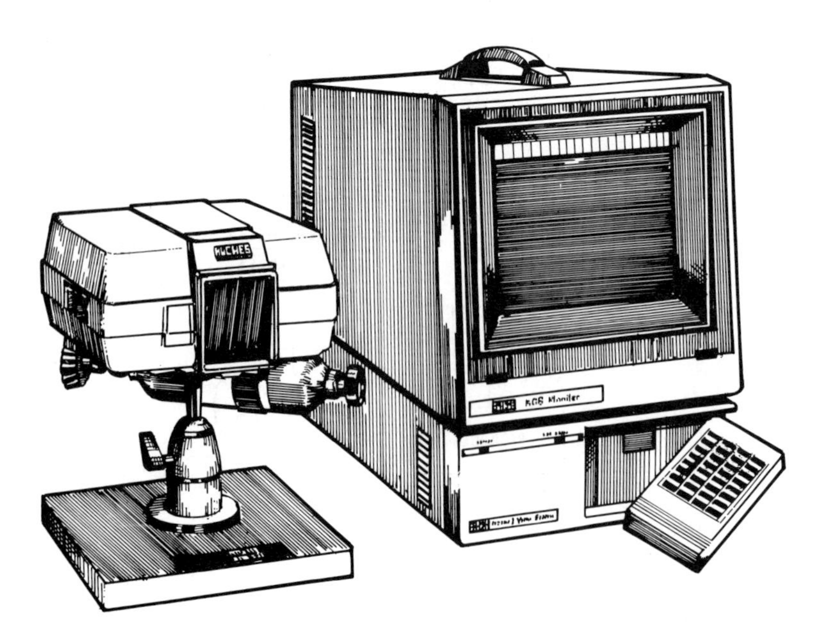

Fig. 4 The Probeye Series 3000 Systems showing the TG head, control, and color monitor (courtesy of Hughes Aircraft Co.).

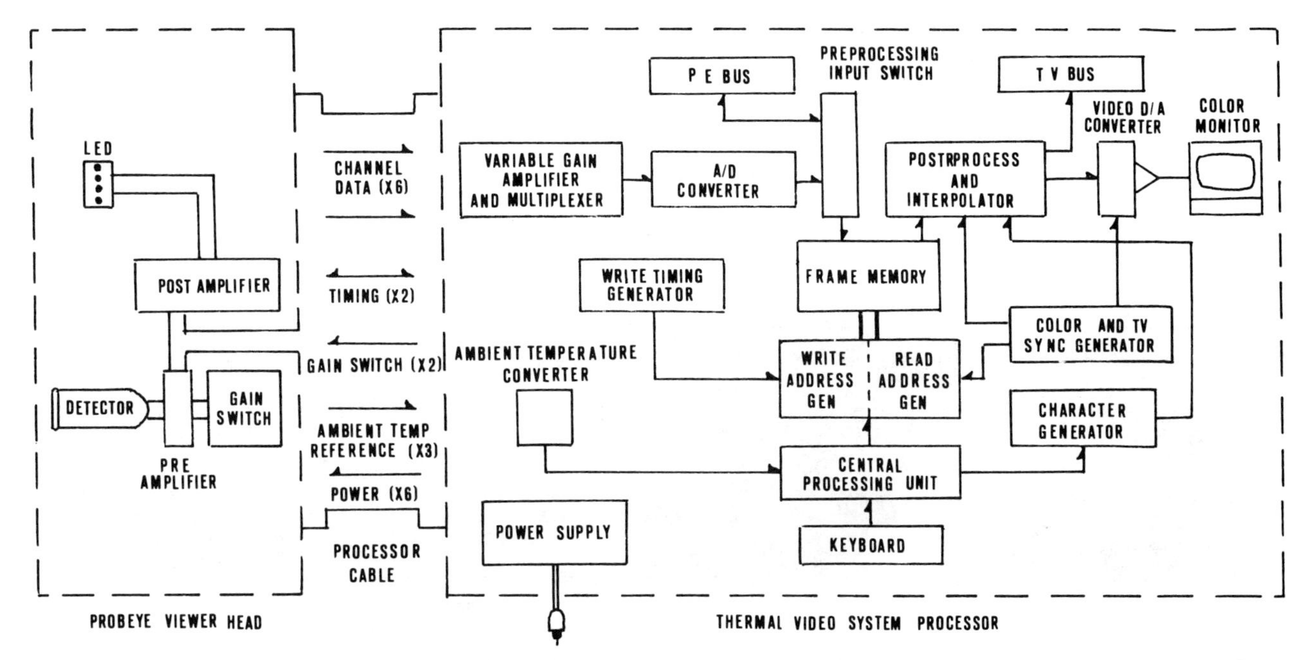

Fig. 5 Schematic of the processor essential elements of the Probeye 3000 System [4] shown in Fig. 4 (courtesy of Hughes Aircraft Co.).

against 15 images per second). The TG frequency can be further reduced whenever the thermal evolution of the observed surface is relatively slow compared with the IR detector response time (this may result in substantial noise reduction).

4. For these reasons it may be convenient in some cases to average thermal images, thus improving the accuracy of temperature measurement.

Typical software packages developed for a very specific application (evaluation of liquid surface temperatures time evolution taken on an orbiting space platform in zero-g conditions) are illustrated in Fig. 6 [5]. In the following sections, examples of typical software packages are given for the three classes of the software categories.

1 Systems Software To make the TG instrumentation applicable to many classes of problems, the systems software must be as versatile as possible to cope with many different problems. With reference to Fig. 2 (or to similar TG systems), systems software must be provided that pilots the TG signal-acquisition sequence and that can be installed either on a dedicated microprocessor or on a microcomputer.

Let us denote by **N** (t) the signal matrix (formed by $I_m \times J_m$ pixels) taken at the instant of time t. Typical systems software should allow different acquisition sequences (i.e., digitalization and recording of the thermal images). For instance,

Periodic acquisition of single images (the time interval Δt must be assigned).

Periodic acquisition of average thermal images. Averaging thermal images over n_A subsequent images yields an improvement of the signal : noise ratio. The acquisition sequence is characterized by the number of images N_A taken and by the time interval Δt between the images.

Nonperiodic acquisition of n images at different times $N(t_k)$. The times Δt_k are to be assigned ($K = 1, n$).

Any combination of the above sequences (i.e., specify times, averaging sequence, and number of acquired images).

The recording may be done on a digital tape recorder, on a VHS tape recorder, or on a floppy disk (via the dynamic memory of the computer).

2 Applications Software The applications software is specific to the problem (see subsequent sections) and is able to extract the parameters relevant for the different applications. Typically the surface temperature $T(I, J, t)$, taken at each target point I, J, at one or more instants of time, is the basic information provided to the application software (the evaluation of **T**(t) from **N**(t), through calibration procedures, is discussed in Sec. V).

Typical tasks of the applications software are

Noise reduction
Digital filtering
Computation of temperature
Smoothing
Computation of isotherms

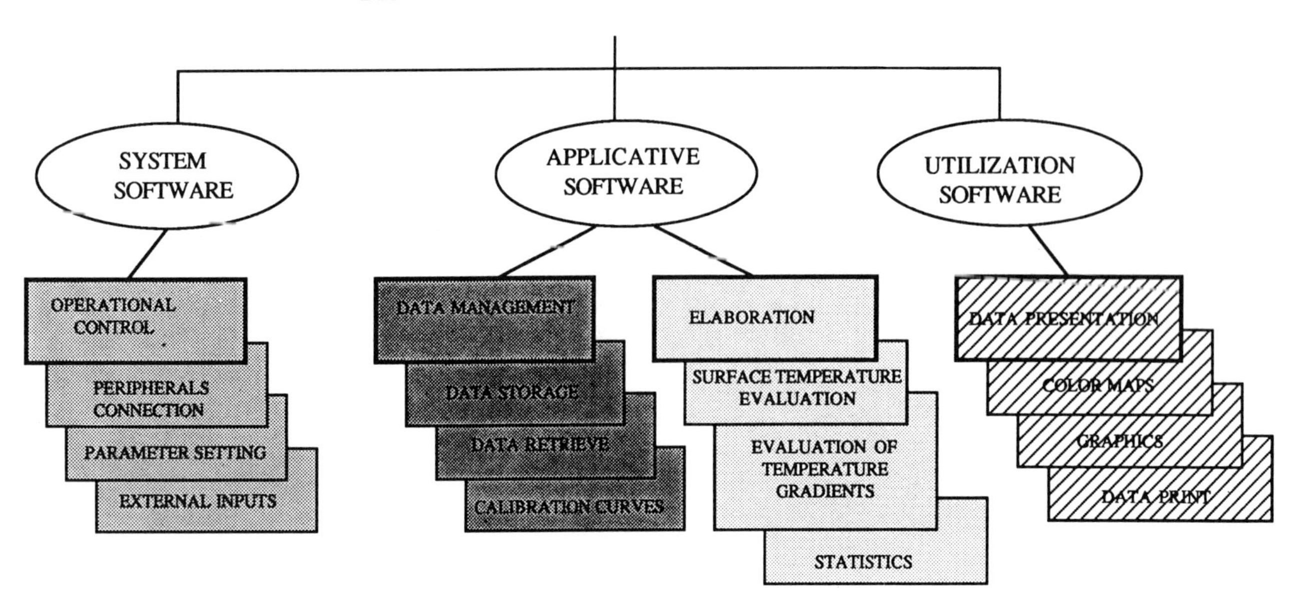

Fig. 6 Software packages developed for the image elaboration and quantitative measurements [5].

- Image measurements (specific to the problems). Typically evaluation of time and space derivates
- Statistical computations (histograms, probability functions, averages, best fitting, etc.)

Most of these functions are similar to those performed by other image-elaboration softwares (e.g., Ref. [6]) and will not be discussed further. Typical software specific of TG systems will be illustrated when dealing with TG applications.

3 Data Display Software Data Display Software consists of those routines used for data presentation. The digital data released by the TG System can be either a matrix of $I_m \times J_m$ numbers representing the digitized pixel signal taken at time t [$\mathbf{N}(t) = N(I, J, t)$] (e.g. for a 12-bit A/D conversion, there will be positive numbers ranging from 0 to 4096) or a matrix of real numbers corresponding to the surface temperature matrix [$\mathbf{T}(t) = T(I, J, t)$]. Software packages are typically available that show on the system monitor

- A preselected temperature matrix (or part of it)
- The matrix of filtered, noise-removed, or transformed temperature values
- The graphics of temperatures along a line, a column, or a diagonal
- Color maps of the temperature levels (e.g., with 16 color levels)
- Curves connecting points with the same value of computed parameters (e.g., temperatures, temperature gradients, time derivatives, etc.)
- Histograms of any computed parameter
- Probability density values

All this information, shown on the screen, can also be typically printed, on demand, on paper (hard copy). Again the software above described is rather general purpose and will not be discussed in detail.

V SURFACE TEMPERATURE MEASUREMENT

Considerations about the stability, accuracy, and resolution of the TG systems that are at the basis of quantitative measurements are reported in Ref. [17]. The first step in all the TG systems applications is the extraction of the surface temperature distribution $T(\xi, \eta, t)$ from the thermographic signal $N(I, J, t)$ and from geometric computations that relate object surface coordinates (ξ, η) to the pixel position (I, J).

The temperature at the surface point $T(I, J, t)$ that corresponds to line I and to column J on the digitized map is related to the signal $N(I, J, t)$ by

$$T = F(N, \varepsilon, \beta, A) \tag{8}$$

where ε is the local surface emissivity of the body (the gray-body assumption is typically made for normal operations), β is the angle between the surface normal n and the TG-view axis, and A denotes the interaction with the ambient temperature and the position of the objects that interact, at the instrument wavelengths, with the surface points.

For objects of limited extensions and of the same type and surface conditions, A and ε can be assumed to be the same for all the surface points. For values not exceedingly large (say, $\beta <= 70°$) one could assume that the calibration made at one surface point could be extended to all points.

A "Ad Hoc" Calibration Curve

The simplest way of obtaining the calibration curve $T = F(N)$ for the test surface is to locate a temperature sensor at a surface point, to allow this temperature T to vary over the range of interest (at the very same ambient conditions that are established during the measurement phase), and to detect the corresponding digitized TG signal N. Inputing the sensor temperature T directly in to the computer and acquiring the TG digital signal leads to the point-by-point calibration curve $T = F(N)$. This procedure is valid only for that specific object surface, observed by the TG from a given position, and at specific ambient conditions.

The calibration and measurement procedures are however more complex if different TG sensitivities are provided or if adjustable TG "windows" (or slots) are available that select a temperature range (for each sensitivity). In this case the transformation of the signal matrix $\mathbf{N}(t)$ into the temperature matrix $\mathbf{T}(t)$ is not immediate because typically the window position is not selectable stepwise (and therefore its position is not known accurately) and because the $T(N)$ calibration curve is highly nonlinear. The calibration and measurement procedures that should be adopted in this case in a laboratory are illustrated below.

The calibration curves are constructed for the surface of interest at a given sensitivity and at constant ambient conditions by looking at a field point where the surface temperature can be changed at will (within the temperature range of interest) and measured by a point sensor. For each "window" position (typically indicated as *black level*) the couple of values of T and N are recorded to construct a calibration curve to cover the entire temperature range. The value of N that corresponds to temperature T is however dependent on the window position.

The calibration curve is built piecewise by patching together different legs corresponding to different window positions (Fig. 7). Starting at window position 1, one records T and N at increased temperature values (starting at $T_{1_{min}}$). When the full window range is covered (e.g., $\Delta T = 10°C$) and the value of N corresponding to $T_{1_{max}} = T_{1_{min}} + \Delta T$ is reached, then the window is shifted to position 2; at this new window position the same $T_{1_{max}}$ then corresponds to a smaller value of $N(N_{2_{min}})$. The two calibration curves are

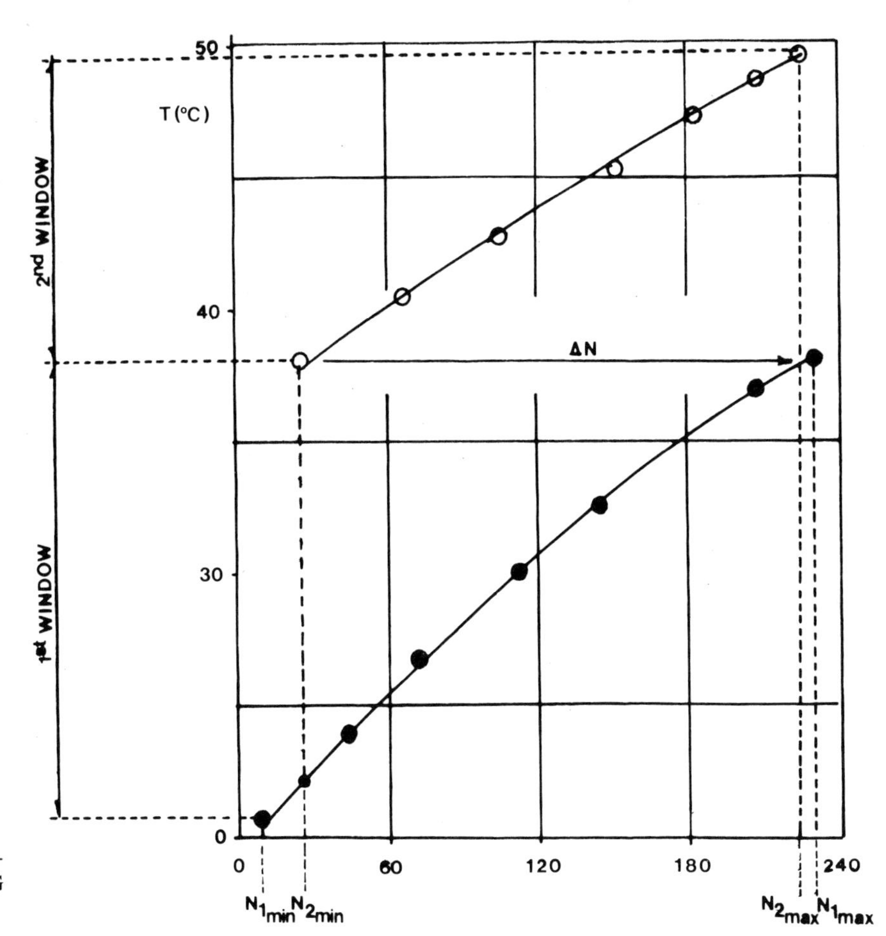

Fig. 7 Construction of an ad hoc calibration curve by a computerized TG system.

then patched by plotting the second curve, shifted by $\Delta N = N_{1_{max}} - N_{2_{min}}$ (see Fig. 7). The maximum value of ΔN is the number corresponding to the available bit/pixel of the A/D process. A complete calibration curve is then obtained covering the temperature range of interest.

The measurement phase needs a reference temperature T_r in the field of view to recognize a point on the calibration curve. By reading the value of N_{ir} corresponding to T_r, taken at the general window position i, one is able to select the proper position r of the calibration curve with respect to the N axis (see Fig. 8). In fact reading on the calibration curve the value of N_r corresponding to T_r allows computation of the value of the shift $\Delta N_{ir} = N_{ir} - N_r$. The point surface temperature corresponding to the value N is then found by computing a "corrected" signal N_c given by

$$N_c = N + \Delta N_{ir}$$

and entering the calibration curve with this new value $T(N_c)$.

B Blackbody Calibration Curve

The surface temperature can also be evaluated in general utilizing the blackbody calibration curve ($\varepsilon = 1, \beta = 0$, no effect of the environment) that relates the signal of the instrument N_B to the blackbody temperature. In this case the signal N_B is proportional to the radiative power (or to the photon flux, if photon detectors are used) collected by the thermograph:

$$N_B(T) = RP_B(T)$$

where R is the instrument response and $P_B(T)$ is the relation between blackbody temperature and power (or photon number) emitted by the surface, derived by Planck's law, Eq. (1).

For a gray body ($\varepsilon < 1$) the signal of the thermograph is a combination of the radiation emitted by the surface and of the radiation from the environment (reflected on the surface). Assuming that the environment is at constant temperature T_a and that it behaves as a blackbody, the TG signal corresponding to a surface point (at which $\beta \simeq 0$) is

$$N = R[\varepsilon P_B + (1 - \varepsilon)P_a] = \varepsilon N_B(T) + (1 - \varepsilon)N_a \quad (9)$$

The emitting surface point temperature can be obtained from Eq. (9) by evaluating the term $N_B(T)$ and by computing T with the help of the blackbody calibration curve. Two

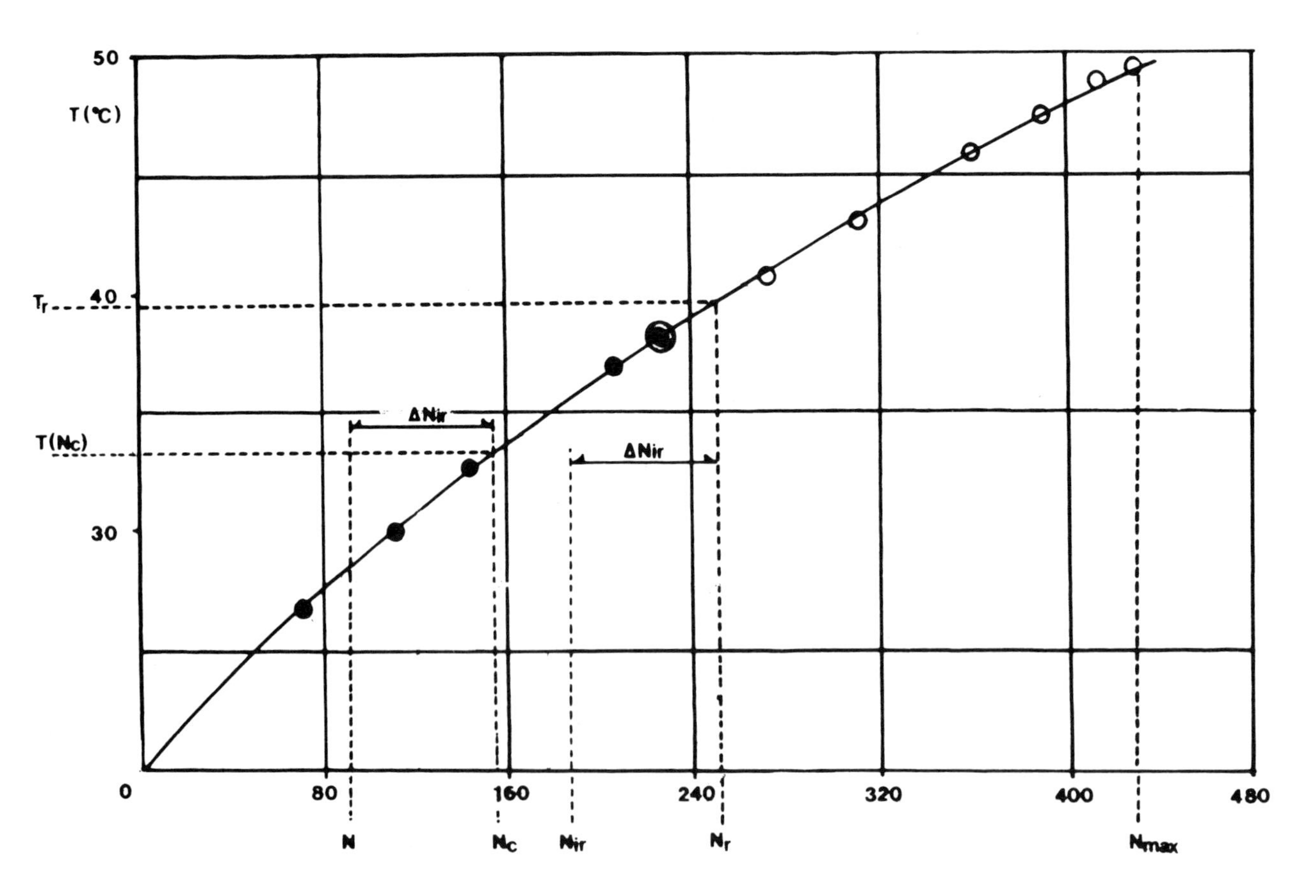

Fig. 8 Measurement of the surface temperature by an ad hoc calibration curve stored in a computerized TG system.

measures are taken at $T = T_a$ (ambient temperature) and at a known reference temperature T_r:

$$N_a = RP_a \qquad N_r = \varepsilon N_B(T_r) + (1 - \varepsilon)N_a \tag{10}$$

Knowledge of the blackbody calibration curve $N_B(T)$ and the evaluation of the differences between the thermographic readings

$$\begin{aligned} N - N_a &= \varepsilon[N_B(T) - N_a] \\ N - N_r &= \varepsilon[N_B(T_r) - N_a] \end{aligned} \tag{11}$$

lead to computation of the signal that the TG would get from a blackbody at the same surface temperature T

$$N_B(T) = N_a + \frac{N - N_a}{N_r - N_a}[N_B(T_r) - N_a] \tag{12}$$

The temperature T can then be obtained from Eq. (12), by measurements of the thermographic signals N_r, N_a, N, by the blackbody calibration curve.

If temperatures T, T_a, T_r are close enough to each other, then a linearization is possible for the calibration curve to get simply

$$T \simeq T_a + \frac{N - N_a}{N_r - N_a}(T_r - T_a) \tag{13}$$

In modern TG systems, however, it is not necessary to assume the above linear dependence since applications software allows accurate, on-line evaluation of the temperature (through appropriate nonlinear calibration curves).

VI FLOW VISUALIZATION

Thermographic equipment is not specifically intended for flow visualization, but in a number of cases it can be used to get information on flow fields. For instance TG can actually detect the position of particles in gas jets, if they emit radiation in the TG wavelength range; these particles (solid or liquid) then play the same role as the visible tracer-particle light scatterers detected by TV cameras in the visible spectrum. Techniques for particle visualization are described elsewhere in this handbook and will not be further discussed.

In the following discussion, therefore, only a number of applications related to the diagnostics of flow fields are presented, together with the specific problems of image processing and computer-assisted methods. The basic idea is to obtain indirect information on flow fields by detecting the thermal behavior of a solid (or liquid) body surface in contact with the flow. In order to single out the parameters of interest that can be measured by a TG system (through appropriate elaborations of the TG signals), the modelings of a number of problems are described. More specifically the following applications are considered:

- Convective heat transfer measurement
- Visualization of separated flow regions
- Measurement of the blood flow in human limbs

For each of these problems, after presentation of the appropriate numerical modelings, experimental results are shown.

A Modeling for Evaluation of the Heat Transfer Coefficient

Correlation is sought of the heat transfer from an object and its surroundings environment with the object surface temperature evolution in time; this will lead to a numerical model that allows heat transfer fluxes to be computed by measurements of surface temperatures taken by TG systems.

1 Basic Equations The three-dimensional unsteady heat transfer problem for a thin wall model is first considered [7, 8]. A local coordinate system is chosen in which n is the axis orthogonal to the surfaces (the inner and outer boundary wall surfaces run parallel), and ξ, η are two orthogonal directions on the plane tangent to the model surface (Fig. 9).

The (constant) thin wall thickness is denoted by s and the two principal radiuses of curvature of the surface by R_1 and R_2. The heat transfer equation, if curvature effects are negligible (in practice, when s/R_1, $s/R_2 << 1$), reads

$$\rho c \frac{\partial T}{\partial t} = k_n \frac{\partial^2 T}{\partial n^2} + K\left(\frac{\partial^2 T}{\partial \xi^2} + \frac{\partial^2 T}{\partial \eta^2}\right) \tag{14}$$

The contributions of the conductive heat transfers along n and in the plane ξ, η tangential to the model surface have been separated; nonisotropic ($k_n \neq k$) materials will be considered as possible candidates for the model wall.

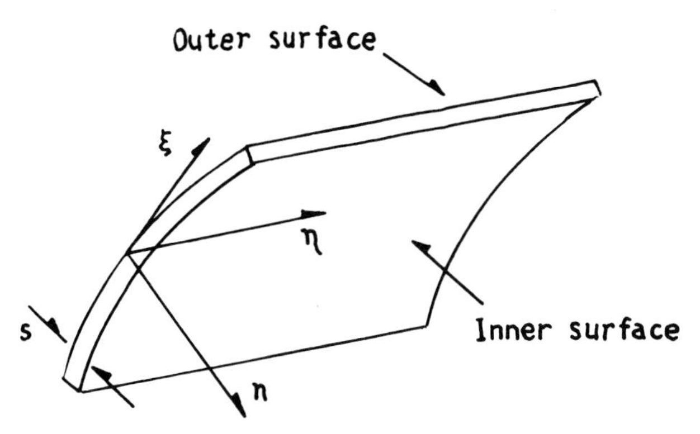

Fig. 9 Model wall coordinate system.

Integration of Eq. (14) across the wall thickness s and the definition of heat fluxes q entering the wall at $n = 0$ and $n = s$

$$q_0 = -k_n \left(\frac{\partial T}{\partial n}\right)_{n=0} \qquad q_s = K_n \left(\frac{\partial T}{\partial n}\right)_{n=s}$$

yields, for material properties that are constant with time, space, and temperature

$$\rho c \int_0^s \frac{\partial T}{\partial t} dn = q_0 + q_s + k \int_0^s \left(\frac{\partial^2 T}{\partial \xi^2} + \frac{\partial^2 T}{\partial \eta^2}\right) dn \quad (15)$$

The main idea of the method is to deal with ranges of values of the wall thickness and with experimental conditions so that the following assumptions can be made:

$$\int_0^s \frac{\partial T}{\partial t} dn = s \frac{\partial T}{\partial t} \simeq s \frac{\partial T_0}{\partial t} \quad (16)$$

$$\int_0^s \left(\frac{\partial^2 T}{\partial \xi^2} + \frac{\partial^2 T}{\partial \eta^2}\right) dn \simeq \left(\frac{\partial^2 T_0}{\partial \xi^2} + \frac{\partial^2 T_0}{\partial \eta^2}\right) s \quad (17)$$

where T_0 denotes the outer surface temperature of the model that can be measured by a TG system. These assumptions are both verified for a thin wall model with a high thermal conductivity along n or when specific experimental conditions are established during the measurements (as examined later).

Assumptions (16) and (17) allow a straightforward application of Eq. (15) for the overall heat transfer flux evaluation when the heat flux q_s at the inside surface ($n = s$) is known:

$$q_0 = q_s - \rho c s \frac{\partial T_0}{\partial t} \left[1 - \frac{\alpha}{\partial T/\partial t}\left(\frac{\partial^2 T_0}{\partial \xi^2} + \frac{\partial^2 T_0}{\partial \eta^2}\right)\right] = q_{c_0} + q_{r_0} \quad (18)$$

where q_{c_0} is the convective heat flux at the outer surface, q_{r_0} is the net heat flux radiated from the model surface, and α is the thermal diffusivity in the plane tangent to model surface. A convective heat transfer coefficient can be defined as:

$$h = \frac{q_{c_0}}{T_0 - T_{aw}} \quad (19)$$

For low-Mach-number regimes the adiabatic wall temperature T_{aw} practically coincides with the asymptotic stream temperature T. The value of q_0 coincides with the convective heat transfer flux if the radiative heat flux can be neglected with respect to the convective one. This condition is fulfilled when

$$q_{c_0} >> \varepsilon\sigma\,(T_0^4 - T_a^4)$$

For a surface temperature T_0 not much different from the ambient temperature T_a, this condition reads

$$h >> 4\sigma\varepsilon T_0^3$$

Simplifications of Eq. (18) can be made whenever (1) the convective heat flux at $n = 0$ is much larger than the heat flux at $n = s$, (2) the model wall material exhibits a rather low thermal diffusivity $\alpha = {}^R(\rho c)$ and the wall surface temperature distribution is almost uniform (the second term in the square bracket can be neglected with respect to 1):

$$\alpha \left(\frac{\partial^2 T_0}{\partial \xi^2} + \frac{\partial^2 T_0}{\partial \eta^2}\right) << \frac{\partial T_0}{\partial t} \quad (20)$$

In this case one would then simply get

$$h \simeq \frac{\rho c s}{T_{aw} - T_0} \frac{\partial T_0}{\partial t} \quad (21)$$

Expression (16) represents in any case the leading term for the heat transfer coefficient evaluation as proposed in Ref. [8]; to evaluate the local value of h it would be necessary to measure the local surface temperature of a preheated model positioned in an air stream at two different times $(t,\ t + \Delta t)$. If these times are close enough for the time derivative to be approximated by its finite difference, then the overall heat transfer coefficient h reads

$$h \simeq \frac{\rho c s}{\Delta t} \frac{T_0(t) - T_0(t + \Delta t)}{T_0\left(t + \dfrac{\Delta t}{2}\right) - T_{aw}} \quad (22)$$

The experimental evaluation of h relies on the accurate computation of the nondimensional ratio between two surface temperature differences that can be measured by the TG system. In particular $T_0(t + \Delta t/2)$ can either be measured by the TG at time $t + \Delta t/2$ or be computed by

$$T_0\left(t + \frac{\Delta t}{2}\right) = \frac{T_0(t) + T_0(t + \Delta t)}{2} \quad (23)$$

2 Implementation of Thermographic Techniques to the Measurement of Convective Flows The evaluation of the heat transfer coefficient h is performed by measuring the thin wall model surface temperatures, by means of a thermograph, at the unsteady thermal conditions that are established after having heated (or cooled) the model above (or below) the airstream temperature (or the adiabatic wall temperature, for compressible regimes) during the thermal evolution toward the airstream temperature. The essential steps in the measurement procedure [7] are the following:

1. Establish the flow field over the model to be tested and measure the surface temperature T_{aw} at the thermal steady state conditions.
2. Heat up (or cool down) the model by appropriate heat sources (convective, radiative, conductive).
3. Establish a thermal steady-state condition during the steady heating or cooling; this step may be necessary to

adjust the heating or (cooling) process for obtaining an almost uniform surface temperature.
4. Shut off the heating (or cooling) source.
5. Record the surface temperature time evolution during the cooling or warming of the model (toward the adiabatic wall temperature).

According to Eqs. (19), (22) computation of the heat transfer calls for accurate measurements of the following quantities:

$$T_{aw} \qquad T_0(\xi, \eta) \qquad \frac{\partial^2 T_0}{\partial \xi^2} \qquad \frac{\partial^2 T_0}{\partial \eta^2} \qquad \Delta t$$

Thermographic measurements are performed at steps 1 and 5; monitoring of the surface temperature at step 3 is often necessary for the appropriate setup of the sources and of the TG system.

The second-order derivatives of the surface temperature with respect to two orthogonal axes ξ, η in the plane tangent to the surface can be computed once the surface temperature measurements are performed by the TG system. The difficulty in the experimental evaluation of these terms arises from the fact that the thermographic systems detect the projection of the surface temperatures over a plane x, y orthogonal to the thermograph viewing axis z; the angles between x, y, z, and n must therefore be known at each surface point in order to compute the values of $\partial^2 T_0/\partial \xi^2$ and $\partial T_0^2/\partial \eta^2$ from the measured values of $\partial^2 T/\partial x^2$ and $\partial^2 T/\partial y^2$.

Equation (18), written at each of the measurement steps 1, 3, and 5, furnishes the necessary information for evaluation of the heat transfer coefficient and for the experimental procedure. These steps are covered in the following discussion.

***a Measurement of* T_{aw}** Step 1 allows measurement of the reference adiabatic wall temperature T_{aw} to be taken at the conditions

$$\frac{\partial}{\partial t} \equiv 0 \qquad q_s = 0 \qquad q_r \simeq 0$$

$$T_0 = T_{aw} + \frac{k_s}{h}\left(\frac{\partial^2 T_0}{\partial \xi^2} + \frac{\partial^2 T_0}{\partial \xi^2}\right) \tag{24}$$

For sufficiently uniform surface temperature and/or for $k_s/hL^2 << 1$ (where L is a characteristic model length), one can assume

$$T_0 \simeq T_{aw}$$

In practice, the adiabatic wall temperature, to which the heat transfer coefficient refers, is constant over the surface so that an average value of T_0 can be computed:

$$\frac{1}{A}\int_A T_0 \, dA \simeq T_{aw} \tag{25}$$

Expression (25) assumes that the conductive heat fluxes, which tend to make the surface temperature more uniform, do not affect the average surface temperature (i.e., their contribution tends to cancel out when averaged over the model area); this is sufficiently true for constant-thickness wall models.

b Measurement of the Temperature Distribution During Heating Step 3 permits checking of the steady-state conditions $\partial/\partial t = 0$; $T_0 = T_0^s$ to be achieved under the adjustable heat flux q_{source}:

$$q_0 = q_s + ks\left(\frac{\partial^2 T_0^s}{\partial \xi^2} + \frac{\partial^2 T_0^s}{\partial \eta^2}\right) + q_{source} \tag{26}$$

The source term can be created by radiative internal and/or external heat sources, by convective internal heat transfer (e.g., by hot air passing inside the model), or by electrical heating of the model walls.

The possibility of adjusting the heat intensity over the surface to obtain a surface temperature distribution as uniform as possible is welcome because the in-plane conductive heat fluxes are then negligible. To obtain these conditions one should ensure larger heat source fluxes at those surface points where the heat transfer coefficient h_0 to the airstream is larger.

c Measurement of the Cooling Rate Measurements of the local surface temperature as a function of time are performed at step 5, and the value of h_0 is computed by Eq. (18). Precision of the evaluation of the term enclosed in brackets in Eq. (18) strongly depends on the thermal unsteadiness of the process (the local value of $\partial T_0/\partial t$ during this step should not vanish at any point of the model surface). The implication is that $\Delta T_0 > \Delta T_{min}$, ΔT_0 is the difference in surface temperatures taken at two instants of time, and ΔT_{min} is the minimum temperature difference detectable by the thermographic system; this condition is necessary (albeit not sufficient) for accuracy of the method.

The fundamental parameter to be identified is the time interval during which the measurements are to be taken. This choice is relevant also for validity of the assumptions (16) and (17) made in deriving Eq. (18); these assumptions, in fact, may hold if the temperature–time derivatives are computed at appropriate time intervals. To prove this point, it is sufficient to follow the wall temperature evolution right after the source is shut off. As shown by numerical tests, after a suitable time the values of $\partial T_s/\partial t$, and $\partial T_0/\partial t$ practically coincide on both sides of the wall, even though T_0 during the wall cooling by the airstream is lower than T_s. In practice, for thin walls, after a few seconds, the temperature distribution at the inside surface appears similar to that of the outside surface so that the temperature–time derivatives attain the same values on both the external and internal surfaces: assumption (16) and, to some extent, also assumption (17) are then verified.

d Software for Image Elaboration Software is needed for the thermographic image elaboration to display appropriately the parameters extracted from the surface temperature. For instance, two subroutines are necessary to compute the local heat transfer coefficient:

1. Evaluation of the local temperature–time derivative. The difference of the values of the surface temperature taken at time $t + \Delta t$ and at time t is computed at each point (I, J) and its value divided by Δt:

$$\frac{\partial T_0}{\partial t} \simeq \frac{T_0(I, J, t + \Delta t) - T_0(I, J, t)}{\Delta t} \tag{27}$$

2. Evaluation of $\nabla^2 T_0$ at each point of the thermographic image:

$$(\nabla^2 T_0)_t \simeq \frac{T_0(I, J - 1) + T_0(I, J + 1) - 2T_0(I, J)}{(\Delta\eta)^2} \frac{T_0(I - 1, J) + T(I + 1, J) - 2T_0(I, J)}{(\Delta\xi)^2} \tag{28}$$

In order to properly evaluate the $\nabla^2 T_0$ in a plane tangent to the observed surface, the equation of the surface $z = F(x, y)$ must be known (where z is the coordinate along the TG viewing axis). A specific software [9] derives, for all the TG pixels, their distance along the z axis (i.e., the $F(x, y)$ equation from an analytical or from a point-by-point description of the surface). Once $z = F(x, y)$ has been derived, the two curves [intersections between the body surface and (z, x) or (z, y) planes] can be computed:

$$Z_\xi = f(\xi) \quad \text{at y = const}$$
$$Z_\eta = g(\eta) \quad \text{at x = const}$$

At the surface point (i, j) two angles (γ, β) are then defined

$$tg(\gamma)_{i,j} = f'_{i,j} = \frac{(Z_\xi)_{i,j} - (Z_\xi)_{i-1,j}}{(x_{i,j} - x_{i-1,j}}$$
$$tg(\beta)_{i,j} = g'_{i,j} = \frac{(Z_\eta)_{i,j} - (Z_\eta)_{i,j-1}}{y_{i,j} - y_{i,j-1}} \tag{29}$$

that allow the transformations

$$\Delta\xi \rightarrow \Delta x \qquad \Delta\eta \rightarrow \Delta y$$

to be performed for use in Eq. (17):

$$(\Delta\xi)_{i,j} = \frac{(\Delta x)_{i,j}}{\cos \gamma_{i,j}} \qquad (\Delta\eta)_{i,j} = \frac{(\Delta y)_{i,j}}{\cos \beta_{i,j}} \tag{30}$$

This description of the algorithm necessary for TG ''image elaboration'' points out that the specific problems encountered in the elaboration of thermographic data are quite different from those of more conventional flow visualizations.

B Experimental Results on Natural and Forced Convection Heat Transfer

In order to show the potential and the capabilities of the thermographic technique, as implemented on the typical TG systems described in Sec. IV and based on the modeling described in Sec. VI.A, three applications are described here for which theoretical, numerical, or other experimental results were available for comparison. The repeatibility and the accuracy of the use of thermography in flow problems of increasing complexity is demonstrated.

1 Natural Convection over Flat Plates [10] Two flat plates, 2 mm thick and measuring 40 × 40 cm, were used for these tests. One plate was glass and the other Plexiglas; both were painted black on both sides. These plates were preheated between two metallic plates (toaster-type heater) to about 20–25°C above the ambient temperature. The plates were subsequently positioned vertically and observed by the thermograph during natural convection cooling. The area observed by the TG system (described in Sec. IV.B) was about 5 × 10 cm and covered the plate leading edge (at the bottom). The background (at ambient temperature) that appeared in the field of view was also continuously monitored.

The experimental procedure is similar to the one described for convective flows [4]. Two thermal images $\mathbf{N}(t)$, $\mathbf{N}(t + \Delta t)$ are recorded $\Delta t = 4s$, from which the temperature matrices are extracted $\mathbf{T}(t)$, $\mathbf{T}(t + \Delta t)$ through the calibration curve; the heat fluxes are then computed and graphs are printed (in almost real time) that show the values of the heat transfer coefficient along the flow direction ($x = 0$ at the leading edge). Three-dimensional fields have also been observed for rectangular plates with nonuniform temperature distribution and for plates of different geometries and inclined at an angle with respect to the vertical. The heat transfer coefficient is then computed at each point of the plate (I, J) to construct the matrix $\mathbf{h}$, and is then visualized by conventional graphs (line, color, or 3-D prospectic graphs). Typical computer outputs are shown in Figs. 10 and 11, together with the theoretical and numerical results.

The theoretical value of the Nusselt number heat flux on a vertical flat plate at constant temperature is

$$\mathrm{Nu}(x) = \frac{q(x)x}{k_a(T_0 - T_a)} = Cx^{3/4}\left(\frac{\partial T^+}{\partial n^+}\right)_{n^+=0} \tag{31}$$

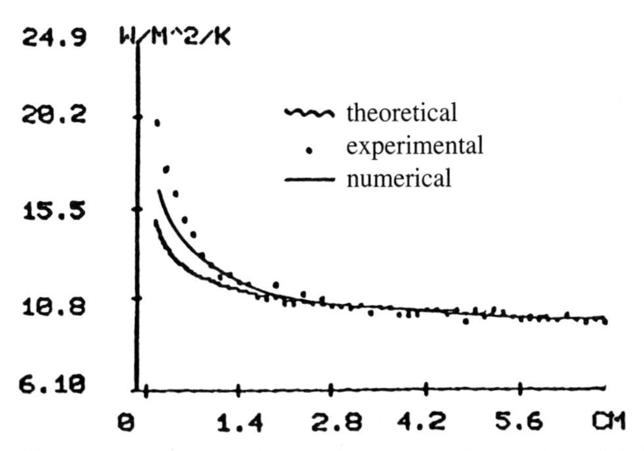

Fig. 10 Distribution of the heat transfer coefficient along a glass flat plate (s = 2 mm).

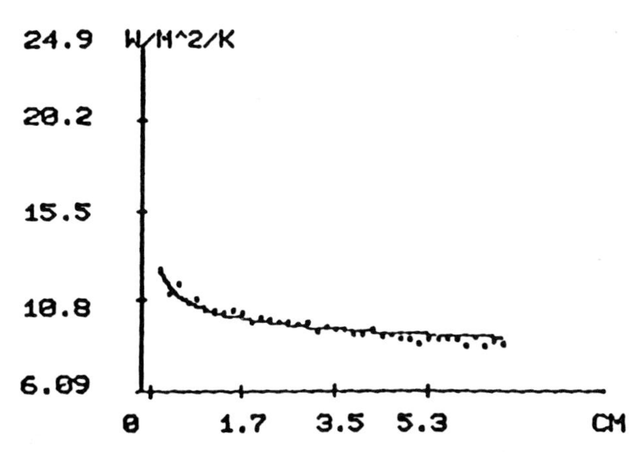

Fig. 11 Distribution of the heat transfer coefficient along a cellulose acetate flat plate (s = 0.15 mm).

where $q(x)$ is the convective heat flux, k_a is the thermal conductivity of the air, and T^+ and n^+ are the nondimensional temperature and coordinate orthogonal to the surface respectively. A total heat transfer coefficient has been computed [in this case the TG coordinates (x, y) coincide with the surface coordinates (ξ, η)], and the radiative cooling has been subtracted for evaluation of the convective heat transfer coefficient [conduction within the plate is practically negligible, upon assumption (20)]:

$$\mathrm{Nu}(x) = \frac{x}{k_a} \frac{1}{T_0(\xi, \eta, t) - T_a} \left[\frac{\rho c s}{2} \frac{\partial T_0}{\partial t} - \varepsilon\sigma \left(T_0^4(\xi, \eta, t) - T_a^4\right) \right] \tag{32}$$

The two plots of Figs. 10, 11 refer to two different thicknesses of the plates (2 and 0.15 mm). As expected, the results with the thin foil are more accurate near the leading edge because it more closely approximates the flat plate geometry assumption. However a number of difficulties arise when using very thin foils:

Poor mechanical rigidity (the foil tends to bend)

Strong cooling of the foil, which means foil temperatures very close to the ambient temperature (that is critical for the computation of h near the leading edge, experiencing the highest cooling rate)

The need to take thermograms in very short time intervals

The good correlation with the theoretical results show the applicability and the accuracy of the technique.

2 Forced Convection on Flat Plates [7] Flat plates, at a zero angle of attack and at different Reynolds numbers, were tested. Models made of black-painted window glass 2 mm thick and Plexiglas 1 mm thick were used. During the heating, of the plate (by radiative lamps) the open tunnel airflow did not impinge on the flat plate model; this procedure allowed an almost uniform temperature distribution of the flat plate to be reached before the thermographic observation (before the airflow was established over the plate). Typical computer output is shown in Fig. 12; the local Nusselt number Nu_x was plotted versus the abscissa

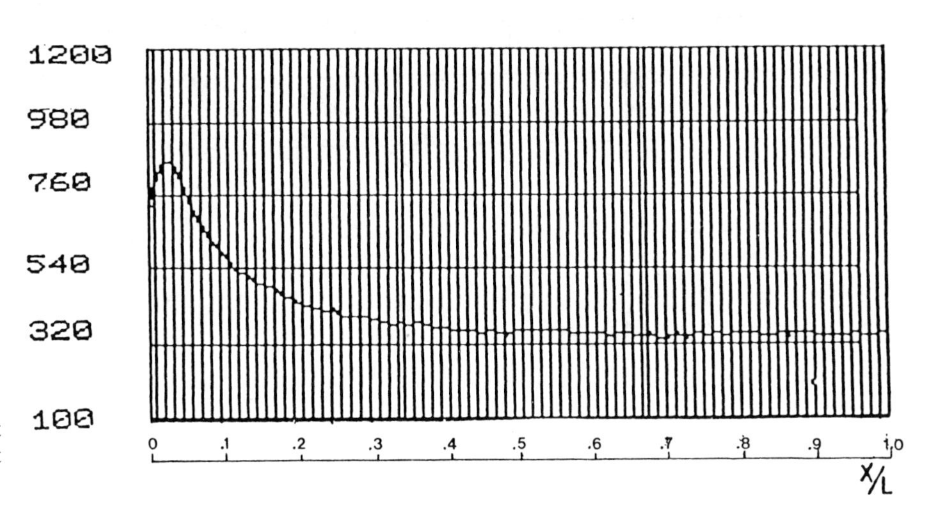

Fig. 12 Distribution of the Nusselt number along a Plexiglas flat plate at V = 12 m/s (Re = 2.1×10^5).

x for an asymptotic velocity of $V = 12$ m/s corresponding to $Re_L = 2.1 \times 10^5$.

As already discussed for the natural convection cases, a poor correlation of the experimental measurements was again found near the leading edge (because of the bluntness of the plate of finite thickness, which altered the flow field at the leading edge).

The experimental data are compared with the classical analytical solution for laminar flow:

$$Nu_x = 0.332\,(P_r)^{1/3} \quad Re_x^{1/2} \tag{33}$$

where $Re_x = Vx/\nu$. The experimental values of Nu_x were computed by

$$Nu_x = \frac{q(x)x}{k_a(T_w - T_a)} \tag{34}$$

and compared with experimental points taken by other experimenters. The comparison is shown in Fig. 13, where four different repeated tests are examined at different airflow velocities, $4 < V < 12$ m/s; the accuracy of the method appears good and its repeatibility excellent. The tests at high Reynolds number show a gradual transition toward the turbulent limits, in agreement with that found by other authors.

3 Forced Convection on Cylinders and Spheres [8]

In the previous cases, the transformation from surface coordinates (ξ, η) to coordinates (x, y), when orienting the TG viewing axis orthogonal to the flat plate (if no distortions are induced by the IR lens system) is immediately performed through two scale factors, i.e., the distances between two contiguous pixels along x and y. In the following example, simple geometries are considered that need the software described in Sec. IV.C for evaluation of the second order derivatives term in Eq. (18).

Nonaerodynamic geometries like spheres and cylinders have been tested in an open-jet wind tunnel. For the case of the sphere a quasispherical lamp bulb was placed in an airflow at different velocities. This case appears to be quite critical for the check of the heat transfer coefficient over the surface (due to separation and reattachment regions over the sphere). The radiative heating source, in this case, is simply the lamp filament.

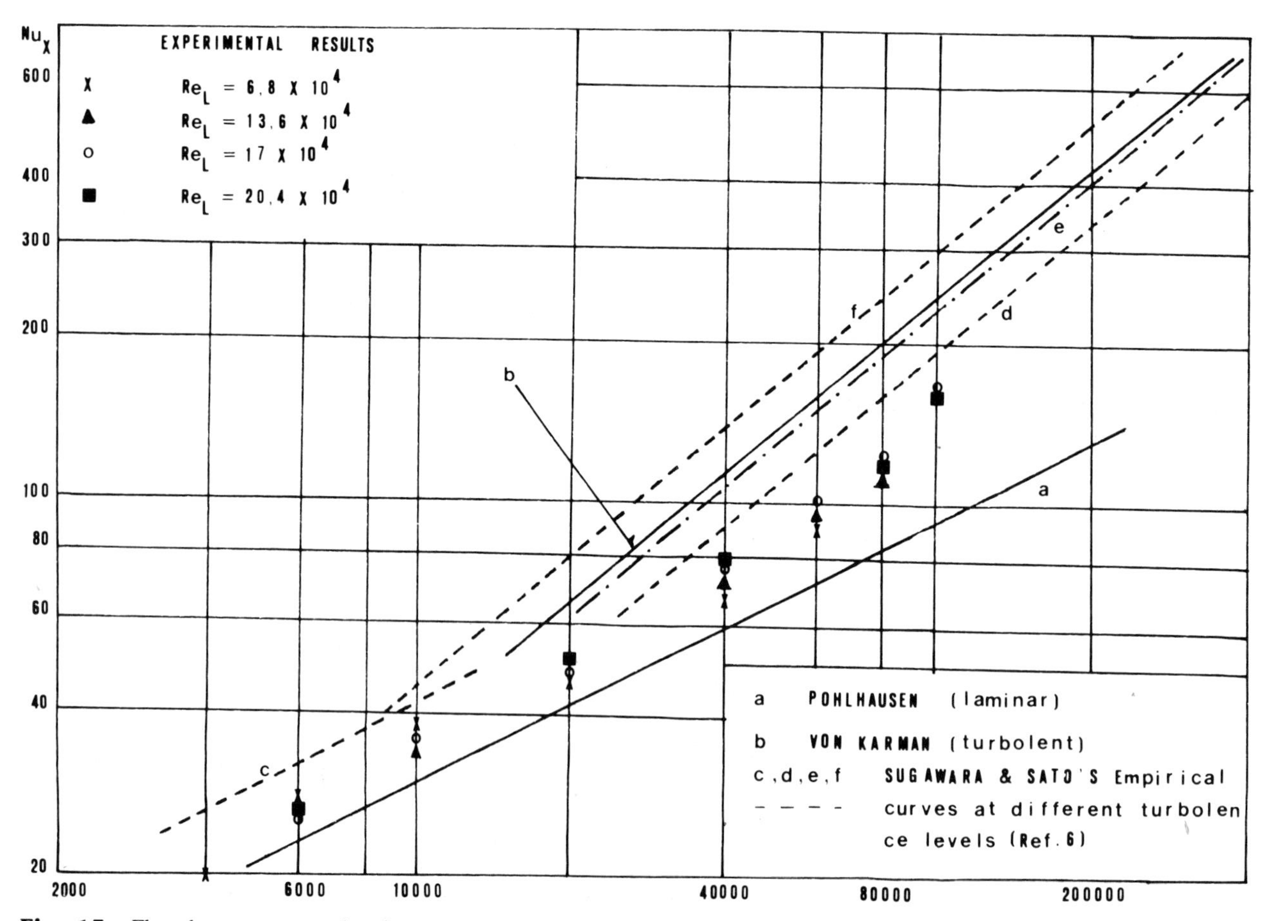

Fig. 13 Flat plate at zero angle of attack: Experimental data are compared with theoretical values and with other experimental results.

Different experimental protocols were followed. The preliminary results, entirely satisfactory, are reported in Fig. 14, where the Nusselt numbers are shown as a function of the angle $\vartheta = 0$ at the front stagnation point through $\vartheta = 180$ at the rear stagnation point). The plot shown is that appearing directly on the screen of the microcomputer system.

The results compare favorably with the experimental data reported in the literature at the same Reynolds numbers. The turbulence level was not measured; however, it does not affect the results by more than 25%, according to the data reported in literature. The position of the separation zone (at about 85° downstream of the front stagnation point) is quite evident, and the method appears particularly suited for the investigation and the exact localization of the separation and reattachment lines.

The results for the cylinder are shown in Fig. 15 for different Reynolds numbers. Thin-wall Plexiglas cylinders were heated by passing hot air into them [7]; the cooling phase was then observed by a TG system. Two thermal maps, taken at two instants of time (t and $t + \Delta t$), were elaborated to yield the temperature matrices $T(t)$ and $T(t + \Delta t)$, from which the Nusselt number was computed by Eq. (18). The results compared with other experiments show the applicability of the methodology to complex shapes.

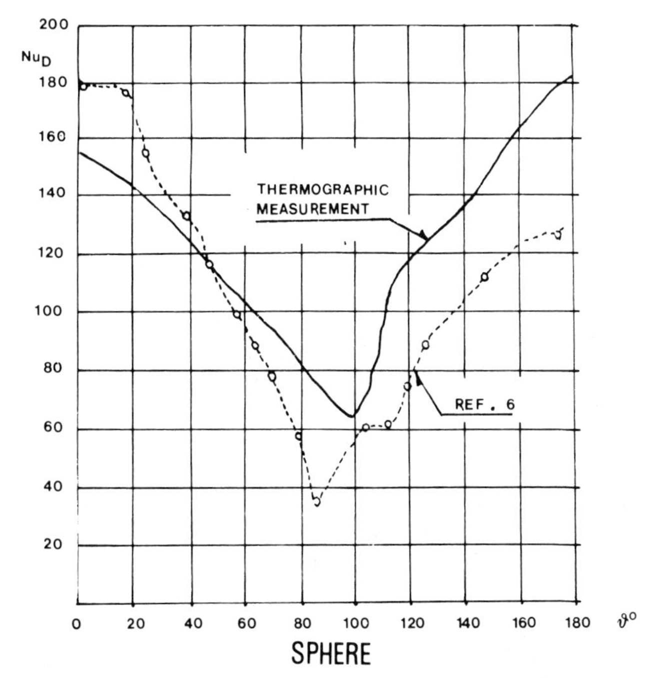

Fig. 14 Comparison between the experimental results and other experiments at $Re_D = 5E + 04$. The difference may be due to different turbulent levels in the air stream.

C Experimental Procedure for Flow Visualization over Airfoil [9]

In aerodynamics it is of great interest to ''visualize'' flow separation lines, extension of separation bubbles, and the transition from laminar to turbulent flow. One of the greatest difficulty occurs when these phenomena take place over a 3-D body. In these cases it is common to utilize oil-film techniques, which are complex, very qualitative, and somewhat invasive. The strict correlation between the above phenomena and the convective heat transfer at the body surface have led to extending the TG unsteady technique to visualization of the flows over complex 3-D aerodynamic and nonaerodynamic bodies [8], to detect the surface points at which (1) the flow separates from the surface, (2) the flow reattaches, and (3) the transition from laminar to turbulent boundary layer takes place.

Aerodynamic profiles (or wing sections) at low Reynolds numbers (where these phenomena are more critical) and 3-D wings have been tested in wind tunnels, showing the applicability of the method and its usefulness when intrusive techniques are not suitable because they alter the surface geometry. The value of the Nusselt number over the model surface gives an indication of the velocity gradient at the surface $\partial V/\partial_n$. In all cases in which the Reynolds analogy between the heat transfer and the tangential stress

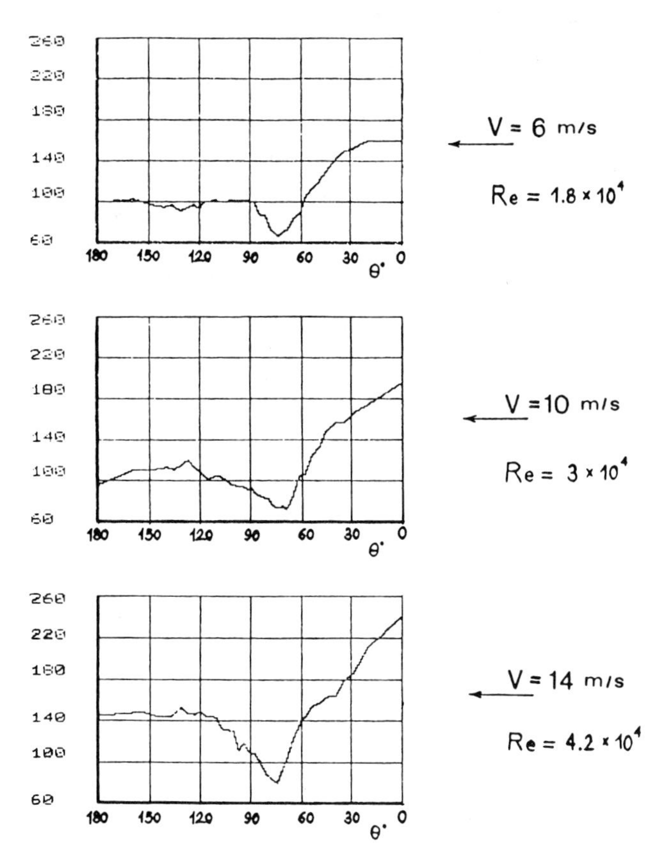

Fig. 15 Distribution of the Nusselt number for the cylinder versus the position angle (from the front stagnation point).

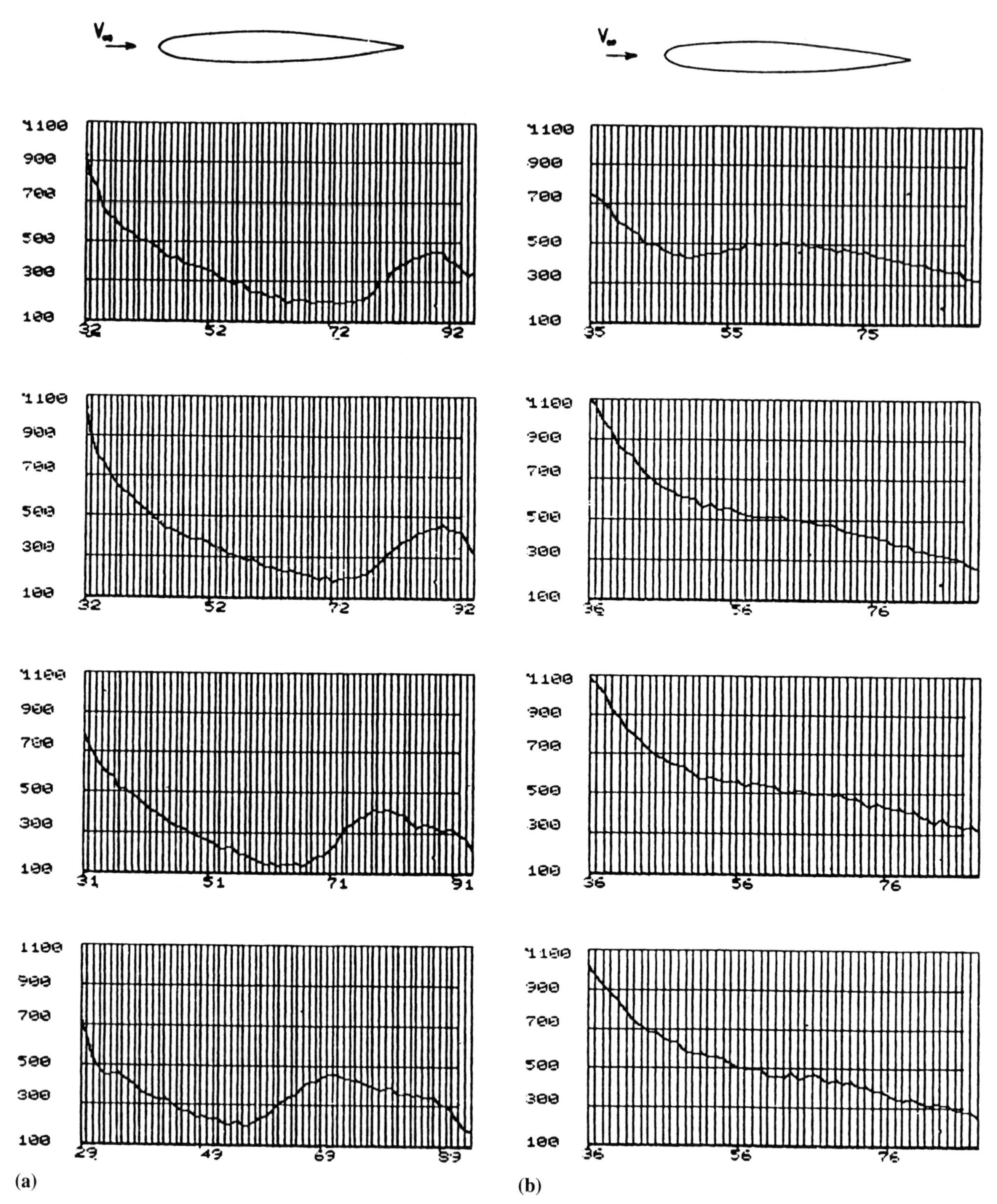

Fig. 16 Distribution of the Nusselt number along the upper surface of the NACA 0012-64 airfoil for different angles of attack at Re = 8.8*E* + 04: (*a*) α = 0–9°; (*b*) α = 12–21°.

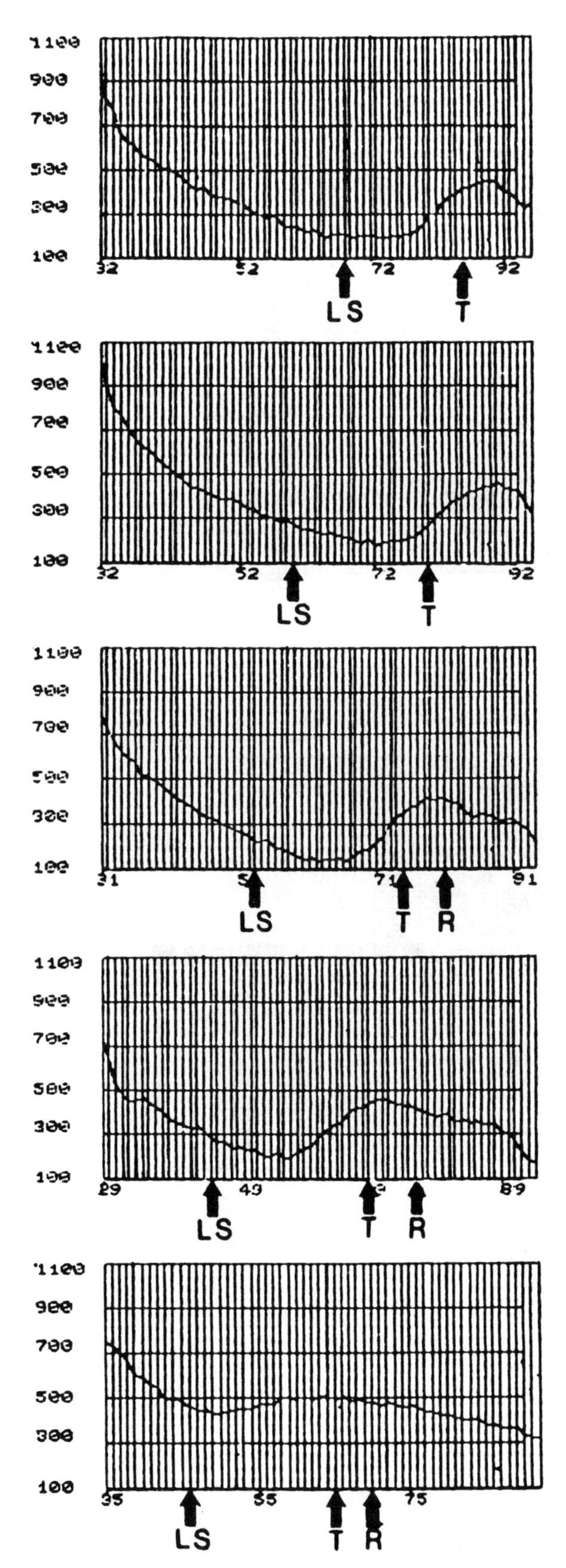

Fig. 17 Characteristic laminar separation *LS*, transition *T*, and reattachment *R*-points for the Wortmann FX 63-137 airfoil at RE = 8*E* + 04, compared with the distribution of the Nusselt number for the airfoil under study.

holds, the Nusselt number can be related to the velocity gradient at the surface by

$$\mathrm{Nu} = \frac{1}{\mathrm{Re}_L} t_\alpha \left| \frac{\partial V}{\partial n} \right|_{n=0}$$

where Re_L is the Reynolds number based on the characteristic length L, t_α is the thermal characteristic time $t_\alpha = L^2/\alpha$, and α is the fluid thermal diffusivity.

The value of the velocity gradient at the surface can therefore be related to the location of the separation points (where this gradient vanishes), to points where transition from laminar to turbulent regime takes place (at which the velocity gradient suddenly increases), or to reattachment points (where again a discontinuity in the velocity gradient at the surface takes place). Typical results for Nusselt numbers along the wing section (NACA 0012-64) are shown at different angles of attack $\alpha = 0$–$21°$ in Figs. 16*a* and 16*b*.

The location of laminar separation *LS*, transition *T*, and reattachment *R* points is quite evident on the Nusselt-versus-chord plots. The low Reynolds number regime justifies the behavior of the separated region displacement as a function of the angle of attack. In Fig. 17 the characteristic points *LS*, *T*, *R* for the Wortmann FX 63-137 airfoil [9] at Re = 8*E* +04 are shown in comparison with the distribution of the Nusselt number for the airfoil under study. The geometry of the Wortmann FX 63-137 and the NACA 0012-64 airfoils is reported in Fig. 18. Figures 19*a*, *b* show the Nusselt number footprint over the entire wing at different angles of attack.

The darker regions correspond to high values of Nu (as shown by the gray scale beneath the maps). At a high angle of attack ($\alpha > 12°$) the separation region at the two wing extremities are established, as can be visualized on the plots of Fig. 17. The results obtained are in qualitative agreement with those reported in the literature [16] that were measured at a steady state by imposing a "uniform" heat flux at the model surface. The difficulty of controlling the uniformity of the heat flux (by Joule heating), the invasive and discrete number of the temperature sensors used, and the approximations made in neglecting conductive and radiative heat

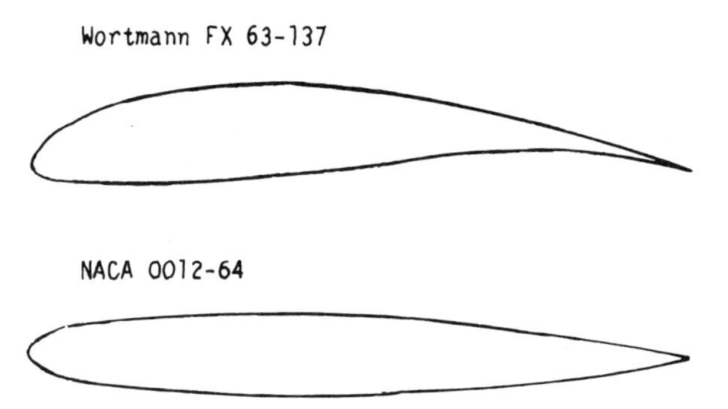

Fig. 18 Geometry for the Wortmann FX 63-137 and the NACA 0012-64 airfoils.

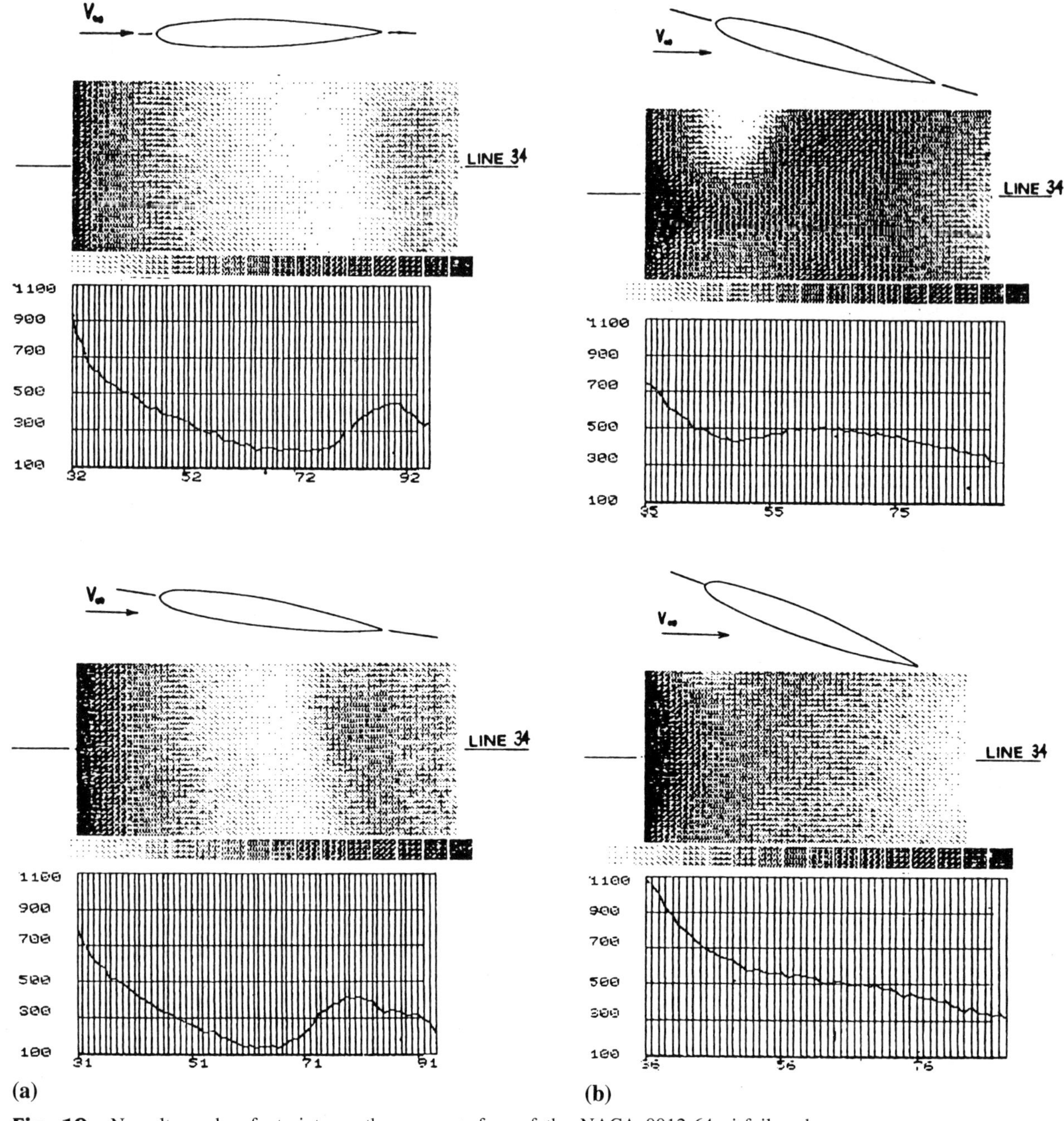

Fig. 19 Nusselt number footprints on the upper surface of the NACA 0012-64 airfoil and distribution of the Nusselt number along the airfoil centerline (line 34): (*a*) At low angles of attack; (*b*) at high angles of attack.

transfer contributions leave some doubts as to the accuracy of the steady-state methods against the method described in Sec. VI and the technique adopted for the evaluation of the heat transfer coefficients.

VII OTHER APPLICATIONS

In this section we discuss a number of applications of the TG techniques that witness the potential of those systems, when duly equipped with a computer that has peripherals for data elaboration and display.

A Computerized Thermography for the Detection of Blood Flows in Limbs [11]

The blood perfusion rate (bpr) is defined as the blood mass flow rate that perfuses the unit mass of the tissue (at the microvascular level). The bpr is, in many instances, an important diagnostic parameter that can be directly related to the pathologies that cause insufficient blood supply to the peripheral tissues. A number of noninvasive methods are being used for the evaluation of blood circulation in limbs (plethysmography, photoplethysmography); all of them are qualitative and indirect methods that give answers that strongly depend on the morphology of the limbs.

The problem is to define and measure an objective parameter able to quantify the blood perfusion rate in the peripheral tissues of the limbs (e.g., fingers and toes). The protocol adopted consists of cooling the fingers (by water immersion or by fan cooling) to a temperature T that is close to the ambient temperature. The thermographic examination proposed consists of measurement of the skin temperature performed at two instants of time: (1) after the skin cooling, (2) during the thermal recovery, after the cooling action has ceased; the increase in the skin temperature under these conditions can be correlated with the blood perfusion rate, as shown in [11].

B Application to Nondestructive Testing [12] [15]

The thermographic method can be applied to nondestructive testing whenever a defect in a structure is characterized by thermophysical properties different from the structure without defects. In this case, if a heat flux is passed through the region that has the defect, a temperature field distortion will result that can be observed on the structure surface. The best detection of a defect depends on the choice of (1) the surface area to be observed; (2) the external thermal stimulus that creates the heat flow; (3) the time of observation at which the surface temperature distribution is evaluated.

Even though it is difficult to make general statements, one may safely claim that the surface area to be thermally observed should be as close to the location of the defect as possible, for the defect to appear on the TG monitor as a high-contrast thermal footprint on the surface. Furthermore, the defect should behave as a "large" obstacle to the heat flow.

The external thermal stimuli may consist of heat addition (or subtraction) from the specimen, which occurs either at the specimen boundary surfaces (e.g., by convective or radiative heat transfer) or inside the specimen (e.g., by radio frequency heating). Another methodology relies on monitoring the heat generated at the defect site by mechanical movements due to the characteristics of the defect itself (friction at the delamination regions, plastic deformation, etc.). The observation time is the time at which the surfaces temperature distribution $T_s(x, y, z, t)$ is measured after the thermal stimuli are first applied (at $t = 0$). Let us denote by t_c the characteristic time (which depends on the experimental conditions) needed for the steady thermal conditions to be attained (applying steady boundary conditions). The values of the ratio $z = t/t_c$ indicates the nondimensional time "distance" from the initial temperature distribution. Two experimental procedures exist according to the relative value of the observation time with respect to the characteristic time:

Thermal steady conditions	$\tau >> 1$
Thermal unsteady conditions	$\tau < 1$

It is worth stressing here that the reliability of the nondestructive control methods substantially improves when evaluation is performed by comparing measurements taken on defected and undefected specimen regions. For instance, the case of a fiber-reinforced composite honeycomb structure is shown in Fig. 20. The structures that were tested

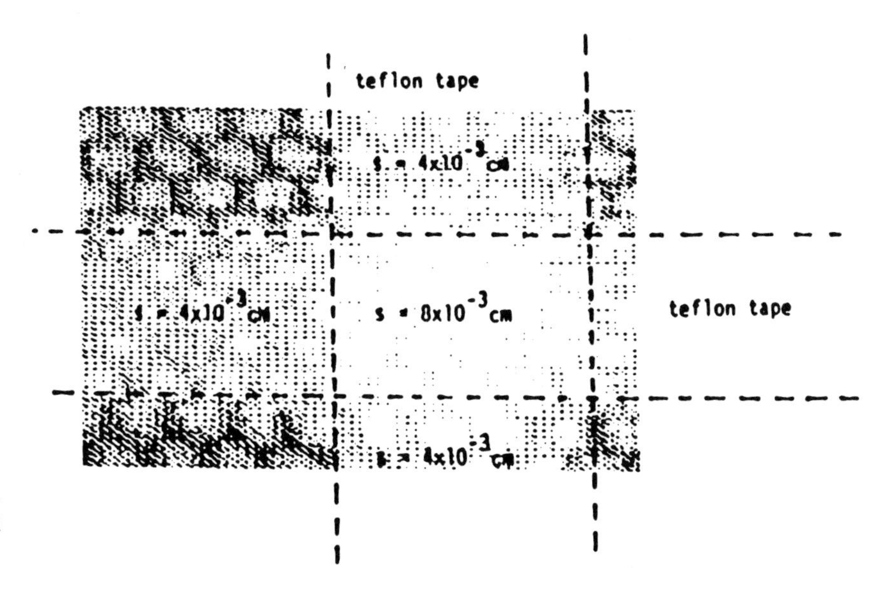

Fig. 20 Semiquantitative plot of the surface temperature on a simulated defected zone.

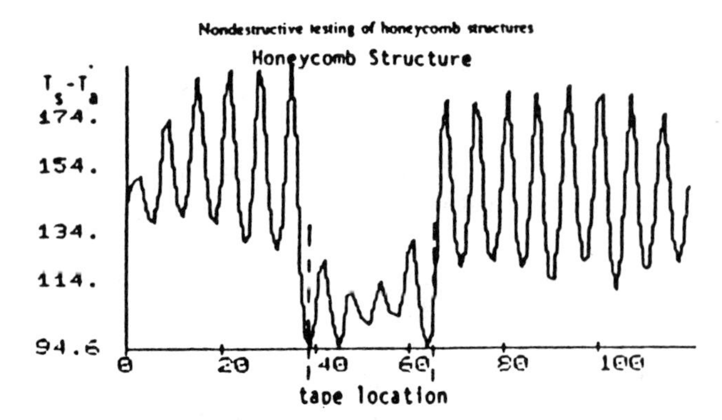

Fig. 21 Experimental surface temperatures along x.

are fiberglass–aluminum honeycomb structures having a height of 0.9–1.8 cm and a plate thickness of 3.5×10^{-2} cm. The defected bonding between the honeycomb and the skin was simulated by inserting, at the manufacturing stage, a Teflon tape of $s = 4 \times 10^{-3}$ cm thickness between the honeycomb and the plate, thus simulating a bonding defect. Tapes run along the x and y directions and at their crossing simulate a defect of 8×10^{-3} cm thickness. The temperature was measured at different times for the two honeycombs with $H = 0.9$ and 1.8 cm and for the two defect sizes $s = 4 \times 10^{-3}$ and 8×10^{-3} cm.

The evaluation was performed by reading the plot of the surface temperature along a line; typical experimental temperatures are shown in Fig. 21. The peaks correspond to the honeycomb bondings; the temperature depression, due to the included defect, is visible in the middle part of the plot.

The thermographic images clearly show the possibility of detecting small defects in the honeycomb structure. Figures 20 and 21 show clearly, in a qualitative and a semiquantitative mode, the presence of a single and a double defect. The procedure is very helpful when not only the presence, the location, and the extension (along the observation surface) of defects are to be detected, but also their size in a direction orthogonal to the thermographic viewing axis.

C Breast Cancer Detection [13]

The unsteady thermographic method of breast cancer diagnosis is aimed at better accuracy of tumor detection and at reduction of diagnostic errors (false positives as well as false negatives). If heat transfer from the tissue to the skin depended on the tissue conduction alone (tumor site near the skin), then the skin temperature would directly yield information on the tumor location, size and shape. The deeper the tumor, the greater the chance that a substantial percentage of heat is transported from the tumor to the skin surface by the blood; this means poor definition of the hot spot on the skin and a poor correspondance between the position of the hot spot and the tumor itself.

When air is blown on the breast (in order to eliminate any residual thermal effect of clothing) the surface temperature tends towards the ambient air temperature. If a thermographic picture is taken after prolonged cooling, very little difference is found between skin areas close to a hot tumor and those far away from it. The hot spots reappear gradually after the ventilation is stopped. Instead of waiting for thermalized conditions to be established, the evolution of the surface temperature toward the thermalized conditions can be followed by the thermograph. The relevant parameter is how rapidly the surface temperature, which is a function of the skin point, and of the time, tends to the new equilibrium value. It can be shown that the average temperature increase of the skin can be related to the metabolic heat production, and that the difference between temperature–time derivatives at two skin locations (e.g., at the two controlateral breasts) can be correlated with the tumor overheating (which depends on the tumor activity or malignancy).

D Liquid Surface Temperature Measurement [14]

Another application relevant to flow visualization, is the use of a thermographic system onboard the Spacelab for microgravity experimentation with fluids. More specifically, the need arises of measuring free-surface temperatures at the boundary of uncontained and contained liquid specimens. Large drops and large liquid bridges (or floating zones) may be created in a microgravity environment (established onboard orbiting platforms like the Spacelab); surface-tension-induced flows can then be studied with thermographic equipment included as diagnostic tools in the multiuser facilities of the European Space Agency (ESA).

A relationship may also be established, under certain conditions, between the surface temperature distribution and the flow field; in these cases measurement of a surface temperature distribution may lead to the evaluation of flow field characteristics.

REFERENCES

1. Baker, L., and MacDonald, R. S., II, Electro-optical Remote Sensors with Related Optical Sensors, in *Manual of Remote Sensing,* pp 325–365 American Society of Photogrammetry, Falls Church, Va., 1975.
2. Burton, B., *ABC's of Infrared,* Foulsham—Sams, Technical Books.
3. Monti, R., Thermographic System for Material Science Experiments, Final Rep. Contract 4091/79/NL/HP (SC), Techno Syst. Rep. TS-2-80, October 1980.

4. *Thermal Video Systems (TVM 3000),* Operational Manual, Hughes Aircraft Co., 1987.
5. Monti, R., Surface Temperature Measurement, Phase 2 Report, Contract Estec-CISE on Fluid Physics Instrum. Study, Techno Syst. Rep. TS-1-84, January 1984.
6. Kendall, P. J., *Image Processing Software: A Survey,* in *Progress in Pattern Recognition,* ed. L. N. Kanal and A. Rosenfeld, North-Holland Publishing Co., Amsterdam, 1981.
7. Monti, R., Thermography, *Flow Visualization and Digital Image Processing,* Proc. Lect. Ser., von Karman Institute for Fluid Dynamics, Rhode-Saint-Genèse, Belgium, June 9–13, 1986.
8. Monti, R., Measurement of the Local Heat Transfer Coefficient by Noninvasive Techniques in Aerodynamics, *VIII Cong. Naz. AIDAA, Torino,* September 1985; *Aerotecn. Miss. Spaz.,* vol. 65, no. 1/2, pp. 55–62, June 1986.
9. Monti, R., and Zuppardi, G., Computerized Thermographic Technique for the Detection of Boundary Layer Separation, *Aerodynamics Data Accuracy and Quality: Requirements and Capability in Wind Tunnel Testing, Naples,* Sept. 28–Oct. 1, 1987, AGARD/FDP Symp. AGARD CPP 429N.30.
10. Monti, R., and Mannara, G., Determinazione Sperimentale dei Coefficienti di Trasmissione del Calore con Metodi Termografici, *Aerotecn. Miss. Spaz.,* vol. 61, no. 2, pp. 92–105, 1981.
11. Monti, R., and Mannara, G., Computerized Thermography for the Detection of Blood Flow in Limbs, *Aerotecn. Miss. Spaz.,* vol. 62, no. 2, pp. 77–82, June 1983.
12. Monti, R., and Mannara, G., Valutazione Sperimentale dei Flussi Convettivi all'Interno di Pannelli Solari, *Aerotecn. Miss. Spaz.,* vol. 61, no. 1, pp. 21–30, March 1982.
13. Monti, R., Spadaro, G., Buonomo, R., and Piscitelli, B., Unsteady Thermography: Software and Hardware Development for Computer Aided Cancer Diagnosis, Configurazioni e Applicazioni dei Sistemi a Minicomputer: Tecnologie dei Sistemi di Informatica, Olivetti, Ivrea, September 1979.
14. Monti, R., Breadboarding of the Thermographic System, Techno System Rep. TS-11-86, October 1986.
15. Monti, R., and Mannara, G., The Computerized Thermography for NDT in Aerospace Applications, 40th European Conf. NDT, London, September 13–18, 1987.
16. Ota, T., Nishiyama, H., et al., Heat Transfer and Flow around an Elliptic Cylinder, *Int. J. Heat Mass Transfer* vol. 27, no. 10, pp. 1771–1779, 1984.
17. Ohman, C., Measurement versus Imaging in Thermography, or What Is Resolution, *Proc. Fifth Infrared Information Exchange, New Orleans,* October 1985.

Chapter 22

Flow Solutions

Robert E. Smith and Robert A. Kudlinski

I INTRODUCTION

During the past 20 years an area of research called computational fluid dynamics (CFD) has emerged along with the development of high-speed digital computers. This area of research is the simulation of fluid motion through the numerical solution of governing partial differential equations and associated boundary conditions [1, 2]. Computational fluid dynamics complements physical experiments as a means of studying fluid motion and its effects on solid and porous boundaries. Computer-generated imagery (CGI) [3, 4] is the electronic storage and presentation of pictorial information (images); CGI also includes the acquisition and/or creation of digital electronic images. Techniques for visualizing CFD solutions using CGI are the subjects of this chapter.

Geometry described by ordered sets of points called grids [5] is the common ingredient relating CFD and CGI. Images of scalar variables (density, temperature, etc.) defined on grid surfaces or lines, and traces displaying vector quantitites (velocities, vorticity, etc.) relative to a grid are the visual manifestations for computer-generated flow fields. The concepts presented in this chapter are directed at computed flow-field solutions, but they can be applied by any system of scalar and vector quantities defined on a structured grid.

Using CGI to show solutions of mathematical and numerical phenomena has a long history dating from the 1960s [6, 7]. In the early days, relatively slow computer speeds and low resolution of visual-display systems made CGI time consuming and expensive for routine use. However, present-day computational speeds and firmware (i.e., graphics processors) incorporated into workstations make CGI a viable tool for the analysis of complex computational models.

In this chapter, we describe the basic concept of color coding scalar variables associated with a grid for CGI and two algorithms that have been used at the NASA Langley Research Center for creating images of flow field variables. Finally, we discuss the use of ''state-of-the-art'' workstations for flow-field visualization and of hard-copy devices.

II COMPUTATIONAL FLUID DYNAMICS AND GRIDS

Computational fluid dynamics is the study of fluid flow using numerical methods to solve differential equations that are postulated from the conservation of the mass, momentum, and energy of a fluid in motion. For a viscous-compressible fluid flow, the Navier–Stokes equations are the model differential equations of motion [8, 9]. At the present time, turbulence is modeled by algebraic and/or differential equations that are coupled with the conservation equations and the fluid viscosity. If the flow has chemical reactions, additional conservation equations must be included. Subsets of the Navier–Stokes equations can be solved, assuming that certain aspects of the flow, i.e., the Euler equations for inviscid compressible flow, are not significant. Most often, compromises must be made based on the available computational power (speed and memory) and the pertinent physics.

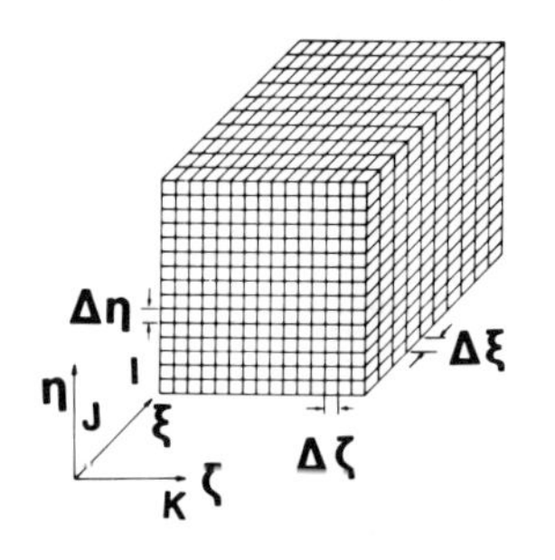

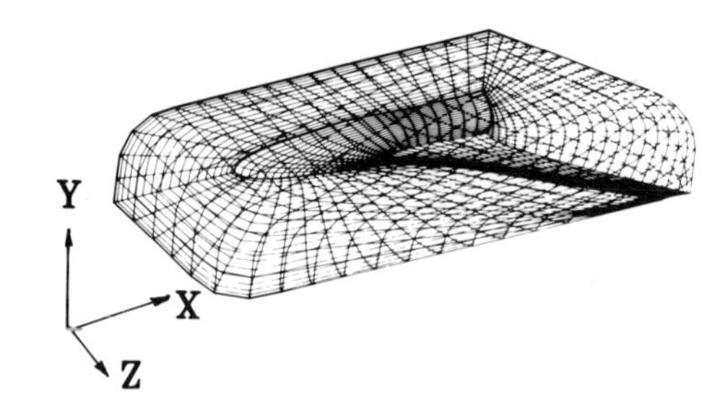

Fig. 1 Computational domain and physical domain.

The most used solution approach in CFD is to discretize the physical domain where the solution is to be obtained into an organized set of points called a grid [5, 10]. The physical grid is transformed into a uniform rectangular grid referred to as the computational domain (Fig. 1). Correspondingly, the equations of motion are transformed from the physical domain, where they are defined, to the computational domain, where they are discretized into algebraic relations between grid points that can be used to compute the solution (density, velocities, pressure, temperature, etc.). A computer-generated flow field consists of a physical grid and the flow variables defined at the grid points. Computer-generated flow-field visualization is the presentation of the physical grid and flow variables in pictorial form.

Grids used in CFD are generally boundary conforming (fit the surface of objects in a fluid) [5] and should be as orthogonal as possible. Also, the grid must conform to the physics of the problem by being finely spaced where there are high gradients. Two types of grids can be used, structured and unstructured grids [11]. In this discussion, only structured grids are considered.

Structured grids have a one-to-one correspondence to some curvilinear coordinate system. That is, a boundary-fitting grid in the physical domain maps uniquely to a rectangular computational grid. For our purposes here, the computational grid can be thought of as a rectangular grid with unit spacing between grid points and coordinate directions that correspond to indices I, J, and K (Fig. 1). Representing the physical grid coordinates and solution variables in index notation (i.e., $X(I, J, K)$, $Y(I, J, K)$, $Z(I, J, K)$, Pressure(I, J, K), etc.) provides a convenient theoretical and practical means of describing a grid and the solution at the grid points. For a given grid point in a structured grid, the position and solution at neighboring grid points are automatically available through increasing or decreasing the indices. Holding one index fixed while letting the remaining two indices vary provides an easy reference to grid surfaces and the solutions on grid surfaces. Also, since the grids are boundary fitted, one or more grid surfaces can correspond to solid boundaries in the physical domain.

Using an index representation, a physical grid and the solution at the grid points can be thought of as a block. For complex geometries of objects submersed in a fluid, it often requires several blocks to represent the geometry of the object and the flow domain. Note that the grid is a collection of discrete points and that the solution is a set of variables prescribed at the grid points. The image displays of solutions described here are general, and any system that can be fitted into the array format just presented can be depicted in an image as described in the following sections.

III COMPUTER-GENERATED IMAGERY AND FLOW-FIELD VISUALIZATION

A digital image is a matrix of pixels [3, 4], where each pixel is assigned a digit code (integer value) that informs a visual-display device (raster CRT and associated hardware) to present at each pixel location either a shade of gray or a color. Computer-generated imagery is the creation and presentation of digital images, which can depict physical observations, mathematical constructions, or a combination of the two. The hardware for visualization of digital images is not the primary concern here, but since so much of the current technology depends on firmware, this topic is covered in Sec. V.

The objective of this section is to describe sample techniques for the creation of digital images depicting computed flow-field data. Basically, the technique involves creating images that show the geometry of a problem such as boundary surfaces and scalar information (pressure, temperature, etc.) or vector information (velocity, vorticity, etc.) relative to the geometry. The principle of creating a digital image of a flow-field variable is to transform selected grid points from the object space to the image plane (Fig. 2) and fill the void between points. The transformation can be either orthographic or perspective [3, 4], and the voids are filled through interpolation of values at the grid points. Generally speaking, the grid points define one or more surfaces, and at least one surface is a boundary surface. The surface can be coded to represent the magnitude of a scalar variable or merely the reflectance from one or more light sources. Further supplementary quantities, such as contour lines, streaklines (which represent the paths of particles interpolated

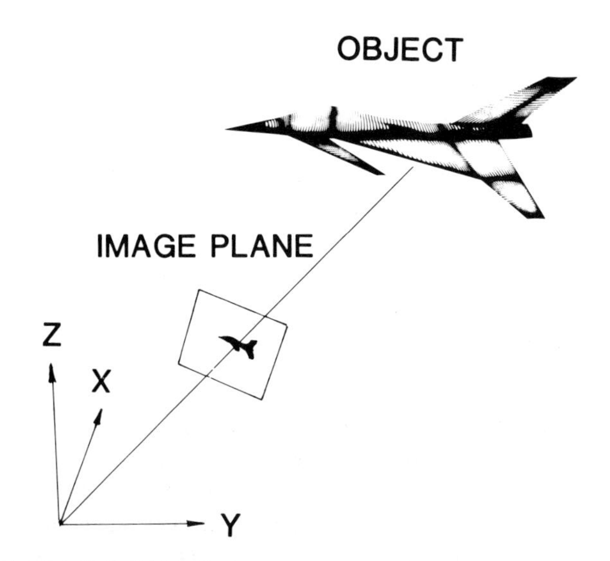

Fig. 2 Relation between image space and object space.

through the velocity field), or line vectors (whose lengths represent the magnitude of some vector quantity), can be transformed to the image plane and displayed.

IV SCALAR VARIABLE PRESENTATION

A color digital image is a color-coded matrix of pixels in an image plane depicting an object space (Fig. 2). Each row of pixels in an image is referred to as a *raster line*. The principle of creating a digital image of a flow-field variable is to choose grid surfaces from the physical domain where one wants to visualize a variable. Associated with the grid points describing the surfaces is the scalar variable. For instance, a boundary grid surface may be $X(I, 1, K)$, $Y(I, 1, K)$, and $Z(I, 1, K)$ and the scalar variable $P(I, 1, K)$. The three-dimensional coordinates describing the surface are transformed from the object space to the image plane (not to be confused with the computational coordinate system for CFD computation) through an orthogonal or perspective transformation [3, 4]. The coordinates in the image plane, corresponding to the three-dimensional surface, are $x(i, j)$ and $y(i, j)$, and the scalar variable is $P(I, J, K)$, which maps into $P(i, j)$ and is invariant under the transformation. The objective is to fill in the space between the image-space coordinates with values of P and code these quantities for display. Two algorithms are presented for this purpose. We describe them as raster line priority image generation (RLPI), and element priority image generation (EPI) [12]. The RLPI algorithm is useful primarily for two-dimensional planar solutions (original data described in a planar surface) whereas the EPI algorithm is more general, in accordance with the surface geometry previously discussed. Both algorithms are presented here for conceptual purposes; both have been incorporated into software at the NASA Langley Research Center. The RLPI algorithm requires a relatively small computer memory because each raster line of information can be disposed of as it is computed. The EPI algorithm is general and amendable for the removal of hidden portions of a surface from view. A general-purpose software package called RASLIB has been developed around the EPI algorithm. Before we describe the image algorithms in detail, we present the process of color coding an individual pixel.

A Color Coding

A digital raster scan display device can be thought of as a matrix of pixels where color is assigned to pixels with a component from each of the primary colors (red, green, and blue, or RGB). The totality of possible colors that can be displayed can be thought of as an ordered set of discrete points uniformly spaced in a 3-D Cartesian coordinate system (color cube) [12]. Each axis of the color cube corresponds to a primary color (Fig. 3). In order to relate a scalar variable to a visual composite of the three primary colors, a set of ordered integers is associated with a subset of the points in the color cube. For a 256-color scheme with integers 0–255, a nominal choice for association is shown in Fig. 4; 7 edges and all 8 vertices of the cube are mapped in uniform increments onto the 256 integers. Increasing integers start with the color black, move to blue, to cyan, to green, to yellow, to red, to magenta, and finally to white. This is only one of virtually an infinite number of possible associations. The variable to be displayed in color is scaled between 0 and 255, and the value at a pixel location is interpolated from surrounding values at image space grid points.

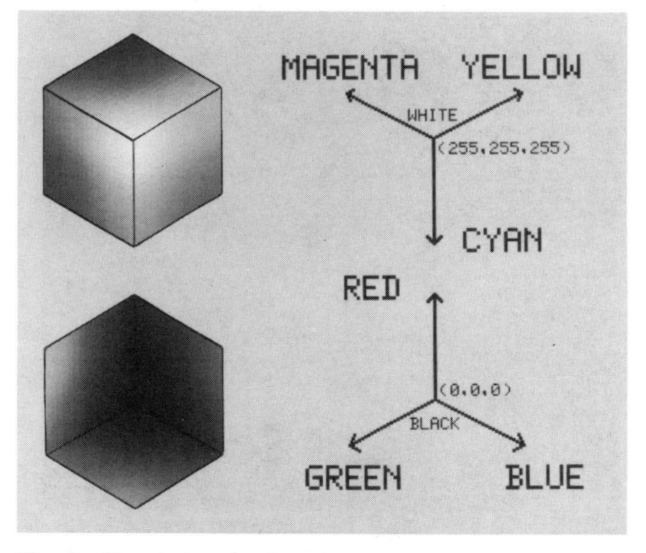

Fig. 3 Description of color cube.

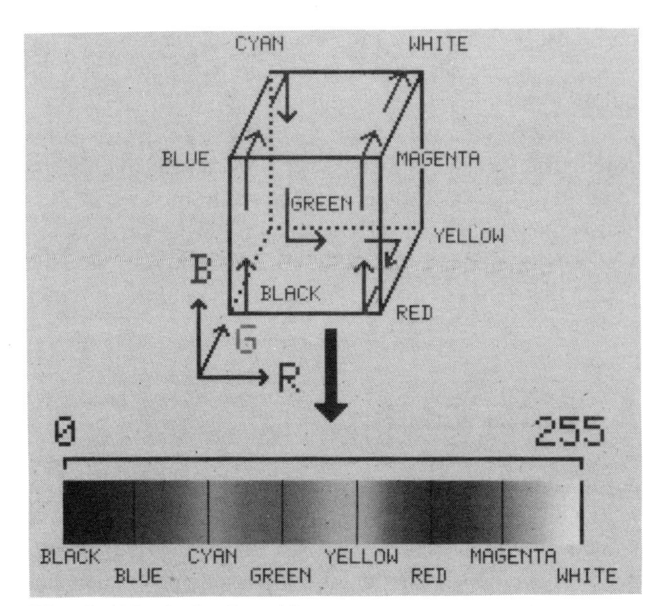

Fig. 4 Nominal color table.

B Raster Line Priority Image Generation (RLPI)

The transformation between the object space and the image plane is either an orthographic projection or a perspective view. The relationship between neighboring node points is invariant under the transformation. However, if the node points in the object space grid are planar and not rotated relative to the image plane, the transformation is merely a rescaling of the physical coordinates to pixel coordinates. For each pixel along a raster line in the field of view from top to bottom, a search is performed to find the quadrilateral of image node points in which the pixel is located. When the correct quadrilateral is found, a bidirectional interpolation is performed to evaluate the dependent variable at the pixel location.

Bicubic interpolation is used to produce a smooth representation when there are relatively few points. Bilinear interpolation is faster and more appropriate when there is a large number of closely spaced points. A linear interpolation into a color-code table designates the color to be displayed at the pixel position. The color code for each pixel along a raster line is stored, and the process proceeds to the next raster line. After all the raster lines for a picture have been processed, the stored digital image is displayed on a CRT. Examples of flow-field visualization obtained with this technique are shown in Figs. 5 and 6.

In Fig. 5, the display is the density about a shuttlelike configuration in a supersonic flow field. Also shown in Fig. 6 is the pressure distribution from a simulated firing of the shuttle solid-rocket booster.

C Element Priority Image Generation (EPI)

The second algorithm is used primarily for three-dimensional grids and corresponding flow fields where grid surfaces are not planar. The transformation into the image plane does not alter the relationship between grid points,

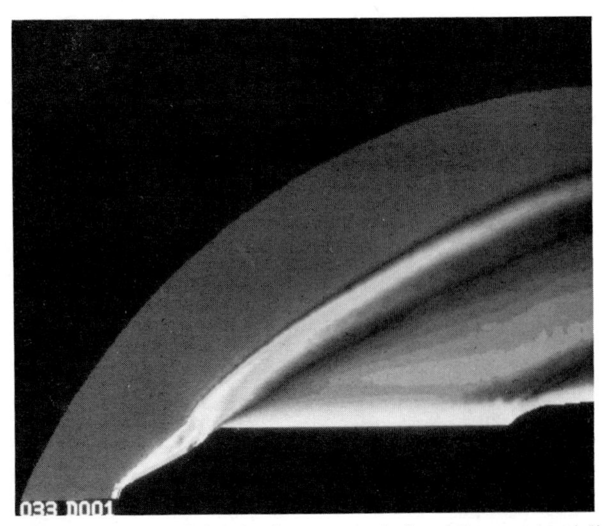

Fig. 5 Image of density in supersonic flow about a shuttlelike configuration.

Fig. 6 Image of pressure from a simulated firing of the shuttle solid-rocket booster.

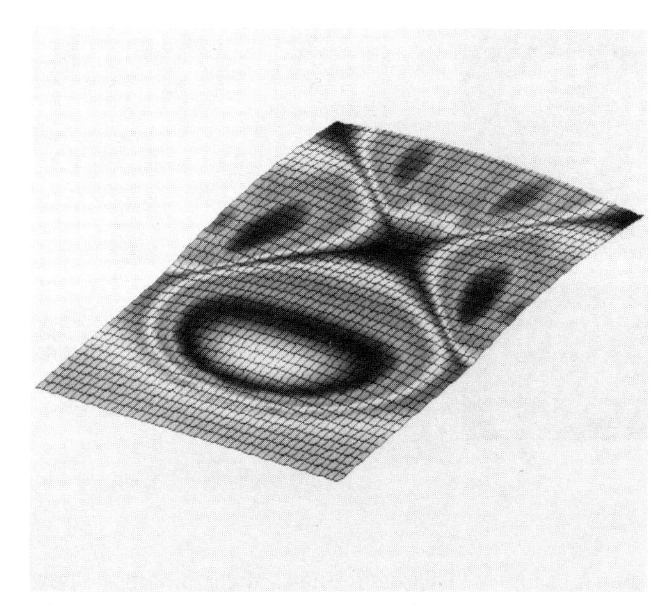

Fig. 7 Image of temperature on a blanket of spherical protuberances (a thermal protection system).

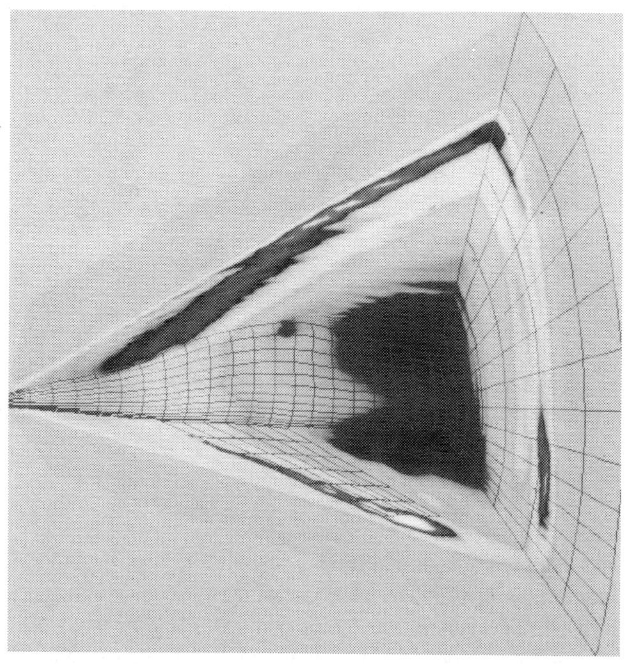

Fig. 8 Image of pressure distribution on four surfaces in a supersonic flow field about the forward part of a fighter aircraft configuration.

but certain parts of a surface may be hidden by other parts. The fill of the region between grid points is different from the previous algorithm in that it is performed element by element rather than by raster lines (an element can spread over several raster lines). This algorithm is essentially the one found in Ref. [13] except for color coding of the pixels.

In the image plane, a surface of grid points is organized into triangular elements according to the neighboring relationship in the object space. The elements are sorted according to decreasing distance from an observer position in the object space. Starting with the most distant element, the pixels within the bounds of the element are identified, and the values of the dependent variable associated with the pixels are obtained by bilinear interpolation from surrounding grid point values. The corresponding color is interpolated from the color-code table. The procedure is applied to the next farthest element from the observer position until the nearest element in the object space has been processed. Pixels corresponding to more than one element in the object space are coded according to the position closest to the observer. In this manner, hidden parts of a surface are removed from the image. The pixels are finally organized into raster lines for visualization.

When there are multiple surfaces in an image, additional logic must be included to assure that the part of one surface hidden by another surface is not visible. Using a color code to describe a variable on a curved surface does not always convey to an observer the concept of curvature. It is often advantageous to display grid lines in addition to the color code for the scalar variable. Grid lines are displayed in a color chosen to contrast with the surface colors. In a similar manner, velocity vectors, contour lines, and streaklines can be simultaneously displayed on an image. Two examples are presented that are processed with the element priority algorithm. The first example is the temperature distribution from the supersonic viscous flow over a blanket of spherical protuberances on a flat plate (Fig. 7). The grid lines are displayed on the boundary surface, and the color code is the one described previously. The second example is the display of pressure on the forward part of a fighter aircraft configuration including a fuselage, canard, and part of the wing (Fig. 8). Pressure is displayed on four grid surfaces, and a grid is superimposed on the fuselage, canard, and back surface.

RASLIB is a general-purpose set of subroutines that employ the EPI algorithm.[1] A general software package for creating images that runs on several general-purpose computers is called MOVIEBYU [14]. It has a long history dating from the 1960s and is the forerunner of current high-performance workstations.

V WORKSTATION-INTERACTIVE FLOW VISUALIZATION

It has been an objective of computer graphics to dynamically visualize an object domain consisting of surface grids and related variables. The requirement is to compute images

[1]Further information about RASLIB can be obtained by contacting the Flight Software and Graphics Branch, MS 125A, NASA, Langley Research Center, Hampton, VA 23665.

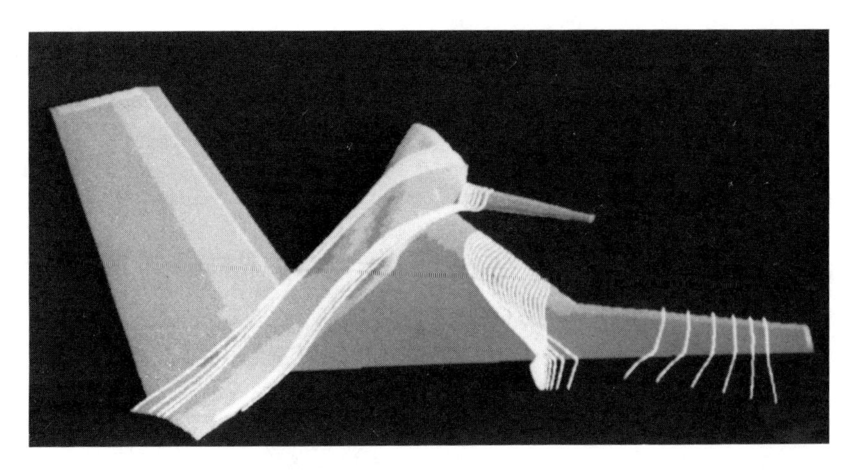

Fig. 9 Streaklines in a flow about a fighter aircraft configuration.

and display them in real time in response to user input. The functions performed include transformations on the surface grids, image generation (color coding), clipping, and hidden-surface removal. The approach has been to develop firmware, graphics processors in which circuitry performs the redundant mathematical operations for image generation. This is in contrast to the total use of software in a general-purpose computer to perform these computations. However, such equipment must be associated with a general-purpose processor for data storage and preprocessing. Graphics processors [15] have been in existence since the early 1970s, but the early equipment was very bulky and expensive. Today, equipment to perform the dynamic graphics function are compact (9 ft^3 plus CRT screen) and relatively inexpensive. Of the wide spectrum of graphics processors presently available, we briefly describe three that we are using at the NASA Langley Research Center for the visualization of computed flow fields: the IRIS workstation, the TEKTRONIX 4129, and the Professional Graphics System installed in a personal computer.

The IRIS workstation is used extensively by NASA for flow-field visualization. There have been several upgrades since its introduction, but the characteristic parts are a general-purpose processor, graphics processor, high-capacity storage disk, display monitor, and mouse. (Performance characteristics are not quoted here because they vary with the model.) This type of workstation is becoming a primary tool for flow-field visualization, but a certain amount of software is necessary to use it for specific types of applications. The achievement of real-time response depends on the number of grid points in the object space.

PLOT3D, a graphics program [16] especially designed for the display of two- and three-dimensional flow fields, runs on the IRIS workstation. Authored by Dr. Peter Buning of the NASA Ames Research Center, PLOT3D manipulates and prepares flow-field data for input into IRIS's geometry processor for dynamic rotation and translation. An example of a single frame obtained from PLOT3D is shown in Fig. 9. This figure shows streaklines in the flow field about a fighter aircraft configuration under study at the NASA Langley Research Center.

Another graphics device used at the NASA Langley Research Center for flow-field visualization is the TEKTRONIX 4129 in conjunction with a MICROVAX minicomputer. The two devices operate together as a workstation but do not allow dynamic rotations and translations like the IRIS. New images are rapidly computed from cursor input. Here again, firmware is used to transform object space coordinates to the image plane, clip the image, and remove hidden parts of surfaces. Preprocessing of object space data such as computing normal vectors relative to a surface must be done on a general-purpose machine such as the MICROVAX. A program called PLOTGRD for the display of computed flow-field data that runs on a MICROVAX and TEKTRONIX 4129 has been written by Mary Adams and William Von Ofenheim of the NASA Langley Research Center. Figure 10 shows the pressure field on selected surfaces about the same fighter aircraft configuration just discussed, but obtained from the TEKTRONIX 4129.

At the low-cost end of the graphics processors is the Professional Graphics System in the form of a standard-sized board that fits into an IBM PC XT, AT, or compatible, and a specialized monitor. It is possible to dynamically display only very simple object spaces with this system. On the other hand, high-quality images that have been created on a general-purpose system (for instance with RASLIB described previously) can be stored on a PC disk and displayed on the Professional Graphics System.

A major factor in the development of graphics workstations has been the reduction in cost and physical size. The next step for computed flow-field solutions is the coupling of workstations with the supercomputers where the equations of motion are being solved. This step will allow dynamic interaction with the solution process as well as the display and will ultimately minimize the time required for flow-field analysis.

Fig. 10 Pressure distribution on grid surfaces about a fighter aircraft configuration.

VI HARD-COPY DEVICES

Regardless of the ability of an electronic device to produce high-quality color images, the effectiveness of showing the results through publications and presentations depends on the quality of the hard-copy device used. The technology of color hard-copy devices has lagged developments in color workstations. The availability of quality color hard-copy devices at a reasonable cost has been very limited, and many vendors must make a great effort to fill the void [17].

In choosing a hard-copy device, several criteria must be addressed to determine their priorities for a given situation. These criteria include image quality, copy speed, ease of use, and cost. Image quality includes resolution, which should be at least equal to the screen resolution of the workstation. Image quality also is determined by the purity of colors, the uniform filling of areas, and the sharpness of edges and lines. Additionally, the copier must have a correct aspect ratio so that, for instance, circles appear round rather than oval. Copy speed is also a primary consideration as often as the workstation is locked up until the hard copy is finished. This problem can be minimized through the choice of workstations with a spooling capability to store the display and control the hard-copy device in the background. Ease of use is of primary concern in unattended workstation systems that support several users who may be unfamiliar with hard-copy devices. Finally, determination of cost must include both initial capital expense for the device and the per-copy cost of using it.

With these considerations, there are primarily four types of color hard copy devices to be considered: impact printers, xerographic devices, ink jets, and photographic devices. These four categories of devices offer tradeoffs on image quality, speed, cost, and ease of use.

Impact printers offer low cost at the expense of low image quality. Xerographic devices are high in capital costs and provide good image quality and the ability to produce transparencies. Ink jets offer average costs with good image quality at the expense of lower reliability and considerable routine maintenance. Photographic devices offer the highest image quality along with the ability to produce transparencies at a high capital and per-copy cost. Copy speed and ease of use vary greatly within each type, being generally related to hardware cost.

VII CONCLUSIONS

Computed flow-field visualization is basically the application of computer-generated imagery to a particular well-suited data structure; the same processes can be applied to other engineering disciplines. It is neither possible nor feasible to cover all the aspects of computational fluid dynamics and its relation to CGI for computed flow-field visualization in a short chapter, particularly since this is also a rapidly changing relationship. We have presented a snapshot of the relation between the disciplines to motivate the reader to search further for more details.

REFERENCES

1. Roach, P. J., *Computational Fluid Dynamics,* Hermosa Publishers, Albuquerque, 1972.

2. Anderson, D. A., Tannehill, J. C., and Pletcher, R. H., *Computational Fluid Mechanics and Heat Transfer,* Hemisphere, Washington, D.C., 1984.
3. Newman, W. M., and Sproull, R. F., *Principles of Interactive Computer Graphics,* McGraw-Hill, New York, 1973.
4. Foley, W. M., and Van Dam, A., *Fundamentals of Interactive Computer Graphics,* Addison-Wesley, Reading, Mass., 1982.
5. Thompson, J. F., Warsi, Z. U. A., and Mastin, C. W., *Numerical Grid Generation, Foundations and Applications,* North-Holland Publishing, New York, 1985.
6. Van Dam, A., Computer-driven Displays and Their Use in Man/Machine Interaction, in *Advances in Computers,* ed. F. Z. Alt and M. Rubinoff, vol. 7, pp. 239–290, Academic, New York, 1966.
7. Sutherland, I. E., Computer Displays, *Sci. Am.,* pp. 36–81, June 1970.
8. Batchelor, G. K., *An Introduction to Fluid Dynamics,* Cambridge Univ. Press, New York, 1967.
9. Schlichting, H., *Boundary Layer Theory,* McGraw-Hill, New York, 1968.
10. Smith, R. E., Two-Boundary Grid Generation for the Solution of the Three-dimensional Compressible Navier-Stokes Equations, NASA Tech. Memo. 83123, May 1981.
11. Smith, R. E., and Eriksson, L.-E., Algebraic Grid Generation, *Comput. Meth. Appl. Mech.,* vol. 64, Oct., pp. 285–300, 1987.
12. Smith, R. E., Speray, D., and Everton, E. L., Visualization of Computer-Generated Flow Fields, in *Flow Visualization III: Proceedings of the Third International Symposium on Flow Visualization,* ed. W.-J. Yang, pp. 190–197, Hemisphere, Washington, D.C., 1985.
13. Newell, M. E., Newell, R. G., and Sancha, T. L., A Solution to the Hidden Surface Problem, *Proc. ACM Nat. Mtng.,* Boston, Aug. 14–16, 1972.
14. Christianson, H., and Stephenson, M., *MOVIE.BYU Training Text,* Community Press, Provo, Utah, 1986.
15. *Line Drawing System Model 1: System Reference Manual,* Evans and Sutherland Computer Corp., Salt Lake City, Utah, 1971.
16. Buning, P. G., and Steger, J. L., Graphics and Flow Visualization in Computational Fluid Dynamics, *Proc. AIAA 7th Comp. Fluid Dyn. Conf.,* Cincinnati, July 1985.
17. *NCGA's Comput. Graph. 87 Eight Ann. Conf. Exposit., Proc.,* vol. 1, National Computer Graphics Association, Philadelphia, March 1987.

Chapter 23

Flow-Field Survey Data

Allen E. Winkelmann

I INTRODUCTION

This chapter describes a new computer-driven color video display technique for the presentation of flow-field survey data. In this technique, data sets taken in wind tunnel flow-field surveys are processed through a computer, which color codes the data and then displays the results on a color video monitor. This technique allows one to present and interpret vast quantities of data, which heretofore would have been difficult or impossible to comprehend. Depending on the probe sensor used in a test, the data may be qualitative or quantitative; in either case, the multicolored data displays constitute a new form of flow visualization.

The presentation of data obtained in surveys of complex three-dimensional flow fields has always been a rather difficult task. The problem is compounded by the fact that large quantities of data are required to properly map a 3-D region, or "volume," of the flow field. The general approach to this problem is to survey the flow-field volume in a series of 2-D planes; a composite of these planes, or "sheets," of data can then be used to gain an understanding of the entire 3-D flow region.

Flow-field survey techniques generally involve a point-by-point measurement scheme. A matrix of measurement points is required to cover a 2-D survey plane. This type of data is commonly presented in the form of contour lines, the shading of regions, or the use of arrows to represent the direction and magnitude of velocity components. The construction of contour lines or the shading of regions of the survey plane requires interpolation schemes to fill in areas between discrete data points.

When mapping flow regions that may contain small-scale flow phenomena such as secondary vortices, a very large number of discrete data points are required so as not to miss the core of the small vortex. An alternate approach to the discrete point-by-point measurement scheme is to continuously scan the flow field with a slowly moving probe. By repeatedly scanning the probe across the flow field and spatially incrementing the probe by a small amount at the end of each scan, one can obtain a 2-D sheet of high-spatial-resolution data. Although standard contour plots could be used to display these data sheets, a more effective and simple way is to color code the data and display it directly as scan lines on a color video monitor.

The general tendency in conducting flow-field survey experiments is to take data that are strictly quantitative. These data are presented as velocity plots, pressure contour maps, etc. To obtain quantitative data in complex 3-D flow fields, one is required to use multicomponent probes (e.g., 7-hole pressure probes or 3-channel laser velocimeter systems). Surveys made in a 3-D flow field using a single-element hot-wire probe result in highly qualitative data sets, i.e., the data taken with the probe is largely meaningless in the quantitative sense. However, as the hot-wire probe encounters various regions of the 3-D flow such as vortex cores or shear layers, a definite change in probe output occurs. This relative change in probe output can be used to map (in a qualitative sense) the entire 3-D flow field.

The technique of high-resolution scanning and the use of color to display data levels was first introduced by Crowder [1, 2]. This approach involves mounting a set of light-emitting diodes (LEDs) on the back of a Pitot probe, scanning the flow field with the probe, and lighting the various colored LEDs (typically three) in response to the probe-transducer output. By taking a time-elapse photograph as the probe sweeps back and forth across the wind tunnel, a multicolored data display is obtained showing the variations of total pressure in the flow field.

Inspired by this work, a computer-driven color video display technique for flow-field survey data was developed in early work by Winkelmann and Tsao [3–6] at the University of Maryland. A technique for the color video display of rake data obtained in a mixer nozzle was developed at about the same time by Anderson et al. [7]. The general technique has recently been further developed by a number of research workers, most notably by Cogotti [8] for use in wake surveys behind land vehicles.

The color video display technique for flow-field surveys is another example of the use of color coding for the presentation of vast quantities of complex data. Color-coded video display techniques have been in use for years in satellite imagery, infrared thermography, and medical diagnostic applications. The use of color shading provides an added "degree of freedom" that greatly expands one's comprehension of large data sets.

In the following sections, the computer-driven color video display technique is described in detail, and typical applications of the method are included.

II DESCRIPTION OF TECHNIQUE

A sketch of a typical setup for conducting flow-field surveys is shown in Fig. 1. In this test, a two-degree-of-freedom traversing rig was used to survey the flow field just downstream of a rectangular planform wing mounted on struts at the exit of a free-jet wind tunnel. By repeatedly scanning across the flow field in the spanwise direction and vertically incrementing the probe by a small amount at the end of each scan, one can obtain a sheet of high-spatial-resolution data at a given station behind the model. Typically 100 scans are used to obtain what in essence is a "slice" of data across a region in the flow. Various types of probes can be used, ranging from simple 1-element hot-wire probes to the more complex 7-hole pressure probe.

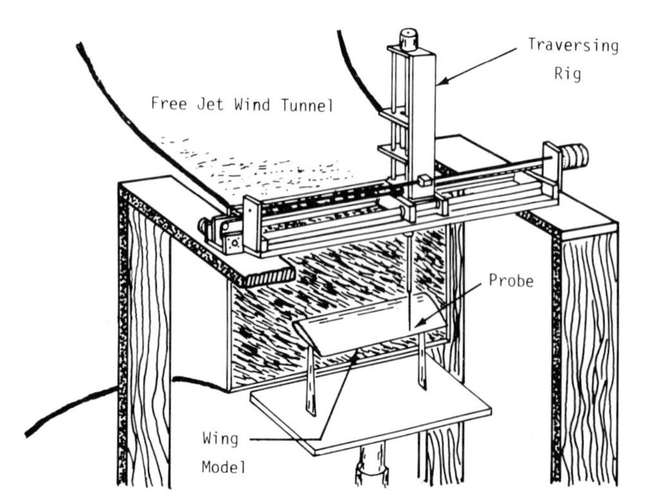

Fig. 1 Typical setup for conducting flow field surveys.

Figure 2 shows a flow chart of the data-acquisition and -processing scheme used in the flow-field imaging technique. The following discussion covers specific details and considerations for the major steps in this flow chart.

A Probes

Since this technique is generic, virtually any type of probe can be used, depending on the type of flow-field survey being conducted. The use of a single-element hot-wire probe in the study of a highly separated flow field (e.g., behind a wing at high angle of attack) yields data that are strictly qualitative, i.e., no hard numerical values (velocities) can be assigned to the data. On the other hand, if our "probe" is a 3-channel laser velocimeter, then with proper data processing, we can obtain hard quantitative data, i.e., the complete 3-D velocity field.

Simple probes such as Pitot tubes, pitch probes, and single-element hot-wire probes are well suited for the color video display technique. Each of these probes produces a characteristic response when encountering vortices, shear layers, etc. With some experience, one can readily identify specific flow structures encountered in any type of flow field. The rms output levels of a hot-wire probe, for example, clearly indicates shear layers where the turbulence levels are very high. The output from a pitch probe shows large and very rapid changes in the vicinity of a vortex. A miniature Pitot tube is very effective in pinpointing the location of the vortex core of even small, secondary vortices.

B Traversing Devices

To survey a flow field, one needs a precision mechanical device to move the probe around. The type of traversing rig used depends on the specific test being conducted. For example, in Fig. 1, the traversing mechanism moves the probe spanwise and vertically (in an "x–y" fashion) and produces a rectangular sheet of data. In the wind tunnels where models are sting mounted, it may be more convenient to mount a traversing device on the sting and survey the flow field in circular arcs (e.g., when surveying the flow field downstream of a cone model). Obviously a number of different traversing rig designs can be used, depending on the test geometry and the type of data desired.

The choice of the traversing mechanism has a definite effect on the subsequent data processing required. The

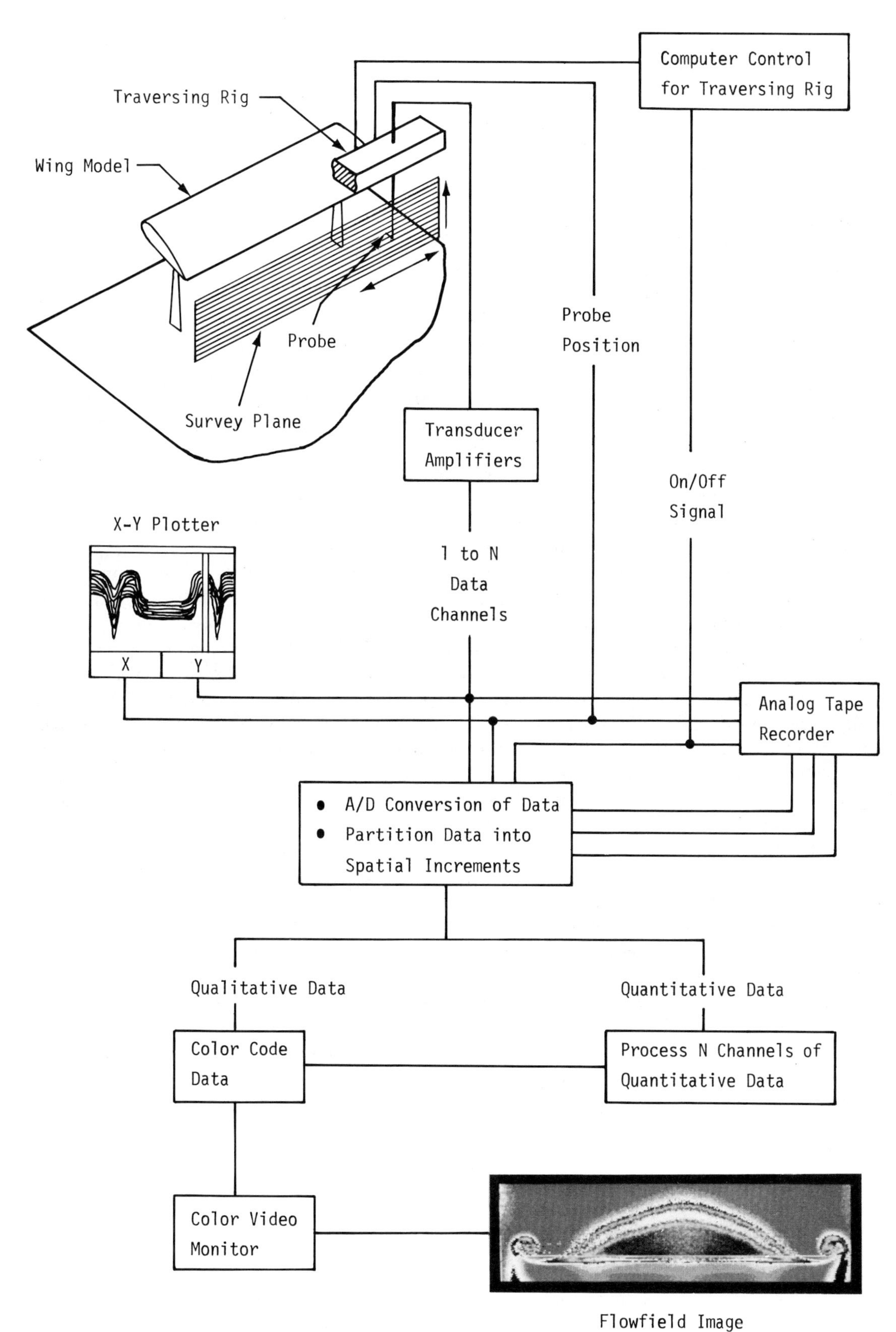

Fig. 2 Data-acquisition, -processing, and -display for flow-field imaging technique.

"x–y" mechanism in Fig. 1 scans the data sheet in the same manner as a video image is projected (scanned), i.e., in the natural raster scan mode. Hence, each scan across the data plane with the probe is displayed on the video screen as a single scan line. Taking data in this manner can simplify data processing since the computer deals with only one complete scan line at a time. On the other hand, if data were taken with a probe that was surveyed in circular arcs, then each scan of the probe would cut across several video scan lines (since the video display is rectangular). In this case, some additional processing and sorting of data would be required.

The traversing rig shown in Fig. 1 was custom designed and fabricated using a number of commercially available components. The main carriage of the traversing rig, which moves back and forth on Thomson linear bearings, is driven by a lead screw connected to a Superior Electric stepper motor. The system is configured such that 4800 counts on the stepper motor drive is equivalent to a traversing motion of 2.54 cm. A Winfred Berg Min-E-Pitch chain is connected to the main carriage and looped around a series of sprockets, one of which is connected to a 10-turn potentiometer, which is used as the position-monitoring device. Vertical motion is supplied by a second traversing stage, also driven by a small stepper motor. The vertical-drive system is configured such that 9600 half-counts are equivalent to a 2.54-cm motion. Since flow-field surveys are conducted by vertically incrementing the probe by a small constant amount at the end of each scan, no position readout system is required on the vertical axis. Although the system shown in Fig. 1 has only two-degrees-of freedom, a third traversing stage could readily be added to provide traversing motion in the chordwise direction.

The time required to complete a high-resolution flow-field survey depends on the speed of the traversing rig, the time response of the probe, and the rate at which the data can be handled. A slower traversing motion results in a better-averaged picture of the flow field. The maximum scanning speed of the traversing rig just described is 1.7 cm/s; however, most flow-field surveys conducted with this rig have been at slower speeds of 0.3 to 1.2 cm/s. A typical survey consisting of 100 vertical scans that are each 40 cm in length can take from 1 to 4 hours to complete. More recent work at the University of Maryland, with a faster traversing rig and a Pitot probe with a pressure transducer embedded in the probe tip, allows complete wake surveys to be taken in about 10 min. The time to complete a flow-field survey could be reduced dramatically by scanning with a multiprobe rake. However, care must be exercised when using a probe rake to avoid flow-field interference effects.

The traversing rig shown in Fig. 1 can be positioned to survey various planes perpendicular or parallel to the free-stream flow direction. Positioning of the traversing rig relative to the model is easily accomplished by using standard machine shop alignment techniques. The entire data-acquisition system (including the traversing rig) should be under the control of a small computer. For setup and alignment purposes, the traversing-rig drive system should be capable of operating in a manual mode.

C Data Acquisition, Processing and Display

The flowchart in Fig. 2 shows the data-acquisition and -processing scheme used to eventually display our data on a color video monitor. From the probe and traversing device, the various data channels are (1) probe/transducer output, 1 to N channels, depending on the probe being used, (2) probe position (x, y, z), and (3) an on–off signal to the computer. Additional channels may be needed to monitor other parameters that may vary during a test (e.g., recording the angle of attack of an oscillating wing model). If data are being taken at a specific station (i.e., x = constant in Fig. 1) and if each scan line is incremented vertically the same amount (i.e., Δz = constant), then the probe position can be recorded merely as the variation in y (from the multiturn potentiometer). In other words, the x and z positions of the probe are implicit, and we need to record only y.

Prior to conducting a flow-field survey, one must decide on the resolution required for the test, i.e., what vertical increment Δz will be used. Since the data will be eventually displayed on a video monitor, Δz must be related in some way to the picture elements (pixels) used to make up a color video display (typical color video monitors have 512 pixels per scan line). As an example, suppose a 51.2-cm-wide flow-field survey is to be made behind a wing model. If a 512-resolution monitor is used, then each pixel is equivalent to a 1-mm increment. A survey consisting of 100 scans (with Δz = 1 mm) would cover a plane measuring 51.2 cm wide by 10.0 cm high. By using Δz = 2 mm, one could cover a 51.2- by 20.0-cm plane in the same 100 scans. In this case an individual data "point" on the monitor would consist of a 1- by 2-pixel array.

An X–Y plotter is very useful as an auxiliary monitoring device. By operating the traversing rig in the manual mode, one can use the plotter to "hunt" for the core of a vortex. Finding the locations in the flow that produce minimum and maximum probe-output levels is essential in setting the transducer amplifier gains. The plotter may also be used to determine whether the traversing rig is scanning slowly enough to allow a probe to "follow" a rapid change in the flow. If the plotter trace shows a lead–lag effect when going through a vortex, then the traversing rig may be moving too fast.

Data obtained during a test either can be recorded on analog tape for subsequent processing or can be fed directly to the computer. The use of an analog tape is strongly advised to allow one to reprocess the data and to avoid the pitfalls of an unreliable or multiuser computer facility. The analog data are first converted to a digital signal by processing the transducer and position channels through the analog–digital convertor (ADC). Some computer ADC systems are configured to take a set number of data points in a certain time interval (the time interval starts when the "On" signal is received). An external clock can be used to set the sampling rate and hence adjust the time interval to be equal to or slightly greater than the time required to complete a flow-field scan. A more desirable data-acquisition system allows one to use an "Off" signal to stop the ADC at the end of a scan. This feature is needed if the time to complete a scan varies throughout a test. Variable scan times occur if the scan length varies (e.g., surveying around the contour of a model). One may also wish to reduce the traversing speed (hence lengthening the scan time) in the vicinity of vortices and shear layers, where the gradients in pressure and velocity are very high. At the end of each scan, a small time interval (5–10 s) is required to dump the data from the ADC to a disk or tape for storage. During this same time interval, the vertical traversing stage increments the probe upward.

As noted earlier, each scan line is divided into a number of small spatial increments, the size of which are determined by the geometry of the survey plane. The data from a given scan (e.g., 10, 240 digitized points per transducer) are grouped together and averaged into the spatial window they correspond to. The computer software required to sort the data is quite simple and straightforward. For example, suppose the digitized positions of the probe were given as 0 to 10, 000. If the scan line is divided into 200 spatial increments, then each increment has a width equivalent to an ADC number of 50. Consider a datum point (y_i, p_i), where y_i = datum point location and P_i = digitized probe output. Suppose $y_i = 536$ for this datum point. Dividing y_i by 50, using a computer function "INT, " which truncates to an integer, and adding 1 results in INT(536/50) + 1 = 11, i.e., our datum point belongs in the eleventh spatial window. All datum points P_i that belong to this window are added together and divided by the number of points to get an average data level (PA) for the spatial window. The average data are stored in an array with the window number used as the array counter; in our example the average data would be PA(11).

If the data are taken with a single-element probe such as a Pitot tube, then the next step in data processing is to color code the averaged data in each pixel array. For example, if the voltage level of the data ranges from −10 to +10 V and 20 colors are used, then each color represents a range of 1 V. The computer software to color code the data can consist of simple "IF-THEN"-type statements. If data are taken using a multichannel laser velocimeter system or a 7-hole pressure probe, then further computer processing is required before the data are color coded. With these probes, our flow-field images are now quantitative displays of total velocity, cross-flow velocity, total pressure, etc.

The final step in Fig. 2 is to display the results on a color video monitor. Typical computer processors work with three binary data files (red, green, and blue) and are able to load the buffer memory of a color monitor with the data image. Colors are made by combining these three primary colors (red, green, and blue). On computers with one color level (intensity), eight colors can be displayed on the monitor. On systems employing more color-intensity levels, a very large number of colors are possible. Although a color graphics terminal may be able to display thousands of shades of colors, only a small fraction can be used for displays where it is possible for one to point to a color in the data and find it on the color-code chart. On the video monitor, this number is at most on the order of 30 colors. By the time the displays are photographed and photocopied for reports, this number is reduced to perhaps about 20. Colors used for data displays can be selected solely on the basis of which combination seems to give the most graphic and colorful displays. Colors arranged in the natural color spectrum (i.e., shades of red to shades of blue) give a very nice display. In some cases, it may be appropriate to select a color code that helps one associate the data display with the conditions existing in the test. For example in Ref. [7], red was used to indicate regions of high temperature and blue to indicate regions of low temperature in a thermocouple survey of a jet engine mixer nozzle. It has been suggested that colors such as red, which are associated with a warning or alarm, should be used to code the specific data levels that are of particular interest (e.g., coding low-velocity separated flow regions in red such that they stand out in a data display).

The display quality of a color video monitor has a direct bearing on the number of spatial increments that can be used per scan line. Quality here refers to how well the monitor's CRT is aligned for convergence. If a white line is to be drawn on the screen, poor alignment for convergence (misconvergence) will make the line appear as shifted, smeared-together lines of red, green, and blue. The problem is most noticeable in the corner regions of a screen. Although a monitor may be specified as having a 512 × 390 display, one may have to use 2- by 2-pixel arrays in order to "see" the true color intended, i.e., one could display only 256 spatial windows per scan on this type of monitor.

The video screen can be readily photographed with a well-aligned 35-mm camera. Because the video picture is refreshed 60 times per second, a long exposure time is required to avoid the appearance of the refresh bar (an exposure too short will show the refresh bar as a shadowy diagonal region extending across the screen). Good results can be obtained using a low-ISO film (e.g., ISO 100) and an exposure time of 1 s. The f-stop can be adjusted to the correct setting by using the built-in (through-the-lens) light meter. Since larger monitors have a curved screen, a certain amount of picture distortion is to be expected. Distortion due to the camera lens can be reduced by using a long-focal-length lens (e.g., 200 mm in place of the standard 50 mm). Very-high-resolution, undistorted images can be obtained with a film recorder. In these devices, a monochrome screen is used together with a red–green–blue filter wheel. The three binary files are displayed and photographed one at a time with the associated color filter; a triple-exposed color photograph is the end result. This approach virtually eliminates convergence problems since a pixel is always illuminated at exactly the same spot for each of the three colors. The CRT tubes used in film recorders are quite small; hence, the screen is completely flat and no image distortion occurs. High-resolution film recorders of this type can produce images with 4096 × 4096 pixels.

Visual ''clues''—outlines of the wind tunnel model, projections of wing trailing edges, etc.—can be added to a display to help convey where the data survey has been made. A reticle, or gridlike pattern, can be added to the display to help determine the specific locations of certain features of the flow field [8]. New CAD/CAM-related software and hardware with zoom and pan capabilities allow one to ''roam around'' a large data set and then zero in on and magnify a specific region of interest. Besides the ''linear'' color coding described earlier, one can use variable scaling, which accentuates various regions in the flow field.

Use of the color video display technique poses a serious problem when the results are to be printed in a publication generally restricted to black-and-white reproduction. It has been suggested that one solution to this problem is to use clearly contrasted bands of a gray scale to display data for black-and-white publication. It is to be hoped that the cost of color printing will someday be low enough that reports can provide the full impact of these truly beautiful color video displays.

III TYPICAL APPLICATIONS

The color video display technique has been used at the University of Maryland to study the flow fields produced by a wide variety of wind tunnel models ranging from delta wings to scale models of aircraft carriers. Data have been taken with a number of different probes, including Pitot tubes, hot-wire probes, ptich/yaw probes, split-film probes, 5- and 7-hole pressure probes, and a laser velocimeter system [4–6]. Most of the testing to date has involved post-processing of data recorded on analog tape. A real-time system, used principally for undergraduate instructional labs, has been developed that allows one to observe the developing color display as the test is under way. The results of several test series of strictly qualitative data are outlined in the following sections.

A Separated Flow on a Finite Wing

Figure 3 shows a composite of various types of data obtained in flow-field surveys just downstream of a partially stalled rectangular planform wing (as sketched in Fig. 1). The finite wing model had a Clark Y-14 airfoil section with chord $c = 15.2$ cm and aspect ratio $AR = 3$. The model was tested at an angle of attack of $\alpha = 25.4°$ and a Reynolds number based on chord of $Re_c = 360{,}000$. Oil-flow tests indicated the formation of a large mushroom-shaped trailing-edge stall cell [9]. Surverys were made 2.54 cm downstream of the trailing edge of the model using a single-element hot-wire probe, a Pitot tube, and a pitch (Conrad) probe. A total of 100 vertical scans were completed, with a vertical scan increment of 1.5 mm. Each data display in Fig. 3 took about 90 min to complete at scanning rates of 1.5 cm/s. The data were processed through three separate computer systems for final display on a VAX/Grinnel image-processing system [6].

Unfortunately, the black-and-white reproduction of these data sets prevents one from clearly discussing these results in terms of the specific colors existing in the displays. In the remainder of this section, reference is made to color levels in the data in an attempt to convey to the reader what can be seen in actual color displays.

The data displays are shown as if viewed from downstream of the model. The white line is the horizontal projection of the trailing edge. The highly qualitative data displays clearly show the outline of the separation bubble formed downstream of the mushroom-shaped stall cell. The tip vortices are also clearly shown in each display. The color-bar chart has 20 colors ranging from shades of red–yellow at the minimum transducer output levels, through shades of green and blue, and finally to shades of pink at the maximum transducer output. The data ''images'' from the hot-wire and Pitot-tube data are quite similar in appearance. The pitch probe data pinpoint the location of the tip vortices by showing rapidly changing color patterns at the vortex core region. Additional information on studies of the mushroom stall cells and more extensive flow-field survey work can be found in Refs. [5, 9].

Figure 4 shows the results of a chordwise flow-field survey on the centerplane of a fully stalled rectangular planform wing (Clark Y-14 airfoil section, $c = 15.2$ cm, $AR = 4$, $\alpha = 28.4°$, $Re_c = 480,000$) [5]. The traversing rig was operated manually to allow positioning of the probe tip as close to the wing surface as possible. A bright light was used to cast a long shadow of the probe tip on the wing surface; the probe was then positioned close to the surface

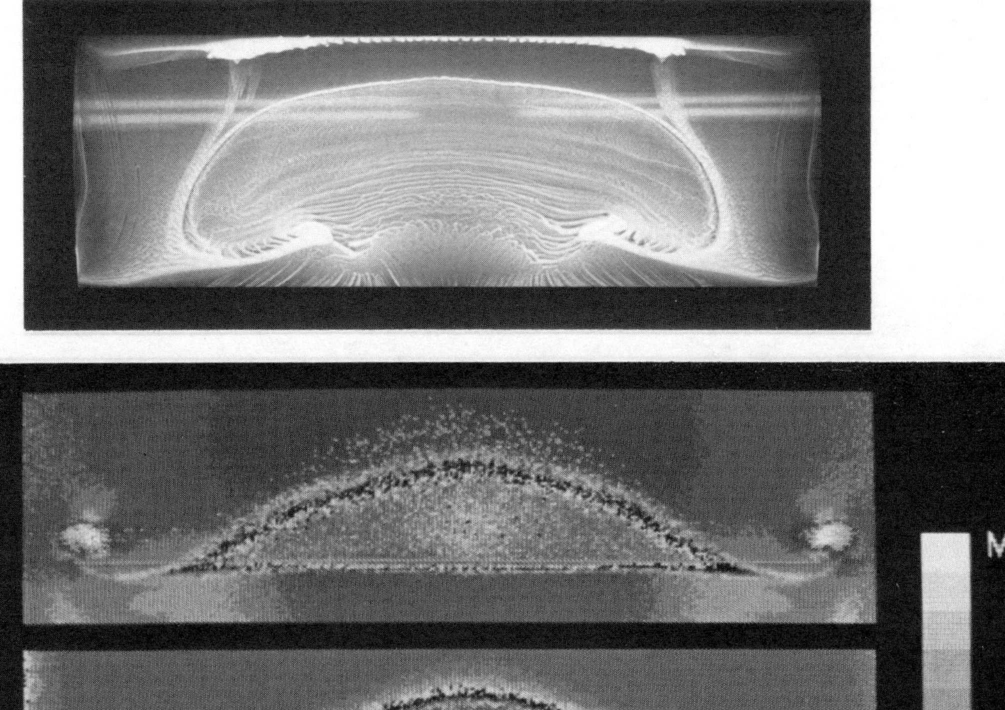

Fig. 3 Surface oil flow and color video displays of flow-field surveys behind a partially stalled wing.

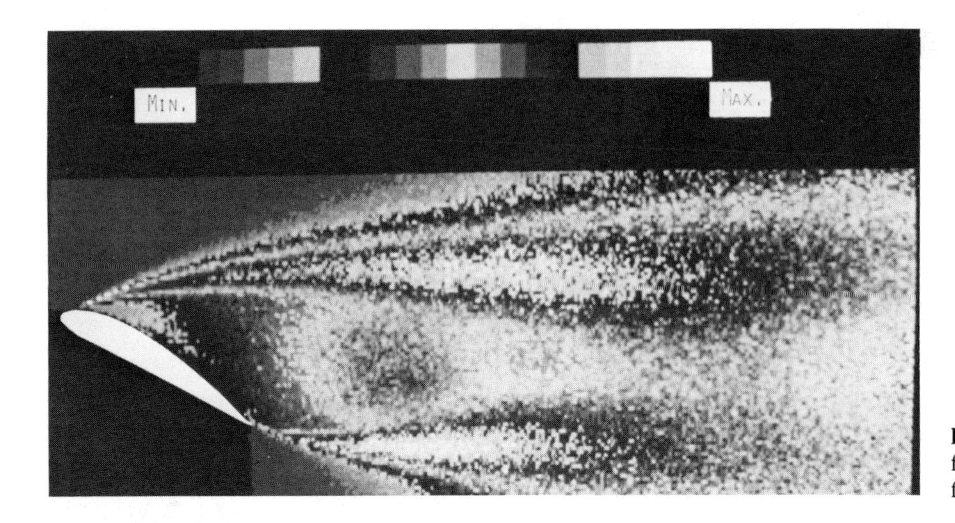

Fig. 4 Video display of a hot-wire flow-field survey behind a fully stalled finite wing.

by observing the relative location between the shadow and the probe tip.

The hot-wire rms output shown in this display clearly indicates the shear layer between the free-stream flow and the separated wake region. The shear layer appears as the highest rms levels in shades of pink. The free-stream flow is displayed as the lowest rms levels in dark red.

B Wake Measurements behind an Oscillating Wing

Figure 5 shows the location of a flow-field survey conducted behind an oscillating semi-infinite wing model. The model (NACA 0012 airfoil section, c = 15.2 cm, span b = 38 cm) was mounted on a wall extension of the free jet wind tunnel shown in Fig. 1. The wing model was oscillated at an angle of attack between 10 and 20° at a frequency of

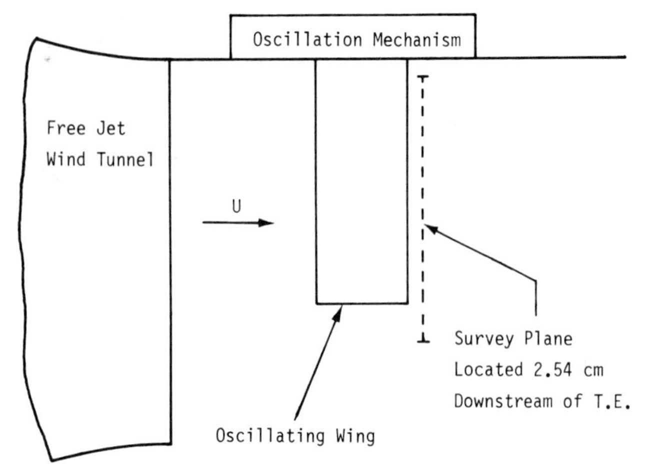

Fig. 5 Location of flow-field survey behind an oscillating wing (*top view*).

7.5 Hz and tested at a Reynolds number based on chord of Re_c = 360, 000. A single-element hot-wire probe was used to scan the flow field at a scan rate of 0.15 cm/s. The instantaneous wing angle was monitored by a rotary variable differential transformer (RVDT) and recorded along with the hot-wire output and traversing rig position.

Conditional sampling techniques were used to "extract" data images at selected angles of attack from the vast data file generated during the 5.5-hr test. A data window of 1° was used in this process (i.e., for some angle θ, all data within ±0.5° of θ were averaged into one pixel array). The "instantaneous data images" displayed in Fig. 6 show attached flow at the lowest 10° angle and fully separated flow at the highest 20° angle (again, the displays are shown as if viewed from downstream of the model, with the white line being the horizontal projection of the trailing edge). The data at +15° for increasing angle of attack show a small wake (i.e., attached flow) while the data at −15° for decreasing angle of attack show a large wake (i.e., fully separated flow). When lift is measured on an oscillating wing, the data plot of lift versus angle of attack contains a large hysteresis loop. The wake images at +15° and −15° correspond respectively to points on the upper and lower portions of the lift hysteresis loop. Variations in the tip vortex structure during the pitch cycle are readily apparent in these displays. By extracting images at 1° intervals over the entire 10 to 20° pitch cycle, it is possible to produce an animated movie film that shows the cyclically changing flow field behind the oscillating wing.

C Pitot Probe Measurements behind a Yawed Delta Wing

Figure 7 shows the location of flow-field surveys conducted downstream of a yawed delta wing. The delta wing (c = 40.6 cm, AR = 1.5) was tested at an angle of attack of α

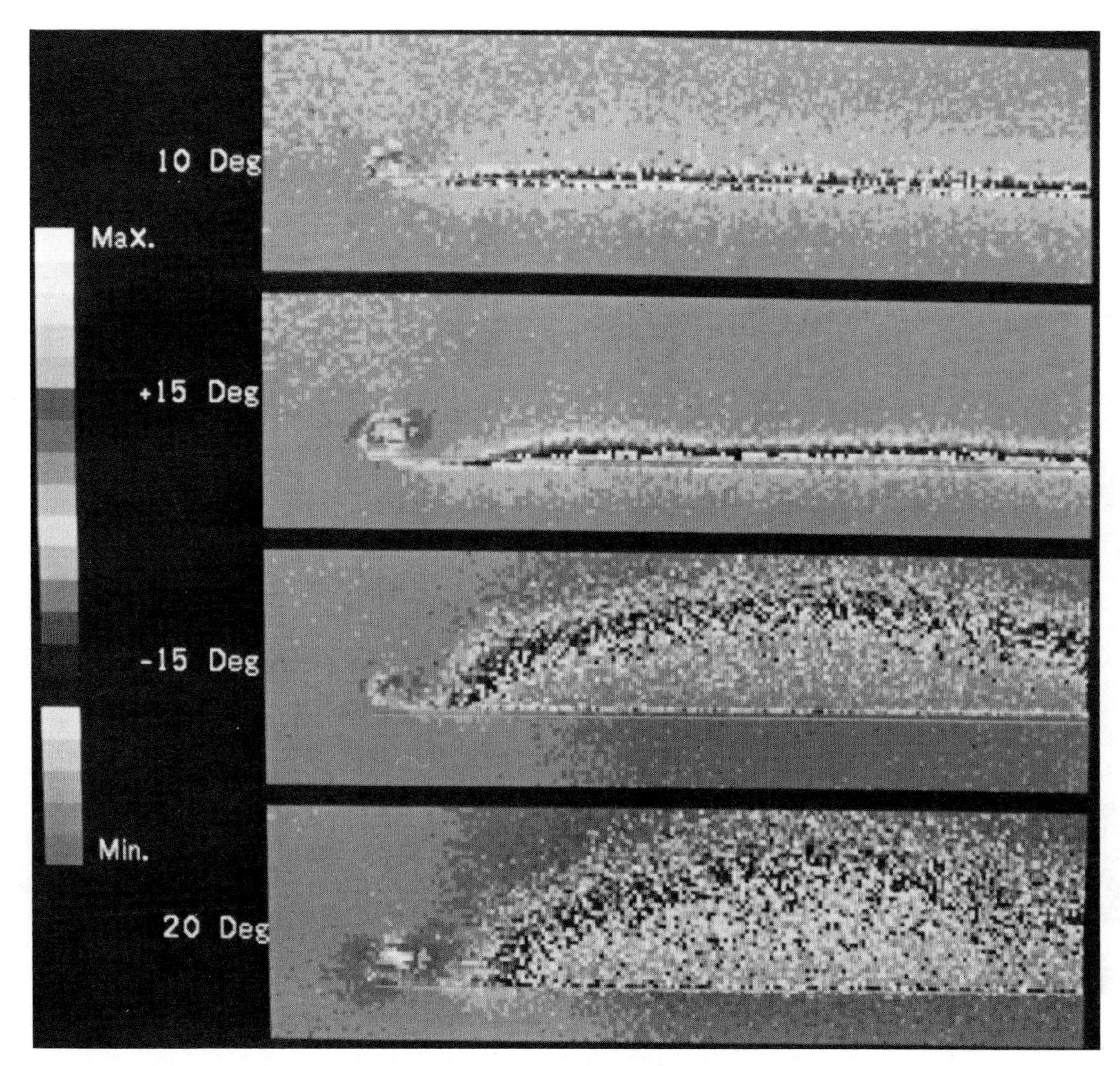

Fig. 6 Video display of hot-wire surveys (DC level) behind an oscillating wing.

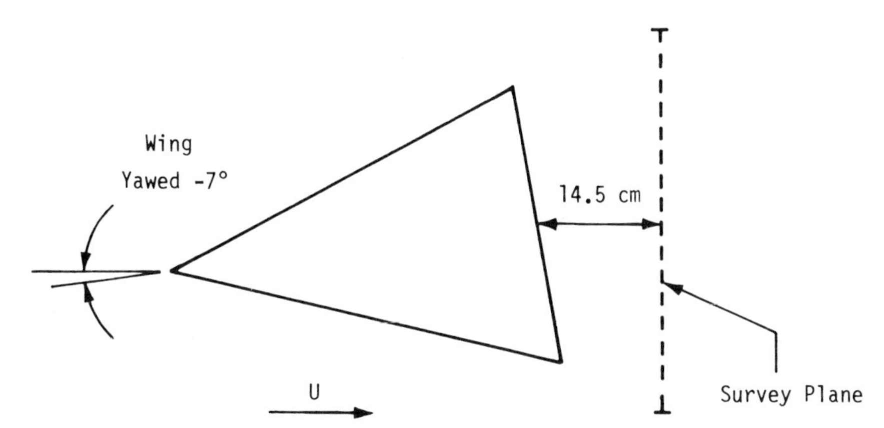

Fig. 7 Location of flow-field surveys behind a yawed delta wind (*top view*).

(a)

(b)

(c)

Fig. 8 Video displays of Pitot tube surveys behind a yawed delta wing: (*a*) Overall survey; (*b*) detailed survey of the boxed region in the overall survey shown in (*a*); (*c*) reprocessed data of the boxed region in the detailed survey shown in (*b*).

= 23°, a yaw angle of $\beta = -7°$, and a Reynolds number based on chord of $Re_c = 960,000$. A sharp-lipped Pitot tube with an outside diameter of 0.5 mm was used to obtain very fine detail of the flow.

The first test consisted of 96 vertical scans with a vertical movement of 1.59 mm and a scan length of 40.64 cm. Slow scanning rates of 0.4 cm/s resulted in a test duration of 3 hr. Each scan line was divided into 256 spatial windows, each consisting of a 2- by 2-pixel array. The data from this test, shown in Fig. 8*a*, were displayed using the 8 basic colors of an HP2627A graphics terminal; the colors were "recycled" such that 24 color levels were utilized. Again the displays are shown as if viewed from downstream of the model, and the black line crossing the picture is the horizontal projection of the trailing edge.

Because the delta wing is yawed 7°, the primary vortex on the right side is known to burst on the upper surface of the wing at a point about one-half chord from the apex (as observed in smoke-flow tests). The data display shows the burst vortex pattern as a large diffuse area. The vortex system on the left side of the wing is far from bursting and shows a very tight structure. The primary vortex on each side interacts with a smaller counter-rotating vortex flow which is referred to by Hummel [10] as the trailing vortex.

The area contained in the box on the left side of the overall data image was resurveyed in a second test. The boxed area, measuring 12.7 by 8.38 cm, was scanned 165 times with vertical scanning increments of 0.51 mm. The data obtained in this survey are shown in Fig. 8*b*. In processing the data for this display, the 8 basic colors were recycled such that 29 color levels were used. Because of the very large suction pressure produced by the primary vortex, most of the color levels were used up in the near vicinity of the primary vortex core. Only a few color levels were used to display the trailing vortex region. Further detail of the trailing vortex was obtained by selectively reprocessing the data in the 3.15- by 3.15-cm box surrounding the trailing vortex. The display obtained for the boxed area is shown in Fig. 8*c*. A total of 24 color levels were used in reprocessing the data. The mosaic pattern formed by the 2- by 4-pixel arrays is quite apparent in this display of 62 scan lines.

D Special Display Techniques

The data shown in the previous examples are "2-D displays," which require one to "imagine" exactly where the data are relative to the model. In the original method developed by Crowder [1, 2], double-exposure photographic techniques were used to "overlay" the wake image behind and alongside of the model in true perspective. There are two basic methods to "overlay" computer-generated data displays on an image of the model. One approach is to generate an image of the model on the video monitor using CAD/CAM-related software. The data images can then be overlain on this computer-generated model. By using computer graphics techniques such as transparency, the displays can be made to look quite realistic. The second overlaying technique is use of special effects photography. An example of this method is shown in Fig. 9. The same delta wing used in the yawed delta wing tests was aligned to zero yaw and pitched to 18°. Surveys were taken at four stations above the wing, equally spaced apart. This display clearly shows the primary and secondary vortex flow on the upper surface of the wing.

Fig. 9 Pitot tube flow-field surveys above a delta wing.

The use of high-resolution video graphics terminals, film recorders, and three-dimensional (stereographic and holographic) display techniques will certainly open a new era in the presentation of flow-field survey data. Three-dimensional images of a flow field could be constructed after one had taken multiple "slices" of data at different stations behind a model. Higher-resolution color video screens would allow one to use colored arrowheads to display the magnitude of velocity (as a color level) and the flow direction in a plane.

REFERENCES

1. Crowder, J. P., Hill, E. G., and Pond, C. R., Selected Wind Tunnel Testing Developments at Boeing Aerodynamics Laboratory, AIAA Paper 80-0458, 1980.
2. Crowder, J. P., Quick and Easy Flow Field Surveys, *Astronaut. Aeronaut.*, vol. 10, pp. 38–39, 1980.

3. Winkelmann, A. E., and Tsao, C. P., An Experimental Study of the Flow on a Wing with a Partial Span Drooped Leading Edge, AIAA Paper 81-1655, 1981.
4. Winkelmann, A. E., and Tsao, C. P., A Color Video Display Technique for Flowfield Surveys, AIAA Paper 82-0611-CP, 1982.
5. Winkelmann, A. E., and Tsao, C. P., A Color Video Display Technique for Flowfield Surveys, *AIAA J.*, vol. 23, no. 9, pp. 1381–1386, 1985.
6. Winkelmann, A. E., and Tsao, C. P., Flow Visualization using Computer-generated Color Video Displays of Flowfield Survey Data, in *Flow Visualization III: Proceedings of the Third International Symposium on Flow Visualization*, ed. W.-J. Yang, pp. 175–179, Hemisphere, Washington, D.C., 1985.
7. Anderson, B. H., Putt, C. W., and Giamati, C. C., Application of Computer Generated Color Graphic Techniques to the Processing and Display of Three-Dimensional Fluid Dynamic Data, *ASME Comput. Flow Predict. Fluid Dynam. Exp.*, pp. 65–72, presented at the Winter Annual Meeting of the ASME, Washington, DC, Nov. 15–20, 1981.
8. Cogotti, A., Car-Wake Imaging using a Seven-Hole Probe, SAE Paper 860214, presented at the SAE International Congress and Exposition, Detroit, Michigan, Feb. 24–28, 1986.
9. Winkelmann, A. E. and Barlow, J. B., A Flow Field Model for a Rectangular Planform Wing beyond Stall, *AIAA J.*, vol. 18, pp. 1006–1007, 1980.
10. Hummel, D., On the Vortex Formation Over a Slender Wing at Large Angles of Incidence, AGARD CP-247, High Angle of Attack Aerodynamics, pp. 15–1 to 15–17, Jan. 1979.

Chapter 24

Computer-Aided Flow Visualization

C. R. Smith

I INTRODUCTION

Computer-aided flow visualization is a rather broad category that refers to the use of digital computer processing in combination with various display devices to enhance our ability to visually understand fluid flow behavior. This description clearly broadens the scope of what is considered "conventional" flow visualization (i.e., two-dimensional images of an optically detectable medium or property translating in an experimental flow field) to include the three-dimensional reconstruction of space- and time-dependent behavior, the visual display of reduced quantitative property data, and the generation of "flow visualization" images from analytically generated data. This chapter reviews a number of general techniques that have been developed to facilitate computer-aided flow visualization. As the reader will note, most techniques discussed are adaptations from the well-developed general fields of image processing and computer-aided design/graphics. Since the techniques employed for computer-aided flow visualization are essentially generic to these other fields, the present section covers only general approaches. For specific details of the computational mechanics and appropriate hardware/software systems, the reader should consult the literature in the appropriate field [e.g., 1–4]. The reader is also referred to Chaps. 16–19 in this handbook, which address the details of image acquisition and processing in more depth.

The overall objectives of computer-aided flow visualization are, at the lowest level, to improve image quality and highlight certain aspects of images (e.g., edges). At higher levels, the objectives of computer-aided flow visualization are to provide methods that can illustrate

1. the "flow" properties, either in the field of view or over a time sequence;
2. the "structure" of a flow field (i.e., characteristic patterns).

A more recent use of computer-aided flow visualization is to help make the connection between conventional flow visualization, which provides primarily qualitative information, and the quantitative information derived from empirical probe measurements and computational fluid dynamics. For example, recent advances now allow quantitative, statistical data to be obtained directly from the deformation, concentration, and relative motion of visual media. In addition, the behavior of flow visualization media is being simulated using data bases generated by large, multidimensional computational fluid dynamic programs. The combination of these types of approaches provides mechanisms for the more direct interpretation and comparison of both flow visualization results and quantitative data bases.

In a broad sense, computer-aided flow visualization can be roughly separated into three areas:

1. *image enhancement:* modification to enhance certain aspects of the original image;
2. *image reconstruction:* reconstruction of higher-dimensional images from lower-dimensional images (i.e., 3-D from 2-D);

3. *image synthesis:* creation of visual images from reduced quantitative data or analytical simulations.

The first two areas are expansions of conventional flow visualization techniques; the third area utilizes computer-generated images to provide a much broader visual mechanism for either rapid evaluation of multidimensional data sets or comparisons between parallel experimental and analytical studies. The following sections highlight the relevant aspects of these areas.

II IMAGE ENHANCEMENT

Clearly, in order to perform any type of computer image enhancement or manipulation, the quality of the original image should be as good as possible. However, in many cases, the image-acquisition process is impaired by inadequate lighting, the wrong shutter speed or depth of focal field, lens irregularities, or poor film or image-detector resolution. In these cases, appropriate enhancement of the image can help improve its clarity. In the majority of cases, however, the purpose of image enhancement is to extract some characteristic property of the image or of the visual medium being viewed (e.g., the concentration gradients in a mixing process).

In order to be able to computationally manipulate an image, the image must first be transferred from the original format (i.e., film or video) to a computational array in computer memory. Normally, images are converted to an array of $N \times N$ dimensions, with each member of the array representing the numerical "gray-level" intensity of the pixel at that particular location in the image. Generally, the level of the intensity varies from 0 to 255, with 0 being black and 255 indicating white. Color pictures are composed of three overlapping arrays, with each array representing the intensity contributions of the constituent colors (e.g., red, green, and blue for conventional video) of the total image. For the purposes of this discussion, only gray-level images are considered.

The primary methods for image transfer to computer memory are by use of either a direct digitizing camera, or a digitizing board interfaced with a video source. Digitizing cameras generally employ a CCD detection tube, which provides a direct digital image. The image is fed directly to a digital display device and/or to a computer for storage. The size of the array that these direct-input devices can convert is limited, with the maximum being typically 200×200 pixels. A modified version of a digitizing camera is also employed in a slow-scan mode for image transfer from film. Cameras configured for slow-scan image acquisition can input arrays up to 4096×4096, but are quite slow, requiring on the order of minutes to perform the image transfer. The second mode of image transfer is the use of a digitizing board, interfaced with either a computer or an image-processing system, to convert a conventional analog video image to a digital array. This can be done either directly, from a video camera, or indirectly, from a videotaped image. This latter approach is very attractive, since it allows the review and preselection of images, but it is often complicated by problems in phase synchronizing the video picture from the VTR with the digitizing board.

The variety of visualization media with which image enhancement can be effectively employed includes smoke, dye, hydrogen bubbles, particles, and density variations, etc. For purposes of illustration, the image-processing approaches discussed in this section are demonstrated using dye and hydrogen bubble flow visualization pictures obtained in a low-speed water flow. Note that the images shown have already been digitized. The first image, shown in Fig. 1*a*, employs surface dye injection into a left-to-right laminar flow over a plate. The injection of dye through the surface initiates a hairpinlike flow structure made visible by the dye. This image shows both a side view (upper part) and plan view (lower part) of the hairpin structure, obtained with two separate cameras and a split-screen mixer in the recording process. The second image, Fig. 1*b*, is a hydrogen bubble timeline visualization of a turbulent boundary layer. In this image, the flow is left to right across a flat plate (at the bottom of the picture), with the hydrogen bubble wire oriented vertically and bubble time lines generated periodically.

A Filtering

Filtering is a general term implying a spatial or frequency transformation of the image intensities in order to enhance or deemphasize certain features of the image. This can be done to sharpen the intensity discontinuities between image regions or, conversely, to soften the discontinuities. The approach employed clearly depends on the information desired. For example, if the object(s) of interest are large, then the image can be blurred to delete small-intensity discontinuities while retaining the boundaries of the large objects. This is a low-pass filtering process, accomplished by use of a local averaging of pixel intensity; the degree of filtering and orientation are controlled by the size and orientation of the averaging window and by the averaging algorithm employed. Figure 2*a* shows the effect of a symmetric, 29- by 29-pixel, boxcar (equal weighting) filter applied to Fig. 1*a*. Alternatively, one may want to emphasize the smaller objects or discontinuities in an image (high-pass filtering). This can be done by such techniques as subtraction of an image, which has been filtered over a finite averaging window, from the original image; the resulting image will emphasize the scales of the order of the original averaging filter. An example of this high-pass-filter approach is shown in Fig. 2*b*, where the image in Fig. 2*a* has been subtracted from Fig. 1*a*. The limit on this high-pass filtering is the subtraction of the average of the intensities

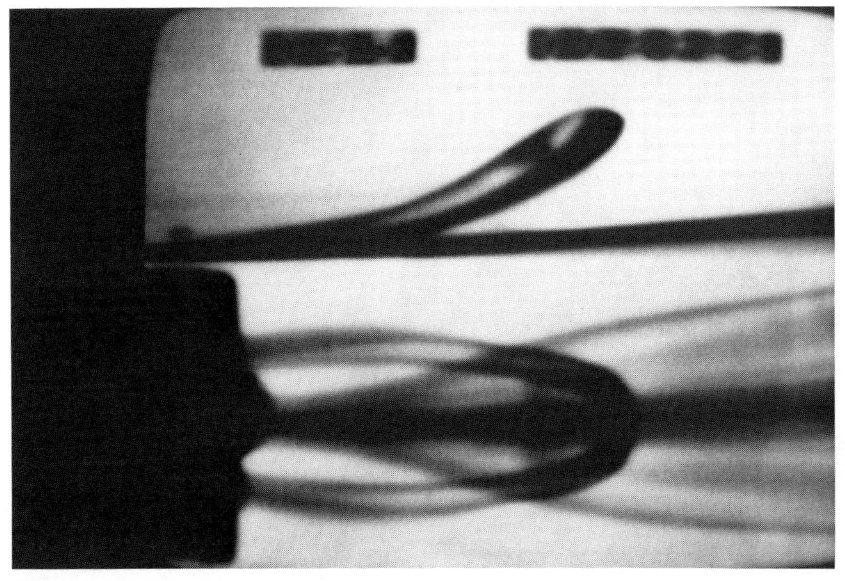

(a)

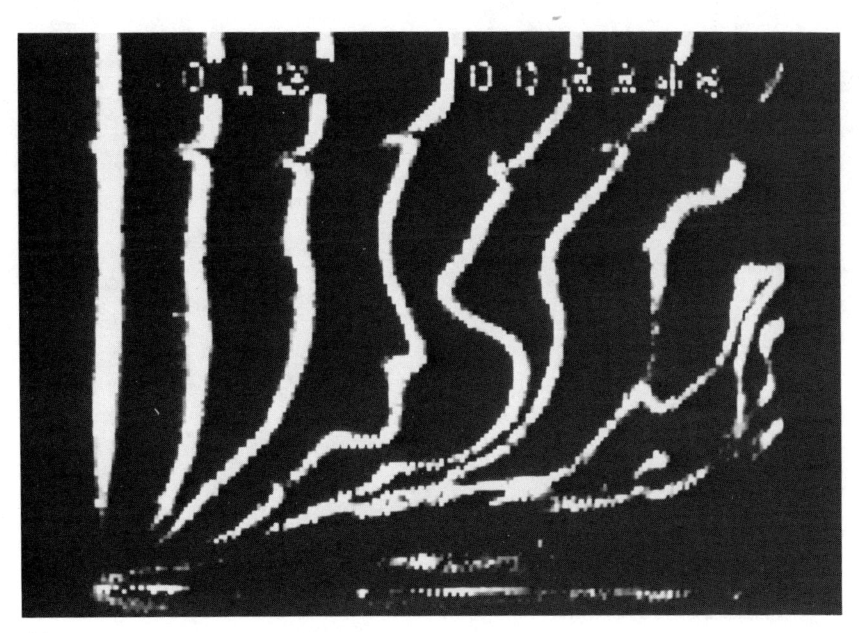

(b)

Fig. 1 Test images that illustrate image-enhancement techniques: (*a*) Surface dye visualization of a hairpin vortex formation in a laminar boundary layer (flow left to right); (*b*) hydrogen bubble timeline visualization of the wall region of a turbulent boundary layer (flow left to right).

(a)

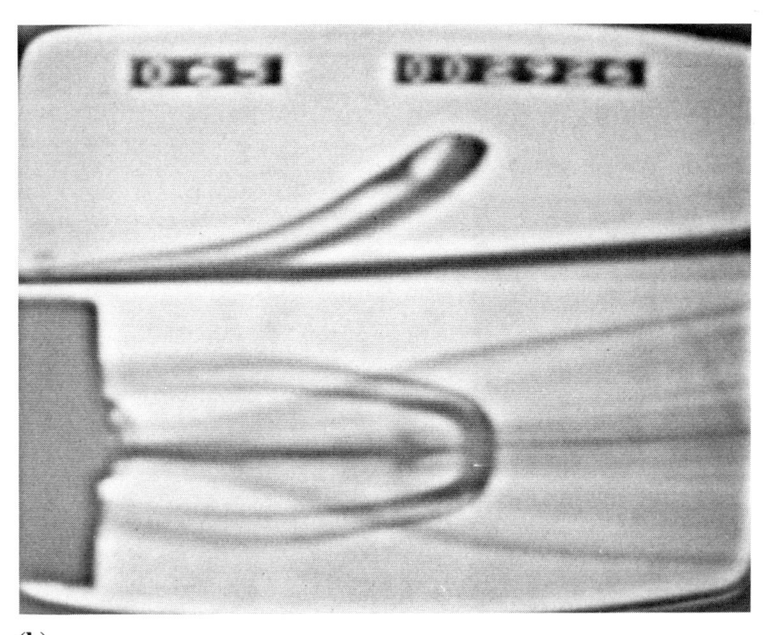

(b)

Fig. 2 Image-filtering examples: (*a*) Boxcar averaging filter (29 × 29 pixels) applied to Fig. 1*a*; (*b*) high-pass filter generated by subtraction of Fig. 2*a* from Fig. 1*a*. (Figure continues.)

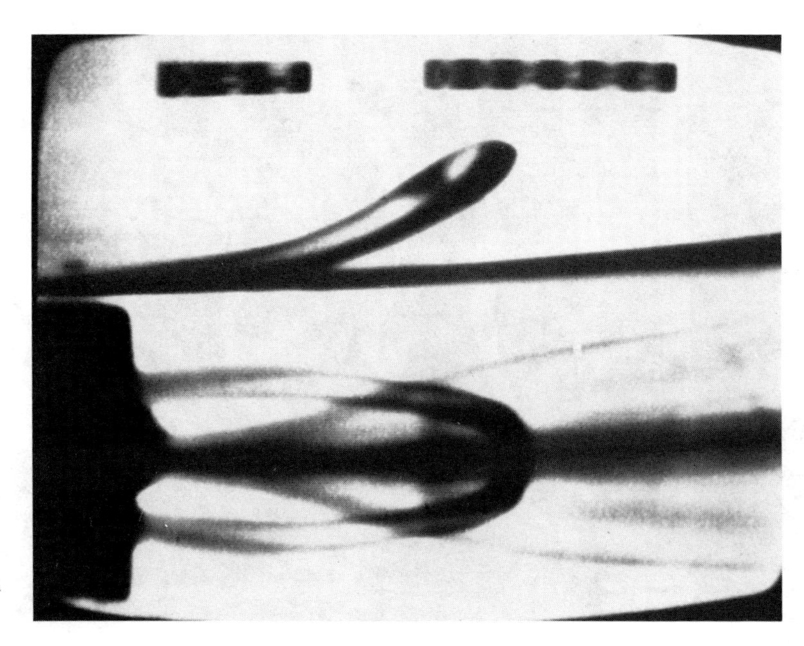

Fig. 2 (continued) (*c*) Histogram equalization of Fig. 1*a*.

over the entire image from the original image. This latter technique is often employed to help compensate for spatially varying illumination.

Histogram equalization is another filtering-type technique that may prove useful when there is a nonuniform distribution of intensities over the intensity range of the image. In this situation, numerical techniques can be effectively employed to equalize the distribution of intensities within the image. An intensity histogram constructed from the total number of pixels comprising the original image normally displays a nonuniform distribution with intensity magnitude; the process of equalization redistributes the pixel intensities uniformly over the range of intensities. The result is an image with expanded contrast, which usually improves the discrimination of many image features. Figure 2*c* is an example of histogram equalization applied to Fig. 1*a*. For details of image filtering, the reader is referred to Refs. [3, 4].

B Thresholding

Thresholding is probably the easiest of all image-enhancement processes to apply. This process searches the original image for selected pixel intensity limits or ranges, which are then reset to some prescribed intensity value. The result may be a segmentally shaded image [5–11], if ranges are established, or a high-contrast black-on-white image, if a limiting value is set. The results of a limiting threshold process applied to Fig. 1*b* is shown in Fig. 3*a*, with white indicating regions in which the intensity magnitudes exceeded the threshold value and black indicating regions where the intensity magnitudes fell below the threshold. Often when a series of upper and lower bounds are set as thresholds, the range of intensity values covered by different ranges is often best discriminated by use of a specified sequence of colors, with each color indicating a particular range of intensities. This process of "coloring" a gray-level image is commonly known as pseudocoloring. Examples of this type of process for both original and synthesized images (see Sec. IV) are numerous. An example for a synthesized image is shown in Figs. 9*c*,*d* and 10.

C Edge Detection

The process of edge detection is employed to locate the supposed edges within an image. Mathematically, this implies the location of locally strong gradients or discontinuities in the image-intensity field. The location of an edge is usually done by establishing a mathematical edge operator, which is applied over a small spatial extent in order to establish a magnitude describing the severity of this change and the direction of the maximum gray-level change [12–14]. Because certain aspects of a flow visualization image may not have uniformly discrete edges, it is often unclear how to specify whether a local edge corresponds to the boundaries of a particular flow characteristic within an image. This often requires the educated input of the investigator and possible adaptation or "tuning" of the edge operator for each type of image evaluated. An additional difficulty encountered in the use of edge operators is the presence of image "noise"; since edges are high-spatial-frequency phenomena, edge detectors are normally very sensitive to high-frequency noise (e.g., snow on a video screen). In order to establish consistent edges, it is often

(a)

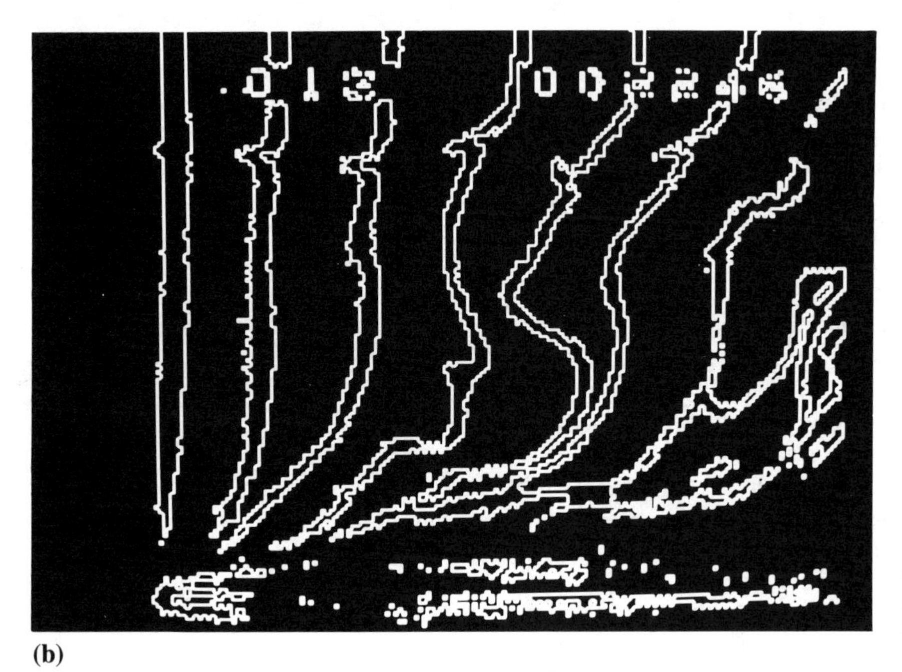

(b)

Fig. 3 Edge establishment: (*a*) Application of a limiting threshold process to Fig. 1*b;* (*b*) simple edge-detection algorithm applied to Fig. 3*a*.

necessary and advisable to filter extraneous noise from the image before attempting the application of an edge operator. One simple technique for edge detection is to create a thresholded image and then to establish the dividing surface between the thresholded region and the remainder of the image. Figure 3*b* is an example of the edges detected for the thresholded image in Fig. 3*a*. Clearly, the edges established by this simple approach strongly depend on the threshold level selected, indicating the need for proper ''education'' of the edge-detector algorithm for the particular type of image considered. Again, it is not appropriate to cover all the possibilities for edge detection; for a review of the classes and procedural details of edge-detector algorithms, the reader is referred to Refs. [1–4].

D Gradient Techniques

From a flow visualization point of view, the boundaries of visualization characteristics may be effectively illustrated by displaying the image gradients [15]. This approach is particularly useful for illustrating regions with strong inten-

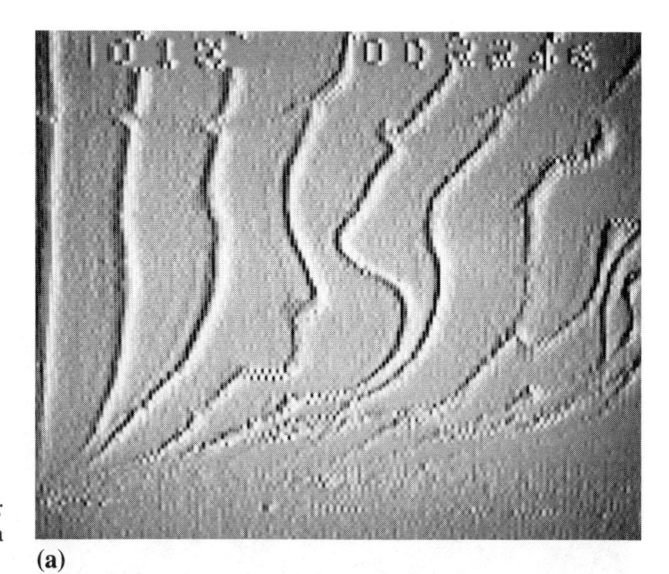

(a)

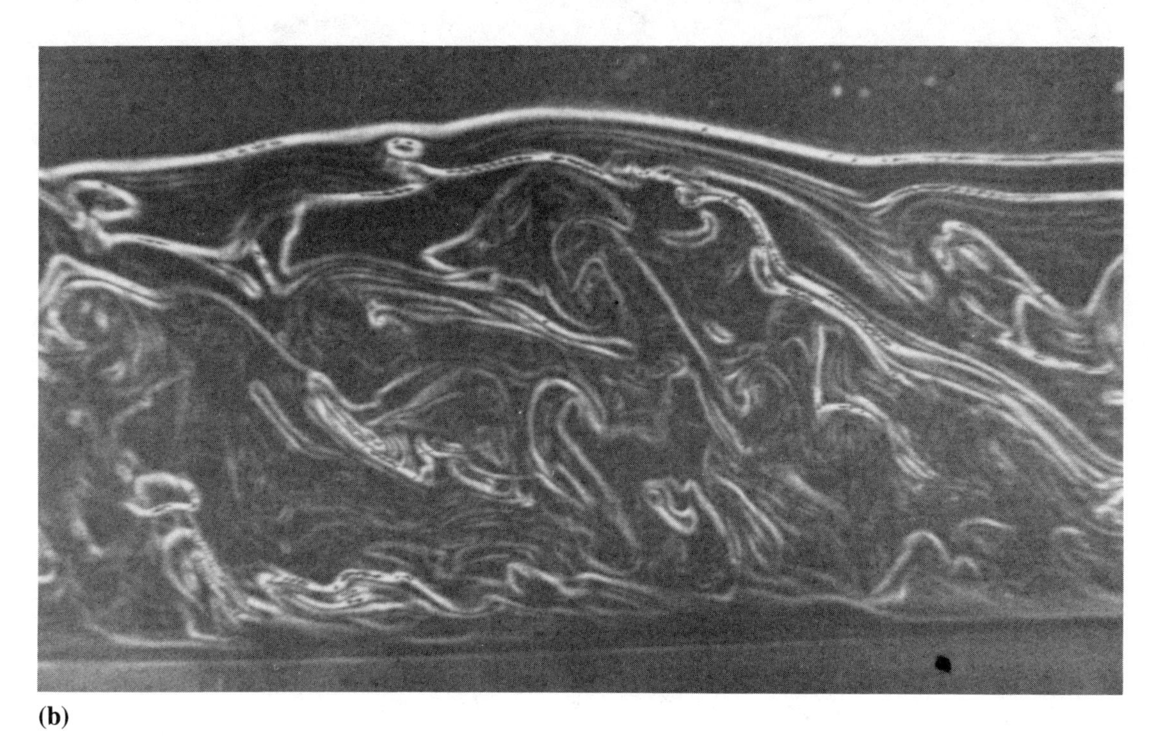

(b)

Fig. 4 Gradient examples: (*a*) Horizontal gradient of Fig. 1*b*; (*b*) maximum gradient of generalized smoke visualization of a turbulent boundary layer (courtesy of J. L. Balint, et al. [15]).

sity variations, such as for visualizations employing smoke, dye, and hydrogen bubbles. The basic approach is to numerically differentiate the image pixel array, displaying the resultant image as a gray-level image. Since differentiated images have both positive and negative gradients, the general method for display is to have white and black represent the maximum positive and negative gradients respectively, let neutral gray represent the zero gradient, and employ linear gray-level shading for variations between the extremes. The horizontal gradient of Fig. 1*b* is shown in Fig. 4*a* using this type of display scheme. This kind of presentation illustrates the derivative of the intensity of the image, allowing the viewer to interpret the location and strength of the gradients as reflecting either edges of objects or characteristic changes within the image. The method for calculating the derivative depends on the type of image required. The derivative may be obtained using either a simple two-point linear approximation at each point or local fitting and differentiation of a higher-order polynomial curve; the gradient may also be displayed for either coordinate direction (vertical or horizontal) or as the maximum gradient (normally the root mean square of the two direc-

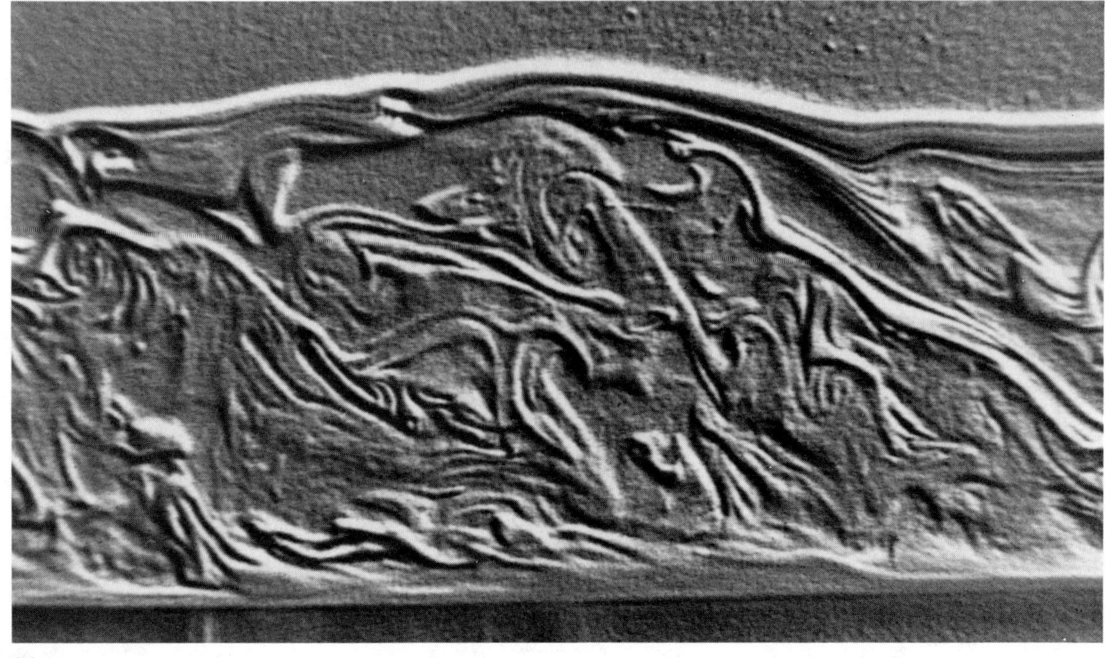

(a)

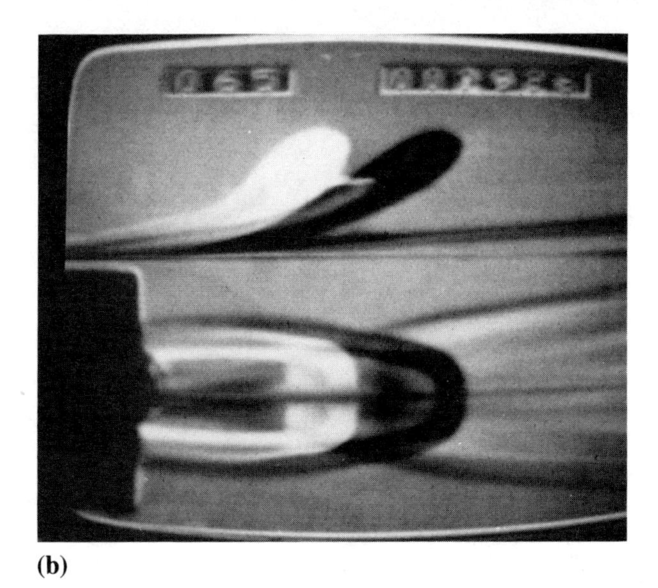

(b)

Fig. 5 Image-differencing examples: (*a*) ''Bas relief'' effect achieved by image shifting, inversion, and addition (courtesy of J. L. Balint, et al. [15]); (*b*) image subtraction of time-sequence images (from the sequence of Fig. 1*a*).

tional derivatives). An example of the maximum gradient is shown in Fig. 4*b* for a generalized smoke visualization of a turbulent boundary layer [15].

E Differencing Techniques

A number of useful enhancement techniques employ the process of image differencing (subtraction of one image from another) in order to highlight or remove certain aspects of an image. Recall that subtraction of a filtered image from the original image is employed as a high-pass type of image filter (see Fig. 2*b*). Other examples of differencing with flow visualization applications are positive–negative addition and sequential scene differencing. Positive–negative addition is a process that has been successfully employed with smoke visualization [15] to enhance the interior details of a complicated concentration gradient. Basically, the positive and the negative of the same image are diagonally displaced a small distance (e.g., 2 pixels) and then added. The resultant image has a ''bas relief'' appearance, which emphasizes the interior characteristics of the flow structure. An application of this technique is shown in Fig. 5*a* (note that Figs. 4*b* and 5*a* are enhancements of the same original image). Sequential scene differencing involves the subtraction of one image in a temporal sequence from another in order to illustrate the temporal changes in the flow visualization medium. Figure 5*b* illustrates the effect of subtracting a later sequential image from the dye visualization image shown in Fig. 1*a*. Note that the effect is to generate highly contrasting regions that reflect the *differences in the pictures,* which can be quantified.

F Line Reduction

For an image in which clearly defined lines of visualization medium are present, such as pulsed hydrogen bubble timelines or smoke streaklines, it is often desirable to "reduce" the original image lines to single-pixel-width lines. Reduction of the image in this fashion facilitates determination of quantitative displacement information, such as time-of-flight velocity, from the spacing and/or curvature of the reduced lines [5, 16, 17]. Methods for line reduction generally employ either a local gradient criterion or use of some mathematical process for iteratively reducing a thresholded image. The use of the local gradient commonly establishes the reduced lines as an "edge" of the original image lines. Note that either the positive or the negative edge (gradient) of the original line may be selected, depending on which is better defined or more appropriate for the information desired. Alternatively, one may wish to consider using the gradient zero crossing, where the lines of visual medium display a maximum (or minimum) in intensity [5, 16]. Figure 6*a* shows the result of a zero-crossing approach employed for the identification of lines in the image shown in Fig. 1*b*. Another line-reduction approach is to "thin" a thresholded image of the original image, as shown in Fig. 3*a*. The broad, thresholded regions of constant intensity are thinned by a process of iterative reduction of the minimum dimension of the thresholded region. Each iteration reduces the threshold regions over the entire image in single pixel increments until all thresholded regions are reduced to single pixel lines. Figure 6*b* is the result of the application of such a thinning process to the thresholded image in Fig. 3*a*.

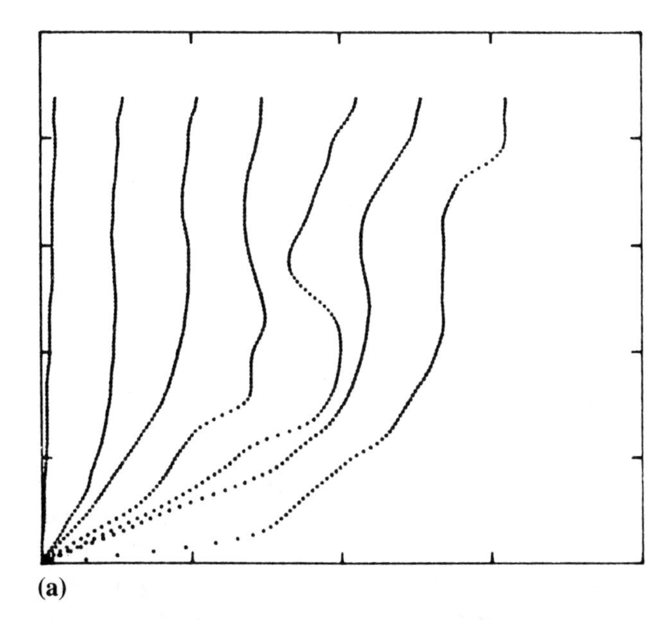

(a)

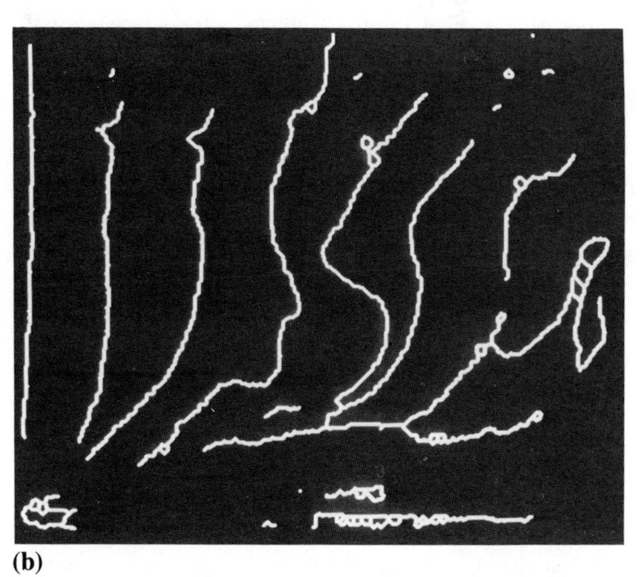

(b)

Fig. 6 Line-reduction examples: (*a*) Lines established by gradient zero-crossing approach applied to Fig. 1*b;* (*b*) lines established by thinning of thresholded image of Fig. 1*b* (i.e., Fig. 3*a*).

III IMAGE RECONSTRUCTION

Image reconstruction deals with the use of time-sequence, phase-averaged, and/or multiple-view images to reconstruct the original time-dependent three-dimensional images from two-dimensional pictures. Most of the techniques employed in these approaches rely on the use of standard graphics techniques interfaced with computer-graphics hardware. Basically, three different approaches are discussed here: phase averaging of single views, multiview reconstruction, and integration of multiple light slices. The use of holography and tomography is not considered; the reader is referred to Chaps. 15 and 20 in this handbook for excellent discussions of these topics. Note that only the highlights of general techniques are touched on here.

A Phase-averaged Reconstruction

When a flow is periodic or responds in a characteristic, repeatable manner to a controlled, phase-locked input, a process of phase-averaged image reconstruction can be employed [17]; the approach is similar to the phase-averaging and signal-education techniques employed with single-sensor probe measurements in periodic and artificially stimulated flows. The technique employs a phase-reference signal to correlate a set of images in time and/or space. After identification of the appropriate aspect of the visualization medium in each image (the introduction of which must also be phase referenced), the phase reference is then employed to allow the geometric characteristics of the image to be graphically reconstructed. This technique has the advantages of requiring only one viewing camera, and allowing reconstruction of the behavior over a broad three-dimensional field. Figure 7 shows the reconstruction of three-dimensional timeline surfaces (i.e., time surfaces) using phase-referenced hydrogen bubble timeline visualization in the wake of a segmented cylinder [17]. This application uses a series of phase-referenced images taken for 17 different planes along the axis of the segmented cylinder, extending from the mid-section of the small diameter to the mid-section of the larger diameter, with symmetry assumed about the mid-sections. The images have been phase referenced to the periodic excitation of the cylinder. An example of a typical digitized image with reduced bubble timelines appears at the bottom of Fig. 7. The corresponding three-dimensional time surfaces are generated by graphically connecting the corresponding phase-locked timelines in each image plane. The results of this time-surface reconstruction (done using conventional Unigraphics computer-aided design hardware and software) are shown in Fig. 7. In concept, this approach is particularly attractive for periodic unsteady flow, where phase referencing of both the images and the introduction of the visual medium is possible. In addition, this technique should be applicable for flow events where controlled stimulation is employed.

B Multiview Reconstruction

This technique makes use of simultaneous, multiple views of a visualization medium to allow geometric reconstruction of local three-dimensional behavior [18–20]. Theoretically, only two different views are necessary to establish the coordinate locations of any objective in the viewed flow field. However, this is the case only when clear, unambiguous, one-to-one correspondence of the viewed characteristics can be established in both views. In addition, one must be very careful to assure proper camera alignment and scale calibration as well as to compensate for uncertainties due to distortions caused by lens nonuniformities, depth-of-field variations, and image resolution. The key point here is to be aware of the limitations of the initial images and to recognize that the reconstruction process amplifies any distortions or uncertainties in the original images. Often, the major difficulty is in establishing image pairs for which clear one-to-one correspondence can be established. Often this requires modification of the normal method in which one visualizes a particular flow. Figure 8 illustrates the application of dual-view image reconstruction to illustrate the 3-D behavior of a hydrogen bubble line near the bounding surface in a turbulent boundary layer [18]. Figure 8*a* illustrates the manner in which the original images were obtained. Simultaneous top (plan) and end views of bubble lines generated transverse and parallel to the boundary were taken; recording was done using a split-screen video display and recording system, as shown. A series of temporal images were then digitized, line reduced in both views, and geometric correlation of the three-dimensional coordinates obtained. Because the generation of multiple bubble lines creates an ambiguous image in the end view, the reconstruction process was performed only for the generation of single bubble lines. The results of this reconstruction process are shown in Fig. 8*b,* which displays the temporal deformation of a single bubble line after its generation. Note that once the geometric data have been reconstructed, the lines or surfaces can be manipulated for viewing from any perspective using conventional computer-aided design techniques. In addition, if a dynamic display system is available, the original dynamics of the visualization process may be recreated by rapid, sequential redisplay of the reconstructed temporal images. Because the three-dimensional information is available, this process not only allows the process dynamics to be reviewed but also provides the capability for optimizing the viewing perspective.

C Multiple-Light-Slice Reconstruction

The use of laser light sheets has allowed the development of methods for illumination and imaging of flow visualization media in thin "sheets" (see Chap. 14). If the light sheet is placed perpendicular to the movement of the visualization medium [21, 22] or if the light sheet is "swept" rapidly through the medium in a sequence of planes [23], the photographic result is a series of images of planar segments of the visualization medium. If a Galilean transformation on the segments obtained for passage through a fixed light sheet is performed or a scanning and image-acquisition rate that far exceeds the convection rate of the visualization medium is employed, these light sheet visualization techniques yield a series of images that can be computationally "stacked" upon one another to reconstruct the three-dimensional geometric shape of the visualization medium (see Ref. [21] for several excellent examples). This technique is quite similar to the medical techniques employed for reconstruction of three-dimensional physiological displays using sequential CAT-scan images. Note that one of the limitations of this technique is dependence on the method and location of introduction of the visual medium employed. For instance, the geometric dispersion of dyes and smoke depends on the integrated history from the point of introduction of the medium. Thus, the derived geometric shape of the visualization medium will be a function of the

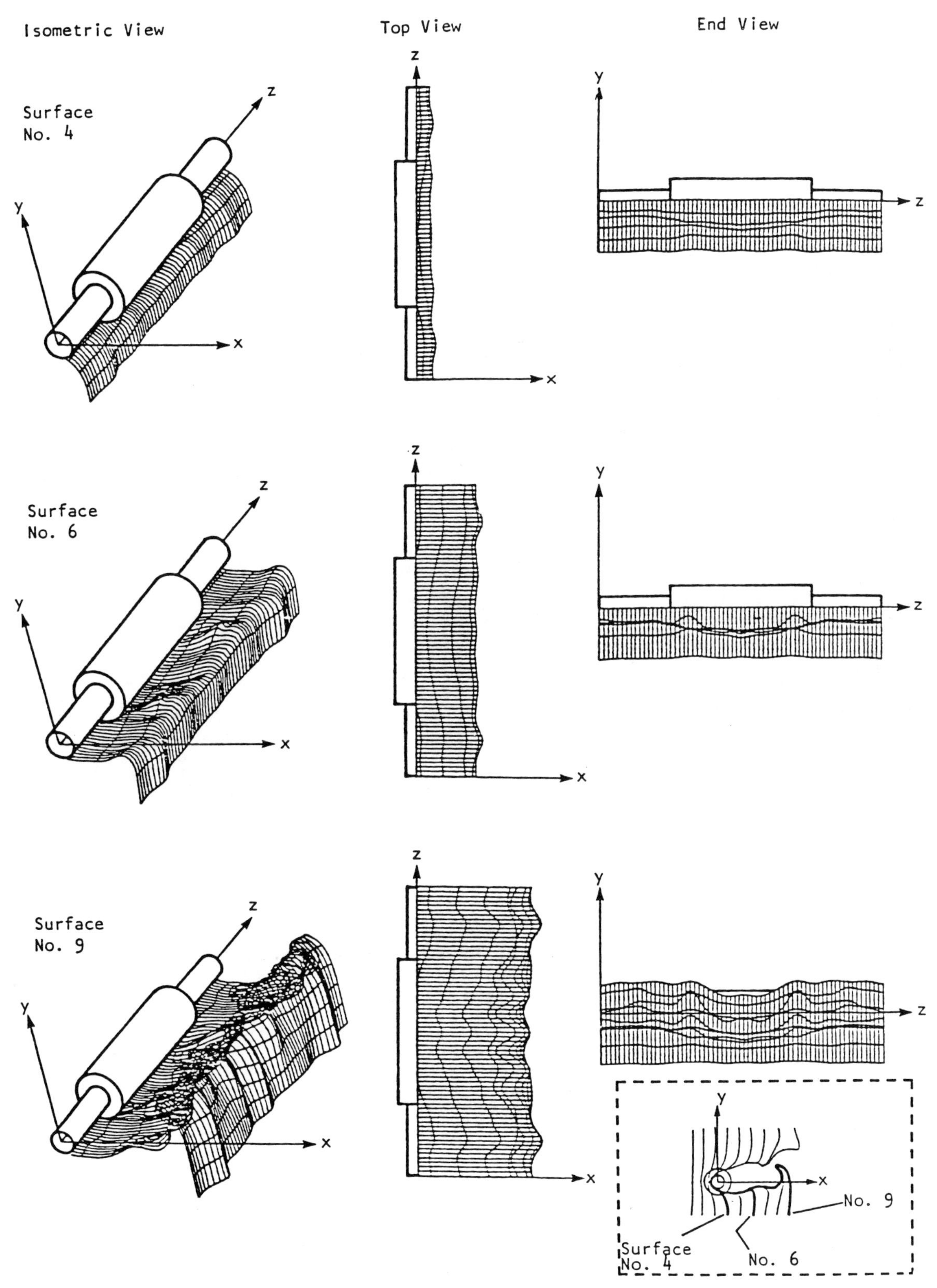

Fig. 7 Isometric, top, and end views of phase-averaged reconstruction of three-dimensional hydrogen bubble timeline surfaces for vortex shedding behind a circular cylinder. Surfaces are obtained by connecting the timelines for a series of 17 phase-referenced image planes similar to the single image plane shown in the lower right-hand corner (courtesy of D. O. Rockwell [17]).

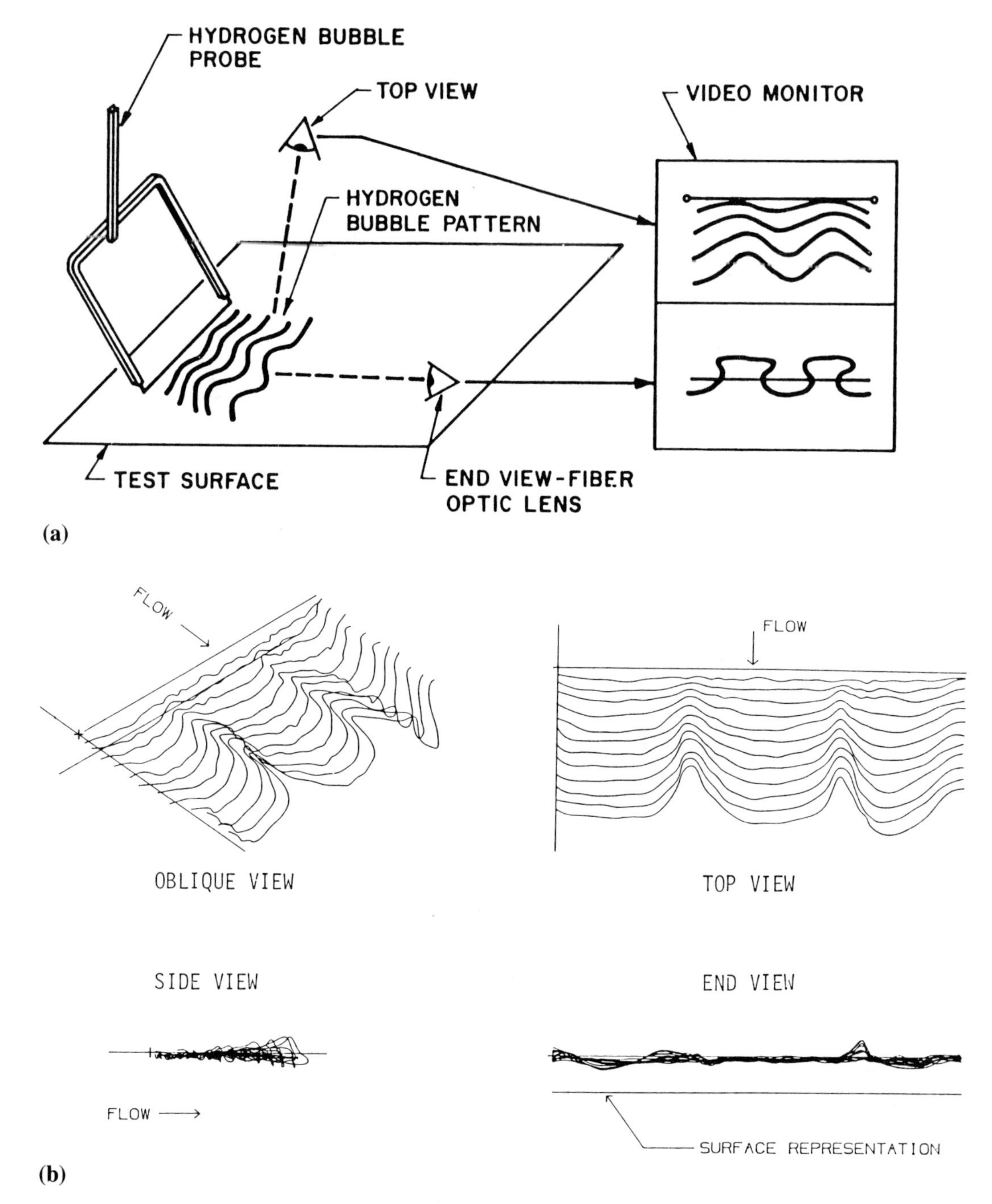

Fig. 8 Dual-view reconstruction of 3-D temporal behavior of a single hydrogen bubble timeline near the wall of a turbulent boundary layer: (*a*) Method of dual-view observation; (*b*) reconstructed 3-D image from four different perspectives [18].

integrated history of the flow behavior, not the result of the local flow conditions. Because of the assumptions of a Galilean transformation for the fixed-light-sheet technique [21] and the occurrence of some relative motion of the visual medium during the rapid-light-scan technique [23], some computer corrections to the image segments must be done. In addition, depending on the clarity and quality of the visualization techniques employed, some enhancement (e.g., noise filtering, histogram equalization, and thresholding were employed by Garcia and Hesselink [21]) of the images must be done prior to computationally stacking the images to reconstruct the final three-dimensional image. For details of the processing and presentation procedure involved, the reader is referred to Refs. [21–23].

IV IMAGE SYNTHESIS

One of the largest areas in which computer-aided techniques have had an impact has been in the synthesis of flow visualization and flow visualization–type images from other than direct, empirical visualization methods. These computer-aided methods fall into two broad categories:

1. creation of "images" of flow properties other than fluid displacement and density variations (the conventional properties that are commonly visualized), and
2. analytical simulations of conventional flow visualization techniques.

The following section reviews some of the typical approaches and applications that have been employed in these two areas.

A Experimental Synthesis

Modern fluid dynamic experiments have become very expansive, often generating substantial quantitative data bases that cover significant regions of a flow field and often including temporal as well as spatial dependency. These data bases may represent spatially time-averaged data [7, 28–30], phase-referenced averages [17, 26, 27], multiple-probe measurements [25], or quantitative evaluation of flow visualization images [5, 6, 10, 11, 16, 24, 31–33], but in all cases the data bases are large and require display techniques that facilitate rapid understanding as well as effective cross comparison with other data bases.

One common mode of presentation is to create a 2- or 3-D "image" of a selected flow-field property by using the measured spatial–temporal distribution of the property magnitude to create intensity arrays, which are then displayed using graphics techniques. In two dimensions, these arrays may be visualized using either gray-level shading or pseudocolor displays of selectively thresholded images. Figure 9 (see colorplate between pages 388 and 389) illustrates several basic display techniques. Figure 9*a* is a series of temporal velocity profiles obtained in the wall region of a turbulent boundary layer using quantitative evaluation of a sequence of hydrogen bubble visualization images (Fig. 1*b* is a representative image from this sequence). The comparable display of the same information using gray-level shading is shown in Figure 9*b*, with velocity magnitude represented by intensity level (white = high, black = low). Note how the shaded image in Fig. 9*b* allows one to easily recognize velocity patterns in either time (abscissa) or space (ordinate). The same information in Fig. 9*b* is redisplayed in Fig. 9*c* using pseudocoloring (5 colors) of the intensity array. Variations in the velocity are again easily observable.

The limitation of this latter approach is the establishment of a coloring sequence that makes logical sense to the untrained observer. It is often not obvious what magnitude (or range of magnitudes) a particular color or color hue represents in a pseudocolored image. There are many schemes, such as spectral ordering of the colors or dark-to-light ordering, but there is no standard. In general, it is left to the discretion and taste of the image creator to choose the sequencing that makes the most sense for the particular image displayed. When quantitative interpretation is of particular concern, it is important that a color bar be displayed in conjunction with the image. An example of an accompanying color bar is shown in Fig. 10 (see colorplate between pages 388 and 389), which illustrates the use of pseudocoloring for display of the time-averaged concentration of smoke (obtained by image averaging) in a round jet [31]. For a more detailed discussion of color perception and the choice of pseudocoloring schemes, the reader is referred to Refs. [1–4, 35].

When dealing with the display of quantitative intensity arrays, particular difficulty is encountered when the data may take on both positive and negative values. This situation is often encountered when dealing with the display of property variations about a mean value. Gray-level shading becomes quite awkward, since some median gray level must be taken as the neutral (mean) value of the property, with shadings higher and lower implying positive or negative excursions from that central value. However, the visual interpretation of such images is not readily obvious. A much more effective technique is to incorporate the advantages of both color and shading to provide clear discrimination between positive and negative magnitudes. Figure 9*d* employs two-color shading to represent the velocity-fluctuation data for the velocity-field data shown in Figs. 9*b* and 9*c*. In this image, red implies values in excess of the local time mean and green represents values in deficit. The shading (i.e., intensity) of the color indicates the magnitude of the excursion, with the mean values appearing as black. From Fig. 9*d* the time–space patterns for the velocity behavior are immediately obvious, revealing a generally repetitive temporal pattern of substantial extent in the spatial direction.

B Analytical Synthesis

Theoretical and computational fluid dynamic analyses frequently give rise to voluminous amounts of information that require creative methods of presentation in order to be effectively assimilated. Through the use of computer-display and -manipulation techniques, these large analytical data bases can be effectively displayed using techniques that are essentially identical to computer-aided flow visualization approaches employed with empirically generated data [8, 34]. Because modern computational data bases can now be generated in three dimensions [34–37], it is necessary to be able to effectively present space–time data in a three-dimensional format. A particularly attractive method of

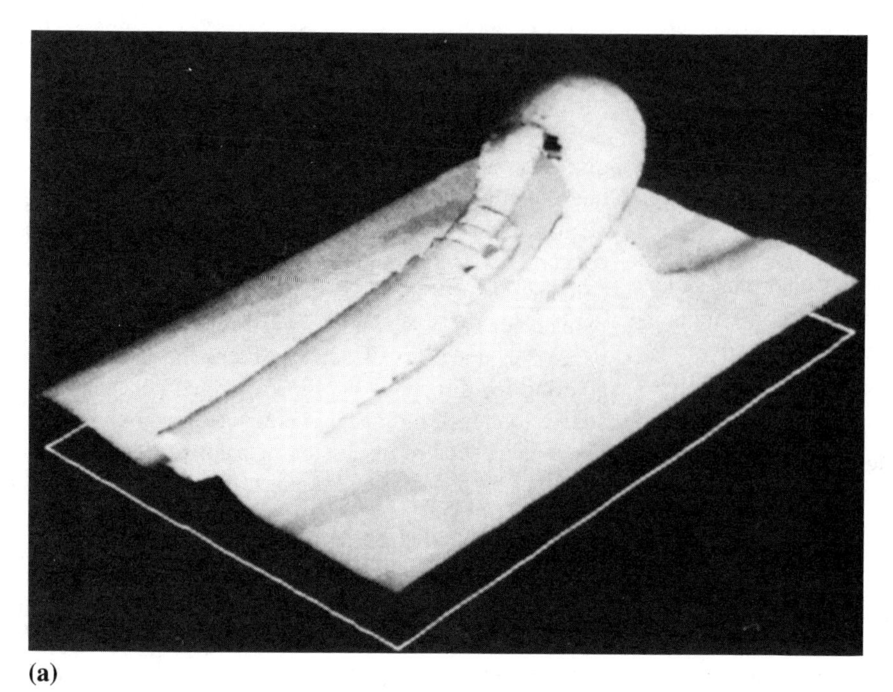

Fig. 11 Perspective views of surfaces and lines of constant enstrophy (one-half of the squared vorticity) for a vortical flow structure simulation in three dimensions: (*a*) Gray-level display of a surface of constant enstrophy; (*b*) (see colorplate between pages 388 and 389) a three-color display of lines of constant enstrophy; Relative values are high = red, intermediate = yellow, blue = low (courtesy of J. Kim [36]).

three-dimensional display in the use of graphical surface-generation techniques to create complex isoproperty surfaces. Perspective display of these constant property surfaces, such as the one shown in Fig. 11*a* for a vortical flow structure simulation in three-dimensions [36], allows rapid spatial evaluations of behavior to be done. By varying the value of the isoproperty surface, the three-dimensional distribution of a property over a discrete region can be rapidly understood.

An alternative to presentation of a series of isoproperty images is the use of color to highlight a series of constant-property surfaces [34–36]. In some cases, contour lines rather than surfaces are employed, so that different property levels and concentrations can be clearly discriminated. An example of a three-dimensional, color contour–line display is shown in Fig. 11*b* (see colorplate between pages 388 and 389) for the same data base as represented in Fig. 11*a*. Note that although the general shape of the flow property illustrated is similar, the interpretation of the different intensity levels is possible in Fig. 11*b*, whereas Fig. 11*a* illustrates only one intensity level. Clearly, the availability of a dynamic display system that allows interactive image rotation greatly facilitates the visual evaluation of images such as Fig. 11.

C Analytical Simulation

The availability of large flow-field data bases makes the analytical simulation of empirical-type flow visualization techniques possible, thus providing another mechanism for direct comparison of analytical/computational and empirical results [37–39]. To perform an analytical simulation, conventional techniques of empirical flow visualization are computationally imitated through the computational re-creation of streaklines, pathlines, and timelines using a computed flow-field data base. The general approach is to computationally track a series of particles introduced within the computed flow field. Depending on the number of particles being tracked, the computational time involved can be quite substantial. Figure 12 shows an example of a simulated visualization of a turbulent channel flow using hydro-

Fig. 12 Simulated hydrogen bubble timeline visualization for a three-dimensional Navier–Stokes computation of a turbulent channel flow (flow left to right) (courtesy of J. Kim).

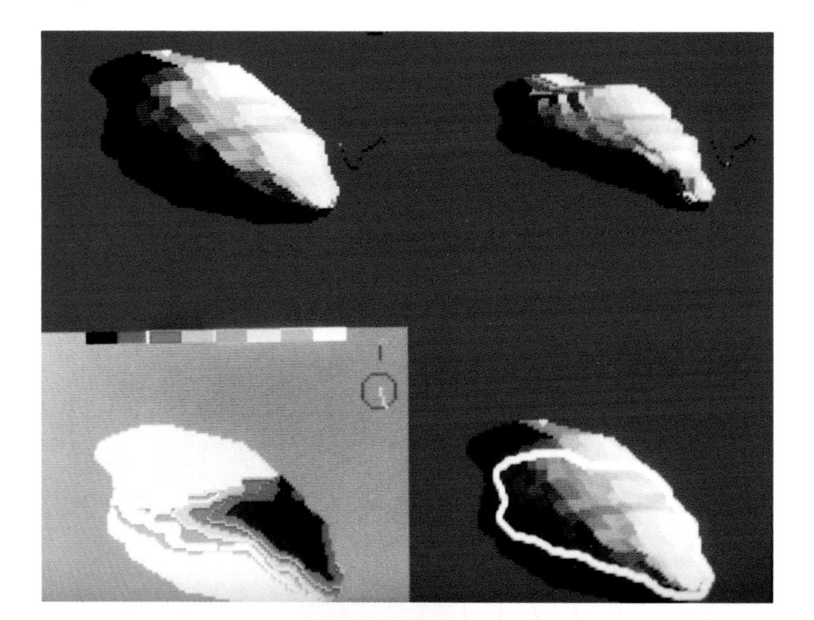

Fig. 28 Three-dimensional left ventricles and percentage shortening image obtained from biplane cineangiocardiograms.

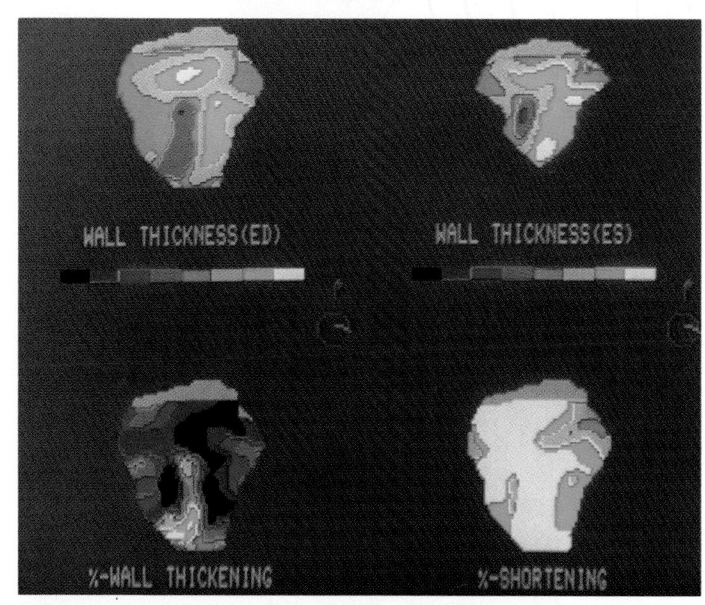

Fig. 29 Functional images of three-dimensional left ventricle wall thickness, percentage wall thickening, and percentage shortening obtained from echocardiograms.

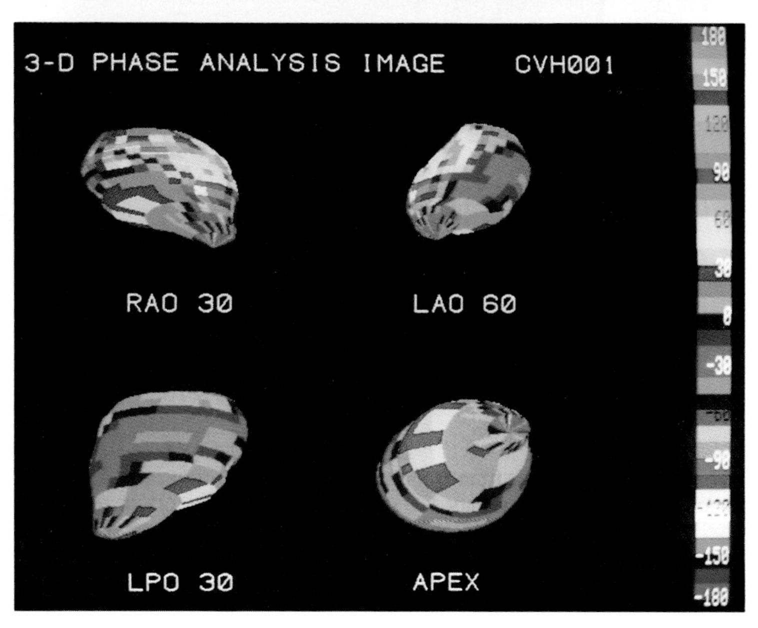

Fig. 30 Color-mapping display of propagation of the contraction and relaxation of the left ventricle.

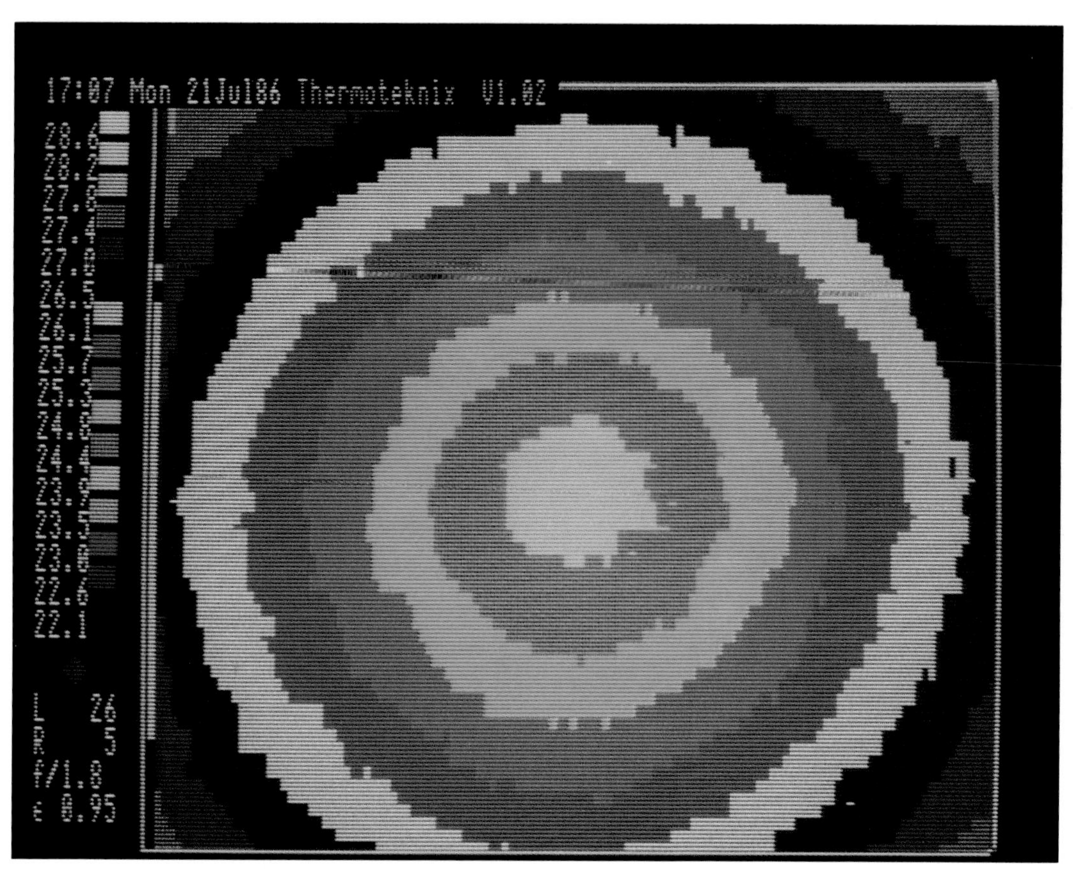

Fig. 17 Color thermogram of the temperature map of a heated plate cooled by an air jet.

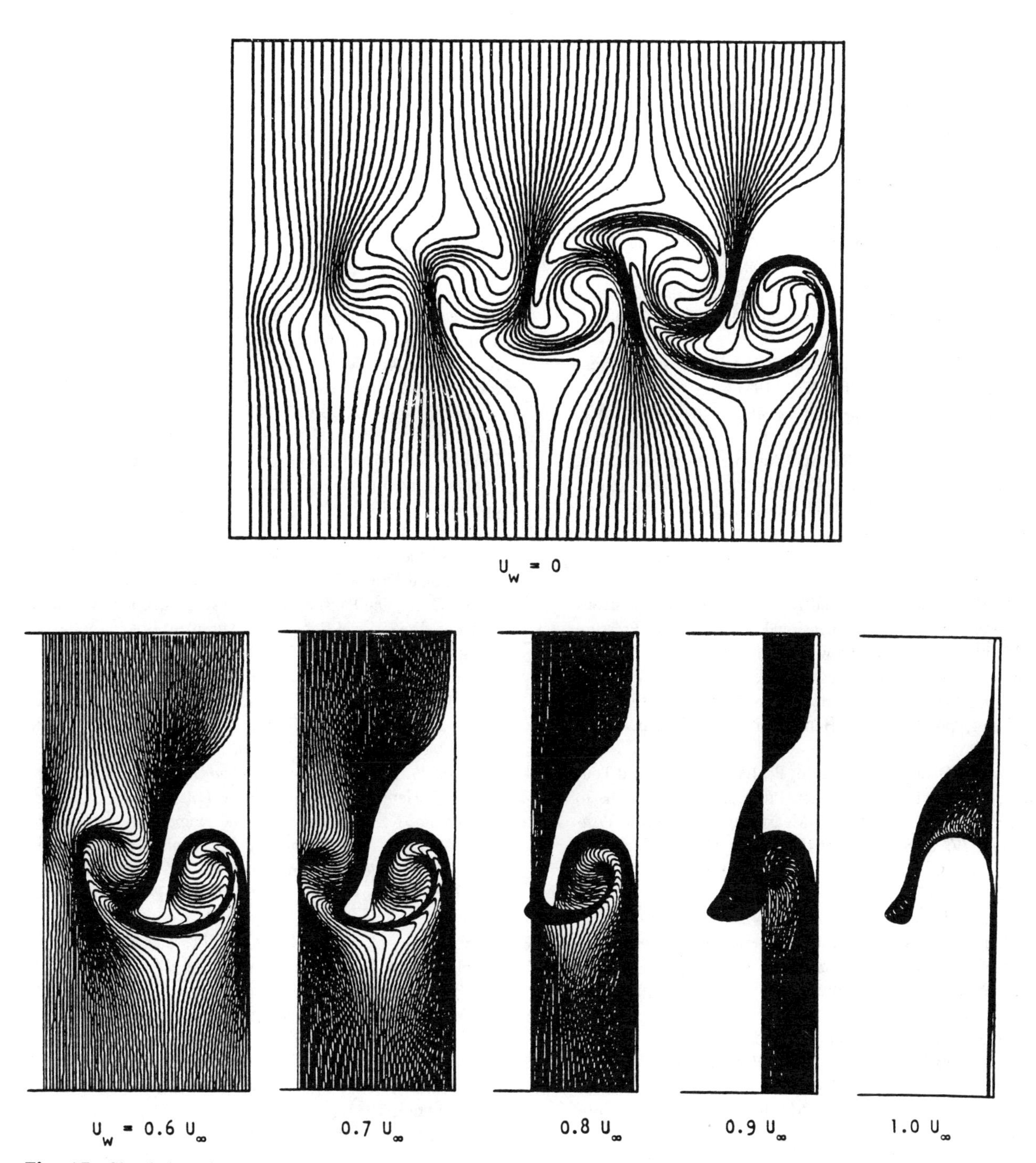

Fig. 13 Simulation of hydrogen bubble timeline visualization using concepts of inviscid stability theory. Top image shows simulated timelines initiated from a stationary bubble wire located at left of the image for a neutral, spatially periodic vorticity field translating left to right. Bottom images show the effect of translating the wire relative to the free stream (courtesy of D. Rockwell et al. [38]).

gen bubble timelines. The utility of such simulations is that direct empirical–computational comparisons are possible, greatly facilitating the proper interpretation of the relationship of the flow visualization images with the flow-field dynamics (compare Fig. 12 to a similar empirical image in Fig. 1*b*). Figure 13 is another example of a simulated flow visualization. This simulation is based on the idealized flow field created by a series of translating, inviscid vortices [38, 39]. As is clearly shown in Fig. 13, the resultant simulated hydrogen bubble flow visualization image is highly dependent on the integrated effect of the flow on the visualization medium. This simulation indicates that the apparent scales in the flow visualization image cannot be directly interpreted, since the scales that appear in the simulation clearly do not represent the actual flow process.

REFERENCES

1. Ballard, D. H., and Brown, C. M., *Computer Vision*, Prentice-Hall, Englewood Cliffs, N.J., 1982.
2. Foley, W. M., and Van Dam, A., *Fundamentals of Interactive Computer Graphics*, Addison-Wesley, Reading, Mass., 1982.
3. Andrews, H. C., and Hunt, B. R., *Digital Restoration*, Prentice-Hall, Englewood Cliffs, N.J., 1977.
4. Pratt, W. K., *Digital Image Processing*, Wiley-Interscience, New York, 1978.
5. Lu, L. J., and Smith, C. R., Application of Image Processing of Hydrogen Bubble Flow Visualization for Evaluation of Turbulence Characteristics and Flow Structure, in *Flow Visualization IV: Proceedings of the Fourth International Symposium*, ed. C. Véret, pp. 247–252, Hemisphere, Washington, D.C., 1987.
6. Hiller, B., Cohen, L. M., and Hanson, R. K., Simultaneous Measurements of Two-dimensional Velocity and Pressure Fields in Compressible Flow through Image-intensified Detection of Laser-induced Fluorescence, in *Flow Visualization IV: Proceedings of the Fourth International Symposium*, ed. C. Véret, pp. 173–178, Hemisphere, Washington, D.C., 1987.
7. Biscos, Y., Bismes, F., Hebrard, P., Lavergne, G., and Toulouse, G., Digital Video Image Processing Applications to Drop Size and Concentration Measurements, in *Flow Visualization IV: Proceedings of the Fourth International Symposium*, ed. C. Véret, pp. 253–260, Hemisphere, Washington, D.C., 1987.
8. Dewey, C. F., and Patera, A. T., Visualizing the Flow of Fluid and Heat as Predicted by Numerical Computations, in *Flow Visualization IV: Proceedings of the Fourth International Symposium*, ed. C. Véret, pp. 269–274, Hemisphere, Washington, D.C., 1987.
9. Carlomagna, G. M., and de Luca, L., Heat Transfer Measurements by Means of Infrared Thermography, in *Flow Visualization IV: Proceedings of the Fourth International Symposium*, ed. C. Véret, pp. 611–616, Hemisphere, Washington, D.C., 1987.
10. Brandt, A., Hydrodynamic Flowfield Imaging, in *Flow Visualization III: Proceedings of the Third International Symposium on Flow Visualization*, ed. W.-J. Yang, pp. 279–284, Hemisphere, Washington, D.C., 1985.
11. Berger, C., Bourgeois, M., Lavergne, G., Lempereur, C., and Mathe, J. M., Shear Flow Patterns Analysed by Video Systems, in *Flow Visualization IV: Proceedings of the Fourth International Symposium*, ed. C. Véret, pp. 281–287, Hemisphere, Washington, D.C., 1987.
12. Keffer, J. F., Shokr, M., and Kawall, J. G., A Digital Image Processing/Clustering Technique for Identifying Organized Wake Structures, in *Flow Visualization IV: Proceedings of the Fourth International Symposium*, ed. C. Véret, pp. 217–220, Hemisphere, Washington, D.C., 1987.
13. Hernan, M. A., Parikh, P., and Sarohia, V., Digital Image Processing Application to Spray and Flammability Studies, in *Flow Visualization III: Proceedings of the Third International Symposium on Flow Visualization*, ed. W.-J. Yang, pp. 269–273, Hemisphere, Washington, D.C., 1985.
14. Gennero, C., and Mathe, J. M., Real-Time Edge Extraction Application to the Study of Vortex Cores, in *Flow Visualization III: Proceedings of the Third International Symposium on Flow Visualization*, ed. W.-J. Yang, pp. 285–289, Hemisphere, Washington, D.C., 1985.
15. Wallace, J. M., Balint, J. L., Ladhari, F., and Morel, R., Applications of Image Processing Analysis to the Study of Turbulent Boundary-layer Structure, in *Flow Visualization III: Proceedings of the Third International Symposium on Flow Visualization*, ed. W.-J. Yang, pp. 249–253, Hemisphere, Washington, D.C., 1985.
16. Lu, L. J., and Smith, C. R., Image Processing of Hydrogen Bubble Flow Visualization for Determination of Turbulence Statistics and Bursting Characteristics, *Exp. Fluids*, vol. 3, pp. 349–356, 1985.
17. Ongoren, A., Chen, J., and Rockwell, D., Multiple Time–Surface Characterization of Time-dependent Three-dimensional Flows, *Exp. Fluids*, vol. 5, 1987. In press.
18. Smith, C. R., and Paxson, R. D., A Technique for Evaluation of Three-dimensional Behavior in Turbulent Boundary Layers using Computer Augmented Hydrogen Bubble-Wire Flow Visualization, *Exp. Fluids*, vol. 1, pp. 43–49, 1983.
19. Doi, J., and Miyake, T., Three-dimensional Analysis of Time-varying Tuft Behavior by Its Successive Geometric Shape Modeling, in *Flow Visualization IV: Proceedings of the Fourth International Symposium*, ed. C. Véret, pp. 229–234, Hemisphere, Washington, D.C., 1987.
20. Yamamoto, K., Kawamata, S., and Nomoto, A., Visualization of Surface Wave by Inclined Grid Moiré Topography, in *Flow Visualization IV: Proceedings of the Fourth International Symposium*, ed. C. Véret, pp. 791–796, Hemisphere, Washington, D.C., 1987.
21. Agüi Garcia, J. D., and Hesselink, L., 3-D Reconstruction of Fluid Flow Visualization Images, in *Flow Visualization IV: Proceedings of the Fourth International Symposium*, ed. C. Véret, pp. 235–240, Hemisphere, Washington, D.C., 1987.
22. Jimenez, J., Cogollos, M., and Bernal, L. P., A Perspective View of the Plane Mixing Layer, *J. Fluid Mech.*, vol. 152, pp. 125–143, 1985.
23. Lynch, M. K., Miller, P., Lewis, C., and Nosenchuck, D. M., Visualization of Active Turbulent Boundary Layer Control using a Scanning Laser Sheet, 38th Ann. Mtng. Am. Phys. Soc., Div. of Fluid Dynam., Tucson, 1985.
24. Kobayashi, T., Saga, T., and Segawa, S., Some Considerations on Automated Image Processing of Pathline Photographs, in *Flow Visualization IV: Proceedings of the Fourth International Symposium*, ed. C. Véret, pp. 241–246, Hemisphere, Washington, D.C., 1987.
25. Corke, T. C., Way, J. L., and Nagib, H. M., Wave Number Analysis of Visualized Flowfields, in *Flow Visualization III: Proceedings of the Third International Symposium on Flow Visualization*, ed. W.-J. Yang, pp. 221–230, Hemisphere, Washington, D.C., 1985.
26. Izutzu, N., Ishii, Y., Oshima, K., and Oshima, Y., Time-

like Visualization of Vortical Flowfield, in *Flow Visualization IV: Proceedings of the Fourth International Symposium,* ed. C. Véret, pp. 425–430, Hemisphere, Washington, D.C., 1987.

27. Baird, J. P., Gai, S. L., and Reynolds, N. T., Flow behind a Rearward Facing Step in Hypersonic High-Enthalpy Flow—Some Interferometric Studies, in *Flow Visualization IV: Proceedings of the Fourth International Symposium,* ed. C. Véret, pp. 549–553, Hemisphere, Washington, D.C., 1987.
28. Winkelmann, A. E., and Tsao, C. P., Flow Visualization using Computer-generated Color Video Displays of Flow Field Survey Data, in *Flow Visualization III: Proceedings of the Third International Symposium on Flow Visualization,* ed. W.-J. Yang, pp. 175–179, Hemisphere, Washington, D.C., 1985.
29. Sakata, K, Shindoh, S., and Yanagi, R., Experimental Flow Analysis with Computer Graphics on Film Cooling Flowfield, in *Flow Visualization III: Proceedings of the Third International Symposium on Flow Visualization,* ed. W.-J. Yang, pp. 203–207, Hemisphere, Washington, D.C., 1985.
30. Crowder, J. P., and Beck, H. M., Electronic Wake Imaging Systems, in *Flow Visualization III: Proceedings of the Third International Symposium on Flow Visualization,* ed. W.-J. Yang, pp. 208–212, Hemisphere, Washington, D.C., 1985.
31. Balint, J.-L., Ayrualt, M.,and Schon, J. P., Qualitative Investigation of the Velocity and Concentration Fields of Turbulent Flows Combining Visualization and Image Processing, in *Flow Visualization III: Proceedings of the Third International Symposium on Flow Visualization,* ed. W.-J. Yang, pp. 254–258, Hemisphere, Washington, D.C., 1985.
32. Hesselink, L., Penders, J., Jaffey, S. M., and Dutta, K., Quantitative Three-dimensional Flow Visualization, in *Flow Visualization III: Proceedings of the Third International Symposium on Flow Visualization,* ed. W.-J. Yang, pp. 295–300, Hemisphere, Washington, D.C., 1985.
33. Iwasaki, T., and Tanaka, H., On the Flow Visualization and Turbulent Measurement on the Ripple Model, in *Flow Visualization III: Proceedings of the Third International Symposium on Flow Visualization,* ed. W.-J. Yang, pp. 681–685, Hemisphere, Washington, D.C., 1985.
34. Hasan, G., and Graham, J., Computer Generated Graphics Display of Navier–Stokes Flow Solutions, in *Flow Visualization III: Proceedings of the Third International Symposium on Flow Visualization,* ed. W.-J. Yang, pp. 180–184, Hemisphere, Washington, D.C., 1985.
35. Smith, R., Speray, D., and Everton, E., Visualization of Computer-generated Flow Fields, in *Flow Visualization III: Proceedings of the Third International Symposium on Flow Visualization,* ed. W.-J. Yang, pp. 190–197, Hemisphere, Washington, D.C., 1985.
36. Kim, J., Evolution of a Vortical Structure Associated with the Bursting Event in a Channel Flow, in *Turbulent Shear Flows,* vol. 5, pp. 221–233, Springer-Verlag, Berlin, 1987.
37. Rogollo, R. S., and Moin, P., Numerical Simulation of Turbulent Flows, *Ann. Rev. Fluid Mech.,* vol. 16, pp. 99–137, 1984.
38. Rockwell, D., Atta, R., Kramer, L., Lawson, R., Lusseyran, D., Magness, C., Sohn, D., and Staubli, T., Flow Visualization and Its Interpretation, AGARD Sym. Aerodynam. Related Hydrodynam. Stud. Water Facil., preprint AGARD-CPP-413, 1986.
39. Lusseyran, D., Gursul, I., and Rockwell, D., On Interpretation of Flow Visualization in Unsteady Shear Flows. (In preparation.)

Part 4
Applications

Chapter 25

Aerospace and Wind Tunnel Testing

G. S. Settles

I INTRODUCTION

Most flow visualization is done in wind tunnels. The focus of this chapter is primarily on flow visualization undertaken during routine testing in wind tunnels and related aerospace facilities. Research applications and facilities dedicated to flow visualization are treated elsewhere in this handbook.

The testing of aircraft configurations, missiles, and engines comprise the most common activities in routine aerospace testing. The facilities are often large and expensive to operate, so the time and resources available for flow visualization are limited and in competition with other testing activities. A survey of flow visualization techniques for such "practical" wind tunnel testing was carried out by the present author in 1981 [1, 2], with the assistance of the Boeing Aerodynamic Laboratory. Much of the content of this chapter is based on that study.

Most of the techniques of flow visualization described earlier are applicable to aerospace testing, including many examples of the traditional surface, tracer, and optical techniques. Those that are most often used satisfy certain criteria of simplicity, ruggedness, and adaptability which make them useful outside a controlled laboratory environment. Further, restricted visual access and safety regulations in some aerospace facilities limit the range of applicable visualization methods. As a consequence of these constraints, an arsenal of "standard" flow visualization methods was developed and has been available for many years. The application of these methods and the interpretation of the results are part of the learning process for aerospace testing personnel. Newer or more radical flow visualization techniques sometimes require extended periods of occasional use before being accepted in the standard repertoire.

The range of flow visualization applications in aerospace testing is very diverse. While traditional wind tunnel testing applications are prevalent, there are many others as well. These include flight testing as well as testing in engine test cells and stands, ballistic ranges and tracks, and acoustic and environmental chambers. Flow regimes cover the full range from incompressible to hypervelocity, and may include cryogenic, ambient, and high-enthalpy thermal environments.

Flow visualization is typically applied to assess aerodynamic performance, to identify and locate certain flow phenomena, and to troubleshoot flow problems. Boundary-layer separation is a prime candidate for both identification and correction. The wide range of other visualized phenomena in aerospace testing includes low-speed wing stall, shock stall and buffeting, vortex shedding and bursting, excrescence flows, control—surface interactions, wakes, heat transfer, store separation, aerodynamic interference, exhaust plume shape and shock structure, boundary-layer transition, bow shocks, and many others.

The goals of this chapter are to describe at least the primary applications of flow visualization in aerospace testing, to provide a brief practical background for the use and interpretation of the various techniques in this context, and to guide the reader to sources of more detailed information.

II WIND TUNNEL TESTING

General information on wind tunnel testing and the applications of flow visualization therein is widely available, e.g. [3–8]. Both the applications and methods employed can be grouped according to flow regimes as follows.

A Low-Speed Applications

Below a Mach number of about 0.3, wind tunnel flows are essentially incompressible. Thus the surface flow visualization and tracer techniques are primarily useful, while optical methods are used only on occasions when significant temperature or gas-concentration differences occur.

Surface flow visualization by way of pigmented or fluorescent oil flow is routinely carried out in wind tunnel tests of aerospace vehicles and components. Many examples can be found in the literature. The resulting patterns are photographed and examined for evidence of phenomena of interest, such as flow separation, boundary layer transition, etc.

With experience, test engineers can obtain limited information about the overlying flowfield from the "skeleton" of skin-friction lines in a surface flow pattern. Some local flow patterns are easily identified, e.g., the "horseshoe" vortex that accompanies boundary layer separation at a wing–body junction (see Fig. 1). However, the surface flow over an entire aircraft model often becomes very complex and difficult to interpret. A further difficulty arises when unsteady flows are present; these are not clearly revealed by oil-flow patterns, which have a very low frequency response.

No simple rules of thumb exist for the interpretation of complex surface flow patterns on wind tunnel models. However, several recent investigators [10–13] have applied the principles of topology as an interpretational aid. Briefly, topology governs the number and type of singular points that, connected by lines of flow separation or attachment, form the framework of a surface flow pattern. Typical examples of the detailed interpretation of such patterns in V/STOL [14], fighter aircraft [15], missile [16], and Space Shuttle [17] wind tunnel tests can be found in the literature.

Recipes for oil-flow mixtures for wind tunnel testing vary widely. Some basic principles for these mixtures are given by Maltby [6]. A popular recipe [18] in the United States involves a mixture of fluorescent dye and lubricating oil. Lighter or heavier petroleum distillates are added, usually by trial and error, to achieve the proper viscosity for a given application. This mixture is sprayed or painted on the leading edges of wind tunnel models prior to a test. Using ultraviolet illumination, the results are photographed during the test.

There are, of course, many variations on this theme. For example, in complex flows the use of various oil-paint colors, thinned to the proper consistency and spotted on the model, helps one follow surface streamlines. Dry pigment spots can be applied in a matrix on the model surface, then washed with a solvent during the test, forming streaks [19]. Where repeated patterns were required during a single test, Crowder [20] devised a porous oil dispenser for the leading edge of a model wing.

Smoke tracers are routinely used in low-speed wind tunnel testing [21–23]. Depending in part on the wind tunnel configuration, an upstream smoke rake, a smoke "wand," or the smoke-wire technique can be employed; $TiCl_4$, painted on models just prior to testing, often yields beautiful results though its fumes are hazardous. A difficulty lies in the absence in many ordinary wind tunnels of the special design features (such as installed smoke rakes, large nozzle contraction ratios, and special flow conditioning) that contribute to effective smoke visualization in smoke tunnels.

Tufts of thread or yarn, taped or glued to wind tunnel models, are traditional in low-speed testing. Concern over possible flow interference has led to the development of fluorescent minitufts [24, 25], which are minute strands of nylon monofilament treated with fluorescent dye and observed under ultraviolet illumination. Off-the-surface tuft installations (on pylons [26] or grids [27]) have also been used, but have largely been superseded by such developments in flowfield visualization as the light sheet.

Light sheet visualization [28, 29] has become very popular in wind tunnel testing, especially for the study of wakes and vortex shedding [30]. The flow is seeded, either locally or in its entirety, with an aerosol, which may be smoke [21], water fog [21, 30], dust, neutrally bouyant soap bubbles [31], etc. Mie scattering of a thin planar light sheet by the aerosol renders visible a "slice" of the flow field. Vortex locations about aircraft and missile models at angles of attack to the flow are readily observed by this method.

Visible lasers in the power range of ≥ 1 W are very useful for such work. A suitable light sheet is produced by passing the laser beam through a cylindrical lens or a simple glass rod. A portable version of this equipment has recently been reported [32]. Further, recent methods of digital image analysis are available [33] to extract quantitative data from light-sheet images, as well as to correct distortions caused by viewing the light sheet obliquely through existing wind tunnel windows.

Unfortunately, the application of light sheet visualization in routine wind tunnel testing can cause problems. Visual access is usually limited, safety regulations may prevent the use of powerful lasers, and the seeding material may not be easy to introduce, and may lead to tunnel contamination.

Other methods related to light sheet visualization include the computer-graphics presentation of probe data in the form of flow-field slices [34], and Crowder's "wake-imaging system" (WIS) [35–37]. In the WIS, a flowfield slice is built up on a time-exposed photograph through successive

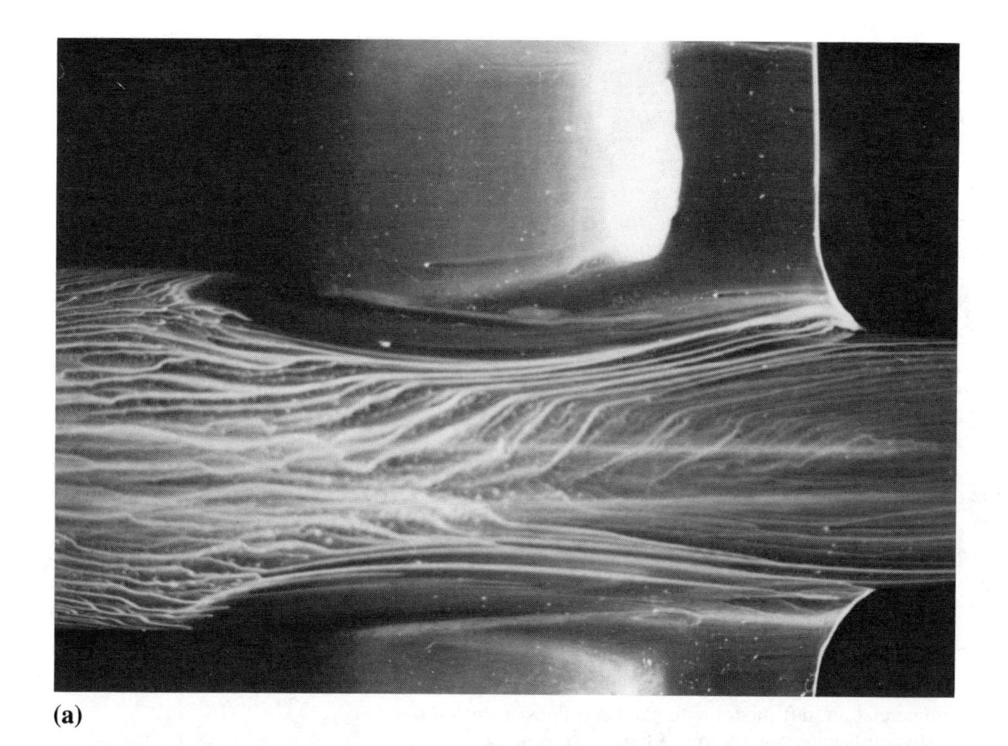

(a)

(b)

Fig. 1 Surface oil-flow photos showing wing–body junction of wind tunnel sailplane model (flow from left to right): (*a*) Natural transition; (*b*) artificial transition (from M. D. Maughmer [9]).

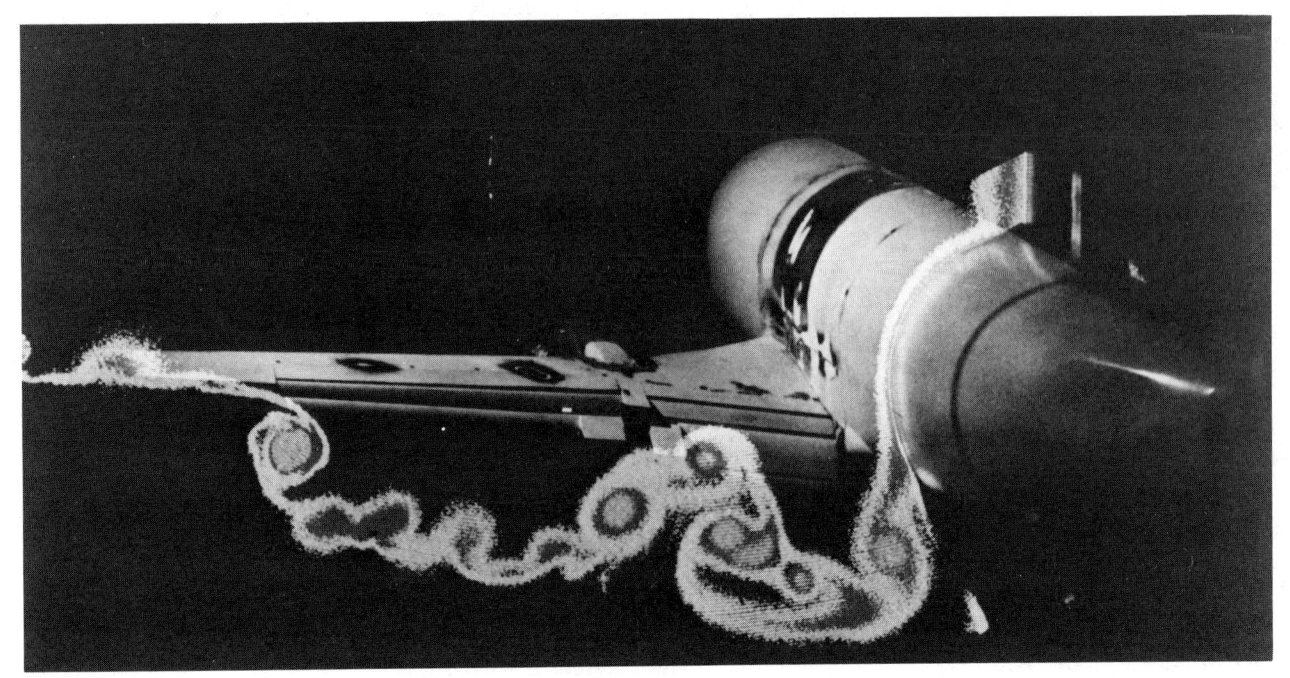

Fig. 2 Wake Pitot pressure map of commercial aircraft model with extended flaps, revealed by the WIS technique (from J. P. Crowder, Boeing Commercial Airplane Co.).

passes of a probe through the flowfield. The flow quantity measured by the probe modulates a signal light mounted on the probe, which is recorded on the film. Results, such as that shown in Fig. 2, are quite striking. They are somewhat similar to light sheet images, but do not require flow seeding. Proper visual access to the test section and extended wind tunnel runs with low-light conditions are required, however.

The WIS technique has been used primarily to image pressure distributions in the wakes of wind tunnel models. Other sensors, such as a yaw head or hot wire, can also be used. As with light sheet visualization, WIS images help one to decipher complex 3-D flows in wind tunnel testing by revealing such flows as a series of 2-D slices. A full-color example of the combined use of WIS and minitufts in low-speed testing of the Japanese Space Shuttle is given in Ref. [38].

Transition visualization on wind tunnel models is especially important in checking Reynolds number scaling and interpreting drag data. Time-honored indicators include "china clay," fluorene, and azobenzine [5, 6], which reveal the different rates of mass exchange in laminar versus turbulent boundary layers. Liquid crystals [39] also showed early promise in indicating transition when accompanied by a significant temperature difference, which is not usually the case in low-speed testing. More recent liquid crystal research has apparently produced mixtures that are primarily shear sensitive, though results are available only from transonic flight tests [40, 41] to date.

B Transonic and Supersonic Applications

The onset of compressibility does not preclude the use of most of the flow visualization methods just described. However, it does eliminate some possibilities while enabling others. The smoke-wire and neutrally bouyant bubble techniques appear strictly limited to low speeds, for example. On the other hand, optical instrumentation can now be used.

Optical instrumentation is available at most high-speed wind tunnel facilities in the form of a Z-type schlieren system. Such a system can produce either schlieren or shadowgram flow images [42–44] and can be modified for holographic interferometry [45, 46]. The optical characteristics of these systems are similar the world over, though details of mounting, alignment, and access vary. In the West, schlieren instruments are not generally available off the shelf, but are made to order by optical contractors [47]. The Soviet Bloc countries, on the other hand, often use a standardized instrument known as the IAB-451 [48].

Holder and North [42, 44] constitute the best practical references on the details of schlieren and shadowgraph alignment, sensitivity, and measuring range. Those interested in mastering these techniques should also consult Schardin's classic work [49].

Some routine problems encountered in the use of schlieren equipment with high-speed wind tunnels include poor alignment, overranging, and knife-edge jitter caused by tunnel vibrations. Careful alignment according to estab-

Fig. 3 Schlieren photo of transonic flow over early developmental Space Shuttle Orbiter model (from G. S. Settles, NASA Ames Research Center).

lished rules [44] is critical to the quality of a schlieren image. This should be done with a low ambient light level and with full access to the optical path. The most useful tool for schlieren alignment is a sheet of white paper, held in the light beam and used to trace its progress. The common error of poor focusing of the image can be corrected by placing a glass rod or candle flame in the center of the test section and adjusting the final lens and image plane for a sharp image.

Both overranging and jitter can be cured to some extent by the use of a graded filter in place of a sharp knife edge [44]. Vibration isolation of the schlieren components is also sometimes required. Contrary to popular opinion, lasers do not make good light sources for routine schlieren work, and should be avoided in favor of traditional white-light sources.

The use of color coding in schlieren flow visualization is straightforward and usually helpful. It can be accomplished in a variety of ways [50, 51]. An example is shown in Fig. 3.

Transonic testing with perforated wind tunnel walls blocks the use of ordinary schlieren or shadow equipment. In such cases, optical flow visualization is usually not done. However, the problem can be circumvented using special equipment, such as the multiple-source schlieren system [52].

Often the test model itself blocks the optical view of a phenomenon of interest. A common example is the transonic shock–boundary layer interaction on the wing of an aircraft model, where the fuselage gets in the way of optical observation. Three techniques that solve this problem are planview shadowgraphy [2], reflection of the schlieren beam from specularly polished model surfaces [53], and observation of the shock against a background grid applied directly to the offending fuselage [2]. With somewhat more expense, holographic interferometry will also accomplish this purpose.

Holographic interferometry [45, 46, 54–58], like its predecessor, Mach–Zehnder interferometry, is somewhat beyond the class of rough-and-ready wind tunnel flow visualization techniques. Nonetheless, it is extremely powerful for such purposes as quantitative density measurements, 3-D flow imaging, seeing through poor optical components, and visualizing the unsteadiness of the flow. The simplest arrangement replaces the light source of an existing schlieren system with a pulsed ruby or YAG laser, with the reference beam routed around the wind tunnel while the object beam passes through the test section. Some remarkable wind tunnel visualizations have been achieved, but holographic interferometry has yet to be widely accepted as a standard wind tunnel tool. A typical example is shown in Fig. 4.

Finally, a variety of other specialized optical techniques has seen occasional wind tunnel use. These include schlieren interferometry [44], moiré deflectometry [59], and conical shadowgraphy [60, 61]. Combined images of shadowgraph, schlieren, oil flow, and thermal paints have also been obtained [62, 63] as aids to flow interpretation. General discussions of optical and nonoptical flow visualization practice in two large wind tunnel labs are given in Refs. [64, 65]. Good examples of the application of optical methods in the study of base flows are shown in Ref. [66].

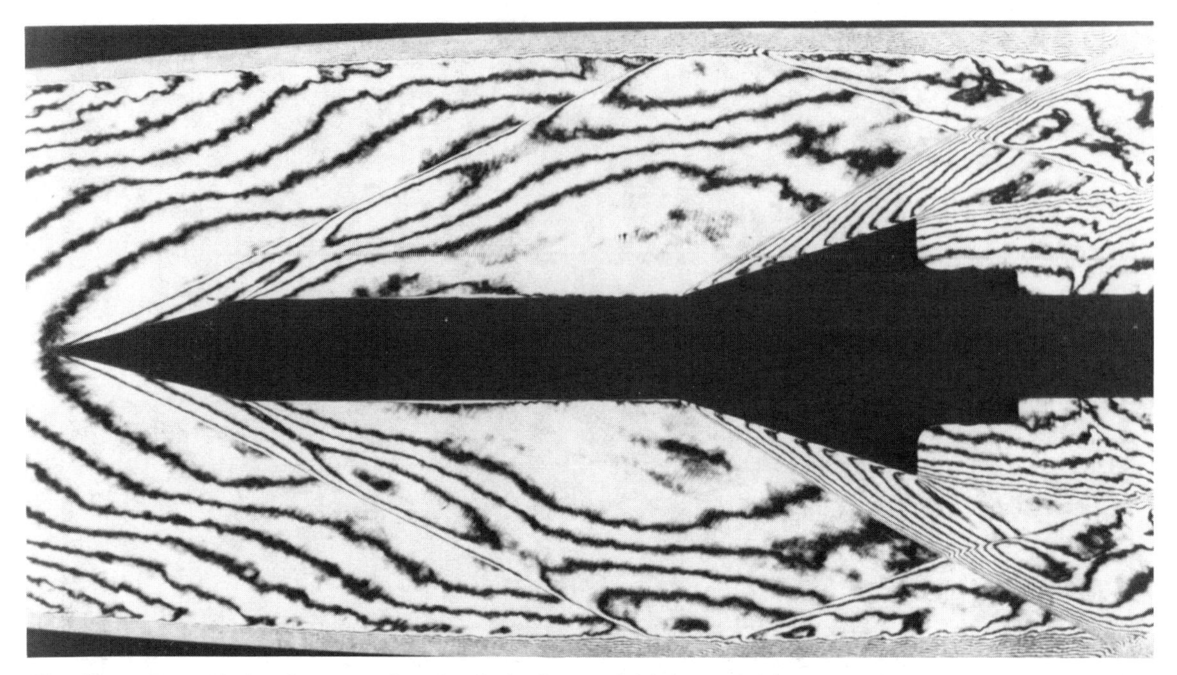

Fig. 4 Holographic interferogram of cone-cylinder-flare model being tested in supersonic flow (from A. G. Havener, U.S. Air Force Wright Aeronautical Laboratories).

Smoke tracers have been applied successfully up to transonic speeds in routine facilities [67], and at supersonic speeds in specially designed tunnels [68]. However, the primary application of smoke and other aerosol tracers in high-speed testing is in conjunction with light sheet observation.

Light sheet visualization was apparently first attempted in a high-speed flow [28], and is now an accepted and useful tool [43, 61, 62, 65, 69–73]. Seeding in high-speed tunnels depends on the tunnel type. Continuous closed-return tunnels can be seeded with moisture or smoke, which is allowed to circulate many times and disperse thoroughly. Uniform seeding is more problematic in intermittent facilities. However, by circumventing the air drier, enough moisture often can be obtained to do light sheet work.

The amount of moisture in the airstream is subject to certain limits. If insufficient, then the scattering by the light sheet will not be properly observable. Too much moisture, on the other hand, obscures the contrast of the light sheet and may also cause condensation shocks that alter the flow properties. Guidelines for moisture in the airstream are given in Refs. [28] and [70].

Many of the problems just discussed are characteristic of global seeding of the flow, and can be avoided by local seeding. For example, a small amount of liquid injected through a pressure tap is converted to an aerosol by the unsteadiness of turbulent boundary layer separation. Separation bubbles and free shear layers can be readily observed by light sheet in this way [61, 71].

As mentioned earlier, suitable light sheets are easy to produce via laser (either by spreading the beam through a cylindrical lens or sweeping it with a scanning mirror [69]). It is much more difficult to achieve the same light sheet quality with a noncoherent source. Nonetheless, safety concerns in wind tunnel testing still sometimes force the latter approach. Where lasers can be used, the smaller, cheaper helium–neon lasers are frequently not powerful enough. Argon–ion lasers of ≥ 1 W of power are very useful in this case. Further, a powerful pulsed laser can produce enough intensity in a nanosecond-range pulse to enable light sheet visualizations. Thus, the fluctuating motions of high-speed flows can be "frozen" [71].

Visual access for light sheet work may be more of a problem in high-speed than in subsonic tunnels. Windows on two adjacent sidewalls of the test section are helpful, but not usually available. When the light sheet is oriented perpendicular to the flow, the optimum viewing position is along the wind tunnel centerline. Since high-speed flow deals harshly with camera equipment, periscopes [70] and endoscopes [72] have been used to extract such views from the flow.

For further information, the reader is referred to McGregor's pioneering work [28] and the recent work of Snow and Morris [70].

Surface flow visualization in high-speed testing is in widespread use. Oil-flow techniques [74] do not vary significantly from those discussed earlier, except that the viscosity of the mixture is tailored to the particular testing conditions (usually by trial and error). A problem sometimes occurs in that tunnel window locations do not allow proper photography of oil-flow results during a test, and the pattern is destroyed by the shock-wave system upon tunnel shutdown. One solution involves the use of a pigment–carrier fluid mixture, which dries during the test [61]. Kerosene is a suitable carrier fluid for this purpose in many cases. Lampblack is often used as a pigment, though colored chalks produce more easily interpreted patterns [75]. A dry, powdery pattern is left on model surfaces after a run. Instead of photographing this, one can lift it off the model using large squares of matte adhesive tape. When pressed on white paper, these surface-flow "traces" reveal vivid, full-scale views of surface streamline patterns. The locations and angles of these streamlines can then be measured with accuracy, providing quantitative data from the flow visualization [61, 75].

Other methods in transonic and supersonic wind tunnel testing include the minituft [73] and WIS [76] methods discussed earlier, and spark tracing [77], which has been used occasionally. Flow visualization in high-enthalpy environments and that involving molecular excitation of the test gas will be discussed in Sec. II.C.

Cryogenic testing imposes some severe constraints on flow visualization [20, 78–81]. Briefly, transonic cryogenic wind tunnels, of which the National Transonic Facility is an archetype, attempt to reach flight-scale Reynolds numbers by way of cold (high-density, low-viscosity) flows. Typical problems include extremely limited optical test-section access, strong thermal gradients in the optical path, frost on the windows, and very thin boundary layers on highly polished models.

The latter problem may preclude the use of conventional surface flow visualization methods in cryogenic testing. Any liquid present on model surfaces could seriously interfere with extremely thin boundary layers. For that matter, no conventional surface tracers would remain fluid under cryogenic conditions.

The use of optical methods is more promising, and has been the subject of intensive research at the NASA Langley Research Center [79–81]. Thermal noise along the optical path was found to be a major problem. However, the most recent results [81] show that this occurs primarily outside the test section, where it can be dealt with.

C Hypersonic Applications

Above Mach 5, the wind tunnel testing environment tends toward much lower flow densities, higher enthalpies, and shorter testing times. Traditional surface flow visualization methods are still usable [82, 83], however, and some have even been developed for millisecond-range shock tunnel facilities [84].

So little hypersonic testing was done during 1970–1985 that many of the newer flow visualization methods have not been tried in this flow regime to date. Optical methods, methods involving molecular excitation of the test gas, and surface thermal-mapping indicators are traditional, and are covered briefly in the following discussion.

Optical instrumentation for hypersonic wind tunnel testing does not differ in many respects from that already discussed for the transonic–supersonic range. Shadowgraph, schlieren [85], color schlieren [86], and holographic interferometry [87] visualizations have all been done successfully in hypersonic flows. An example is shown in Fig. 5. The major difficulty lies in low flow-density levels, which can violate the sensitivity limits of standard optics [85, 88]

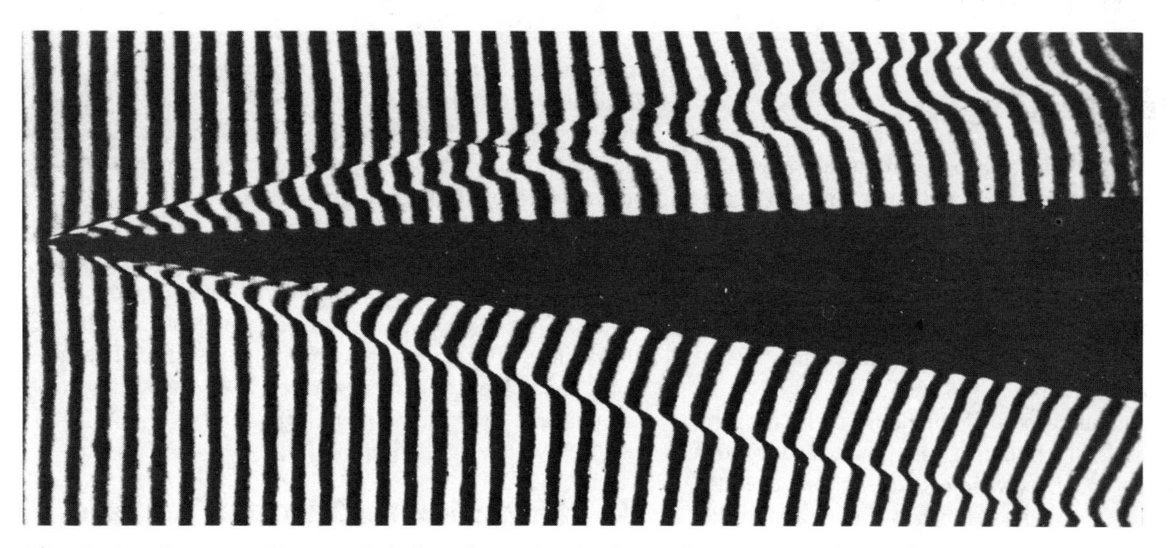

Fig. 5 Interferogram of hypersonic helium flow at Mach 13 over flat plate, revealing merging and strong interaction of shock wave and boundary layer (from Princeton University Gas Dynamics Laboratory).

in extreme cases. When this occurs, one can take advantage of the low-density conditions by resorting to molecular excitation.

Molecular-excitation methods, such as the glow discharge and electron beam methods, are classical flow visualization techniques [89]. An electrical discharge or an electron beam is passed through the flow, causing the excitation of outer electron levels and possibly the dissociation or ionization of test-gas molecules. Spontaneous photon emissions occur as these excited molecules relax, producing a visible glow. Results depend on the test-gas composition, density, and method of excitation. A variety of flow visualization examples, including some recent ones in hypersonic flows, can be found in Refs. [82, 90–92].

Thermal mapping indicators are applied to the surfaces of hypersonic wind tunnel models to give an indication of heat flux in high-enthalpy testing. Early versions of this technique used temperature-sensitive paints [63, 83, 93, 94] and phase-change coatings [95]. These were successful, but sometimes required long run times and had restricted indicating ranges. Infrared thermography [96, 97] has been used more recently with success, though it has limited spatial resolution and requires special wind tunnel windows. Thermographic phosphors [97, 98] are the latest development in this technique. These phosphor coatings emit temperature-dependent fluorescence under ultraviolet illumination.

III OTHER AEROSPACE TESTING

For the present purposes, flow visualization applications in aerospace testing other than in wind tunnels are discussed under the general headings of engine, propeller and rotor, ballistic, and flight testing. Again, many of the techniques already discussed remain applicable in these testing regimes.

A Engine Testing

Jet and rocket engines for aerospace propulsion are tested in engine test stands, test cells, and environmental chambers designed to simulate the altitude, enthalpy, mass flux, and Mach number conditions encountered in flight. These facilities often impose access restrictions, which complicate the task of performing flow visualization. Consequently, the literature contains few examples of novel flow visualizations in engine testing.

However, a unique aspect of this topic is the property of engine exhausts, which, unlike cold flows, tend to often be self-luminous and thus visible to the naked eye. An example of a rocket exhaust is shown in Fig. 6, and further examples of both rocket and jet engine exhausts can be found in Refs. [99, 100]. These supersonic exhaust plumes emit visible radiation according to their temperature and

Fig. 6 Space Shuttle main engine undergoing test firing, revealing natural visibility of Mach disk (from NASA National Space Technology Laboratories).

sometimes due to continuing combustion outside the nozzle. Mach disks, which cause abrupt temperature increases, are especially visible (Fig. 6). Even in some cases where the radiation is not visible, infrared thermography can still be used for flow visualization [101].

The visualization of flows in turbomachinery cascades has been attempted using shadowgraph, schlieren, color schlieren, and holographic interferometry techniques [102–105], as well as surface streak techniques and smoke [106]. This rotating environment imposes special restrictions, and holographic interferometry has been used to great advantage in overcoming them [104, 105].

B Propeller and Rotor Testing

Oil flow and fluorescent minitufts can be applied directly to propeller blades and observed during testing via strobe

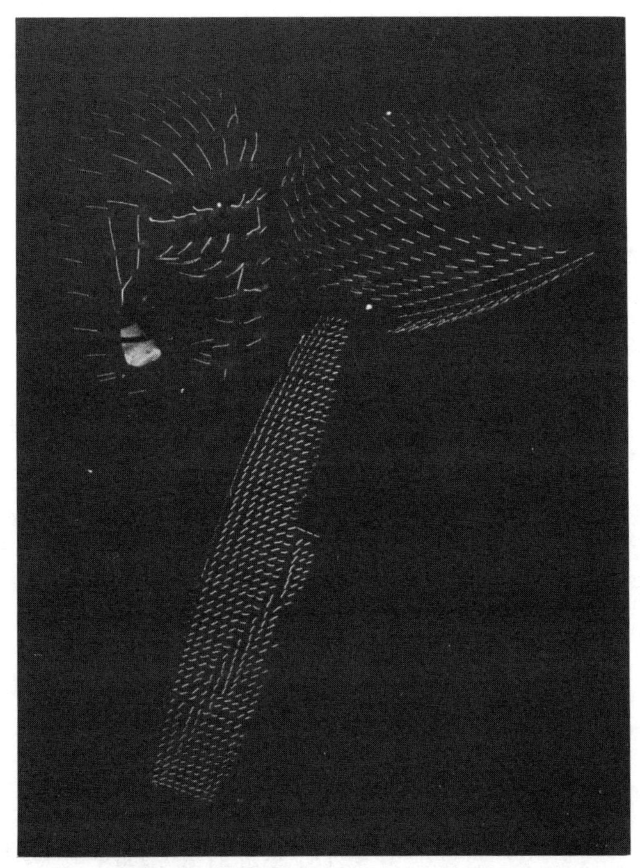

Fig. 7 Fluorescent minitufts on light twin-engine aircraft propeller, photographed in flight (from J. P. Crowder, X-Aero Co.).

lighting [107, 108]. An example, due to Crowder, is shown in Fig. 7. However, caution is advised: centrifugal forces on these indicators can overcome the aerodynamic forces, with unpredictable results. In particular, an oil-flow line of separation with superimposed centrifugal force may not bear much resemblance to the familiar separation line in nonrotating flows.

Helicopter rotors have been studied in test stands using dust [109], shadowgraphy [110], holographic interferometry [111], and tomography [112]. The visualizations of transonic tip flows and blade–vortex interactions are topics of current interest.

C Ballistic Testing

The visualization of flows about models fired along ballistic ranges obviously requires both short exposures and good synchronization. Spark shadowgraphy [113–115] has been a baseline approach to this problem for many years; an example is shown in Fig. 8. Both monochrome and color [116] schlieren techniques have also been used.

Typically, a ballistic range is fitted with a number of stations including a light source, one or more reflectors or

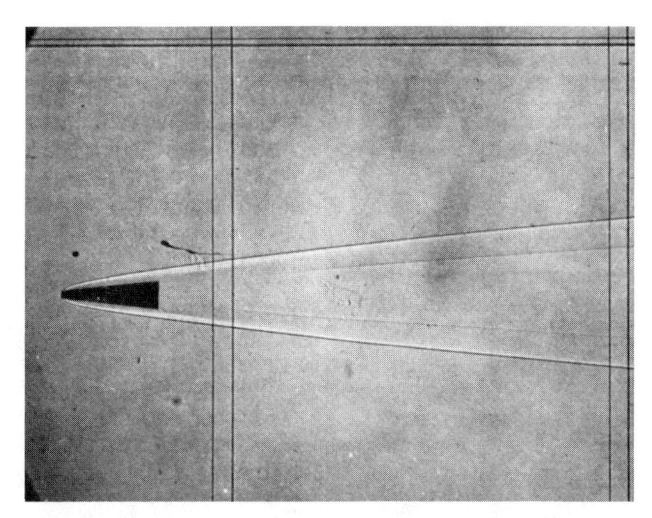

Fig. 8 Shadowgram of blunted cone in free flight at Mach 16 in a hypervelocity ballistic range (from NASA Ames Research Center).

Fresnel lenses, a camera, and an electronic triggering apparatus. Xenon flashtubes of ≤ 1 μs duration have replaced the early spark gaps as white-light sources.

Hypersonic and hypervelocity range testing with ablating, self-luminous models involves an extraneous-light problem in shadowgraph visualizations. This can be solved by pulsed-laser illumination coupled with a narrow-bandpass filter over the camera lens. Laser front lighting against a grid background has also been tried successfully, as has X-ray shadowgraphy [114, 115]. Finally, a high-speed framing camera has been used to obtain multiple shadowgrams of projectiles in flight [115].

D Flight Testing

Compared to ground-test facilities, flight tests of aerospace vehicles tend to be more expensive, more dangerous, and less accessible. Here, in particular, flow visualization efforts must overcome their reputation for consuming valuable test time.

Tufts, and more recently minitufts [117], are most often used in flight tests. A more stable modification of the tuft principle, known as the flowcone [118, 119], has also been used.

Oil-flow techniques [120] are reasonably simple to apply in flight. As mentioned earlier, recent developments in liquid-crystal technology now provide effective shear-sensitive boundary layer transition indications in flight tests [40, 41]. Surprisingly, even some of the more sophisticated flow visualization approaches, such as the WIS [119] and light sheet [121], have been used successfully in flight.

Finally, some special flow phenomena are naturally visible in flight. These include sunlight-illuminated shadowgrams of shocks on transonic wings [2, 119] and the

Fig. 9 Visible condensation above the wing of the Concorde Supersonic Transport during landing (from Air France).

condensation of moisture in transonic or vortical regions about flight vehicles. An example of the latter is shown in Fig. 9.

REFERENCES

1. Settles, G. S., New Flow Visualization Techniques for Practical Aerodynamic Testing, Rep. submitted to Aerodynamic Laboratory, Boeing Commercial Airplane Co., April 1981.
2. Settles, G. S., Flow Visualization Techniques for Practical Aerodynamic Testing, in *Flow Visualization III: Proceedings of the Third International Symposium on Flow Visualization*, ed. W.-J. Yang, pp. 306–315, Hemisphere, Washington, D.C., 1985.
3. Rae, W. H., Jr., and Pope, A. Y., *Low-Speed Wind Tunnel Testing*, 2d ed., Wiley, New York, 1984.
4. Pope, A. Y., and Goin, K. L., *High-Speed Wind Tunnel Testing*, Wiley, New York, 1965.
5. Pankhurst, R. C., and Holder, D. W., *Wind-Tunnel Technique*, Pitman, London, 1952.
6. Maltby, R. L., ed., Flow Visualization in Wind Tunnels using Indicators, AGARDograph 70, 1962.
7. Hunter, W. W., and Foughner, J. T., Jr., eds. Flow Visualization and Laser Velocimetry for Wind Tunnels, *Proc. NASA Workshop*, NASA CP-2243, 1982.
8. Settles, G. S., Modern Developments in Flow Visualization, *AIAA J.*, vol. 24, no. 8, pp. 1313–1323, 1986.
9. Maughmer, M. D., An Experimental Investigation of Wing Fuselage Integration Geometries, AIAA/AHS/ASEE Aircraft Design, Systems, and Operations Mtg., St. Louis, Missouri, Sept. 1987, AIAA Paper 87-2937.
10. Hunt, J. C. R., Abell, C. J., and Woo, H., Kinematical Studies of the Flows around Free or Surface-mounted Obstacles. Applying Topology to Flow Visualization, *J. Fluid Mech.*, vol. 86, pp. 179–200, 1978.
11. Tobak, M., and Peake, D. J., Topology of Three-dimensional Separated Flows, *Ann. Rev. Fluid Mech.*, vol. 14, pp. 61–85, 1982.
12. Hornung, H., and Perry, A. E., Some Aspects of Three-dimensional Separation: Part I. Streamsurface Bifurcations, *Z. Flugwiss. Weltraumforschung*, vol. 8, no. 2, pp. 77–87, 1984.
13. Perry, A. E., and Hornung, H., Some Aspects of Three-dimensional Separation: Part II. Vortex Skeletons, *Z. Flugwiss. Weltraumsforschung*, vol. 8, no. 3, pp. 155–160, 1984.
14. Kao, H. C., Burstadt, P. L., and Johns, A. L., Flow Visualization and Interpretation of Visualization Data for Deflected Thrust V/STOL Nozzles, AIAA 22d Aerosp. Sci. Mtng., Reno, Nev. AIAA Paper 84-1012, 1984.
15. Řezníček, S. G., and Flores, J., Strake-generated Vortex Interactions for a Fighter-like Configuration, AIAA 25th Aerosp. Sci. Mtng., Reno, Nev., AIAA Paper 87-0589, 1987.
16. Stoy, S. L., Dillon, J. L., and Roman, A. P., Correlation and Analysis of Oil Flow Data for an Air-breathing Missile Model, AIAA 23d Aerosp. Sci. Mtng. Reno, Nev., AIAA Paper 85-0452, 1985.
17. Ericsson, L. E., and Reding, J. P., Flow Visualization, A Diagnostic Tool for Space Shuttle Aero-Analysis, *ICIASF'83 Record, IEEE*, pp. 1-14, 1983.
18. Miller, D. S., Boeing Commercial Airplane Co., private communication, March 29, 1967.
19. Langston, L. S., and Boyle, M. T., Surface Streamline Flow Visualization, *NASA Tech Briefs*, vol. 8, pp. 372–373, Spring 1984.
20. Crowder, J. P., Surface Flow Visualization Using Indicators, NASA CP-2243, pp. 37–46, 1982.
21. Maltby, R. L., and Keating, R. F. A., Smoke Techniques for Use in Low Speed Wind Tunnels, in *Flow Visualization in Wind Tunnels using Indicators*, ed. R. L. Maltby, AGARDograph 70, pp. 1–74, 1962.
22. Mueller, T. J., Smoke Visualization in Wind Tunnels," *Astronaut. Aeronaut.*, vol. 21, pp. 50–54, 62, 1983.
23. Mueller, T. J., Recent Developments in Smoke Flow Visualization, in *Flow Visualization III: Proceedings of the*

Third International Symposium on Flow Visualization, ed. W.-J. Yang, pp. 30–40, Hemisphere, Washington, D.C., 1985.
24. Crowder, J. P., Hill, E. G., and Pond, C. R., Selected Wind Tunnel Testing Developments at the Boeing Aerodynamics Laboratory, AIAA 11th Aerodyn. Test. Conf., Colorado Springs, AIAA Paper 80-0458-CP, 1980.
25. Crowder, J. P., Fluorescent Minitufts for Surface Flow Visualization, in *Flow Visualization II: Proceedings of the Second International Symposium*, ed. W. Merzkirch, pp. 663–668, Hemisphere, Washington, D.C., 1982.
26. Petrov, A. V., Certain Types of Separated Flow over Slotted Wings, *Fluid Mech. Sov. Res.*, vol. 7 no. 5, pp. 80–89, 1978.
27. Bird, J. D., Tuft-Grid Surveys at Low Speeds for Delta Wings, NASA Tech. Note D-5045, February 1969.
28. McGregor, I., The Vapor-Screen Method of Flow Visualization,'' *J. Fluid Mech.*, vol. 11, no. 4, pp. 481–511, 1961.
29. Véret, C., Flow Visualization by Light Sheet, in *Flow Visualization III: Proceedings of the Third International Symposium on Flow Visualization*, ed. W.-J. Yang, pp. 106–112, Hemisphere, Washington, D.C., 1985.
30. Bisplinghoff, R. L., Coffin, J. B., and Haldeman, C. W., Water Fog Generation System for Subsonic Flow Visualization, *AIAA J.*, vol. 14, no. 8, pp. 1133–1135, 1976.
31. Hale, R. W., Tan, P., and Ordway, D. E., Experimental Investigation of Several Neutrally-Bouyant Bubble Generators for Aerodynamic Flow Visualization, Sage Action, Inc., Ithaca N.Y., Rep. SAI-RR-6901, August 1969.
32. Koga, D. J., Abrahamson, S. D., and Eaton, J. K., Development of a Portable Laser Sheet, *Exp. Fluids*, vol. 5, no. 3, pp. 215–216, 1987.
33. O'Hare, J. E., Digital Image Analysis for Aerodynamic Testing, *ISA Proc. 27th Symp.*, vol. 18, pt. 1, 1981.
34. Winkelmann, A. E., and Tsao, C. P., A Color Video Display Technique for Flowfield Surveys, *AIAA J.*, vol. 23, pp. 1381–1386, 1985.
35. Crowder, J. P., Quick and Easy Flow-Field Surveys, *Astronaut. Aeronaut.* vol. 18, pp. 38–39, 1980.
36. Stremel, P. M., Experimental Investigations to Fully Map the Flow Field around a Wind Tunnel Model of a Transport Airplane, SAE Turb. Powered Exec. Aircr. Mtng., Phoenix SAE Paper 800630, April 1980.
37. Crowder, J. P., and Beck, H. M., Electronic Wake Imaging System, in *Flow Visualization III: Proceedings of the Third International Symposium on Flow Visualization*, ed. W.-J. Yang, pp. 208–212, Hemisphere, Washington, D.C., 1985.
38. Covault, C., Japan Pursues Space Shuttle Advanced Technology Work, *Aviat. Week Space Technol.* vol. 125, p. 84, July 21, 1987.
39. Klein, E. J., Liquid Crystals in Aerodynamic Testing, *Astronaut. Aeronaut.*, vol. 6, pp. 70–73, 1968.
40. Holmes, B. J., Gall, P. D., Croom, C. C., Manuel, G. S., and Kelliher, W. C., A New Method for Laminar Boundary Layer Transition Visualization in Flight: Color Changes in Liquid Crystal Indicators, NASA Tech. Memo 87666, January 1986.
41. NASA Testing Laminar Flow Experiment, *Aviat. Week Space Technol.*, vol. 126, p. 135, April 27, 1987.
42. Holder D. W., and North R. J., Optical Methods for Examining the Flow in High-Speed Wind Tunnels. Part 1. Schlieren Methods, AGARDograph 23, 1956.
43. Volluz R. J., Wind Tunnel Instrumentation and Operation, U.S. Navy NAVORD Rep. 1488, vol. 6, sec. 20, 1961.
44. Holder, D. W., and North, R. J., Schlieren Methods, Br. Nat. Phys. Lab., NPL Notes Appl. Sci. 31, 1963.
45. Beamish, J. K., Gibson, D. M., Sumner, R. H., Zivi, S. M., and Humberstone, G. H., Wind Tunnel Diagnostics by Holographic Interferometry, *AIAA J.*, vol. 7, no. 10, pp. 2041–2043, 1969.
46. Trolinger, J. D., Flow Visualization Holography, *Opt. Eng.*, vol. 14, pp. 470–481, September–October 1975.
47. Optical contractors in the United States include, for example: J. Unertl Optical Co., 3551–55 East St., Pittsburgh, Pa. 15214, and Muffoletto Optical Co., 6100 Everall Ave., Baltimore, MD. 21206.
48. Vasil'ev L. A. *Tenevye Metody*, Izdatel' stvo Nauka, Moscow, 1968, English transl.: *Schlieren Methods*, Israel Prog. for Sci. Transl. Ltd., New York, 1971.
49. Schardin H., Schlierenverfahren und Ihre Anwendungen, *Ergeb. Exakten Naturwiss.*, vol. 20, pp. 303–439, 1942; English transl.: NASA TTF-12731, 1970.
50. Settles G. S., Color Schlieren Optics—A Review of Techniques and Applications, in *Flow Visualization II: Proceedings of the Second International Symposium*, ed. W. Merzkirch, pp. 749–759, Hemisphere, Washington, D.C., 1982.
51. Settles, G. S., Colour-Coding Schlieren Techniques for the Optical Study of Heat and Fluid Flow, *Int. J. Heat Fluid Flow*, vol. 6, no. 1, pp. 3–15, 1985.
52. Kantrowitz, A., and Trimpi, R. L., A Sharp-Focusing Schlieren System, *J. Aeronaut. Sci.*, vol. 17, no. 5, pp. 311–315, 1950.
53. Moseley, G. W., Peterson, Jr., J. B., and Braslow, A. L., An Investigation of Splitter Plates for the Aerodynamic Separation of Twin Inlets at Mach 2.5, NASA Tech. Note D-3385, 1966.
54. Wuerker, R. F., ''Holographic Interferometry,'' *Jpn. J. Appl. Phys.*, vol. 14, pp. 203–212, 1975.
55. Vest, C. M., *Holographic Interferometry*, Wiley, New York, 1979.
56. O'Hare, J. E., and Strike, W. T., Holographic Interferometry and Image Analysis for Aerodynamic Testing, AEDC-TR-79-75, September 1980.
57. Trolinger, J. D., Holographic Flow Visualization: State-of-the-Art Overview, NASA CP-2243, pp. 149–158, 1982.
58. Bachalo, W. D., and Houser, M. J., Optical Interferometry in Fluid Dynamics Research, *Opt. Eng.*, vol. 24, pp. 455–461, May/June 1985.
59. Stricker, J., and Kafri, O., A New Method for Density Gradient Measurements in Compressible Flows, *AIAA J.*, vol. 20, pp. 820–823, 1982.
60. Love, E. S., and Grigsby, C. E., A New Shadowgraph Technique for the Observation of Conical Flow Phenomena in Supersonic Flow and Preliminary Results Obtained for a Triangular Wing, NACA Tech. Note 2950, 1953.
61. Settles, G. S., and Teng, H.-Y., Flow Visualization Methods for Separated Three-dimensional Shock Wave/ Turbulent Boundary Layer Interactions, *AIAA J.*, vol. 21, pp. 390–397, 1983.

62. Sedney, R., Kitchens, C. W., Jr., and Bush, C. C., The Marriage of Optical, Tracer, and Surface Indicator Techniques in Flow Visualization, U.S. Army BRL Rep. 1763, 1975.
63. Creel, T. R., and Hunt, J. L., Photographing Flow Fields and Heat-Transfer Patterns Simultaneously, *Astronaut. Aeronaut.*, vol. 10 pp. 54–55, 1972.
64. Véret, C., Philbert, M., Surget, J., and Fertin, G., Aerodynamic Flow Visualization in the ONERA Facilities, in *Flow Visualization*, ed. T. Asanuma, pp. 335–340, Hemisphere, Washington, D.C., 1979.
65. Corlett, W. A., Operational Flow Visualization Techniques in the Langley Unitary Plan Wind Tunnel, NASA CP-2243, pp. 65–74, 1982.
66. Mueller, T. J., The Role of Flow Visualization in the Study of Afterbody and Base Flows, *Exp. Fluids*, vol. 3, no. 1, pp. 61–70, 1985.
67. Stahl, W., Zur Sichtbarmachung von Stromungen um Schlanke Deltaflugel bei hohen Unterschallgeswindigkeiten, German DFVLR Rep. AVA-70-A-45, 1970.
68. Batill, S. M., Nelson, R. C., and Mueller, T. J., High Speed Smoke Flow Visualization, U.S. Air Force Rep. AFWAL-TR-81-3002, March 1981.
69. Porcar, R., and Prenel, J. P., Supersonic Flow Visualization by Light Scattering, in *Flow Visualization II: Proceedings of the Second International Symposium*, ed. W. Merzkirch, pp. 433–438, Hemisphere, Washington, D.C., 1982.
70. Snow, W. L., and Morris, O. A., Investigation of Light Source and Scattering Medium Related to Vapor-Screen Flow Visualization in a Supersonic Wind Tunnel, NASA Tech. Memo 86290, December 1984.
71. Bonnet, J. P., and Chaput, E., Large-scale Structures Visualization in a High Reynolds Number Turbulent Flat-plate Wake at Supersonic Speed, *Exp. Fluids*, vol. 4, pp. 350–356, 1986.
72. Philbert, M., Faleni, J. P., and Deron, R., Air Intake Flow Visualization, in *Flow Visualization IV: Proceedings of the Fourth International Symposium*, ed. C. Véret, pp. 753–758, Hemisphere, Washington, D.C., 1987.
73. Pittman, J. L., Experimental Flowfield Visualization of a High Alpha Wing at Mach 1.62, AIAA 25th Aerosp. Sci. Mtng., Reno, Nev., AIAA Paper 87-0329, 1987.
74. Loving, D. L., and Katzoff, S., The Fluorescent-Oil Film Method and Other Techniques for Boundary-Layer Flow Visualization, NASA Memo 3-17-59L, March 1959.
75. Lu, F. K., Settles, G. S., and Horstman, C. C., Mach Number Effects on Conical Surface Features of Swept Shock Boundary-Layer Interactions, AIAA 19th Fluid Plasma Dyn. Lasers Conf., Honolulu, AIAA Paper 87-1365, June 1987.
76. Crowder, J. P., Transonic Applications of the Wake Imaging System, in *Flow Visualization and Laser Velocimetry for Wind Tunnels*, ed. W. W. Hunter and J. T. Foughner, Jr., NASA CP-2243, pp. 109–116, 1982.
77. Frungel, F., High-Frequency Spark Tracing and Application in Engineering and Aerodynamics, *Proc. 12th Int. Cong. High Speed Photog.*, Toronto, *SPIE* Proc., vol. 97, pp. 291–301, 1976.
78. Haldeman, C. W., Suggested Modification of Fog Flow Visualization for Use in Cryogenic Wind Tunnels, *Proc. 1st Int. Symp. Cryogen. Wind Tunnels*, Univ. of Southampton, England, pp. 21.1–21.3, 1979.
79. Rhodes, D. B., and Jones, S. B., Flow Visualization in the Langley 0.3-Meter Transonic Cryogenic Tunnel and Plans for the National Transonic Facility, NASA CP-2243, pp. 117–132, 1982.
80. Burner, A. W., and Goad, W. K., Flow Visualization in a Cryogenic Wind Tunnel using Holography, in *Flow Visualization III: Proceedings of the Third International Symposium on Flow Visualization*, ed. W.-J. Yang, pp. 444–448, Hemisphere, Washington, D.C., 1985.
81. Snow, W. L., Burner, A. W., and Goad, W. K., Improvement in the Quality of Flow Visualization in the Langley 0.3-Meter Transonic Cryogenic Tunnel, NASA Tech. Memo 87730, August 1986.
82. Duller, C. E., An Investigation of Flow Visualization Techniques in Helium at Mach Numbers of 15 and 20, NASA Tech. Note D-1769, April 1963.
83. Rhudy, R. W., Flow Visualization Techniques for Use in Hypersonic Wind Tunnels, AEDC-TDR-64-108, October 1964.
84. Li, S.-H., Yang, Z.-Q., and Wang, X.-R., Oil-Flow Technique and Application in Hypersonic Shock Tunnel, in *Flow Visualization IV: Proceedings of the Fourth International Symposium*, ed. C. Véret, p. 901, Hemisphere, Washington, D.C., 1987.
85. North, R. J., and Stuart, C. M., Flow Visualization and High-Speed Photography in Hypersonic Aerodynamics, Br. Nat. Phys. Lab., NPL Aero Rep. 1029, July, 1962.
86. Cash, R. F., and Catley, A. M., Separation Measurements on a Delta Wing in a Shock Tunnel at $M = 8.6$ using Monochrome and Colour Schlieren Photography, Br. Nat. Phys. Lab., NPL Aero Note 1097 (ARC 32 399), 1970.
87. Havener, A. G., Holographic Visualizations of Viscous Interactions in Hypersonic Flows, *SPIE Proc.*, vol. 788, paper no. 10, May 1987.
88. Merzkirch, W. F., Sensitivity of Flow Visualization Methods at Low-density Flow Conditions, *AIAA J.*, vol. 3, no. 4, pp. 794–795, 1965.
89. Merzkirch W. F., *Flow Visualization*, Academic, New York, 1974.
90. Kunkel, W. B., and Hurlbut, F. C., Luminescent Gas Flow Visualization in a Low-Density Supersonic Wind Tunnel, WADC Tech. Rep. 57-441, September, 1957.
91. Kimura, T., and Nishio, M., Visualizing Method of Shock Wave by Electric Discharge using Point-Line Electrodes, in *Flow Visualization III: Proceedings of the Third International Symposium on Flow Visualization*, ed. W.-J. Yang, pp. 473–477, Hemisphere, Washington, D.C., 1985.
92. Nishio, M., and Kimura, T., Visualization of Shock Wave and Streamline Around Hypersonic Vehicles by using Electrical Discharge, in *Flow Visualization IV: Proceedings of the Fourth International Symposium*, ed. C. Véret, pp. 185–190, Hemisphere, Washington, D.C., 1987.
93. Stainback, P. C., A Visual Technique for Determining Qualitative Aerodynamic Heating Rates on Complex Configurations, NASA Tech. Note D-385, October 1960.
94. Kafka, P. G., Gaz, J., and Yee, W. T., Measurement of Aerodynamic Heating of Wind-Tunnel Models by Means of

Temperature-Sensitive Paint, *AIAA J. Spacecr.*, vol. 2, no. 3, pp. 475–477, 1965.

95. Jones, R. A., and Hunt, J. L., Use of Fusible Temperature Indicators for Obtaining Quantitative Aerodynamic Heat-Transfer Data, NASA Tech. Rep. R-230, February 1966.
96. Bandettini, A., and Peake, D. J., Diagnosis of Separated Flow Regions on Wind-Tunnel Models using an Infrared Camera, *ICIASF'79 Record IEEE*, pp. 171–185, 1979.
97. Matthews, R. K., Nutt, K. W., Wannenwetsch, G. D., Kidd, C. T., and Boudreau, A. H., Developments in Aerothermal Test Techniques at the AEDC Supersonic/Hypersonic Wind Tunnels, *Prog. Astronaut. Aeronaut.* vol. 103, pp. 373–392, 1986.
98. The Phosphor Thermography Center Brochure, Martin Marietta Energy Systems, Inc., PO Box P, Oak Ridge, Tenn. 37381.
99. Rocketdyne Fires Prototype Orbital Transfer Vehicle Engine, *Aviat. Week Space Technol.*, vol. 125, p. 87, June 16, 1986.
100. Pratt & Whitney PW1128 EMD Undergoes Tests, *Aviat. Week Space Technol.*, vol. 124, cover, October 28, 1985.
101. Hilton, M., Quantitative Thermography in Aero-Engine Research and Development, Rolls-Royce Rep. PNR-90021, 1980.
102. Kaspar, J., Colour in High-Speed Photography by Modified Schlieren Apparatus, *Proc. 8th Int. Cong. High-Speed Photog.*, pp. 357–358, Wiley, New York, 1968.
103. Véret, C., Review of Optical Techniques with Respect to Aero-Engine Applications. AGARD LS-90, Paper 2, 1977.
104. Boldman, D. R., Buggele, A. E., and Decker, A. J., Three-dimensional Shock Structure in a Transonic Flutter Cascade, *AIAA J.* vol. 20, pp. 1146–1148, 1982.
105. Bryanston-Cross, P. J., High Speed Flow Visualization, *Prog. Aerosp. Sci.*, vol. 23, pp. 85–104, 1986.
106. Hansen, A. G., Herzig. H. Z., and Costello, G. R., A Visualization Study of Secondary Flows in Cascades, NACA Tech. Note 2947, May 1953.
107. Stefko, G. L., Paulovich, F. J., Greissing, J. P., and Walker, E. D., Propeller Flow Visualization Techniques, NASA CP-2243, pp. 75–90, 1982.
108. Crowder, J. P., In-Flight Propeller Flow Visualization using Fluorescent Minitufts, NASA CP-2243, pp. 91–95, 1982.
109. Taylor, M. K., A Balsa-Dust Technique for Airflow Visualization and its Application to Flow through Model Helicopter Rotors in Static Thrust, NACA Tech. Note 2220, 1950.
110. Parthasarathy, S. P., Cho, Y. I., and Back, L. H., Wide-Field Shadowgraph Flow Visualization of Tip Vortices Generated by a Helicopter Rotor, AIAA 18th Fluid Dynam., Plasmadynam. Lasers Conf., Cincinnati, AIAA Paper 85-1557, 1985.
111. Kittleson, J. K. and Yu, Y. H., Reconstruction of a three-dimensional, Transonic Rotor Flow Field from Holographic Interferogram Data, AIAA 23d Aerosp. Sci. Mtng., Reno, Nev., AIAA Paper 85-0370, 1985.
112. Hesselink, L., and Snyder, R., Three-dimensional Tomographic Reconstruction of the Flow around a Revolving Helicopter Rotorblade: A Numerical Simulation, in *Flow Visualization III: Proceedings of the Third International Symposium on Flow Visualization*, ed. W.-J. Yang, pp. 213–220, Hemisphere, Washington, D.C., 1985.
113. Clemens, P. L., and Hendrix, R. E., Development of Instrumentation for the VKF 1000-Ft Hypervelocity Range, in *Advances in Hypervelocity Techniques*, ed. A. M. Krill, pp. 245–277, Plenum, New York, 1962.
114. Henderson, W. F., Robertson, G. W. Jr., and Hill, J. W., Investigation of Shadowgraphs for Use with Highly Ablating, Self-luminous, Ballistic Projectiles, AEDC-TR-71-225, 1971.
115. Hendrix, R. E., and Dugger, P. H., High-Speed Photography in the Aeroballistic Range and Track Facilities of the von Karman Gas Dynamics Facility, *SPIE Proc.* vol. 97, pp. 238–247, 1976.
116. Berger, A. G., Sheetz, N. W. Jr., and Cords, P. H., Temperature-Control Techniques and Instrumentation for Viscous Flow Investigations in a Ballistics Range, AIAA Paper 68-384, 1968.
117. Crowder, J. P., Fluorescent Minitufts for Flow Visualization on Rotating Surfaces, in *Flow Visualization III: Proceedings of the Third International Symposium on Flow Visualization*, ed. W.-J. Yang, pp. 55–59, Hemisphere, Washington, D.C., 1985.
118. Crowder, J. P., and Robertson, P. E., Flow Cones for Airplane Flight Test Flow Visualization, in *Flow Visualization III: Proceedings of the Third International Symposium on Flow Visualization*, ed. W.-J. Yang, pp. 60–64, Hemisphere, Washington, D.C., 1985.
119. Crowder, J. P., Flow Visualization Techniques Applied to Full Scale Vehicles, in *Flow Visualization IV: Proceedings of the Fourth International Symposium*, ed. C. Véret, pp. 15–24, Hemisphere, Washington, D.C., 1987.
120. Anon., Visualization System Depicts F/A-18 High Angle of Attack Airflow Patterns, *Aviat. Week Space Technol.* vol. 126, no. 18, pp. 50–51, May 4, 1987.
121. Lamar, J. E., Bruce, R. A., Pride, J. D. Jr., Smith, R. H., and Brown, P. W., In-Flight Flow Visualization of F-106B Leading-Edge Vortex Using the Vapor-Screen Technique, AIAA 3d Flight Test. Conf., Las Vegas, Nev., AIAA Paper 86-9785, 1986.

Chapter 26

Water Tunnel Testing

H. Werlé

I TYPES OF WATER TUNNEL SETUPS

A wide variety of hydraulic setups are used for flow visualizations, differing essentially in the arrangement of test section and in operating principle. The following three main types can be distinguished [1]:

- vertical, open-circuit tunnels, usually operating by means of gravitational draining, such as those of ONERA (Fig. 1) [2], AFWAL (the United States), and IAI (Israel);
- tunnels with vertical or horizontal test sections, operating in closed circuit with a motor-driven pump or fan; they are the most common, and we mention first those of NAE (Ottawa) [3], NASA Ames [4], NASA Dryden [5], Lockheed (Marietta, Georgia), MBB (Ottobrun, West Germany) [6], and Bertin (France) [7];
- canals with a free surface, usually without water circulation, in which the model is towed by a trolley, such as that of DFVLR (Göttingen, West Germany) [8] and MATRA (Velizy, France), [9].

II EXAMPLES OF APPLICATIONS

A classic example in the aeronautical domain is the upper-surface vortex flow that develops around a Concorde aircraft (1/140 scale model) in engine simulation and during the landing configuration (Fig. 2*a,b*) [10].

A second example of this type is the flow around an Airbus aircraft model at small incidence with only tip and flap vortices (Fig. 2*c*) [10].

To evoke unsteady flows, the hovering test in the TH3 tunnel of an isolated three-blade helicopter rotor reveals the helicoidal-shaped tip vortices (Fig. 2*e*). But the most complex example of an aeronautical vehicle is probably a motorized helicopter model in forward flight with rotation of the main and tail rotors. In this case (Fig. 2*d*), the series of vortices emitted by the blades is modified by the fuselage and the air intake and gas exhaust simulation [11].

A last example, in the domain of space, concerns an Ariane launcher model. The visualizations reveal the formation (Fig. 2*f*) and the asymmetric shape (Fig. 2*g*) of the vortices observed at high angles of attack [12].

An example of a contribution to building aerodynamics is the flow around a fixed structure, such as a skyscraper model, in the wind (Fig. 3*a,b*), characterized by the three-dimensional shape of the wake and the relative stability of the separated zone issued from the wedges [13].

In the case of a tubular array having cylinders disposed in staggered files, we can observe wake instabilities. Vortices issued from the first row run round the elements of the second row and so on (Fig. 3*c*) [14].

Another example of internal flow and unsteady regime is the case of the three-dimensional axial turbomachine model of Fig. 3*f*. The colored streaks emitted upstream show the appearance of a rotating stall moving around the annular section of an off-design regime [15].

Among the models tested during work done for industry, vehicles are of primary importance. This is because of the advantage offered by the realistic simulation of the ground by means of a floor moving at the same speed as the flow [10, 13]. This type of vehicle is represented here by a car model (Fig. 3*d*) and a TGV high-speed train model (Fig. 3*e*).

Fig. 1 Overall view of the three vertical water tunnels in the hydrodynamic visualization laboratory at the Châtillon center of ONERA [2].

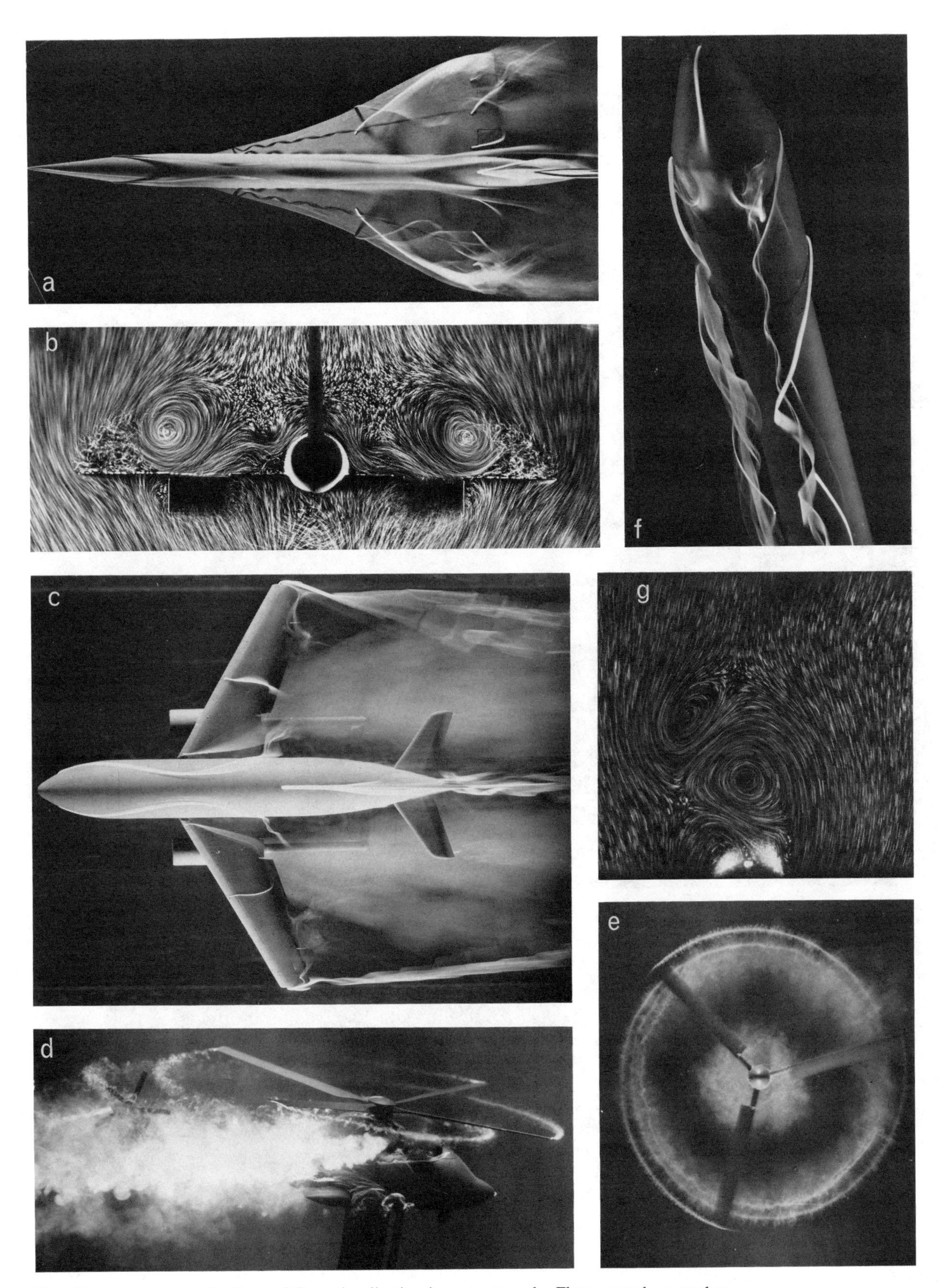

Fig. 2 Aerospace applications of flow visualization in water tunnels. Flow around a complete aircraft with jet engine simulation: (*a,b*) Concorde model at 12° incidence (upper surface and cross section at trailing edge); (*c*) airbus model at 2.5° incidence (upper surface). Flow around a helicopter rotor: (*d*) Motorized helicopter model in forward flight (tunnel TH2); (*e*) hovering isolated rotor (tunnel TH3). Ariane launcher model with incidence: (*f*) Formation of the main upper surface vortices on the round nose; (*g*) asymmetric vortices observed by high angles of attack (cross section at half length).

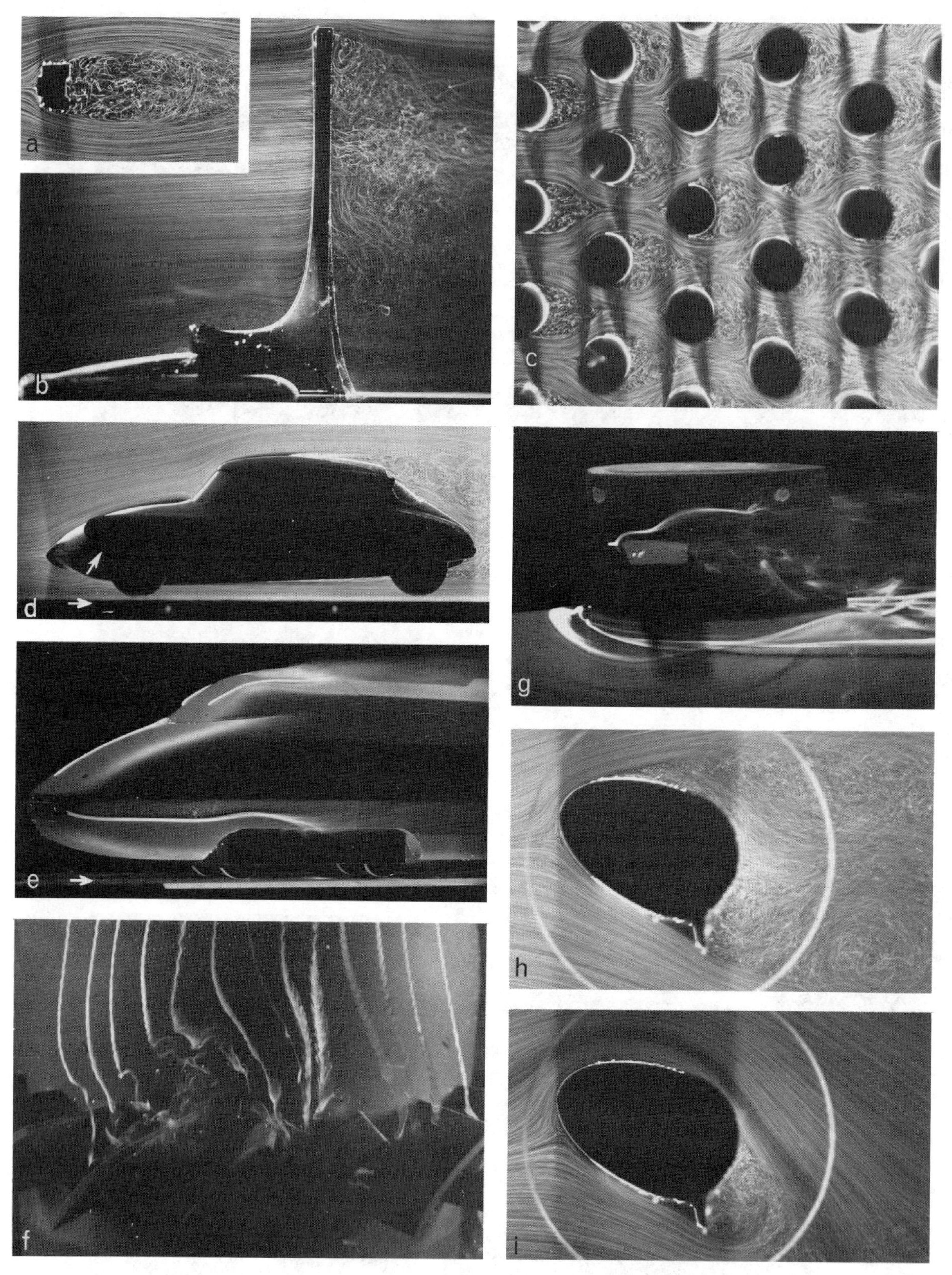

Fig. 3 Nonaerospace applications of flow visualization in water tunnels: (a,b) Flow around a building tower model (cross section and medium section); (c) flow in a tubular array (medium section). Flow around vehicles over a ground simulated by a moving belt: (d) Car model with low-mounted air-intake simulation (medium section); (e) TGV high-speed train model. (f) Rotating separation in an axial flow turbomachine model (tunnel TH3); (g) vortices around the conning tower of a submarine model; (h,i) suction effect of the flow around a turbine sail model at 25° incidence.

The last application examples concern the naval domain. The case of a submarine model is illustrated by Fig. 3*g*. The dye emissions reveal the horseshoe vortices developing around the conning tower and their diving planes in neutral position [13, 16].

Another naval application is the turbine sail invented by Professor Malavard and tested at sea by Commander Jacques Cousteau. The sail consists of a mast with elliptical profile, and the gas tracers bring to light the suction effect of the flow around the model (Fig. 3*h,i*) [16, 17].

REFERENCES

1. Werlé, H., Flow Visualization Techniques for the Study of High Incidence Aerodynamics, AGARD VKI Lect. Ser. 121, March 1982.
2. Werlé, H., The New Hydrodynamic Visualization Laboratory of the Aerodynamics Division, *Rech. Aerosp.,* no. 5, pp. 1–23, 1982.
3. Dobrodzicki, G. A., Flow Visualization in the NAE Water Tunnel, NRC, NAE LR-557, February 1972.
4. Olsen, J., and Liu, H. T., The Construction and Operation of a Water Tunnel in Application to Flow Visualization Studies of an Oscillating Airfoil, Flow Res. Rep. 13-CR 114696, May 1973.
5. Owen, F. K., and Peake, D. J., Vortex Breakdown and Control Experiments in the Ames-Dryden Water Tunnel, Monterey, AGARD CCP 413, Paper 2, October 1986.
6. Stieb, R., and Gross, V., The Water Tunnel–A Helpful Simulation Facility for the Aircraft Industry (MBB), ICIASF Cong., September 1981.
7. Chezlepretre, B., and Brocard, Y., Qualification d'un Tunnel Hydrodynamique pour des Pesées de Maquettes Aéronautiques, Monterey, AGARD CCP 413, Paper 3, October 1986.
8. Bippes, H., and Colak-Antic, P., Der Wasserschleppkanal der DFVLR, *Z. Flugwiss,* vol. 21, pp. 113–120, 1973.
9. Perinelle, J., and Lupieri, A., Présentation d'un Bassin Hydrodynamique. Etude d'un Missile aux Grandes Incidences, Monterey, AGARD CCP 413, Paper 8, October 1986.
10. Werlé, H., Le Tunnel Hydrodynamique à Visualisation au Service de la Recherche Aérospatiale, ONERA Publ. 156, 1974.
11. Werlé, H., Visualisation des Écoulements Tourbillonnaires Tridimensionnels, Rotterdam, AGARD CP 42, Paper 8, April 1983.
12. Werlé, H., Tourbillons de Corps Fuselés aux Incidences Elevées, *Aeronaut. Astronaut.,* vol. 6, pp. 3–22, 1976.
13. Werlé, H., Applications Aérospatiales, Industrielles et Maritimes de la Visualisation Hydrodynamique des Écoulements, ATMA Sess., 1975.
14. Werlé, H., Sur l'Écoulement autour d'un Faisceau Tubulaire, Rev. Fr. Mec., vol. 41, pp. 7–19, 1972.
15. Werlé, H., and Gallon, M., Ecoulement dans une Maquette Hydraulique de Turbomachine Axiale, *Rech. Aerosp.,* vol. 5, pp. 267–288, 1977.
16. Werlé, H., Possibilités d'Essai Offertes par les Tunnels Hydrodynamiques à Visualisation de l'ONERA dans les Domaines Aéronautique et Naval, Monterey, AGARD CCP 134, Paper 14, October 1976.
17. Werlé, H., Hydrodynamic Visualization of the Flow around a Streamlined Cylinder with Suction: Cousteau–Malavard Turbine Sail Model, *Rech. Aerosp.,* vol. 4, pp. 29–38, 1984.

Chapter 27

Internal Flows

K. C. Cheng

I INTRODUCTION

Flow visualization has contributed greatly to the development of modern fluid dynamics and has been utilized by such pioneers as O. Reynolds (1842–1912), L. Prandtl (1875–1953), G. I. Taylor (1886–1975), and others. Flow visualization reveals the whole flow and temperature fields and always plays an important role in understanding flow phenomena. A study of complex fluid flow patterns also provides a basis for physical-intuitive reasoning. Such complex phenomena as the transition from laminar to turbulent flow in a tube leading to a sudden increase in pressure drop, Karman vortices causing periodic forces on a cylinder in a cross flow, and the drop of lift due to flow separation from an airfoil when the angle of attack on an airplane wing exceeds a limiting value can be understood from flow visualization photographs obtained by various techniques.

For complex fluid flows, it is generally difficult to understand the whole flow field by using a probe for measurement of flow velocity, direction, and other quantities. Observation of the whole flow field is obviously an advantage, and flow visualization represents a powerful tool complementing the probe method. Once the flow phenomenon is understood, one can then carry out the measurements for a further detailed study. The motion of fluids is usually invisible, and various techniques have been developed to make fluid flows visible. Flow visualization techniques reveal streamlines, streaklines, particle paths, timelines, flow directions, flow separation, limiting streamlines, density distributions, temperature distributions, vortices, and stress distributions [1].

The purpose of this chapter is to review examples of applications of flow visualization techniques to internal flow problems with an emphasis on the physical understanding of complicated flow phenomena. A brief historical review of the applications of flow-field visualization in internal flows is presented. This is followed by examples of a class of flow phenomena revealed by flow visualization photographs. The applications of flow visualization techniques to selected internal flow problems are also described, using flow photographs and quoting references according to categories.

II HISTORICAL PERSPECTIVE

Internal fluid flows occur in many fluid-handling devices and turbomachines and have great technical importance. Internal flow is an old subject in practical hydraulics. Around 1500, Leonardo da Vinci (1452–1519) presented many sketches depicting flow patterns, vortices, separation and reattachment phenomena. Some examples [2–4] are shown in Fig. 1 together with results from flow visualization and computer simulation. Internal flows were studied by such well-known figures as, for nonviscous flow, Bernoulli (1738), Euler (1766), and d'Alembert (1744) and, for viscous flow, Hagen (1839, 1854), Poiseuille (1828,

This chapter was prepared using an operating grant from the Natural Sciences and Engineering Research Council of Canada. The author wishes to thank Professors Yasuki Nakayama (Tokai University), Eisuke Outa (Waseda University), Tadashi Sakaguchi (Kobe University), Kazuyoshi Takayama (Tohoku University), Koji Kikuyama (Nagoya University), and Dr. Toyoaki Yoshida (National Aerospace Laboratory), Japan, for kindly providing technical papers and visual results. The author is indebted to Miss Betty-Ann Brenneis for her skillful typing of the manuscript.

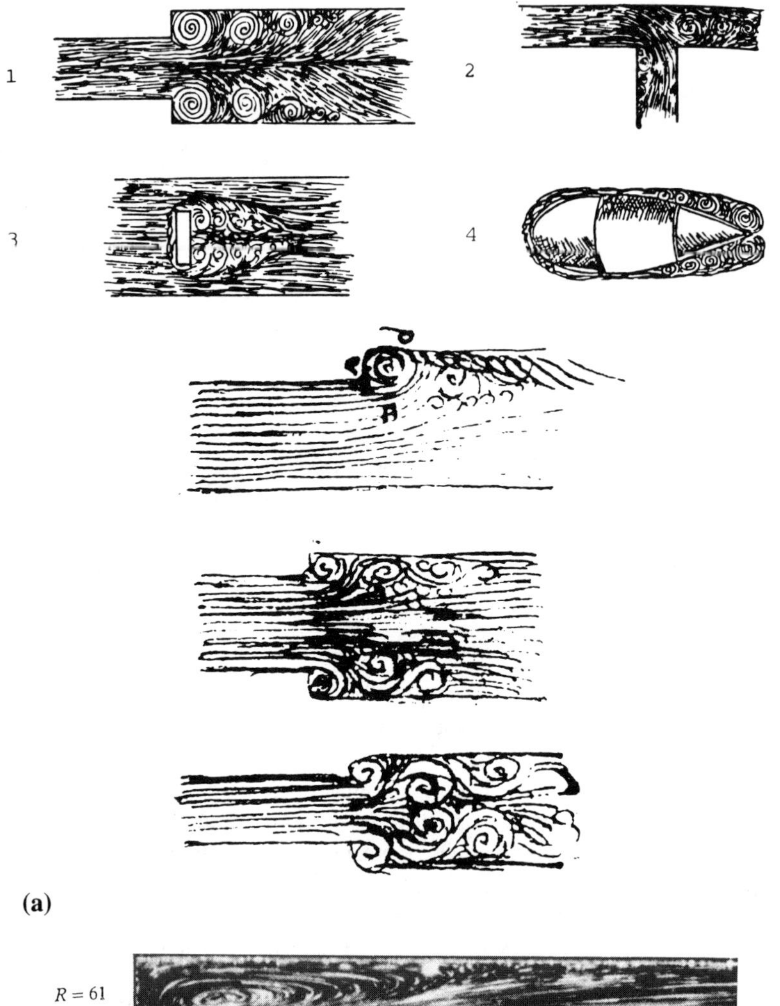

Fig. 1 (*a*) Leonardo da Vinci's sketches [2(a)–2(c)]; (*b*) flow visualization and numerical solution: top, Re = 60; bottom, Re = 200 [3]; (Figure continues.)

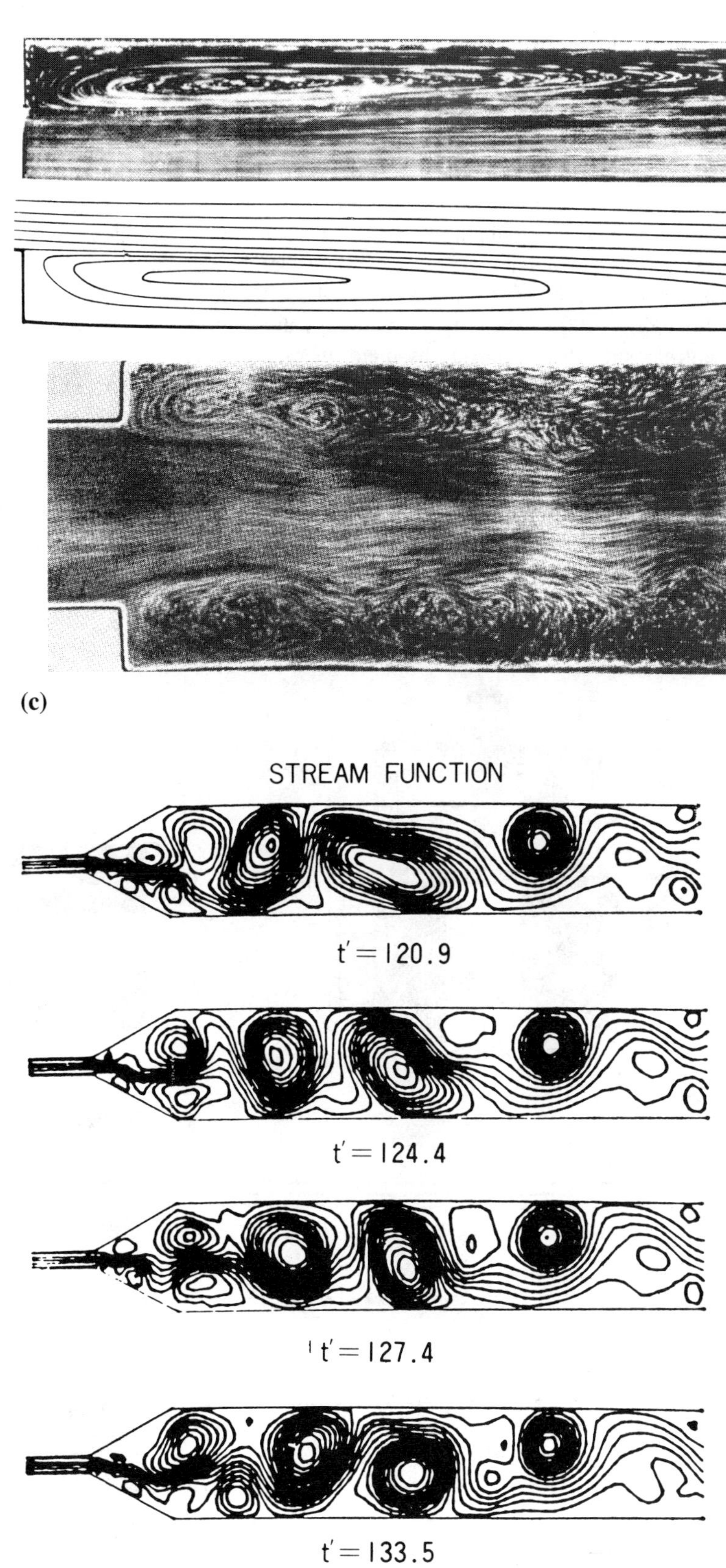

Fig. 1 (continued) (*c*) flow in a pipe with sudden enlargement [4]; (*d*) computer simulation [2(d)].

1841), Hagenbach (1833–1910), and Weisbach (1806–1871) among many others. Hagen clearly showed the existence of two different modes of flow.

Osborne Reynolds (1883) [5] conducted famous flow visualization experiments to study the transition processes in long glass-tube water flow by observing the behavior of dye filaments introduced into the flow. Reynolds's sketches of the transition from laminar to turbulent flow are best known; he determined the critical Reynolds number for pipe flow at about Re = 2000. Reynolds's original dye-injection experiment has since been repeated by many investigators using different flow visualization techniques. The photographs obtained by Dubs [6] are shown in Fig. 2 together with Reynolds's sketches. Lanchester's sketch of wing-tip vortices based on observations may have provided an important hint to Prandtl's physical model development for the three-dimensional aerofoil theory. Also, von Karman's observation of Hiemenz's repeated experiments on cross flow around a cylinder in a water channel with alternating vortices in the wake led to his development of the theory of the Karman vortex street in 1911 [7].

The development of Prandtl's boundary layer concept [8] in 1904 was apparently based on his physical-intuitive reasoning from a series of flow visualization experiments on flow separation phenomena using the surface-floating method in a rather simple water channel [9]. It is significant to observe that 12 photographs depicting flow separation are shown in a pioneer paper on boundary layer [8] by Prandtl. Prandtl's books [10–12] are illustrated with many beautiful and revealing original photographs of fluid motion. This represents quite a contrast to Lamb's famous book on hydrodynamics. Prandtl's 1925 development of the mixing-length theory in turbulent flow was inspired by his flow visualization experiments [13]. The onset of a turbulent spot was clearly shown by Prandtl in Fig. 2 of [14], reproduced in Fig. 3.

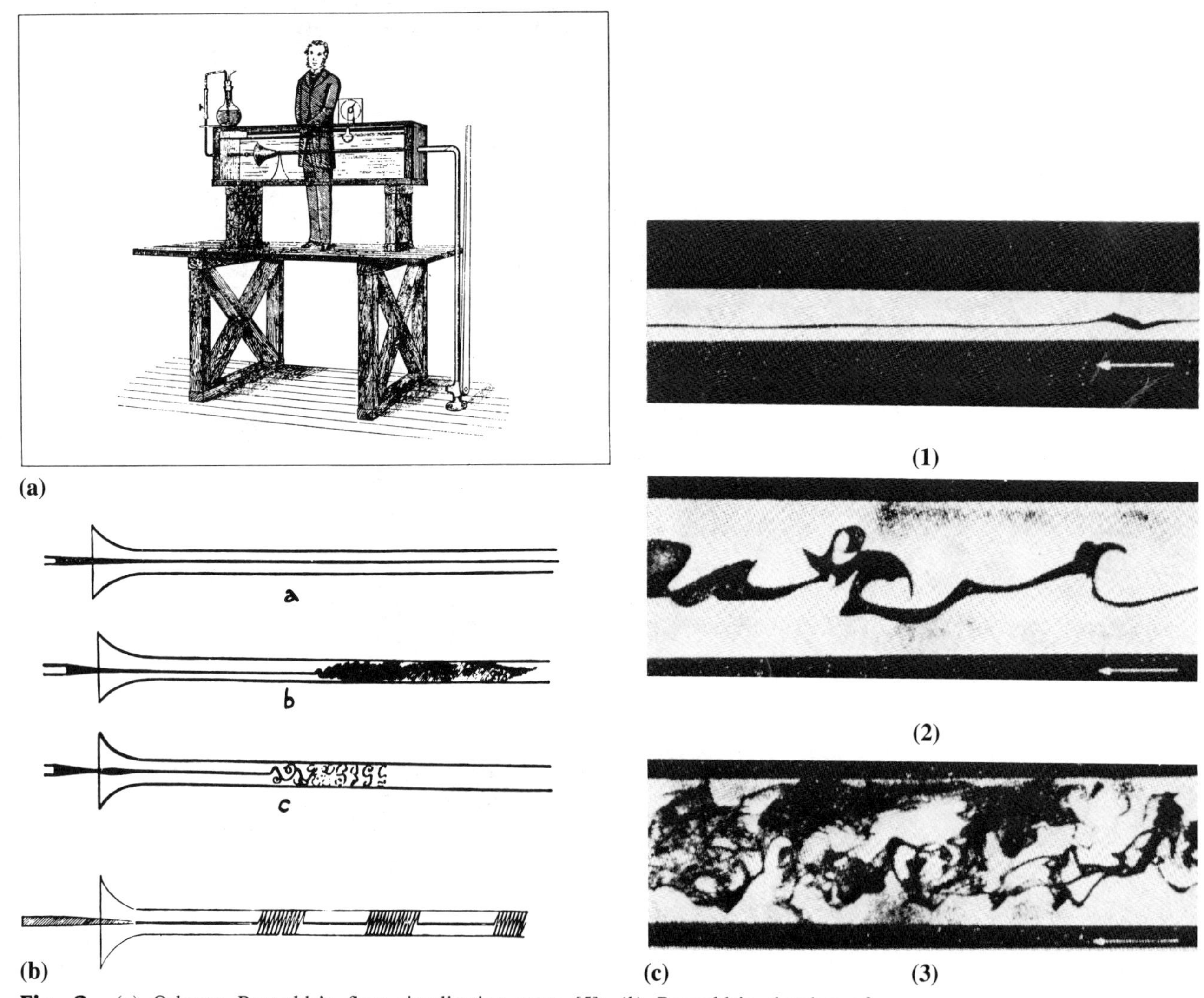

Fig. 2 (*a*) Osborne Reynolds's flow visualization setup [5]; (*b*) Reynolds's sketches of flow observation [5]; (*c*) the Reynolds experiment by Dubs: (1) Re = 1150, (2) Re = 1760, (3) Re = 2520 [6].

Fig. 3 Development of turbulent spots for flow along a flat plate [14].

Mach and Salcher [15] obtained a schlieren photograph of a brass bullet in supersonic flight through air in 1887; the photograph is reproduced in [16]. Ernst Mach (1836–1916) was the first person to use the schlieren method for the visual observation of supersonic flow [7]. Optical methods such as shadowgraph, schlieren, and the Mach–Zehnder interferometer contributed greatly to research in gas dynamics and aerodynamics during the period 1900–1955.

The tracer method of flow visualization by injection of foreign particles such as aluminum powders was used first by Ahlborn [17] in 1902; most of Prandtl's flow photographs were obtained by this method. Kline and his co-workers [18–20] (1959, 1967) carried out flow visualization studies using dye injection and the hydrogen bubble technique for the turbulent boundary layer on a flat plate. They revealed the coherent structure in the viscous sublayer. Their work clearly marked the beginning of a new era in research on coherent structure in turbulent shear flow after the phenomenological and statistical studies revealed mainly by hot-wire measurements in the turbulent boundary layer. It is of interest to note that the phenomenological studies of turbulent flow started with Boussinesq more than a hundred years ago, but active research started with Prandtl's development of the boundary layer concept in 1904 and the mixing-length theory in 1925. Important works were subsequently carried out by G. I. Taylor, Th. von Karman (1881–1963), R. G. Deissler (1921–), and many others.

The concept of coherent structure started with rather simple flow visualization experiments. Hama, in [21], revealed the streaklines (see Fig. 4) near the flat plate by using the dye-injection method. Since turbulent flow was considered to be irregular then, his experimental technique was questioned. However, the flow visualization photographs [18] (see Figs. 5 and 6) obtained by Kline and his co-workers using the hydrogen bubble technique conclusively proved the existence of a coherent turbulent structure. The low-speed streaks formed in the region very near the wall interact with the outer portions of the flow through a process of gradual "lift-up"; then there is sudden oscillation, bursting, and ejection. Flow visualization clarifies the flow struc-

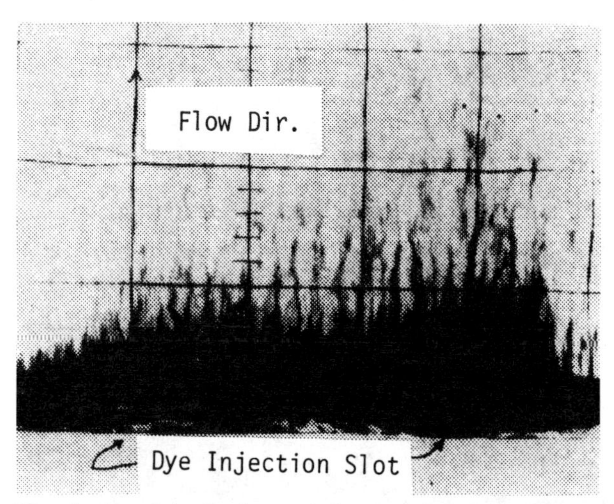

Fig. 4 Dye injection into turbulent boundary layer [21].

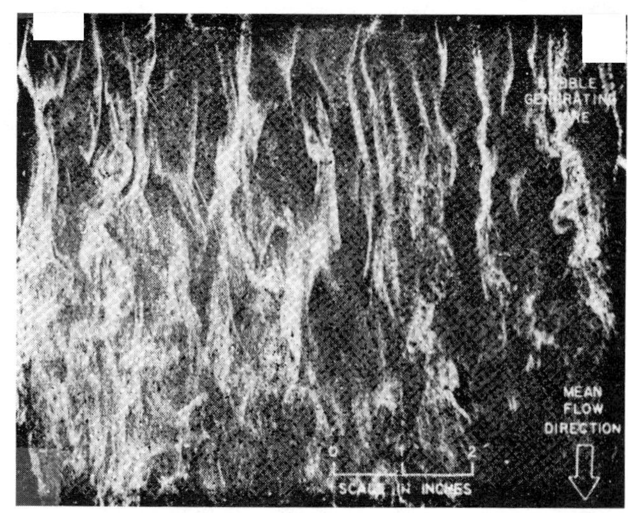

Fig. 5 Eddy structure in sublayer region $y^+ = 8$ ($U_\infty = 0.43$ fps) of turbulent boundary layer on a flat plate (hydrogen bubble method) [18(b)].

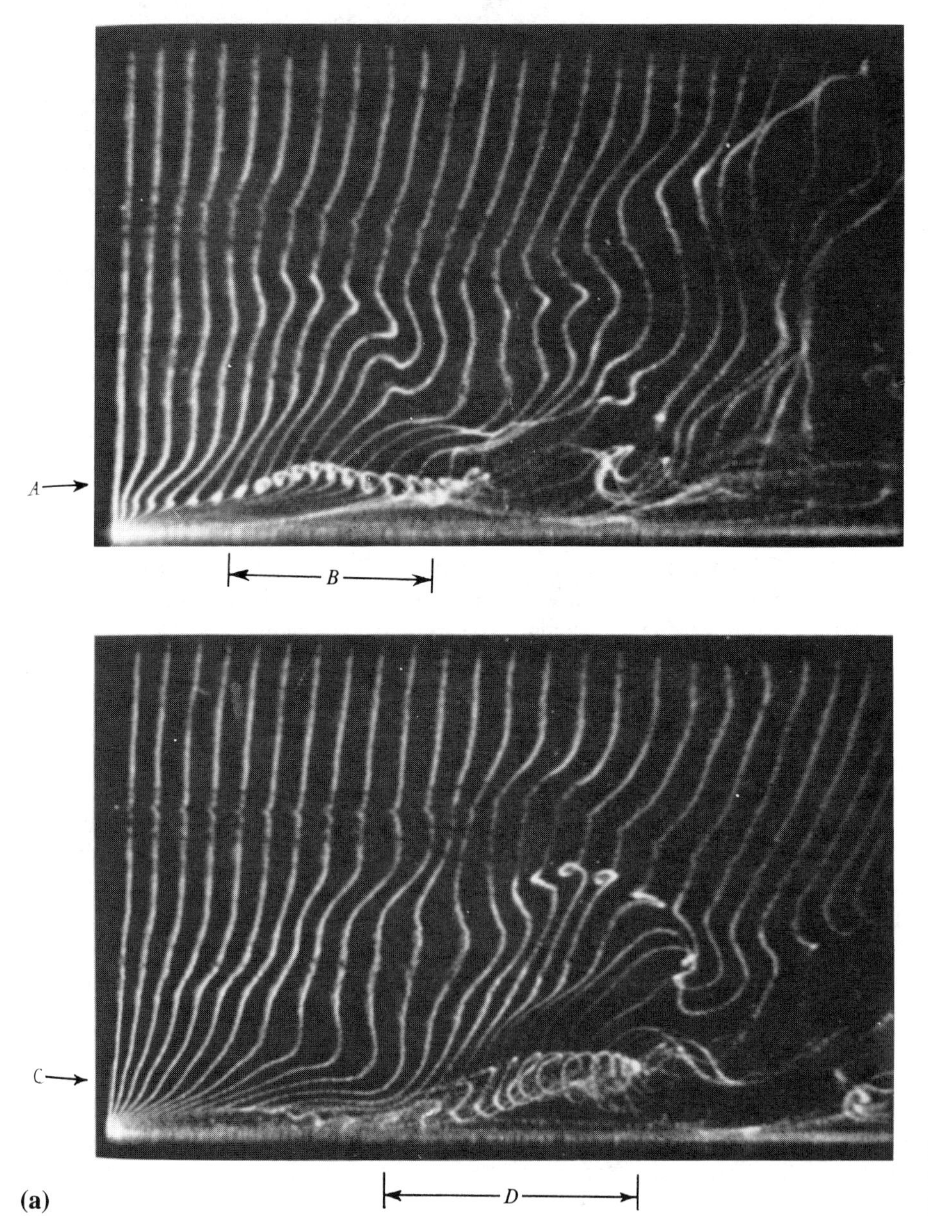

Fig. 6 (*a*) Formation of streamwise vortex motion during bursting. (Figure continues.)

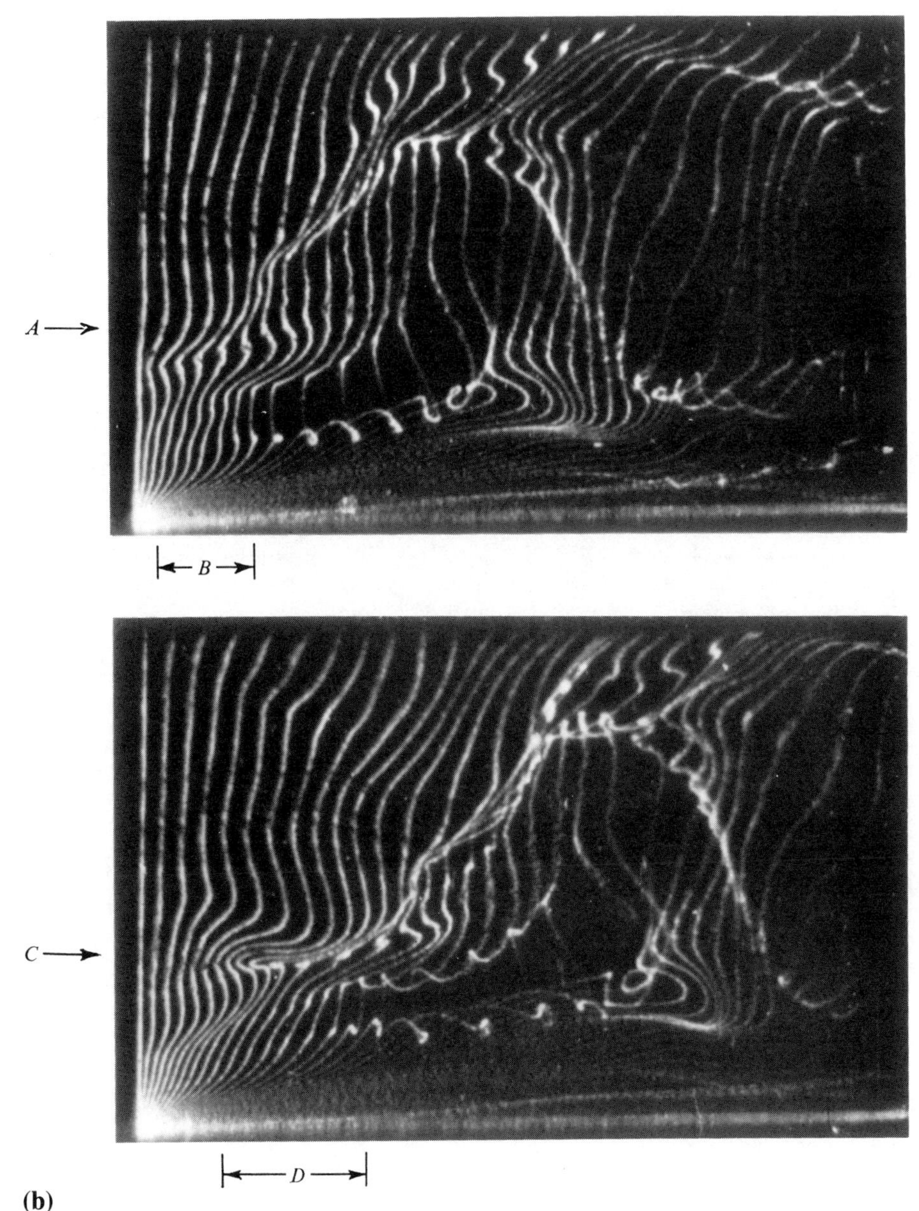

Fig. 6 (continued) (*b*) formation of transverse vortex during bursting [19(b)].

ture near the wall and paves the way for the current active research on coherent structures in turbulent flow. Many recent photographic results on instability and turbulence are shown in [16]. By examining the photographs in [10–12, 22, 23], one also gains some historical perspective on the development of flow visualization in experimental fluid mechanics. The most recent results can be found in [24–27].

It is clear that flow visualization has from early times played an important part in understanding flow structure and the development of theory. Review of the history of fluid mechanics from the viewpoint of flow visualization will no doubt shed considerable insight into the development of theories in fluid mechanics, since observation of flow phenomenon always precedes the development of theory.

III FLOW PHENOMENA IN INTERNAL FLOWS AND FLOW VISUALIZATION TECHNIQUES

A Flow Visualization Techniques for Internal Flows

A wide variety of experimental techniques are used in flow visualization. These range from the foreign tracer-injection method to the nonintrusive optical methods for compressible flow. The techniques are used not only to obtain flow pictures but also to give quantitative data about the details of the flow. Details of visualization techniques are usually buried in papers (journals or proceedings) or reports. Flow visualization techniques are described in books [28–30] or chapters in books [11, 31–44]. The principles, procedures, limitations, and evaluation techniques pointing out the main sources of error can be found in books. For convenience, a list of flow visualization techniques is given in Table 1 [30], and various techniques will be referred to later. Methods 22–25 are based on density change and methods 27–29 are based on fluid boundary.

B Classical Experiments

Many internal flow situations occur in fluid dynamics, and some classical examples are considered first. In this category, the classical experiments of Reynolds in 1883 [5] are best known; details can be found in [45]. His dye-injection experiments demonstrated conclusively that flow in a tube changes from laminar to turbulent over a given region of the tube and that the flow regime depends on Reynolds number (law of similarity). Prandtl presented 12 photographs in the 8 pages of his pioneer paper in 1904 [8], 12 photographs for flow around a cylinder and in diverging channels in a chapter (39 pages) of a handbook [46], and 24 photographs in his classic book [10]. Among the 69 photographs showing fluid motion near a solid wall for fluids (air, water) of small viscosity in [11], Figs. 39 and 40 for turbulent flow in an open channel are considered to have inspired Prandtl's concept of mixing length [13]. Bradshaw [13] notes that almost any method of flow visualization contradicts Prandtl's idea of momentum exchange via lumps of fluid. He speculates that the concept of mixing length may have originated from his rather simple flow visualization photographs based on Ahlborn's surface floating method using aluminum powders. It is also of interest to note that Prandtl used 33 photographs and a small water channel in one of his benchmark publications [47] to illustrate the flow phenomena.

Other classical examples of flow visualization experiments include Bénard cells [48] and Taylor vortices [49] for the onset of hydrodynamic instability phenomena and Schmidt's photographs illustrating a series of free convection phenomena by the schlieren method [50]. Photographs of Bénard convection can be found in [51, 52] and those of Taylor vortex flow in [52] among many others.

C Hydrodynamic Instability Phenomena due to Body Forces

When body forces such as buoyancy, centrifugal, or Coriolis forces act in a direction normal to the main flow in

Table 1 Classification of Flow Visualization Techniques [30]

I. Surface tracing methods	
1. Oil film	2. Oil dot
3. Mass transfer method	3.1. Soluble chemical film
3.2. Sublimation	3.3. Evaporation (drying)
4. Electrolytic etching	5. Thermosensible paint
6. Pressure-sensitive paper	
II. Tuft methods	
7. Surface tuft	8. Tuft grid
9. Depth tuft	10. Tuft stick
III. Tracer injection methods	
11. Streakline	12. Pathline
13. Timeline injection	14. Particle suspension
15. Surface floating	
IV. Dye (tracer) production (or chemical reaction tracer) methods	
16. Chemical reaction	17. Electrolytic
18. Photochemical	
V. Electrical control methods	
19. Hydrogen bubble	20. Spark tracing
21. Smoke wire	
VI. Optical methods	
22. Shadowgraph	23. Schlieren
24. Mach–Zehnder	25. Holography
26. Laser speckle photograph	27. Stereophotography
28. High-speed camera	29. Moiré
30. Mirage	31. Streaming birefringence
32. Spark discharge	33. Light luminescence
34. Laser-induced fluorescence	35. Ultraviolet absorption
VII. Others	
36. Thermography	37. Computed tomography
38. Acoustic intensity	39. Computer-aided visualization
40. Computational fluid dynamics	

channels, convective instability or a hydrodynamic instability problem may arise in the form of longitudinal vortices. Well-known examples are longitudinal vortices in a horizontal-plane Poiseuille flow heated from below [53, 54], Taylor vortices in two concentric infinite cylinders with the inner cylinder rotating and the outer cylinder at rest [49], Dean vortices in a fully developed laminar flow in curved parallel-plate channels [55], and longitudinal vortices in plane Poiseuille flow subjected to a spanwise rotation [56]. Bénard cells occur in a horizontal fluid layer heated from below [57], and the Bénard problem is also of considerable interest in the field of meteorology. Görtler vortices occur in the laminar boundary layer along a concave wall [58].

Flow visualization has played an important role in the understanding of thermal and centrifugal instability phenomena since the beginning of this century. Thermal and hydrodynamic instability problems are well covered in monographs [59–62] and review articles [51, 52, 61–66], where excellent flow visualization photographs can also be found. Probably the most striking examples of visualization data are Bénard cells and Taylor vortices; some examples are shown in Fig. 7. Figure 7*a* shows a plan of the convection cells in a silicone oil with regular hexagons as the predominent polygons [67]. Figure 7*b* shows G. I. Taylor's famous 1923 photograph of the flow between concentric circular cylinders with the inner cylinder rotating while the outer cylinder was at rest. The toroidal vortices now bear his name. Coles [68] carried out comprehensive experiments on the Taylor problem and presented many visualization data in the postcritical regime. One series of photographs is reproduced in Fig. 7*c*. Based on flow visualization experiments, Coles [68] proposed the concept of transition by spectral evolution to describe flows in which the laminar flow evolves through a succession of gradually more complex states to reach turbulent flow. Apparently, similar phenomena involving linear and nonlinear processes also occur in the Bénard convection problem, but the flow visualization experiments in the supercritical regime for the Taylor problem are relatively easier than for the Bénard problem. Because of the relative simplicity, flow visualizations for Couett flow and the Bénard problem will continue to receive attention in future studies involving hydrodynamic stability and turbulent flow.

Longitudinal vortices caused by the unstable vertical distribution of density similar to that in Bénard convection in a horizontal-plane Poiseuille flow heated from below with constant upper and lower wall temperatures are shown in Fig. 8*a* [54] at Re = 8.3 and Rayleigh number Ra = 4160. The experimental critical Rayleigh number for the onset of

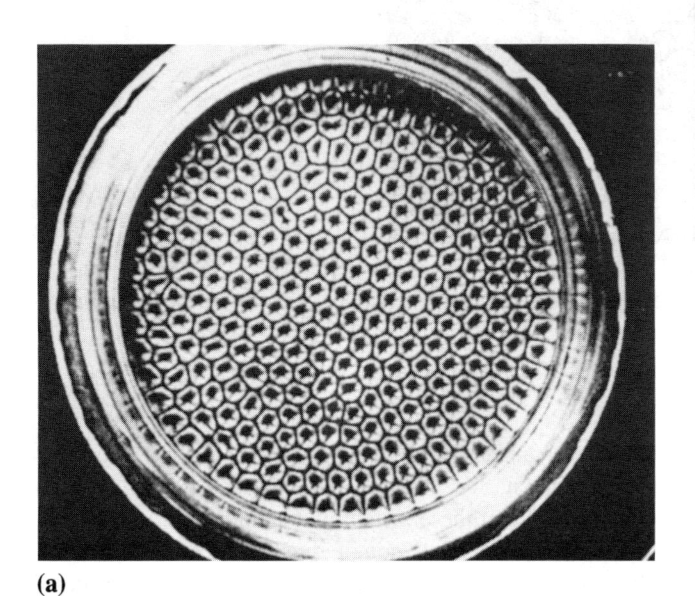

(a)

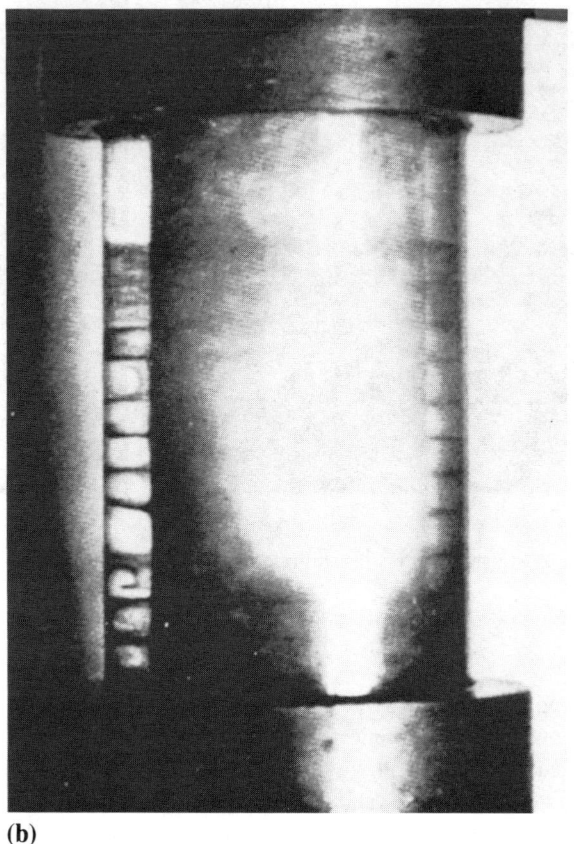

(b)

Fig. 7 (*a*) Bénard cells under an air surface [67]; (*b*) Taylor vortices between rotating cylinders [49]; (Figure continues.)

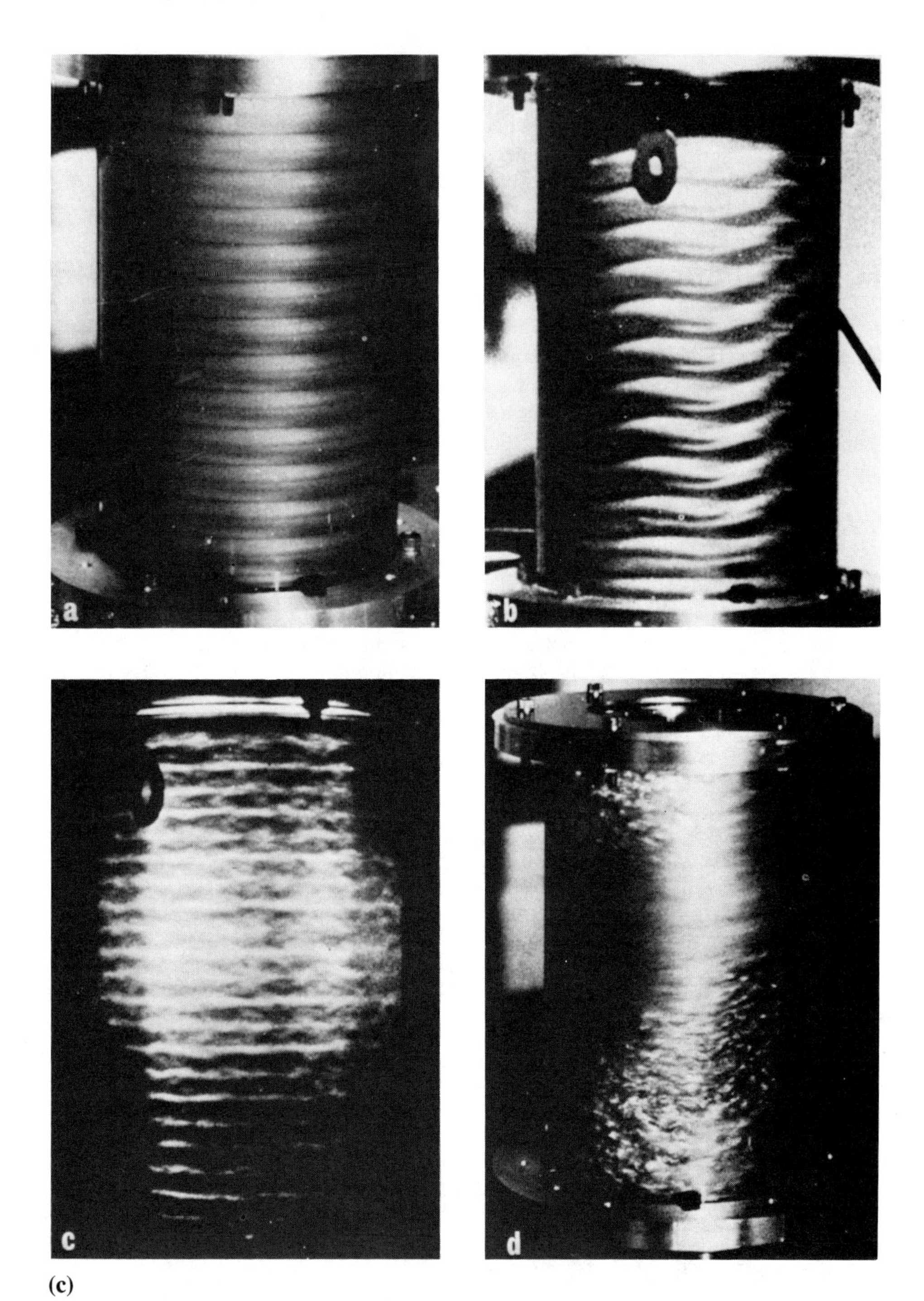

(c)

Fig. 7 (continued) (*c*) instability of rotating Couette flow in postcritical regime [68].

secondary flow is $Ra_c = 1807$, for this case, and the deviation from the theoretical value 1708 is 6%. Dean vortices caused by centrifugal instability in a curved rectangular channel with aspect ratio 12 and curvature ratio $a/R = 0.25$ are shown in Fig. 8*b* [69] at Dean number $(u_m a/\nu)(a/2R)^{1/2} = 69.6$, where R = radius of curvature and a = channel height. Both visualization photos in Fig. 8 were obtained by the smoke-injection method. Flow visualization data for centrifugal and thermal instability problems reveal that these instability phenomena are analogous [70–72]. Qualitatively longitudinal vortices in horizontal Blasius flow heated from below [73, 74] can be considered to be analogous to Görtler vortices in the flow along concave walls [75, 76].

Fig. 8 (*a*) Secondary flow due to thermal instability in horizontal-plane Poiseuille flow heated from below at Re = 8.3, Re = 4160 [54]; (*b*) Dean vortices in a curved rectangular channel [69, 93].

D Secondary Flow Patterns in Curved, Heated, and Rotating Channels

When body forces such as buoyancy, centrifugal, or Coriolis forces act in a direction normal to the main flow in ducts or channels, secondary flow occurs in a cross section normal to the main flow. This class of problems belongs to the boundary-value problems, in contrast to the eigenvalue problems considered in Sec. III.C. Some examples include secondary flow caused by buoyancy forces in heated or cooled horizontal tubes, centrifugal forces in curved pipes, and Coriolis forces in tubes where the main flow is rotating about an axis perpendicular to the tube axis. Trefethen [77] points out that secondary flow phenomena in bent tubes, heated horizontal tubes, and rotating radial tubes are analogous. Because of the practical importance in applications, the secondary-flow problems relating to heat and mass transfer have been studied both theoretically and experimentally by many investigators in the past. Selected secondary-flow patterns for flow in curved and heated pipes obtained recently by the smoke-injection method are presented for reference.

Secondary-flow patterns revealed by flow visualization and numerical solution for fully developed laminar flow in a curved square channel are shown in Fig. 9 [78]. At higher Dean numbers, the centrifugal instability problem (onset of Dean vortices in the form of an additional pair of vortices near the center of the outer concave wall) also occurs in a curved square channel. The same problem also occurs in a curved circular pipe; the results are shown in Fig. 10 [79]. For combined free- and forced-laminar convection in horizontal rectangular channels with uniform wall heat flux, the convective instability problem (onset of longitudinal vortices) occurs at higher ReRa [80]. Secondary streamlines and isotherms from numerical solution for the case of a square channel showing convective instability phenomena are shown in Fig. 11 [80]. Figures 10 and 11 clearly show the analogy between the two physically different instability problems, and flow visualization provides considerable physical insight. Visualization photos showing two or three pairs of vortices in a square channel are presented in [81]. Visualization data in the thermal-entrance region of a horizontal rectangular channel for convective instability phenomena are presented in [82, 83]. Visualization results in secondary flow patterns at the exit of an isothermally heated curved pipe (180° bend) in vertical and horizontal planes are shown in [84].

The developing secondary-flow patterns in the entrance region of a curved pipe with parabolic entrance velocity profile are shown in Fig. 12 for Dean number K = 364 and 520 [71]. The results can be compared with those obtained by flow visualization, measurements, and numerical solutions shown in [85] for Dean number K = 333. Unsteady flow in curved pipes is important for studying blood flow in the human arterial systems and heat and mass transfer in engineering systems. Secondary-flow patterns for oscillating viscous flow in a curved pipe obtained from experimental [86] and theoretical [87] investigations are shown in Fig. 13. The generation of liquid aerosols for the visualization of oscillatory flows in a curved rectangular channel is shown in Fig. 8 in [88]. Other secondary-flow patterns in curved pipes or ducts and in heated channels are reviewed in [72]. Flow visualization of the flow field caused by body forces was reviewed by Nakayama et al. [89], giving examples for rotating fluids and natural convection phenomena in high-temperature and extremely low-temperature equipment.

In a flow field with significant body-force effect, the flow is usually three dimensional, and phenomena such as the existence of dual solutions and bifurcation may occur, depending on the relative magnitude of the body force, inertia, and viscous forces present. For this class of problems, the

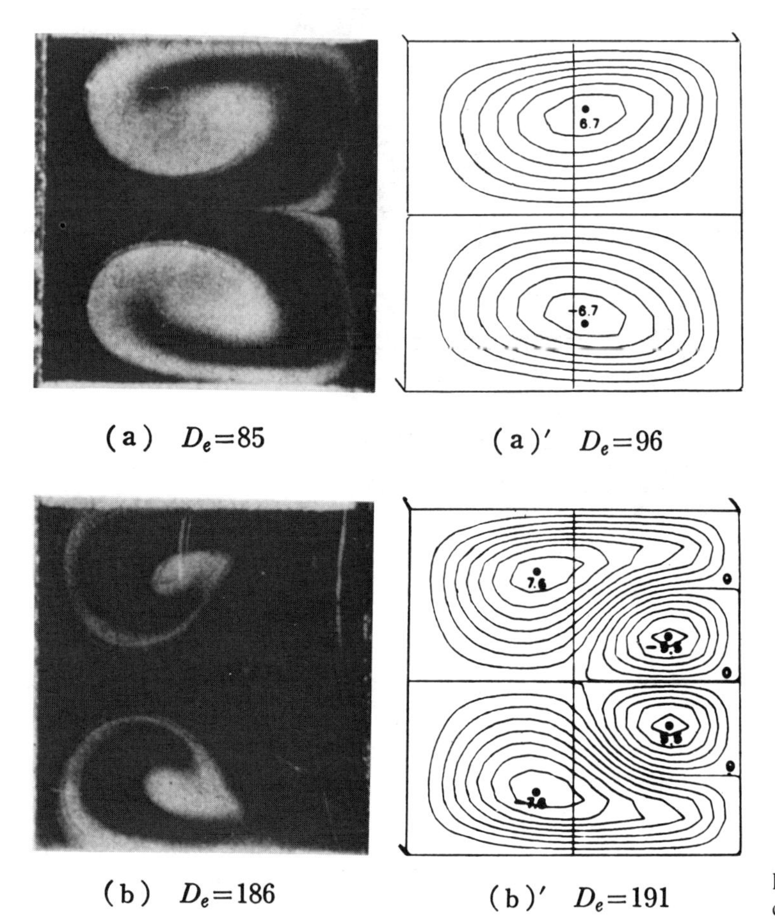

Fig. 9 Secondary-flow patterns in a curved square channel [78].

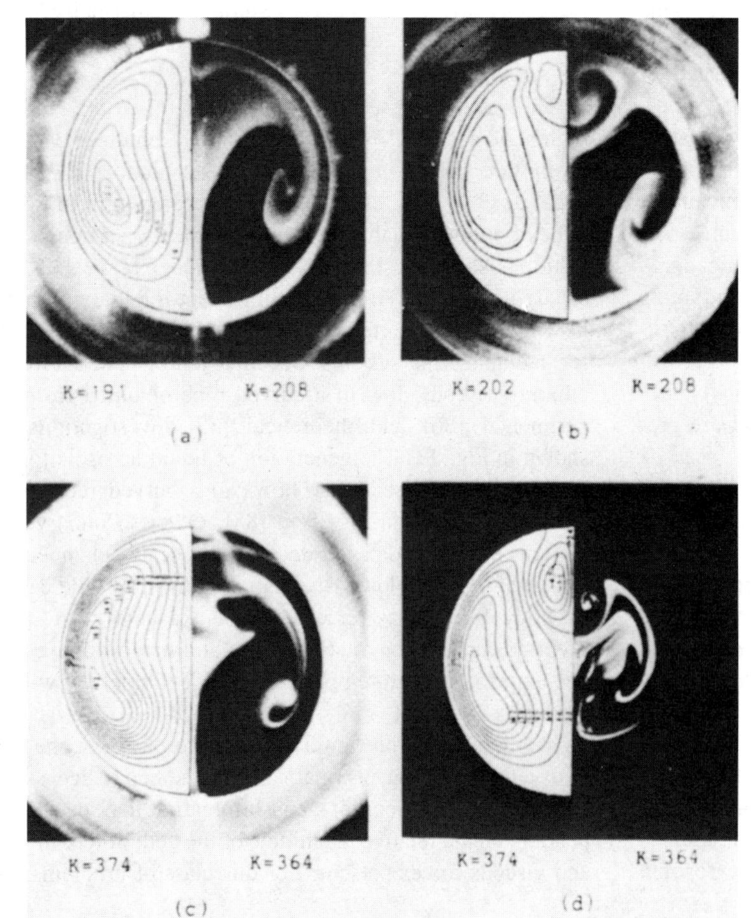

Fig. 10 Secondary-flow patterns in curved tubes [79].

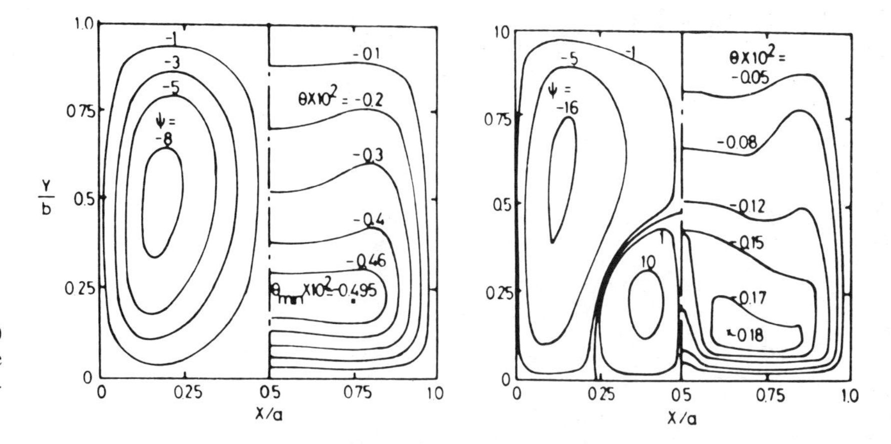

Fig. 11 Secondary streamlines (left) and isotherms (right) for convective instability in a horizontal square channel [80].

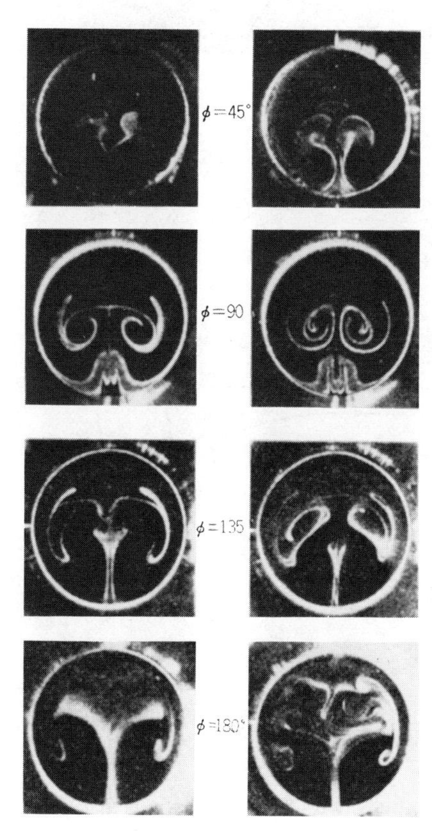

Fig. 12 Developing secondary-flow patterns in the entrance region of a curved circular pipe [71].

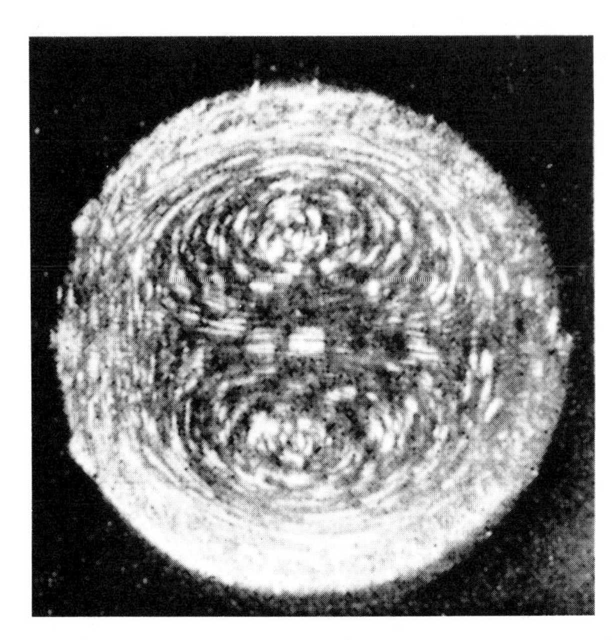

(a)

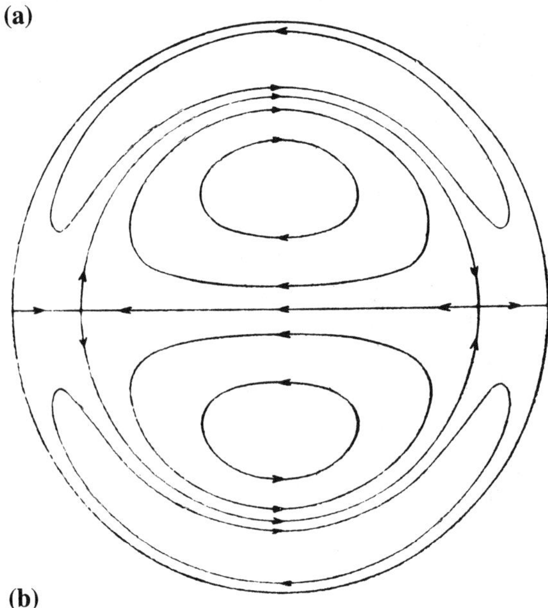

(b)

Fig. 13 Secondary flow patterns for oscillating flow in curved pipe: (*a*) Experiment [86]; (*b*) theory [87].

numerical solution of the Navier–Stokes equations also provides the flow pattern. Flow visualization can be used in confirming the predictions of the numerical solution.

It is useful to observe the analogy among various secondary-flow problems caused by different body forces; the governing physical parameters are listed here for reference:

$$\text{Taylor number} \quad \mathrm{Ta} = \left(\frac{Ud}{\nu}\right)\left(\frac{d}{R}\right)^{1/2}$$

$$\text{Görtler number} \quad \mathrm{G}_{\delta} = \left(\frac{U\delta}{\nu}\right)\left(\frac{\delta}{R}\right)^{1/2}$$

$$\text{Dean number} \quad \mathrm{K} = \mathrm{Re}\left(\frac{a}{R}\right)^{1/2}$$

$$\text{Rayleigh number} \quad \mathrm{Ra} = \frac{g\beta\,\Delta T h^3}{\nu\alpha}$$

$$\text{Ekman number} \quad \mathrm{E} = \frac{\mu}{\rho\Omega L^2}$$

$$\text{Coriolis parameter} \quad \frac{\mathrm{Re}}{E}$$

$$\text{Rotating Grashof number} \quad \mathrm{Gr}_r = \frac{R\Omega^2\beta\,\Delta T L^3}{\nu^2} \tag{1}$$

where Re = Reynolds number, ΔT = temperature difference, **U** = characteristic velocity, d, h, L, and δ = characteristic lengths, a = radius of tube, g = gravitational acceleration, R = radius of rotation or radius of curvature, β = coefficient of thermal expansion, μ = viscosity, ρ = density, Ω = angular velocity, ν = kinematic viscosity, and α = thermal diffusivity.

Akiyama and his co-workers [90–98] carried out a series of flow visualization studies on secondary-flow problems involving centrifugal or buoyancy forces. Further studies are reported in [99–125]. The developing axial velocity profiles in a U bend of square cross section are visualized by the laser-induced fluorescence method in [105, 107, 108]; the visual results are shown in Fig. 14*a* [108]. Figures 14*b,c* show developing velocity profiles revealed by timelines in a curved circular pipe using the hydrogen bubble method [85].

Secondary flow in curved pipes delays the onset of transition from laminar to turbulent flow; the problem was originally studied experimentally by G. I. Taylor [126] in 1929. When fully developed turbulent flow enters a coiled tube, the relaminarization phenomenon occurs. Flow visualization study secondary flow in the relaminarization process in curved pipes apparently has not been carried out so far; when it is, it should provide considerable insight into the relaminarization mechanism. The relaminarization phenomenon also occurs in channels involving other body forces such as buoyancy and the Coriolis forces; it also remains to be investigated in the future. The relaminarization of fluid flows in general was reviewed in [127], and the visual study of laminarization of turbulent flow in an axially rotating pipe is reported in [128, 129].

E Photographs of Internal Flow Phenomena

Visual results for the internal flow phenomena are scattered in many sources (journals, proceedings, reports, and books). A collection of photographic results is available in book form [16, 130–134]. Visual results can also be found in proceedings [24–27, 135–137] and books on flow visualization [28–30]. The *Journal of the Flow Visualization Society of Japan* has published visual results in color and in black and white regularly as frontispieces since its inception in 1981. In the 1960s the Educational Development Center produced 16-mm sound films and 8-mm film loops in cassettes for the National Committee for Fluid Mechanics Films. The subject listing of film loops is given in [131], and the films are distributed by the Encyclopaedia Britannica Educational Corporation in the United States and many other countries. The topics of the 16-mm sound films are discussed in 21 chapters in [131], and the purposes and references for the 8-mm silent film loops are also given in [131]. The following 8-mm silent film loops on internal flows are available: (1) propagating stall in airfoil cascade, (2) incompressible flow through area contractions and expansions, (3) flow from a reservoir to a duct, (4) flow patterns in venturis, nozzles, and orifices, (5) flow regimes in subsonic diffusers, (6) wide-angle diffuser with suction, (7) flow through right-angle bends, (8) flow through ported chambers, (9) flow through tee-elbow, (10) secondary flow in a bend, and (11) secondary flow in a teacup.

The literature on visualization techniques and visual results is now very extensive, and only selected examples of visual results are mentioned here from the viewpoint of understanding basic physical phenomena. Reynolds's celebrated 1883 visual investigation of flow instability in a transparent tube was documented in sketches rather than photographs. The visual results repeating Reynolds's dye experiment a century later, using his original apparatus at the University of Manchester, are presented on page 61 in [16]. Reynolds's experiment has since been repeated by many investigators, and the visual results can be found, for example, in [6, 138–140]. Turbulent slug in pipe flow was studied by E. R. Lindgren, from 1957 to 1960, who used optical methods to study the flow of a birefringent liquid (bentonite sol), and is discussed by Coles [141]. Reynolds's visual experiments using color bands in glass tubes revealed two modes of flow (laminar and turbulent) and intermittency in turbulent shear flow.

At this point, it is interesting to note Kline's observation on a selective history of the role of visual studies of coherent structures of turbulent boundary layers [142]. Reynolds concluded that turbulence is more or less "a random fluctuation on the basic mean-flow" based on his visual studies with dye injection at the tube center. Apparently, this view of turbulence prevailed until the 1950s, when coherent structure in turbulent boundary layers was revealed conclusively by flow visualization studies with markers used directly in the wall layers [142]. Kline [142] noted that in Reynolds's experiment, the dye was injected in the center of a pipe, and he viewed the markers from the reference frame of laboratory coordinates. Such a marking location and view do give an appearance of relatively steady random fluctuations superposed on a near flow [142]. Reynolds's conclusions were consistent with the marking

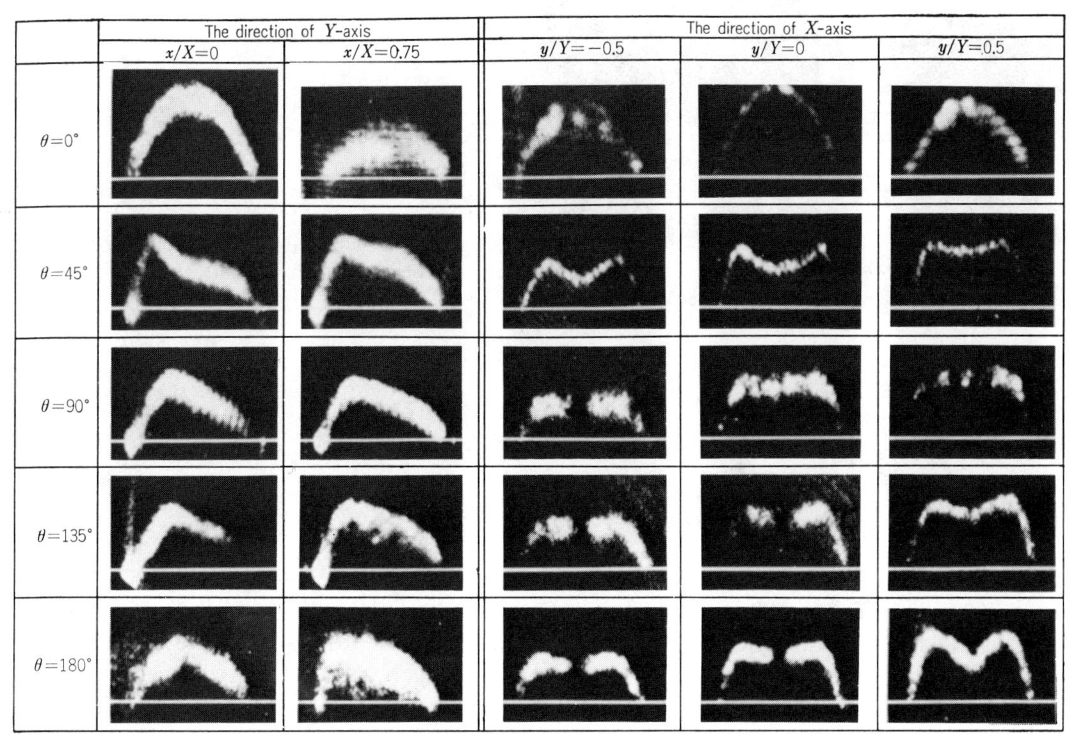

(a)

Fig. 14 (*a*) Developing velocity profiles in a curved rectangular channel (U-bend) by the laser-induced fluorescence method, $\theta = 0°$, start of bending; x, y = coordinates in a cross section [108]. (Figure continues.)

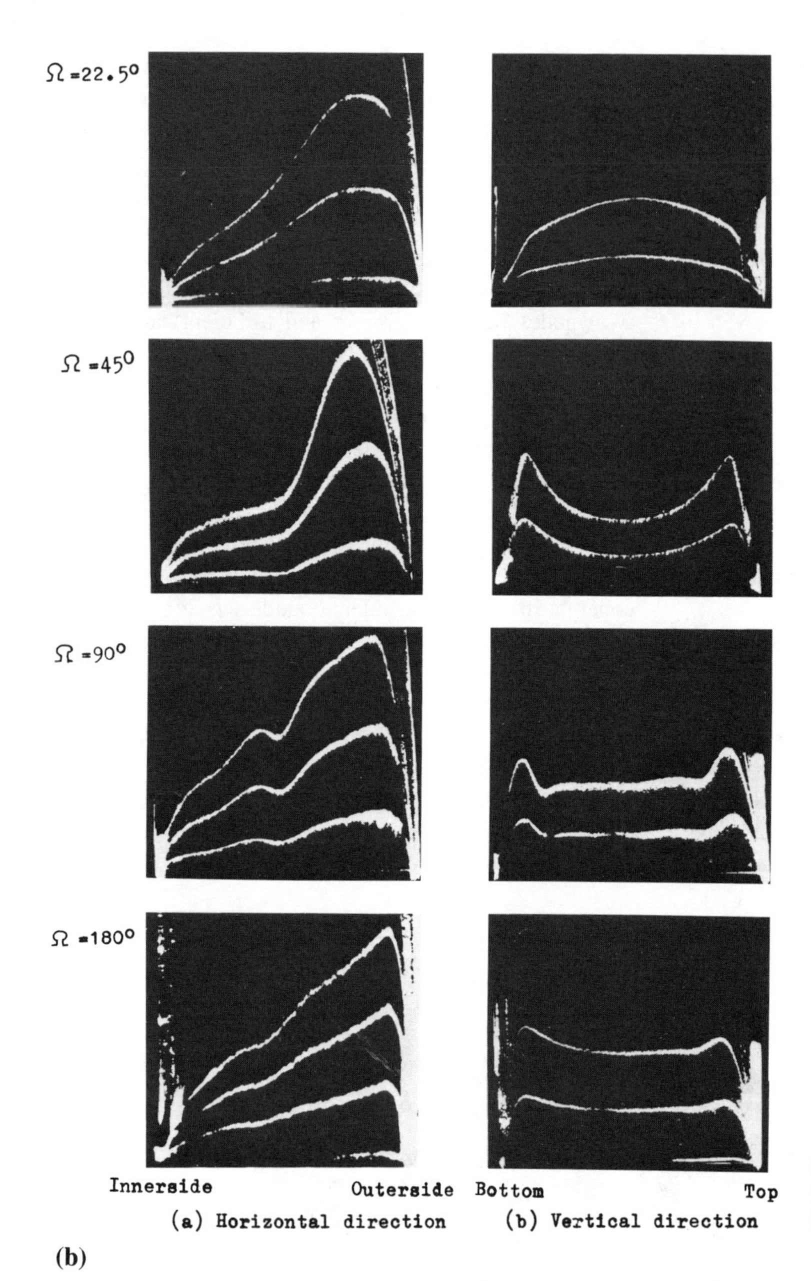

Fig. 14 (continued) (*b*) developing velocity profiles (timeline of main flow) in a curved circular pipe, Re = 1000, De = 333 [85]. (Figure continues.)

method and the reference frame he employed. It is seen that interpretation of the visual data is very important.

The two photographs for turbulent flow in an open channel that motivated Prandtl's development of the mixing-length theory for turbulent flow are shown in Fig. 15 [11]. Similar photographs for turbulent flow of water in a channel, visualized by a camera moving at various velocities in the flow direction, are also shown in [22, 143]. Prandtl used flow visualization extensively, and one further example is shown in Fig. 16 [144]. Figure 16 shows flow-separation phenomena in a diverging channel. Kline [142] noted that the Prandtl–Ahlborn pictures, taken in the late 1920s with a traversing camera, show clear evidence of quasicoherent structure in turbulent flow.

Many revealing photographs for internal-flow patterns involving abrupt contraction or expansion in a conduit, two-

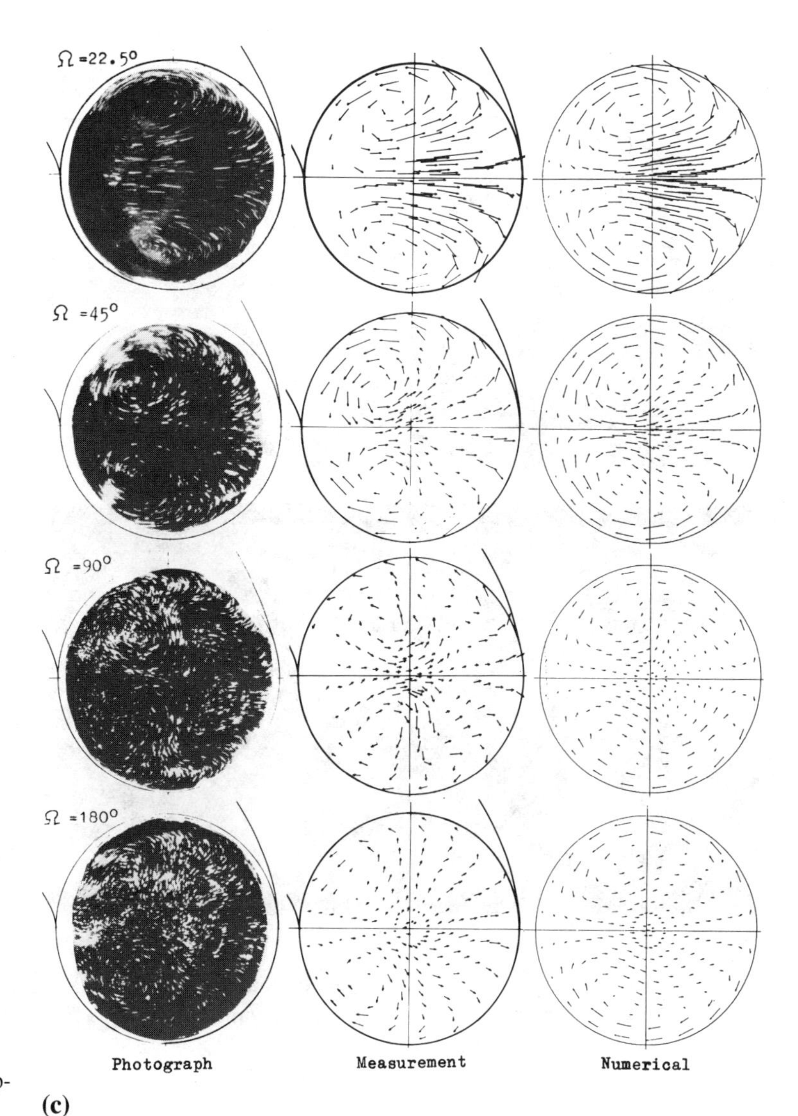

Fig. 14 (continued) (*c*) secondary-flow development, Re = 1000, De = 333 [85].

dimensional plate orifice, rounded orifice, venturi, throat, nozzle, diffuser, and others can be found in [16, 132, 138, 145].

The developing laminar flow (velocity profiles) in a circular tube and in a parallel-plate channel visualized by the hydrogen bubble method is shown in [132]. Clear pictures of the transition downstream of Tollmien–Schlichting waves, turbulent spots, and the structure of the turbulent boundary layer along a flat plate are shown in [16, 22, 132]. Visualization of basic flow phenomena is useful for physical understanding. Many examples involving jet, cavitation, non-Newtonian flow, and shock waves [146] for internal flow can also be found in [16, 132]. The velocity profile of Blasius flow (boundary layer) on a flat plate revealed by the chemical-reaction tracer method is shown as Fig. 30 in [16]. Two earlier flow visualization photos obtained by dye

(a)

(b)
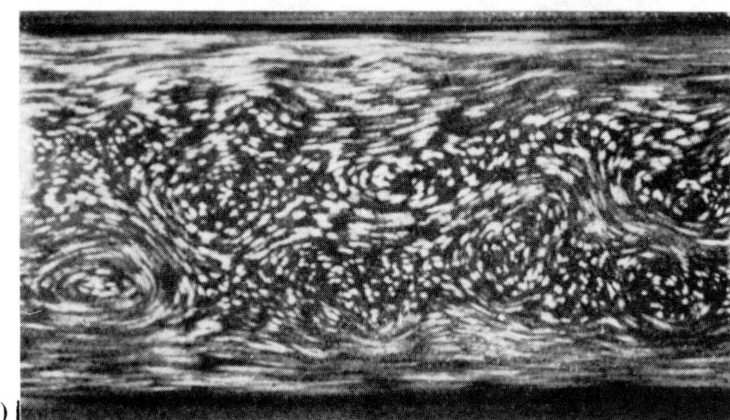

Fig. 15 Turbulent flow in an open channel: (*a*) Camera speed is about equal to the water near the wall; (*b*) camera speed is that of the water in the center [11].

injection for the vortex deformation in forced boundary layer flow with tripped transition [147] and by smoke threads for the natural transition to turbulence of free-convection boundary layer [148] are shown for reference in Figs. 17*a,b* respectively.

IV EXAMPLES OF APPLICATIONS

Internal flows can be classified into those in (1) pipes and valves; (2) equipment (heat exchangers, electronic equipment, steam generators, nuclear reactors, combustion equipment, reactors, injectors, separators); (3) furnaces (boilers, industrial furnaces); (4) chambers or rooms (ventilation, drying); (5) tanks; (6) towers (spray, distillation, reaction, bubble); (7) fluid machinery (pumps, fans, blowers, compressors, turbomachinery); (8) heat engines; (9) fluidic amplifiers; (10) biological systems; and (11) others. The internal flow problems are further complicated by vortex flow, jets, convective flow, particle flow, mixed-phase flow, and magnetic flow. It is obvious that the subject of internal flow is very extensive, and in many cases flow visualization studies were also carried out as part of experimental investigations involving measurements by probes. Selected examples of visualization applications are considered next.

A Internal Flow in Turbomachinery and Engines

Because of rotation, curvature effect, high velocity, complicated flow passage, and other factors, internal flow in fluid machinery represents one of the most complicated fluid mechanics problems; one cannot attempt a theoretical analysis from the inlet to the outlet of the rotating machinery in the same way as the entrance flow problem for internal flow in ducts or channels. The secondary flow caused by centrifugal and Coriolis forces is also a complicating factor. Thus, measurements, flow visualization, and computer simulation are useful tools in understanding the flow phenomena inside the rotating machinery.

Various visualization techniques for turbomachinery have been used in the past, and review articles [149–156] are available. Recent visualization results obtained by the spark-tracing method [156] are shown in Fig. 18, where the relative flow patterns of a centrifugal blower impeller along the shroud plane, the intermediate plane, and the hub plane are shown for large, design, and small flow rates. It is seen that at a large flow rate the main flow moves fast along the negative pressure side and a large flow-separation region occurs on the pressure side. At a design flow rate, the flow-separation region becomes small and the main flow moves almost uniformly along the outlet edge. At a small

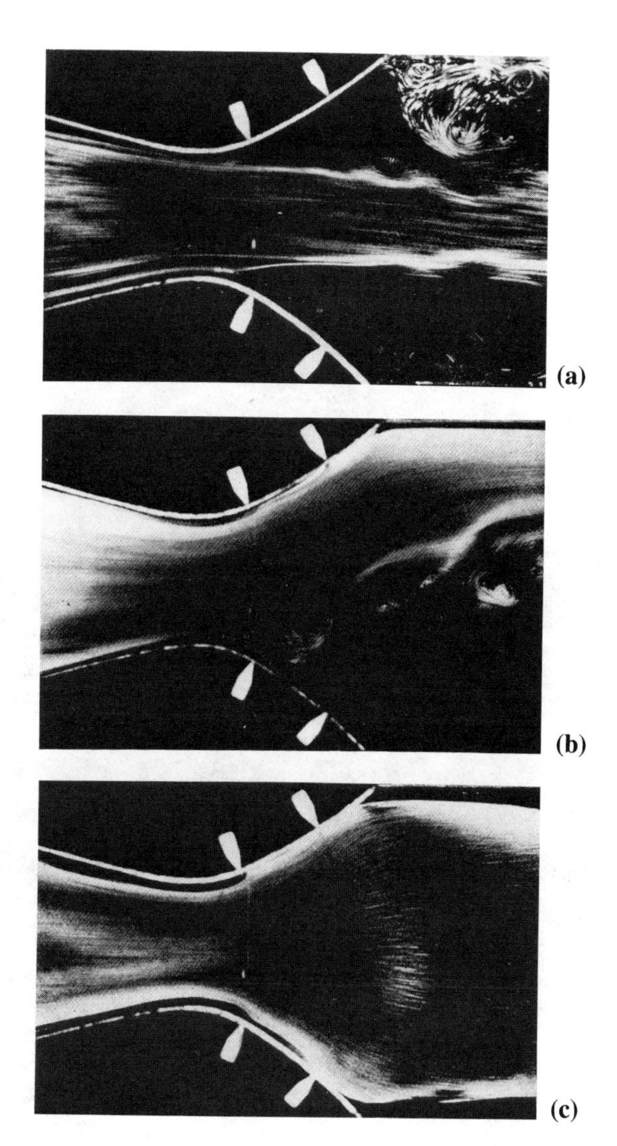

Fig. 16 Flow-separation phenomena in a sharply diverging channel: (*a*) No suction; (*b*) suction on upper side; (*c*) suction on both sides [11, 144].

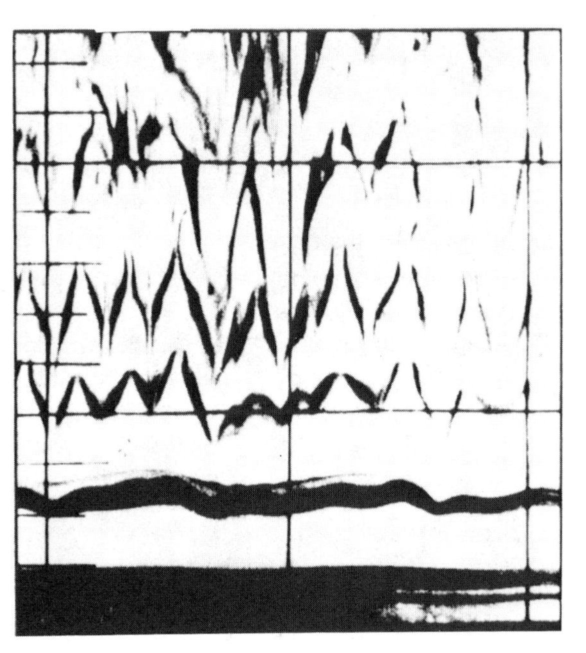

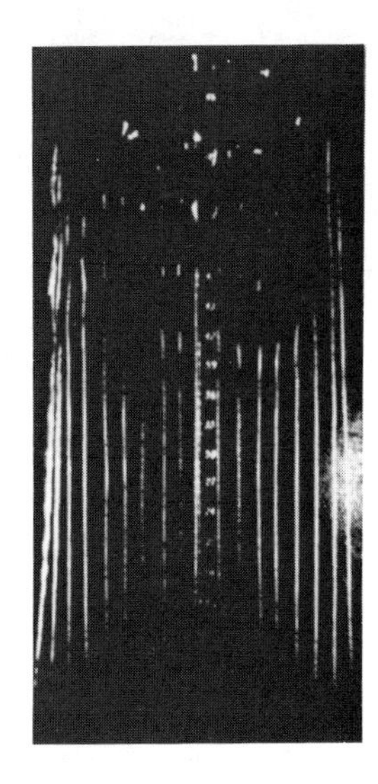

Fig. 17 Visualization of transition from laminar to turbulent flow: (*a*) Dye injection in forced boundary layer flow with tripped transition; (*b*) natural transition in free-convection boundary layer made visible by smoke threads [148].

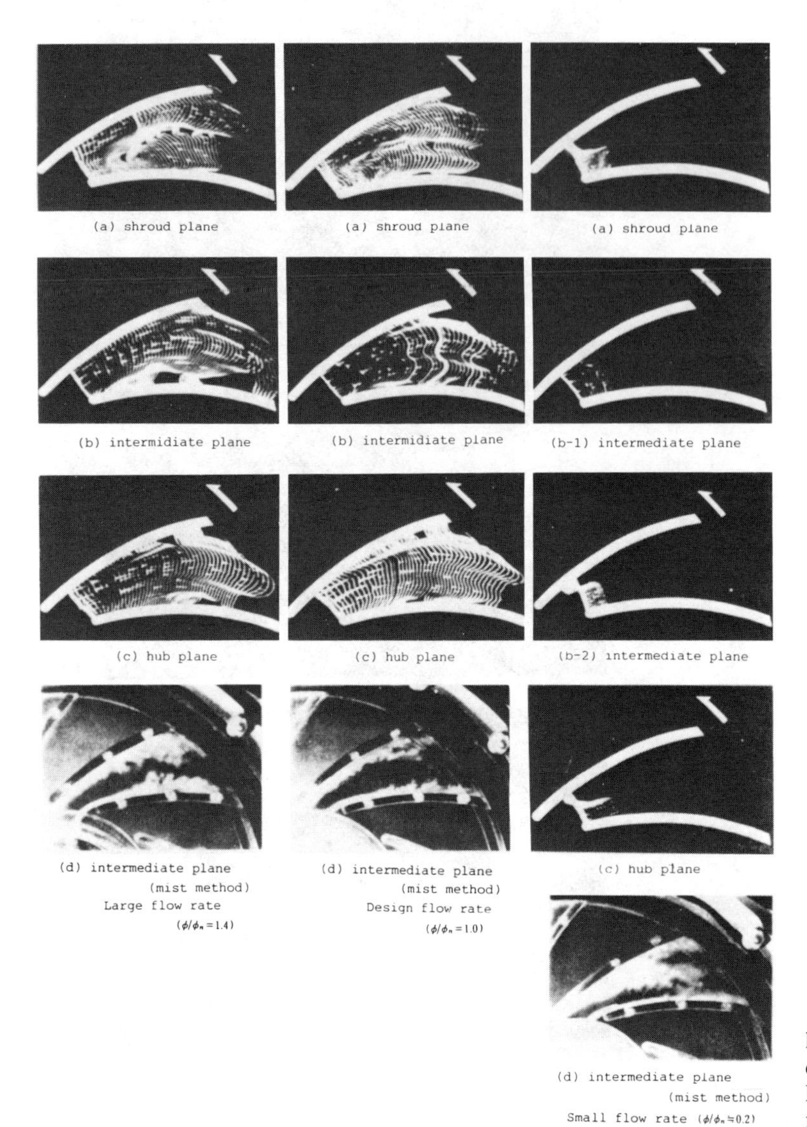

Fig. 18 Effects of flow rate on relative velocity distribution in centrifugal blower impeller: (*a*) Large flow rate; (*b*) design flow rate; (*c*) small flow rate [156].

flow rate, the main flow moves slowly along the pressure side, and a back flow may or may not occur on the negative pressure side. Thus it is seen that an unstable surging develops. The details of the flow visualization technique are given in [156]. The spark-tracing method was first applied to visualize flow patterns in rotating centrifugal impellers in 1966 by Fister [157] and subsequently was used by other investigators [156, 158–161].

The fluid motion within the cylinder of an internal-combustion engine is reviewed in [162] using visual results. Flow visualization in the Space Shuttle's main engine is discussed in [163], and optical methods in combustion research are reviewed in [164–167]. Flow visualizations in explosion, detonation, and shock waves in internal flow can be found in [24–27, 168–174].

B Flow inside Valves

Internal flow inside valves is rather difficult to visualize, and one example [175] is considered here. Acoustic noise of intermittent and discrete tone characters is frequently generated in a cage-type valve by the throttling of a compressible fluid [176]. Multiple supersonic jets collide at the center of the cage, and the disturbances of the collision act as sources of cross-sectional wave motion in the cage cavity. Thus the frequencies of the tone can be determined by the acoustic eigenmodes of the cavity. A control valve may be regarded as an energy converter. Within a certain range of pressure ratios, a strong discrete tone of several kilohertz (kHz) can be generated, and the noise level may increase considerably. The flow patterns inside the cage-type control valve with an 80-mm inlet dia. (jet dia. = 7 mm) are shown in Fig. 19 for pressure ratio P_1/P_2 = 2–15 with the corresponding pressure fluctuation levels (PFL) shown for reference. The experimental apparatus for flow visualization using a multiframing shadowgraph technique is similar to that shown in [176, 177].

At a low pressure ratio P_1/P_2 = 4–5, two jets collide near the center of the cage, but at a higher pressure ratio the collision occurs near the outlet of one jet. The pressure-fluctuation level inside the tube also changes with the collision pattern of the two jets. Photo A shows a flow pattern when the flow is in a subsonic region and the collision of the jets occurs at the center. For photos B and C, two opposing jets pass each other and vibrations are generated. For photo D (pressure ratio ≈ 5) the collision plane may be located near the lower part or near the center, and the hysteresis phenomenon arises. The PFL value changes rather smoothly when the pressure ratio becomes greater than 5, and the positioning of the collision plane is irregular. The flow pattern inside the valve is closely related to noise-generation characteristics. The details of the flow patterns for flow structures A, B, and C are shown in Fig. 20*a,b,c* respectively. The eigenmode frequencies of the cage cavity can be estimated by solving the wave equation in cylindrical coordinates [175]. The discrete noise frequency relating to flow pattern B in Fig. 20 is found to coincide with the frequency of modes (n = 2, m = 1), 5.8 kHz, where the indices n and m represent angular and radial modes respectively. The flow patterns for one cyclic oscillation of supersonic flow relating to wave mode (2,1) are shown in Fig. 21. For this case, the disturbances from the jet system behave as the quadrupole due to the feature of the collision, and thus the mode (2,1) can be sustained. This mode occurs at a relatively high pressure ratio (2–7). Other examples of flow visualization in valves can be found in [178–186]; one further example is shown in Fig. 22 [180].

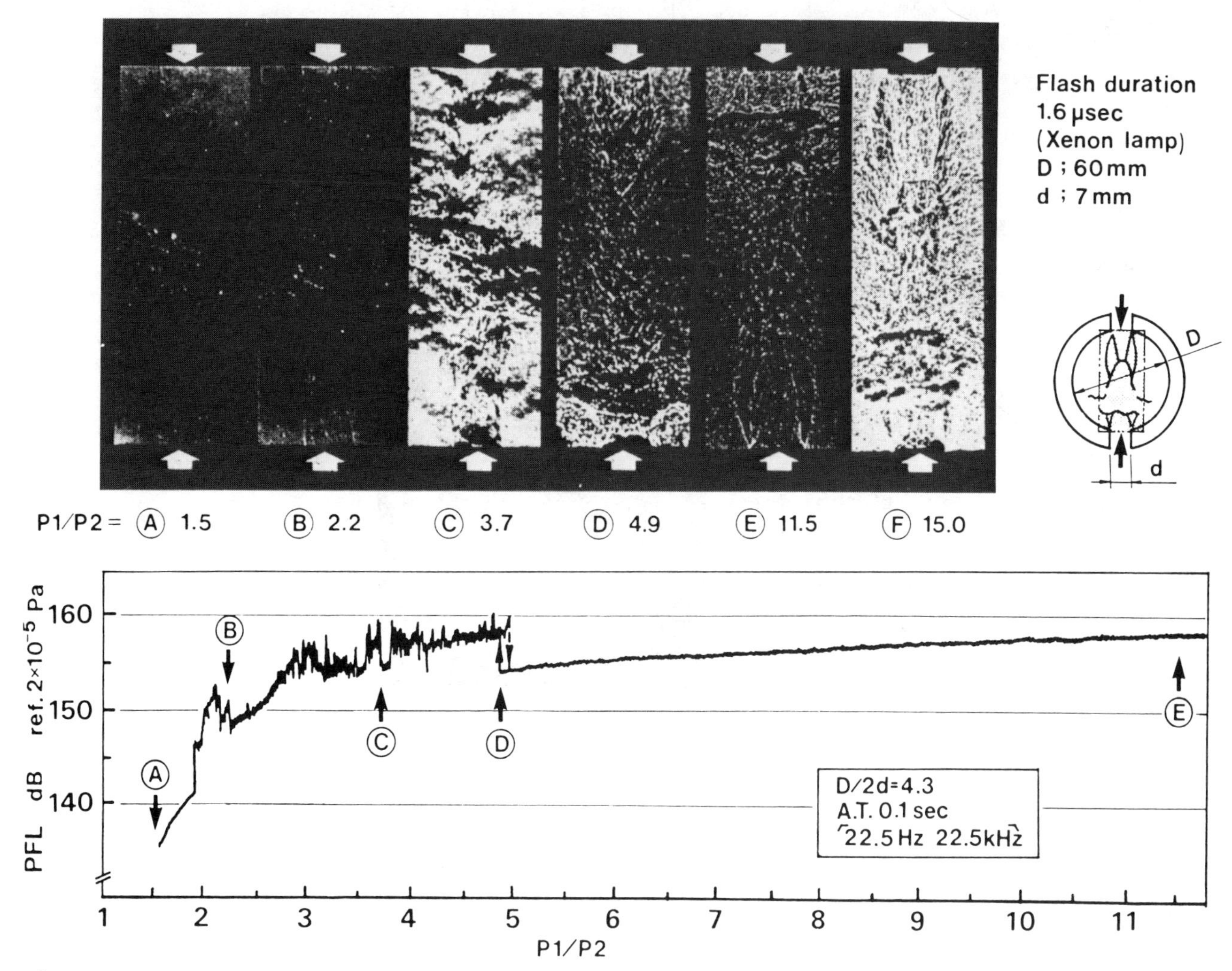

Fig. 19 Flow patterns and pressure-fluctuation level (PFL) inside a cage type valve as a function of pressure ratio P_1/P_2 [175].

FLOW STRUCTURE Ⓐ

Large eddies are formed along the jet boundary.

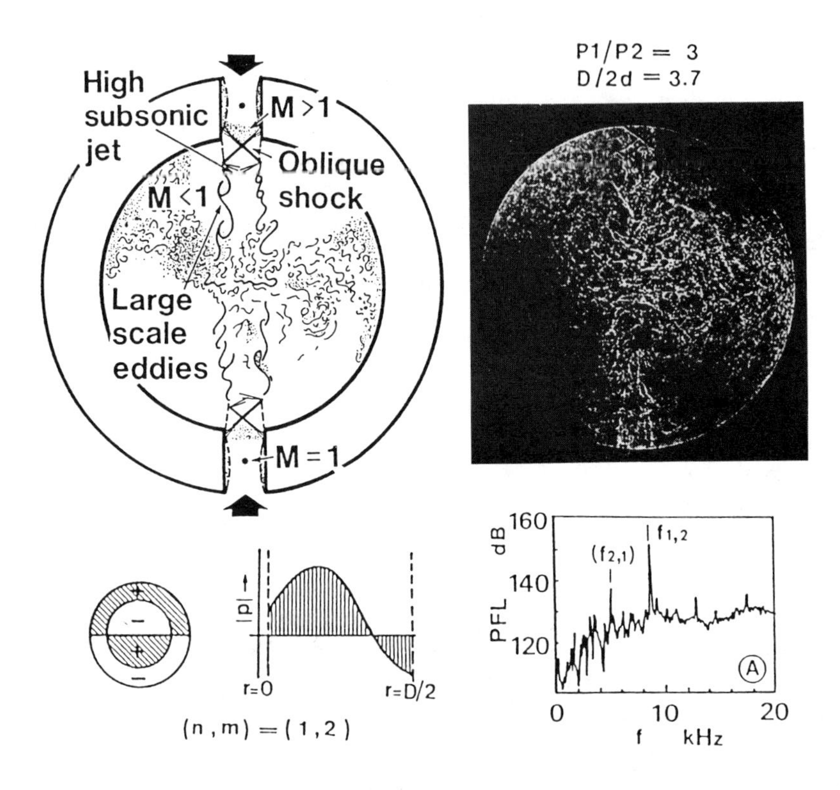

FLOW STRUCTURE Ⓒ

Stable supersonic structure of a one-sided collision

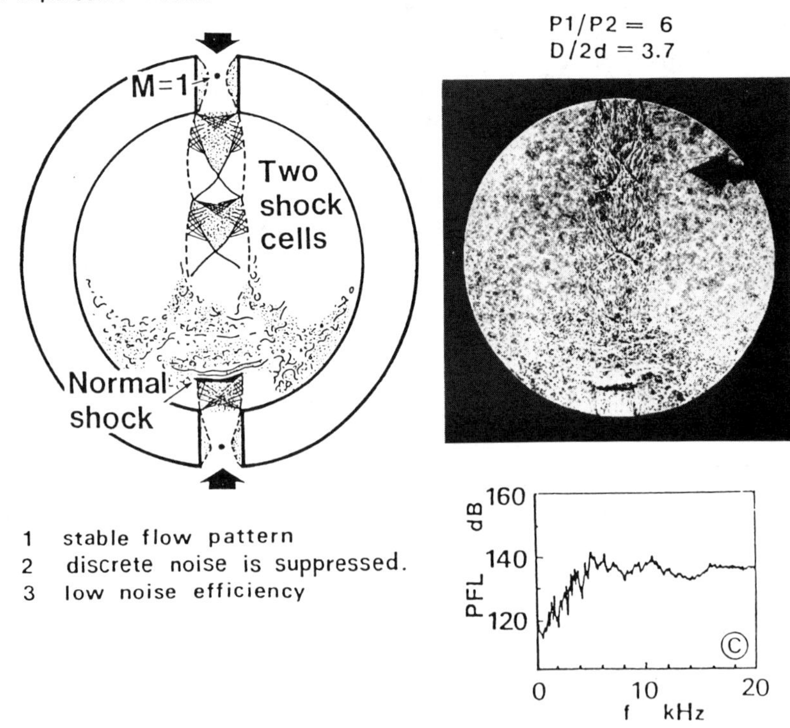

Fig. 20 Flow pattern and structure (*a, b*) [175]. (Figure continues.)

FLOW STRUCTURE Ⓑ

Colliding parts of the jets brush past each other.

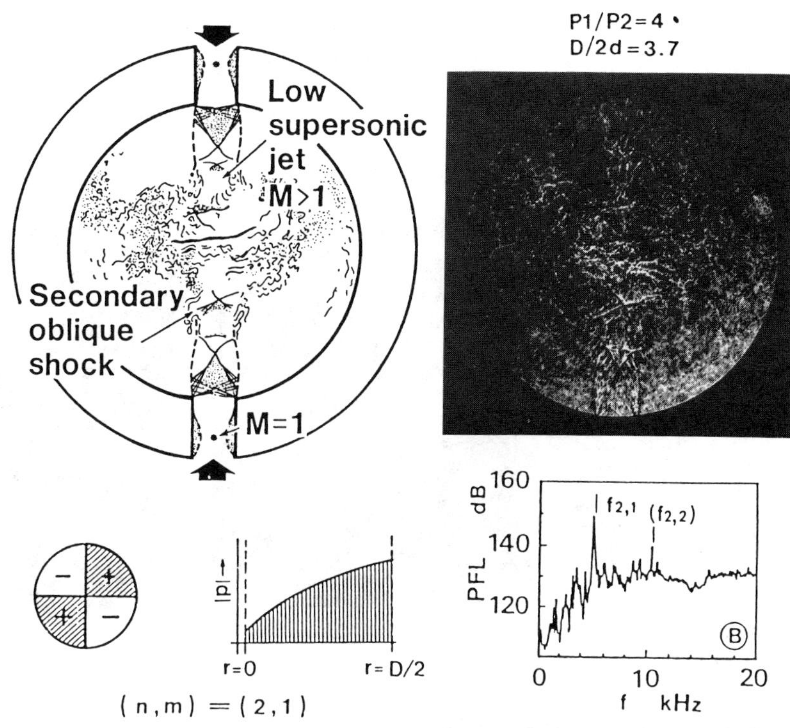

Fig. 20 (c) (continued)

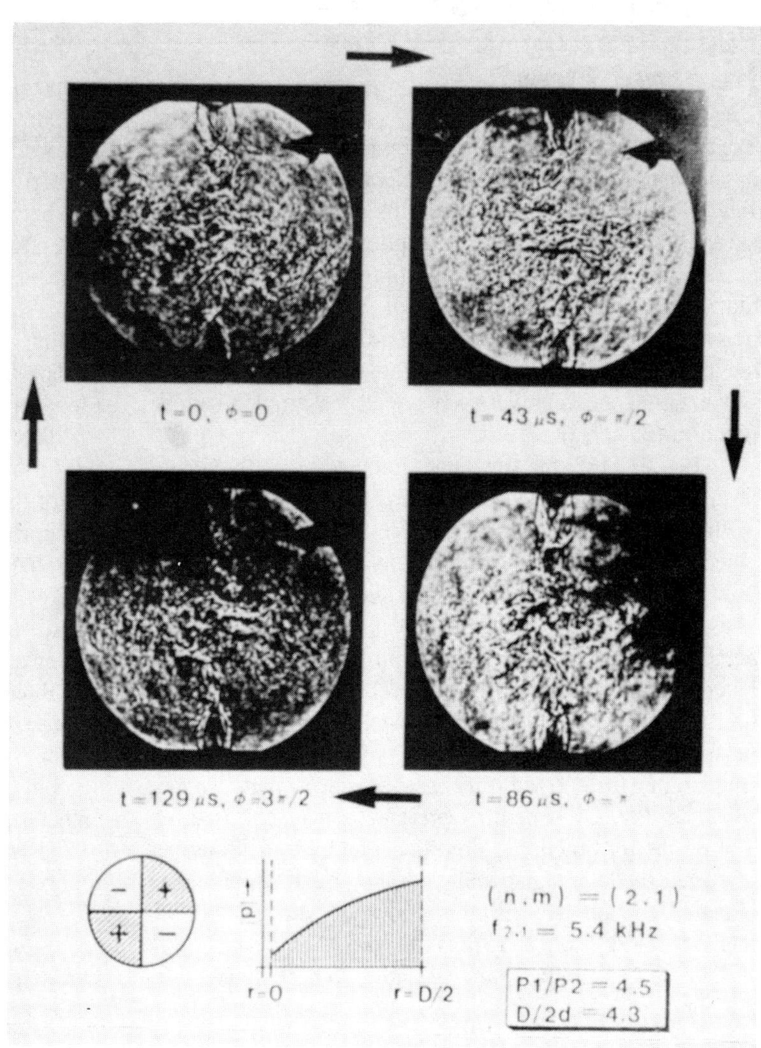

Fig. 21 Wave mode (2,1) and the related oscillation of supersonic jets [175].

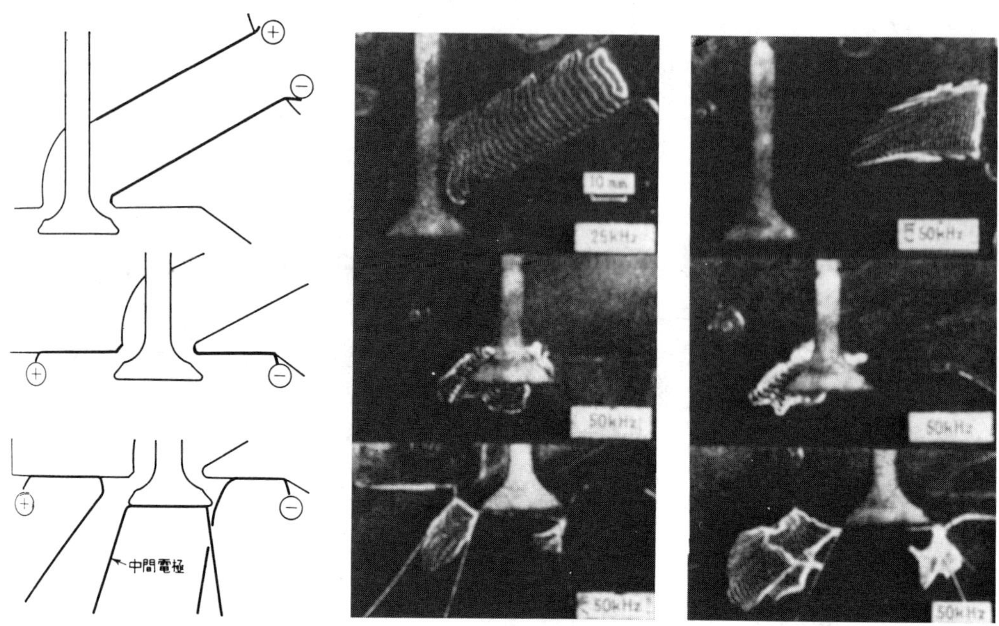

Fig. 22 Flow around suction port [180].

C Computer-aided Flow Visualization (CAFV) for Internal Flow

The flow pattern and the temperature field can be visualized by (1) visualization experiments, (2) numerical solution of Navier–Stokes and energy equations and numerical simulation of turbulent flow, (3) application of the image-processing method to visual results (data) to obtain quantitative data, and (4) application of computer graphics to measurement data by a probe method. The applications of the image-processing method and computer graphics to flow visualization are relatively new; recent examples can be found in [25–27].

In Fig. 23 [188] the streaklines in the turbulent boundary layer on a flat plate visualized by the hydrogen bubble method at $y^+ = 9.6$ [19, 187] are compared with the computed streaklines from the numerical simulation of turbulent flow. One is impressed by the striking similarity of streaklines obtained by the visualization experiment and the computation. It is noted that the region $0 \leq y^+ \leq 7$ is the innermost layer. Kobayashi et al. [189] showed that turbulent flow in a parallel-plate channel with regular rectangular turbulence promoters can be predicted by large eddy simulation (LES). The timelines of tracer particles are shown in Fig. 24, where t represents dimensionless time. The results for timelines, streaklines, and Strouhal number of vortex shedding agree well with the visualization results obtained by the dye-injection and the hydrogen bubble methods [189]. Moin and Kim [190] showed that large eddy simulation can predict bursting and sweep phenomena in the boundary layer.

The application of computer graphics to image display is considered next. Temperature field displays for surface-temperature measurements with an infrared camera for actual aircraft turbine vanes, with film cooling from multiple holes with mainstream velocity of 33 m/s, are shown in Fig. 25 [191] for mass velocity (density × velocity) ratio (M) between coolant and mainstream $M = 0.25$, 0.50, and 0.75. The original photos are in color, and the white elliptical spots represent coolant injection holes with the maximum fluid temperature. It is seen that at $M = 0.50$, the cooling effectiveness is the highest, and at $M = 0.75$, the coolant penetrates through the mainstream boundary layer. Flow separation occurs at the vane surface.

Mass transfer near the injection holes of the air-cooled gas turbine blades can be surveyed experimentally with the use of the depth-measurement method of naphthalene sublimation [192]. A row of three injection holes with a space of 3 diameters is inclined at an angle of 35° and mounted flush with a test flat plate. The mainstream boundary layer was always turbulent, and the mass flow rate of the injectant was changed to provide laminar and turbulent jets. The two- and three-dimensional images for distribution of the ratio of mass transfer coefficients with and without film cooling are shown in [192] for a range of injection parameters (mass velocity ratio) $M = 0.369 \sim 2.001$. The visualization results for $M = 0.51$ are shown in Fig. 26. The image display of the pressure-distribution data obtained by Pitot tube at the exit section of a gas-turbine vane (annular) cascade is shown in Fig. 27 [193]. The perspective displays by a 3-D processor and plotter are shown in Fig. 28. The unsteady two-dimensional temperature fields of a gas-turbine cooling blade using computer graphics are shown in Fig. 29 [194].

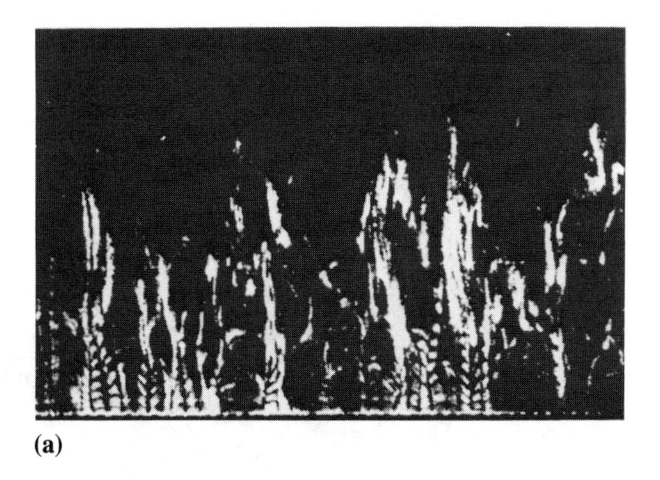

Fig. 23 Structure of turbulent boundary layer illustrated by combined-time-streak markers at $y^+ = 9.6$: (*a*) Experiment [19, 187], (*b*) calculation [188].

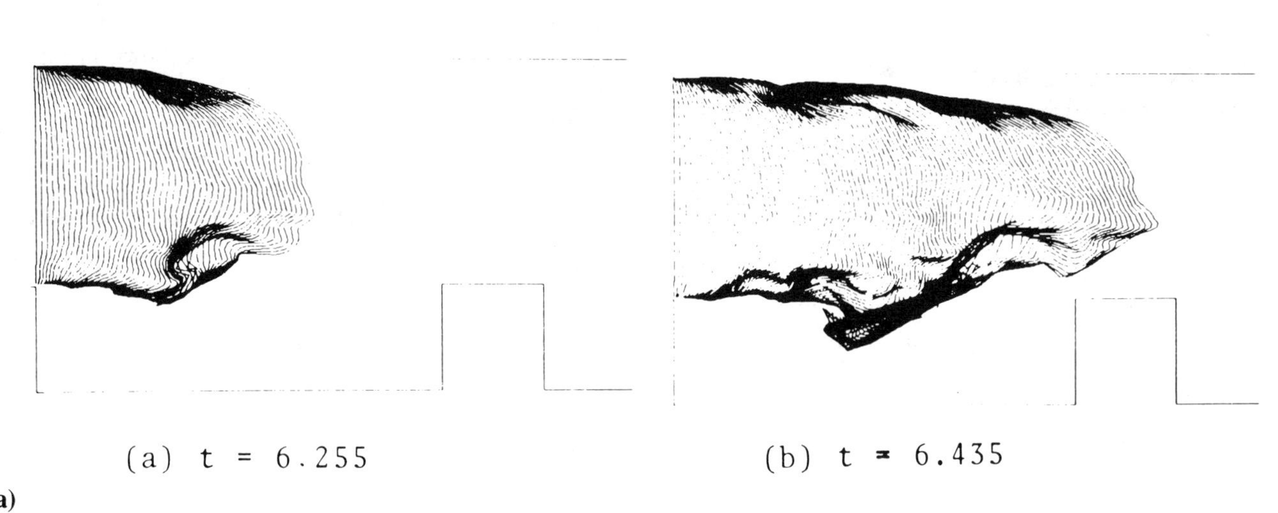

(a)

Fig. 24 (*a*) Time-dependent structure of timelines (computation, Re = 1.1×10^4). (Figure continues.)

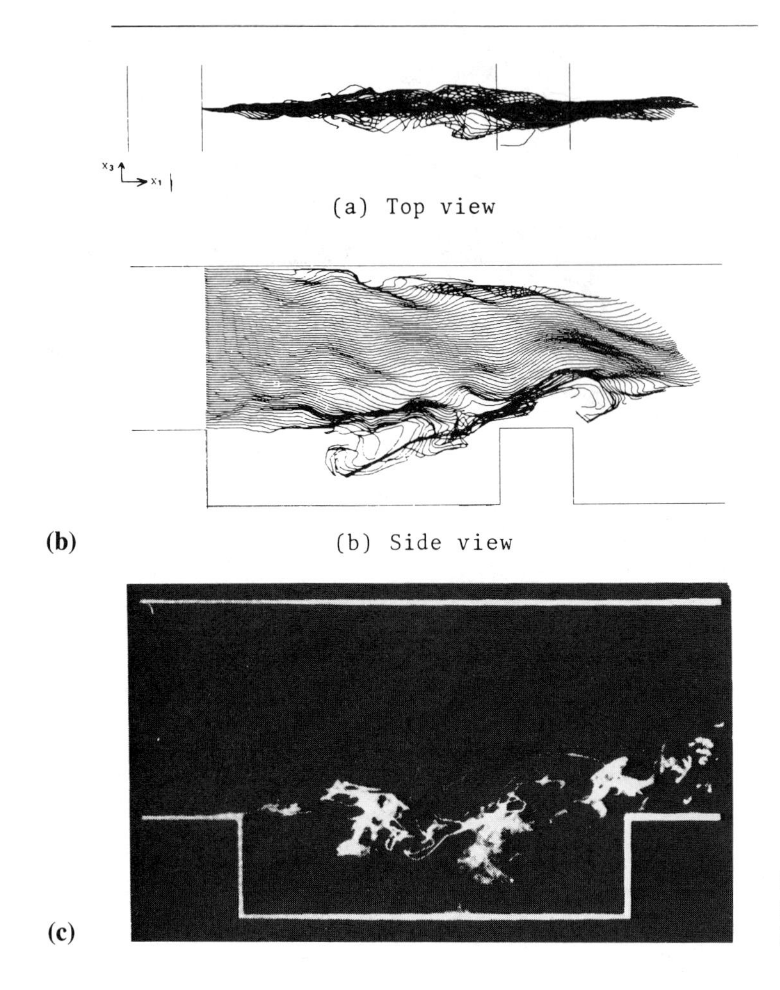

Fig. 24 (continued) (*b*) formation of streaklines (computation, $t = 6.605$); (*c*) comparison of streaklines [189].

The numerical simulation of turbulence is of current interest. A numerical study of turbulent flow in a parallel-plate channel was first reported by Deardorff [195]. Recent works [190, 196–198] reveal bursting and sweep phenomena numerically, and constant lines for velocity fluctuation, pressure, and vorticity are presented.

D Mixed-Phase Flow Phenomena

The understanding of such complicated phenomena as boiling, two-phase flow, evaporation, and condensation in internal flows depends greatly on flow visualization. Flow visualization indeed plays an important role in understanding the mechanisms and in formulating the physical models for these phenomena. Flow visualizations in boiling, two-phase flow, and condensation are well reviewed in [199–203]. Flow visualization photos can also be found in books [204–206]. Apparently, the visualization of vapor bubbles generating on fine wires led Nukiyama [207] to the discovery of maximum and minimum boiling heat-flux rates, and he obtained a substantially complete boiling curve in 1934. Visualization studies are used in the design of two-phase flow equipment such as boilers, steam generators, heat exchangers, nuclear-fission reactor cores, refrigerators, heat pumps, turbines, compressors, pumps, heat pipes, thermosyphons, cooling systems for nuclear-fusion reactor blankets and for immersion cooling of electronic components. One example of the flow pattern for gas–liquid–solid (aluminum particle–air–water) three-phase flow in a vertical pipe (20 mm ID) is shown in Fig. 30 [208, 209].

Computer tomography using X-ray [210, 211], γ-ray [212], and neutrons [213] for flow visualization in two-phase flow now reaches a practical stage. References for flow visualizations in multiphase flow or two-phase flow can be found in [24, 25, 27].

E Heat Transfer

Eckert [214] recently pointed out the need for a collection of photographs illustrating various heat transfer processes and temperature fields for conduction and radiation, as well as for velocity and temperature fields in convection, obtained by computer simulation similar to the book *An Album of Fluid Motion* by Van Dyke [16]. Literature reviews on

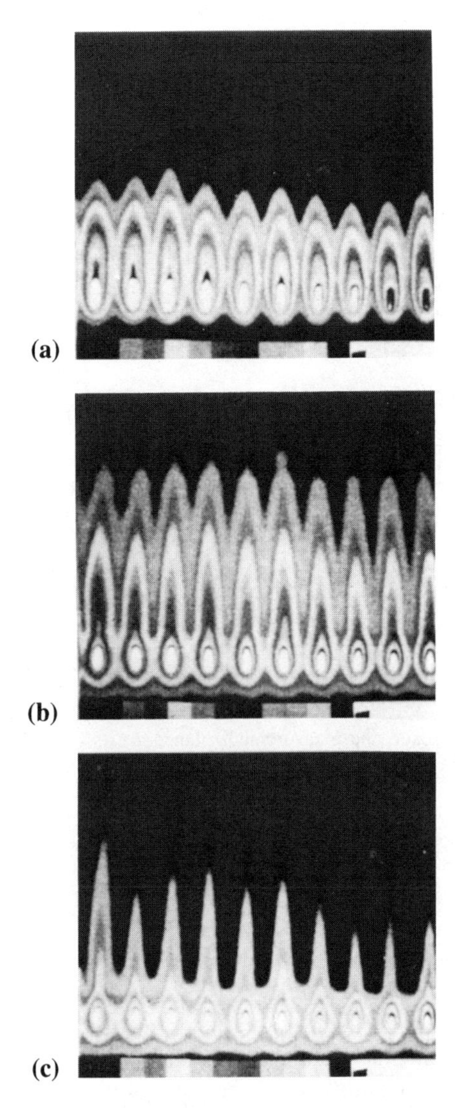

Fig. 25 Temperature distribution: (*a*) $M = 0.25$; (*b*) $M = 0.50$, (*c*) $M = 0.75$ [191].

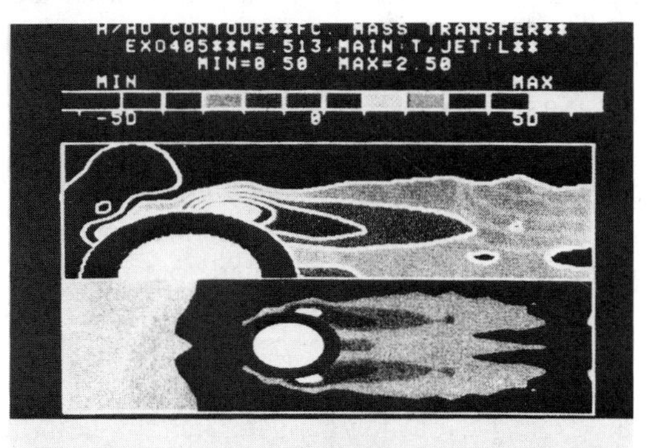

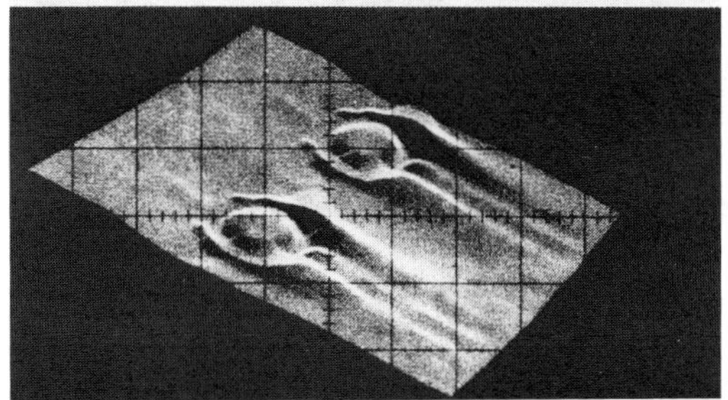

Fig. 26 Two- and three-dimensional images of mass transfer coefficient distributions [192].

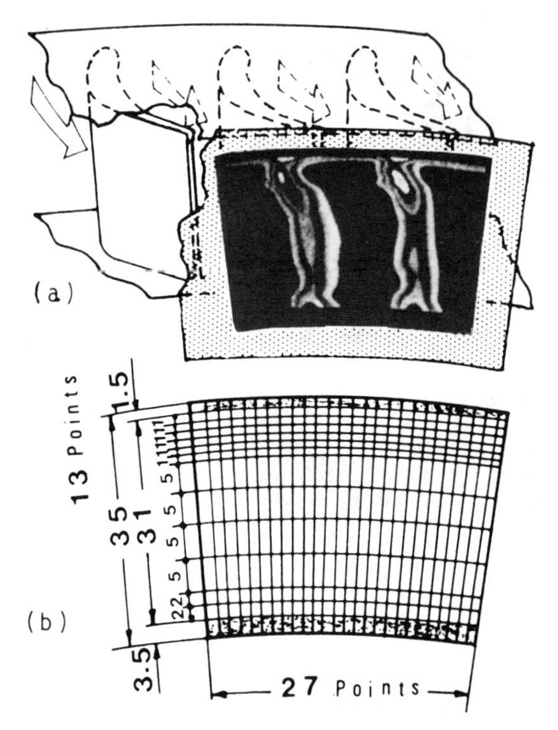

Fig. 27 (*a*) Vane cascade and measurement plane; (*b*) its grid dimension [193].

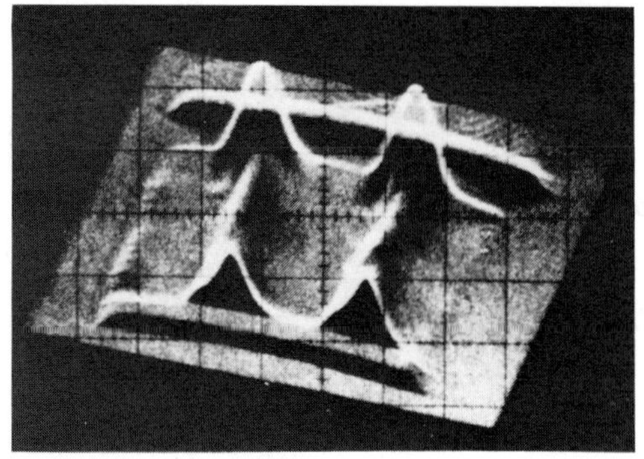

(a)

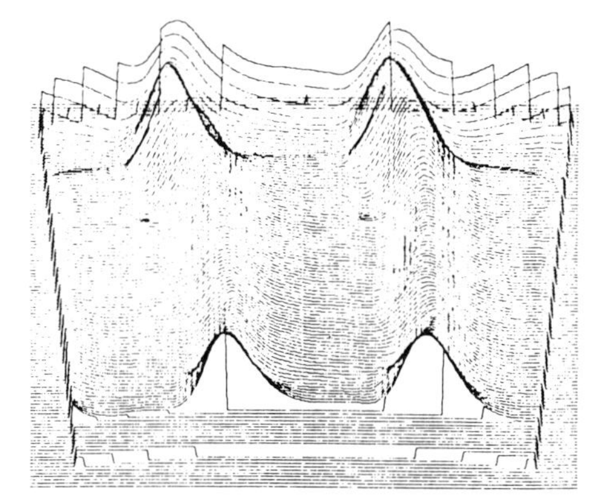

(b)

Fig. 28 Perspective display by (*a*) 3-D processor, (*b*) plotter [193].

flow visualization for convective (forced and natural) heat transfer are available [89, 215, 216]. When heat transfer is associated with a convective fluid flow, visualization of heat transfer employs essentially the methods of flow visualization.

The schlieren pictures of temperature fields for natural convection near heated bodies obtained by Schmidt [50] are well known. The first applications of the Mach–Zehnder interferometer to heat transfer were made by Kennard [217] in 1932 and by Eckert and Soehngen [218] in 1948. The first interference photographs of transition from laminar to turbulent flow in a free-convection boundary layer on a vertical flat plate were also obtained by Eckert and Soehngen. The photographs can be found in heat transfer books [219, 220]. It is of interest to observe that optical methods (shadowgraph, schlieren, and interferometer) were used widely in free-convection studies. Optical methods are reviewed in heat transfer [202, 221–223] and in combustion research [164–167].

The examples of flow visualization in heat transfer (forced and natural) for internal flow are too numerous to be listed here, and only a few examples will be mentioned. Figure 31 shows the growth of the dendritic ice across the cross section of a short length of a horizontal water pipe under transient cooling conditions [224]. Figure 32 shows the stages in the freezing of a water pipe without main flow. Figure 31*a* shows supercooling near the top to −3°C just before the onset of ice nucleation. Figure 31*b–d* shows the growth sequence of ice dendrites until they completely block the cross section. The total time required for the completion of this dendritic ice-growth phase is only about 30 s. The photos in Fig. 31*e,f* showing the inward-growing annular ice growth were taken 3 and 18 hr later, respectively, after the whole liquid temperature field reached practically 0°C. It is seen that a water pipe without a main flow can be completely blocked by dendritic ice formation, instead of by the commonly understood annular ice growing inward. Figure 33 shows the formation of an ice band in a cooled water pipe with main flow at Re = 3025, based on pipe diameter at several values of cooling ratio θ [225]. These two examples [226] clearly show the importance of flow visualization in understanding the freezing mechanism in a freezing pipe.

Streak photographs for transformer oil in a narrow vertical rectangular cavity with aspect ratio 15 and two vertical walls maintained at uniform but different temperatures are shown in Fig. 34 [227] for Ra = 5.0×10^9. Figure 34*b* shows a close-up view of tertiary flow. Examples of the flow patterns and the temperature field for the Bénard problem visualized by a numerical solution can be found in [228–230] among many others. The applications of liquid crystal in heat transfer experiments are discussed in [231, 232], and two photos are shown in Figs. 35 and 36. Two

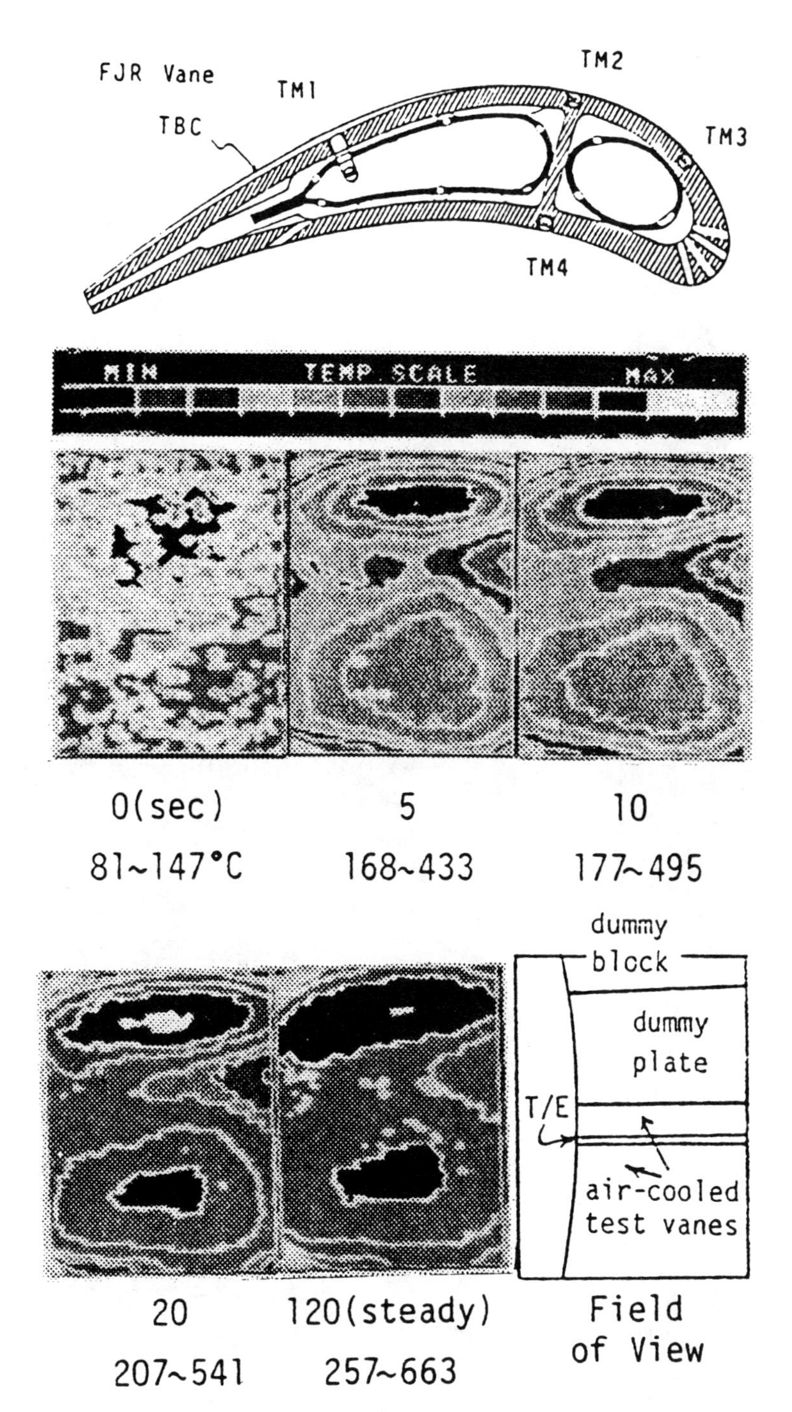

Fig. 29 Image display of unsteady 2-D temperature fields [194].

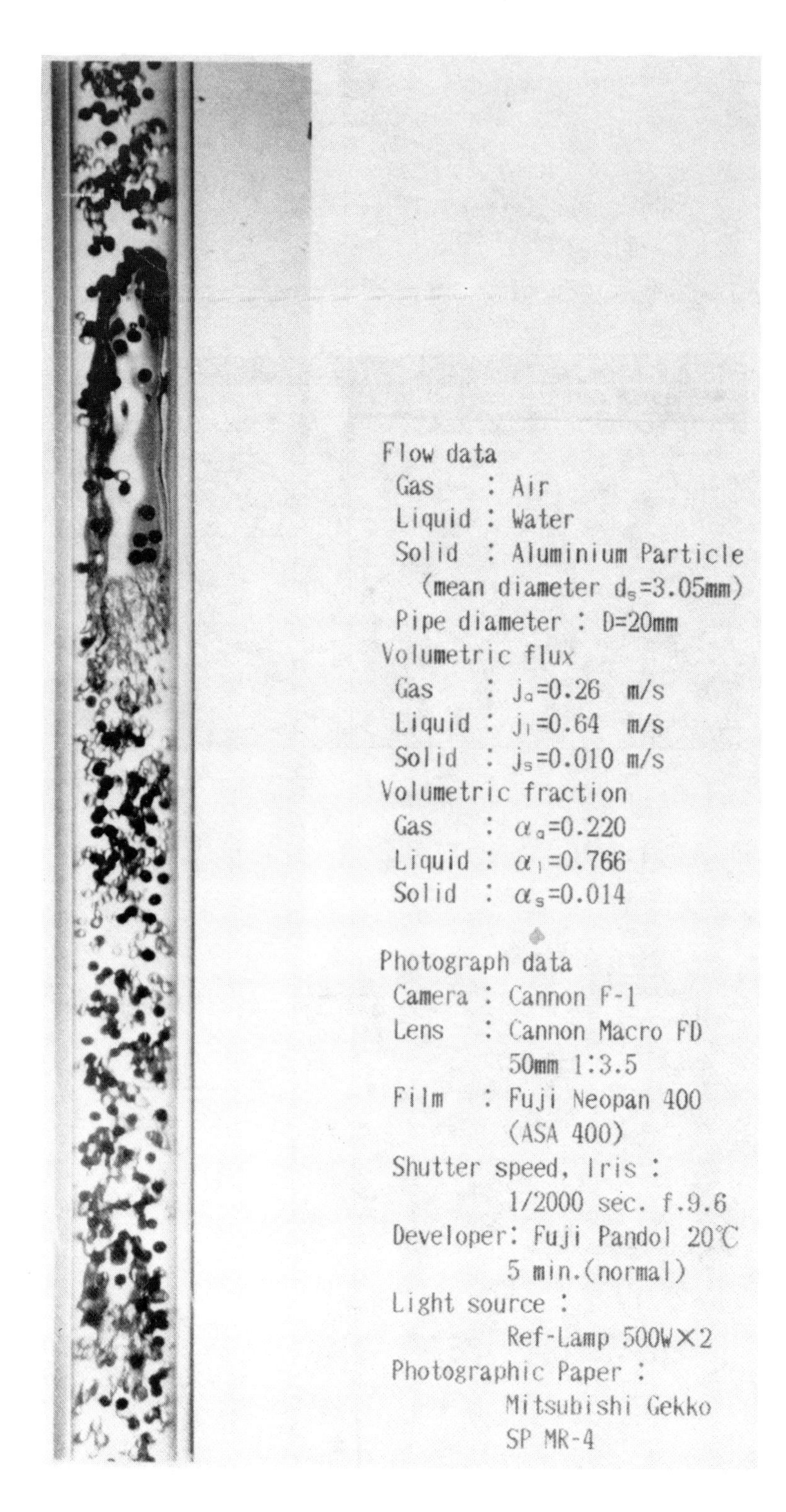

Fig. 30 Gas–liquid–solid three-phase flow (air, water, aluminum particles (mean dia. = 3.05 mm), pipe dia. = 20 mm) [208].

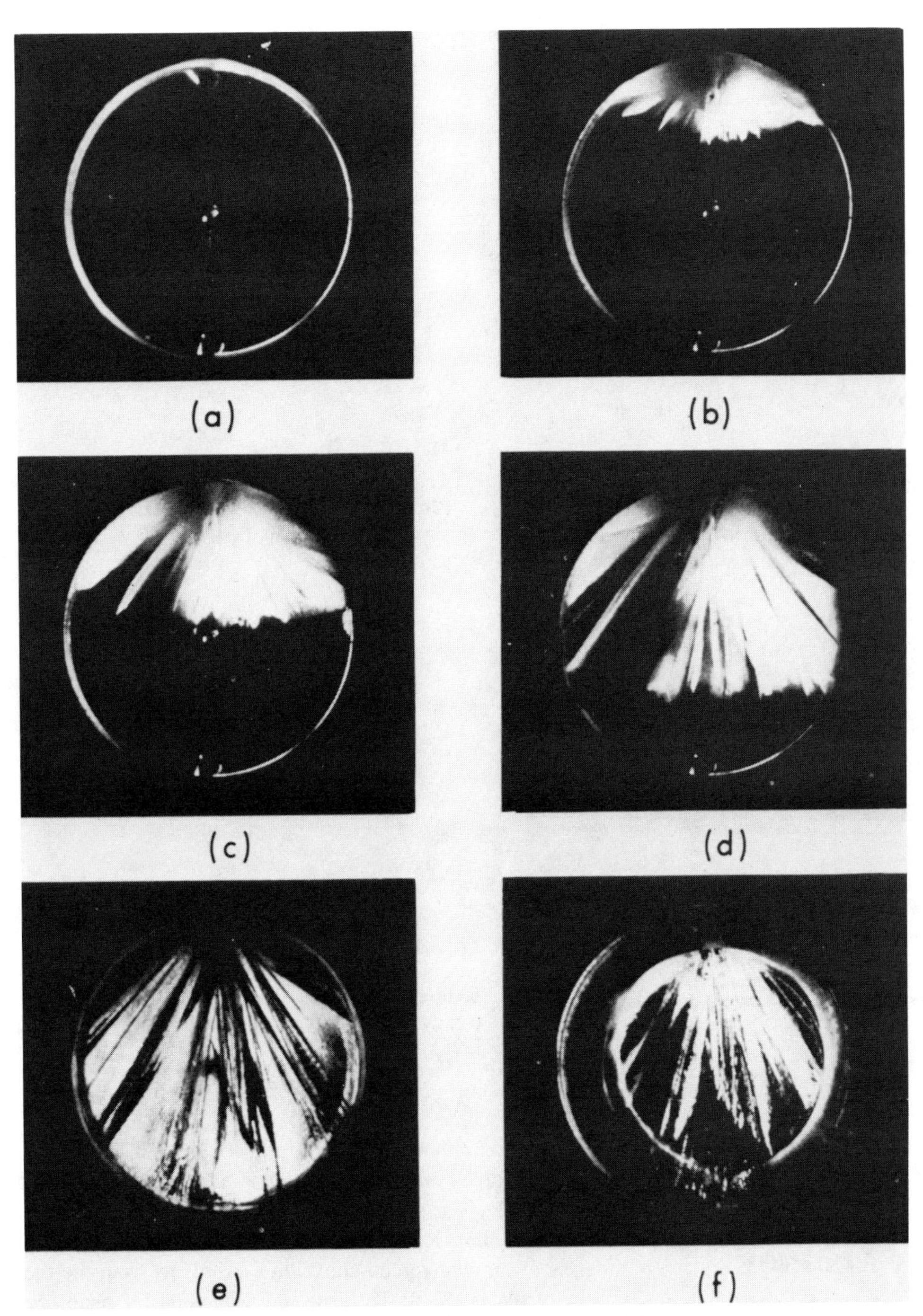

Fig. 31 The growth of ice in a cooled pipe without main flow [224].

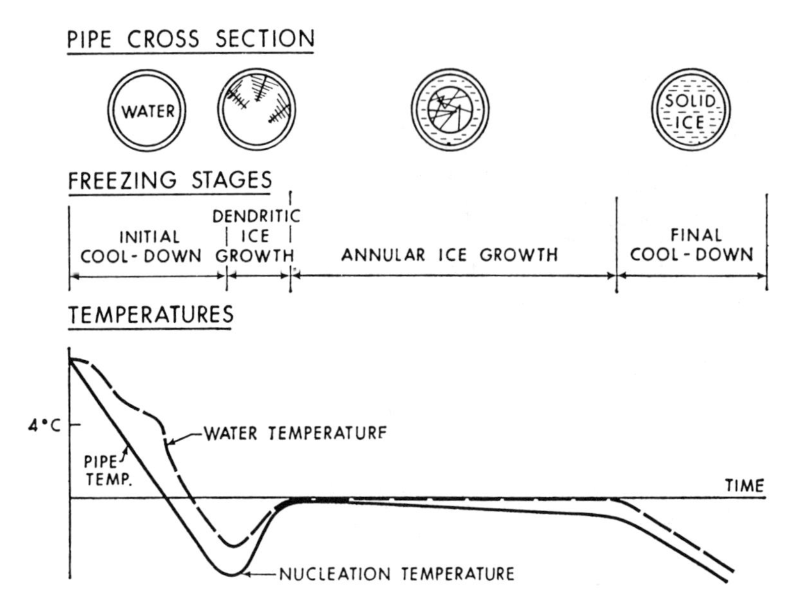

Fig. 32 The stages in the freezing of a water pipe that has no main flow through it.

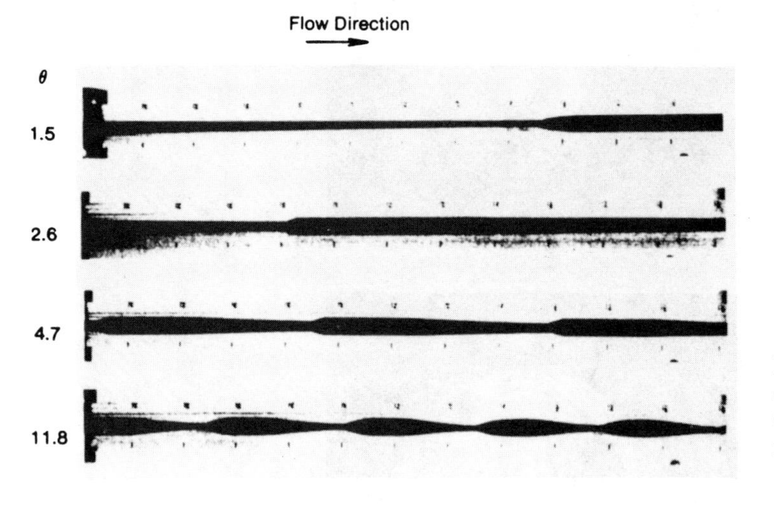

Fig. 33 The steady-state ice-band structure in a cooled pipe for various nondimensional temperatures θ. In each photo the dark line of varying thickness is the unfrozen flow channel in the pipe, Re = 3025 [225].

examples of interference fringes for natural convection in horizontal cylinders are shown in Fig. 37 [233].

F Visualization by Analogy or Simulation

Very often, different physical phenomena are governed by the same or similar differential equations and the associated boundary conditions, and the observation of the analogy leads to considerable physical insight when visual results are available for one problem. The experimental analogic method for heat-conduction problems using the fluid flow analogy and the membrane (soap-film) analogy is discussed in [234]. It is interesting to observe that a dilated soap film [234] represents many physical problems. The relationship of heat transfer to fluid mechanics and other branches of science is discussed by Jakob [235] and by Eckert [236].

An observation of the analogy between fully developed laminar forced convection in ducts and simply supported thin flat plates of the same geometrical shape under uniform lateral loading leads immediately to exact solutions of a class of laminar forced-convection problems from the theory of plates [237]. Thus the moiré method for plates can be utilized to solve fluid mechanics and heat transfer problems [238]. The application of hydraulic analogy to flow visualization is discussed in [29, 239, 240], and analog methods are described in [241, 242]. Steady two-dimensional potential flow can be visualized by the Hele–Shaw method [16]. Visualization of free convective flow in Hele–Shaw cells is reported in [243].

Eckert [244] discusses analogies to heat transfer processes involving conduction, radiation, and convection. The mass transfer analogy for a constant property fluid can be used for flow visualization, and examples for the flow of air across aligned and staggered tube banks [245] are

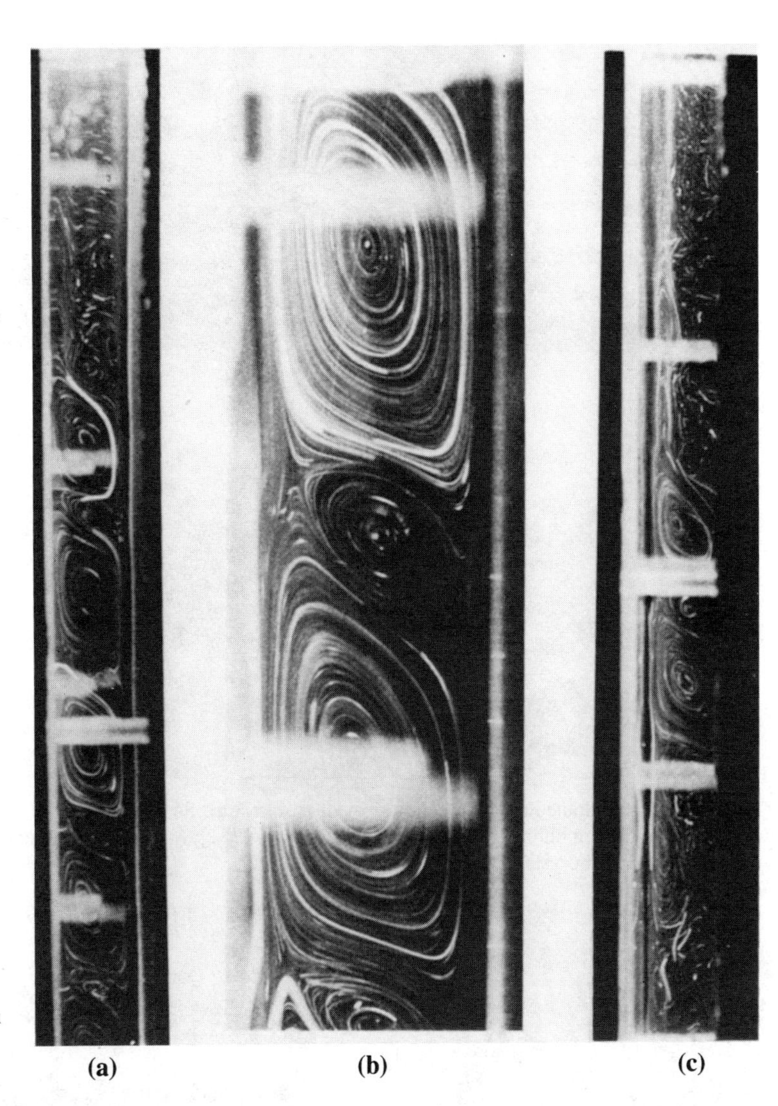

Fig. 34 Streak photographs for transformer oil in a rectangular cavity with aspect ratio 15 [227].

shown in [244, 246]. One example is shown in Fig. 38 [245]. The observation of analogies between thermal (Bénard cells) and viscous (Taylor–Görtler vortices, Dean vortices) instabilities [70, 247] leads to physical understanding of different phenomena. The analogy is confirmed by the similarity of the flow patterns revealed by the visualization results.

The dispersion of pollutants in a neutrally stable atmospheric boundary layer can be simulated by the plume from a source of a neutrally buoyant salt–water–ethanol mixture in a water channel. Other analogies are discussed in [71].

V CONCLUDING REMARKS

Flow visualization experiments have been performed by Reynolds (1883) and Prandtl (1904) in understanding flow structure and phenomena. Flow visualization has played an important role in the development of modern fluid mechanics. In the 1950s Kline and his co-workers revealed conclusively the existence of coherent structures in the fluid layer near the wall by using relatively simple visual methods; their pioneering visual research apparently marked the beginning of a new era in turbulence research after about 50 years of probe measurements. Such important flow phenomena as the transition from laminar to turbulent flow, flow separation, stall, surging, Karman vortices, laminar and turbulent boundary layers, shear flow, wake flow, wing-tip vortices, secondary flows, impinging jets, instabilities (Tollmien–Schlichting waves, Taylor–Görtler vortices, Bernard cells), vortex wings, Coanda effect, shock waves, turbulence spots, vortices, compressible flows, unsteady flows, and many others were clarified by visual experiments. It is indeed instructive to contrast the history of visual research with the evolution of turbulence research through phenomenological, statistical, and structural periods over the past 100 years or so.

The dissemination of scientific and technical information related to all aspects of flow visualization started with the first flow visualization symposium organized by the ASME in 1960, and a series of annual symposia on flow visualization in Japan starting in 1973. The series of international

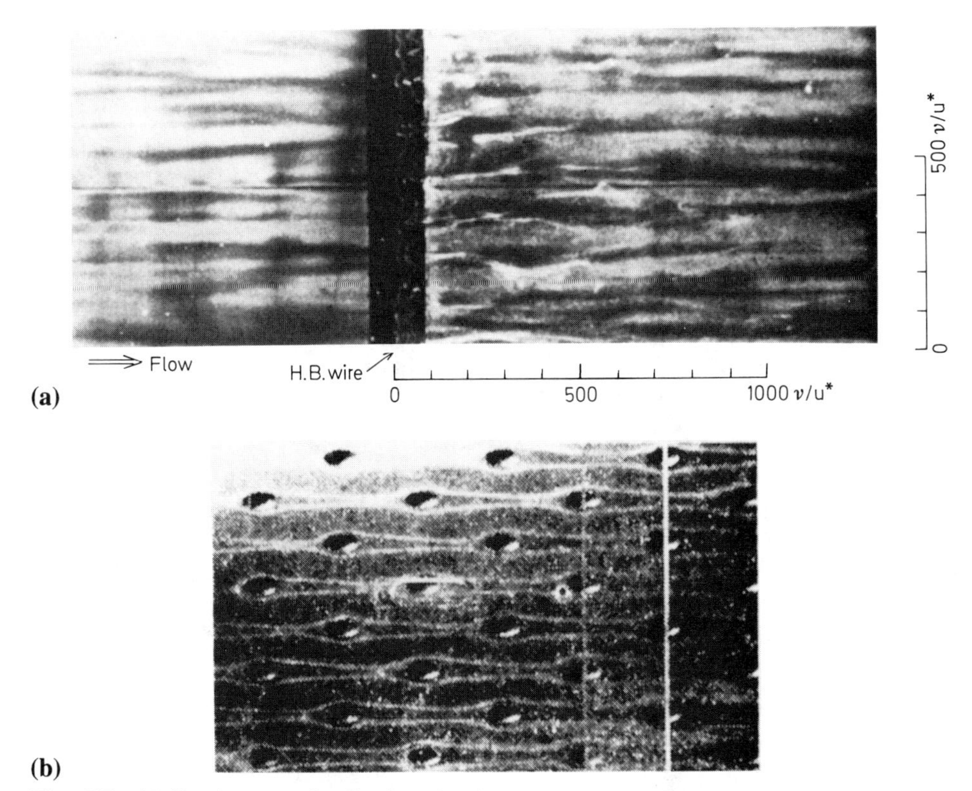

Fig. 35 (*a*) Simultaneous visualization of wall temperature fluctuation (liquid crystal sheet) and flow field (a hydrogen bubble wire is located at $y^+ = 6$) [231]; (*b*) surface temperature distribution by cholesteric liquid crystal [232].

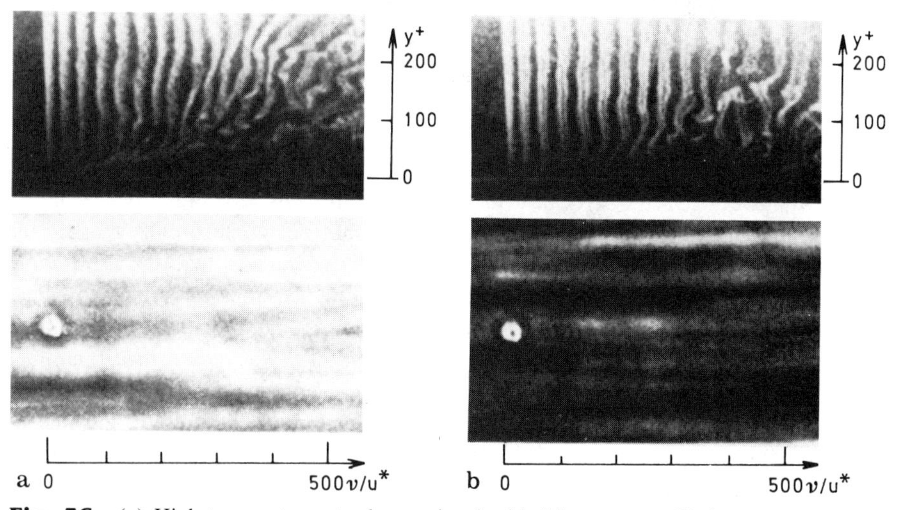

Fig. 36 (*a*) High-temperature streak associated with lift-up event; (*b*) low-temperature streak associated with sweep event [231].

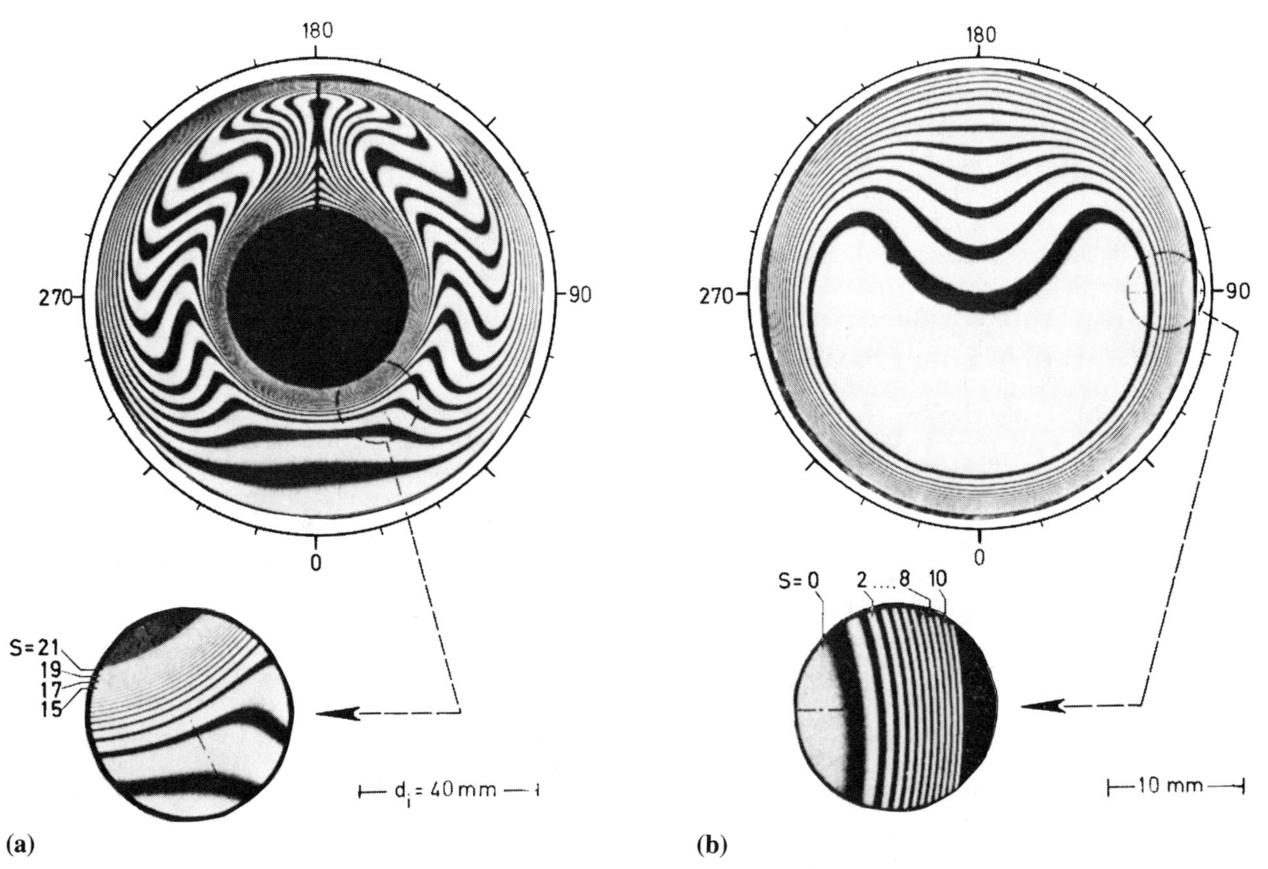

Fig. 37 (*a*) Natural convection between two horizontal coaxial cylinders; (*b*) natural convection in a horizontal cylinder heated uniformly from the outside (photos from Technische Universität München [233]).

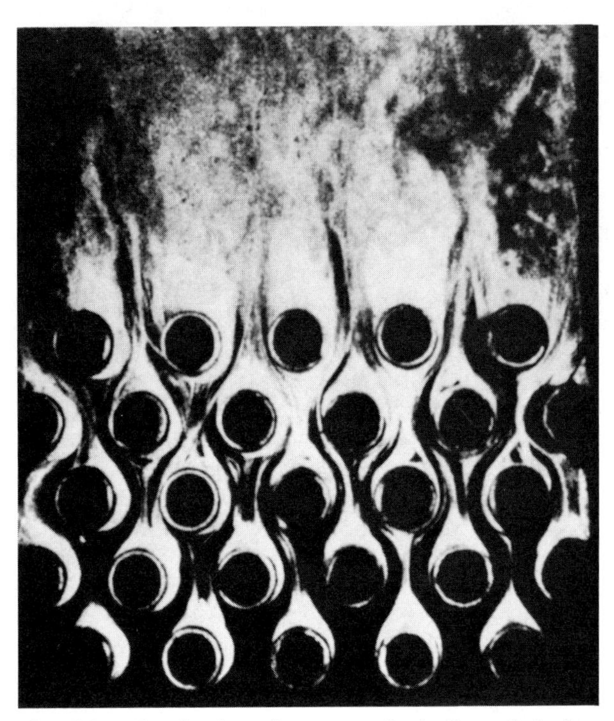

Fig. 38 Visualization of mass transfer in flow of air through staggered tube bank [245].

symposia on flow visualization (Tokyo, 1977; Bochum, 1980; Ann Arbor, 1983; Paris, 1986) and the publication of the *Journal of Flow Visualization of Japan,* starting in 1981, have accelerated dissemination of the worldwide literature on flow visualization. In the past, flow visualization experiments were also performed as part of experimental investigations on flow and heat transfer problems, and the visual results were reported extensively in various journals and proceedings. The literature on the visual results relating to the various internal flows is extensive at present, and an exhaustive listing of technical papers is not possible within the scope of this article. The recent development of flow visualization techniques and the visual results for internal flows can be found in the publications mentioned above. The present chapter deals only with selected topics in internal flows, and the technical papers quoted are based on materials available at hand.

The role of flow visualization in revealing flow patterns and in providing physical insight into complex flow phenomena will remain the same in the future. It is also clear that the visual method, the probe method, and the computer method will be the main tools in flow and heat transfer research involving internal flows. The difficulty of interpreting visual results has been pointed out repeatedly in the literature and will remain a challenge in future studies. Flow visualization has the unsurpassed advantages over the probe methods of providing the whole flow and temperature (concentration) fields. Quantitative data from visual results can be obtained by applying image processing, computer graphics, and theories in fluid mechanics, and the distributions of velocity, pressure, density, and temperature can be presented. Recently flow visualization of the pressure field has become possible [248]. Visualization of the density field at high speed is quite advanced and common in contrast to the visualization of streamlines. In this respect, it is of interest to observe that the upper limits of the numerical simulation and the visualization by the smoke method for the secondary flow in curved pipes, for example, are about the same at present.

Extensive literature citation relating to applications of various flow visualization techniques to internal flow problems can be found in [30], *Journal of Fluid Mechanics*, and the *Journal of the Flow Visualization Society of Japan*, among others. Listing of the application examples to internal flow problems according to flow visualization techniques should be most useful and remains to be done in the future. It is also useful to present a summary of the visual results clearly illustrating the present understanding of the transition process from the onset of instability to the fully developed turbulent flow together with the numerical and the probe results for pipe flow, flow on a flat plate, and hydrodynamic instability problems in channels involving Taylor–Görtler vortices, Bénard cells, and Dean vortices. Flow visualization experiments of such phenomena as relaminarization and bifurcation should be performed in conjunction with numerical solutions.

Flow visualization is the most useful experimental tool in obtaining information on flow structure. Recently, many new techniques such as the laser technique, holographic interferometry, digital image processing, computer-generated color graphics, computed tomography, and nuclear magnetic resonance have been developed [1]; the applications of these techniques to internal flow problems are expected in the future. Flow visualization will continue to play a key role in our understanding of the complex flow phenomena. It would be interesting and informative if the history of fluid mechanics and heat transfer were also reviewed from the viewpoint of flow visualization. The utilization of flow visualization up to 1934 can be found in a classic work [249].

REFERENCES

1. Taneda, S., Flow Field Visualization, in *Theoretical and Applied Mechanics*, ed. F. I. Niordson and N. Olhoff, pp. 399–410, Elsevier, New York, 1985.
2. (a) Flachsbart, O., *Handbuch der Experimentalphysik*, vol. 4, Pt. 2, p. 8, 1932.
   (b) Lugt, H. J., *Vortex Flow in Nature and Technology*, Wiley, New York, 1983.
   (c) Truesdell, C., *Essays in the History of Mechanics*, p. 73, Springer-Verlag, New York, 1968.
   (d) Fromm, J. E., Numerical Solutions of Two-Dimensional Stall in Fluid Diffusers, *Phys. Fluids*, suppl. II, p. 113, 1969.
3. Macagno, E. O., and Hung, T. K., Computational and Experimental Study of a Captive Annular Eddy, *J. Fluid Mech.*, vol. 28, pp. 43–64, 1967.
4. Truckenbrodt, E., *Fluidmechanik*, vol. 1, p. 262, Springer-Verlag, Berlin, 1980.
5. Reynolds, O., An Experimental Investigation of the Circumstances Which Determine Whether the Motion of Water Shall Be Direct or Sinuous and of the Law of Resistance in Parallel Channels, *Philos. Trans. R. Soc. London*, vol. 174A, pp. 935–982, 1883.
6. Dubs, W., Über den Einfluss Laminarer und Turbulenter Strömung auf das Röntgenstreubild von Wasser und Nitrobenzol. Ein Röntgenographischer Beitrag zum Turbulenz-Problem, *Helv. Phys. Acta*, vol. 12, pp. 167–228, 1939.
7. von Karman, T., *Aerodynamics: Selected Topics in the Light of Their Historical Development*, pp. 67–73, Cornell University Press, Ithaca, N.Y., 1954.
8. Prandtl, L., Grenzschichten in Flüssigkeiten mit Kleiner Reibung, *Verh. III Int. Math. Kongr., Heidelberg*, pp. 484–491, 1904; *Ges. Abh.* vol. 2, pp. 575–584, 1961.
9. Schlichting, H., Some Developments in Boundary Layer Research in the Past Thirty Years, *J. R. Aeronaut. Soc.*, vol. 64, pp. 64–80, 1960.
10. Prandtl, L., *Abriss der Strömungslehre*, Vieweg, Braunschweig, 1931.
11. Prandtl, L., and Tietjens, O. G., Applied Hydro- and Aeromechanics, pp. 279–306, Dover, New York, 1934.
12. Prandtl, L., *Fuhrer durch die Strömungslehre*, Vieweg, Braunschweig, 1965.
13. Bradshaw, P., Possible Origin of Prandtl's Mixing-Length Theory, *Nature*, vol. 249, pp. 135–136, 1974.
14. Prandtl, L., Neuere Ergebnisse der Turbulenzforschung, *Z VDI*, vol. 77, pp. 105–114, 1933; see also *Coll. Works*, vol. 2, pp. 819–845, Springer-Verlag, Berlin, 1961.
15. Mach, E., and Salcher, P., Sitzungsber. *Wien Akad. Wiss.*, II 95, pp. 764–780, 1887.
16. Van Dyke, M., *An Album of Fluid Motion*, Parabolic Press, Stanford, Calif., 1982.
17. Ahlborn, F., On the Mechanism of Hydrodynamic Drag, abh. a.d. Geb.d. Naturwiss, vol. 17, pp. 8–37, 1902; *Jahrb Shiffbautech. Ges.*, Vol. 5, pp. 417–453, 1904. In German.
18. (a) Kline, S. J., and Runstalder, P. W., Some Preliminary Results of Visual Studies of the Flow Model of the Wall Layers of the Turbulent Boundary Layer, *J. Appl. Mech.*, vol. 26, pp. 166–170, 1959.
    (b) Schraub, F. A., Kline, S. J., Henry, J., Runstadler, P. W. and Little, A., Use of Hydrogen Bubbles for Quantitative Determination of Time-dependent Velocity Fields in Low-Speed Water Flows, *J. Basic Eng.*, vol. 87, pp. 429–444, 1965.
19. (a) Kline, S. J., Reynolds, W. C., Schraub, F. A., and Runstadler, P. W., The Structure of Turbulent Boundary Layers, *J. Fluid Mech.*, vol. 30, pp. 741–773, 1967.

(b) Kim, H. T., Kline, S. J., and Reynolds, W. C., The Production of Turbulence near a Smooth Wall in a Turbulent Boundary Layer, *J. Fluid Mech.*, vol. 50, pp. 133–160, 1971.
20. Kline, S. J., Observed Structure Features in Turbulent and Transitional Boundary Layers, in *Fluid Mechanics of Internal Flow*, ed. G. Sovran, pp. 27–79, Elsevier, New York, 1967.
21. Brodkey, R. S., Coherent Structures in Turbulent Shear Flows, *J. Jpn. Soc. Mech. Eng.*, vol. 83, no. 736, pp. 245–252, 1980.
22. Schlichting, H., *Boundary Layer Theory,* McGraw-Hill, New York, 1979.
23. Tritton, D. J., *Physical Fluid Dynamics,* Van Nostrand Reinhold, New York, 1977.
24. Asanuma, T., ed., Flow Visualization, Hemisphere, Washington, D.C., 1979.
25. Merzkirch, W., ed., *Flow Visualization II: Proceedings of the Second International Symposium,* Hemisphere, Washington, D.C., 1982.
26. Yang, W.-J., ed., *Flow Visualization III: Proceedings of the Third International Symposium on Flow Visualization,* Hemisphere, Washington, D.C., 1985.
27. Véret, C., ed., *Flow Visualization IV: Proceedings of the Fourth International Symposium,* Hemisphere, Washington, D.C., 1987.
28. Řezníček, R., Visualisace Proudeni "Flow Visualization", Academia, Praha, 1972.
29. Merzkirch, W., Flow Visualization, Academic, New York, 1974.
30. Asanuma, T., ed., *Handbook of Flow Visualization,* 2d ed., Asakura Shoten, Tokyo, 1986. In Japanese.
31. Pankhurst, R. C., and Holder, D. W., *Wind-Tunnel Technique,* Pittman, London, 1965.
32. Bradshaw, P., *Experimental Fluid Mechanics,* 2d Ed., Pergamon, Oxford, 1970.
33. Kirschner, J. M., ed., *Fluid Amplifiers,* McGraw-Hill, New York, 1966.
34. Dean, R. C., ed., *Aerodynamic Measurements,* Massachusetts Institute of Technology, Gas Turbine Laboratory, Cambridge, Mass., 1953.
35. Chang, P. K., *Control of Flow Separation,* Hemisphere, Washington, D.C., 1976.
36. Nakayama, Y., *Fluid Mechanics,* Yokendo, Tokyo, 1979. In Japanese.
37. Nakayama, Y., *Fluid Measurements,* JSME, App. A, pp. 301–308, 1985.
38. Holder, D. W., and North, R. J., Schlieren Methods, AGARDograph 23, pt. I, 1956.
39. Wood, G. P., Interferometer Methods, AGARDograph 23, pt. II, 1953.
40. Ladenburg, R. W., Lewis, B., Pease, R. N., and Taylor, H. S., *Physical Measurements in Gas Dynamics and Combustion,* Princeton Univ. Press, Princeton, N.J., 1954.
41. Eckert, E. R. G., and Goldstein, R. J., eds., *Measurements in Heat Transfer,* 2d ed., chap. 5, Hemisphere, Washington, D.C., 1976.
42. Goldstein, R. J., ed., *Fluid Mechanics Measurements,* Hemisphere, Washington, D.C., 1983.
43. Maltby, R. L., Flow Visualization in Wind Tunnels using Indicators, AGARDograph 70, 1962.
44. Howarth, L., ed., *Modern Developments in Fluid Dynamics, High Speed Flow,* vol. 2, pp. 578–605, Oxford Univ. Press, Oxford, England, 1956.
45. McDowell, D. M., and Jackson, J. D., *Osborne Reynolds and Engineering Science Today,* Manchester Univ. Press, Manchester, England, 1970.
46. Prandtl, L., Einfuhrung in die Grundbegriffe der Strömungslehre, *Handbuch der Experimentalphysik,* vol. 4, pt. 1, pp. 3–41, Akademische, Leipzig, 1931.
47. Prandtl, L., The Generation of Vortices in Fluids of Small Viscosity, *Wilbur Wright Mem. Lect., R. Aeronaut. Soc.,* vol. 31, pp. 720–743, 1927.
48. Bénard, H., Tourbillons Cellulaires dans une Nappe Liquide, *Rev. Gen. Sci.,* vol. 11, pp. 1261–1271, 1309–1328, 1900.
49. Taylor, G. I., Stability of a Viscous Liquid Contained between Two Rotating Cylinders, *Philos. Trans.,* vol. 223A, pp. 289–343, 1923.
50. Schmidt, E., Schlierenaufnahmen des Temperaturfeldes in der Nahe Wärmeabgebender Korper, *Forsch. Geb. Ingenieurwes.,* vol. 3, pp. 181–189, 1932.
51. Koschmieder, E. L., Bénard Convection, *Adv. Chem. Phys.,* vol. 26, pp. 177–212, 1974.
52. Koschmieder, E. L., Stability of Supercritical Bénard Convection and Taylor Vortex Flow, *Adv. Chem. Phys.,* vol. 32, pp. 109–133, 1975.
53. Nakayama, W., Hwang, G. J., and Cheng, K. C., Thermal Instability in Plane Poiseuille Flow, *J. Heat Transfer,* vol. 92, pp. 61–68, 1970.
54. Akiyama, M., Hwang, G. J., and Cheng, K. C., Experiments on the Onset of Longitudinal Vortices in Laminar Forced Convection between Horizontal Plates, *J. Heat Transfer,* vol. 93, pp. 335–341, 1971.
55. Dean, W. R., Fluid Motion in a Curved Channel, *Proc. R. Soc. London,* vol. 121A, pp. 402–420, 1928.
56. Lezius, D. K., and Johnston, J. P., Roll-Cell Instabilities in Rotating Laminar and Turbulent Channel Flow, *J. Fluid Mech.,* vol. 77, pp. 153–175, 1976.
57. Bénard, H., Les Tourbillons Cellulaires dans une Nappe Liquide, *Rev. Gen. Sci. Pur. Appl.,* vol. 11, pp. 1261–1271, 1309–1328, 1900.
58. Görtler, H., Über eine Dreidimensionale Instabilitat Laminarer Grenzschichten an Konkaven Wanden, *Nachr. Ges. Wiss. Göttingen, Math.-Phys. Kl.* 1, pp. 1–26, 1940.
59. Chandrasekhar, S., *Hydrodynamic and Hydromagnetic Stability,* Oxford Univ. Press, Oxford, England, 1961.
60. Greenspan, H. P., The Theory of Rotating Fluids, Cambridge Univ. Press, Cambridge, England, 1969.
61. Turner, J. S., Buoyancy Effects in Fluids, Cambridge Univ. Press, Cambridge, England, 1973.
62. Drazin, P. G., and Reid, W. H., Hydrodynamic Stability, Cambridge Univ. Press, Cambridge, England, 1981.
63. Stuart, J. T., Hydrodynamic Stability, in *Laminar Boundary Layers,* ed. L. Rosenhead, pp. 492–579, Oxford Univ. Press, Oxford, England, 1963.
64. Segel, L. A., Non-linear Hydrodynamic Stability Theory and Its Applications to Thermal Convection and Curved Flows, in *Non-Equilibrium Thermodynamics, Variational*

Techniques and Stability, ed. Donnelly, R. J., Herman, R., and Prigogine, I., pp. 165–197, Univ. of Chicago Press, Chicago, 1966.

65. Stuart, J. T., Stability of Laminar Flows, *Fluid Dyn. Trans.,* vol. 5, pt. I, pp. 267–287, 1971.
66. Ostrach, S., Laminar Flows with Body Forces, in *Theory of Laminar Flows,* ed. F. K. Moore, pp. 528–718, Princeton Univ. Press, Princeton, N.J., 1964.
67. Koschmieder, E. L., and Pallas, S. G., Heat Transfer through a Shallow, Horizontal Convecting Fluid Layer, *Int. J. Heat Mass Transfer,* vol. 17, pp. 991–1002, 1974.
68. Coles, D., Transition in Circular Couette Flow, *J. Fluid Mech.,* vol. 21, pp. 385–425, 1965.
69. Cheng, K. C., Nakayama, J., and Akiyama, M., Effect of Finite and Infinite Aspect Ratios on Flow Patterns in Curved Rectangular Channels, in *Flow Visualization,* ed. T. Asanuma, pp. 181–186, Hemisphere, Washington, D.C., 1979.
70. Zierep, J., Analogies between Thermal and Viscous Instabilities, in *Convective Transport and Instability Phenomena,* ed. J. Zierep and H. Oertel, pp. 25–37, G. Braun, Karlsruhe, 1982.
71. Cheng, K. C., Flow Visualization of Secondary Flow in Curved, Heated and Rotating Channels and Hydrodynamic Instability Problems, *J. Flow Visual. Soc. Jpn.,* vol. 6, no. 20, pp. 2–9, 1986.
72. Cheng, K. C., and Nakayama, Y., Flow Visualization of Secondary Flow in Curved, Heated and Rotating Channels, and Internal Flow in Rotating Machinery, Yang, W.-J., Ed., Hemisphere Publishing Corp., pp. 115–136, 1987.
73. Gilpin, R. R., Imura, H., and Cheng, K. C., Experiments on the Onset of Longitudinal Vortices in Horizontal Blasius Flow Heated from Below, *J. Heat Transfer,* vol. 100, pp. 71–77, 1978.
74. Takimoto, A., Hayashi, Y., and Matsuda, O., Thermal Instability of Blasius Flow over a Horizontal Flat Plate, *Heat Transfer–Jpn. Res.,* vol. 12, pp. 19–33, 1983.
75. Ito, A., The Generation and Breakdown of Longitudinal Vortices along a Concave Wall, *J. Jpn. Soc. Aeronaut. Space Sci.,* vol. 28, pp. 327–333, 1980.
76. Aihara, Y., and Koyama, H., Secondary Instability of Gortler Vortices: Formation of Periodic Three-dimensional Coherent Structure, *Trans. Jpn. Soc. Aeronaut. Space Sci.,* vol. 27, pp. 78–94, 1981.
77. Trefethen, L., Fluid in Radial Rotating Tubes, *Proc. 9th Int. Cong. Appl. Mech.,* vol. 2, pp. 341–350, 1957.
78. Akiyama, M., Kikuchi, K., Suzuki, M., Nishiwaki, I., Cheng, K. C., and Nakayama, J., Numerical Analysis and Flow Visualization on the Hydrodynamic Entrance Region of Laminar Flow in Curved Square Channels, *Trans. JSME,* vol. 47, no. 422B, pp. 1960–1970, 1981.
79. Cheng, K. C., and Yuen, F. P., Flow Visualization Studies on Secondary Flow Patterns in Straight Tubes Downstream of a 180 Deg. Bend and in Isothermally Heated Horizontal Tubes, *J. Heat Transfer,* vol. 109, pp. 49–54, 1987.
80. Chou, F. C., and Hwang, G. J., Combined Free and Forced Laminar Convection in Horizontal Channels for High ReRa, *Can. J. Chem. Eng.,* vol. 62, pp. 830–836, 1984.
81. Chou, F. C., and Hwang, G. J., Experiments on Combined Free and Forced Laminar Convection in Horizontal Square Channels, *Proc. ASME–JSME Thermal Eng. Jt. Conf.,* vol. 4, pp. 135–141, 1987.
82. Hwang, G. J., and Liu, C. L., An Experimental Study of Convective Instability in the Thermal Entrance Region of a Horizontal Parallel-Plate Channel Heated from Below, *Can. J. Chem. Eng.,* vol. 54, pp. 521–525, 1976.
83. Incropera, F. P., Knox, A. L., and Maughan, J. R., Mixed-Convection Flow and Heat Transfer in the Entry Region of a Horizontal Rectangular Duct, *J. Heat Transfer,* vol. 109, pp. 434–439, 1987.
84. Cheng, K. C., and Yuen, F. P., Flow Visualization Experiments on Secondary Flow Patterns in an Isothermally Heated Curved Pipe, *J. Heat Transfer,* vol. 109, pp. 55–61, 1987.
85. Akiyama, M., Hanaoka, Y., Cheng, K. C., Urai, I., and Suzuki, M., Visual Measurements of Laminar Flow in the Entry Region of a Curved Pipe, in *Flow Visualization III: Proceedings of the Third International Symposium on Flow Visualization,* ed. W.-J. Yang, pp. 526–530, Hemisphere, Washington, D.C., 1985.
86. Bertelsen, A. F., An Experimental Investigation of Low Reynolds Number Secondary Streaming Effects Associated with an Oscillating Viscous Flow in a Curved Pipe, *J. Fluid Mech.,* vol. 70, pp. 519–527, 1975.
87. Lyne, W. H., Unsteady Viscous Flow in a Curved Pipe, *J. Fluid Mech.,* vol. 45, pp. 13–31, 1971.
88. Griffin, O. M., Ramberg, S. E., Votaw, C. W., and Kelleher, M. D., The Generation of Liquid Aerosols for the Visualization of Oscillatory Flows, *ICIASF '73 Record, IEEE,* pp. 133–139, 1973.
89. Nakayama, W., Torii, T., and Ogata, H., Flow Visualization of Flow Fields with Body Forces, *J. Flow Visual. Soc. Jpn.,* vol. 2, no. 5, pp. 11–17, 1982.
90. Urai, I., Akiyama, M., Suzuki, M., and Nishiwaki, I., Visualization of Low Speed Water Flow by using Tracer Injection, Electrochemical and Two-Color Electrochemical Methods, *4th Symp. Flow Visual., ISAS Univ. of Tokyo,* pp. 93–96, 1976.
91. Urai, I., Akiyama, M., Suzuki, M., and Nishiwaki, I., Flow Pattern of the Mixed Laminar Convection in Rectangular Channels with Heated Vertical Side-Wall, *8th Symp. Flow Visual., ISAS, Univ. of Tokyo,* pp. 141–146, 1980; see also *7th Symp. Flow Visual., ISAS Univ. of Tokyo,* pp. 75–80, 1979.
92. Urai, I., Akiyama, M., Suzuki, M., and Nishiwaki, I., Flow Pattern of the Mixed Convection in Rectangular Channels with Two Heated Side-Walls, *J. Flow Visual. Soc. Jpn.,* vol. 1, no. 2, pp. 71–76, 1981.
93. Akiyama, M., Cheng, K. C., Urai, I., Suzuki, M., and Nishiwaki, I., Effect of Curvature on the Hydrodynamic Instability in Curved Rectangular Channels, *J. Flow Visual. Soc. Jpn.,* vol. 2, no. 6, pp. 553–558, 1982.
94. Akiyama, M., Hanoaka, Y., Urai, I., Suzuki, M., and Cheng, K. C., Visual Study of Laminar Flow in the Entrance Region of a Curved Pipe, *J. Flow Visual. Soc. Jpn.,* vol. 2, no. 6, pp. 617–622, 1982.
95. Akiyama, M., Takamura, S. I., Suzuki, M., Nishiwaki, I., Nakayama, J., and Cheng, K. C., Flow Visualization of the Entrance Region for Viscous Flow in Coiled Circular Pipes, in *Flow Visualization II: Proceedings of the Second International Symposium,* ed. W. Merzkirch, pp. 221–226, Hemisphere, Washington, D.C., 1982.
96. Cheng, K. C., Inaba, T., and Akiyama, M., Flow Visuali-

zation Studies of Secondary Flow Patterns and Centrifugal Justability in Curved Circular and Semicircular Pipes, in *Flow Visualization III: Proceedings of the Third International Symposium on Flow Visualization,* ed. W.-J. Yang, pp. 531–536, Hemisphere, Washington, D.C., 1985.

97. Akiyama, M., Kikuchi, K., Nakayama, J., Suzuki, M., Nishiwaki, I., and Cheng, K. C., Two-Stage Developments of Entry Flow with an Interaction of Boundary-Wall and Dean's Instability Type Secondary Flows (Hydrodynamic Entry Region Problem of Laminar Flow in Curved Rectangular Channels), *Trans. JSME,* vol. 47, no. 421B, pp. 1705–1714, 1981.
98. Akiyama, M., Endo, Y., Suzuki, M., and Nishiwaki, I., Numerical Experiment on the Elliptic Nature of Developing Flow in a 180° Bend Square Duct, *J. Flow Visual. Soc. Jpn.,* vol. 4, no. 14, pp. 181–184, 1984.
99. Cheng, K. C., and Yuen, F. P., Flow Visualization Studies on Secondary Flow Pattern for Mixed Convection in the Thermal Entrance Region of Isothermally Heated Inclined Pipes, Fundamentals of Forced and Mixed Convection, ASME, HTD, vol. 42, pp. 121–130, 1985.
100. Cheng, K. C., and Mok, S. Y., Flow Visualization Studies on Secondary Flow Patterns and Centrifugal instability Phenomena in Curved Tubes, in *Fluid Control and Measurement,* ed. M. Harada, Vol. 2, pp. 765–773, Pergamon, New York, 1986.
101. Cheng, K. C., and Yuen, F. P., Flow Visualization Studies on Transient Secondary Flow Development in the Thermal Entrance Region of an Isothermally Heated Horizontal Pipe, in *Flow Visualization IV: Proceedings of the Fourth International Symposium,* ed. C. Véret, pp. 623–628, Hemisphere, Washington, D.C., 1987.
102. Karashima, H., Nakayama, Y., and Mizushima, J., Flow Analysis in a Curved Pipe using Image Processing Technique, *J. Flow Visual. Soc. Jpn.,* vol. 2, suppl., pp. 15–18, 1982.
103. Sugiyama, S., Hayashi, T., and Yamazaki, K., Flow Characteristics in the Curved Rectangular Channels (Visualization of Secondary Flow), *Bull. JSME,* vol. 26, pp. 964–969, 1983.
104. Muto, H., Comparison between Flow Patterns in Vertically Standing 180-Degree Bends of Rectangular Ducts with Two Strong Curvatures, *J. Flow Visual. Soc. Jpn.,* vol. 4, no. 14, pp. 279–284, 1984.
105. Sato, J., and Ohba, K., Visualization of Flow Velocity Profile in a Curved Duct using Laser Induced Fluorescence Method, *J. Flow Visual. Soc. Jpn.,* vol. 5, suppl., pp. 25–28, 1985.
106. Tomoto, Y., Sugihara, S., Maruyama, M., and Muto, H., A Study of Flow in Vertically Standing 180° Bend Rectangular Section with Strong Curvature, *J. Flow Visual. Soc. Jpn.,* vol. 6, no. 22, pp. 331–336, 1986.
107. Ohba, K., and Sato, M., Visualization and Measurement of Flow Velocity Profile in a U-Bend of Square Cross-Section Using a Laser Induced Fluorescence Method, *J. Flow Visual. Soc. Jpn.,* vol. 6, no. 22, pp. 377–382, 1986.
108. Ohba, K., Flow Visualization by Laser Induced Fluroescence Method, *J. Flow Visual. Soc. Jpn.,* vol. 7, no. 25, pp. 77–83, 1987.
109. Sugiyama, S., Hayashi, T., and Furukawa, H., Flows in Curved Rectangular Channels (1st Report: Visualization of Secondary Flow in the Developing Region of Laminar Flow), *Trans. JSME,* vol. 52, no. 484B, pp. 3905–3911, 1986.
110. Muto, H., A Study of Flow in Vertically Standing 180° Bend with Strong Secondary Motion (First Part: Flow Visualization), *J. Flow Visual. Soc. Jpn,* vol. 7, no. 24, pp. 39–45, 1987.
111. Hirai, S., Takagi, T., and Tanaka, K., Flow Pattern Transition in an Annulus with a Concentric Rotating Inner Cylinder, *J. Flow Visual. Soc. Jpn.,* vol. 4, no. 14, pp. 209–212, 1984.
112. Wang, C. C., and Coney, J. E. R., An Investigation of Adiabatic Spiral Vortex Flow in Wide Annular Gaps by Visualization and Digital Analysis, *Int. J. Heat Fluid Flow,* vol. 3, pp. 39–44, 1982.
113. Gu, Z. H., and Fahidy, T. Z., Characteristics of Taylor Vortex Structure in Combined Axial and Rotating Flow, *Can. J. Chem. Eng.,* vol. 63, pp. 710–715, 1985, Also see pp. 14–21.
114. Kelleher, M. D., Flentie, D. L., and McKee, R. J., An Experimental Study of the Secondary Flow in a Curved Rectangular Channel, *J. Fluids Eng.,* vol. 102, pp. 92–96, 1980.
115. Koseff, J. R., and Street, R. L., Visualization Studies of a Shear Driven Three-dimensional Recirculating Flow, *J. Fluids Eng.,* vol. 106, pp. 21–29, 1984.
116. Maughan, J. R., and Incropera, F. P., Secondary Flow in Horizontal Channels Heated from Below, *Exp. Fluids,* vol. 5, pp. 334–343, 1987.
117. Incropera, F. P., Buoyancy Effects in Double-Diffusive and Mixed Convection Flows, in *Heat Transfer 1986: Proceedings of the Eighth International Heat Transfer Conference, San Francisco, August 1986,* ed. C. L. Tien, V. P. Carey, and J. K. Ferrell, vol. 1, pp. 121–130, Hemisphere, Washington, D.C., 1986.
118. Winoto, S. H., Visualization of Naturally Developing Goertler Vortices, in *Flow Visualization III: Proceedings of the Third International Symposium on Flow Visualization,* ed. W.-J. Yang, pp. 408–412, Hemisphere, Washington, D.C., 1985.
119. Winoto, S. H., Visualization of Longitudinal Vortices in Curved Rectangular Channel Flow of High Aspect Ratio, *Int. Symp. Phys. Numer. Flow Visual., ASME, FED,* vol. 22, pp. 1–7, 1985.
120. Wimmer, M., Experiments on a Viscous Fluid Flow between Concentric Rotating Spheres, *J. Fluid Mech.,* vol. 78, pp. 317–335, 1976.
121. Zierep, J., Instabilitaten in Stomungen Zaher, Warmeleitender Medien, *Z. Flugwiss. Weltraumforschung,* vol. 2, pp. 143–150, 1978.
122. Buhler, K., and Oertel, H., Thermal Cellular Convection in Rotating Rectangular Boxes, *J. Fluid Mech.,* vol. 114, pp. 261–282, 1982.
123. Sawatzki, O., and Zierep, J., Das Stromfeld im Spalt zwischen Zwei Kozentrischen Kugelflachen, von Denen die Innere Rotiert, *Acta Mech.,* vol. 9, pp. 13–35, 1970.
124. Krishnamurti, R., On the Transition to Turbulent Convection: Part 1. The Transition from Two- to Three-dimensional Flow, *J. Fluid Mech.,* vol. 42, pp. 295–307, 1970. Part 2. The Transition to Time-dependent Flow, vol. 42, pp. 309–320, 1970.

125. Krishnamurti, R., Some Further Studies on the Transition to Turbulent Convection, *J. Fluid Mech.*, vol. 60, pp. 285–303, 1973.
126. Taylor, G. I., The Criterion for Turbulence in Curved Pipes, *Proc. R. Soc. London,* vol. 124A, pp. 243–249, 1929.
127. Narasimha, R., and Sreenivasan, K. R., Relaminarization of Fluid Flows. *Adv. Appl. Mech.*, vol. 19, pp. 221–309, 1979.
128. Cannon, J. N., and Kays, W. M., Heat Transfer to a Fluid Flowing inside a Pipe Rotating about Its Longitudinal Axis, *J. Heat Transfer,* vol. 91, pp. 135–139, 1969.
129. Nishibori, K., Kikuyama, K. and Murakami, M., Laminarization of Turbulent Flow in the Inlet Region of an Axially Rotating Pipe, *JSME Int. J.*, vol. 30, no. 260, pp. 255–262, 1987.
130. Shapiro, A. H., *Shape and Flow: The Fluid Dynamics of Drag,* Doubleday, New York, 1961.
131. National Committee for Fluid Mechanics Films, *Illustrated Experiments in Fluid Mechanics (the NCFMF Book of Film Notes),* MIT Press, Cambridge, Mass., 1972.
132. JSME, Visualized Flow, Fluid Motion in Basic and Engineering Situations Revealed by Flow Visualization, Pergamon Press, Oxford, 1988.
133. Flow Visualization Society of Japan, *Photographic Journal of Flow Visualization,* no. 1, 1984; no. 2, 1985; no. 3, 1986; no. 4, 1987. In Japanese.
134. Journal of the Flow Visualization Society of Japan, *Fantasy of Fluid Flow,* Kodansha, Tokyo, 1986. In Japanese.
135. ASME, Symposium on Flow Visualization, ASME symp. volume, 1960.
136. ISAS, University of Tokyo, Symposium on Flow Visualization: 1st—1973; 8th—1980.
137. Journal of the Flow Visualization Society of Japan, Symposium on Flow Visualization: 9th—1981, 15th—1987.
138. Rouse, H., *Fluid Mechanics for Hydraulic Engineers,* Dover, New York, frontispiece, pp. 169–175, 1961.
139. Flugge, S., and Truesdell, C., eds., Handbuch der Physik, vol. 8/1: Fluid Dynamics I, pp. 111, 352, Springer-Verlag, Berlin, 1959.
140. Tietjens, O., *Stromungslehre,* vol. 2, p. 112, Springer-Verlag, Berlin, 1970.
141. Coles, D., Interfaces and Intermittency in Turbulent Shear Flow, in *CNRS, The Mechanics of Turbulence,* pp. 229–250, Gordon and Breach, New York, 1964.
142. Kline, S. J., The Role of Visualization in the Study of the Structure of the Turbulent Boundary Layer, in *Coherent Structure of Turbulent Boundary Layers,* eds. C. R. Smith and D. E. Abott, pp. 1–27, Lehigh Univ., Bethlehem, PA, 1978.
143. Eckert, E. R. G., Heat Transfer, in *Osborne Reynolds and Engineering Science Today,* Manchester Univ. Press, Barnes & Noble, New York, pp. 160–175, 1970.
144. Prandtl, L., The Generation of Vortices in Fluids of Small Viscosity, *J. R. Aeronaut. Soc.*, vol. 31, pp. 720–743, 1927.
145. Eck, B., *Technische Strömungslehre,* Springer-Verlag, Berlin, 1966.
146. Prandtl, L., *Gasbewegung, Handwörterbuch der Naturwissenschaften,* vol. 4, pp. 544–560, Jena: Fischer, 1913.
147. Hama, F. R., Lang, J. D., and Hegarty, J. C., On Transition from Laminar to Turbulent Flow, *J. Appl. Phys.*, vol. 28, pp. 388–394, 1957.
148. Eckert, E. R. G., Mass, Momentum, and Heat Transfer; A Survey of the Field, in *Recent Advances in the Engineering Sciences,* pp. 132–154, McGraw-Hill, New York, 1958.
149. Akashi, K., Application of Flow Visualization to Machinery, *J. Jpn. Soc. Mech. Eng.*, vol. 81, pp. 663 669, 1978.
150. Bryanston-Cross, P. J., High Speed Flow Visualization, *Prog. Aeros. Sci.*, vol. 23, pp. 85–104, 1986.
151. Ohki, H., Yoshinaga, Y. and Tsutsumi, Y., Visualization of Relative Flow Patterns in Centrifugal Impellers, in *Flow Visualization III: Proceedings of the Third International Symposium on Flow Visualization,* ed. W.-J. Yang, pp. 723–727, Hemisphere, Washington, D.C., 1985.
152. Fabri, H., Flow Visualization Techniques for Radial Compressor, VKI Lect. Ser. 66, 1974.
153. Paulon, J., Optical Measurements in Turbomachinery, AGARDograph 207, pp. 123–139, 1975.
154. Sieverding, C., Starken, H., Lichtfuss, H. and Schimming, P., AGARDograph 207, pp. 1–76, 1975.
155. Kraft, H., Nonsteady Flow in the Turbine, Recent Work and Thinking, ASME Symp. Unsteady Flow, 68-FE-41, 1968.
156. Nakayama, Y., Yamamoto, T., Aoki, K. and Ohta, H., Measurement of Relative Velocity Distribution in Centrifugal Blower Impeller, *Bull. JSME,* vol. 28, no. 243, pp. 1978–1985, 1985.
157. Fister, W., Sichtbarmachung der Strömungen in Radialverdicterstufen, Besonders der Relativstromung in Rotierenden Laufradern, durch Funkenblitze, *Brennst.-Wärme-Kraft,* vol. 18, pp. 425–429, 1966.
158. Asanuma, T., Tanida, Y., Kurihara, K., On the Measurement of Flow Velocity by means of Spark Tracing Method, in *Flow Visualization,* ed. T. Asanuma, pp. 227–232, Hemisphere, Washington, D.C., 1979.
159. Gallus, H. E., Survey of the Techniques in Computation and Measurement of the Unsteady Flow in Turbomachines, *Proc. 5th Conf. Fluid Mech., Budapest,* vol. 1, p. 335, 1975.
160. Fister, W., Eikelmann, J., and Witzel, U., Expanded Application Programs of the Spark Tracer Method with Regard to Centrifugal Compressor Impellers, in *Flow Visualization II: Proceedings of the Second International Symposium,* ed. W. Merzkirch, pp. 107–120, Hemisphere, Washington, D.C., 1982.
161. Frungel, F., High-Frequency Spark Tracing and Application in Engineering and Aerodynamics, *Proc. 12th Int. Cong. High Speed Photog., Soc. Photo-Optical Inst. Eng.*, vol. 97, pp. 291–301, 1976.
162. Heywood, J. B., Fluid Motion within the Cylinder of Internal Combustion Engines—The 1986 Freeman Scholar Lecture, *J. Fluids Eng.*, vol. 109, pp. 3–35, 1987.
163. Belie, G., Flow Visualization in the Space Shuttle's Main Engine, *Mech. Eng.*, vol. 107, pp. 27–33, 1985.
164. Winter, E. F., Flow Visualization Techniques, *Prog. Combust. Sci. Technol.*, vol. 1, pp. 1–36, 1980.
165. Weinberg, E. J., Geometric-optical Techniques in Combustion Research, *Prog. Combust. Sci. Technol.*, pp. 111–144, 1960.

166. Weinberg, E. J., *Optics of Flames,* Butterworths, London, 1963.
167. Weinberg, E. J., Optical Methods in Combustion Research, in *Flow Visualization II: Proceedings of the Second International Symposium,* ed. W. Merzkirch, pp. 3–14, Hemisphere, Washington, D.C., 1982.
168. Oppenheim, A. K. and Soloukhin, R. I., Experiments in Gasdynamics and Explosions, *Ann. Rev. Fluid Mech.,* vol. 5, pp. 31–58, 1973.
169. Glass, I. I., *Shock Waves and Man,* Univ. of Toronto, Toronto, 1974.
170. Takayama, K., et al., Cinematographic Study of Disintegration of Human Calculi by Underwater Shock Wave Focusing, in *Flow Visualization IV: Proceedings of the Fourth International Symposium,* ed. C. Véret, pp. 803–808, Hemisphere, Washington, D.C., 1987.
171. Takayama, K. and Sekiguchi, H., Formation and Diffraction of Spherical Shock Waves in a Shock Tube, Rep. Inst. High Speed Mech., Tohoku Univ., vol. 43, pp. 89–119, 1981.
172. Takayama, K. and Itoh, K., Unsteady Drag Over Circular Cylinders and Aerofoils in Transonic Shock Tube Flows, Rep. Inst. High Speed Mech., Tohoku Univ., vol. 51, pp. 1–41, 1986.
173. Takayama, K. and Ben-Dor, G., Reflection and Diffraction of Shock Waves Over a Circular Concave Wall, Rep. Inst. High Speed Mech., Tohoku Univ., vol. 51, pp. 43–87, 1986.
174. Henckels, A. and Takayama, K., A Study of Shock Wave Propagation Phenomena in PMMA and Glass, Rep. Inst. High Speed Mech., Tohoku Univ., vol. 52, pp. 13–28, 1986.
175. Okutsu, R., Oomoto, W., Outa, E. and Machiyama, T., Discrete Tone Noise Generation Mechanism in Cage-Guided Control Valve, Nippon Inst. Tech., Japan, NITL-86015-M, 1986.
176. Okutsu, R., Oomoto, W., Outa, E., and Machiyama, T., Collision of Multiple Supersonic Jets Related to PIP Noise Generation in Cage Valve, in *Fluid Control and Measurement,* ed. M. Harada, vol. 1, pp. 533–538, Pergamon, New York, 1986.
177. Okutsu, R., Kuramochi, S., Outa, E., and Machiyama, T., Noise and Vibration Induced by Throttling of High Pressure Compressible Fluid, *Bull. JSME,* vol. 28, pp. 837–845, 1985.
178. Kitamura, M., Yamazaki, K., Okada, Y., and Hashimoto, K., Visualizing Flows in Butterfly Valves, in *Flow Visualization III: Proceedings of the Third International Symposium on Flow Visualization,* ed. W.-J. Yang, pp. 537–541, Hemisphere, Washington, D.C., 1985.
179. Kimura, T. and Ogawa, K., Visualization of Flashing Phenomena around a Butterfly Valve, *J. Flow Visual. Soc. Jpn.,* vol. 5, suppl., pp. 15–18, 1985.
180. Nagao, F., et al., Air Flow Measurement using Repeated Discharge, *Trans. JSME,* vol. 37, no. 296, pp. 797–806, 1971.
181. Nakano, M., Outa, E., and Tajima, K., Noise and Vibration Induced by Oscillating Supersonic Flow in a Pressure Reducing Gas Valve, *Symp. Flow-induced Vibr., ASME,* vol. 4, pp. 87–103, 1984.
182. Woo, Y. R., Williams, F. P., Faughan, P. D. and Yoganathan, A. P., Pulsatile Flow Visualization studies with Aortic and Mitral Mechanical Valve Prostheses, *Chem. Eng. Commun.,* vol. 47, pp. 23–48, 1986.
183. Nakano, M., Outa, E., and Tajima, K., Flow Patterns of Supersonic Cold Air Flow Discharging through Conical Valve Model, *6th Symp. Flow Visual., ISAS Univ. of Tokyo,* pp. 121–126, 1978.
184. Nakayama, Y., Aoki, K. and Ohta, H., Flow Visualization in Artificial Heart, *8th Symp. on Flow Visual., ISAS Univ. of Tokyo,* pp. 31–34, 1980.
185. Heuser, G., and Kohler, J., The Use of Flow Visualization Techniques in the Design of Blood Pumps, in *Flow Visualization II: Proceedings of the Second International Symposium,* ed. W. Merzkirch, pp. 599–603, Hemisphere, Washington, D.C., 1982.
186. Swanson, W. M., and Clark, R. E., Streaming Birefringent Flow Qualitative Evaluation of Prosthetic Valves, Flow Visualization in *Flow Visualization II: Proceedings of the Second International Symposium,* ed. W. Merzkirch, pp. 605–609, Hemisphere, Washington, D.C., 1982.
187. Kline, S. J., Observed Structure Features in Turbulent and Transitional Boundary Layers, in *Fluid Mechanics of Internal Flow,* ed. G. Sovran, pp. 27–79, Elsevier, New York, 1967.
188. Kobayashi, T., Numerical Simulation of Turbulent Flow—Dream of Numerical Wind Tunnel, *J. Jpn. Soc. Mech. Eng.,* vol. 88, no. 799, pp. 644–647, 1985.
189. Kobayashi, T., Kano, M., and Saga, T., Numerical Experiment of the Flow Around Two-dimensional Steps by Large Eddy Simulation, *J. Flow Visual. Soc. Jpn.,* vol. 4, no. 14, pp. 337–340, 1984.
190. Moin, P., and Kim, J., Numerical Investigation of Turbulent Channel Flow, *J. Fluid Mech.,* vol. 118, pp. 341–377, 1982.
191. Yoshida, T., Sakata,K., Mimura, F., and Shindoh, S., Surface Temperature Measurements with Infrared Camera, *J. Flow Visual. Soc. Jpn.,* vol. 2, no. 5, pp. 382–390, 1982.
192. Yoshida, T., and Mimura, F., A Fundamental Research on Film Cooling of Air Cooled Gas Turbine Blades, *J. Flow Visual. Soc. Jpn.,* vol. 3, no. 10, pp. 205–210, 1983.
193. Usui, H., and Yoshida, H., Image Display of Exit Flow in Gas Turbine Vane Cascade, *J. Flow Visual. Soc. Jpn.,* vol. 4, no. 14, pp. 349–352, 1984.
194. Yoshida, T., Kumagai, T., and Mimura, F., Unsteady Temperature Analysis of Air-Cooled Turbine Vanes, *1983 Tokyo Int. Gas Turb. Cong.,* vol. 1, p. 51, 1983.
195. Deardorff, J. W., A Numerical Study of Three-dimensional Turbulent Channel Flow at Large Reynolds Numbers, *J. Fluid Mech.,* vol. 41, pp. 453–480, 1970.
196. Horiuti, K., and Kuwahara, K., Study of Incompressible Turbulent Channel Flow by Large Eddy Simulation, in *Lecture Notes in Physics,* vol. 170, pp. 260–265, Springer-Verlag, New York, 1982.
197. Kano, M., and Kobayashi, T., Numerical Analysis of Turbulent Plane Couette Flow at Low Reynolds Number by Large Eddy Simulation, *Trans. JSME,* vol. 53, no. 488, pp. 1199–1206, 1987.
198. Mizunuma, H., and Kato, H., Experimental Investigation of Transition in Plane Poiseuille Flow (1st Report, Newton-

ian Fluids), *Trans. JSME,* vol. 53, no. 488, pp. 1214–1223, 1987.
199. Hsu, Y. Y., Simoneau, R. J., Simon, F. F. and Graham, R. W., Photographic and Other Optical Techniques for Studying Two-Phase Flow, Two-Phase Flow Instrumentation, *ASME,* pp. 1–23, 1969.
200. Akiyama, M., Flow Visualization in Convective Heat Transfer: 4. Boiling and Two-Phase Flow, *J. Flow Visual. Soc. Jpn.,* vol. 2, no. 5, pp. 359–365, 1982.
201. Tanasawa, I., Flow Visualization in Convective Heat Transfer: 5. Condensation, *J. Flow Visual. Soc. of Jpn,* vol. 2, no. 5, pp. 365–373, 1982.
202. Mayinger, F., and Panknin, W., Holography in Heat and Mass Transfer, in *Heat Transfer 1974: Proceedings of the Fifth International Heat Transfer Conference, Tokyo, September 1974,* vol. 6, pp. 28–43, Hemisphere, Washington, D.C., 1974.
203. Delhaye, J. M., Optical Methods in Two-Phase Flow, *Proc. Dynamic Flow Conf.,* Marseille, France and Baltimore, MD, pp. 321–343, 1978.
204. Mayinger, F., *Strömung und Wärmeübergang in Gas-Flussigkeits-Gemischen,* Springer-Verlag, Berlin, 1982.
205. Hsu, Y. Y., and Graham, R. W., *Transport Processes in Boiling and Two-Phase Systems,* chap. 12, Hemisphere, Washington, D.C., 1976.
206. JSME, Boiling Heat Transfer, 1965.
207. Nukiyama, S., Netsu "Heat". Collection of Nukiyama's Papers. Yokendo, Tokyo, 1969.
208. Sakaguchi, T., Personal communication.
209. Sakaguchi, T., et al., Estimation of Volumetric Fraction of Each Phase in Gas–Liquid–Solid Three-Phase Flow, *Proc. ASME–JSME Thermal Eng. Joint Conf.,* vol. 5, pp. 373–380, 1987.
210. Ikeda, T., Kotani, K., and Maeda, Y., Preliminary Study on Application of X-Ray CT Scanner to Measurement of Void Fractions in Steady State Two-Phase Flows, *J. Nucl. Sci. Technol.,* vol. 20, p. 1, 1983.
211. Reimers, P., and Goebbels, J., New Possibilities of Nondestructive Evaluation by X-Ray Computer Tomography, *Mater. Eval.,* vol. 41, p. 732, 1983.
212. Tsumaki, K., et al., Measurement of Void Fraction Distribution by Gamma-Ray Computer Tomography, *J. Nucl. Sci. Technol.,* vol. 21, p. 315, 1984.
213. Richards, W. J., McClellan, G. C., and Two, D. M., Neutron Tomography of Nuclear Fuel Bundles, *Mater. Eval.,* vol. 40, p. 1263, 1982.
214. Eckert, E. R. G., Engineering Education in Heat Transfer—A Contribution to Its History and Thoughts of Future Trends, ASME Paper 84-WA/HT-31, 1984.
215. Merzkirch, W., Visualization of Heat Transfer, in *Heat Transfer 1982: Proceedings of the Seventh International Heat Transfer Conference, München, Federal Republic of Germany, September 1982,* ed. U. Grigull, E. Hahne, K. Stephan, and J. Straub, vol. 1, pp. 91–102, Hemisphere, Washington, D.C., 1982.
216. Kamotani, Y., Flow Visualization in Natural Convection, *ICIASF'81 Record (Int. Cong. on Inst. Aerosp. Simul. Facil.), IEEE,* pp. 188–191, 1981.
217. Kennard, R. B., An Optical Method for Measuring Temperature Distribution and Convective Heat Transfer, *J. Res. Nat. Bur. Stand.,* vol. 8, p. 787, 1932.
218. Eckert, E. R. G., and Soehngen, E., Studies on Heat Transfer in Laminar Free Convection with the Zehnder–Mach Interferometer, USAF Tech. Rep. 5747, 1948.
219. Jakob, M., *Heat Transfer,* vol. 1, Wiley, New York, 1956.
220. Eckert, E. R. G., and Drake, R. M., *Heat and Mass Transfer,* McGraw-Hill, New York, 1959.
221. Eckert, E. R. G., and Goldstein, R. J., eds., *Measurements in Heat Transfer,* 2d ed., chap. 5, Hemisphere, Washington, D.C., 1976.
222. Goldstein, R. J., ed., *Fluid Mechanics Measurements,* chap. 8, Hemisphere, Washington, D.C., 1983.
223. Hauf, W., and Grigull, U., Optical Methods in Heat Transfer, *Advances in Heat Transfer,* ed. J. P. Hartnett and T. F. Irvine, Jr., vol. 6, pp. 133–366, Academic, New York, 1970.
224. Gilpin, R. R., The Effects of Dendritic Ice Formation in Water Pipes, *Int. J. Heat Mass Transfer,* vol. 20, pp. 693–699, 1977.
225. (a) Gilpin, R. R., The Morphology of Ice Structure in a Pipe at or near Transition Reynolds Numbers, AIChE Symp. Ser. No. 189, Vol. 75, pp. 89–94, 1979.
(b) Gilpin, R. R., Ice Formation in a Pipe Containing Flows in the Transition and Turbulent Regimes, *J. Heat Transfer,* vol. 103, pp. 363–368, 1981.
226. Cheng, K. C., and Gilpin, R. R., Freezing Mechanism in Some Solidification Problems, in *Flow Visualization II: Proceedings of the Second International Symposium,* ed. W. Merzkirch, pp. 497–502, Hemisphere, Washington, D.C., 1982.
227. Seki, N., Fukusako, S., and Inaba, H., Visual Observation of Natural Convective Flow in a Narrow Vertical Cavity, *J. Fluid Mech.,* vol. 84, pp. 695–704, 1978.
228. Fromm, J., Solutions to Nonlinear Incompressible Flow Problems through a Finite Difference Method, *Fluid Dyn. Trans.,* vol. 3, Polish Scientific Publishers, Warszawa, Poland, pp. 169–191, 1967.
229. Grotzbach, G., Direct Numerical Simulation of Laminar and Turbulent Bénard Convection, *J. Fluid Mech.,* vol. 119, pp. 27–53, 1982.
230. Cheng, K. C., and Ou, J. W., Convective Instability and Finite Amplitude Convection in the Thermal Entrance Region of Horizontal Rectangular Channels Heated from Below, in *Heat Transfer 1982: Proceedings of the Seventh International Heat Transfer Conference, München, Federal Republic of Germany, September 1982,* ed. U. Grigull, E. Hahne, K. Stephen, and J. Straub, vol. 2, pp. 189–198, Hemisphere, Washington, D.C., 1982.
231. Iritani, Y., Kasagi, N., and Hirata, M., Heat Transfer Mechanism and Associated Turbulence Structure in the Near-Wall Region of a Turbulent Boundary Layer, in *Turbulent Shear Flows vol. 4,* Springer-Verlag, New York, pp. 223–234, 1985.
232. Kasagi, N., Hirata, M., and Kumada, M., Studies of Full-Coverage Film Cooling (1st Report, Cooling Effectiveness of Thermally Conducting Wall), *Trans. JSME,* vol. 48B, no. 430, pp. 1146–1155, 1982.
233. Brun, E. A., ed., *Modern Research Laboratories for Heat*

and Mass Transfer, UNESCO Press, Paris, pp. 49–53, 1975.
234. Schneider, P. J., *Conduction Heat Transfer,* chap. 13, Addison-Wesley, Reading, Mass., 1955.
235. Jakob, M., Relationship of Heat Transfer to Mechanics and Other Branches of Science, *Proc. 1st U.S. Nat. Cong. Appl. Mech.,* ASME, pp. 687–692, 1951.
236. Eckert, E. R. G., A Pioneering Era in Convective Heat Transfer Research, in *Heat Transfer 1982: Proceedings of the Seventh International Heat Transfer Conference, München, Federal Republic of Germany, September 1982,* ed. U. Grigull, E. Hahne, K. Stephan, and J. Straub, vol. 1, pp. 1–8, Hemisphere, Washington, D.C., 1982.
237. Marco, S. M. and Han, L. S., A Note on Limiting Laminar Nusselt Number in Ducts with Constant Temperature Gradient by Analogy to Thin-Plate Theory, *Trans. ASME,* vol. 77, pp. 625–630.
238. Cheng, K. C., Analog Solution of Laminar Heat Transfer in Noncircular Ducts by Moiré Method and Piont-Matching, *J. Heat Transfer,* vol. 88, pp. 175–182, 1966.
239. Klein, E. J., Interaction of a Shock Wave and a Wedge: An Application of the Hydraulic Analogy, *AIAA J.,* vol. 3, pp. 801–808, 1965.
240. Werlè, H., Hydrodynamic Flow Visualization, *Ann. Rev. Fluid Mech.,* vol. 5, pp. 361–382, 1973.
241. Shapiro, A. H., Free Surface Water Table, in *Physical Measurements in Gas Dynamics and Combustion,* ed. R. W. Ladenburg, *High Speed Aerodynamics and Jet Propulsion,* vol. 9, pp. 309–321; also see pp. 322–340. Princeton Univ. Press, Princeton, N.J., 1954.
242. Hoyt, I. W., The Hydraulic Analogy for Compressible Gas Flow, *Appl. Mech. Rev.,* vol. 15, pp. 419–425, 1962.
243. Koster, J. N., and Müller, U., Visualization of Free Convective Flow in Hele–Shaw Cells, in *Flow Visualization III: Proceedings of the Third International Symposium on Flow Visualization,* ed. W.-J. Yang, pp. 738–742, Hemisphere, Washington, D.C., 1985.
244. Eckert, E. R. G., Analogies to Heat Transfer Processes, in *Measurements in Heat Transfer,* 2d Ed., ed. E. R. G. Eckert and R. J. Goldstein, pp. 397–423, Hemisphere, Washington, D.C., 1976.
245. Lohrisch, W., Bestimmung von Warmeubergangszahlen durch Diffusionsversuche, *Forschung. Geb. Ingenieurwes.,* no. 322, p. 46, 1929.
246. Jakob, M., *Heat Transfer,* vol. 1, pp. 605–613, Wiley, New York, 1955.
247. Oertel, H., Thermal Instabilities, in *Convective Transport and Instability Phenomena,* eds. J. Zierep and H. Oertel, pp. 3–24, G. Braun, Karlsruhe, 1982. See also Koschmieder, E. L., *Rayleigh–Bénard Convection and Taylor Vortex Flow,* pp. 39–54,
248. Nakayama, Y., Aoki, K., and Ohta, H., Visualization of Pressure Distribution due to Impact Accompanying Collapse of Cavity on the Vanes of Mixed Flow Pump-Turbine, *Int. Symp. Phys. Numer. Flow Visual.,* ASME, FED, vol. 22, pp. 101–108, 1985.
249. Goldstein, S., ed., *Modern Developments in Fluid Dynamics,* vols. 1,2, Oxford Univ. Press, Oxford, England, 1938.

Chapter 28

Vortices

Peter Freymuth

I INTRODUCTION

Much has been learned about vortices by means of flow visualization. Particularly successful have been the vortex-tagging techniques whereby smoke is released in air [1–2] or dye or fine bubbles are released in water [3–6] from vorticity-generating solid surfaces. When vorticity separates, the tagging fluid separates with it and makes the vortex development visible. In contrast, smoke-rake and smoke-wire techniques as well as streamline techniques do not visualize vortices directly; their application is not considered in this chapter.

Vortex tagging is often considered to be a streakline technique. This is approximately correct when the separation points of the vortex filaments do not change their position in space. For moving bodies, in particular for pitching and plunging airfoils, the classification as a streakline technique is incorrect. It seems best to consider vortex tagging as a class of its own.

Vortex tagging appears most useful in unsteady flows at modest Reynolds numbers where vortices can be visualized near bodies. Far downstream, differences in viscous and smoke or dye diffusion render visualization results questionable. In three-dimensional visualizations, in particular in turbulent flows, the problem of credibility is further compounded by differences in vortex stretching and smoke or dye stretching. This chapter largely emphasizes the visualization of two-dimensional vortices; three-dimensional vortex patterns are described briefly.

The photographic cooperation of W. Bank and F. Finaish is greatly appreciated. This work has been supported by AFOSR contract F49 620-84-C-0065.

Vortex-tagging techniques foremost reveal vortex shapes or patterns [7]. These patterns, viewed as a secret code of fluid motion, are a source of inspiration and understanding.

II CONCEPTUAL ROOTS OF VORTEX STRUCTURES

Vortex structures as spiritual and possibly observational concepts can be traced back thousands of years [8, 9]. The ancient Greek gravestone [8] shown in Fig. 1 (bottom left) depicts, among other ornaments, a clockwise and a counterclockwise rotating simple spiral vortex, with the vortex filament entering from one side only and terminating at the vortex center. Such starting vortices are generated in many unsteady flows around bluff bodies. Figure 2 (top left) shows a large clockwise rotating starting spiral embedded in other detail. The spiral was generated behind the leading edge of a pitching airfoil (from 0 to 60°) in a steady flow from left to right.

The entrance stone to a grave in Ireland dating back to the bronze age [8] and reproduced in Fig. 1 (top) shows vortices produced from a folded filament such that there is no filament termination at the vortex center. Such vortices are often generated in the midst of a vortex filament due to folding of the filament that enters the vortex from two sides. An example of this filament instability is shown in Fig. 2 (top center), where the trailing-edge vortices in the lower part of the picture have been generated by filament folding. The filament enters each vortex from two sides.

Figure 1 (bottom right) shows an Irish bronze relief (eighth century A.D.) [8] depicting the crucifixion and incorporating rather complex vortex structures. For instance,

Fig. 1 Three ancient representations of vortex patterns [8].

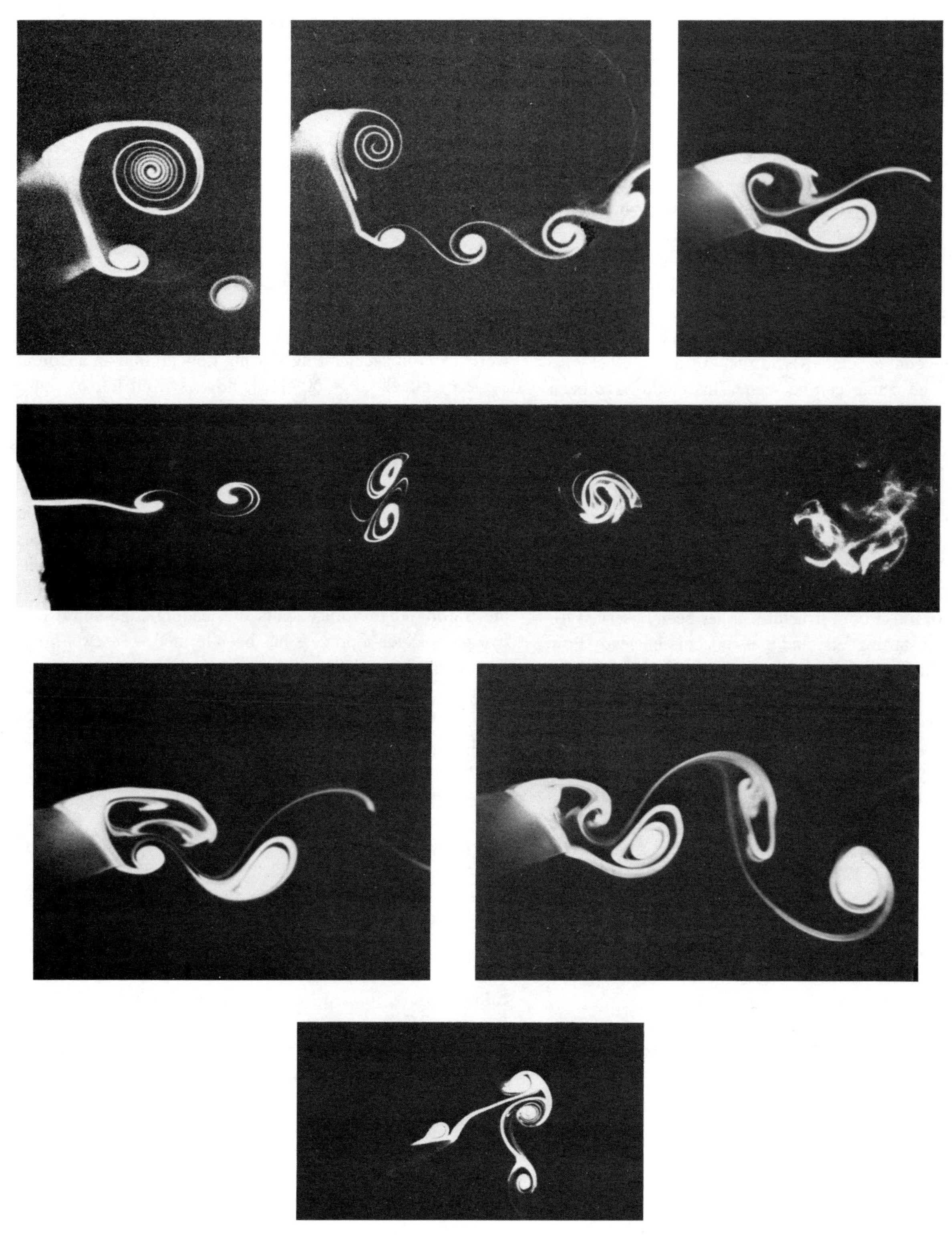

Fig. 2 Modern visualization of vortex patterns.

the two large vortices below the chest are each composed of two paired vortices with spiral arms. Above these are structures consisting of three paired or merged vortices. Merging, pairing, coalescence, fusion, or condensation of vortices with the same sense of rotation is a ubiquitous process of two-dimensional vortex dynamics [7, 10–14, 17] that can involve many vortices. Figure 2 (2nd row) shows a modern example of vortex pairing in the shear layer of a laminar jet [14]. To the left of the frame center two vortices in the process of pairing can be discerned, to the right of the center two vortices have fully paired, and to the extreme right a paired vortex decays turbulently.

Consider the ornaments above the head in Fig. 1 (bottom right). Three counter-clockwise structures and a clockwise structure in the center can be seen. This could have been facilitated by the clockwise vortex which split a large counter-clockwise structure into three vortices. While such a process has not yet been observed in modern times, Fig. 2 (top right) shows splitting of a vortex into two parts. This has been observed in accelerating starting flow around an airfoil at high angle of attack [15, 16].

The two counter-rotating vortices below the chest in Fig. 1 (bottom right) could be interpreted as a counter-rotating vortex pair or mushroom structure. It has been observed for a long time in starting jets, and it also occurs in steady flow over an oscillating airfoil, as shown in Fig. 2 (bottom). On a large scale, the arrangement of vortices into vortex streets of complicated symmetries is also evident in Fig. 1 (bottom right).

Developing vortices have spiral shapes but the simplest vortex is circular. Such a structure is often attained by a tendency of noncircular vortices to become rounded in the absence of strong external deformation. This tendency is termed *aggregation* [7] or *axisymmetrization* [17] and is further aided by viscous diffusion. An example is shown in the frames of Fig. 2 (third row). The left frame depicts a spiral with a droplike deformation. The right frame shows a rounding of this vortex at a later time, with the round vortex being close to the right edge of the frame.

It should be mentioned that many variations on the one-sided starting spiral structure have been found, as a consequence of filament instability. Figure 3 shows a collection of spiral variations as we found them in the starting flows around airfoils at various conditions [18]. In contrast, two-sided or folded spirals are small, and pattern variation is limited, since a vortex filament is shared by many vortices in this case. In some visualizations, notably in starting jets, smoke or dye is inadvertently introduced beyond the vortical region. In such a case an irrotational smoke or dye filament may mimic a vortical filament, and a starting spiral may look like a folded spiral. A prominent example of such pseudofolding is seen in the starting spirals of Fig. 76 in Ref. [19], where many other examples of the vortex-tagging technique can also be found.

III VORTEX INTERACTIONS

Vortices do not exist in isolation. They interact with other vortices and with solid surfaces. A vocabulary to describe many of these interactions has been assembled [7] and slightly extended [20] recently. The purpose of this section is to furnish examples of the various vortex interactions in time sequence.

A Vortex–Vortex Interactions

Figure 4 represents a time sequence of movie frames. They were taken in accelerating starting flow around an airfoil at angle of attack $\alpha = 60°$ [18]; flow is from left to right. Movie frames are ordered into columns from top to bottom, starting with the left column. Time between adjacent frames is $\Delta t = 1/64$ s. The Reynolds number for this flow is Re $= a^{1/2}c^{3/2}/\nu = 350$, where a is the constant flow acceleration, c is the airfoil chord length, and ν is the kinematic viscosity of the airflow. In column 1 spiral starting vortices develop near the leading and trailing edges of the airfoil. In column 2 the spirals interfere with each other such that the trailing-edge vortex splits the leading-edge spiral into two parts. Such a process has been termed *vortex splitting*. In column 3, the left remnant of the split vortex coalesces with the next vortex, which forms near the leading edge, i.e., the process of merger of vortices of the same sense of rotation. The so-augmented leading-edge vortex then deforms the trailing-edge vortex into a droplike shape, referred to as *vortex squeezing*, before it nips off a very small part of the trailing-edge vortex, i.e., the process of vortex nipping (which is a weak form of vortex splitting). In the last three frames of column 3, the drop shape returns to a more rounded form, i.e., the process of aggregation or axisymmetrization.

Another vortex interaction, which we termed *ornamentation*, is displayed [21] in Fig. 5. There an array of small folded spirals, formed from a filament that leaves the leading edge of an airfoil at $\alpha = 80°$, align themselves into a large composite starting spiral of ornamental quality. Ho and Huang [12] termed a similar but not identical process *collective interaction*. In their case small vortices interacted to form a composite folded spiral rather than a composite starting spiral. Composite structures appear the most pronounced at intermediate Reynolds numbers prior to the onset of turbulence.

Another basic vortex interaction is the marching of a pair of vortices rotating in opposite directions on a straight or curved path, depending on their relative strengths. An example is shown in Fig. 6, where a periodically pitching airfoil exposed to steady wind generates among other vortices a pair or mushroom behind the trailing edge. The airfoil pitches at a reduced frequency $K = \pi f c/U_\infty = 0.52$,

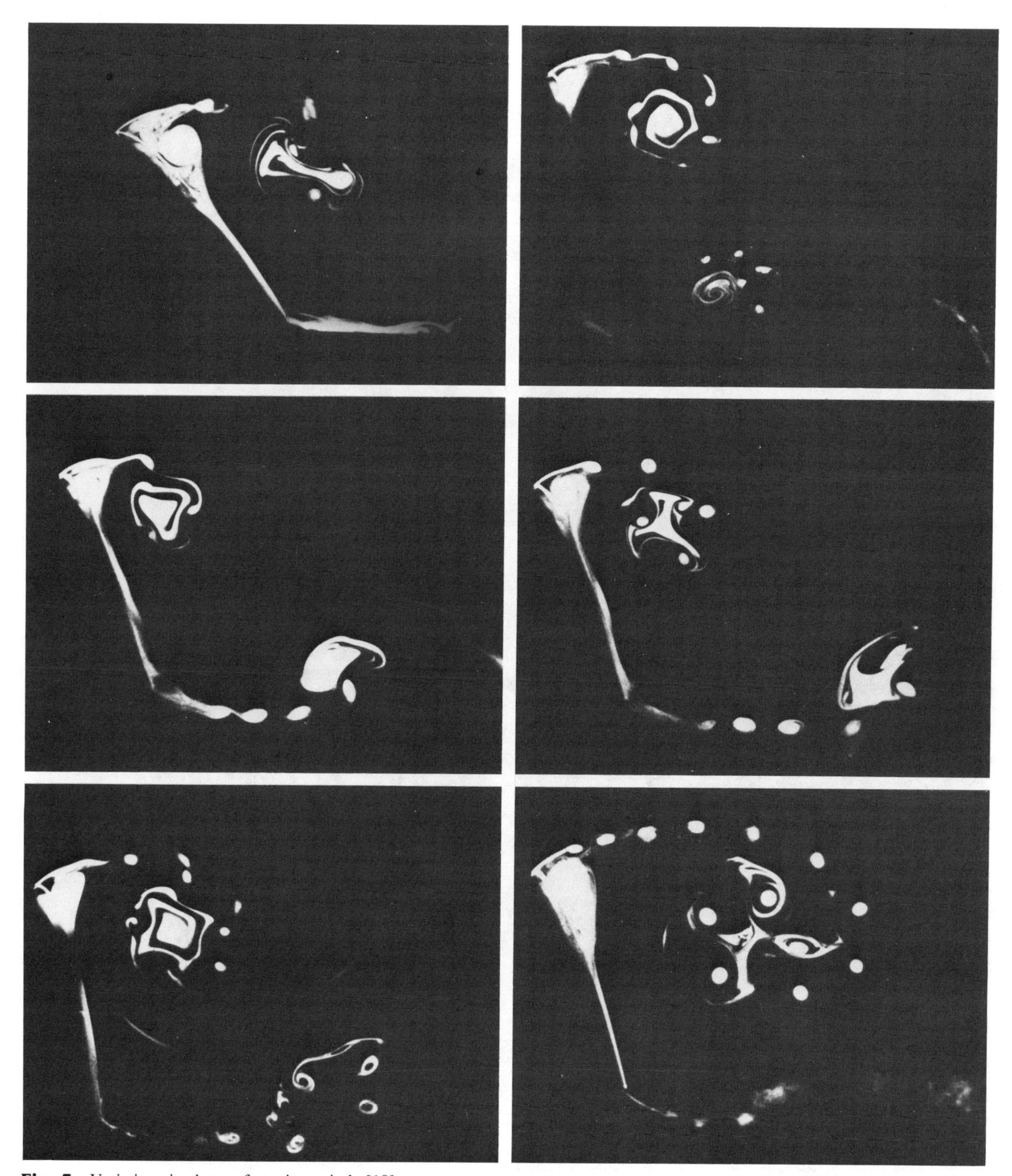

Fig. 3 Variations in shape of starting spirals [18].

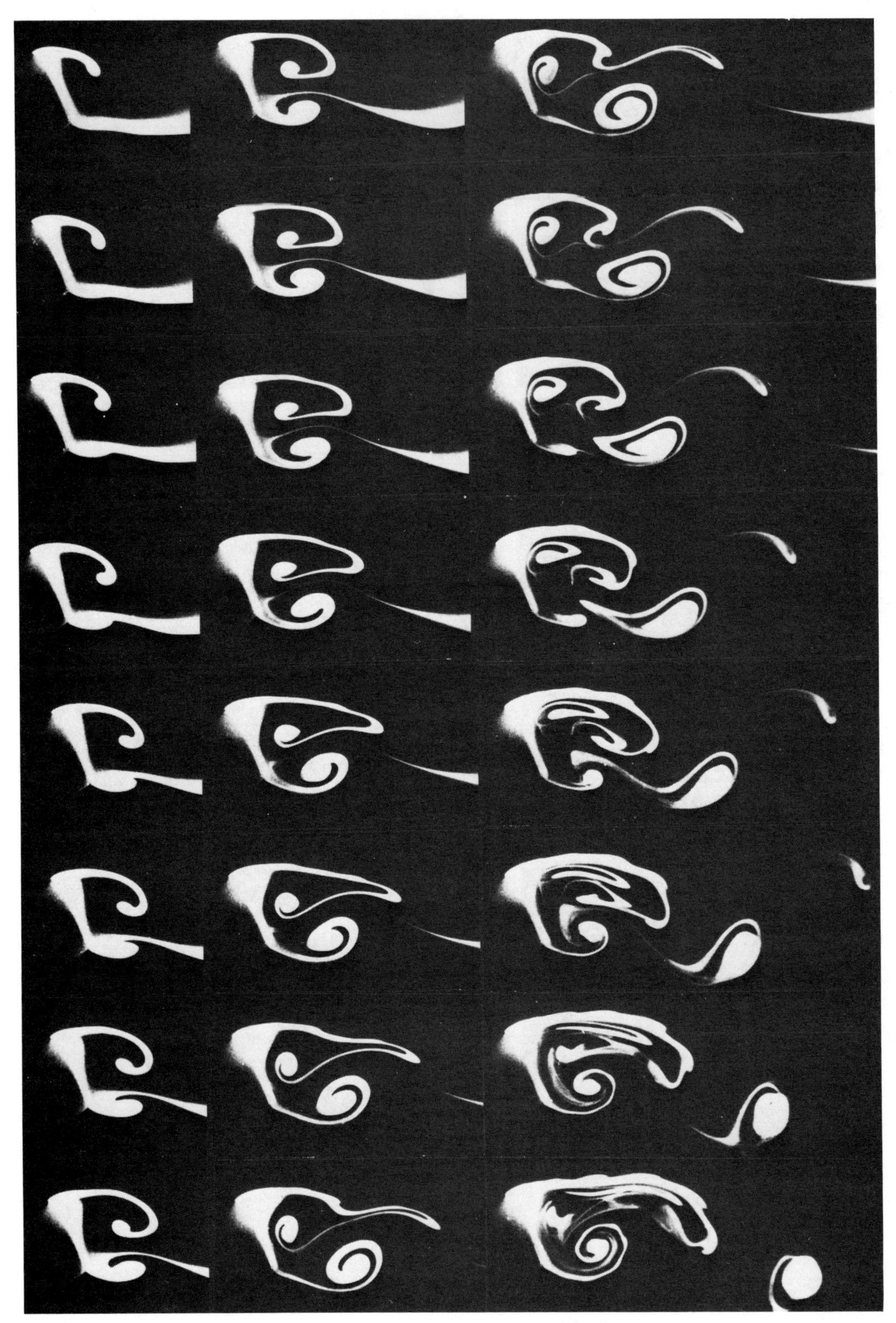

Fig. 4 Sequence of accelerated starting flow around an airfoil at $\alpha = 60°$ [18].

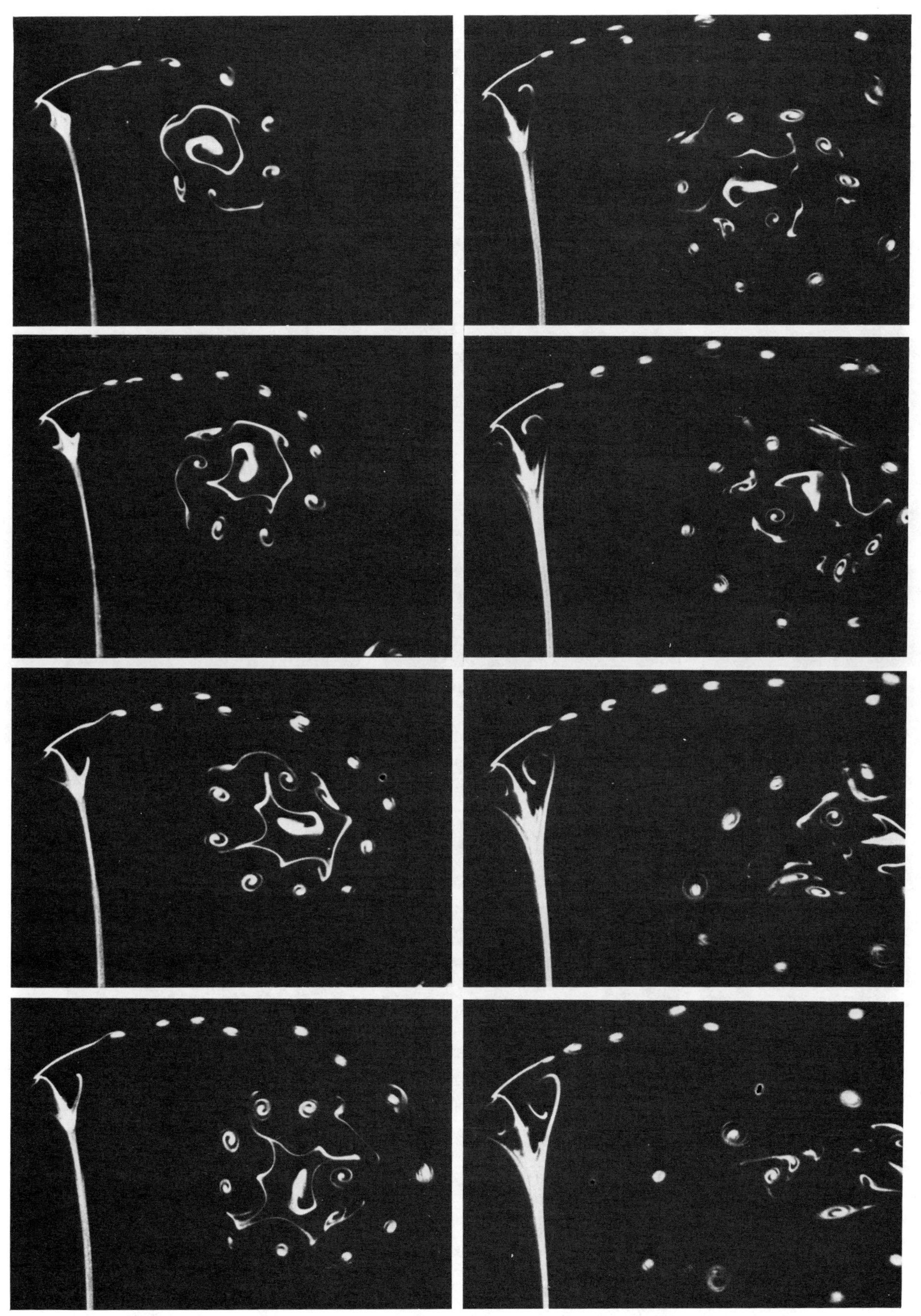

Fig. 5 From a sequence of accelerated starting flow around an airfoil, $\alpha = 80°$, Re = 5200 [21].

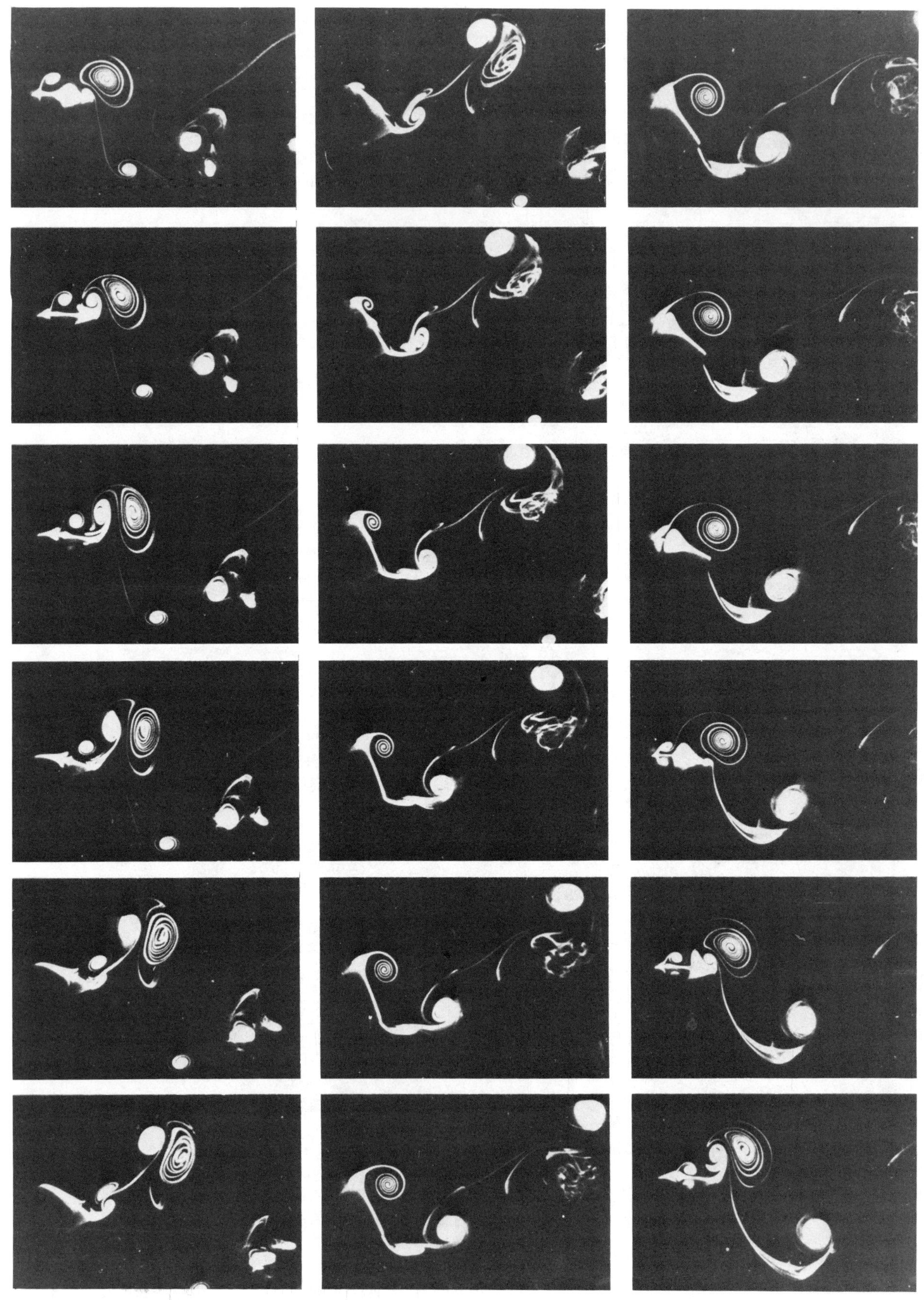

Fig. 6 Periodically pitching airfoil in steady flow, 30 ± 30°, Re = 1700, $K = 0.52$, $\Delta t = 1/32$ sec., $U_\infty = 61$ cm/s, $c = 5.1$ cm. The pitching axis is $c/4$ from the leading edge.

where f is the pitching frequency and U_∞ is the wind tunnel speed. The Reynolds number is $\mathrm{Re} = U_\infty c/\nu = 1700$, and pitching is between angles 0 and 60°. Pairs of counterrotating vortices have been visualized for some time, for instance a starting–stopping vortex pair of an airfoil [22] by streamline visualization and the starting jet [23] by a schlieren technique. More recently mushrooms have been created by an oscillating jet impinging on a wedge [24] and by a flat plate that starts oscillating [6]. The process of mushroom formation has recently been termed *binding* by Zabusky [17]. On a more global scale an array of counterrotating vortices can move into an irrotational flow [6–8] i.e., the formation of vortex streets.

Many of the two-dimensional vortex interactions also occur in turbulent flows. Turbulence is easily identified in vortex-tagging visualizations by the fuzziness in the appearance of vortices and by a decrease in two-dimensional detail as a consequence of turbulent diffusion.

B Vortex–Surface Interactions

Solid surfaces are foremost the source of vortex filaments from which vortices develop. Once developed, the vortices can interact with solid surfaces to create secondary and higher order vortices as illustrated in Fig. 7. The sequence of six frames in Fig. 7 (top) shows in the third frame of column 1 a clockwise-rotating primary vortex and a counter-clockwise secondary vortex to the left of it. The secondary vortex acts as a barrier to the vortex filament, redirecting it into another primary vortex. This process we termed *filament redirection*. In the top frame of column 2 the second primary vortex induces another secondary vortex close to the leading edge. Turbulence blurs two-dimensional detail in subsequent frames. Figure 7 (bottom) shows a marginally visualized primary vortex to the right and a secondary and a tertiary vortex. There is a hint of a quartary vortex nested in the left corner of the tertiary vortex. The tertiary vortex also acts as a barrier for the secondary vortex filament and redirects it into a second secondary vortex located to the right of the first secondary vortex.

Vortex filaments can spring from locations on the wing surface other than the leading and trailing edges. In Fig. 8 (column 1) weak separation occurs also near mid-chord, from which it moves downstream. Such additional separations we term *multiple separations*. When the separation tongue reaches the trailing edge, it elicits vortex formation from the trailing edge. In column 2 the leading edge vortex becomes turbulent. When it reaches the trailing edge again, the interaction of vortex elicitation occurs.

Another interaction, which we termed *vortex shredding*, is illustrated in Fig. 9 and involves a secondary vortex sucked into the primary vortex (a composite spiral in this case) and stretched out.

A process of vortex pinching has been found [20] behind a fence on a flat plate in accelerating starting flow. Vortices of the same sense of rotation move over each other close to the flat surface and in the process flatten each other.

Vortices that have been created by separation can attach to solid surfaces downstream, as has been shown by Ziada and Rockwell [25]. Vortices moving over a wedge can attach themselves to either side of it or be severed, depending on impact position.

Finally, under favorable conditions, a separated vortex filament can reattach to the surface from the front. For instance, for an oscillating airfoil shown in the sequence of Fig. 6 (first column) vortices over the airfoil are flushed downstream, and the filament reattaches from the front.

III THE CHOREOGRAPHIC ORGANIZATION OF VORTEX PATTERNS

In the previous section we identified various vortex interactions such as pairing, splitting, mushroom formation, secondary-vortex generation, multiple separation, and others. There is an almost infinite variety of contexts in which these interactions can arise and succeed each other. We term a particular realization of interactions *the choreography of the pattern development*. This section is devoted to a study of the choreographic variety of vortical interactions. Having developed the vocabulary of vortex interactions in the previous section, we now proceed to use them as elements of a descriptive language.

Let us commence by studying the choreographic variety of interaction of a primary vortex with its secondary vortex. Figure 9 already showed the shredding of a secondary vortex by its primary vortex. Another possibility is an escape of the secondary vortex by slipdown over the trailing edge and merger with a trailing-edge vortex; this can be observed nicely in Fig. 6 (column 1). Figure 10 shows a sequence for an airfoil pitching periodically at high rate ($K = 3.1$) in steady flow ($\mathrm{Re} = 12{,}000$) between 0 and 60°. In column 1 a secondary vortex develops to the left of the primary leading-edge vortex. This secondary vortex is of sufficient strength to form a mushroom with the primary vortex instead of being shredded. This mushroom lifts off the airfoil in column 2. These three basic variations in the fate of a secondary vortex as consequences of its interaction with the primary vortex have been observed in many other situations.

Let us consider some choreographic aspects of the marching of a vortex pair or mushroom. Figure 11 sequences a periodically pitching airfoil at reduced frequency $K = 2.4$ and $\mathrm{Re} = 12{,}000$; pitching is $\pm 30°$. The two vortices generated behind the trailing edge (column 2) have a relative position that propels them downstream, i.e., the pitching airfoil generates thrust. On a larger scale an in-

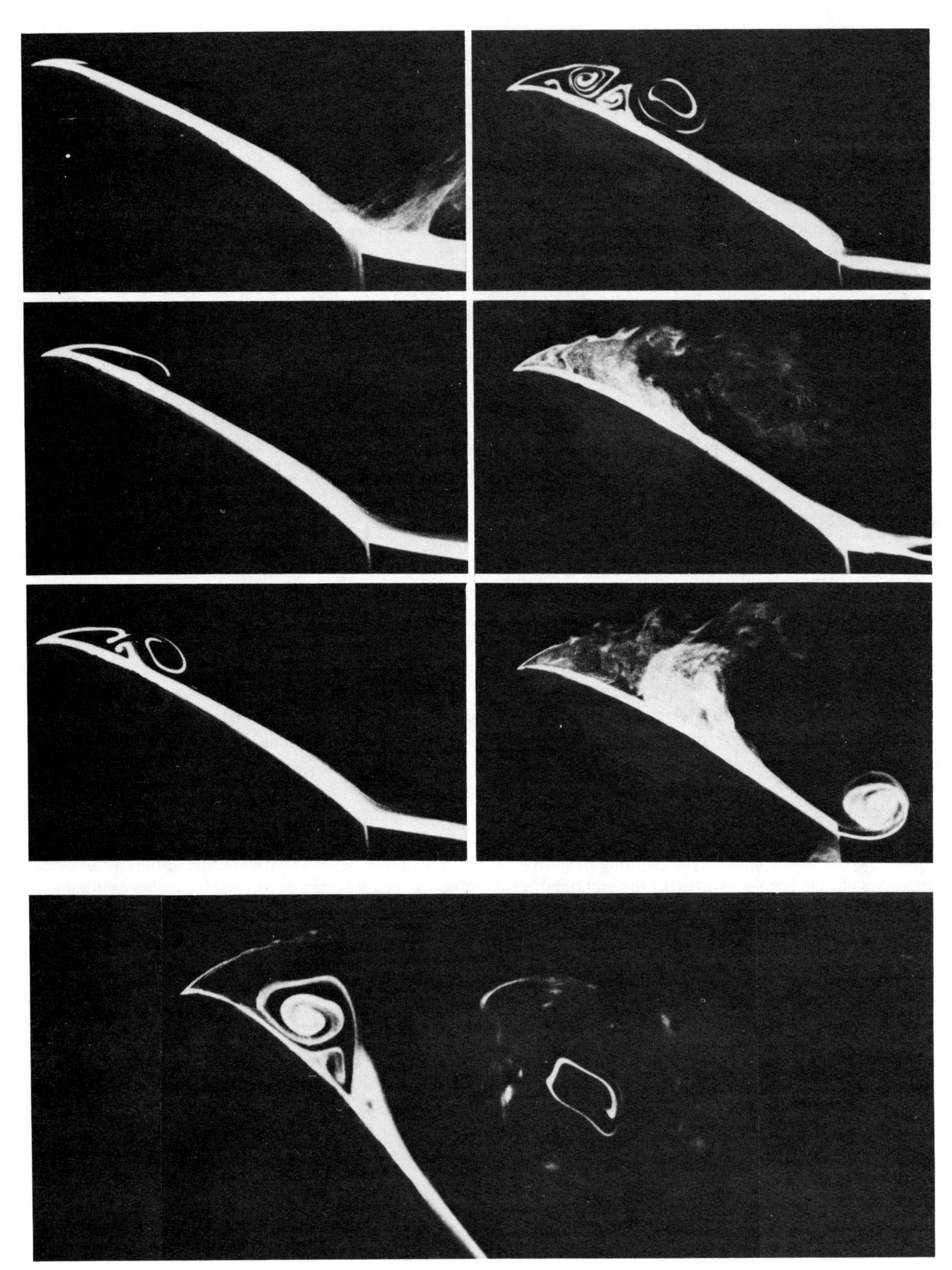

Fig. 7 Accelerated starting flow over airfoil: Top six frames, $\alpha = 26°$, Re = 5200; bottom frame, $\alpha = 40°$, Re = 5200 [21].

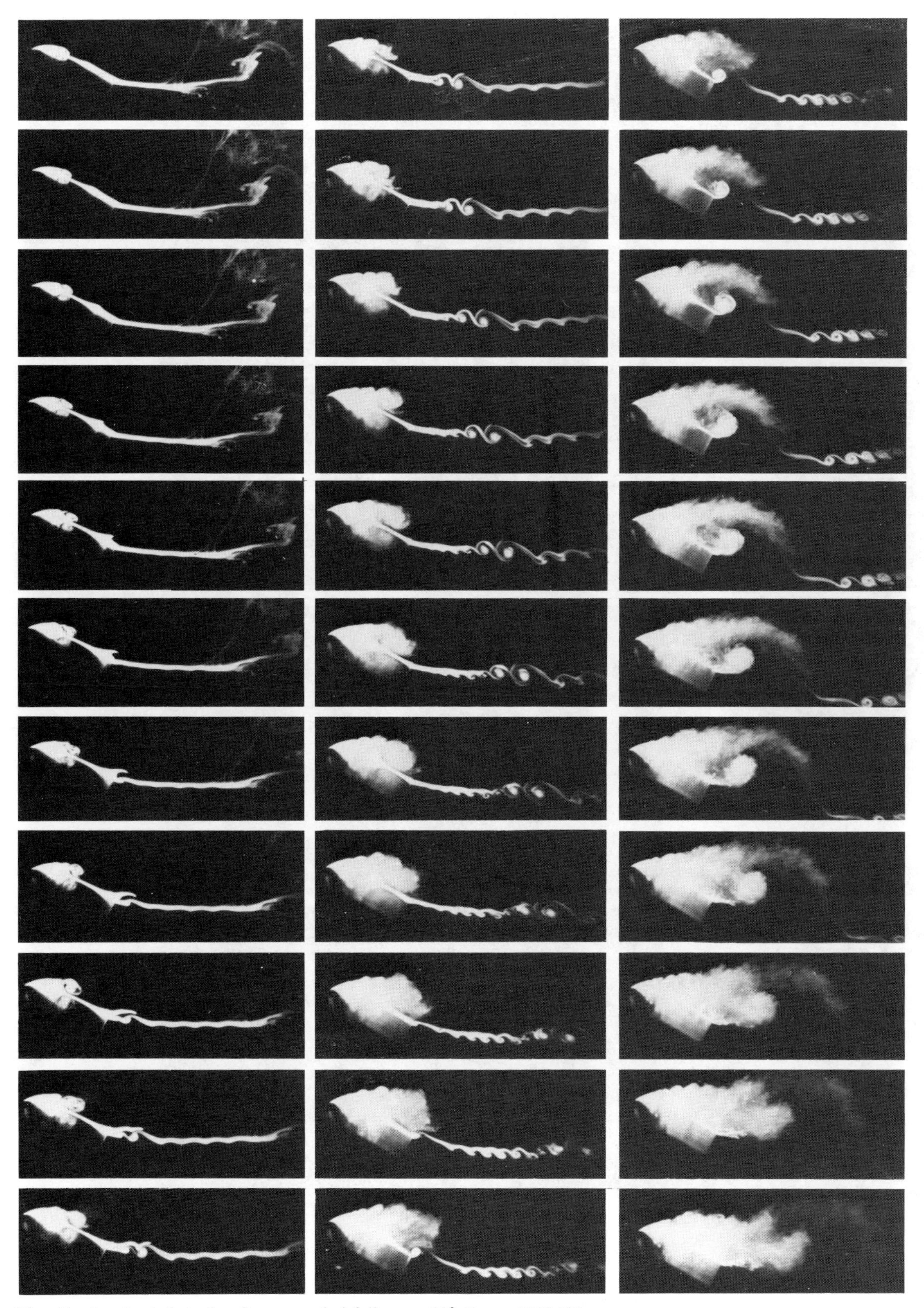

Fig. 8 Accelerated starting flow around airfoil, $\alpha = 20°$, Re = 5200 [7].

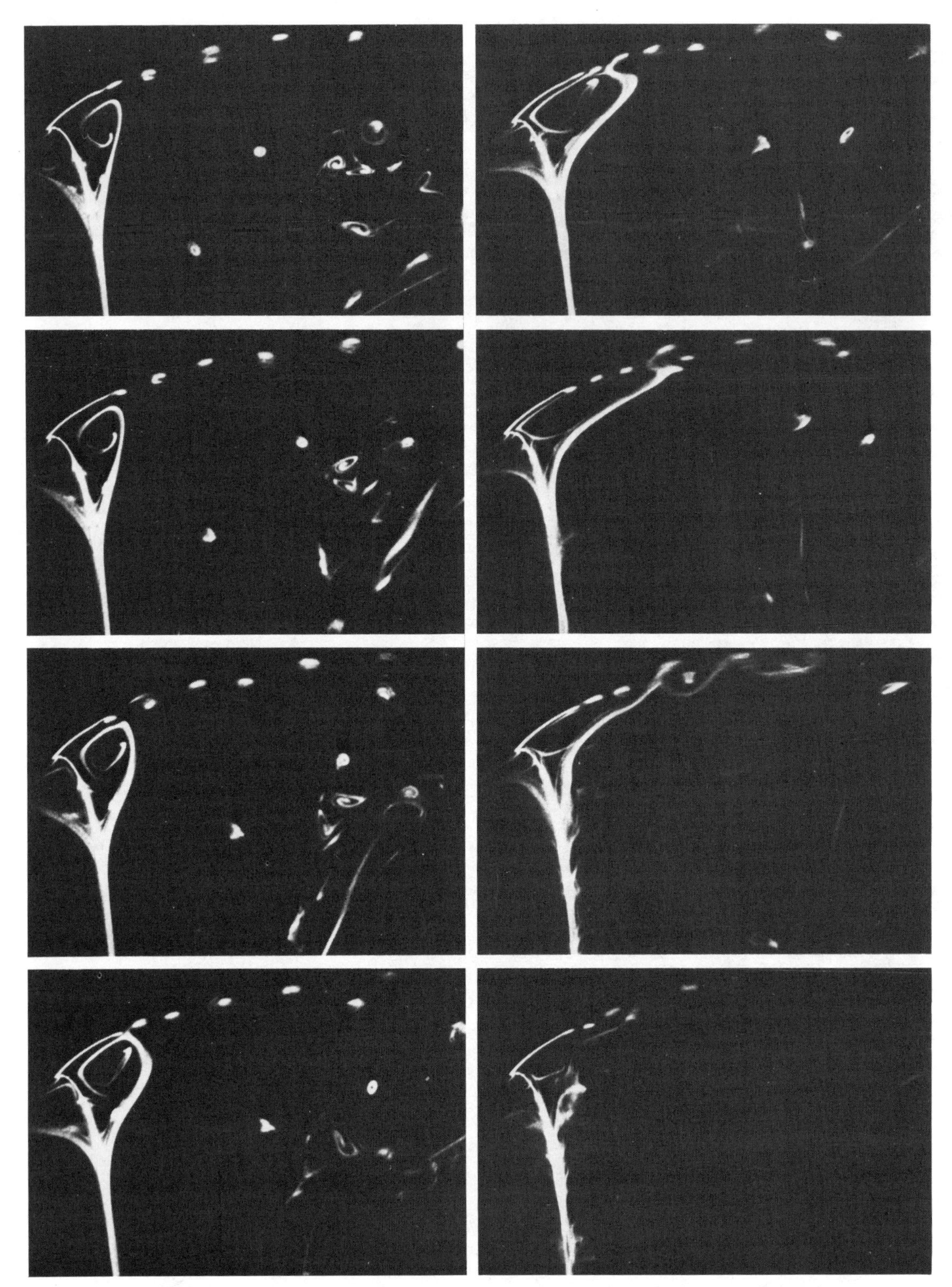

Fig. 9 From a sequence of accelerated starting flow around an airfoil, $\alpha = 80°$, Re = 5200 [21].

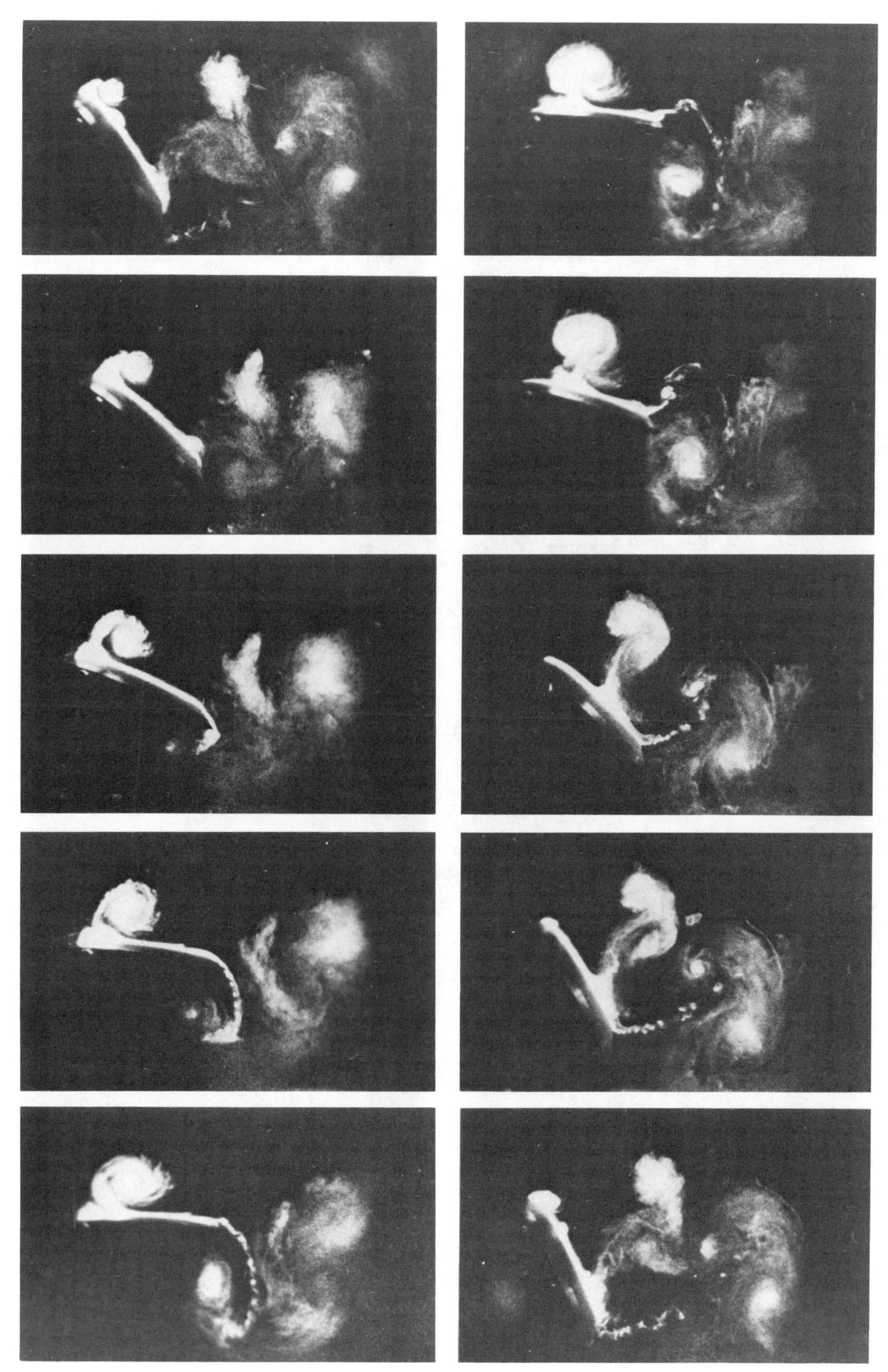

Fig. 10 Periodically pitching airfoil in steady flow, 30 ± 30°, Re = 12,000, K = 3.1, c = 35.6 cm, U_∞ = 61 cm/s, Δt = 1/16 s.

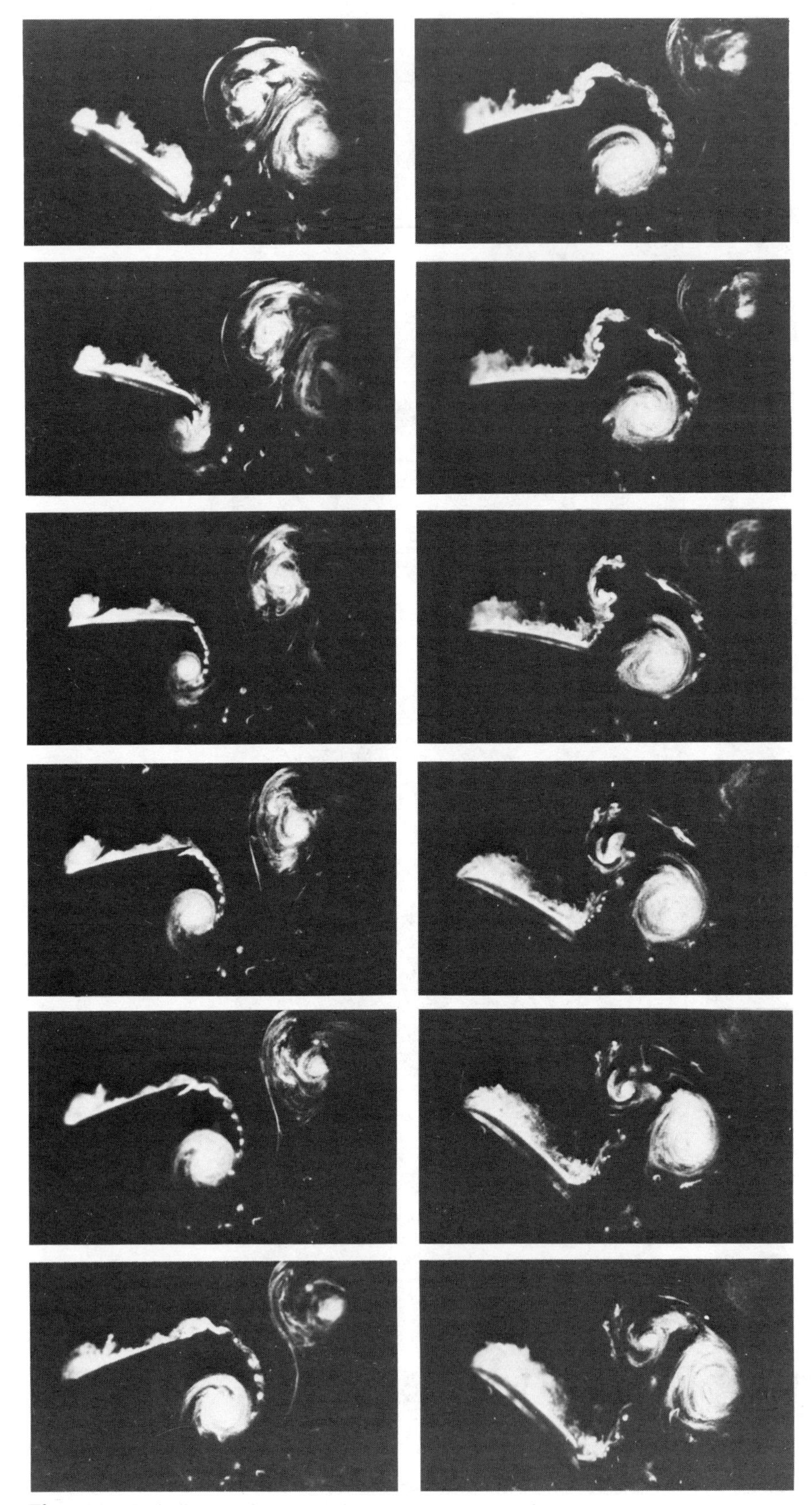

Fig. 11 Periodically pitching airfoil in steady flow, 0° ± 30°, Re = 12,000, K = 2.4, c = 35.6 cm, U_∞ = 61/s, Δt = 1/32 s.

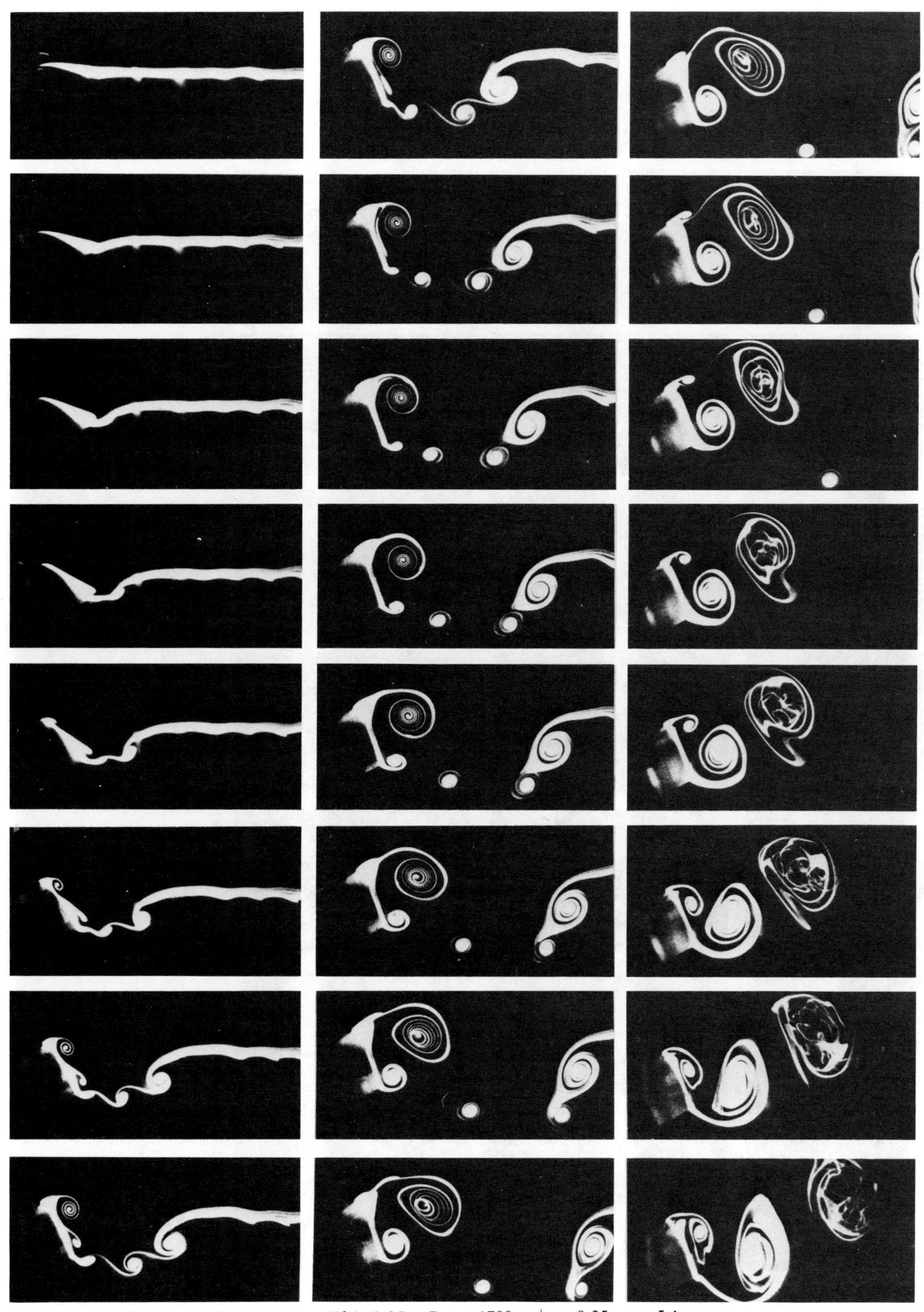

Fig. 12 Airfoil in single pitch from 0 to 60° in 0.25 s, Re = 1700, $\alpha^+ = 0.35$, $c = 5.1$ cm, $U_\infty = 61$ cm/s, $\Delta t = 1/32$ s.

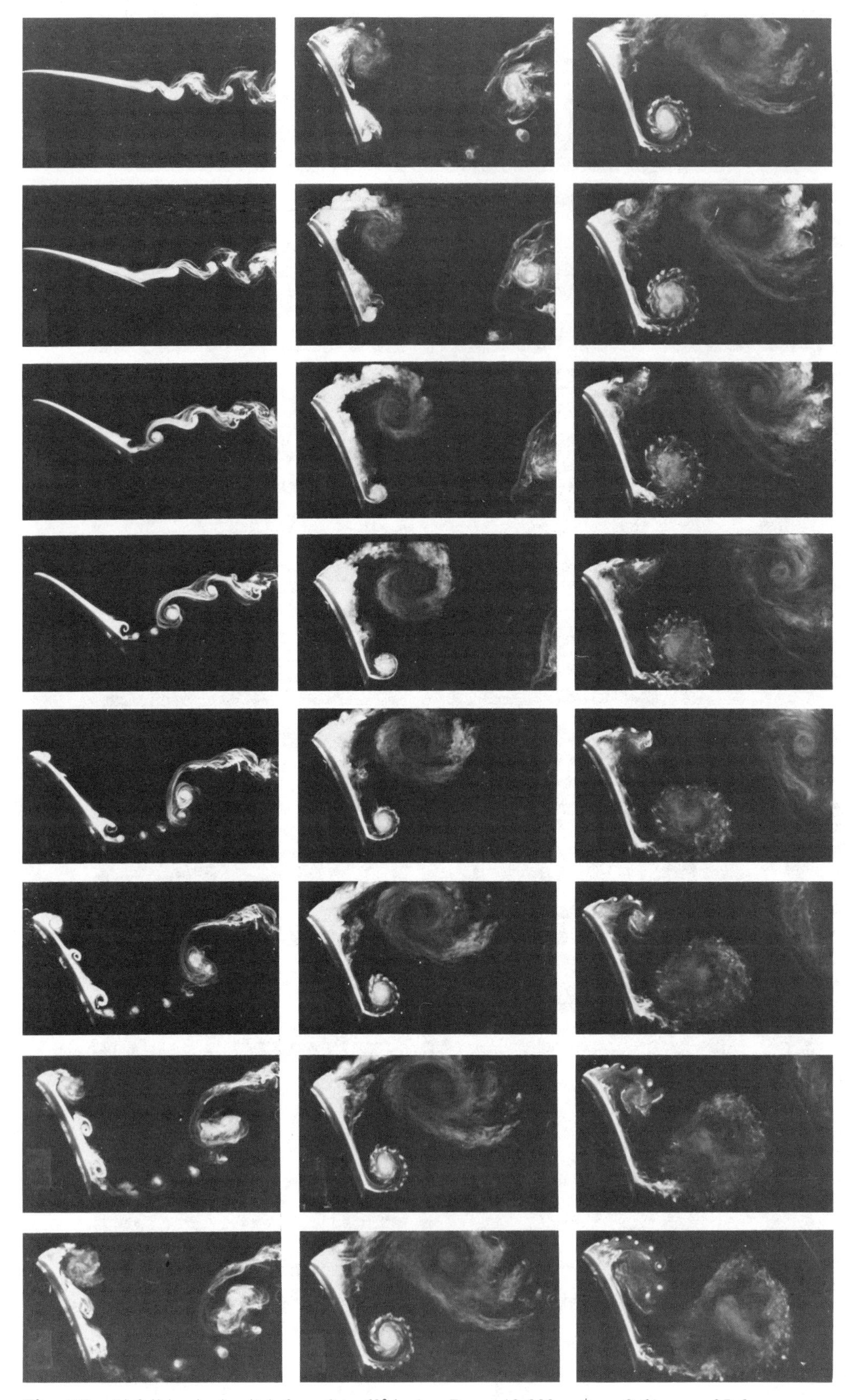

Fig. 13 Airfoil in single pitch from 0 to 60° in 1 s, Re = 12,000, $\alpha^+ = 0.61$, $c = 35.6$ cm, $U_\infty = 61$ cm/s, $\Delta t = 1/8$ s.

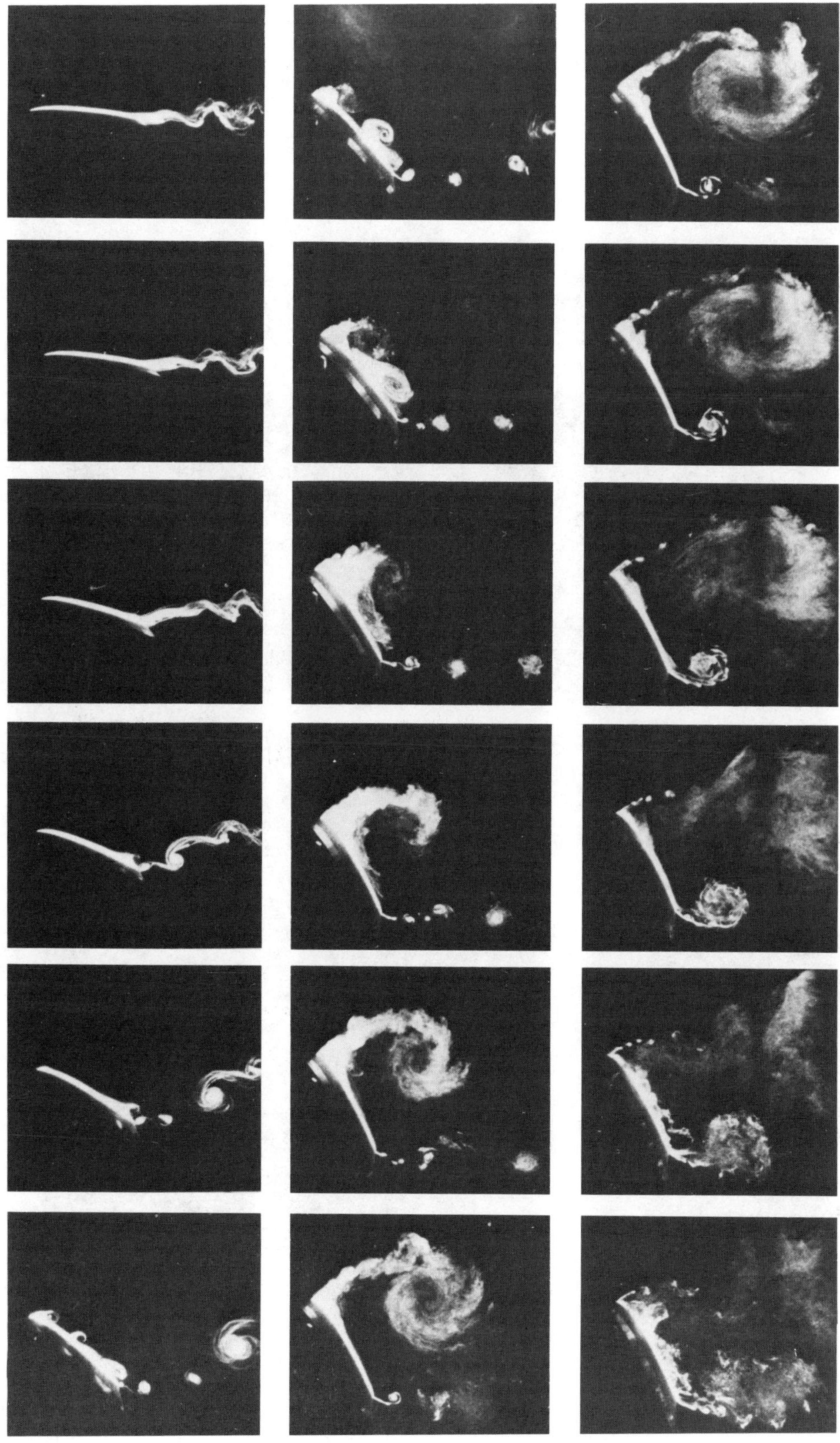

Fig. 14 Airfoil in single pitch from 0 to 60° in 2 s, Re = 12,000, $\alpha^+ = 0.3$, $c = 35.6$ cm, $U_\infty = 61$ cm/s, $\Delta t = 1/4$ s.

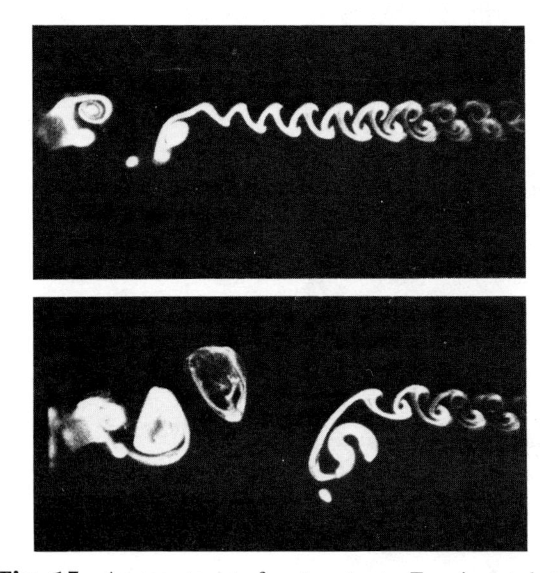

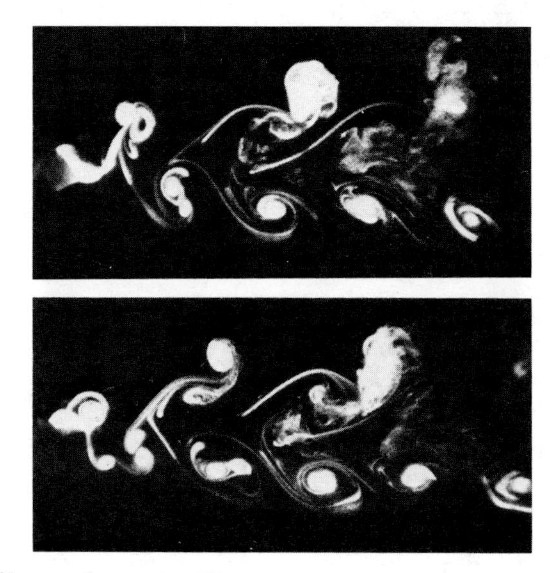

Fig. 15 An assortment of vortex streets: Top, inverted street of a jet [8]; second row, street in the wave of a circular cylinder; third and fourth rows, left, two frames from a sequence for airfoil in single pitch 0 to 60°; third and fourth rows, right, two frames for airfoil in periodic pitch 30 ± 30°.

verted vortex street exists where counter-clockwise rotating vortices are located above clockwise rotating vortices, in contrast to the regular drag-inducing vortex street behind a circular cylinder in steady flow. Figure 6 shows mushroom formation in column 1 from a leading-edge primary vortex that has rolled over the trailing edge and from the merged secondary and trailing-edge vortices. Simultaneous formation of two mushrooms is shown in Fig. 10, one near mid-chord and one behind the trailing edge.

Mushroom formation by pitching airfoils should have

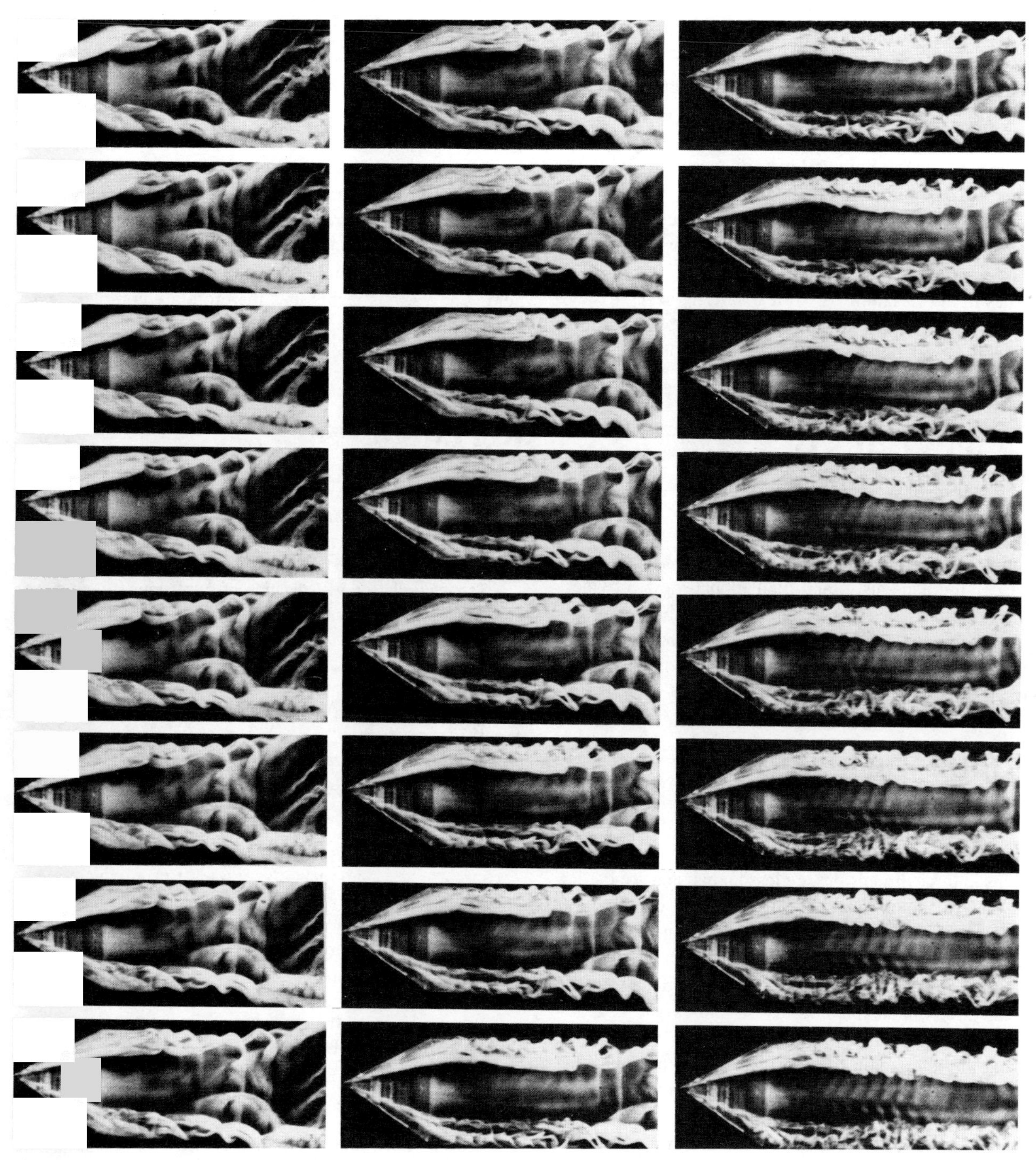

Fig. 16 Accelerated flow over a delta wing, $\alpha = 10°$, Re = 5200, spanwise view.

some relevance to bird- and insect-flight propulsion. Even more relevance can be expected from investigations of simultaneously pitching and plunging airfoils, which could serve as a crude two-dimensional model of wing flapping.

Let us consider the choreographic variability of multiple separations over a wing. In Fig. 8 a separation developed near mid-chord and moved downstream. In Fig. 12 (column 1) a separation developed below mid-chord in addition to the leading-edge separation. The separating vortex below mid-chord in this case moved upstream rather than downstream and merged with the primary vortex. The sequence of Fig. 12 depicts an airfoil in single pitch from 0 to 60° within 0.25 s. The pitching process is finished after the first column. Other relevant parameters are Re = 1700 and $\alpha^+ = \dot{\alpha}c/U_\infty = 0.35$, where $\dot{\alpha}$ is the pitch rate.

Even more choreographic variability is possible in cases where three separations occur over the airfoil. Figure 13 sequences an airfoil in single pitch from 0 to 60°, Re =

12000, $\alpha^+ = 0.6$. Column 1 shows the separation in three places. In column 2 the mid-chord separation moves upstream and merges with the leading-edge vortex. Thereafter, also the separation close to the trailing edge moves upstream and merges with the leading-edge vortex. In Fig. 14 for the same airfoil but at a lower pitch rate $\alpha^+ = 0.3$, a merger of the mid-chord and trailing-edge separations occurs prior to their merger with the leading-edge vortex. It is clear that many more variations are possible under different circumstances.

Studies in the variability of vortex splitting have already been presented [16]. Many other choreographic aspects can be considered but an exhaustive development is not intended.

IV TAPESTRIES OF VORTEX PATTERNS

In the previous section we focused on particular vortical interactions and their variability. Usually many interactions take place simultaneously, and their global view provides stunning tapestries that border on being an abstract ornamental art. Figure 15 shows a collection of vortex streets obtained under different conditions. The study of global patterns is still in its infancy.

V THREE-DIMENSIONAL VORTICES

The statement by Saffman and Baker [26] that vortex interactions in three dimensions are less known and are harder to analyze still holds. The development of a detailed vocabulary and its use in a descriptive language would be premature at this point.

Again the vortex-tagging techniques have much to offer to reveal three-dimensional vortex structures and interactions. Vortex rings have been visualized by Maxworthy [27], and the interaction of two rings has been studied by Yamada and Matsui [28]. We have visualized combined wing-tip and starting-vortex systems [29–31] as well as transition [32]. Lugt [9] has given a rather systematic account of three-dimensional vortices.

A single figure may suffice to demonstrate the capability of vortex tagging of three-dimensional vortex systems. Figure 16 sequences accelerating starting flow over a delta wing in the form of an equilateral triangle at $\alpha = 10°$, Re = 5200. The conical leading-edge vortices can be seen and their connection to a number of starting vortices shed from the trailing edge. In addition, columns 2 and especially 3 show formation of vortical laces emanating from the trailing-edge corners. Undoubtedly visualization of three-dimensional structures will be further improved and the application range be broadened in the future.

VI CONCLUSION

We have demonstrated the application of vortex-tagging techniques to the visualization of vortices. These techniques are far superior to smoke-rake and smoke-wire techniques as well as to streamline techniques for this specific purpose. Smoke-wire, smoke-rake, and other streakline techniques on the other hand are the best suited to visualizing the potential flow that surrounds vortices, and the streamline technique has a dominating role in visualizing the velocity field. The various techniques therefore are complementary and cannot be substituted for each other.

REFERENCES

1. Mueller, T. J., Flow Visualization by Direct Injection, in *Fluid Mechanics Measurements,* ed. R. J. Goldstein, pp. 307–375, Hemisphere, Washington, D.C., 1983.
2. Freymuth, P., Bank, W., and Palmer, M., Use of Titanium Tetrachloride for Visualization of Accelerating Flow around Airfoils, in *Flow Visualization III: Proceedings of the Third International Symposium on Flow Visualization,* ed. W.-J. Yang, pp. 99–105, Hemisphere, Washington, D.C., 1985.
3. Taneda, S., Visual Study of Unsteady Separated Flows around Bodies, *Prog. Aerosp. Sci.,* vol. 17, pp. 287–348, 1977.
4. Werlè, H., Visualization Hydrodynamique d'Ecoulements Instationaires, ONERA. Note Tech. 180, 1971.
5. McAlister, K. W., and Carr, L. W., Water Tunnel Experiments on Oscillating Airfoil at Re = 21,000, NASA Tech. Memo 78446, March 1978.
6. Oshima, Y., and Oshima, K., Vortical Flow Behind an Oscillating Airfoil, IUTAM, pp. 357–368, 1980.
7. Freymuth, P., The Vortex Patterns of Dynamic Separation: A Parametric and Comparative Study, *Prog. Aerosp. Sci.,* vol. 22, pp. 161–208, 1985.
8. Schwenk, T., *Das Sensible Chaos,* Verlag Freies Geistesleben, Stuttgart, 1961 (a 5th edition was published in 1980).
9. Lugt, H. J., *Vortex Flow in Nature and Technology,* Wiley, New York, 1983.
10. Roshko, A., Structure of Turbulent Shear Flows: A New Look, *AIAA J.* vol. 14, pp. 649–664, 1976.
11. Winant, D. C., and Browand, F. K., Vortex Pairing, the Mechanism of Turbulent Mixing Layer Growth at Moderate Reynolds Numbers, *J. Fluid Mech.,* vol. 63, pp. 237–255, 1974.
12. Ho, C.-M., and Huang, L.-S., Subharmonics and Vortex Merging in Mixing Layers, *J. Fluid Mech.,* vol. 119, pp. 443–473, 1982.
13. Ho, C.-M., and Huerre, P., Perturbed Free Shear Layers, *Ann. Rev. Fluid Mech.,* vol 16, pp. 365–424, 1984.
14. Freymuth, P., On Transition in a Separated Laminar Boundary Layer, *J. Fluid Mech.,* vol. 25, pp. 683–704, 1966.
15. Freymuth, P., Bank, W., and Palmer, M., First Experimental Evidence of Vortex Splitting, *Phys. Fluids,* vol. 27, pp. 1045–1046, 1984.

16. Freymuth, P., Bank, W., and Palmer, M., Further Experimental Evidence of Vortex Splitting, *J. Fluid Mech.*, vol. 152, pp. 289–299, 1985.
17. Zabusky, N. J., Visualizing Mathematics: Evolution of Vortical Flows, Physica D, vol. 18D, pp. 15–25, 1986.
18. Freymuth, P., Bank, W., Palmer, M., Vortices around Airfoils, *Am. Sci.*, vol. 72, pp. 242–248, 1984.
19. Van Dyke, M., *An Album of Fluid Motion,* Parabolic Press, Stanford, Calif., 1982.
20. Finaish, F., Freymuth, P., and Bank, W., Starting Flow over Spoilers, Double Steps and Cavities, *J. Fluid Mech.*, vol. 168, pp. 383–392, 1986.
21. Freymuth, P., Bank, W., and Finaish, F., Surveying Unsteady Flows by Means of Movie Sequences–A Case Study. *ASME Int. Symp. Phys. Numer. Flow Visual.*, ed. M. L. Billet et al., pp. 27–37, 1985.
22. Prandtl, L., and Tietjens, O. G., Applied Hydro- and Aeromechanics, p. 301, Dover, New York, see 1934.
23. Garside, J. E., Hall, A. R., and Townend, D. T. A., *Nature,* vol. 152, p. 748, 1943.
24. Kaykayoglu, R., and Rockwell, D., Unstable Jet-edge Interaction, *J. Fluid Mech.*, vol. 169, pp. 125–149, 1986.
25. Ziada, S., and Rockwell, D., Vortex–Leading Edge Interaction, *J. Fluid Mech.*, vol. 118, pp. 79–107, 1982.
26. Saffman, P. G., and Baker, G. R., Vortex Interactions, *Ann. Rev. Fluid Mech.*, vol. 11, pp. 95–122, 1979.
27. Maxworthy, T., Some Experimental Studies of Vortex Rings, *J. Fluid Mech.*, vol. 81, pp. 465–495, 1977.
28. Yamada, H., and Matsui, T., Preliminary Study of Mutual Slip-through of a Pair of Vortices, *Phys. Fluids,* vol. 21, pp. 292–294, 1978.
29. Freymuth, P., Visualizing the Combined System of Wing Tip and Starting Vortices, *TSI Flow Lines,* premier issue, May 1986.
30. Freymuth, P., Finaish, F., and Bank, W., The Wing Tip Vortex System in a Starting Flow, *Z. Flugwiss. Weltraumforschung,* vol. 10 pp. 116–118, 1986.
31. Freymuth, P., Finaish, F., Bank, W., Visualization of Wing Tip Vortices in Accelerating and Steady Flow, *J. Aircr.*, vol. 23, pp. 730–733, 1986.
32. Freymuth, P., Finaish, F., and Bank, W., Three-dimensional Vortex Patterns in a Starting Flow, *J. Fluid Mech.*, vol. 161, pp. 239–248, 1986.

Chapter 29

Explosive Flows: Shock Tubes and Blast Waves

John M. Dewey

I INTRODUCTION

Because of the transitory nature of explosive events, high-speed flow visualization has been essential for the understanding of such phenomena, dating from the pioneering work in the last century of Ernst Mach [1] and Dvorak [2] in Germany, and Boys [3] in England. The rapid expansion of detonation products or the released gases from a ruptured pressurized container produce a shock wave that is characterized at its leading edge by an almost instantaneous increase of the ambient gas density, pressure, velocity, and temperature. The change of gas density results in a large gradient of refractive index so that shock waves can be easily visualized using shadowgraph and schlieren photography, and the varying density fields can be mapped by interferometry. The movement of gases within shock and blast waves can be followed by photography of smoke tracers introduced into the ambient gas before the arrival of the shocks. Because shock waves travel supersonically, any flow visualization technique used to study them must involve high-speed photography with very short exposure times and high framing rates.

The author is grateful for the assistance of D. J. McMillin and Martin Gastell in preparation of this chapter. The information contained here is available as a result of the long-term collaboration with colleagues from Ernst-Mach-Institut, Germany; Tohoku University, Japan; the University of Sydney and the National University, Australia; the Ballistic Research Laboratory and Denver Research Institute, United States; and the University of Toronto Institute for Aerospace Studies and the Defence Research Establishment Suffield, Canada.

II SHOCK TUBES

A shock tube is a relatively simple device for producing high-speed gas flows in the laboratory [4–11]. The essential features of a shock tube with windows for flow visualization are shown in Fig. 1. The tube is divided into a high-pressure chamber and an expansion chamber, which may be partially evacuated, separated by a plastic or metal diaphragm. Various combinations of gases and temperatures can be used in the two chambers so that when the diaphragm is ruptured, a shock wave is produced with the desired characteristics. Shock tubes vary in size from those a few meters long and centimeters in diameter, to large blast simulators, which may be hundreds of meters long and tens of meters in diameter. A window section for flow visualization is usually placed at least 20 tube diameters from the diaphragm to permit the damping of transverse waves, which are generated during the process of diaphragm rupture. A tube with a circular cross section has a preferred strength-to-weight ratio, but a rectangular cross section is needed to accommodate plane windows for flow visualization. Most tubes incorporating a flow visualization facility are therefore rectangular. Plane windows can be used in a circular cross-section tube with an appropriately designed transition section before the windows, but this may have some adverse effects on the flow being studied.

The windows in a shock tube, in addition to having the necessary optical properties and flatness, e.g., quarter wavelength per centimeter, for schlieren and interferometric

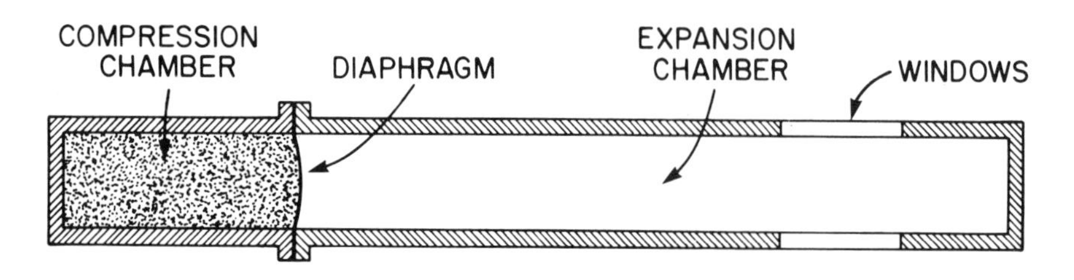

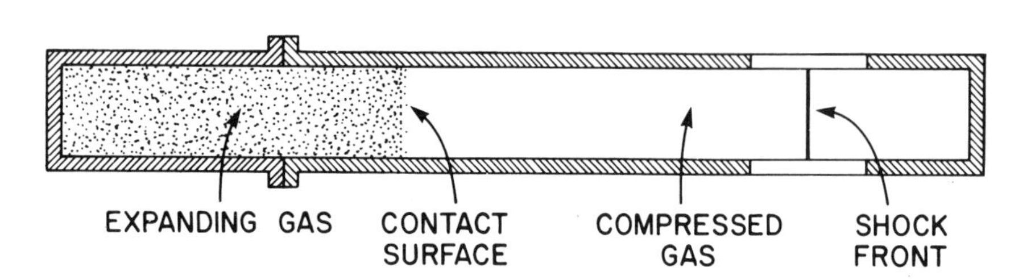

Fig. 1 Plan views of a shock tube. In the upper diagram a high-pressure gas in the expansion chamber is separated by a plastic or metal diaphragm from the low-pressure gas in the expansion chamber. In the lower diagram the diaphragm has been ruptured, and the high-pressure gas expands, producing a pressure wave with a supersonic shock front that can be photographed as it passes between the two windows. A diffused and turbulent contact surface exists between the expanding and the compressed gases.

studies, must also be strong enough to withstand the high pressure created behind a shock, particularly if the shock wave reflects from a model in the tube or the closed end of the tube, as this produces a pressure more than twice that behind the incident wave. Careful selection and mounting of windows in a shock tube is therefore of great importance. This question is addressed in Refs. [12–14]. Great care must be taken to avoid nonuniform stresses in the glass. The edges of windows must be bevelled to accurately fit the mounting frame, and a torque wrench should be used to ensure uniform pressures from the bolts of the frame that clamps the window to the shock tube. For high-pressure shocks, circular windows are preferred to avoid the stresses that build at the corners of a rectangular window. It should always be assumed that a shock tube window may fail, and suitable precautions must be taken to protect operators from both flying glass and the intensity of sound released from a failed window, which may damage eardrums. Glass from a shattered window will easily penetrate 2-cm-thick plywood. Safety screens can be made using 3 cm of foam plastic backed with 2 cm of plywood and a 0.5-cm steel plate. In some laboratories wire or rope blast mats are used. Extremely high pressures on the order of gigapascals may also affect the optical properties of glass [14].

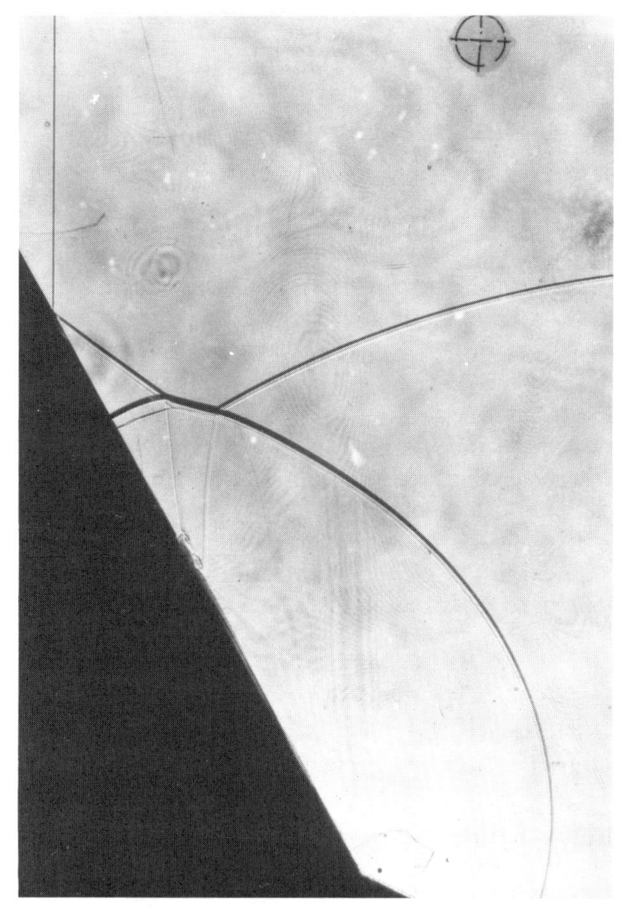

Fig. 2 Direct-contact shadowgraph of a plane shock wave interacting with a double wedge. The shadowgraph was made by pressing the film against the outside surface of the shock-tube window. The light source was a single pulse from a ruby laser expanded by a diverging lens from the principal focus of a parabolic mirror to produce a parallel beam perpendicular to the windows. The fiducial marker at the top of the photograph was 6 mm in diameter and attached to the surface of the window (courtesy of the University of Victoria Shock Studies Laboratory).

A Shadow, Schlieren, and Interferometry

Shadowgraph, schlieren, and interferometric photography are the most common flow visualization techniques used for the study of shock-tube flows, and the fundamentals of these methods are described in Chaps. 12–14 of this handbook. Some other references that specifically describe the application of these techniques for shock tubes are [5, 8, 12, 15–26].

The simplest technique for shock visualization is undoubtedly direct-contact shadow photography [12], in which the photographic film is pressed against the outside surface of the window opposite the light source. This is the least sensitive of all shadowgraph and schlieren methods, but for some purposes this insensitivity may be advantageous in that it produces very clear images of only shock fronts, as shown in Fig. 2. There is little or no optical distortion in such an image, and measurements can be made directly from the film without scaling. If the image from a direct shadowgraph is enlarged (Fig. 3), the diffraction pattern can be observed and analyzed to identify very accurately the positions of the shock fronts [27]. Examples of the use of conventional shadow photography for studying shock-tube flows are given in [16, 17].

Schlieren is a more sensitive technique than shadow photography in that the deflection of light is proportional to the density gradient $\partial\rho/\partial x$ rather than its second derivative $\partial^2\rho/\partial x^2$. Schlieren may therefore reveal features of the flow, such as the density gradient across a contact surface between two gases with different physical properties, that would not be seen in a shadowgraph (Fig. 4). On the other hand, a schlieren system can be made so sensitive that it reveals very slight gradients and weak transverse shocks that detract from the primary features of the flow being visualized. If the primary density gradients being visualized are in one direction, a slit aperture can be used, but a circular aperture should be used for two-dimensional flows as it will be equally sensitive to gradients in all directions.

Schlieren photography has been used in shock-wave research primarily to identify the positions of shock fronts, contact surfaces, and vortices. Limited use has been made of colored schlieren, in which the aperture is replaced by a multicolored filter [28–31] to better visualize small variations of the gas density. Schlieren can be made quantitative such that an actual measure of the density gradient is obtained. These methods usually use photodetectors rather than photography and perhaps should not be considered as flow visualization techniques [32–37]. An alternative is to use Moiré schlieren [38–41], in which the single slit of a schlieren system is replaced by two grids of equally spaced transparent and opaque strips on opposite sides of the window section such that under ambient conditions no light passes through the combination. A refraction of light by a density gradient permits light to pass the system and visualizes the gradient. If the plates are rotated relative to one another or have a slightly different grid spacing, a series of fringes is produced that are shifted by a change of the refractive index. This shift can be calibrated so as to provide a quantitative measure of the change of refractive index.

Another valuable adaptation of shadowgraph, schlieren, and interferometric methods is streak photography, in which the image of the slit is focused onto a rapidly moving film on a rotating drum [9, 42–44]. The slit produces a shadow or schlieren image, or an interferogram of a narrow one-dimensional cross section of the flow. As this sweeps across the film, it produces an x–t diagram of the movement of shocks and other density gradients in the direction of the

Fig. 3 High-resolution contact shadowgraph of a Mach reflection showing the diffraction patterns produced by the shocks. Analysis of the fringes can be used to identify the positions of the shocks with great accuracy (courtesy of the University of Victoria Shock Studies Laboratory).

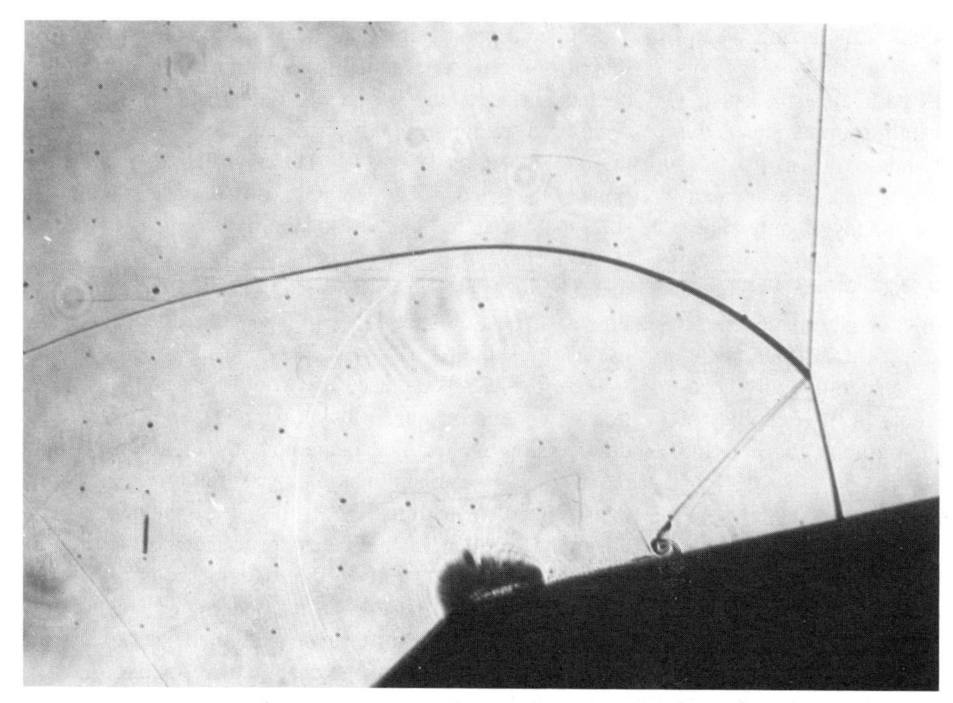

Fig. 4 Schlieren photograph of a plane shock interacting with a multiple-faceted wedge. The contact surfaces, vortices, and weak compressions and rarefaction waves are clearly visible. This photograph is one of a series taken at 25 kHz using a multiple-pulse ruby laser and recorded with a simple rotating-mirror camera. Fiducial marks on a 1-cm grid permit the correction of any optical distortions (courtesy of the University of Victoria Shock Studies Laboratory).

slit. From such an image, direct measurements of the velocities and the interaction of these features can be made. In a further adaptation of this method, Ben-Dor and Takayama [45] used curved slits corresponding to the surfaces of curved wedges in order to make precise measurements of the transition of shock waves from regular to Mach reflection on these surfaces.

B Interferometry and Holography

Interferometry is the most accurate flow visualization method to determine the density of a gas in a shock-wave flow, since the fringe shift is directly related to the density. At one time, interferometry required the use of a Mach–Zehnder type of system [46, 47], which was expensive and required skilled adjustment [5, 7, 8, 48, 49]. In recent years the development of high-speed holography has greatly simplified the applicability of this technique [50–63]. Figure 5 shows an infinite-fringe interferogram made using such a double-exposure holographic method, in which the fringes trace the lines of constant refractive index and thus the isopycnics, since in most flows it may be

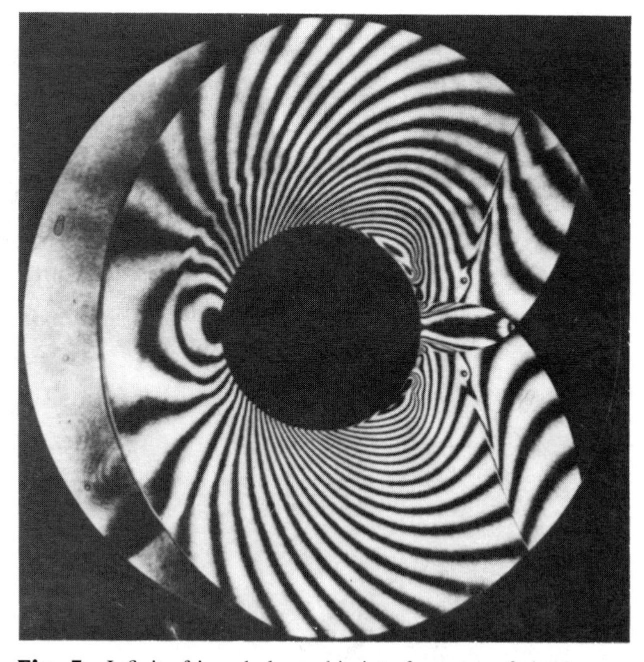

Fig. 5 Infinite-fringe holographic interferogram of shock-wave interactions. Each fringe is a contour of constant refractive index and is therefore an isopycnic. This interferogram was taken using a system similar to that shown in Fig. 6 (courtesy of Professor K. Takayama, Tohoku University, Japan).

assumed that the Gladstone–Dale constant does not change [64, 65].

In double-exposure holographic interferometry two holograms are superimposed on the same film to form an interference pattern. The first, or reference, hologram is taken before the shock wave reaches the window section, and the second, or object, hologram is taken of the event being studied. The interval between the holograms can be a few microseconds or a much longer period, depending on the capabilities of the pulsed-laser system, but must be short enough to ensure that there are no changes to the ambient conditions in the gas ahead of the shock between the two exposures. Figure 6 shows a double-pass arrangement for obtaining double-exposure holographic interferograms. In this arrangement a high-quality stainless steel mirror is used behind the second window of the shock tube. This has the advantages of doubling the sensitivity of the system and providing the convenience of having all the optical components on one side of the shock tube. High-quality front-surface mirrors should be used, but other components need not be of exceptional quality since imperfections that are the same for both exposures will not influence the interferogram. The object- and reference-beam distances should be the same, within one or two centimeters.

The fringes obtained by double-pulse holographic interferometry are lines of constant refractive index. The refractive index of a gas is related to its density by the Gladstone–Dale equation

$$n - 1 = k\rho, \qquad (1)$$

where n is the refractive index, K is the Gladstone–Dale constant for a given medium and wavelength, and ρ is the density. The difference in refractive index Δn between adjacent fringes is therefore related to a change of density $\Delta\rho$ by

$$\Delta n = K\,\Delta\rho \qquad (2)$$

For constructive interference the optical-path difference must be an integral number N of wavelengths so that

$$\Delta n\, d = N\lambda \qquad (3)$$

where d is the distance the light travels between the windows (twice the width of the shock tube for the double-pass system illustrated in Fig. 6) and λ is the wavelength of the light used to create the hologram. The density difference between adjacent fringes is therefore

$$\Delta\rho = \frac{\lambda}{dK} \qquad (4)$$

Focused holographic interferograms can be created with a Q-switched ruby laser and reconstructed with any monochromatic or broad spectrum source.

Holography has also been used to study three-dimensional shock flows [66–68], such as those around a conical body. This technique has the advantage that the hologram itself can be used subsequently to produce shadowgraphs, schlieren photographs, or interferograms.

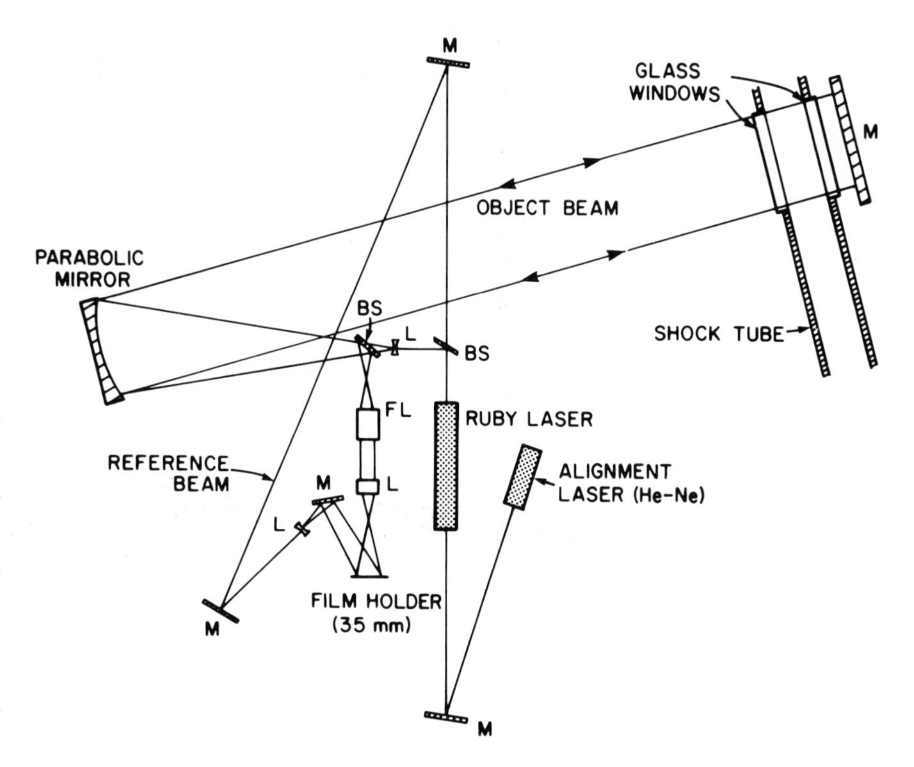

Fig. 6 Double-pass, double-pulse holographic interferometer for producing infinite-fringe interferograms such as that shown in Fig. 5. The double-pass system, using a large mirror behind the second window of the shock tube, doubles the sensitivity of the system and permits all the optical components to be on the same side of the shock tube. The total length of the object beam should equal that of the reference beam, within a few centimeters: M, front surface mirror; L, lens; FL, focusing lens system; BS, 50/50 beam splitter.

C Particle Tracer Techniques

Two particle tracer techniques have been used for the study of shock-tube flows. One method traces the flow by using injected smoke particles, and the other follows the movement of the luminous path of a spark.

Muirhead [69] first demonstrated how a stable laminar jet of tobacco smoke could be formed across a 5-cm-dia. shock tube in the ambient gas ahead of the shock. High-speed photography of the subsequent movement of the smoke showed that it behaved as an excellent flow tracer without any measurable inertia during the passage of the shock (Fig. 7). Subsequently Dewey and Whitten [70], using smoke tracers generated in this manner, made a complete calibration of a shock-tube flow. For this work it was found possible to simultaneously form up to 15 stable laminar smoke jets across a 25-cm-dia. shock tube. Tobacco smoke was again used, and the jets were drawn into the tube through 0.15-cm-dia. orifices, using a vacuum of about 18 cm of water. The smoke tracers were illuminated with a 1000-W photolamp against a matt black background placed behind the second window. The movement of the tracers was photographed at 5000 frames per second using a Hycam 16-mm camera. By changing the positions of the smoke-injection and window sections of the shock tube, the smoke tracers could be initiated at any position in the expansion chamber of the tube, and their movement could be recorded at any other position. This required multiple experiments, and it was necessary to ensure that shocks of the same strength were produced on each occasion. The first movement of each smoke jet identified the time of arrival of the incident shock and of the reflected shock from the closed end of the tube, so that the shock velocity could be determined at any point. The movement of the tracers gave the particle velocity in the flow, and the change of the distance between adjacent jets gave the average density of the gas in that region. From these three measurements all the physical properties of the flow could be determined at any point in the shock tube. More sophisticated methods for analysis of particle tracer trajectories in shock flows have now been developed and will be discussed later.

An advantage of the technique just discussed is that relatively low-quality windows can be used. A disadvantage is that the movement of the tracers can be measured accurately only at right angles to the smoke jets, and the method therefore has limited use for the study of two-dimensional flows such as the oblique reflection of a shock wave.

This difficulty was overcome when it was noted that the

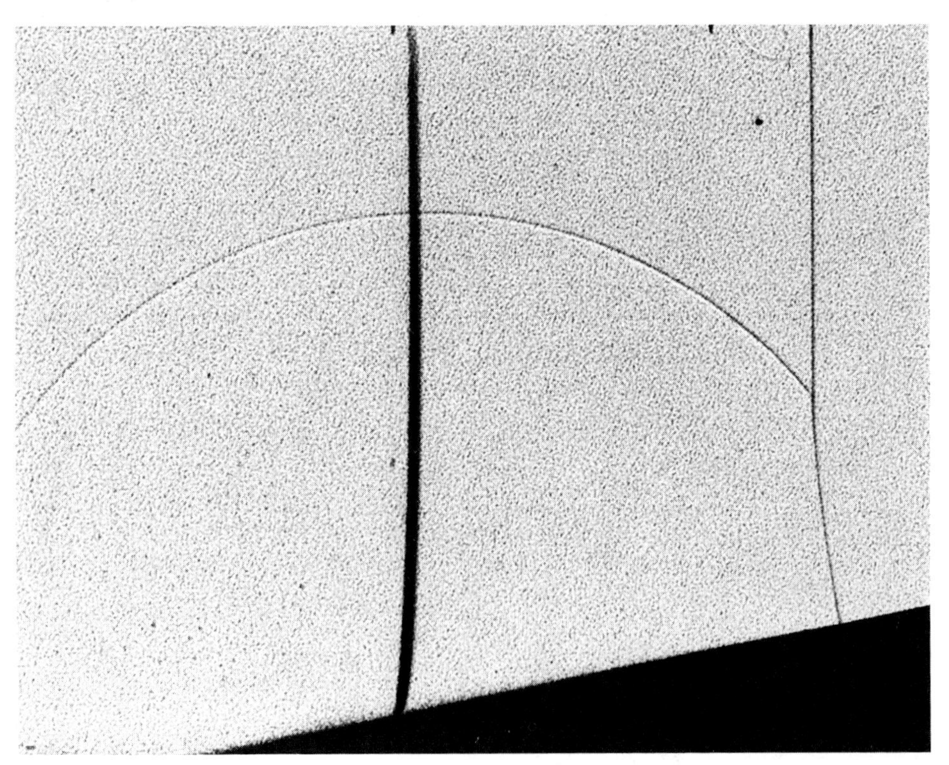

Fig. 7 Shadowgraph of the Mach reflection of a plane shock and a laminar jet of tobacco smoke used as a flow tracer. The initial position of the smoke jet was just in front of the leading edge of the wedge. This shadowgraph is one of a sequence of 24 taken using a Cranz–Schardin multiple-spark system at a frequency of 17 kHz, with individual exposure times of less than 1 μs. The vertical dimension of the shock tube was 11 cm (courtesy of the Ernst-Mach-Institut, Freiburg/Br., Germany).

laminar smoke jets were created by an appropriate pressure gradient across the input nozzle, and without the need for exit nozzles. Dewey and Walker [71] were therefore able to replace the second window in a shock tube with a front-surface stainless steel mirror and to visualize the shocks by a double-pass schlieren system (Fig. 8). Laminar smoke jets were injected into the ambient gas ahead of the shock through a grid of more than 700 holes in the mirror, spaced 1 cm apart and each 0.33 mm in diameter. The smoke jets were created about 0.5 s before the shock tube was fired so that they did not quite reach the window on the other side of the tube before the arrival of the shock. The smoke jets, when viewed end on by the schlieren system, appeared as individual point tracers (Fig. 9). The two-dimensional movement of these tracers was photographed using a ruby laser, pulsed at a frequency of about 20 kHz. This technique was subsequently used to make accurate and detailed studies of shock reflections [72]. In the subsequent analysis advantage was taken of the images of the 700 holes, whose positions were accurately known, so that optical distortions could be corrected at every point in the field of view.

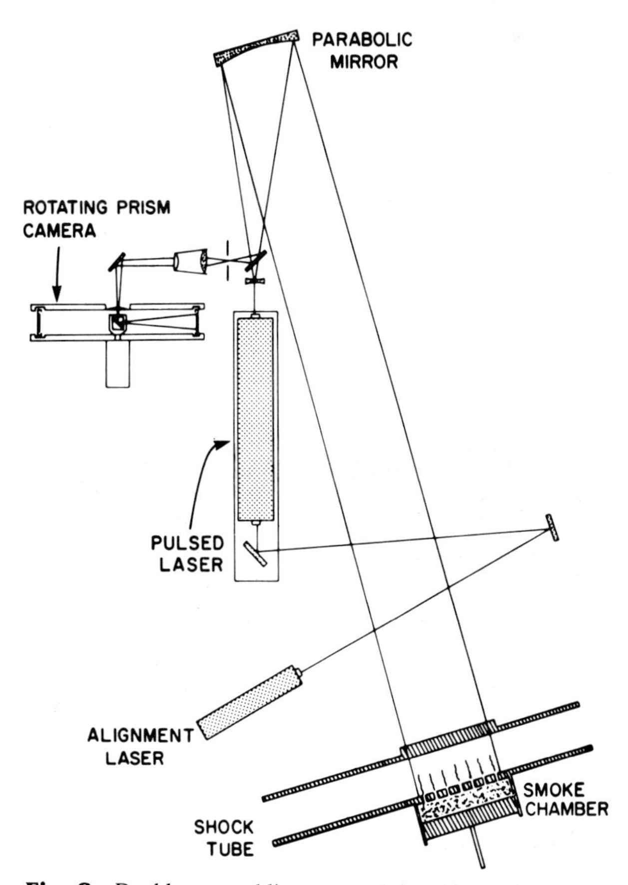

Fig. 8 Double-pass schlieren system in which the second window of the shock tube has been replaced by a front-surface stainless steel mirror. A grid of 0.33-mm-dia. holes through the mirror permits the injection of laminar smoke jets that act as flow tracers. Multiple pictures can be taken using a pulsed ruby laser. A continuous He–Ne laser is used to align the optical system and to focus the rotating-prism camera.

Because of the extremely rapid acceleration that takes place when a gas is traversed by a shock front, it is necessary to use very small particle tracers to study shock-tube flows. Tracer materials are discussed in Chap. 5 of this handbook, and a useful graphic of tracer particle sizes is given by Somerscales [73]. Tobacco smoke works extremely well but leaves a tar residue that requires regular cleaning of the injection orifices. In the work reported above [71], ammonium chloride was found to be very satisfactory if first bubbled through water to remove all but the smallest particles. Care must be taken to ensure that any gas introduced with the tracers is not significantly different in density or sound speed from the gas being studied. In order to visualize two-dimensional flows in a shock tube, Dewey [74] produced small heated cells of gas that could be visualized with a schlieren system. When these cells were traversed by a shock, the heated gas moved more rapidly than the ambient gas, and the cells formed stable vortices that traveled for long periods at this enhanced velocity, and so were not valid flow tracers. The same phenomenon occurs if a gas such as helium, with a lower density and higher velocity of sound, is introduced as a tracer into a heavier gas. The validity of a tracer should be evaluated by comparing its observed velocity immediately behind a shock with the Rankine–Hugoniot value calculated from the shock speed [74]. A good discussion of the acceleration of micron-sized particles by shock waves is given by von Stein and Pfeifer [75].

Smoke tracer techniques have not been successful in shock-tube flows in which the working gas is significantly below atmospheric pressure due to the high rate of diffusion of the smoke particles. A tracer technique that can be used in gases at reduced pressures involves the formation of a series of sparks behind the shock to form luminous ionized paths that can be photographed for some milliseconds [76–78]. This technique is also discussed in Chap. 5 of this handbook. Spark frequencies of up to 70 kHz can be obtained, and because the paths of the spark are self-luminous a series of up to 50 paths can be recorded on a single photograph, making possible the use of a very simple photographic system. However, for the reasons given in the previous paragraph, the sparks must be triggered after the passage of the shock in a region where the fluid accelerations are relatively slow. Also, as with linear smoke jets, this method visualizes only the component of displacement perpendicular to the paths of the sparks.

III BLAST WAVES

Blast waves in air are produced by the sudden release of energy from sources such as a chemical detonation, a nuclear explosion, or the rupture of a pressurized vessel. Blast

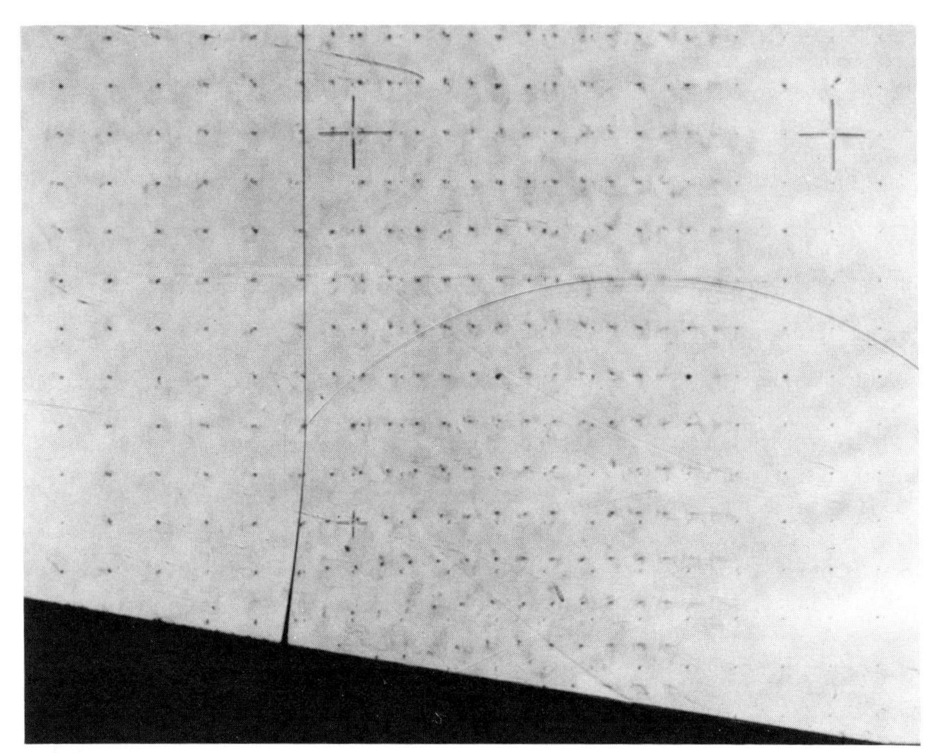

Fig. 9 Schlieren photograph of a Mach reflection and a grid of smoke tracers produced by the system illustrated in Fig. 8. On the right side the grid of holes through the mirror can be seen. On the left side the laminar jets of smoke are viewed end on. The moving smoke tracers can be seen to the right of the shock in the center of the picture. The holes through the mirror are 1 cm apart and act as a complete grid of fiducial marks. The shadowgraph is one of a sequence at a frequency of 24 kHz and individual exposure times of approximately 50 ns (courtesy of the University of Victoria Shock Studies Laboratory).

waves are characterized by a supersonic shock front followed by an exponential-type decay of the physical properties of the gas [79–84]. Flow visualization techniques are invaluable tools for studying blast waves from sources at the microscale, such as a spark, a laser discharge, or the detonation of a sub-gram-sized chemical charge, to those at very large scale such as a nuclear explosion with an energy release equivalent to millions of tons of TNT.

Small-scale explosions can be created in a laboratory so that the flow visualization methods described above for shock-tube flows can also be used, and in fact may be simplified since no windows are required between the camera and the event under study. However, blast waves from free-field explosions are three-dimensional, which complicates the interpretation of interferometric measurements, even if spherical symmetry can be assumed. Shadow photography is therefore the most useful technique for providing exact measures of the position, and thus the velocity, of blast-wave shock fronts (Fig. 10).

For large-scale explosions both shadow and particle tracer photogrammetry have been used. For shadow photography to be successful a suitable background is required. This may occur naturally, such as a bright sky against which relatively strong shocks can be visualized or a sharp contrast in background intensity such as the outline of hills against the sky. For consistent results an artificial background, such as a series of smoke trails or large screens painted with

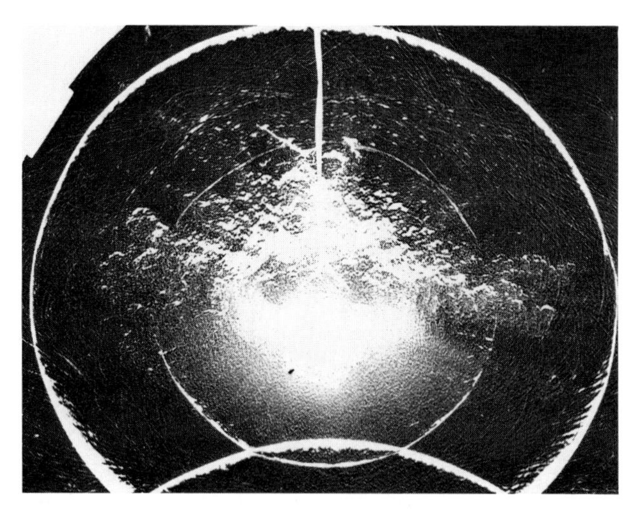

Fig. 10 Shadowgraph of the blast wave from a hexogen explosive charge of approximately 1 *g* detonated 8.3 cm above a solid reflecting surface. This photograph is one of a sequence of 24 taken at 100 kHz using a Cranz–Schardin system (courtesy of the Ernst-Mach-Institut, Freiburg/Br., Germany).

Fig. 11 Shock fronts from exploding charges visualized against a striped background. The stripes were painted on canvas screens approximately 15 m high. The experiment involved the simultaneous detonation of two 450-kg charges, one at a height of 8 m and the other at 24 m so that the mutual reflection of the two blast waves could be compared with the reflection from the ground of the blast wave from the lower charge. This photograph is one of a sequence taken at 3 kHz using a WF5 (Fastax) half-frame camera (courtesy of the Defence Research Establishment Suffield, Canada).

black-and-white stripes, is normally created (Fig. 11). Figure 12 shows a tracing of the shock fronts from a sequence of photographs, such as that in Fig. 11, taken at a frequency of 3 kHz. For visualizing complex shock structures, such as those that occur when an intense blast wave from an air-burst explosion is reflected from the ground, Wisotski [85] has successfully used Scotchlite screens illuminated by a pulsed ruby laser and photographed with a high-speed framing camera (Fig. 13).

Care must be taken in interpreting shadowgraphs of

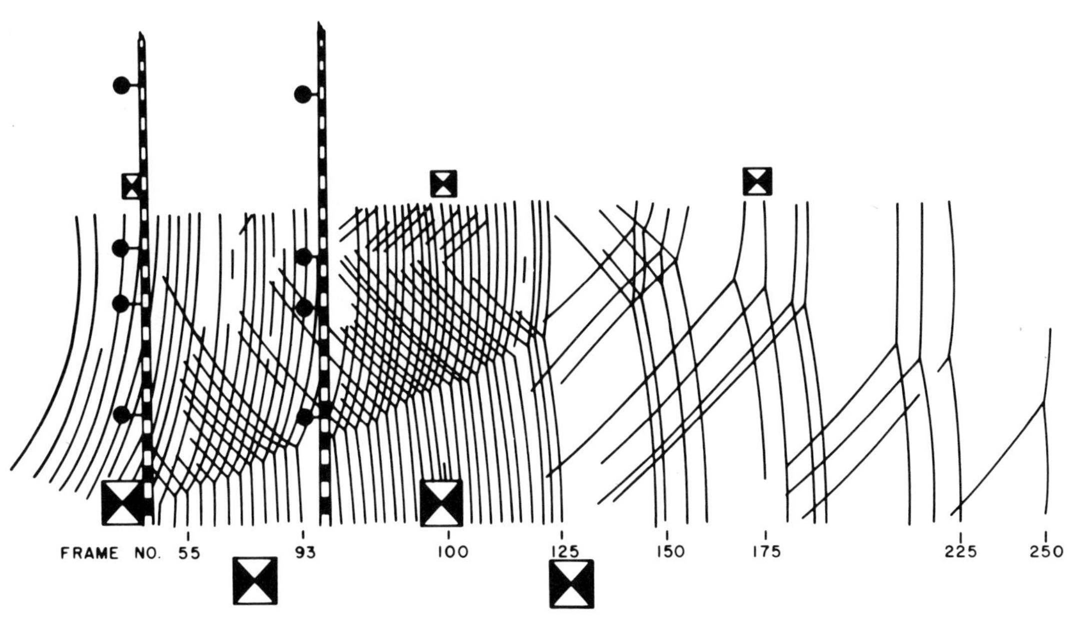

Fig. 12 Tracings of shock fronts made from a sequence of photographs such as that in Fig. 11. The vertical black-and-white scales and the fiducial marks are carefully surveyed relative to the camera. The two fiducial marks at the bottom of the figure were situated approximately midway between the object plane and the camera so that the orientation of the camera could be calculated.

Fig. 13 Large-scale pulsed-laser shadowgraph of the complex Mach reflection region produced by the detonation of a 450-kg charge. The shocks are visualized against a Scotchlite background at a framing rate of approximately 30 kHz, and the fiducial marks are 0.6 m above the ground. The dark "bench" in the lower right-hand section of the photograph is the side view of a toroidal vortex (courtesy of J. Wisotski, Denver Research Institute, Colorado).

spherical blast waves because there is no fixed plane of measurement parallel to the film plane. Figure 14 shows a plan view of the spherical shock front of a blast wave. The line of sight from the camera is tangential to the spherical surface so that the point being visualized traces a circular path whose diametric end points are the charge center and the camera. The radius of the shock is not therefore linearly related to the apparent radius measured in the film plane [86, 87].

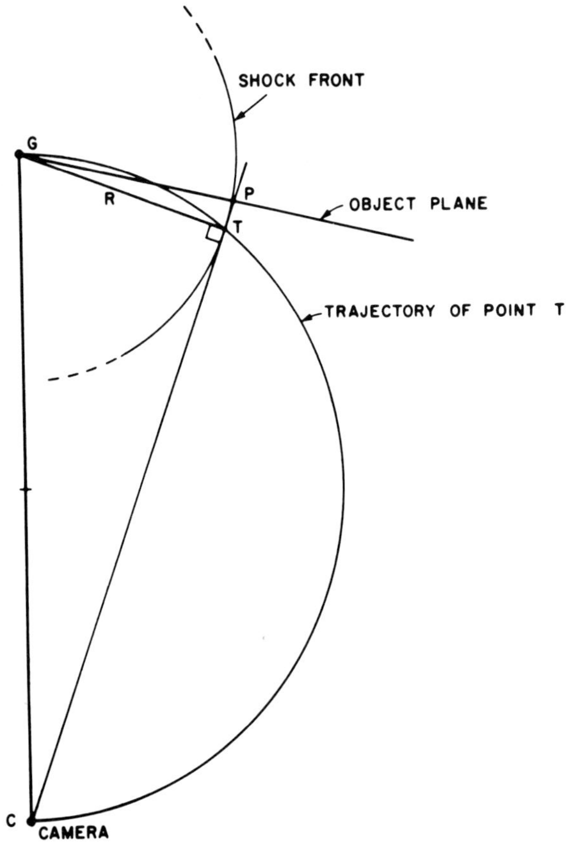

Fig. 14 Plan view of a spherical shock front centered at G. The object plane through G is parallel to the film plane of the camera, and the refractive image of the shock will appear to be at point P in the object plane. The line of sight from the camera to P is tangential to the shock at T, which is also the intersection of CP and the circle with CG as diameter. GT is the actual radius of the shock.

Particle tracer photogrammetry of blast waves requires a mechanism for creating smoke trails or puffs immediately ahead of the incident shock. If spherical symmetry can be assumed, smoke trails released by mortars can be used [86], formed in a vertical plane containing the charge center (Fig. 15). For flows that are two-dimensional in the object plane of the camera, such as those that result from the reflection from the ground of the blast wave from an air-burst explosion, a grid of smoke puffs is required [88] (Fig. 16). In order for the smoke tracers to be photographed, there must be a suitable contrast between the smoke and the background. This can be achieved by using smoke of appropriate color, for example, red or black against the sky and white against dark hills. Alternatively an artificial background of black smoke can be created by burning oil, against which white smoke is well contrasted [86].

White smoke trails can be produced by releasing chlorosulfonic acid, ammonium chloride, or titanium tetrachloride from a projectile fired by a mortar [86]. In order to obtain a uniform release of the smoke agent, the exit ports must be at the head of the projectile. The projectile will be decelerated by air drag, and inertia will force the smoke agent to the forward end. Release ports at the trailing end of the projectile will not produce uniform smoke trails. Carbon black is an excellent particulate flow tracer for use against a light background. It can be released in a trail from an open ended PVC tube fired from a mortar, or released as a puff from a suspended plastic bag and dispersed by an electric squib. White trails and puffs for use against dark backgrounds can be formed in the same way, using fumed silica or titanium dioxide, for example. An 8 to 1 mixture by volume of fumed silica and titanium dioxide has been found to have high visibility and good properties for flow tracing.

Fig. 15 450-tonne TNT explosion. The refractive image of the hemispherical shock can be seen against the cloudy-sky background in the upper left part of the picture. On the right-hand side a series of white-smoke trails, which acted as flow tracers in the blast wave, are visualized against a black-smoke background produced by burning crude oil in a shallow trench (courtesy of the Defence Research Establishment Suffield, Canada).

IV CAMERAS AND LIGHT SOURCES

Cameras and light sources for shock-wave photogrammetry are characterized by the need for high framing rates and short exposure times in order to freeze the supersonic shocks. Until the advent of the pulsed laser, a spark source was the most economical method of achieving the required intensity and short duration [12]. By 1929 Cranz and Schardin [89] had developed a multiple-spark system for photographing shock waves, in which 24 separate sparks and optical systems were used to record 24 images on a single photographic plate, with frequencies up to 50 kHz. With modern triggering mechanisms there is virtually no limit to the frequency at which sparks can be initiated. The

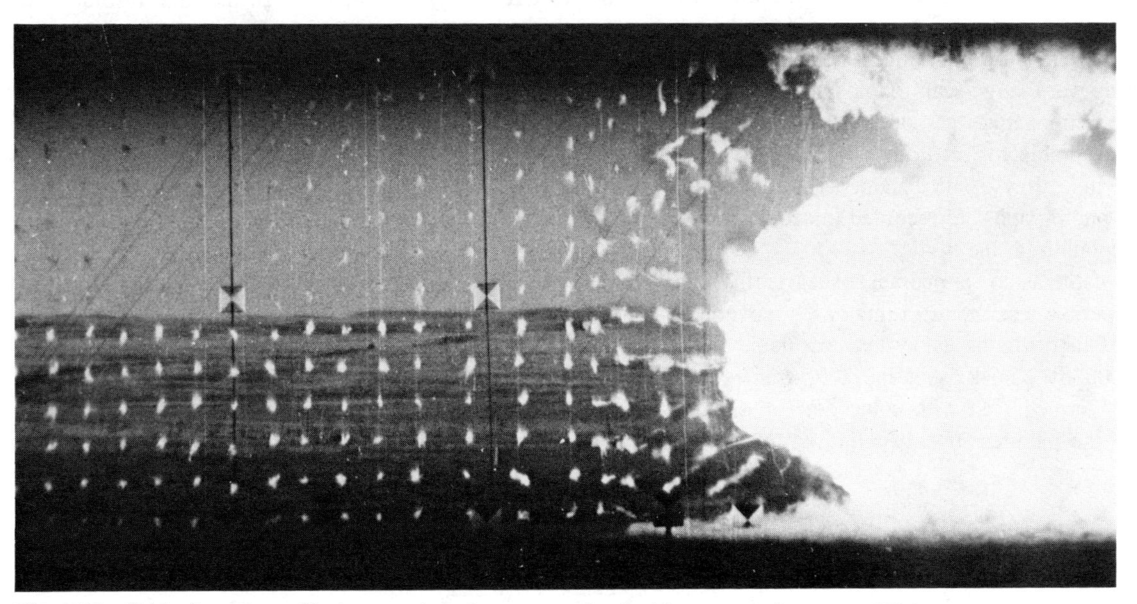

Fig. 16 Grid of smoke puffs that acted as flow tracers in the blast waves from two 450-kg charges detonated simultaneously at heights of 8 and 24 m. This photograph is one of a sequence taken at 3 kHz with a WF5 (Fastax) camera (courtesy of the Defence Research Establishment Suffield, Canada).

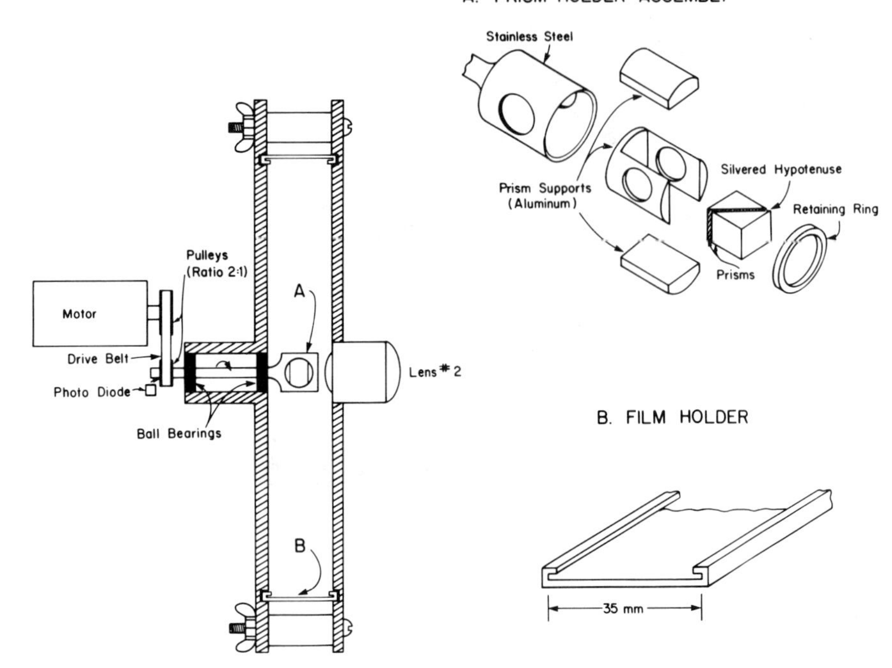

Fig. 17 Simple rotating-prism camera that can be used in conjunction with a pulsed-laser light source (courtesy of the University of Victoria Shock Studies Laboratory).

Cranz–Schardin system is still widely used, and recent developments to the technique are described in Refs. [90–92]. Because a separate optical system is used for each picture, individual parallax corrections must be made if accurate measurements are required from the sequence of photographs.

A variety of high-speed framing cameras are available for use with continuous-light sources or for photographing self-luminous events such as the fireball from an explosion. These cameras use films ranging from 16 to 70 mm and with framing rates from a few hundred to several million frames per second. The slower framing rates are adequate for photographing very large-scale explosive events, but only the highest framing rates and appropriately short exposure times are suitable for recording small-scale explosions and shock-tube flows. These cameras have the advantage that the photographs are recorded in sequence and with correct orientation on the film.[1]

Normal video cameras do not operate at sufficiently high framing rates to permit accurate recording of the movement of shock waves. One of the fastest systems available is the Spin Physics 2000 by Kodak,[2] but the 2000-frames-per-second frequency is achieved with some deterioration of the resolution of the pictures. If only a limited number of small-format photographs is required, an image-converter camera can be used.[3] Such a camera can operate at frequencies of 10^{10} frames per second with picosecond time resolution, producing up to 20 frames, each about 1 cm wide and with a spatial resolution of up to 10 line pairs per millimeter. The image in such a device is formed on a photocathode that releases electrons. These are directed by deflection plates to form a series of images on a single photographic plate. Image-converter cameras are versatile and relatively easy to use, but lack the high resolution of more conventional cameras.

Shock-wave photogrammetry in the laboratory has been greatly simplified by the availability of pulsed lasers [22–24, 27, 93–96]. The most common system is a ruby laser, switched by a Pockels cell. High light intensities are possible at frequencies up to 100 kHz, and with exposure times of a few nanoseconds. These characteristics permit the use of relatively simple, continuous-writing, rotating-mirror cameras that do not require fast-acting shutters, since the exposure time is controlled by the laser pulses [97, 98]. In the simplest cameras, such as that shown in Fig. 17, no attempt is made to uniformly orient the pictures on the film, but this is not a major disadvantage for work in which each frame of a sequence is to be individually analyzed.[4]

[1]Some of the companies producing high-speed cameras and light sources suitable for shock- and blast-wave photography are Cordin (Beckman and Whitley products), 2230 South 3270 West, Salt Lake City, Utah 84119; Red Lake Labs, Inc. (Hycam and Fastax), 2991 Corvin Dr., Santa Clara, Calif. 95051; Impulsphysik GmbH, Sulldorfer, Landstrasse 400, 2000 Hamburg 56, Federal Republic of Germany.

[2]Spin Physics (Kodak), 925 Page Mill Rd., Palo Alto, Calif. 94034.

[3]Imacon Cameras, Hadland Photonics Ltd., New House Laboratories, Bovington, Hemel Hempstead, Herts, United Kingdom.

[4]A camera of this type, the HISPIN 35, is produced by Vivitro Systems, Inc., 1900 Fort St., Victoria, B.C., Canada V8R1J8.

V ANALYSIS

High-speed photography of shock and blast waves is often used qualitatively in order to understand the phenomenology of an event. On the other hand, a large amount of quantitative information is available from an accurately timed photographic sequence. Most other measurement techniques are able to record only a single physical property, such as pressure, temperature, or velocity, in a shock-wave flow at a single position, and the gauge itself may be intrusive, affecting the flow being measured. Photogrammetry, on the other hand, is almost entirely nonintrusive, even when flow tracers are used, and measurements can be made throughout a large volume of the flow. It will be shown in this section that a simple measurement of position from a photograph can be used to determine all the physical properties of most shock-wave flows. In addition to measurement of the position of such features as shock fronts, contact surfaces, flow tracers, and interference fringes, the degree of exposure of film has been used to calculate the temperature of self-luminous objects such as an explosive fireball [99].

In order to make accurate measurements of the positions of shocks, density discontinuities, and flow tracers, it is necessary to correct for optical distortion throughout the field of view. Such corrections are particularly important with schlieren, which is usually an off-axis system. The best way to make these corrections is to place an accurate grid of fiducial markers in the object plane [72]. This grid can be photographed simultaneously with the flow field or immediately before or after the experiment, and before any optical components have been moved. The calibration must include the playback optics such as the projector of a digitizer or an enlarger. If the measurements are made from photographic prints, this can also cause distortion due to the nonuniform drying of the print paper. High-quality photogrammetry paper should be used to reduce this error. Fiducial markers need not be placed in the object plane if their positions are accurately surveyed relative to the camera and the object plane. Fiducial markers incorporated within a high-speed camera itself may not be reliable for making accurate measurements in the object plane, as even with sprocket cameras there is some wandering of the image, and a very small movement in the image plane introduces a large measurement error in the object plane.

Some difficulties arise in the interpretation and analysis of interferograms of multiple shocks, such as a Mach reflection. This is because there is a large gradient of the refractive index through a shock front, and it is normally not possible to unambiguously identify the same fringe on both sides of a shock. In the case of a plane incident shock, the shock's strength, and thus the density jump across the shock, can be determined from the shock velocity, as described below. As soon as the density associated with one fringe has been identified, the density associated with other fringes can be determined using eq. (4). However, the strength of reflected shocks may be more difficult to assess, and this calls for some judgment in interpretation of the interferogram [48, 49]. This problem has been overcome by using a broad-spectrum light source to produce colored interferograms in which each fringe can be clearly identified, even when passing through multiple shocks [100]. Reference [100] also describes a static calibration of such an interferometry system to compensate for irregularities in the fringes produced by imperfections in the components of the interferometer or the shock-tube windows.

The position of a shock front measured from a sequence of shadow or schlieren photographs permits the shock velocity to be calculated. This is a useful measurement in that a knowledge of the shock-front Mach number used in the Rankine–Hugoniot relationships [5, 7, 8], gives an accurate measure of all the physical properties of the gas immediately behind the shock. The reliability of this method is such that it is often used to calibrate other measurement devices such as pressure transducers. This method is most useful for shocks with Mach numbers in the range of about 1.05 to 3. For shocks stronger than Mach 3 real gas effects become important and must be taken into consideration in using the Rankine–Hugoniot relationships. In the case of very weak shocks the dependency of the physical properties behind the shock on the Mach number is small so that the shock velocity must be known with great accuracy.

In the field of view of a shock-tube window section most shocks travel with constant velocity so that the velocity can be measured from two accurately timed photographs. If a sequence of such photographs is available, the displacement perpendicular to the shocks can be measured along orthogonal curves [72], and the results can be fitted by least squares and differentiated to give the shock velocity. The Mach number of the shock can then be calculated in terms of the speed of sound in the ambient gas ahead of the shock, but this requires an accurate measure of the temperature of that gas. The appropriate relationships are as follows:

$$M_s = \frac{V_s}{a_0} \tag{5}$$

where M_s is the shock Mach number, V_s is the normal shock velocity, and a_0 is the speed of sound in the gas ahead of the shock.

$$a_0 = a_s \frac{T_0}{T_s} \tag{6}$$

where a_s is the known speed of sound in the gas at absolute temperature T_s, and T_0 is the absolute temperature of the ambient gas ahead of the shock.

The ratios of the physical properties of the gas immediately behind the shock to the corresponding property in the ambient gas, in terms of the shock Mach number, are

$$\frac{P}{P_0} = \frac{2\gamma M_s^2 - \gamma + 1}{\gamma + 1} = \frac{7M_s^2 - 1}{6} \quad \text{if} \quad \gamma = \frac{7}{5} \tag{7}$$

$$\frac{\rho}{\rho_0} = \frac{(\gamma + 1)M_s^2}{(\gamma - 1)M_s^2 + 2} = \frac{6M_s^2}{M_s^2 + 5} \quad \text{if} \quad \gamma = \frac{7}{5} \tag{8}$$

$$\frac{T}{T_0} = \frac{(2\gamma M_s^2 - \gamma + 1)[(\gamma - 1)M_s^2 + 2]}{(\gamma + 1)^2 M_s^2}$$

$$= \frac{4(7M_s^2 - 1)(M_s^2 + 5)}{49M_s^2} \quad \text{if} \quad \gamma = \frac{7}{5} \tag{9}$$

and

$$\frac{u}{a_0} = \frac{2(M_s^2 - 1)}{(\gamma + 1)M_s} = \frac{5(M_s^2 - 1)}{6M_s} \quad \text{if} \quad \gamma = \frac{7}{5} \tag{10}$$

where P is the static pressure, ρ is the density, T is the absolute temperature, u is the particle velocity, γ is the ratio of specific heats, and the subscript 0 refers to the state of the ambient gas. Other physical properties of the gas such as the dynamic pressure and the specific energy can be calculated from the properties defined above.

In the case of a spherical blast wave the shock speed decays as the shock front expands. From a sequence of photographs of such an expanding shock the radius R of the shock can be calculated using the procedure illustrated in Fig. 14. The time t of each photograph relative to the initiation of the explosion, should be known from timing marks recorded on the film. The normal method of determining the shock speed from these values of R and t is to make a least squares fit to an appropriate function. Dewey [101] suggests a function of the form

$$R = A + Ba_0t + C \ln(1 + a_0t) + D\sqrt{\ln(1 + a_0t)} \tag{11}$$

where A, B, C, and D are the fitted coefficients and the other symbols are as defined above. This function works well for data covering a large range of R, from close to the explosive source out to distances at which the shock speed has fallen to Mach 1.1 or lower. For shorter ranges of data close to the charge

$$R = A + B\,a_0t + C \ln(1 + a_0t) \tag{12}$$

is recommended, and for weaker shocks Sadek and Gottlieb [102] recommend

$$R = A + a_0t + B\sqrt{\ln(1 + a_0t)} \tag{13}$$

The derivatives of these functions give the shock velocity V_s in terms of R and t, and this can be used in Eqs. (5)–(10) to give the peak values of the physical properties of the blast waves.

As a blast wave expands and decreases in strength, it leaves the air in a state of radially decreasing entropy. As a result there is no single valued functional relationship between the physical properties of the flow within the wave. This means, for example, that a measurement of pressure does not permit a determination of the gas density. It has been shown, however [86, 101], that a knowledge of the time-resolved particle trajectories in the blast wave is sufficient to determine all the physical properties. The particle trajectories can be obtained by high-speed photography of smoke tracers established just before the arrival of the shock front. An analytical method to do this is given in [101], in which a three-dimensional surface is described by a set of orthogonal polynomials fitted to the observed particle trajectories. The disadvantage of this technique is that the order of the polynomials used is arbitrary, and the solution may not lead to the correct thermodynamic relationship between the calculated physical properties. An improved technique is described by Lau and Gottlieb [103], who used numerical modeling in conjunction with photogrammetric observations. In a further development of this method, a particle trajectory close to an explosive charge, obtained by photography of a flow tracer, is used to define a piston path in a numerical model. This piston generates a blast wave, and the model is used to calculate the expected particle trajectory of an observed flow tracer. The piston path is iteratively adjusted to match the calculated to the observed trajectory in an optimum manner. This technique, using only the data obtained from the high-speed photography of the flow tracers, gives an excellent measure of all the physical properties in the blast wave, and even permits reasonable extrapolation beyond the range of the observed flow tracers.

VI FUTURE DEVELOPMENTS

Future developments using flow visualization to study shock-tube and blast-wave flows will undoubtedly extend the use of high-powered pulse lasers with high-frequency and short-duration pulses. This will permit multiple high-speed holography using framing or continuous-writing cameras. Because of the necessity for high framing rates and photographs with high resolution, it appears unlikely that the traditional film will be replaced by direct video recording in the immediate future. However, the advantages of video recording for storage, high-speed random retrieval, and image enhancement would appear to justify the transfer of film images to video form using cameras with a resolution of at least 1024 × 1024 pixels. These frames can be stored digitally on optical disks or tapes [104]. In addition to the possibilities for image enhancement, digital video will simplify measurements of position, color, and intensity and provide the opportunity for automatic analysis of flow visualization records [105].

REFERENCES

1. Mach, E., Ueber den Verlauf der Funkenwellen in der Ebene und im Raume, Oesterreichische Akademie der Wissenschaften, Vienna, *Mathematisch-naturwissenschaftliche Klasse, Sitzungsberichte,* vol. 77, no. 3, pp. 819–838, 1878.
2. Dvorak, V., Ueber eine Neue Einfache Art der Schlierenbeobachtung, *Ann. Phys. Chem.,* vol. 9, pp. 502–512, 1880.
3. Boys, C. V., On Electric Spark Photographs; or, Photography of Flying Bullets, etc., by the Light of the Electric Spark, *Nature,* vol. 47, no. 1219, pp. 440–446, 1893.
4. Bleakney, W., Weimer, D. K., and Fletcher, C. H., The Shock Tube: A Facility for Investigations in Fluid Dynamics, *Rev. Sci. Instrum.*, vol. 20, no. 11, pp. 807–815, 1949.
5. Gaydon, A. G., and Hurle, I. R., *The Shock Tube in High-Temperature Chemical Physics*, Chapman and Hall, London, 1963.
6. Gaydon, A. G., Shock Tubes, *Proc. R. Inst. GB*, vol. 43, no. 202, pp. 427–437, 1970.
7. Wright, J. K., *Shock Tubes*, Methuen, London, 1961.
8. Bradley, J. N., *Shock Waves in Chemistry and Physics*, Methuen, London, 1962.
9. Glass, I. I., The Design of a Wave Interaction Tube, Univ. of Toronto, Inst. for Aerosp. Studies, UTIA Rep. 6, 1950.
10. Glass, I. I., Martin, W., Patterson, G. N., A Theoretical and Experimental Study of the Shock Tube, Univ. of Toronto, Inst. for Aerosp. Studies, UTIA Rep. 2, 1953.
11. Glass, I. I., Shock Tubes: Part 1. Theory and Performance of Simple Shock Tubes, Univ. of Toronto, Inst. for Aerosp. Studies, UTIA Rev. 12, 1958.
12. Holder, D. W., and North, R. J., Schlieren Methods, *Notes on Applied Science*, no. 31, Her Majesty's Stationery Office, London, 1963.
13. Bitondo, D., and Lobb, R. K., Design and Construction of a Shock Tube, Univ. of Toronto, Inst. for Aerosp. Studies, UTIA Rep. 3, 1950.
14. Chhabildas, L. C., Survey of Diagnostic Tools used in Hypervelocity Impact Studies, *J. Imp. Eng.*, vol. 5, 1987.
15. Glass, I. I., Beyond Three Decades of Continuous Research at UTIAS on Shock Tubes and Waves, Univ. of Toronto, Inst. for Aerosp. Studies, UTIAS Rev. 45, 1981.
16. Shine, A. J., An Optical Study of a Shock Wave Interacting with a Change in Cross Section, *Trans. ASME, Ser. D*, vol. 93, no. 2, pp. 329–331, 1971.
17. Sturtevant, B., and Kulkarny, V. A., The Focusing of Weak Shock Waves, *J. Fluid Mech.*, vol. 73, no. 4, pp. 651–671, 1976.
18. Takayama, K., Honda, M., and Onodera, O., Shock Propagation along 90 Degree Bends, Rep. Inst. High Speed Mech., Tohoku Univ., Japan, vol. 35, no. 299, pp. 83-111, 1976.
19. Wu, J. H. T., and Ostrowski, P. P., Shock Propagation in a 90 Degree Bend, *Can. Aeronaut. Space J.*, vol. 22, no. 5, pp. 230–242, 1976.
20. Deckker, B. E. L., and Yang, A. W., The Glancing Collision of Two Shock Waves in a Branched Duct, *Inst. Mech. Eng., Thermodyn. Fluid Mech. Group Proc.*, vol. 189, 29/75, pp. 293–303, 1975.
21. Deckker, B. E. L., and Weekes, M. E., The Unsteady Boundary Layer in a Shock Tube, *Inst. Mech. Eng., Thermodyn. Fluid Mech. Group Proc.*, vol. 190, 11/76, pp. 287–296, 1976.
22. Dewey, J. M., and Walker, D. K., A Multiply Pulsed Double-Pass Laser Schlieren System for Recording the Movement of Shocks and Particle Tracers within a Shock Tube, *J. Appl. Phys.*, vol. 46, no. 8, pp. 3454–3458, 1975.
23. Zakharenkov, Yu. A., and Shikanov, A. S., Multiframe Ultrahigh-Speed Schlieren Photography in a Ruby Laser Beam, *Instrum. Exp. Tech.*, vol. 17, no. 5, pt. 2, pp. 1445–1448, 1975.
24. Oppenheim, A. K., Urtiew, P. A., and Weinberg, F. J., On the Use of Laser Light Sources in Schlieren-Interferometer Systems, *Proc. R. Soc. London, Ser. A,* vol. 291, pp. 279–290, 1966.
25. Hiang, P.-Y., Bershader, D., and Wray, A., Optical Studies of Shock Generated Transient Supersonic Base Flows, in *Shock Tubes and Waves*, ed. C. E. Treanor and J. G. Hall, pp. 200–208, SUNY Press, Albany, N.Y., 1981.
26. Dugger, P. H., and Handrix, R. E., Laser Photography: A Role at the AEDC, *Opt. Spectra*, vol. 9, no. 5, pp. 32–34, 1975.
27. Julian, R. J., On Fringes Observed When Making Shadowgraphs of Shocks using Monochromatic Light from a Ruby Laser, University Microfilms, Ann Arbor, Mich., 1970.
28. Holder, D. W., and North, R. J., A Schlieren Apparatus Giving an Image in Colour, *Nature*, vol. 169, no. 4298, p. 466, 1952.
29. North, R. J., A Colour Schlieren System using Multicolour Filters of Simple Construction, Br. Nat. Phys. Lab., NPL Aero Rep. 266, 1954.
30. Howes, W. L., Rainbow Schlieren and Its Applications, *Appl. Opt.*, vol. 23, no. 14, pp. 2449–2459, 1984.
31. Howes, W. L., Rainbow Schlieren vs. Mach–Zehnder Interferometer: A Comparison, *Appl. Opt.* vol. 24, no. 6, pp. 816–821, 1985.
32. Ahlborn, B., and Humphries, C. A. M., Quantitative Schlieren Densitometer Employing a Neutral Density Wedge, *Rev. Sci. Instrum.,* vol. 47, no. 5, pp. 570–573, 1976.
33. Bander, J. A., and Sanzone, G., An Improved Schlieren System for the Measurement of Shock-Wave Velocity, *Rev. Sci. Instrum.*, vol. 45, no. 7, pp. 949–951, 1974.
34. Bryanston-Cross, P. J., Camus, J.-J., and Richards, P. H., Intensity Correlation Measurements on a Schlieren Image, *Opt. Acta*, vol. 29, no. 9, pp. 1161–1166, 1982.
35. Davidson, G. P., and Emmony, D. C., A Schlieren Probe Method for the Measurement of the Refractive Index Profile of a Shock Wave in a Fluid, *J. Phys. E: Sci. Instrum.*, vol. 13, no. 1, pp. 92–97, 1980.
36. Ellingson, W. A., and Hall, J. L., Development of a Quantitative Schlieren System for Studying Shock-Waves, *Proc.*

4th Int. Cong. Instrum. Aerosp. Simul. Facil., Rhode-Saint-Genèse, Belgium, pp. 201–208, 1971.
37. Kiefer, J. H., and Lutz, R. W., Simple Quantitative Schlieren Technique of High Sensitivity for Shock Tube Densitometry, *Phys. Fluids,* vol. 8, no. 7, pp. 1393–1394, 1965.
38. Mortensen, T. A., An Improved Schlieren Apparatus Employing Multiple Slit Gratings, *Rev. Sci. Instrum.*, vol. 21, no. 1, pp. 3–6, 1950.
39. Waddell, P., A Large Field Retroreflective Moiré-Schlieren System, *Electro-Optics/Laser Int. Conf.*, Brighton, England, pp. 74–79, 1982.
40. Stricker, J., and Kafri, O., Moiré Deflectometry, a New Method for Density Gradient Measurement in Compressible Flows, TECHNION—Israel Institute of Technology, Israel, TAE Rep. 436, 1981.
41. Kafri, O., and Glatt, I., Moiré Deflectometry: A Ray Deflection Approach to Optical Testing, *Opt. Eng.*, vol. 24, no. 6, pp. 944–960, 1985.
42. Christiensen, A. B., and Isbell, W. M., Technique for Streak Camera Writing Rate Calibration using a Pulsed Laser, *Rev. Sci. Instrum.*, vol. 37, no. 5, pp. 559–561, 1966.
43. Tanner, L. H., The Optics of Laser Streak Interferometry, *J. Sci. Instrum.*, vol. 44, no. 9, pp. 725–730, 1967.
44. Tyburczy, J. A., Blayney, J. L., Miller, W. F., and Ahrens, T. J., Streak Camera Recording of Shock Wave Transit Times at Large Distances using Laser Illumination, *Rev. Sci. Instrum.*, vol. 55, no. 9, pp. 1452–1454, 1984.
45. Ben-Dor, G., and Takayama, K., Streak Camera Photography with Curved Slits for the Precise Determination of Shock Wave Transition Phenomena, *Can. Aeronaut. Space J.*, vol. 27, no. 2, pp. 128–134, 1981.
46. Zehnder, L., Ein Neuer Interferenzrefraktor, *Z. Instrumentenkd.*, vol. 11, pp. 275–278, 1891.
47. Mach, L., Ueber ein Interferenzrefraktometer, *Z. Instrumentenkd.*, vol. 12, pp. 89–93, 1892.
48. Ben-Dor, G., and Whitten, B. T., Interferometric Techniques and Data Evaluation Methods for the UTIAS 10 cm × 18 cm Hypervelocity Shock Tube, Univ. of Toronto, Inst. for Aerosp. Studies, UTIAS Tech. Note 208, 1979.
49. Ben-Dor, G., Whitten, B. T., and Glass, I. I., Evaluation of Perfect and Imperfect Gas Interferograms by Computer, *Int. J. Heat Fluid Flow*, vol. 1, pp. 77–91, 1979.
50. Tanner, L. H., Some Laser Interferometers for Use in Fluid Mechanics, *J. Sci. Instrum.*, vol. 42, no. 12, pp. 834–837, 1965.
51. Abramson, N., "Interferometric Holography" without Holograms, *Laser Focus*, vol. 4, no. 23, pp. 26–28, 1968.
52. Witte, A. B., and Wuerker, R. F., Laser Holographic Interferometry Study of High-Speed Flow Fields, AIAA 4th Aerodyn. Test. Conf., Cincinnati, AIAA Paper 69-347, 1969.
53. Jones, A. R., Schwar, M. J. R., and Weinberg, F. J., Generalizing Variable Shear Interferometry for the Study of Stationary and Moving Refractive Index Fields with the use of Laser Light, *Proc. R. Soc. London, Ser. A*, vol. 322, pp. 119–135, 1971.
54. Tanner, L. H., A Holographic Interferometer and Fringe Analyzer, and their use for the Study of Supersonic Flow, *Opt. Laser Technol.*, vol. 4, no. 6, pp. 281–287, 1972.
55. Barker, L. M., Laser Interferometry in Shock Wave Research, *Exp. Mech.*, vol. 12, no. 5, pp. 209–215, 1972.
56. Zien, T.-F., Ragsdale, W. C., and Spring, W. C. III, Quantitative Determination of Three-dimensional Density Field by Holographic Interferometry, *AIAA J.*, vol. 13, no. 7, pp. 841–842, 1975.
57. Armstrong, W. T., and Forman, P. R., Double-pulsed Time Differential Holographic Interferometry, *Appl. Opt.*, vol. 16, no. 1, pp. 229–232, 1977.
58. Surget, J., Two Reference Beam Holographic Interferometry for Aerodynamic Flow Studies, *Appl. Holography and Optical Data Processing*, P. Soc. Photo. Opt. Inst., Pergamon, N.Y., pp. 183–192, 1977.
59. Kolyshkina, L. L., Shevtsov, V. D., Arkhipov, V. V., Use of Diffraction Displacement Interferometer in Shock-Tube Research, *Soviet Phys.—Tech. Phys.*, vol. 25, no. 10, pp. 1283–1285, 1980.
60. Toyota, T., and Nishida, M., Diagnostics of Shock Tube Flows by Laser Interferometry, *Kyoto Univ. Fac. Eng. Mem.*, vol. 44, p. 3, pp. 410–429, 1982.
61. Takayama, K., Application of Holographic Interferometry to Shock Wave Research, *Industrial Applications of Laser Technology*, Geneva, Switzerland, *SPIE Proc.*, vol. 398, pp. 174–180, 1983.
62. Takayama, K., Onodera, O., and Ben-Dor, G., Holographic Interferometric Study of Shock Transition over Wedges, *High Speed Photography*, Strasbourg, France, *SPIE Proc.*, vol. 491, pp. 976–983, 1984.
63. Bachalo, W. D., and Houser, M. J., Optical Interferometry in Fluid Dynamics Research, *Opt. Eng.*, vol. 24, no. 3, pp. 455–461, 1985.
64. Anderson, J. H. B., An Experimental Determination of the Gladstone-Dale Constants for Dissociating Oxygen, Univ. of Toronto, Inst. for Aerosp. Studies, UTIAS Tech. Note 105, 1967.
65. Alpher, R. A., and White, D. R., Optical Refractivity of High-Temperature Gases. I. Effects Resulting from Dissociation of Diatomic Gases, *Phys. Fluids*, vol. 2, no. 2, pp. 153–161, 1959.
66. Trolinger, J. D., Flow Visualization Holography, *Opt. Eng.*, vol. 14, no. 5, pp. 470–481, 1975.
67. Burner, A. W., and Goad, W. K., Combined Single Pulse Holography and Time-resolved Laser Schlieren for Flow Visualization, NASA Tech. Memo 83109, 1981.
68. Brooks, R. E., Heflinger, L. O., Wuerker, R. F., and Briones, R. A., Holographic Photography of High-Speed Phenomena with Conventional and Q-Switched Ruby Lasers, *Appl. Phys. Lett.*, vol. 7, no. 4, pp. 92–94, 1965.
69. Muirhead, J. C., Smoke Streams as Tracers in Shock Tube Flows, *J. Appl. Phys.*, vol. 30, no. 5, p. 789, 1959.
70. Dewey, J. M., and Whitten, B. T., Calibration of a Shock Tube Flow by Analysis of the Particle Trajectories, *Phys. Fluids*, vol. 18, no. 4, pp. 437–445, 1975.
71. Dewey, J. M., and Walker, D. K., A Multiply Pulsed Double-Pass Laser Schlieren System for Recording the Movement of Shocks and Particle Tracers within a Shock Tube, *J. Appl. Phys.*, vol. 46, no. 8, pp. 3454–3458, 1975.
72. Dewey, J. M., and McMillin, D. J., Observation and Analysis of the Mach Reflection of Weak Uniform Plane Shock

Waves. Part 1. Observations, *J. Fluid Mech.*, vol. 152, pp. 46–66, 1985.
73. Somerscales, E. F. C., Measurement of Velocity: Tracer Methods, in *Methods of Experimental Physics*, ed. R. J. Emrich, vol. 18, pt. A, pp. 1–240, Academic, New York, 1981.
74. Dewey, J. M., A Preliminary Investigation of Particle Trajectory Analysis for Studying the Interaction of a Shock Wave with a Structure, Ernst-Mach-Institut, Freiburg/Br., West Germany, Rep. 2/73, 1973.
75. von Stein, H. D., and Pfeifer, H. J., Acceleration of Micron-sized Particles in Shock Tubes, in *Recent Developments in Shock Tubes*, ed. D. Bershader and W. Griffith, pp. 804–814, Stanford Univ. Press, Stanford, Calif., 1973.
76. Bomelburg, H. J., Herzog, I., and Weske, I., The Electric Spark Method for Quantitative Measurements in Flowing Gases, *Zt. Flugwiss. Weltraumforschung*, vol. 7, no. 11, p. 322, 1959.
77. Kyser, J. B., Tracer-Spark Technique for Velocity Mapping of Hypersonic Flow Fields, *AIAA J.*, vol. 2, no. 2, pp. 393–394, 1964.
78. Matsuo, K., Ikui, T., Yamamoto, Y., and Setoguchi, T., Measurements of Shock Tube Flows Using a Spark Tracer Method, in *Flow Visualization,* ed. T. Asanuma, pp. 157–162, Hemisphere, Washington, D.C., 1979.
79. Baker, W. E., *Explosions in Air,* Wilfred Baker Eng., San Antonio, Tex., 1973.
80. Held, M., Blast Waves in Free Air, *Propell. Explos. Pyrotech.,* vol. 8, pp. 1–7, 1973.
81. Glass, I. I., *Shock Waves and Man,* Univ. of Toronto Press, Toronto, 1974.
82. Glasstone, S., and Dolan, P. J., *The Effects of Nuclear Weapons,* 3d ed., U.S. Department of Defense and Energy Research and Development Administration, Washington, D.C., 1977.
83. Kinney, G. F., and Graham, K. J., *Explosive Shocks in Air,* 2d ed., Springer-Verlag, New York, 1985.
84. Boyer, D. W., An Experimental Study of the Explosion Generated by a Pressurized Sphere, *J. Fluid Mech.,* vol. 9, no. 3, pp. 401–429, 1960.
85. Wisotski, J., Ultra High-Speed Ruby Laser Photographic Light System, *6th Symp. Int. Appl. Militair. Simul. Souffle, Cahors, France,* pp. 1.13–1.13.25, 1979.
86. Dewey, J. M., The Air Velocity in Blast Waves from TNT Explosions, *Proc. R. Soc. London, Ser. A,* vol. 279, pp. 366–385, 1964.
87. Dewey, J. M., McMillin, D. J., and Classen, D. F., Photogrammetry of Spherical Shocks Reflected from Real and Ideal Surfaces, *J. Fluid Mech.*, vol. 81, no. 4, pp. 701–717, 1977.
88. Dewey, J. M., and McMillin, D. J., An Analysis of the Particle Trajectories in Spherical Blast Waves Reflected from Real and Ideal Surfaces, *Can. J. Phys.,* vol. 59, no. 10, pp. 1380–1390, 1981.
89. Cranz, C., and Schardin, H., Kinematographie auf ruhendem Film und mit extrem hoher Bildfrequenz, *Z. Phys.*, vol. 56, pp. 147–183, 1929.
90. North, R. J., A Cranz–Schardin High-Speed Camera for Use with a Hypersonic Shock Tube, Br. Nat. Phys. Lab., NPL Aero Rep. 399, 1960.
91. deLeeuw, J. H., Glass, I. I., and Heuckroth, L. E., A High-Speed Multi-Source Spark Camera, Univ. of Toronto, Inst. for Aerosp. Studies, UTIAS Tech. Note 26, 1962.
92. Conway, J. C., An Improved Cranz-Schardin High-Speed Camera for Two-dimensional Photomechanics, *Rev. Sci. Instrum.,* vol. 43, no. 8, pp. 1172–1174, 1972.
93. Rowlands, R. E., and Taylor, C. E., Pulsed Laser High-Speed Photography, *Proc. Int. Cong. Instrum. Aerosp. Simul. Facil., Polytech. Inst. Brooklyn, N.Y.,* 1969.
94. Rowlands, R. E., and Wentz, J. L., A Low-voltage Pockels Cell Having High-Repetition Rates and Short Exposure Durations, *Proc. 9th Int. Cong. High Speed Photog., Denver,* pp. 51–57, 1970.
95. Oertel, F. H., Jr., Transient Experiments using a Multiple-Pulse Laser Light Source, *Rev. Sci. Instrum.,* vol. 48, no. 10, pp. 1256–1261, 1977.
96. Oppenheim, A. K., and Kamel, M. M., Laser Cinematography of Explosions, Int. Cent. Mech. Sci., Courses and Lect., Udine, Italy, CISM 100, 1972.
97. Walker, D. K., Scotten, L. N., and Dewey, J. M., Construction and Evaluation of a Simple and Inexpensive Rotating Prism Camera for Recording the Movement of Shocks and Particle Tracers within a Shock Tube, *Proc. 15th Int. Cong. High Speed Photog. Photonics, San Diego,* pp. 125–130, 1982.
98. Eisfeld, F., Rotating Mirror Camera for Use in Holographic High Speed Interferometry, *Proc. 15th Int. Cong. High Speed Photog. Photonics, San Diego,* pp. 621–625, 1982.
99. Dudziak, W., SBM Data for Nuclear Detonation Photography, Defense Nuclear Agency Rep. 4132F, 1976.
100. Dewey, J. M., Heilig, W., Reichenbach, H., and Walker, D. K., The Analysis of Coloured Interferograms of Shock Waves, in *Flow Visualization III: Proceedings of the Third International Symposium on Flow Visualization,* ed. W.-J. Yang, pp. 585–589, Hemisphere, Washington, D.C., 1985.
101. Dewey, J. M., The Properties of a Blast Wave Obtained from an Analysis of the Particle Trajectories, *Proc. R. Soc. London, Ser. A,* vol. 324, pp. 275–299, 1971.
102. Sadek, H. S. I., and Gottlieb, J. J., Initial Decay of Flow Properties of Planar, Cylindrical and Spherical Blast Waves, Univ. of Toronto, Inst. for Aerosp. Studies, UTIAS Tech. Note 244, 1983.
103. Lau, S. C. M., and Gottlieb, J. J., Numerical Reconstruction of Part of an Actual Blast-Wave Flow Field to Agree with Available Experimental Data, Univ. of Toronto, Inst. for Aerosp. Studies, UTIAS Tech. Note 251, 1984.
104. Dewey, J. M., and Racca, R. G., A Feasibility Study of a Facility for the Transfer of Air Blast Photogrammetry Films to Randomly Accessible Video Storage, Univ. of Victoria, Contract Rep, Tech Reps, Inc., Albuquerque, 1986.
105. Racca, R. G., and Dewey, J. M., A Method for Automatic Particle Tracking in a Three-dimensional Flow Field, *Exp. Fluids,* vol. 6, pp. 25–32, 1988.

Chapter 30

Interferometry in Heat and Mass Transfer on Earth and in Space

J. Koster and R. Owen

I BACKGROUND

In heat and mass transfer, nonintrusive interferometric methods such as Mach–Zehnder and holography have been particularly valuable, as they permit the direct measurement of such parameters as thermal and concentration profiles in transparent liquid layers. Experiments in crystal growth, convective flow, transient heat transfer, electrodeposition, immiscible mixing, and separation have all made extensive usage of interferometric methods.

A Relevant Optical Measurement Techniques

Optical flow visualization techniques like schlieren, shadowgraph, and interferometers are commonly used. Apart from the assessments in Part 2 of this handbook, extensive descriptions of these optical technologies are given by Goldstein [32], Merzkirch [65], Hauf and Grigull [39], Vest [97], and others.

1 Color Schlieren Color schlieren techniques have generally been useful in the analysis of complex refractive fields. They have advantages that make them quite suitable for materials-processing model fluid experiments. With color schlieren it is possible to make a reasonably complete analysis by simply scanning the schlieren photograph and identifying the colors. The traditional schlieren advantages of simplicity and sensitivity are still retained. The color system proposed for low-gravity research is based on an improved version of a high-efficiency color-filter method developed by Howes [41, 42].

2 Mach–Zehnder Interferometer The Mach-Zehnder interferometer was the major technique used in heat transfer at the beginning of interferometry applications [25, 27]. The main advantage of the technique is the separation of the object beam through the test cell with the reference beam passing outside the test cell [32, 33, 91, 102]. In the beginning, the light sources were conventional ones, which made alignment of the interferometer a tremendous task. Light waves interfere only if they are coherent, meaning that the optical distance of the two light paths have to be exact within the wavelength of the light at the plane of interference. Coherent monochromatic laser-light sources eased the alignment enormously.

3 Holographic Interferometry With the advent of the laser and the development of holography, the technique of holographic interferometry was developed [40]. Two methods can be distinguished: double-exposure and real-time interferometry. In double-exposure interferometry, two light waves create an interferogram at a reference time t_0 stored in the holographic plate. A second interferogram is superimposed and recorded at a later time t_1 onto the first hologram. Both holographic patterns, stored on one plate, are subsequently reconstructed simultaneously by laser light. The variation of the refractive index fields between exposures gives rise to an interference pattern. The visualized fringe pattern is correlated to the change of the density field during that time change. A disadvantage is that this technique can be used best particularly for steady-state experiments as no optical information is available on the density field between the two measuring times t_0 and t_1 (black-box approach). This limitation was eliminated when real-time holographic interferometry was developed.

When the real-time technique is applied, the hologram, processed at reference time t_0, stores the phase information of the refractive index field at that time in the hologram. The reconstructed initial-reference density field is continuously superimposed onto the refractive index field in the test chamber existing at a later time t_1, thus creating a real-time interferogram. The interference of reference and actual index field thus allows a continuous evaluation of the density field in a transparent liquid layer.

Usually the application of interferometry to heat and mass transfer is limited to two-dimensional flow and temperature fields. However there was a demand for evaluating three-dimensional density fields. Three-dimensional tomography was known from the fields of X-ray analysis, electron microscopy, and radio astronomy. Sweeney and Vest [94] investigated the possibility of measuring three-dimensional temperature fields by using multidirectional interferometric data. Other techniques were to follow [60, 89].

4 Two-wavelength Interferometry As refractive index fields are given by two parameters, local temperature and concentration, interferometry can be used to evaluate the temperature and concentration fields. This is possible by using a two-wavelength interferometric Mach–Zehnder method as proposed by Ross and El-Wakil [86, 87]. Mayinger and Panknin [62] applied the technique to a holographic interferometer.

The method is based on the simultaneous recording of two double-exposure holographic interferograms on one photographic plate, using different wavelengths and reference angles so that the reconstructed interferograms can be spatially separated. Each interferogram is reconstructed separately by each wavelength, and each contains information on sample temperature and concentration profiles. Since the corresponding sample refractive-index profiles are a function of wavelength, it is possible to evaluate temperature and concentration profiles from differences between the two interferograms.

5 Differential Interferometry The optical component that allowed the development of this technique was the Wollaston prism [71]. In the first Wollaston prism each light ray is split into two polarized rays, which diverge by a small amount. The second polarizer superimposes the light rays again and brings them to interference at an image plane defined by the optics. The two light beams entering the test cell, separated by a small distance given by the angle of the Wollaston prism, experience a different optical pathlength due to a difference of the local refractive index. The fringe shift obtained is thus a function of the density difference between the locations of the two polarized light rays traversing the liquid layer. The differential interferometer thus measures the density gradient in the plane of beam separation. The sensitivity of the interferometer can be set by the angle of the Wollaston prisms. The interpretation of visualized density gradients is difficult in comparison to a Mach–Zehnder interferogram, especially to the nonuser of such a system. The advantages of the differential interferometer are its variable resolution and its lower sensitivity to vibrations.

B Basics

Consider a homogeneous transparent liquid system in which the index of refraction depends on the density field according to the Lorenz–Lorentz relation [39]. The density in turn may be governed by local temperature and concentration distributions.

The physics of boundary layers is important in many investigations of heat and mass transfer. Grigull [37] has analyzed the optical characteristics of boundary layers. At a solid boundary, the no-slip condition allows evaluation of the heat or mass transfer to the body by using the conduction equation. In case of heat conduction this is

$$\frac{q}{A} = -K \left.\frac{dT}{dx}\right|_{\text{wall}}$$

where q/A is the heat flux per area, K is the thermal conductivity and dT/dx the temperature gradient at the surface of the body. The heat transfer is expressed by

$$\frac{q}{A} = h(T_w - T_1)$$

with unit thermal conductance h and the temperature difference between wall and liquid. Thus the interferogram can be used to determine the local heat transfer, also expressed as the Nusselt number [43]:

$$\text{Nu} = h\frac{L}{K} = \left(\frac{dT}{dx}\right)_{\text{wall}} \frac{L}{T_w - T_1} = \frac{L}{\delta}$$

where L is a characteristic length of the problem and δ the thickness of the thermal boundary layer. With a reference temperature T_w or T_1, the temperature gradient and the heat transfer coefficient can be evaluated. Thermal boundary layer thicknesses can now be evaluated locally [49, 50].

A similar evaluation can be done for density changes due to changing concentrational fields.

C Low-Gravity Research Objectives

The environment of outer space offers unique opportunities for the study of heat and mass transfer under weightlessness. Effects such as buoyancy-driven convection, hydrostatic pressure, sedimentation, and flotation either become quite minor or simply cease to exist. However, a careful assessment of the importance of secondary effects, such as surface tension gradients, must be considered as they may generate unexpected flow patterns and results. It is anticipated that low-gravity research will have sweeping applications in the understanding of materials processing and that the promise of new and improved materials from space is actual and close to realization.

The major incentive in going into space is to avoid strong gravity-determined effects. Since weak effects may influence the flow and temperature pattern, it is of the greatest importance to avoid any perturbation by any type of sensor. Optical methods meet this requirement as they are nonintrusive data-gathering techniques.

We shall examine some of the major space flight experiments that have used interferometric measurement as a major support system. The space experiments covered include those flown on Spacelab 3 and on the West German D-1 Spacelab mission. Ground-based preparatory experiments include aircraft studies, data reduction and evaluation, supportive laboratory hardware development, and future experiments.

D Aims

In the following study we concentrate on interferometric techniques. Interferograms are sometimes used as qualitative information but are of great value in guiding intuition for further research, and sometimes for quantitative studies in heat transfer.

II CONVENTIONAL EARTH-BASED APPLICATIONS

A Horizontal, Vertical, and Inclined Plates

An early application of Mach–Zehnder interferometers to heat transfer problems was investigation of the heat transfer from vertical [26] and horizontal [30] plates in order to assess the results of Nusselt and Jurgens [70] and Weise [101]. Goldstein and Eckert [35] applied Mach–Zehnder interferometry to measure heat transfer coefficients of a uniformly heated vertical plate. Their investigation also gave valuable information on transient effects in thermal boundary layers, which at that time was a major problem in nuclear reactor technology. The problem of the step function change in surface temperature was investigated by Schetz and Eichorn [90]. Their results closely agreed with theoretical results. Aung and O'Regan [2] applied the then-newly developed holographic technology to precise heat transfer measurements. Their results of heat transfer measurement on a heated isothermal vertical flat plate agreed well with existing theories.

Polymeropoulos and Gebhart [82] used interferometry to visualize temperature waves induced by a perturbation of the thermal boundary layer of a vertical flat plate. The stability of the thermal boundary layer was analyzed as a function of the frequency of a mechanical disturbance. Zinnes [107] investigated the vertical flat plate with arbitrary heating distribution in its surface. The results obtained by holographic double-exposure interferometry indicated that natural convection in the fluid is greatly influenced by the plate–fluid conductivity ratio.

A horizontal cylinder was investigated by Kennard [46] to evaluate debated theories. Reisbig and Reifel [84] applied a double-exposure laser holographic system to the measurement of complex temperature fields in a gas.

Black and Carr [9] and Black and Norris [10] introduced the concept of a differential interferometer to measure heat transfer coefficients. Later they investigated free convection from a horizontally vibrating isothermal cylinder using this technique [16].

Black and Norris [11] investigated the local heat transfer for free convection from an inclined vertical plate. They used a differential interferometer to show that thermal waves traverse the surface of the heated plate and cause significant variations in the local heat transfer coefficient.

B Plumes and Line Heat Sources

The heated plume above a line source of heat was studied by Pera and Gebhart [80] using a Mach–Zehnder interferometer. Interactions between laminar thermal natural convection plumes generated by line and point sources were investigated by Pera and Gebhart [81]. Adjacent plane plumes were found to interact more strongly than axisymmetric plumes. Free boundaries were found to be affected by adjacent surfaces.

C Interfaces

Kapur and Macleod [45] used holographic interferometry to investigate mass transfer rates at solid–liquid interfaces, yielding detailed data for the spatial variation of the coefficients. They concluded that the holographic technique is about one order of magnitude better than previously used techniques, due to higher speed of operation, convenience, and field view.

An optical study of mass transfer and interfacial turbulence in a liquid–liquid system was done by Nakaike et al. [68]. The concentration profile of propionic acid near the interface was measured by microinterferometric and Mach–Zehnder interferometric methods.

D Boiling Phenomena

Bubble growth phenomena in nucleate boiling can also be investigated interferometrically [3, 79]. The temporal and spatial temperature distributions associated with boiling can be determined [4, 61]. Since these phenomena are transient processes, high-speed film techniques have to be applied.

E Concentration

Holographic interferometry can also be used to observe concentration fields in an electrodialysis cell [88]. Mass transfer through selective membranes can thus be studied. Sanchez and Clifton [88] used the holographic technique to show that if the limiting current is reached, the concentration at the interface of membrane and liquid tends to be zero at the outlet of the cell.

Santoro et al. [89] used optical tomography to analyze an off-axis turbulent methane–air free jet, to determine the mean methane concentration throughout the mixing region. Mean concentration measurements were obtained in several locations, and comparisons with other technical results show good agreement, demonstrating the capability of optical tomography.

A special tomographic interferometer to investigate convection and mass transfer problems was studied by Mayinger and Lubbe [60]. The volume under investigation was viewed in different directions. The refractive index field was calculated, giving the local and current concentration field information. Such a technique is useful for mixing and mass transfer processes that influence the production rates of chemical reactions.

Heterodyne holographic interferometry was used by Farrell et al. [29] to measure specie concentration and temperature measurements in dilute binary gas mixtures. The specific application was the study of gas boundary layers adjacent to a vertical heated surface through which a second gas was ejected. This technique allows further understanding of degassing, ignition, and combustion phenomena.

F Combustion

Interferometry was applied early in combustion research [100]. An interferometer used for such an application has to satisfy stringent requirements due to the high amount of heat generated, which can distort optical components. Since the temperature gradients in flames are high, the light deflection is large, which has to be considered in the error analysis. On top of this problem, many flames are three-dimensional, requiring special evaluation.

Ross and El-Wakil [86, 87] considered evaluating the chemistry of flames by introducing the new concept of two-wavelength interferometry.

G Rayleigh–Bénard Convection

Farhadieh and Tankin [28] used a Mach–Zehnder interferometer to study two-dimensional Bénard convection. They observed the reversal of the temperature profile across convection cells at high Rayleigh numbers that characterizes the temperature difference between the cold and hot boundaries. In general, their evaluation of the Nusselt number agreed with existing theories.

Goldstein and Chu [34] investigated the basic stability behavior of a horizontal layer of air and found that the critical Rayleigh number for the onset of convection agreed well with theoretical predictions. The interferograms qualitatively showed the periodic convection roll pattern.

Chu and Goldstein [17] investigated the turbulent convection in a horizontal layer of water. The overall Nusselt number of the heat transfer was evaluated. It was shown that thermals, released from the boundary layers, contribute extensively to the heat transfer. Unsteady natural convection in a nonuniformly stratified liquid layer has also been studied by Viskanta and Behnia [98] using a Mach–Zehnder interferometer. The theoretical predictions of a turbulence model overpredicted the mixed layer depth.

Time-dependent thermal convection is a topic of major interest for understanding of the transition to turbulence and chaos. Flow visualization is very important in this research as the knowledge of the flow pattern gives some information about the cause of turbulent flow in unstable thermal liquid layers. Bergé and Dubois [8, 21, 22] investigated the stability of thermal boundary layers in small convection boxes and observed the oscillators responsible for the time-dependent flow. Koster and Müller [54, 56] did similar investigations in Hele–Shaw cells, which are discussed in the next section.

Bühler et al. [14] and Oertel and Bühler [71] introduced a more simplified and advanced differential interferometer to investigate convective flow. Bühler and Oertel [15] applied this technique to the investigation of thermal cellular convection in rotating rectangular boxes. The stability regions of different flow configurations in a rectangular container could thus be studied as a function of the Taylor number, which characterizes the rotation. Kirchartz et al. [48] used differential interferometry to investigate the influence of shear flows on Rayleigh–Bénard convection in a rectangular box. A basic flow was generated with tilting of the box, upon heating from below and cooling from above. The stability of such a system could be verified experimentally and theoretically.

In most theoretical analysis the heat transfer from a boundary of a plate or container is averaged over a certain length. Kim and Viskanta [47] studied the local wall conductance on natural convection in square cavities. They determined the temperature profile in the liquid and the local heat transfer at the walls.

An experimental study of the convection between two vertical flat plates was performed in early work by Oshima [72] utilizing a Mach–Zehnder data-gathering technique. At high Rayleigh numbers they observed the boundary layer–type flow, and at even higher temperature difference the flow became turbulent and wavy motions occurred.

H Hele–Shaw Convection

The Hele–Shaw cell is a special type of geometry. The flow equations of such cells have some similarities with the flow through porous media. The geometry of a Hele–Shaw cell is a slender vertical gap where the spacing between the walls is much smaller than any other characteristic length. Koster and Müller [49–56] investigated the stability of liquid layers confined in such vertical cells heated from below. Two kinds of cavities were studied: Hele–Shaw cells extended in one horizontal direction, celled the Hele–Shaw gap, and those with larger vertical than horizontal extension, called the Hele–Shaw slot.

The transition to time-dependent flow was of major interest. In Hele–Shaw gaps the unstable horizontal thermal boundary layer was shown to generate secondary vortices that drive the time-dependent flow (Fig. 1). The optical analysis revealed that thermal boundary layer thickness is locally different and time dependent (Fig. 2) [50]. The flow pattern in Hele–Shaw slots exhibits complicated structures and transitions between different flow patterns (Fig. 3). Symmetry of the existing flow pattern is considered to be of major importance to the time structure of the oscillations. Nonsteady end effects observed in a special Hele–Shaw cell indicated that complicated oscillation patterns develop only when there is a coupling between the oscillators [51].

I Convection with Phase Transition

It has been observed that natural convection shapes the interface of a material undergoing a liquid–solid phase transition. In many materials-processing techniques it is important to understand the formation of the morphology of the interface. Diaz and Viskanta [19] as well as Dietsche and Müller [20] investigated the stability of such systems and investigated the onset of the convection and the hysteresis at this threshold. The geometry also has a major effect on the flow pattern and thus on the shape of the interface.

J Thermohaline Convection

Concern with the environment has fostered the study of heat and mass transfer in oceans and seas. Freezing or evaporation at the surface often results in the development of unstable liquid layers. ''Salt fingers'' may develop, which are thought to determine the salinity distribution in oceans. Other appearances of this type of convection are in crystal growth from melts and solutions, casting, purification of materials, phase-change storage, and solar collectors. Viskanta and his collaborators [5–7, 36, 58, 59, 99] used Mach–Zehnder interferometry to investigate these problems in great detail.

K Crystal Growth

Holographic interferometry can also be applied to fluid flow and heat transfer phenomena in crystal growth. Williams and Peterson [103] applied interferometry to measure temperature profiles in an air-cooled, vertical quartz-walled metalorganic chemical vapor deposition (MOCVD) reactor for better understanding of the growth process. Fluid dynamics and thermal gradients greatly affect uniformity of epitaxial layers, impurity incorporation, and interfacial quality and sharpness. An understanding of gas flow patterns and thermal gradients becomes very useful when designing reactor cells.

L Nuclear Energy

In nuclear energy research, holographic interferometry was introduced to investigate heat transfer in rod bundle heat transfer channels [79]. It must be considered that in these cases the integration length of the optical path is parallel to the heating walls that could induce three-dimensional heat advection. Such cooling liquids are also subject to forced turbulent flow, which again alters the local refractive index field. This technique has also been proposed to control whisker processes [79].

Boyd [12] used a Mach–Zehnder interferometer to study natural convection in an annulus with irregular boundaries. The inner cylinder was hexagonal, while the outer cylinder was circular. This configuration models a fuel assembly in a fast-breeder nuclear reactor. The controlled heat transfer in these subchannels is important for the operation and safety of the reactor.

Other applications include heat transfer problems in liquid layers with internal heat generation, as this can occur during nuclear reactor accidents. Using model fluids, Steinberner and Mayinger [93] measured the heat transfer coefficients in semicircular containers that model nuclear core catcher designs.

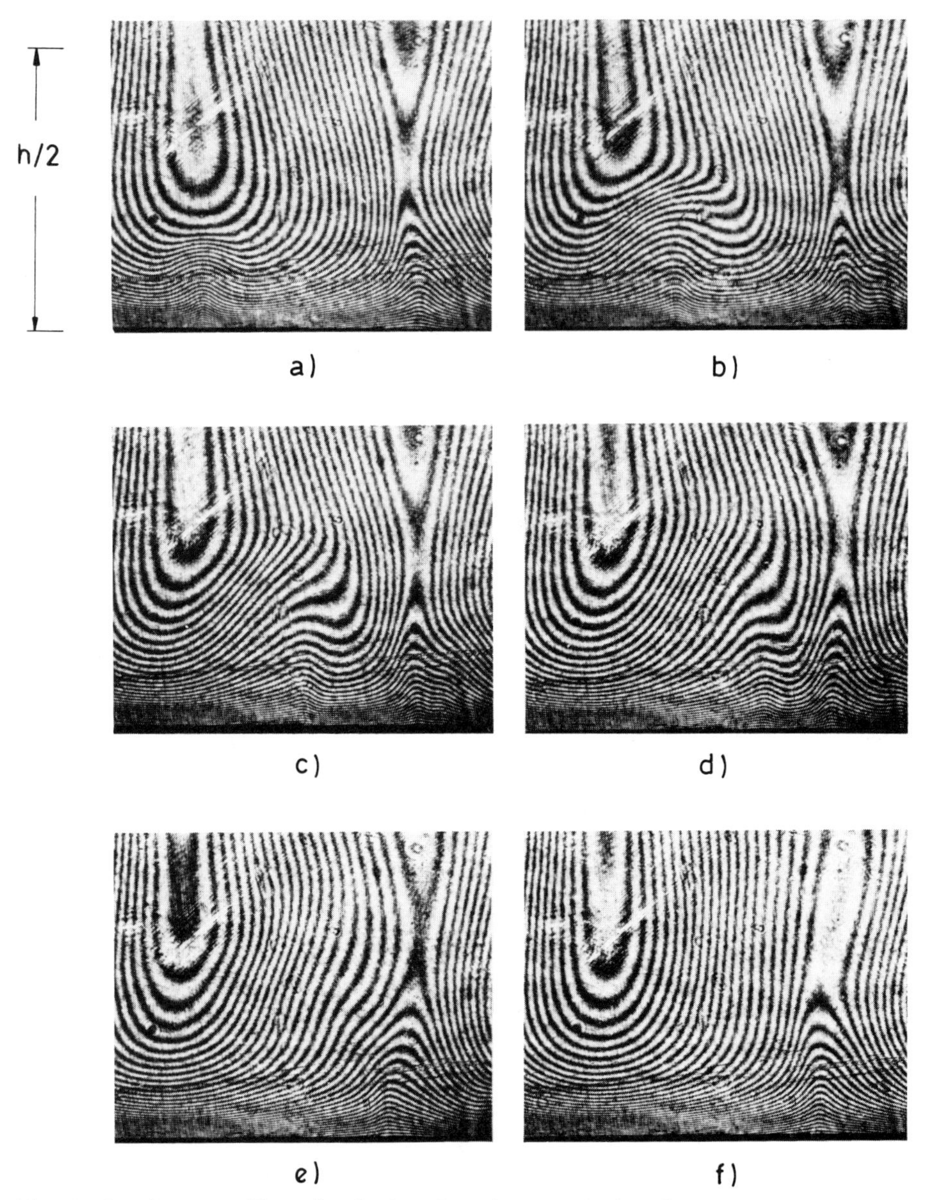

Fig. 1 Interferograms illustrating the time-dependent perturbation of the thermal boundary layer. The period is 240 s (from Koster and Müller [54]).

A problem related to the safety of reactors is the basic understanding of convection from a horizontal heated plate facing downward. Hatfield and Edwards [38] studied this problem with a rectangular plate in air, water, and oil, and compared the experimental results with theoretical correlation functions.

In laser-induced inertial confinement fusion, understanding of the physical process taking place in the plasma atmosphere surrounding the target is important, as is the location where the energy is absorbed. Attwood et al. [1] used interferometry to investigate the electron density around irradiated microballoons.

M Solar Energy

Meyer et al. [66] used a Mach–Zehnder interferometer to study heat transfer between parallel plates tilted at various angles. Honeycomb cells were introduced between plates, and these arrays were used as solar collectors. Knowledge of the loss of heat of these collectors is important in order to assess their efficiency.

A problem often appearing in solar energy is the heating or cooling of a liquid layer from above. Behnia and Viskanta [5–7, 59] investigated this problem with Mach–Zehnder interferometry and shadowgraph techniques. They

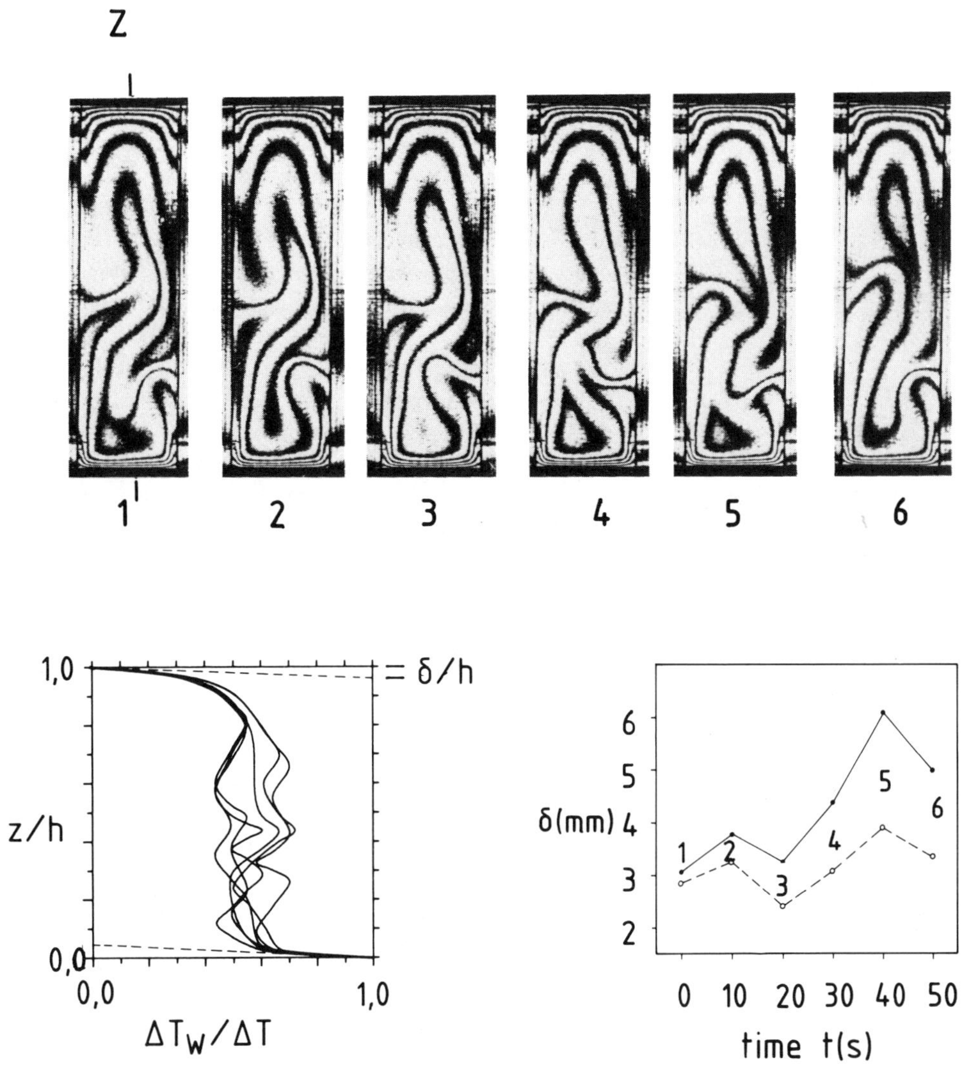

Fig. 2 One period of an asymmetric oscillation and the relevant centerline temperature profiles. The thermal boundary layer thickness is different in each interferogram (from Koster [49]).

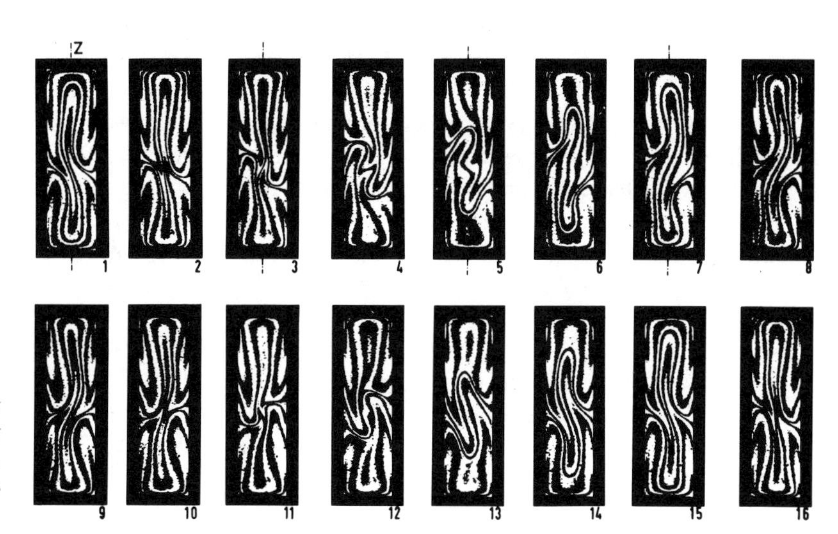

Fig. 3 One period of a periodic oscillatory convective flow in a high-conductivity Hele–Shaw slot of aspect ratio $h/b = 3.5$. The time interval between two interferograms is 10 s (from Koster and Müller [56]).

observed two flow regimes when the liquid layer was cooled by airflow. A turbulent mixing zone developed underneath the surface, extending to the lower stratified layer. If the liquid layer was also radiatively heated, the absorption of radiation influenced the flow and thermal structure in the liquid layer.

III LOW-GRAVITY RESEARCH APPLICATIONS

A Low-Gravity-related Ground-based Research

Low-gravity research and development utilizing optical interferometric techniques include ground-based support studies for the casting and solidification technology (CAST) experiment; Spacelab 3 data reduction and evaluation, using data from the holographic unit flown on that mission; and studies of immiscible liquids, protein crystal–growth experiments, and hardware system development.

1 The CAST Experiment The CAST experiment is a study aimed at determining some of the effects of the low-gravity environment on metal casting. It makes use of a transparent material (ammonium chloride) that closely models metal solidification; the solidification process can be observed optically. The test sample for this experiment will be flown in the holographic Fluid Experiments System (FES) [106].

Specific scientific objectives are to determine the influence of gravity on the fluid flow and nucleation that occur during casting as well as to study the solidification and coarsening processes of dendrite arms and their subsequent influence on the grain structure in casting. The approach taken is to directionally solidify a metal model material under low-gravity and one-gravity conditions in order to determine the thermal and concentrational profiles ahead of a growing interface during supercooling [63]. This experiment will make use of test flights on the KC–135 aircraft.

2 Spacelab 3 Data Reduction and Evaluation Considerable laboratory effort is currently being mounted to reduce and evaluate the data from the Spacelab 3 FES holographic unit. A special-purpose Holographic Ground System allows the holograms to be evaluated in a variety of modes, including schlieren, shadowgraph, Mach–Zehnder interferometric, and holographic interferometric systems [95, 104].

An important component in the analysis of the FES holograms is the Automated Holographic Data Reduction System [92, 96]. This system digitizes the holographic images and automatically produces a refractive index map based on the original holographic interferograms. This mapping can then be utilized to accurately determine the concentration of solutes and temperature profiles in the FES fluid sample.

3 Protein Crystal Growth The major bottleneck in the widespread application of protein crystallography has been the lack of large, high-quality protein crystals suitable for structural analysis. There are definite indications that the reduction in convective and sedimentation effects that occurs in a low-gravity environment will allow larger samples of these extremely delicate crystals to be grown [13, 18].

In support of this research, ground-based optical systems are being used to specify sample characteristics and growth parameters. Holographic systems are being used to study the details of protein crystals in the laboratory; refractive methods are being used to measure temperature and concentration profiles in hanging growth drops, and image-analysis techniques are being used to size and characterize drops.

4 Immiscibles Space experiments have shown that the low-gravity environment has a strong effect on the formation of binary alloys whose components have a liquid-phase miscibility gap. For this reason, model transparent immiscible-fluid systems are being studied optically to gain a better understanding of phase-separation processes in binary miscibility-gap metal alloys [105]. Holography microscopy is used to study the growth of minute droplets in the diethylene glycol–ethyl salicylate immiscible system. Sequential holograms are taken over several months of very slow growth. The holographic system is capable of recording particle densities up to 10^7 particles/cm^3 and can resolve particles as small as 2–3 μm in diameter throughout the entire test cell volume.

Immiscible systems have also been studied on board the KC-135 aircraft using standard photographic techniques as well as shadowgraph and schlieren methods. This study concentrated on meniscus shapes and the concentration profile of the succinonitrile–water solution as a function of gravity levels. There was a clear enhancement of meniscus curvature upon removal of the hydrostatic pressure in low gravity. The interface between the two immiscible phases flattened during the 2-*g* pullout, indicative of density differences between the two, and remained so for some time into low gravity, a relaxation phenomenon that is not yet understood.

In many materials science experiments it is necessary to evaluate the temperature distribution within the liquid. One application is the liquid floating zone that is extensively studied for crystal-growth purposes. Monti and Russo [67] investigated the possibility of studying floating zones with schlieren and differential interferometers.

5 Hardware System Development A number of laboratory optical systems are being developed that would aid and support the study of heat and mass transfer in low-gravity conditions. These include a compact color schlieren system [83] for use as a Shuttle mid-deck locker package; possible space station applications; and a two-color holographic system [23] that has been used on the KC-135 aircraft and that may also find a place in the low-gravity research module of the space station.

A test two-color holographic system has been flown on the NASA KC-135 low-gravity simulation aircraft by Ecker [24], and further tests and development efforts are being conducted to determine the potential usefulness of this technique for space station low-gravity research.

B Aircraft Research and Development

In special cases, it is possible to simulate the weightlessness in aircraft flying Keplerian parabolic trajectories. The NASA KC-135 aircraft is able to simulate low gravity to within 10^{-2} *g* for periods of approximately 25 s. A typical parabola profile is shown in Fig. 4. A 2- to 3-hr mission will involve 30 to 40 parabolas. Such experiments are much cheaper to perform than actual space flight experiments and are more easily repeated. Experiments in crystal growth, transient heat transfer, and electrodeposition are discussed in the following sections.

Mach–Zehnder interferometers and holographic units have been developed that are able to withstand the rigors of flight [73, 74]. These systems have been used to conduct a variety of experiments in materials processing on board the KC-135 aircraft.

1 Crystal Growth A saturated solution of NH_4Cl–H_2O ammonium chloride has been used as a transparent metal-model material in solidification experiments, and extensive use has been made of Mach–Zehnder interferometry. The aircraft studies included growth morphology, dendrite remelting and coarsening, and fluid instabilities [44, 76]. Changes in the fluid-sample temperature profile and concentrational plume development were clearly shown by the interferograms.

Another KC-135 crystal-growth experiment using interferometry involved a cloud physics study on ice-crystal growth in a saline solution [108]. Past experiments that attempted to duplicate the growth of ice in clouds by studying solution-grown ice crystals have been plagued by gravity-driven convective effects that arose during growth. Mach–Zehnder interferometry was used to study temperature and concentration profiles within the cell and to verify convective damping during nucleation and growth. Preliminary results indicate much less ice crystal morphological dependence on *g* level than was anticipated.

2 Transient Heat Transfer The study of transient heat input into a confined compressible fluid is a problem in the design of engines and nuclear power plants as well as in a variety of chemical engineering processes. Optical techniques offer significant improvements over traditional methods. Optical approaches allow direct measurement of the temperature field, and they are nonintrusive and able to perform high-speed data gathering.

In transient heat transfer, one enters regimes of large time-varying, nonuniform refractive index gradients. For these extreme conditions, many classical techniques, such as schlieren and Mach–Zehnder interferometry, which make use of collimated light beams, are no longer useful for quantitative studies. Although techniques have been developed for evaluating interferometric data in boundary lay-

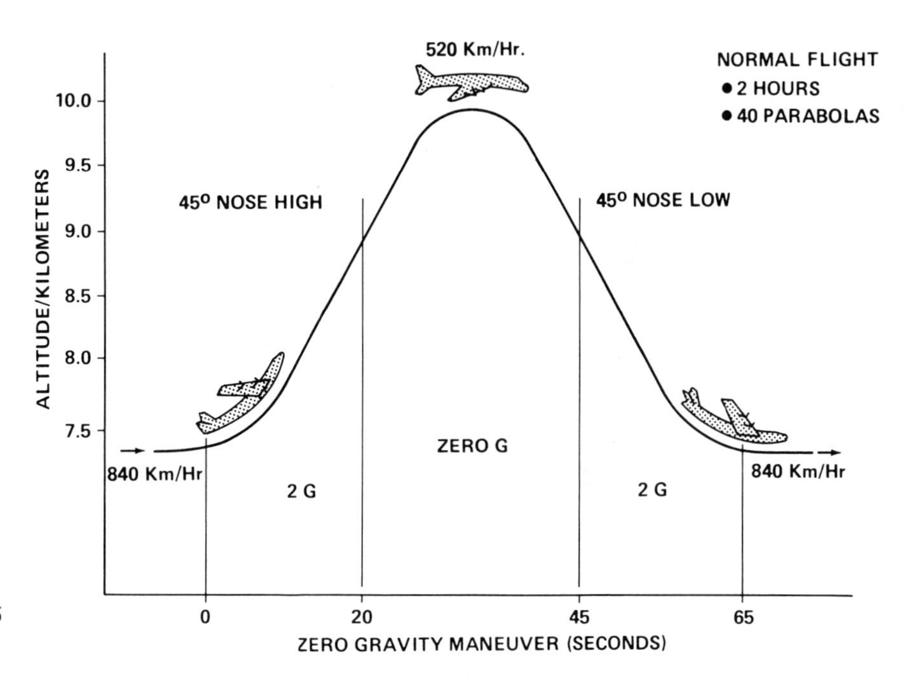

Fig. 4 Flight parabola of KC-135 low-gravity simulation aircraft.

ers for electrodeposition samples [64], these methods are not adequate for the extremely small (0.2 mm) high-density boundary layers of interest in this study.

The main source of error in the classical optical techniques studied is the use of collimated light beams; these errors result from focusing effects and from nonlinear refractive index gradients. Since holographic methods permit the use of a diffuse light source in the object beam, holographic interferometry was applied to the measurement of transient heat transfer boundary layer temperature profiles. It was found that the application of holographic microscopy using diffusive light allowed considerably more of the boundary layer to be measured than was possible with classical optical techniques. A spatial resolution of 0.02 mm and a time resolution of 1 ms were achieved [75].

It is expected that aircraft holographic measurements taken on board the NASA KC-135 low-gravity simulation aircraft will determine the extent to which gravity-driven convection is responsible for the observed divergence between theory and experiment.

3 Electrodeposition The main purpose of the KC-135 electrodeposition experiment was to examine concentration gradients and flow in plating solutions subjected to a low-gravity environment [85]. These observations were aimed at determining the extent to which larger neutral particles that cannot be readily suspended on earth can be co-deposited in low gravity with metals, possibly improving desirable material properties such as tensile strength, wear, and hardness. $CoSO_4$ was selected as the deposition system because of its transmissivity at He–Ne wavelengths and its low toxicity.

A Mach–Zehnder interferometer was again used to examine sample concentration profiles. Results from repeated comparisons of electrodeposition under varying gravity loads showed apparent differences in concentration profiles formed under low- and high-gravity conditions.

C Space Flight Experiments

While there have been numerous low-gravity experiments since the beginning of the Space Age [69], no major interferometric systems were flow in space until the era of the Space Shuttle. To date, the principal space interferometric systems used for low-gravity research were flown on two Spacelab missions, Spacelab 3 and the West German D-1 Spacelab mission.

1 Crystal Growth One of the experiments conducted on Spacelab 3 involved the solution growth of triglycine sulfate (TGS) crystals. This material is useful as an infrared thermal detector and has interesting ferroelectric properties as detailed elsewhere [57, 77]. Common crystallographic defects such as inclusions and dislocations have been proven to adversely affect the electrical characteristics of TGS and determine its quality as a sensor. It has been conjectured that some of these defects might be caused by gravity-driven convection [57, 77]. The purpose of the experiment in space was to study the basic crystallization

MET 79:44:26

MET 84:44:30

MET 86:57:53

MET 87:13:02

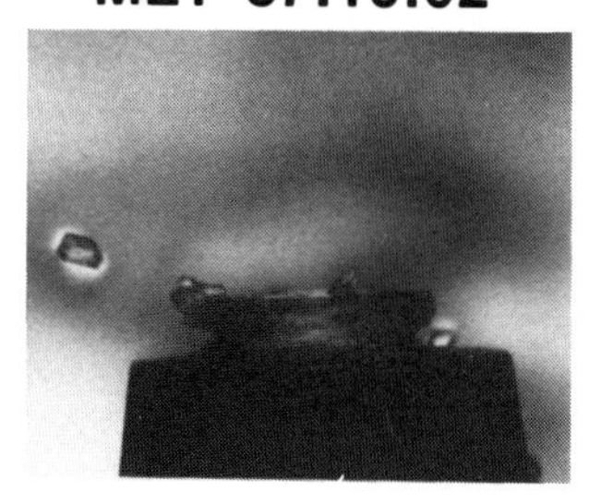

MET 89:43:34

Fig. 5 Double-exposure holograms taken of a triglycine sulfate crystal growing in the Fluid Experiment System during the Spacelab 3 mission. MET, mission elapsed time (hr : min : s) (courtesy of W. K. Witherow, NASA Marshall Space Flight Center).

growth process in a convection-free environment. Since the melt solution is transparent, this process can readily be optically observed.

Test crystals were grown in the Fluid Experiment System (FES). Reconstruction of the holograms takes place in a special-purpose facility [95]. The single-exposure holograms with diffuser are used to study the crystal profile and can be examined microscopically. The single-exposure holograms without diffuser can be used to obtain shadowgraph, schlieren, and interferometric data. Double-exposure holograms were recorded with a 100-s delay between exposures. Examples of reconstructed images from the flight holograms are shown in Fig. 5.

Some results have been described by Owen et al. [78]. The optical images from the test cells clearly showed that the mass transport of solution to the crystal surface was by diffusion. Some of the crystals grew freely suspended in the solution. Growth rate and concentration profiles of the seed and around the seed, and of the freely suspended crystallites could be determined. Mach–Zehnder interferograms created from the flight holograms are shown in Fig. 6.

MET 111:37:46

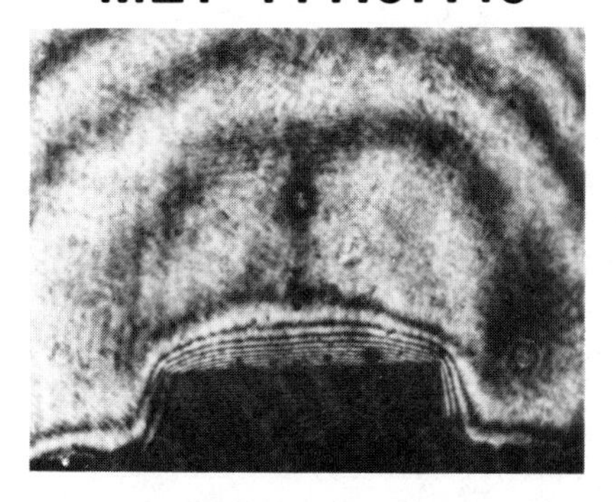

MET 117:33:01

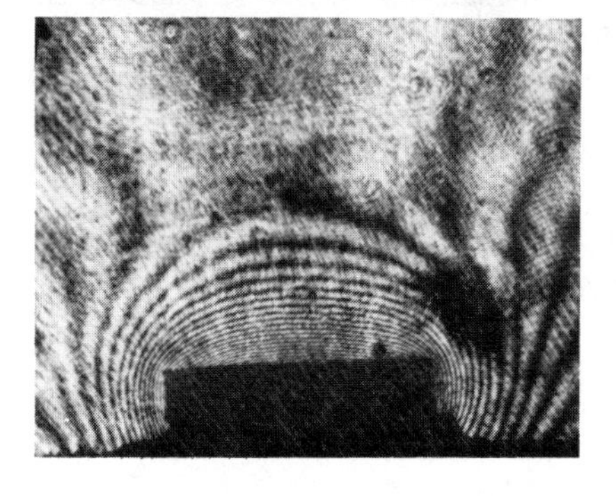

MET 125:16:29

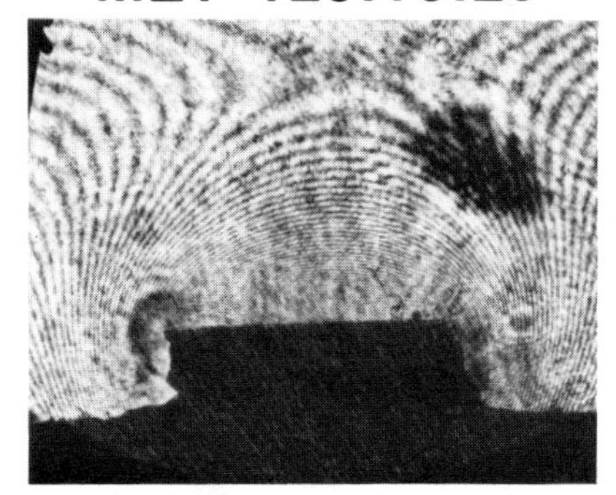

Fig. 6 Mach–Zehnder interferograms reconstructed from holograms of a triglycine sulfate crystal growing in the Fluid Experiment System during the Spacelab 3 mission. MET, mission elapsed time (hr : min : s) (courtesy of W. K. Witherow, NASA Marshall Space Flight Center).

2 HOLOP One of the major experiments on the West German D-1 Spacelab mission was the Holographic Optics Laboratory (HOLOP), which was used to examine a variety of transparent media [31]. The system followed three-dimensional reconstruction of refractive-index variations within a liquid sample volume. These variations could then be interpreted in terms of density and composition of the material of interest. If the sample contained particles, it was possible to determine their distribution and velocities, thereby establishing fluid flow velocities.

Three samples were examined by the HOLOP during the D-1 mission. These included chemical waves in a liquid, observed via the motion of gas bubbles; changes of the density distribution inside a fluid sample when approaching its critical point; and bubble motions induced by temperature gradients.

REFERENCES

1. Attwood, D. T., Sweeney, D. W., Auerbach, J. M., and Lee, P. H. Y., Interferometric Confirmation of Radiation-Pressure Effects in Laser-Plasma Interactions, *Phys. Rev. Lett.*, 40, pp. 184–187, 1978.
2. Aung, W., and O'Regan, R., Precise Measurement of Heat Transfer using Holographic Interferometry, *Rev. Sci. Instrum.*, vol. 42, pp. 1755–1759, 1971.
3. Beer, H., Interferometry and Holography in Nucleate Boiling, in *Boiling Phenomena,* ed. S. van Stralen and R. Cole, vol. 2, pp. 821–843, Hemisphere, Washington, D.C., 1979.
4. Beer, H., Interfacial Velocities and Bubble Growth in Nucleate Boiling, *Int. Symp. Two-Phase Systems, Haifa,* 1971.
5. Behnia, M., and Viskanta, R., Free Convection in Thermally Stratified Water Cooled from Above,'' *Int. J. Heat Mass Transfer,* vol. 22, pp. 611–623, 1979.
6. Behnia, M., and Viskanta, R., Natural Convection flow Visualization in Irradiated Water Cooled by Air Flow over the Surface, in *Natural Convection in Enclosures,* ed. K. E. Torrance and I. Catton, ASME, HTD, vol. 8, pp. 17–26, 1980.
7. Behnia, M., and Viskanta, R., Heat Transfer in Irradiated Shallow Layers of Water, in *Heat Transfer and Thermal*

Control, ed. A. L. Crosbie, *Prog. Astronaut. Aeronaut.,* vol. 78, pp. 110–129, 1981.

8. Bergé, P., and Dubois, M., Study of Unsteady Convection Through Simultaneous Velocity and Interferometric Measurements, *J. Phys. (Paris), Lett.,* vol. 40, pp. 505–509, 1979.
9. Black, W. Z., and Carr, W. W., Application of a Differential Interferometer to the Measurement of Heat Transfer Coefficients, *Rev. Sci. Instrum.,* vol. 42, pp. 337–340, 1971.
10. Black, W. Z., and Norris, J. K., Interferometric Measurement of Fully Turbulent Free Convective Heat Transfer Coefficients, *Rev. Sci. Instrum.,* vol. 45, pp. 216–218, 1974.
11. Black, W. Z., and Norris, J. K., The Thermal Structure of Free Convection Turbulence from Inclined Isothermal Surfaces and Its Influence on Heat Transfer, *Int. J. Heat Mass Transfer,* vol. 18, pp. 43–50, 1975.
12. Boyd, R. D., An Experimental Study of the Natural Convection in a Horizontal Annulus with Irregular Boundaries, in *Natural Convection in Enclosures,* ed. K. E. Torrance and I. Catton, *ASME, HTD,* vol. 8, pp. 89–96, 1980.
13. Bugg, C. E., "The Future of Protein Crystal Growth, *J. Cryst. Growth,* vol. 76, pp. 535–544, 1986.
14. Bühler, K., Kirchartz, K. R., and Oertel H., Jr., Steady Convection in a Horizontal Fluid Layer, *Acta Mech.,* vol. 31, pp. 155–171, 1979.
15. Bühler, K., and Oertel H., Jr., Thermal Cellular Convection in Rotating Rectangular Boxes, *J. Fluid Mech.,* vol. 114, pp. 261–282, 1982.
16. Carr, W. W., and Black, W. Z., Interferometric Flow Visualization of Free Convection from a Horizontally Vibrating Isothermal Cylinder, *Int. J. Heat Mass Transfer,* vol. 18, pp. 583–587, 1975.
17. Chu, T. Y., and Goldstein, R. J., Turbulent Convection in a Horizontal Layer of Water, *J. Fluid Mech.,* vol. 60, pp. 141–159, 1973.
18. DeLucas, L. J., Suddath, F. L., Snyder, R., Broom, M. B., Pusey, M., Yost, V., Herren, B., Carter, D., Nelson, B., Meehan, E. J., McPherson, A., and Bugg, C. E., Preliminary Investigations of Protein Crystal Growth using the Space Shuttle, *J. Cryst. Growth,* vol. 76, pp. 681–693, 1986.
19. Diaz, L. A., and Viskanta, R., Visualization of the Solid–Liquid Interface Morphology Formed by Natural Convection During Melting of a Solid from Below, *Int. Comm. Heat Mass Transfer,* vol. 11, pp. 35–43, 1984.
20. Dietsche, C., and Müller, U., Influence of Bénard Convection on Solid–Liquid Interfaces, *J. Fluid Mech.,* vol. 161, pp. 249–268, 1985.
21. Dubois, M., and Bergé, P., "Experimental Evidence For The Oscillators in a Convective Biperiodic Regime," *Phys. Lett.,* vol. 76A, pp. 53–56, 1980.
22. Dubois, M., and Bergé, P., Instabilities de Couche Limite dans un Fluide en Convection, *J. Phys. (Paris),* vol. 42, pp. 167–174, 1981.
23. Ecker, A., A Two Wavelength Holographic Technique for Simultaneous Measurement of Temperatures and Concentration during the Solidification of Two Component Systems, *Proc. AIAA 25th Aerosp. Sci. Mtng., Reno, Nev.,* January 12–15, 1987.
24. Ecker, A., personal communication, 1987.
25. Eckert, E. R. G., Drake, R. M., Jr., and Soehngen, E., Manufacture of a Zehnder–Mach Interferometer, Wright-Patterson AFB, Tech. Rep. 5721, ATI-34235, 1948.
26. Eckert, E. R. G., and Soehngen, E. E., Studies on Heat Transfer in Laminar Free Convection with the Mach–Zehnder Interferometer, Air Force Tech. Rep. 5747, ATI-44580, 1948.
27. Eckert, E. R. G., and Soehngen, E., Interferometric Studies on the Stability and Transition to Turbulence of a Free Convection Boundary Layer, *Proc. Gen. Discuss. Heat Transfer, Inst. Mech. Engs., ASME, London,* 1951.
28. Farhadieh, R., and Tankin, R. S., Interferometric Study of Two-dimensional Bénard Convection Cells, *J. Fluid Mech.,* vol. 66, pp. 739–752, 1974.
29. Farrell, P. V., Springer, G. S., and Vest, C. M., Heterodyne Holographic Interferometry: Concentration and Temperature Measurements in Gas Mixtures, *Appl. Opt.,* vol. 21, pp. 1624–1627, 1982.
30. Faw, R. E., and Dullforce, T. A., Holographic Interferometry Measurement of Convective Heat Transport beneath a Heated Horizontal Circular Plate in Air, *Int. J. Heat Mass Transfer,* vol. 25, pp. 1157–1166, 1982.
31. Feuerbacher, B., Hamacher, H., and Naumann, R. J., *Materials Sciences in Space,* pp. 281–283, Springer-Verlag, New York, 1986.
32. Goldstein, R. J., Optical Systems For Flow Measurement, in *Fluid Mechanics Measurements,* ed. R. J. Goldstein, pp. 377–422, Hemisphere, Washington, D.C., 1983.
33. Goldstein, R. J., Optical Techniques for Temperature Measurement, in *Measurements in Heat Transfer,* ed. E. R. G. Eckert and R. J. Goldstein, pp. 241–293, Hemisphere, Washington, D.C., 1976.
34. Goldstein, R. J., and Chu, T. Y., Thermal Convection in a Layer of Air, *Prog. Heat Mass Transfer,* vol. 2, pp. 55–75, 1969.
35. Goldstein, R. J., and Eckert, E. R. G., The Steady and Transient Free Convection Boundary Layer on a Uniformly Heated Vertical Plate, *Int. J. Heat Mass Transfer,* vol. 1, pp. 208–218, 1960.
36. Grange, B. W., Viskanta, R., and Stevenson, W. H., Interferometric Observation of Thermohaline Convection During Freezing of Saline Solution, *Lett. Heat Mass Transfer,* vol. 4, pp. 85–92, 1977.
37. Grigull, U., Einige Optische Eigenschaften Thermischer Grenzschichten, *Int. J. Heat Mass Transfer,* vol. 6, pp. 669–679, 1963.
38. Hatfield, D. W., and Edwards, D. K., Edge and Aspect Ratio Effects on Natural Convection from the Horizontal Heated Plate Facing Downwards, *Int. J. Heat Mass Transfer,* vol. 24, pp. 1019–1024, 1981.
39. Hauf, W., and Grigull, U., Optical Methods in Heat Transfer, in *Advances in Heat Transfer,* ed. J. P. Hartnett and T. F. Irvine, Jr., vol. 6, pp. 133–366, Academic, New York, 1970.
40. Heflinger, L. O., Wuerker, R. F., and Brooks, R. E., Holographic Interferometry, *J. Appl. Phys.,* vol. 37, pp. 642–649, 1966.
41. Howes, W. L., Rainbow Schlieren and Its Applications, *Appl. Opt.,* vol. 23, no. 14, pp. 2449–2460, 1984.
42. Howes, W. L., Rainbow Schlieren vs. Mach–Zehnder In-

terferometry: A Comparison, *Appl. Opt.*, vol. 24, pp. 816–821, 1985.
43. Jakob, M., *Heat Transfer*, vol. 1, p. 493, Wiley, New York, 1959.
44. Johnston, M. H., and Owen, R. B., Optical Observations of Unidirectional Solidification in Microgravity, *Metall. Trans.*, vol. 14A, pp. 2163–2167, 1983.
45. Kapur, D. N., and Macleod, N., The Determination of Local Mass-Transfer Coefficients by Holographic Interferometry-I, *Int. J. Heat Mass Transfer*, vol. 17, pp. 1151–1162, 1974.
46. Kennard, R. B., An Optical Method for Measuring Temperature Distribution and Convective Heat Transfer, *J. Res. Nat. Bur. Stand.*, vol. 8, pp. 787–805, 1932.
47. Kim, D. M., and Viskanta, R., Study of the Effects of Wall Conductance on Natural Convection in Differently Oriented Square Cavities, *J. Fluid Mech.*, vol. 144, pp. 153–176, 1984.
48. Kirchartz, K. R., Srulijes, J., and Oertel, H., Jr., "Steady and Time-dependent Rayleigh-Bénard Convection under the Influence of Shear Flows, *Adv. Space Res.*, vol. 3, pp. 19–22, 1983.
49. Koster, J. N., Heat Transfer in Vertical Gaps, *Int. J. Heat Mass Transfer*, vol. 25, pp. 426–428, 1982.
50. Koster, J. N., Time Dependent Heat Transfer in Hele–Shaw Slots, *Int. Comm. Heat Mass Transfer*, vol. 12, pp. 159–167, 1985.
51. Koster, J. N., Ehrhard, P., and Müller, U., Nonsteady End Effects in Hele–Shaw Cells, *Phys. Rev. Lett.*, vol. 56, pp. 1802–1804, 1986.
52. Koster, J. N., and Müller, U., Free Convection in Vertical Slots, *Natural Convection in Enclosures*, ed. K. E. Torrance and I. Catton, *ASME, HTD*, vol. 8, pp. 27–30, 1980.
53. Koster, J. N., and Müller, U., Time-dependent Convection in Vertical Slots, *Phys. Rev. Lett.*, vol. 47, pp. 1599–1602, 1981.
54. Koster, J. N., and Müller, U., Free Convection in Vertical Gaps, *J. Fluid Mech.*, vol. 125, pp. 429–451, 1982.
55. Koster, J. N., and Müller, U., Visualization of Free Convective Flow in Hele–Shaw Cells, in *Flow Visualization III: Proceedings of the Third International Symposium on Flow Visualization*, ed. W.-J. Yang, pp. 738–742, Hemisphere, Washington, D.C., 1985.
56. Koster, J. N., and Müller, U., Oscillatory Convection in Vertical Slots, *J. Fluid Mech.*, vol. 139, pp. 363–390, 1984.
57. Kroes, R. L., and Reiss, D., The Pyroelectric Properties of TGS for Application in Infrared Detection, NASA Tech. Memo 82394, January 1981.
58. Lewis, W. T., Incropera, F. P., and Viskanta, R., Interferometric Study of Stable Salinity Gradients Heated from Below or Cooled from Above, *J. Fluid Mech.*, vol. 116, pp. 411–430, 1982.
59. Lewis, W. T., Incropera, F. P., and Viskanta, R., Interferometric Study of Mixing Layer Development in a Laboratory Simulation of Solar Pond Conditions, *Sol. Energy*, vol. 28, pp. 389–401, 1982.
60. Mayinger, F., and Lubbe, D., Ein Tomographisches Messverfahren und Seine Anwendung auf Mischvorgange und Stoffaustausch, *Warme Stoffübertrag.*, vol. 18, pp. 49–59, 1984.
61. Mayinger, F., Nordman, D., and Panknin, W., Holographische Untersuchungen zum Unterkühlten Sieden, *Chem. Ing. Tech.*, vol. 46, p. 209, 1974.
62. Mayinger, F., and Panknin, W., Holography in Heat and Mass Transfer, in *Heat Transfer 1974: Proceedings of the Fifth International Heat Transfer Conference, Tokyo, September 1974*, vol. 6, pp. 28–43, Hemisphere, Washington, D.C., 1974.
63. McCay, M. H., Univ. of Tennessee Space Institute, Tullahoma.
64. Meharnon, F. R., Muller, R. H., and Tobias, C. W., Light-Deflection Errors in the Interferometry of Electrochemical Mass Transfer Boundary Layers, *J. Electrochem. Soc.*, vol. 122, pp. 59–64, 1975.
65. Merzkirch, W., *Flow Visualization*, Academic, New York, 1974.
66. Meyer, B. A., El-Wakil, M. M., and Mitchell, J. W., An Interferometric Investigation of Heat Transfer in Honeycomb Collector Cells, in *SUN, Mankind's Future Source of Energy*, ed. F. De Winter and M. Cox, vol. 2, pp. 956–959, Pergamon, New York, 1978.
67. Monti, R., and Russo, G. P. Nonintrusive Methods for Temperature Measurements in Liquid Zones in Microgravity Environments, *Acta Astronaut.*, vol. 11, pp. 543–551, 1984.
68. Nakaike, Y., Tadenuma, Y., Sato, T., and Fujinawa, K., An Optical Study of Interfacial Turbulence in a Liquid–Liquid System, *Int. J. Heat Mass Transfer*, vol. 14, pp. 1951–1961, 1971.
69. Naumann, R. J., and Herring, H. W., *Materials Processing in Space: Early Experiments*, NASA SP-443, 1980.
70. Nusselt, W., and Jürgens, W., Das Temperaturfeld über Einer Lotrecht Stehenden Geheizten Platte, *Z. VDI*, vol. 72, p. 597, 1928.
71. Oertel, H., Jr., and Bühler, K., A Special Interferometer used for Heat Convection Investigations, *Int. J. Heat Mass Transfer*, vol. 21, pp. 1111–1115, 1978.
72. Oshima, Y., Experimental Studies of Free Convection in a Rectangular Cavity, *J. Phys. Soc. Jpn.*, vol. 30, pp. 872–882, 1971.
73. Owen, R. B., Optical Measurements and Tests Performed in a Low-Gravity Environment *Opt. Eng.*, vol. 20, pp. 634–638, 1981.
74. Owen, R. B., Interferometry and Holography in a Low-Gravity Environment, *Appl. Opt.*, vol. 21, pp. 1349–1355, 1982.
75. Owen, R. B., Giarratano, P. J., and Arp, V. D., The Use of Holography to Measure Heat Transfer in High Gradient Systems, Opt. Soc. Am. Ann. Mtng., Seattle, Oct. 19–24, 1986; Abstract: *J. Opt. Soc. Am. A*, vol. 2A, ser. 2, no. 12, p. P113, 1986.
76. Owen, R. B., and Johnston, M. H., Optical Observations of Unidirectional Solidification and Related Fluid Parameters in Microgravity, *Opt. Lasers Eng.*, vol. 5, pp. 95–108, 1984.
77. Owen, R. B., and Kroes, R. L., Holography on the Spacelab 3 Mission, *Opt. News*, vol. 11, pp. 12–16, 1985.
78. Owen, R. B., Kroes, R. L., and Witherow, W. K., Results and Further Experiments using Spacelab Holography, *Opt. Lett.*, vol. 11, pp. 407–409, 1986.
79. Panknin, W., Einige Techniken und Anwendungen der Hol-

ographischen Durchlichtinterferometrie, *CZ-Chem. Tech.*, vol. 3, pp. 219–225, 1974.
80. Pera, L., and Gebhart, B., On the Stability of Laminar Plumes: Some Numerical Solutions and Experiments, *Int. J. Heat Mass Transfer,* vol. 14, pp. 975–984, 1971.
81. Pera, L., and Gebhart, B., Laminar Plume Interaction, *J. Fluid Mech.,* vol. 68, pp. 259–271, 1975.
82. Polymeropoulos, C. E., and Gebhart, B., Incipient Instability in Free Convection Laminar Boundary Layers, *J. Fluid Mech.,* vol. 30, pp. 225–239, 1967.
83. Poteet, W. M., and Owen, R. B., Compact Field Color Schlieren System for Use in Microgravity Materials Processing, *Opt. Eng.,* vol. 25, pp. 841–845, 1986.
84. Reisbig, R. L., and Reifel, A. J., The Application of Interferometric Holography for Measuring Complex Temperature Fields in a Gas, *Adv. Thermal Conduct., 13th Int. Conf. Thermal Conduct.,* ed. R. L. Reisbig and H. J. Sauer, Jr., pp. 310–317, 1973.
85. Riley, C., and Coble, H. D., "Electrodeposition in Microgravity," Final Rep., NASA Contract NAS8-33812, June 1982.
86. Ross, P. A., Application of Two-Wavelength Interferometry to the Study of a Simulated Drop of Fuel, Ph.D. dissertation, University of Wisconsin, Madison, 1960.
87. Ross, P. A., and El-Wakil, M. M., A Two-Wavelength Interferometric Technique for the Study of Vaporation and Combustion of Fuels, *Prog. Astronaut. Rocketry, vol. 2,* pp. 265–298, 1960.
88. Sanchez, V., and Clifton, M., Determination du Transfert de Matiere par Interferometrie Holographique dans un Motif Elementaire d'un Electrodialyseur, *J. Chim. Phys.,* vol. 77, pp. 421–426, 1980.
89. Santoro, R. J., Semerjian, H. G., Emmerman, P. J., and Goulard, R., Optical Tomography for Flow Field Diagnostics, *Int. J. Heat Mass Transfer,* vol. 24, pp. 1139–1150, 1981.
90. Schetz, J. A., and Eichorn, R., Natural Convection with Discontinuous Wall Temperature Variations, *J. Fluid Mech.,* vol. 18, pp. 167–176, 1964.
91. Shardin, H., Theorie und Anwendung des Mach-Zehnderschen Interferenz-Refraktometers, *Z. Instrumentenkd.,* vol. 53, pp. 396, 424, 1933.
92. Spectron Development Laboratories, Inc.
93. Steinberner, U., and Mayinger, F., Untersuchung von Temperatur-grenzschichten mit Hilfe der Holographischen Interferometrie, Zweite Fachtagung der Fachgruppe "Thermound Fluiddynamik" der Kerntechnischen Gesellschaft, Hannover, February 28–March 2, 1977.
94. Sweeney, D. W., and Vest, C. M., Measurement of Three-dimensional Temperature Fields above Heated Surfaces by Holographic Interferometry, *Int. J. Heat Mass Transfer,* vol. 17, pp. 1443–1454, 1974.
95. TAI Corporation, Huntsville, Ala.
96. Trolinger, J. D., Tan, H., Modarress, D., and Salehpour, A., Automated Holographic Data Reduction System Manual, Spectron Development Lab., SDL 87-51041, March 1987.
97. Vest, C., *Holographic Interferometry,* Wiley, New York, 1979.
98. Viskanta, R., and Behnia, M., Experimental and Analytical Study of Heat Transfer and Mixing in Thermally Stratified Buoyant Flows, *Int. J. Heat Mass Transfer,* vol. 25, pp. 847–861, 1982.
99. Viskanta, R., Karalis, A., and Behnia, M., Effects of a Barrier on Temperature Structure and Mixing in Thermally Stratified Water Cooled from Above, *Wärme Stoffübertrag.,* vol. 11, pp. 229–239, 1978.
100. Weinberg, F. J., *Optics of Flames,* Butterworths, London, 1963.
101. Weise, R., Warmeubergang durch Freie Konvektion an Quadratischen Platten, *Forsch. Ingenieurwes.,* vol. 6, pp. 281–292, 1935.
102. Wilkie, D., and Fisher, S. A., Measurement of Temperature by Mach–Zehnder Interferometry, *Proc. Inst. Mech. Eng.,* vol. 178, pp. 461–470, 1963.
103. Williams, J. E., and Peterson, R. W., The Application of Holographic Interferometry to the Visualization of Flow and Temperature Profiles in a MOCVD Reactor Cell, *J. Cryst. Growth,* vol. 77, pp. 128–135, 1986.
104. Witherow, W. K., Reconstruction Techniques of Holograms from Spacelab 3, *Appl. Opt.,* vol. 26, pp. 2465–2473, 1987.
105. Witherow, W. K., and Facemire, B. R., Optical Studies of a Binary Miscibility Gap System, *J. Colloid Interface Sci.,* vol. 104, pp. 185–192, 1985.
106. Wuerker, R. F., Heflinger, L. O., Flannery, J. V., and Kassel A., Holography on Space Shuttle, *Recent Adv. Hologr., SPIE Proc.,* vol. 215, pp. 76–84, 1980.
107. Zinnes, A. E., The Coupling of Conduction with Laminar Natural Convection from the Vertical Flat Plate with Arbitrary Surface Heating, *J. Heat Transfer,* vol. 92, no. 1, pp. 528–535, 1970.
108. Lord, A., Hallett, J., Keller, Z., Owen, B., Influence of Convective Motions on the Growth of Crystal in Solution, Proc. 7th Conf. Crystal Growth, p. 8, 1984, Fallenleaf Lake, California, May 31–June 3, 1983.

Chapter 31

Double-Diffusive Convection

Josef Tanny and A. B. Tsinober

I INTRODUCTION

Double-diffusive convection can arise in a fluid that contains opposing gradients of two buoyancy components with different molecular diffusivities. Although the fluid is initially statically stable, in certain conditions the fluid can become unstable, due to the release of potential energy stored in the unstably distributed component, and a convective motion is produced (for reviews see [1–4]).

In order to describe the physical mechanism of double-diffusive convection, let us imagine a fluid in which vertical gradients of temperature and salinity are present. The temperature and salinity increase with depth in such a way that the resulting density also increases with depth, and the fluid is statically stable. If a fluid particle is displaced for any reason, some distance upward, it reaches a colder and less salty surrounding. Since the coefficient of thermal diffusivity is much larger than that of the salt diffusivity, the particle loses heat, but almost no salt, and soon reaches a level of neutral buoyancy. Due to its motion the particle overshoots this level, continues to lose heat, and eventually reaches a level where its density is greater than that of its surroundings. A downward motion starts, and the particle is set into an oscillatory motion. Under certain conditions these oscillations may be amplified to produce unstable motion. An interface formed in this situation is called a ''diffusive interface,'' since all transport across it takes place by molecular diffusion. If the gradients are distributed inversely so that the temperature and the salinity decrease with depth and the net density increases with depth, a small upward displacement of a fluid particle produces a steady convective motion known as ''salt fingers'' [5, 6].

In the two examples just described, only vertical gradients were present in the fluid. The existence of horizontal gradients in the fluid, in addition to the basic vertical gradients, may also lead to unstable motions. If a statically stable stratified fluid contains opposing horizontal gradients of two diffusive components, a small lateral displacement of a fluid particle produces an unbalanced buoyancy force that brings the fluid into an unstable steady convective motion [7, 8].

Like the well-known single-component natural convection, double-diffusive convection stems from density variations in the fluid. These variations may be due to temperature or solute-concentration gradients existing in the fluid. A necessary condition for double-diffusive convection is that the difference between the coefficients of molecular diffusion of the two components be large enough. Therefore, besides the heat–salt system, many laboratory experimental investigations utilize salt–sugar or heat–sugar systems, or any other combination of two appropriate components.

It should be noted that normally double-diffusive convection exists in low–Reynolds number flows. For large Reynolds numbers the flow may become turbulent and the difference between the coefficients of molecular diffusion of the two components may play a minor role in the flow. In certain situations, when laminar flow exists outside the boundaries of a turbulent region, double-diffusive convection is also significant. This property of double-diffusive convection implies that the phenomenon may be found mainly in fluids with relatively large viscosity (i.e., liquids and not gases) and involves relatively small velocities in laminar flow. These limitations should be kept in mind when one is choosing a visualization technique for double-diffusive convection.

As mentioned earlier, density differences are the basic cause of double-diffusive convection. For this reason flow visualization of these phenomena utilized, in many circumstances, optical methods that are capable of detecting variations in the refractive index, due to local density gradients existing in the fluid. Of these methods the most widely used is the shadowgraph technique, mainly because of its relative simplicity. The interferometry and schlieren methods were rarely used due to the complexity involved in their operation. The "color polarigraph," a simple and very efficient optical technique suggested by Ruddick [9], is also described.

Another group of methods widely used involves marking the fluid. The markers are usually water-soluble dyes, which are illuminated by a wide beam or a thin sheet of light. Fluorescent dyes, which provide high contrast between dye and background, are mostly used, but simple soluble dyes of different kinds are also used. Marking the fluid by an electrochemical method using a pH indicator is also done.

The third group of methods consists of adding particles to the fluid and photographing them with relatively long time exposures. In this method a particle streakline picture is obtained, from which details of the flow field can be extracted. Here, special care should be taken regarding the difference between the densities of the particles and the fluid.

This chapter is divided according to the three groups of methods just described. In Sec. II the optical methods are described; Secs. III and IV are devoted to the marker and particle methods respectively. Section V is a short summary.

II OPTICAL METHODS

The classical optical methods, namely, the shadowgraph, schlieren, and interferometry techniques, utilize the fact that the refractive index of a fluid is a function of its density. This means that a broad beam of light passing through a fluid in which the density is not uniform is refracted at different angles at different locations, generating an image that consists of brighter and darker regions.

These three methods differ in their sensitivity to variations in density. Interferometry is sensitive to changes in the density, the schlieren technique is sensitive to changes in the first spatial derivative of the density, and the shadowgraph is sensitive to changes in the second spatial derivative of density. It is thus obvious that the shadowgraph and schlieren techniques are used mainly to obtain qualitative information whereas quantitative data on the density field can be extracted from the interferometer images.

A The Shadowgraph Technique

When a linear stable-solute gradient is heated uniformly from below, convective layers are formed successively from the bottom of the tank. These layers are well mixed and are separated by relatively thin-density interfaces across which large density gradients exist. Figure 1 [10] is a typical shadowgraph in which the location of the horizontal interfaces is easily identified. The sequence of photographs demonstrates the formation of new layers at the top and the merging of the bottom layer with adjacent layers above it.

When the stable-solute gradient is heated from the side, it is also split into a system of convective layers separated by sharp density interfaces. The shadowgraph in Fig. 2 shows double-diffusive layers formed when heat is applied uniformly at the side wall of a tank containing a linear salt gradient. This shadowgraph was produced in our laboratory using a 1-W argon laser. The laser beam was expanded onto a spherical mirror, 28 cm in diameter at its focal length, thus generating a parallel circular beam of the same diameter. The beam passed horizontally through the test fluid and then was allowed to fall on a square back-projection screen behind which a camera was located.

These two examples show that besides the qualitative information obtained from the shadowgraphs, some quantitative data on the thickness of the layers and their rate of development may be extracted from such pictures. It should be emphasized that since the shadowgraph (and in fact the schlieren and interferometry too) provides an integrated image, such data can be extracted only from two-dimensional and, in certain situations, from axisymmetric flows. Moreover, small departures from two-dimensionality may cause significant errors so that care should be taken in processing shadowgraphs to obtain quantitative data.

Another phenomenon visualized by shadowgraph is the formation of salt fingers. An interface consisting of vertical fingers separating convecting layers of sugar solution above salt solution is shown in Fig. 3 [11]. This picture was obtained by shining a collimated light beam horizontally through the tank and onto a screen. The fingers, formed according to the mechanism discussed in the introduction, are quite uniform and vertical; again, only information on their size and orientation can be obtained.

The shadowgraph was also used to obtain a top (instead of side) view of salt fingers. An example is shown in Fig. 4 [12]. Here the tank was illuminated from below by a 500-W projection lamp. Another plan-view shadowgraph, shown in Fig. 5 [13], is that of a diffusive interface.

Huppert and Turner [14] observed an ice block melting in a salinity gradient; the shadowgraphs they obtained (Fig. 6) also showed the rate of melting of the ice block and the shape of the block in different stages during this process.

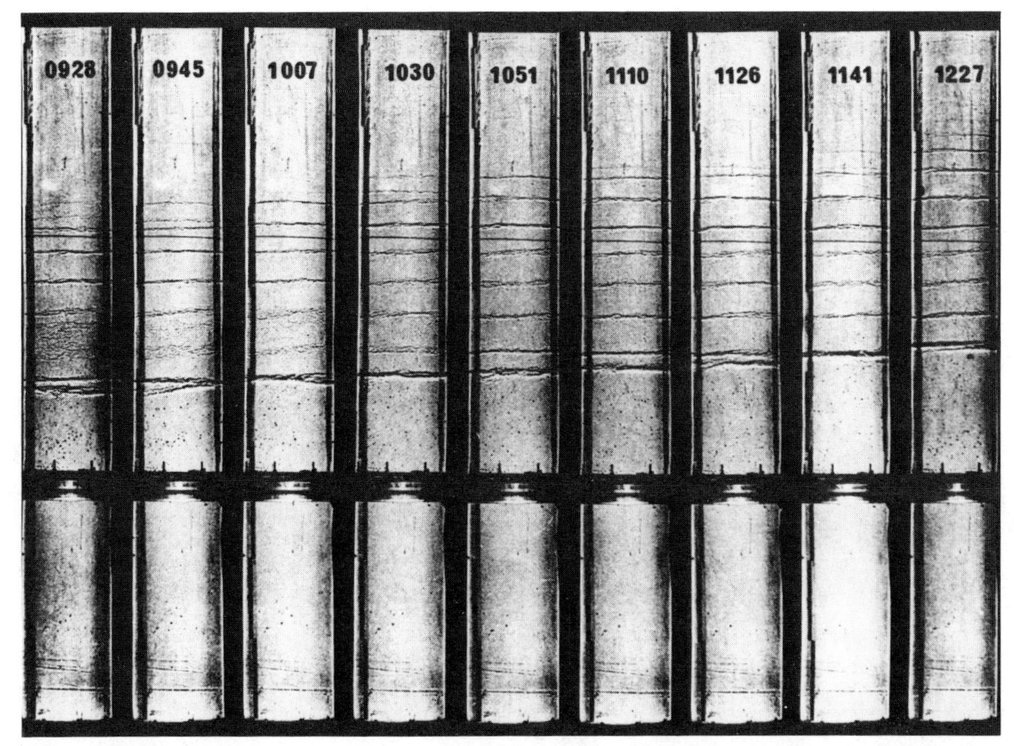

Fig. 1 Shadowgraph showing the development of layers when heat is applied at the bottom of a stable salt gradient (from Huppert and Linden [10]).

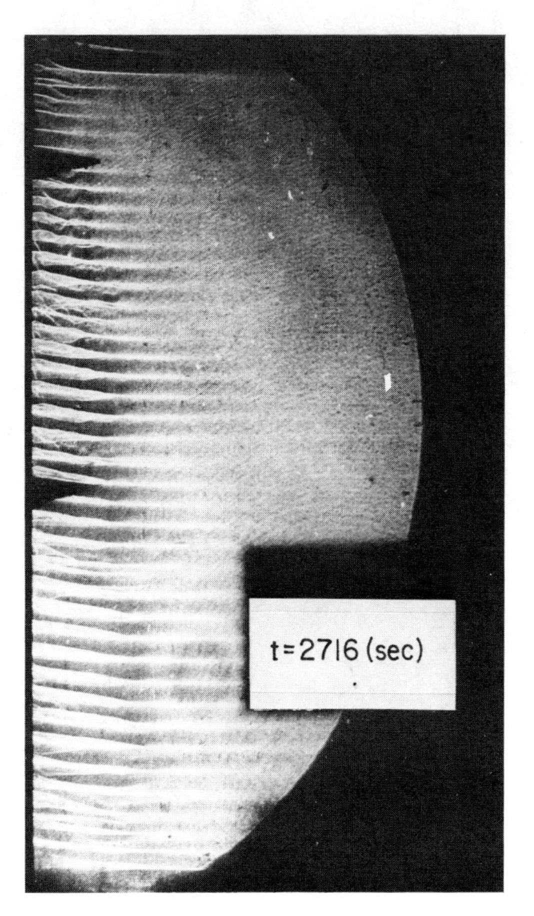

Fig. 2 Shadowgraph of layers formed by heating uniformly a stable salinity gradient from a side wall. *t*, time from initiation of heating.

Fig. 3 Shadowgraph of a thick layer of vertical salt fingers with convecting layers of sugar solution from above and salt solution from below (from Shirtcliffe and Turner [11]).

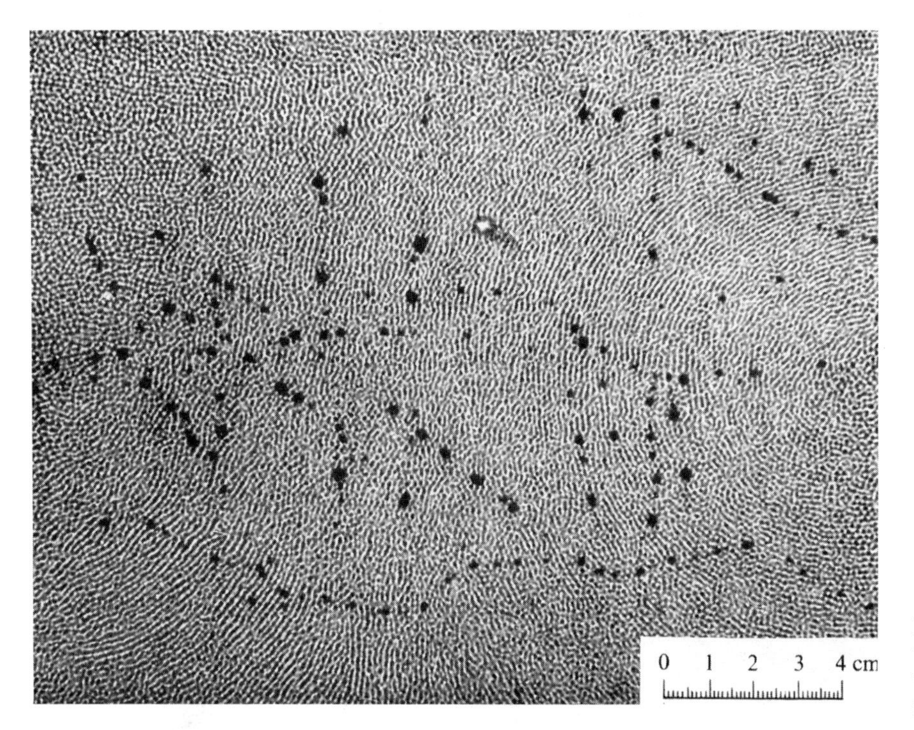

Fig. 4 Plan view shadowgraph of vertical salt fingers; the black dots are water droplets (from Chen and Sandford [12]).

Fig. 5 Plan view shadowgraph of a sugar–salt diffusive interface (from Linden and Shirtcliffe [13]).

Fig. 6 Shadowgraph of an ice block melting in a salinity gradient. Here the shadowgraph provides information on the layered structure as well as on the form of the melting block (from Huppert and Turner [14]).

The crystallization of a linear gradient of Na_2CO_3 cooled from the side is shown by the shadowgraph in Fig. 7, obtained by Chen and Turner [15]. Like the case of the melting ice blocks, the shadowgraph provides some information on the crystallization process.

Griffiths [16] visualized double-diffusive convection in a Hele–Shaw cell in order to model convection in a porous medium. In Fig. 8 is seen a shadowgraph of a diffusive interface separating two solute layers. This shadowgraph also shows the unsteady motion, with plumes intermittently appearing and breaking away from the edge of the interface.

Shirtcliffe and Turner [11] used the schlieren technique in order to study the cell structure of salt fingers. The same fingers shown in the shadowgraph in Fig. 3 were inspected by a schlieren apparatus that was oriented to produce a plan view, as shown in Fig. 10. Instead of a knife edge, Shirtcliffe and Turner used a small aperture having an adjustable diameter. With a large aperture the depth of focus was small, and it was possible to distinguish between different horizontal plane sections through the field of vertical fingers.

B The Schlieren Technique

The schlieren method is somewhat more complicated than the shadowgraph but has the advantage of high resolution and the possibility of obtaining quantitative data on the density field, with some limitations due to diffraction effects. However, in the two examples described in this section, the schlieren technique was utilized only to obtain information on the structure of the flow; no specific data on the density field were extracted.

Thangam and Chen [17] used the schlieren technique to visualize salt fingers in the surface discharge of heated saline jets. Figure 9*a* is a schlieren top view of a hot saline solution discharged on a stable solute gradient. The focal plane is about 3 cm below the water surface. Figure 9*b* is a shadowgraph side view of the same flow, presented for comparison.

C The Interferometry Technique

Of the three classical optical techniques, interferometry is the most complicated to operate but has the great advantage of enabling one to obtain quantitative data on the density field in the flow. Since in double-diffusive convection density variations are due to two components, say, temperature and salinity, it is necessary to know the distribution of one component in conjunction with the density measurements in order to obtain a complete knowledge of the flow.

We know of only one attempt to utilize this technique in double-diffusive convection, the work done by Lewis et al. [18]. They used the Mach–Zehnder interferometer in conjunction with temperature measurements by thermocouples to obtain the salinity distribution in a stable-salinity gradient heated from below or cooled from above.

Fig. 7 Shadowgraph showing crystallization process in a fluid with constant solute gradient being cooled from the side (from Chen and Turner [15]).

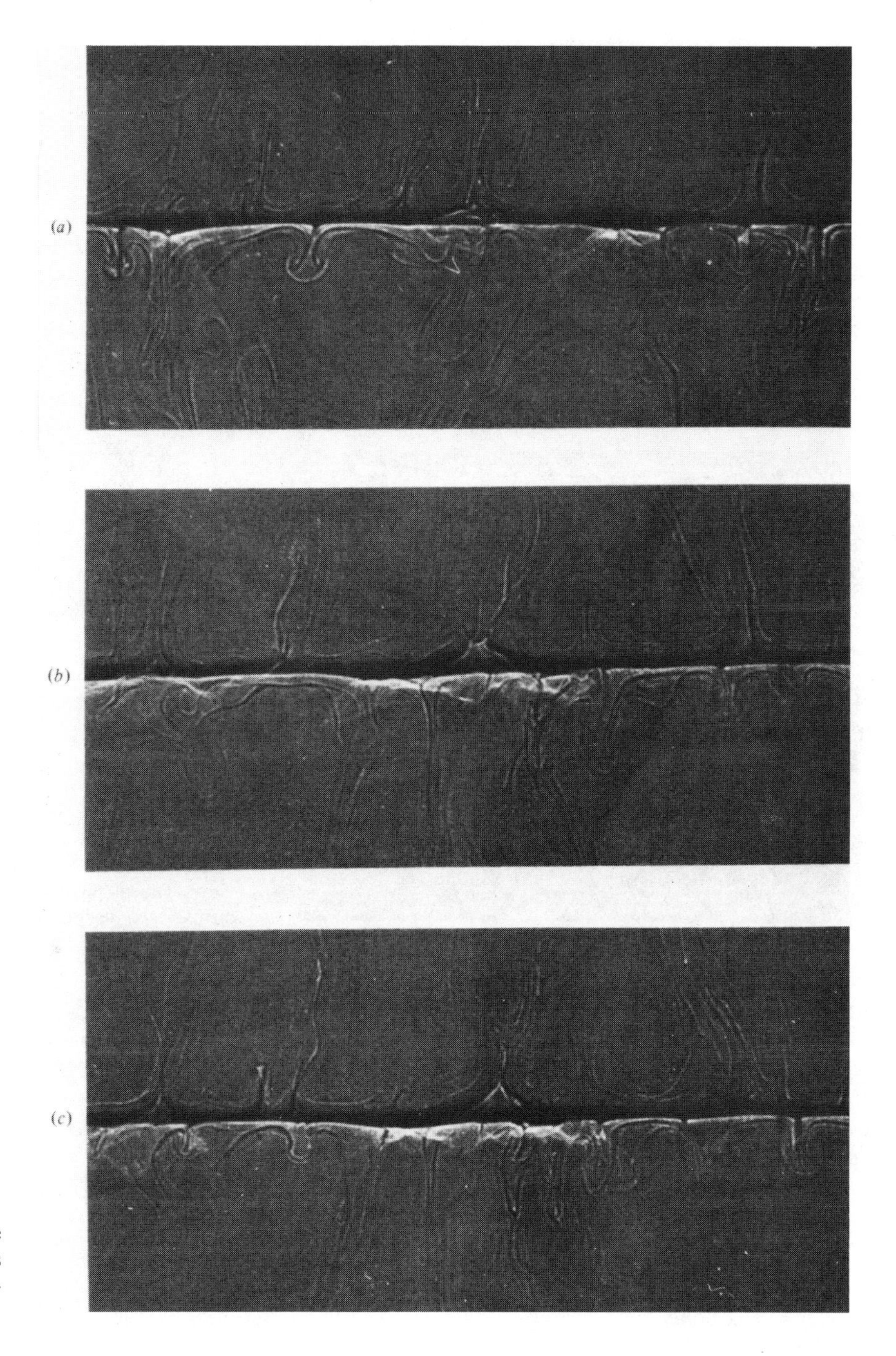

Fig. 8 Shadowgraphs of a two-solute interface in a Hele–Shaw cell; frames are at about 1-min intervals (from Griffiths [16]).

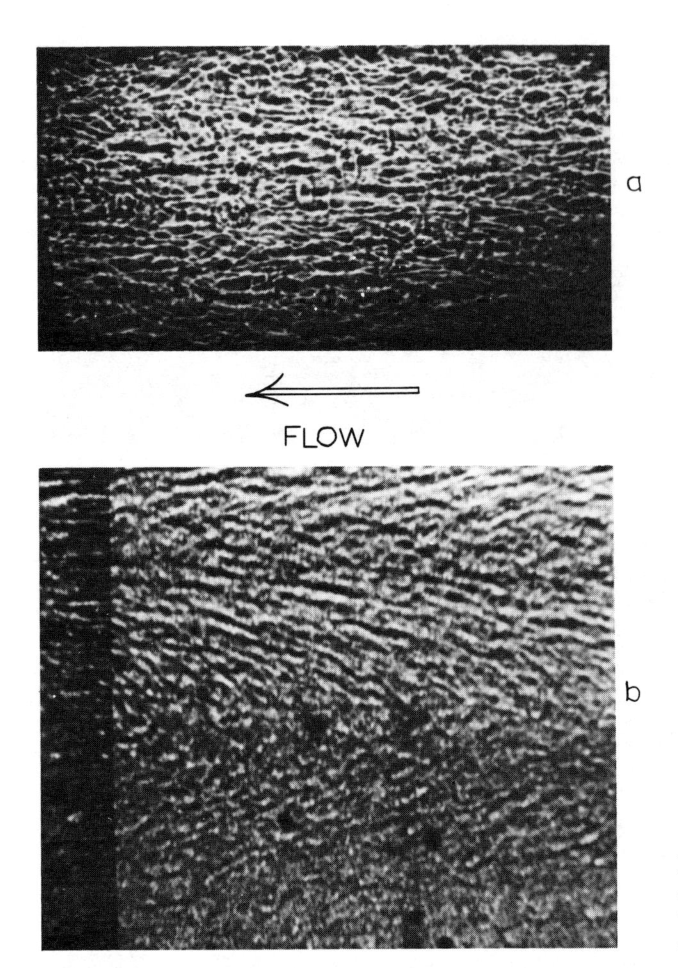

Fig. 9 Two-dimensional heated saline jet discharged on a stable solute gradient: (*a*) Schlieren top view; (*b*) shadowgraph side view. The pictures show the salt fingers formed due to double-diffusive convection (from Thangam and Chen [17]).

The light source was a 25-mW helium–neon laser (wavelength = 632.8 nm) with a beam collimated and split into two components, one passing through the test tank and the other along a reference path. A set of interferograms of a salt-stratified water layer heated from below is shown in Fig. 11.

Lewis et al. [19] also used the interferometry technique to study development of the mixing layer in a laboratory-simulated solar pond.

D The Color Polarigraph

This simple and efficient technique, suggested by Ruddick [9], is used for determining the two-dimensional distribution of sugar concentration. The method is based on the fact that optical rotation is almost linearly proportional to sugar concentration and almost unaffected by salt. Thus a double-diffusive system consisting of sugar and salt gradients is most appropriate for investigation by this technique. The experimental tank is photographed through a Polaroid analyzing filter, and an image is obtained that is color coded according to the concentration of sugar. Quantitative results can be obtained by using a calibration tank; the estimated accuracy is 1 wt.%.

An example of the resulting picture is shown in Fig. 12 (originally in color). Figure 12*a* is a polarimeter photograph showing a sugar gradient at the left and a salt gradient at the right; the two gradients are separated by a removable vertical barrier. Figure 12*b* is a polarimeter photograph 17 min 20 s after the barrier was removed, showing how the sugar solution had flowed to the right in a series of interleaving layers. The calibration tank is at the bottom; the numbers (in cm) correspond to the optical pathlength of the sugar solution.

Ruddick used three polarigraphs like that shown in Fig. 12*b*, each taken with the Polaroid analyzer rotated at a different angle, in order to build up a relatively complete map of sugar concentration.

III MARKER METHODS

Marking the fluid has usually been done by dyeing a portion of the fluid with a soluble dye and tracing the motion of

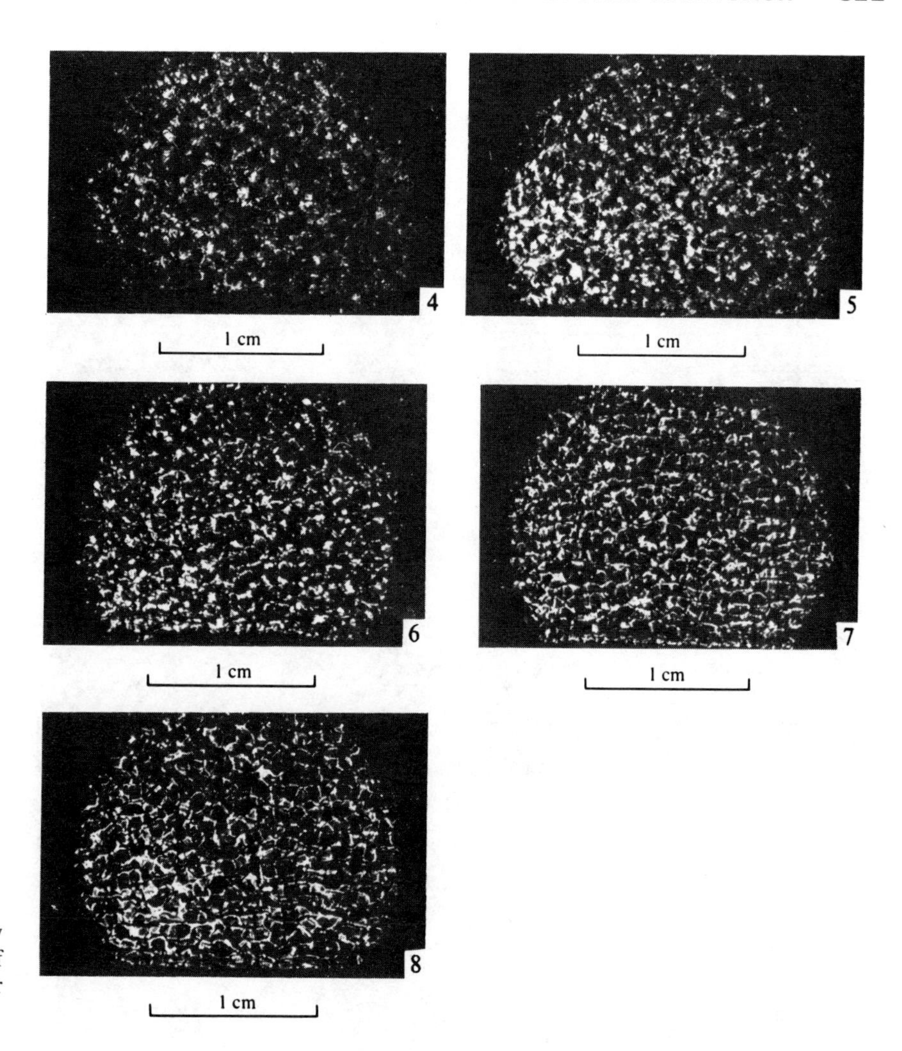

Fig. 10 Sequence of schlieren plan-view photographs showing the cell structure of salt fingers (from Shirtcliffe and Turner [11]).

the dyed fluid. Fluorescent dyes are preferred since they provide high-contrast pictures, but other dyes have also been used.

A Dyeing Techniques

Tsinober et al. [20] used Rhodamine B to visualize the flow due to a point source of heat in a stable salinity gradient. The vertical plane containing the point source was illuminated by a thin sheet of laser light produced by a 3-W argon laser. The fluid in this sheet of light was tagged by dropping small particles of the fluorescent dye. The lines of dye, which were initially vertical, deformed due to the flow, thus illustrating the structure of the flow above and around the heat source.

In Fig. 13 [20] a comparison is shown between the same flow visualized by shadowgraph (Fig. 13*a*) and by dye tracing (Fig. 13*b*). It is shown that the vortices in the double-diffusive layers are observed only in Fig. 13*b*, a result that indicates a significant advantage of the dyeing technique over the shadowgraph, which provides an integrated image with a serious lack of detail.

Another dye widely used is fluorescein. The melting of an ice block in a salinity gradient, mentioned earlier, was visualized by Huppert and Turner [14] using this dye. Figure 14 shows an ice block melting into a salinity gradient with fluorescein frozen initially into the block. By this procedure it was possible to observe the motion of the melted water or the mixture of melted water–ambient water, as distinct from the ambient water.

The fluorescein was also used to observe salt fingers, as shown in Fig. 15 [3]. This picture was produced by A. J. Faller and H. Stommel in 1960 and seems to be the first visualization of double-diffusive convection. Here a dyed salt solution was poured at the top of a stable temperature gradient, and the illumination was made through a slit from below.

Thorpe et al. [7] used potassium permanganate to visualize the formation of layers when a stable salinity gradient was heated from a side wall. Figure 16 shows the cells formed when a stable solute gradient was heated differen-

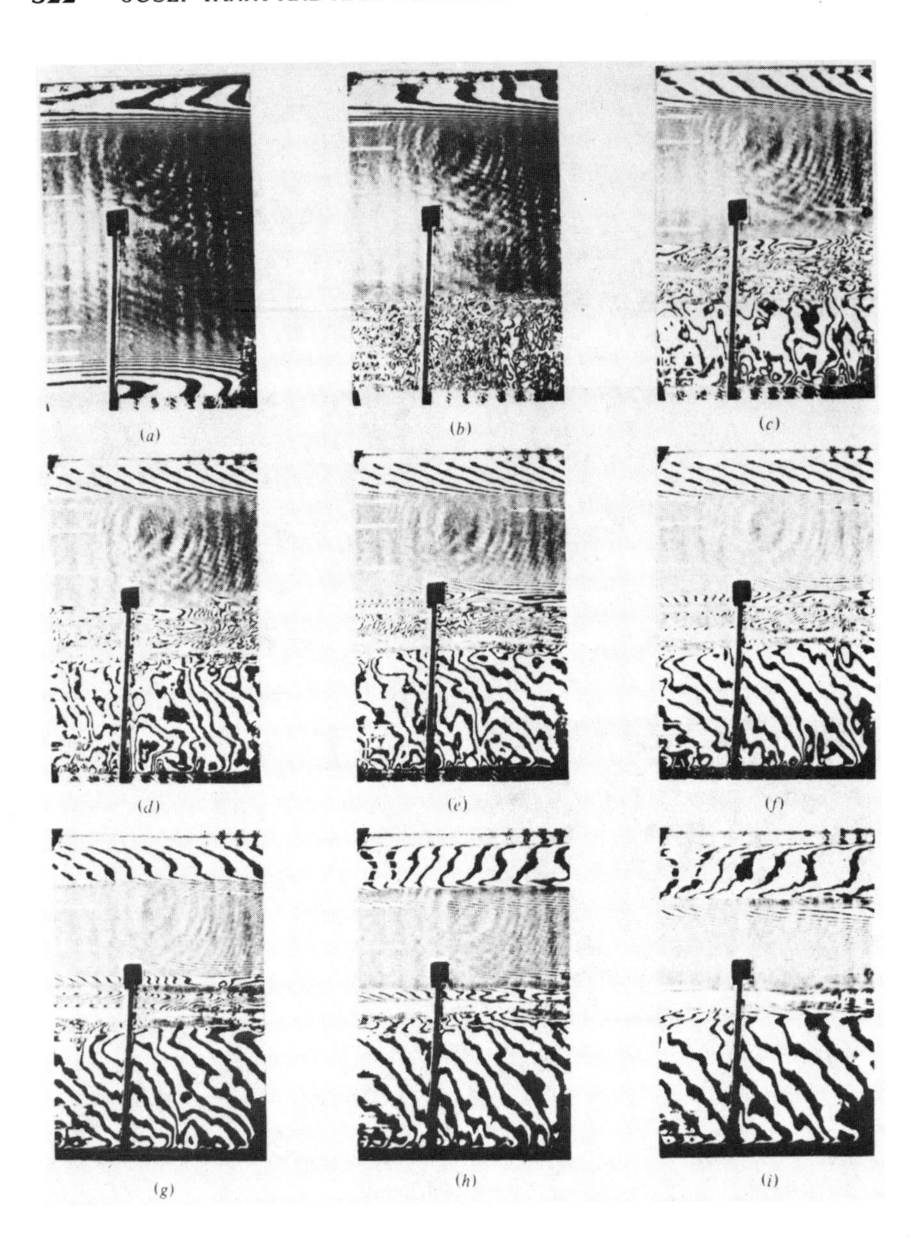

Fig. 11 Sequence of interferograms of a stable salt gradient heated from below: t = (*a*) 0 min, (*b*) 2 min, (*c*) 5 min, (*d*) 15 min, (*e*) 30 min, (*f*) 45 min, (*g*) 60 min, (*h*) 80 min, (*i*) 105 min (from Lewis et al. [18]).

tially in a vertical narrow slot. Here small crystals of potassium permanganate were suspended on thin wires in the slot in order to trace the motion. When the same phenomenon was investigated in a wide tank where only one side wall was heated, the visualization was performed by dropping small crystals of the same dye. The formation and development of the layers visualized by this technique are shown in Fig. 17. The illumination was produced by a photoflood light located behind the tank.

Care must be taken in interpreting dye-traced images, especially when mixing occurs. In this case the dyed region did not represent the location of the fluid originally dyed but the location of the mixture of dyed and ambient fluid.

B The Thymol Blue Technique

This technique, suggested by Baker [21], utilizes local pH variations to observe flows in liquids to which an amount of analytical indicator with light acidic color has been added. When the pH level is varied by electrolysis, the color of the indicator is changed, allowing observation of the flow. Hart [22] used this technique to study the stability of a linear stable-solute gradient heated differentially in a vertical narrow slot. Figure 18 shows the convective cellular motion at supercritical conditions. The local variation in pH was produced by electrolysis near the vertical wire at the middle of the slot.

(a)

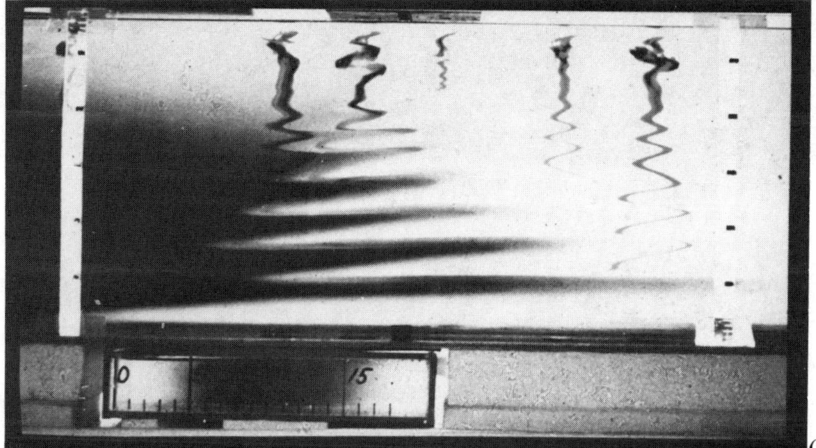

(b)

Fig. 12 Polarimeter photographs. In (*a*) a sugar gradient is on the left and a salt gradient on the right; the gradients are separated by a barrier. The picture in (*b*) was taken about 17 min after the barrier was removed and shows the interleaving layers of sugar solution flowing into the salt gradient (from Ruddick [9]).

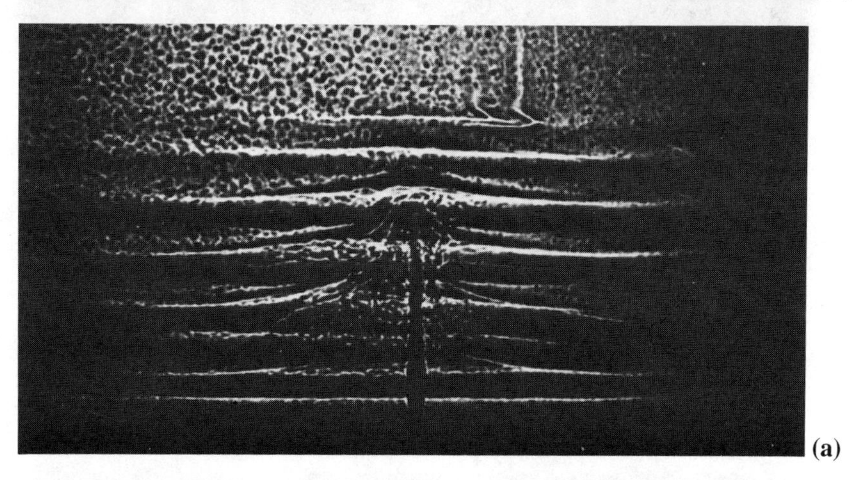

(a)

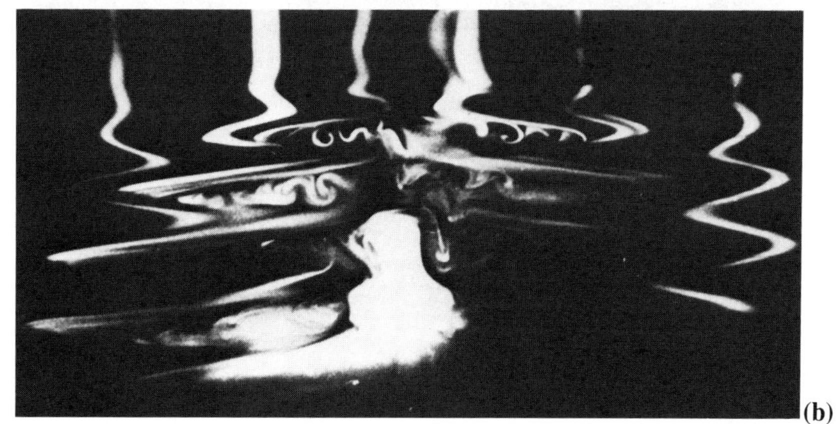

(b)

Fig. 13 Comparison of two visualization techniques: (*a*) Shadowgraph; (*b*) fluorescent dye. The flow is due to a point source of heat in a stable salinity gradient; the two photographs were taken simultaneously (from Tsinober et al. [20]).

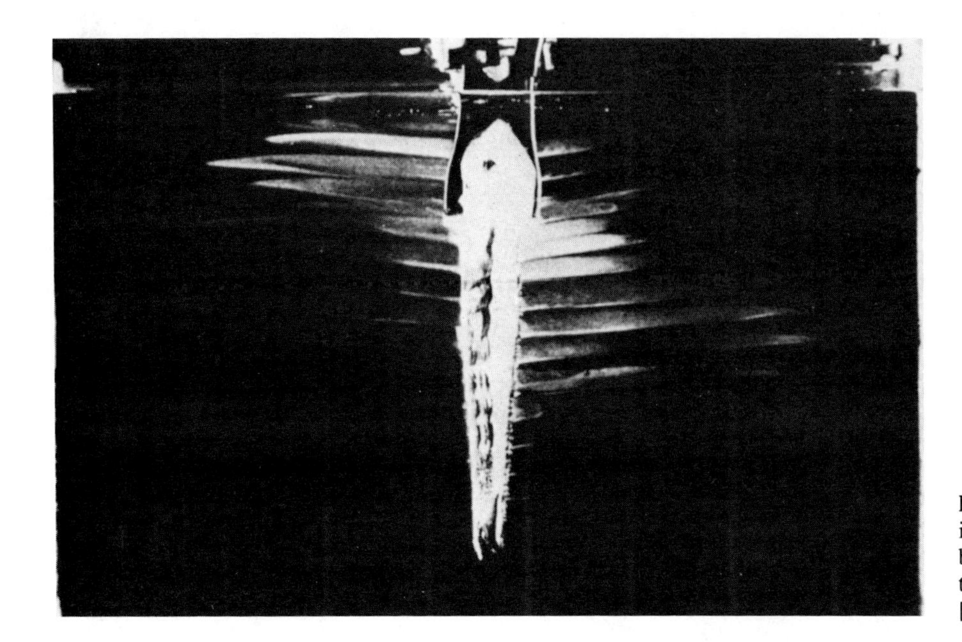

Fig. 14 Ice block melting into a salinity gradient; the flow was made visible by fluorescein frozen initially into the block (from Huppert and Turner [14]).

Fig. 15 Field of salt fingers formed by setting up a stable temperature gradient and pouring a little salt solution on top. The fingers were made visible by dyeing the salt solution with fluorescein (from Huppert and Turner [3]).

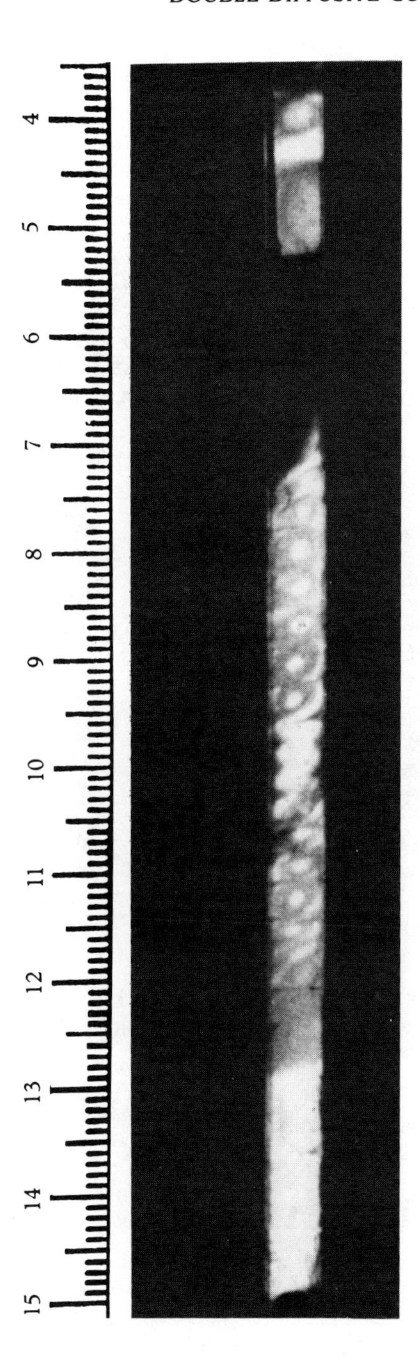

Fig. 16 Cells formed when a stable solute gradient was heated differentially in a vertical narrow slot; visualization was by potassium permanganate (from Thorpe et al. [7]).

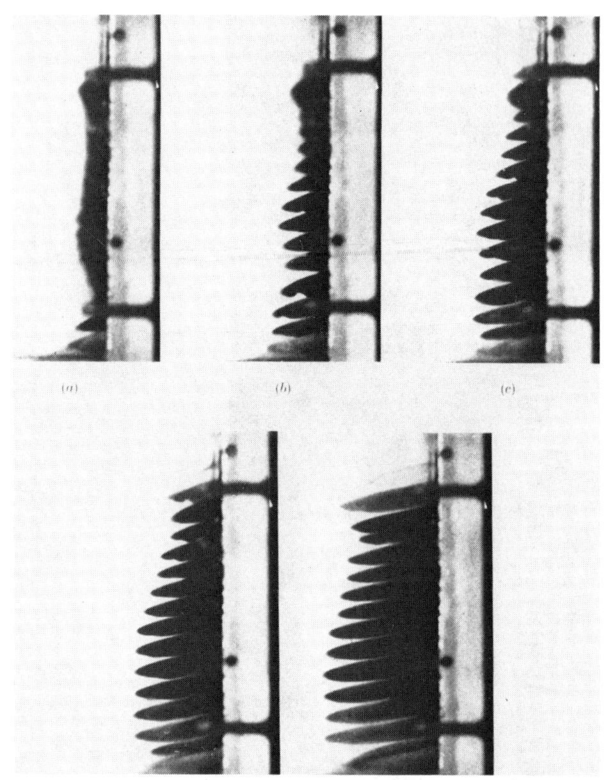

Fig. 17 Visualization of double-diffusive layers formed due to side wall heating of a stable solute gradient. The fluid near the heated wall was tagged by dropping a particle of potassium permanganate (from Thorpe et al. [7]).

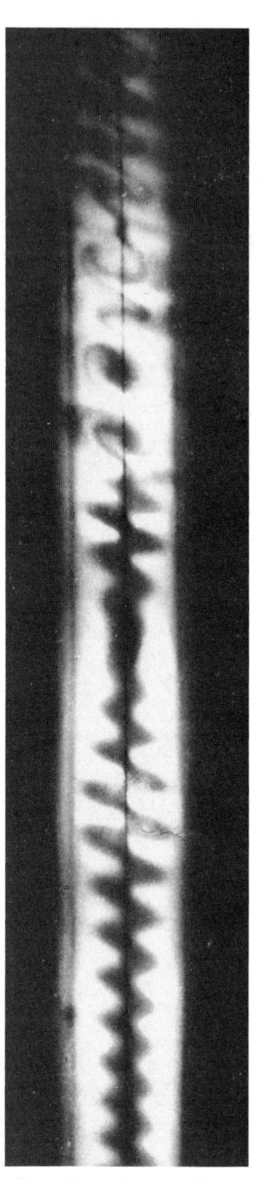

Fig. 18 Thymol blue technique used to determine the moment of cells formation when a stable solute gradient is heated through the vertical walls of a narrow slot (from Hart [22]).

The advantage of this technique is that marking of the fluid is temporary; when the pH returns to its basic level, the indicator returns to its initial color. On the other hand, tracing by this method is limited to the vicinity of metallic surfaces or wires where electrolysis is possible.

IV PARTICLE METHODS

Visualization using particles added to the fluid is very efficient in obtaining detailed data on the flow field. In particular, velocity-field and related quantities can be extracted from particle streakline photographs. In addition, qualitative information on the structure of the flow is available from such pictures.

Utilizing particles in double-diffusive convection is limited due to the low velocities involved and the density variations present in the fluid. Care should therefore be taken when choosing particles to visualize these convective flows.

Chen et al. [8] used aluminum particles to visualize flow in the developed layers formed by heating a stable salinity gradient from a side wall. The fluid containing the aluminum powder was illuminated from the top by a slit light source about $\frac{1}{4}$ in. wide. The photograph, shown in Fig. 19, indicates the shearing motion at the interfaces separating the layers and the circulating motion within the layers.

Huppert and Linden [10] used aluminum particles to observe the structure of the flow in a stable solute gradient heated uniformly from below. The streak photograph in Fig. 20 was obtained by illuminating the tank using a vertical slit of light 1 cm thick. The exposure time was 10 s.

Tsinober et al. [20] used another type of particle to visualize the flow due to a point source of heat in a stable salinity gradient. An example is shown in Fig. 21. Small PVC particles (about 4 μm dia., specific gravity 1.4) were dyed with a yellow fluorescent dye. The fluorescence was excited by a 3-W argon laser operating in the blue line. The flow was photographed through a yellow optical filter (absorbing blue, transmitting yellow), and an extremely enhanced contrast between the streaks and the background was achieved. In spite of the relatively large density of the

Fig. 19 Aluminum-particle flow visualization showing convective motion in developed layers formed due to side wall heating of a stable salt gradient (from Chen et al. [8]).

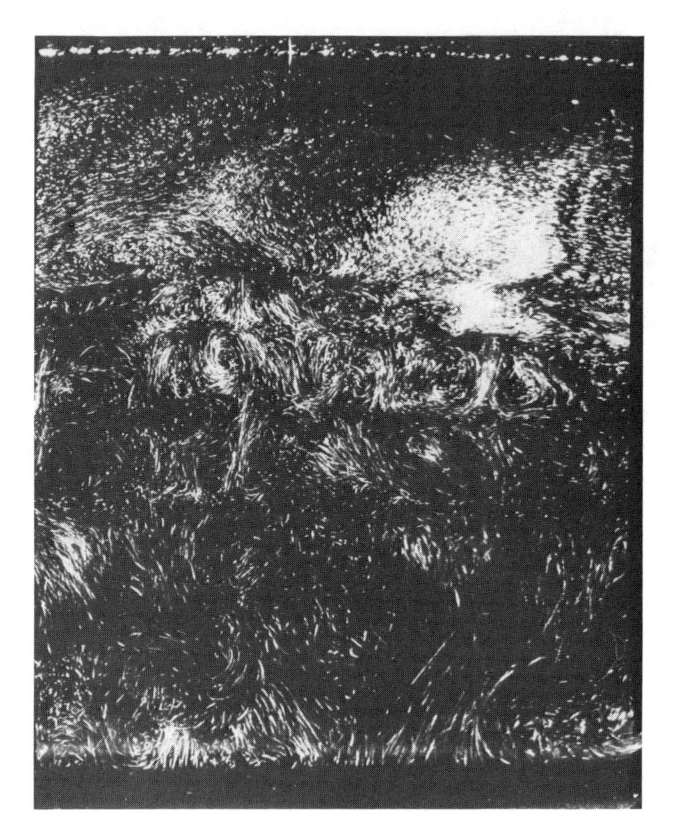

Fig. 20 Visualization of a stable solute gradient heated from below, using aluminum particles suspended in the fluid (from Huppert and Linden [10]).

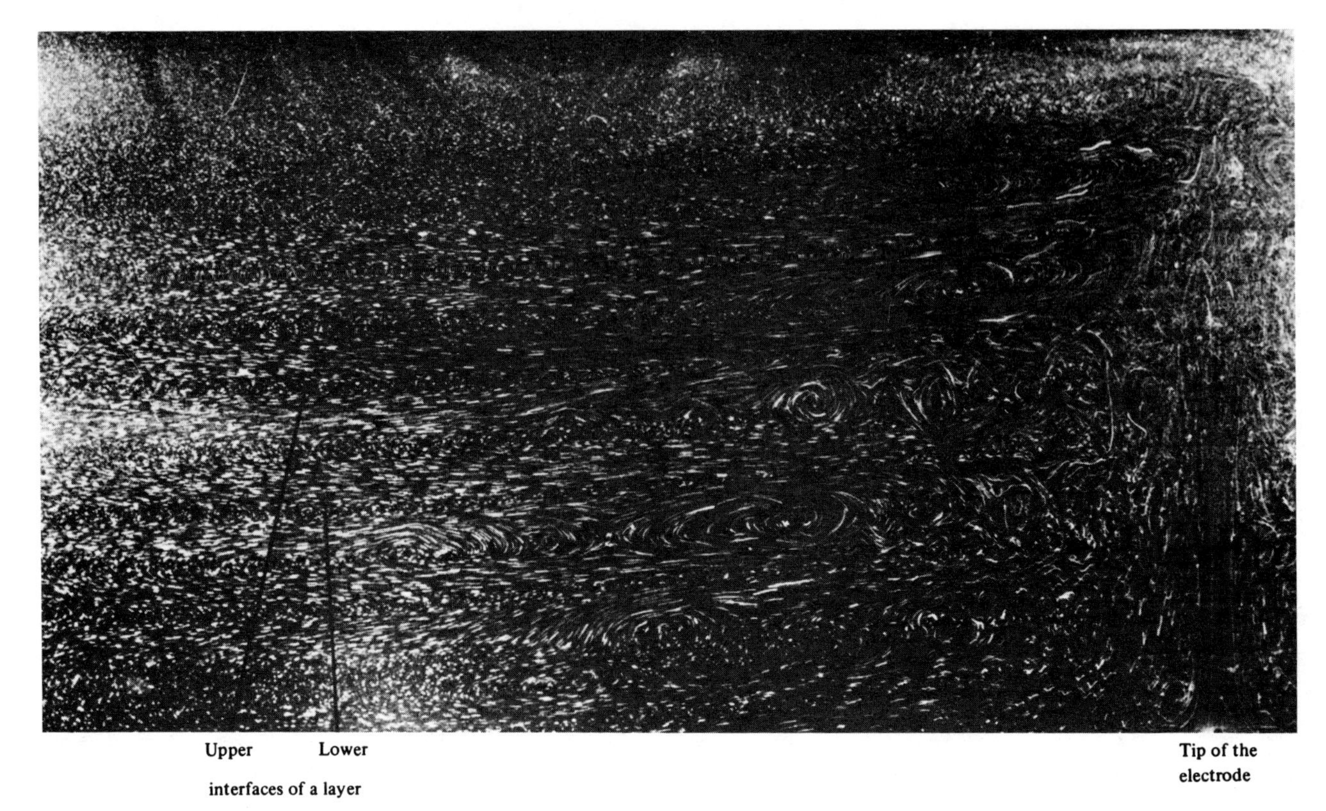

Fig. 21 Particle-streak photograph of the flow due to a point source of heat in a stable salt gradient. The particles are made of PVC and dyed by a fluorescent yellow dye (from Tsinober et al. [20]).

particles with respect to the fluid, their sinking velocity was negligible, due to their very small diameter; in fact they were neutrally buoyant.

V SUMMARY

This short review of flow visualization in double-diffusive convection aims to give the reader a general first insight into the variety of visualization techniques used thus far in this relatively new field of research. It seems that the variety of techniques utilized is a direct consequence of the variety of phenomena existing in double-diffusive convection.

Although this review does not include all visualization experiments reported in the literature, all the basic techniques are described here; it is hoped that this description will be helpful to anyone applying an appropriate method for any phenomenon in double-diffusive convection.

REFERENCES

1. Turner, J. S., Double Diffusive Phenomena, *Ann. Rev. Fluid Mech.,* vol. 6, pp. 37–56, 1974.
2. Turner, J. S., Buoyancy Effects in Fluids, Cambridge Univ. Press, Cambridge, England, 1979.
3. Huppert, H. E, and Turner, J. S., Double Diffusive Convection, *J. Fluid Mech.,* vol. 106, pp. 299–329, 1981.
4. Turner, J. S., Multicomponent Convection, *Ann. Rev. Fluid Mech.,* vol. 17, pp. 11–44, 1985.
5. Stommel, H., Arons, A. B., and Blanchard, D., An Oceanographical Curiosity: The Perpetual Salt Fountain, *Deep-Sea Res.,* vol. 3, pp. 152–153, 1956.
6. Stern, M. E., The "Salt Fountain" and Thermohaline Convection, *Tellus,* vol. 12, pp. 172–175, 1960.
7. Thorpe, S. A., Hutt, P. K., and Soulsby, R., The Effect of Horizontal Gradients on Thermohaline Convection, *J. Fluid Mech.,* vol. 38, no. 2, pp. 375–400, 1969.
8. Chen, C. F., Briggs, D. G., and Wirtz, R. A., Stability of Thermal Convection in a Salinity Gradient due to Lateral Heating, *Int. J. Heat Mass Transfer,* vol. 14, pp. 57–65, 1971.
9. Ruddick, B. R., The "Colour Polarigraph"—A Simple Method for Determining the Two-dimensional Distribution of Sugar Concentration, *J. Fluid Mech.,* vol. 109, pp. 277–282, 1981.
10. Huppert, H. E., and Linden, P. F., On Heating a Stable Salinity Gradient from Below, *J. Fluid Mech.,* vol. 95, no. 3, pp. 431–464, 1979.
11. Shirtcliffe, T. G. L., and Turner, J. S., Observations of the Cell Structure of Salt Fingers, *J. Fluid Mech.,* vol. 41, no. 4, pp. 707–719, 1970.
12. Chen, C. F., and Sandford, R. D., Sizes and Shapes of Salt Fingers near the Marginal State, *J. Fluid Mech.,* vol. 78, no. 3, pp. 601–607, 1976.

13. Linden, P. F., and Shirtcliffe, T. G. L., The Diffusive Interface in Double Diffusive Convection, *J. Fluid Mech.*, vol. 87, no. 3, pp. 417–432, 1978.
14. Huppert, H. E., and Turner, J. S., Ice Blocks Melting into a Salinity Gradient, *J. Fluid Mech.*, vol. 100, no. 2, pp. 367–384, 1980.
15. Chen, C. F., and Turner, J. S., Crystallization in a Double Diffusive System, *Geophys. Res.*, vol. 85, pp. 2573–2593, 1980.
16. Griffiths, R. W., Layered Double Diffusive Convection in Porous Media, *J. Fluid Mech.*, vol. 102, pp. 221–248, 1981.
17. Thangam, S., and Chen, C. F., Salt Finger Convection in the Surface Discharge of Heated Saline Jets, Geophys. *Astrophys. Fluid Dynam.*, vol. 18, pp. 111–146, 1981.
18. Lewis, W. T., Incropera, F. P., and Viskanta, R., Interferometric Study of Stable Salinity Gradients Heated from Below or Cooled from Above, *J. Fluid Mech.*, vol. 116, pp. 411–430, 1982.
19. Lewis, W. T., Incropera, F. P., and Viskanta, R., Interferometric Study of Mixing Layer Development in a Laboratory Simulation of Solar Pond Conditions, *Sol. Energy*, vol. 28, no. 5, pp. 389–401, 1982.
20. Tsinober, A. B., Yahalom, Y., and Shlien, D. J., A Point Source of Heat in a Stable Salinity Gradient, *J. Fluid Mech.*, vol. 135, pp. 199–217, 1983.
21. Baker, D. J., A Technique for the Precise Measurement of Small Fluid Velocities, *J. Fluid Mech.*, vol. 26, pp. 573–575, 1966.
22. Hart, J. E., Finite Amplitude Sideways Diffusive Convection, *J. Fluid Mech.*, vol. 59, no. 1, pp. 47–64, 1973.

Chapter 32

Infrared Thermography in Heat Transfer

Giovanni Maria Carlomagno and Luigi de Luca

I BASIC PRINCIPLES OF INFRARED THERMOGRAPHY

Thermography is a measurement technique of thermal maps. Accurate quantitative analysis of thermal images acquired in real time is an essential performance requirement of a thermographic system. The technology of modern infrared (IR) thermography can currently attain this goal to a considerable extent.

Basically the infrared scanning radiometer (IRSR) detects the electromagnetic energy radiated in the IR spectral band from an object (whose temperature has to be measured) and converts it into an electronic video signal. In particular, starting from the object, IR energy is first radiated through a medium (typically the atmosphere). It then enters the sensing system, passing through a lens, a scanning mechanism, an aperture (or a filter), and finally impinges on an IR detector, which transduces the radiation into an electrical signal.

Monochromatic radiation intensity E_λ, emitted by a surface having an absolute temperature T, is given by Planck's law:

$$E_\lambda = \varepsilon_\lambda E_{\lambda 0} = \frac{\varepsilon_\lambda C_1}{\lambda^5 (e^{C_2/\lambda T} - 1)} \tag{1}$$

where ε_λ is the spectral hemispherical emittance, C_1 and C_2 are the first and second radiation constants respectively, λ is the wavelength of the radiation being considered, and $E_{\lambda 0}$ is the blackbody monochromatic radiation intensity.

By integrating Planck's law over the entire spectrum the total radiation intensity E is obtained (Stefan–Boltzmann law):

$$E = \varepsilon \sigma T^4 \tag{2}$$

where ε is the total hemispherical emittance (or emissivity coefficient) and σ is the Stefan–Boltzmann constant.

Measurements made by means of spectral radiometers or pyrometers are generally based on Eq. (1). Infrared scanners are often classed as total-radiation radiometers, although Eq. (2), does not apply, since their detectors sense radiation in a limited band width (window) of the IR spectrum.

In fact, IRSRs typically perform measurements in two different windows of the IR band: the short-wave (SW) window and the long-wave (LW) one. In the first case the radiometer generally uses silicon optics with a coating having a transmittance peak at a radiation wavelength of about 5 μm; the detector is typically manufactured from indium–antimonide (InSb) and gives a spectral response between 3.5 and 5.6 μm. A broad-band coating of the optics of an SW system increases the relative response of the scanner by bringing the lower edge of the SW window down to 2 μm, the short-wave broad (SWB) band. This is an advantage when spectral filtering is required, e.g., reading the temperature of semitransparent objects, target signature, and laser applications. However, broad-band systems are most susceptible to atmospheric losses.

In the LW band, the instrument uses germanium optics with a peak coating at about 10 μm; the detector is typically manufactured from mercury–cadmium–telluride (HgCdTe), which gives a spectral response between 8- and 14-μm wavelengths. Mercury–cadmium–telluride detectors can be used also in the SW window.

The choice of an appropriate spectral band (SW or LW) depends on different factors. Some surfaces have a higher emissivity factor in the SW window; moreover, low-cost detectors and/or thermoelectrically cooled detectors are also available in this band. However, in spite of the relatively high atmospheric transmission coefficient, usually the SW region requires some compensation while performing high-accuracy measurements at viewing distances greater than 1 m. On the other hand, the LW region exhibits a very low coefficient of atmospheric absorption, except in the case of very high water-vapor content. Due to a higher thermal contrast or sensitivity in this window, higher overall system performances can be achieved.

The detector is the core of the IR thermographic system. The most frequently used detectors are the so-called photon detectors, in which the release or transfer of electrons is directly associated with photon absorption. The main characteristic of photon detectors is that they have a very short response time (of the order of microseconds) but they require cooling well below the ambient temperature to allow for rapid scanning and high sensitivity. Therefore, the sensor is frequently located in the wall of a Dewar chamber, which is cooled by liquid nitrogen (LN_2). To increase the operating time of the radiometer, a demand-flow Joule–Thompson cryostat using high-pressure nitrogen (or argon) gas or a closed-cycle cryogenic refrigerator can also be employed. An alternative method has recently been introduced. Based on a thermoelectrical (Peltier) cooling effect, it eliminates the use of detector cooling agents. This, however, generally results in a sensitivity loss of the radiometer.

Since the sensing element is zero-dimensional (practically a point), in order to have it receive energy from different parts of the field of view (i.e., to scan the object), a proper electromechanical scanning mechanism must be used. This scanning mechanism may consist of moving mirrors, refractive elements (such as prisms), or a combination of the two. For two-dimensional imaging such a mechanism allows object scanning in both vertical and horizontal directions. Infrared scanning radiometers that scan the object in only one direction (one-dimensional IRSRs) are also available. They are convenient when measuring temperatures of objects moving in a direction perpendicular to the scanning one or to have very high scanning speed in the study of fast transient phenomena (e.g., heat transfer in shock tunnels).

Before reaching the detector, IR energy passes through specially designed lenses. Only IR-transmitting glasses such as germanium, silicon, or sapphire can be used. Moreover, lenses must be coated with a proper material to allow for maximum IR transmission. The use of filters also allows one to see through certain atmospheres or to measure the surface temperature of objects, such as glass or plastic, or even of flames. Low-pass, high-pass, band-pass/reject, and attenuating filters are available.

The overall performance of an IR imaging system is conventionally measured by the amount of useful and accurate information that can be acquired per unit of time. This can be expressed by means of the following parameters: thermal sensitivity, or equivalent random noise level; scan speed, or update rate of the scanning mechanism; image resolution, or number of independent measurement data points that compose the image; intensity resolution, or number of intensity levels that allow one to resolve fine temperature differences.

Energy is radiated from objects by means of individual photons; the timing between the emission of photons is random. The random emission of photons produces a variation in signal intensity, referred to as thermal noise. The sensitivity of an IRSR system is generally expressed by the noise equivalent temperature difference (NETD), the temperature difference between two images resulting in a signal equal to the random background noise of the camera. Noise equivalent temperature difference is usually given at a specified object temperature, namely, 30°C. Typical values of NETD range from 0.07 to 0.5°C for commercially available systems (see Table 1).

The scan speed of the system is the rate at which complete thermal images are updated by the scanning mechanism. The total field of view is scanned into a certain number of horizontal scanning lines by synchronized horizontal and vertical motors. Accordingly, the scan speed is usually expressed through two parameters: the scan rate per line and the scan rate per field. Depending on the line interlace used to update the whole imaging frame, the rate at which completely interlaced picture frames are updated is finally obtained (frame frequency). Low scan speeds offer the advantages of low cost, but result in images that severely distort during transients or when the scanner and/or the object move. High-performance imaging radiometers are characterized by a high scan rate and full compatibility with the frame rate of standard television. Typical values are reported in Table 1.

Image resolution is the capability of a thermal imaging system to detect and accurately measure the temperature of small portions (slits of reduced width) of the object surface, with the term *small* referring to the size of the total image. Resolution is generally determined by characteristics of the detector such as size and response time. Typically, an IRSR detects radiation using a detector of finite size; therefore such finiteness limits the image resolution. Electronics generally digitizes the image into pixels smaller than the resolution elements corresponding to the detector size. Pixel digitization therefore does not reduce system performance, but it must be noted that the pixels do not usually contain as many true image resolution elements.

Generally, image resolution is defined either by means of the instantaneous field of view (IFOV) of the detector,

Table 1 Specifications of Some Commercially Available IRSRs

	Agema			Hughes		Inframetrics	
	Model 782	**Model 870**	**Model 880**	**Model 3000**	**Model 4000**	**Model 600**	**Model 522**
Detector	InSb or HgCdTe	HgCdTe	HgCdTe	10 elements InSb	6 elements InSb	HgCdTe	HgCdTe
Spectral response (μm)	2–5.6 3.5–5.6 8–12	2–5	8–12	2–5.6	2–5.6	8–12 3–5 3–12	8–12 3–5 3–12
Cooling	LN_2	Peltier	LN_2	Argon gas	Argon gas	LN_2	LN_2
NETD (°C)	0.1	0.1	0.07	0.1–0.5	0.1–0.5	0.1	0.1
Field frequency (Hz)	25	25	25	20	20	50	50
Frame frequency (Hz)	25/4	25/4	25/4	20/2	20/2	50/2	50/2
Interlace	4:1	4:1	4:1	2:1	2:1	2:1	2:1
IFOV (mrad)	2.3	2.3	2.0	2.18	2.18	2.0	3.0
IFOV lens (°)	12 × 12	12 × 12	20 × 20	10 V × 15 H	7.5 V × 15 H	15 V × 20 H	14 V × 18 H
Horizontal IFOVs	100	100	175	120	120	175	100
Vertical IFOVs	100	100	100	80	60	130	80
Intensity levels	256	256	256	256	256	128	64

or by the number of instantaneous fields of view (IFOVs) that subtend the complete field of view (FOV). Theoretically IFOV is the ratio of the detector width over the focal length of the lens; however, this is only an ideal design parameter and does not describe the performance of the actual system. In practice, it is useful to define as an actual IFOV the smallest thermically detectable width of the object that can be viewed by the sensor for a fixed value of contrast (or modulation). Generally the 50% point of the slit response function is considered [1], but care has to be taken because often resolution is defined differently depending on the main function to which it should apply. A distinction should always be made between resolution in a measurement situation and resolution in an imaging situation. Conventionally, IFOV is measured in mrad.

When a lens of reduced total FOV is used, the image resolution is increased since the number of resolution elements (IFOVs) on the image is an invariant while the single IFOV is reduced. Generally telescopes, expander lenses, and supplementary close-up microscope lenses are available to alter FOV (and consequently IFOV). Table 1 shows typical values of the image resolution for different IR imaging systems. Due to continuous improvement, technical specifications from different manufacturers are subject to change.

The intensity resolution (commercially referred to as dynamic range) defines the ability of a thermographic system to resolve temperature differences with respect to the temperature-measuring range. Dynamic range can be expressed by means of the number of gray shades (or digital levels of intensity) used to encode the thermal image.

Typical temperature ranges of IRSRs span from about −20 to 800°C, and can be extended up to 1500–2000°C by using filters.

II THE IR IMAGING SYSTEM

The block diagram in Fig. 1 shows a typical configuration of the entire IR thermal-imaging system, including hardware for data acquisition and digital image processing.

Infrared radiation emitted by the model is amplified and converted into an electronic video signal that is displayed on the monitor of the driving unit. This houses all the operational controls of the IR camera.

Thermal images can be continuously recorded by a videotape recorder (VTR). After the test, the recorded data can be replayed in playback mode and digitized by the analog–digital converter.

Fully computerized IR imaging technology also allows on-line automated direct-digitizing and -recording processing. Viewing and processing images in real time permit scanner adjustments during the test, thus ensuring the accuracy of the posttest analysis.

After analog–digital conversion, thermal data are fed into a computer for more sophisticated quantitative analysis [2].

In relating the radiation detected by IRSR to the radiation emitted by the three main sources (object, surroundings, and atmosphere), image-processing software typically takes into account, among others, the following parameters: thermal level, thermal range, surface emissivity coefficient, lens type, viewing distance, and air and ambient temperatures. The basic computer display mode is generally a colored image where each color is related to a temperature interval and the isotherm is visualized as the boundary line between two colors. Standard image-processing software usually can provide thermal profiles and gradients across the surface, temperature frequency histograms in a given area, temperature differences from different images, and

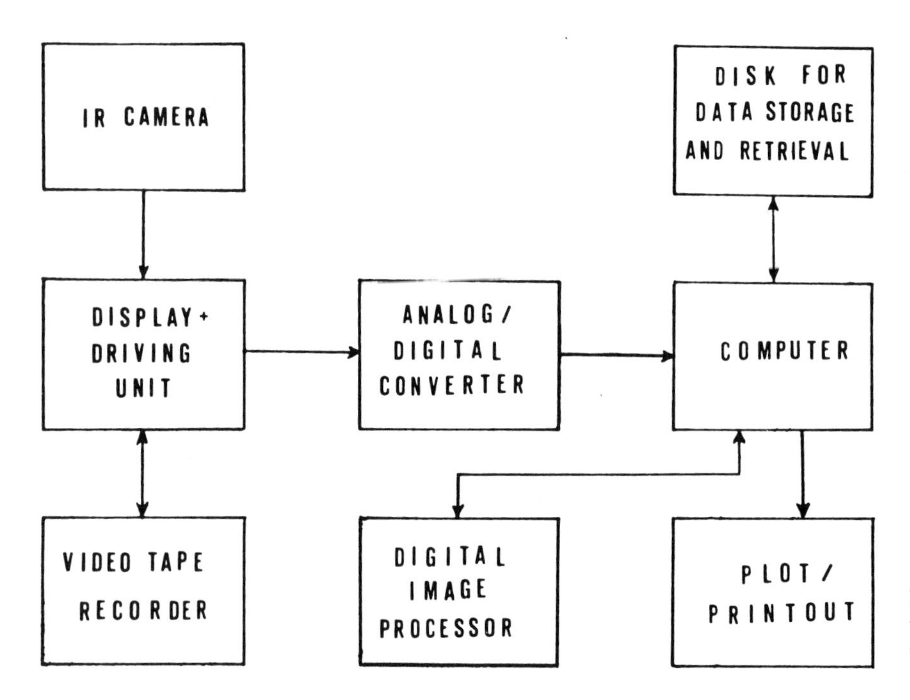

Fig. 1 Schematic representation of a typical IR imaging system configuration.

image filtering and zooming. Dedicated software can be developed to reduce data in final form for specific applications.

Finally, thermal images can be stored on a hard disk or diskette for future retrieval and/or printed out on a color graphics printer.

III THE USE OF IRSR TO MEASURE CONVECTIVE HEAT FLUXES

Measurement of heat flux rates and/or convective heat transfer coefficients from a surface to a stream may be more difficult to perform than other common thermo-fluid dynamic quantities.

Usually, measuring heat fluxes involves measuring temperatures. In the ordinary techniques [3–8], where temperature is measured by thermocouples, resistance temperature detectors (RTDs), or pyrometers, the sensor yields the local heat flux at a single point (or the space-averaged one); hence, the sensor itself can be considered as zero-dimensional. This limitation makes experimental work particularly troublesome whenever temperature and/or heat flux fields exhibit high spatial gradients. The use of encapsulated liquid crystals (LCs) [9, 10] represents a first step toward overcoming the limitations of zero-dimensional techniques. An LC could in principle be considered a two-dimensional sensor, since it permits the visualization of lines of constant temperature. However, for quantitative measurements, it is necessary to use LCs that change color within a small temperature range and/or to choose a given boundary of color change. Hence, in practice, LCs can visualize one isotherm at a time; moreover, the LC working range is relatively limited (from -40 to 285°C [10]).

The IRSR constitutes a true two-dimensional temperature sensor since it allows the performance of accurate measurement of surface temperature maps even in the presence of high spatial-temperature and/or heat-flux gradients (the spatial resolution depending on the optics employed). In the recent past IRSR has been used to a large extent for qualitative analyses including nondestructive testing of materials, energy conservation, plant maintenance, and control and optimization of processes involving heat transfer [11]. Indeed, the potential of IRSR still seems little exploited as far as its quantitative utilization is concerned, especially for convective heat transfer measurement. In particular, IR thermography can be fruitfully employed to measure convective heat fluxes, in both steady and transient techniques [12]. In this context, note that the IR radiometer can be considered intrinsically to be a thin-film sensor, since it measures skin temperatures. Since the thermal map obtained by means of currently available computerized thermographic systems is formed through a large amount of IFOVs, then IRSR can be regarded as a two-dimensional array of thin films. Unlike standard thin films, however, which have a response time of the order of microseconds, the typical response time of IRSR is of the order of 10^{-1} s (see Table 1).

Using IRSR as a temperature sensor in convective heat transfer measurement appears advantageous, from several points of view, compared to standard sensors. In fact, as already mentioned, IRSR is a fully two-dimensional sensor; it allows the evaluation of errors due to tangential conduction and radiation, and it is noninvasive. This last charac-

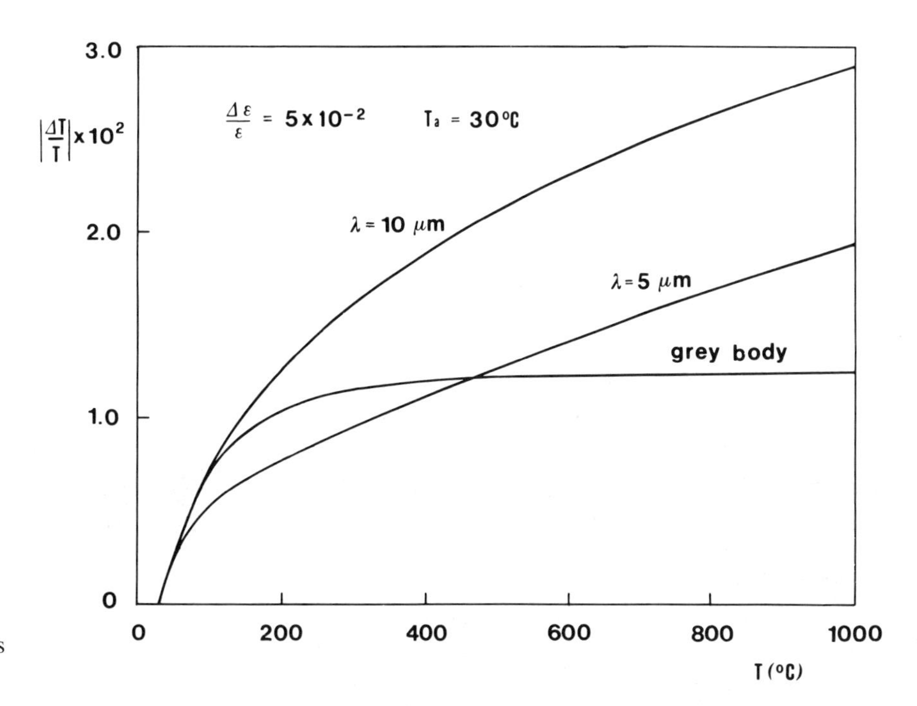

Fig. 2 Temperature-reading error as a function of target temperature.

teristic also eliminates the conduction errors through the thermocouple's or RTD's wires.

Accuracy of the measurements depends on knowledge of the emissivity coefficient of the surface viewed by an IR camera. However, even if the emissivity is set in error, the corresponding error of the temperature measurement is generally less than the emissivity-setting error. In fact, if it is assumed that the radiating surface is a gray body and that the IR detector uniformly absorbs the whole electromagnetic spectrum, the relation linking the temperature-reading error $\Delta T/T$ to the emissivity-error $\Delta\varepsilon/\varepsilon$ is

$$\frac{\Delta T}{T} = -\frac{1}{4}\frac{\Delta\varepsilon}{\varepsilon}\left[1 - \left(\frac{T_a}{T}\right)^4\right] \tag{3}$$

where T is the true target temperature and T_a is the ambient temperature.

Since IR detectors are generally sensitive in a relatively narrow window of the IR band, their spectral response showing a marked peak at a certain wavelength, it is more useful to discuss the case of a monochromatic spectral response. The requested relation in this case is

$$\frac{\Delta T}{T} = 1 - \frac{\lambda T}{C_2}\ln\left(1 + \frac{1 + \Delta\varepsilon/\varepsilon}{A}\right) \tag{4}$$

where

$$A = \frac{\lambda^5}{C_1}\left[E_{\lambda 0}(T) + \frac{\Delta\varepsilon}{\varepsilon}E_{\lambda 0}(T_a)\right] \tag{5}$$

Assuming that $\Delta\varepsilon/\varepsilon = 0.05$ and $T_a = 30°C$, Fig. 2 shows the plots of Eq. (3) for the gray body and those of Eq. (4) for the two cases of $\lambda = 5$ and 10 μm, which represent the wavelengths of the peak spectral response respectively in the SW and LW windows. For gray-body behavior, the temperature-reading error is always smaller than $(\Delta\varepsilon/\varepsilon)/4$, a value reached only at relatively high temperatures. For the monochromatic case, however, $\Delta T/T$ depends strongly on λ and tends to $\Delta\varepsilon/\varepsilon$ for $T \to \infty$.

The following three sections analyze the potential of IR imaging systems to measure convective heat-flux rates, as well as the ranges of applicability of the most commonly used types of heat-flux sensors when temperature is measured by means of IRSR.

IV FREQUENCY RESPONSE OF ONE-DIMENSIONAL HEAT-FLUX SENSORS

Heat-flux sensors generally consist of slabs with a known thermal behavior, whose temperature is measured at fixed points [3–8]. The equation for heat conduction in solids applied to a proper sensor model yields the relationship by which measured temperature is correlated to convective heat transfer rate.

The most commonly used heat-flux sensors are the so-called one-dimensional ones, where the heat flux to be measured is assumed to be normal to the sensing-element surface, and the temperature gradient components parallel to the plane of the slab are negligible. In practice, the slab surfaces may be also curved, but the curvature is neglected if the layer affected by the input heat flux is rather small compared with the local radius of curvature of the slab.

Strictly speaking, there is another type of one-dimensional sensor, the Gardon gauge, in which the heat flux normal to the sensor surface is correlated with a radial temperature difference, along the direction parallel to the plane of the slab [3].

In the following, only ideal one-dimensional sensors are considered. The term *ideal* means that the thermophysical properties of the sensor material are assumed independent of the temperature and that the presence of the actual temperature-sensing element is not considered.

The one-dimensional sensor models considered here are

1. **Thin-film sensor.** A very thin resistance thermometer classically measures the surface temperature of a "thermally" thicker slab on which it is mounted. Heat flux is inferred from the theory of heat conduction in a semi-infinite one-dimensional wall. The surface film is made thin so as to have negligible heat capacity and thermal resistance compared to the slab. Surface temperature can also be measured by means of IR thermography [12].
2. **Thick-film sensor.** The slab is used as a calorimeter; heat flux is inferred from the time rate of change of mean slab temperature. This temperature is usually measured by using the slab as a resistance thermometer.
3. **Wall calorimeter sensor,** or thin-skin method. The slab is made thermally thin (so that its temperature can be considered to be constant across the thickness) and is used as a calorimeter. Heat flux is typically inferred from the time rate of change of the rear-face temperature, usually measured by a thermocouple. The use of IR thermography in this method is also practicable [12].
4. **Gradient sensor.** In this sensor the temperature difference across the slab thickness is measured. Considering a heat transfer process, the heat flux is computed by means of the temperature gradient across the slab. The temperature difference is usually measured by a thermopile of very-thin-ribbon thermocouples or by two thin-film resistance thermometers. As an alternative, IRSR can be applied [12].

Figure 3 shows the generalized model representing the sensor, where L is the slab thickness and the distance x is measured from the lower surface. Temperature is denoted by T (with subscripts 1 or 2 referring to the upper or the lower surface respectively). The heat flux to be measured Q_1, which crosses the upper surface, is assumed in the form of a simple harmonic function:

$$Q_1(t) = |Q_1| \exp(i\omega t) \tag{6}$$

where ω and t are the circular frequency and the time respectively.

The four types of one-dimensional sensors just described involve the following temperature measurements:

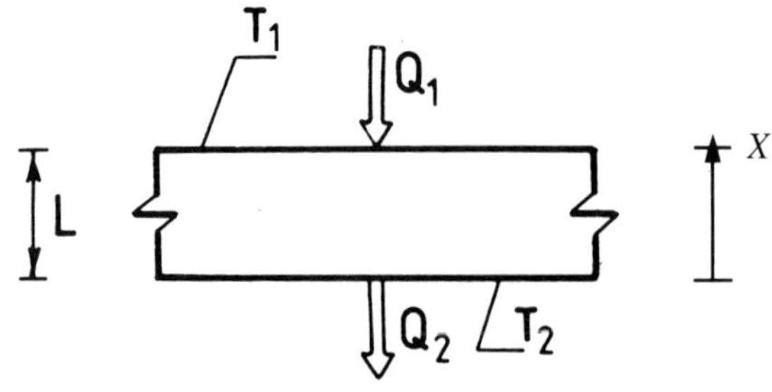

Fig. 3 Sketch of the generalized one-dimensional heat-flux sensor.

Thin film. Temperature to be measured is that of the heated surface T_1.

Thick film. Mean temperature $\overline{T} = \int_0^L T\,dx/L$ is measured.

Wall calorimeter. Temperature T_2 of the lower face of the slab is measured.

Gradient sensor. The temperature difference $\Delta T = T_1 - T_2$ is measured.

The frequency response of T_1, T_2, $\overline{T}$, and ΔT when the heat flux (6) is applied on the upper face of the slab, is studied in Ref. [13]. To give a unitary solution to the problem, two boundary conditions on the lower surface are considered: perfectly insulated surface (i.e., adiabatic, $Q_2 = 0$) or constant surface temperature (i.e., in contact with a heat sink, $T_2 = 0$). In any case, at a given point x the harmonic solution for T has the form

$$T(x, t) = |T(x)| \exp\{i[\omega t + \Phi(x)]\} \tag{7}$$

where $\Phi(x)$ is the phase angle by which T lags Q_1. The frequency response of the various sensors is reported in Figs. 4 and 5, where the amplitude factor $k|T|/(L|Q_1|)$

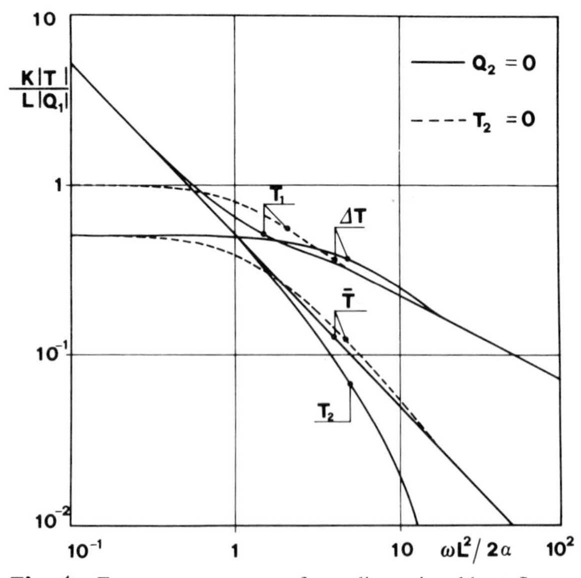

Fig. 4 Frequency response of one-dimensional heat-flux sensors (amplitude factor) [13].

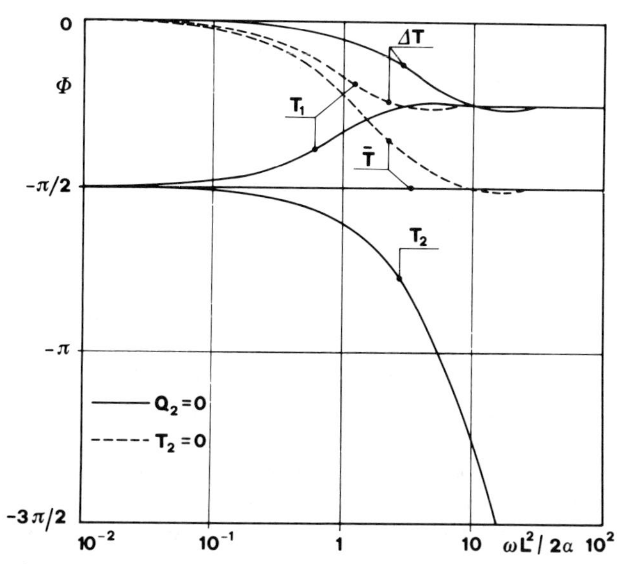

Fig. 5 Frequency response of one-dimensional heat-flux sensors (phase lag) [13].

(k being the thermal conductivity of the slab) and the phase lag Φ are plotted, respectively, against the dimensionless frequency $\omega L^2/2\alpha$ (α being the thermal diffusivity of the material of the sensor).

Comparing the different sensors, it is evident that for devices of the same thickness and material, the gradient sensor has the flattest response. Consequently, this sensor represents the best choice when the heat flux to be measured contains relatively low frequencies, and a good frequency response is of prime importance. Figures 4 and 5 also show that the wall calorimeter is applicable at intermediate heat-flux frequencies; the thin- and thick-film sensors are applicable at higher frequencies, with the thin-film sensor being generally more sensitive.

In the following two sections the use of IRSR as a temperature-measuring device for the different types of sensor is discussed. In this context, on the basis of the frequency response results, it appears useful to group the different sensors and the corresponding measuring techniques in two classes, namely quasisteady and transient techniques.

V APPLICATION OF IRSR TO QUASISTEADY TECHNIQUES

One of the classical methods of measuring convective heat-flux rates in steady conditions is the gradient-sensor technique. In steady conditions heat flux is inferred by the relationship

$$Q_1 = k\frac{T_1 - T_2}{L} \tag{8}$$

which justifies the name *gradient sensor*.

The gradient-sensor method can also be used in the unsteady regime. In particular, if the functions $T_1(t)$ and $T_2(t)$, which describe the time–temperature histories, do not contain harmonic components with frequencies greater than $0.1\ \alpha/L^2$, the heat flux can be inferred from the relation [7]:

$$Q_1(t) = k\frac{T_1(t + L^2/3\alpha) - T_2(t - L^2/6\alpha)}{L} \tag{9}$$

which for $L^2/\alpha \to 0$ in particular reduces to Eq. (8).

One of the major problems arising when IRSR is used as a temperature sensor in the gradient-sensor technique lies in the inability of the IR camera to look at more than one surface of the sensor. This difficulty can be bypassed either by scanning the wall surfaces one at a time or by viewing both of them by means of mirrors; the first technique is more advantageous because it allows the performance of a higher spatial resolution. If Q_1 is not relatively large, an alternative method consists of viewing by IR camera the upper sensor surface while keeping the lower at a known temperature T_2 (e.g., with a boiling liquid or a condensing vapor). The high heat transfer coefficient obtained in these cases allows T_2 to coincide with the phase-change temperature of the fluid.

When the heat flow across the sensor is not truly one-dimensional, the use of the IR radiometer in the gradient-sensor technique gives the possibility of measuring heat-flux variations over the exchange surface as well as evaluating tangential conduction errors.

Another very practical technique in quasisteady conditions is the thin-disk sensor (Fig. 6), otherwise known as the thermal window or Gardon foil gauge, in which a radial-temperature difference is related to the (normal) heat flux to be measured. Assuming that the disk edge is in contact with a heat sink and the input heat flux Q (entering the upper surface of the disk, the lower one being adiabatic) is uniform, Q is proportional to the temperature difference $T_0 - T_R$ between the center and the edge of the disk:

$$Q = 4kL\frac{(T_0 - T_R)}{R^2} \tag{10}$$

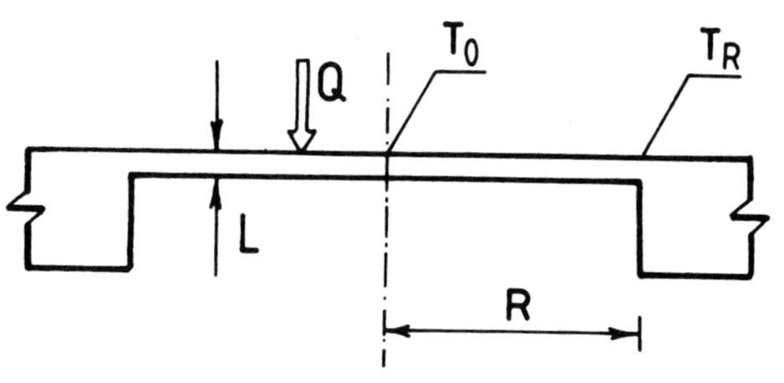

Fig. 6 Sketch of the thin-disk sensor.

where L and R are, respectively, the thickness and the radius of the disk. Equation (10) is valid in steady-state conditions.

Use of the Gardon gauge can be also extended to quasi-unsteady regime. Under the assumptions stated above, if the functions $T_0(t)$ and $T_R(t)$ do not contain harmonic components with frequencies greater than $4\alpha/R^2$ heat flux Q can be obtained by the formula [7]:

$$Q(t) = \rho c L \left[\frac{3}{4}\frac{dT_0}{dt} + \frac{1}{4}\frac{dT_R}{dt} + 4\alpha \frac{T_0(t) - T_R(t)}{R^2} \right] \quad (11)$$

where ρ and c are, respectively, the density and the specific heat coefficient of the disk material.

Use of IR thermography in the thin-disk method appears advantageous compared to the standard techniques because IRSR allows checking for uniformity of heat-flux rate over the gauge. Moreover, it is in general possible to use thinner disks because in IRSR application the error due to conduction through the thermocouples, which usually measures the temperature at the center of the disk, is suppressed. This allows design of sensors with smaller diameters, i.e., having higher spatial resolution.

It has to be kept in mind, however, that the thin-disk method is intrinsically a zero-dimensional heat-flux sensor (i.e., it measures the heat flux at a given point). Therefore, in order to measure heat-flux distribution, it is necessary to have an array of such sensors, which may not be practical or convenient, especially from the point of view of the spatial resolution.

Within the class of steady-state techniques to measure convective heat fluxes, another method where the application of IRSR seems to be very effective is the heated-thin-foil technique. This method consists of heating a thin metallic foil by Joule effect and measuring the heat transfer coefficient h from the foil to a stream flowing on it. The convective heat transfer coefficient is inferred from the relationship

$$h = \frac{Q_0 - Q_1}{T_w - T_{aw}} \quad (12)$$

where Q_0 is the Joule heating per unit area; Q_1 is the heat loss, including radiation, free convection, and lateral conduction; T_w and T_{aw} are the local wall temperature of the foil and the local adiabatic wall temperature of the stream respectively. Use of the IR radiometer allows visualization of the map of constant-temperature lines and thus the measurement of heat-flux distribution over the foil. In fact, if the constant heat-flux boundary condition (including Q_1) is realized, each isotherm translates into a locus of a constant heat transfer coefficient.

Under the assumption that the Biot number $\mathrm{Bi} = hL/k$ (where L is the thickness of the foil) is very small compared to unity, temperature can be considered practically uniform across the foil thickness. In this case the surface of the foil to be viewed can also be chosen as the one opposite the exchange surface.

In evaluating the behavior of the heated-thin-foil sensor in unsteady regime, note that the response of the sensor follows the exponential law, the time constant being $-L^2/(\alpha \mathrm{Bi})$.

Application of IRSR to the heated-thin-foil technique seems advantageous compared to the use of LCs because it simultaneously visualizes the whole map of isotherms. Use of thermocouples in this technique could yield substantial conduction errors owing to the very thinness of the foil.

At higher Mach numbers, in order to take into account compressibility effects, the adiabatic wall temperature must also be measured. As T_{aw} distribution may be assumed to coincide with the temperature exhibited by the foil when Joule heating is suppressed, each test run may be composed of two parts. First, when T_{aw} is measured, the cold image of the foil is obtained. Then the surface temperature distribution T_w is measured when the electric current is heating the foil (hot image is recorded). The temperature difference $T_w - T_{aw}$ distribution is directly obtained by subtracting the two thermal images of the foil, the cold from the hot. The possibility of digitizing thermal image and processing data by computer makes this procedure possible simultaneously over the whole foil, whereas standard thermometric measurement would be more difficult to perform.

Whatever technique is used, T_{aw} can be directly measured by realizing the exchange surface of insulating material, provided the radiative conductance of the surface is negligible compared to the convective conductance of the stream.

VI APPLICATION OF IRSR TO TRANSIENT TECHNIQUES

With regard to transient techniques, as already pointed out, IRSR can be considered a two-dimensional array of thin films whose temperature can be correlated to heat-flux rate by using the one-dimensional semi-infinite wall model. By assuming this latter to be isothermal at initial time $t = 0$, a suitable formula to evaluate heat flow from the measured film temperature is [7]

$$Q_1(t) = \frac{k}{2\pi\sqrt{\alpha}} \left[\frac{2T_1(t)}{\sqrt{t}} + 2m\sqrt{t} - \int_0^{t-\delta} \frac{T_1(p) - T_1(t) + (t-p)m}{(t-p)^{3/2}}\, dp \right] \quad (13)$$

where $m = [T_1(t) - T_1(t - \delta)]/\delta$, with δ being a real number such that the function $T_1(p)$ can be assumed linear within the interval $t - \delta < p < t$.

In practice, a heat-flux sensor is a slab of finite thickness L; hence the thin-film model is applicable only for relatively small measurement times (i.e., there is a lower limit to the frequencies at which the sensor gives valid results). On a quantitative basis, if t_M is the measurement time, it has to be verified (Figs. 4 and 5):

$$t_M < \frac{L^2}{2\alpha} \tag{14}$$

In this case the boundary condition on surface 2 is irrelevant since the assumption of semi-infinite wall is valid.

Since the IR radiometer has a typical response time of the order of 0.1 s, much larger than typical thin films, by taking into account the result enforced by formula (14), due to relatively high response time of IRSR, severe limitations arise. In particular, the use of IRSR requires relatively thick sensors and/or materials of low thermal diffusivity.

Another very practical transient technique is the thin-skin one, where the model wall behaves as an ideal calorimeter. In this method the sensor, practically a thin plate, is modeled as an ideal calorimeter (isothermal across its thickness), heated on one face and thermically insulated on the other. If the heat flux Q_1 is constant in time, the relation linking Q_1 to the time rate of change of T (sensor temperature) is

$$Q_1 = \rho c L \frac{dT}{dt} \tag{15}$$

If the heat flow is varying in time and does not contain harmonic components with frequencies greater than $2\alpha/L^2$, the following equation can be used [7]:

$$Q_1(t) = \rho c L \frac{dT_2(t + L^2/6\alpha)}{dt} \tag{16}$$

where T_2 is the adiabatic face temperature, usually measured.

In the case of convection heat transfer, in order to calculate the heat transfer coefficient, it is necessary to know the temperature T_1 of the heated surface. This latter can be correlated to temperature T_2 by the formula

$$T_1(t) = 3T_2\left(t + \frac{L^2}{6\alpha}\right) - 2T_2(t) \tag{17}$$

Heat flux may also be correlated to temperature T_1 by means of the formula

$$Q_1(t) = \rho c L \frac{dT_1\,(t + L^2/3\alpha)}{dt} \tag{18}$$

Both Eqs. (17) and (18) are valid in the same range of frequencies of Eq. (16). Analysis of the frequency response results (Figs. 4 and 5) shows that within the assumption made, it is preferable to measure the heated-surface temperature instead of the insulated one, because at higher frequencies T_2 becomes smaller than T_1.

The use of IRSR in the wall-calorimeter technique seems advantageous because one can measure the temperature on either side of the model wall. Further, in this technique the possibility of evaluating tangential conduction errors, due to edge effects and/or nonuniform heat flow, is very valuable.

VII MEASURING HEAT TRANSFER IN WIND TUNNEL FACILITIES

Heat transfer rates to aerodynamic shapes are often measured in wind tunnels. For high-Mach-number flows (high enthalpy), the model is typically at ambient temperature, so it can be suddenly exposed to the stream; the increase of the wall temperature versus time is measured. For low-Mach-number flows, a temperature difference must generally be created between model and stream; e.g., a common method consists in preheating the model and studying the subsequent convective cooling by the stream. If the thermal properties of the wall are known, on the basis of the results reported in the previous sections, the study of the heat conduction in solids can be used to determine a relationship between the heat transfer rate and the measured wall temperature history.

Often it is desirable to obtain a quick survey of the distribution of the heat transfer rate in complicated shapes. In this case, the use of classic sensors makes instrumentation of the model time consuming and expensive. In certain cases moderate accuracy can be tolerated, so alternative methods to measure surface-temperature distribution have been developed.

The use of paints that change color at certain temperatures has been described in Refs. [14–17]. The temperature at which the paints change color depends on the rate of change of the temperature, on the pressure, and in some cases also on the humidity absorbed by the paints. It is therefore usual to calibrate the whole technique with experiments on hemispheres for which the results are known. In any case, including when liquid crystals are used [9], only one isotherm at a time can be detected. Moreover, it is necessary to paint the model before each test run. Another method [18] makes use of a thin opaque coating that melts to a clear liquid at given temperatures. In this case, care must be taken to keep the liquid from flowing along the surface at the risk of creating accumulations that may disturb the flow.

It is important to remember that surface-temperature sensing techniques for wind tunnel facilities can be classed, in general, either as mounted or unmounted (remote or non-invasive sensing). Models having surfaces unsuitable for attachment of a sensor require remote methods using pyrometers and/or radiometers.

A wide field measurement with good spatial resolution, good thermal response range and high frequency response is mandatory whenever the experimenter does not know the optimum location or is unable to mount surface sensors. The technology of IRSR currently available fills this need. The use of IR thermography allows not only measurement of the distribution of local heat transfer coefficients but also visualization of the local flow condition, i.e., in particular the location of the boundary layer transition and the location and extent of the flow-separation region. Moreover, IRSR constitutes a rapid, accurate technique for determining thermal loads as a function of the model geometry (e.g., when the flow around an aircraft or space vehicle is simulated).

A typical experimental arrangement showing the model and camera installation in a wind tunnel is shown in Fig. 7 [19]. Proper lens focal length has to be chosen as a function of the dimensions of the tunnel and/or the model, or of the desired spatial resolution [20].

The optical-access window is probably one of the main problems to be solved when IRSR is used in a wind tunnel facility, since neither standard glass nor quartz can be used for the wavelengths for which the IR detector is sensitive. In the case of test runs simulating supersonic or hypersonic flow regimes, the choice of the window material is determined by both optical and mechanical considerations. In fact, in these cases the window must resist the stresses that originate in it because of the temperature and pressure differences existing between the wind tunnel and the environment around it. The temperature differences across the window can be minimized, as shown in Fig. 7, by having it in a remote position.

Figure 8 gives the transmittance of some of the most-used materials. Different technical solutions are also possible: i.e., zinc sulfide Zn S; calcium fluoride CaF_2; cadmium telluride CdTe and potassium bromide KBr.

Germanium's primary transmission range is from 2 to 15 μm, making it useful for IR laser applications, particularly in the LW band. It is opaque in the visible region, and because of a high surface reflectivity of 36% (related to its high refractive index, which is greater than 4), Ge should be antireflection (AR) coated. Silicon Si is similar to germanium but with greater resistance to mechanical and thermal shocks. Its use in the LW region requires extensive calibration due to the strong variation of the transmission coefficient. Arsenic trisulfide As_2S_3 has a transmission range from 0.5 to 13 μm but is quite soft and brittle. Its coefficient of thermal expansion is very similar to that of aluminum. Zinc selenide, ZnSe with a transmission range from 0.58 to 15 μm, is useful for IR applications in both SW and LW bands, permitting visual alignment. An AR coating is required to decrease the 17% single-surface reflection loss. Calcium fluoride CaF_2 [22] and magnesium fluoride MgF_2 [21] have excellent transmission over a broad spectral range. The former is usable from 0.15 to 9 μm, and the latter from 0.11 to 7.5 μm. Both of these materials are slightly water soluble. Their low index of refraction

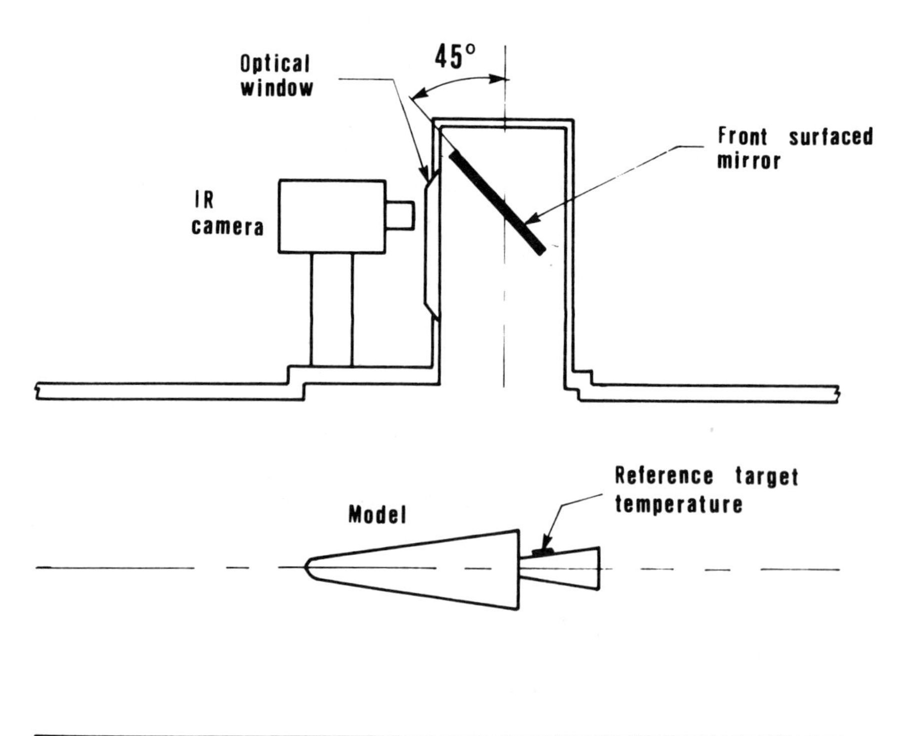

Fig. 7 Typical model and IR camera installation in a wind tunnel [19].

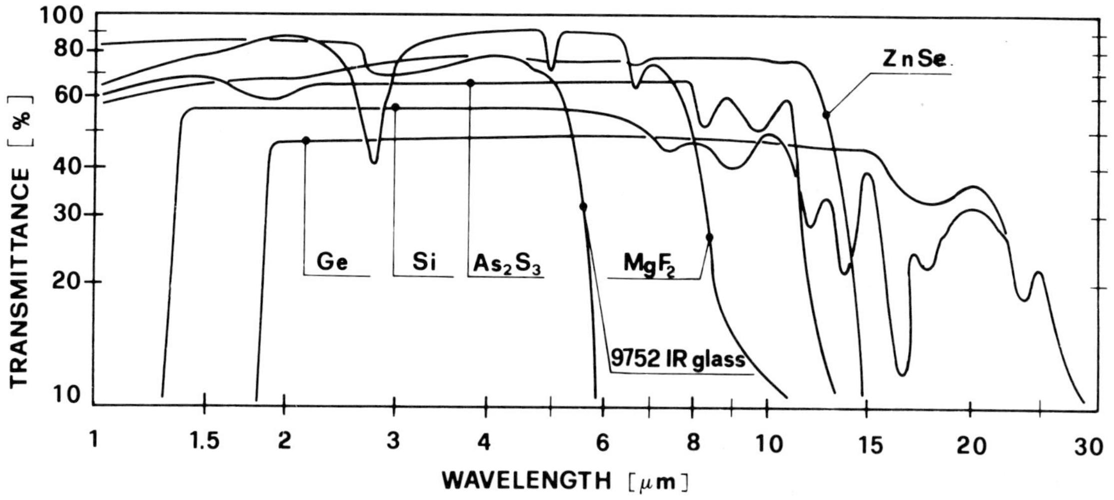

Fig. 8 Transmittance of some IR optical materials (thickness, 2 mm).

allows them to be used without AR coatings; MgF_2 is more durable than CaF_2, and both are sensitive to thermal shock. Sapphire Al_2O_3 [20], also known as 9752 IR glass, has an extremely hard surface, is chemically inert, and is insoluble except at very high temperatures. It exhibits high transmittance (all the way from 0.15 to 6 μm), which makes it useful in the SW band and allows visual alignment. Because of its great strength, the sapphire window can safely be made much thinner than windows of other material. The sapphire window is therefore useful even at wavelengths very close to its transmission limits. Because of sapphire's exceptionally high thermal conductivity, thin windows made of it can be effectively cooled by forced air or other methods.

A simple solution for the optical-window problem, described in [23], makes use of a thin plastic foil supported by a perforated force-carrying metal sheet. If the sheet is far away from the focal plane of the camera, it decreases the IR radiation intensity without interfering with the optical quality of the picture. In the specific application where this system was tested, in which IRSR was used to measure the heat transfer rate on a paraboloid in a hypersonic blowdown wind tunnel, the foil had to support a pressure difference of about 1 bar. Holes of 8-mm dia. were used. No difficulties due to heating of the foil were encountered, as it was far away from the hot jet boundary.

When testing conditions do not impose any particular restriction, such as pressure differences, more simple solutions can be considered. The simple thin plastic (e.g., vinylidene chloride/vinyl chloride copolymer) foil may often be used as an IR window in room-temperature low-speed wind tunnel environments [24].

A block diagram of the entire recording and data-reduction system, as used by Bynum et al. [19] to obtain model surface temperature and heat transfer coefficient distributions, is shown in Fig. 9.

To circumvent the problem of measuring absolute radiation from the model surface, a relative method of measurement is frequently adopted [19, 21, 24], requiring known temperatures within the scanned field of view. Reference thermocouples are embedded in the subject surface at fixed points, and the rate of the strip chart used to read the reference thermocouple output can be synchronized with the tape-footage indicator of the video recorder of the thermal images. This allows continuous direct comparison of the intensity level and the subject temperature in the zones near the reference thermocouples. The scanner aperture setting, thermal range, and thermal level can also be noted at each test point.

To achieve a high accuracy of absolute temperature measurement, modern IRSRs include a microprocessor controlled measuring system. One or more miniature reference sources are built into the scanner and, by scanning these references, the gain and level of the system can be precisely controlled.

Calibration of the IRSR is generally accomplished with a blackbody calibration source mounted with the aperture on the centerline of the wind tunnel test section and in the center of the field of view of the camera. The radiation from the source is viewed with the camera through all system optical components, thus negating corrections for transmission and reflection losses.

In order to enhance the thermal-image detection of the IR system it may be necessary to increase the emissivity coefficient of the surface to be measured, especially when the model is constructed of metallic material. The use of a thin coat of flat black paint raises the emissivity coefficient to 0.95–0.98 [21, 24]. To minimize interference and collimate radiation from the test model's hot surface, measuring the temperature of the tunnel walls in order to evaluate their influence on the measurements is recommended. In any case, it may be useful to keep the tunnel walls near the

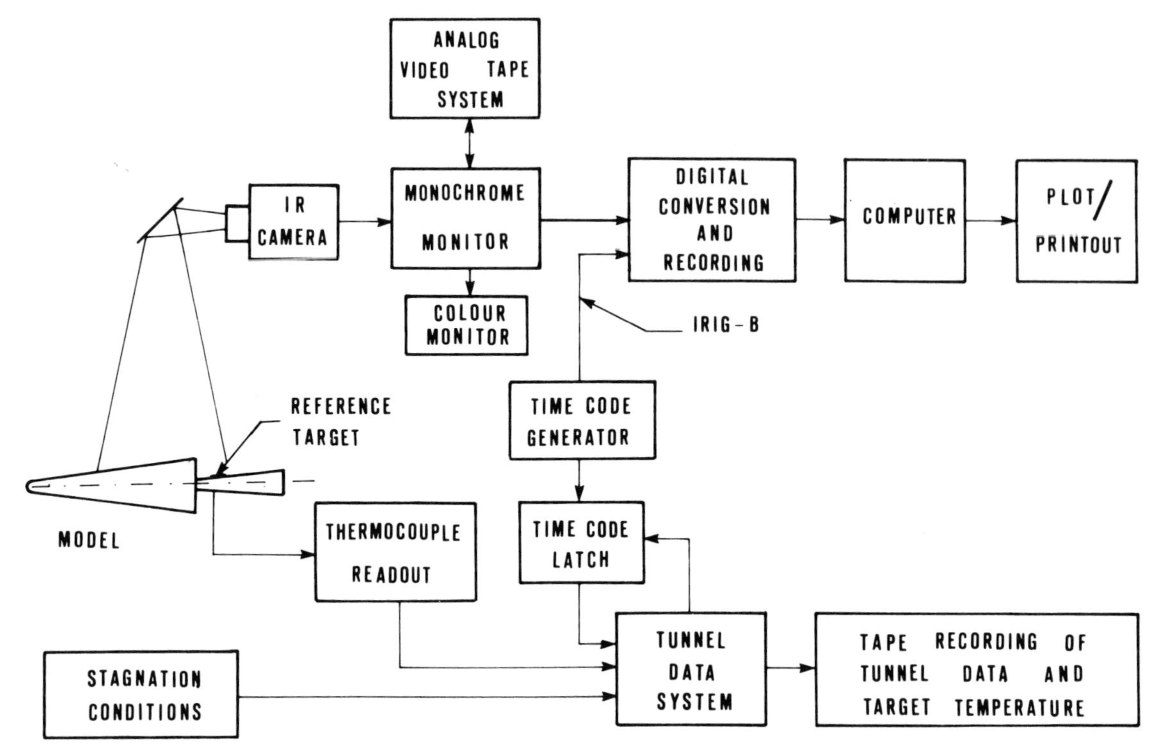

Fig. 9 Data-recording and digitizing system for heat transfer measurements in a wind tunnel [19].

test section very cool compared with the model surface and/or to coat them, using selective coatings.

It is also recommended that the optical-access window be AR coated also on the side facing the camera so as to avoid reflection of infrared radiation coming from the environment.

When particular model shapes such as cones or cylinders are tested, certain regions at the edge of the model are often viewed with a great angle of incidence measured from the normal. Such regions may appear cooler or hotter than they actually are since all materials exhibit a dependence of the emissivity on the angle of incidence (directional emissivity). In particular, electric conductors and nonconductors exhibit two different behaviors. For conductors, the emissivity first increases and then decreases as the emission angle approaches 90°. Nonconductors, on the other hand, have a practically constant emissivity for angles of 60° or less. Beyond 60° the emissivity falls off rapidly. If a constant emissivity is assumed in thermal-image acquisition, the interpretation of the heat transfer contour maps therefore needs proper corrections.

Thin-film and thin-skin models are generally employed in wind tunnel testing.

A Wind Tunnel Applications

The earliest application of an IR imaging system to measure and map aerodynamic heating parameters is probably that reported in Ref. [23], in which an IR camera was used to measure heat transfer rates on a paraboloid in a hypersonic blow-down wind tunnel. The data-acquisition system made use of an AGA Thermovision camera with an InSb detector cooled with liquid nitrogen, utilizing an SWB band between 2 and 5.4 μm. The thin-film data-reduction technique was used to determine the heat transfer data. The measured distribution of the Stanton number St $= h/\rho_\infty V_\infty c_p$ along the surface of a paraboloid at Mach 7.1 agrees in a satisfactory way with similar results obtained using melting coating (Fig. 10). Both are about 30% below the results from Ref. [25]. The agreement between the two methods indicates common sources of error, the main one being the uncertainty of the physical properties of the wall material. Another source of error is the nonuniform wall-temperature distribution.

The thin-film method was also selected in Ref. [19], which describes the IR system used to acquire and reduce the heating data, and the results of the measurement and mapping of aerodynamic heating parameters in the AEDC-VKF continuous wind tunnels. Heat transfer data were obtained on a cone and on a hemisphere-cylinder model at zero angle of attack for Mach 8 and free-stream Reynolds numbers ranging from 2.4×10^6 to 3.5×10^6. The IR camera used was an AGA Thermovision Model 680, with an InSb detector sensitive to the 2- to 5-μm wavelength band.

Figure 11 reports the longitudinal centerline Stanton number distribution obtained from a 6° cone with laminar

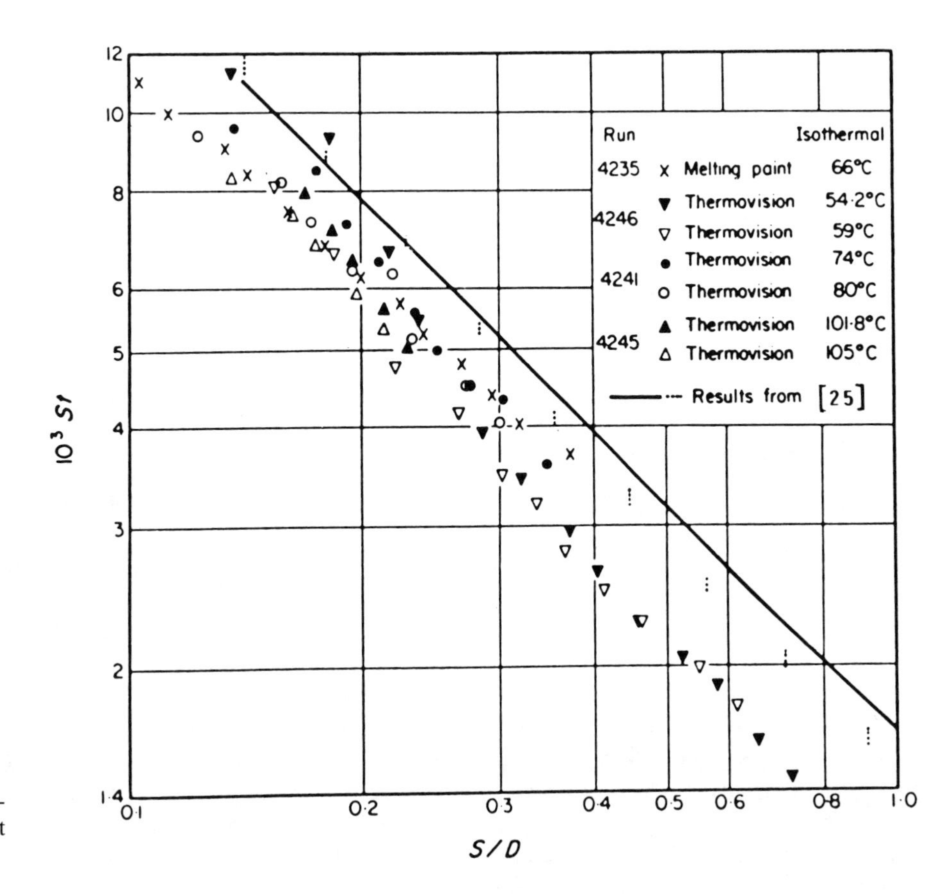

Fig. 10 Distribution of Stanton number along the surface of a paraboloid at M = 7.1 and Re = 3.2 × 10^5 [23].

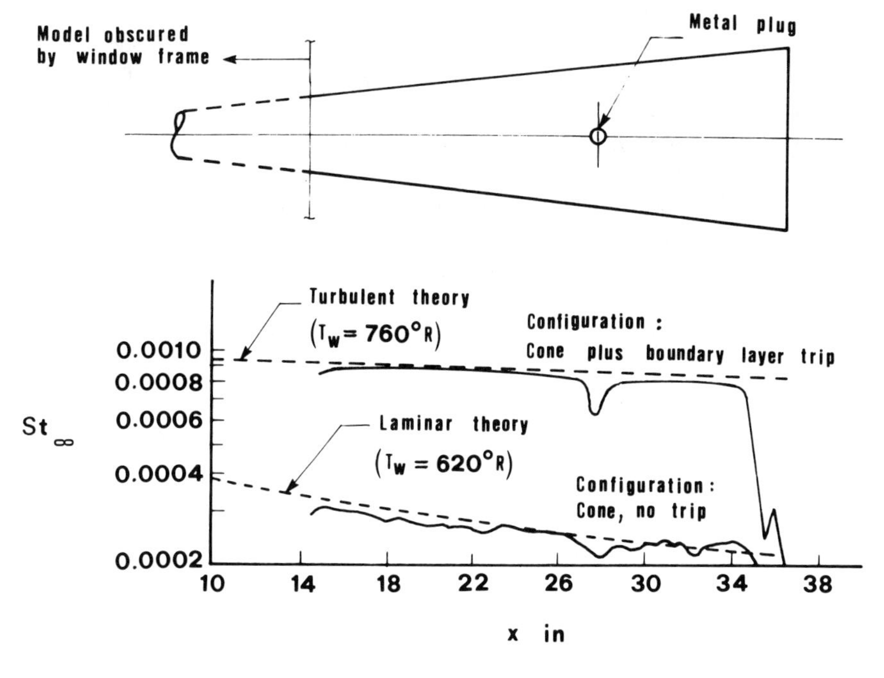

Fig. 11 Longitudinal centerline Stanton number distribution on a 6° cone [19].

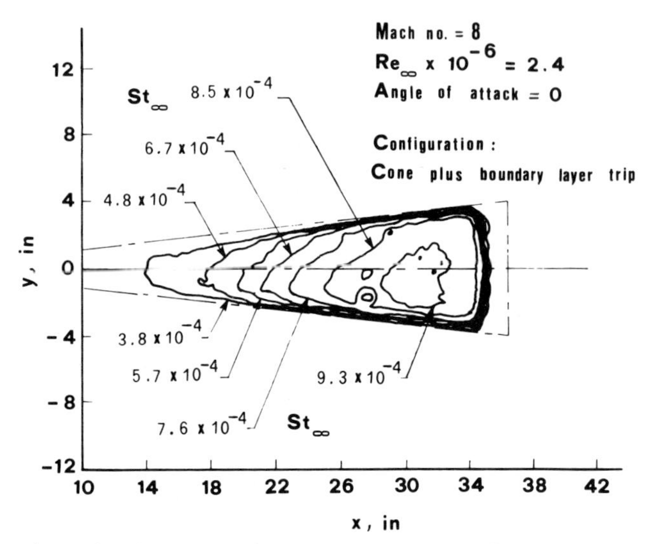

Fig. 12 Stanton number contour map on a 6° cone with a transitional boundary layer [19].

and turbulent boundary layers. The measured distributions are in reasonably good agreement with the calculated values. A typical Stanton contour map is presented in Fig. 12. Since these data were obtained at zero angle of attack, the isotherms would be expected to appear as straight lines normal to the body axis. These data, indeed, show an apparent gradual change in heat transfer at the edge of the model, from the lowest contour value to the higher ones. This is attributed mainly to the directional emittance of the model surface.

Similar measurements performed in hypersonic flow on a slender 5° half-angle cone, to evaluate the accuracy and test procedure of an IR thermal-mapping system, are described in Ref. [26].

An example of the application of the thin-skin method to measuring convective heat transfer, where surface temperature is detected by means of IRSR, is reported in Ref. [21]. The IRSR (Hughes Series 4300, working in the SWB band) was applied to determine the surface heat transfer coefficient on a full-scale model of a double wedge, simulating a jet vane used in guided-missile thrust vector control. Preliminary tests were performed in a subsonic wind tunnel to evaluate the feasibility of using IR imaging system vice thermocouples. The model was preheated, then cooled by means of forced convection to ambient conditions.

Surface-temperature readings were taken both by IRSR and by thermocouples. The results obtained from the thermal-imaging system were in satisfactory agreement with the results obtained with thermocouples. In general, IRSR results showed an average uncertainty band contained within 7%, which is of the same order of thermocouple results. The size of the uncertainty band is, however, found to increase as the temperature-sensitivity setting of IRSR is widened.

The application of IRSR in the thin-skin technique is also described in Ref. [27], where local values of the Nusselt number over a flat plate and spherical bodies are reported.

The IRSR has also been evaluated as a diagnostic tool for aerodynamic research. The results reported in the literature characterize the system's capability of performing a variety of experimental investigations, such as temperature transients, air-velocity distributions, capture of vortices, boundary layer flows, separated flows, and wakes [28]. In particular, the IR technique has been applied to the detection of boundary layer transition in wind tunnel settings [29, 30]. Since turbulent flow evinces an increase in skin friction, heat transfer rates are greater in turbulent flow than in laminar flow. Transition, then, can be detected by an increase in the measured heat transfer coefficient.

The use of the IR camera to detect boundary layer transition is relatively simple for the case of kinetic heating, i.e., at high Mach numbers [29]; at low Mach numbers, however, the kinetic heating is not large enough to achieve good sensitivity to the transition effect; an active heating technique must therefore be used. In Ref. [31] the surface of a natural laminar-flow airfoil was heated radiatively by an external source and the temperature decay was monitored when heating ceased. The developed measurement technique consisted of three prime elements: a laser heating source, an infrared camera for data acquisition, and a video recorder for data storage. A laser beam was scanned over the airfoil profile to heat its surface a few degrees above ambient. The IR camera then measured the temperature of the airfoil surface as a function of time; the temperature history was stored in the VTR, and after digitization an iterative algorithm was used to extract the heat transfer coefficient.

Figure 13 depicts the heat transfer coefficient as a function of position. In this case, transition was artificially induced at 3.2 cm from the leading edge by placing a thin line of grit on the airfoil. The upper curve, representing the

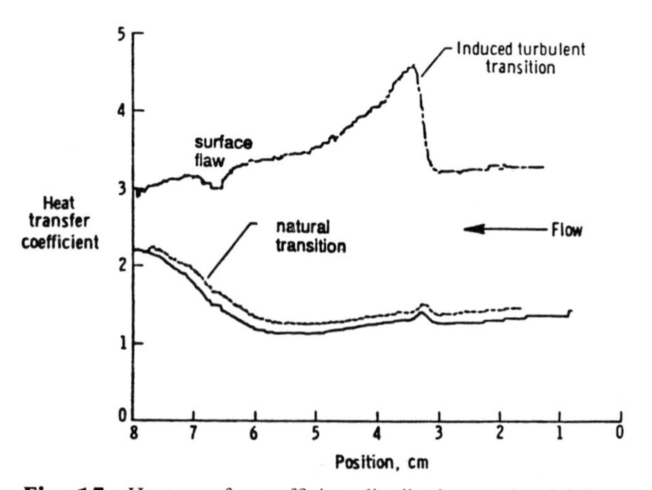

Fig. 13 Heat transfer coefficient distribution on the airfoil surface; positions are measured from the leading edge [31].

case of high wind velocity (80 mph), clearly shows the change in the heat transfer coefficient between fully laminar and fully turbulent flow. For this velocity, hot-film sensor measurements show that turbulent conditions exist directly behind the artificial trip point but not before it. Results obtained with LCs sensitive to shear stress indicate a transition at 3.1 cm from the leading edge. The two lower curves of Fig. 13 were obtained at a smaller wind velocity, with two different input heat fluxes. The wind velocity in this case was not sufficiently high to induce transition at the trip, as verified also by the hot-film sensors.

The heated thin foil technique was used in Ref. [32] to detect boundary layer transiton from laminar to turbulent flow. Experimental tests were carried out in a Gottingen type wind tunnel at Reynolds number Re = 259,000 and turbulence level of the free stream equal to about 0.9%. A rectangular model wing (0.18 m chord) with a Gottingen 797 cross-section was used during the tests. The model was made of a thin layer of fiberglas epoxy over polyurethane foam and had two 5 mm thick copper ribs at its extremities. The upper surface of model was coated with stainless steel that was glued by means of epoxy resin. The coating foil was 155 mm wide and 0.030 mm thick, and started 13 mm away from the leading edge. The foil covered the whole span of the model and was electrically connected to the copper ribs. The upper surface of the foil was blackened by means of a very thin film of paint that led to an emissivity factor of approximately 0.95. The thermal data were reduced by using the procedure described in Sec. V. In the present case T_{aw} practically coincided with the free stream static temperature. Typical heat transfer fluxes were about $7 \cdot 10^{-2}$ W/cm^2. The angle of attack of the wing was varied from -12 to 20°.

Figure 14*a* and Fig. 14*b* show the distribution of the Nusselt number on the upper surface of the model for various angles of attack. These data refer to the section along the center span of the wing. As the angle of attack is increased from -12 to 20°, the evolution of the boundary layer from completely laminar, to partially, and then almost fully turbulent is clearly apparent. Even at the highest angle of attack, the Gottingen 797 airfoil did not show separation of the turbulent boundary layer in the portion of upper surface covered by the metallic foil. The above interpretations were confirmed by tuft visualization of the flow over the wing.

A particular application of IR thermography is related to the development of advanced gas-turbine combustion systems. An experimental study of several gas-turbine combustor-liner cooling concepts was conducted in a heat transfer rig capable of simulating advanced turbine-engine-operating conditions [20]. An AGEMA Model 782 SWB scanner and a digital image-processing system were employed to acquire hot-surface-temperature distributions on the candidate liner cooling samples. The subject surfaces were successfully imaged through a high-velocity flow of

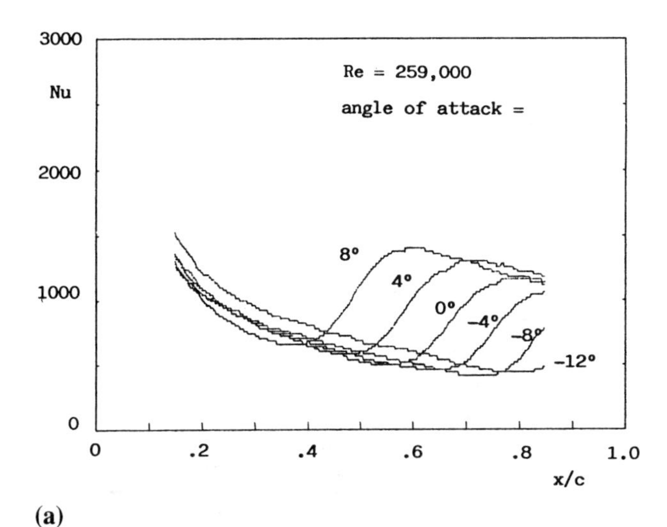

(a)

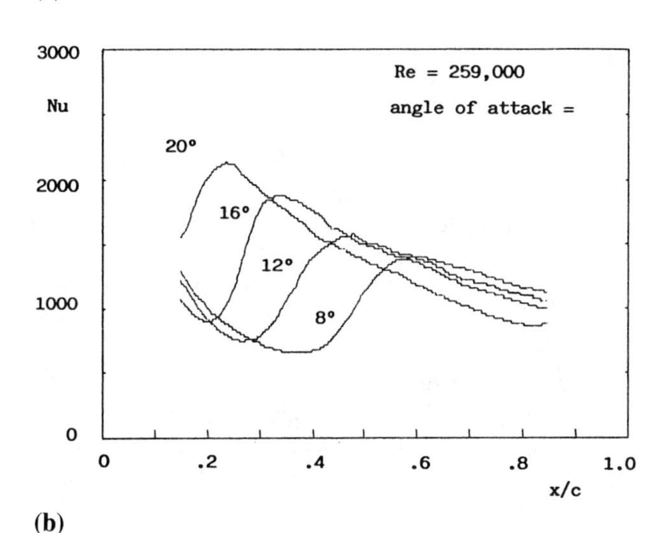

(b)

Fig. 14 Chordwise distributions of the Nusselt number (based on wing chord) over the wing upper surface for various angles of attack α: (a) $-12° < \alpha < 8°$; (b) $8° < \alpha < 20°$ [32].

luminous combustion products at gas temperatures approaching 2600°F. The test panel surfaces were imaged through a glass spectral filter, a sapphire window, and the flow of combustion products in a highly reflective, ceramic-lined duct. In view of this complex optical system, a relative method of temperature measurement was chosen, requiring a known temperature within the scanned field of view. Oxidation of the panel surface and the concurrent changes in emissivity also suggested the use of this relative technique. Accordingly, two reference thermocouples were embedded in the test surface. In addition to developing a valuable pool of wall-cooling design information, performance data acquired by IR thermography were used to benchmark temperature-prediction computer codes. Satisfactory agreement was obtained between the code predictions and the thermography measurements.

The IR technique used to measure local film-cooling effectiveness distribution on large-scale models of turbine

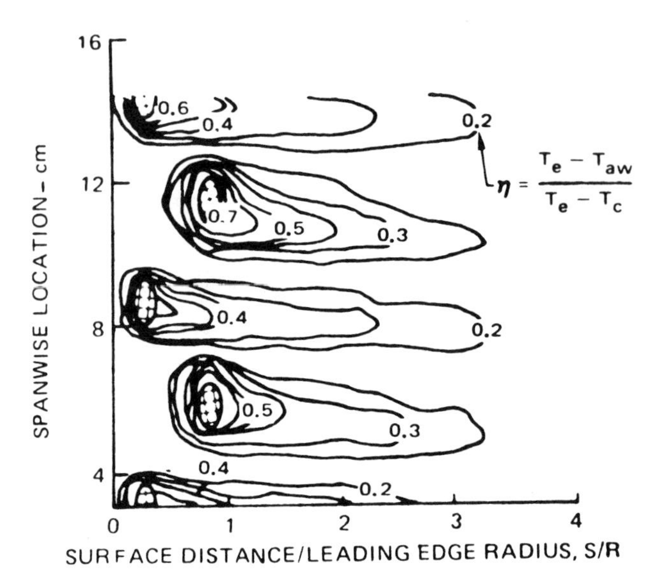

Fig. 15 Isoeffectiveness distribution for airfoil leading-edge cooling configuration [24].

blade and vane sections was also developed for use in subsonic, room-temperature wind tunnel environments [24]. Highly detailed effectiveness distributions were obtained on near-adiabatic cast rigid-foam test models using a Bofors IRSR system. The film cooling was simulated by air heated to a temperature approximately 28°C hotter than the mainstream air.

A sample plot of local effectiveness contours is presented in Fig. 15. The data are presented in a plane view of the curved leading-edge model with the coolant holes shown cross-hatched. For an adiabatic wall the film-cooling effectiveness parameter is defined as

$$\eta = \frac{T_e - T_{aw}}{T_e - T_c} \tag{19}$$

where T_e is the free-stream hot-gas temperature, T_{aw} is the measured adiabatic wall temperature, and T_c is the coolant temperature. The reported results, showing large streamwise and spanwise variations in film-cooling effectiveness, agree qualitatively with similar results in the literature.

A survey on application of the IR scanner to measurement of surface temperatures of large panels for aerodynamic testing in wind tunnels, performed at NASA Langley Research Center, is described in Ref. [22]. An InSb photovoltaic liquid nitrogen–cooled detector, sensitive at the 2 to 5.6 μm spectral range, was used for sensing. Measured data were compared to surface thermocouples and to analytical results. Agreement to better than 5% was reached.

The application of IRSR has been considered very valuable in design and evaluation of the performance parameters of the thermal protection system (TPS) of space vehicles. In fact, heat transfer measurement of the surface of a space vehicle is of great interest because it can also provide information on the chemical and mechanical state of the fluid medium (e.g., stagnation enthalpy, degree of dissociation, laminar or turbulent state).

A space vehicle is subjected to three distinct environments during a typical mission: ascent, in orbit, and reentry. Each produces unique constraints on the design of the TPS, especially when coupled with the reuse requirement. In fact, design of the TPS must consider the long-term effects that will accumulate over the lifetime of the vehicle and the number of its missions. Failure of the TPS is probably the most serious problem, because of the gradual accumulation of damage due to thermal loads as a function of exposure time. This damage is virtually undetectable under most operating conditions. These considerations result in an urgent requirement for more in-depth studies related to hypersonic flows, especially as far as re-entry problems are concerned, including test cases with regard to aerothermodynamic problems of space vehicles' TPS design.

Infrared thermography was tested as a useful tool in designing the TPSs of space vehicles (namely, the Space Shuttle) [33–35]. References [34, 35] discuss the feasibility of remote high-resolution IR imagery of the lower surface of the Shuttle (IRIS) Orbiter during re-entry, to obtain accurate measurements of aerodynamic heat transfer. Using available technology, such images can be taken from an existing aircraft-telescope system (C141 AIRO) flying parallel to the orbiter re-entry ground track. These images can have a spatial resolution of 1 m or better and a temperature resolution of 2.5% between temperatures of 800 and 1900 K. An IR acquisition system has been designed using a telescope with a 6.25-cm aperture and a single indium–antimonide IR detector together with an image-plane system making use of 600 indium–antimonide detectors disposed in two arrays.

VIII HEAT TRANSFER FROM A PLATE TO IMPINGING JETS

One method of measuring convective heat transfer where IRSR application seems to be very effective is the heated-thin-foil technique (Sec. V). A very interesting test of the capabilities of the IR scanner as a temperature sensor in this technique is the cooling of a heated plate by impinging air jets. In this problem the capabilities of IRSR are strongly stressed since large variations of the heat transfer coefficient are present on the plate.

Relatively recent research on momentum, heat and mass transfer to impinging flows led to a multitude of experimental data obtained by different techniques. A certain spread of results is present, even for the much simpler flow regime of a single jet. Moreover, empirical correlations of the average heat transfer coefficient could often be better explained if more in-depth knowledge of the spatial distributions were available. As mentioned in Sec. V, the use of ordinary zero-dimensional heat transfer sensors (e.g., the Gardon gauge [36, 37]) is limited from the point of view of spatial resolution. Encapsulated LCs have recently been used [10], but they can visualize just one isotherm at a time.

In the present section the use of IRSR is discussed. Advantages related to the use of IRSR in the heated-thin-foil technique have been described in Sec. V. The literature includes a few papers dealing with the measurement by IRSR of the heat transfer from a plate to impinging jets.

Buchlin [38] presented some preliminary results of single, round, normal, and oblique jets and arrays of three co-linear jets. In the following discussion, extensive data obtained by the present authors [39, 40] are reported.

Figure 16 shows a sketch of the experimental apparatus with the IR camera installation. The apparatus consisted of a vertical stainless steel thin foil (150 mm wide, 500 mm long, and 0.040 mm thick), heated by passing an electric current through it and cooled by air jets perpendicular to it. Cooling air, supplied by a compressor, went through a pressure-regulating valve and a heat exchanger, then entered the stagnation chamber, where pressure and temperature were measured. The stagnation temperature was equal to that of the ambient air. Surface-temperature distribution was measured by viewing the rear face of the foil with the IR camera.

The IR thermal-imaging system included an Agema Thermovision camera 782 SW; an A/D convertor data link; a computer BMC IF800 with color display; a black-and-white hard-copy printer; a Panasonic NV100 analog video recorder, and a calibration unit.

According to the desired spatial resolution, a 7 or 20° lens was used. For the 7° lens, the spatial resolution of the thermal image obtained on the computer display was about 1.2 pixel/mm at a viewing distance of 1 m. Higher spatial

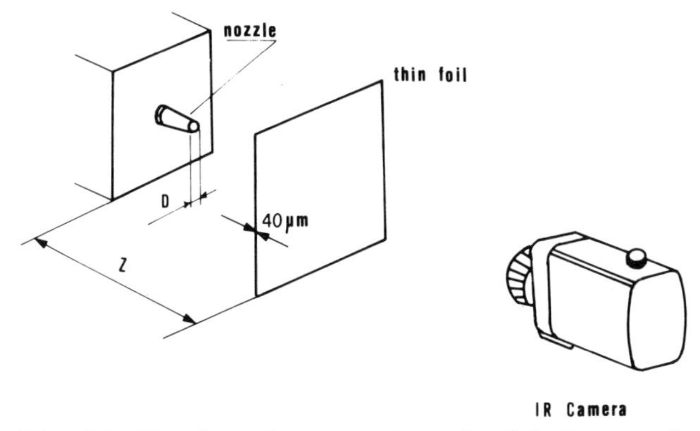

Fig. 16 Experimental apparatus for study of the heat transfer between a jet and a heated plate.

resolution was obtained by using extension rings. For example, 12- and 21-mm extension rings coupled with a 7° lens gave a resolving power of 2.1 and 3.0 pixel/mm respectively. Nozzle exit velocity was varied to investigate the influence of the Reynolds number of jets in the range of 15,000–60,000. Nozzle diameter equaled 3, 5, and 10 mm, and the dimensionless nozzle-to-plate distance Z/D ranged from 2 to 32. Truncated cone-shaped nozzles with well-rounded inlet sections were used.

To enhance the thermal-image detection, the foil surface was coated with a very thin film of black paint. Ribbon thermocouples checked the foil temperature at fixed reference points.

The adiabatic wall temperature also had to be measured in the tests. According to the procedure described in Sec. V, T_{aw} distribution was assumed to coincide with temperature distribution due to the jets on the foil surface without Joule heating. The convective heat transfer h was then inferred from Eq. (12), the local temperature difference $T_w - T_{aw}$ being performed by software.

Dedicated software was used to check the geometric position of the jet axis on the foil (i.e., stagnation point) in the case of a single jet (axisymmetric physical situation). It was also possible to obtain temperature (or recovery factor, or Nusselt number) profiles along any direction, as well as to compute the average heat transfer coefficient on a circular ring or on a given area. To minimize background noise, in addition to the numerical filtering process, the average of a number of images could be performed.

Figure 17 (see colorplate between pages 388 and 389) shows a typical color thermogram. It refers to the measured temperature map (warm image) for a single jet obtained by means of a nozzle with an exit diameter $D = 5$ mm and a dimensionless nozzle-to-plate distance $Z/D = 6$. The Reynolds number of the jet $Re = VD/\nu$ (where V is the nozzle exit velocity and ν the kinematic viscosity of air) was 28,000, and the heat flux was $Q_0 = 5.9 \times 10^{-2}$ W/cm^2. The surface of the foil shown was about 100×100 mm^2. On the left of the thermogram the scale defining the color-temperature (in °C) correspondence is also shown. This picture reproduces clearly the amount of detailed information that can be detected by the IR imaging system. Note that measurement was done by setting the thermal range of the camera at $R = 5$, which corresponds to a range of measured temperature of about 7°C; L represents the thermal level fixed during the calibration stage, and the lens aperture was set at $f = 1.8$.

As an example of perspective view of output data, the relief map of the Nusselt number for the single-jet case is reported in Fig. 18. Test conditions are Re $= 28,000$, $D = 10$ mm, $Z/D = 2$, $Q_0 = 0.144$ W/cm^2. The Nusselt number is defined as Nu $= hD/k$, where the physical properties are evaluated at the film temperature.

Figure 19 shows the variation of the recovery factor at

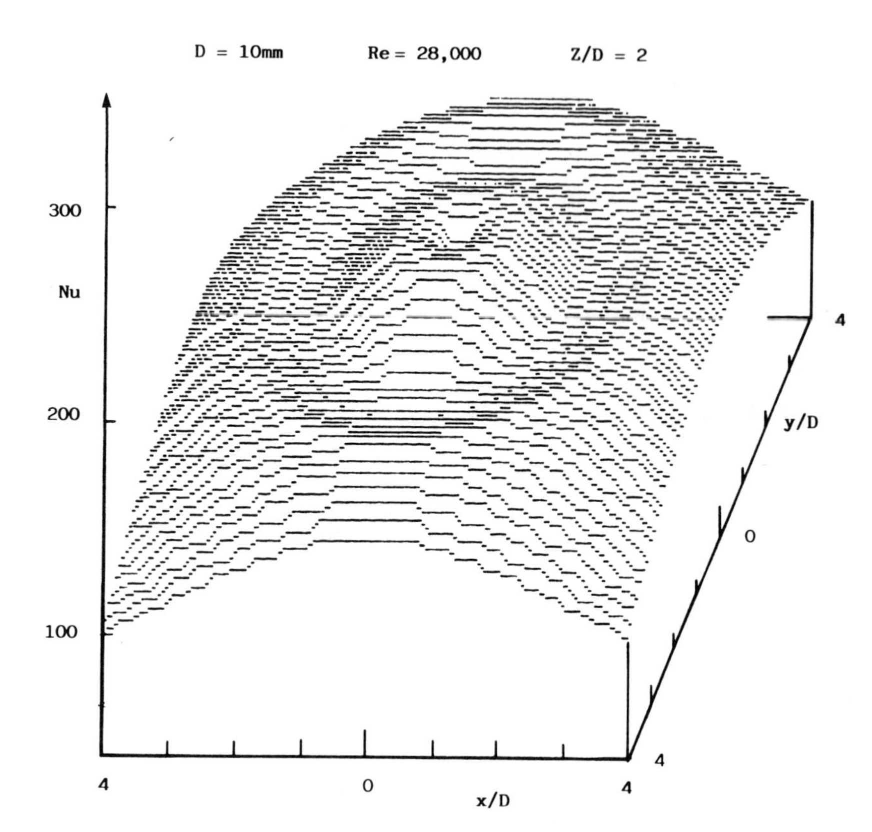

Fig. 18 Perspective view of Nusselt number distribution over a heated plate cooled by a jet.

stagnation point F_0 for the single jet. Recovery factor F is defined as

$$F = \frac{T_{aw} - T_s}{T_0 - T_s} \tag{20}$$

where T_s and T_0 are static and total temperatures of the jet at nozzle exit. F_0 is practically constant until Z/D exceeds

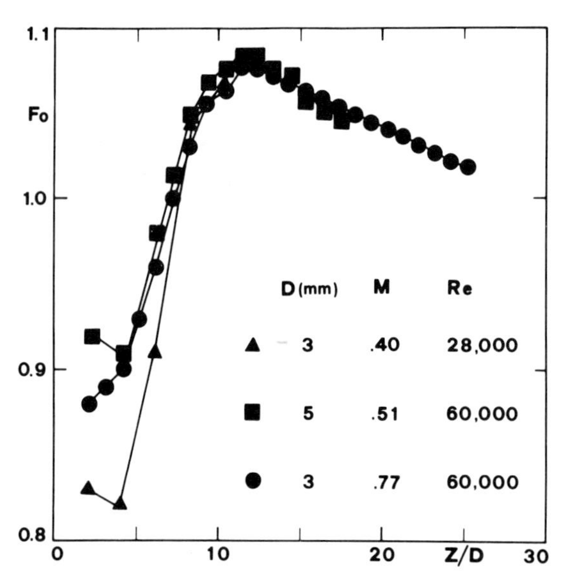

Fig. 19 Recovery factor at the stagnation point of a jet as a function of the dimensionless nozzle-to-plate distance [39].

4 because mixing does not take place in the potential core of the jet; for the lower Re number, F_0 is smaller, while for Re = 60,000, it seems practically independent of the nozzle diameter. For $Z/D > 4$, F_0 is independent of the Re number; it increases with increasing Z/D due to entrainment into the jet of ambient air, which has the same stagnation temperature as the issuing jet. Beyond $Z/D = 12$ the recovery factor decreases because the velocity of the jet is decreasing; consequently its stagnation temperature approaches its static temperature, which is reaching that of the ambient air due to the mixing.

These results practically coincide with those of Ref. [36] for $Z/D > 5$ and agree only qualitatively with those of Refs. [41, 42], the present values of F_0 being lower.

In Fig. 20 the radial distributions of the local recovery factor are reported for three significant values of the normal distance $Z/D = 2$, 4, 8, and Re = 28,000, $D = 5$ mm, $T_0 = 19.2$°C. Profiles are obtained by averaging the thermographic data over each circumference of given radius r/D. For the smaller Z/D a local central maximum and a local minimum, at about $r/D = 1.5$, are present. This minimum can be attributed to the vortex rings in the shear layer surrounding the jet when, at small Z/D distance, the impingement occurs within the potential core region [42]. At $Z/D = 8$ the minimum disappears, while a central maximum close to unity is present. For all distances, F approaches unity (not shown) far away from the stagnation region $r/D > 4$. Present data are in good qualitative agreement with those of Ref. [42], but generally lower; however,

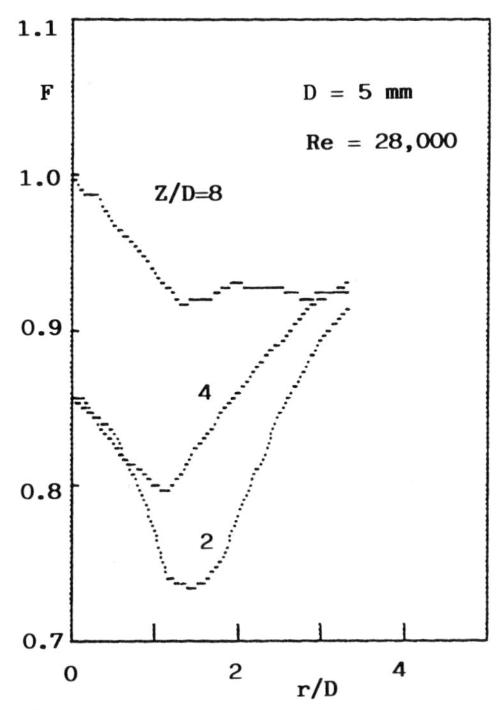

Fig. 20 Local recovery factor over the plate as a function of the dimensionless distance from jet axis.

the spatial resolution obtained with IRSR seems at least 10 times greater than that of Ref. [42]; tangential conduction errors should be lower in the present case.

Figure 21 shows Nusselt number values at the stagnation point Nu_0 as a function of Z/D for the various tested Reynolds numbers and nozzle diameters. As a general trend, all the curves confirm a maximum value of Nu_0 at Z/D between 6 and 7 and practically coincide with those of Ref. [36]. However, analysis of similar data drawn from Refs. [42–45] shows that quantitative differences in the heat transfer coefficient exist, especially in the stagnation region. This discrepancy can be attributed at least partly to differences in the nozzle geometry used in the different studies [40, 45]. Moreover, consistent improvements in the design of (standard) transducers used in different studies helped to produce more reliable measurements; e.g., it has been found [44] that improving contact between the transducer and the support plate and reducing thermocouple wire conduction errors increased the measured heat transfer. This may explain why present data obtained by IRSR (a fully noninvasive instrument) are generally higher than those of other studies. The considerably better spatial resolution of IRSR must also be considered.

In Fig. 22 the radial distributions of the local Nu number are reported for different values of Z/D and Re = 28,000, D = 10 mm, Q_0 = 0.144 W/cm^2. In this case Nu profiles are also obtained by averaging data over circular rings. For distances equal to 2 and 4 diameters, the radial distribution of Nu substantially agrees with the results of Ref. [36]: the inner peak occurs at about $r/D = 0.5$ while the outer one (occurring at $r/D = 2$) is reduced and has the form of an annular "hump." For $Z/D = 6$ the central minimum disappears and is replaced by a marked peak similar to that reported in Refs. [10, 41, 42]. Results of Ref. [36] showed a bell-shaped curve, probably due to the lower spatial resolution near the stagnation point. Such a bell-shaped curve is recovered in the present data at the highest values of Z/D.

Preliminary tests were also performed for the case of multiple jets issuing from nozzles laid out in an in-line 3 × 3 square array, with nozzle spacing S = 30 mm.

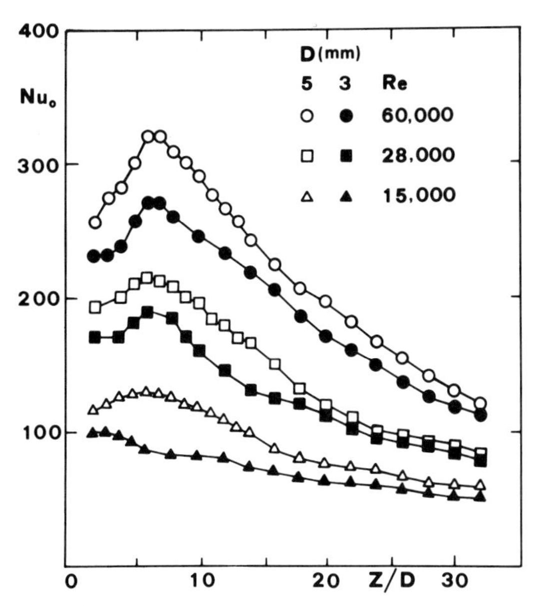

Fig. 21 Nusselt number values at the stagnation point as a function of nozzle-to-plate distance [39].

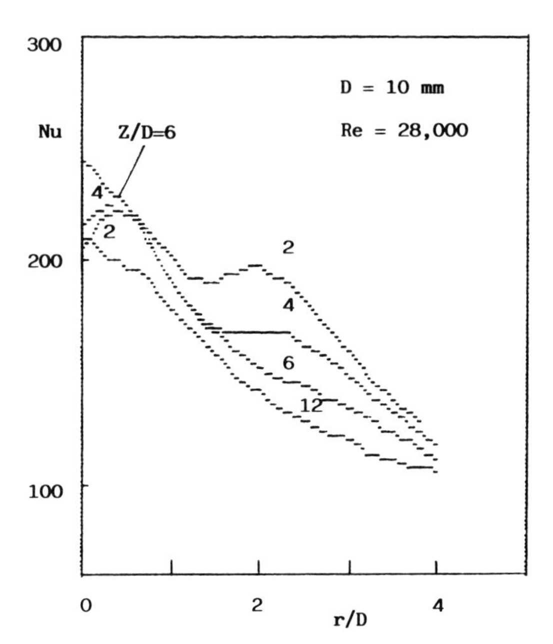

Fig. 22 Radial distributions of local Nusselt number over the plate.

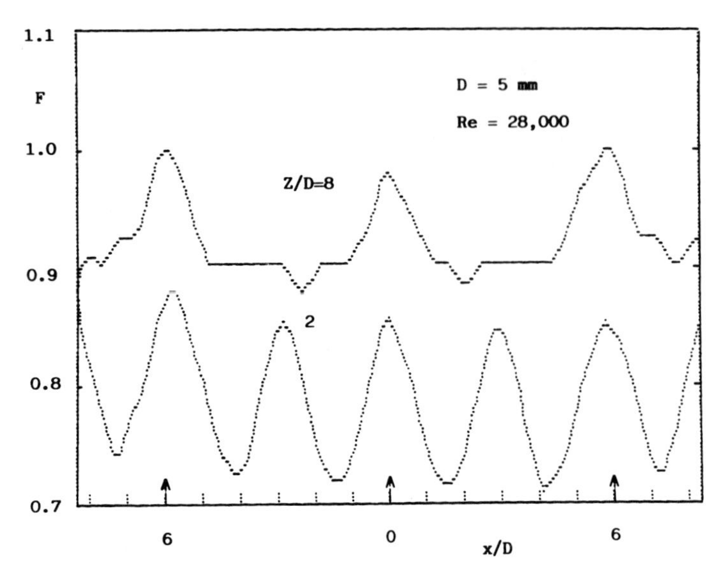

Fig. 23 Spatial distributions of local recovery factor along the central row of nozzles in a square array of 3 × 3 nozzles.

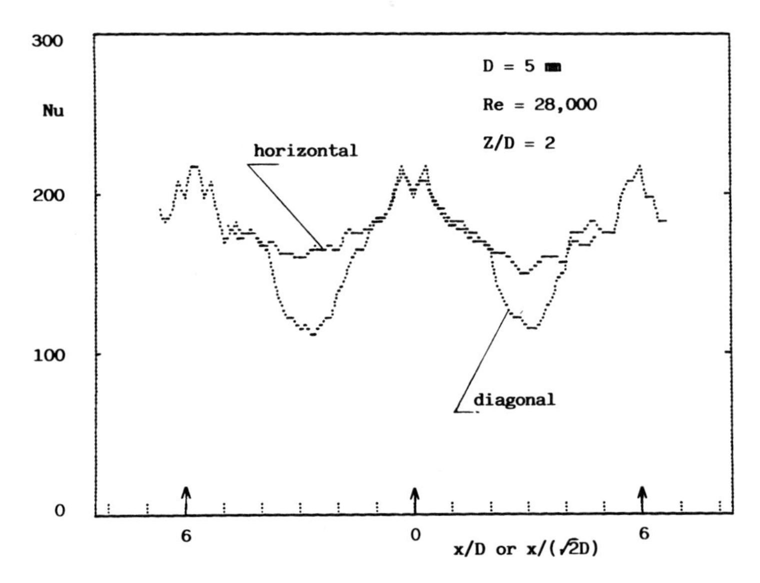

Fig. 24 Distributions of local Nusselt number over the plate along the central row and a diagonal line in a square array of 3 × 3 nozzles.

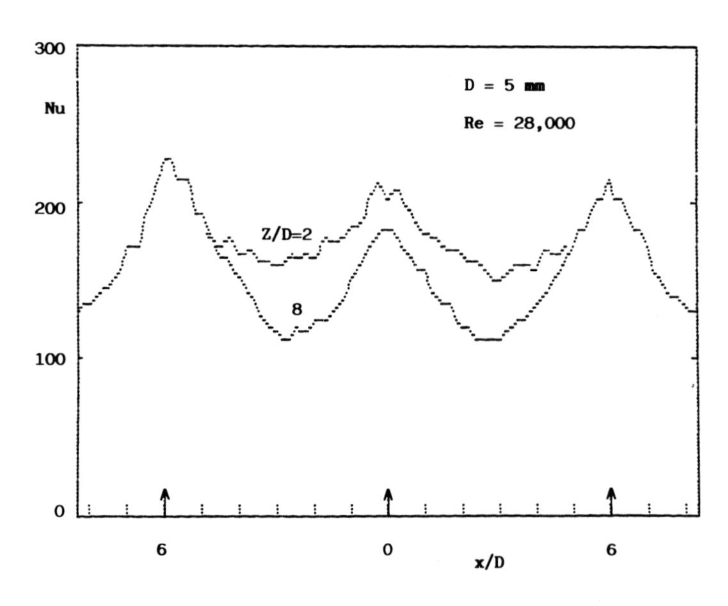

Fig. 25 Distributions of local Nusselt number over the plate along the central row of nozzles in a square array of 3 × 3 nozzles.

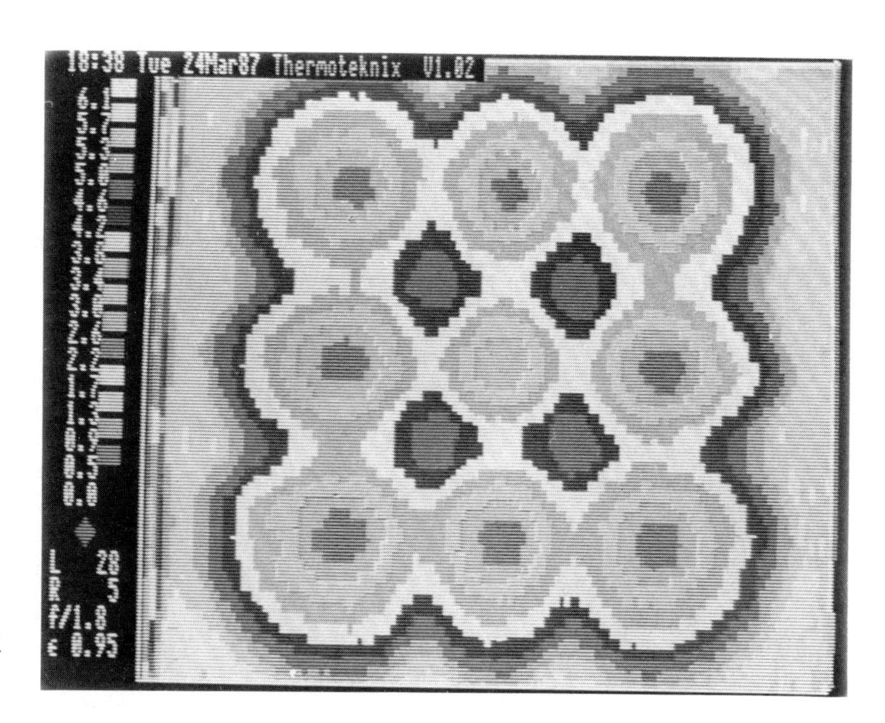

Fig. 26 Thermogram of the temperature-difference (between hot and cold images) map of a heated plate cooled by a square array of 3 × 3 nozzles.

Figure 23 shows the spatial distributions of the recovery factor recorded along the horizontal central row of nozzles for Re = 28,000, D = 5 mm (S/D = 6), T_0 = 18.2°C, and nozzle-to-plate distances Z/D = 2 and 8. Arrows indicate jet axes. For the smaller Z/D, local maxima occur at the stagnation point of each jet and midway between the centers of two neighboring jets, while local minima are present between them. The strong interaction between impinging jets (due to small Z/D and S/D) does not allow the recovery factor to attain unity midway between the stagnation regions of two neighboring jets. For Z/D = 8, local minima near the stagnation points and maxima midway between jets disappear, while maxima are present opposite the nozzles; the central peak is lower than the outer ones, due to interaction between jets before their impingement [37].

Figure 24 shows the spatial distributions of the local Nusselt number along the horizontal row of nozzles and along a diagonal. The horizontal profile was obtained using a 7° lens with an extension ring of 12 mm. Test conditions are the same as those reported in Fig. 23: Z/D = 2 and Q_0 = 0.256 W/cm^2. The heat transfer rate shows a peak opposite the nozzles, while minima occur at midpoints of the horizontal row and of the diagonal. The latter are much lower than the former [36]. In Fig. 25, the spatial distribution of Nu along the central row is plotted for two normal distances Z/D = 2 and 8. Other test conditions are the same as in Fig. 24. At Z/D = 8, a strong interaction between jets occurs before impingement; as a result the peak Nu numbers are no longer equal [37]. These results are shown globally in Fig. 26, where the temperature-difference map $T_w - T_{aw}$ is reported for Z/D = 8. The figure also represents the map of constant heat transfer coefficients. Minima of temperature difference (resp. maxima of the heat transfer coefficient) are found opposite the centers of the nozzles; maxima of $T_w - T_{aw}$ (resp. minima of h) occur at the center of each square formed by the four neighboring jet centers (secondary stagnation points). Furthermore, due to jet interaction, the local minimum value of the central jet is greater than values of minima of the outer jets.

NOMENCLATURE

Bi	Biot number, hL/k
c	specific heat coefficient or wing chord (Sect. VII)
c_p	specific heat of test gas at constant pressure
C_1, C_2	Planck's constants
D	nozzle diameter
E	total radiation intensity
E_λ	monochromatic radiation intensity
$E_{\lambda 0}$	blackbody monochromatic radiation intensity
F	recovery factor, $(T_{aw} - T_s)/(T_0 - T_s)$
F_0	recovery factor at stagnation point
h	convective heat transfer coefficient
k	thermal conductivity
L	sensor thickness or thermal level
M	Mach number
Nu	Nusselt number, hD/k for jet or hc/k for wing

Nu_0 Nusselt number at stagnation point of jet
Q heat flux rate
Q_1 heat loss per unit area and time
Q_0 Joule heating power per unit area
r radial distance from geometrical jet center
R radius of the Gardon foil gauge or thermal range
Re Reynolds number of jet, VD/ν
S nozzle spacing
St Stanton number, $h/(\rho_\infty V_\infty c_p)$
t time
t_M measurement time
T temperature
$\bar{T}$ mean temperature, $\int_0^L T\,dx/L$
T_{aw} adiabatic wall temperature
T_s static temperature of jet
T_w wall temperature
T_0 total temperature of jet or temperature at the center of the Gardon foil gauge
T_R temperature at the edge of the Gardon foil gauge
x, y spatial coordinates
V velocity
Z normal nozzle-to-plate distance

Greek Symbols

α thermal diffusivity
ΔT temperature difference $\Delta T = T_1 - T_2$ or temperature-reading error, Eqs. (3) and (4)
ε total hemispherical emittance (emissivity coefficient)
ε_λ spectral hemispherical emittance
η film-cooling effectiveness, $\eta = (T_e - T_{aw})/(T_e - T_c)$
λ radiation wavelength
ν kinematic viscosity
ρ density
σ Stefan–Boltzmann constant
Φ phase angle
ω circular frequency

Subscripts

a ambient
c coolant
e free-stream hot gas
1 sensor, heated face
2 sensor, rear face
∞ free-stream conditions

REFERENCES

1. Ohman, C., Measurement versus Imaging in Thermography, or What Is Resolution?, *Proc. 5th Infrared Inform. Exch. (IRIE), New Orleans,* vol. 2, pp. 65–70, 1985.
2. Carlomagno, G. M., and De Luca, L., Computer Aided Application of Infrared Scanning Radiometer to Measure Heat Fluxes, *The Computer in Experimental Fluid Dynamics,* Proc. Lect. Series, von Karman Institute Rhode-Saint-Genèse, Belgium, pp. 43–45, 1985.
3. Gardon, R., A Transducer for the Measurement of Heat Flow Rate, Trans. *J. Heat Transfer,* vol. 82, pp. 396–398, 1960.
4. Vidal, R. J., Transient Surface Temperature Measurements, CAL rep. 114, pp. 1–55, 1962.
5. Scott, C. J., Transient Experimental Techniques for Surface Heat Flux Rates, *Measurements Techniques in Heat Transfer,* AGARDograph 130, pp. 309–328, 1970.
6. Willeke, K., and Bershader, D., An Improved Thin-Film Gauge for Shock-Tube Thermal Studies, *Rev. Sci. Instrum.,* vol. 44, pp. 22–25, 1973.
7. Baines, D. J., Selecting Unsteady Heat Flux Sensors, Instr. Contr. Syst., pp. 80–83, 1972.
8. Thompson, W. P., Heat Transfer Gages, in *Methods of Experimental Physics,* ed. L. Marton and C. Marton, vol. 18B, pp. 663–685, Academic, New York, 1981.
9. Cooper, T. E., Field, R. J., and Meyer, J. F., Liquid Crystal Thermography and Its Application to the Study of Convective Heat Transfer, *J. Heat Transfer,* vol. 97, pp. 442–450, 1975.
10. Goldstein, R. J., and Timmers, J. F., Visualization of Heat Transfer from Arrays of Impinging Jets, *Int. J. Heat Mass Transfer,* vol. 25, pp. 1857–1868, 1982.
11. *Proc. 5th Infrared Inform. Exch. (IRIE), New Orleans,* vols. 1–2, 1985.
12. Carlomagno, G. M., and De Luca, L., Termografia all'Infrarosso nella Determinazione dei Coefficienti di Scambio Termico, *Proc. 3d Nat. Cong. UIT Heat Transfer, Palermo,* C125–133, 1985.
13. Carlomagno, G. M., and De Luca, L., La Risposta dei Sensori di Flusso Termico in Regime Instazionario, *Proc. 41th Nat. Cong. ATI, Napoli,* vol. 6, pp. 141–152, 1986.
14. Sartell, R. J., and Lorenz, G. C., A New Technique for Measurement of Aerodynamic Heating Distributions on Models of Hypersonic Vehicles, *Proc. 1964 Heat Transfer Fluid Mech. Inst.,* Stanford Univ. Press, Stanford, Calif., 1964.
15. Kafka, P. G., Gaz, J., and Yee, W. T., Measurement of Aerodynamic Heating of Wind Tunnel Models by Means of Temperature Sensitive Paint, *J. Spacecr. Rockets,* vol. 2, p 475, 1965.
16. Ceresuela, R., Betremieux, A., and Caders, J., Mesure de l'Echauffement Cinétique dans les Souffleries Hypersoniques au Moyen de Peinture Thermosensibles, *Rech. Aerosp.,* vol. 109, pp. 13–19, 1965.
17. Baker, S. S., and Matthews, R. K., Demonstration of the Thermographic Phosphor Heat-Transfer Technique as Applied to Aerodynamic Heating of External Stores, AEDC Rep. TR-73-128, 1973.
18. Jones, R. A., and Hunt, J. L., Use of Fusible Temperature Indicators for Obtaining Quantitative Aerodynamic Heat Transfer Data, NASA Tech. Rep. R-230, 1966.

19. Bynum, D. S., Hube, F. K., Key, C. M., and Diek, P. M., Measurement and Mapping of Aerodynamic Heating with an Infrared Camera, AEDC Rep. TR-76-54, pp. 1–33, 1976.
20. Myers, G., Van der Geest, J., and Rodrigue, A., Thermography and the Development of Advanced Gas Turbine Combustion Systems, *Proc. 5th Infrared Inform. Exch. (IRIE), New Orleans,* vol. 2, pp. 23–30, 1985.
21. Spence, T. M., Applications of Infrared Thermography in Convective Heat Transfer, Master's thesis, Naval Postgraduate School, Monterey, Calif., 1986.
22. Kantsios, A. G., Infrared Scanners for Temperature Measurement in Wind Tunnels, *Proc. 3d Infrared Inform. Exch. (IRIE), St. Louis,* pp. 149–154, 1976.
23. Thomann, H., and Frisk, B., Measurement of Heat Transfer with an Infrared Camera, *Int. J. Heat Mass Transfer,* vol. 11, pp. 819–826, 1967.
24. Blair, M. F., and Lander, R. D., New Techniques for Measuring Film Cooling Effectiveness, *J. Heat Transfer,* vol. 97, pp. 539–543, 1974.
25. Thomann, H., Heat Transfer Measurements at M = 7.17—First Step of a Comparison between Tunnel Test and Free Flight, Aeron. Res. Inst. Sweden Rep. 110, 1967.
26. Noble, J. A., and Boylan, D. E., Heat Transfer Measurements on a 5-deg Sharp Cone using Infrared Scanning and On-Board Discrete Sensor Techniques, AEDC Rep. TSR-78-V51, pp. 1–26, 1978.
27. Monti, R., Measurement of the Local Heat Transfer Coefficient by Non-Invasive Techniques in Aerodynamics, *Aerotecn. Miss. Spaz.,* vol. 65, pp. 55–62, 1986.
28. Gartenberg, E., Roberts, A. S., and Selby, J. B., Infrared Surface Imaging as a Flowfield Diagnostic Tool, *Proc. 12th Int. Con. Inst. Aerosp. Simul. Fac., Williamsburg, Va.,* pp. 343–349, 1987.
29. Bouchardy, A., and Durand, G., Processing of Infrared Thermal Images for Aerodynamic Research, *Proc. SPIE Applic. Digital Image Process. Conf., Geneva,* pp. 304–309, 1983.
30. Quast, A., Detection of Transition by Infrared Image Technique, *Proc. 12th Int. Cong. Instrum. Aerosp. Simul. Fac.,* Williamsburg, Va., pp. 125–134, 1987.
31. Heath, D. M., Winfree, W. P., Carraway, D. L., and Heyman, J. S., Remote Noncontacting Measurements of Heat Transfer Coefficients for Detection of Boundary Layer Transition in Wind Tunnel Tests, *Proc. 12th Int. Cong. Instrum. Aerosp. Simul. Fac., Williamsburg, Va.,* pp. 135–139, 1987.
32. Carlomagno, G. M., De Luca, L., Buresti, G., and Lombardi, G., Characterization of Boundary Layer Conditions in Wind Tunnel Tests through IR Thermography Imaging, in Applications of Infrared Technology, ed. T. L. Williams, Proc. SPIE, vol. 599, pp. 23–29, 1988.
33. Bradley, P. F., An Experimental Investigation to Determine the Effect of Window Cooling by Mass Injection for the Shuttle Infrared Leeside Temperature Sensing (SILTS) Experiment, NASA LaRC, Rep. NASA Tech. memo 80170, pp. 1–44, 1979.
34. Nutt, K. W., Space Shuttle Orbiter SILTS Pod Flow Angularity and Aerodynamic Heating Tests, AEDC Rep. TSR-79-V70, 1979.
35. Chocol, C. J., Infrared Imagery of Shuttle (IRIS), Task 1, Rep. NASA-CR-152123, pp. 1–151, 1977.
36. Gardon, R., and Cobonpue, J., Heat Transfer between a Flat Plate and Jets of Air Impinging on It, *Proc. 2d Int. Heat Transf. Conf., New York,* pp. 454–460, 1962.
37. Gardon, R., and Akfirat, J. C., Heat Transfer Characteristics of Impinging Two-dimensional Air Jets, *J. Heat Transfer,* vol. 88, pp. 101–108, 1966.
38. Buchlin, J. M., Thermography Application to Heat Transfer of Impinging Jets, *Flow Visualization and Digital Image Processing,* Proc. Lect. Ser. von Karman Institute, Rhode-Saint Genèse, Belgium, pp. 13–22, 1986.
39. Carlomagno, G. M., and De Luca, L., Heat Transfer Measurements by Means of Infrared Thermography, in *Flow Visualization IV: Proceedings of the Fourth International Symposium,* ed. C. Véret, pp. 611–616, Hemisphere, Washington, D.C., 1987.
40. Carlomagno, G. M., De Luca, L., and Meola, C., Misure Termografiche dello Scambio Termico tra una Lastra Piana e Getti d'Aria Incidenti su Essa, *Proc. 41th ATI Nat. Cong., Napoli,* vol. 6, pp. 153–160, 1986.
41. Goldstein, R. J., and Behbahani, A. I., Impingement of a Circular Jet with and without Cross Flow, *Int. J. Heat Mass Transfer,* vol. 25, pp. 1377–1382, 1982.
42. Goldstein, R. J., Behbahani, A. I., and Heppelmann, K. K., Streamwise Distribution of the Recovery Factor and the Local Heat Transfer Coefficient to an Impinging Circular Air Jet, *Int. J. Heat Mass Transfer,* vol. 29, pp. 1227–1235, 1986.
43. Martin, H., Heat and Mass Transfer between Impinging Gas Jets and Solid Surfaces, in *Advances in Heat Transfer,* ed. J. P. Hartnett and T. F. Irvine, Jr., vol. 13, pp. 1–60, Academic, New York, 1977.
44. Hrycak, P., Heat Transfer from Round Impinging Jets to a Flat Plate, *Int. J. Heat Mass Transfer,* vol. 26, pp. 1857–1865, 1983.
45. Obot, N. T., Majumdar, A. S., and Douglas, W. J. M., The Effect of Nozzle Geometry on Impingement Heat Transfer under a Round Turbulent Jet, ASME Paper 79-WA/HT-53, 1979.

Chapter 33

Gas Turbine Disk Cooling Flows

J. R. Pincombe

I INTRODUCTION

Improvements in the performance and efficiency of modern gas turbine engines have been made by increasing the compression ratios and maximum temperatures at which they operate. These increases have been made possible by the use of new materials and improvements in cooling technology. The cooling is accomplished, as illustrated in Figs. 1*a, b*, by allowing air bled from the compressor stages to circulate around the hot components. One of the problems of great importance to the gas turbine designer is the heat transfer resulting from the flow of air circulating in the various chambers or cavities formed by the turbine disk assembly of the rotor. In some cases these cavities are sealed at their inner and outer radii; in other cases they may be open to allow the entry and exit of cooling air. Under "real" engine conditions these cooling flows are usually turbulent and may be subject to significant buoyancy effects.

In order to gain a better understanding of these flow structures a research program using flow visualization and laser Doppler anemometry (LDA) was undertaken. The approach was to construct a simplified experimental rig that incorporated the essential features of the more complex coolant-flow arrangements to be found in a gas turbine. For this a cylindrical cavity, formed by two annular disks and a peripheral shroud, was used.

By having either one disk or both disks rotating, and by varying the flow entry and exit positions, it was possible to model a number of coolant-flow geometries.

Figure 2*a* shows a simplified model of an air-cooled turbine disk rotating close to the stationary engine casting. This arrangement takes the form of a partially shrouded rotating disk. The "cooling air" enters the cavity axially through a central hole in the stationary disk and leaves the cavity radially via the axial clearance between the rotating disk and the stationary shroud. In this case, as well as removing heat from the disk, the coolent flow is also used to seal the cavity (that is, to prevent ingress of hot gas into the cavity) from the mainstream turbine flow.

Figure 2*b* shows the experimental model used to study the flow between co-rotating compressor disks. In this case both disks, which rotate about a common axis, are sealed at their peripheries by a cylindrical shroud. To simulate cooling air on its way to the turbine blades a central axial flow enters the cavity through one disk ("upstream disk") and leaves axially through a central hole in the other disk ("downstream disk"). Thus the arrangement forms a rotating cylindrical cavity with axial throughflow of air.

A rotating cylindrical cavity with a radial outflow of air, as shown in Fig. 2*c*, is used to simulate the cooling flow between co-rotating turbine disks. In this case the flow enters the cavity through a central hole in one of the disks and leaves the cavity radially via a series of holes in the peripheral shroud.

The aspect ratios (length to diameter) of the cylindrical cavities, used to study the respective flow geometries just described, were less than unity. Research has also been undertaken in rotating cavities with l/d ratios much greater than unity (better described as rotating tubes). As shown in

I wish to acknowledge those who have contributed to the work reported here. In particular I would like to thank Dr. J. M. Owen for the encouragement and advice he has given over the years. I would also like to thank the Science and Engineering Research Council and Rolls-Royce Ltd. for funding the research.

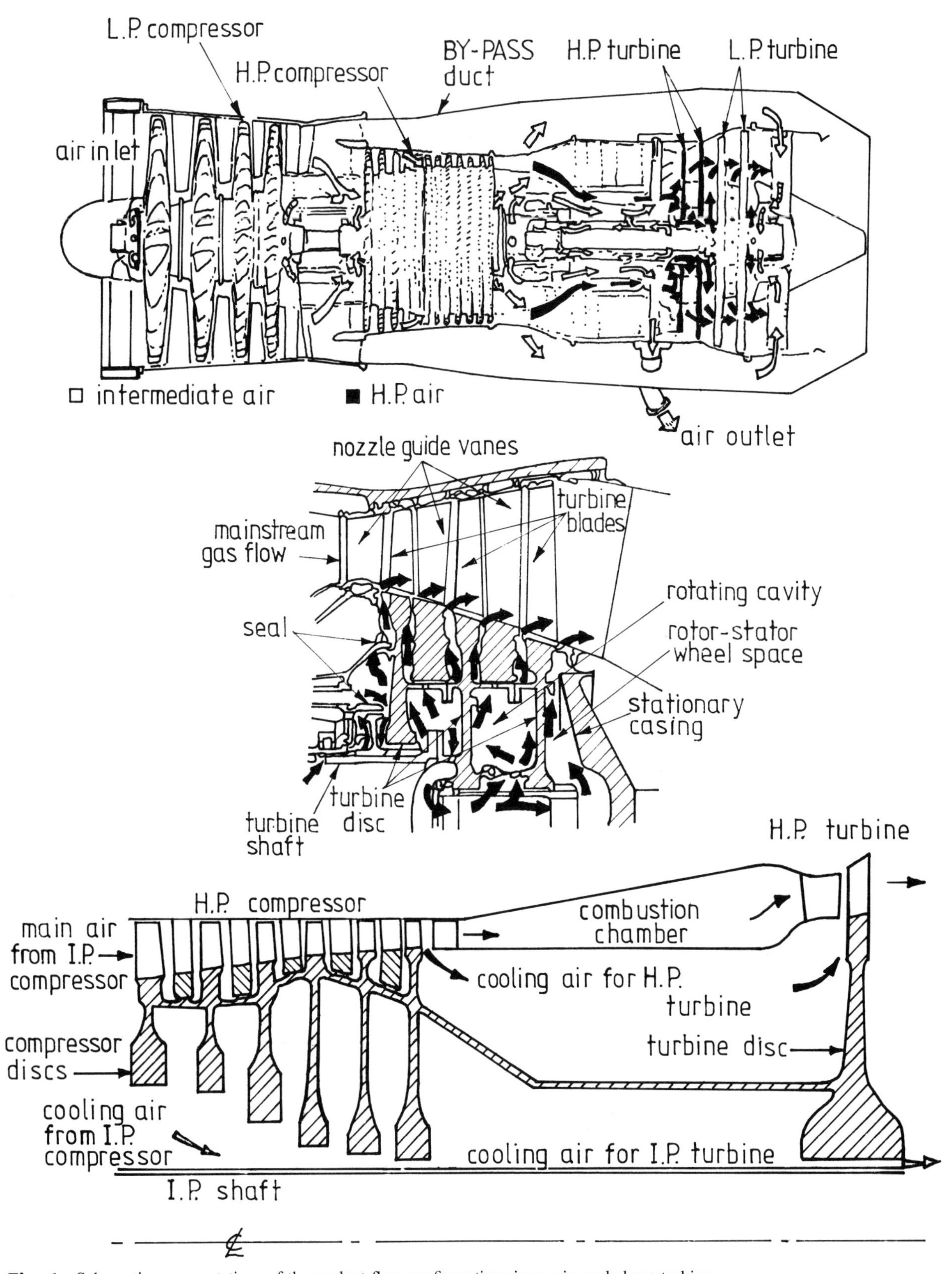

Fig. 1 Schematic representations of the coolant-flow configurations in an air-cooled gas turbine.

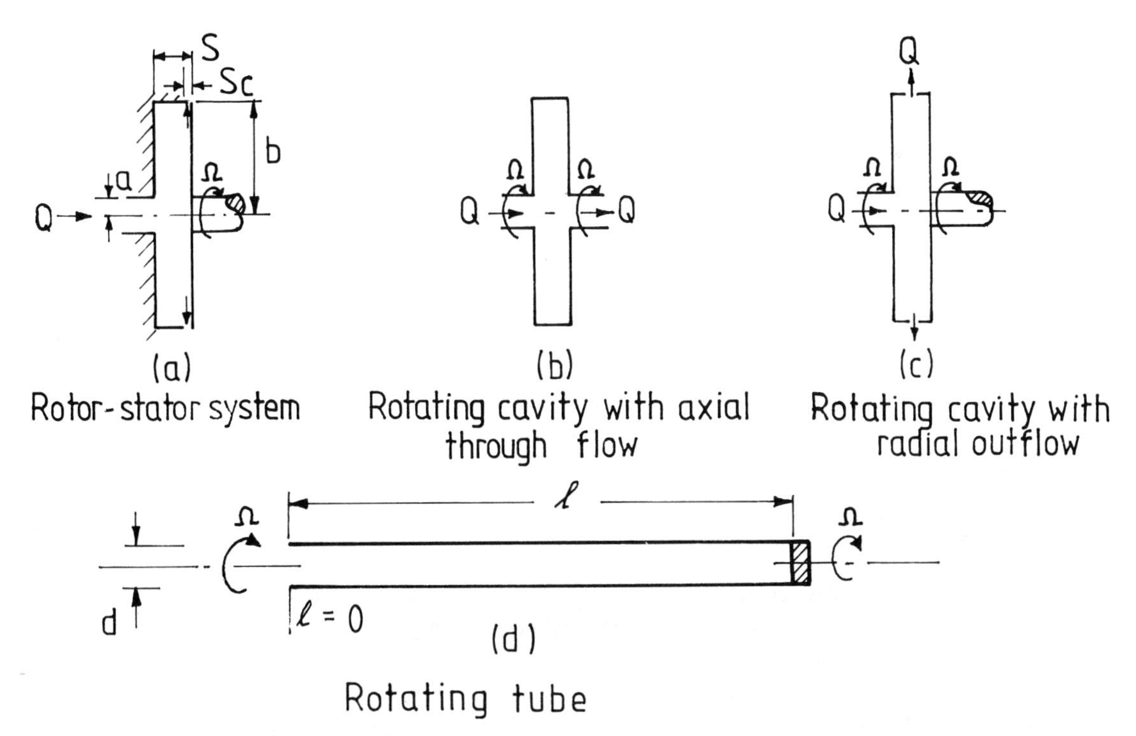

Fig. 2 Schematic drawings of rotating cavity-flow geometries.

Fig. 2*d*, the basic arrangement is a rotating tube with one end sealed and the other end open. While the engineering application of this arrangement is less well defined than those of the geometries previously described, flow visualization has revealed the existence of an interesting type of flow phenomenon.

The experimental techniques that were applied to study the flow structure, in the rotating cavity geometries described above, were flow visualization using sheet illumination and smoke [1] to provide a qualitive picture over the full flow field, and LDA [2] to provide quantitative measurements of velocity at point locations within the flow field. The flow visualization was used mainly to aid interpretation of the LDA measurements. The choice of flow visualization technique was limited by the class of flow (isothermal, incompressible) and the fact that optical access was through rotating windows. However, the technique used, as will be discussed in the following sections, proved to be very effective in advancing understanding of the nature of the rotating flows under investigation. The optical program also complemented a computational fluid dynamics study and a separate experimental heat transfer investigation carried out in the Thermo-Fluid Mechanics Research Centre at the University of Sussex.

Details of the experimental apparatus are given in the next section, and are followed by a discussion of the experimental results obtained for each of the cases mentioned above.

II EXPERIMENTAL APPARATUS

A brief description of the experimental apparatus is given here; further information can be obtained from [3].

A The Rotating-Disk Systems

For the rotor–stator system shown schematically in Fig. 2*a*, the rotor was a plane aluminum disk of radius, $b = 190$ mm, and the stator and shroud were made from acrylic (transparent thermoplastic). The gap s between the rotor and the stator was 19 mm ($G = 0.1$, where $G \equiv s/b$), and the shroud clearance s_c was maintained constant at 1.9 mm ($G_c = 0.01$ where $G_c \equiv s_c/b$). The rotor was driven up to 4000 rev/min ($\mathrm{Re}_\phi \simeq 10^6$, $\mathrm{Re}_\phi \equiv \Omega b^2/\nu$) by means of a variable-speed electric motor, and the coolant flow rate was supplied through the inlet pipe ($a = 19$ mm) at rates up to 0.06 m^3/s ($C_w = 2 \times 10^4$, $C_w \equiv Q/\nu b$).

For the rotating cavity with axial throughflow shown in Fig. 2*b*, the rig comprised two acrylic disks, of radius $b =$ 190 mm and a cylindrical shroud. The radius of the rotating inlet and outlet pipes was $a = 19$ mm, and the spacing between the disks could be varied from $G = 0.267$ to $G = 0.533$. The cavity was rotated up to $\mathrm{Re}_\phi = 4 \times 10^5$, and axial Reynolds numbers of $\mathrm{Re}_z \simeq 10$ ($\mathrm{Re}_z \equiv 2\overline{W}a/\nu$, with $\overline{W}$ being the bulk-average velocity in the inlet pipe) were produced.

The rotating cavity with radial outflow shown in Figure 2*c* was based on the rig just described. The only difference was that the acrylic shroud contained thirty 12.7-mm-dia. holes equi-spaced in the mid-axial plane between the disks.

To study the flow structure inside a rotating tube with one end sealed and the other end open, a glass tube of length l = 397 mm and internal diameter d = 14.5 mm (giving l/d = 27.4) was inserted into a metal tube (with an internal diameter sufficient to allow entry of the glass tube but tight enough to provide a good mechanical coupling between them) mounted on bearings, which could be rotated via pulleys by an electric motor up to speeds of 5000 rev/min. This arrangement allowed optical access to the glass tube in the overhanging sections (of equal length) on each side of the metal tube.

B Flow Visualization Apparatus

Illumination of the cavity was achieved using a 2-W argon–ion laser. As shown in Fig. 3, the illuminating laser was used in conjunction with a cylindrical and a collimating lens to produce "slit illumination" in a plane through the axis of rotation (hereafter referred to as "the r–z plane"). The focal lengths of the lenses were chosen to produce a beam that, viewed normal to the r–z plane, was slightly wider than the cavity width s; viewed normal to the r–θ plane, the beam was brought to a focus at the axis of rotation of the cavity. The resulting "sheet" of light that was, less than 1 mm thickness.

For photography, a camera (operating in the "aperture-preferred" mode) was arranged with the axis of its lens normal to the illuminated plane. With an f-1.8 lens and ASA 400 (film uprated in processing to 1600 ASA), a minimum exposure time of 1/60 s was required. For flow visualization, the coolant air was "seeded" by means of a smoke generator. This vaporized oil and the resulting smoke were driven off with carbon dioxide. The size of the oil particles was approximately 0.8 μm dia. For LDA, the air was seeded by means of a "microfog lubricator," which generated silicon oil particles of approximately 2 μm dia.

For flow visualization in a rotating tube, a similar optical arrangement, shown in Fig. 3, was used in conjunction with an argon–ion laser. In this case the collimated light sheet (of width just greater than the glass tube diameter) was aligned along the length of the tube to illuminate the r–z plane. The smoke generator just described was used to pulse smoke around the open end of the rotating tube. The camera, arranged with its optical axis normal to the plane of illumination, was used to record the progression of the smoke down the tube.

C The Laser Doppler Anemometer

For measurements in the rotating-cavity rig shown in Figs. 2*b, c*, the LDA optics were arranged in a forward-scatter real-fringe mode, "looking through" the disk. This is illustrated in Fig. 3, where the incident beams are arranged to detect the radial component of velocity; by rotating the beams through 90° it was possible to measure the tangential component. Alternatively by arranging the optics looking through the shroud it was possible to measure the axial and tangential components of velocity.

The transmitting optics comprised a 5-mW HeNe laser (wavelength 632.8 nm), a rotating diffraction grating, steering prisms, and transmitting lens. The receiving optics comprised an f-1.8, 50-mm lens, a 50-μm pinhole aperture, and a photomultiplier tube, all of which were contained in a single housing.

The diffraction grating was a bleached radial grating

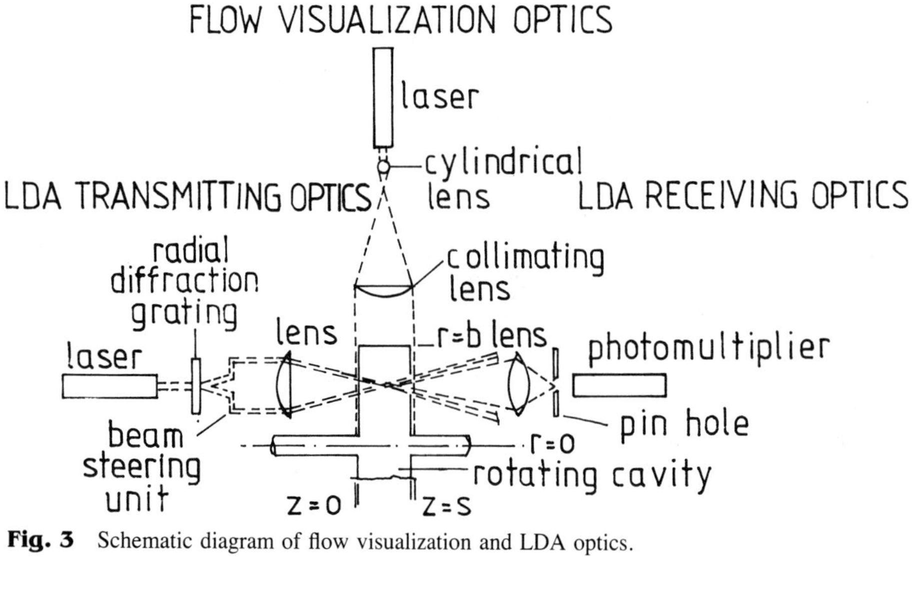

Fig. 3 Schematic diagram of flow visualization and LDA optics.

with 21,600 lines on an effective diameter of 133 mm. Approximately 50% of the incident light was transmitted in the first-order beams, and a rotational speed of 2570 rev/min produced a frequency shift of 1.85 MHz. The grating could be yawed about its optical axis, allowing measurements to be made in either of two orthogonal directions. The emergent first-order beams were reflected by steering prisms to create two parallel beams, with a separation distance of 50 mm. Using a transmitting lens of 200-mm focal length, the probe volume was approximately 0.25 mm dia. and 2 mm long, and the fringe spacing was approximately 2.5 μm. The Doppler signal from the photomultiplier was processed by a tracking filter, which had an upper frequency limit of 15 MHz. The tracker output voltage, which was proportional to the magnitude of the velocity component being measured, was measured using a time-domain analyzer which produced a true-time average of the mean velocity.

III THE FLOW AROUND AN AIR-COOLED TURBINE DISK

A Introduction

In this case (see Fig. 2*a*) the coolant enters the cavity through a central hole in the stationary disk and leaves the cavity via the peripheral gap between the stationary shroud and the rotating disk. As mentioned in Sec. I, the coolant serves to prevent the ingress of the hot mainstream gases (through the clearance gap) into the cavity; it is of primary importance to establish the minimum coolant flow rate required to prevent ingress [4].

B The Flow Structure

Using the argon–ion laser (at an output power of approximately 1 W) in conjunction with the optical arrangement (described in Sec. II.B) to illuminate the r–z plane of the experimental rig (described in Sec. II.A), and injecting smoke into the coolant flow upstream of the cavity, it was possible to observe the flow structure between the rotor and the stator. A series of photographs was taken of the progression of the smoke through the cavity. The photographs that appear in Figs. 4 and 5 were chosen to provide the maximum information on the flow structure. It should be noted that only the top half of the cavity is shown; the coolant flow enters axially near the bottom left-hand side and leaves near the top right-hand side of the photographs. The camera shutter (operating at 5 frames per second) was not synchronized with the disk rotation.

Figures 4 and 5 show the smoke patterns obtained for four values of coolant flow rate, C_w, in the cavity with rotational Reynolds number $Re_\phi = 5 \times 10^4$, axial spacing $G = 0.1$ and a gap clearance $G_c = 0.01$. Figures 4*a–c* show the progression of the smoke through the cavity for $C_w = 200$. Referring to Fig. 4*a*, it can be seen that the flow from the axial jet (entering the cavity at $z = 0$) impinged on the rotating disk (at $z = s$) where it moved radially outward in a thin boundary layer, attached to the rotor, and left the cavity (at $r = b$) through the axial clearance between the shroud and the rotor.

Smoke can also be seen near the stator, where it was observed that the direction of the fluid motion was radially inward. The photograph has captured the smoke when it had moved inward to reach $r/b = 0.6$. The overall flow structure comprises a single recirculation zone with outflow on the rotor and inflow on the stator. Further tests were conducted in which the smoke was introduced into the air (outside the cavity) surrounding the shroud; this revealed that for the condition shown in Figs. 4*a–c*, some of the fluid circulating in the cavity was in fact coming from outside the cavity through the peripheral clearance between the shroud and the rotor. This ingress of fluid was caused by the disk rotation's reducing the pressure in the cavity below that of the ambient level outside the cavity.

Figure 4*b* reveals a series of vortex instabilities in the inflow region on the stator. Close inspection of the smoke pattern also reveals that for $r/b < 0.6$, the boundary layer on the rotor was not entraining flow from the core of fluid between the rotor and the stator; for $r/b > 0.6$, however, there is evidence (by the way in which the smoke has streaked) that the boundary layer on the rotor is entraining fluid and behaving like a "free-disk" boundary layer [5]. The final photograph in this series, Fig. 4*c* shows that at this flow rate, the change from an impinging boundary layer to an entraining boundary layer occurs without disturbance to the flow on the rotor.

The flow structure shown in Figs. 4*d–f* for $C_w = 280$ was broadly unchanged with unsteady radial inflow on the stator and entrainment of fluid into the boundary layer on the rotor for $r/b > 0.6$. There is, however, evidence of instability in the impinging (nonentraining) jet boundary layer on the rotor at $r/b \approx 0.42$.

Figures 5*a–c* show the progression of the smoke for $C_w = 400$. In this case the impinging boundary layer on the rotor appears to extend almost to the shroud; however, the localized instability reported above for $C_w = 280$ increased in intensity and scale extending out to about half the cavity width at $r/b \approx 0.42$, as can be seen in Fig. 5*c*. The unsteadiness of the flow on the stator also increased in scale and intensity.

A further increase in coolant flow rate, as shown in Figs. 5*d–f* for $C_w = 490$, caused a marked increase in the instability of the flow in the rotor boundary layer; in fact the disturbances from the rotor extend right across the cavity to the stator, dividing the cavity into two regions: an inner recirculation zone and an outer recirculation zone. At higher flow rates, rapid mixing of the smoke made it difficult to

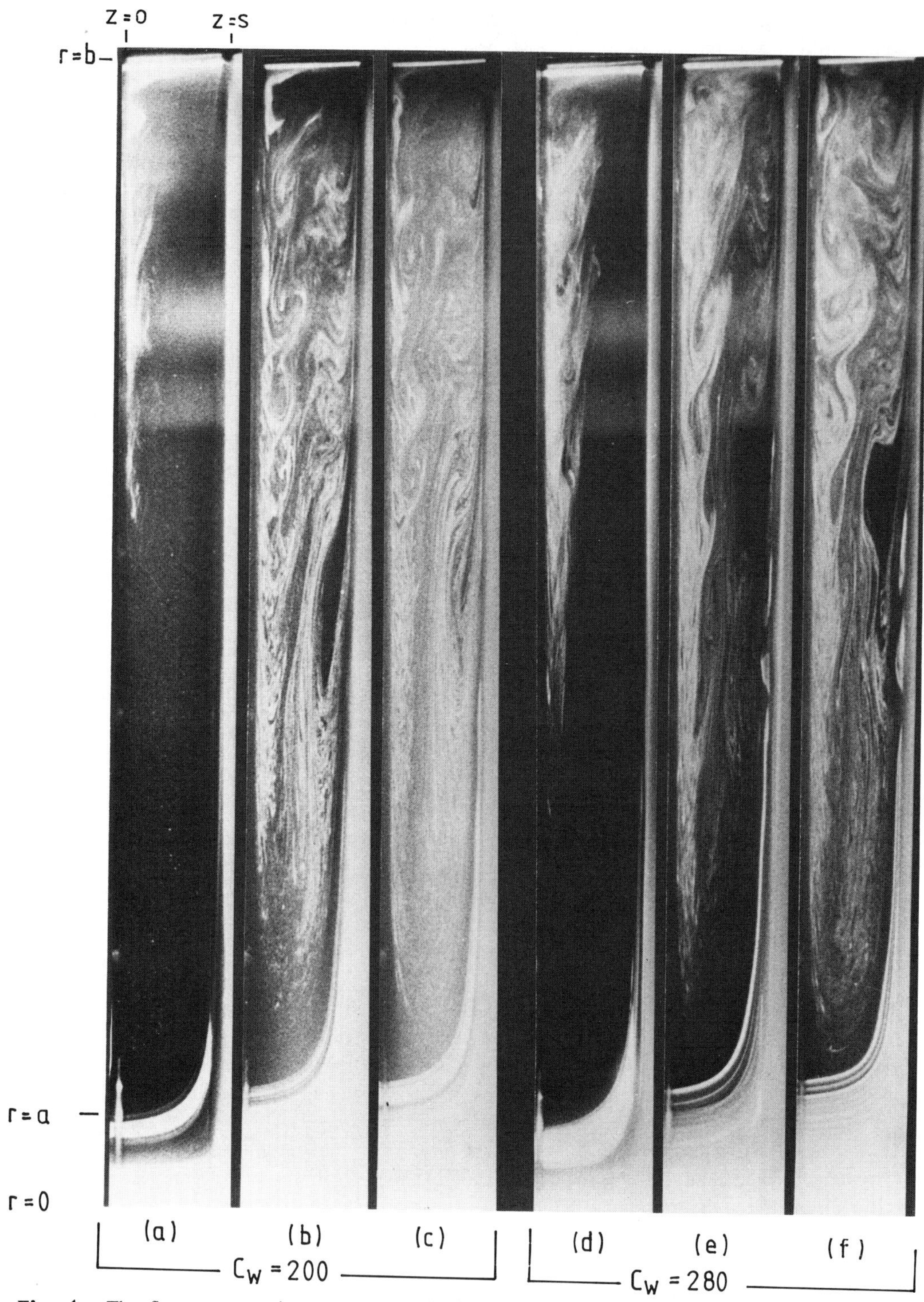

Fig. 4 The flow structure in a rotor–stator cavity for $Re_{\phi} = 5 \times 10^4$, $G = 0.1$, and $G_c = 0.01$.

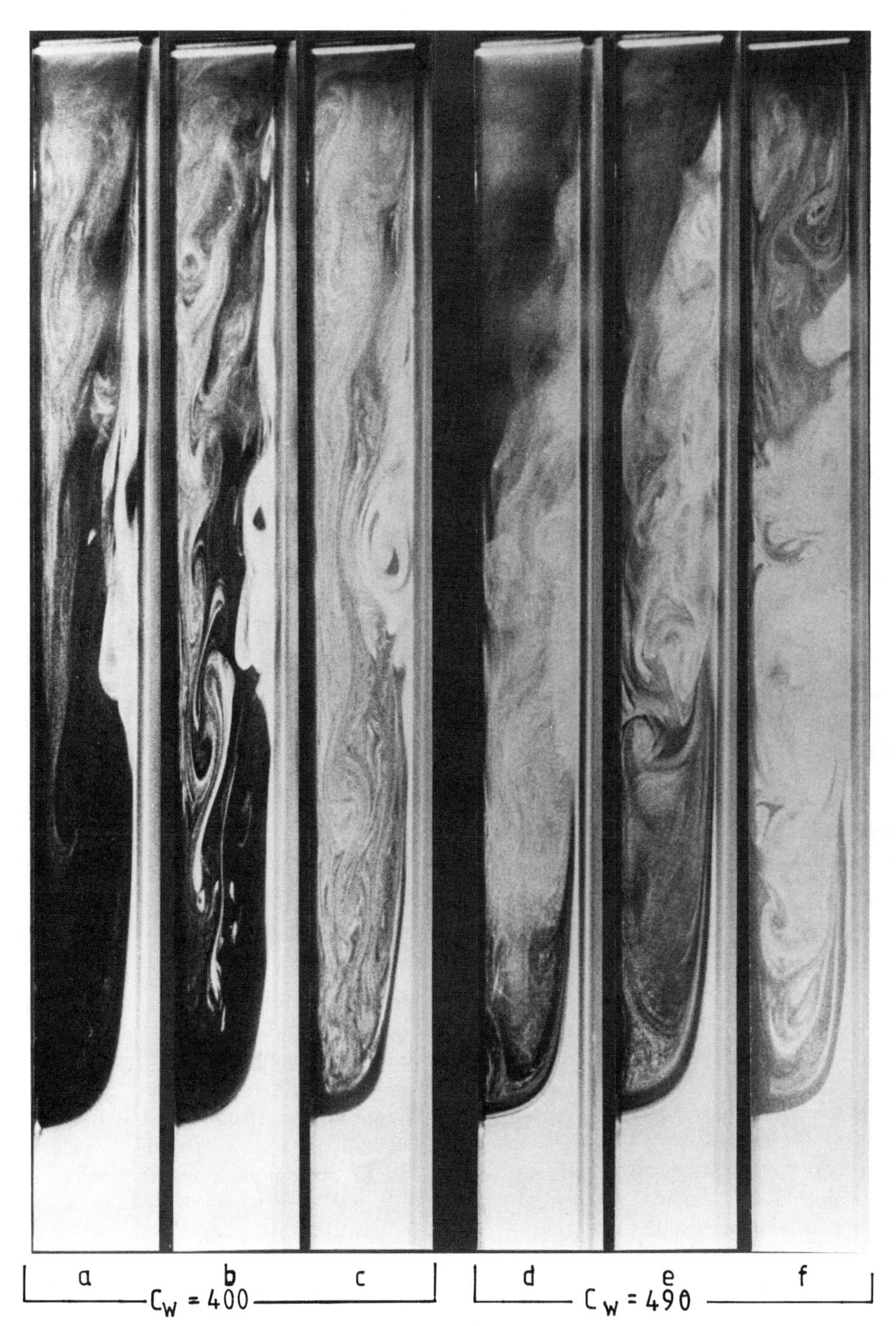

Fig. 5 The flow structure in a rotor–stator cavity for $Re_\phi = 5 \times 10^4$, $G = 0.1$, and $G_c = 0.01$.

discern the flow structure in detail; however, the overall impression was that the flow structure remained the same.

By using peripheral seeding it was possible to quantify, for a particular rotational Reynolds number and gap clearance, the minimum flow rate C_w to prevent the ingress of flow into the cavity through the clearance gap. To do this smoke was continually pulsed into the air surrounding the shroud, and the coolant flow was progressively increased from zero until no smoke entered the cavity; this was repeated for a range of values of Re_θ. The correlation obtained using flow visualization for $G_c = 0.01$ (the value used for the photographs shown in Figs. 4 and 5) was

$$C_{w_{\min}} = 0.0075 \, \mathrm{Re}_\phi \tag{1}$$

Equation (1) predicts that for $\mathrm{Re}_\phi = 5 \times 10^4$ the cavity was sealed when $C_w = 375$. Therefore for Fig. 4 ingress was present for the two flow rates shown; however, no ingress was present for the two flow rates shown in Fig. 5.

No detailed velocity measurements using LDA were undertaken for this flow geometry. Some measurements of the tangential component of velocity midway between the rotor and the stator were obtained in a rig similar to that described above, and these measurements were in general consistent with those reported in [6].

IV THE FLOW AROUND AIR-COOLED, CO-ROTATING COMPRESSOR DISKS

A Introduction

In this case (see Fig. 2*b*) flow enters axially through a central hole in one disk and leaves axially through a central hole in the other disk.

Flow visualization revealed that over certain ranges of axial Rossby number ε_z ($\varepsilon_z = \overline{W}/\Omega a$), the central axial jet ceased to be axisymmetric and precessed violently about the axis of rotation. This behavior, which was attributed to vortex breakdown (an abrupt change in the structure of a swirling axial flow), caused a large increase in the circulation within the cylindrical cavity [7].

A review of breakdown phenomena in swirling flows is given in [8].

B Flow Structure in a Stationary Cavity ($\varepsilon_z = \infty$)

By illuminating the r–z plane and injecting smoke into the flow upstream of the cavity, it was possible to observe the flow structure. In this case, because the flow could be nonaxisymmetric, the complete cavity was included in the field of view covered by the camera.

Transition from laminar to turbulent flow in the inlet pipe was found to be delayed well beyond the traditionally accepted value for a circular pipe of $\mathrm{Re}_z \approx 2300$, in fact the onset of fully turbulent flow could be delayed up to $\mathrm{Re}_z = 2 \times 10^4$. By inserting a trip upstream of the cavity, transition could be made to occur at $\mathrm{Re}_z \approx 2300$.

The effect of transition in the axial jet on the flow structure in a stationary cavity is shown in Fig. 6, for $\mathrm{Re}_z = 0.5 \times 10^4$ and $G = 0.533$. Figs. 6*a–c* show a laminar jet entering the cavity through a hole in the upstream disk (on the left) and leaving the cavity through a hole in downstream disk (on the right). The smoke reveals that the jet was mainly axisymmetric with well-defined boundaries. There is also evidence of a weak nonaxisymmetric secondary recirculation inside the cavity. (This asymmetry was caused by buoyancy effects due to small temperature differences between the incoming air and the walls of the cavity.)

In contrast to this are the smoke patterns shown in Figs. 6*d–f* for a turbulent jet (the flow tripped). It can be seen that the jet boundaries were irregular, and there was a significant increase of entrainment of air from the jet to the cavity. This caused a relatively powerful axisymmetric toroidal vortex to be generated in the cavity, with the vortex center at $r/b \simeq 0.8$ and $z/s \simeq 0.5$.

C Flow Structure in a Rotating Cavity

The effect of rotation on the turbulent flow structure in a cavity with a gap ratio of $G = 0.533$ is shown in Fig. 7. There appeared to be no discernible change in the behavior of the powerful toroidal vortex generated by a turbulent axial jet (and illustrated in Fig. 6*f* until $\varepsilon_z \lesssim 100$, where an occasional precession of the main jet about the central axis was seen. As the Rossby number was further reduced, the jet precession occurred more regularly and eventually became continuous. This form of spiral breakdown is shown in Figs. 7*a–c* for $\varepsilon_z = 29$. The amplitude of the jet's precission increased to reach a maximum at $\varepsilon_z \approx 21$, whereafter, at higher rotational speeds, the jet suddenly stopped precessing. If, for $\varepsilon_z < 21$, the Rossby number were increased (by speed reduction or by a flow-rate increase), the jet would not resume precessing until $\varepsilon_z \approx 23$.

For $\varepsilon_z < 21$, the jet appeared to be axisymmetric with occasional oscillations of the jet boundaries; this will be referred to as mode-IIa breakdown. Occasional excursions of the jet into the cavity were observed, and for $\varepsilon_z < 10$, an inner core of smoke with imprecise boundaries was formed, as shown in Figs. 7*d*, *e* for $\varepsilon_z = 5$.

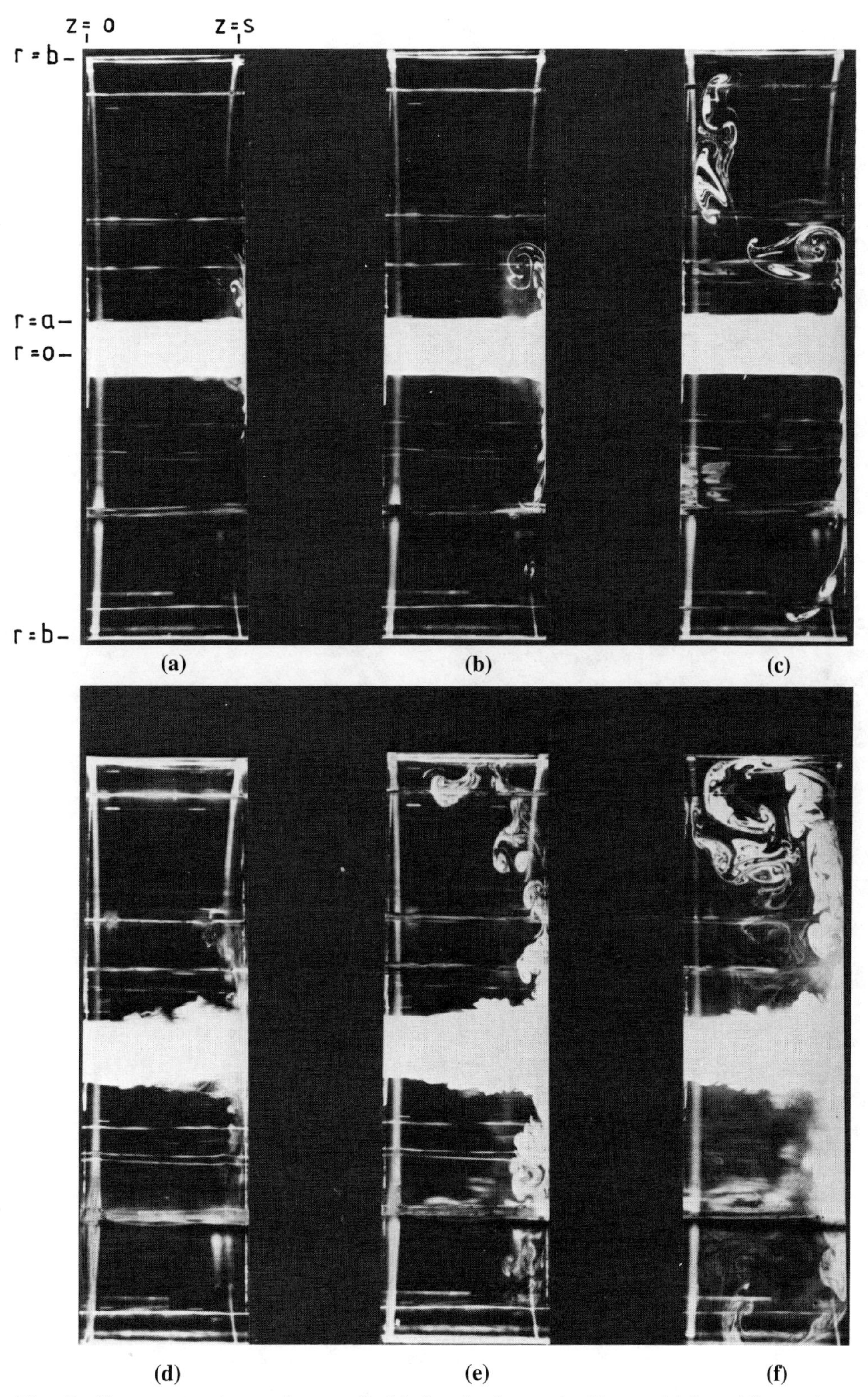

Fig. 6 Flow structure in a stationary cylindrical cavity ($\varepsilon_z = \infty$) with an axial throughflow and $G = 0.533$, for $Re_z = 0.5 \times 10^4$; (*a–c*) Flow with a laminar central jet; (*d–f*) flow with a turbulent jet.

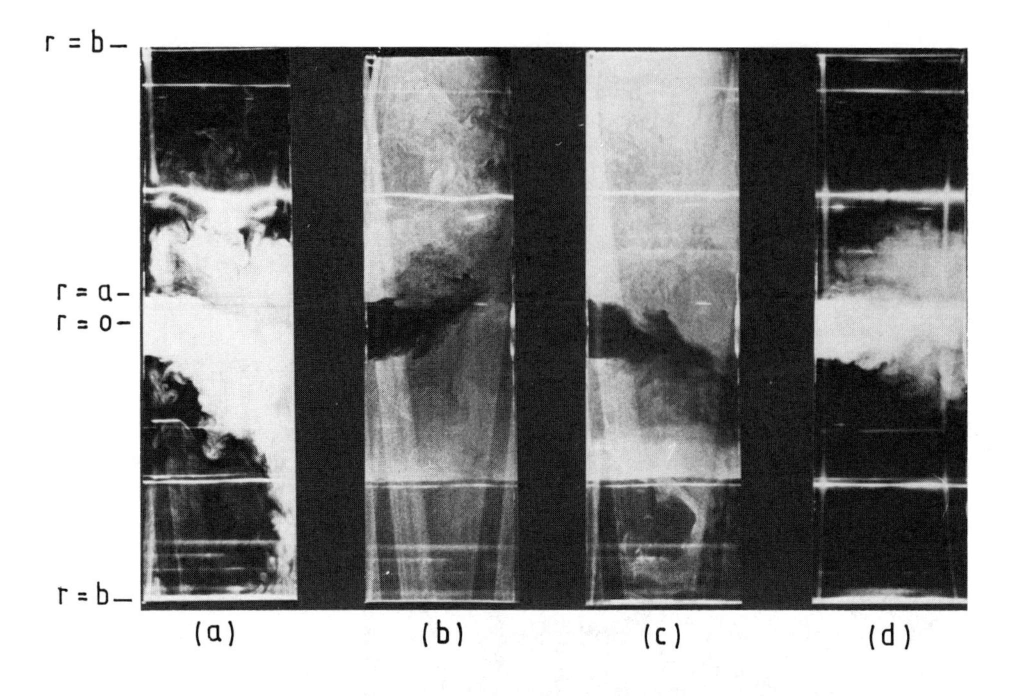

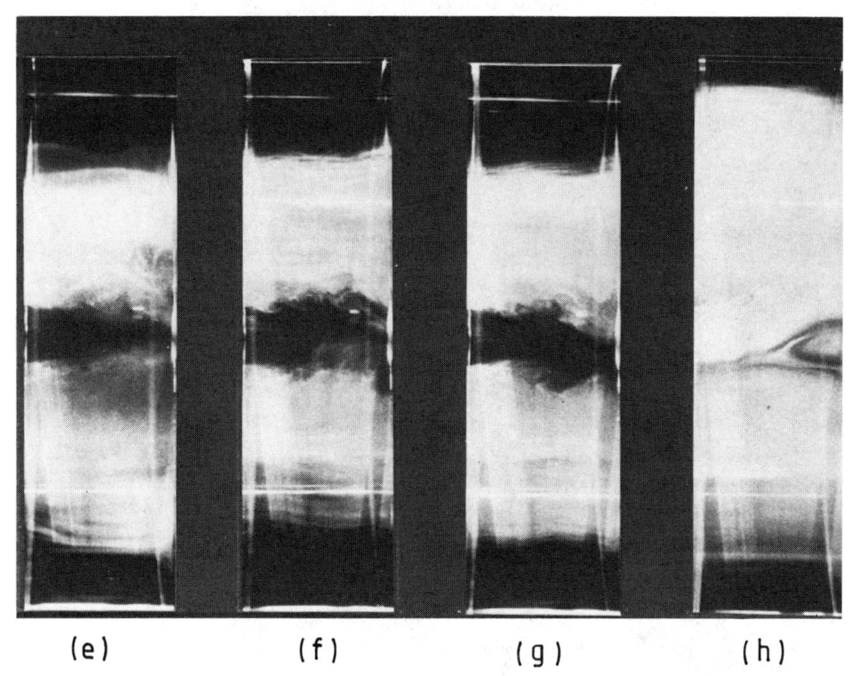

Fig. 7 Flow structure in a rotating cylindrical cavity with an axial thoughflow (turbulent central jet) and $G = 0.533$, for $\mathrm{Re}_z = 1 \times 10^4$; (*a–c*) $\varepsilon_z = 29$; (*d, e*) $\varepsilon_z = 5$; (*f, g*) $\varepsilon_z = 2.0$; (*h*) $\varepsilon_z = 1.25$.

At $\varepsilon_z \approx 2.6$, the jet became nonaxisymmetric again, as revealed in Figs. 7*f*, *g*. This mode of behavior indicated that a second regime of spiral breakdown, similar to that observed at the higher Rossby number, had been identified.

At $\varepsilon_z \approx 1.5$, signs of intermittent reverse flow were observed on the center line at the downstream end of the jet, and the axial flow appeared to separate around an apparent obstruction. The effect can be seen in Fig. 7*h*. This axisymmetric mode of breakdown caused a significant increase in the shedding air from the jet to the cavity, resulting in an increase in the size of the smoke core.

The occurrence of the intermittent reverse flow reached a maximum at $\varepsilon_z \approx 1.0$. For $\varepsilon_z \simeq 1$ the smoke core appeared to reduce in size. As the Rossby number was reduced, the smoke core continued to shrink.

These results were for a turbulent jet; in the case of

rotating cavity with laminar jet, the dramatic spiral mode of breakdown observed for $100 > \varepsilon_z > 21$ was not present. However at low values of Rossby number $\varepsilon_z < 2$, the modes of behavior were similar to that of a turbulent jet.

Recent tests conducted in a heated cavity suggest that instabilities in the axial jet play a significant role in increasing the heat transfer from the hot disk to the air in the cavity.

As was stated in Sec. I, flow visualization was used mainly as a tool to aid interpretation of the LDA measurements. While it would be inappropriate to discuss these measurements in detail, the velocity data presented in Fig. 8 are complementary to the smoke patterns shown in Figs. 6 and 7. In particular it is interesting to note the significant reduction of the axial component of velocity W [normalized to the centerline value $W(0)$, measured with the cavity stationary], shown in Fig. 8*a*, near the centerline when the central axial jet is precessing (over the range of Rossby number $21 < \varepsilon_z < 100$). Note how it almost returns to the value obtained for the stationary cavity when the jet regains its axisymmetry (for $\varepsilon_z < 21$). Also, in Fig. 8*b*, how increasing the rotational speed of the cavity suppresses the radial motion associated with the strong toroidal vortex, described above. (The measured radial component of velocity U is normalized to the bulk average velocity $\overline{W}$ in the inlet pipe.) The effect of rotation speed on the tangential component of velocity, V_ϕ is shown in Figs. 8*c*, *d* (where the measured values have been normalized to Ωb and Ωr respectively). There is a change of the radial distribution from that of a free vortex to that of a forced vortex.

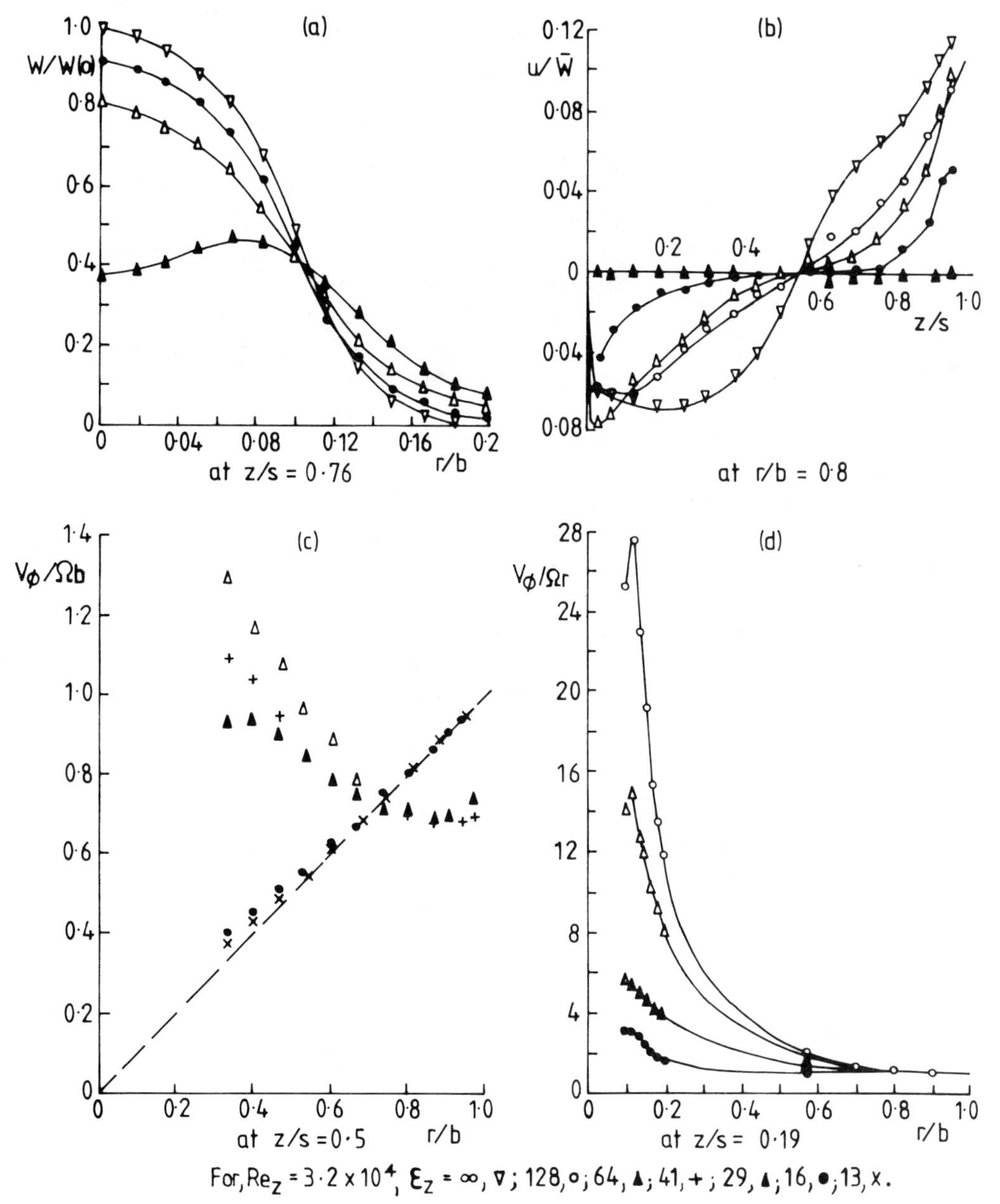

Fig. 8 Velocity measurements inside a rotating cavity with an axial throughflow and $G = 0.533$.

V THE FLOW AROUND AIR-COOLED, CO-ROTATING TURBINE DISKS

A Introduction

In this case (see Fig. 2*c*) the coolant flow enters at the center of the cavity, moves radially outward, and exits via a series of holes in the peripheral shroud. Two inlet geometries were studied: (1) a radial inlet, where the flow enters the cavity radially through a fine-mesh gauze at $r = a$; (2) an axial inlet, where the flow enters the cavity axially through a central hole in one disk. For these tests the axial spacing was $G = 0.27$.

The overall flow structure observed for both inlet geometries was qualitatively similar to that predicted by [9] for a rotating cavity with a laminar radial flow (from a uniform source to a uniform sink). The flow structure comprised a source region, Ekman layers with a $Q/2$ (Q being the total volume flow rate entering the cavity) on each disk, a sink layer, and an inviscid central-core region. A rotating cylindrical cavity with a radial throughflow (outflow and inflow; laminar and turbulent) has also been the subject of extensive theoretical and experimental study [10, 11].

B Flow Structure with a Radial Inlet

By illuminating the r–z plane and pulsing the smoke generator, it was possible to observe the flow structure inside the rotating cavity. A sequence of photographs was taken of the progression of a pulse of smoke through the cavity, from the time it entered the cavity at $r = a$ to the time it had been convected to all the regions of interest. Because the flow structure was mainly axisymmetric, the field of view of the camera was limited to the top half of the cavity; the camera shutter operation (at 5 frames per second) was not synchronized with the rotation of the cavity. Figure 9 shows the progression of smoke through a rotating cavity with a radial outflow for $C_w = 79$ and $\mathrm{Re}_\phi = 2.5 \times 10^4$. In this case the flow entered the cavity radially through a fine cylindrical mesh at $r = a$. Figure 9 relates to the time from the smoke's arrival at the inlet to its departure from the cavity through the perforated shroud at $r = b$. In Fig. 9*a* the smoke had entered the source region (mainly on the left-hand side of the cavity, which it reached first), where the flow divided equally before feeding into the Ekman layers on each disk, as can be seen in Figs. 9*b*, *c*. Figure 9*d* shows the instant when the smoke had been convected through the Ekman layers and the sink layer (on the shroud) and had just left the cavity through the holes in the shroud.

Rotating the cavity caused the pressure inside the cavity to be lower than that of its surroundings. Consequently, at this low flow rate there was evidence of an ingress of fluid from outside the cavity through the holes in the shroud: smoke that left the cavity at $r = b$ subsequently re-entered. The inflow created an ingress region, which was several times the size of the sink layer, and this compound outer region is visible in Figs. 9*e*, *f*.

Ekman-layer theory predicts that the radial component of velocity has a decaying sinusoidal profile in the axial direction and that there should hence be evidence of inflow within the Ekman layers. Close inspection of Fig. 9*f* revealed a trail of smoke adjacent to the outflow region on each disk. In Figs. 9*g*, *h* this smoke moved radially inward (in fact, in Fig. 9*h* it had almost reached back to the source region). Also in Fig. 9*h* there is evidence of the next outflow region (the two lobes of smoke penetrating radially into the central core) in the Ekman layers; however this radial motion was very weak.

C Flow Structure with Axial Inlet

Figures 10*a–d* show the smoke patterns for $C_w = 79$ and $\mathrm{Re}_\phi = 5 \times 10^4$ for the case where the flow entered axially, from the left. Comparison with the radial-inlet case reveals that apart from the flow in the source region, the overall flow structures are the same. Smoke streaks in the central axial flow show that some (approximately half) of the incoming air, immediately upon entry into the cavity, is disgorged from the jet and has been entrained on to the upstream disk; the rest moves axially towards the downstream disk, where it is entrained and fed into the Ekman layer on that disk. Hence in the source region the flow adjusts from a predominantly axial motion to that of an Ekman spiral on each disk. Observation of the radial extent of the source region, for a range of conditions, revealed that the size of the source region increased with increasing C_w and decreased with increasing Re_ϕ; the Ekman-layer thickness also decreased with increasing Re_ϕ. These effects can be seen by comparing Fig. 9 for $\mathrm{Re}_\phi = 2.5 \times 10^4$ and Figs. 10*a–d* for $\mathrm{Re}_\phi = 5 \times 10^4$. The effect of increasing the flow rate C_w can be seen by comparing Figs. 10*a–d* with those shown in Figs. 10*e–h* for $C_w = 316$. In this case all the flow in the central axial jet reached the downstream disk, where it impinged to form a wall jet boundary layer extending to $r/b \approx 0.5$. At this radius some of the flow disgorged from the disk moved radially inward and across the cavity to be entrained on the upstream disk. At radius $r/b \approx 0.7$, Ekman-layer flow was established, with the flow being divided equally on both disks. Close inspection of the figure reveals the existence of cellular instabilities in the Ekman layers; these are similar to those reported by [12]. It can also be seen that at this flow rate, ingress through the shroud was suppressed.

It is interesting to compare the Ekman boundary layer,

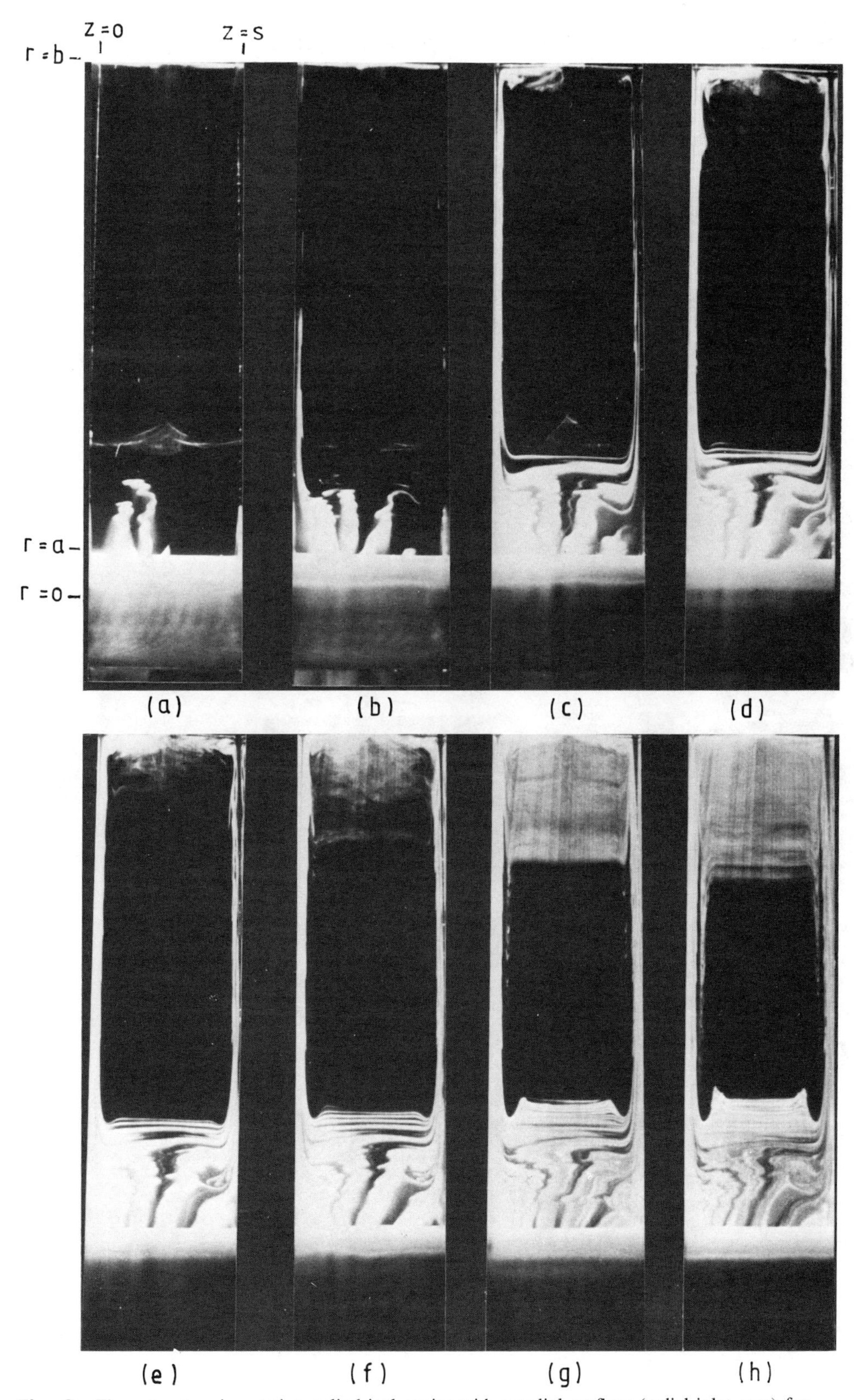

Fig. 9 Flow structure in rotating cylindrical cavity with a radial outflow (radial-inlet case) for $C_w = 79$, $\mathrm{Re}_\phi = 2.5 \times 10^4$, and $G = 0.27$.

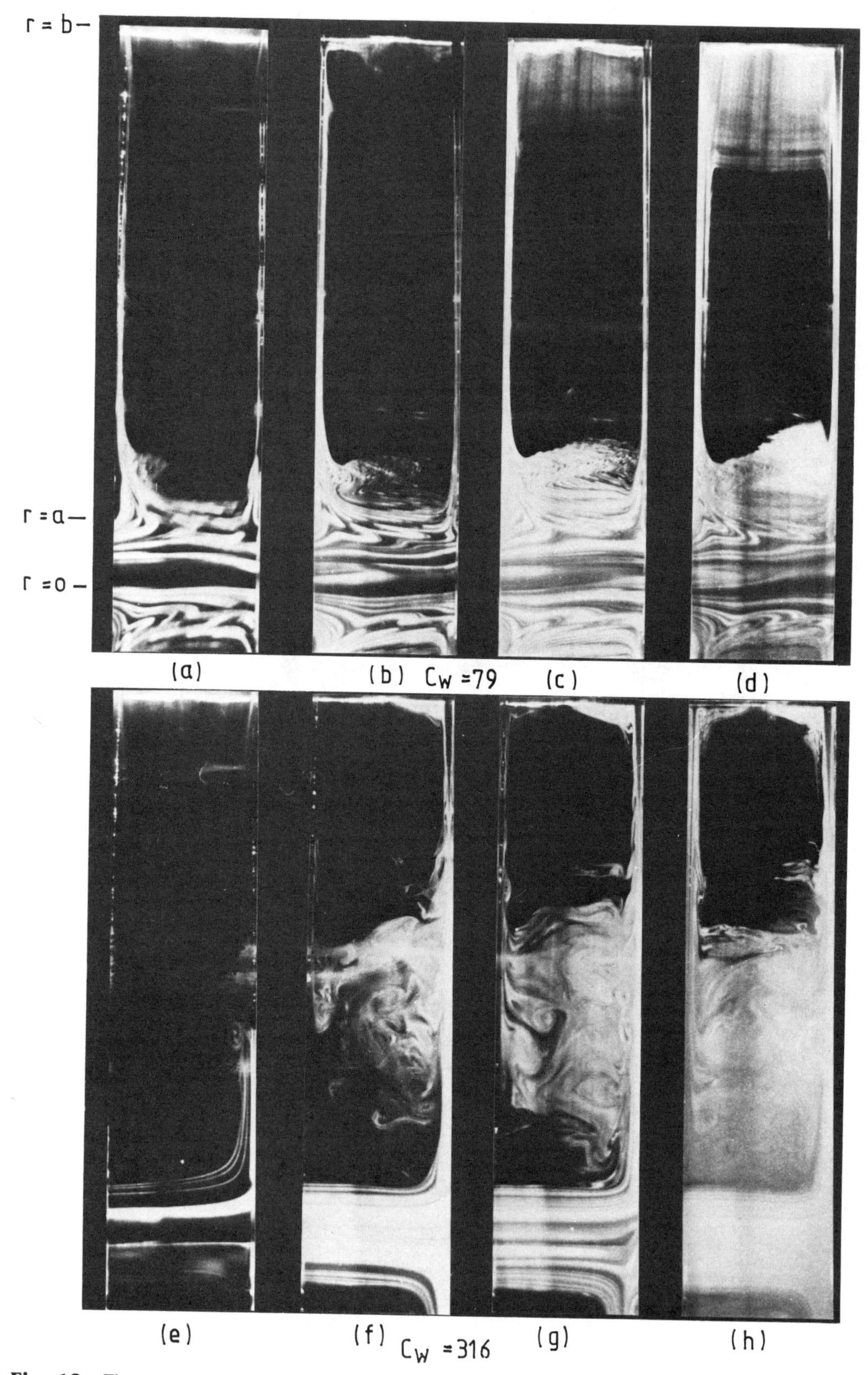

Fig. 10 Flow structure in a rotating cylindrical cavity with a radial outflow (axial-inlet case) for $Re_\phi = 5 \times 10^4$ and $G = 0.27$.

in which the volume flow rate remains constant with increasing radius, with the "free-disk" boundary layer (observed on the rotating disk in the rotor–stator geometry, as described in Sec. III), where flow rate, (through entrainment) increases with radius. Giving careful consideration to the smoke patterns shown in Figs. 10*a–d*, it is apparent that both types of boundary layer are present. In fact three types of boundary layer are present: (1) a wall jet on the downstream disk for $r/b < 0.4$, (2) Ekman layers on both disks for $r/b > 0.6$, and (3) a free-disk entrainment boundary layer on the upstream disk for $r/b < 0.6$. Clearly a complex redistribution of the flow takes place in the source region for this flow rate and inlet geometry.

The effect of increasing C_w further is shown in Figs. 11*a–d* for $C_w = 632$ and $Re_\phi = 5 \times 10^4$. In this case the source region extends to $r/b \simeq 0.85$; in fact the source region has almost merged with the sink layer on the shroud.

The effect of increasing Re_ϕ is shown in Figs. 11*e–h* for $C_w = 632$ and $Re_\phi = 20 \times 10^4$. It can be seen that the source region has reduced in size to $r/b \simeq 0.55$. Also there is evidence of bursts of high-spatial-frequency (with a period much less than the Ekman-layer thickness) disturbances that penetrate into the central core and are similar to those reported by [13]. These disturbances are believed to mark the onset of turbulent Ekman-layer flow. The transition to turbulent flow in the Ekman layers has been predicted and measured by [10] as occurring at a radial Reynolds number of $Re_r \approx 180$ (where $Re_r = Q/2\pi \nu r$).

This flow geometry has been modeled numerically in [14]. Figures 12*a, b* show the comparison between the streamline computations and flow visualization for $C_w = 331$ and $Re_\phi = 2.5 \times 10^4$. The similarity between theory and experiment is quite remarkable. (It should be noted that the mirror-image reflections from the walls of the cavity have not been removed in this photograph.)

As with the axial-throughflow case, flow visualization was used to aid interpretation of LDA velocity measurements. To illustrate this, some velocity data are presented in Fig. 13, for both inlet geometries, for $Re_\phi = 5.0 \times 10^5$. While it is inappropriate to discuss these in detail, the salient facts are presented in the following discussion and can be compared with the preceding text and associated figures. Also shown in Figure 13 are the laminar-Ekman-layer flow predictions of [10].

Figure 13*a* shows the axial distribution of the radial component of velocity (normalized to the bulk average radial velocity $\overline{U}$ where $\overline{U} = Q/2\pi rs$) for two values of flow rate ($C_w = 190$ and 603) at radius $r/b = 0.6$. In the low-flow-rate case the LDA measurements are consistant with the theoretical predictions: there is radial inflow in the Ekman layers, and zero radial velocity in the central-core region. This is in marked contrast to the high-flow-rate case, where the source region extends beyond the radial location of the probe $r/b = 0.6$ (see Figs. 11*a–d*, where the radial outflow near the downstream disk (at $0.8 < z/s < 1.0$) corresponds to the wall jet; the inflow at $0.2 < z/s < 0.8$ corresponds to the flow disgorging from the downstream disk and moving inward before being entrained on the upstream disk.

Figure 13*b* shows a more detailed axial distribution of the normalized radial component of velocity in the Ekman layer on the upstream disk for two flow rates ($C_w = 135$ and 603). The lower flow rate is in good agreement with the laminar-Ekman-layer theory whereas, at the higher flow rate, the measurements are in good agreement with the free-disk boundary layer theory.

Figure 13*c* shows the axial distribution of the tangential component of velocity V_ϕ (normalized with respect to Ωr) for two flow rates ($C_w = 190$ and 603). At the lower flow rate the measurements for both the radial-inlet and the axial-inlet geometries are in good agreement with the laminar-Ekman-layer theory. For the high-flow-rate case there is a significant difference between the measured swirl components for the two inlet geometries. The tangential velocity for the radial inlet is close to zero, as predicted by the theory, whereas for the axial inlet there is a significant increase in the swirl of the core of fluid within the source region.

Finally, Fig. 13*d* shows the radial distribution of the tangential component of velocity at $z/s = 0.5$ for $C_w = 251$. In the Ekman-layer region (for $r/b > 0.5$) the measurements are in good agreement with the theory, for both inlet cases; in the source region (for $r/b < 0.5$) the measurements depart from the theory and there is a significant difference between the results for the two inlet cases.

These complex flow structures, which have been identified using flow visualization and quantified using LDA, have since been used to interpret heat transfer measurements conducted for radial outflow [15].

VI THE FLOW IN A ROTATING TUBE

A Introduction

During an investigation of the flow structure in a sealed, heated, rotating cylinder (with $l/d \approx 1$) it was noticed that if a plug, used to seal a small hole in one disk, were left out, smoke in the laboratory was observed to enter through the hole, travel with a steady axial motion toward the other disk, and return in an annular layer back toward the hole (even under isothermal conditions).

The effect was investigated further [16], and the cylinder was replaced with a tube that had one end sealed and the other end open to the atmosphere. It was found that over the practical length-to-diameter ratios that were possible (up to 40), the smoke would always reach the sealed end if the tube were rotated fast enough.

While some work appears to have been conducted on the flow structure in a rotating tube with a superimposed

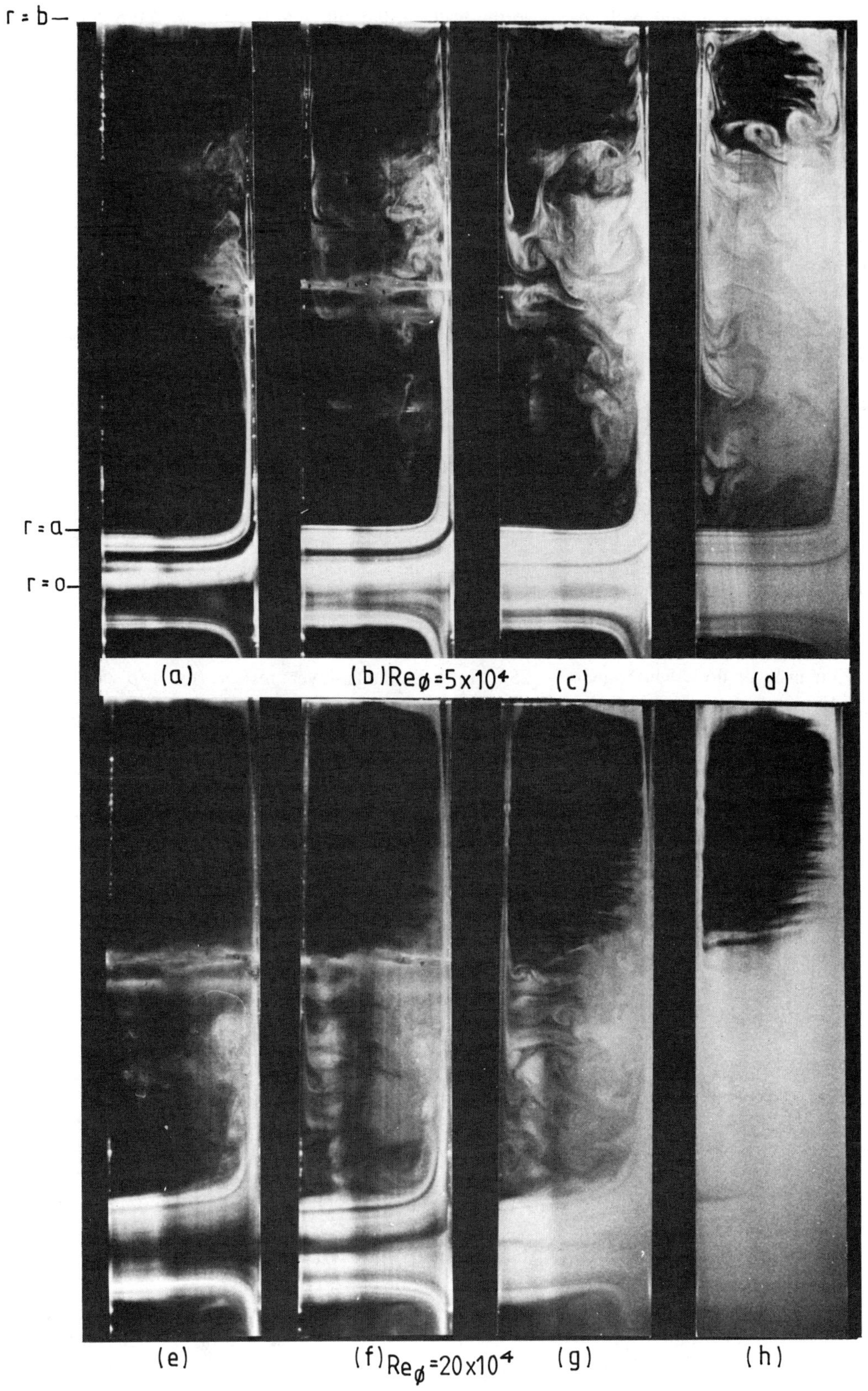

Fig. 11 Flow structure in a rotating cylindrical cavity with a radial outflow (axial inlet) for $C_w = 632$ and $G = 0.27$.

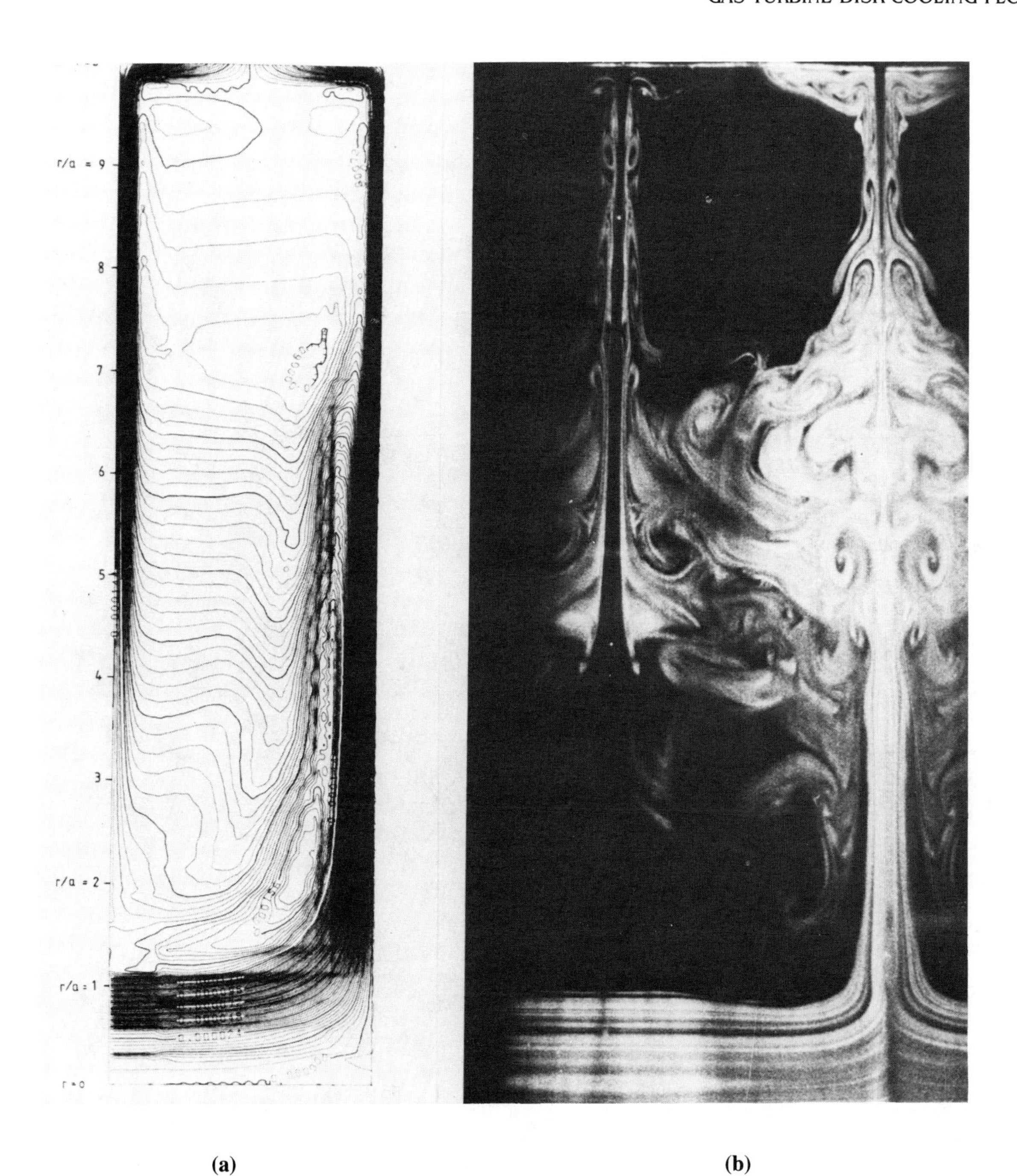

(a) (b)

Fig. 12 Comparison between the streamline computations obtained by [14] and flow visualization, for the axial-inlet case with $Re_\phi = 2.5 \times 10^4$; (*a*) streamlines in the *r–z* plane predicted by the numerical method for $C_w = 331$; (*b*) smoke pattern observed experimentally in rig 1 for $C_w = 316$.

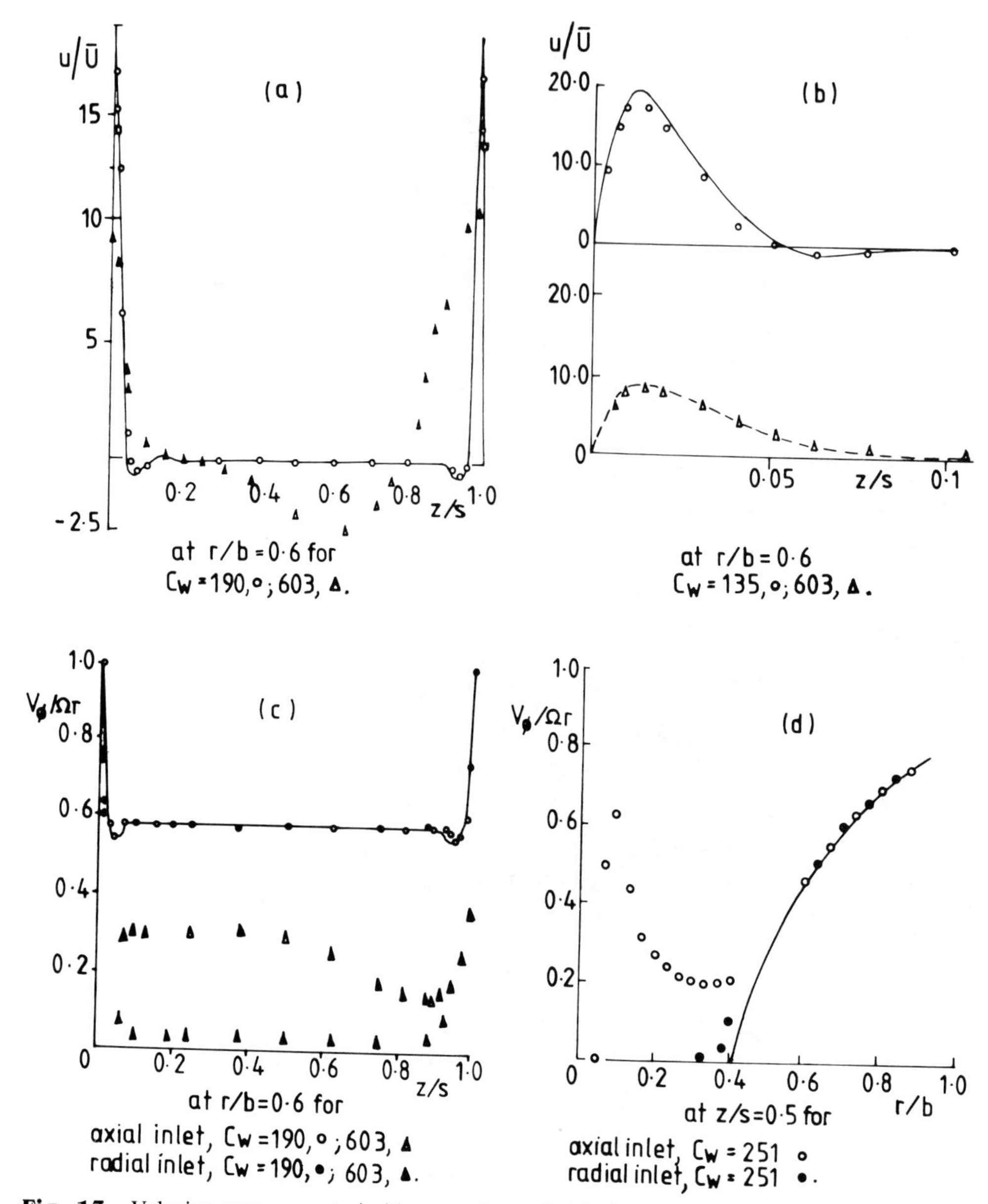

Fig. 13 Velocity measurements inside a rotating cylindrical cavity with a radial outflow for $Re_\phi = 5 \times 10^4$ and $G = 0.27$ ———, Ekman-layer theory [10]; ———, free-disk theory [10].

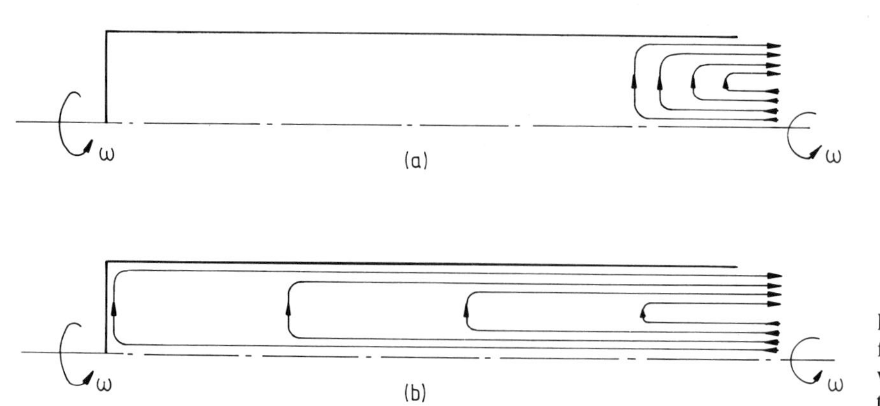

Fig. 14 Schematic diagram of the flow structure in (*a*) a long rotating tube with one end sealed; (*b*) a short rotating tube with one end sealed.

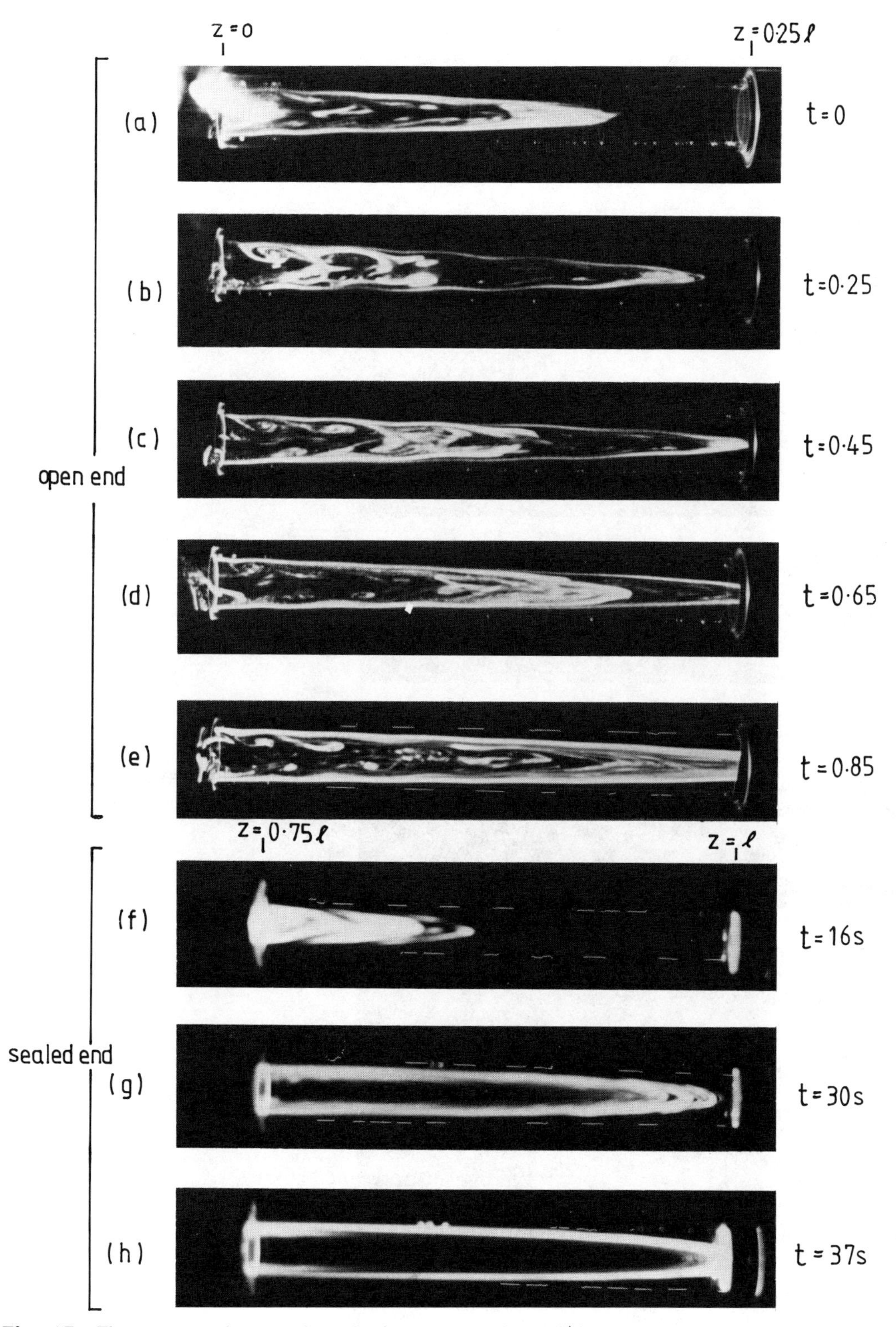

Fig. 15 Flow structure in a rotating tube for $Re_\phi = 440$ and $l/d = 27.4$; (*a–e*) Open end; (*f–h*) sealed end.

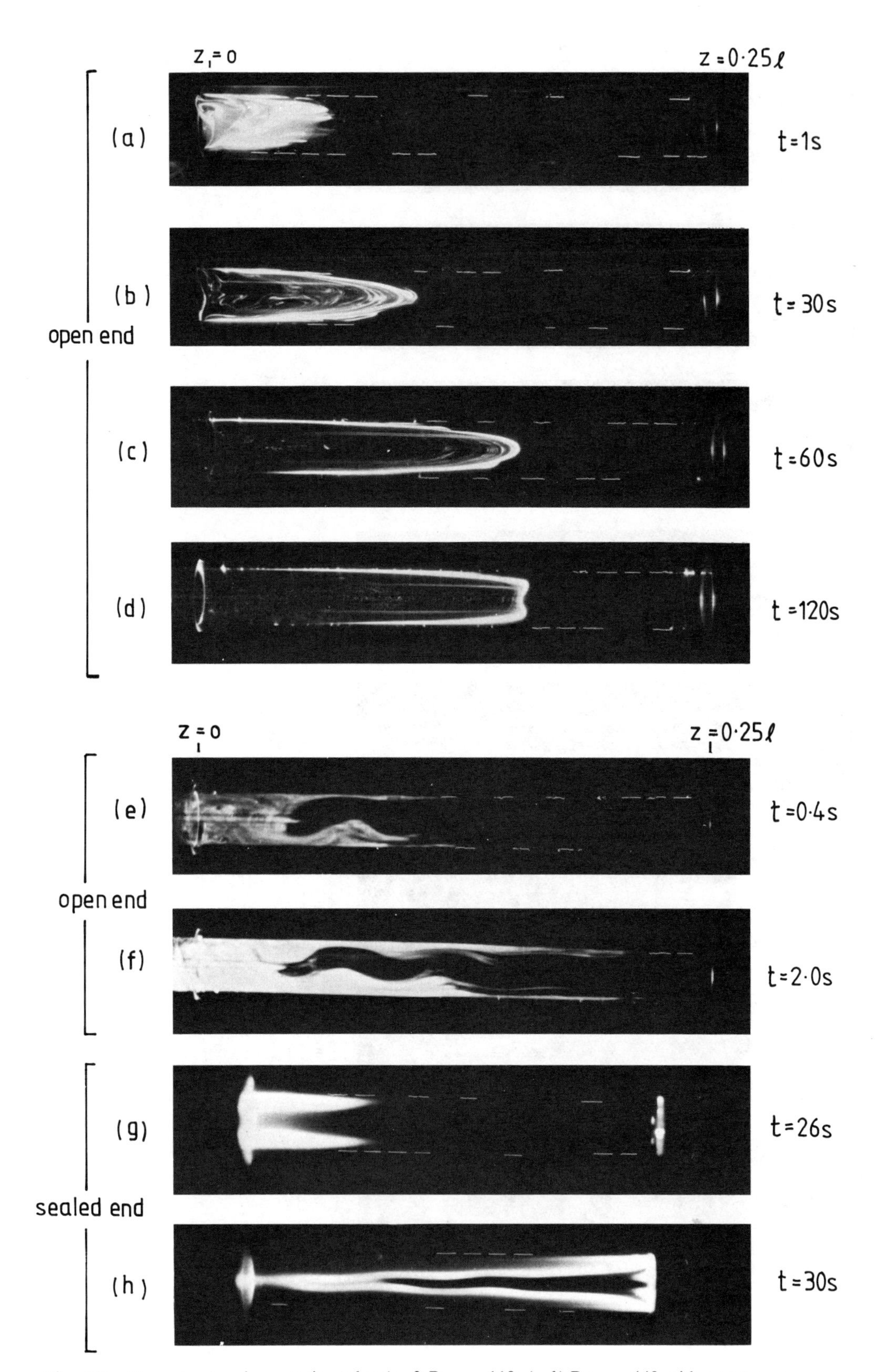

Fig. 16 Flow structure in a rotating tube; (*a–d*) $Re_{\phi} = 110$; (*e–h*) $Re_{\phi} = 440$ with a transverse flow.

axial flow, surprisingly little appears to have been published on a partially sealed rotating tube.

B Flow Structure in a Rotating Tube

Schematic representations of the flow structure observed in an open-ended rotating tube are shown in Figs. 14*a*, *b* for two cases: (1) where the axial flow fails to penetrate the full length of the tube, referred to as the long-rotating-tube condition; (2) where the axial flow reaches the sealed end and impinges on the inner face of the sealing plug, referred to as the short-rotating-tube condition.

The flow in an open-ended rotating tube is illustrated by the series of photographs, arranged sequentially in Fig. 15, for a glass tube with $l/d = 27.4$ and $Re_\phi = 440$. Figures 15*a*–*e* show the axial progression of the smoke near the open end ($0 < z/l < 0.3$, with z being measured axially from the open end); Figs. 15*f*–*h* show the flow near the sealed end ($0.7 < z/l < 1$).

Figure 15*a* shows the penetration of smoke immediately after it had been released (at time $t = 0$) into the air surrounding the open end; the smoke has entered the tube and has created a "front" with a distinct profile. In Fig. 15*b* ($t = 0.2$ s) the smoke front has moved axially to the right, toward the sealed end; unseeded, clear air has moved into the tube behind the front. Away from the open end, the counter-flowing annular boundary layer appears as the clear region on the (radial) outside of the smoke front. Careful inspection of Figs. 15*b*–*e* (which were all taken at intervals of 0.2 s) reveals the presence of what appear to be spatially periodic spiral vortices on the interface between the central column and the annular layer.

Figure 15*f* ($t = 16$ s) shows the smoke front near the sealed end (which, owing to surface scatter, appears as a white region) of the tube at $z/l \simeq 0.87$; in Fig. 15*g* ($t = 30$ s) the front has nearly reached $z/l = 1$. In Fig. 15*h* ($t = 37$ s) smoke can be seen to have moved radially outward in a thin boundary layer on the sealed rotating end and to be about to enter the returning annular layer. The almost stationary smoke front virtually delineates the boundry between the central column and the annular layer, but the instabilities that were observed near the open end are not visible near the sealed end of the tube.

Below a certain rotational speed, it was observed that the central column would never reach the sealed end of the tube. This condition is illustrated by the series of photographs shown in Figs. 16*a*–*d*, taken at $Re_\phi = 110$ near the open end at $t = 1$, 30, 60 and 120 s, respectively. In Fig. 16*c* the central smoke front has moved to $z/l \simeq 0.19$; in Fig. 16*d*, taken 60 s later, there is no evidence of any significant further axial movement. The change from Figs. 16*c*–*d* in the shape of the profile of the smoke front is attributed to weak radial convection (in addition to molecular diffusion) of the smoke-filled air into the annular layer.

Another effect (discovered accidentally!) was that the direction of flow in the rotating tube could be reversed by blowing a jet of air, in a direction normal to the axis of rotation, across the open end of the tube. Under these conditions, the flow in the central column moves toward the open end, and the flow in the annular layer moves toward the sealed end of the tube. This effect is illustrated in Figs. 16*e*, *f*, which show, for $Re_\phi = 440$, the flow near the open end at $t = 0.4$ and 2.0 s, respectively; Figs. 16*g*, *h* show the flow near the sealed end, where the smoke moves radially inward, at $t = 26$ and 30 s, respectively. The speed of the external transverse flow was approximately 8 m/s.

The sheet-illumination flow visualization technique has been successfully applied to the study of the gas turbine disk coolant-flow geometries described here. It has advanced understanding of flow behavior and proved to be invaluable in interpretation of the laser Doppler anemometry measurements.

REFERENCES

1. Merzkirch, W., *Flow Visualization*, Academic, New York, 1974.
2. Durst, F., Melling, A., and Whitelaw, J. H., *Principles and Practice of Laser Doppler Anemometry*, 2d ed., Academic, New York, 1981.
3. Pincombe, J. R., Optical Measurements of the Flow inside a Rotating Cylinder, D. Phil. thesis, University of Sussex, Brighton, England.
4. Phadke, U.P., Aerodynamic Aspects of the Sealing of Rotor-Stator Systems in Gas Turbines, D. Phil. thesis, University of Sussex, Brighton, England.
5. Cochran, W. G., The Flow due to a Rotating Disc, *Proc. Cambridge Philos. Soc.* vol. 30, p. 365, 1934.
6. Daily, J. W., Ernst, W. D., and Asbedian, V. V. Enclosed Rotating Disks with Superposed Throughflow. MIT Hydrodynamic Lab., Rep. 64, 1964.
7. Owen, J. M., and Pincombe, J. R., Vortex Breakdown in a Rotating Cylindrical Cavity, *J. Fluid Mech*, vol. 90, p. 109, 1979.
8. Leibovich, S., The Structure of Vortex Breakdown, *Ann. Rev. Fluid Mech.*, vol. 10, p. 221, 1946.
9. Hide, R., Source-Sink Flows in a Rotating Fluid, *J. Fluid Mech.* vol. 32, p. 737, 1968.
10. Owen, J. M., Pincombe, J. R., and Rogers, R. H., Source-Sink Flow inside a Rotating Cylindrical Cavity, *J. Fluid. Mech.*, vol. 155, p. 233, 1985.
11. Owen, J. M., and Pincombe, J. R., Velocity Measurements inside a Rotating Cylindrical Cavity with a Radial Outflow of Fluid, *J. Fluid Mech.*, vol. 99, pp. 111, 1980.
12. Faller, A. J., An Experimental Study of the Instability of the Laminar Ekman Boundry Layer, *J. Fluid Mech.* vol. 15, p. 560, 1963.

13. Tatro, P. R., and Mollo-Christensen, E. L., Experiments on Ekman Layer Stability, *J. Fluid Mech.*, vol. 28, p. 531, 1967.
14. Chew, J. W., Owen, J. M., and Pincombe, J. R., Numerical Predictions for Laminar Source-Sink Flow in a Cylindrical Cavity, *J. Fluid Mech.*, vol. 143, p. 451, 1984.
15. Long, C. A., and Owen, J. M., The Effect of Inlet Conditions on Heat Transfer in a Rotating Cavity with a Radial Outflow of Fluid, *J. Turbomach.*, vol. 108, p. 145, 1986.
16. Owen, J. M., and Pincombe, J. R., Rotationally-induced Flow and Heat Transfer in Circular tubes, Univ. of Sussex, Brighton, England, Rep. TFMRC/32, 1981.

Chapter 34

Coronary Arteries

Hani N. Sabbah, Fareed Khaja, James F. Brymer,
Thomas M. McFarland, and Paul D. Stein

I INTRODUCTION

The predilection of particular sites of the arterial system to atherosclerosis has provoked speculations that mechanical forces related to the local dynamics of blood flow might be a factor leading to the development of atheroma. Several fluid dynamic fa :tors have been implicated in atherogenesis; they include flow separation, turbulence, and wall shear. Injury to the vascular endothelium by high fluid shear stresses has been suggested as a focus for the subsequent development of atheroma at the site of injury [1]. Other investigators [2], however, suggested that exposure of the vascular endothelium to low fluid shear stresses may be conducive to the formation of atherosclerosis by adversely affecting the mass transfer of lipids across the arterial wall. Both hypotheses have provided an impetus for assessment of the characteristics of blood flow in arteries with the use of both in vivo [3–6] and in vitro [7–15] preparations.

Although considerable work on the characteristics of blood flow in large arteries has been accomplished, in vivo studies of flow patterns in human coronary arteries are lacking because of limitations of instrumentation and access. We recently described a method to visualize patterns of flow in the coronary arteries of patients that utilizes routine diagnostic coronary arteriograms. The approach is useful in the qualitative evaluation of wall shear and flow separation in human coronary arteries [16–19].

II METHOD OF VISUALIZATION

Coronary arteriography refers to the selective injection of a radiopaque constrast medium (injectate containing organically bound iodine) into the coronary arteries of patients. It is used as a tool for the diagnosis of coronary artery disease. Typically, the tip of a catheter is introduced into the ostium of the coronary artery, and the contrast medium is injected to opacify the lumen while at the same time, a 35-mm cine at 30–60 frames per second is obtained to document the process.

To show the pattern of flow in the coronary artery, the tip of the catheter was withdrawn from the ostium while continuing to obtain 35-mm cine at 30 frames per second as the artery is cleared of contrast material. Two distinct observations with respect to the patterns of flow in the coronary arteries were made by observation of the characteristics of clearing of the artery. These were (1) a qualitative assessment of relative differences of blood velocity, and therefore wall shear, between the inner and outer wall of the artery; (2) a qualitative evaluation of sites of stasis that may reflect sites of flow separation.

III STUDIES OF THE RIGHT CORONARY ARTERY

Differences of blood velocity between the inner and outer wall of the right coronary artery were determined qualitatively from selective right coronary arteriograms performed in patients during diagnostic cardiac catheterization (Fig. 1). Selective coronary arteriograms of 30 patients with angiographically normal right coronary arteries were reviewed retrospectively. Arteriograms of the right coronary artery were obtained in a 30 to 45° left anterior oblique projection during the injection of meglumine and sodium diatrizoates (Renografin 76). Arteriograms were recorded on 35-mm film at 30 frames per second. The onset of systole was designated by a mark on the appropriate frame of film coincident with the R wave of the electrocardiogram. The difference of blood velocity between the inner wall (nearest to the myocardium) and outer wall (nearest to the epicardial surface) of the right coronary artery was assessed on the basis of the rate of clearance of contrast material after the tip of the injecting catheter was withdrawn from the coronary ostium.

In every patient, the contrast material was cleared more rapidly along the outer wall of the right coronary artery than along the inner wall (Fig. 1) [16]. This rapid clearing of contrast material along the outer wall was particularly prominent in the mid-right coronary artery, where the vessel curved around the right border of the heart. In all instances, a slower clearing of contrast material occurred along the inner wall, which represented the surface of the vessel in close proximity to the myocardium. The contrast material along the inner wall persisted 2 to 6 cardiac cycles after the outer wall of the artery had cleared. Definition of the inner wall of the right coronary artery, therefore, could be appreciated several cardiac cycles after the outer wall had cleared of contrast material. Figure 1 shows frames from a coronary arteriogram that depict the typical sequence of clearance of contrast material from the right coronary artery.

IV STUDIES OF THE LEFT ANTERIOR DESCENDING CORONARY ARTERY

Differences of blood velocity, and therefore wall shear, between the inner and outer wall of the left anterior descending coronary artery were evaluated qualitatively from selective arteriograms performed in 21 patients with angiographically normal left anterior descending coronary arteries. In 18 patients, the left anterior descending coronary artery formed nearly a direct linear conduit with the left main coronary artery. In the remaining 3 patients, the origin of the left anterior descending coronary artery arose from the left main coronary artery at an acute angle. In 2 patients, a large intermediate coronary branch was present. Arteriograms of the left anterior descending coronary artery were obtained in multiple views to exclude the presence of coronary artery disease. However, only the 90 to 105° left lateral projection was used for the qualitative assessment of velocity and wall shear. In this projection, the inner wall of the left anterior descending coronary artery (the wall in contact with the myocardium) and outer wall (the wall farthest from the myocardium) can be visualized as the left anterior descending coronary artery curves over the border of the heart (Fig. 2).

As with the right coronary artery, in every patient the contrast material was cleared more rapidly along the outer wall of the left anterior descending coronary artery than along the inner wall [17]. Slower clearance of contrast material occurred in all instances along the inner wall, which represented the surface of the vessel in direct contact with the myocardium. Contrast material along the inner wall persisted two to six cardiac cycles after the outer wall of the artery had completely cleared. Figure 2 depicts a typical sequence of the rate of clearance of contrast material from the left anterior descending coronary artery. The slower clearance of contrast material along the inner wall suggests a lower velocity and shear rate along this wall. The presence of a large intermediate coronary artery in two patients or of an acute angle between the left anterior descending coronary artery and the left main coronary artery in three patients did not alter the distribution of velocity and shear rate in the left anterior descending coronary artery in comparison to other patients who did not possess this anatomy.

Observations of the rate of clearance of contrast material from the right coronary artery and left anterior descending coronary artery in humans qualitatively indicate a lower velocity along the inner wall than along the outer wall. This suggests that the velocity profile at various sites along the right coronary artery and left anterior descending coronary artery may be skewed toward the outer wall. Skewing of the velocity profiles is likely to arise from the curvature of these vessels as each one courses over the border of the heart. This is consistent with flow dynamics in curved pipes [20]. As such, shear rate would be expected to be lower along the inner wall than along the outer wall of these coronary arteries.

V EVIDENCE OF FLOW SEPARATION IN THE LEFT AND RIGHT CORONARY ARTERIES

Observations of the clearance of contrast medium, by the method discussed earlier, was used in 7 patients with mild focal atherosclerotic plaques of either the left or right coronary artery. The atherosclerotic plaques in each patient caused a 30 to 50% narrowing of the lumen diameter and were in all instances located on the inner wall of the vessel. In each of these 7 patients, a region of slow clearance of contrast material was present just distal (downstream) from the plaque (Fig. 3). Contrast material persisted 3 to 6 cardiac cycles in this region even though the remainder of the vessel (distal and proximal) had cleared completely. These localized sites of slow clearance of contrast material just distal to a mild obstruction suggest the presence of flow separation, which is typically a region of low wall shear. The latter is compatible with the dynamics of flow downstream from an obstruction.

Observations based on the visualization of flow in the coronary arteries of patients suggested the presence of a skewed velocity profile. This is compatible with direct

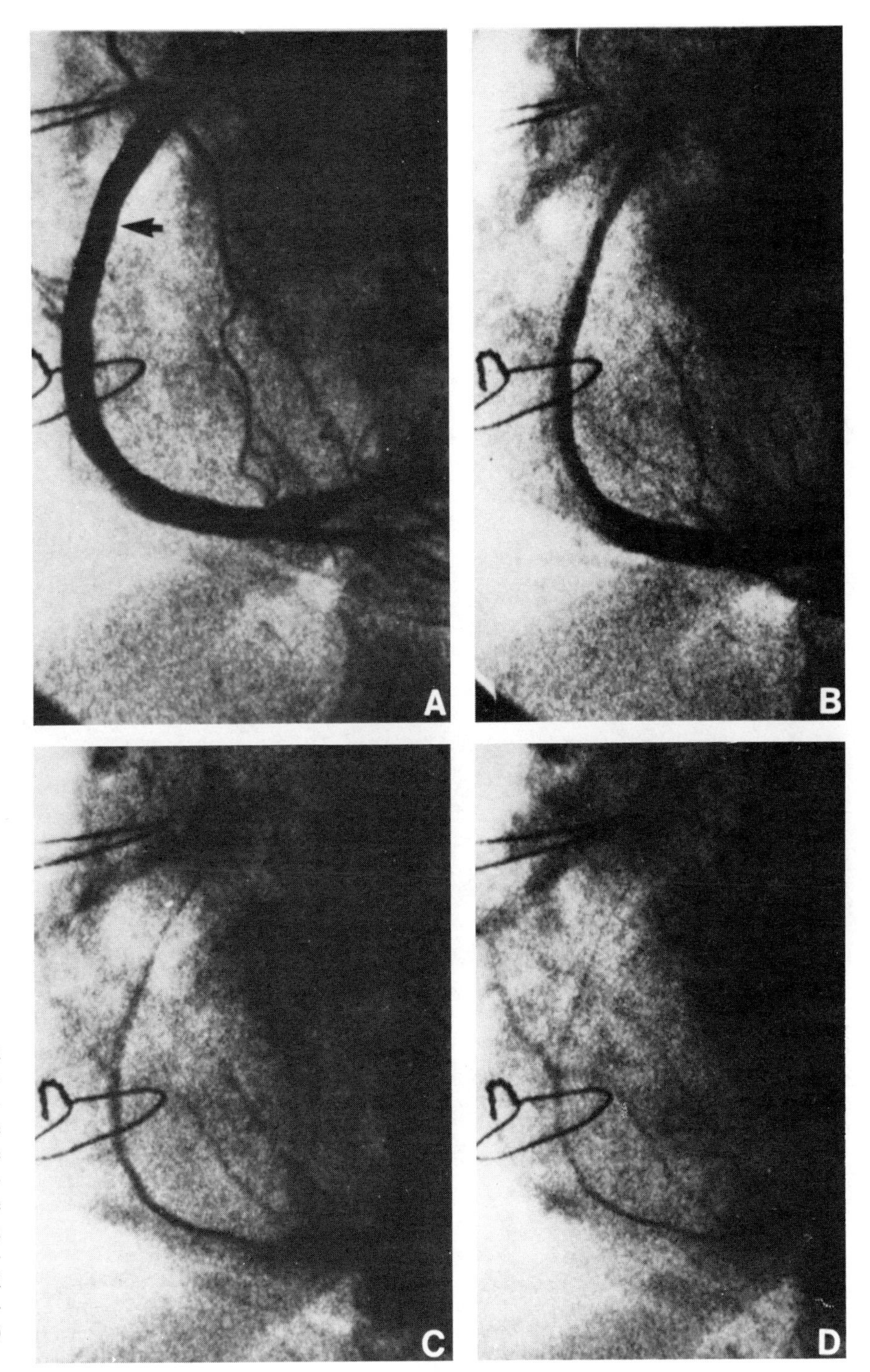

Fig. 1 Right coronary arteriogram from a patient with a porcine bioprosthetic valve in the mitral position; the sequence was taken every 1 s from the moment of withdrawal of the tip of the catheter: (*A*) The artery completely opacified; arrow indicates inner wall of the vessel; (*B*) clearing of the outer wall; (*C*, *D*) the persistence of contrast material along the inner wall [reprinted from the American Journal of Cardiology (vol. 53, pp. 1008–1012, 1984) with permission].

measurements of blood velocity in the coronary arteries of horses. Nerem et al. [6] used hot-film anemometry to measure blood velocity in the left coronary artery of horses. Their results demonstrated skewed profiles of velocity; velocity profiles in the left main, left anterior descending, and left circumflex coronary arteries, in general, were skewed toward the outer wall.

Theoretically, secondary flows are expected to develop in curved vessels as a result of centrifugal forces. Evaluation of the patterns of flow in the coronary arteries of humans utilizing clinical coronary arteriograms, however, was not helpful in the evaluation of secondary flows. To evaluate the nature of secondary flows in the epicardial coronary arteries, we developed a flow-through mold of the coronary arteries of a pig; in our model secondary flows could be visualized using dye-streamer techniques.

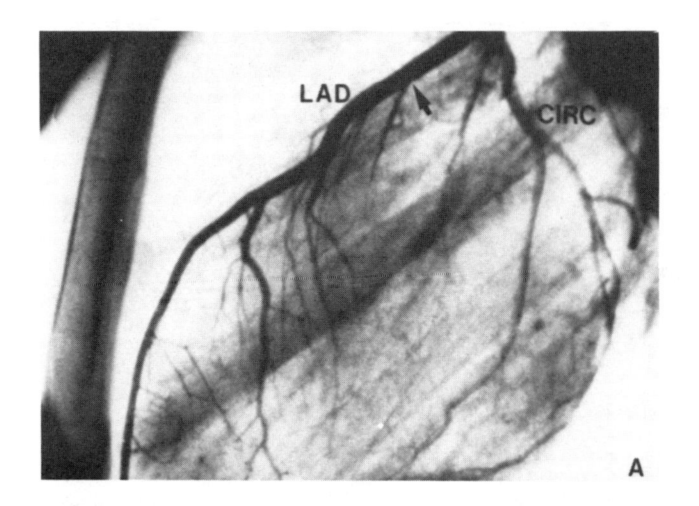

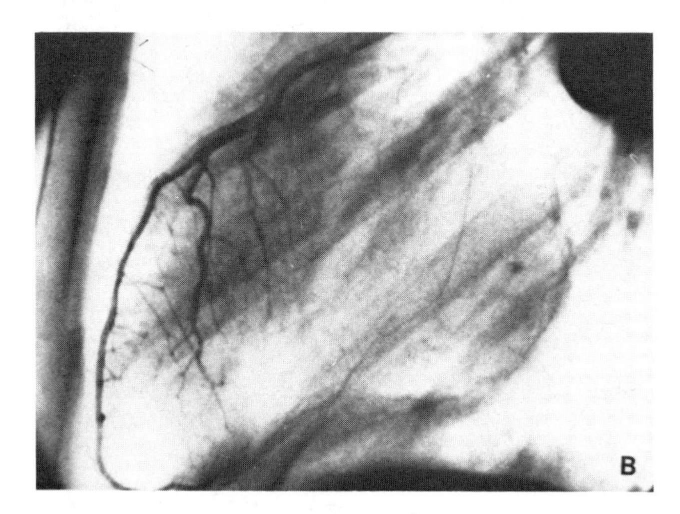

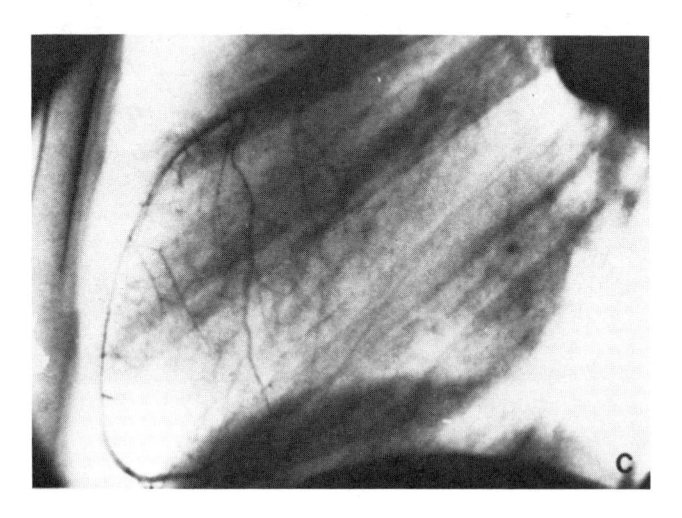

Fig. 2 Sequence of three frames of coronary arteriograms showing delayed clearance of contrast material along the inner wall of the left anterior descending coronary artery (LAD): (*A*) The LAD completely opacified; arrow indicates inner wall of the vessel (CIRC, circumflex coronary artery); (*B*) clearing of the outer wall; (*C*) the persistence of contrast material along the inner wall [reprinted from the American Heart Journal (vol. 112, pp. 453–458, 1986) with permission].

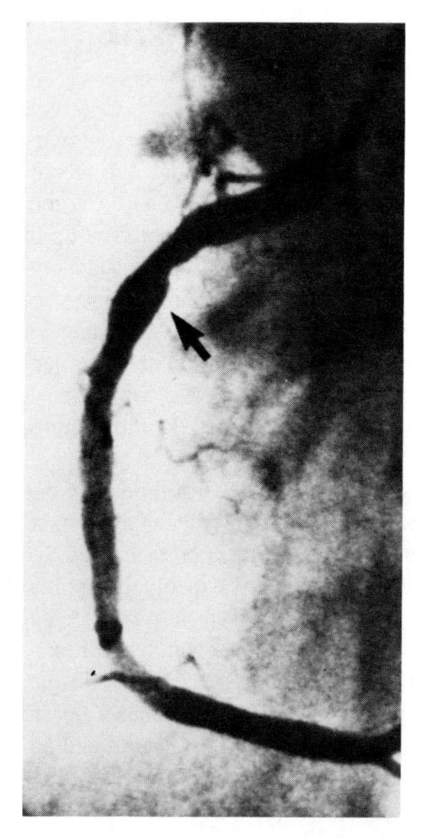

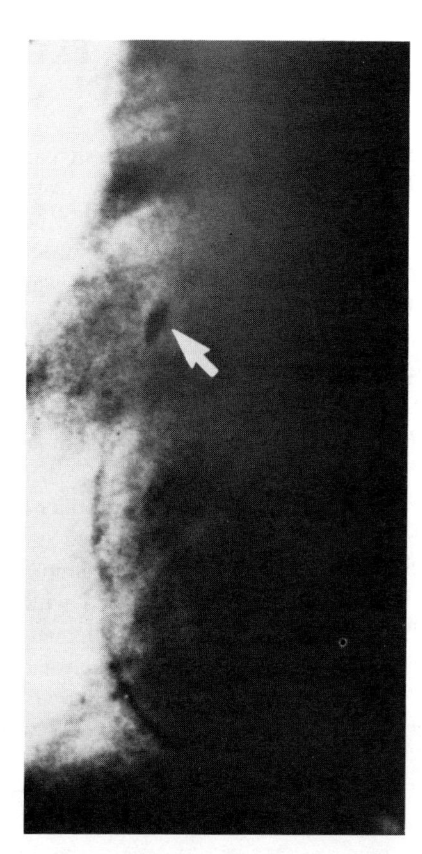

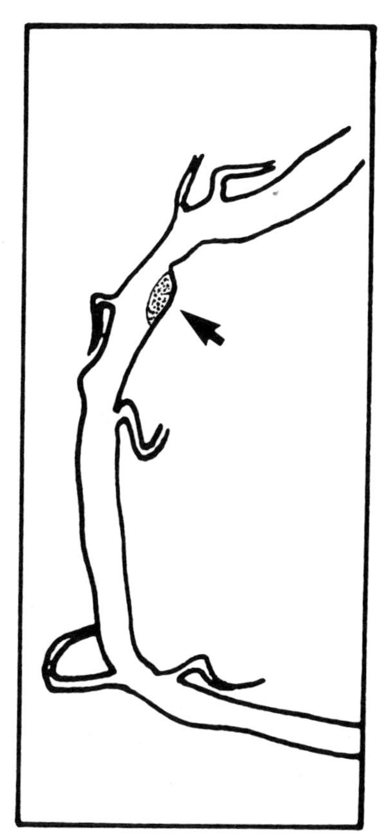

Fig. 3 Two frames from a right coronary arteriogram of a patient with coronary atherosclerosis showing delayed clearance of contrast material (arrow) from a region just distal to an atherosclerotic plaque: (*Left*) Artery is completely opacified; (*center*) delayed clearing distal to the atherosclerotic plaque (white arrow); (*right*) drawing of the artery; the shaded area represents the site of delayed clearance of contrast material.

VI CONSTRUCTION OF THE CORONARY MOLD

An intact fresh heart of a pig, which included a 7-cm segment of the ascending aorta, was obtained at a local slaughterhouse. After the heart was flushed thoroughly with saline, the aortic valve was closed by suturing together the free margins of the cusps. The ascending aorta was cannulated, and the heart was allowed to float in a container of saline to minimize deformation. A liquid casting plastic[1] was injected into the aorta at a pressure of 100 mm Hg. The perfusion pressure (100 mm Hg) was maintained until the plastic hardened. A cast, which duplicated in detail the geometry of the lumen of the coronary arteries, was obtained by immersing the heart in a 50% solution of potassium hydroxide until the tissue was dissolved (usually 5 days). The cast included the left coronary artery and its branches, the proximal 1 cm of the right coronary artery, the sinuses of Valsalva and a 3-cm segment of the ascending aorta [8]. Fine (1-mm-dia.) branches of the left coronary artery were carefully removed from the cast to minimize the marked difficulties encountered in preparing a flow-through mold of such fine vessels. The flow-through mold of the coronary cast was manufactured commercially from clear acrylic resin using a modified method of the lost-wax technique. Because the leaflets of the aortic valve were sewn closed prior to preparing the cast, the resulting mold was suited only for studies of flow during diastole, which is the period of greatest coronary flow.

Flow through the mold of the left coronary artery was studied during steady flow and during pulsatile flow in a pulse-duplicating system that simulated the magnitude and phasic pattern of coronary flow [8, 21, 22]. In both instances flow rates were adjusted to simulate rest and exercise [8].

VII FLOW IN THE SINUSES OF VALSALVA

The patterns of flow in the sinuses of Valsalva (three small dilatations at the root of the aorta [8]) were studied during

[1]Ward's, Rochester, N.Y.

steady flow [23]. Flow was visualized with streamers of red dye injected through a needle with multiple holes positioned transversely across the root of the aorta approximately 10 cm from the aortic valve. Motion pictures were obtained at 24 frames per second on 16-mm film. Resting conditions were evaluated with a mean steady flow in the left coronary artery of 110 ml/min and in the right coronary artery of 80 ml/min. Conditions of high flow were evaluated with a steady flow of 340 ml/min in the left coronary artery and 310 ml/min in the right coronary artery. The test fluid was a mixture of glycerin and saline with a viscosity of 0.04. Poise. The red dye was mixed with glycerin to achieve a density comparable to that of the circulating fluid.

Flow in the sinuses of Valsalva, as it entered the coronary artery, was always laminar [23]. During levels of steady flow that were comparable to the magnitude of flow at rest, laminar streamers gently curved into the orifices of the coronary arteries. Some streamers followed the course of the closed valves, curved over the valves, and superiorly into the coronary arteries (Figs. 4, 5). With high flow, as might occur with exercise, flow remained laminar in the sinuses of Valsalva, coursed down the valves, curved superiorly, and spiraled into the ostia (Fig. 6).

VIII FLOW PATTERNS IN THE CORONARY ARTERIES

The overall pattern of flow in the mold of the coronary arteries was assessed during steady flow and during pulsatile flow. A number 5 French end-hole catheter (1.67 mm OD) was introduced 5 cm proximal to the mold. The tip of the catheter was placed within the sinuses of Valsalva approximately 1 cm from the ostium of the left coronary artery. The catheter was used to inject a single streamer of red dye to visualize the patterns of flow in the coronary artery [8]. The fluid used in the system was a mixture of glycerin and saline with a viscosity of 0.04 Poise and a density of 1.16 g/cm^3. The red dye was mixed with glycerin to produce a density equal to that of test fluid. The dye was injected at rates of 0.4 to 3.0 ml/min using a constant-infusion pump. The higher rates of injection of the red dye were used with higher coronary flows to obtain optimal visualization. The patterns of flow in the left anterior descending coronary artery and the left circumflex coronary artery and their branches were photographed with a 35-mm camera. Two views were photographed at planes 90° to each other. The views were equivalent to a lateral and anterior–posterior projection [8].

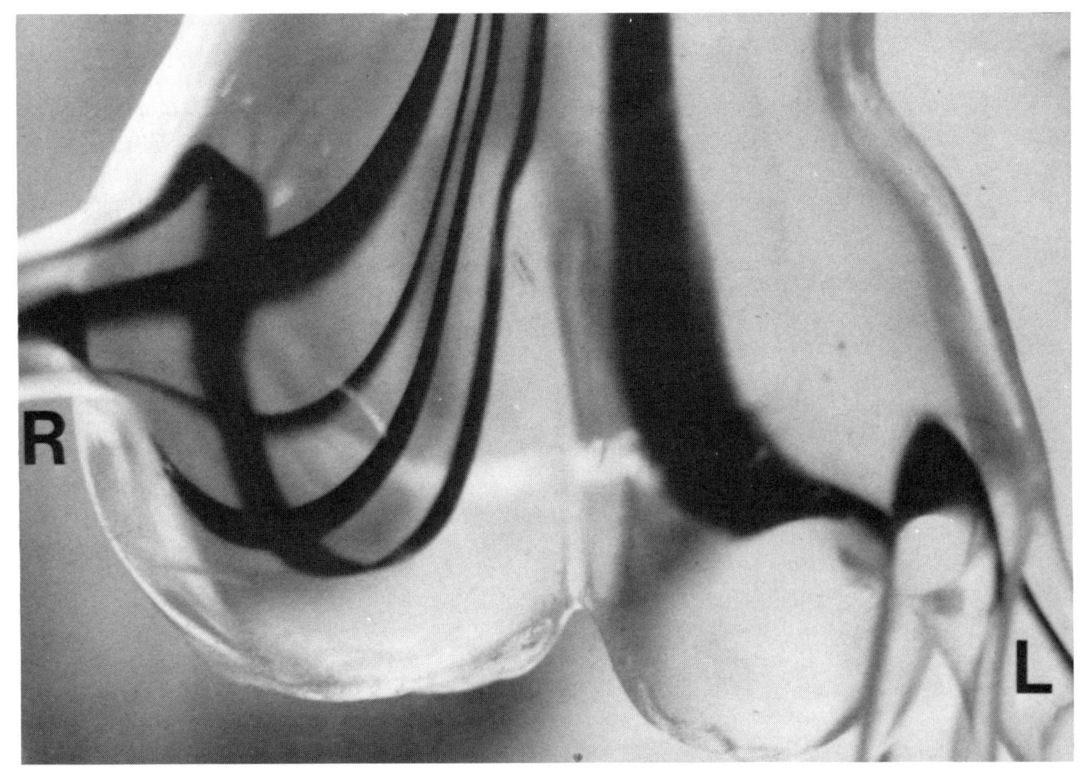

Fig. 4 Flow pattern in the sinuses of Valsalva during steady flow under conditions that simulate rest. L, Ostium of the left coronary artery; R, ostium of right coronary artery.

Fig. 5 Flow pattern in the sinuses of Valsalva during steady flow under conditions that simulate rest (photograph was taken at 45° relative to the photograph in Fig. 4). R, Ostium of right coronary artery.

Fig. 6 Flow pattern in the sinuses of Valsalva during steady flow under conditions that simulate exercise. L, Ostium of left coronrary artery; R, ostium of right coronary artery.

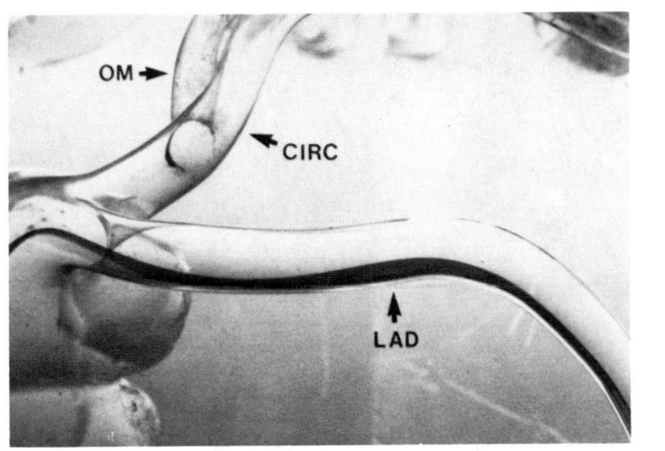

FLOW = 100 ml/min

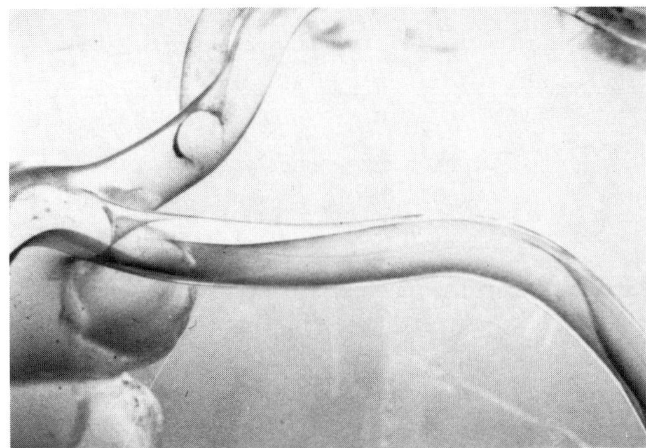

FLOW = 400 ml/min

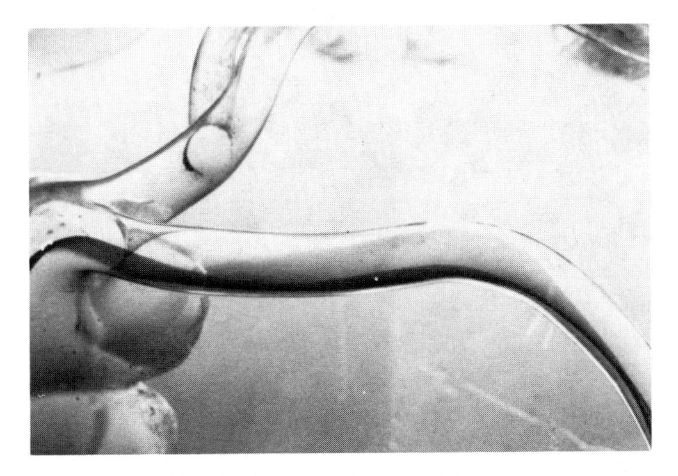

FLOW = 200 ml/min

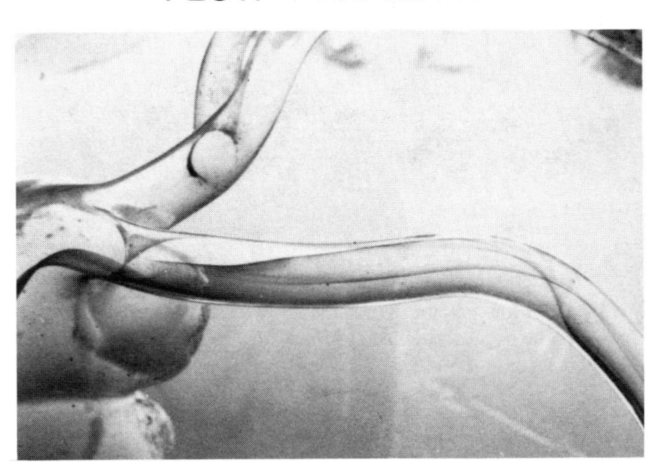

FLOW = 500 ml/min

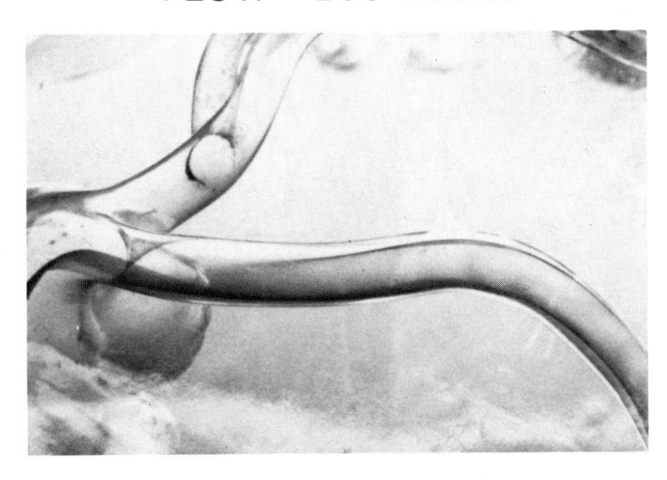

FLOW = 300 ml/min

Fig. 7 Secondary flow in the left anterior descending coronary artery (LAD) during steady flow. Patterns of secondary flow are shown at coronary flows of 100, 200, 300, 400, and 500 ml/min. CIRC, Left circumflex coronary artery; OM, obtuse marginal branch of CIRC [reprinted from the Journal of Biomechanical Engineering (vol. 106, pp. 272–279, 1984) with permission].

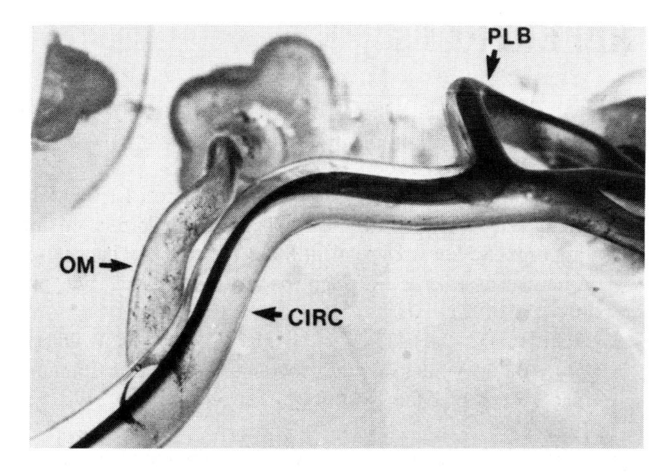

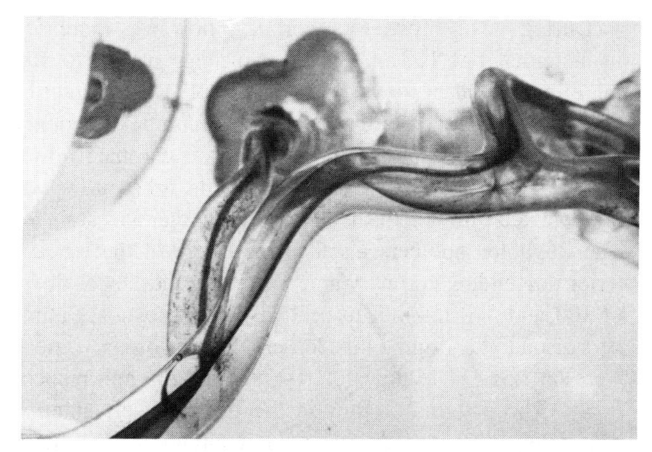

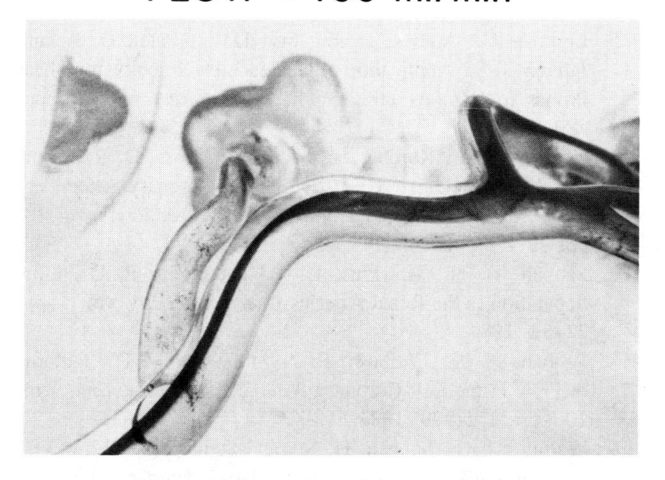

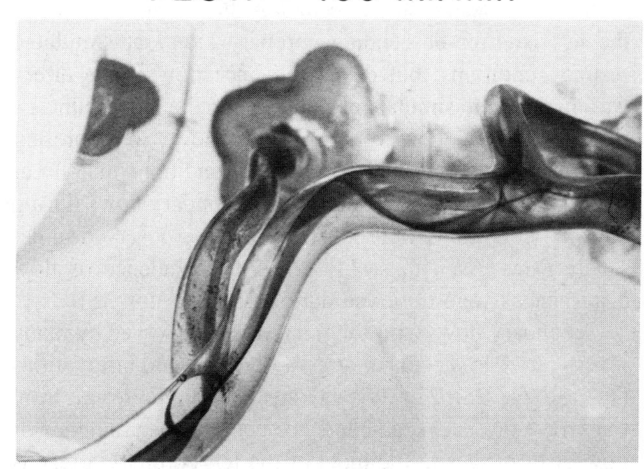

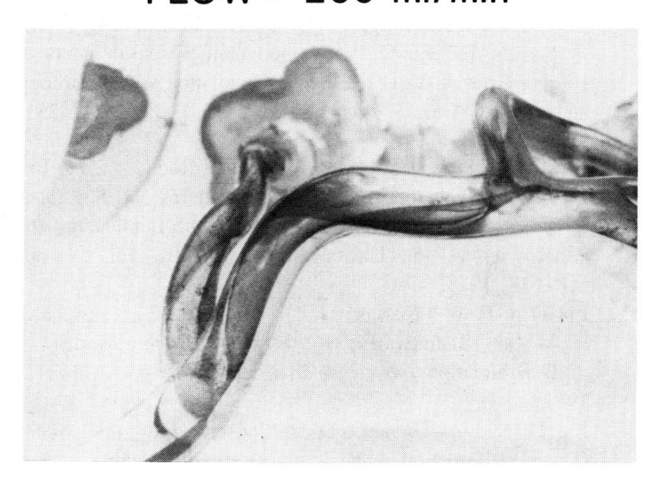

Fig. 8 Secondary flow in the left circumflex coronary artery (CIRC) during steady flow. The patterns of secondary flow are shown at coronary flows of 100, 200, 300, 400, and 500 ml/min. OM, obtuse marginal branch; PLB, first posterior lateral branch [reprinted from the Journal of Biomechanical Engineering (vol. 106, pp. 272–279, 1984) with permission].

During steady flow, total coronary flow was increased in increments of 100 ml/min over a range of 100 to 500 ml/min. Secondary flows developed in both the left anterior descending coronary artery and left circumflex coronary artery and became more prominent as left coronary artery flow was increased [8]. Turbulence or disturbed flow was not detected throughout the range of flow that was studied. The Reynolds numbers distal to the origin of the left anterior descending coronary artery were 57 and 284 at flows of 100 and 500 ml/min respectively. The Reynolds numbers distal to the origin of the left circumflex coronary artery were 65 and 321 at flows of 100 and 500 ml/min respectively. The patterns of flow in the proximal left anterior descending coronary artery and left circumflex coronary artery are shown in Figs. 7 and 8.

During pulsatile flow, secondary flows were absent in the left anterior descending coronary artery at simulated resting conditions, but developed once flow in the artery was increased to simulate exercise [8]. In the left circumflex coronary artery, however, secondary flows were present both at resting conditions and during conditions that simulated exercise [8]. The nature of the secondary flows during pulsatile conditions were identical to those observed under steady flow. As with steady flow, no turbulence or flow disturbances were observed during pulsatile flow [8].

Secondary flows (spiraling) have been observed by many investigators in models of arterial branches and bifurcations [11–13, 24, 25, 27, 28] and are thought to result from transverse pressure gradients. The absence of turbulence or flow disturbances in the mold of the left coronary artery is consistent with observations made by others in vivo [26, 27]. Nerem et al. [26], using hot-film velocity sensors in the coronary arteries of horses, showed that coronary flow was generally free of disturbances. Others, using a pulsed Doppler velocity meter in the coronary arteries of ponies, came to the same conclusion [27]. In both of these studies, however, large-amplitude, low-frequency oscillations were observed in the velocity patterns in the coronary arteries.

To date, flow visualization in the coronary arteries has proven to be a useful approach to understanding some of the characteristics of flow in these vessels. Clearly, more studies are needed to further understand the overall patterns of flow in the epicardial coronary arteries. Techniques such as streaming birefringence may be more effective in vitro than dye streamers and are currently under investigation. The difficulties of in vivo flow visualization in the coronary arteries or other blood vessels remain. One approach currently under consideration in this laboratory is use of the arteriographic technique in conjuction with computer image analysis to enhance some aspects of the flow that may not be optically apparent.

REFERENCES

1. Fry, D. L., Acute Vascular Endothelial Changes Associated with Increased Blood Velocity Gradients, *Circ. Rev.*, vol. 22, pp. 165–197, 1968.
2. Caro, C. G., Fitz-Gerald, J. M., and Schroter, R. C., Atheroma and Arterial Wall Shear; Observation, Correlation and Proposal of a Shear Dependent Mass Transfer Mechanism for Atherogenesis, *Proc. R. Soc. London, Ser. B*, vol. 177, pp. 109–159, 1971.
3. Stein, P. D., Sabbah, H. N., Anbe, D. T., and Walburn, F. J., Blood Velocity in the Abdominal Aorta and Common Iliac Artery of Man, *Biorheology*, vol. 16, pp. 249–255, 1979.
4. Nerem, R. M., and Seed, W. A., An in Vivo Study of Aortic Flow Disturbances, *Cardiovasc. Res.*, vol. 6, pp. 1–14, 1972.
5. Ling, S. C., Atabek, H. B., Fry, D. L., Patel, D. J., and Janicki, J. S., Application of Heated-Film Velocity and Shear Probes to Hemodynamic Studies, *Circ. Res.*, vol. 23, pp. 789–801, 1968.
6. Nerem, R. M., Rumberger, J. A., Jr., Gross, D. R., Muir, W. W., and Geiger, G. L., Hot Film Coronary Artery Velocity Measurements in Horses, *Cardiovasc. Res.*, vol. 10, pp. 301–313, 1976.
7. Sabbah, H. N., Hawkins, E. T., and Stein, P. D., Flow Separation in the Renal Arteries, *Arteriosclerosis*, vol. 4, pp. 28–33, 1984.
8. Sabbah, H. N., Walburn, F. J., and Stein, P. D., Patterns of Flow in the Left Coronary Artery, *J. Biochem. Eng.*, vol. 106, pp. 272–279, 1984.
9. Walburn, F. J., Sabbah, H. N., and Stein, P. D., Flow Visualization in a Mold of an Atherosclerotic Human Abdominal Aorta, *J. Biomech. Eng.*, vol. 103, pp. 168–170, 1981.
10. Friedman, M. H., Bargeron, C. B., Hutchins, G. M., Mark, F. F., and Deters, O. J., Hemodynamic Measurements in Human Arterial Casts and Their Correlation with Histology and Luminal Area, *J. Biomech. Eng.*, vol. 102, pp. 247–251, 1980.
11. El Masry, O. A., Feuerstein, I. A., and Round, G. F., Experimental Evaluation of Streamline Patterns and Separated Flows in a Series of Branching Vessels with Implications for Atherosclerosis and Thrombosis, *Circ. Res.*, vol. 54, pp. 608–618, 1978.
12. LoGerfo, F. W., Nowak, M. D., Quist, W. C., Crawshaw, H. M., and Bharadvaj, B. K., Flow Studies in a Model Carotid Bifurcation, *Arteriosclerosis*, vol. 1, pp. 235–241, 1981.
13. Ku, D. N., and Giddens, D. P., Pulsatile Flow in a Model Carotid Bifurcation, *Arteriosclerosis*, vol. 3, pp. 31–39, 1983.
14. Roach, M. R., Scott, S., and Ferguson, G. G., The Hemodynamic Importance of the Geometry of Bifurcations in the Circle of Willis (Glass Model Studies), *Stroke*, vol. 3, pp. 255–267, 1972.

15. Gutstein, W. H., Schneck, D. J., and Marks, J. O., In Vitro Studies of Local Blood Flow Disturbance in a Region of Separation, *J. Atheroscler. Res.*, vol. 8, pp. 381–388, 1968.
16. Sabbah, H. N., Khaja, F., Brymer, J. F., Hawkins, E. T., and Stein, P. D., Blood Velocity in the Right Coronary Artery: Relation to the Distribution of Atherosclerotic Lesions, *Am. J. Cardiol.*, vol. 53, pp. 1008–1012, 1984.
17. Sabbah, H. N., Khaja, F., Hawkins, E. T., Brymer, J. F., McFarland, T. M., Van Der Bel-Kahn, J., Doerger, P. T., and Stein, P. D., Relation of Atherosclerosis to Arterial Wall Shear in the Left Anterior Descending Coronary Artery of Man, *Am. Heart J.*, vol. 112, pp. 453–458, 1986.
18. Sabbah, H. N., Khaja, F., Brymer, J. F., McFarland, T. M., Hawkins, E. T., Marzilli, M., and Stein, P. D., In Vivo Patterns of Flow in the Coronary Arteries of Man, ASME, publication pp. 109–110, 1984.
19. Sabbah, H. N., Khaja, F., Brymer, J. F., McFarland, T. M., and Stein, P.D., Patterns of Blood Flow Distal to Mild Atherosclerotic Plaques in the Coronary Arteries of Patients, *Clin. Res.*, vol. 32, p. 781A, 1984.
20. Schlichting, H., *Boundary Layer Theory*, 6th ed., pp. 589–590, McGraw-Hill, New York, 1968.
21. Sabbah, H. N., and Stein, P.D., Hemodynamics of Multiple versus 50 Percent Coronary Artery Stenoses, *Am. J. Cardiol.*, vol. 50, pp. 276–280, 1982.
22. Sabbah, H. N., and Stein, P. D., Effect of Aortic Stenosis on Coronary Flow Dynamics: Studied in an *In Vitro* Pulse Duplicating System, *J. Biomech. Eng.*, vol. 104, pp. 221–225, 1982.
23. Stein, P. D., Sabbah, H. N., and Walburn, F. J., Patterns of Flow in the Sinuses of Valsalva, in *Flow Visualization III: Proceedings of the Third International Symposium on Flow Visualization*, ed., W.-J. Yang, pp. 115–119, Hemisphere, Washington, D.C., 1985.
24. Karino, T., and Motomlya, M., Flow Visualization in Isolated Transparent Natural Blood Vessels, *Biorheology*, vol. 20, pp. 119–127, 1983.
25. Brech, R., and Bellhouse, B. J., Flow in Branching Vessels, *Cardiovasc. Res.*, vol. 7, pp. 593–600, 1973.
26. Nerem, R. M., Rumberger, J. A., Jr., Gross, D. R., Hamlin, R. L., and Geiger, G. L., Hot-Film Anemometer Velocity Measurement of Arterial Blood Flow in Horses, *Circ. Res.*, vol. 34, pp. 193–203, 1974.
27. Wells, M. K., Winter, D. C., Nelson, A. W., and McCarthy, T. C., Blood Velocity Patterns in Coronary Arteries, *J. Biochem. Eng.*, vol. 99, pp. 36–38, 1977.

Chapter 35

Cardiovascular Systems

Ajit P. Yoganathan, H.-W. Sung, E. F. Philpot, Y.-R. Woo, S. T. McMillan, A. Jimoh, and A. J. Ridgway

I INTRODUCTION

The new modalities of noninvasive cardiac imaging, such as color Doppler flow mapping (CDFM) and magnetic resonance imaging (MRI), are providing cardiologists and cardiac surgeons with the ability to observe (i.e., visualize) blood-flow patterns in the heart and great vessels [1]. These noninvasive techniques are being used on an ever-increasing basis to diagnose and treat a variety of cardiac defects and abnormalities, such as (1) valvular stenosis; (2) valvular regurgitation (incompetence); (3) prosthetic valve function; (4) ventricular septal defects; (5) atrial septal defects; and (6) other congenital heart defects [2]. Interpretation of the images produced by these noninvasive techniques is not trivial, and in many instances is quite complex. In addition, the CDFM instruments commercially available at present are fraught with artifacts, pitfalls, and aliasing problems, and they have inadequate spatial and frequency resolution.

In order to better understand the flow patterns created in the presence of different cardiac defects and abnormalities, a series of well-defined in vitro laboratory-bench flow visualization studies were conducted. These experiments were conducted in mock circulation loops that mimicked the pressure and flow wave forms observed in the human left and right heart. An overview of these studies is presented in this chapter. In addition, examples of in vitro color Doppler flow mapping studies conducted in our laboratory are also described.

This work was supported by contracts from the Food and Drug Administration, and grants from the American Heart Association–Georgia Affiliate, the Whitaker Foundation, the Rich Foundation, Medtronic Blood Systems, Baxter Edwards Laboratories, Johnson and Johnson Cardiovascular, Medical Inc., St. Jude Medical, and Shiley Laboratories.

II APPARATUS AND METHODOLOGY

A Mock Circulation Loops

Experiments were conducted in the Georgia Tech left- and right-heart pulse-duplicator systems. Details of the pulse-duplicator systems and the aortic, mitral, and pulmonic flow chambers can be found elsewhere [3–5]. Figures 1 and 2 are schematics of the left- and right-heart mock circulation loops and the different flow chambers used. The flow chambers were constructed from Plexiglas and/or glass for optical clarity. The experiments were performed under simulated physiologic conditions, for example: (1) heart rate (HR) of 70 beat/min; (2) systolic duration of about 300 ms in the aortic and pulmonic valve studies, and diastolic time of about 550 ms in the mitral valve studies; (3) mean aortic (or pulmonic) pressure of 90 to 100 (or 15 to 20) mm Hg; (4) cardiac outputs (CO) in the range of 2.5 to 7.5 liter/min. An aqueous glycerine solution with a viscosity of 3.5 cP was used as the blood analog fluid.

B Flow Visualization (i.e., Streak Photography) Technique

A 7-mW He–Ne laser was used as a light source. The laser beam was diverged into a plane of light by directing it

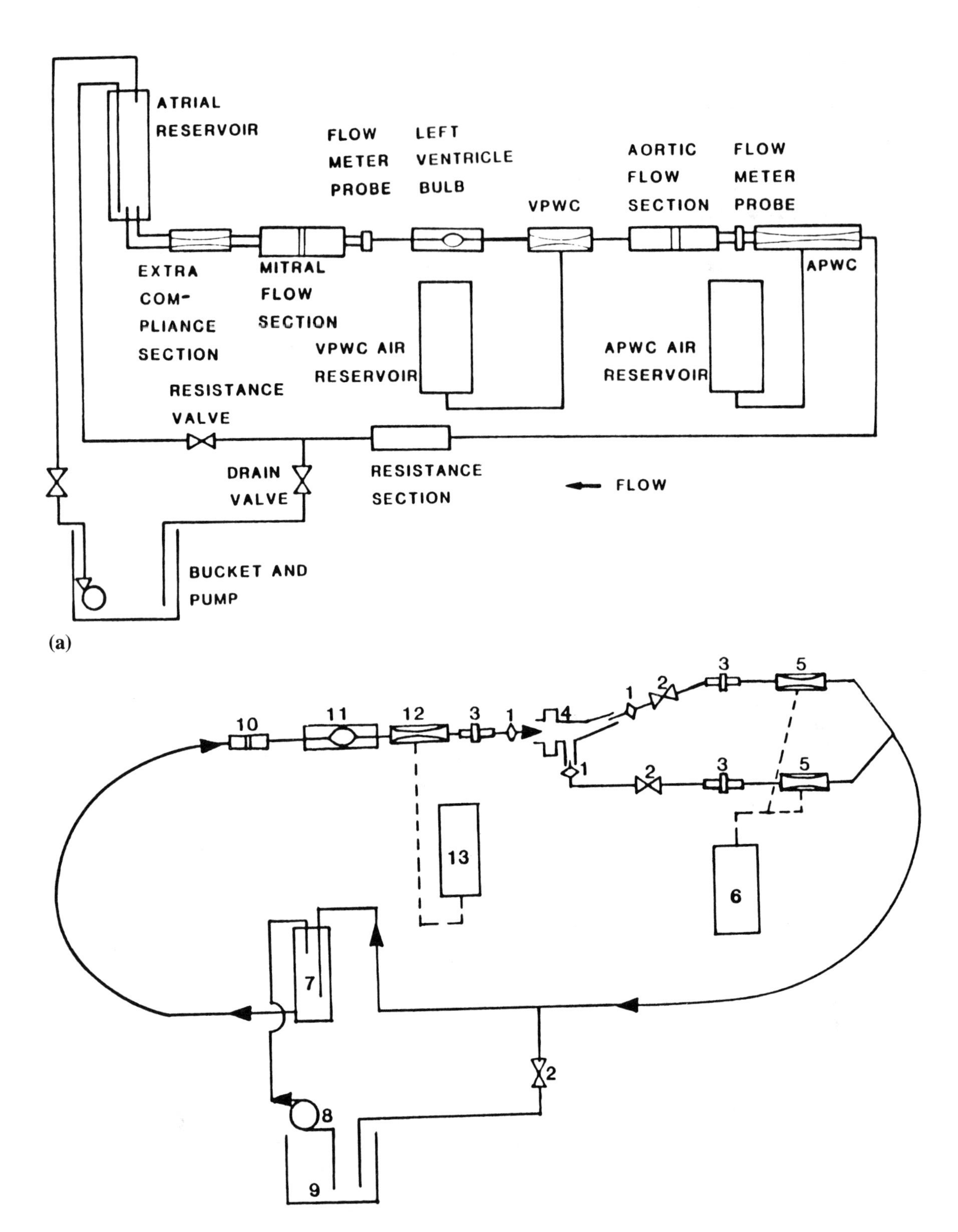

1. pressure taps
2. Valve
3. Flow Meter Probe
4. Model
5. Pulmonic Pressure Wave Control (PPWC)
6. PPWC Air Reservoir
7. Atrial Reservoir
8. Pump
9. Bucket
10. Straight Valve Section
11. Right Ventricular Bulb
12. Ventricular Pressure Wave Control (VPWC)
13. VPWC Air Reservoir

(b)

Fig. 1 Schematics of pulse-duplicator systems: (*a*) Left heart; (*b*) right heart.

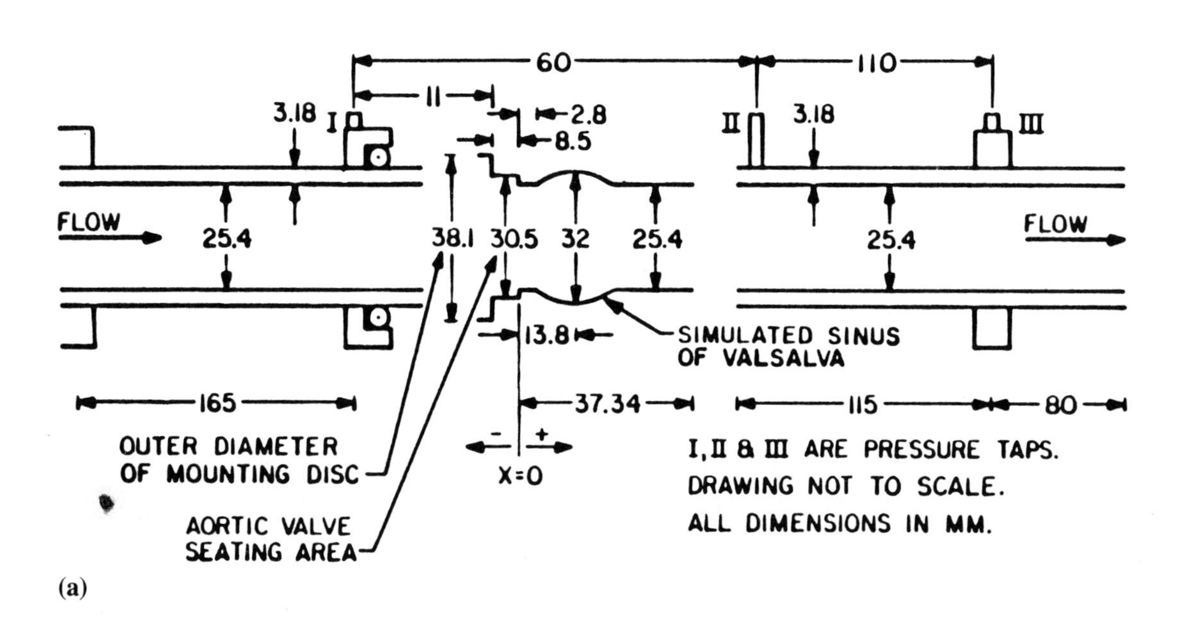

(a)

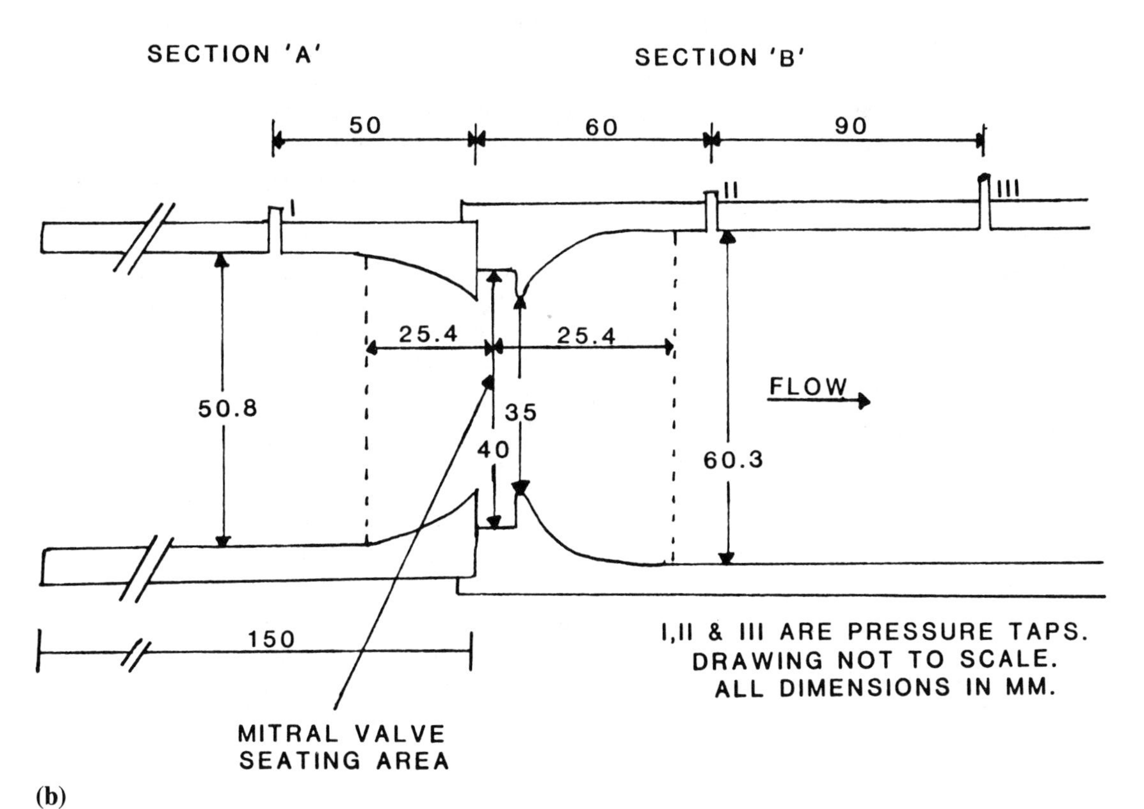

(b)

Fig. 2 Schematics: (*a*) Aortic flow chamber; (*b*) mitral flow chamber. (Figure continues.)

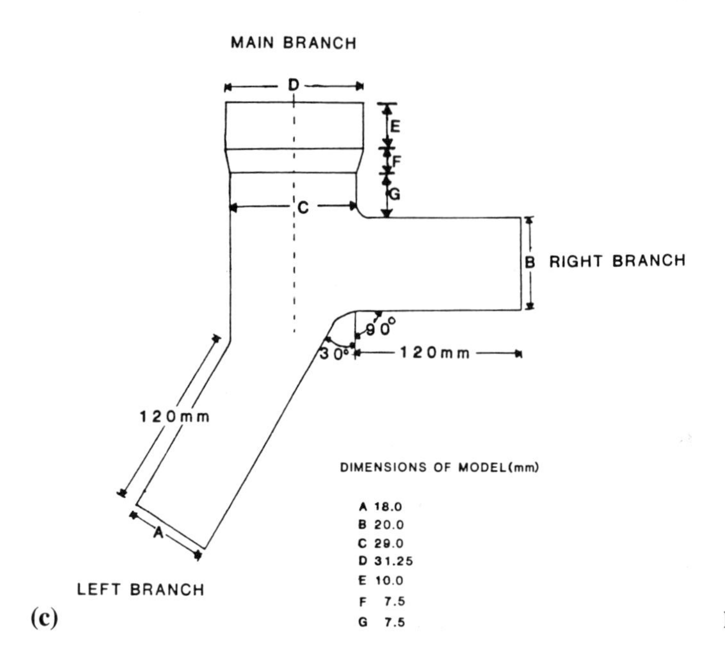

Fig. 2 (continued) (c) pulmonary artery model.

through a glass rod 1 cm in diameter; the plane of light was about 1.2 mm wide. The width of the plane of light could be controlled by two variables: (1) the diameter of the glass rod and (2) the distance between the rod and the flow section of interest. As the diameter of the glass rod was decreased, the angle of divergence increased, permitting a shorter distance between the laser and the flow section, for a desired width of the plane of light. The intensity of the light was inversely proportional to the width. A schematic of the optical setup used is shown in Fig. 3*a*.

In addition, the setup could be adjusted such that the sheet of light could pass through various cross sections of the flow channel, thereby facilitating flow visualization at different cross sections. Such cross-sectional variations are useful in studying any asymmetry in the flow fields.

Amberlite[1] particles, about 100 μm in diameter, were added to the blood analog fluid to act as tracers to observe the flow patterns. The density of the particles (1.07 g/cm^3) was close to that of the blood analog fluid used (1.05 g/cm^3), so that buoyancy effects were not a problem. The Amberlite particles were illuminated by the laser light, and the flow patterns were photographed against a dark background with either Tri-X or Ilford XP-1 film (ASA 400) using a Cannon A-1 camera. The camera *f* stop was generally set at 3.5 and the shutter speed at 1/15, 1/30, or 1/60 s.

In these experiments the camera was synchronized with the timer on the pulse-duplicator system (i.e., mock circulation loop). By using an electronic time-delay circuit, it was possible to take photographs of the flow field at any given instant during the heart cycle. The instant the photograph was taken, an electronic pulse signal was emitted from the stroboscope connection on the camera body. Figure 3*b* is a schematic drawing of the photography technique employed under pulsatile-flow conditions. The electronic pulse was recorded on a dual-beam storage oscilloscope (Tektronix T912) together with either the aortic, mitral, or pulmonic flow or pressure curve. Subsequently a Polaroid photograph of the stored electronic pulse signal and the aortic, mitral, or pulmonic flow curve was taken, thereby recording the instant the photograph was taken with respect to the cardiac cycle.

In addition to the still photographs, video films of the pulsatile flow fields were made using a Panasonic WV3240 video camera. The temporal and spatial resolution of the video films were, however, inferior compared to the still photographs.

C Color Doppler Flow Mapping Technique

Color Doppler flow mapping (CDFM) may represent one of the most important advances in noninvasive cardiac imaging technology. The technique offers, for the first time, true spatial relationships of cardiac blood flow. Color Doppler flow mapping utilizes multigate and multiple-sample volume ultrasound Doppler technology in an attempt to display the actual movement of blood within the heart [1, 6]; CDFM has been implemented in both phased-array and mechanical systems, using colors to indicate the direction and ''magnitude'' of blood velocity. Two of the three primary colors are used to differentiate flow direction. Blood flow toward the ultrasound transducer is generally displayed as shades of red, while blood flow away from the transducer is displayed as shades of blue. Color brightness increases in proportion to blood flow velocity, as shown in

[1]Rohm and Hass Corp., Pa.

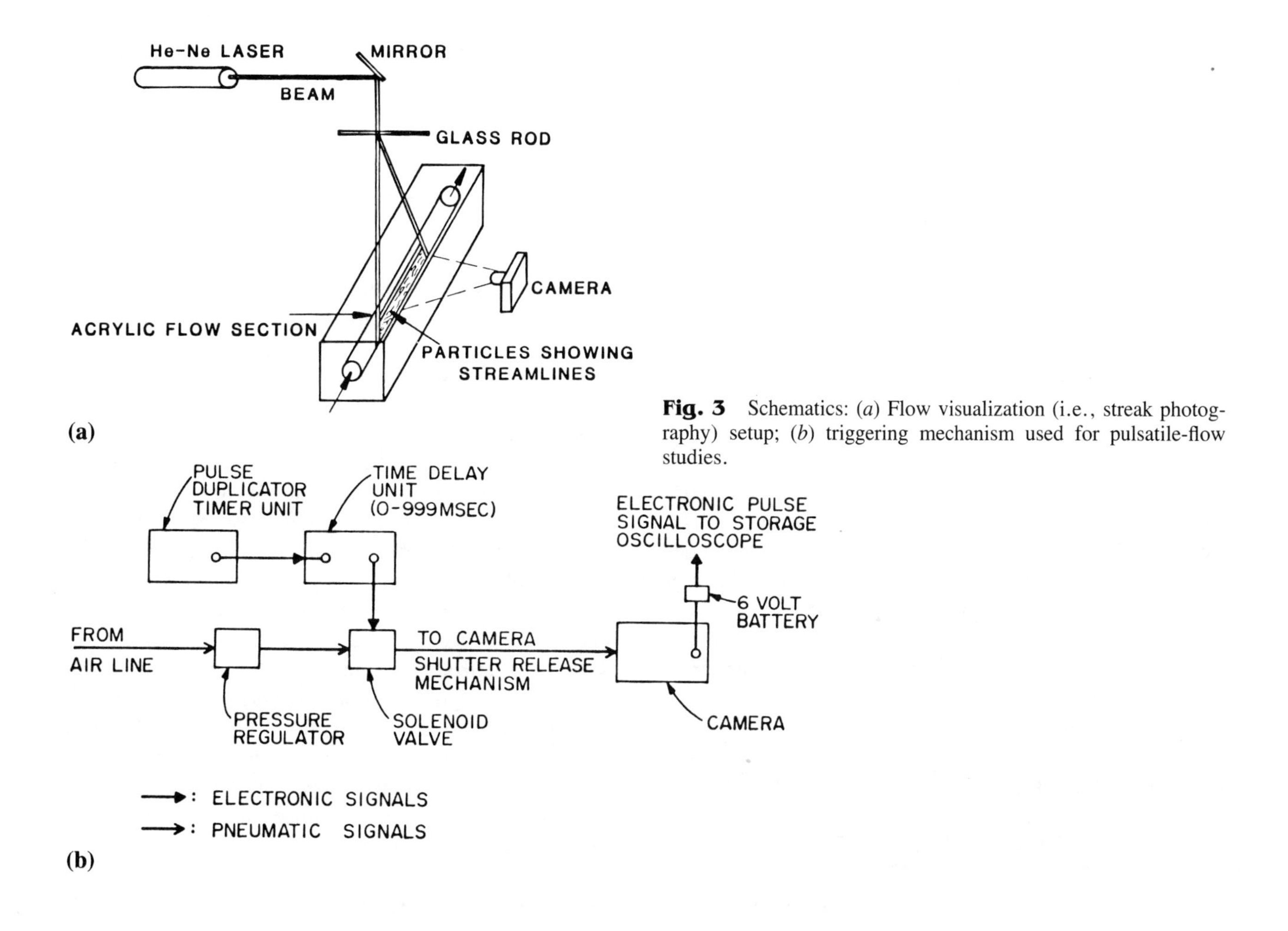

Fig. 3 Schematics: (*a*) Flow visualization (i.e., streak photography) setup; (*b*) triggering mechanism used for pulsatile-flow studies.

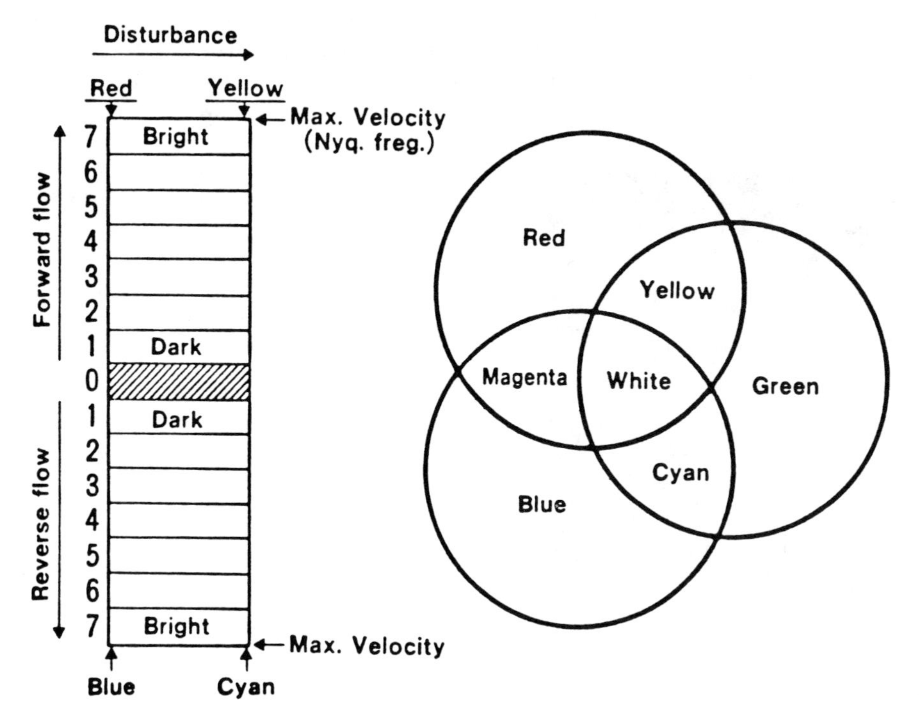

Fig. 4 Display of flow velocity by color in CDFM instruments.

Fig. 4. The third primary color, green, represents flow variance, and is added proportionally to the directional colors (red and blue) to highlight regions of disturbed (not necessarily turbulent) flow [7]. The mixing of red and green creates shades of yellow, while the mixing of blue and green create shades of cyan. Therefore, disturbed flow patterns in the forward flow direction are yellow, while those in the reverse direction are cyan. Generally, disturbed and/or turbulent flow fields, such as high velocity jets, exhibit mosaic color patterns.

Color-encoded two-dimensional Doppler studies were conducted with an Aloka SSD 880 system and a Toshiba SSH65 system. Color Doppler flow patterns were imaged in real time and in slow-motion gating using either the mitral, aortic, or pulmonic flow trace from an electromagnetic flowmeter. Two carrier frequencies (2.5 and 3.5 MHz) and three pulsed repetition frequencies (4, 6, and 8 kHz) were used for a total of five maximal resolvable velocities without aliasing from 46 to 122 cm/s. All the CDFM studies were conducted with the transducer placed parallel to the bulk flow direction [5]. For example, in the mitral and aortic flow chambers, the ''apical'' and ''supersternal notch'' views, respectively, were used.

III RESULTS AND DISCUSSION

A Aortic Stenosis

Examples of the flow visualization photographs obtained at peak systole with varying degrees of aortic stenosis are shown in Fig. 5. Flow through the ''normal'' aortic valve (valve leaflet opening area 5.0 cm^2) was relatively undisturbed and evenly distributed across the aorta. No regions of flow separation, secondary motion, or jet-type flow were observed. For cases of the mild (3.0 cm^2), moderate (1.0 cm^2), and severely (0.5 cm^2) stenotic aortic valves certain general trends were observed. For example, it was observed that the fluid generally exited from the valve as an asymmetric central jet. The jet broadened progressively as it traveled downstream by entraining the surrounding fluid. The asymmetric and angulated behavior of the aortic stenosis jets is clearly seen in Fig. 5. However, the size of the jet formed and the distal location where the jet hit the aorta varied with the degree of stenosis. As the valve became more stenotic, the diameter of the jet at the base of the aorta decreased and the flow field became more disturbed and chaotic (see Fig. 5). Furthermore, the annular region of flow separation surrounding the aortic stenosis jet and the turbulence intensity of the jet-type flow field increased as the degree of stenosis increased. The photographs show only two-dimensional fluid flow behavior, but physical observation of the fluid flow through the Plexiglas model allowed a three-dimensional sketch to be described.

The CDFM studies revealed flow phenomena qualitatively similar to that observed during the flow visualization studies. Color Doppler images were recorded to outline the inner highest velocity layered core of the jet, which was usually surrounded by an area of turbulence as observed in the flow visualization studies. As the degree of stenosis increased, the degree of aliasing and the amount of variance in the jet-type flow field increased. In addition, the length of proximal acceleration region at the inlet of the valve increased with increasing degrees of stenosis. Therefore, proximal acceleration is an estimate of the severity of the degree of valvular stenosis.

Unfortunately, it is not possible to show the color flow maps obtained with the varying degrees of aortic stenosis in this publication. The high velocities and turbulence levels created by the moderately and severely stenotic valves made *quantitative* interpretation of the color Doppler images virtually impossible. Furthermore, the CDFM experiments showed that, in present day-color flow-mapping systems, the amount of flow imaged and the area of variance changes with gain, pulse repetition frequency (PRF), and other imaging factors that relate to tissue distortion and attenuation

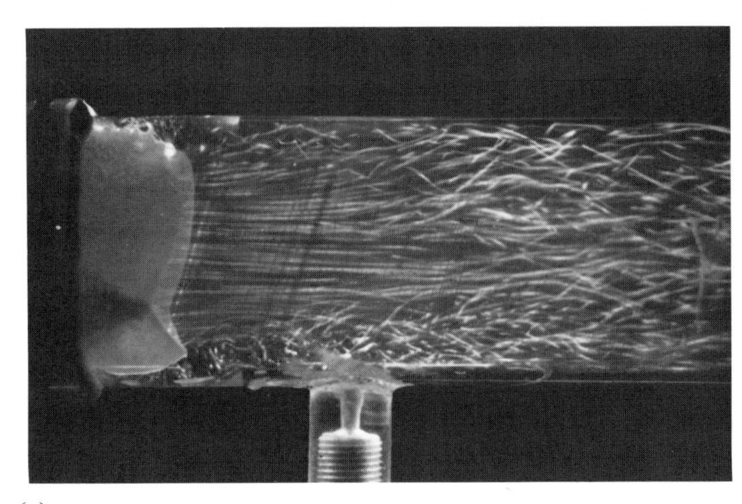

(a)

Fig. 5 Flow visualization photographs of aortic valves at peak systole: (*a*) ''Normal.'' (Figure continues.)

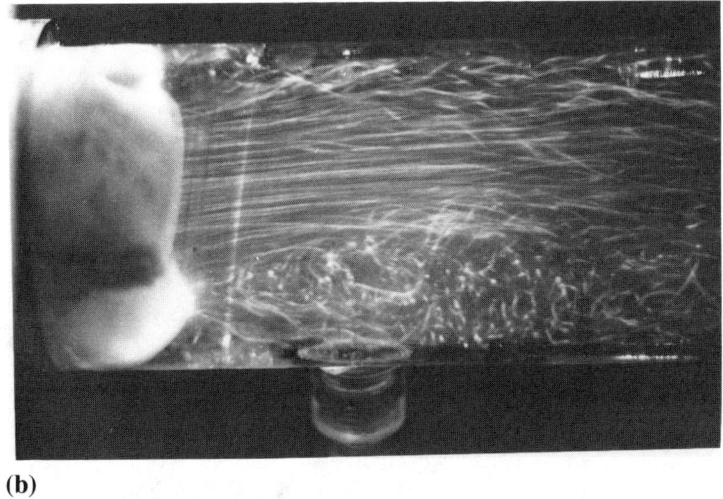

(b)

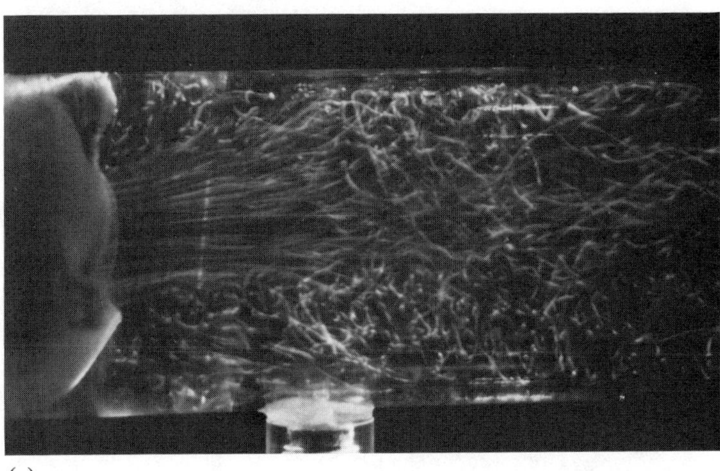

(c)

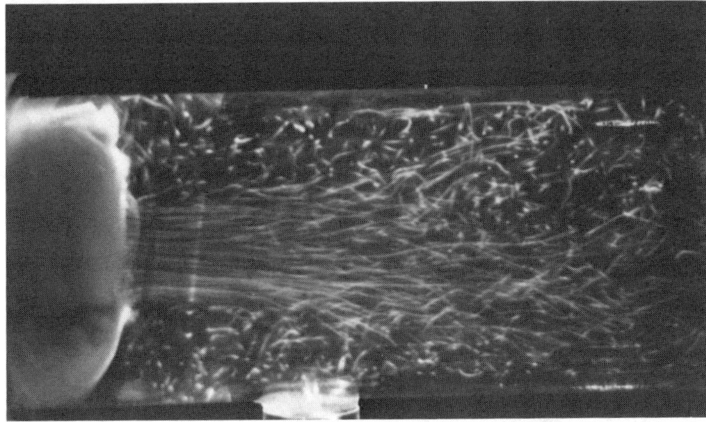

(d)

Fig. 5 (continued) (*b*) mildly stenotic; (*c*) moderately stenotic; (*d*) severely stenotic.

or, in our model, the Plexiglas interfaces themselves. The size of a measured jet varies with dynamic range and the degree to which flow information is filled in by the variance indicator and varies between instruments. The color Doppler flow maps were, however, very valuable in identifying the directions (i.e., the angulation) of the aortic stenosis jets.

B Pulmonic Stenosis

In vitro pulsatile flow visualization and CDFM studies were conducted in an adult-sized pulmonary artery model to observe the effects of valvular pulmonic stenosis on the flow fields of the main, left, and right pulmonary arteries. Trileaflet valves with leaflet opening areas of 5.0, 3.0, 1.0, and 0.5 cm^2 were used to simulate "normal," mild, moderate, and severely stenotic "pulmonic" valves, respectively [8, 9]. The flow patterns (see Figs. 6 and 7) revealed that as the degree of stenosis increased, the jet-type flow created by the valve became narrower, and it impinged on the far (distal) wall of the left pulmonary artery farther downstream from the junction of the bifurcation. This in turn led to larger regions of disturbed turbulent flow, as well as helical-type secondary flow motions in the left pulmonary artery, compared to the right pulmonary artery (see Fig. 8). The flow field in the main pulmonary artery also became more disturbed and turbulent, especially during peak systole and the deceleration phase.

For a given set of physiological experimental conditions, as the degree of valvular "pulmonic" stenosis increased, (1) the velocity and turbulence intensity of the jet in the main pulmonary artery (MPA) increased, (2) the intensity of the jet along the distal wall of the left pulmonary artery (LPA) increased more rapidly than the intensity of the jet along the distal wall of the right pulmonary artery (RPA), (3) the amount of turbulence in the flow field also increased, especially in the MPA and LPA, (4) the sizes of the regions of low axial flow and secondary flow were also observed to increase, especially in the MPA and LPA, (5) the region of disturbed flow in the LPA increased and extended farther downstream to a greater extent than in the RPA, and (6) the intensity of the helical motion increased in both the LPA and RPA, especially in the LPA.

C Aortic Incompetence

The left-heart pulse duplicator was used to provide a completely controllable system for study of aortic incompetence jet morphologies as a function of hemodynamic extremes [10]. Bioprosthetic valves with 1-, 3-, or 5-mm-dia. holes (or defects) punched into one or more leaflets were used to simulate aortic incompetence. The system was first used to calibrate the limits of resolution of color Doppler imaging. Next, to define which jet features reliably predict the defect size or the regurgitant fraction (RF), and which are primarily influenced by instantaneous hemodynamic variables, the jets' maximal length, width, proximal width, and temporal pattern of color variance were measured during in-

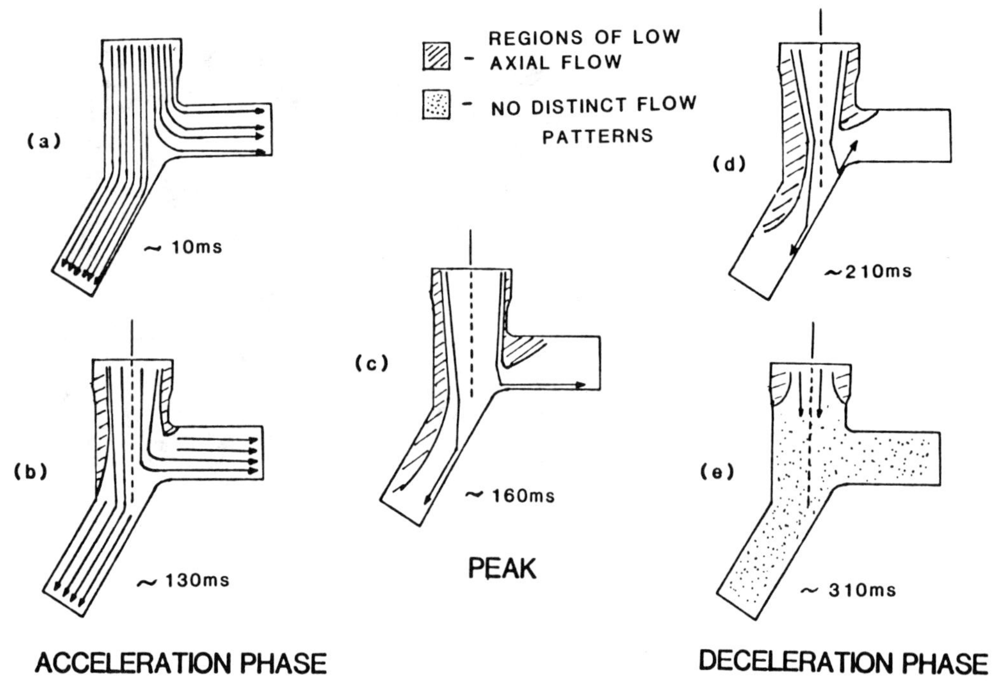

Fig. 6 Schematic of flow patterns observed during systole in the pulmonary artery model.

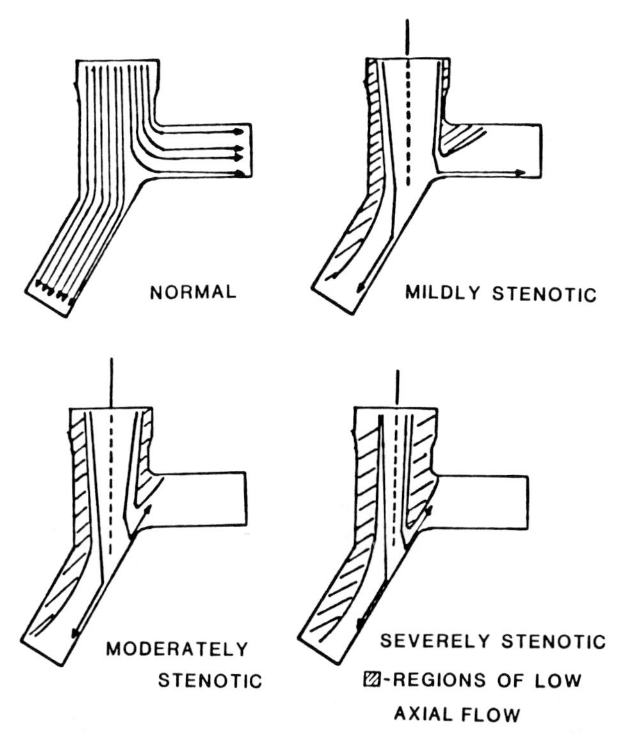

Fig. 7 Schematic of flow patterns observed at peak systole with varying degrees of pulmonic stenosis.

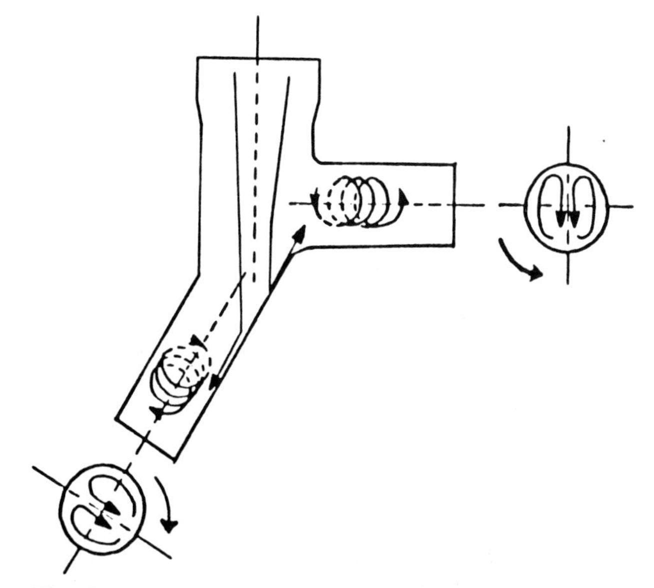

Fig. 8 Schematic of observed helical secondary flow motion in the pulmonary artery model.

dependent variations in the heart rate (HR), cardiac output (CO), and pressure gradient across the incompetent valve.

Net forward flow could be detected (in this system) with a CO as low as 0.5 liter/min. Aortic incompetence (AI) could be detected without difficulty through the 5- and 3-mm defects, as well as through the 1-mm pinhole, provided that the CO exceeded 1.5 liter/min and the mean aortic pressure (MAP) was at least 75 mm Hg. The dimensions of the AI jet, and therefore the sensitivity of the CDFM instruments for detecting AI, were most dependent on the pressure gradient across the valve. This was particularly relevant for the 1-mm pinhole. The sensitivity of color imaging for detecting AI was compromised by tachycardia (HR $\geq$ 110) or a low CO ($\leq$3.5 liter/min).

During the evaluation for lateral resolution, the two jets originating from defects placed in separate leaflets were always easily differentiated. The two jets were separated laterally by up to 20 mm on the color Doppler image. Furthermore, even the two jets originating from two defects placed only 3 mm apart on one leaflet could be differentiated regardless of the hemodynamic settings. However, when subjected to a significantly elevated MAP (>150 mm Hg), the maximum width of the two jets increased and obscured the fact that these were two distinct jets.

The temporal profile of color variance (signifying disturbed flow) often provided additional information useful in grading the severity of AI, particularly in the presence of elevated pressure gradients. With small defects, such color variance persisted throughout the jets' duration, whereas with the 5-mm defects, color variance was limited to very early diastole; the jet then became a homogeneous red or orange color for the remainder of diastole. Streak photography (i.e., flow visualization) was used to validate the use of these color changes in predicting flow patterns. Qualitative comparisons confirmed that when the color image predicted flow variance, streak photography demonstrated significantly disturbed flow, whereas monochromatic homogeneous color jets correctly predicted undisturbed laminar flow.

Analysis of AI jet morphology (from any size defect) was optimal at slower heart rates. This was particularly relevant when imaging jets from small defects (1 mm) during a low MAP (<100 mm Hg) and/or a CO $\leq$ 3.5 liter/min. It was observed that jets from small defects taper to a width significantly less than their maximal downstream width, whereas jets from larger defects have a more cylindrical shape. This distinction was true regardless of extremes in CO and MAP. Figure 9 shows that jet depth is clearly greater for jets arising from a larger defect (5 > 3 > 1 mm). However, it also demonstrates that the jet depth has a very strong dependence on the driving pressure gradient. The jets' maximal width was an inadequate predictor of either the defect size or the RF. Figure 10 demonstrates the poor correlation found. The minimal proximal width of the jets (measured within the first centimeter immediately below the valve plane) was predictive of the size of the defect, and the measured RF (see Fig. 11). The dimension was 100% specific in differentiating the three defect sizes, provided that aortic end diastolic pressure was $\geq$80 mm Hg.

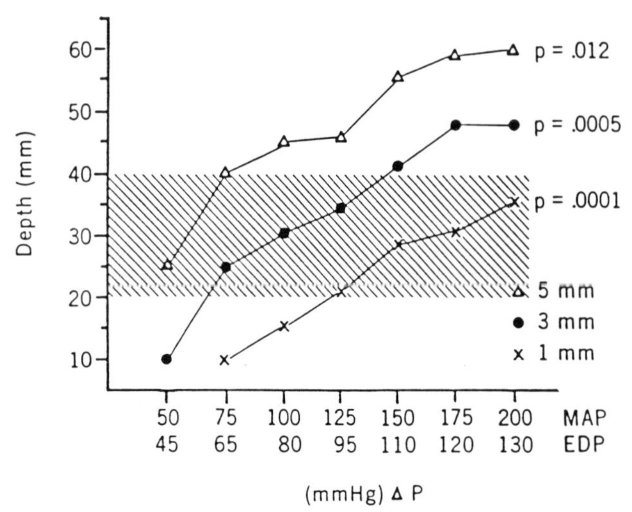

Fig. 9 Plot of regurgitant jet depth versus mean aortic pressure for 1-, 3-, and 5-mm size defects.

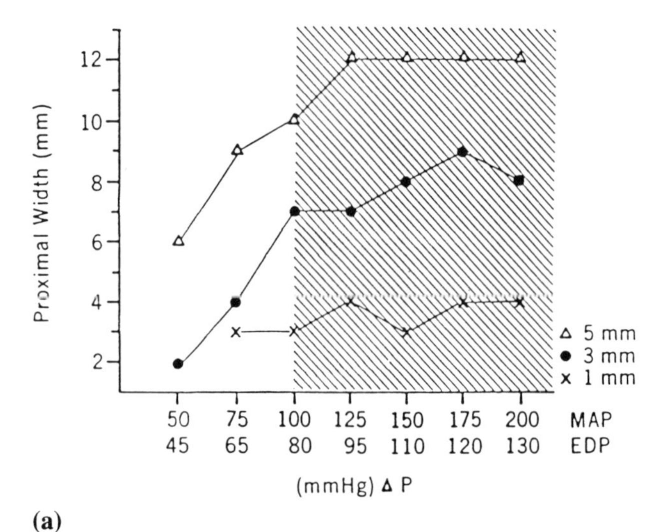

(a)

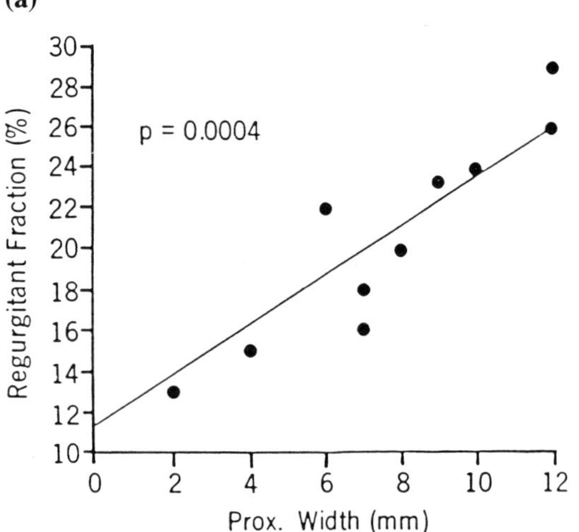

(b)

Fig. 11 (*a*) Plot of the minimal proximal jet width versus the pressure drop across the valve during diastole; (*b*) plot of the measured regurgitant fraction versus the minimal proximal jet width.

D Prosthetic Mitral Heart Valves

Examples of the in vitro flow visualization and CDFM results obtained in the mitral flow chamber at peak diastole downstream of the various generic valve designs are presented in this section. All figures presented are examples of the flow fields observed at peak diastole. More detailed results can be found elsewhere [3, 11, 12].

1 Ball and Cage Valve At early peak diastole, the forward flow emerging from the valve was jetlike. The flow separated from the ball about one quarter of the way around it, with a 45° angle to the axial direction (see Fig. 12). A region of flow separation, upstream of this jet, could be observed adjacent to the valve sewing ring. A well-defined vortex (22 mm in length) existed in this region. A turbulent wake could be observed distal to the ball. This area of turbulence was caused by the vortices shed from the distal surface of the ball as a result of boundary layer separation. The wake extended 45 mm downstream of the sewing ring. The region immediately downstream of the cage apex appeared stagnant. At late peak diastole, the jet separated from the ball about one third of the way around it. The jet then curved upward and impinged on the wall of the flow channel 50 mm downstream of the sewing ring. A region of flow separation existed between the jet and the sewing ring. The vortex structure that appeared in this region at early peak diastole had broken down into small vortices. The turbulent wake distal to the ball extended up to 50 mm downstream of the sewing ring. The region immediately downstream of the cage apex appeared stagnant.

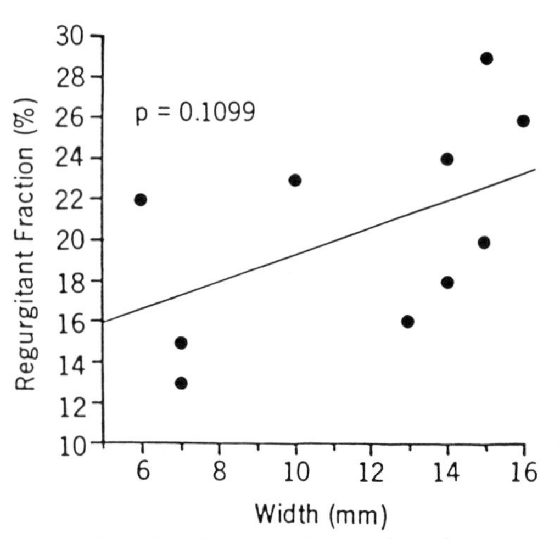

Fig. 10 Plot of measured regurgitant fraction versus maximal jet width for 3- and 5-mm defects.

The CDFM studies revealed two narrow jets of similar size oriented along the walls of the left ventricular (i.e., mitral flow chamber) model. These jets had similar peak

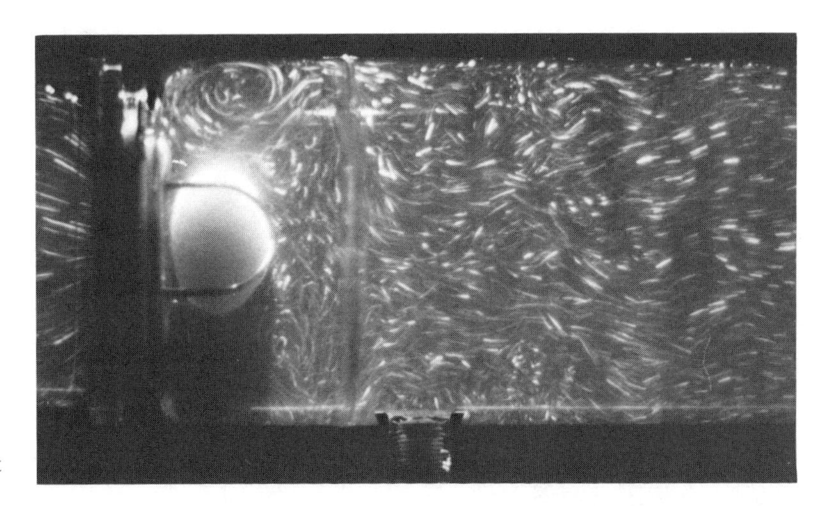

Fig. 12 Flow visualization photograph at peak diastole with ball and cage valve design.

velocities. A small wake region was observed immediately distal to the cage. The CDFM studies also showed two small annular regions of reverse flow immediately adjacent to the walls of the LV model. In addition, variance was detected along the edges of the annular forward flow jet.

2 Caged-Disk Valve At early peak diastole, a very thin circumferential jet emerged from the valve at a 20° angle to the occluder. The jet impinged on the wall 16 mm downstream from the sewing ring. A doughnut-shaped flow-separation region was formed between the jet and the valve sewing ring. The forward flow then traveled along the wall in a thin layer. It started to diverge 48 mm downstream of the sewing ring and immediately occupied the whole flow channel. Reverse flow occurred in the center of the flow channel up to 50 mm downstream of the sewing ring, and impinged on the downstream face of the occluder. Thus, a large doughnut-shaped vortex existed downstream of the valve as well, which appeared as a pair of symmetric vortices in the glance of the laser light. A small region of stagnation existed adjacent to the downstream face of the occluder. At late peak diastole, the flow field was qualitatively similar to that at early peak diastole (see Fig. 13).

Similar results were obtained from the CDFM experiments. Two narrow jets of similar peak velocity were observed adjacent to the walls of the left ventricular flow chamber. A large region of flow reversal, a result of flow separation from the occluder, was detected just distal to the valve cage. Variance was observed (the presence of yellow in the color map) along the edges of the jetlike flow field. Very little or no variance was, however, observed in the central wake region.

3 Tilting-Disk Valve At early peak diastole, the forward flow emerging from the major orifice gained more momentum and was jetlike. The jet exited the major orifice at a 40° angle to the axial direction, impinged on the flow channel wall 25 mm downstream of the sewing ring and then traveled along the wall (see Fig. 14). The region of flow separation between the jet and the flow channel wall showed a well-defined vortex about 11 mm in diameter. The forward flow from the minor orifice had lower velocity

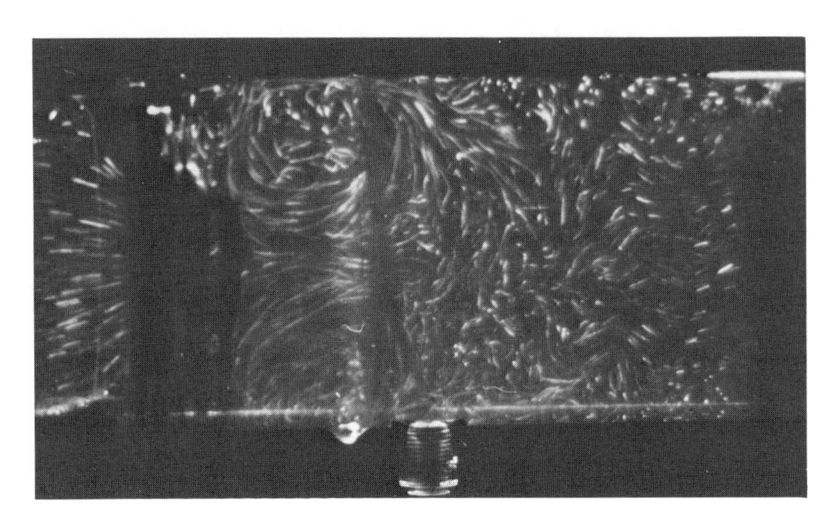

Fig. 13 Flow visualization photograph at peak diastole with caged-disk valve design.

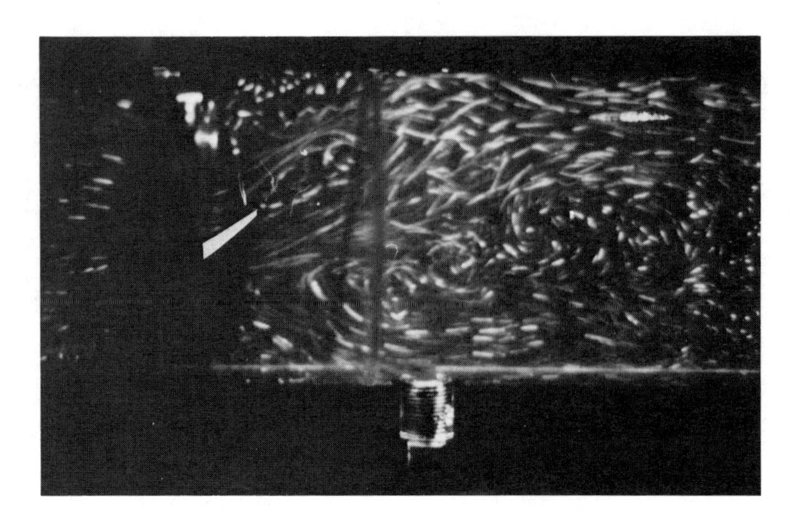

Fig. 14 Flow visualization photograph at peak diastole with tilting-disk valve design.

(shorter streaks) than the major orifice jet, and dissipated about 33 mm downstream of the sewing ring. The region between the forward flow through the minor orifice and the major orifice appeared relatively stagnant.

At late peak diastole, the major orifice jet formed a 30° angle to the axial direction as it emerged from the valve. The jet impinged on the wall about 48 mm downstream of the sewing ring. The region of flow separation between the jet and the flow channel wall extended farther downstream than observed at early peak diastole, since the point of jet impingement moved farther downstream. The vortex structure, which was observed at early peak diastole in this separation region, had broken down. Forward flow covered the whole minor orifice region with higher velocity than that at early peak diastole. This forward flow split into two parts 20 mm downstream from the sewing ring with a region of flow separation between. The upper part of the forward flow joined the major orifice jet.

The CDFM studies also revealed two jetlike flow fields: one from the major orifice and the other from the minor orifice. Peak velocities detected for both jets were approximately of the same magnitude. A small region of flow reversal was observed between the two jets. In addition, CDFM indicated a region of reverse flow along the wall of the left ventricular flow chamber adjacent to the minor orifice of the valve. Variance in the color spectrum was observed along the edges of both jets. Little or no variance was detected in the regions of reverse flow.

4 Bileaflet Valve At early peak diastole, the flow emerging from the valve formed a jet-type flow field in the central part of the flow channel with a doughnut-shaped flow-separation region around it. There appeared to be a pair of symmetric vortices on either side of the jet (see Fig. 15). These vortices were well defined and were located 25 mm downstream of the sewing ring in the central orifice region. In the lateral (i.e., side) orifices, the sizes of the vortices were the same as those in the central orifice but were located 30 mm downstream of the sewing ring. At late peak diastole, diminished forward flow (shorter streaks) relative to early peak diastole, could be observed through the central orifice. The vortex adjacent to the sewing ring increased in size to about 60 mm in length and 30 mm in width. The forward flow through the lateral orifices formed jets similar to those at early peak diastole. The vortices around the jets were about the same size as those at early

Fig. 15 Flow visualization photograph at peak diastole with bileaflet valve design.

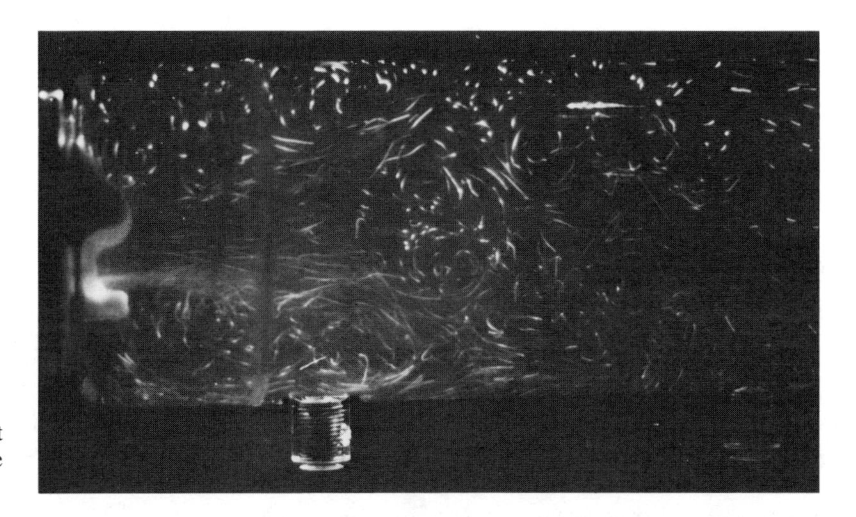

Fig. 16 Flow visualization photograph at peak diastole with porcine trileaflet valve design.

peak diastole, while the center of the vortices migrated 35 mm downstream of the sewing ring.

Three jet-type flow fields, two from the lateral orifices and one from the central orifice, were also observed in the color Doppler flow map. These jets (70–90 cm/s) were centrally located and were surrounded by annular regions of flow reversal. Variance was observed along the edges of the three jets.

5 Trileaflet Porcine Valves At early peak diastole, a jetlike flow emerging from the valve occupied the center part of the flow channel. The jet traveled axially 45 mm downstream of the sewing ring and then suddenly moved upward at a 40° angle to the axial direction (see Fig. 16). A region of flow separation existed on either side of the jet. The annular region between leaflets and flow channel, about 10 mm in width, adjacent to the sewing ring, appeared relatively stagnant. The separation region above the jet was not disturbed very much, and no high-velocity recirculation flow could be observed. The separation region below the jet was quite disturbed with high-velocity recirculation flow. A well-defined vortex could be seen in this region with a diameter of about 30 mm. At late peak diastole, the general features of the flow field were the same as that at early peak diastole. The region of stagnation above the jet adjacent to the sewing ring diminished. A low-velocity vortex, with a diameter of about 25 mm, existed in this region. The region beneath the jet adjacent to the sewing ring was still relatively stagnant.

The CDFM studies also revealed a relatively high-velocity aliased asymmetric jet in the center of the flow chamber. Regions of flow separation were detected adjacent to the valve leaflets; these regions were attributed to vortices originating from the forward flow jet. Variance was clearly observed along the edges of the forward flow jet.

6 Trileaflet Pericardial Valve At early peak diastole, a high-velocity jet emerged from the valve. The jet

Fig. 17 Flow visualization photograph at peak diastole with trileaflet pericardial valve design.

exited the valve axially, started to curl downward about 34 mm downstream of the sewing ring, and impinged on the flow channel wall 45 mm downstream (see Fig. 17). Flow separated from the downstream edges of the valve leaflets. The annular region adjacent to the sewing ring and around the outflow faces of the valve leaflets appeared relatively stagnant. A large vortex having a diameter of about 40 mm existed above the jet, 45 mm downstream of the sewing ring. At late peak diastole, the jet emerging from the valve had a much higher velocity than that at early peak diastole. The jet maintained an axial direction and did not move downward as it did at early peak diastole. The annular region of stagnation around the outflow faces of the leaflets extended up to 30 mm downstream of the valve sewing ring.

A broad central flow field was also observed in the color Doppler flow map. Regions of reverse and separated flow, which were attributed to vortices originating from the forward flow jet, were detected adjacent to the valve leaflets. Variance was observed along the edges of the forward flow jet.

REFERENCES

1. Sochurek, H., Medicine's New Vision, *Nat. Geog.*, vol. 171, pp. 2–41, 1987.
2. Omoto, R., *Color Atlas of Real-Time Two-dimensional Echocardiography,* Shindan-To-Chiryo, Tokyo, 1984.
3. Woo, Y.-R., *In Vitro* Velocity and Shear Stress Measurements in the Vicinity of Prosthetic Heart Valves, Ph.D. dissertation, Georgia Institute of Technology, Atlanta, 1984.
4. Philpot, E. F., *In Vitro* Flow Visualization and Pressure Measurement Studies in the Pulmonary Artery, M.S. thesis, Georgia Institute of Technology, Atlanta, 1984.
5. Valdes-Cruz, L. M., Yoganathan, A. P., Tomizuka, F., Woo, Y.-R., and Sahn, D. J., Studies *in Vitro* of the Relationship between Ultrasound and Laser Doppler Velocimetry and Applicability to the Simplified Bernoulli Relationship, *Circulation,* vol. 73, pp. 300–308, 1986.
6. Nanda, N. C., *Doppler Echocardiography,* Igaku-Shoin, New York, 1985.
7. Omoto, R., and Kasai, C., Basic Principles of Doppler Color Flow Imaging, *Echocardiography,* vol. 3, pp. 463–474, 1986.
8. Philpot, E. F., Yoganathan, A. P., Woo, Y.-R., Sung, H.-W., Franch, R. H., Sahn, D. J., and Valdes-Cruz, L. M., *In Vitro* Pulsatile Flow Visualization Studies in a Pulmonary Artery Model, *J. Biomech. Eng.,* vol. 107, pp. 368–375, 1985.
9. Yoganathan, A. P., Ball, J., Woo, Y.-R., Philpot, E. F., Sung, H.-W., Franch, R. H., and Sahn, D. J., Steady Flow Velocity Measurements in a Pulmonary Artery Model With Varying Degrees of Pulmonic Stenosis, *J. Biomech.,* vol. 19, pp. 129–146, 1986.
10. Switzer, D. F., Yoganathan, A. P., Nanda, N. C., Woo, Y.-R., and Ridgway, A. J., *In Vitro* Calibration of Color Doppler Flow Mapping during Extremes of Hemodynamic Conditions: A Foundation for a Reliable Quantitative Grading System for Aortic Incompetence, *Circulation,* vol. 75, pp. 837–846, 1987.
11. Woo, Y.-R., Williams, F. P., Faughnan, P. D., and Yoganathan, A. P., Pulsatile Flow Visualization Studies with Aortic and Mitral Mechanical Heart Valve Prostheses, *Chem. Eng. Commun.,* vol. 47, pp. 23–48, 1986.
12. Jones, M., McMillan, S. T., Eidbo, E. E., Woo, Y.-R., and Yoganathan, A. P., Evaluation of Prosthetic Heart Valves by Doppler Flow Imaging, *Echocardiography,* vol. 3, pp. 513–525, 1986.

Chapter 36

Medicine and Biology

Wen-Jei Yang

I INTRODUCTION

A Content of the Chapter

The first part of this chapter covers conventional techniques, including direct and projection imaging. Direct imaging is achieved with the use of laser Doppler velocimetry and tracers. Projection imaging, also referred to as machine vision, includes radiography, sonography, and thermography. The second part of the chapter deals with the digitized version of flow visualization, consisting of video imaging and computer graphics. Video imaging refers to the celebrated "machine's new vision," e.g., incredible machines that can peep into the human body to detect the flow and metabolic phenomena as well as the functioning of various organs. Computer graphics refers to graphics and color graphics displays of computed or measured field phenomena. Figures 1 and 2 summarize the content of this chapter.

The methods of flow visualization have provided significant contributions not only to the investigation of normal and morbid anatomy but also to the exploration of normal and pathological physiology. In clinical applications, they serve as a diagnostic guide in surgery, as a means of determining the activity of blood-borne pharmacologic agents in normal and malignant cells, as an aid in the diagnosis and localization of tumors, and as a help in establishing the in vivo topography of the vascularization (angiogenesis) of a growing malignant tumor.

Most chapters in the book deal with physical systems while the present one concerns biological systems. It is important to recognize the distinct difference between physical and biological (in particular, human body) systems: the former can be readily exposed or investigated in vitro (through modeling), while the latter requires body-scanning systems to peep into the human body without the trauma of surgery.

B Historical Background

In 1615 the principle of blood circulation was conceived by William Harvey, who observed the circulation of blood in a frog lung by means of a microscope. By the impulse of heart beats, he saw the movement of blood through very small arteries, and thus, pulmonary circulation was introduced.

In 1800 Sir William Herschel discovered infrared rays. The direct impact of infrared radiation on medical examination has been minimal; however, the development of a thermal, or infrared, camera in recent years has made it possible to visualize a heat map of the body's surface through the detection of small amounts of infrared radiation emitted from the human body. It is a noninvasive means of producing a two-dimensional heat map of a living body.

The X-ray, discovered by W. C. Roentgen in 1895, has been developed as a seeing machine, which penetrates the body, producing a view of the tissues within. In 1972 computer tomography (CT) scanners were developed to convert X-ray pictures into digital computer codes in order to make high-resolution video images. The breakthrough required the mathematical equations of A. M. Cormack and the technical ingenuity of G. N. Hounsfield [1], two 1979 Nobel laureates of medicine.

The basic discovery of nuclear magnetic resonance (NMR) was made in the United States by two groups working independently, one led by F. Bloch [2] and the other

First Generation (Qualitative)

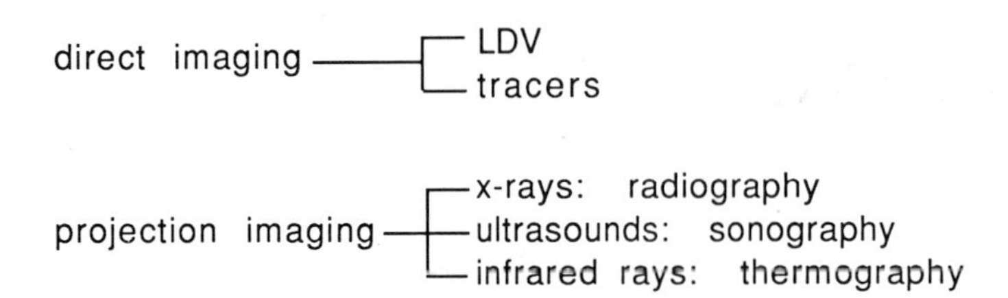

Second Generation (Digitalization)

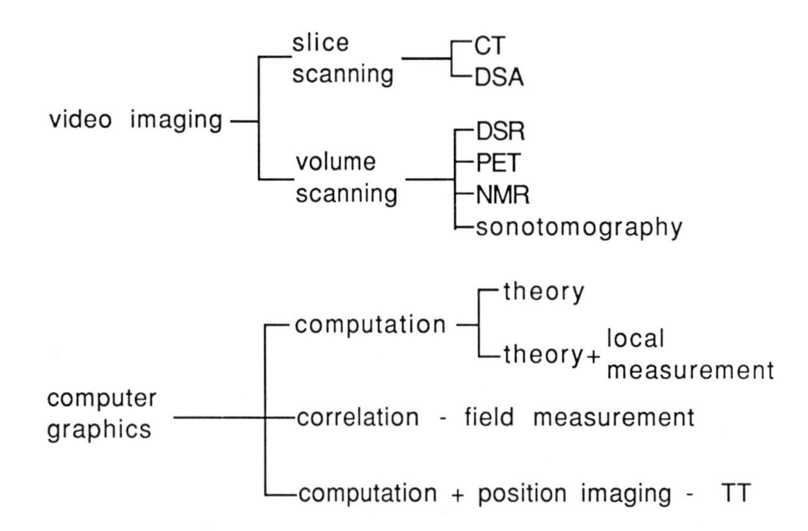

Fig. 1 Classification of flow visualization techniques in medicine and biology.

by E. M. Purcell [3]. Their results were made public in 1946. For this discovery, Bloch and Purcell were jointly awarded the Nobel Prize for Physics in 1952.

The Doppler effect is a change in the frequency of echo signals that occurs whenever there is a relative motion between the sound source and the reflector. It was discovered in 1842 by the Austrian physicist Christian Johann Doppler [4], who applied the principle to shifts in red light from double stars. Later in the same decade, Bays Ballot applied the principle to sound. Sonography, a method of mapping a physical or biological system structure, is based on the technique of sonar (sound navigation ranging), based on work by the French physicist Paul Langevin in 1917. Satomura [5] first applied the Doppler technique to determine

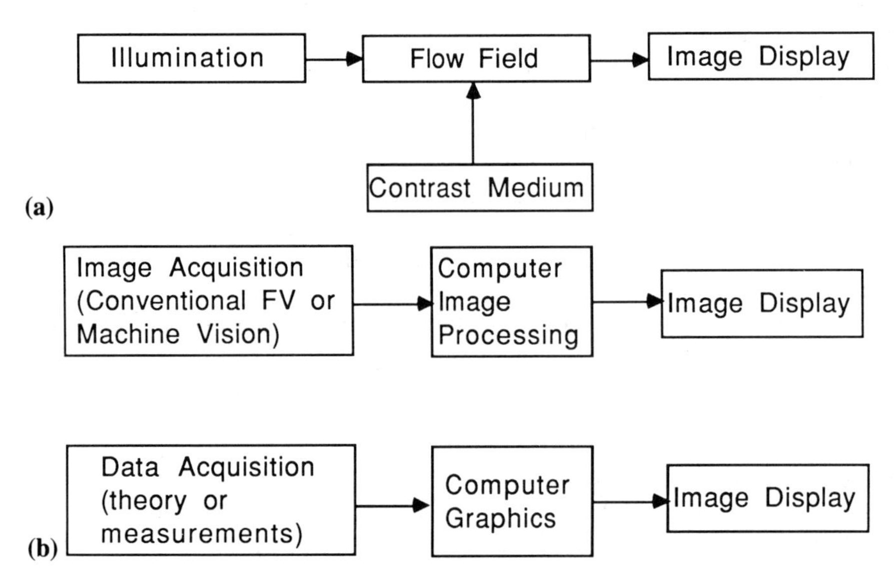

Fig. 2 Flow visualization in medicine and biology; *(a)* Conventional, and *(b)* computer-assisted.

blood velocity in 1956, while the use of sonography in obstetrics was first investigated by Donald et al in 1958 [6].

II CONVENTIONAL FLOW VISUALIZATION TECHNIQUES

The flow systems in medicine and biology include vascular vessels and excretory passages ranging from regular to micro size. In order to facilitate recognition of the anatomical morphology, function, and necrosis of the flow systems, contrast media are needed for image intensification. They are nothing but dyes, which include stains, fluorescein, and radiopaque and radioactive agents. These contrast media can be in the form of gases (not common), liquids, or fine particulates suspended in solution. The image is intensified through tinting, staining, fluorescence, interaction between X-rays and radioactive contrast media, and radioisotope labeling. Observation is with the unaided eye or through an optical device. The quality of visualization can be enhanced by means of appropriate vision magnification or image intensification. Photography and radiology (when interactions between X-rays and radioactive contrast media are involved) are commonly employed in either a single or sequential recording. Because of potential hazards to living organisms as a result of long exposure to radioactive contrast media, other diagnostic methods are being constantly sought and developed in an effort to replace radiology in medicine. The use of ultrasounds and infrared rays are examples of this attempt.

As Fig. 1 indicates, the conventional flow visualization techniques can be divided into direct and projection imaging according to the methods of imaging. The former is achieved through the use of tracers and laser Doppler velocimetry (LDV). The latter, also referred to as machine vision, includes radiography, sonography, and thermography by means of X-rays, ultrasound, and infrared rays, respectively. Contrast media, including tracers, are employed in the direct-imaging methods and in radiography.

A Laser Doppler Velocimetry

Laser Doppler velocimetry can be used to determine the dynamic characteristics of blood flow through vessels [7]. The flow of blood is different from ordinary fluid flow in that (1) the blood flow in arteries varies with time, e.g., pulsatile, (2) blood vessels are characterized by a complicated geometry and viscoelastic properties, and (3) blood is a highly concentrated suspension of red blood cells and thus non-Newtonian in rheological property. Hence, studies of blood flow dynamics are quite difficult compared with nonbiological fluid flows. However, recent progress in laser optics has made it possible to apply the Doppler method to determine blood flow dynamics related to the pathogenesis of vascular diseases such as arteriosclerosis or thrombosis.

The blood flow in arteries has been measured in vivo using a hot-film anemometer or a pulse Doppler ultrasonic velocimeter. Recently, LDV has become a useful tool in blood flow measurements. Figure 3 is a schematic of an LDV system [7]. A laser is split into two beams, which intersect at their waist point in a blood vessel with an angle of $\pm\theta$ against the normal of the vessel axis. When the two beams are scattered at a crossed probing area by blowing erythrocytes, they independently undergo a small phase shift of $\pm\Delta f/2$ due to the Doppler effect. The scattered light with Doppler frequency shifts is received by a photomultiplier, producing beat signals (Doppler's burst signals). The signals from the photomultiplier can be displayed on an oscilloscope and/or processed by a signal analyzer for recording. The distance between the two interference fringes d is given by

$$d = \frac{\ell}{2 \sin (\theta/2)}$$

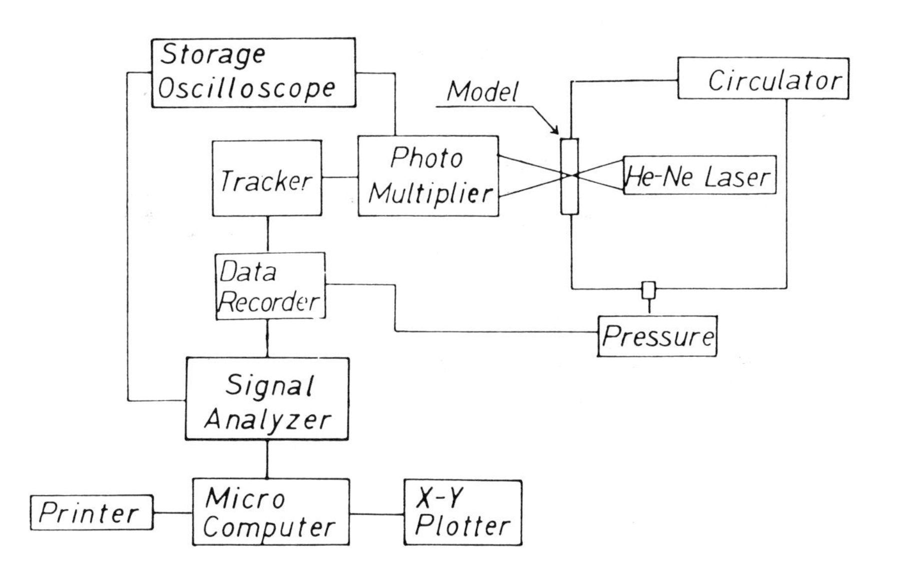

Fig. 3 Schematic of an LDV system [7].

where ℓ denotes wavelength. The flow velocity v and the period of beat signals τ are related by

$$\tau = \frac{d}{v}$$

In biomedical applications, the direct use of LDV is limited to in vitro experiments and exposed vascular vessels due to the penetration capability of laser light. However, the introduction of optical fibers into LDV provides excellent accessibility to small vessels such as epicardial arteries and veins as well as coronary vessels with high spatial resolution. Figure 4 is a schematic of an LDV system that has an optical fiber [7]. The light is split into dual beams. One beam is focused onto the entrance of an optical fiber, while the other is used as the reference beam. The beam emitted from the fiber tip is scattered by flowing red blood cells, and the back-scattered light from the Doppler shift signals is collected by the same fiber and transmitted back to its entrance. A frequency shifter is placed on the path of the reference beam to distinguish the reverse flow from the forward one. An optical heterodyning (generating beats) is produced by mixing the reference beam with the back-scattered beam having Doppler shift signals. The signals received by a photodetector (e.g., an avalanche photodiode) are processed by a spectrum analyzer to determine the Doppler shift frequency Δf, which is related to the blood flow velocity by

$$\Delta f = 2nv \frac{\cos \theta}{\lambda}$$

where n is the refractive index of blood, θ is the angle of incidence from the beam to the blood stream, and λ is the wavelength of the laser light.

B Contrast Media

Contrast media in medicine and biology can be in the liquid or gas state or fine particulates suspended in solution. A contrast medium is used in visualization to intensify the image in contrast to its surrounding tissues (1) through tinting, smearing or staining (dyes), (2) as the direct result of absorption of radiation of different wavelengths (fluorescein), (3) through interactions with X-rays (radioactive contrast media), and (4) by tagging a radioactive tracer to immunoglobulins (radioisotope labeling).

Dyes were conventionally employed for visualizing blood vessels in tissue sections. For example, the staining technique has facilitated recognition of small blood vessels in various sections. Stains, i.e., dyes regularly used for blood, are of the so-called Leishman type. They are essentially mixtures of red, blue, and purple values made by mixing in the solution eosin (a yellowish-red acid dye) with methylene blue, methyl violet, methylene azure, and other basic analine dyestuffs to make a polychrome stain. Numerous other compounds, such as eosinates, are also available. When blood is stained by one of these variants of the original Leishman mixture, red cells turn pink, and white cells, containing nuclei and granules, become distinctly colorful. Nuclei exhibit various shades from royal purple to bright magenta, granules show red, rose, blue, or purple in pale-blue cell bodies, and platelets are pink and purple. Through the microscope, these colors appear bright and dark by transmitted light. The Groat–Jenner stain, a modification of the Leishman stain, can be used for photographing blood.

Luxol Fast Blue stain is an alcohol-soluble amine salt of sulfonated copper phthalocyanine developed by Kluver and Barrera [8] as a myelin stain. It gives an intense bright-blue coloring of myelin, which is resistant to light, heat, acids,

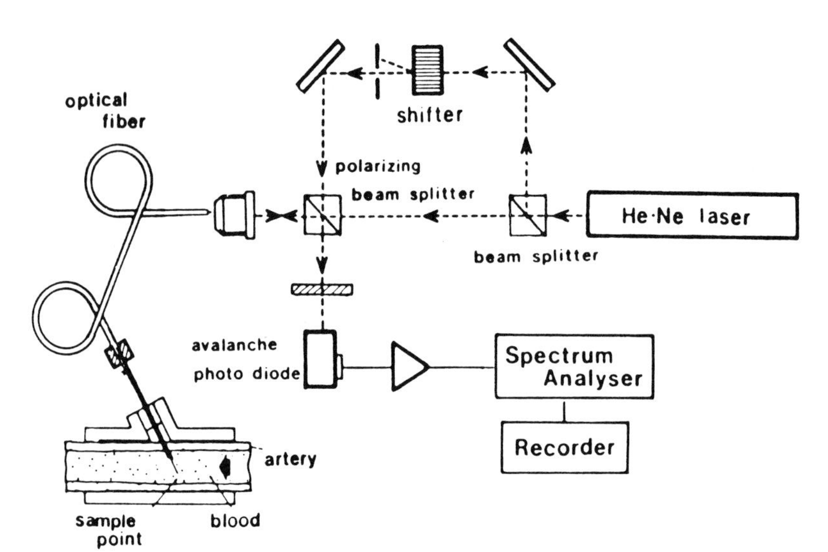

Fig. 4 Schematic of LDV system with an optical fiber [7].

and alkalis. Margolis and Pickett [9] used the stain in combination with the periodic acid–Schiff (PAS) reaction and hematoxylin, thus obtaining maximum contrast, to study the nervous system in humans and experimental animals. The PAS-positive vascular endothelium gave good contrast with blue-stained erythrocytes and hematoxylin-stained nuclei. Hematoxylin and eosin were also used [10].

Stained smears are recorded photomicrographically in natural color or by transference of these colors into black-and-white halftone contrasts. All conventional histological techniques of using dyes to show blood vessels in tissue sections have some important shortcomings: they cannot provide information about the speed of blood flow, and furthermore, they often give rather poor contrast between blood vessels and other tissue elements.

Fluorescein is a dye that produces fluorescene under radiation of different wavelengths.

Fluorescene, a kind of luminescence, is an emission of radiation energy resulting from absorption of the incident radiation. When incident photons give energy to an acceptor molecule A, some of the electrons within the molecule are raised to a higher energy level. The electrons may then return to their ground level with the emission of photons equal to the energy-level difference. The fluorescence is usually less energetic than the incident photons. Where the acceptor A is nonfluorescent, the transferred energy is not re-emitted; this process is referred to as quenching:

$$h\nu_1 + A \rightarrow A^* \qquad A^* \rightarrow A + h\nu_2 \qquad \text{(fluorescence)}$$

or

$$A^* \rightarrow A + \text{thermal energy} \qquad \text{(quenching)}$$

Where h is the Planck constant, ν is the frequency of the radiation emitted or absorbed by the molecule, and A^* is A in the excited state. According to Stokes's law, the fluorescent light always has a longer wavelength than that of the exciting radiation. Radiation throughout the range from the near infrared to γ rays is capable of exciting fluorescence in suitable substances. However, attention is confined to visible fluorescent light excited by ultraviolet or visible light, since the majority of medical and biochemical applications fall within this category.

Medical applications of fluorescence include the following: (1) fluorescent roentgen ray screens (fluoroscopy) and roentgen ray photography; (2) clinical diagnosis and therapy by ultraviolet fluorescence (through fluorescence of body substances); (3) fluorescence and photosensitization; (4) fluorescence of drugs, hormones, enzymes, and vitamins; (5) fluorophotometry or fluorometry; (6) the fluorescence microscope; (7) the phosphorescence microscope; and (8) medicolegal aspects of fluorescence.

A part of the body, for example, the kidney, the stomach, or a blood vessel, may attenuate radiation to no greater or lesser extent than the tissues surrounding it, thus the radiograph would yield no separate visualization of the organ. To overcome this difficulty, the body cavity (kidney, stomach, or blood vessel) can be filled with a contrast medium consisting of a material with an atomic number and density different from the surrounding tissues.

The contrast media currently used in clinical diagnostic radiology are of two basic types: highly radiolucent and densely radiopaque. They enhance radiographic contrast by virtue of their differential absorption of X-rays. Enhancement of contrast improves diagnostic visualization of the size, shape, and structure of organs, of spaces in and surrounding organs, and the course and patency of certain vessels and ducts.

The radiolucent-type media have atomic numbers and densities less than those of the surrounding structures. Air, oxygen, and carbon dioxide are in this category. They produce a clear outline of the gastrointestinal tract, of cavities such as the ventricles and cisternae of the brain, and the subarachnoid spaces of the spinal cord. Gaseous media can also be injected into joint cavities and into retroperitoneal connective tissue spaces.

There are many media of the radiopaque type, with higher atomic numbers and densities than those of body tissues. They are all, except barium sulfate, iodine-containing compounds. The increased attenuation of these media is due largely to increased photoelectric attenuation. Some are soluble in water, and others are used as a solid or are suspended in an aqueous or oily vehicle. The contrast agent can be administered directly into a luminal or cavitary region; they are commonly instilled into the tracheobronchial tree (bronchography), the common bile duct (cholangiography), the excretory tracts of the urinary system (retrograde ureteropyelography, cystography, and urethrography), the ductal tracts of the salivary glands and seminal vesicles (sialography and seminal vesiculography), the genital system (ureterosalpingography), and various sinus tracts and fistulae. They can be injected into the spinal subarachnoid space (myelography) and selectively into certain sections of the circulatory system for direct opacification of arteries, veins, cardiac chambers, and lymphatics (arteriography, venography, angiocardiography, and lymphangiography). The lumen of the gastrointestinal tract can be outlined by using both barium sulfate suspensions and the iodine compounds.

Some radiopaque media are introduced indirectly to the organ to be examined. A material having a high atomic number is combined with another substance to produce a compound that the organ will absorb from the bloodstream. Following intravascular or oral administration, the medium is transported via the bloodstream to any organ that selectively concentrates and excretes the medium, thus imparting radiopacity to the organ and its excretory passages, e.g., by the hepato-biliary system (cholecystography and cholangiography) and by the renal system (excretory urography). Sodium tetraiodophenolphthalein is the contrast medium often used. After oral administration, it is secreted by the liver and is then concentrated by the gallbladder.

The absorption of X-rays is governed by the formula

$$A = C\lambda^3 Z^4 + 0.2$$

Absorption A thus varies with the third power of the wavelength λ and the fourth power of the atomic number Z; C denotes a constant, while 0.2 is the factor to account for scattering effects, depending on the wavelength at which the observation is made.

Besides having a different mass-attenuation coefficient, the contrast medium must have other essential features: low systemic and local tissue toxicity, minimal pharmacodynamic action, selective localization, ease of administration, and prompt and complete elimination.

Radiopaque media for angiography are water soluble or water miscible and are used for contrast in X-ray filming. They can be classified into two groups: iodinated organic contrast media and inorganic contrast media [11]. These media have permitted the visualization not only of the heart and blood vessels but also of other viscera that are non-opaque to X-rays. The media used for angiographic procedures throughout the past three decades include Diodrast, Neo-Iopax, Urokon, Miokon, Hypaque, Renografin, Ditriokon, Conray, Angi-Conray, and Isopaque [12]. None of the opaque media currently in use is entirely satisfactory. In general, the diatrizoate drugs (Hypaque and Renografin) have been widely accepted and are the media of choice in most situations.

Movement of certain drugs and antibodies in the vascular system, tissue sections, or within cells can be visualized by means of radioisotope labeling (also known as tapping) under ultraviolet light illumination. This method is used for visualization, identification, and localization of various antigens, protein hormones and many other substances [13, 14]. Fluorescein, labeled antiserum (i.e., a fluorescein tagged to an antibody), which retains its immunologic reactivity, is used as a marker. The most generally used fluorescent material is fluorescein isothiocyanate.

The use of a labeling technique to visualize antigen in tissue sections was first perfected by Coons [13] during the 1950s. At present, immunofluorescent antibody producers are performed as one of three basic variations: (1) direct, (2) indirect, or "sandwich," and (3) complement fluorescence [15].

C Radiography

The morphology and function of vascular and excretory systems can be observed by the naked eye or with the aid of an optical instrument. Appropriate vision magnification or image intensification may be required to enhance the quality of visualization. Commonly, photography and radiology are used in either a single or sequential recording.

A major portion of the vascular system, such as fine arteries, arterioles, and capillaries, is too small to be visualized by standard angiographic technique. Some form of magnification is required in order to delineate these structures. Three methods are available to magnify vision or to produce a magnified image: (1) photographic or optical magnification, (2) direct geometric magnification, and (3) electronic magnification [16]. The microscope is, of course, the basic instrument used to magnify the range of vision; microphotography and microradiography are the processes of recording magnified images.

The process involved in making the fluoroscopic image brighter is known as image intensification or image amplification. Three types are in use: (1) The *image intensifier* is a vacuum type, which converts the light image from a fluoroscopic type of screen into an electron image. The electrons thus produced are then accelerated, focused, and made to form a smaller and much brighter light image. (2) In the *electro-optical image intensifier,* the large fluoroscopic screen image is transferred by the mirror optical system to the input phosphor of the image-intensifier tube. Another optical system transfers the output phosphor image of the intensifier tube, without a change in size, to a cine camera. (3) In the *solid-state image intensification,* an alternating voltage is applied between the two outer electrodes. Due to this alternating voltage, emission of light from the electroluminescent layer will increase when the photoconductor is irradiated.

Radiology is due to interactions between X-rays and matter; the transmitted beam emerging from the subject contains information. The X-ray beam has a more or less uniform intensity before the attenuation processes (including unmodified scattering, photoelectric absorption, Compton effect, and pair effect) takes place. When it leaves the subject, the beam contains an image made up of regions of different intensities. Figure 5 depicts the components of a conventional radiography system, consisting of an X-ray tube, film–screen combination, film processor, and light box.

D Thermography

Any use of radiation (electromagnetic wave phenomenon) for visualization, both fluoroscopy and radiography, is accompanied by hazards, particularly when the time of exposure is prolonged. For this reason, new diagnostic methods are being constantly sought and developed in an effort to replace radiology in medicine. Total replacement may seem unlikely in the near future but efforts have been directed toward two new means of visualization: thermogra-

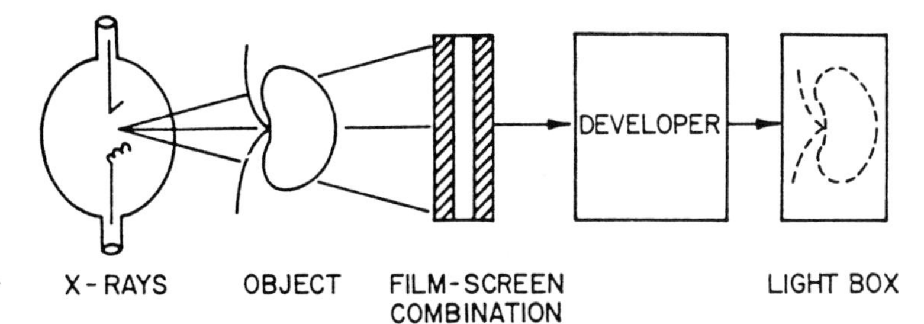

Fig. 5 Components of a conventional radiography system.

phy and ultrasonics. Thermography is discussed here, while the use of ultrasonics is presented later, under Sonography (see Sec. III.G).

Thermography, or infrared thermography, is a method that gives a photographic display of point-to-point temperature differences of the surfaces of the body. It is used in medical infravision for mapping the temperature of the body surface [17, 18]. This is a noncontact-type thermometry, which is unmatched by any other method by virtue of its high spatial and thermal resolution.

Extremely low-energy, infrared rays are constantly emitted from the body surface, as the most important means of dissipating excess energy in order to maintain a constant body temperature. An infrared radiometer or infrared camera (Fig. 6) is used to detect the intensity of infrared radiation. The intensity is then converted into electrical signals; after being enhanced by an amplifier, the signals are displayed on a cathode ray tube, producing a two-dimensional image of body-surface temperature. A temperature difference of 0.1°C can be detected. Quantitative results can be obtained by placing the photographic film in a reader and evaluating the amount of reflected light from various points on the picture as compared to the gray scale.

As an aid in visualization, the skin can be photographed after being treated with a phosphor, since upon irradiation with ultraviolet light, it emits visible light at different intensities depending on temperature. This is very useful because many pathological conditions, such as tumors, produce local changes of temperature. The thermographic technique is also used in the detection of drug effects on the peripheral circulation. In botanical applications, thermography records the thermal image of plant leaves as a guide in the study of heat and mass transfer between the leaves and the ambient.

III COMPUTER-ASSISTED FLOW VISUALIZATION TECHNIQUES

The new generation of flow visualization techniques is the combination of imaging devices and computers for three-dimensional mapping of flow field or structures. Often referred to as machine vision in medicine, it consists of an image-acquisition system, an image processor, and a display, as shown in Fig. 16 at the end of the chapter. The difference in body-scanning systems has led to five different medical vision methods: computed tomography (CT); digital subtraction angiography (DSA); magnetic resonance imaging (MRI), which is also called nuclear magnetic resonance (NMR); radioisotope imaging (RII); and sonography (SONO). Radioisotope imaging includes positron emission tomography (PET) and single photon emission computed

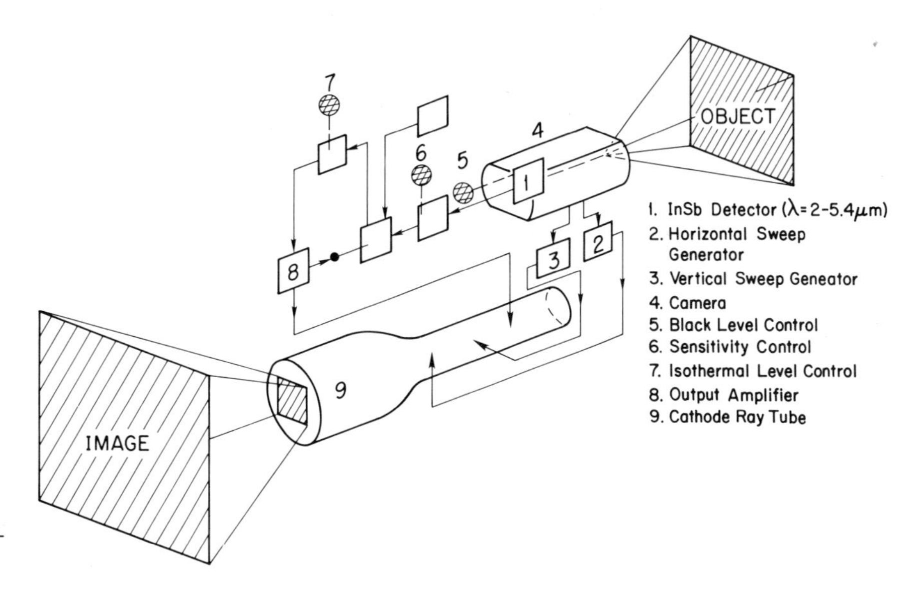

Fig. 6 Schematic diagram of a thermographic device.

tomography (SPECT). Both MRI and RII belong in nuclear medicine. The X-ray is employed in body scanning in both CT and DSA. The CT, SONO, DSA, RII, and MRI techniques are used for morphography and diagnosis, while CT, SONO, DSA, and RII visualize vascular networks.

In the image-acquisition system, the image flow, namely, an analog signal, is converted into a digital signal by means of an analog–digital convertor (ADC). With the digital image as input, the image processing is performed by a computer, which also provides memory and storage. The main functions of image processing include edge enhancement, segmentation of objects, identification of objects, smooth spatial averaging, 3-D reconstruction, densitometry, and real-time imaging. The digital information is then reconverted into the analog signal (an image) for display. Thermotomography (TT) provides a three-dimensional view of body temperature, which is evolved from a heat map of two dimensions obtained by thermography (TG).

A Image Acquisition

1 Computed Tomography (CT) Using a thin, fan-shaped X-ray beam, a CT scanner produces a cross-sectional view of tissues within the body by detecting the X-ray absorption of the examined tissues or organs as a function of space and time. The CT machines view a slice of the body from many angles by revolving an X-ray tube around the body.

2 Magnetic Resonance Imaging (MRI) The magnetic fields are varied sequentially in three directions: first, in the Z direction along the body axis from head to toe; next, on the plane in the Y direction from the top of the plane to the bottom; and finally, on the plane from left to right in the X direction. Detecting the signal resulting from proton resonance, a computer locates each voxel and assign it a spot on the video screen.

3 Positron Emission Tomography (PET) A radioisotope solution emits positrons, which strike crystals in a ring of detectors around the body, causing the crystals to light up. A computer records the location of each flash and plots the source of radiation, translating those data into an image.

4 Sonography (SONO) Using a wedge-shaped ultrasound beam, a SONO scanner produces an image from echoes received from the interior structures of the human body. The SONO scanner views a slice of the structure from many angles by varying the positions of the scanner around the structure.

5 Digital Subtraction Angiography (DSA) First, a picture of the system is made by a digital X-ray scanner. Next, as a contrast medium is injected through a catheter into the system, a second X-ray image is made, showing the agent flowing through the blood vessels in the system.

B Digital Image Processing [19, 20]

An image, such as a photographic record, can be transmitted electronically from one place to another. First, the analog signal is converted into a digital signal by means of an analog–digital convertor (ADC), a transducer that can quantize the continuous analog signal into discrete bundles. These bundles represent the gray tone scale values of the digitized signal.

An ADC consists of comparators, which are electronic devices used to compare two electronic input signals, the reference signal and analog signal to be quantized. The summed comparator voltages are converted into a binary number as is the output of the ADC. Since it is far simpler to construct electronic devices that performs binary arithmetic rather than decimal arithmetic, binary storage and image processing are employed in most computer manipulations. In general, if a code consisting of N bits is used, then 2^N possible gray levels can be encoded. Therefore, an 8-bit code, called a byte, can encode 256 (namely, 2^8) gray levels. The finer the spacing (with a larger number of bits), the less the error in the quantization process. This means that a finite number of bits should be used in the digitization process for error-free quantization. In practice, an 8-bit digitizer would be the minimum used, and more bits than eight would be likely to result in little overall improvement.

Once an image or a series of images have been digitized and stored, software is used to perform the desired arithmetic manipulations, called digital processing. The resulting image information can be printed out from the computer in the numbers associated with each pixel. However, it is a far more common practice to reconvert information into an analog image to be displayed on a cathode ray tube or made into a film transparency. A digital signal is reconverted to analog form by transforming binary numbers into video voltage levels. A device for performing this function is called a digital–analog convertor (DAC). Figure 7 depicts different scopes of image processing in radiology.

C Computed Tomography

In 1972 a CT device for medical applications was introduced by a group of researchers at Electronic Musical Instruments, Ltd., in England [1]. In principle, CT scanners convert X-ray pictures into digital computer codes to produce high-resolution video images. The computer graphics used are similar to those employed in reassembling pictures beamed back from distant space probes. Computed tomog-

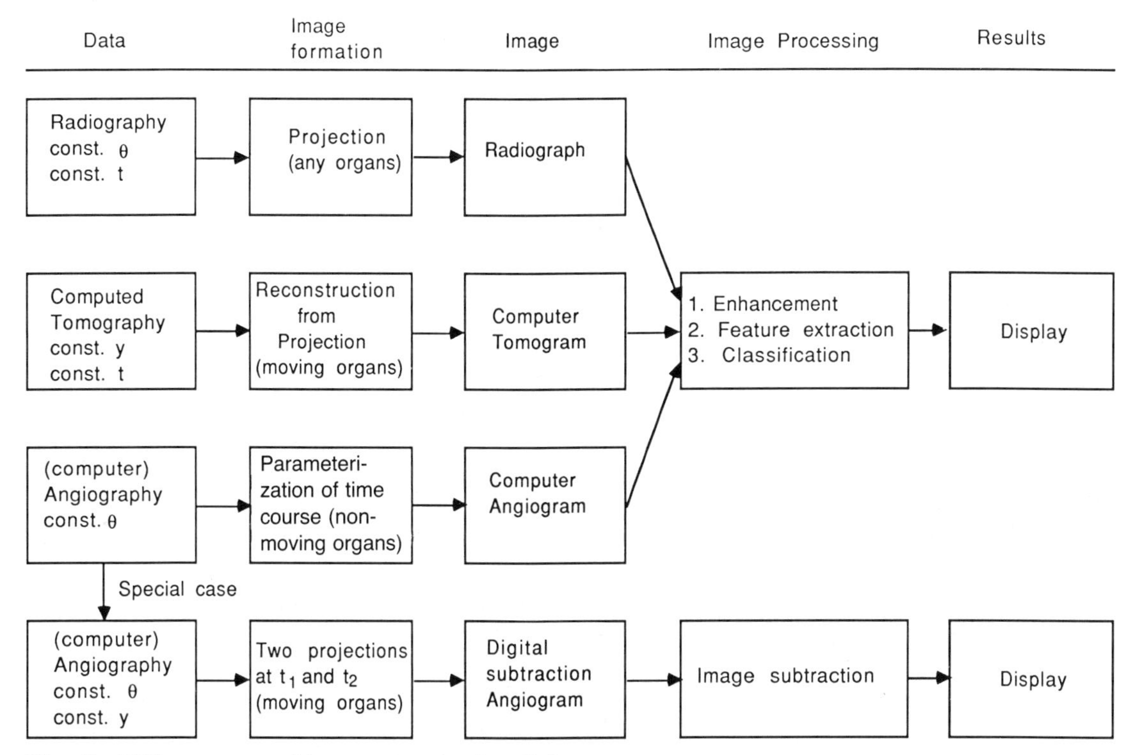

Fig. 7 Different scopes of image processing in radiology.

raphy scans can depict bone structures in fine detail as well as small differences between normal and abnormal tissues in the brain, lungs, and other organs.

The physical components of a fourth-generation CT system are schematically depicted in Fig. 8. A collimated beam of X-rays is passed through an object at a large number of angular orientations. The transmitted X-ray photons are intercepted by a suitable detector system, and a current is generated which can be sampled and digitized. The transmission data are fed into a computer system, which utilizes an algorithm to reconstruct an image of the anatomical structures within a transverse section. This image is dis-

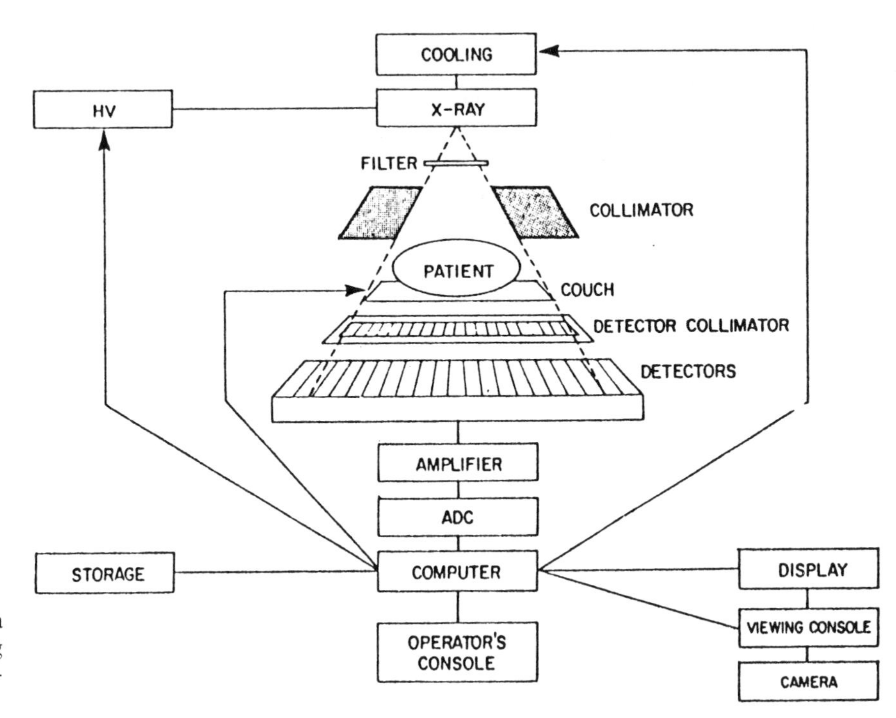

Fig. 8 Schematic diagram of a fourth-generation CT system showing the relationships of various components.

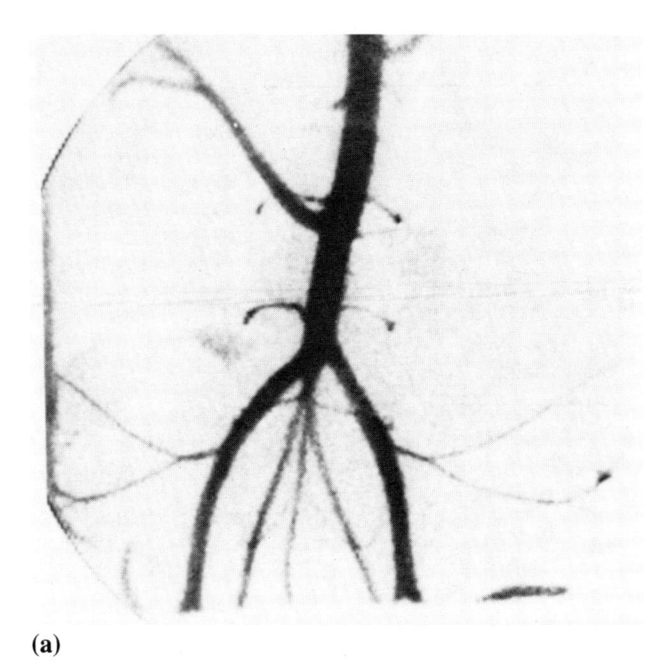
(a)

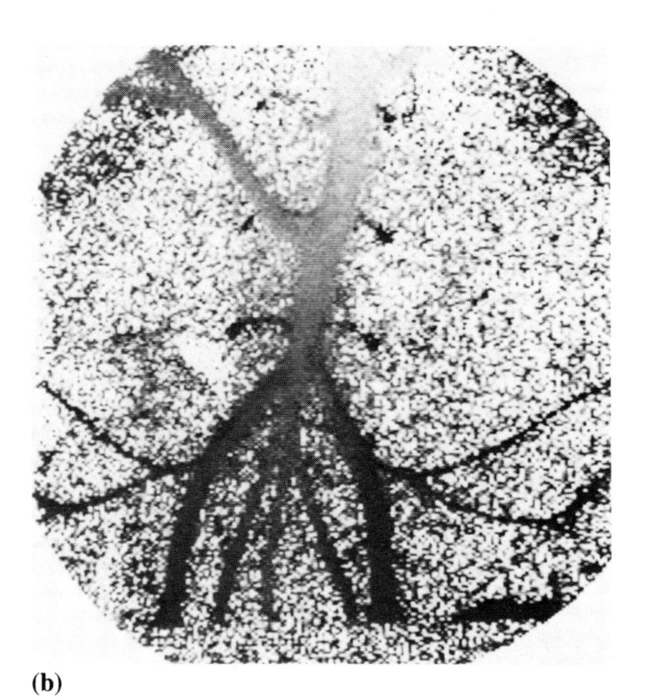
(b)

Fig. 9 Computed tomography images of the abdominal aorta of a pig: (*a*) Depth parameter image; (*b*) time parameter image [46].

played on the screen of a CRT, which can then be photographed. Figure 9 shows CT images of the abdominal aorta of a pig; (*a*) is the depth parameter image indicating the regional thickness of vessels by gray intensities, and (*b*) is the time parameter image. Increasing gray levels indicate the temporal delay of contrast flow through the vessels after the start of contrast injection. The gray scatter surrounding the vessels is due to erroneous time values as calculated from the image noise, which should be masked out for consecutive image analysis. Numerous publications are available on CT (e.g., [21–31]).

In current medicine, CT scanning is established foremost as a standard medical visual technique. It is used for the body scan only to a certain extent. For example, CT can be used in angiography (or arteriography, by injecting an X-ray attenuating agent into the bloodstream and imaging through X-ray techniques), cerebral pneumography (by injecting air into the spinal canal and visualizing through X-ray techniques), and radionuclide brain scanning (injecting a radioactive material into the bloodstream and viewing through a γ camera), to detect blood vessel and other abnormalities.

D Digital Subtraction Angiography (DSA) [32–34]

The combined applications of computers with standard radiologic equipment have resulted in the development of electronic, or so-called filmless, imaging. Digital subtraction angiography (DSA), one application of computed angiography, has become of particular value in providing a clean, clear view of the central vascular system or the blockage of blood flow by narrow vessels.

Figure 10 is a schematic diagram of a basic DSA unit. It consists of an under-table X-ray tube and an above-table chain of image-intensifier tube, TV, and computer processing system. The DSA procedure consists of injection into the blood vessels of a contrast agent containing iodine that is opaque to X-rays. This is accomplished by threading a catheter through a blood vessel in the arm or groin to a desired location, for example, a coronary artery. Before injection of the contrast medium, an X-ray image is made and stored in a computer. After injection, a second image is made highlighting the flowing blood as revealed by the

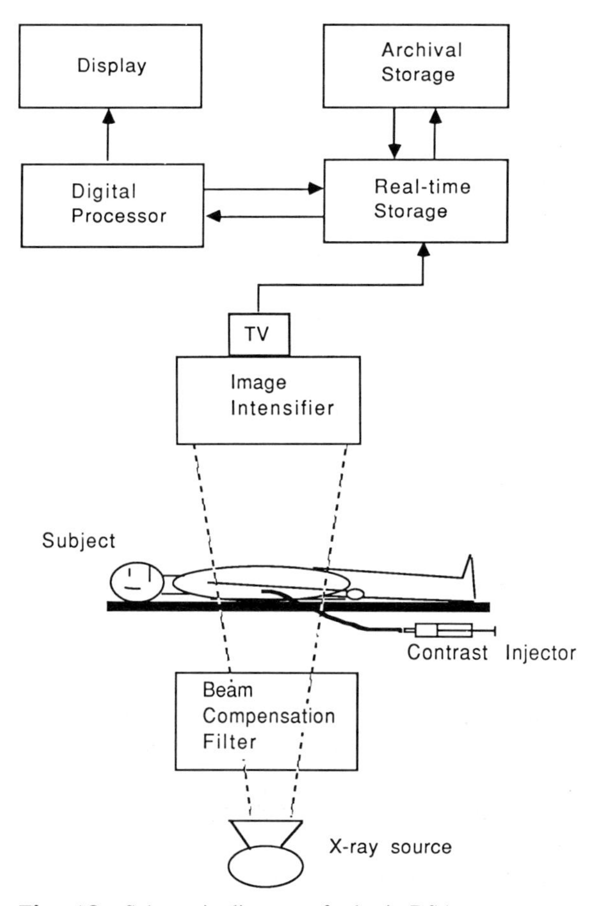

Fig. 10 Schematic diagram of a basic DSA system.

medium. The computer then subtracts the first image from the second, leaving only what has changed, namely blood vessels containing the contrast agent. For example, an X-ray picture of the heart is made first. Next, a contrast substance is injected through a catheter into the coronary arteries. A second X-ray image is made showing the substance flowing through the heart's vessels. A computer subtraction of the images reveals the coronary arteries.

Addition, subtraction, and averaging of the digital data results in images with adequate signal–noise ratios to permit imaging of vascular structures with markedly lower contrast densities than previously required using standard angiography (conventional screen-film techniques). Computer subtraction of unneeded background information improves the conspicuousness of the opacified vessels to permit detection of vascular structures containing a concentration of no more than 1 to 3% of contrast medium. This improved visualization is possible even with intravenous peripheral injections or a reduced amount of contrast medium given intra-arterially. With either method of contrast-medium administration, DSA has become an excellent means of anatomic demonstration of the heart and great vessels with decreased morbidity and at lower cost. The retrieval of digital data over selected areas of vascular systems also

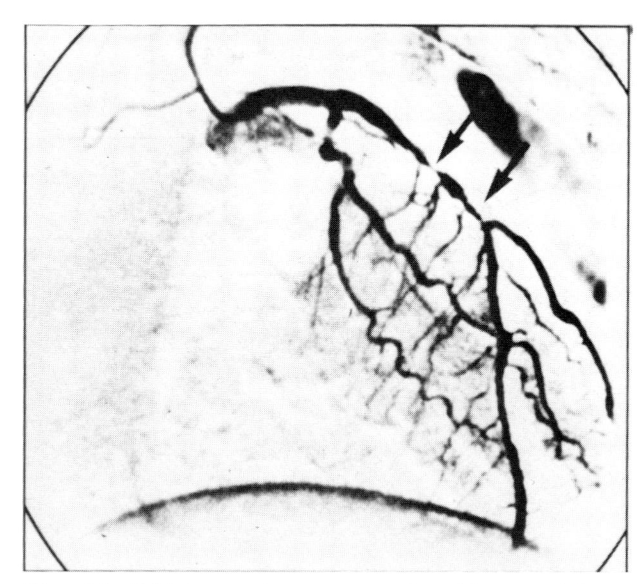

Fig. 11 DSA image of coronary arteries with sequential stenoses [33].

produces functional assessment of the circulation. Quantification of the time of appearance, peak concentration, and clearance of the injected contrast medium results in reasonable estimates of cardiovascular dynamics, which include ejection fraction, wall motion, and shunt analysis.

Figure 11 depicts the digital coronary angiography with injection of dilute contrast into the left coronary artery and imaging in the right anterior oblique projection. Sequential stenoses in the left anterior descending artery are indicated by arrows.

E Emission Computed Tomography Radioisotope Imaging

Blood flow can be imaged by tracing amounts of radioisotopes in the flow stream. Both single photon emission computed tomography (SPECT) and positron emission tomography (PET) are the methods used [35–38]. The latter is more versatile and is also able to measure metabolism, revealing how well the body is working. Both PET and SPECT depict the distribution of blood into tissue (which absorbs a radioisotope, such as N-13 ammonia or technetium 99-m, from the blood), but PET does so with greater accuracy. The use of radioactive tracers is well suited to investigations of stroke, epilepsy, schizophrenia, and Parkinson's disease.

In a PET setup, a small, low-energy cyclotron creates a radioisotope that is tagged on body substances. The radioactive solution emits positrons wherever it flows. When the positrons collide with electrons to annihilate each other, a burst of energy is released in the form of two γ rays shoot-

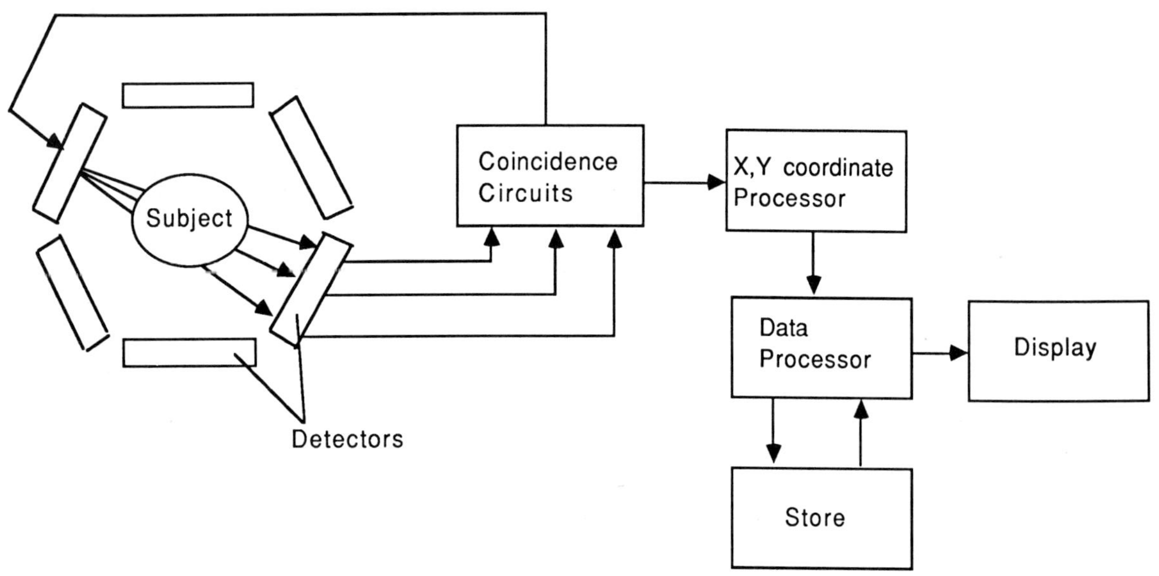

Fig. 12 Schematic diagram of a PET imaging system.

ing in opposite directions. A ring of detectors in the PET scanner detect the γ rays (Fig. 12). A computer records the location of radiation and translates the data into an image. By tracing the radioactive substance, one can determine areas of abnormal cell activity. Unlike PET, which is generally equipped with a cyclotron, SPECT uses commercially available radioisotopes.

F Magnetic Resonance Imaging [39–41]

Magnetic resonance imaging (MRI) is also known as nuclear magnetic resonance (NMR). The MRI scanner surrounds the human body with powerful electromagnets, which create a magnetic field as much as 60,000 times stronger than that of the earth (Fig. 13). The field aligns the nuclei of hydrogen atoms, which are normally pointing in random directions, in the direction of the field's poles. These aligned hydrogen atoms vibrate at a specific frequency. The stronger the magnetic field, the greater the frequency. When the protons are excited by a radio pulse from the scanner that has the same frequency as their wobbling, they are knocked out of alignment and spiral back into place, emitting a faint radio signal of their own. A computer translates these faint signals into an image of the area scanned. The image discloses changing densities of hydrogen atoms and their interactions with surrounding tis-

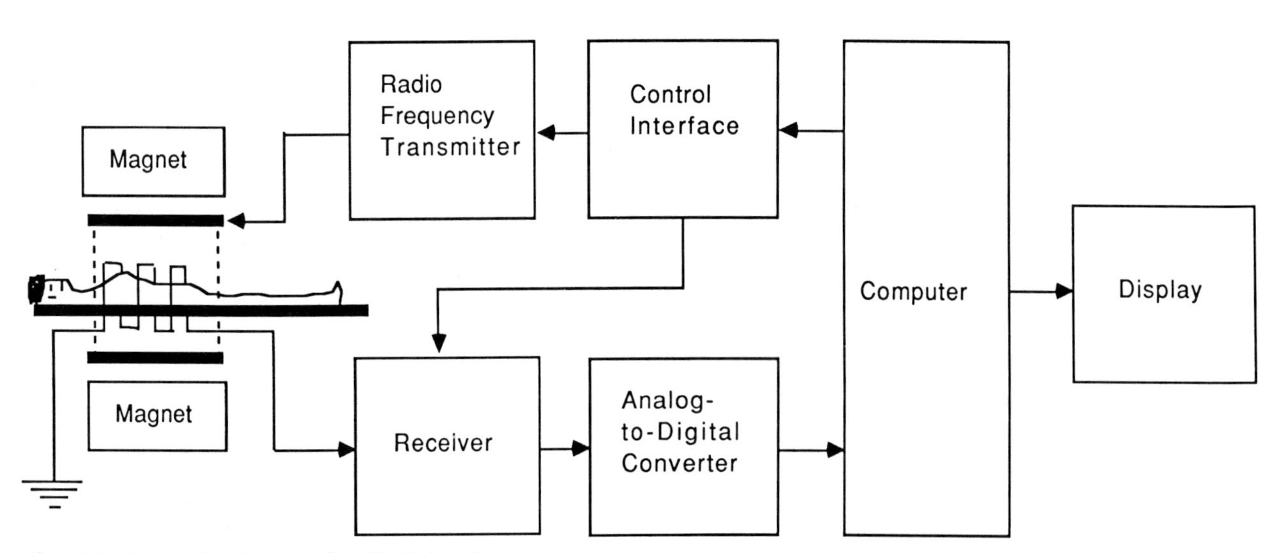

Fig. 13 Schematic diagram of an NMR imaging system.

sues. It makes distinctions between tissues, since hydrogen reflects water content.

Hydrogen is selected as the basis for MRI scanning for two reasons: its abundance in the body and its prominent magnetic qualities. Other elements such as sodium and phosphorus may also serve the purpose.

6 Sonography [42–45]

The sound range audible to the human ear is between 20 and 20,000 Hz, beyond which it is classified as ultrasound (US). The main medical applications of US include (1) hyperthermia at the frequency ν of 0.8 to 1.0 MHz with the intensity I of 0.5 to 4.0 W/cm^2. One advantage of US in diagnosis is that it can reflect off small objects by virtue of its short wavelength. The wavelengths of audible sound waves are too large for good resolution of biological structures. A second merit of US is that it can be condensed into a narrow beam more easily than audible sound and can be transmitted through biologic tissue with a wide margin of safety.

Ultrasound energy is generated by a piezoelectric element, which in most cases is a quartz crystal. With application of an electric voltage, piezoelectric crystals can change shape and produce vibrations that propagate away in the form of alternating compressions and rarefactions. The piezoelectric element also acts as a receiver when US waves are reflected from tissue interfaces. The echo is then converted into an electric signal, which can be amplified and displayed on a CRT or oscilloscope (Fig. 14).

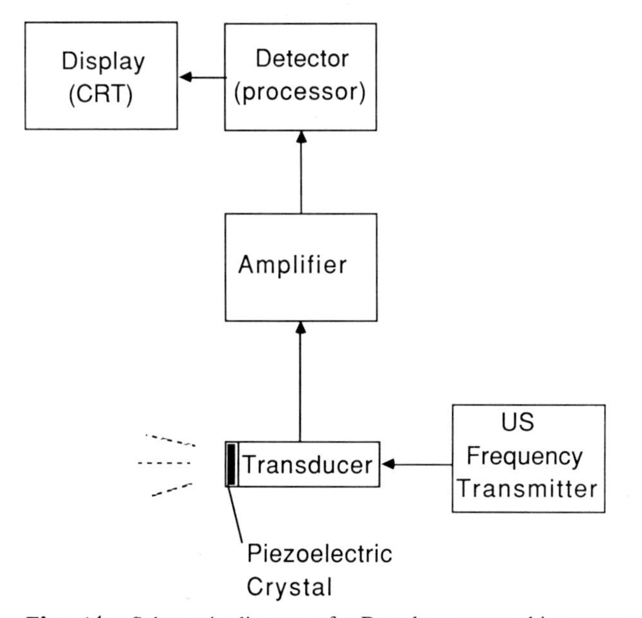

Fig. 14 Schematic diagram of a Doppler sonographic system.

Sonography is one of the simplest and cheapest of the five medical imaging techniques. There are three kinds of sonography: imaging, Doppler sonography, and their combination, called duplex sonography. In the standard imaging sonography with a pulse-echo transmitter, a transducer with a piezoelectric crystal is placed in contact with the area of the body to be investigated. Ultrasound waves penetrate the body, strike internal structures, and reflect back to the surface where the echo is then converted into electric signals. A B-mode image processor in the detector produces the target location, size, shape, and even texture for display on a screen. The standard imaging sonography has been used to outline nonmoving anatomical structures such as tissue, organ, and fetal morphology, typically in obstetrics and cardiology.

The Doppler scanners are used for mapping moving structures, namely, flow and movement patterns. A change in the frequency of echo signals, called the Doppler effect, occurs whenever there is relative motion between the sound source and the reflector. The Doppler devices in the transducer may be of continuous-wave (CW), pulse, or duplex type. The CW-type Doppler scanner has two crystals in the transducer, one for transmitting the US beam and the other for receiving the reflected echo. The transmitter continuously excites the transducer with sinusoidal electric signals. The returning signal is compared with the transmitted signal in the detector to determine the Doppler frequency shift.

In a pulsed Doppler instrument, the transducer is excited with a short-duration burst from a pulsed transmitter. Reflected echo signals are detected by the same transducer, amplified in the receiver, and applied to the detector, where the Doppler frequency shift is obtained. In comparison with a CW Doppler, a pulsing circuit is added, consisting of a clock, two range gates, and an adjustable delay for the length and location of the gates. By gating the receiver for a short duration at a specified time following each transmission of the pulse, Doppler signals originating only from a particular depth are selected for display.

A duplex scanner is a real-time B-mode scanner with built-in Doppler capabilities. In other words, it is a combination of a pulse-echo scanner and a Doppler instrument. In typical applications, the pulse-echo B-mode image intensifies the areas where flow will be examined using the Doppler instrument. The operator selects the direction of blood flow with respect to the US beam; this angle must be known to evaluate flow velocity from the frequency of the Doppler signal. Figure 15 depicts a Doppler sonographic image of a carotid bifurcation.

The Doppler instrument can show the flow and eddying (at stenosis or bifurcation) of blood as it courses through the heart, veins, and arteries. It can also be applied to detect arterial and venous flow in the extremities and to measure blood pressure in hypotensive and surgical patients.

Details on echocardiography are available in Chap. 18 of this handbook.

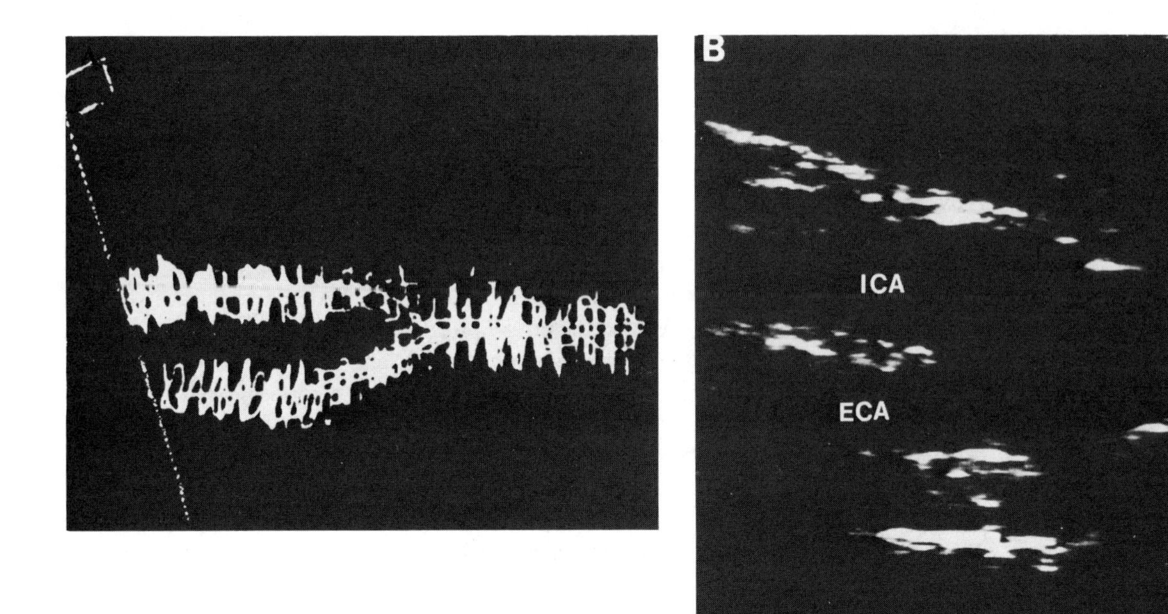

Fig. 15 Doppler SONO image of a carotid bifurcation [47].

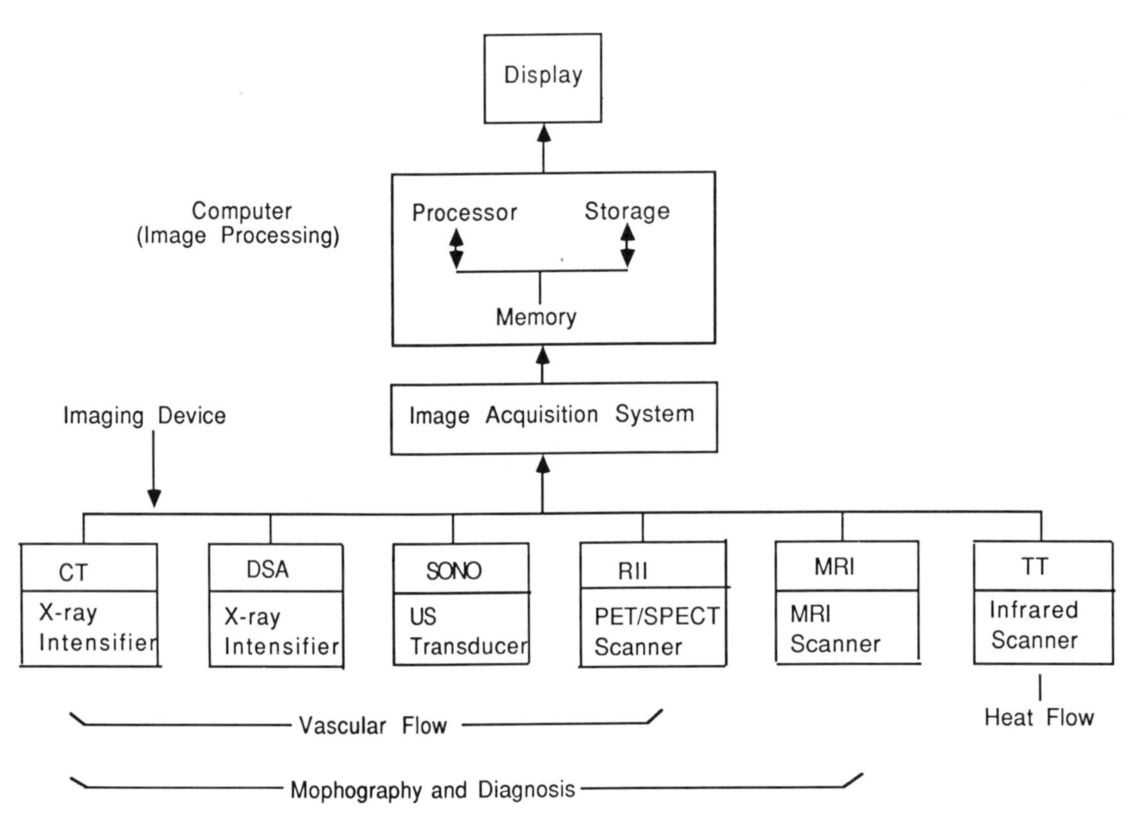

Fig. 16 Machine vision in medicine: the marriage of imaging devices and a computer.

H Thermotomography

Computer-assisted thermography, called thermotomography, is in its infancy. The purpose of this thermal vision is to construct a three-dimensional body temperature in the area being scanned. A computer is installed between the scanner and the display unit, as shown in Fig. 16. The computer is equipped with a program, which solves the bioheat equation to generate three-dimensional temperature distribution using the scanned thermogram (two-dimensional surface temperature) as the input.

REFERENCES

1. Hounsfield, G. N., Computerized Transverse Axial Scanning (Tomography): Part I. Description of System. *Br. J. Radiol.,* vol. 46, pp. 1016–1022, 1973.
2. Bloch, F., Hansen, W. W., and Packard, M. E., Nuclear Induction, *Phys. Rev.,* vol. 69, p. 127, 1946.
3. Purcell, E. M., Torrey, H. C., and Pound, R. V., Resonance Absorption by Nuclear Magnetic Moments in a Solid, *Phys. Rev.,* vol. 69, pp. 37–38, 1946.
4. Doppler, C., Uber das Farbige Licht der Doppelsterne. Abhandlungen der Koniglishen Bohmischen Gesellschaft der Wissenschaften., vol. 2, p. 465, 1842.
5. Satomura, S., A Study on Examining the Heart with Ultrasonics: I. Principles; II. Instruments, *Jpn. Circ. J.,* vol. 20, p. 227, 1956.
6. Donald, J., MacVicar, J., and Brown, T. G., Investigation of Abdominal Masses by Pulsed Ultrasound, *Lancet,* vol. 1, pp. 1188–1194, 1958.
7. Takahei, T., *The Application of Laser Doppler Velocimetry,* Power Co., Tokyo, 1984.
8. Kluver, H., and Barrera, E., A Method for the Combined Staining of Cells and Fibers in the Nervous System, *J. Neuropathol. Exp. Neurol.,* vol. 12, p. 400, 1953.
9. Margolis, G., and Pickett, J. P., New Applications of Luxol Fast Blue Myelin Stain: Myelo-angio-cytoarchitectonic Method; Myelin–Neuroglia Method; Myelin-Fat Method; Myelin–Axis Cylinder Method, *Lab. Invest.,* vol. 5, p. 459, 1956.
10. Endrich, B., Reinhold, H. S., Gross, J. F., and Intaglietta, M., Tissue Perfusion Inhomogeneneity during Early Tumor Growth in Rats, *J. Nat. Cancer Inst.,* vol. 62, p. 387, 1979.
11. Strain, W. H., The Radiopaque Media: Nomenclature and Chemical Formulas, in *Abrams Angiography,* ed. H. L. Abrams, 3rd ed., vol. 1, chap. 3, Little, Brown, Boston, pp. 41–77, 1971.
12. Abrams, H. L., The Opaque Media: Physiologic Effects and Systemic Reactions in *Abrams Angiography,* ed. H. L. Abrams, 3rd ed., vol. 1, chap. 2, Little, Brown, Boston, pp. 15–39, 1983.
13. Coons, A. H., Fluorescent Antibody Methods, in *General Cytochemical Methods,* ed. J. F. Danielli, vol. 1, pp. 399–422, Academic, New York, 1958.
14. Nairn, R. C., ed., *Fluorescent Protein Tracing,* E. & S. Livingstone, London, 1962.
15. Friedman, H., Immunologic Techniques with Pharmocologic Application, in *Animal and Clinical Pharmacologic Techniques in Drug Evaluation,* ed. P. E. Siegler and J. H. Moyer III, vol. 2, chap. 50, Year Book Publishers, Chicago, 1967.
16. Greenspan, R. H., Magnification Angiography, in *Angiography,* ed. H. Abrams, 2d ed., vol. 1, chap. 8, Little, Brown, Boston, 1971.
17. Atsumi, K., ed. *Medical Thermography,* Univ. of Tokyo Press, Tokyo, 1973.
18. Engel, J. M., Flesch, U., and Stuttgen, G., eds., *Thermological Methods,* VCH, Weinheim, West Germany, 1985.
19. Macvski, A., *Medical Imaging Systems,* Prentice-Hall, Englewood Cliffs, N.J., 1983.
20. Reinhardt, E., and Tan, S. C., *Medical Image Processing,* vol. 593, Int. Soc. Opt. Eng., Bellingham, Wash., 1985.
21. Waggener, R. G., and McDavid, W. D., Computed Tomography, *Adv. Biomed. Eng.,* vol. 7, pp. 65–100, 1979.
22. Ell, P. J., and Holman, B. L., eds., *Computed Emission Tomography,* Oxford Univ. Press, Oxford, England, 1982.
23. Mancuso, A. A., and Hanafee, W. N., *Computed Tomography and Magnetic Imaging of the Head and Neck,* Williams & Wilkins, Baltimore, 1985.
24. Brooker, M. J., *Computed Tomography for Radiographers,* MTP Press, Boston, 1986.
25. Federle, M. P., and Brant-Zawadzki, M., eds., *Computed Tomography in the Evaluation of Trauma,* Williams & Wilkins, Baltimore, 1986.
26. Herman, G. T., *Image Reconstruction from Projections: The Fundamentals of Computerized Tomography,* Academic, New York, 1980.
27. Lee, J. K. T., Sagel, S. S., and Stanley, R. J., eds., *Computed Body Tomography,* Raven, New York, 1983.
28. Gonzalez, C. F., Grossman, C. B., and Palacios, E., *Computed Brain and Orbital Tomography,* Wiley, New York, 1976.
29. James, A. E., Anderson, J. D., and Higgins, C. B., Digital Image Processing in Radiology, Williams & Wilkins, Baltimore, 1985.
30. Erickson, J. J., and Rollo, F. D., *Digital Nuclear Medicine,* Lippincott, Philadelphia, 1983.
31. Brody, W. R., *Digital Radiology,* Raven, New York, 1984.
32. Kruger, R. A., and Riederer, S. J., *Basic Concepts of Digital Subtraction Angiography,* G. K. Hall, Boston, 1984.
33. Van Breda, A., and Katzen, B. T., *Digital Subtraction Angiography,* Slack, Thorofare, N.J., 1986.
34. Moodie, D. S., and Yiannikas, J., *Digital Subtraction Angiography of the Heart and Lungs,* Grune & Stratton, Orlando, Fla., 1986.
35. Croft, B. Y., *Single-Photon Emission Computed Tomography,* Year Book Medical Publishers, Chicago, 1986.
36. Reivich, M., and Alavi, A., eds., *Positron Emission Tomography,* Liss, New York, 1985.
37. Phelps, M. E., Mazziotta, J. C., and Schelbert, H. R., eds., *Positron Emission Tomography and Autoradiography Principles and Applications for the Brain and Heart,* Raven, New York, 1986.
38. Heiss, W. D., and Philps, M. E., eds., *Positron Emission Tomography of the Brain,* Springer-Verlag, New York, 1983.

39. Bradley, W. G., Adey, W. R., and Hasso, A. N., *Magnetic Resonance Imaging of the Brain, Head and Neck,* Aspen Systems, Rockville, Md., 1985.
40. Bangert, V., Nuclear Magnetic Resonance Tomography, NMR Scanner Techniques and the Theory of Image Reconstruction, VDI-Verlag, Düsseldorf, 1982.
41. Kaufman, L., Crooks, L. E., and Margulis, A. R., eds., *Nuclear Magnetic Resonance Imaging in Medicine,* Igaku-shoin, New York, 1981.
42. Callen, P. W., *Ultrasonography in Obstetrics and Gynecology,* Saunders, Philadelphia, 1983.
43. Giglia, R. V., Mayden, K. L., and Gleicher, N., *A Practical Guide to Real-Time Office Sonography in Obstetrics and Gynecology,* Plenum, New York, 1986.
44. Arvan, S. B., *Echocardiography: An Integrated Approach,* Churchill Livingstone, New York, 1983.
45. Goldberg, S. J., *Doppler Echocardiography,* Lea & Febiger, Philadelphia, 1984.
46. Heintzen, P. H., and Brennecke, R., *Digital Imaging in Cardiovascular Radiology,* G. Thieme Verlag, Stuttgart, New York; Thieme-Stratton, New York, 1983.
47. Zwiebel, W. J., ed. *Introduction to Vascular Ultrasonography,* 2d ed., Grune & Stratton, Orlando, Fla., 1986.

Chapter 37

Indoor Environments

G. S. Settles

I INTRODUCTION

Flow visualization in indoor environments is generally connected with heating, ventilation, and air conditioning (HVAC) work. Other applications include airborne contamination control and fire safety studies. As with most other applications of flow visualization, it is used here primarily to gain a phenomenological understanding of airflow patterns. Such understanding is useful in maintaining proper ventilation patterns and rates, efficient energy distribution for thermal comfort and economy, and effective removal of harmful fumes and contaminants.

Indoor environments involve some special difficulties in flow visualization. For example, the array of different ventilation schemes in buildings and residences is quite diverse and difficult to classify. Further, some applications (especially those in clean environments) preclude use of some of the most powerful flow visualization methods. Finally, problems may arise from the large scales of typical indoor environments.

II MODEL TESTING AND SCALING

While full-scale flow visualization studies have been done indoors many times, this is often expensive or impractical in large buildings and industrial plants. Further, it may be useful to conduct scale-model tests to verify ventilation patterns during design of new buildings.

When this is done, careful attention must be given to proper scaling of the model flow with respect to its full-scale analog. Thorough discussions of this issue can be found, e.g., in Refs. [1, 2]. Briefly, geometric similarity between model and full-scale environments is almost always required. Dynamic similarity is also required, but is achievable in varying degrees, depending on the particular circumstances of the simulation. For example, indoor airflows solely involving natural convection can be simulated properly as long as the Rayleigh number,

$$\mathrm{Ra} \equiv \frac{\beta\ \Delta T g L^3 \rho^2 c_p}{\mu k}$$

for model and full-scale flows is matched. (Here, $\beta \equiv$ expansion coefficient, $\Delta T \equiv$ temperature difference, $g \equiv$ gravity, $L \equiv$ characteristic length, $\rho \equiv$ density, $c_p \equiv$ specific heat at constant pressure, $\mu \equiv$ viscosity, and $k \equiv$ thermal conductivity.) Provided that the Rayleigh numbers of both full-scale and model flows are large, i.e., several million or more, the simulation becomes insensitive to Rayleigh number, and an exact match is not required.

Similarly, forced convection in isothermal indoor airflows is governed solely by the familiar Reynolds number,

$$\mathrm{Re} \equiv \frac{\rho V L}{\mu}$$

where V denotes the air speed. Again, for Reynolds numbers above a few thousand, both full-scale and model flows are turbulent and relatively insensitive to Reynolds number.

However, when mixed natural and forced convection occurs, the simulation requirements are more stringent. In addition to Rayleigh number similarity, the Archimedes number,

$$\mathrm{Ar} \equiv \frac{\beta\, \Delta T g L}{V^2}$$

must also be matched between model and full-scale flows. Further, within local isothermal forced flows such as ventilation jets, a consideration of Reynolds number scaling is required. It seems unlikely that these several scaling criteria could all be simultaneously matched in a model experiment. Baturin [1] cites some relevant practical examples including cases of ventilation for foreign-gas removal. Here, though comprehensive similarity was not obtained, the Archimedes number was regarded as the primary criterion.

Provided that reasonable dynamic similarity is observed as just described, the experimenter is free to choose a model working fluid other than air. While a few examples using heavy gases or exotic liquids can be found, water models [2–8] have been quite popular for the study of indoor environmental flows. The high density of water relative to that of air permits conveniently sized models of 1/6 or smaller scale to be used. Further, water has advantages over air for flow visualization in that neutrally buoyant tracer particles are easier to produce. Dyes, colored milk, immiscible liquid droplets, plastic pellets, and even air bubbles have been used for this purpose. Finally, optically accessible models of rooms and buildings can be conveniently built from clear acrylic plastic sheets.

III ROOM VENTILATION

The airflows in ordinary rooms of residences and businesses contain a multitude of particles ranging from submicron to several millimeters in size. These include smoke, dust, textile fibers, pollen, and flakes of human skin. Mie scattering of light by these particles may render room airflow patterns visible. For this to occur, a brilliant light beam and a low background illumination level are required. Such conditions often occur naturally, e.g., when a shaft of sunlight enters a room. Flow visualization is sometimes used to study the motion of these particles and to aid in their reduction or elimination.

More often, however, flow visualization is applied to room environments in connection with HVAC studies. Figure 1 shows an example from a model study by Grigull [9], in which Mach–Zehnder interferometry was used to reveal temperature differences in the airflow.

Tracer flow visualization is traditional in HVAC studies, and many examples can be found in the literature (e.g., [4, 10–12]). In addition to the usual means of generating smoke, such as tobacco, heated kerosene, and $TiCl_4$, more exotic tracers such as methaldehyde particles [11], ethylenediaminetetraacetic acid [13], and $MgCO_2$ powder [14] have been used. While these tracers are often mixed with the ventilation inlet air, they can also be introduced through rakes or smoke wands. General or light-sheet illumination can be used, depending on flow complexity.

Fig. 1 Interferogram of scale-model flow, depicting air exchange through a window in summer (outside air warmer) (from U. Grigull [9]).

IV SOLAR HEATING

A subclass of residential HVAC studies, solar heating, has received particular attention beginning in the energy-conscious 1970s. Flow visualization has been applied in several model studies [15–19] in attempts to better understand the airflow and heat transfer mechanisms involved.

The "Trombe wall," which is typical of passive-solar room-heating schemes, admits sunlight through a large window followed immediately by a thermal storage device (stacked, water-filled cans, masonry wall, etc.). A thermosiphon effect is created by a narrow air gap between the window and the heated wall, presumably leading to an even temperature distribution in the room.

A critical issue in this process is the character of the airflow and heat transfer in the air gap. Several studies have focused on this flow [15, 16, 18] using tracers such as smoke or methaldehyde for flow visualization. Briefly, the air gap contains a buoyancy-driven channel flow, which may form subsidiary recirculating cells under some conditions.

Color schlieren [17] and differential interferometry [19] techniques have been applied to scale-model flows to visualize the airflow patterns in solar-heated rooms. Despite some exaggerated architechtural claims to the contrary, thermal stratification was observed to occur in these solar-heated rooms as it does in ordinary rooms (see Fig. 2 from Ref. [17]). Partitions [19] or ceiling fans can be used to

Fig. 2 Schlieren photograph of thermal stratification in a scale-model room with Trombe wall solar heating (from [17]).

control this stratification. In principle, however, thermal air currents derived from a solar source should not be expected to behave differently from those due to conventional room-heating means.

V INDUSTRIAL VENTILATION

In addition to thermal conditioning, industrial environments often require special ventilation considerations for removal of hazardous fumes or dust. Particular attention to ventilation is required in industries in which legislation has mandated minimum ventilation levels or maximum contaminant concentrations in order to protect workers. The principles of industrial HVAC are beyond the scope of the present chapter; they can be found, for example, in Ref. [1].

Such environments are usually too large to permit full-scale flow visualization studies, so models are traditionally used. Both air [20–22] and water [3, 5–7, 23] have been used as working fluids. Some water models [6, 23] employ free-surface flow visualization due to tracer particles sprinkled on the surface, thus yielding only two-dimensional flow information.

Three-dimensional water models [5–7] have been helpful in designing effective ventilation schemes for metal-ore processing plants and blast furnaces, wherein strong local heat sources, cold walls, and external winds complicate the issue. Thermal stratification [7] has been used advantageously to isolate contaminants above head level until they can be removed through roof vents. Locally injected, colored water dyes with either general or light-sheet illumination provide effective visualizations.

Local removal of hazardous fumes from the vicinity of chemical vats calls for special ventilation schemes, such as the push–pull ventilator [22]. Here, an air jet is forced tangentially across the liquid surface, capturing the fumes by turbulent entrainment. The airflow is removed by suction at the opposite side of the vat. A computational solution of the Navier–Stokes equations yielded visualizations of this flow, such as the one shown in Fig. 3. Color schlieren observations of a scale-model flow were used to verify the computations.

Color schlieren [24] and smoke [20] have been used to study methane stratification in mine-ventilation scale models. Once again, the explosion hazard in these cases makes effective ventilation a serious issue. In Ref. [24], quantitative readings of methane concentrations were extracted from the schlieren visualizations.

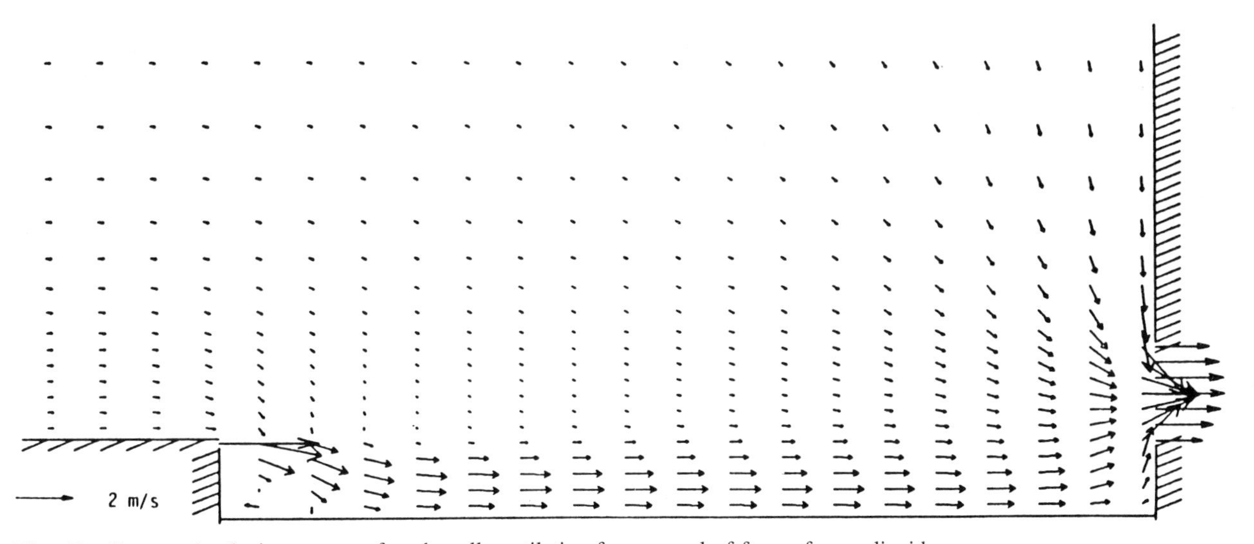

Fig. 3 Computed velocity vectors of push–pull ventilation for removal of fumes from a liquid-filled vat (from [22]).

VI CLEAN-ROOM TECHNOLOGY

The rapid growth of the microelectronics industry is demanding manufacturing environments of extraordinary cleanliness. Airborne particulates are especially insidious in causing defects that limit the yield of functional microelectronics. Modern clean-room facilities attempt to limit these airborne particles through high-performance filters, controlled airflows, and limited access to contamination sources. To date, these attempts have been only partially successful. Flow visualization studies [8, 13, 14, 25–35] are helping to provide an understanding of clean-room airflow phenomena, from which further improvements should result. Much of the following discussion applies as well to hospital operating theaters, where the prevention of disease spread by airborne contamination is a major concern.

A typical clean-room airflow is introduced through high-performance (HEPA) filters at the ceiling and removed through wall vents or (better) a perforated floor. A typical downflow of about 0.5 m/s is thus generated. In clean-room parlance, such airflows are termed vertical laminar flows (VLF). However, recent flow visualization studies [14, 32, 35] indicate that they are neither laminar nor vertical, especially when manufacturing benches and tools are in place.

As before, model studies [8, 14, 29] are useful in this regard. For example, Akabayashi et al. [14] used an air model with magnesium carbonate seeding and laser light-screen illumination. They observed a variety of flows about clean-room equipment and operators, and arrived at some helpful guidelines. In particular, vortical flow separation from obstacles or sharp edges is to be avoided, as this creates recirculation zones in which particles are trapped.

Several full-scale clean-room studies have also been done, primarily using smoke for flow visualization [13, 25, 28, 30, 32]. The work of Sodec and Veldboer [32] was especially productive. They observed flow disturbances from ceiling lights, heat sources, hot walls, perforated and unperforated work surfaces, and various floor venting arrangements. While some solutions were presented, it remains to be seen whether or not the conflicting demands of maintenance access and cost will allow proper fixes to be made in production clean rooms.

Smoke and similar tracers are somewhat self-defeating in full-scale clean-room tests, since they amount to contamination sources. To avoid this problem, optical instruments such as the schlieren system have been used by the present author and colleagues [31, 35–37]. A sensitive schlieren system can detect minor natural or induced thermal gradients in clean-room flows without causing contamination. A fixed laboratory-based instrument of 1-m aperture [35] has been used to study flows in open and enclosed VLF workstations with operators present (see Fig. 4). More recently, a portable 32-cm-aperture schlieren instrument was developed and qualified for clean-room use [31]. Noncontaminating flow studies in a state-of-the-art clean room are now in progress.

Fig. 4 Schlieren photograph showing flow separation and recirculation about a clean-room worker at an open workstation with downflow at 0.5 m/s (from G. S. Settles and A. J. Smits).

Other noncontaminating flow visualization techniques include the use of ''streamers'' and visible fog from liquid nitrogen. Streamers can be conveniently made from audio-cassette tape. No further details on these methods were found in the literature.

The human body is a critical issue in clean-room work, since it is a prime contamination source. Special garments attempt to contain this contamination with varying degrees of success. For example, it has been observed [33, 36] that the breath plume from a human cough has sufficient inertia to penetrate the surrounding airflow up to 1 or 2 m distant (see Fig. 5). Standard clean-room face coverings prevent this, but do not prevent local airborne contamination [33]. Apparently full-helmet-type facial enclosures are required for this purpose.

Finally, a number of computational studies of clean-room flows have been carried out [13, 25, 27, 29, 34, 35]. To the extent that these computations accurately predict the flow, various types of flow visualization can be extracted from the results. Examples can be found in Fig. 3 and in the cited references.

VII LARGE-FIELD OPTICAL VISUALIZATION

Studies on the spread of residential and building fires require large models or full-scale buildings, since the phenomena do not scale readily. Though such fires are naturally luminous, additional flow visualization aids are sometimes required. Weinberg and his colleagues [38, 39] are major contributors in this area, having produced several new laser-based techniques.

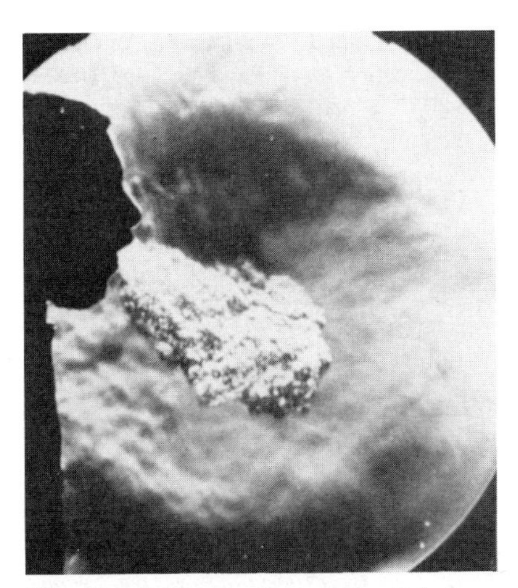

Fig. 5 Schlieren photograph of the contaminated plume generated by a human cough (from G. S. Settles [36]).

In general terms, optical visualization methods are traditionally limited in aperture due to the cost and clumsiness of large precision mirrors. The 1-m-aperture schlieren instrument used to produce Figs. 4 and 5 is near the upper practical limit in this regard. Larger, simpler instruments capable of imaging full-scale rooms would be quite useful in visualizing indoor environmental flows.

In fact, several such instruments are available. Given a sufficient pathlength, the divergent-beam, infinitesimal-shear interferometer [39, 40] can image large fields without expensive optical elements. Moiré interferometry [39] also has this potential. Finally, as demonstrated by Wuerker [41], diffuse-reflection holographic interferometry can also image large fields indoors. Figure 6 shows an example, intended primarily to reveal a parabolic microwave reflector, but imaging as well the thermal plume from the human subject present in the field of view. To implement this for indoor flow visualization, one needs a pulsed laser of significant output (0.5 J or more per pulse) and reasonably long coherence length. While such a laser is expensive, the large optical element required is simply the wall of the room, painted with retroreflective paint to improve the optical gain. A holographic plate is exposed twice, once with airflow disturbances present and once without. Upon reconstruction, only the difference in the flow between the two pulses is revealed. The poor optical quality of the wall, which did not change between pulses, is not seen in the holographic interferogram.

REFERENCES

1. Baturin, V. V., *Fundamentals of Industrial Ventilation,* 3d ed., pp. 12–24, Pergamon, New York, 1972.
2. Winter, E. F., Flow Visualization Techniques, in *Progress in Combustion Science and Technology,* ed. J. Ducarme, M. Gerstein, and A. Lefebvre, vol. 1, pp. 1–36, Pergamon, New York, 1960.

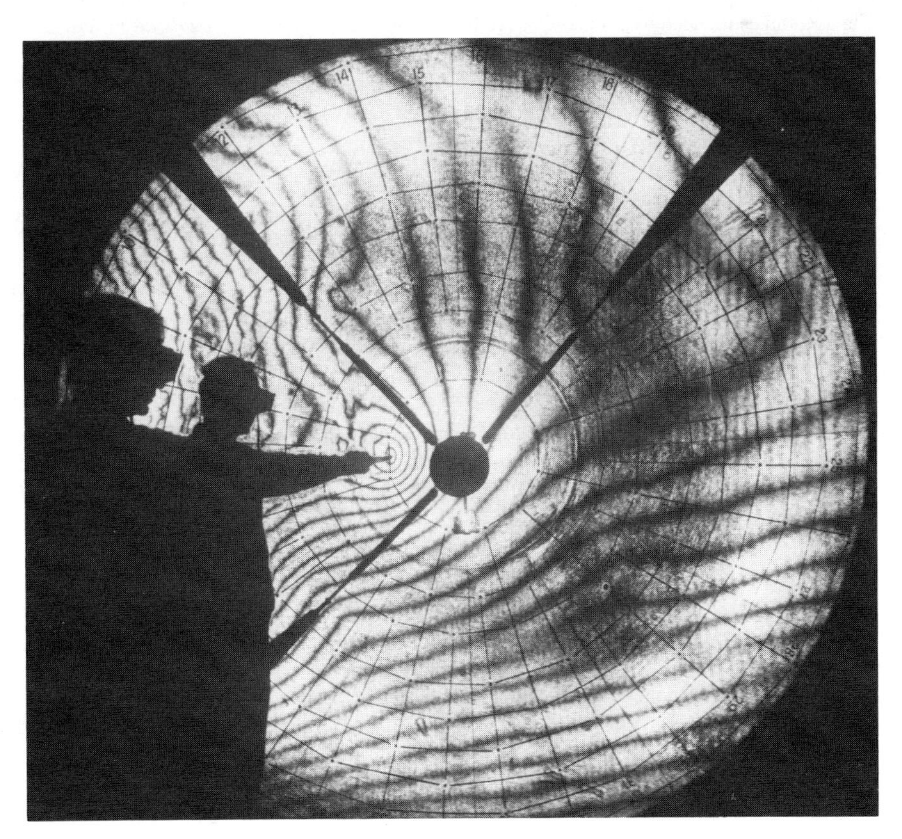

Fig. 6 Holographic interferogram of human subject touching microwave antenna, revealing thermal plume from the body (from R. F. Wuerker, TRW Inc. [41]).

3. Gibson, D. J., Scale Model Flow Testing, *Res./Dev.*, pp. 24–30, November, 1976.
4. Akiyama, M., Suzuki, M., and Nishiwaki, I., Transport Phenomena of Ventilating Flows in a Rectangular Room, in *Flow Visualization*, ed. T. Asanuma, pp. 135–141, Hemisphere, Washington, D.C., 1979.
5. Sucker, D., and Boenecke, H., Practical Application of Fluid-Dynamic Models on Power and Process Engineering in the Iron and Steel Industry, in *Flow Visualization II: Proceedings of the Second International Symposium,* ed. W. Merzkirch, pp. 185–192, Hemisphere, Washington, D.C., 1982.
6. Sucker, D., and Boenecke, H., Improvement of Flow, Heat Transfer, and Combustion in Steel Plants by Means of Physical Modeling Techniques, in *Flow Visualization III: Proceedings of the Third International Symposium on Flow Visualization,* ed. W.-J. Yang, pp. 797–802, Hemisphere, Washington, D.C., 1985.
7. Skaret, E., Industrial Ventilation—Model Tests and General Development in Norway and Scandinavia, *Proc. 1st Int. Symp. Ventil. Contam. Cont., Toronto,* October 1–3, 1985, ed. H. D. Goodfellow, Elsevier Chem. Eng. Monogr., 24, pp. 19–31, 1986.
8. Shi, N., The Water Model Experimental Equipment for Research on Airflow Field of Laminar Clean Room, *Proc. 8th Int. Symp. Contam. Contr., Milan,* pp. 859–868, 1986.
9. Grigull, U., Visualization of Heat Transfer, *Proc. 4th Int. Heat Transfer Conf., Versailles,* vol. 9, pp. 7–21, 1970.
10. Nottage, H. B., Slaby, J. G., and Gojsza, W. P., A Smoke Filament Technique for Experimental Research in Room Air Distribution, *ASHVE Trans.*, vol. 58, p. 399, 1952.
11. Euser, H., Hoogendoorn, C. J., and Van Ooijen, H., Airflow in a Room as Induced by Natural Convection Streams, *Energy Conservation in Heating, Cooling, and Ventilating Buildings: Heat and Mass Transfer Techniques and Alternatives,* ed. C. J. Hoogendoorn and N. H. Afgan, vol. 1, pp. 259–270, Hemisphere, Washington, D.C., 1978.
12. Sandberg, M., Blomqvist, C., and Sjoberg, M., Efficiency of General Ventilation Systems in Residential and Office Buildings—Concepts and Measurements, *Proc. 1st Int. Symp. Ventil. Contam. Contr., Toronto,* Oct. 1–3, 1985, ed. H. D. Goodfellow, Elsevier Chem. Eng. Monogr. 24, pp. 323–332, 1986.
13. Hayashi, T., Okonogi, T., Takemura, M., Proposal of Air Supply Method for Clean Tunnel System, *Proc. 8th Int. Sym. Contam. Contr., Milan,* pp. 332–339, 1986.
14. Akabayashi, S., Murakami, S., Kato, S., and Chirifu, S., Visualization of Air Flow around Obstacles in Laminar Flow Type Clean Room with Laser Light Sheet, *Proc. 8th Int. Symp. Contam. Contr., Milan,* pp. 691–697, 1986.
15. Robert, J. F., Puebe, J. L., and Trombe, F., Experimental Study of Passive Air-cooled Flat-Plate Solar Collectors: Characteristics and Working Balance in the Odeillo Solar Houses, *Energy Conservation in Heating, Cooling, and Ventilating Buildings: Heat and Mass Transfer Techniques and Alternatives,* ed. C. J. Hoogendoorn and N. H. Afgan, vol. 2, pp. 761–781, Hemisphere, Washington, D.C., 1978.
16. Okeke, B. O., Visualization of Natural Convection in Flat-Plate Solar Collectors, Ph.D. dissertation, Mississippi State University, 1979.
17. Carlson, A. B., Harrje, D. T., and Settles, G. S., An Optical Study of Thermal Convection in a Passive Solar Heated Room, ASME Paper 80-C2/Sol-1, August 1980.
18. Linthorst, S. J. M., Schinkel, W. M. M., and Hoogendoorn, C. J., Natural Convection Flow in Inclined Air-filled Enclosures of Small and Moderate Aspect Ratio, in *Flow Visualization II: Proceedings of the Second International Symposium,* ed. W. Merzkirch, pp. 93–97, Hemisphere, Washington, D.C., 1982.
19. Chou, X., Li, Y.-Z., and Shu, J.-Z., Flow Visualization in the Model of a Solar House by Means of a Differential Interferometer, in *Flow Visualization IV: Proceedings of the Fourth International Symposium,* ed. C. Véret, pp. 691–696, Hemisphere, Washington, D.C., 1987.
20. Renner, K., and Wesley, R., Visualization of Mixing and Removing Methane Layerings—Problems of Mine Ventilation, in *Flow Visualization II: Proceedings of the Second International Symposium,* ed. W. Merzkirch, pp. 193–200, Hemisphere, Washington, D.C., 1982.
21. Gamble, S. L., and Irwin, P. A., Application of Scale Model Techniques in the Design of Ventilation Systems for Railway Locomotive Maintenance Shops and Stations, *Proc. 1st Int. Symp. Ventil. Contam. Contr., Toronto,* October 1–3, 1985, ed. H. D. Goodfellow, Elsevier Chem. Eng. Monogr. 24, pp. 111–120, 1986.
22. Heinsohn, R. J., Yu, S.-T., Merkle, C. L., Settles, G. S., and Huitema, B. C., Viscous Turbulent Flow in Push–Pull Ventilation Systems, *Proc. 1st Int. Symp. Ventil. Contam. Contr.*, ed. H. D. Goodfellow, Toronto, October 1–3, 1985, Elsevier Chem. Eng. Monogr. 24, pp. 529–566, 1986.
23. Řezníček, R., Flow Visualization in Czechoslovakia and Some Countries in Central and Eastern Europe, in *Flow Visualization,* ed. T. Asanuma, pp. 19–28, McGraw-Hill, New York, 1979.
24. Phillips, H., A Three-Colour Quantitative Schlieren System, *J. Sci. Instrum.*, vol. 1, no. 2, 1968.
25. Broyd, T. W., Deaves, D. M., and Oldfield, S. G., The Use of Computational Modeling Techniques for Cleanroom Design, *Proc. Semicon/Europa 1983, Semiconductor Process. Equip. Symp., Zurich,* pp. 324–347, 1983.
26. Caplan, K. J., Research and Development Trends and Needs, *Proc. 1st Int. Symp. Ventil. Contam. Contr., Toronto,* pp. 1–17, 1986.
27. Deaves, D. M., and Malam, D., Advanced Analysis Techniques for the Optimum Design of Clean Rooms, *J. Environ. Sci.*, vol. 28, pp. 17–20, September/October, 1985.
28. Dai, J.-L., An Alternative Design of Class II Laminar Flow Biosafety Cabinet, *Proc. 8th Int. Symp. Contam. Contr., Milan,* pp. 631–640, 1986.
29. Kato, S., Murakami, S., and Chirifu, S., Study on Air Flow in Conventional Flow Type Clean Room by Means of Numerical Simulation and Model Test, *Proc. 8th Int. Symp. Contam. Contr., Milan,* pp. 781–791, 1986.
30. Oshige, K., Newly Developed Changing Room for Clean Room Garments, *Proc. 8th Int. Symp. Contam. Contr., Milan,* pp. 662–671, 1986.
31. Settles, G. S., and Via, G., A Portable Schlieren Optical System for Clean Room Applications, *Proc. 8th Int. Symp. Contam. Contr., Milan,* pp. 381–392, 1986.
32. Sodec, I. F., and Veldboer, W., Influences on the Stability

of Airflow in Clean Rooms, *Proc. 8th Int. Symp. Contam. Contr., Milan,* pp. 869–885, 1986.

33. Sullivan, G., and Trimble, J., Evaluation of Face Coverings, *Microcontamination,* vol. 4, no. 5, pp. 64–70, 1986.
34. Toshigami, K., Kanayama, H., and Yashima, S., Finite Element Analysis of Air Flow and Advection-Diffusion of Particles in Clean Rooms, *Proc. 8th Int. Symp. Contam. Contr., Milan,* pp. 278–285, 1986.
35. Settles, G. S., Huitema, B. C., McIntyre, S. S., and Via, G. G., Visualization of Clean Room Flows for Contamination Control in Microelectronics Manufacturing, in *Flow Visualization IV: Proceedings of the Fourth International Symposium,* ed. C. Véret, pp. 833–838, Hemisphere, Washington, D.C., 1987.
36. Settles, G. S., and Kuhns, J. W., Visualization of Airflow and Convection Phenomena about the Human Body, *Bull. Am. Phys. Soc.,* vol. 29, no. 9, p. 1515.
37. Settles, G. S., Colour-Coding Schlieren Techniques for the Optical Study of Heat and Fluid Flow, *Int. J. Heat Fluid Flow,* vol. 6, no. 1, pp. 3–15, 1985.
38. Weinberg, F. W., *Optics of Flames,* Butterworths, London, 1961.
39. Popovich, M. M., and Weinberg, F. J., Laser Optical Methods for the Study of Very Large Phase Objects, *Exp. Fluids,* vol. 1, pp. 169–178, 1983.
40. Weinberg, F. J., and Wong, W. W.-Y., Optical Studies in Fire Research, *Proc. 16th Int. Symp. Combust.,* pp. 799–807, 1977.
41. Wuerker, R. F., Holographic Interferometry, *Jpn. J. Appl. Phys.,* vol. 14, pp. 203–212, 1975.

Chapter 38

Agriculture

Junta Doi

I INTRODUCTION

The role of flow visualization in agriculture is similar to that in industry except that the objects of visualization are mostly biological and natural phenomena. Almost all methods that have been used in various applications are employed in this field. In the experiments on agricultural problems, however, so many parameters are involved in the phenomena investigated that the initial and boundary conditions are not definitely determined. Therefore, repetitive observations with pointwise measurement devices such as the Pitot tube or a hot-wire probe sensor are not always effective. Instead, instantaneous flow-pattern observation with visualization techniques works best.

Following are some typical examples reported mainly in agricultural journals over the last two decades. Most of the examples are relevant to flow problems; however, the last section is related particularly to the flow of soils around working tools.

II BOUNDARY LAYER AROUND PLANT LEAVES

The energy transfer to and from a plant involves solar radiation; thermal infrared radiation from the ground, surrounding surfaces, and the atmosphere; thermal emission by the plant; and free convection in still air and forced convection in wind. This energy transfer creates the temperature difference between the plant and the surrounding air. This temperature difference, though a slight quantity, can be visualized by optical methods.

There are some reports on the observation of plant leaves by the schlieren photograph. Gates et al. [1, 2] observed the phenomenon of free convection from leaves of broad-leaved and coniferous trees in still air. Their purpose was to observe the convection from plants and to estimate, for various types of plants, the quantity of heat carried away by free convection. They used a schlieren system with a small plane mirror and a parabolic mirror with a diameter of 20 cm and a radius of curvature of about 3.3 m. In these experiments all the sample stems were cut from the tree and placed in water until just prior to viewing in the laboratory. The leaves were irradiated from above with visible and infrared heat radiation from an illumination lamp. In order to ensure the quantitative measurement, a fine thermocouple was used in the vicinity of the convection plumes and on the leaf surface.

The investigators recorded the schlieren field with a movie camera, delivering a slight disturbance to the plume by blowing a small puff of air into the base of the plume so as to obtain the rate at which the plume rose. They also observed the plume motion by injecting small bursts of cigar smoke in the same base. They reported that both systems worked very well and gave nearly identical results. In these experiments, the surface temperature of the leaf was also measured by a radiometer in addition to the thermocouple. The experimenters concluded that the thermocouple generally gave a surface temperature of a degree or more lower than the radiometer due to a lack of contact between the surface and the thermocouple; the radiometric temperature was taken to be the correct surface temperature.

When the leaves were irradiated in sunlight or equivalent light, the share of the energy carried away by transpiration

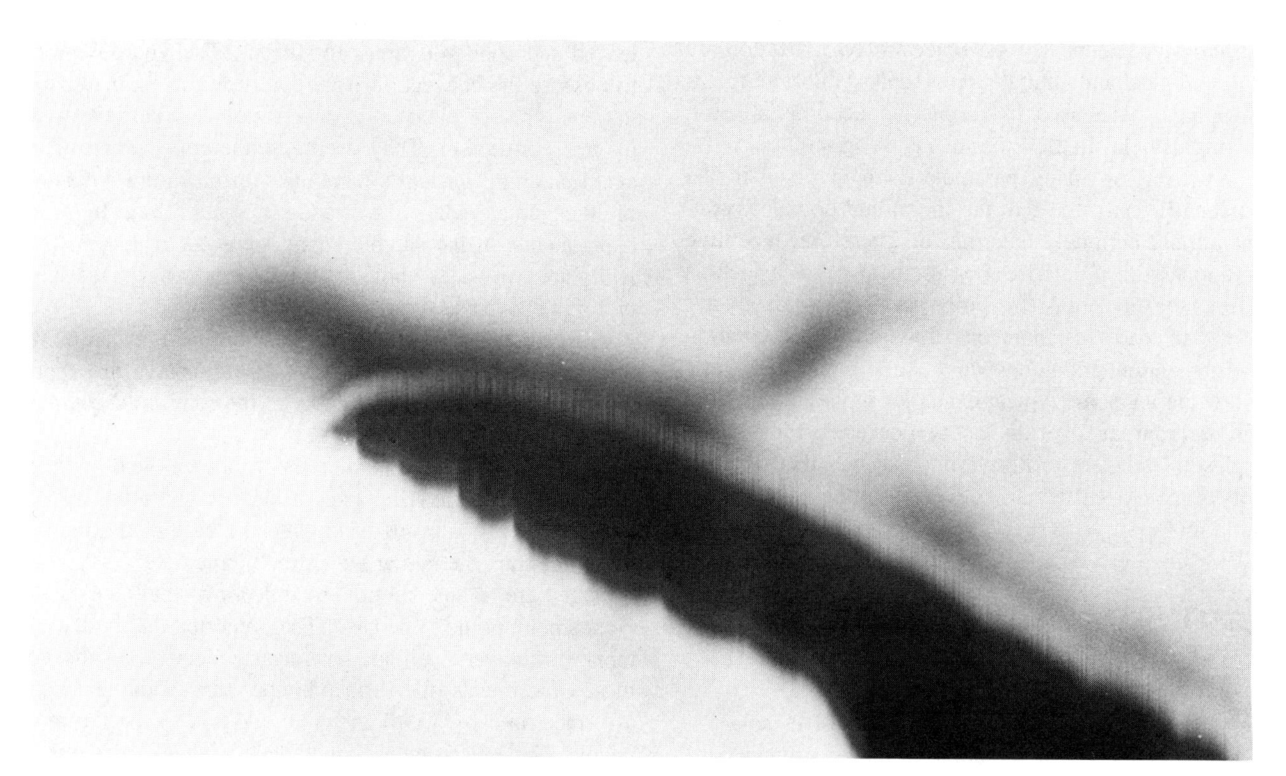

Fig. 1 Boundary layer around a plant leaf (top) and its separation by a draft (bottom), observed with a schlieren method.

increased. In the same time, the gas exchange of CO_2 and O_2 occurred according to the progress of photosynthesis. When the light intensity decreased, the reverse process of respiration was dominant, and the exchange of O_2 and CO_2 took place. Doi [3, 4] showed photographs of boundary layer developed on the broad leaf of a gloxinia (*Sinningia speciosa*) in the same condition as the field. They used a schlieren system with a pair of parabolic mirrors of 30 cm dia. Figure 1 (top) shows the boundary layer developed under illumination with a halogen lamp at about 0.17 W/cm^2 of energy in still-air conditions. A plume of thick layers is observed in the schlieren field, and it rises along the direction of the leaf. If the front end of the leaf is down, the shape of the plume is different from that seen in this photograph. It can be observed that this boundary layer blocks the free exchange of these gases and is disadvantageous to photosynthesis. This state is, however, advantageous to suppression of the respiratory process and to reduction of consumption of photosynthesized products.

In Fig. 1 (bottom) this boundary layer is separating from the leaf surface at the moment of a small draft. As mentioned previously, a thick and stationary boundary layer is disadvantageous to photosynthesis. However, in fact some wind or draft always exists in the atmosphere, and the stem, branches, and leaves move and flutter in the wind, preventing development of the thick and stationary layer. This observation agrees with the fact that adequate wind is necessary for efficient photosynthesis.

III FLOW AROUND A FISH

An important problem in physiological studies of fish behavior is the question of how fish swim and what the fluid mechanics of their propulsion are. The problem has two major aspects. One is the physiological activity in relation to fishery, and the other is an engineering application. The latter is called *biomimetics*, an application aimed at manufacturing artificial fish or developing new mechanisms having fishlike propulsion.

A major portion of the fish's propulsive force is thought to be generated by the wiggling motion at the rear of its body and around its tail. As a result, the angle of attack of the fish's head and main body to the direction in which it is moving is very large in comparison with the mechanical propulsion of ships or submarines. Doi and Miyake [5] investigated a fish's behavior in a tank, showing figures of the three-dimensional fish position, head direction, velocity or trajectory, and acceleration by means of time-to-time tracking. The figures were constructed using successive three-dimensional models with three simultaneous video images. From these figures it was observed that the fish's motion is essentially three-dimensional. Flow visualization studies of fish propulsion, however, are done in two-dimensional form because of technical limitations. These studies are done by observing direct multiplex photographs using aluminum powder on the water surface [6]; dyes, hydrogen bubbles generated in swimming pools; [7]; and liquid tracers such as milk and Chinese black ink [8].

Abe et al. [8] used three types of visualization apparatus, shown in Figs. 2, 3, to observe eddies made by a fish's motion. In Fig. 2 the liquid tracer, a mixture of condensed milk and Chinese black ink, was poured into the water adjacent to the upper test section in which the swimming fish and eddies were observed in a laminar flow condition. Observations of the fish and eddies were made in a shallow rectangular pool with diffused back lighting (from two directions), as shown in Figs. 3*a*, *b*. In these cases the sugared water of 30% concentration and Chinese ink composed the liquid tracer, which was poured from a container along a thin glass rod in such a manner that it did not disturb the water. It is reported that with these apparatuses, clear

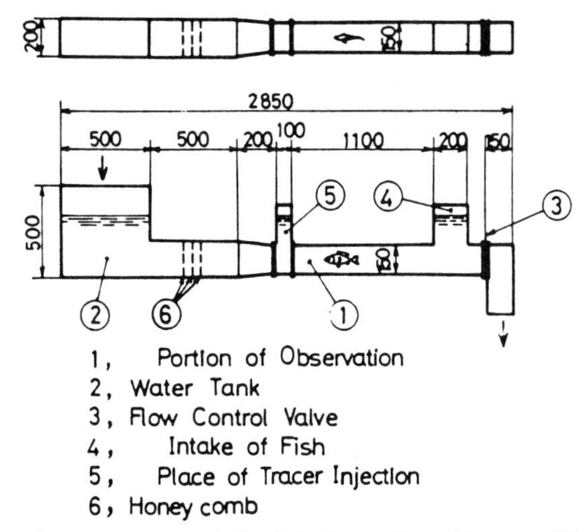

Fig. 2 Water tank for fish observation with tracer [8].

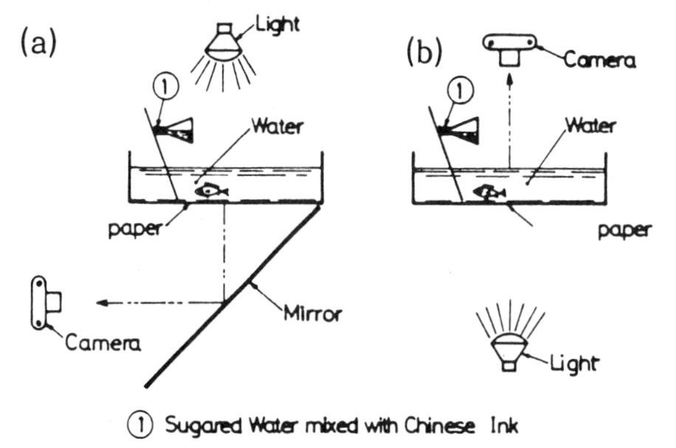

Fig. 3 Swimming pools for visualization with tracer [8].

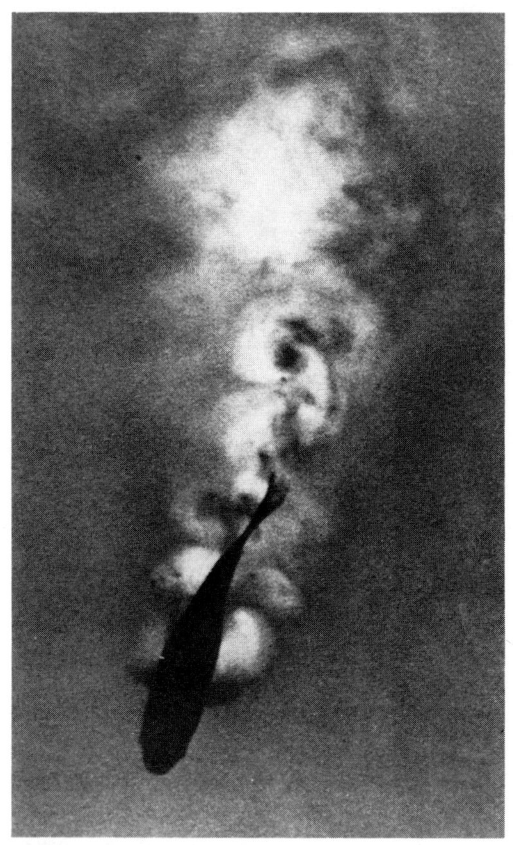

Fig. 4 Eddies observed by sugared water and mixed Chinese ink [8].

photographs of eddies generated by the swimming fish could be taken, as seen in Fig. 4 [9].

The hydrogen bubble method was applied to observations of flows around a fish (carp) by Nakayama and Narasako [7]. Their apparatus is shown in Fig. 5, and an example of the visualized picture is given in Fig. 6. The apparatus was installed in a circulation tank, and the water-stream velocity was recorded at 0.122–0.208 m/s. To avoid passing electric shock through the swimming fish, they set four anodes, one in each corner of the pool. It was observed that the flow did not separate along the fish's body; however, bubbles were diverged near the tail, due to its wiggling motion. From the hydrogen bubble method it can be observed that the high-voltage application for bubble generation is dangerous to the swimming fish and that the fish itself has a tendency to avoid the bubbles when swimming.

IV AGRICULTURAL METEOROLOGY

Flow visualization techniques have wide areas of application in the field of agricultural meteorology. Most of the techniques are the same as those applied to general meteorology; the difference is that the targets exist near the earth's surfaces.

One example, the measurement of air movements in a forest clearing, is presented here. Bergen [10] reported a

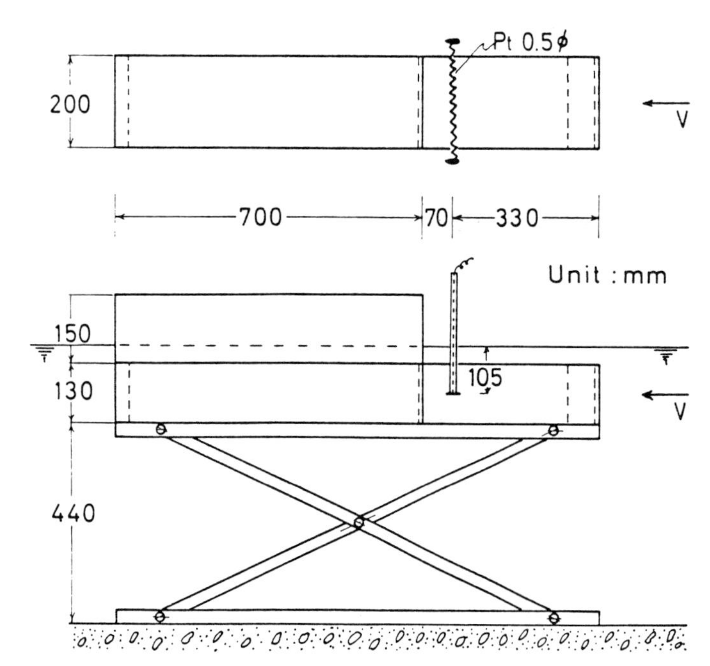

Fig. 5 Swimming pool used for visualization with hydrogen bubbles [7].

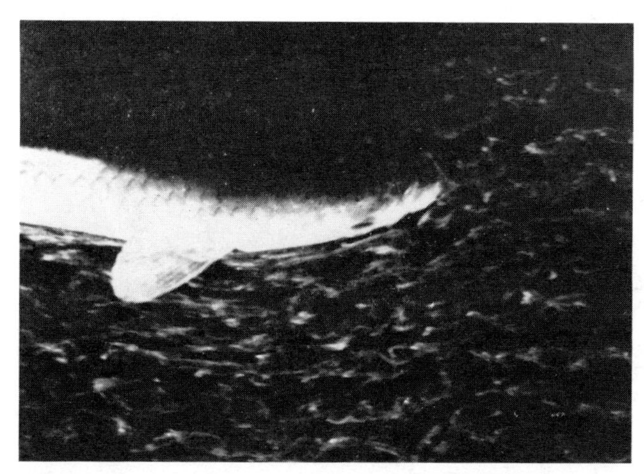

Fig. 6 Flow around a fish observed with hydrogen bubbles [7].

Fig. 7 Breakdown of flow in a forest clearing (U.S. Forest Service photo [10]).

cinematic observation of the behavior of multiple smoke plumes in a stand of lodgepole pine. Measurements were made against a tower grid, with simultaneous recordings of the wind speed and direction above the canopy. The observation site was a 10- by 50-m rectangular clearing in the pine stand, which had a density of 200 stems per 100 m^2. Tree height averaged 10 m, and tree diameter at breast height averaged 10.4 cm.

In this experiment two types of smoke were used: candles lit by fuse produced white smoke, and mechanically fired grenades produced yellow smoke. Grenades were set at the levels of 8 and 5 m, and candles were set at the levels of 2 and 0.5 m. This placement gave good contrast between the colors of the upper and lower pairs of smoke plumes. Smoke-drift patterns were recorded with color film at a speed of 10.6 frames per second. The results are shown in Fig. 7.

It was reported that the buoyancy of the smoke plumes seemed to cause serious error because both the candles and the grenades burned at high temperatures. It is doubtful that the apparent trajectories indicate true vertical displacement of the airflow. Horizontal velocities were estimated from frame-to-frame displacement of plumes, but vertical velocities were computed by continuity. It was concluded that air movement in the forest clearing alternated between an unseparated flow and a central vortex motion similar to that found in a square notch. The transition involved strong circulation along the crosswind axis of the clearing, and the frequency of the alternation was in fair agreement with the vortex-shedding frequency of a flat plate having the same vertical dimensions as the live crowns.

V WINDBREAKS

The frictional drag and obstacle reduction created by trees, solid barriers, or walls can cause diversions in airflow. This effect brings noticeable changes not only in the reduced pressure on vegetables and trees but also in the temperature and humidity of the air and in snowdrift formation. Another benefit is soil protection from wind erosion.

The type and layout of a windbreak have a perceptible effect on the resulting velocity and pattern of the airflow and on the area of protection. Wind tunnel experiments on the effect of a barrier shape on the resultant flow have been studied in various scale models. Most visualization techniques applied for this purpose are smoke and tuft methods. Figure 8 is an example of a two-dimensional model experiment with tuft observations by Yamamoto [11]. In this experiment a 50-mm tuft of woolen yarn was used in the arrangement of a 50- by 100-mm matrix located in the central plane of a windbreak model.

Moysey and McPherson [12] demonstrated flow patterns around porous windbreak models. They showed airflow patterns behind a model having some solid and some porous barriers. They observed that the solid windbreak model created a vortex behind it, extending several model heights leeward. With the 35%-porous model, the vortex was broken up by air coming through the openings. They concluded that windbreaks with porosities in the range of 15 to 30% provide better shelter than solid windbreaks, particularly at distances of 1 to 8 barrier heights leeward.

VI FLOW OUTSIDE AGRICULTURAL BUILDINGS

Flow visualization studies on agricultural buildings have two major aspects. One is the role of windbreaks with reference to the object and area of protection. Housing layout for residential colonies and its influences on windbreak performance were tested in the field and in modeling experiments [13]. In these studies the effects of the layout of multiple housing, trees, and hedges on airflow patterns were

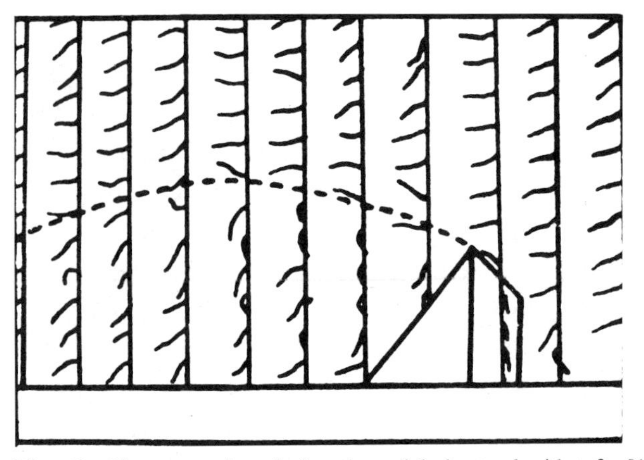

Fig. 8 Flow around a windbreak model observed with tufts [11].

investigated using modeling experiments with smokestreak observations.

Another point of consideration is how to avoid and protect the agricultural facilities from wind damage, which is a serious problem for small, simple, and light buildings such as greenhouses. Wind tunnel testing was carried out on models of greenhouses with gabled, semicircular, and semielliptical roofs [14]. Smoke observations were also made in these studies. In Fig. 9 an example of flow patterns around two parallel greenhouse models is shown.

VII FLOW INSIDE AGRICULTURAL BUILDINGS

In agricultural buildings bioclimatic characteristics are studied in both analytical and experimental treatments for problems of air movement and temperature distribution. Most of the efforts are concerned with design criteria and microclimate control in buildings housing livestock. The ideal ventilation system provides an optimum environment at the level of the livestock. Intensive investigations concerning ventilation have been executed for this purpose. For these investigations it is undesirable to use actual livestock houses because costs are high and too many variables exist to be considered and measured. Most of the investigations, therefore, are made with the use of scale models; however, because of the difficulties in similitude, results may require verification on a full scale. Factors affecting the ventilating system are ambient wind, fan speed, inlet-air temperature, internal obstructions and divisions, inlet and outlet locations, livestock arrangement, and internal sources of heat. In any type of experiment, measurement of airflow patterns or flow visualization is essential.

Three kinds of measurement have been reported regarding the observation of airflow in livestock buildings. One is an experiment with a scale model in a water table [15]; air flowing around buildings can be considered incompressible and able to be simulated using an incompressible fluid such as water. However, most of the experiments were carried out in air using smoke observations of the general pattern of air movements [13, 16, 17].

The third method of observation is the use of soap-film bubbles [18–21], which are suitable for use in the real building housing livestock because they are nontoxic. In the bubble experiments, helium was introduced in the bubbles to yield a particle near neutral buoyancy. It was reported that the maximum terminal velocity in still air was

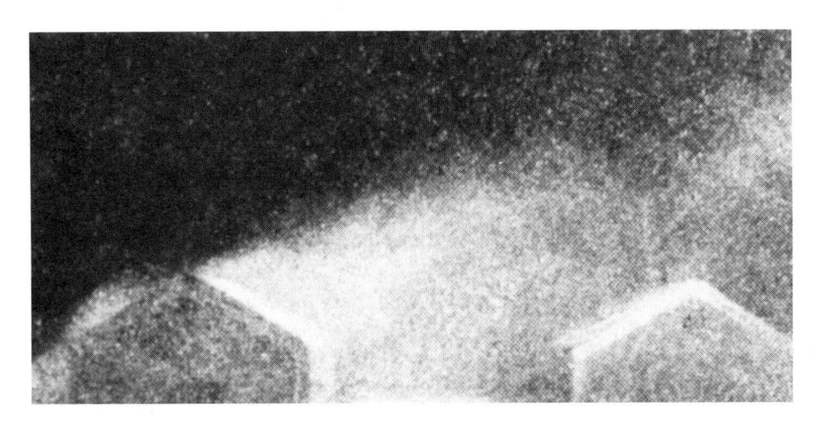

Fig. 9 Flow around two parallel greenhouse models observed with smoke [14].

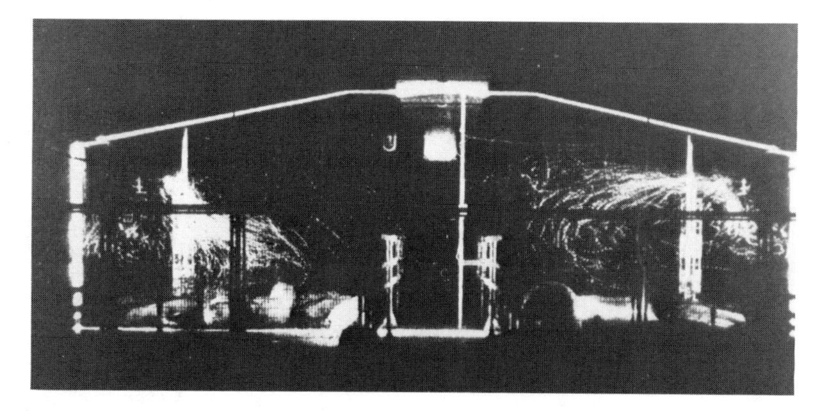

Fig. 10 Flow inside a livestock building observed with bubbles illuminated by a light sheet [21].

about 0.02 m/s, and a satisfactory bubble lifetime was obtained when the relative humidity of the ambient air exceeded 55%. This humidity level resembles that of many livestock buildings [18, 20]. An airflow pattern visualized by Boon [21] is shown in Fig. 10. In this experiment, liquid-film bubble generators were positioned in the building section so as to inject bubbles in the central plane, where they were illuminated by a light sheet from modified slide projectors. The floor of the model building was 3.30 m long by 7.60 m wide, and the height was 1.87 m to the eaves and 2.43 m at the ridge. This building section was surrounded above and on two sides by an outer shell, in which air was controlled by fans, heaters, and refrigerators.

VIII SPRAYERS AND SPRINKLERS

Atomizers have been used for the application of pesticides and insecticides in hand-held, vehicle-attached, or aircraft-mounted applicators. The technical background of atomization in agriculture is the same as of a combustion chamber. Flow visualizations play great roles in these studies. Dombrowski [22] reported shadowgraph pictures in his study of drop formation from single-orifice fan-spray nozzles. Figure 11 shows a liquid sheet atomized close to the nozzle by aerodynamic wave motions. Lake [23] and Frost [24] used high-speed photography to investigate direct-drop, ligament, and sheet formation in rotary atomizers.

Another application of the flow visualization technique is the observation of droplets in irrigation sprinklers. Using a spray-type sprinkler jet breakup, a cylindrical jet of water impinging perpendicularly onto a smooth and serrated plate was described using photographic information [25].

Flow visualization investigations of agricultural aircraft have been made from the viewpoint of efficient application of liquid and powder. In a crop-spraying aircraft, which generally is heavily loaded and has to fly slowly, a strong swirling motion near each wing tip is generated. The spray near each tip is swept into the swirling flows, and some of it is thrown into the air and blown away from the target area. Efficient wing-tip design was developed using an aircraft fitted with wool tufts [26].

IX FOOD TECHNOLOGY

In the field of food technology, many problems are to be solved with respect to fluid dynamics and thermodynamics. Some of these problems have been investigated with visualization techniques. Denk and Stern [27] reported the flow in an alcoholic-fermentation-model experiment in a cylindroconical tank. Flow patterns of CO^2 bubbles inside this axisymmetric fermenter were thought to be also axisymmetric; however, it was found by light-sheet observation that the bubble flow was quasisteady or unsteady, not axisymmetric.

Fig. 11 Distribution of spray sheet visualized with shadowgraph [22].

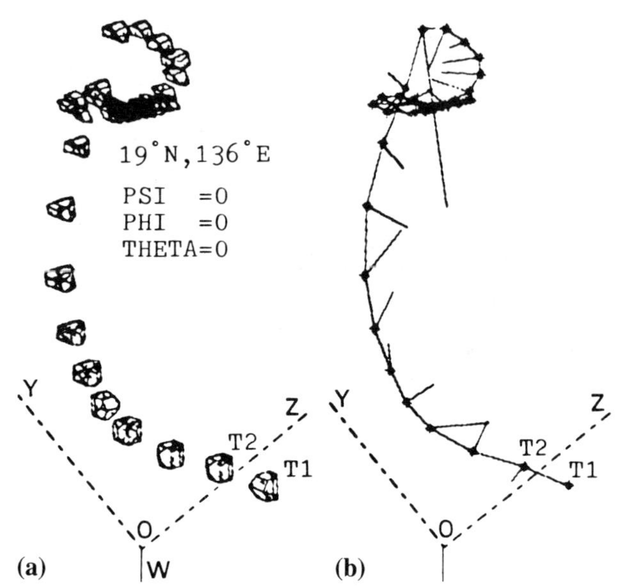

Fig. 12 Flow inside an agitation tank observed by 3-D shape tracking of a tracer particle: (*a*) Position of polyhedron (*b*) trajectory and acceleration.

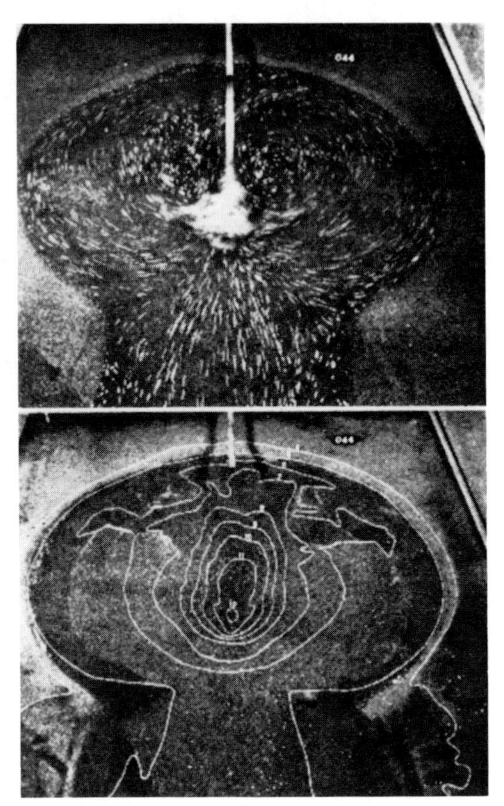

Fig. 13 Flow around a scored hole in a sand bed observed with confetti (top) and contours of the scored bed outlined with white yarn (bottom) [31].

As a technique applicable to food technology, three-dimensional flow visualization in an agitating tank was reported by Chang and Tatterson [28]. In this report, a standard stereoscopic-motion-picture technique was used to track neutrally buoyant tracer particles, resulting in three-dimensional particle paths that describe the various flow systems evolving in time. Doi et al. [29] developed a system that measures the shape of a target particle with three simultaneous video-image inputs and reconstructs its geometric modeling in a three-dimensional solid-model form. In Fig. 12 an example is illustrated in a wire-frame representation. In this system using reconstructed models, the position, the path or trajectory, the velocity, and the acceleration are calculated in addition to its time-to-time varying shape. This method will be applicable in understanding the behavior of a bubble or a similar target that moves while changing its size, shape, and direction.

X SOIL AND WATER

Irrigation, drainage, and soil conservation are fundamental concerns in agriculture; water movement on the surface and in the subsurface of soil is a very basic problem in agricultural fields. Flow visualization is used in these studies. Water drops from rain or irrigation sprinklers cause soil erosion. The impacting and splashing of raindrops were observed using a high-speed camera. The photograph shows that a drop landing on wet soil causes a crown similar to that of Edgerton's milkdrop splash [30]. The flow onto a soil surface scores a hole. Robinson [31] observed the flow patterns and the scored-hole configurations that resulted when flow was discharged from a pipe above the model bed of various sizes of sand. Confetti was used on the water surface to show the flow patterns. Fig. 13 shows this example and contours of the scored bed with white yarn.

The so-called seepage, infiltration, and percolation phenomena are related to water movements in subsoils and to water storage, drainage, and transpiration through plant roots. Water movement in subsurface soil was studied using a two-phase flow model of air and water in a porous medium. Mechrez et al. [32] investigated this using a two-dimensional sandbox and glass beads, photographing the results with a movie camera through a glass window. Chen and Wada [33] investigated the phenomena using a 25-mm-dia. cylindrical tube packed with quartz particles about 600–700 μm in diameter; they observed the sections of different heights by illuminating a light sheet using an argon–ion laser. They showed images of horizontal and vertical cross sections.

XI SOIL AND GRAIN

Tillage, soil loosening, leveling, mole drainage, and trenchless pipe laying are fundamentals in agricultural op-

Fig. 14 Streamlines of soil particles and soil wedge formation in front of a moving tool bar [37].

erations, and all these operations emphasize soil movement. Soil movement around a tillage tine or a plow is analogous to the external flow around a rocket or a ship. Investigations of these problems, in relation to cutting, shearing, compressing, mixing, turning, and elevating, involve the mechanics of soil failure around the areas of working tools, resistance against the tools, density changes, and flow patterns. Most of the soil-movement studies have been carried out by observing the movements of soil inside glass-sided bins [34–40]. Investigations of field conditions are also carried out using subsurface tracers such as colored beads [41].

In the soil-bin experiment, the glass side is assumed to be the axis of symmetry of the tine in the direction of travel. To observe the trajectory of deformation, strings made of wet paper tissue were laid in the soil along the glass side of the soil bin at different depths, and the glass was lightly greased to prevent the test material from sticking to it when clay soil was tested [34]. Resistance of the test material to the glass side was considered to be negligible for other kinds of soil. Krause [37] observed that a wedge developed in front of a tool bar. This wedge and the streamline of soils, photographed through the vertical glass side, are shown in Fig. 14. Godwin and Spoor [38] examined horizontal soil failures by pushing vertical tines of different widths horizontally into the soil beneath a horizontal sheet of laminated glass. They showed a plan photograph at 0.5-s exposure showing the soil-particle trajectories, indicating the particle trajectory as having forward, sideways, and backward components.

Another investigation of granular materials involves the flow of grain. Pneumatic conveyers are widely used to transport grain. Observations of the flow of air and grain around different types of suction nozzles were done by taking photographs through glass-sided models to obtain the optimal design criteria [42]. Flow of granular materials has also been investigated in food-processing research.

XII CONCLUDING REMARKS

Some examples of flow visualization studies applied to agriculture have been introduced. In addition to these examples, there are many applicable problems such as external flow around individual plants and livestock or the microclimate problems in agriculture, the internal flow of plants, flow through standing crops and trees, flow in irrigation channels, flow from charged and noncharged sprayers, flow in grain drying, and flow in food processing. These applications are, by nature, similar to those in industry. However, most of the phenomena involved in agriculture are scarcely reproducible so that a great deal of observation depends, more than ever, on flow visualization that gathers the pattern information gradually.

REFERENCES

1. Gates, D. M., and Benedict, C. M., Convection Phenomena from Plants in Still Air, *Am. J. Bot.*, vol. 50, no. 6, pp. 563–573, 1963.
2. Tibbals, E. C., Carr, E. K., Gates, D. M., and Kreith, F., Radiation and Convection in Conifers, *Am. J. Bot.*, vol. 51, no. 5, pp. 529–538, 1964.
3. Doi, J., Boundary Layer of a Plant Leaf, Photographic *J. Flow Visual.*, no. 2, pp. 30–31, 1985. In Japanese.
4. Doi, J., Boundary Layers on a Plant Leaf in *Fantasy of Flow*, ed. Flow Visualization Society of Japan, p. 132, Kodansha, Tokyo, 1986.
5. Doi, J., and Miyake, T., Measurement of Three-dimensional Behavior, *2d Symp. Image Sensing Technique in Industry*, Tokyo, pp. 31–34, 1987.
6. Hertel, H., *Struktur-Form-Bewegung*, Krausskopf-Verlag, Mainz, West Germany, 1963.
7. Nakayama, H., and Narasako, Y., On the Thrusting-Occurrence-Possibility by the Movement of the Scale Plates, *Mem.*

Fac. Fish., Kagoshima Univ., vol. 27, no. 1, pp. 173–182, 1978. In Japanese.
8. Abe, J., Yoshinaga, A., and Hatanaka, H., Visualization of Eddies Formed by Movement of a Fish, *J. Flow Visual. Soc. Jpn.*, vol. 1, no. 2, pp. 89–92, 1981. In Japanese.
9. Abe, J., Gold Fish Swimming in *Fantasy of Flow*, ed. Flow Visualization Society of Japan, pp. 130–131, Kodansha, Tokyo, 1986.
10. Bergen, J. D., Air Movement in a Forest Clearing as Indicated by Smoke Drift, *Agric. Meteorol.*, vol. 15, no. 2, pp. 165–179, 1975.
11. Yamamoto, R., Mechanism of Lessening of Protective Effect of the Shelter-Hedge when Approached by Oblique Wind, *J. Agric. Meteorol. Jpn.*, vol. 19, no. 3, pp. 103–108, 1964. In Japanese.
12. Moysey, E. B., and McPherson, F. B., Effect of Porocity on Performance of Windbreaks, *Trans. ASAE*, vol. 9, pp. 74–76, 1966.
13. Olgray, V., *Design with Climate*, Princeton Univ. Press, Princeton, N.J., 1963.
14. Nakazaki, A., Tamai, S., Kuwabara, T., and Hara, K., On the Wind Pressure Distribution about Two Houses of the Same Type Adjacent to Each Other, *Trans. Jpn. Soc. Irrig. Drain. Reclam. Eng.*, no. 47, pp. 49–55, 1973. In Japanese.
15. Kelly, T. G., Dodd, V. A., and Ruan, D. J., Ventilation and Air Flow Patterns in Climatic Calf Houses, *J. Agric. Eng. Res.*, vol. 33, pp. 187–203, 1986.
16. Zeisig, H. D., Luftfuhung in Stallen, *Glundlag. Landtech.*, vol. 19, no. 3, pp. 79–84, 1969.
17. Turnbull, J. E., and Coates, J. A., Temperatures and Air-Flow Patterns in a Controlled-Environment, Cage Poultry Building, *Trans. ASAE*, vol. 14, pp. 109–113, 120, 1971.
18. Moulsley, L. J., and Boothroyd, D. N., A Device for Producing Small Bubbles for Use in the Visualization of Air Movement, *J. Agric. Eng. Res.*, vol. 16, pp. 364–367, 1971.
19. Carpenter, G. A., Moulsley, L. J., and Randall, J. M., Ventilation Investigations using a Section of a Livestock Building and Air Flow Visualization by Bubbles, *J. Agric. Eng. Res.*, vol. 17, pp. 323–331, 1972.
20. Randall, J. M., The Prediction of Airflow Patterns in Livestock Buildings, *J. Agric. Eng. Res.*, vol. 20, pp. 199–215, 1975.
21. Boon, C. R., Airflow Patterns and Temperature Distribution in an Experimental Piggery, *J. Agric. Eng. Res.*, vol. 23, pp. 129–139, 1978.
22. Dombrowski, N., Some Flow Characteristics of Single-Orifice Fan-Spray Nozzles, *J. Agric. Eng. Res.*, vol. 6, pp. 37–44, 1961.
23. Lake, J. R., An Apparatus for Photographing Sprays, *J. Agric. Eng. Res.*, vol. 24, pp. 215–218, 1979.
24. Frost, A. R., Rotaty Atomization in the Ligament Formation Mode, *J. Agric. Eng. Res.*, vol. 26, pp. 63–78, 1981.
25. Kohl, R. A., and DeBoer, D. W., Drop Size Distributions for a Low Pressure Spray Type Agricultural Sprinkler, *Trans. ASAE*, vol. 27, pp. 1836–1840, 1984.
26. Parkin, C. S., and Spillman, J. J., The Use of Wing-Tip Sails on a Spraying Aircraft to Reduce the Amount of Material Carried Off-Target by a Cross-Wind, *J. Agric. Eng. Res.*, vol. 25, pp. 65–74, 1980.
27. Denk, V., and Stern, R., Examples of Flow Visualization in Food Technology, in *Flow Visualization II: Proceedings of the Second International Symposium*, ed. W. Merzkirch, pp. 404–408, Hemisphere, Washington, D.C., 1982.
28. Chang, T. P., and Tatterson, G. B., An Automated Analysis Method for Complex Three Dimensional Mean Flow Fields, in *Flow Visualization III: Proceedings of the Third International Symposium on Flow Visualization*, ed. W.-J. Yang, pp. 236–243, Hemisphere, Washington, D.C., 1985.
29. Doi, J., Miyake, T., and Asanuma, T., Three-dimensional Flow Analysis by On-Line Particle Tracking, in *Flow Visualization III: Proceedings of the Third International Symposium on Flow Visualization*, ed. W.-J. Yang, pp. 14–18, Hemisphere, Washington, D.C., 1985.
30. Palmer, R. S., Waterdrop Impact Forces, *Trans. ASAE*, vol. 8, pp. 69–70, 72, 1965.
31. Robinson, A. R., Model Study of Scour from Cantilevered Outlets, *Trans. ASAE*, vol. 14, pp. 571–576, 581, 1971.
32. Mechrez, E., Kessler, A., and Rubin, H., Optical Visualization of Oil Spill Penetration into the Capillary Zone of Groundwater Aquifers, in *Flow Visualization IV: Proceedings of the Fourth International Symposium*, ed. C. Véret, pp. 827–831, Hemisphere, Washington, D.C., 1987.
33. Chen, J. D., and Wada, N., Visualization of Immiscible Displacement in a Three-dimensional Transparent Porous Medium, *Exp. Fluids*, vol. 4, pp. 336–338, 1986.
34. Tanner, D. W., Further Work on the Relationship between Rake Angle and the Performance of Simple Cultivation Implements, *J. Agric. Eng. Res.*, vol. 5, pp. 307–325, 1960.
35. Koolen, A. J., Mechanical Behavior of Soil by Treatment with a Curved Blade having a Small Angle of Approach, *J. Agric. Eng. Res.*, vol. 17, pp. 355–367, 1972.
36. Hettiaratchi, D. R. P., and Reece, A. R., Boundary Wedges in Two-dimensional Passive Soil Failure, *Geotechnique*, vol. 25, pp. 197–220, 1975.
37. Krause, R., The Most Important Phenomena Subsoiling in Dry Sand, *J. Terramech.*, vol. 12, pp. 119–130, 1975.
38. Godwin, R. J., and Spoor, G., Soil Failure with Narrow Tines, *J. Agric. Eng. Res.*, vol. 22, pp. 213–228, 1977.
39. De Albuquerque, J. C. D., and Hettiaratchi, D. R. P., Theoretical Mechanics of Sub-surface Cutting Blades and Buried Anchors, *J. Agric. Eng. Res.*, vol. 25, pp. 121–144, 1980.
40. Harrison, H. P., Soil Reactions from Laboratory Studies with an Inclined Blade, *Trans. ASAE*, vol. 25, pp. 7–12, 17, 1982.
41. Spoor, G., and Fry, R. K., Soil Disturbance Generated by Deep-working Low Rake Angle Narrow Tines, *J. Agric. Eng. Res.*, vol. 28, pp. 217–234, 1983.
42. Allen, J. R., Baxter, R. I., and Willis, A. H., Experiments on Intake Nozzles for Pneumatic Conveying of Grain under Suction, *J. Agric. Eng. Res.*, vol. 9, pp. 329–334,1964.

Chapter 39

Building Aerodynamics

W. Frank

I INTRODUCTION

In order to design buildings with the right dimensions it is necessary to know very exactly all details concerning the airflow around and through buildings. Due to the complexity of air motion, an exact mathematical description does not yet exist, and therefore, predetermination of all aerodynamic processes cannot be made. In this situation, experimental investigations of the flow field around scaled-down models in wind tunnels are useful. These investigations answer many questions and solve many problems in building aerodynamics, e.g., the wind loads on a building, the air-conditioning and ventilation of skyscrapers, the spreading of polluted exhaust gases, the wind noise, the heat balance within the building.

The shape and the size of a building strongly influence the flow in its vicinity. As a result, wind velocities and gusty winds in the ground vicinity are possible, leading to extreme wind loading on facade elements, protruding roofs, etc.; such conditions can be dangerous to life, as reported by Penwarden [1] and Melbourne and Joubert [2].

In the following discussion a simple technique is described for simulating the natural earth-surface boundary layer in the test section of a wind tunnel. In this boundary layer the three-dimensional flow pattern around a building model is visualized by a combined smoke and laser beam sweeping technique and by a surface-coating method (using petroleum and soot).

The influence of the variation of the model shape to the flow field is demonstrated through the example of a high building with an additional roof protruding in a position perpendicular to the vertical walls of the model.

The experimental results are then compared with the results of a two-dimensional calculation.

II EXPERIMENTAL ARRANGEMENTS

The experiments were conducted in a low-speed wind tunnel of the Prandtl type with an opened test section and a circular-nozzle-exit cross section of 1.80 m dia. The natural earth surface boundary layer was simulated in the following way: a system of several wire screens having different heights and widths of meshes was located downstream of the exit nozzle. The wire screens were so arranged that the wind velocity above the ground plate, on which the models were mounted, increased with increasing distance from the plate. By this method the uniform oncoming flow was transformed into a thick shear layer flow. The shape of the velocity profile could be changed by the arrangements of the wire screens.

Different lucid models with a rectangular ground plan and sharp edges, $b = 200$ mm wide and $l = 100$ mm deep, were constructed. The heights of the bodies were $h_1 = 400$ mm and $h_2 = 200$ mm. The protruding roof was arranged around the whole building and had a depth of 50 mm. The experiments were carried out with freestream velocities from 4 to 12 m/s. The Reynolds numbers, referred to as the width b of the bodies, were $Re \sim 10^5$.

For the flow visualization a focused laser beam was guided over an oscillating mirror parallel to the freestream direction and perpendicular to the ground plate. With the aid of a smoke generator, white smoke was blown into the test section. Thus the flow field on a thin light sheet parallel to the oncoming flow could be made visible with the oscillating laser beam.

The streamlines of the flow field could be made visible directly by an oil-wire method. For that purpose, a thin heated wire was suspended in a position perpendicular to

Fig. 1 Visualization of the streaklines on the ground plate around a model by means of a petroleum–soot surface-coating method. Flow is from left to right.

the ground plate in the test section. Along this wire, small drops of oil flowed down and were burned, generating a dense, white smoke, which was released to the flow. The density of this smoke could be regulated by varying the rate of oil drops. The flow pattern on the ground plate of the test section was visualized by means of a surface-coating method. For this purpose the surface of the plate was coated with a specially prepared mixture of petroleum and soot. The airstream caused the mixture to flow along the surface of the plate. The black soot particles indicated the streamline pattern of the flow in the immediate neighborhood of the plate. This technique allowed observation, in particular, of the lines of separation and reattachment of the flow.

III RESULTS

Figures 1–10 give an impression of the flow conditions that can arise from the flow around high buildings.

Figure 1 shows the streaklines on the ground plate made visible by the petroleum–soot surface-coating method. We can observe very distinctly the horseshoe vortex around the model, the flow separation along the side walls, and the strong pair of vortices in the separation zone behind the building. In Fig. 2 the calculated streamlines at a certain distance from the ground plate are plotted for the same Reynolds number. We see that the numerical result agrees well with the experiment.

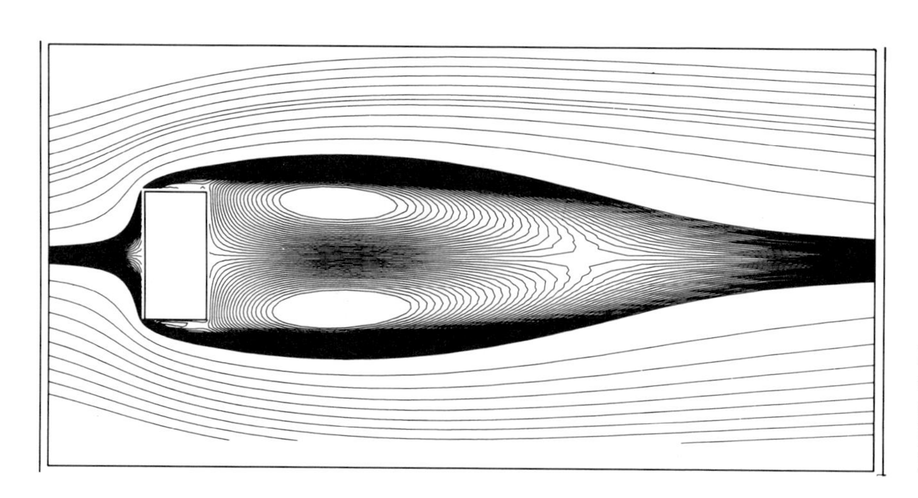

Fig. 2 Calculated streamlines at a certain distance from the ground plate, using the same conditions as the experiment reported in Fig. 1. Flow is from left to right.

Fig. 3 Visualization of the vortex system in front of a skyscraper model with a protruding roof; a combined smoke and laser beam sweeping technique was used. Flow is from right to left.

Figure 3 shows the flow pattern in front of a skyscraper model with a protruding roof in the vicinity of the ground. The flow direction is from right to left. The horseshoe vortex and the strong vortex under the front roof are made visible in a thin light sheet parallel to the oncoming flow by a combined smoke and laser beam sweeping technique [3]. In Fig. 4 we see the calculated flow field for the same case.

If the model of a lower building is located in front of the skyscraper model, the flow field is strongly influenced by this front building. Figure 5 shows the flow pattern for this case visualized by means of the oil-wire method. Due to the front model a very-low-pressure region exists between the two models [3]. This leads to an increase in the intensity of the vortex located under the protruding roof. Now this vortex has a diameter that corresponds to the distance of the roof from the ground (Figs. 5 and 6). Due to the high vorticity, the exhaust gases sucked in from the two buildings (Fig. 6) become concentrated under the roof. This is a very important finding on the aerodynamics of high buildings and their influence on the biological sphere around them. In Fig. 7 the numerically calculated streamlines for the two models are plotted.

If a passage exists under the building, it strongly influ-

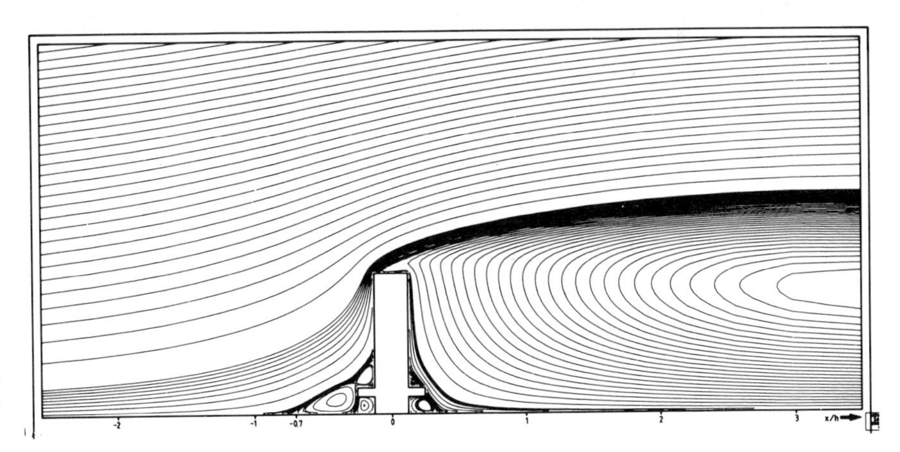

Fig. 4 Calculated streamlines around the model shown in Fig. 3. Flow is from left to right.

Fig. 5 The flow field between two building models visualized by means of the oil-wire method. Flow is from right to left.

Fig. 6 Concentration of exhaust gases in front of a skyscraper model with a lower building in front. Flow is from right to left.

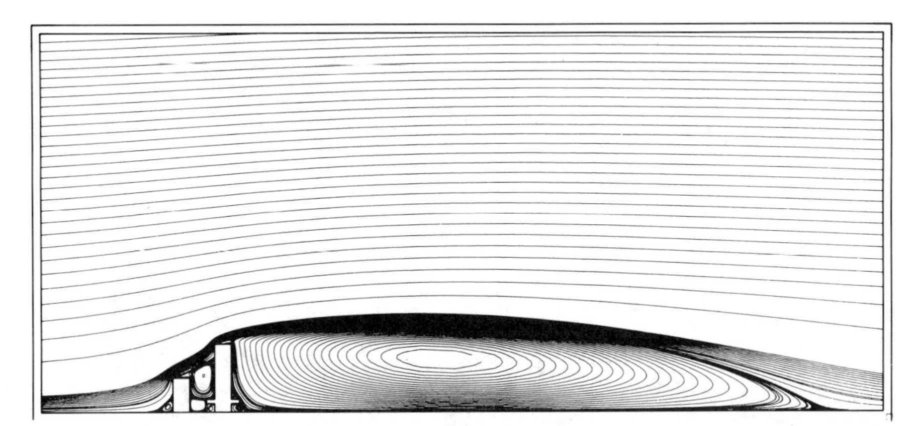

Fig. 7 Calculated streamlines for two building models shown in Fig. 6. Flow is from left to right.

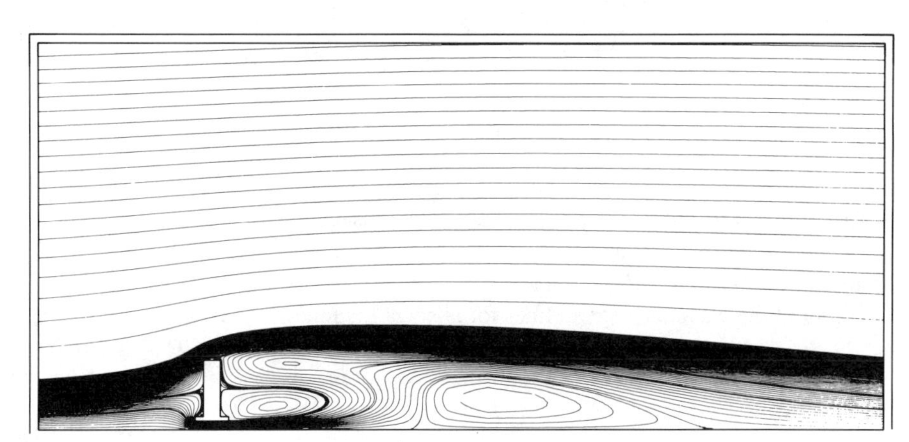

Fig. 8 Calculated streamlines for a building with a passage. Flow is from left to right.

Fig. 9 Flow through the passage and in the wake of the building visualized by blowing smoke into the test section. Flow is from right to left.

Fig. 10 Flow pattern around a skyscraper model with a passage visualized by blowing smoke into the test section. Flow is from right to left.

ences the flow field behind the body, as can be seen in Figs. 8–10. Due to the high overspeed through the passage, a great separation bubble develops in the vicinity of the ground in the wake of the building.

In Fig. 8 the calculated streamlines for a Reynolds number of $Re = 10^5$ are plotted.

In Figs. 9 and 10 the flow through the passage and in the wake of the building is visualized by blowing smoke into the test section. We see that the calculated flow pattern agrees well with the experimental result.

REFERENCES

1. Penwarden, A. D., Acceptable Wind Speeds in Towns, Building Res. Est. CP-1/74, 1974.
2. Melbourne, W. H., and Joubert, P. N., Problems of Wind Flow at the Base of Tall Buildings, *Proc. 3rd Int. Conf. Wind Effects on Buildings and Structures, Tokyo,* pp. 105–114, 1971.
3. Frank, W., The Flow Field in the Neighbourhood of High Buildings, in *Flow Visualization III: Proceedings of the Third International Symposium on Flow Visualization,* ed. W.-J. Yang, pp. 553–557, Hemisphere, Washington, D.C., 1985.

Chapter 40

Land Vehicles

A. Cogotti

In recent years, the aerodynamics of land vehicles have become increasingly important for reasons of fuel economy, comfort, top-speed performance, and side-wind response. Flow visualization techniques play an important role in the aerodynamic development of these vehicles, even more so than for other types of vehicles. This is because land vehicles can often be regarded as "bluff bodies," that is to say, they generate flow fields where large areas of separated flow exist.

By their nature, these flow fields are rather complex and difficult or even impossible to investigate with such techniques as numerical codes. Flow visualization techniques are therefore the most helpful and practical way to understand flow fields around land vehicles.

I THE MAIN CHARACTERISTICS OF FLOW VISUALIZATION TECHNIQUES FOR LAND VEHICLES

A characteristic peculiar to land vehicles is that they usually run at speeds in the low subsonic region. Top speeds up to 100 m/s can be found in racing cars, some high-level sports cars, and high-speed trains. However, typical velocities of most land vehicles range from 15 to 60 m/s; therefore, only a few of existing flow visualization techniques (which in most cases have been developed for aeronautical testing) are suitable for these "low" speeds.

A second characteristic peculiar to land vehicles is that, whenever possible, full-scale models are used to test them. This is almost always the case for passenger cars and motorcycles. Industrial vehicles, buses, and trains are often tested as small-scale models; however, as they are tested in the same full-scale wind tunnels that have been designed and built to test passenger cars, the size of their scale models is also quite large.

The flow visualization techniques applied for land vehicles must therefore be appropriate for use in large facilities. People are present during the tests, precluding the use of any toxic materials. Also any materials likely to produce significant soiling, contamination, or corrosion to the facility itself must be avoided. These prerequisites reduce the number of types of material that can be used as tracers or as seeding and therefore the relevant techniques based on these materials.

A third characteristic demanded relative to flow visualization techniques for land vehicles is that preparation and execution time must be as short as possible. This is necessary because full-scale wind tunnel test time is rather expensive, and only rarely do aerodynamic studies have the highest priority in development of land vehicles (due to higher priority being given to other factors such as aesthetics, functionality and production costs). In order to be really helpful, flow visualization techniques must not require long periods of wind tunnel run time.

In conclusion, and as a general rule, flow visualization techniques for land vehicles must be quick to conduct and suitable for low flow velocities in large facilities. Contamination of the environment must also be minimized.

II TYPES OF LAND VEHICLES TESTED IN WIND TUNNELS

The land vehicles developed in wind tunnels and therefore making use of the available flow visualization techniques include the following:

- Cars (passenger, sport, racing)
- Motorcycles (touring, racing)
- Bicycles
- Industrial vehicles including buses
- Trains
- Experimental land vehicles (basic bodies, record vehicles)

These are only a few of the many individual types of land vehicles developed in wind tunnels.

Passenger cars are, by far, the land vehicles most studied in wind tunnels. Good aerodynamic characteristics are a must for every new car model, for both technical and commercial reasons. To achieve good aerodynamics in new vehicles, great attention must be paid to details during the aerodynamic development phase, and an ever-increasing part of the relevant wind tunnel time is spent on flow visualization tests.

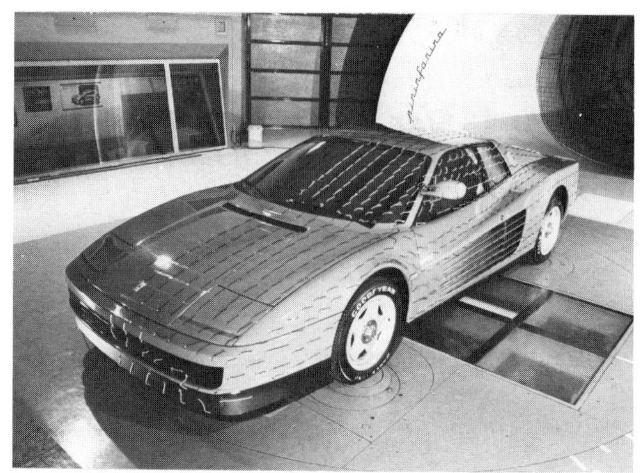

Fig. 1 Surface visualization by yarn tufts: Ferrari Testarossa [1]

Fig. 2 Surface visualization by yarn tufts: F1 racing car.

III EXAMPLES OF THE MAIN TECHNIQUES USED ON LAND VEHICLES

Investigations by flow visualization techniques are typically carried out both on the vehicle surface and in the flow field near the vehicle.

A Techniques Used to Investigate the Vehicle Surface

Various techniques are used on the vehicle surface to detect

- Areas of flow separation
- Streamline direction

A complete description of these techniques is reported in Part A of Ref. [1].

Among the techniques used the most are:

- Yarn tufts and, sometimes, fluorescent minitufts
- Oil, both as droplets and as a continuous film

Examples of surface visualization by yarn tufts on various kind of land vehicles are reported in Figs. 1–6. Examples of minituft visualization are shown in Fig. 7. Visualization by oil droplets is shown in Figs. 8 and 9. Visualization by a continuous oil film are shown in Figs. 10 and 11.

Fig. 3 Surface visualization by yarn tufts: Motorcycle KZ 125, Gilera.

Fig. 4 Surface visualization by yarn tufts: Racing motorcycle.

Fig. 5 Surface visualization by yarn tufts: Scale model of Bus [1]

Fig. 6 Surface visualization by yarn tufts: Low-drag (C_d = 0.049) basic body, 1979, Istituto Motorizzazione Politecnico Torino.

Fig. 7 Minituft visualization: Full-scale passenger car tested on the road [1].

Fig. 8 Visualization by oil droplets: Low-drag (C_d = 0.049) basic body, 1979, Istituto Motorizzazione Politecnico Torino.

Fig. 9 Visualization by oil droplets: Low-drag (C_d = 0.205) car body, 1978, CNR Pininfarina.

Fig. 10 Visualization by continuous oil film: Full-size CNR PFT model

Fig. 11 Visualization by continuous oil film: Small-scale truck cabin [1].

B Techniques Used to Investigate the Flow Field Near the Vehicle

The flow field close to the vehicle can be investigated by

- Streamers
- Planar Surveys

1 Streamers In the case of streamers, several types of tracers can be used to find the streamlines around the vehicle body (see also part B of Ref. [1]).

Examples include the following:

- Single and multiple wands
- Smoke lines
- Water stream
- Liquid nitrogen-water stream
- Helium bubbles

Wands (Fig. 12) are simple but are only approximate streamline indicators.

Smoke (Fig. 13) gives some contamination and soiling problems; furthermore it mixes quickly in the regions where flow is turbulent. Indications provided by smoke therefore leave much to be desired for flow visualization around land vehicles. An exception is smoke injected inside a separation bubble, which can easily provide a clear image of the size and shape of the bubble.

Water steam (Figs. 14 and 15) can be used to replace smoke. It is safe and clean, but much less visible than smoke.

Liquid nitrogen–water steam (Fig. 16) is also a clean, safe tracer. It is very clearly visible and from this point of view, it is superior to smoke or water steam. It seems however that this tracer is not as practical as was first thought for use as a routine visualization technique.

Helium bubbles (Figs. 17–22) are clean, safe, and relatively easy to use, the main problem being setting up an

Fig. 12 Multiple wands over a convertible car [2].

Fig. 13 Smokelines over a small-scale truck model [1].

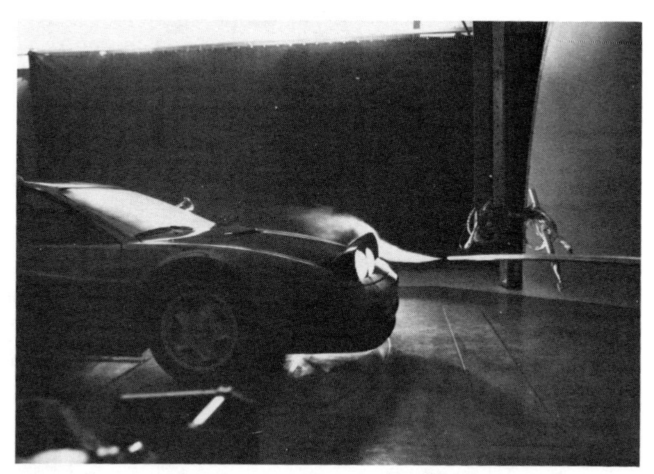

Fig. 14 Water steam over a pop-up headlamp.

Fig. 15 Water steam under a passenger car air dam [2].

Fig. 16 Liquid nitrogen-water steam over a full-scale car [1].

Fig. 17 Helium bubbles in front of a Fiat 124 notchback.

Fig. 18 Helium bubbles in front of a Spyder Europa Pininfarina.

Fig. 19 Helium bubbles in front of an Alfa Romeo 164.

Fig. 20 Helium bubbles under the Alfa Romeo 164 air dam.

Fig. 21 Helium bubbles in front of a Giltera KZ 125 motorcycle. Pilot upright.

Fig. 22 Helium bubbles in front of a Gilera KZ 125 motorcycle. Pilot lowered.

appropriate lighting system. The bubbles have the capability of showing the fine details of the flow field with each bubble as a single unit. The bubbles do not ever merge with each other even in conditions of high turbulent flow.

Considering all aspects, the author considers helium bubbles to be the best technique for streamline visualization of land vehicles when tested in large, low-speed wind tunnels.

2 Flow Field Planar Surveys The techniques in this category are gaining increasing acceptance and popularity (see also Part C of Ref. [1]). As a general rule they provide information on a given flow field by the reconstruction, point by point, of the local values of pressures, velocities, turbulence, etc. Therefore, by their nature, these techniques should be regarded as "second generation." Instead of being based on some types of tracers (tufts, smoke, etc.) that show the air path, the surveys are produced and visualized by studying and recording the velocity vectors, point by point, in the flow field.

In order to achieve this, planar surveys heavily utilize electronic equipment and accurate, fast data-acquisition systems, processors, and computer graphics. The primary task of this electronic equipment is to speed up the complete process (data acquisition, processing, and presentation) such that a large amount of information on the flow field is gathered in a relatively short time.

Some examples of results which have been obtained by these techniques, are reported here.

a Maps of Total Pressure Coefficients These maps are usually made by using a Kiel probe, which measures total pressures correctly up to local flow angles of $\pm 50°$ without requiring any alignment to the approaching flow (see Refs. 3–6 and Figs. 23 and 24).

b Maps of Total and Static Pressures, Velocities, and Cross Flows. These maps can be generated by using various types of multihole probes. In the literature examples of 5-, 6-, 7-, 14-hole probes can be found. These probes have different capabilities, which span from $\pm 30°$ for 5-hole probes, through $\pm 70°$ for the 7-hole probes, to $\pm 180°$ for the 14-hole probes.

Figures 25–27 show examples of pressure and velocity maps made by a 7-hole probe [7]. Figures 28–33 show pressure and velocity maps recently made by a new 14-hole probe [2, 8, 9].

From these maps the following information can easily be found:

- Size and shape of the vehicle wakes
- Pressure and velocity levels in the flow field, both outside and inside the wake
- Presence of vortices inside the wake and, from the cross-flow map, their direction of rotation.

Furthermore, it is possible to process these data in order to generate maps of vorticity and local drag.

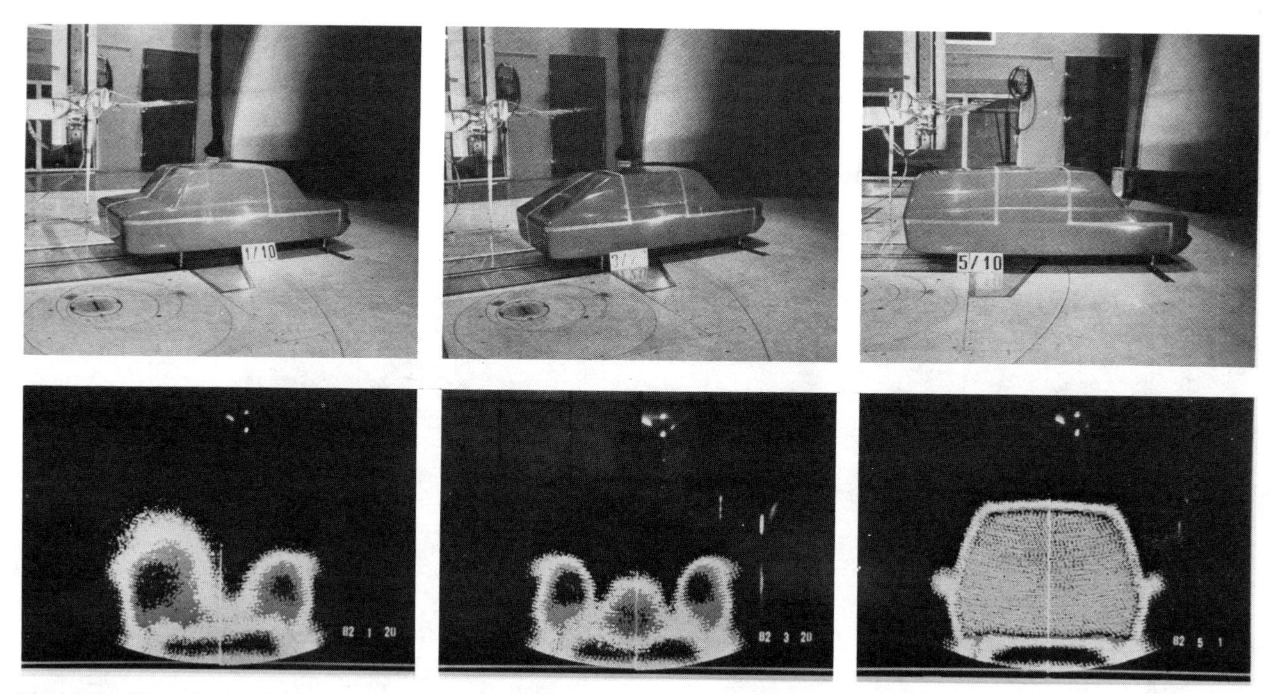

Fig. 23 Wake image behind a small-scale CNR model having three different rear ends—Visualization with the LED Technique [2], 1982.

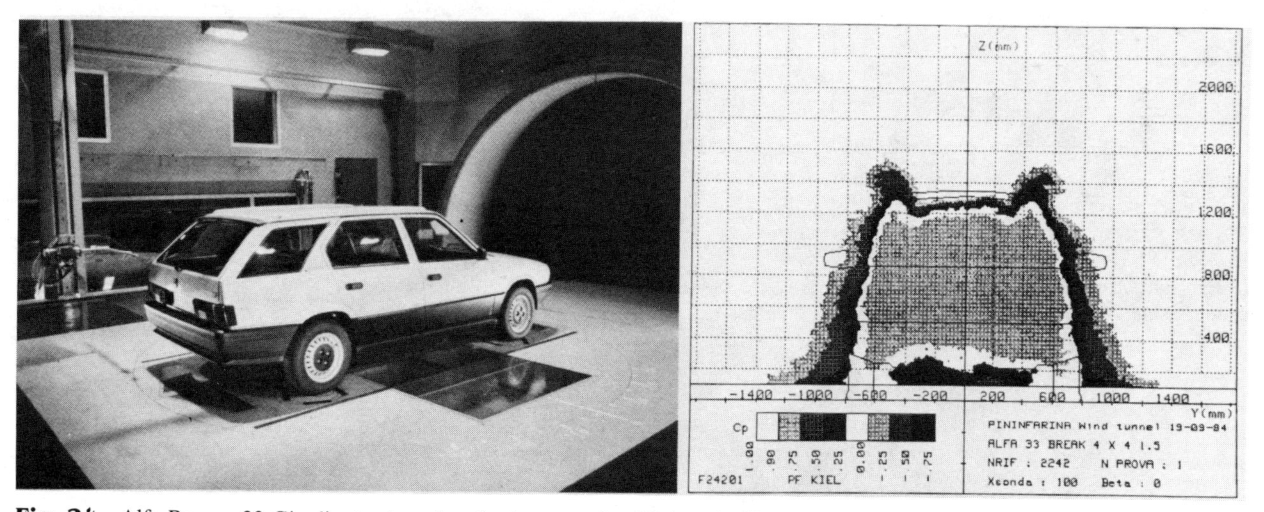

Fig. 24 Alfa Romeo 33 Giardinetta 4 × 4 and relevant wake: Kiel probe/Computer generated graphics [2], 1984.

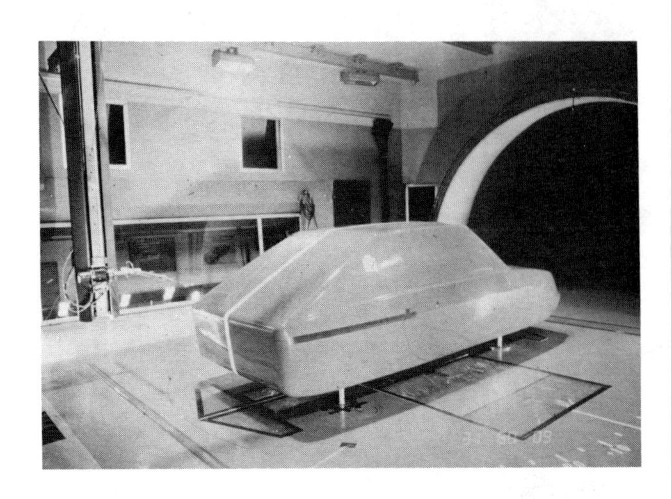

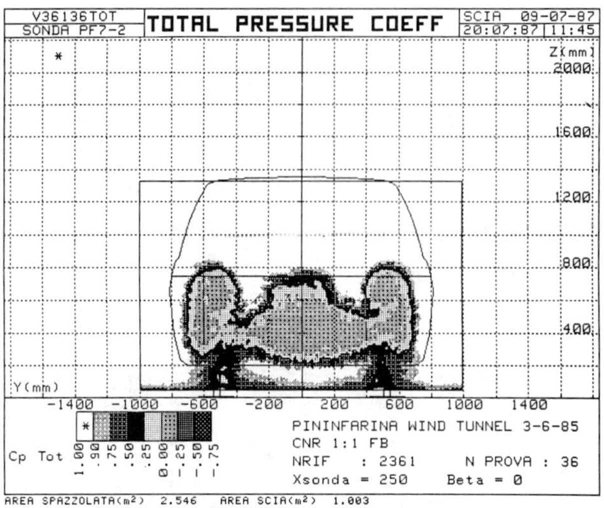

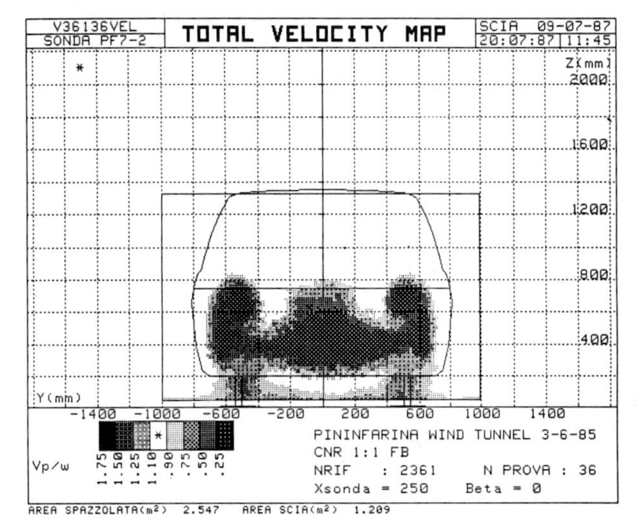

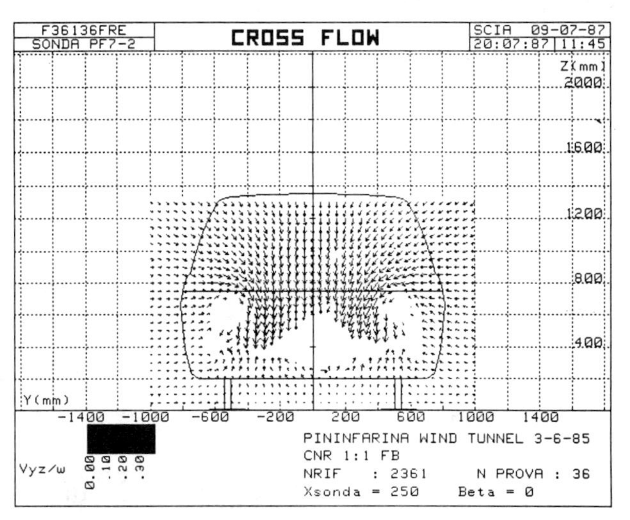

Fig. 25 Full-size CNR model, fastback shape: pressure and velocity maps made 0.25 m behind the model using 7-hole probe/computer-generated graphics [1], 1985.

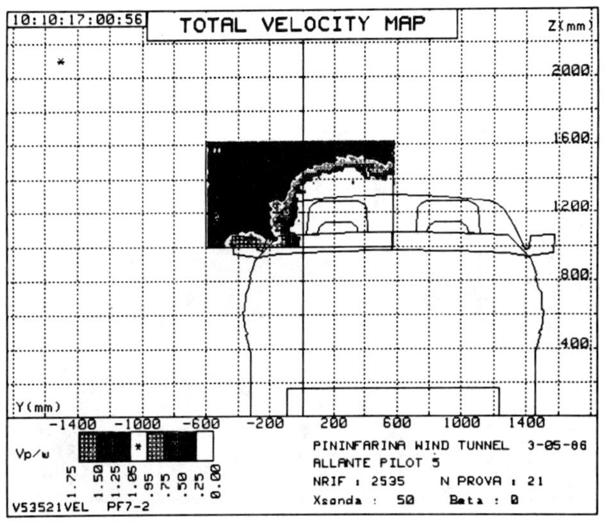

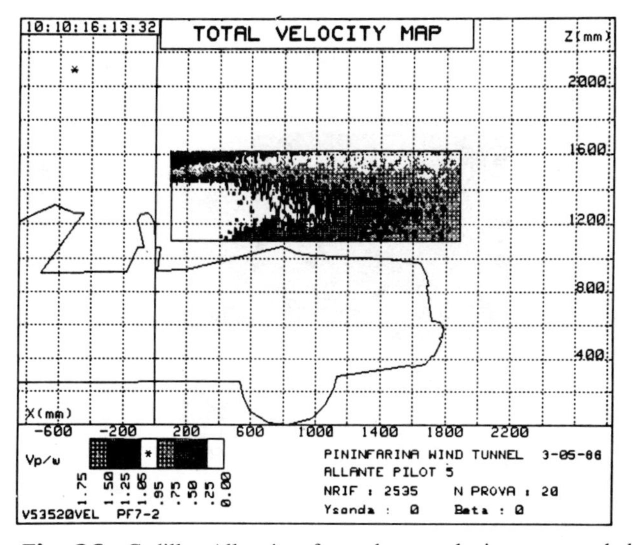

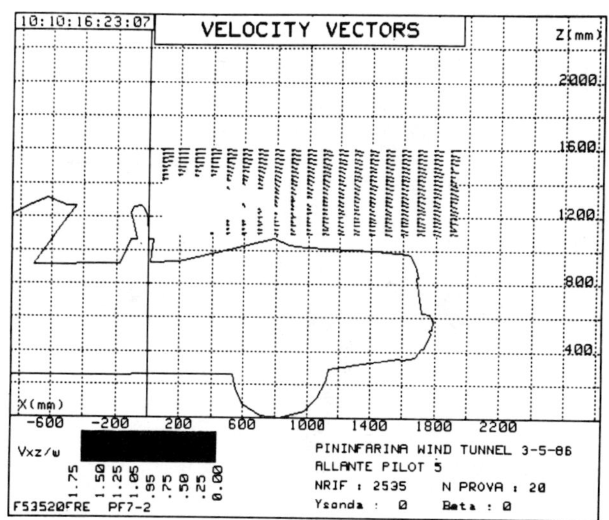

Fig. 26 Cadillac Allanté, soft top down: velocity maps made behind the driver's headrest using 7-hole probe/computer-generated graphics [2], 1986.

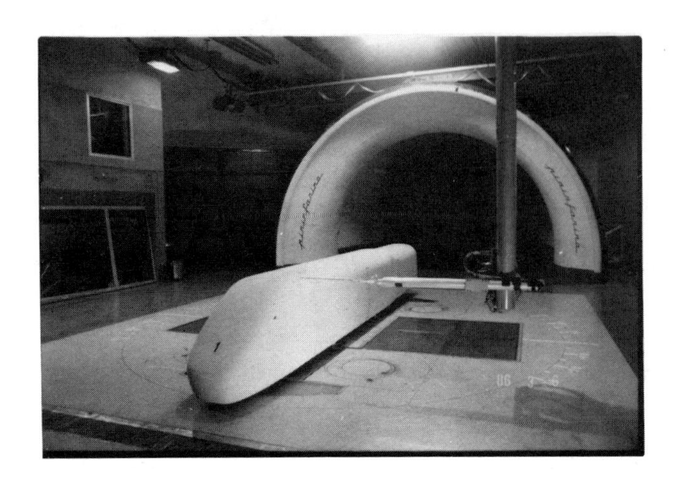

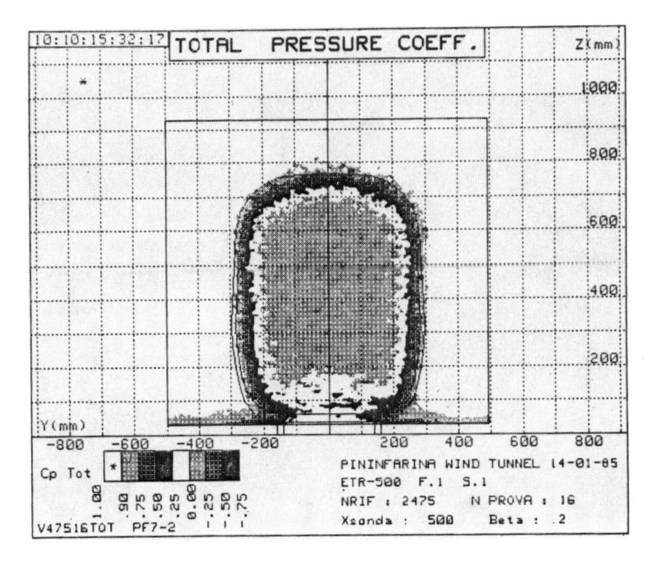

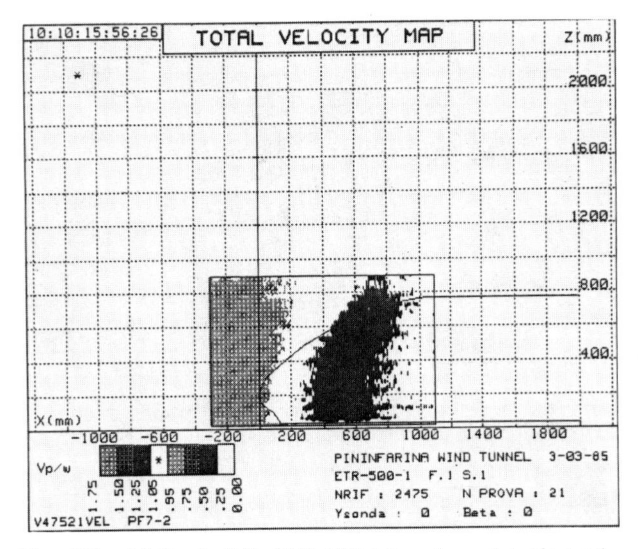

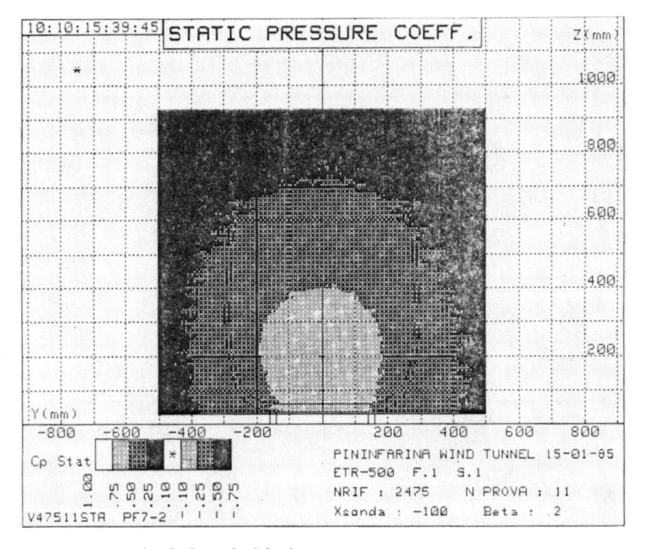

Fig. 27 (*a*) Breda C.F. ETR 500 1:5 scale model; (*b*) total pressure map made 0.5 m behind the model; (*c*) velocity map made at the model side, (*d*) static pressure map made 0.1 m in front of the model using 7-hole/computer-generated graphics [2], 1985.

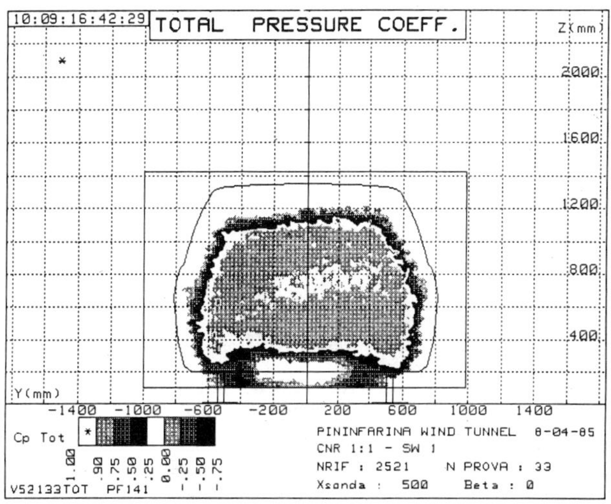

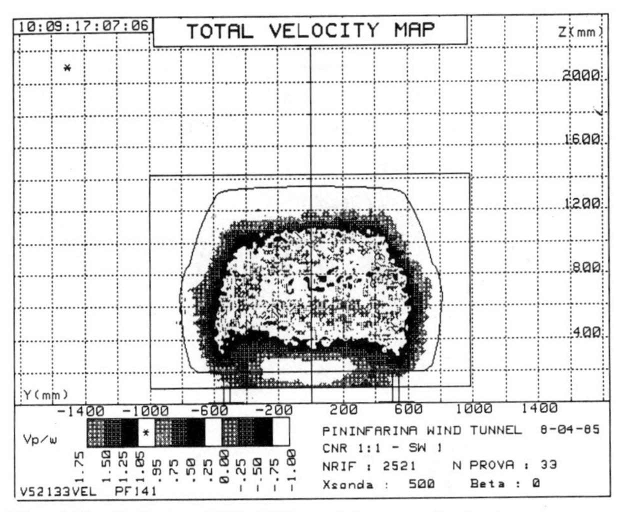

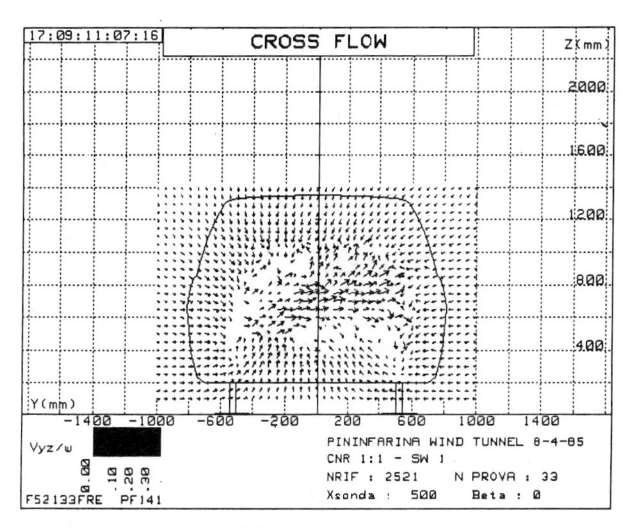

Fig. 28 Full-size CNR-PFT model, squareback shape: pressure and velocity maps made 0.5 m behind the model using 14-hole probe/computer-generated graphics [8], 1986.

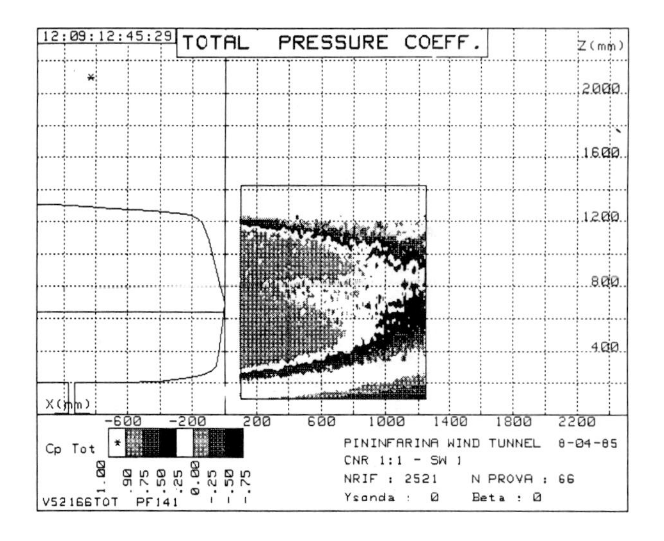

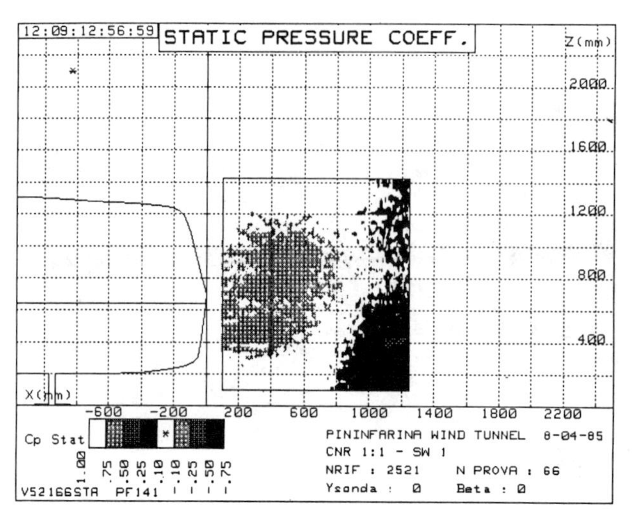

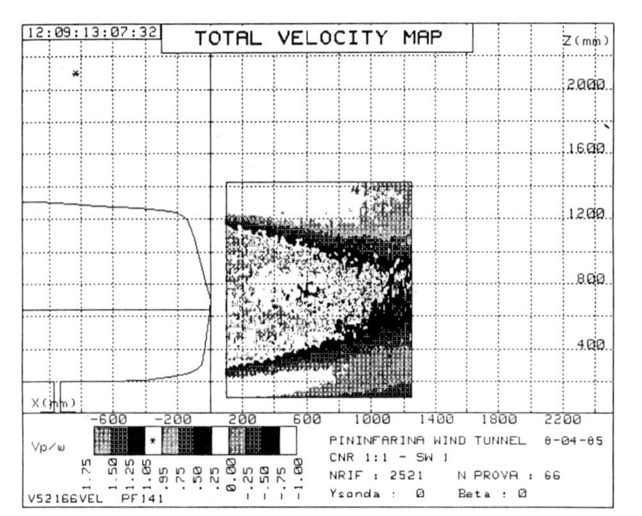

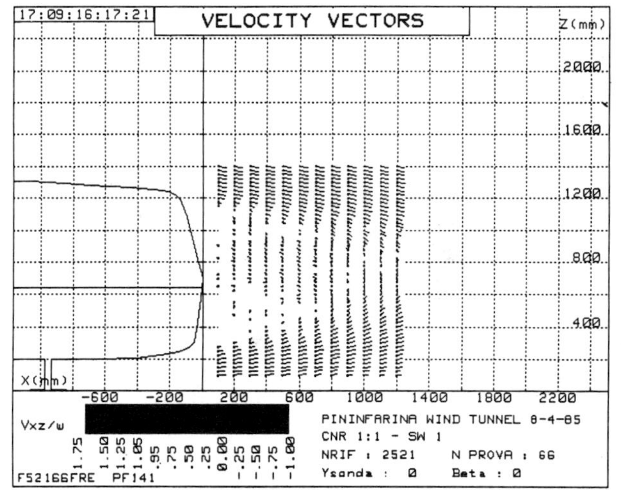

Fig. 29 Full-size CNR-PFT model, squareback shape: pressure and velocity maps made along the center line vertical longitudinal plane (XZ) behind the model using 14-hole probe/computer-generated graphics [8], 1986.

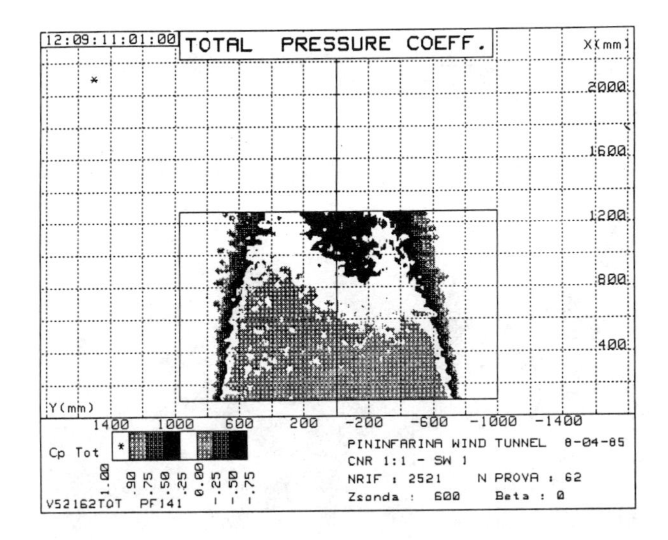

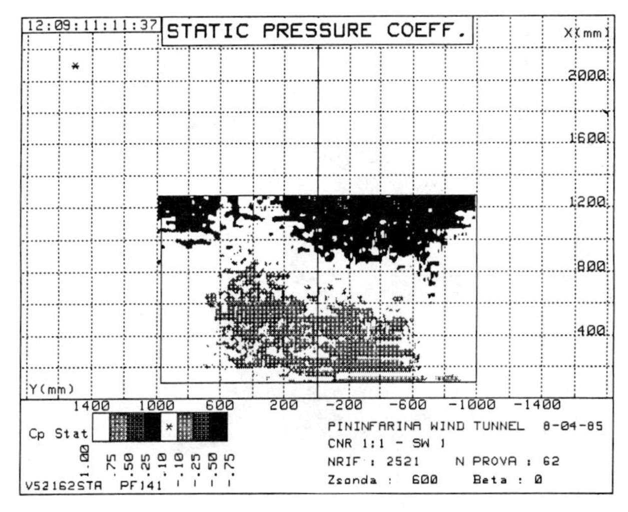

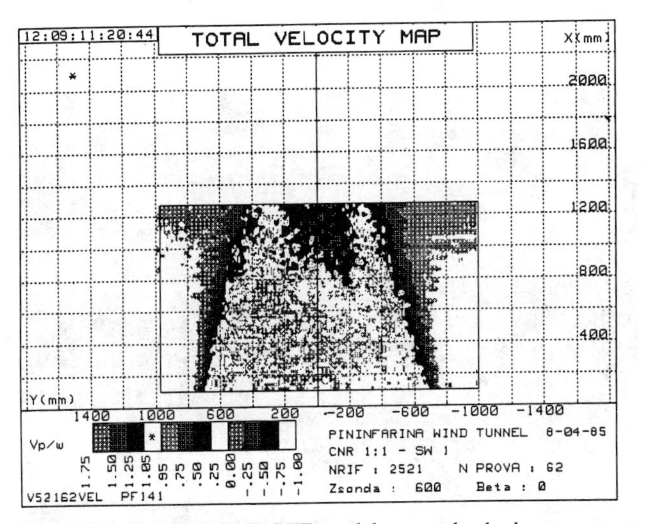

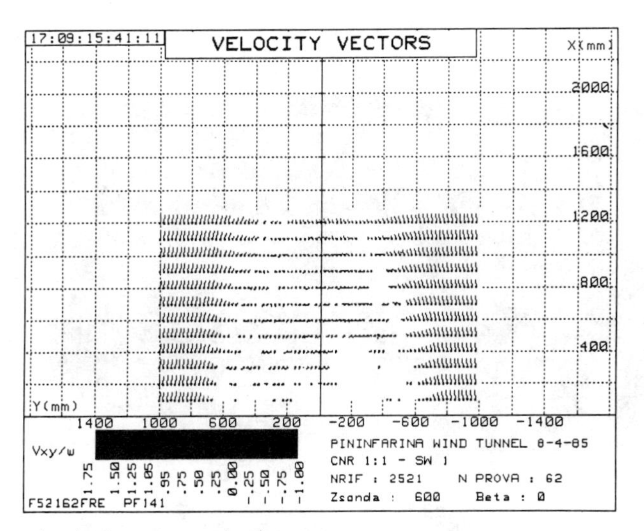

Fig. 30 Full-size CNR-PFT model, squareback shape: pressure and velocity maps made along a horizontal plane (XY) behind the model, at z = 0.6 m from the ground, using 14-hole probe/computer-generated graphics [8], 1986.

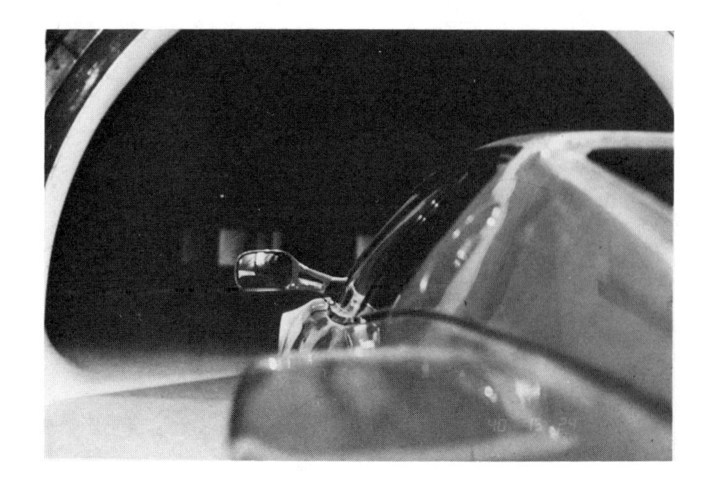

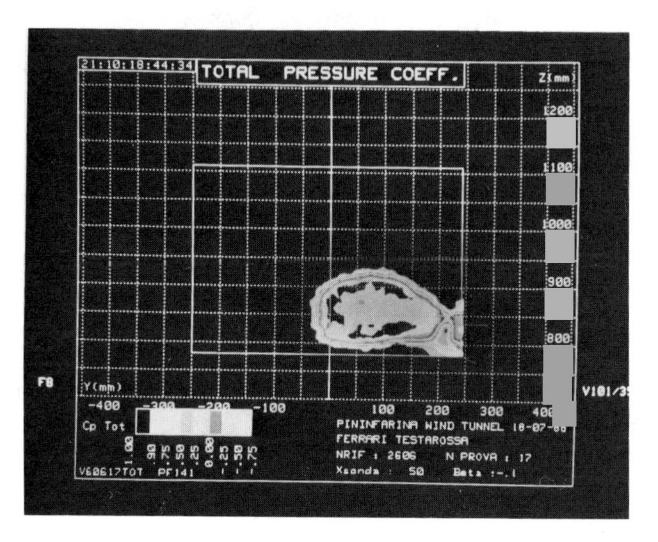

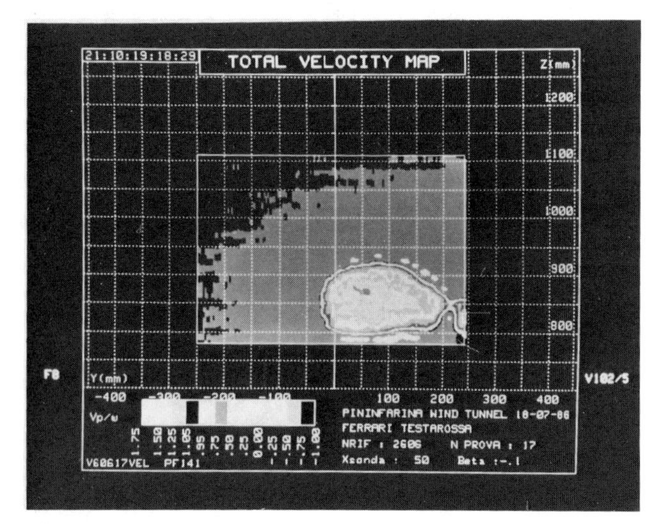

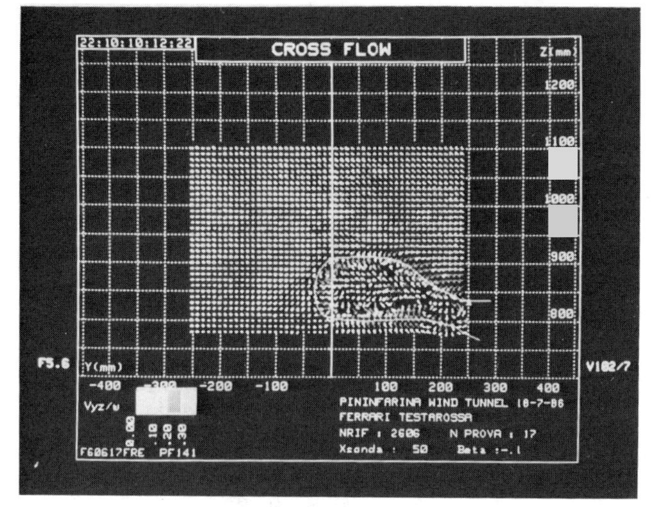

Fig. 31 Ferrari Testarossa, with pressure and velocity maps made 0.05 m behind the rear-view mirror using 14-hole probe/computer-generated graphics, 1986.

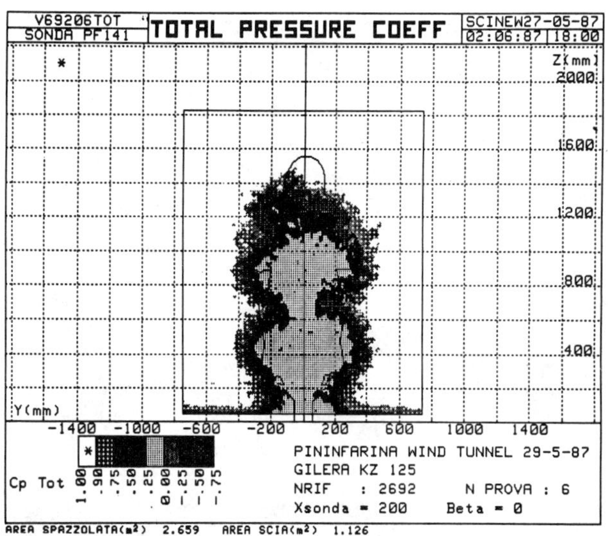

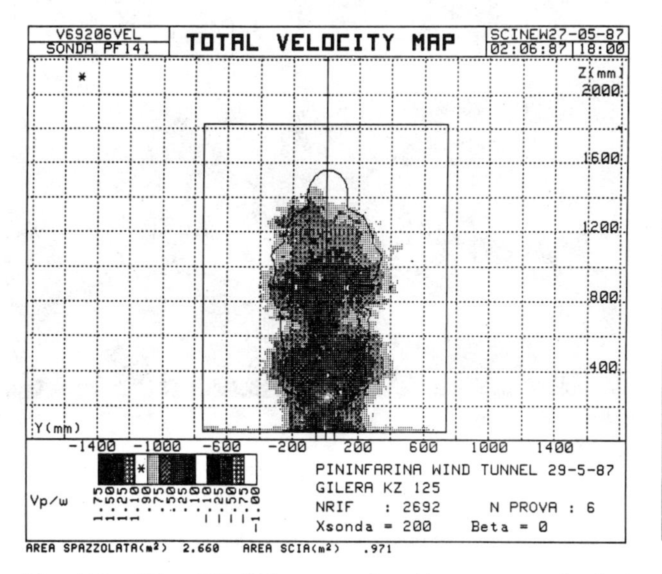

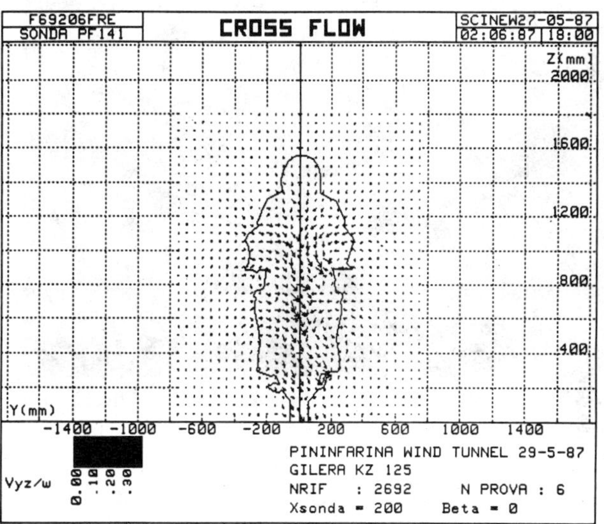

Fig. 32 Gilera KZ 125 motorcycle with pressure and velocity maps made 0.2 m behind the motorcycle using 14-hole probe/computer-generated graphics, 1987.

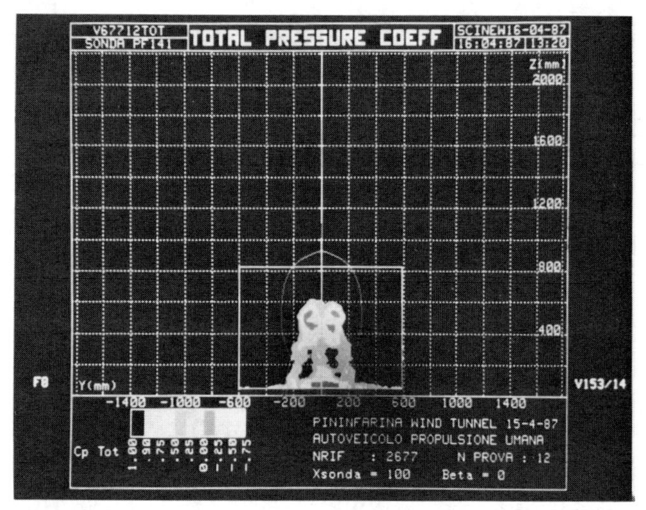

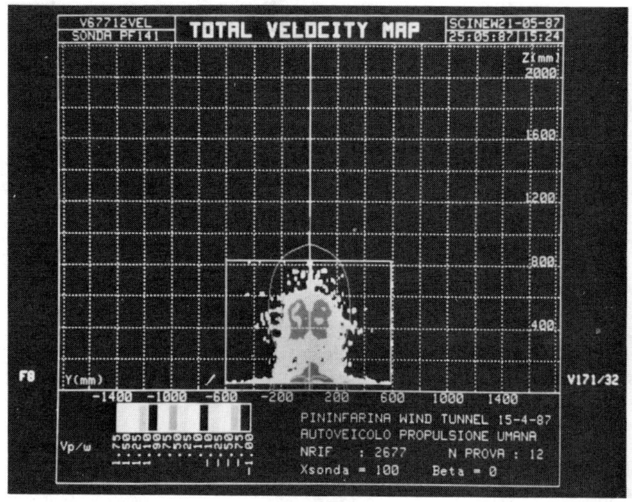

Fig. 33 Human-powered Vehicle made by Istituto Motorizzazione Politecnico Torino
Pressure and velocity maps using 14-hole probe/computer-generated graphics, 1987.

IV CONCLUSIONS

A number of flow visualization techniques are available today for investigating the flow fields around land vehicles and solving the problems that may arise during the aerodynamic development of these vehicles.

Each of these techniques has its own peculiarities. This means that depending on the type of flow investigation to be carried out as well as the available wind tunnel time, one specific technique may be more appropriate than another.

It seems very likely that in the near future, due to the continuous progress in a number of fields, such as electronics, holography and lasers, powerful new techniques will become available to allow even more detailed surveying of land vehicle flow fields.

REFERENCES

1. Aerodynamic Flow Visualization Techniques and Procedures, SAE Inform. Rep. HS J1566, January, 1986.
2. Cogotti, A., Recent Advances in Flow Field Mapping Techniques, *Montecarlo Autotechnologies,* January 1987.
3. Crowder, J. P., Quick and Easy Flow Field Survey, *Astronaut. Aeronaut.,* October 1980.
4. Cogotti, A., A Passenger Car Wake Survey using Coloured Isopressure Maps, in *Flow Visualization III: Proceedings of the Third International Symposium on Flow Visualization,* ed. W.-J. Yang, pp. 668–675, Hemisphere, Washington, D.C., 1985.
5. Cogotti, A., Wake Surveys of Different Car-Body Shapes with Coloured Isopressure Maps, Detroit, SAE Paper 840299, February 1985.
6. Cogotti, A., Pininfarina New Measurement Techniques in the Wind Tunnel, *Montecarlo Autotechnologies,* February 1985.
7. Cogotti, A., Car-Wake Imaging using a Seven-hole Probe, Detroit, SAE Paper 860214, February 1986.
8. Cogotti, A., Flow Field Surveys behind Three Squareback Car Models using a New Fourteen-Hole Probe, Detroit, SAE Paper 870243, February 1987.
9. Cogotti, A., A Strategy for Optimum Surveys of Passenger-Car Flow Fields. Detroit, SAE Congress, February 1989.

Index